THE OXFORD HANDBOOK OF

NANOSCIENCE AND TECHNOLOGY

The Oxford Handbook of Nanoscience and Technology

Volume I of III

Basic Aspects

Edited by

A.V. Narlikar

Y.Y. Fu

OXFORD

UNIVERSITY PRESS

OXFORD
UNIVERSITY PRESS

Great Clarendon Street, Oxford, OX2 6DP,
United Kingdom

Oxford University Press is a department of the University of Oxford.
It furthers the University's objective of excellence in research, scholarship,
and education by publishing worldwide. Oxford is a registered trade mark of
Oxford University Press in the UK and in certain other countries

Published in the United States of America by Oxford University Press
198 Madison Avenue, New York, NY 10016, United States of America

British Library Cataloguing in Publication Data
Data available

Library of Congress Cataloging in Publication Data
Data available

ISBN 978–0–19–953304–6

Preface

Wolfgang Pauli is known to have remarked, "God made solids, but surfaces were the work of the Devil." This Handbook deals with the Devil's work. As the size of the material is reduced, surfaces acquire increasing importance, and indeed override the bulk when one of the dimensions of the material shrinks to nanometers. Simultaneously, at the nanoscale, quantum effects come into play and the properties of matter confined to nanodimensions are dramatically changed. Nanoscience and nanotechnology are all about relating and exploiting the above phenomena for materials having one, two or three dimensions reduced to the nanoscale. Their evolution may be traced to three exciting happenings that took place in a short span from the early to mid-1980s with the award of Nobel prizes to each of them. These were the discovery of the quantum Hall effect in a two-dimensional electron gas, the invention of scanning tunnelling microscopy (STM) and the discovery of fullerene as the new form of carbon. The latter two, within a few years, further led to the remarkable invention of the atomic force microscope (AFM) and, in the early 1990s the extraordinary discovery of carbon nanotubes (CNT), which soon provided the launch pad for the present-day nanotechnology. The STM and AFM have emerged as the most powerful tools to examine, control and manipulate matter at the atomic, molecular and macromolecular scales and these functionalities constitute the mainstay of nanotechnology. Interestingly, this exciting possibility of nanolevel tailoring of materials was envisioned way back in 1959 by Richard Feynmen in his lecture, "There's plenty of room at the bottom."

During the last 15 years, the field of nanoscience and technology has expanded internationally and its growth has perhaps been more dramatic than in most other fields. It has been transformed into an intense and highly competitive research arena, encompassing practically all disciplines that include theoretical and experimental physics, inorganic, organic and structural chemistry, biochemistry, biotechnology, medicine, materials science, metallurgy, ceramics, electrical engineering, electronics, computational engineering and information technology. The progress made in all these directions is truly spectacular. In this edited Handbook of Nanoscience and Technology, we have attempted to consolidate some of the major scientific and technological achievements in different aspects of the field. We have naturally had to follow a selective rather than exhaustive approach. We have focused only on those topics that are generally recognized to have had a major impact on the field. Inherent in this selection process is the risk of some topics inadvertently getting overemphasized, while others are unavoidably left out. This is a non-trivial problem especially in the light of the great many developments that have taken place in the field. However, a great diversity of important developments is represented in this Handbook and helps us overcome some of these risks.

The present Handbook comprises 3 volumes, structured thematically, with 25 chapters each. Volume I presents fundamental issues of basic physics, chemistry, biochemistry, tribology, etc. at the nanoscale. Many of the theoretical papers in this volume are intimately linked with current and future nanodevices, molecular-based materials and junctions (including Josephson nanocontacts) and should prove invaluable for further technology development. Self-organization of nanoparticles, chains, and nanostructures at surfaces are further described in detail. Volume II focuses on the progress made with a host of nanomaterials including DNA and protein-based nanostructures. This volume includes noteworthy advances made with the techniques of improved capability used for their characterization. Volume III highlights engineering and related developments, with a focus on frontal application areas like Si-nanotechnologies, spintronics, quantum dots, CNTs, and protein-based devices, various biomolecular, clinical and medical applications. The other prominent application areas covered in this chapter are nanocatalysis, nanolithography, nanomaterials for hydrogen storage, nanofield emitters, and nanostructures for photovoltaic devices. This volume concludes the Handbook with a chapter that analyses various risks that are associated in using nanomaterials.

We realize that the boundaries separating a few of the topics of the above three volumes are somewhat shadowy and diffuse. Some chapters of Volumes II and III could have also provided a natural fit with Volume I. For instance, some of the novel molecular devices of Volume III could have alternatively been included in the realm of basic studies that form a part of Volume I.

The three volumes together comprise 75 chapters written by noted international experts in the field who have published the leading articles on nanoscience and nanotechnology in high-profile research journals. Every chapter aims to bring out *frontiers and advances* in the topic that it covers. The presentation is technical throughout, and the chapters in the present set of 3 volumes are not directed to the general and popular readership. The set is not intended as a textbook; however, it is likely to be of considerable interest to final-year undergraduates specializing in the field. It should prove indispensable to graduate students, and serious researchers from academic and industrial sectors working in the field of nanoscience and technology from different disciplines like physics, chemistry, biochemistry, biotechnology, medicine, materials science, metallurgy, ceramics, electrical, electronics, computational engineering, and information technology. The chapters of the three volumes should provide readers an analysis of the state-of-the-art technology development and give them an opportunity to engage with the cutting edge of research in the field.

We would like to thank all the contributors for their splendid and timely cooperation throughout this project. We are grateful to Dr Sonke Adlung for being most cooperative and considerate and for his important suggestions to help us in our efforts, and acknowledge with thanks the efficient assistance provided by Ms April Warman, Ms Phaedra Seraphimidi and Mr Dewi Jackson. Special thanks are due to Mrs Emma Lonie and Ms Melanie

Johnstone for commendably coordinating the proof correction work with over 200 contributors. One of us (AVN) thanks the Indian National Science Academy, New Delhi for financial assistance in the form of a Senior Scientist fellowship and the UGC-DAE Consortium for Scientific Research, Indore, for providing infrastructural support. He thanks the Consortium Director, Dr Praveen Chaddah, and the Centre Director (Indore), Dr Ajay Gupta, for their sustained interest and cooperation. He further acknowledges with thanks the technical assistance provided by Mr Arjun Sanap, Mr D. Gupta, Dr N.P. Lalla, Mr Suresh Bharadwaj and Mr U.P. Deshpande on many occasions. He is particularly grateful to his wife Dr Aruna Narlikar for her invaluable help, patience, and support throughout, and especially for her useful suggestions on many occasions during the course of the present project. He acknowledges the commendable technical support of his daughter Dr Amrita Narlikar at Cambridge, and also of Dr Batasha who remains a close and valued friend of the family. YYF extends his thanks to the National Natural Science Foundation of China (Contracts No. 60776053 and No. 60671021), and the National High Technology Research and Development Program of China (Program 863 and Contract No. 2007AA03Z311) for financial support. He remains indebted to his father, who passed away many years ago, for his invaluable guidance, advice and help to build his life and career, and to his mother, wife and son, for their sustained patience and support.

A.V. Narlikar
Y.Y. Fu

November 2008

Contents

List of Contributors

Aydin, K. Nanotechnology Research Center, Bilkent University, Ankara, Bilkent 06800, Turkey. aydin@fen.bilkent.edu.tr, koraydin@gmail.com

Bertel, Erminald Institute of Physical Chemistry, Universität Innsbruck, Innrain 52a A - 6020 Innsbruck, Austria. Erminald.Bertel@uibk.ac.at

Biswas, S.K. Department of Mechanical Engineering, Indian Institute of Science, Bangalore 560 012, India. skbis@mecheng.iisc.ernet.in

Blase, X. Insitut Néel, CNRS, and Université Joseph Fourier BP 166, 38042 Grenoble cedex 9, France. xavier.blase@grenoble.cnrs.fr

Blunt, M. O. School of Physics and Astronomy, University of Nottingham, Nottingham NG7 2RD, UK. matthew.blunt@nottingham.ac.uk

Boero, Mauro Center for Computational Sciences, University of Tsukuba, Tennodai 1-1-1, Tsukuba, Ibaraki 305-8577, Japan, and Institut de Physique et Chimie des Matériaux de Strasbourg (IPCMS), 23 rue du Loess, F-67037, Strasbourg, France. boero@comas.frsc.tsukuba.ac.jp

Burton, J. D. Nebraska Center for Materials and Nanoscience and Department of Physics and Astronomy, 116 Brace Lab., University of Nebraska-Lincoln, NE, 68588-0111, USA. jdburton1@gmail.com

Datta, Supriyo Purdue University School of Electrical and Computer Engineering, Electrical Engineering Building, 465 Northwestern Ave., West Lafayette, Indiana 47907-2035, USA. datta@ecn.purdue.edu

Fernández-Serra, M.-V. Department of Physics and Astronomy, and New York Center for Computational Science, Stony Brook University (SUNY), Stony Brook, NY 11794-3800, USA. maria.fernandez-serra@stonybrook.edu

Golizadeh-Mojarad, Roksana Purdue University School of Electrical and Computer Engineering, Electrical Engineering Building, 465 Northwestern Ave., West Lafayette, Indiana 47907-2035, USA. rgolizad@ecn.purdue.edu

Graupner, Ralf Technical Physics, University of Erlangen-Nürnberg Erwin-Rommel-Strasse 1, 91058 Erlangen, Germany. ralf.graupner@physik.uni-erlangen.de

Hauke, Frank Institute for Organic Chemistry, University of Erlangen-Nürnberg Henkestrasse 42, 91054 Erlangen, Germany. frank.hauke@chemie.uni-erlangen.de

Kadowaki, Kazuo Institute of Materials Science, University of Tsukuba, Tsukuba, Ibaraki 305-8573, Japan. kadowaki@ims.tsukuba.ac.jp

Kanda, Akinobu Institute of Physics, University of Tsukuba, Tsukuba, Ibaraki 305-8571, Japan. kanda@lt.px.tsukuba.ac.jp

Kirczenow, George Department of Physics, Simon Fraser University, Burnaby, B.C., Canada V5A 1S6, and Canadian Institute for Advanced Research, Nanoelectronics Program, Toronto, Canada. kirczeno@sfu.ca

Král, Petr Department of Chemistry, MC111, University of Illinois at Chicago, 845 West Taylor Street, Chicago, IL60607, USA. pkral@uic.edu

Lanzara, Alessandra Department of Physics, University of California, Berkeley, Berkeley, CA 94720, USA, and Materials Sciences Division, Lawrence Berkeley National Laboratory, Berkeley, California 94720, USA. alanzara@lbl.gov

Li, Qunxiang Hefei National Laboratory for Physical Sciences at the Microscale, University of Science and Technology of China, Hefei, Anhui 230026, P. R. China. liqun@ustc.edu.cn

Lin, Nian Department of Physics, The Hong Kong University of Science and Technology, Clear Water Bay, Kowloon, Hong Kong. phnlin@ust.hk

Martin, C. P. School of Physics and Astronomy, University of Nottingham, Nottingham NG7 2RD, UK. ppxcpm1@nottingham.ac.uk

Menzel, Alexander Institute of Physical Chemistry, Universität Innsbruck, Innrain 52a A - 6020 Innsbruck, Austria. alexander.menzel@uibk.ac.at

Moriarty, P. School of Physics and Astronomy, University of Nottingham, Nottingham NG7 2RD, UK. Philip.Moriarty@nottingham.ac.uk

Nikolić, Branislav K. Department of Physics and Astronomy, University of Delaware, Newark, DE 19716-2570, USA. bnikolic@physics.udel.edu

Ootuka, Youiti Institute of Physics, University of Tsukuba, Tsukuba, Ibaraki 305-8571, Japan. ootuka@lt.px.tsukuba.ac.jp

Ozbay, Ekmel Nanotechnology Research Center, Bilkent University, Ankara, Bilkent 06800, Turkey. ozbay@bilkent.edu.tr

Ozkan, G. Nanotechnology Research Center, Bilkent University, Ankara, Bilkent 06800, Turkey. gozkan@bilkent.edu.tr

Pauliac-Vaujour, E. School of Physics and Astronomy, University of Nottingham, Nottingham NG7 2RD, UK. ppxev@nottingham.ac.uk

Peeters, François M. Departement Fysica, Universiteit Antwerpen, Groenenborgerlaan 171, B-2020 Antwerpen, Belgium. francois.peeters@ua.ac.be

Rubio, Angel Nano Bio Spectroscopy Group and European Theoretical Spectroscopy Facility (ETSF), and Departamento de Fisica de Materiales, Unidad de Materiales Centro Mixto CSIC-UPV/EHU, Universidad del Pas Vasco, Edificio Korta, Avd. Tolosa 72, E-20018 Donostia, Spain. angel.rubio@ehu.es

Sanvito, Stefano School of Physics and CRANN Trinity College, Dublin 2, Ireland. sanvitos@tcd.ie

Segal, Dvira Department of Chemistry, University of Toronto, Toronto, Canada M5S 3H6. dsegal@chem.utoronto.ca

Sergeenkov, Sergei Departamento de Fisica, Universidade Federal da Paraiba Cidade Universitaria, Caixa Postal 5008 58051-970 Joao Pessoa, PB Brazil. sergei@fisica.ufpb.br

Shapiro, Moshe Department of Chemical Physics, Weizmann Institute of Science, Israel, and Department of Chemistry, University of British Columbia, 2036 Main Mall, Vancouver, Canada V6T 1Z1. mshapiro@chem.ubc.ca

Souma, Satofumi Department of Electronics and Electrical Engineering, Kobe University, 1-1 Rokkodai, Nada, Kobe 657-8501, Japan. ssouma@harbor.kobe-u.ac.jp

Stafström, Sven Department of Physics, Chemistry and Biology, IFM Linköping University, SE-581 83 Linköping, Sweden. sst@ifm.liu.se

Stamenova, Maria School of Physics and CRANN, Trinity College, Dublin 2, Ireland. tsonevam@tcd.ie

Stannard, A. School of Physics and Astronomy, University of Nottingham, Nottingham NG7 2RD, UK. ppxas1@nottingham.ac.uk

Stepanow, Sebastian Max-Planck-Institute for Solid State Research, Heisenbergstrasse 1, D-70569 Stuttgart, Germany. s.stepanow@fkf.mpg.de

Šuvakov, Milovan Department of Theoretical Physics, Jožef Stefan Institute, Box 3000, SI-1001 Ljubljana, Slovenia. milovan.suvakov@ijs.si

Tadić, Bosiljka Department of Theoretical Physics, Jožef Stefan Institute, Box 3000, SI-1001 Ljubljana, Slovenia. bosiljka.tadic@ijs.si

Tateno, Masaru Center for Computational Sciences, University of Tsukuba, Tennodai 1-1-1, Tsukuba, Ibaraki 305-8577, Japan. tateno@ccs.tsukuba.ac.jp

Thiele, Uwe School of Mathematics, Loughborough University, Leicestershire LE11 3TU, UK. u.thiele@lboro.ac.uk

Thygesen, Kristian Center for Atomic-scale Materials Design (CAMD), Department of Physics, Technical University of Denmark, DK-2800 Kgs, Lyngby, Denmark. thygesen@fysik.dtu.dk

Tsymbal, Evgeny Nebraska Center for Materials and Nanoscience and Department of Physics and Astronomy, 116 Brace Lab., University of Nebraska-Lincoln, NE, 68588-0111, USA. tsymbal@unl.edu

Unge, Mikael Department of Physics, Chemistry and Biology, IFM, Linköping University, SE-581 83 Linköping, Sweden. unge.mikael@gmail.com

Vancea, Ioan School of Mathematics, Loughborough University, Leicestershire LE11 3TU, UK. I.Vancea@lboro.ac.uk

van Houselt, Arie Physical Aspects of Nanoelectronics and Solid State Physics Groups, MESA+ Institute for Nanotechnology University of Twente, P.O. Box 217, 7500AE, Enschede, The Netherlands. a.vanhouselt@tnw.utwente.nl

van Ruitenbeek, J. M. Kamerlingh Onnes Laboratory, Leiden Institute of Physics, Leiden University, P.O. Box 9506, 2300 RA Leiden, The Netherlands. ruitenbeek@Physics.LeidenUniv.nl

Watanabe, Kazuyuki Department of Physics, Tokyo University of Science, 1-3 Kagurazaka, Shinjuku-ku, Tokyo 162-8601, Japan. kazuyuki@rs.kagu.tus.ac.jp

Watanabe, Satoshi Department of Materials Engineering, The University of Tokyo, Hongo 7-3-1, Bunkyo-ku, Tokyo 113-8656, Japan. watanabe@cello.t.u-tokyo.ac.jp

Winkler, Adolf Institute of Solid State Physics, Graz University of Technology, Petersgasse 16, A-8010 Graz, Austria. a.winkler@tugraz.at

Yamamoto, Takahiro Department of Materials Engineering, The University of Tokyo, Hongo 7-3-1, Bunkyo-ku, Tokyo 113-8656, Japan. takahiro@cello.t.u-tokyo.ac.jp

Yang, Jinlong Hefei National Laboratory for Physical Sciences at the Microscale, University of Science and Technology of China, Hefei, Anhui 230026, P. R. China. jlyang@ustc.edu.cn

Zandvliet, Harold J.W. Physical Aspects of Nanoelectronics and Solid State Physics Groups, MESA+ Institute for Nanotechnology, University of Twente, P.O. Box 217, 7500AE Enschede, The Netherlands. H.J.W.Zandvliet@tnw.utwente.nl

Zârbo, Liviu P. Department of Physics, Texas A&M University, College Station, TX 77843-4242, USA. liviuzarbo@physics.tamu.edu

Zhou, Shuyun Department of Physics, University of California, Berkeley, California, 94720, USA. SZhou@lbl.gov

Nanoelectronic devices: A unified view

Supriyo Datta

1.1   Introduction

Since "everyone" has a computer these days and every computer has nearly a billion field effect transistors (FETs) working in concert, it seems safe to say that the most common electronic device is an FET, which is basically a resistor consisting of an active region called the channel with two very conductive contacts at its two ends called the source and the drain (Fig. 1.1). What makes it more than just a resistor is the fact that a fraction of a volt applied to a third terminal called the gate changes the resistance by several orders of magnitude. Electrical switches like this are at the heart of any computer and what has made computers more and more powerful each year is the increasing number of switches that have been packed into one by making each switch smaller and smaller. For example, a typical FET today has a channel length (L) of $\sim$50 nm, which amounts to a few hundred atoms!

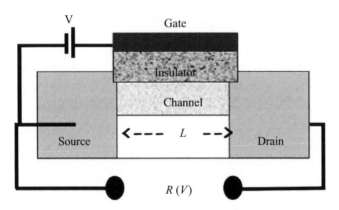

Fig. 1.1 Schematic representing a field effect transistor (FET), which consists of a channel with two contacts (labelled "source" and "drain"), whose resistance R can be controlled through a voltage V applied to a third terminal labeled the "gate", which ideally carries negligible current.

Nanoscale electronic devices have not only enabled miniature switches for computers but are also of great interest for all kinds of applications including energy conversion and sensing. The objective of this chapter, however, is not to discuss specific devices or applications. Rather it is to convey the conceptual

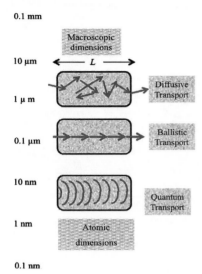

0.1 mm

Macroscopic
dimensions

10 µm

L

Diffusive
Transport

1 µ m

Ballistic
Transport

0.1 µm

10 nm

Quantum
Transport

1 nm

Atomic

dimensions

0.1 nm

Fig. 1.2 As the length "*L*" of the channel in Fig. 1.1 is reduced the nature of electronic transport from one contact to the other changes qualitatively from diffusive to ballistic to quantum.

framework that has emerged over the last twenty years, which is important not only because of the practical insights it provides into the design of nanoscale devices, but also because of the conceptual insights it affords regarding the meaning of resistance and the essence of all non-equilibrium phenomena in general.

This new conceptual framework provides a unified description for all kinds of devices from molecular conductors to carbon nanotubes to silicon transistors covering different transport regimes from the diffusive to the ballistic limit (Fig. 1.2). As the channel length L is reduced, the nature of electronic transport changes qualitatively. With long channels, transport is *diffusive*, meaning that the electron gets from one contact to another via a random walk, but as the channel length is reduced below a mean free path, transport becomes *ballistic*, or "bullet-like". At even shorter lengths the wave nature of electrons can lead to *quantum* effects such as interference and tunnelling. Historically our understanding of electrical resistance and conduction has progressed top-down: from large macroscopic conductors to small atomic-scale conductors. Indeed thirty years ago it was common to argue about what, if anything, the concept of resistance meant on an atomic scale. Since then there has been significant progress in our understanding, spurred by actual experimental measurements made possible by the technology of miniaturization. However, despite this progress in understanding the flow of current on an atomic scale, the standard approach to the problem of electrical conduction continues to be top-down rather than bottom-up. This makes the problem of nanoscale devices appear unduly complicated, as we have argued extensively (Datta 2005, 2008). The purpose of this chapter is to summarize a unified bottom-up viewpoint to the subject of electrical conduction of particular relevance to nanoelectronic devices.

The viewpoint we wish to discuss is summarized in Fig. 1.3(a): Any nano-electronic device has an active "channel" described by a Hamiltonian [H] that also includes any potential U due to other charges, external (on the electrodes) or internal (within the channel). The channel communicates with the source and drain (and any additional contacts) that are maintained in local equilibrium with specified electrochemical potentials. The communication between the channel and the contacts is described by the self-energy matrices $[\Sigma_1]$ and $[\Sigma_2]$ (Caroli *et al.* 1972). Finally, there is a self-energy matrix $[\Sigma_s]$ describing the interaction of an individual electron with its surroundings, which unlike $[\Sigma_{1,2}]$ has to be calculated self-consistently. Each of these quantities is a matrix whose dimension ($N \times N$) depends on the number of basis functions (N) needed to represent the channel. How these matrices are written down varies widely from one material to another and from one approach (semi-empirical or first principles) to another. But once these matrices have been written down, the procedure for calculating the current and other quantities of interest is the same, and in this chapter we will stress this generic procedure along with its conceptual underpinnings.

The schematic model of Fig. 1.3(a) includes both the diffusive (Fig. 1.3(b)) and the ballistic (Fig. 1.3(c)) limits as special cases. In the ballistic limit, the flow of electrons is controlled by the contact terms $[\Sigma_1]$ and $[\Sigma_2]$, while the interactions within the channel are negligible. By contrast, in the diffusive

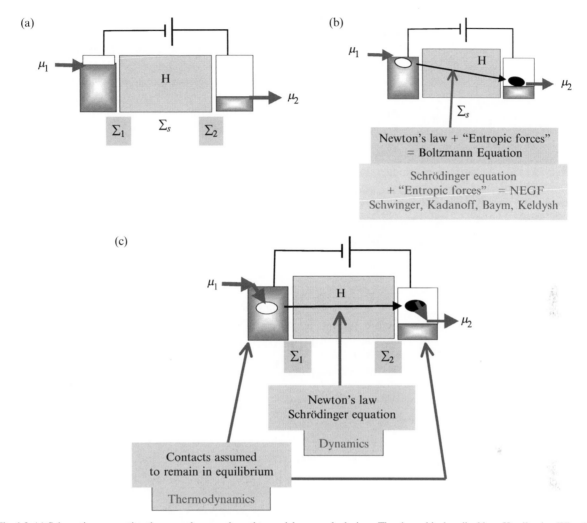

Fig. 1.3 (a) Schematic representing the general approach used to model nanoscale devices: The channel is described by a Hamiltonian [H], while the communication between the channel and the contacts is described by the self-energy matrices $[\Sigma_1]$ and $[\Sigma_2]$. The self-energy matrix $[\Sigma_s]$ describes the interaction of an individual electron with its surroundings. (b) In traditional long devices it is common to ignore the contacts, while (c) in the coherent limit a "Landauer model" neglecting incoherent interactions within the channel is more appropriate (adapted from Datta 2005).

limit, the flow of electrons is controlled by the interactions within the channel described by $[\Sigma_s]$ and the role of contacts ($[\Sigma_1]$ and $[\Sigma_2]$) is negligible. Indeed, prior to 1990, transport theorists seldom bothered even to draw the contacts. Note that there is an important distinction between the Hamiltonian matrix [H] and the self-energy matrices $[\Sigma_{1,2,s}]$. The former is Hermitian representing conservative dynamical forces, while the latter is non-Hermitian and helps account for the "entropic forces". Let me elaborate a little on what I mean by this term.

Consider a simple system like a hydrogen atom having two energy levels separated by an energy $\varepsilon_2 - \varepsilon_1$ that is much larger than the thermal energy $k_B T$ (Fig. 1.4). We all know that an electron initially in the upper level ε_2 will

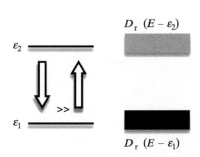

Fig. 1.4 A system with two energy levels $[\varepsilon_1]$ and $[\varepsilon_2]$, coupled to a reservoir whose corresponding density of states are $D_r(E - \varepsilon_1)$ and $D_r(E - \varepsilon_2)$. The downward transition rate from ε_2 to ε_1 far exceeds the upward transition rate ε_1 to ε_2 although the Schrodinger equation would have predicted them to be equal. The unidirectionality arises from entropic forces as discussed in the text.

lose energy, possibly by emitting a photon, and end up in the lower level ε_1, but an electron initially in the lower level ε_1 will stay there forever. Why? This tendency of all systems to relax unidirectionally to the lowest energy is considered so "obvious" that only a novice student would even raise the question. But it is important to recognize that this property does not follow from the Schrödinger equation. Hamiltonians are always Hermitian with $|H_{12}| = |H_{21}|$. Any perturbation that takes a system from ε_2 to ε_1 will also take it from ε_1 to ε_2. The unidirectional transfer from ε_2 to ε_1 is the result of an entropic force that can be understood by noting that our system is in contact with a reservoir having an enormous density of states $D_r(E_r)$ that is a function of the reservoir energy E_r (Feynman 1972). Using E to denote the total energy of the reservoir plus the system, we can write the reservoir density of states as $D_r(E - \varepsilon_1)$ and $D_r(E - \varepsilon_2)$ corresponding to the system energy levels ε_1 and ε_2, respectively (Fig. 1.4).

The ratio of the downward to the upward transition rate is given by

$$\frac{R_{2 \to 1}}{R_{1 \to 2}} = \frac{D_r(E - \varepsilon_1)}{D_r(E - \varepsilon_2)}.$$

Why is the downward rate far greater that the upward rate: $R_{2 \to 1} \gg R_{1 \to 2}$? Simply because for all normal reservoirs, the density of states is an increasing function of the reservoir energy so that with $(E - \varepsilon_1) \gg (E - \varepsilon_2)$, we have $D_r(E - \varepsilon_1) \gg D_r(E - \varepsilon_2)$. We call this an entropic force because the density of states is related to the entropy through the Boltzmann relation $(S = k_B \ln \Omega)$:

$$\frac{D_r(E - \varepsilon_1)}{D_r(E - \varepsilon_2)} = \exp\left(\frac{S(E - \varepsilon_1) - S(E - \varepsilon_2)}{k_B}\right) \approx \exp\left(\frac{\varepsilon_2 - \varepsilon_1}{k_B}\frac{dS}{dE}\right).$$

Noting that the temperature T is defined as $1/T = dS/dE$, we can write

$$\frac{R_{2 \to 1}}{R_{1 \to 2}} = \frac{D_r(E - \varepsilon_1)}{D_r(E - \varepsilon_2)} = \exp\left(\frac{\varepsilon_2 - \varepsilon_1}{k_B T}\right), \tag{1.1}$$

so that with $\varepsilon_2 - \varepsilon_1 \gg k_B T$, $R_{2 \to 1} \gg R_{1 \to 2}$ and the system relaxes to the lower energy, as "everyone" knows.

The point I am trying to make is that the Schrödinger equation alone is not enough even to describe this elementary behavior that we take for granted. Like numerous other phenomena in everyday life, it is driven by entropic forces and not by mechanical forces. Clearly, any description of electronic devices, quantum or classical, must incorporate such entropic forces into the dynamical equations. Over a century ago, Boltzmann showed how to combine entropic forces with Newton's law, and his celebrated equation still stands as the centerpiece in the transport theory of dilute gases, though it was highly controversial in its day and its physical basis still provokes considerable debate (see for example, McQuarrie 1976). The non-equilibrium Green's function (NEGF) formalism we use in this chapter, originating in the work of Martin and Schwinger (1959), Kadanoff and Baym (1962) and Keldysh (1965), can be viewed as the quantum version of the Boltzmann equation: it combines entropic forces with Schrödinger dynamics.

What makes both the Boltzmann and the NEGF formalisms conceptually challenging is the intertwining of dynamical and entropic forces. By contrast, the ballistic limit leads to a relatively simple model with dynamical and entropic processes separated spatially. Electrons zip through from one contact to the other driven purely by dynamical forces. Inside the contacts they find themselves out of equilibrium and are quickly restored to equilibrium by entropic forces, which are easily accounted for simply by legislating that electrons in the contacts are always maintained in local equilibrium. We could call this the "Landauer model" after Rolf Landauer who had proposed it in 1957 as a conceptual tool for understanding the meaning of resistance, long before it was made experimentally relevant by the advent of nanodevices. Today, there is indeed experimental evidence that ballistic resistors can withstand large currents because there is negligible Joule heating inside the channel. Instead the bulk of the heat appears in the contacts, which are large spatial regions capable of dissipating it. I consider this separation of the dynamics from the thermodynamics to be one of the primary reasons that makes a bottom-up viewpoint starting with ballistic devices pedagogically attractive.

Our objective is to present the complete NEGF–Landauer model for nanodevices (Fig. 1.3(a)) that incorporates the contact-related insights into the classic NEGF formalism following Datta (1989,1990) and Meir and Wingreen (1992). I will summarize the complete set of equations (Section 1.2), present illustrative examples (Section 1.3) and conclude with a brief discussion of current research and unanswered questions (Section 1.4). I have written extensively about the NEGF-Landauer model in the past (Datta 1995, 2005, 2008) and will not repeat any of the detailed derivations or discussions. Neither will I attempt to provide a balanced overview of the vast literature on quantum transport. My purpose is simply to convey our particular viewpoint, namely the bottom-up approach to nanoelectronics, which I believe should be of interest to a broad audience interested in the atomistic description of non-equilibrium phenomena.

1.2   The NEGF–Landauer model

Figure 1.5 shows a schematic summarizing the basic inputs that define the NEGF–Landauer model. The channel is described by a Hamiltonian $[H_0]$ while the communication between the channel and the contacts is described by the self-energy matrices $[\Sigma_1]$ and $[\Sigma_2]$. The self-energy $[\Sigma_s]$ and the potential $[U]$ describe the interaction with the surroundings and have to be determined self-consistently as we will explain shortly. Each of these quantities is a matrix whose dimension $(N \times N)$ depends on the number of basis functions (N) needed to represent the channel. $[H_0]$ and $[U]$ are Hermitian, while $[\Sigma_{1,2,s}]$ have anti-Hermitian components

$$\Gamma_{1,2,s} = i\,[\Sigma_{1,2,s} - \Sigma_{1,2,s}{}^{+}].$$

All contacts (Fig. 1.5 shows two, labelled source and drain) are assumed to remain in local equilibrium with electrons distributed according to specified

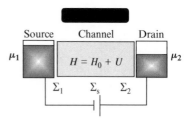

Fig. 1.5 Schematic summarizing the basic inputs that define the NEGF–Landauer model widely used to model nanoscale devices.

Fermi functions

$$f_{1,2}(E) = \frac{1}{1 + \exp\left(\frac{E-\mu_{1,2}}{k_B T_{1,2}}\right)}.$$

Given these inputs, we can calculate any quantity of interest such as the density of states or the electron density or the current using the equations summarized in Section 1.2.1. But first let me briefly mention a simplified version (Fig. 1.6) that can be obtained from the full NEGF–Landauer model with appropriate approximations as described in Section 1.2.2. The inputs to this model are the density of states, $D(E - U)$ that floats up or down according to the local potential U, along with escape rates $\gamma_{1,2,s}$ that are simple numbers representing the same physics as the anti-Hermitian part $[\Gamma_{1,2,s}]$ of the self-energy matrices. Despite the simplifications that limit its applicability, this model has the advantage of illustrating much of the essential physics of nanoelectronic devices (Datta 2005, 2008).

For example, in Section 1.2.2 we obtain the following equation

$$I(E) = \frac{q}{\hbar} \frac{\gamma_1 \gamma_2}{\gamma_1 + \gamma_2} D(E) \left(f_1(E) - f_2(E)\right) \quad \text{(same as eqn(1.8))}$$

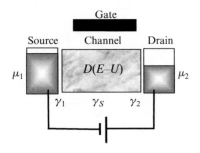

Fig. 1.6 Schematic representing the independent-level model for nanoscale devices that can be viewed as a simple version of the general model of Fig. 1.5 with matrices replaced by ordinary numbers.

for the current per unit energy as a special case of the general matrix equations. However, this equation can be obtained from elementary arguments without invoking any advanced concepts, as I do in an undergraduate course on nanoelectronics that I have developed (see Chapter 1 of Datta 2005). The point I want to make about eqn (1.8) is that it illustrates the basic "force" that drives the flow of current : $f_1(E) - f_2(E)$. Contact 1 tries to fill the states in the channel according to $f_1(E)$, while contact 2 tries to fill them according to $f_2(E)$. As long as $f_1(E) \neq f_2(E)$, one contact keeps pumping in electrons and the other keeps pulling them out, leading to current flow. It is easy to see that this current flow is restricted to states with energies close to the electrochemical potentials of the contacts. For energies E that lie far below μ_1 and μ_2, both $f_1(E)$ and $f_2(E)$ are approximately equal to one and there is no steady-state current flow. Although this conclusion appears obvious, it is not necessarily appreciated widely, since many view the electric field as the driving force, which would act on all electrons regardless of their energy. But the real driving force is the difference between the two Fermi functions, which is sharply peaked at energies close to the electrochemical potentials.

Once we recognize the role of $f_1(E) - f_2(E)$ as *the driving force*, thermoelectric effects are also easily understood. If both contacts have the same electrochemical potential μ, but different temperatures, we have a driving force $f_1(E) - f_2(E)$ that changes sign at $E = \mu$, leading to a thermoelectric current whose sign depends on whether the density of states $D(E)$ is increasing or decreasing around $E = \mu$. The molecular Seebeck effect predicted from this argument (Paulsson and Datta 2003) seems to be in good agreement with recent experimental observations (Reddy *et al.* 2007). This viewpoint also provides a natural explanation for phenomena like the Peltier effect that form the basis for thermoelectric refrigerators (Shakouri 2006). We mentioned earlier that in the Landauer model all the Joule heat is dissipated in the two contacts. But if a conductor has a non-zero density of states only above the electrochemical potentials (Fig. 1.7) then an electron in order to transmit has to first absorb heat from contact 1, thus cooling this contact.

In order for electrons to flow in the direction shown we must have $f_1(E) > f_2(E)$, which requires

$$\frac{E - \mu_1}{T_1} < \frac{E - \mu_2}{T_2}.$$

Noting that $E - \mu_1$ represents the heat removed from contact 1 and $E - \mu_2$ represents the heat released to contact 2, we recognize this as a statement of the Carnot principle.

What I am trying to illustrate here is the clarity with which many key concepts can be understood within the bottom-up approach. In this chapter we will not discuss this version any further. Instead we will focus on the full matrix version (Fig. 1.5).

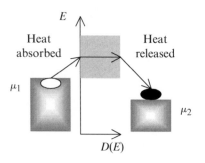

Fig. 1.7 "Peltier effect": If a conductor has a non-zero density of states only above the electrochemical potentials then an electron in order to transmit has to first absorb heat from contact 1, thus cooling this contact.

1.2.1 Summary of equations

A derivation of the basic equations of the NEGF–Landauer method can be found in Datta 2005 both from a one-electron Schrödinger equation (see Chapter 9) and from a second quantized formalism (see Appendix). Here, we will simply summarize the equations without derivation. Our notation is a little different from the standard NEGF literature as explained in Datta (1995, see Chapter 8):

$$G^n \equiv -iG^<, G^p \equiv +iG^>,$$

$$\Sigma^{\text{in}} \equiv -i\Sigma^<, \Sigma^{\text{out}} \equiv +i\Sigma^>,$$

$$G \equiv G^R, G^+ \equiv G^A.$$

In quantum transport we have a matrix corresponding to each quantity of interest from which the desired information can be extracted. For example, we have a *spectral function* whose diagonal elements give us the local density of states (times 2π), an electron and a hole *correlation function* whose diagonal elements give us the electron and hole density per unit energy (times 2π) and a *current operator* $[I^{\text{op}}]$ whose trace gives us the current. The following equations allow us to calculate these quantities.

(1) **Spectral function [A(E)]** is obtained from

$$G(E) = [EI - H_0 - U - \Sigma_1 - \Sigma_2 - \Sigma_s]^{-1} \qquad (1.2a)$$

$$A(E) = i[G - G^+]. \qquad (1.2b)$$

(2) **Electron and hole correlation functions [G^n (E) and G^p (E)]** are obtained from

$$[G^n(E)] = [G\Gamma_1 G^+] f_1 + [G\Gamma_2 G^+] f_2 + [G\Sigma_s^{\text{in}} G^+] \qquad (1.3a)$$

$$[G^p(E)] = [G\Gamma_1 G^+] (1 - f_1) + [G\Gamma_2 G^+] (1 - f_2)$$

$$+ [G\Sigma_s^{\text{out}} G^+]. \qquad (1.3b)$$

It can be shown that $A = G^n + G^p$, as we would expect since the density of states should equal the sum of the electron and hole densities.

(3) **Current operator, I_i at terminal 'i'** per unit energy is obtained from

$$I_i^{\mathrm{op}}(E) = \frac{iq}{h}\left([\Gamma_i G^+ - G\Gamma_i]f_i - [\Sigma_i G^n - G^n\Sigma_i^+]\right).\qquad(1.4\mathrm{a})$$

The charge current per unit energy (to be integrated over all energy for the total current) is obtained from the trace of the current operator:

$$I_i(E) = \frac{q}{h}\left(\mathrm{Trace}\,[\Gamma_i A]f_i - \mathrm{Trace}[\Gamma_i G^n]\right),\qquad(1.4\mathrm{b})$$

while the coherent component of the current can be calculated from the relation

$$I_{\mathrm{coh}}(E) = \frac{q}{h}\,\mathrm{Trace}\,[\Gamma_1 G\Gamma_2 G^+](f_1(E) - f_2(E)),\qquad(1.4\mathrm{c})$$

where the quantity $\bar{T}_{\mathrm{coh}}(E) \equiv \mathrm{Trace}\,[\Gamma_1 G\Gamma_2 G^+]$ is called the "transmission". Equation (1.4c) only gives the *coherent* part of the current while eqn (1.4b) gives us the full current, the coherent plus the incoherent.

Note that the current operator from eqn (1.4a) can be used to calculate other quantities of interest as well. For example, the spin current could be obtained from Trace $[\vec{S}\,I_i^{\mathrm{op}}]$, where $\vec{S}$ is an appropriate matrix representing the spin.

Equations (1.2) through (1.4) involve three quantities $[U]$, $[\Sigma_s]$ and $[\Sigma_s^{\mathrm{in}}]$ that describe the interactions of an individual electron with its surroundings. These quantities are functions of the correlation functions ($[G^n, G^p]$) and have to be calculated *self-consistently*. The actual function we use embodies the physics of the interactions as we will outline below. But let us first neglect these interactions and try to get a physical feeling for eqn (1.2) through (1.4), by applying them to a particularly simple problem.

1.2.2   Independent level model

Equations (1.2) through (1.4) provide a general approach to the problem of quantum transport, with inputs in the form of $(N \times N)$ matrices. The Hamiltonian matrix $[H]$ has N eigenstates and a simple approach is to treat each eigenstate separately and add up the currents as if we have N independent levels in parallel. We call this the "independent level model", which would be precisely correct if the self-energy matrices were also diagonalized by the transformation that diagonalizes $[H]$. This is usually not the case, but the independent level model often provides good insights into the basic physics of nanoscale transport.

Consider a channel with a single energy eigenstate in the energy range of interest. We can use this eigenstate as our basis to write all input parameters as (1×1) matrices or pure numbers:

$$[H] = \varepsilon, [\Sigma_1] = -i\gamma_1/2, [\Sigma_2] = -i\gamma_2/2, [\Gamma_1] = \gamma_1, [\Gamma_2] = \gamma_2.$$

Neglecting all interactions and setting each of the quantities $[U]$, $[\Sigma_s]$ and $[\Sigma_s^{in}]$ to zero, we have from eqn (1.2) for the

$$\text{Green's function } G = \frac{1}{E - \varepsilon + i\,(\gamma_1 + \gamma_2)/2}, \qquad (1.5a)$$

$$\text{and the Spectral function } A = \frac{\gamma}{(E - \varepsilon)^2 + (\gamma/2)^2},$$

$$\text{where } \gamma \equiv \gamma_1 + \gamma_2. \qquad (1.5b)$$

The density of states is equal to $A/2\pi$, showing that the energy level is broadened around the energy level ε. Equation (1.5) gives the occupation of this broadened level

$$\text{Electron correlation function } G^n = \frac{\gamma_1 f_1 + \gamma_2 f_2}{(E - \varepsilon)^2 + (\gamma/2)^2}, \qquad (1.6a)$$

or the lack of occupation thereof

$$\text{Hole correlation function } G^p = \frac{\gamma_1(1 - f_1) + \gamma_2(1 - f_2)}{(E - \varepsilon)^2 + (\gamma/2)^2}. \qquad (1.6b)$$

The electron and hole density per unit energy are given by $G^n/2\pi$ and $G^p/2\pi$, respectively, and as expected, $A = G^n + G^p$.

Finally, the current can be calculated from eqn (1.4b) or eqn (1.4c)

$$I(E) = \frac{q}{h} \frac{\gamma_1 \gamma_2}{(E - \varepsilon)^2 + (\gamma/2)^2} \, (f_1(E) - f_2(E)). \qquad (1.7)$$

Using eqns (1.7) and (1.5b) we can write

$$I(E) = \frac{q}{\hbar} \frac{\gamma_1 \gamma_2}{\gamma_1 + \gamma_2} \, D(E) \, (f_1(E) - f_2(E)), \qquad (1.8)$$

where $D(E) = A(E)/2\pi$ is the broadened density of states associated with the level.

Now, if we superpose the results from N levels we still have exactly the same equation for the current. It is just that $D(E)$ now represents the total density of states rather than just the part associated with a particular level. Indeed one can include a self-consistent potential U into this model simply by letting the density of states float up or down, $D(E-U)$ and this approach (Fig. 1.6) has proved quite successful in providing a simple description of nanoscale transistors (Rahman *et al.* 2003). Elastic and inelastic interactions can also be included straightforwardly into this model (Datta 2008). However, we will not discuss this model further in this chapter. Instead we will focus on the full matrix version.

1.2.3   Self-consistent potential, $[U]$

The potential $[U]$ represents the potential that an individual electron feels due to the other electrons and as such we expect it to depend on the electron density or more generally the correlation function $[G^n]$. In semi-empirical theories the Hamiltonian $[H_0]$ often includes the potential under equilibrium conditions, so that $[U]$ itself should account only for the deviation $[\delta\, G^n]$ from equilibrium. How $[U]$ is related to $[G^n]$ or to $[\delta\, G^n]$ depends on the approximation used, the simplest being the Hartree approximation, which is equivalent to using the Poisson equation or classical electrostatics. More sophisticated approaches using many-body perturbation theory or density-functional theory will include corrections to account for exchange and correlation. We will not go into this any further, except to note that there are examples of devices whose current–voltage characteristics cannot be described within this approach no matter how sophisticated our choice of "U". These devices seem to require models that go beyond the framework described here (see concluding section).

1.2.4   Intrachannel interactions: $[\Sigma_s]$ and $[\Sigma_s^{in}]$

As I mentioned earlier, the classic NEGF formalism like much of the pre-mesoscopic literature on transport theory paid no attention to the contacts except perhaps for tunneling devices, as in Caroli *et al.* (1972). Instead it was focused on the quantities $[\Sigma_s]$ and $[\Sigma_s^{in}]$ and provided systematic prescriptions for writing them down using diagrammatic perturbation theoretic treatment to treat interactions (Danielewicz 1984). In the self-consistent Born approximation (SCBA) we can write for any interaction involving an exchange of energy ε

$$\left[\Sigma_s^{in}(E)\right]_{ij} = D_{ijkl}(\varepsilon)\left[G^n(E-\varepsilon)\right]_{kl} \tag{1.9a}$$

$$\left[\Sigma_s^{out}(E)\right]_{ij} = D_{lkji}(\varepsilon)\left[G^p(E+\varepsilon)\right]_{kl}, \tag{1.9b}$$

where summation over repeated indices is implied. $[\Sigma_s]$ is obtained as follows: Its anti-Hermitian component is given by $\Gamma_s(E) = \Sigma_s^{in}(E) + \Sigma_s^{out}(E)$, while the Hermitian part is obtained by finding its Hilbert transform. Note that we have ordered the indices differently: our D_{ijkl} is conventionally labeled D_{ikjl} (see, for example, Chapter 10 of Datta 2005).

The "scattering current" is given by (cf.eqn (1.4b))

$$I_s(E) = \frac{q}{h}\left(\text{Trace}\,[\Sigma_s^{in}A] - \text{Trace}[\Gamma_s G^n]\right) \tag{1.10a}$$

$$= \frac{q}{h}\left(\text{Trace}\,[\Sigma_s^{in}G^p] - \text{Trace}[\Sigma_s^{out}G^n]\right), \tag{1.10b}$$

and it can be shown that $\sum_i I_i(E) + I_s(E)$ is assured to equal zero at all energies, as required for current conservation. Making use of eqns (1.9a,1.9b) we can write eqn (1.10) in the form

$$I_s(E) = \frac{q}{h}\sum_{i,j,k,l} D_{ijkl}(\varepsilon)\,G_{kl}^n(E-\varepsilon)\,G_{ji}^p(E) - D_{lkji}(\varepsilon)\,G_{kl}^p(E+\varepsilon)\,G_{ji}^n(E),$$

$$\tag{1.10c}$$

which can be integrated to show that $\int dE\, I_s(E) = 0$, as we would expect since there is no net exchange of electrons with the scatterers. However, $\int dE\, E\, I_s(E) \neq 0$, indicating the possibility of energy exchange. This equation can be understood in semi-classical terms if we assume that the electron and hole matrices are both purely *diagonal*:

$$I_s(E) \rightarrow \frac{q}{h} \sum_{i,k} D_{iikk}(\varepsilon) \, G_{kk}^n(E - \varepsilon) \, G_{ii}^p(E) - D_{kkii}(\varepsilon) \, G_{kk}^p(E + \varepsilon) \, G_{ii}^n(E).$$

This is essentially the standard scattering term in the Boltzmann equation if we associate the D tensor with the scattering probabilities: $D_{iikk}(\varepsilon) \rightarrow S_{ik}(\varepsilon)$. We know from the Boltzmann treatment that if the entity (like phonons) with which the electrons interact is in equilibrium with temperature T_s, then in order to comply with the laws of thermodynamics, we must have $S_{ik}(\varepsilon) = S_{ki}(-\varepsilon) \, \exp(-\varepsilon/k_B T_s)$. The corresponding relation in quantum transport

$$D_{ijkl}(\varepsilon) = D_{lkji}(-\varepsilon) \, \exp(-\varepsilon/k_B T_s) \tag{1.11}$$

is more subtle and less appreciated. Note, however, that neither the semi-classical nor the quantum restriction is operative, if the interacting entity is not in equilibrium.

If we assume *elastic interactions ($\varepsilon = 0$), along with the equilibrium condition (eqn (1.11))*, then we can write

$$\left[\Sigma_s^{in}(E)\right]_{ij} = D_{ijkl} \left[G^n(E)\right]_{kl} \text{ and } \left[\Sigma_s^{out}(E)\right]_{ij} = D_{ijkl} \left[G^p(E)\right]_{kl}, \tag{1.12}$$

so that $[\Gamma_s(E)]_{ij} = D_{ijkl} \, [A(E)]_{kl}$, and $[\Sigma_s]$ can be related directly to $[G]$:

$$[\Sigma_s]_{ij} = D_{ijkl} \, [G]_{kl}. \tag{1.13}$$

This simplifies the calculation by decoupling eqn (1.2) from (1.3) but it is important to note that eqns (1.12) and (1.13) are valid only for elastic interactions with scatterers that are in equilibrium.

As mentioned above, the NEGF formalism provides clear prescriptions for calculating the tensor $[[D]]$ starting from any given microscopic interaction Hamiltonian. Alternatively, we have advocated a phenomenological approach whereby specific choices of the form of the tensor $[[D]]$ give rise selectively to phase, momentum or spin relaxation and their magnitudes can be adjusted to obtain desired relaxation lengths for these quantities as obtained from experiment. For example, the following choice

$$D_{ijkl} = d_p \, \delta_{ik}\delta_{jl}, \tag{1.14a}$$

d_p being a constant, leads to pure phase relaxation. This is equivalent to writing $[\Sigma_s]$ and $[\Sigma_s^{in}]$ as a constant times $[G]$ and $[G^n]$, respectively:

$$[\Sigma_s]_{ij} = d_p \, [G]_{ij} \text{ and } \left[\Sigma_s^{in}\right]_{ij} = d_p \, [G^n]_{ij}. \tag{1.14b}$$

I will present a concrete example showing that this choice of the tensor $[[D]]$ indeed relaxes phase without relaxing momentum. But one can see the reason intuitively by noting that the SCBA (eqn (1.9)) effectively takes electrons out of the channel and feeds them back with a randomized phase similar in concept to the Buttiker probes widely used in mesoscopic physics (Datta 1989; Hershfield 1991). A constant multiplier as shown in eqn (1.14b) suggests that the electrons are fed back *while preserving the initial correlation function exactly*. We thus expect no property of the electrons to be relaxed except for phase.

Another choice

$$D_{ijkl} = d_m \, \delta_{ij} \, \delta_{ik} \, \delta_{jl} \tag{1.15a}$$

that we will illustrate is equivalent to writing

$$[\Sigma_s]_{ij} = d_m \, \delta_{ij} \, [G]_{ij} \text{ and } \left[\Sigma_s^{in}\right]_{ij} = d_p \delta_{ij} \, [G^n]_{ij}. \tag{1.15b}$$

Unlike the phase relaxing choice (eqns (1.23)), this choice feeds back only the diagonal elements. In a real-space representation this leads to momentum relaxation in addition to phase relaxation, as we will see in Section 1.3.

A choice that leads to pure spin relaxation is

$$D_{abcd} = d_s \, \vec{\sigma}_{ac} \bullet \vec{\sigma}_{db}, \tag{1.16a}$$

where we have used a separate set of indices (a,b,c,d instead of i,j,k,l) to indicate that these are spin indices. The tensor has the same form as that for pure phase-relaxing interactions (eqn (1.23)) as far as the indices other than spin are concerned. Here, $\vec{\sigma}$ denotes the Pauli spin matrices and eqn (1.16a) is equivalent to writing

$$[\Sigma_s] = d_s \, ([\sigma_x][G][\sigma_x] + [\sigma_y][G][\sigma_y] + [\sigma_z][G][\sigma_z]) \quad \text{and}$$

$$\left[\Sigma_s^{in}\right] = d_s \, ([\sigma_x][G^n][\sigma_x] + [\sigma_y][G^n][\sigma_y] + [\sigma_z][G^n][\sigma_z]). \tag{1.16b}$$

It is straightforward to show that Trace $[\Sigma_s^{in}\vec{\sigma}] = - \, d_s$ Trace $[G^n\vec{\sigma}]$, indicating that this choice for the tensor $[[D]]$ feeds back a spin equal to $- \, d_s$ times the original spin, thus leading to spin relaxation.

In the next section we present a few examples to give the reader a flavor of how these equations are applied. More examples, especially those involving spin are discussed in Chapter 3 (Golizadeh-Mojarad *et al.*) and Chapter 24 (Nikolic *et al.*) in this volume.

1.3   A few examples

1.3.1   Single-moded channel

Consider first a one-dimensional single-band tight-binding model with a nearest-neighbor Hamiltonian of the form

$$\begin{bmatrix} \varepsilon & -t & 0 & 0 & \cdots \\ -t & \varepsilon & -t & 0 & \cdots \\ 0 & -t & \varepsilon & -t & 0 & \cdots \end{bmatrix}, \tag{1.17}$$

which can be represented schematically as shown in Fig. 1.8. In principle, the Hamiltonian should also include the potential due to any external voltages applied to the electrodes, but for our examples we will neglect it assuming it to be small. We will also ignore the self-consistent potential $[U]$.

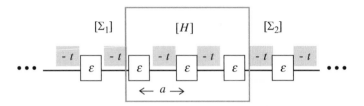

Fig. 1.8 One-dimensional single-band tight-binding model with site energy ε and nearest-neighbor overlap "$-t$" having a dispersion relation of the form $E = \varepsilon - 2t\cos ka$, a being the nearest-neighbor distance.

Let us treat just one site as the channel $[H_0] = \varepsilon$ and the rest of the semi-infinite wire on either side as self-energies that are given by (Caroli *et al.* 1972)

$$[\Sigma_1] = -t\exp^{ika} \text{ and } [\Sigma_2] = -t\exp^{ika},$$

so that $[\Gamma_1] = 2t\sin ka$ and $[\Gamma_2] = 2t\sin ka$, where "ka" is related to the energy by the dispersion relation $E = \varepsilon - 2t\cos ka$.

From eqn (1.2),

$$[G] = \frac{1}{E - \varepsilon + 2t\exp^{ika}} = \frac{-i}{2t\sin ka}.$$

From eqn (1.4),

$$I(E) = (q/h)\,(f_1(E) - f_2(E)), \tag{1.18}$$

as long as $\varepsilon - 2t < E < \varepsilon + 2t$. Outside this energy range, "ka" is imaginary, making $[\Sigma_1]$ and $[\Sigma_2]$ purely real and hence $[\Gamma_1] = [\Gamma_2] = 0$.

From eqn (1.18) we obtain for the total current

$$I = (q/h)\int dE\,(f_1(E) - f_2(E)) = (q/h)\,(\mu_1 - \mu_2).$$

Since $\mu_1 - \mu_2 = qV$ this shows that a one-dimensional ballistic wire has a conductance equal to the quantum of conductance: $I/V = q^2/h$.

Note that the single-band tight-binding Hamiltonian in eqn (1.17) can alternatively be viewed as a discrete version of a one-dimensional effective mass Hamiltonian of the form $-\frac{\hbar^2}{2m}\frac{\partial^2}{\partial x^2}$, if we set $t = \hbar^2/2ma^2$, $\varepsilon = 2t$. Any potential $U(x)$ can be included in eqn (1.17) by adding $U(x = x_i)$ to each diagonal element (i, i). The continuum version has a dispersion relation $E = \hbar^2 k^2/2m$, while the discrete version has a dispersion relation $E = 2t(1 - \cos ka)$. The two agree reasonably well for $ka < \pi/3$, with energies in the range $0 < E < t$.

1.3.2   Conductance quantization

Figure 1.9 shows the transmission versus energy calculated for a rectangular conductor of width 102 nm using the model described below. Note the discrete integer steps in the transmission as the energy increases and new subbands or transverse modes come into play. The discrete integer values for the transmission lead to low bias conductance values that are approximate integer multiples of the conductance quantum. This quantization of the conductance in multimoded wires, discovered experimentally in 1988 (van Wees *et al.* 1988;

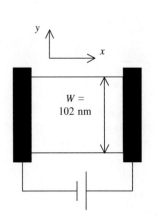

Energy (eV)

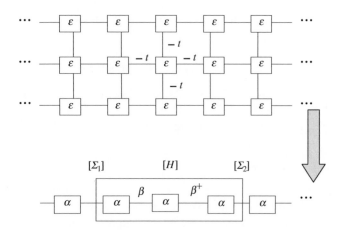

Fig. 1.9 Transmission versus energy for a rectangular conductor of width 102 nm modelled with a single-band tight-binding model with $t = \hbar^2/2ma^2 \approx 0.04\,\text{eV}$, $\varepsilon = 4t$, m = $0.25 \times$ free electron mass, $a = 2\,\text{nm}$.

Fig. 1.10 Single-band tight-binding model on a square lattice with site energy ε and nearest-neighbor overlap "$-t$" having a dispersion relation of the form $E = \varepsilon - 2t \cos k_x a - 2t \cos k_y a$, a being the nearest-neighbor distance. Conceptually, we can lump each column into a single matrix α, with neighboring columns coupled by a matrix β to the left and β^+ to the right.

Wharam *et al.* 1988) serves as a good benchmark for any theory of quantum transport.

1.3.2.1 *Model details*

The rectangular conductor is modelled with a single-band tight-binding model with $t = \hbar^2/2ma^2 \approx 0.04\,\text{eV}$, $\varepsilon = 4t$, $m = 0.25 \times$ free electron mass and $a = 2\,\text{nm}$ (Fig. 1.10). Conceptually, we can lump each column of the square lattice into a single matrix α, which is essentially the one-dimensional Hamiltonian from the last section, (eqn (1.17)). Neighboring columns are coupled by a matrix β to the left and β^+ to the right. In this example, $\beta = \beta^+ = -t\,[I]$, $[I]$ being the identity matrix, but in general β need not equal β^+.

The overall Hamiltonian is written as

$$\begin{bmatrix} \alpha & \beta^+ & 0 & 0 & \cdots \\ \beta & \alpha & \beta^+ & 0 & \cdots \\ 0 & \beta & \alpha & \beta^+ & 0 & \cdots \end{bmatrix}. \tag{1.19}$$

The contact self-energies are given by $\Sigma_1 = \beta g_1 \beta^+$ and $\Sigma_2 = \beta^+ g_2 \beta$, where g_1 and g_2 are the surface Green's functions for the left and right contacts, respectively (they are the same in this example, but need not be in general). These surface Green's functions can be obtained by solving the matrix quadratic equations

$$[g_1]^{-1} = \alpha - \beta g_1 \beta^+ \text{ and } [g_2]^{-1} = \alpha - \beta^+ g_2 \beta. \qquad (1.20)$$

These can be solved iteratively in a straightforward manner but this can be time consuming for wide conductors and special algorithms may be desirable. If the matrices α and β can be simultaneously diagonalized then a faster approach is to use this diagonal basis to write down the solutions to eqn (1.20) and then transform back. In this basis the multimoded wire decouples into separate single-moded wires. However, this simple decoupling is not always possible since the same unitary transformation may not diagonalize both α and β.

1.3.3   Ballistic Hall effect

Figure 1.11 shows another interesting result, namely the Hall resistance normalized to the resistance quantum (h/e^2) as a function of the magnetic field (applied along the z-direction) calculated for a rectangular conductor of width $W = 102$ nm. Note the plateaus in the Hall resistance equal to the inverse of integers 2, 3, 4, etc. representing the quantum Hall effect. This calculation is done using essentially the same model as in the last example, but there are two additional points that need clarification.

The first point is that the magnetic field $\vec{B} = B \hat{z}$ enters the Hamiltonian through the phase of the nearest-neighbor coupling elements as shown in Fig. 1.12. The second point is the concept of a *local electrochemical potential* that we have used to obtain the Hall voltage. Our calculations are done at a single electron energy E and at this energy we assume the Fermi functions $f_1(E)$ and $f_2(E)$ to equal one and zero respectively. At all points "i" within

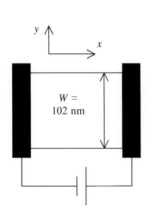

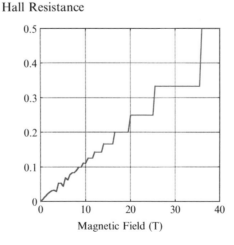

Normalized
Hall Resistance

Magnetic Field (T)

Fig. 1.11 Hall resistance (= Hall voltage/current) normalized to the resistance quantum (h/e^2) versus magnetic field (applied along the z-direction) calculated for a rectangular conductor of width $W = 102$ nm. Note the plateaus in the Hall resistance equal to the inverse of integers 2, 3, 4, etc. representing the quantum Hall effect. Electron energy = $t \sim 0.04$ eV.

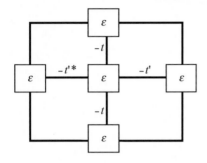

Fig. 1.12 The magnetic field $\vec{B} = B\hat{z}$ represented through a vector potential $\vec{A} = -By\hat{x}$, appears in the single-band tight-binding model in the phase of the coupling elements along $x : t' = t\ \exp(+iBya)$

the channel, the occupation lies between 0 and 1, and it is this occupation that we call the local electrochemical potential and estimate it from the ratio of the local electron density to the local density of states (McLennan *et al.* 1991):

$$\mu(i) = G^n(i, i)/A(i, i). \tag{1.21}$$

Figure 1.13 shows a plot of this local electrochemical potential μ across the width of the conductor. At zero magnetic field, μ is constant ($= 0.5$) and develops a slope as the field is increased. The oscillations arise from coherent interference effects that usually get washed out when we sum over energies or include phase-relaxation processes. Here, we have estimated the Hall voltage simply by looking at the difference between μ at the two edges of the conductor and the Hall resistance in Fig. 1.11 is obtained by dividing this transverse Hall voltage by the current.

1.3.4   "Potential" drop across a single-moded channel

An instructive example to look at is the variation of the electrochemical potential (defined by eqn (1.21)) across a scatterer in a single-moded wire modelled with a tight-binding model as described in Section 1.3.1. As expected, the potential drops sharply across the scatterer (Fig. 1.14), but a purely coherent calculation usually yields oscillations arising from interference effects (see Fig. 1.14(a)). Such oscillations are usually muted if not washed away in room-temperature measurements, because of strong phase relaxation. Much of the phase relaxation arises from electron–electron interactions, which to first order do not give rise to any momentum relaxation. Such processes could be included by including an interaction self-energy of the form shown in eqn (1.23) and indeed it suppresses the oscillations (Fig. 1.14(b)). The momentum relaxing

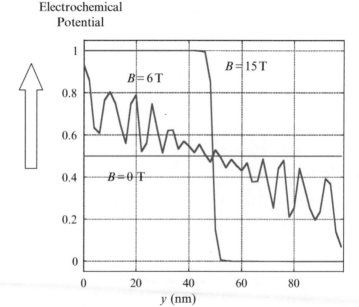

Fig. 1.13 Profile of the local electrochemical potential (defined in eqn (1.21)) across the width of the conductor at three different values of the magnetic field. Electron energy $= t \sim 0.04$ eV.

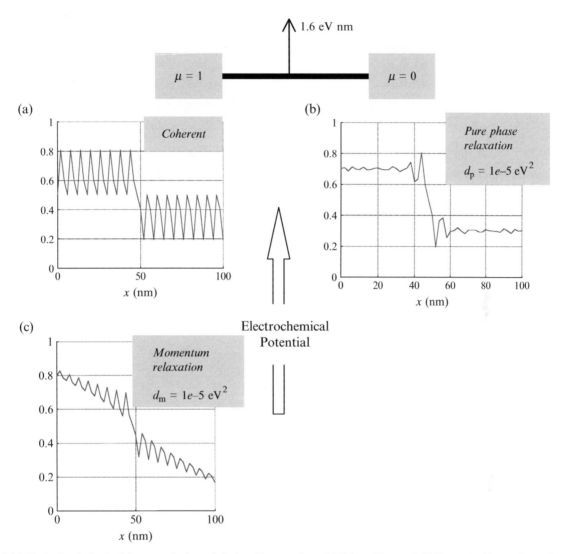

Fig. 1.14 Electrochemical potential across a single-moded wire with one scatterer. (a) Coherent transport, (b) Transport with pure phase relaxation, (c) Transport with momentum relaxation. Electron energy $= t \sim 0.04\,\text{eV}$.

interaction shown in eqn (1.15) also suppresses oscillations, but it leads to an additional slope across the structure (Fig. 1.14(c)) as we would expect for a distributed resistance.

1.3.5 "Potential" drop across a single-moded channel including spin

Another interesting example is the variation of the electrochemical potential for the up-spin and down-spin channels across a single-moded wire connected to antiparallel ferromagnetic contacts assumed to have a coupling to the majority spin that is $(1 + P)/(1 - P)$ times the coupling to the minority

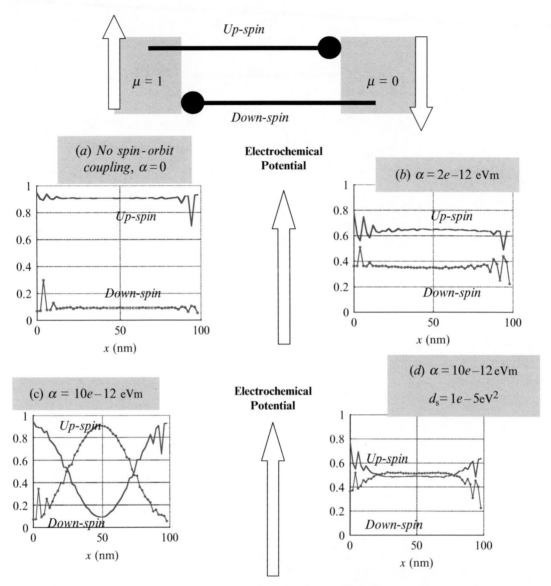

Fig. 1.15 Electrochemical potential for the up-spin and down-spin channels across a single-moded wire connected to antiparallel ferromagnetic contacts assumed to have a coupling to the majority spin that is $(1 + P)/(1 - P)$ times the coupling to the minority spin ($P = 0.95$): Ballistic conductor with (a) weak spin-orbit coupling, (b) weak spin-orbit coupling, (c) strong spin-orbit coupling and finally (d) a conductor with spin relaxation in addition to strong spin-orbit coupling. All calculations include pure phase relaxation ($d_p = 1e - 5\,\mathrm{eV}^2$), which reduce oscillations due to multiple spin-independent reflections.

spin ($P = 0.95$). The up-spin channel is strongly coupled to the contact with $\mu = 1$ and weakly coupled to the contact with $\mu = 0$, with the roles reversed for the down-spin channel. Consequently, the electrochemical potential for the up-spin channel is closer to 1, while that for the down-spin channel is closer to 0 (Fig. 1.15(a)). The difference is reduced when we introduce a little spin-orbit coupling (Fig. 1.15(b)), but with strong spin-orbit coupling the potential actually oscillates back and forth. This oscillation is the basis for many "spin transistor" proposals (for a recent review see Bandyopadhyay and Cahay 2008), but it should be noted that we are assuming a contact efficiency (95%) that is considerably better than the best currently available. Also, our calculations include pure phase relaxation ($d_\mathrm{p} = 1e - 5\,\mathrm{eV}^2$) to account for electron–electron interactions. These processes reduce any oscillations due to multiple spin-independent reflections. Finally Fig. 1.15(d) shows the effect of spin-relaxing processes (eqn (1.16)) in equalizing up-spin and down-spin electrochemical potentials.

1.3.5.1   *Model details*

A brief explanation of how we include spin-orbit coupling into the single-band tight-binding or effective mass equation described in Section 1.3.2.1. Conceptually each "grid point" effectively becomes two grid points when we include spin explicitly and so the site energy becomes $\varepsilon[I]$, $[I]$ being a (2×2) identity matrix and the nearest-neighbor coupling elements become $-t[I]$. Spin-orbit coupling modifies these coupling elements as shown in Fig. 1.16, which add to the usual $-t[I]$ (not shown). It is straightforward to show that this Hamiltonian leads to a dispersion relation

$$E = (\varepsilon - 2t\ \cos ka)[I] + \frac{\alpha}{a}\ ([\sigma_x]\ \sin k_y a - [\sigma_y]\ \sin k_x a), \qquad (1.22a)$$

which for small "ka" reduces to the effective mass-Rashba Hamiltonian (Bychkov and Rashba 1984):

$$E = \frac{\hbar^2 k^2}{2m}[I]\ +\ \alpha\,([\sigma_x]\,k_y - [\sigma_y]\,k_x). \qquad (1.22b)$$

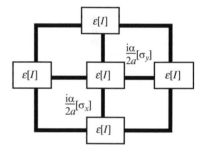

Fig. 1.16 Rashba spin-orbit Hamiltonian on a discrete lattice.

1.4   Concluding remarks

A central point that distinguishes our approach based on the NEGF–Landauer method is the explicit acknowledgement of the important role played by the *contacts*, a role that was highlighted by the rise of mesoscopic physics in the late 1980s. Indeed we are arguing for a bottom-up approach to electronic devices that starts from the coherent or Landauer limit where there is a clear separation between the role of the channel and the contact. The channel is governed by purely dynamical forces, while the contacts are held in local equilibrium by entropic forces. This separation provides a conceptual clarity that makes it very attractive pedagogically. In general, dynamic and entropic processes are intertwined and even the channel experiences entropic forces like the contacts, as long as it has degrees of freedom such as phonons that can be excited. One could say that contacts are not just the physical ones at the ends of the conductor described by $[\Sigma_{1,2}]$. Abstract contacts of all kinds described by $[\Sigma_\mathrm{s}]$ are distributed throughout the channel.

Usually all these contacts are assumed to be held in equilibrium by entropic forces. In practice, it is not uncommon for contacts, especially "nanocontacts," to be driven out-of-equilibrium. This is true of physical contacts made to nanotransistor channels, as well as abstract contacts like the non-itinerant electrons in nanomagnets driven by spin-torque forces or the nuclear spins in semiconductors driven by the Overhauser effect. Such out-of-equilibrium "contacts" can be included straightforwardly into the model we have described by coupling the NEGF–Landauer model to a dynamic equation describing the out-of-equilibrium entity, like the Bloch equation for isolated spins or the Landau–Lifshitz–Gilbert (LLG) equation for nanomagnets (see for example, Salahuddin and Datta 2006 and references therein).

The real conceptual problem arises when we allow for the possibility of correlations or entanglement. This can be understood from a simple example. Consider a channel with just two spin-degenerate levels (Fig. 1.17) biased such that contact 1 wants to fill both levels and contact 2 wants to empty them. If both contacts are equally coupled, we would expect each level to be half-filled:

$$f_{up} = 0.5 \quad \text{and} \quad f_{dn} = 0.5.$$

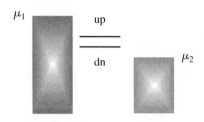

Fig. 1.17 A channel with two spin-degenerate levels biased so that contact 1 wants to fill each level and contact 2 wants to empty them both. Assuming both contacts to be equally coupled to the channel, we would expect each state to be half-filled at steady state.

This is exactly what we would get if we applied the methods discussed in this chapter to this simple problem.

Now, if we ask for the probability that the up-spin level is filled and the down-spin level is empty $P(10)$ we can write it as $f_{up}(1 - f_{dn})$. We can write the probabilities of all four possibilities as

$$P(00) = (1 - f_{up})(1 - f_{dn}), \quad P(01) = (1 - f_{up}) f_{dn}$$
$$P(10) = f_{up}(1 - f_{dn}), \quad P(11) = f_{up} f_{dn}. \tag{1.23}$$

In this case, this yields $P(00) = P(01) = P(10) = P(11) = 1/4$.

However, if the electrons are strongly interacting then the energy cost of occupying both levels can be so high that the state (11) has zero probability. Indeed, it can be shown that under these conditions $P(00) = P(01) = P(10) = 1/3$ and $P(11) = 0$.

The point I want to make is that there is no possible choice of f_{up} and f_{dn} that when inserted into eqn (1.23) will lead to this result! Since $P(11) = 0$ we must have either f_{up} or f_{dn} equal to zero, so that $P(01)$ or $P(10)$ would have to be zero. There is no way to obtain non-zero values for both $P(01)$ and $P(10)$, while making $P(11)$ equal zero.

This is an example of a "strong correlation" where the dynamics of individual electrons is so correlated by their interaction that it is inaccurate to view each electron as moving in a mean field due to the other electrons. This "Coulomb blockade" regime has been widely discussed (see for example, Likharev 1999, Beenakker 1991, Braun *et al.* 2004, Braig and Brouwer 2005) and it can have an important effect on the current–voltage characteristics of molecular scale conductors (Muralidharan 2006) if the single-electron charging energy is well in excess of the broadening as well as the thermal energy.

My purpose, however, is not to talk about Coulomb blockade in particular. I use this example simply to illustrate the meaning of correlation and the conceptual issues it raises. One can no longer "disentangle" different electrons. Instead one has to solve a multi-electron problem and a complete transport theory is not yet available in such a multiparticle framework. This is true not just for correlated electrons, but for electrons correlated to other entities such as nuclear spins as well. Any interaction generates correlations, but the standard approach in transport theory is to neglect them following the example of Boltzmann who ignored them through his assumption of "molecular chaos" or "Stohsslansatz", leading to the increase of entropy characteristic of irreversible processes. Exactly how such multiparticle correlations are destroyed will hopefully become clearer as more delicate experiments are conducted leading to the next level of understanding in transport theory involving "correlated contacts". In the meantime there are many electronic devices for switching, energy conversion and sensing that can be analyzed and designed (see for example, Lake *et al.* 1997 and Koswatta *et al.* 2007) using the conceptual framework that has emerged in the last twenty years, starting from the Boltz-mann (semi-classical dynamics) or the NEGF description (quantum dynamics) appropriate for weak interactions, and extending them to include the contact-related insights from the Landauer model. The distinguishing feature of this framework is the explicit acknowledgement of contacts, leading naturally to a bottom-up approach, which we believe can be very powerful both for teaching and research.

This work was supported by the NSF-sponsored Network for Computational Nanotechnology (NCN) and the Intel foundation, see *Electronics from the Bottom Up: A New Approach to Nanoelectronic Devices and Materials* (2008).

References

This is a very limited set of references directly related to the viewpoint and discussion in this chapter. It is by no means comprehensive or even representative of the vast literature on quantum transport.

Bandyopadhyay, S., Cahay, M. *Introduction to Spintronics* (Taylor & Francis, 2008).

Beenakker, C.W. *J. Phys. Rev. B* **44**, 1646 (1991).

Braig, S., Brouwer, P.W. *Phys. Rev. B* **71**, 195324 (2005).

Braun, M., Koenig, J., Martinek, J. *Phys. Rev. B* **70**, 195345 (2004).

Bychkov, Y.A., Rashba, E.I. *J. Phys. C* **17**, 6039 (1984).

Caroli, C., Combescot, R., Nozieres, P., Saint-James, D. *J. Phys. C: Solid State Phys.* **5**, 21 (1972).

Danielewicz, P. *Ann. Phys.* NY, **152**, 239 (1984).

Datta, S. *Phys. Rev. B* **40**, 5830 (1989).

Datta, S. *J. Phys.: Condens. Matter* **2**, 8023 (1990).

Datta, S. *Electronic Transport in Mesoscopic Systems* (Cambridge University Press, 1995).

Datta, S. *Quantum Transport: Atom to Transistor* (Cambridge University Press, 2005).

Datta, S. *"Nanodevices and Maxwell's Demon", Lecture Notes in Nanoscale Science and Technology.* Vol. 2, *Nanoscale Phenomena: Basic Science to Device Applications.* ed. Z.K. Tang and P. Sheng, (Springer, arXiv:condmat0704.1623 2008).

Electronics from the Bottom Up: A New Approach to Nanoelectronic Devices and Materials. http:nanohub.org/topics/ElectronicsFromTheBottomUp (2008).

Feynman, R.P. *Statistical Mechanics. Frontiers in Physics* (Addison-Wesley 1972).

Hershfield, S. *Phs. Rev. B* **43**, 11586 (1991).

Kadanoff and Bay, *Quantum Statistical Mechanics. Frontiers in Physics Lecture Notes* (Benjamin/Cummings, 1962).

Keldysh, L.V. *Sov. Phys. JETP* **20**, 1018 (1965).

Koswatta, S.O., Hasan, S., Lundstrom, M.S., Anantram, M.P., Nikonov. Dmitri P. *IEEE Trans. Electron Devices* **54**, 2339 (2007).

Lake, R., Klimeck, G., Bowen, R.C., Jovanovich, D. *J. Appl. Phys.* **81**, 7845 (1997).

Likharev, K. *Proc. IEEE* **87**, 606 (1999).

Martin, P.C., Schwinger, J. *Phys. Rev.* **115**, 1342 (1959).

McLennan, M.J., Lee, Y., Datta, S. *Phys. Rev. B* **43**, 13846 (1991).

McQuarrie, D.A. *Statistical Mechanics* (Harper and Row, 1976).

Meir, Y., Wingreen, N.S. *Phys. Rev. Lett.* **68**, 2512 (1992).

Muralidharan, B., Ghosh, A.W., Datta, S. *Phys. Rev. B* **73**, 155410 (2006).

Paulsson, M., Datta, S. *Phys. Rev. B* **67**, 241403(R) (2003).

Rahman, A., Guo, J., Datta, S., Lundstrom, M.S. *IEEE Trans. Electron Devices* **50**, 1853 (2003).

Reddy, P., Jang, S.Y., Segalman, R., Majumdar A. *Science* **315**, 1568 (2007).

Salahuddin, S., Datta, S. *Appl. Phys. Lett.* **89**, 153504 (2006).

Shakouri, Ali. *Proc. IEEE* **94**, 1613 (2006).

van Wees, B.J., van Houten, H., Beenakker, C.W.J., Williamson, J.G., Kouwenhoven, L.P., van der Marel, D., Foxon, C.T. *Phys. Rev. Lett.* **60**, 848 (1988).

Wharam, D.A., Thornton, T.J., Newbury, R., Pepper, M., Ahmed, H., Frost, J.E.F., Hasko, D.G., Peacock, D.C., Ritchie, D.A., Jones, G.A.C. *J. Phys. C* **21**, L209 (1988).

Electronic and transport properties of doped silicon nanowires

2

M.-V. Fernández-Serra and X. Blase

2.1 Introduction

The ability of the semiconductor industry on scaling down the size of electronic components and cramming more and more transistors in silicon chips is coming to an end. Throughout the past years it was still possible to follow Moore's Law, but we have started to reach the limits of the lithographic process, a top-down technology. The alternative, a bottom-up approach is based on the assembly of well-defined nanoscale building blocks to achieve functional devices. Silicon nanowires (SiNWs), together with carbon nanotubes (CNTs), belong to the most promising nanostructures to become the fundamental nanocomponents of these bottom-up-based devices. The synthesis of semiconducting nanowires has renewed the interest that the discovery of carbon nanotubes generated. Due to their compatibility with silicon-based technology, silicon nanowires are being extensively studied and numerous experiments have already characterized some of their structural (Menon *et al.* 2004; Wu *et al.* 2004) and electronic (Cui *et al.* 2000; Yu *et al.* 2000; Cui and Lieber 2001) properties. Currently synthesized SiNWs, used in devices such as field effect transistors (Cui *et al.* 2003) (FET), are passivated by an oxide layer or by hydrogen (Ma *et al.* 2003) and their conduction properties are modified by the type and concentration of the dopants (Cui *et al.* 2000; Ma *et al.* 2001; Lew *et al.* 2004).

Theoretical studies have been undertaken to characterize the electronic structure of SiNWs under a large number of different structural considerations. A fundamental interest related to wire-specific bulk or surface reconstructions explains why most studies to date have focused on the properties of undoped and unpassivated wires (Ismail-Beigi and Arias 1998; Zhao and Yakobson 2003; Kagimura *et al.* 2005; Ponomareva *et al.* 2005; Rurali and Lorente 2005; Singh *et al.* 2005). Important surface (Rurali and Lorente 2005) and core

(Kagimura *et al.* 2005) reconstructions have been observed, leading to novel properties such as an intrinsic metallic or semi-metallic character mediated by dispersive surface states (Kobayashi 2004; Rurali and Lorente 2005).

Until very recent theoretical studies have lagged behind experimental realizations, limited on one side by the availability of microscopic information to build realistic model systems (Ismail-Beigi and Arias 1998; Ponomareva *et al.* 2005; Rurali and Lorente 2005), and on the other by the reduced implementation of algorithmic and methodological procedures to study electronic transport in state-of-the-art *ab-initio* methods. However, the extraordinary theoretical advances in the area of electronic transport in low-dimensional systems (Calzolari *et al.* 2004; Rocha *et al.* 2006) together with the availability of much more powerful supercomputers have inverted this tendency, and more and more experimental studies are relying on existing theoretical calculations to interpret their results (Iwanari *et al.* 2007).

In this chapter we will review a number of theoretical works and methods, dedicated to the analysis of the structure, both atomic and electronic, doping properties and transport characteristic of SiNWs. The aim of the work is to present a perspective of our current understanding on the physical properties of these quasi-one-dimensional systems from a theoretical point of view, showing how quantum-confinement effects and dimensionality effects can intrinsically change the behavior of SiNWs as compared to their bulk and thin film counterparts.

The structure of the chapter is as follows: Section 2.2 will review work done on surface reconstructions and electronic structure of SiNWs as a function of the system doping and passivation. Section 2.3 will analyze the problem of doping in SiNWs. In Section 2.5 we will present the methodology commonly used to analyze the problems of transport, and will show transport characteristics of SiNWs as a function of dopant type, and structure in Section 2.6. We will provide in Section 2.6 an introduction to the chemical functionalization of SiNWs, and conclusions in Section 2.7.

2.2   Electronic structure of silicon nanowires

2.2.1   Unpassivated Si wires

The growth of SiNWs, with diameters below the 5 nm range (Holmes *et al.* 2000; Ma *et al.* 2003; Wu *et al.* 2004), has stimulated the theoretical characterization of Si unidimensional structures, with a crystalline Si core and different growth directions. Relevant questions such as the stability of nanowires as a function of the growth axis and diameter have drawn much attention. (Zhao and Yakobson 2003) used classical force fields (Tersoff 1988) to analyze the surface, edge and bulk contributions to the total energy of SiNWs grown along the $\langle 110 \rangle$ direction, and with different surface plane orientations. One of the structures that was found to be very stable consisted on a SiNW, grown along the $\langle 110 \rangle$ direction with (100) and (111) reconstructed facets at the surface (Fig. 2.1). This structure was obtained after generating six configurations of a $\langle 110 \rangle$-grown SiNW, with different surface reconstructions. These were equilibrated in a series of molecular-dynamics (MD) runs

(a)

(b)

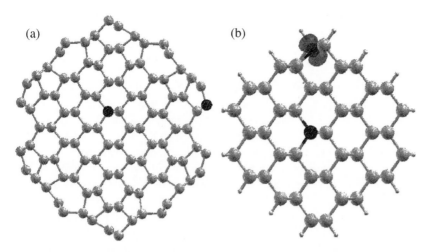

Fig. 2.1 ⟨110⟩-grown wires. (a) Unpassivated, surface-reconstructed SiNW-c (in black, the bulk and most stable surface dopant positions). (b) H-passivated SiNW-d, with a density isosurface corresponding to the "B + DB" complex deep into the gap (see text). Reprinted with permission from Fernandez-Serra and Blase, Phys. Rev. Lett., 96, 166805 (2006). Copyright 2006 by the American Physical Society.

(time $= 5$ ps, $T = 1000$ K), using a tight-binding model for silicon within the learn-on-the-fly (LOTF) method (Csanyi *et al.* 2004). The lowest-energy configuration was finally relaxed using the same DFT method employed throughout this chapter. This structure is consistent with observed experimental wires, as reported by Ma *et al.* (2003).

Going a step beyond these classical simulations, Rurali and Lorente used density-functional theory (DFT) calculations to study the electronic structure of SiNWs grown along the ⟨100⟩ direction. The surface of these wires can be formed by (100) or {110} facets. The lower surface energy of (100) surfaces makes them the final structural choice, although the even or odd number of atoms in the surface unit cell is equally as relevant, given that dimerization considerably lowers the surface energy, independently of its orientation. What they found is that for two different (100) facet reconstructions, the electronic character of the wire could be either metallic or semiconducting, depending on the periodicity imposed along the c-axis of the wire. We (Fernández-Serra *et al.* 2006b) built a new wire structure, grown along the ⟨110⟩ direction with (100) and (111) reconstructed facets at the surface. This last wire has a radius of 15 Å (instead of 8 Å in the case of Rurali and Lorente's wires) and, again, is found to be metallic with two bands crossing the Fermi level near the X point. These results stress the importance of surface reconstructions in determining the electronic properties of SiNWs. This will be shown to be a key aspect when analyzing structural and transport properties of doped SiNWs in Sections 2.3 and 2.5.

2.2.2 Hydrogen-passivated Si wires

Many of the theoretical studies dealing with transport properties and impurity or dopant effects on the electronic structure of SiNWs are performed on model structures where the dangling bonds at the wire surfaces are saturated with hydrogen atoms. This procedure avoids dealing with surface reconstruction effects and is a more realistic approach to actual SiNW experimental realizations. The structures are carved out of the diamond crystal. An effective nanowire radius R can be defined following Niquet *et al.* (2006):

$$R = \sqrt{\frac{N_{sc}a^3}{8\pi l}}, \qquad (2.1)$$

where R is the radius of a cylinder of length l (unit cell length) comprising N_{sc} Si atoms and with a volume $\Omega = \pi R^2 l$ equal to that occupied by N_{sc} atoms in bulk Si, namely $\Omega = N_{sc}a^3/8$. The SiNWs unit cell length l depends on the growth direction with $l = a(< 001 > $ NWs$)$, $l = a/\sqrt{2}(< 110 > $ NWs$)$, or $l = a/\sqrt{3}(< 111 > $ NWs$)$, a being the Si bulk lattice parameter.

An extensive study of electronic properties of SiNWs is provided by Niquet *et al.* (2006). They show how the bandgap opens with decreasing R due to quantum confinement, and show this effect to scale as $\propto 1/R^2$. However, so-called self-energy corrections, not accounted for in standard *ab-initio* calculations such as density-functional theory (DFT) approaches, also contribute to the rescaling of the bandgap of SiNWs as a function R, and are specially relevant for small radius wires. While self-energy corrections within, e.g. the GW approach amounts to correcting the DFT bandgap of bulk Si by ~ 0.6 eV (Hybertsen and Louie 1986), additional image charge effects, strongly depending on the dielectric mismatch between the Si wire and its environment, contribute to a $1/R$ contribution to the quasi-particle bandgap, a dependence that decays more slowly that quantum confinement. Fully *ab-initio* GW calculations have been performed on ultrathin wires by several groups confirming, the strong dependence of the self-energy correction with diameter (Zhao *et al.* 2004; Bruno *et al.* 2007; Yang *et al.* 2007).

Attempts to use hybrid functionals with the DFT have also been proposed by Rurali *et al.* (2007). Further, many-body excitonic calculations building upon the correct quasi-particle bandgap have been explored by Yang *et al.* (2007) and Bruno *et al.* (2007). We will not focus here on the optical properties.

Such many-body formalisms, devoted to the study of the excited state beyond the ground-state properties, have been recently applied to the crucial problem of the ionization energies of dopants in SiNWs, Again, the effects of quantum confinement and dielectric mismatch have been shown to yield significantly enhanced ionization energies in small-diameter wires (Diarra *et al.* 2007), a result that questions the doping efficiency of standard impurities such as boron or phosphorus in small wires.

2.3   Doping characteristics of SiNWs

One of the main advantages of the use of SiNWs in nanodevices, as opposed to CNTs, stems from their intrinsic semiconducting character that allows conductance to be controlled by doping and application of a gate voltage (Cui *et al.* 2000; Yu *et al.* 2000; Lew *et al.* 2004) in standard FETs. In contrast with CNTs, impurity atoms in SiNWs can be located either in the core or at the surface, highlighting the need to study dopant localization and its related electronic properties. Recent experimental results suggest the segregation of B atoms at the surface of SiNWs (Ma *et al.* 2001; Xiangfeng *et al.* 2003; Zhong *et al.* 2005).

Due to the large surface to bulk aspect ratio, the question of impurity trapping by surface defects at, e.g. the Si/SiO_2 interface becomes crucial in SiNWs. Taking as a lower bound a concentration of $\sim 10^{12}\,cm^{-2}$ for interface dangling bond (DB) defects (Pierreux and Stesmans 2002), one can immediately conclude that with "bulk" impurity concentrations of 10^{18}–$10^{19}\,cm^{-3}$ (Sze 1981), there are as many surface traps as dopants in SiNWs up to a few nm in diameter. Wires in this size range, starting to compete with multiwall CNTs, are currently being synthesized (Ma *et al.* 2003) and simulations (Wang *et al.* 2005) show that they would be optimal for integration in devices with sub-100 Å gate length.

The results presented in this section have been obtained by means of DFT calculations within a generalized gradient approximation to the exchange and correlation functional (Perdew *et al.* 1996). Core electrons are replaced by norm-conserving pseudopotentials (Troullier and Martins 1991) and wavefunctions are expressed as linear combinations of a basis set of strictly localized numerical orbitals (Soler *et al.* 2002).

2.3.1 Unpassivated wires

Unpassivated surface-reconstructed SiNWs might not be of special interest given that it is well known that the surface of the wires is always covered by a thin shell of SiO_2, passivated by hydrogen or they might form other types of core-shell composites. This is the final product obtained after experimental nanowire growth process. However, questions such as the spatial localization of dopants are important to understand transport properties of the wires, and they are relevant even during the growth process, when passivation of the surfaces has not yet taken place. Because of that, understanding the energetics of dopant segregation is important from a thermodynamic point of view and differentiating between passivated and non-passivated wires is relevant.

We first study the segregation of B and P in various unpassivated wires displaying different surface reconstructions and electronic properties, as described in Section (2.1) of the chapter. Three different systems are considered. Nanowires labelled *a* and *b*, grown along the ⟨100⟩ direction, are described by Rurali and Lorente (2005). SiNW-a is metallic without doping, due to its particular surface reconstruction, while SiNW-b is semi-metallic due to a different surface reconstruction associated with a doubling of the SiNW-a unit cell. An additional system (SiNW-c), grown along the ⟨110⟩ direction with (100) and (111) reconstructed facets at the surface, is studied (Fig. 2.1). This last wire has a radius of 15 Å (instead of 8 Å for SiNW-a and -b) and is metallic with two bands crossing the Fermi level near the *X* point. In order to minimize impurity interactions upon "periodic" doping, the cells of SiNW-a and SiNW-c are doubled, yielding, respectively, 114 atoms and 200 atoms systems. In the case of SiNW-a, we have verified that using a triple cell does not change by more than 10% the energies provided below.

We also compare these results to the case of planar semiconducting Si surfaces (Bedrossian *et al.* 1989; Ramamoorthy *et al.* 1999). In agreement with previous work (Ramamoorthy *et al.* 1999), we find a clear tendency for surface

Table 2.1 Segregation energies (in eV, see text). The energy reference is the bulk value

	Si(001)-2 × 1	Wire a	Wire b	Wire c	Wire d
B_{doped}	−0.7	−0.03	−0.14	−0.11	−0.99
P_{doped}	−1.08	−0.98	−1.05	−1.10	−1.62

segregation in the case, e.g. of a Si(001)-2 × 1 surface (Table 2.1). While large unpassivated SiNWs will certainly yield similar results, the modifications of the structural and electronic surface properties in small wires (Kobayashi 2004; Rurali and Lorente 2005) are expected to strongly affect the magnitude, and possibly the sign, of these segregation energies. Further, elastic relaxation effects, previously invoked in the case of planar surfaces (Ramamoorthy *et al.* 1999), may be significantly affected in the limit of small radii.

We report in Table 2.1 the surface to bulk energy difference for the most stable surface positions in wires (a) to (c) (all non-equivalent positions between bulk and surface, including subsurface sites, were tested). Clearly, even in this limit of small metallic or semi-metallic wires, impurities will segregate at the surface, even though the effect is reduced as compared to planar surfaces. Comparing SiNW-a to SiNW-b, the effect is more pronounced with reduced surface metallicity, consistent with the larger value obtained for semiconducting planar surfaces. These results seem to point to a strong electronic contribution to the surface stabilization.

The above results strongly suggest that in all situations, from small to large diameters, both in metallic and semiconducting wires, impurities will segregate at the surface (see also Peeters *et al.* 2006). As will be shown, such a surface segregation will strongly affect the conductance associated with surface states. Further, these results clearly indicate that in all cases, P surface segregation will be more pronounced than B segregation. Therefore, different behavior might be expected upon n- or p-type doping of SiNWs. We will address this point in more detail after studying the case of passivated wires.

2.3.2   Passivated wires

Consistent with the conclusion that surface segregation is mainly driven by electronic effects, in the case of fully H-passivated wires and flat surfaces, we do not see any sizeable energy difference between different positions below the surface. However, as emphasized above, the large surface to bulk aspect ratio significantly accentuates the importance of surface or interface defects that are known to exist at Si/SiO_2 interfaces (Pierreux and Stesmans 2002). All of these defects produce a less-co-ordinated Si surface atom left with at least one dangling bond (DB). To mimic this situation, we generate a 5.6-Å radius H-passivated wire (SiNW-d in Fig. 2.1) grown along the ⟨110⟩ direction. All surface atoms, but one, are saturated with hydrogen. Two unit cells (124 atoms) are used for this segregation study.

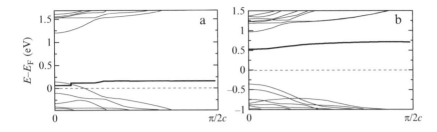

Fig. 2.2 Band structure of B-doped SiNW-d. (a) B in the center of the wire, the system has a degenerate semiconductor character. The black thick line represents the DB state. (b) B at the surface, the structure becomes semiconducting, with an isolated B state deep into the gap (gray line). (Dashed lines indicate the Fermi level.) Reprinted with permission from Fernandez-Serra and Blase, Phys. Rev. Lett., 96, 166805 (2006). Copyright 2006 by the American Physical Society.

Again, we find (Table 2.1) that impurities prefer to segregate at the surface to passivate DB defects. The segregation energy is found to be much larger than in the case of unpassivated surfaces, indicating again the importance of electronic effects. To further understand the effect of such a segregation, we show in Fig. 2.2 the band structures for the surface/bulk doped SiNW-d. For B in a core position, the wire exhibits a degenerate semiconducting behavior (this degenerate behavior is related to the rather large B concentration associated with our supercell size) and the unpassivated DB yields an empty (hardly dispersive) band close to the top of the valence bands. However, upon segregation of B at the surface DB site, the "boron + DB" complex yields an empty level deep into the bandgap (Fig. 2.2(b)) with a clear boron p_z character, B adopting a planar sp^2-like configuration (Fig. 2.1(b)). As a result, the Fermi level rises above the top of the valence bands and the wire becomes truly semiconducting (undoped). *The impurity trapped at the surface is electronically inactive and does not yield a carrier at room temperature*, as the in-gap level is several thousand K away from the band edges.

The segregation energy of P atoms at the surface is, again, significantly larger than that of B. This means that for the same impurity concentration, a much larger fraction of P atoms will be trapped at the surface and electronically inactive. Namely, *the concentration of free carriers will be much larger in the case of B doping than P doping for the same impurity density*. Cui *et al.* (2000) and Yu *et al.* (2000) have reported experimental measurements of conductance in B- and P-doped wires. It is observed that for similar doping levels, B-doped SiNWs present a lower resistance than P-doped ones. In bulk-doped Si, for the same concentration of these dopants, the effect is known to be exactly the opposite (Sze 1981).

In both p- and n-type doping, these results suggest that in SiNWs, surface traps can neutralize a significant fraction of impurities up to a critical concentration c_0 that will depend on the wire radius. Below c_0, the conductivity in wires should increase very slowly with doping percentage and p-doping should be more efficient than n-type (contrary to bulk). Beyond c_0, most impurities will start again to be ionized at room temperature and the relation between impurity and free carrier concentrations will follow that of bulk but with a c_0 shift. As a result, larger dopant concentrations will be necessary in SiNWs as compared to bulk for the same conductivity. This is consistent with the values of resistivity measured in B-doped SiNWs that are found to be higher than expected in bulk Si at similar dopant concentrations (Lew *et al.* 2004).

2.4   Electronic transport

Beyond information related to the density of free carriers, device performances are also strongly related to the mobility (or mean-free path) of these carriers. At low temperature and bias, where acoustic and optical phonon inelastic scattering remain limited, such quantities are governed in particular by the elastic scattering of electrons by defects (inelastic scattering by phonons will not be addressed in this chapter).

While ionized impurities significantly increase the density of carriers at the device's working temperature, they also induce a strong drop of transmission for conducting channels close in energy to the impurity-related bound-state energies. This so-called resonant backscattering effect was clearly evidenced in the case of nanotubes for isolated substitutional impurities (Choi *et al.* 2000; Kaun *et al.* 2002; Latil *et al.* 2004; Adessi *et al.* 2006; Avriller *et al.* 2006).

In this section we will analyze the effect that isolated impurities have on the Landauer transmission along the wires (Datta 1995). Our *ab-initio* transport calculations are performed thanks to the TABLIER code developed by Adessi and Blase (2006) and that exploits the sparse nature of the Hamiltonian and overlap matrices as produced by the well-known SIESTA DFT ground-state package (Soler *et al.* 2002). Our theoretical framework proceeds as follows. We start with a ground-state calculation of the system of interest using the same electronic structure method we have previously described. The locality of the SIESTA basis (Soler *et al.* 2002) and the local density approximation lead to the partitioning of the system into nearest-neighbor sections so that the Hamiltonian H and overlap S matrices are tridiagonal by blocks. This property allows us to calculate with an $O(N)$ scaling the Green's function of the system. The computational scheme is similar to standard tight-binding implementations of transport but with submatrices replacing the scalar on-site and hopping terms. The Green's function of the "device" (see Fig. 2.3) reads

$$\hat{G}_D(\epsilon) = [\epsilon\hat{S}_D - \hat{H}_D - \hat{\Sigma}_L(\epsilon) - \hat{\Sigma}_R(\epsilon)]^{-1}, \tag{2.2}$$

$$\hat{\Sigma}_L(\epsilon) = [\hat{T}_{LD} - \epsilon\hat{S}_{LD}]^\dagger \hat{G}_{00,L}(\epsilon)[\hat{T}_{LD} - \epsilon\hat{S}_{LD}], \tag{2.3}$$

with $\Sigma_L(\Sigma_R)$ the self-energy operator that accounts for the interaction of the device with the left (right) electrode. The bulk "surface" Green's function ($\hat{G}_{00,L}$ for the left electrode), which is the restriction of the semi-infinite electrode Green's function to the electrode subsection in contact with the device area, can then be calculated iteratively as follows:

$$\hat{G}_{00,L}^{(n+1)}(\epsilon) = [\epsilon\hat{S}_L - \hat{H}_L - \hat{U}^\dagger \hat{G}_{00,L}^{(n)}(\epsilon)\hat{U}]^{-1} \tag{2.4}$$

$$\hat{U} = [\hat{T}_{LL} - \epsilon\hat{S}_{LL}]^{-1} \tag{2.5}$$

$$\hat{G}_{00,L}^{(0)}(\epsilon) = [\epsilon\hat{S}_L - \hat{H}_L]^{-1}. \tag{2.6}$$

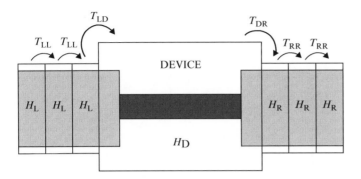

Fig. 2.3 Symbolic representation of the partitioning of the studied systems into device and bulk areas yielding a block-tridiagonal representation of the infinite system Hamiltonian and overlap matrices.

where $T(\epsilon) = \mathrm{Tr}(\hat{\Gamma}_L \hat{G}_D^a \hat{\Gamma}_R \hat{G}_D^r)$ with, e.g. $\hat{\Gamma}_L = \mathrm{Im}(\hat{\Sigma}_L)$. This approach is now standard and is similar, within the linear response, to other implementations using a strictly localized basis (Rocha *et al.* 2006) or Wannier functions (Calzolari *et al.* 2004).

Transport calculations are performed on a 13-Å diameter SiNWs grown along the $\langle 110 \rangle$ direction (Fig. 2.4(a)). The corresponding Kohn–Sham band structure around the forbidden gap is provided in Fig. 2.5(a). Our device consists of a supercell made of 12 wire sections (480 atoms for a length of 47.6 Å), repeated periodically. The on-site and hopping Hamiltonian blocks associated with the two left (right) most sections are used to reconstruct the semi-infinite left (right) electrode Hamiltonian. The remaining central part of the supercell defines the "channel". In a first step, all surface Si atoms are passivated by hydrogen and we study phosphorus substitutional doping. Since phosphorus

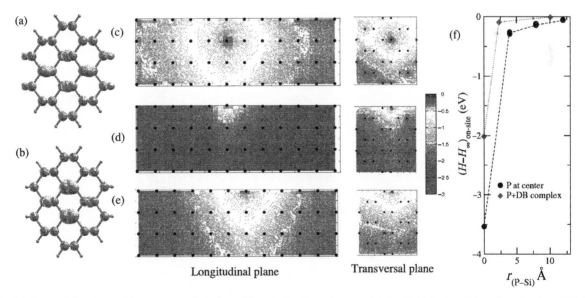

Fig. 2.4 Image of the wire used for transport calculations with an isodensity surface associated with (a) the top of the valence bands and (b) the bottom of the conduction bands. (c–e) Onsite potential 2D maps along transversal and longitudinal planes sectioning the wire: (c) P impurity at the wire center, (d) an isolated DB surface site and (e) the "P + DB" complex. (f) Profile of the on-site potential along a line in the longitudinal plane, for the P impurity at the center (circles) and "P + DB" complex (diamonds). Reprinted with permission from Fernandez-Serra and Blase, Nanoletters, 6, 2674 (2006). Copyright 2006 by the American Chemical Society.

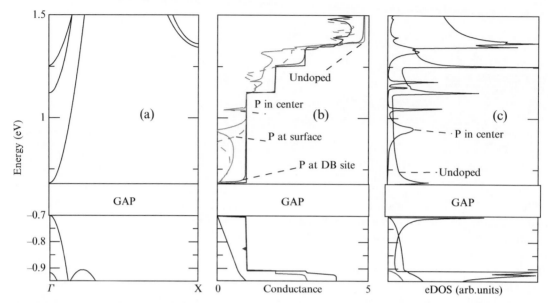

Fig. 2.5 (a) Wire band structure around the gap. Band edges are at $\pm 0.7\,\text{eV}$ (DFT value. The gap is not on-scale on the graph). (b) Conductance (in unit of $G_0 = 2e^2/h$) for the undoped wire, P atom at wire center, P atom at the surface (see text), and P atom at a DB site (P + DB complex). (c) Density of states: undoped wire and P atom in the center. Reprinted with permission from Fernandez-Serra and Blase, Nanoletters, 6, 2674 (2006). Copyright 2006 by the American Chemical Society.

is a donor with respect to silicon (n-type doping), we focus mainly on the conduction bands.

The conductance and electronic density of states (eDOS) of the undoped wire are represented in Figs. 2.5(b) and (c). The eDOS, calculated from the trace of the density operator built from the scattering states, exhibits the expected van Hove singularities at band edges. The structure in plateaus of the transmission shows that our supercell is large enough to insure that the two left-most sections, used to build the infinite leads, are far enough from the impurity to recover the behavior of the perfect (undoped) SiNWs. Upon doping, with phosphorus substituting a central Si atom or a surface Si atom connected to an hydrogen, we observe a clear drop of conductance related to the backscattering of incoming wave packets by the potential well created by the impurity (Figs. 2.4(c,f)). At specific energies on the first conduction plateau, the transmission even goes to zero, that is, the incoming propagating states are entirely backscattered.

In Fig. 2.5(c) the impurity-related local density of states obtained by restricting the trace of the density operator to the phosphorus orbitals (P at the wire center) is plotted. Noticeably, the peaks in the P-related eDOS correspond to the drop of conductance in Fig. 2.5(b). This is an evident signature of the resonant character of the interaction between the impurity-induced bound states and the propagating waves, an effect that can be easily interpreted in terms of first-order perturbation theory. The presence of two peaks on the first conduction plateau, a sharp one just below the second van Hove singularity and a broad one at lower energy, is very similar to what has been observed in nitrogen-doped nanotubes (Choi *et al.* 2000) with the presence

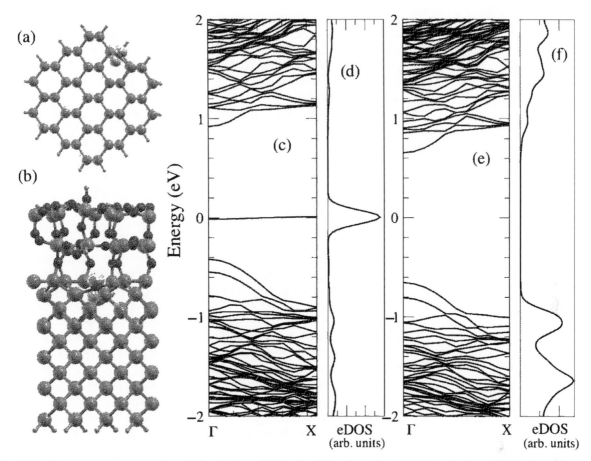

Fig. 2.6 Band structure and corresponding eDOS of the large SiNW with a DB at the surface. (a) Model of the large SiNW with and isodensity surface of charge associated to the DB in-gap band. (b) Model of the Si/SiO$_2$ interface with an isodensity surface of charge associated to the P_b in-gap band. (c) DB isolated, undoped wire. (d) eDOS projected onto the Si DB orbitals. (e) "P + DB" complex structure. (f) eDOS projected onto the P orbitals. Reprinted with permission from Fernandez-Serra and Blase, Nanoletters, 6, 2674 (2006). Copyright 2006 by the American Chemical Society.

of a large *s*-like state and a sharp p-like one. In the valence bands, the impurity located at the center also induces a drop of conductance, but with a smooth energy dependence, free from any resonant character. The same conclusion can be given concerning the projected eDOS. As mentioned above, phosphorus is a donor with respect to silicon and the related bound states are located in the conduction bands. Note that in Fig. 2.5, the P-induced donor in-gap levels cannot be seen, as the eDOS is calculated from the trace of the density matrix built from the propagating states. As in-gap states are strictly localized, not propagating, they cannot be seen in this "transport-related" density of states. This is not the case in Fig. 2.6 where stationary Bloch states are used.

The phosphorus impurity located on a surface silicon atom, and bonded to a passivating hydrogen, also induces a strong drop of conductance, with a zero of transmission on the first conduction plateau, but with resonant energies shifted as compared to the central wire substitutional position (dashed line in

Fig. 2.5(b)). The effects of hydrogen and confinement can certainly explain this radial dependence of the transmission peak structure. It demonstrates in particular that care should be taken when building a model scattering potential for semi-empirical transport studies whenever the impurities come close to the wire surface. The large shift in energy at which the transmission cancels, upon moving the impurity from the wire center to the surface, indicates that a broad drop of conduction on a large energy window is to be expected for a random distribution of dopants.

2.4.1   Role of dangling-bond defect

We turn now to the study of wires in the presence of surface dangling bond (DB) defects, that is in the presence of under-co-ordinated surface Si atoms. It is found that DB defects alone have hardly any direct impact on the conductance, namely the conductance of our wire is unaffected by removing one hydrogen atom, thus leaving an unpassivated surface Si atom. Such a result can be rationalized by analyzing in Fig. 2.4 the evolution away from the defect or impurity site, of the on-site potential energy associated with Si s orbitals; very similar results are obtained for the p-orbital on-site energies. Such on-site energies are directly extracted from the diagonal of the *ab-initio* Hamiltonian. The perturbation associated with a substitutional phosphorus atom in the bulk is found to be large and long ranged (Figs. 2.4(c,f)), explaining that the transmission of bulk states is significantly affected by the impurity as shown above. On the contrary, the perturbing potential generated by the DB defect at the surface is extremely short ranged (Fig. 2.4(d)). As the states around band edges are mainly localized in the core of the wire (Figs. 2.4(a,b)), they are hardly affected by the presence of the dangling bond.

In the presence of dopants, the role of the DB defects becomes, however, significant, as has been shown in the first section of the chapter, because they represent centers of attraction for impurities, energetically being favorable to migrate to these sites, electronically becoming neutralized impurities.

The transmission associated with the "P + DB" complex is shown in Fig. 2.5(b). Clearly, it is nearly indistinguishable from that of the perfect wire. Namely, and in marked contrast with the case of phosphorus in the bulk, the impurity is transparent to the incoming wave packets and the transport is nearly ballistic. We emphasize that it is not just a geometric effect, as phosphorus located at the same surface site, but connected to a hydrogen, yields a significant drop of conductance, as shown above (dashed line). The analysis of the on-site energy profile around the P + DB complex (Figs. 2.4(b,f)) shows that the induced perturbation is weaker and shorter ranged as compared to the one associated with phosphorus in the core. In particular (Fig. 2.4(f)), while the phosphorus on-site energies are $\sim 3.6\,\text{eV}$ below that of unperturbed silicon when the dopant is at the center, they are only $\sim 2.1\,\text{eV}$ below on the P + DB complex. Namely, the potential well associated with the P + DB complex is much shallower.

To better understand this behavior, we present the electronic properties of the DB defect and the P + DB complex. The band structure of the SiNW with a DB defect at the surface is shown also in Fig. 2.6(c). As expected, the DB yields a half-filled band deep into the forbidden gap, consistent with the findings of Singh *et al.* (2006). With P in the core and in the presence of the DB, the extra electron fills the DB level that is now close to the top of the valence band, a situation described in Blomquist and Kirczenow (2006) in the case of hydrogenated silicon nanocrystals. This configuration is, however, energetically less favorable than having the dopant at the surface DB site (Table 2.1). Upon formation of the P + DB complex (Fig. 2.6(e)), the occupied gap state forms a stable lone pair localized on the P atom, leaving the bulk Si wire unperturbed. The phosphorus relaxes in a sp^3-like configuration with the three Si–P backbonds and the lone pair forming the tetrahedral environment. In the case of p-type doping (B, Al), the impurity + DP complex adopts rather an sp^2 configuration as the dangling bond is now completely empty.

The stability of this lone pair can be further evidenced by analyzing the related projected density of states (Fig. 2.6 and Fig. 2.5(c)). The P + DB levels are now found to land mainly deep in the valence bands, a situation that is also reflected in the transmission profile with a drop of conductance no longer in the conduction bands, but on the second plateau in the valence bands (red line, Fig. 2.5(b)). Similar results have been obtained in the case of p-type doping with the removal from the valence bands of any impurity character upon segregation into the surface DB sites.

While the transparency to wave packets propagation of the P + DB complex is a favorable situation, since the ballistic character of the transport in perfect wires is preserved, the removal of the impurity band away from the bandgap means that the dopant is electronically neutralized, namely it cannot yield any free carrier, as the additional electron is now bound to the surface lone pair. The wire with its P + DB complex is a true semiconductor (Fig. 2.6(e)) and hardly any free carriers can exist at room temperature. The impurity is thus completely neutralized upon segregation on the DB site, as it cannot be ionized and is transparent to propagating states.

Recently, a detailed study of phosphorus-doped SiNW-FETs with controlled dopant concentration (Zheng *et al.* 2004) revealed an unexpected increase of conductance with increasing doping percentage, a result at odds with the present conclusions and also with what is observed in bulk silicon. This observation was interpreted in terms of a modification (lowering) of the contact resistance upon doping. The importance of a possible Schottky barrier at the electrode/SiNWs interface was previously emphasized in Landman *et al.* (2000). Whatever the impact of the contact resistance, the increased mobility obtained in the high-doping limit (Si/P = 500 : 1) implies that the SiNW intrinsic conductance is not affected significantly by the dopants, suggesting either that dopants segregate at the growing end of the wire, and/or at the wire surface where a significant fraction is neutralized by DB defects. Other strategies, such as doping of the shell in core/shell structures (Zhong and Stocks 2006) or the use of undoped Ge/Si NWs heterostructures (Xiang *et al.* 2006) have been proposed to reduce as much as possible the backscattering induced by dopants located in the conducting core.

2.5   Multiple impurities and disorder

Markussen *et al.* (2007) have studied the influence of impurity disorder on the wire conductance using a similar method to the one presented above (Tablier code) but generalizing the early studies by Fernández-Serra *et al.* (2006a) to the case of a random distribution of dopants along the SiNWs in order to"bridge the scales" between microscopic and mesoscopic physics. Exploiting the order (N) scaling of the Landauer conductance calculations restricted to the linear regime (no update of the scattering potential upon bias changes), it is nowadays possible to compute the transmission of micrometer-long nanowires. Technically, such long wires are built from the recursive addition of smaller sections (a few nanometers in length) of the kind studied in the previous sections and containing either one impurity, located at various positions between the wire axis and the surface, or no impurities (pristine wire sections). Such a "lego" scheme allows us to "reconstruct" realistically long wires ($L > 100$ nm) and study their conductance in the presence of a random distribution of impurities at a given concentration. The mean conductance $< g >$ can then be calculated by averaging over a large number (200) of different realization of disorder (impurities distribution), allowing us to discriminate between the different regime of transport (quasi-ballistic, diffusive, localization) from the evolution of conductance with length or dopant concentration. Such an analysis leads to the evaluation of the energy-dependent mean-free path (MFP) l_e (diffusive regime) or localization length ξ (localization regime). Indeed, for a wire of length $L \ll \xi$, ξ being the localization length, the resistance increases linearly as $R(L) = R_0 + R_0 L/l_e$. In the localization regime, the resistance increases exponentially and ξ is calculated, for very large L to ensure convergence as:

$$\xi = - \lim_{L \to \infty} \frac{2L}{< lng >}. \tag{2.7}$$

Further, such calculations permit testing of the scaling theories developed in the 1980s concerning the fluctuations (standard deviations) of conductance upon various realization of disorder both in the ballistic and localization regimes. Other scaling relations, such as the so-called Thouless prediction relating the MFP l_e and the localization length ξ as a function of dimensionality and number of conducting channels could also be tested. (For an extensive review in the case of doped nanotubes, including the effect of a magnetic field, see Avriller *et al.* 2007.)

An important outcome of the work by Markussen and colleagues is that the average conductance $< g >$ of a randomly doped wire in the diffusive regime can be extrapolated from the knowledge of the conductance associated with a single impurity averaged over the different positions along the wire diameter. Such a result provides much weight to "single-impurity" calculations that are much less demanding in terms of computer time.

Similar mesoscopic studies have been performed (Markussen *et al.* 2007) to explore the effect of surface disorder using either a random on-site (Anderson-type) surface disorder in the case of unpassivated wires, or by removing randomly one hydrogen atom in the case of passivated SiNWs. As shown by

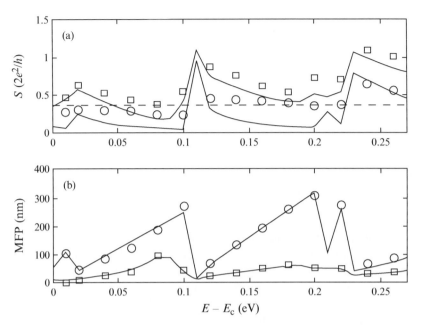

Fig. 2.7 Maximal standard deviation (a) for P-doped wires ($L = 10$ nm), with a homogeneous radial distribution (squares) and a pure surface-doped wire (circles). The solid lines show the standard deviation, s, among the single-dopant transmissions and the pristine wire, and the dashed line marks the UCF value $0.73e^2/h$. Panel (b) shows the MFP vs. energy for the two radial distributions (circles and squares). The lines represent values obtained from the single-dopant transmissions. Reprinted with permission from Markussen *et al.*, Phys. Rev. Lett., 98, 076803 (2007). Copyright 2007 by the American Physical Society.

Fernández-Serra *et al.* (2006b), surface conductance in the case of unpassivated SiNWs is very sensitivive to disorder. In the case of passivated SiNWs, the MFP was shown to be strongly energy dependent, shifts of the Fermi energy of hundreds of meV can cause a transition from the diffusive (Ohmic) regime to the localization regime. Other models of surface disorders, such as a variation of radius along wire length (surface roughness) has been explored within a tight-binding approach and the Kubo–Greenwood approach, leading again to a discussion of the MFP, localization length and Thouless relation as a function of disorder strength (Lherbier *et al.* 2008).

This feature can be exploded in the use of SiNWs as chemical sensor devices, where a so-called "chemical gate", a molecular complex capable of binding to a receptor already bound to the wire surface, can shift the Fermi level of the semiconducting wire. We study this path on the following section, where the functionalization of SiNWs by different chemical linkers is modelled in order to understand how different covalently bounded organic groups can influence the transport characteristic of passivated SiNWs.

2.6 Covalent functionalization of SiNWs

Chemical functionalization offers an important means to manipulate and tailor the properties of carbon nanotubes (CNTs) and nanowires (NWs). From CNTs separation, selection and self-assembling (Dyke and Tour 2004; Balasubramanian and Burghard 2005), to the modification of the electronic properties of the conducting channel, or of the contact resistance, in a CNT or NW-based field-effect transistor (FET), a large body of work devoted to understanding the effect of functionalization has received, at least at the experimental level, considerable attention in the last few years. Experimentally, the use of functionalized semiconducting NWs in rectifying, optical and chemical or biological molecular sensor devices has been explored recently by several groups (Li *et al.* 2000, 2004; Cui *et al.* 2001; Francinelli *et al.* 2004; Hahm and Lieber 2004; Streifer *et al.* 2005; Bunimovich *et al.* 2006; Haick *et al.* 2006; Winkelmann *et al.* 2007).

An important limitation related to the functionalization of CNTs is that the interaction of the side molecules with the sp^2 graphitic network may considerably reduce the mobility of the charge carriers. This is particularly true in the case of covalent functionalization where the conjugated π-network is strongly perturbed (Lee *et al.* 2005). Similar conclusions have been obtained in the case of substitutional doping by, e.g. boron or nitrogen atoms (Choi *et al.* 2000; Adessi *et al.* 2006; Avriller *et al.* 2006; Charlier *et al.* 2007). Even though a specific class of reactions may reduce the negative impact of grafting side chains (Lee and Marzari 2006), it remains that this issue strongly impairs most chemical routes available for the functionalization of CNTs.

In this section we explore the functionalization of SiNWs, following the studies already performed for CNTs, but looking at the most standard organic chains or linkers experimentally sought in the functionalization of Si surfaces. The literature on the covalent functionalization of flat silicon surfaces (Srtoher *et al.* 2000; Bent 2002; Shirahata *et al.* 2005) and SiNWs (Li *et al.* 2000; Francinelli *et al.* 2004; Hahm and Lieber 2004; Bunimovich *et al.* 2006; Haick *et al.* 2006) point to a few important classes of reactions. The most important routes for covalent functionalization are the hydroxylation of H-terminated Si surfaces or the alkylation of halogen-terminated surfaces, leading to the formation of Si–C bonds connecting the silicon surface atoms to saturated alkyl radicals. Such chains are called spacers or linkers. They are usually terminated on the other end by an active chemical group (photoactive function, molecular recognition receptor, etc.) which is thus spatially separated from the SiNWs. Similar reactions can lead to the grafting of carbon-based π-conjugated oligomers such as alkenyl chains or phenyl rings. Other chemical bonds can be formed, such as Si–O through the alkoxylation of hydrogen or halogen-terminated surfaces, or Si-N through the reaction of an amino group on a chlorine-terminated surface (Shirahata *et al.* 2005).

Our system of study is the same one considered in the previous section, a prototype hydrogen-passivated ⟨110⟩ SiNW with a diameter of 13 Å. Since alkyl chains are the most common linkers used to functionalized H-terminated Si wires or surfaces, we first replace one hydrogen by an R $= -(CH_2)_n$-H alkyl radical, with n the number of carbon atoms in the lateral chain (see Fig. 2.8). In the case of a butyl ($n = 4$) lateral chain, both functionalization of the {111} (Figs. 2.2(a,e)) and {100} facets (Figs. 2.9(b,f)) are studied.

The conductance of alkyl-functionalized nanowires is presented in Figs. 2.9(a)–(d) and Figs. 2.9(e) and (f) for the valence and conduction bands, respectively. Black lines indicate the standard plateau of the perfect NW conductance (in units of $G_0 = 2e^2/h$) while thick colored lines indicate the effect of grafting one molecule. As clearly evidenced in Figs. 2.9(e)–(h), *the functionalization has absolutely no effect on the transmission in the conduction bands over a large energy range.* A similar conclusion applies to the first plateau below the Fermi level. While some reduction in transmission can be

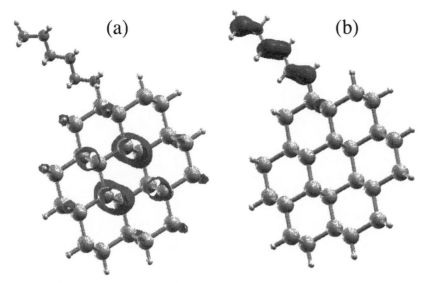

Fig. 2.8 (a) Symbolic representation of the hexane radical (hexyl)-functionalized SiNW. A plot of the integrated charge density for states with energy in the range of Fig. 2.9 is provided (contour at 10% of the maximum density value). (b) Symbolic representation of the alkenyl ($n = 6$) functionalized wire with a plot of the side chain "HOMO" level (see peak at $\sim -0.8\,eV$ in Fig. 2.10(c)). Reprinted with permission from Fernandez-Serra and Blase, Rev. Lett., 100, 046802 (2008). Copyright 2008 by the American Physical Society.

observed at lower an energy (around $\sim -1.17\,eV$), the effect is rather small, especially compared to CNT functionalization (Lee and Marzari 2006) or NW core doping (Fernández-Serra *et al.* 2006a) where a drop of 50% to 100% of the conductance could be observed at resonant energies. The drop of transmission at $\sim -1.01\,eV$ is present in the perfect wire and is related to a small gap opening between two bands (forbidden crossing).

In order to understand these results, we plot in Fig. 2.10(b) the functionalized SiNWs density of states (*e*DOS) projected onto the side chains orbitals in the case of an hexyl radical ($n = 6$) that we compare in Fig. 2.10(a) to the total

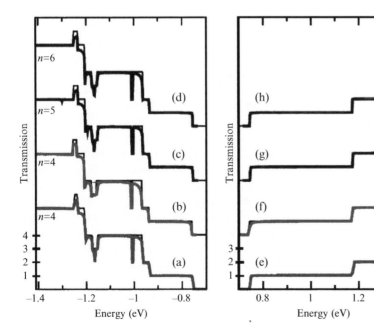

Fig. 2.9 Transmission for alkyl-functionalized SiNWs. The index (n) indicates the number of carbon atoms in the alkyl chain. In (a) and (b), the ($n = 4$) butyl chain is grafted on the {111} and {100} facets of the wire. Reprinted with permission from Fernandez-Serra and Blase, Rev. Lett., 100, 046802 (2008). Copyright 2008 by the American Physical Society.

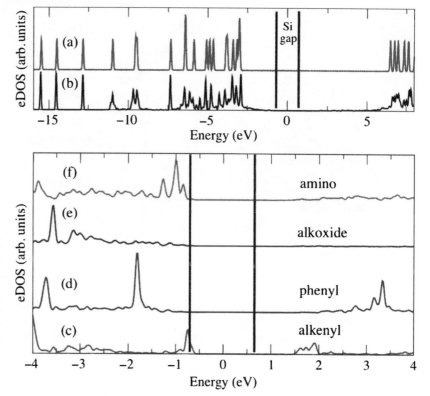

Fig. 2.10 (a) *e*DOS for the hexane ($n = 6$) isolated molecule. (b–f) Local density of states projected on the side-chain orbitals: (b) hexyl, (c) phenyl, (d) alkoxide, and (e) amino radicals. The two thick vertical lines indicate the SiNWs bandgap. Note that the two graphs are not on scale. *e*DOS in the lower panel have been normalized to the number of valence electrons in the side chain for the sake of comparison. Reprinted with permission from Fernandez-Serra and Blase, Rev. Lett., 100, 046802 (2008). Copyright 2008 by the American Physical Society.

*e*DOS of the corresponding isolated hexane chain. The isolated alkane and grafted alkyl-chain *e*DOS have been aligned on the lowest-occupied level at ~ -15.5 eV (well below the Si occupied bands). The bandgap of the SiNWs is indicated by the two vertical thick lines. Clearly, the highest-occupied (HOMO) and the lowest-unoccupied molecular level (LUMO) of the grafted alkyl chains fall well below and above the SiNWs bandgap, respectively. Integrating the charge density over the energy range covered by the conductance plot in Fig. 2.9, we find indeed hardly any contribution from the side chain (see Fig. 2.8(a)). As a result, weak hybridization and no resonant backscattering are expected to take place in the functionalized wire. Identical results apply to the various chains studied above. The bandgap of the infinite alkane chains is $E_g = 8.8$ eV (DFT-LDA value). As such, the present results generalize to longer chains not studied here explicitly.

There exists a large decoupling in energy between alkyl molecular levels and the SiNWs states around the bandgap that insures that no resonant backscattering is to be expected. Furthermore, the analysis of the scattering potential induced by the side chain, defined by the variations of the total electronic potential inside the SiNW upon grafting of the alkyl group, is found to be extremely shallow. The on-site Si-*s* and Si-*p* self-consistent Hamiltonian matrix elements show indeed a maximum variation of ~ 0.1 eV for the Si atom directly connected to the carbon chain, while further away in the SiNW, the on-site energies do not differ by more than 0.01 eV with respect to equivalent Si atoms located far away from the alkyl group along the nanowire. This can be compared with the values obtained in the previous section, where it was shown that either for boron or phosphorus doping the substitutional impurity builds a potential well of the order of a few eVs. As a result, non-resonant scattering is also weak in alkyl-functionalized SiNWs. This shallow scattering potential is found to assume positive values in the SiNW core, explaining that mostly holes (valence channels) are slightly affected.

Functionalizing SiNWs without affecting significantly their conductivity is an important result as it suggests that heavily functionalized SiNWs can be used in transport devices such as FETs. Contrary to nanotubes where most covalent grafting routes severely degrade the carriers mobility, the conductance in SiNWs is expected to remain large even in the limit of a substantial number of grafted linker molecules. The possibility of integrating SiNWs channels with a large number of active side chains per unit surface should significantly enhance the sensitivity of SiNW-based sensors or photoactivated devices. Together with selectivity, this is certainly one of the central issues influencing the future of such devices.

Another important consequence of these findings is that the mechanisms involved in SiNWs-based molecular sensors (Hahm and Lieber 2004) cannot be attributed to a change of carrier mobility upon binding of the side chains. This provides weight to a scenario where adsorbed molecules act as "chemical gates" by transferring charge to the SiNW substrate (Cui *et al.* 2001; Hahm and Lieber 2004), an effect that changes the current either through opening of new conduction channels and/or tuning of the contact Schottky barrier by a change in the Fermi-level position and work function.

In order to sample a large variety of possible linkers, we have extended the study to other, less common functional groups. We provide in Figs. 2.11(a)–(c) the conductance associated with our prototype SiNW functionalized with π-conjugated -$(CH)_n$-H alkenyl chains, with $n = 4$ (Figs. 2.11(a) and (g)) and $n = 6$ (Figs. 2.11(b) and (h)) carbon atoms, a phenyl radical (C_6H_5) (Figs. 2.11(c) and (i)), an $-OC_3H_7$ (alcohol radical) alkoxide group (Figs. 2.11(d) and (j)) and an amino $-NH_2$ group (Figs. 2.11(e) and (k)). Again, there is no reduction of the conductance in the first unoccupied plateaus whatever the functional group. However, in the case of the valence bands, the physics of resonant backscattering reappears, in particular in the case of the conjugated alkenyl (Fig. 2.11(a)) and amino (Fig. 2.11(e)) groups with a drop to zero of the conductance at specific energies (see stars on graph).

This effect again can be clearly seen on the local density of states provided in Figs. 2.10(c)–(f). In the case of the alkenyl group (see Figs. 2.10(c) and 2.11(b)), the molecular HOMO level his just below the top of the valence bands, explaining the drop of conductance in Fig. 2.11(b) in this energy range (see star). Similarly, the states present below the top of the valence bands in the amino case (see Fig. 2.10(f)) can explain the large drop of conductance observed just below the bandgap (see Fig. 2.11(e)). However, in all cases, no molecular derived levels can be seen above the Fermi level for several subbands, the alkenyl case being the less favorable here.

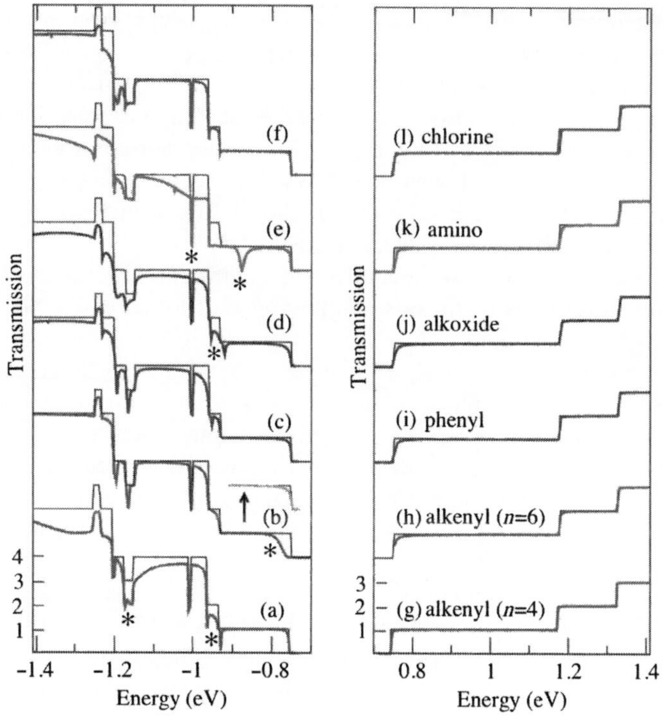

Fig. 2.11 Transmission for SiNWs functionalized with conjugated (a)–(g) -(CH)$_4$-H, (b,h)-(CH)$_6$-H alkenyl radicals, (c) and (i) a phenyl group, (d) and (j) a -OC$_3$H$_7$ alkoxide group, (e) and (k) an amine group, and (f) and (l) a residual chlorine atom. The stars indicate some typical resonant backscattering drop of conductance. The effect of passivating the Si-connected C atom of the -(CH)$_6$-H alkenyl group by a hydrogen is indicated (see arrow). Reprinted with permission from Fernandez-Serra and Blase, Rev. Lett., 100, 046802 (2008). Copyright 2008 by the American Physical Society.

2.7   Conclusions

We have presented in this chapter a summary of some of the most recent theoretical studies of electronic structure and transport on SiNWs. In particular, we have shown how dimensionality and quantum-confinement effects make the physical properties of these Si-based nanotructures very different from their thin-film counterparts. Doping SiNWs has been shown to be much less effective, in terms of providing free carriers to the wires, than expected. Dopants located in the wire core induce a strong resonant backscattering that significantly reduce the mobility of carriers at selected energies. The resonant energies depend on the radial location of the dopants, suggesting a drop of conductance on a large energy window upon random doping. These results make the case for alternative doping strategies such as shell doping in core-shell structures with a properly engineered bandgap alignment between core and shell. In addition, surface dangling-bond defects, the importance of which is enhanced by the large surface/bulk ratio in nanowires, are shown to trap the impurities and neutralize them. Namely, dopants are more stable when located on surface defects where they are transparent to transport. Such results could affect significantly the behavior of small-diameter wires as compared to bulk silicon.

One of the most promising applications of SiNW-based devices is that of serving as electrochemical biosensors. Due to their high surface to volume ratio and tunable electron transport properties SiNWs are potentially capable of detecting the binding of a very few molecules at their surface by changing the intrinsic core conductance or tuning the canal average potential as a "chemical gate". Contrary to the case of nanotube covalent functionalization, the binding of molecules at the surface of NWs can preserve excellent conducting properties if spacers or linkers are used to connect chemically active functional groups to the wire surface. We have shown in particular that the functionalization by alkyl chains, the most standard linkers, hardly affect the conductivity of SiNWs. In particular, the transport remains quasi-ballistic within several subbands below and above the SiNWs bandgap. Functionalization by more active side groups, such as alkenyl, phenyl, amino or alkoxide groups, is shown to be less favorable as resonant backscattering reappears in the valence bands, even though in the conduction bands the conductance is hardly affected. These results are a strong indication that selective functionalization is a viable route for tailoring the properties of SiNWs in (opto)electronic devices and chemical or biological sensors.

References

Adessi, C., Roche, S., Blase, X. *Phys. Rev. B* **73**, 125414 (2006).

Avriller, R., Latil, S., Triozon, F., Blase, X., Roche, S. *Phys. Rev. B* **74**, 121406 (2006).

Avriller, R., Roche, S., Triozon, F., Blase, X., Latil, S. *Mod. Phys. Lett. B* **21**, 1955 (2007).

Balasubramanian, K., Burghard, M. *Small* **1**, 180 (2005).

Bedrossian, P., Meade, R.D., Mortensen, K., Chen, D.M., Golovchenko, J.A., Vanderbilt, D. *Phys. Rev. Lett.* **63**, 1257 (1989).

Bent, S.F. *Surf. Sci.* **500**, 879 (2002).

Blomquist, T., Kirczenow, G. *Nano Lett.* **6**, 61 (2006).

Bruno, M., Palummo, M., Marini, A., Sole, R.D., Ossicini, S. *Phys. Rev. Lett.* **98**, APS, 036807 (2007).

Bunimovich, Y.L., Shin, Y., Yeo, W., Amori, M., Kwong, G., Heath, J. *J. Am. Chem. Soc.* **128**, 16323 (2006).

Calzolari, A., Marzari, N., Souza, I., Buongiorno Nardelli, M. *Phys. Rev. B* **69**, 035108 (2004).

Charlier, J.-C., Blase, X., Roche, S. *Rev. Mod. Phys.* **79**, 677 (2007).

Choi, H., Ihm, J., Louie, S., Cohen, M. *Phys. Rev. Lett.* **84**, 2917 (2000).

Csanyi, G., Albaret, T., Payne, M.C., Vita, A.D. *Phys. Rev. Lett.* **93**, 175503 (2004).

Cui, Y., Duan, X., Hu, J., Lieber, C.M. *J. Phys. Chem. B* **104**, 5213 (2000).

Cui, Y., Lieber, C. *Science* **291**, 851 (2001).

Cui, Y., Wie, Q., Park, H., Lieber, C. *Science* **293**, 1289 (2001).

Cui, Y., Zhong, Z., Wang, D., Wang, W., Lieber, C. *Nano Lett.* **3**, 149–152 (2003).

Datta, S. *Electronic Transport in Mesoscopic Systems* (Cambridge University Press, Cambridge, UK, 1995).

Diarra, M., Niquet, Y.-M., Delerue, C., Allan, G. *Phys. Rev. B (Condens. Matter Mater. Phys.)* **75**, 045301 (2007).

Dyke, C.A., Tour, J.M. *J. Phys. Chem. A* **108**, 11151 (2004).

Fernández-Serra, M., Adessi, C., Blase, X. *Nano Lett.* **6**, 2674 (2006 a).

Fernández-Serra, M.V., Adessi, C., Blase, X. *Phys. Rev. Lett.* **96**, 166805 (2006 b).

Francinelli, A., Tonneau, D., Clément, N., Abed, H., Jandard, F., Nitsche, F., Dallaporta, H., Safarov, V., Gautier, J. *Appl. Phys. Lett.* **85**, 5272 (2004).

Hahm, J.-I., Lieber, C. *Nano Lett.* **4**, 51 (2004).

Haick, H., Hurley, P., Hochbaum, A., Yang, P., Lewis, N. *J. Am. Chem. Soc.* **128**, 8990 (2006).

Holmes, J., Johnston, K., Doty, R., Korgel, B. *Science* **287**, 1471 (2000).

Hybertsen, M.S., Louie, S.G. *Phys. Rev. B* **34**, 5390–5413 (1986).

Ismail-Beigi, S., Arias, T. *Phys. Rev. B* **57**, 11923 (1998).

Iwanari, T., Sakata, T., Miyatake, Y., Kurokawa, S., Sakai, A. *J. App. Phys.* **102**, 114312 (2007).

Kagimura, R., Nunes, R.W., Chacham, H. *Phys. Rev. Lett.* **95**, 115502 (2005).

Kaun, C.-C., Larade, B., Mehrez, H., Taylor, J., Guo, H. *Phys. Rev. B* **65**, 205416 (2002).

Kobayashi, K. *Phys. Rev. B* **69**, 115338 (2004).

Landman, U., Barnett, R.N., Scherbakov, A.G., Avouris, P. *Phys. Rev. Lett.* **85**, 1958–1961 (2000).

Latil, S., Roche, S., Mayou, D., Charlier, J.-C. *Phys. Rev. Lett.* **92**, 256805 (2004).

Lee, Y.-S., Marzari, N. *Phys. Rev. Lett.* **97**, 116801 (2006).

Lee, Y.-S., Nardelli, M.B., Marzari, N. *Phys. Rev. Lett.* **95**, 076804 (2005).

Lew, K.-K., Pan, L., Bogart, T.E., Dilts, S.M., Dickey, E.C., Redwing, J.M., Wang, Y., Cabassi, M., Mayer, T.S., Novak, S.W. *Appl. Phys. Lett.* **85**, 3101 (2004).

Lherbier, A., Persson, M., Niquet, Y.M., Triozon, F., Roche, S. *Phys. Rev. B* **77**, 085301 (2008).

Li, C., He, H.X., Bogozi, A., Bunch, J., Tao, N. *Appl. Phys. Lett.* **76**, 1333 (2000).

Li, Z., Chen, Y., Li, X., Kamins, T., Nuka, K., Williams, R. *Nano Lett.* **4**, 245 (2004).

Ma, D., Lee, C., Lee, S. *Appl. Phys. Lett.* **79**, 2468–2470 (2001).

Ma, D., Lee, C.S., Au, F.C.K., Tong, S.Y., Lee, S.T. *Science* **299**, 1874 (2003).

Markussen, T., Rurali, R., Jauho, A.-P., Brandbyge, M. **99**, APS, 076803 (2007).

Menon, M., Srivastava, D., Ponomareva, I., Chernozatonskii, L.A. *Phys. Rev. Lett.* **70**, 125313 (2004).

Niquet, Y.M., Lherbier, A., Quang, N.H., Fernández-Serra, M.V., Blase, X., Delerue, C. *Phys. Rev. Lett.* **73**, 165319 (2006).

Peelaers, H., Partoens, B., Peeters, F. *Nano Lett.* **6**, 2781–2784 (2006).

Perdew, J.P., Burke, K., Ernzerhof, M. *Phys. Rev. Lett.* **77**, 3865 (1996).

Pierreux, D., Stesmans, A. *Phys. Rev. B* **66**, 165320 (2002).

Ponomareva, I., Menon, M., Srivastava, D., Andriotis, A.N. *Phys. Rev. Lett.* **95**, 265502 (2005).

Ramamoorthy, M., Briggs, E., Bernholc, J. *Phys. Rev. B* **59**, 4813–4821 (1999).

Rocha, A.R., García-Suárez, V.M., Bailey, S., Lambert, C., Ferrer, J., Sanvito, S. *Phys. Rev. Lett.* **73**, 085414 (2006).

Rurali, R., Aradi, B., Frauenheim, T., Gali, A. *Phys. Rev. B* **76**, APS, 113303 (2007).

Rurali, R., Lorente, N. *Phys. Rev. Lett.* **94**, 026805 (2005).

Shirahata, N., Hozumi, A., Yonezawa, T. *Chem. Records* **5**, 145 (2005).

Singh, A.K., Kumar, V., Note, R., Kawazoe, Y. *Nano Lett.* **5**, 2302 (2005).

Singh, A.K., Kumar, V., Note, R., Kawazoe, Y. *Nano Lett.* **6**, 920 (2006).

Soler, J.M., Artacho, E., Gale, J.D., García, A., Junquera, J., Ordejón, P., Sánchez-Portal, D. *J. Phys. Condens. Matter* **14**, 2745 (2002).

Srtoher, T., Cai, W., Zhao, X., Hamers, R., Smith, L. *J. Am. Chem. Soc.* **122**, 1205 (2000).

Streifer, J.A., Kim, H., Nichols, B.M., Hamers, R.J. *Nanotechnology* **16**, 1868–1873 (2005).

Sze, S. *Physics of Semiconductors* (John Wiley & Sons, 1981).

Tersoff, J. *Phys. Rev. B* **37**, 6991–7000 (1988).

Troullier, N., Martins, J.L. *Phys. Rev. B* **43**, 1993 (1991).

Wang, J., Rahman, A., Ghosh, A., Klimeck, G., Lundstrom, M. *Appl. Phys. Lett.* **86**, 093113 (2005).

Winkelmann, C.B., Ionica, I., Chevalier, X., Royal, G., Bucher, C., Bouchiat, V. *Nano Lett.* **7**, 1172 (2007).

Wu, Y., Cui, Y., Huynh, L., Barrelet, C., Bell, D., Lieber, C. *Nano Lett.* **4**, 433 (2004).

Xiang, J., Lu, W., Hu, Y., Yan, H., Lieber, C. *Nature* **441**, 489 (2006).

Xiangfeng, D., Chunming, N., Vijendra, S., Jian, C., Wallace, J., Empedocles, S., Goldman, J. *Nature* **425**, 274 (2003).

Yang, L., Spataru, C.D., Louie, S.G., Chou, M.Y. *Phys. Rev. B* **75**, 201304 (2007).

Yu, J.-Y., Chung, S.-W., Heath, J.R. *J. Phys. Chem. B* **104**, 11864 (2000).

Zhao, X., Wei, C.M., Yang, L., Chou, M.Y. *Phys. Rev. Lett.* **92**, 236805 (2004).

Zhao, Y., Yakobson, B.I. *Phys. Rev. Lett.* **91**, 035501 (2003).

Zheng, G., Lu, W., Jin, S., Lieber, C. *Adv. Mater.* **15**, 1890 (2004).

Zhong, J., Stocks, G. *Nano Lett.* **6**, 128 (2006).

Zhong, Z., Fang, Y., Lu, W., Lieber, C. *Nano Lett.* **5**, 1143 (2005).

3

NEGF-based models for dephasing in quantum transport

Roksana Golizadeh-Mojarad and Supriyo Datta

3.1   Introduction

Although models for coherent quantum transport are fairly well established (Landauer *et al.* 1992), approaches for including incoherent dephasing and spin-flip processes represent an active area of current research. Much of the work is motivated by the basic physics of conductance fluctuation in chaotic cavities (Brouwer *et al.* 1997; Beenakker *et al.* 2005). However, there is also a strong motivation from an applied point of view (see, for example, Venugopal *et al.* 2003). For example, if we calculate the transmission $T(E)$ through a two-dimensional conductor (which could be the channel of a nanotransistor) with a random array of scatterers (due to defects or surface roughness), we see fluctuations as a function of energy that arise from quantum interference. Such fluctuations, however, are seldom observed at room temperature in real devices (when the device length is larger than the phase-relaxation length), because interference effects are destroyed by dephasing processes. Clearly, realistic quantum transport models require including such dephasing processes.

A common way of including incoherent dephasing is through additional Buttiker probes (Buttiker 1988) whose effects can then be modeled within the Landauer–Buttiker framework for coherent transport theory. Recently, this approach has been used to randomize the spin as well (Michaelis *et al.* 2006). Such probes typically also introduce momentum relaxation, since the probes themselves destroy momentum of itinerant electrons by partially reflecting them. With a continuous distribution of Buttiker probes it appears that we do not have the flexibility of adjusting the degree of phase, spin and momentum independently. However, in real devices incoherent dephasing processes arising from "electron–electron" interaction scattering and spin-flip processes arising from interactions with electron/hole spins and nuclear spins destroy only phase or phase and spin but not momentum.

The non-equilibrium Green's function (NEGF) method (see Chapter 1 of this volume) provides a rigorous prescription for including any kind of dephasing mechanisms to any order starting from a microscopic Hamiltonian through an appropriate choice of the self-energy function $\Sigma_s(E)$. In this chapter we describe three simple phenomenological choices of $\Sigma_s(E)$ that allows one to incorporate phase, momentum and spin relaxation independently. Note, we restrict our discussion to elastic dephasing where strictly speaking, no energy is exchanged between the electrons and the dephasing source. Real dephasing mechanisms often involve some energy exchange but this has a negligible effect on the current if the bias and thermal energy ($eV + k_BT$) are small compared to the energy range ε_c, over which the transmission characteristics remain essentially unchanged (see discussion on p. 105 of Datta 1995). Our examples largely correspond to this transport regime with large ε_c, which excludes strong localization. Even if ε_c is small, our elastic dephasing model can be used, but it will not capture effects like hopping that require inelastic interaction.

Finally, we note that our purpose is not to provide a microscopic theory of any specific mechanisms which is already available in the literature (see references cited above). Rather it is to provide an NEGF-based phenomenological model that is comparable to the Buttiker probe model in conceptual and numerical simplicity. Yet it allows one the flexibility of adjusting the degree of momentum phase and spin relaxation independently so as to match experimental values of the momentum-relaxation length L_m, spin-coherence length L_s, and phase-relaxation length L_φ, respectively.

3.2 Dephasing model

The general approach to quantum transport illustrated schematically in Fig. 3.1 provides a general method for modelling a wide class of nanotransistor and spin devices (Datta 1995, 2005. The relevant equations are summarized in Section 1.2.1 of Chapter 1 of this volume).

In NEGF, phase-breaking processes can be described by an additional self-energy Σ_S and in/out-scattering functions $\Sigma_s^{in/out}$ (Danielewicz 1984; Mahan

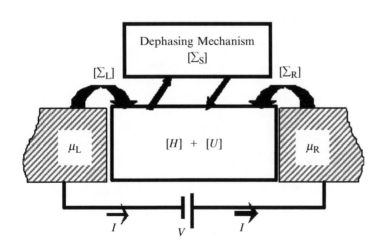

Fig. 3.1 General schematic illustration used for NEGF quantum-transport calculation. Contacts are assumed to remain in local equilibrium with the electrochemical potentials μ_L and μ_R.

1987; Datta 1990; Lake *et al.* 1997). In the first order self-consistent Born approximation, the Σ_S and $\Sigma_s^{in/out}$ due to elastic dephasing processes are given by:

$$\Sigma_s(i, j) = \sum_{k,l} D(i, j, k, l) \, G(k, l), \tag{3.1}$$

$$\Sigma_s^{in/out}(i, j) = \sum_{k,l} D(i, j, k, l) \, G^n(k, l), \tag{3.2}$$

where D is a fourth-order tensor and i, j, k, l can be either spatial or spin indices. We define three different types of elastic dephasing:

$$D(i, j, k, l) = d_m \delta_{ij} \delta_{kl} \delta_{ik} \qquad \text{("Momentum-relaxing" dephasing)} \tag{3.3}$$

$$D(i, j, k, l) = d_s (\vec{\sigma}_{ik} \cdot \vec{\sigma}_{lj}) \qquad \text{("Spin-relaxing" dephasing)} \tag{3.4}$$

$$D(i, j, k, l) = d_p \, \delta_{ik} \delta_{jl}$$
$$\text{for all } i, j, k, l \quad \text{("Pure" dephasing)} \tag{3.5}$$

where d_m, d_s and d_p, constant factors, represent the strength of dephasing. i, j, k and l represent in eqn (3.3) spatial indices, while in eqn (3.4) spin indices. In eqn (3.5) they are both spatial and spin indices. For more details about the definitions above, see Chapter 1, Section 1.2.4.

In the next section using a simple four-probe geometry, we demonstrate that these three choices of D indeed match their definitions. The "momentum-relaxing" dephasing relaxes momentum while conserving spin. The "spin-relaxing" dephasing relaxes spin, but keeps momentum conserved. The "pure" dephasing conserves both spin and momentum. Any linear combination of these three choices of D can be used to adjust momentum-relaxation length L_m, spin-coherence length L_s, and phase-relaxation length L_φ independently as appropriate for a specific problem.

3.3   Effect of different types of dephasing on momentum and spin relaxation

In order to understand the effect of the described dephasings in the last section, we use a particular four-probe geometry. The structure of this four-probe model (Fig. 3.2) is partitioned into the channel and four ferromagnetic contacts with 100% (spin-up/down) polarization. As shown in Fig. 3.2, probe "P_1" injects the spin-up current into the channel, while the rest of the probes are grounded. In this structure, probe "P_1" and "P_3" can only inject/transmit spin-up electrons, while probe "P_2" and "P_4" can just insert/detect spin-down electrons. For simplicity we assume the channel is one-dimensional.

The reason behind using this specific geometry is that by injecting spin-up current to Probe "P_1" and looking at the transmitted currents at the rest of the probes (which are grounded), we can easily investigate the effect of the

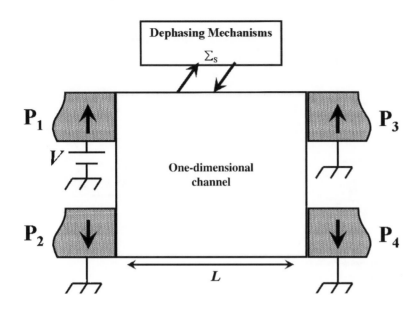

Fig. 3.2 Schematic illustration of four-probe model to understand the effect of different types of dephasing. In this structure, ferromagnetic probes "P_1" and "P_3" can only inject/transmit spin-up electrons, while ferromagnetic probes "P_2" and "P_4" can just insert/detect spin-down electrons. Probe "P_1" injects the spin-up current into the channel, while the rest of the probes are grounded.

different types of dephasing inside the channel on momentum and spin relaxation. For example, if we collect spin-down current at probe "P_4/P_2", we can conclude that the dephasing mechanism inside the channel flips or relaxes spin.

Now we will introduce the different types of dephasing defined in the previous section to the channel of the four-probe model and investigate their effects on momentum and spin relaxations. The details about how to write $[H]$, $[\Sigma_{1,2}]$ and $[\Sigma_s]$ and how to calculate current I for a 1D channel is provided in Chapter 1, Section 1.3.

With "momentum relaxing" dephasing applied into the channel, the detected current at both probes "P_2" and "P_4" is zero (Fig. 3.3(a)); indicating that this choice of D conserves the spin polarization of the injected current (spin-up). However, the transmitted spin-up current at probe "P_3" drops as the channel gets longer (Fig. 3.3(a)) due to the momentum relaxation inside the channel.

By introducing "spin-relaxing" dephasing to the channel, the spin-down current is detected at probe "P_4"; indicating that this model of dephasing indeed relaxes and flips the polarization of the injected spin-up current (Fig. 3.3(b)). Since the sum of the transmitted current into probes "P_3" and "P_4" is equal to the injected current at probe "P_1" independent of the channel length, this choice of D does not relax the momentum (the injected current goes through the channel without changing its momentum and reflecting back). Zero-transmitted current at probe "P_2" also indicates that this type of scattering does not introduce momentum relaxation to the channel, while it does relax the spin.

When "pure" dephasing is applied to the channel of four-probe structure, the zero detected currents at probes "P_2" and "P_4" indicate that this type of dephasing conserves spin (Fig. 3.3(c)). Since the transmitted current at probe

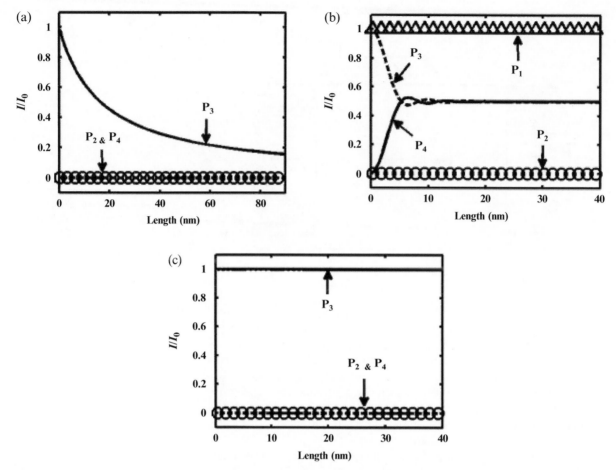

Fig. 3.3 Effect of the different type of dephasings using four-probe geometry. $I_0 = (e/h)(eV)$, where V is the applied bias to probe "P_1". Current from different probes as a function of the channel length (a) "Momentum relaxing" dephasing ($d_m = 0.01\,\text{eV}^2$), (b) "spin relaxing" dephasing ($d_s = 0.001\,\text{eV}^2$), (c) "pure" dephasing ($d_p = 0.001\,\text{eV}^2$). $m = m_e$ (free electon mass), $E_f = 0.43\,\text{eV}$ and $a = 3\,\text{nm}$ where a represents the lattice spacing.

"P_3" is independent of the channel length, one can conclude that this choice of D does not relax momentum. One might wonder what the effect of this type of scattering is on the channel. The answer is that this type of incoherent scattering only relaxes the phase. We will show in the next section that all of the different choices of dephasing discussed here introduce incoherent dephasing to the channel.

3.4   Effect of different types of dephasing on phase relaxation

In this section we show the effect of these three types of incoherent scattering on phase relaxation. Here, we employ the same four-probe configuration

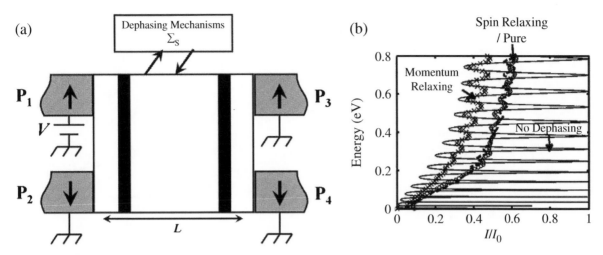

Fig. 3.4 Transmission as a function of energy for a 1D four-probe model with two scatterers as shown in (a). (b) Solid line: without any phase breaking pro-cesses; circled line: with "pure" dephasing; dashed line: with "spin-relaxing" dephasing and crossed line: with "momentum-relaxing" dephasing. Each scatterer is represented by a potential of 0.5 eV at one lattice site ($m^* = m_e$, $L = 12$ nm, $a = 3$ nm, $d_p = 0.001$ eV^2, $d_m = 0.01$ eV^2, $d_s = 0.001$ eV^2).

illustrated in Fig. 3.2, but with two additional scatterers (Fig. 3.4(a)) whose interference in the ballistic regime leads to large oscillations in the current as a function of energy (Fig. 3.4(b) solid line). All of the three types of dephasing randomize phase and destroy oscillations in the current. Although in Fig. 3.4(b) we adjust dephasing factors (d_p, d_m and d_s) to obtain approximately the same phase-relaxation effect from these types of dephasing mechanisms, "momentum-relaxing" dephasing leads to a current that is almost half the current calculated either with "spin-relaxing" dephasing or with "pure" dephasing. This reduction in current indicates that "momentum-relaxing" dephasing adds an additional resistance to the channel. The calculated current with "momentum-conserving" dephasings ("pure" or "spin-relaxing" dephasing) is an average of the oscillating transmission yielded from the coherent dephasing (the areas under the curves shown with solid line and either circled or dash line in Fig. 3.4(b) are the same).

3.5   Calculating L_m, L_s, and L_φ

In this section we show how the momentum-relaxation length L_m, spin-coherence length L_s, and phase-relaxation length L_φ, corresponding to a particular dephasing strength factor d_m, d_s, and d_p can be estimated from a "numerical experiment". Note that by using any linear combination of the three choices of D, we can adjust L_m, L_s, and L_φ independently with the dephasing strength factors d_m, d_s, and d_p.

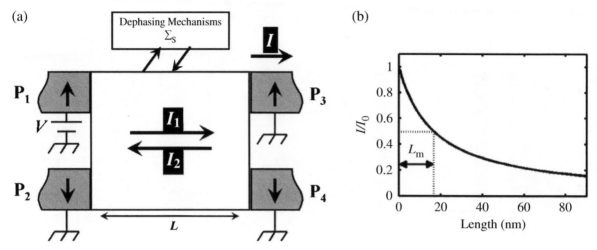

Fig. 3.5 (a) Four-probe model used to estimate L_m. (b) Mean free path L_m estimation from the calculated current I at probe P_3 as a function of the channel length L. ($m^* = m_e$, $a = 3\,\text{nm}$, $d_m = 0.01\,\text{eV}^2$).

3.5.1   Calculating mean-free path L_m

To estimate L_m for a specific d_m, we apply "momentum-relaxing" dephasing with the strength factor of d_m to the channel of a four-probe model (Fig. 3.5(a)), and then we compute the transmitted current I at probe P_3 as a function of the channel length. The mean-free path L_m is approximately equal to the length of the channel where the current I is about half of the current for the zero-length channel [$= I_0 = (e/h)(eV)$] or in other words half of the current of the ballistic channel (Fig. 3.5(b)).

$$I(L_m) = I_0/2. \tag{3.6}$$

We use the following analytical argument to justify what we mentioned above. With "momentum-relaxing" dephasing in the channel, there are two current flows inside the channel, one flowing from probe P_1 into the channel (I_1) and one reflecting back from the channel into probe $P_1(I_2)$ (as shown in Fig. 3.5(a)). The momentum-relaxing mechanism in the channel reflects back a part of the injected current to probe P_1. Hence, these two antipropagating currents, I_1 and I_2, are coupled together by a coupling factor of $\alpha(= 1/L_m)$. The variation of the current I_1 in the channel can be written as

$$\frac{\partial I_1}{\partial x} + \alpha I_1 = \alpha I_2. \tag{3.7}$$

The term αI_1 on the left-hand side of eqn (3.7) represents the decay rate of the current I_1 per unit length, while the term αI_2 on the right-hand side of eqn (3.7) shows the grow rate of the current I_1 per unit length (source term). From the current continuity, we can write

$$I_1(x) - I_2(x) = I, \tag{3.8}$$

and the boundary condition

$$I_1(L) = I, \quad I_2(L) = 0. \tag{3.9}$$

Solving eqn (3.7) by using eqns (3.8) and (3.9), the current I at probe P_3 can be written as

$$I = I_0 \frac{L_m}{L_m + L}. \tag{3.10}$$

From eqn (3.10) it is clear that when the channel length L is equal to the mean-free path L_m, the transmitted current I at probe P_3 is half of the current I_0, as explained above.

3.5.2   Calculating spin-coherence length L_s

To calculate L_s for a specific d_s, we apply "spin-relaxing" dephasing with the strength factor of d_s to the channel of a four-probe model (Fig. 3.6(a)), and then we compute the transmitted spin-down current $I_\downarrow$ at probe P_4 as a function of the channel length. The spin-coherence length L_s is approximately equal to the length of the channel where the current $I_\downarrow$ is about $0.43 (= 0.5[1 - e^{-2}])$ times the current $I_0[= (e/h)(eV)]$ at probe P_1 (Fig. 3.6(b)). Note that since this type of dephasing does not relax momentum the current I_0 at probe P_1 does not change with the channel length [shown in Fig. 3.3(b)] and is the same as the current of the ballistic channel.

$$I_\downarrow(L_s) = 0.5 I_0 (1 - e^{-2}). \tag{3.11}$$

When we apply "spin-relaxing" dephasing in the channel, there are two current flows inside the channel, one spin-up current $I_\uparrow$ flowing from probe P_1 to probe P_3 and one spin-down current $I_\downarrow$ going from probe P_1 to probe P_4 (shown in Fig. 3.6(a)). Since the spin-relaxing dephasing in the channel flips the spin polarization of the currents inside the channel. Therefore, these two copropagating currents, $I_\uparrow$ and $I_\downarrow$, are coupled together by a coupling

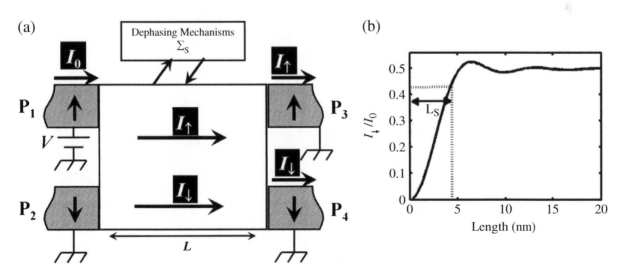

Fig. 3.6 (a) Four-probe model used to estimate L_s. (b) Spin-coherence length L_s estimation from the calculated current $I_\downarrow$ at probe P_4 as a function of the channel length L. ($m^* = m_e$, $a = 3\,\text{nm}$, $d_s = 0.001\,\text{eV}^2$).

factor of $\alpha(=1/L_s)$. The variation of the current $I_\uparrow$ in the channel can be described as

$$\frac{\partial I_\uparrow}{\partial x} + \alpha I_\uparrow = \alpha I_\downarrow. \tag{3.12}$$

This equation follows the same argument as eqn (3.7). From the current continuity, we can write

$$I_\uparrow(x) + I_\downarrow(x) = I_0, \tag{3.13}$$

and the boundary condition

$$I_\uparrow(0) = I_0, \, I_\downarrow(0) = 0. \tag{3.14}$$

Solving eqn (3.12) with eqns (3.13 and 3.14), the spin-down current $I_\downarrow$ at probe P_4 can be written as

$$I_\downarrow = 0.5 I_0 (1 - e^{-\frac{2L}{L_s}}). \tag{3.15}$$

From eqn (3.15) it is clear that when the channel length L is equal to the spin-coherence length L_s, the spin-down current $I_\downarrow$ at probe P_4 is about 0.43 time of the current I_0 as shown above.

3.5.3   Calculating phase-relaxation length L_φ

In order to calculate the phase-relaxation length L_φ, we look at the coherent part of the transmitted current through the channel. The coherent part of the current I_{coh} (which can be calculated from eqn (1.4c), Chapter 1), as a function of the channel length shows how far the coherent part of the current injected at probe 'P$_1$' can transmit inside the channel before its phase becomes randomized by a phase-breaking process (Fig. 3.7(a)).

Using the same structure shown in Fig. 3.4(a), one can validate the estimated L_φ from the transmitted current as a function of energy (Fig. 3.7(b)). As long

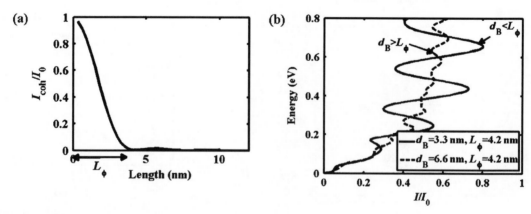

Fig. 3.7 (a) Phase-relaxation length L_φ estimation from the coherent part of the transmitted current I_{coh} as a function of the channel length L. (b) Validating estimated L_φ using four-probe model with two scatterers.

as the spacing between the two scatters is longer than the phase-relaxation length L_φ, the fluctuations in the transmission are destroyed, otherwise the dephasing processes do not remove the oscillations.

NOTE: "Pure" dephasing mimics electron–electron interactions and also helps the NEGF transport model transition to a semiclassical model without scattering.

3.6   Example: "spin-Hall" effect

Any linear combination of these three choices of D can be used to adjust phase, spin- and momentum-relaxation lengths independently as appropriate for a specific problem. In this section we will look at an example considering these three types of dephasing mechanisms. We study the spin accumulation in a two-dimensional conductor due to the intrinsic "spin-Hall" effect in different transport regimes, from ballistic to diffusive using our different choices of dephasing. Note the "spin-Hall" effect is usually associated with an accumulation of spin-up and -down electrons at the opposite edges of a finite-sized sample.

For this example, we start with a two-dimensional conductor (Fig. 3.1) described by a single-band effective-mass Hamiltonian

$$H_0 \equiv (p_x{}^2/2m^*) + (p_y{}^2/2m^*) + U(y), \tag{3.16}$$

with a confinement potential $U(y)$ along with an additional spin-orbit interaction assumed to have the Rashba form

$$H \equiv H_0 I + (\alpha/\hbar)(\sigma_x p_y - \sigma_y p_x), \tag{3.17}$$

where α is the spin-orbit coupling factor and where σ presents the Pauli matrix. We can find the details about how to write $[H]$, $[\Sigma_{1,2}]$ and $[\Sigma_s]$ for a rectangular conductor in matrix format in the finite-difference method in Chapter 1, Section 1.3.1. The information about how to include spin-orbit coupling into the single-band effective-mass equation is provided in Chapter 1, Section 1.3.5.

Without any dephasing mechanism in the sample, the net z-directed spin density per unit applied bias due to the spin-orbit coupling is shown in Fig. 3.8(a) with identical parameters corresponding to the GaAs epilayer used in the recent experiment: $m^* = 0.07\,m_e$ (m_e: free-electron mass), $\alpha = 1.8\,\text{meV nm}$ (as measured), $k_f \cong 10^8\,\text{m}^{-1}$ corresponding to an electron density of $n = k_f{}^3/3\pi^2 \cong 3 \times 10^{16}\,\text{cm}^{-3}$ and $E_f = 5\,\text{meV}$.

The oscillations in the spin density in Fig. 3.8(a) correspond to the Fermi wave number k_f. In order to smooth out these oscillations, we introduce "pure" dephasing to the channel (Fig. 3.8(b)), and we adjust the strength of this dephasing $d_p (= 6 \times 10^{-5}\,\text{eV}^2)$ in a way that $L_\varphi k_f < 1$ as shown in Fig. 3.8(c).

As is clear from Figs. 3.8(a) and (b), even in the ballistic limit, we observe the "spin-Hall" effect. We find that the linear spin (z-directed) accumulation

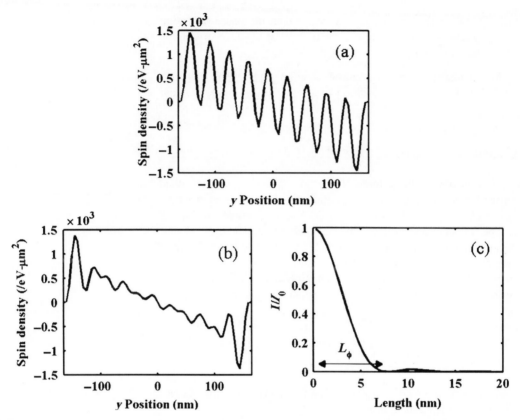

Fig. 3.8 "Spin-Hall" effect in a ballistic sample (a) Without any dephasing. (b) With "pure" dephasing. (c) Phase-relaxation length L_φ estimation from the coherent part of the transmitted current I_{coh} as a function of the channel length L ($d_p = 6 \times 10^{-5}$ eV^2).

across the sample (Figs. 3.8(a) and (b)) can be described approximately by ($-W/2 < y < +W/2$)

$$\tilde{n}_z(y) \approx (k_\alpha^2/2\pi)[M(E_f)/(E_f - E_c)](-2y/W)V, \qquad (3.18)$$

where $k_\alpha \equiv m^*\alpha/\hbar^2$ and M is the number of transverse modes or subbands, which can be estimated from the expression $M = 2 k_f W/\pi$ (E_c: conduction-band edge, V: applied voltage). Equation (3.18) describes our numerical results fairly well and can be justified from a Landauer approach as explained in Golizadeh-Mojarad and Datta (2007). Note eqn (3.18) is valid when $k_\alpha W \ll 2\pi$. The maximum spin accumulation at the edges in the ballistic limit can be written from eqn (3.18) as

$$[\tilde{n}_z]_{max} \approx (m^*/2\pi \hbar^2)(k_\alpha/k_f)(k_\alpha W)V. \qquad (3.19)$$

By introducing "momentum-relaxing" and "spin-relaxing" dephasing gradually to the sample, we will study the spin accumulation in this conductor continuously from the ballistic to the diffusive regime.

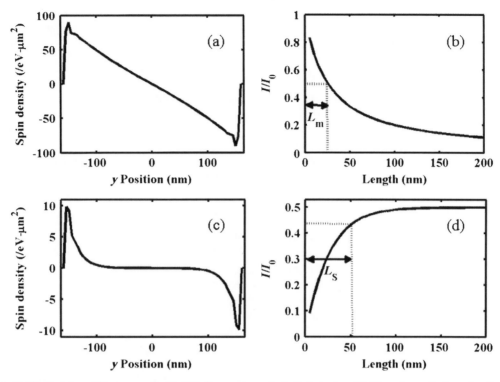

Fig. 3.9 "Spin-Hall" effect in a diffusive sample (a) With "momentum-relaxing" dephasing. (b) Mean free path L_m estimation ($d_m = 1 \times 10^{-3}\,\mathrm{eV}^2$). (c) With "spin-relaxing" dephasing. (d) Spin-coherence length L_s calculation ($d_s = 3 \times 10^{-5}\,\mathrm{eV}^2$).

First, we only add "momentum-relaxing" dephasing and adjust the mean-free path L_m in the channel. As shown in Fig. 3.9(b), the method described in Section 5.1 is used to estimate L_m. With "momentum-relaxing" dephasing, the spin density drops linearly across the channel (Fig. 3.9(a)). The reduction of spin-accumulation density in the diffusive limit when $L_m < W \ll L_s$ (Fig. 3.9(a)) from the ballistic limit (Fig. 3.8(a)) by a factor of ~ 0.06 can be understood as a reduction in applied voltage V by $L_m/(L + L_m) \sim 0.06$ ($L_m \sim 25\,\mathrm{nm}$, $L = W = 350\,\mathrm{nm}$). That is because in a diffusive sample, we should use an effective voltage V_{eff} obtained by scaling down the full voltage V applied across the length L by the factor $L_m/(L + L_m)$

$$V_{\mathrm{eff}} = V L_m/(L + L_m) = F L L_m/(L + L_m) \approx F L_m, \qquad (3.20)$$

where F is the electric field. This can be justified by noting that in a diffusive sample the separation in the electrochemical potentials for $+k$ and $-k$ states is $L_m/(L + L_m)$ times the applied voltage (Datta 1995).

Later, we apply "spin-relaxing" dephasing to the channel and by changing d_s the spin-coherence length L_s is altered to be much smaller than the channel width W. We use the process explained in Section 3.5.2 to calculate L_s (Fig. 3.9(d)). When $L_s \ll W$, the spin density localizes within an L_s near the edges (Fig. 3.9(c)) as observed experimentally (Kato *et al.* 2004; Sih *et al.* 2006). Note since $L_s \ll W$, the width W entering eqn (3.19) should be replaced by an effective width W_{eff} determined by the spin-coherence length L_s as

$$W_{eff} = WL_s/(L_s + W).   \tag{3.21}$$

The reduction in the spin-accumulation density when $L_m, L_s \ll W$ in Fig. 3.9(c) from the ballistic case by a factor of ~ 0.007 can be explained as a product of two factors: a reduction in V_{eff} by ~ 0.06 (mentioned above) and a reduction in W_{eff} by $L_s/(L_s + W) \sim 0.12(L_s = 50\,nm)$. In summary, in the diffusive regime the maximum spin accumulation density due to the intrinsic "spin-Hall" effect can be written as

$$\left[\tilde{n}_z\right]_{max} = (m^*/2\pi\hbar^2)(k_\alpha^2/k_f)W_{eff}V_{eff},   \tag{3.22}$$

which describes our NEGF-based results fairly well.

Although eqn (3.22) is obtained for a 2D structure, it can be applied to the 3D settings in the experiments (Kato *et al.* 2004; Sih *et al.* 2006). Note in the experiments the spin density was measured optically at the surface of a GaAs layer, 2 μm thick. As shown in the experiment (Sih *et al.* (2006)), the experimentally observed spin accumulation extends a distance of $\sim 10\,\mu m (= L_s)^2$ from each edge, while the thickness of the experimental layer is only a fraction of this distance. Hence, the calculated spin accumulation due to the spin-orbit interaction on the surface can penetrate into the 2-μm layer. Using the same parameters as noted in the experiment; $W \approx 50\,\mu m$, $F = 10\,meV/\mu m$, $L_s \approx 10\,\mu m$, $L_m \approx 0.3\,\mu m \ll L \approx 300\,\mu m$ (assuming a mobility of 3–4 m^2/V s); we calculate $V_{eff} \approx 3\,meV$ and $W_{eff} \approx 8\,\mu m$. m^* and k_f are chosen to reflect those in the experiment, as mentioned earlier. We use an $\alpha (= 1.8\,meV\,nm)$ appropriate for confined GaAs layers (Sih *et al.* 2006), assuming it represents the condition at the surface of a bulk sample. From eqn (3.22) we obtain ~ 90 spins per μm^2 or ~ 180 spins per μm^3 (since effective thickness $\sim 0.5\,\mu m$), which is larger than the spin accumulation observed experimentally, suggesting that the experiments in GaAs can be understood without invoking additional effects.

Figure 3.10(a) shows one of the structures studied experimentally by Sih and colleagues (2006) to generate spin-up and -down accumulation in the middle of the device. In this structure, at negative values of x, the only edges are at the two ends, while for positive values of x, there is a notch in the middle with two additional edges. The NEGF-based calculated spin density in both ballistic and diffusive regimes ($L_s = 50\,nm$ and $L_m \sim 25\,nm$), Figs. 3.10(b) and (c), respectively, look very similar in shape to the experimental results (including the "anticipatory" effect ahead of the notch) and also agrees with the sign of the effect. Note that while the experimental structure has $W \sim 50\,\mu m$, the structure studied theoretically has $W \sim 175\,nm$. However, in the theoretical model we have deliberately introduced a large amount of scattering such that $L_s \sim 50\,nm$, making the L_s/W ratio comparable to the experimental setting.

Another interesting structure (Fig. 3.11(a)) studied experimentally (Sih *et al.* 2006) shows the diffusion of the accumulated spin away from the main

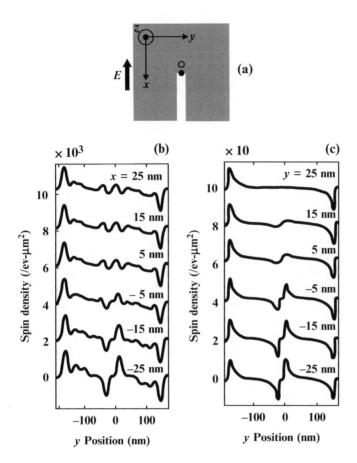

Fig. 3.10 (a) Schematic illustration of a channel splitting into two smaller channels. O indicates the origin. Spin density as a function of transverse position y for different longitudinal positions x from (b) NEGF-based calculation in the ballistic regime averaged over an energy range of $E_f \pm k_B T$ or (c) NEGF-based calculation in diffusive regime ($L_m = 25$ nm, $L_s = 50$ nm).

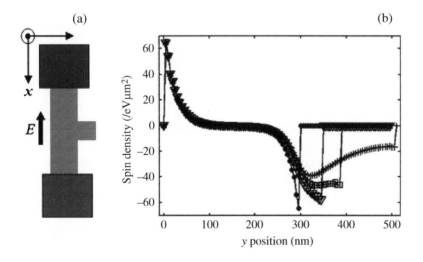

Fig. 3.11 (a) Schematic device geometry. (b) NEGF-based calculated spin density as a function of position y in the diffusive regime for different sidearm lengths ($L_s \sim 100$ nm).

current path into protruding sidearms and these are also in good agreement with our numerical model, especially when the spin-coherence length L_s is adjusted to have the same ratio L_s/W as the experimental structures (Fig. 3.11(b)).

3.7   Summary

In summary, we have proposed an NEGF-based dephasing model with three specific choices of the self-energy Σ_s that provide "momentum-relaxing", "spin-relaxing" and "pure" dephasing. Any linear combination of these three choices can be used to adjust phase-momentum- and spin-relaxation lengths independently, as appropriate for a specific problem. We believe this approach provides a flexibility that is not currently available, while retaining the simplicity of other phenomenological models.

The dephasing approach described here based on NEGF formalism allows one to investigate transport phenomena continuously from the ballistic to the diffusive limit. As an example, in this chapter we examined the spin-Hall effect all the way from the ballistic to diffusive regime and showed how the spin accumulation evolves as momentum and/or spin-relaxation processes are introduced to the system in a controlled way. At the end we presented analytical expressions that describe our numerical results fairly well. We further showed good quantitative agreement with recent experimental observations in GaAs suggesting that the spin-Hall effect in GaAs can be understood in terms of an intrinsic effect driven by the Rashba interaction. For more NEGF-based discussion of spin transport, see Chapter 24 of this volume.

References

Beenakker, C.W.J., Michaelis, B. *J. Phys. A* **38**, 10639 (2005).

Brouwer, P.W., Beenakker, C.W. *J. Phys. Rev. B* **55**, 4695 (1997).

Buttiker, M. *IBM J. Res. Dev.* **32**, 72 (1988).

Danielewicz, P. *Ann. Phys.* **152**, 239 (1984).

Datta, S. *J. Phys.: Condens. Matter* **2**, 8023 (1990).

Datta, S. *Electronic Transport in Mesoscopic Systems* (Cambridge University Press, Cambridge, 1995).

Datta, S. *Quantum Transport: Atom to Transistor* (Cambridge University Press, Cambridge, 2005).

Golizadeh-Mojarad, R., Datta, S. *Phys. Rev. B* **75**, 081301(R) (2007).

Golizadeh-Mojarad, R., Datta, S. cond-mat/0703280 (2007).

Kato, Y.K., Myers, R.C., Gossard, A.C., Awschalom, D.D. *Science* **306**, 1910 (2004).

Lake, R., Klimeck, G., Bowen, R.C., Jovanovic, D. *J. Appl. Phys.* **81**, 7845 (1997).

Landauer. R. *Phys. Scr. T* **42**, 110 (1992).

Mahan, G.D. *Phys. Rep.* **145**, 251 (1987).

Michaelis, B., Beenakker, C.W. *J. Phys. Rev. B* **73**, 115329 (2006).

Sih, V., Lau, W.H., Myers, R.C., Horowitz, V.R., Gossard, A.C., Awschalom, D.D. *Phys. Rev. Lett.* **97**, 096605 (2006).

Venugopal, R., Paulsson, M., Goasguen, S., Datta, S., Lundstrom, M.S. *J. Appl. Phys.* **93**, 5613 (2003).

4 Molecular nanowires and their properties as electrical conductors

George Kirczenow

4.1   Introduction

Since the seminal theoretical proposal of Aviram and Ratner (1974) that certain single molecules may act as current rectifiers, there has been continuing interest in the possibility that individual molecules may be the ultimate nanoelectronic devices; for a review of the early history of this idea and its pre-history see Hush (2003). However, the first indications that this may be more than just a theoretical possibility came in the late 1990s when researchers learned to make molecular nanowires and measure their current–voltage characteristics. This remarkable development was widely recognized as a crucial step towards the realization of single-molecule nanoelectronic devices and stimulated an intense world-wide experimental and theoretical research effort in single-molecule nanoelectronics that has continued to grow through the last decade. During that time much has been learned but many crucial questions remain unanswered and many have very likely not yet been asked.

Given the rapid growth of this area of research and the controversial nature of much of what has been published in the way of both experiments and theory there is at this time a need for a critical assessment of the field clarifying what the key results are, what they mean, what the caveats are, how much confidence one should have in the different experimental and theoretical methodologies that are being used and in the results that have been obtained, and to what extent real contact is being made between the experiments and the theories. The goal of this chapter is to address this need. After ten years of intense research activity, it is no longer feasible to review all of the experimental and theoretical papers that have contributed to this subject. Therefore, this chapter is an attempt to make sense of what has been happening in the field and where it may be going and also to provide an introduction to the subject suitable for researchers new to the field rather than to produce an exhaustive survey. I apologize to the authors of many excellent papers a discussion of which I have not been able to include here.

The chapter is organized as follows: In Section 4.2 molecular nanowires are defined. Then, to clarify the definition a specific example of a molecular nanowire is described, and the idea of molecular nanowire self-assembly is introduced. Common ways in which molecular nano-wires are realized in the laboratory are outlined in Section 4.3 together with some salient comments regarding the relationships between these methodologies, the systems that are produced and some experiments being performed on them. In Section 4.4 one of the major experimental challenges in this field, namely, the current lack of direct experimental measurements of the detailed atomic-scale structure of molecular wires and especially of the atomic geometries of the bonding between molecule and electrodes is discussed in some depth, together with its implications and the efforts being made to address or circumvent this challenge. In Section 4.5 I give a brief overview of the different kinds of molecules, electrodes and linkers out of which molecular nanowires are being or may be constructed and comment on some of the distinctive characteristics of the various systems assembled out of these ingredients. In Section 4.6 the discussion of the theoretical foundations opens with an introduction to and discussion of the Landauer approach to electrical conduction in molecular nanowires. In Section 4.7 the principles and limitations of *ab-initio* and semi-empirical modelling of molecular nanowires in the context of electrical conduction are discussed. In Section 4.8 the contact being made between theory and experiment is addressed by discussing four specific experimental systems and the extent to which their observed behavior has been understood theoretically. Section 4.9 is a summary focusing on key issues for the future development of the field.

4.2   What are molecular nanowires?

A molecular nanowire is an organic molecule with dimensions on the nanometer scale and with two electrodes in its close proximity, one electrode being the electron source from which electrons enter the molecule, the other being the electron drain to which the electrons then flow when a bias voltage is applied across the molecule. The molecule may be attached to one or both of the electrodes through strong chemical bonds or by weak van der Waals forces.

One of the most studied (but still very controversial) examples of a molecular nanowire is the benzenedithiolate molecule contacted with gold electrodes. A schematic representation of one of the many possible atomic geometries of this system is shown in Fig. 4.1. The molecule consists of a six-carbon-atom benzene ring with two sulfur atoms bound to the carbon atoms at opposite sides of the ring replacing hydrogen atoms that occupy those positions in benzene. The sulfur atoms bond strongly to gold and thus act as linkers or "alligator clips" connecting the molecule both structurally and electrically to the gold electrodes. In Fig. 4.1 the sulfur atom on the left bonds to a single protruding gold atom while that on the right is located over a hollow site between three gold atoms. However, many different spatial arrangements of the benzene ring,

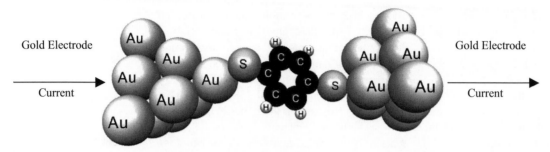

Fig. 4.1 One of many geometries of a molecular nanowire with a benzenedithiolate molecule bridging two gold electrodes of which only a few atoms are shown. An electric current flows through the molecule if a bias voltage is applied between the electrodes.

the gold surface atoms and the sulfur linkers are believed to be possible and are probably being realized in the experiments (Reed *et al.* 1997; Xiao *et al.* 2004a; Dadosh *et al.* 2005; Tsutsui *et al.* 2006, 2008; Venkataraman *et al.* 2006b; Lörtscher *et al.* 2007; Martin *et al.* 2008) on this system. Electrical conduction through these molecular nanowires has been shown theoretically to be sensitive to the details of the atomic geometry (Emberly and Kirczenow 1998a, 2001b,c, 2003; Di Ventra *et al.* 2000a; Kornilovitch and Bratkovsky 2001; Bratkovsky and Kornilovitch 2003; Toher *et al.* 2005; Bratkovsky 2007 and many others), and the measured electrical conductances reported by the different experimental groups have also differed by as much as two orders of magnitude. However, to date it has not been possible to measure the atomic-scale geometry experimentally, as will be discussed further in Section 4.4. Because of this, comparison between theory and experiment continues to be subject to large uncertainties. The gold–benzenedithiolate–gold system will be revisited from time to time throughout this chapter and will be discussed in detail in Section 4.8.2.

Molecular nanowires in which the molecule bonds chemically to both electrodes (as in Fig. 4.1) are very appealing in part because they can be made by *self-assembly*. What is meant by this can be understood in the simplest terms by considering the following example: If a benzenedithiol molecule HS–C_6H_4–SH (similar to the benzenedithiolate shown in Fig. 4.1 but with a hydrogen atom bound to each sulfur atom to form a thiol (SH) group) finds itself in a break junction in a gold wire, a chemical reaction can occur in which the H atoms detach from the thiol groups and the sulfur atoms bond instead to the gold. Thus, if the break in the gold wire is of an appropriate size, structures such as that in Fig. 4.1 with a benzenedithiolate molecule bridging the two gold electrodes should have a non-zero probability of forming spontaneously and, having formed, they should be somewhat robust because of the strong bonding between the gold and the sulfur atoms provided that the geometry of metal electrodes is sufficiently stable, as it can be, for example, for pairs of nanoscale electrodes formed in mechanically controlled break junctions (Reed *et al.* 1997) to be described below.

4.3 Molecular nanowires have been realized in a variety of ways

One approach to the realization of molecular nanowires has been to introduce molecules into a nanoscale gap in a thin metal wire on an insulating substrate, the gap being formed by breaking the wire by stretching it and then returning the broken ends to close proximity with each other (Reed *et al.* 1997; Kergueris *et al.* 1999; Reichert *et al.* 2002; Kiguchi *et al.* 2008; Martin *et al.* 2008; Tsutsui *et al.* 2008). Because metals are ductile the stretching and breaking process can result in the broken ends of the metal wire being sharp on the atomic scale so that it is possible for them to be bridged by just one or a few molecules. Such *mechanically controlled break junctions* have the important virtues of simplicity and stability.

Alternatively, the break in the metal wire may be created by passing an electric current through the wire and thus inducing electromigration of the metal atoms that eventually breaks the wire. The resistance of the wire is monitored as it thins and the gap opens in it (Park *et al.* 2000). An advantage of this approach is that no mechanical apparatus is involved, however, metal nanoparticles can form in or near the gap in the metal wire and can strongly influence or even dominate electrical conduction through the device (de Picciotto *et al.* 2005; van der Zant *et al.* 2006).

Another widely used approach has been to contact a molecule adsorbed on a metal (Bumm *et al.* 1996; Datta *et al.* 1997) or semiconductor (Wolkow 1999) substrate with a scanning tunnelling microscope (STM) or conducting atomic force microscope (AFM) tip directly, or indirectly (Cui 2001) via a metal nanoparticle bound to the molecule. A related method (the STM break junction technique) that is increasingly the experimental technique of choice today is to drive a gold STM tip into a gold substrate with adsorbed molecules (and/or with the tip and substrate immersed in a solution of the molecules) so as to make good metal-to-metal electrical contact between tip and substrate and then to retract the tip until a nanoscale gap is formed between the tip and substrate. Sometimes, the gap is spontaneously bridged by one or a few molecules and this is detected by monitoring the electrical conductance of the tip–substrate junction as the tip is retracted (Xiao *et al.* 2004a,b,c, 2005; Xu *et al.* 2004, 2005; He *et al.* 2005a,b; Chen *et al.* 2006a; Li *et al.* 2006a,b; Venkataraman *et al.* 2006a,b, 2007; Quinn *et al.* 2007a). More elaborate methods are also used in which a molecule bridges a pair of metal nanoparticles by bonding to each of them chemically and then this nanoparticle dimer is electrostatically trapped between a pair of metal electrodes each of which contacts one of the metal nanoparticles electrically (Dadosh *et al.* 2005).

Still other ways are to evaporate (Chen *et al.* 1999) or stamp (Loo *et al.* 2002) a metal film onto a molecular monolayer adsorbed on a nanoscale region of a metal or semiconductor substrate or to bring together two crossed metal wires one of which is coated with molecules (Kushmerick *et al.* 2002a). Various chemical and electrochemical techniques are also being used to produce nanoscale metal–molecular monolayer–metal sandwiches (Cai *et al.* 2004, 2005; Selzer *et al.* 2005; Tang *et al.* 2007). All of these methods yield not a single molecular nanowire but large numbers of molecules between the electrodes. Cartoons are often drawn representing these systems as orderly arrays of molecular nanowires bridging the electrodes electrically in a parallel arrangement. However, this can be misleading: Putative per-molecule electrical conductances of such metal–molecular monolayer–metal sandwiches can differ by orders of magnitude from those of single-molecule junctions for many reasons (Selzer *et al.* 2005). At the simplest level it is important to bear in mind that the fraction of the molecules in the film that make good electrical contact with *both* metal electrodes is not measured directly and may in some cases be very small and that, conversely, even a single defect in the molecular film can dominate electrical conduction through such a device. Nonetheless, some of the most interesting molecular nanoelectronic phenomena (that may eventually lend themselves to technological applications) were observed initially in experiments on such metal–molecular monolayer–metal sandwiches. These phenomena include negative differential resistance, voltage-triggered switching between strongly and weakly electrically conducting states and memory effects (Chen *et al.* 1999, 2000; Collier *et al.* 2000; Lau *et al.* 2004; Cai *et al.* 2005; Gergel-Hackett *et al.* 2006; Majumdar *et al.* 2006). Various explanations have been suggested: It has been proposed that they may be primarily single-molecule properties due to such mechanisms as charging of the molecule (Seminario *et al.* 2000), ligand rotation (Di Ventra *et al.* 2001), changes in the molecule–electrode hybridization or bond angle (Bratkovsky and Kornilovitch 2003; Keane *et al.* 2006; Bratkovsky 2007) or other molecular rearrangements (Collier *et al.* 2000), resonant tunnelling (Cornil *et al.* 2002) or polaronic effects (Galperin *et al.* 2005, 2008; Troisi and Ratner 2006a; Bratkovsky 2007; Yeganeh *et al.* 2007). It has also been proposed that they may be due to or controlled by interactions between several or many molecules (Taylor *et al.* 2003; Gergel-Hackett *et al.* 2006; Majumdar *et al.* 2006; Long *et al.* 2007), or that in some cases they may *not* be inherently molecular effects at all but due instead to the formation of metal filaments between the electrodes (Lau *et al.* 2004; Stewart *et al.* 2004; Beebe and Kushmerick 2007; Bratkovsky 2007; Miao *et al.* 2008), possibly influenced by the Casimir force between electrodes (Katzenmeyer *et al.* 2008). Recently, the experimental situation became clearer when certain molecules with NO_2 groups in monolayer films and similar single molecules were contacted with STM, mechanically controlled break junctions and other probes and negative differential resistance and/or hysteretic voltage-triggered switching were observed in these systems (Fan *et al.* 2004; Blum *et al.* 2005; Keane *et al.* 2006; Kushmerick *et al.* 2006; Lörtscher *et al.* 2006): In these experiments the observed phenomena could be convincingly identified as molecular in origin. However, both voltage-triggered and stochastic switching have also been reported in systems with molecules without NO_2 groups (Danilov *et al.* 2006). Which, if any, of the above mechanisms are responsible for the switching phenomena remains uncertain; there has as yet been no "smoking gun" experiment clarifying this.

A few experiments have also been reported in which oblique evaporation of metal electrodes onto insulating substrates through shadow masks (Kubatkin *et al.* 2003; Danilov *et al.* 2006) or onto different sides of an insulating tip

(Zhitenev *et al.* 2002) while monitoring the conductance of the device has been used to make nanoscale junctions that may have been bridged by as few as a single molecule. Similar methods have also been used to fabricate large numbers of junctions, each bridged by thousands of molecules (Shamai *et al.* 2007).

In some cases a third electrode is also present, separated from the molecule by an insulating layer (Park *et al.* 2000, 2002; Liang *et al.* 2002; Zhitenev *et al.* 2002; Kubatkin *et al.* 2003; Lee *et al.* 2003; Yu *et al.* 2004; de Picciotto *et al.* 2005; Heersche *et al.* 2006; Jo *et al.* 2006; Keane *et al.* 2006; Natelson *et al.* 2006; Goto *et al.* 2007; Henderson *et al.* 2007; Osorio *et al.* 2007a; de Leon *et al.* 2008). The role of this third electrode has been that of a weakly coupled gate, i.e. it has served to modify the electric potential at and around the molecule and thus to influence electrical conduction through the molecule. Thus, systems of this kind are referred to as "single-molecule transistors." In practice, the presence of the insulating layer between the gate and the molecule and screening of the gate potential by the other two conducting electrodes that are usually closer to the molecule than the gate electrode result in the changes in the potential at the molecule that can be imposed by varying the potential applied to the gate being weak, i.e. much smaller than the changes in the bias voltage applied to the gate. This limitation can be overcome by means of electrochemical gating. Here, both the molecular nanowire and the third electrode are immersed in an electrolyte solution and by varying the chemical species in solution and/or the potential applied to the third electrode it is possible to switch molecular nanowires with redox centers between oxidized and reduced states thereby changing the electrical conductances of these wires (Tao 1996; Jäckel *et al.* 2004; Albrecht *et al.* 2005a,b, 2006a,b,c, 2007; Alessandrini *et al.* 2005; Xiao *et al.* 2005, 2006; Xu *et al.* 2005; He *et al.* 2006; Chen *et al.* 2006b; Li *et al.* 2006b, 2007; Haiss *et al.* 2007; Leary *et al.* 2008).

4.4 A major challenge: The atomic-scale geometry is not known

As has already been mentioned, a price is paid for the attractive features of gold-thiol chemistry-based self-assembly, namely, that determining *experimentally* the atomic structures of the molecular nanowires thus formed has not been feasible to date: Introducing a third nanoscale electrode between the two electrodes to which the molecule has bonded and using the third electrode as a scanning probe has not been feasible due to geometrical constraints. Although transmission electron microscopy (TEM) has been used to image the individual atoms of a gold atomic chain bridging two electrodes (Ohnishi *et al.* 1998), this technique has not succeeded to date in imaging the atomic structures of molecular nanowires because these consist of lighter atoms to which TEM is less sensitive. Thus, the atomic geometry of such a molecular wire does not lend itself readily to being measured in this way. Moreover, it is often not even known whether a single molecule or more than one is directly involved in electrical conduction between the electrodes, and different assumptions regarding this can yield theoretical conductance values for the device that differ by orders of magnitude (Emberly and Kirczenow 2001b,c). Given the lack of a direct knowledge of the atomic-scale geometry, it would be desirable to image such molecular wires even with nanometer resolution. Ballistic electron emission microscopy is capable of imaging individual molecules through a thin metal electrode with such resolution (Kirczenow 2007). However, while this technique has recently begun to be applied to molecular films at metal semiconductor interfaces (Li *et al.* 2005), experimental resolution of individual molecules in this way has not yet been achieved.

Many software packages can calculate low-energy atomic geometries of molecular nanowires and the metal contacts to which they bond by using density-functional theory and/or molecular-mechanics models. However, this theoretical approach is limited in that it does not determine which of many possible low-energy configurations is actually realized in a particular experiment. Furthermore, recent experimental work using the STM break junction technique (Xiao *et al.* 2004a; Xu *et al.* 2004; Li *et al.* 2006a; Venkataraman *et al.* 2006b) has yielded convincing evidence that a *different* atomic configuration is realized each time that a new molecular nanowire is formed even though the molecular species involved and the material of the electrodes remain the same (Hu *et al.* 2005; Andrews *et al.* 2008). Thus, a powerful and increasingly used experimental methodology (Xiao *et al.* 2004a,b,c, 2005; Xu *et al.* 2004, 2005; He *et al.* 2005a,b; Chen *et al.* 2006a; Li *et al.* 2006a,b; Venkataraman *et al.* 2006a,b, 2007; Quinn *et al.* 2007a; Yamada *et al.* 2008) is to repeat the process of making a molecular nanowire and measuring its electrical conductance many times so as to gather statistics on electrical conduction for many different atomic geometries of the nanowire. This statistical approach has made it possible to compare the electronic transport properties of nanowires made with different molecular species and thus to identify trends that would otherwise be hidden because of random variations in the nanowire geometry (Venkataraman *et al.* 2006a, 2007; Lindsay and Ratner 2007; Quinn *et al.* 2007a; Hybertsen *et al.* 2008). It represents a major advance in experimental technique. However, it does not determine what the atomic geometries being realized experimentally are. This is a significant limitation of the method since, while it facilitates comparisons between theory and experiment (Lindsay and Ratner 2007; Hybertsen *et al.* 2008), it leaves open the question whether the atomic geometry assumed in calculations is similar to the geometries realized most frequently in the experiments. Moreover, such statistical experiments carried out with the same molecules and electrode materials but under differing conditions or in different ways can yield very different results, as recent statistical studies (for example Xiao *et al.* 2004a and Lörtscher *et al.* 2007) of the benzenedithiol-gold system that yielded the most probable conductances of the molecular nanowires differing by two orders of magnitude have demonstrated.

The uncertainties regarding the geometries of molecular nanowires can be *partly* resolved by studying molecular nanowires in which there is a chemical bond between the molecule and *only one* of the two electrodes; the other electrode can then be the tip of an STM that can image the molecular nanowire. There is a very extensive literature of experimental and theoretical work devoted to STM imaging of molecules on surfaces and interpreting those images. However, for molecular nanowires on metal surfaces the *atomic-level* details of the conformation of the molecule and especially of the bonding to the metal substrate have been difficult or impossible to determine in this way for all but the simplest of molecules. Furthermore, the atomic structure of the STM tip is also not normally determined experimentally, although techniques for doing this are available (Lucier *et al.* 2005). Thus, uncertainties regarding the details of the atomic-level structure are considerable in this case as well. Another complication is that in systems where the molecule is anchored only at one of its ends it can be more susceptible to geometrical fluctuations that result in random switching of the molecular nanowire's electrical conductance (Gaudioso *et al.* 2000; Donhauser *et al.* 2001; Ramachandran *et al.* 2003; Wassel *et al.* 2003; Jung *et al.* 2006; Moore *et al.* 2006; Pitters and Wolkow 2006). However, stochastic switching phenomena are not limited to STM studies; they are common in other types of molecular nanowires as well (Ramachandran *et al.* 2003; Danilov *et al.* 2006).

Another experimental technique that can yield partial structural information about molecular nanowires or at least confirm that a particular type of molecule is in fact involved in some way in electrical conduction between the electrodes is inelastic tunnelling spectroscopy. Here, features of the current–voltage characteristic of a molecular nanowire that are due to the excitation of molecular vibrations (phonons) by electrons as they pass through the nanowire are detected experimentally (Park *et al.* 2000; Kushmerick *et al.* 2004, 2006; Qiu *et al.* 2004; Wang *et al.* 2004, 2008; Yu *et al.* 2004; Cai *et al.* 2005; Djukic *et al.* 2005; Long *et al.* 2006; Naydenov *et al.* 2006; Wang and Richter 2006; Parks *et al.* 2007; Yu *et al.* 2007; de Leon *et al.* 2008; Hihath *et al.* 2008; Kiguchi *et al.* 2008). Typically, only a small fraction of the electrons passing through the molecule (up to a few percent) are involved in the inelastic processes and therefore the latter affect the current I flowing between the electrodes quite weakly. Thus, sensitive experimental methods designed to directly measure the dependence of the differential conductance $G = \mathrm{d}I/\mathrm{d}V$ or of its derivative $\mathrm{d}^2I/\mathrm{d}V^2$ on the applied bias voltage V are used to study them. These measurements yield information regarding the frequencies of some of the vibrational modes of the molecular nanowire that together with theoretical modelling (Bonča and Trugman 1995, 1997; Emberly and Kirczenow 2000a; May 2002; Chen *et al.* 2003, 2005a; Seideman and Guo 2003; Troisi *et al.* 2003; Čížek *et al.* 2004; Galperin *et al.* 2004, 2007a,b, 2008; Pecchia *et al.* 2004, 2007; Sergueev *et al.* 2005; Troisi and Ratner 2005, 2006b; Benesch *et al.* 2006; Jean and Sanvito 2006; Paulsson *et al.* 2006, 2008; Walczak 2006, 2007; Yan 2006; Frederiksen *et al.* 2007; La Magna and Deretzis 2007; Härtle *et al.* 2008; Kula and Luo 2008; Lüffe *et al.* 2008) can be used to identify the molecules involved in the conduction process and may yield some atomic-scale information regarding the molecular nanowires' geometries. The theoretical literature on this topic is extensive and the reader is referred to the recent review of Galperin and colleagues (2007a) for further references.

Clearly, the present, very limited knowledge of the microscopic geometries of molecular nanowires with metal electrodes is a serious impediment to developing a full understanding of the properties of these systems. However, as is explained in Sections 4.5.3 and 4.8.4 the situation is better in this regard for molecular nanowires on silicon.

4.5 Brief overview of molecular nanowire varieties: Different molecules, linkers and electrodes

4.5.1 Molecular species thiol-bonded to gold electrodes

Over the last decade measurements of electrical conduction through numerous different molecules thiol-bonded to one or two gold electrodes have been carried out by many groups and the list continues to grow. Surveys of this literature from various perspectives have been published by Salomon *et al.* (2003), James and Tour (2005), Tao (2006) and Lindsay and Ratner (2007). Here, to help orient readers new to this field, I will mention the main (sometimes overlapping) classes of these molecules together with just a few representative experimental papers: This category of molecules includes thiolates and dithiolates of the strongly insulating saturated hydrocarbon (alkane) chains (Wang *et al.* 2005; Jang *et al.* 2006; Li *et al.* 2006a, 2008; Nishikawa *et al.* 2007), of the somewhat less insulating conjugated hydrocarbon molecules including molecules with one or more aromatic rings (Bumm *et al.* 1996; Reed *et al.* 1997; Reichert *et al.* 2002; Kushmerick *et al.* 2002b; Xiao *et al.* 2004a; He *et al.* 2005a; Beebe *et al.* 2006; Yamada *et al.* 2008), of molecules that include redox centers (Chen *et al.* 1999; Xiao *et al.* 2005), of photochromic molecules whose geometries (and hence their electrical conductances) can be switched optically (Dulic *et al.* 2003; Yasuda *et al.* 2003; He *et al.* 2005b; van der Molen *et al.* 2006; Mativetsky *et al.* 2008), of molecules that include magnetic or non-magnetic transition-metal atoms (Park *et al.* 2002; Yu *et al.* 2004, 2005; Xiao *et al.* 2004c; Mahapatro *et al.* 2008), as well as short thiol-terminated peptide (Xiao *et al.* 2004b,c) and DNA (Xu *et al.* 2004) sequences.

4.5.2 Other linkers

A smaller number of experiments measuring conduction through molecular nanowires bonded to gold electrodes through a variety of non-thiol linkers has also been reported. These linkers including amine (Chen *et al.* 2006a; Venkataraman *et al.* 2006a,b, 2007; Park *et al.* 2007; Martin *et al.* 2008), carboxylic acid (Chen *et al.* 2006a), isonitrile (Beebe *et al.* 2002; Venkataraman *et al.* 2006b), cyanide and isocyanide (Kiguchi *et al.* 2006; Kim *et al.* 2006, 2007), selenol (Patrone *et al.* 2002), aryl (Fan *et al.* 2002), disulfide (Tseng *et al.* 2004), methyl sulfide and dimethyl phosphine (Park *et al.* 2007), and thiocyanate (Yamada *et al.* 2008) groups. Of these the amine-linked molecular nanowires are currently receiving the most attention since they have yielded simpler results in statistical studies of the molecular nanowire conductance than the thiol-linked molecules (Venkataraman *et al.* 2006b). This relative simplicity has been attributed (Venkataraman *et al.* 2006b; Quek *et al.* 2007b) to amine-linked molecules forming fewer different stable bonding geometries with gold surfaces in the presence of a solvent (Martin *et al.* 2008) than thiol-linked molecules and has facilitated systematic *comparative* studies of electrical conduction through a variety of molecules (Venkataraman *et al.* 2006a, 2007; Quinn *et al.* 2007a,b; Hybertsen *et al.* 2008).

4.5.3 Silicon electrodes

Molecular nanowires involving a silicon electrode offer a possible avenue for bridging the divide between the well-established semiconductor electronic technologies and the nascent field of molecular electronics. Thus, the structures and bonding geometries of individual molecules bound to silicon have received considerable attention. The review of Wolkow (1999) provides a clear discussion of this subject as it stood several years ago. More recently, research in this area has focused increasingly on understanding and controlling electrical conduction through the molecular nanowires. For example, experimental observations of negative differential resistance in a variety of molecules bound to silicon and contacted with an STM tip have been reported (Guisinger *et al.* 2004). This data was explained in terms of resonant tunnelling via molecular levels and silicon states at band edges (Rakshit *et al.* 2004, 2005). However, subsequent experimental work (Pitters 2006) pointed to another mechanism, namely, fluctuations of the molecular geometry affecting the time-averaged STM current. Other experiments have suggested different possible approaches to controlling currents in molecule–semiconductor hybrid devices: A recent experimental study found the current through a molecule on silicon to be strongly affected by the Coulomb field of a nearby charged dangling bond on the silicon surface (Piva *et al.* 2005), however, it has been suggested that both an electronic state localized around this defect and dephasing also strongly affected the STM current that was measured experimentally (Raza *et al.* 2008a). It has also been found that the electric current flowing through a surface silicon atom can be strongly affected by subtle changes in a nearby adsorbed molecule (Harikumar *et al.* 2006). Furthermore, electric fields emanating from the electric dipoles and higher multipoles associated with polar molecules bound to a silicon surface can strongly affect conduction not only through those molecules themselves but also that through other nearby molecules and through the silicon surface (Piva *et al.* 2008; Kirczenow *et al.* 2009).

An important difference between molecular nanowires on silicon and those thiol-bonded to gold electrodes is that for molecules bound to a clean or hydrogen-covered silicon surface in ultrahigh vacuum by means of a covalent carbon–silicon bond, determination of the atomic-scale geometry of the molecular nanowire with the help of an STM (see Wolkow 1999) is subject to much less uncertainty than for molecules thiol-bonded to metals in part because the carbon–silicon bond is inherently simpler. Moreover, it is possible to grow well-ordered linear chains of molecules on silicon by self-assembly and to characterize their structures and electronic properties in detail experimentally with STM (Lopinski *et al.* 2000). This raises the possibility of building nanoelectronic devices out of groups of molecules that self-assemble in a controlled way on silicon surfaces. Moreover, experimental and theoretical study of well-characterized groups of identical and different molecules on silicon in close proximity to each other can yield unique insights into the electronic structure and transport properties of molecular nanowires (Kirczenow *et al.* 2005, 2009; Piva *et al.* 2008), as will be discussed in Section 4.8.4.

4.5.4   Transition d-metal electrodes

Theoretical studies have predicted that molecular nanowires thiol-bonded to transition-metal electrodes other than gold should also have interesting electronic transport properties, and spintronic transport properties as well for magnetic electrodes. For magnetic transition d-metal electrodes, thiol-bonded molecular nanowires have been predicted to function as spin valves (Emberly and Kirczenow 2002a; Pati *et al.* 2003; Rocha *et al.* 2005; Dalgleish and Kirczenow 2005a, 2006c; He *et al.* 2006; Sanvito *et al.* 2006; Walczak and Platero 2006; Waldron *et al.* 2006), spin current injectors (Emberly and Kirczenow 2002a; Dalgleish and Kirczenow 2006c), electric and spin current rectifiers (Dalgleish and Kirczenow 2006c), and in some cases to exhibit negative magnetoresistance (Dalgleish and Kirczenow 2005a). However, the most closely related experiments to date have been those of Petta and colleagues (2004) on transport through octanethiol monolayers (*not* single molecules) between nickel electrodes and of Wang and Richter (2006) on octanethiol monolayers between nickel and cobalt electrodes. Magnetoresistance effects were reported suggesting that spin transport through the molecules may be occurring, but a complete understanding of these experimental results is yet to be achieved (Ning *et al.* 2008) although imperfections in the molecular films appear to play a significant role (Petta *et al.* 2004). Measurements of electrical conduction through decanedithiol monolayers between cobalt electrodes have also been carried out and hysteretic current–voltage characteristics were reported by Fan and colleagues (2007). Experiments on spin transport through individual molecular nanowires thiol-bonded to magnetic contacts have still not been reported, although it should be possible to apply statistical STM break junction techniques similar to those used by Xiao and colleagues (2004a) to study transport through molecular wires with gold electrodes to transition d-metal electrodes as well. For some non-magnetic transition d-metal electrodes thiol-bonded to simple alkyl chains negative differential resistance and current rectification have been predicted (Dalgleish and Kirczenow 2006a,b). Some experiments involving electrical conduction through molecular films thiol-bonded to non-magnetic transition d-metal electrodes (for example, Beebe *et al.* 2002; Cai *et al.* 2004, 2005; Tang *et al.* 2007) and through benzene molecules bonded directly to Pt electrodes (Kiguchi *et al.* 2008) have been reported. Recently, negative differential resistance has been reported in a cobalt phthalocyanine molecule on a gold substrate contacted with a Ni STM tip

(Chen *et al.* 2007). However, more experimental and theoretical work in this area is needed.

4.5.5 Molecules weakly coupled to the electrodes: Coulomb blockade and electroluminescence

Electrical conduction through molecules that are *not* bonded chemically to *either* electrode has also received considerable attention. Because for such systems the electronic coupling between the molecule and the electrodes is weak it can be more appropriate to picture the molecule, source and drain electrodes and the gate electrode (if present) as a set of electrostatically coupled, nanoscale capacitors separated by weakly transmitting tunnel barriers, rather than as a nanoscale wire. The charging energy associated with moving an electron between an electrode and the molecule in such systems can be large enough to strongly suppress electrical conduction through the device at low temperatures for certain ranges of bias and gate voltages, a phenomenon known as "Coulomb blockade." This effect has been studied in a variety of mesoscopic and nanoscale systems including semiconductor quantum dots, metallic nanoparticles and carbon nanotubes. It has also been observed experimentally in a variety of molecular systems with metal electrodes, some of them also exhibiting the Kondo effect; this work has been reviewed recently by Selzer and Allara (2006). The molecules present in the systems in which these phenomena have been reported include buckminsterfullerene (C_{60}) (Park *et al.* 2000; Pasupathy *et al.* 2004; Parks *et al.* 2007), saturated and π-conjugated molecules (Kubatkin *et al.* 2003; de Picciotto *et al.* 2005; Osorio *et al.* 2007a,b; Danilov *et al.* 2008), molecules containing transition-metal atoms (Liang *et al.* 2002) and magnetic molecules containing larger numbers of transition-metal atoms (Jo *et al.* 2006; Henderson *et al.* 2007). Coulomb blockade is also observed in systems containing molecules that can bond strongly to the electrodes (as in the case of gold–thiol bonding) but where the charging center in the molecule is weakly coupled *electronically* to both electrodes or the molecule has not bonded strongly to either electrode despite its ability to do so (Park *et al.* 2002; Yu *et al.* 2004, 2005; Heersche *et al.* 2006; Goto *et al.* 2007; Danilov *et al.* 2008). Since gold nanoparticles that may also be present in these junctions can give rise to similar Coulomb-blockade signatures of their own (de Picciotto *et al.* 2005; Goto *et al.* 2006; van der Zant *et al.* 2006) it is essential to determine whether the observed effects are truly due to molecules or nanoscale metallic grains. However, this has not always been done. It can be done by looking for features due to inelastic scattering involving molecular vibrational modes having known energies in the current–voltage characteristics of the devices (Park *et al.* 2000; Yu *et al.* 2004; Osorio *et al.* 2007a; Parks *et al.* 2007) and magnetic and spin excitations in magnetic molecules (Heersche *et al.* 2006; Jo *et al.* 2006).

Electrical conduction through molecular nanowires can result in the emission of light, i.e. electroluminescence (Buker and Kirczenow 2002, 2008; Galperin and Nitzan 2005, 2006; Harbola *et al.* 2006). Experimentally, electroluminescence from STM tips over *clean* metal surfaces has been observed (Berndt *et al.* 1991) and is due to the excitation of plasmons by the electron flux and their subsequent decay with the emission of photons. Electroluminescence was also detected in STM experiments where molecules were present on the metal surface (Smolyaninov and Khaikin 1990; Berndt *et al.* 1993; Poirier 2001; Hoffmann *et al.* 2002), however, opinions differed as to whether the observed photons were coming from the molecules or were again due to the decay of plasmons associated primarily with the STM tip and metal substrate. Theoretical work indicates that light emission from the molecule itself should be favored in systems where the electronic couplings between the molecule and the source and drain electrodes have similar strengths (Buker and Kirczenow 2002). This, however, was *not* the case in the above experiments where the molecule was adsorbed directly to a metal substrate that constituted one of the electrodes, while the other electrode was an STM tip separated from the molecule by a weaker link, namely, a vacuum gap that behaves as a tunnel barrier. In more recent electroluminescence experiments on various porphyrin-like molecules (Qiu *et al.* 2003; Dong *et al.* 2004, 2006; Guo *et al.* 2004, 2005) the molecules have been placed not directly on the metal substrate but on an intervening thin insulating layer that ensures that the couplings between the molecule and the underlying metal electrode and STM tip are *both* weak, satisfying the above theoretical requirement (Buker and Kirczenow 2002) for light emission from the molecule to be favored. The presence of the insulating layer also helps to suppress the excitation of plasmons. The evidence that electroluminescence observed in the more recent experiments was coming from the molecules themselves is persuasive. In these experiments the STM images of the molecules, current–voltage characteristics of the molecular nanowires and the spectra of the emitted photons can be taken simultaneously. With appropriate modelling (Buker and Kirczenow 2005, 2008) this wealth of experimental data can provide uniquely *definitive* information regarding the evolution of the electronic structure of the molecular nanowire under applied bias that, as is explained in Section 4.7, is one of the most challenging unsolved problems in the field of molecular nanoelectronics.

4.5.6   Some related experimental systems

The above brief survey of different types of experimentally studied molecular nanowires is by no means exhaustive. Some of the many related experimental systems that the interested reader may also wish to explore include: Large organic molecules on metal surfaces probed and manipulated by STM reviewed by Rosei and colleagues (2003) and Grill and Moresco (2006), single-molecule amplifiers based on a C60 molecule contacted and deformed by an STM tip (Joachim and Gimzewski 1997), conducting hydrogen-molecule bridges connecting metal electrodes (Smit *et al.* 2002; Djukic *et al.* 2005), molecules bridging gaps in carbon nanotubes (Guo *et al.* 2006), bistable switches based on monolayer films of interlocking molecules such as rotaxanes and catenanes reviewed by Mendes and colleagues (2005) and Saha and Stoddart (2007) and molecular film rectifiers reviewed by Metzger (2003).

4.6   Electrical conduction as a quantum scattering problem

Much of the theoretical work on electrical conduction in molecular nanowires is founded at the most basic level on the Landauer view of transport that regards electrical conduction as a quantum-mechanical scattering problem. The relationship, within this theoretical framework, between the electric current flowing through a nanostructure and the bias voltage applied between the source and drain electrodes connected to the nanostructure was first identified correctly by Economou and Soukoulis (1981) and Fisher and Lee (1981). At low bias voltages, in the linear response regime where the electric current is proportional to the bias voltage, the zero-temperature Landauer formula of Economou and Soukoulis and Fisher and Lee for the two-terminal conductance G of a nanostructure takes the form

$$G \equiv \frac{I}{V} = \frac{e^2}{h} T, \tag{4.1}$$

where I is the electric current flowing through the nanostructure, V is the bias voltage applied between the source and drain, e is the electron charge and h is Planck's constant. T is the *total* electron transmission probability between the electron source and drain at the common Fermi energy E_F of the electrodes and is given by

$$T = \sum_{p,q} T_{pq}. \tag{4.2}$$

Here, T_{pq} is the probability that an electron travelling towards the nanostructure at the Fermi energy in scattering channel q of the source electrode is transmitted through the nanostructure into the outgoing scattering channel p of the drain electrode, the scattering being assumed to be elastic. For semi-infinite quasi-one-dimensional crystalline source and drain electrodes parallel to the z-axis the asymptotic scattering states p and q in eqn (4.2) can be identified with the Bloch states $\psi_{kl} = e^{ikz} u_{kl}(\mathbf{r})$ propagating towards (away from) the nanostructure at the Fermi energy in the source (drain) electrode. Here, $u_{kl}(\mathbf{r})$ is a function of position $\mathbf{r}$ having the same periodicity along the z-axis as the crystalline electrode (see, for example, Peierls 1956). Then, p or q in eqn (4.2) is just the Bloch band label l for the respective electrode. Note that here states with electron spin up and down are assigned separate values of the indices p, q and l.

An elementary derivation of eqn (4.1) proceeds as follows: Consider to start with an ideal infinite quasi-one-dimensional conductor having only a single band of Bloch states, i.e. only one value of the Bloch band index n, and for simplicity also initially ignore the electron spin.

Let the energy of an electron in Bloch state ψ_{kn} be $\varepsilon(k)$. The velocity of an electron in state ψ_{kn} is then $v(k) = d\varepsilon(k)/d(\hbar k)$. Now, suppose that electrons are being fed into the rightward-moving states of the conductor from an ideal electron (source) reservoir at an electrochemical potential μ_S and zero temperature that completely fills the rightward-moving states up to an energy $\varepsilon(k_S) = \mu_S$. Similarly, suppose that the leftward-moving states of

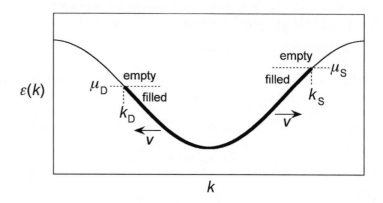

Fig. 4.2 Schematic of the energy ε vs. wave vector k for an ideal one-dimensional conductor with a single band of Bloch states carrying an electric current at zero temperature. The occupied (empty) electron states are indicated by the thick (thin) curve. The right- and left-moving states are filled up to energies μ_S and μ_D that are the electrochemical potentials of the source and drain electron reservoirs that feed electrons to the conductor.

the conductor are filled from an ideal electron (drain) reservoir at a lower electrochemical potential μ_D up to an energy $\varepsilon(k_D) = \mu_D$. This situation is depicted schematically in Fig. 4.2, where the thick (thin) curve shows $\varepsilon(k)$ for the filled (empty) states. The electric current I flowing through the conductor is then proportional to an integral of the electron velocity over the filled electron states, i.e.

$$I = \frac{1}{2\pi} \int_{k_D}^{k_S} e v(k) \mathrm{d}k = \frac{e}{2\pi} \int_{\mu_D}^{\mu_S} \frac{1}{\hbar} \frac{\mathrm{d}\varepsilon}{\mathrm{d}k} \frac{\mathrm{d}k}{\mathrm{d}\varepsilon} \mathrm{d}\varepsilon = \frac{e}{h}(\mu_S - \mu_D) = \frac{e^2}{h} V. \quad (4.3)$$

Here, $V = (\mu_S - \mu_D)/e$ is the electrostatic potential difference between the electron source and drain reservoirs. Notice that the final result in eqn (4.3) is independent of the details of the electronic dispersion $\varepsilon(k)$ in the conductor.

If instead of an ideal conductor through which electrons are transmitted with probability 1, we now consider a similar conductor divided into two semi-infinite pieces (i.e., electrodes) that are bridged by a nanostructure (such as a molecule) through which electrons at the Fermi energy are transmitted with probability T then in the limiting case of small V the current given by eqn (4.3) should be adjusted by a factor T and we thus arrive at eqn (4.1). The generalization of the above argument to more realistic electrodes with multiple overlapping bands of Bloch states populated with both spin-up and-down electrons is straightforward and yields eqn (4.1) with T given by eqn (4.2). Since T is now a *sum* of transmission probabilities over different channels; in general the total transmission probability T may be larger than, smaller than or equal to 1. If the nanostructure connecting the two electrodes is itself a conductor that transmits an integer number N of appropriately defined scattering channels *perfectly*, eqn (4.1) reduces to

$$G = Ne^2/h. \quad (4.4)$$

That is, the conductance of the system is an *integer* multiple of the fundamental conductance quantum e^2/h. Semiconductor nanostructures known as "quantum point contacts" were found experimentally (van Wees *et al.* 1988; Wharam *et al.* 1988) to exhibit conductance features described by eqn (4.4) with reasonable accuracy (typically within $\sim 0.1 e^2/h$) at low temperatures.

These experiments laid to rest the theoretical debate as to *which* Landauer formula is correct: The conductance formula originally proposed by Landauer (1970) was not eqn (4.1) but

$$G_{\text{Landauer}} = \frac{e^2}{h} \frac{T}{1-T} \qquad (4.5)$$

for the one-channel spinless case. Thouless (1981) presented a mathematical argument that supported the original Landauer formula (4.5) and not eqn (4.1). On the other hand, Imry (1986) argued heuristically that eqn (4.1) should be interpreted as the *experimentally observed* two-terminal conductance that is the ratio of the measured current I and the measured potential difference *V between the macroscopic physical source and drain reservoirs that are connected to the electrodes* when conductance measurements are made. He rationalized eqn (4.5) as the ratio of the current to the potential difference not between the electron source and drain reservoirs but across the scatterer (nanostructure) *itself*, i.e. as a quantity that is not measured in two-terminal experiments. The experiments of van Wees *et al.* (1988) and Wharam *et al.* (1988) settled the issue in favor of eqn (4.1) as the appropriate choice for the experimentally measured two-terminal conductance since for perfect transmission in the one-channel case (where $T = 1$) eqn (4.5) predicts $G_{\text{Landauer}} = \infty$, whereas what was observed instead was the quantized conductance described by eqn (4.4) with integer N that follows from eqn (4.1). However, nearly vanishing *four*-terminal *resistances* in other semiconductor devices with two weakly coupled voltage probes have been observed experimentally (de Picciotto *et al.* 2001) and these measurements can be interpreted as approximate realizations of the prediction $1/G_{\text{Landauer}} = 0$ of eqn (4.5) for a perfectly transmitting nanoscale conductor.

A general theory of four-terminal and other multiterminal conductances that generalizes eqn (4.1) to those cases was put forward by Büttiker (1986) and has since then been the standard framework used to interpret experiments such as those of de Picciotto *et al.* (2001). While four-terminal measurements have long been a mainstay of experimental studies of quantum transport in semiconductor nanostructures (for reviews see Beenakker and van Houten 1991; Ulloa *et al.* 1992), contacting individual molecules with four separate electrodes has not as yet been feasible. However, single-molecule transistors in which a third electrode has been present have been realized; see the end of Section 4.3 for an experimental bibliography. In the cases where the third electrode is a gate separated from the molecule by an insulating layer so that there is no direct electron transmission between the gate and the molecule, transport can be regarded as a two-terminal scattering problem, since while the gate potential modifies the scattering between the source and drain electrodes via the molecule there is no direct electron transmission between the molecule and gate. Thus, from a theoretical perspective such molecular transistors are effectively *two*-terminal scattering experiments with the third electrode only modifying the potential profile in the device and affecting conduction in this way. There have been many theoretical predictions of transistor action in single-molecule models of this type (Emberly and Kirczenow 2000b,c; Di Ventra *et al.* 2000b; Lang 2001; Taylor *et al.* 2001a; Bratkovsky and Kornilovitch 2003; Mujica *et al.*

2003; Yang *et al.* 2003; Cornaglia *et al.* 2004, 2005; Ghosh *et al.* 2004; Zhang *et al.* 2004; Chen *et al.* 2005b; Ke *et al.* 2005; Lang and Solomon 2005; Paaske and Flensberg 2005; Donarini *et al.* 2006; Kaun and Seideman 2006; Jiang *et al.* 2006; Su *et al.* 2006; Farajian *et al.* 2007; Galperin *et al.* 2007b; Malyshev 2007; Peng and Chen 2007; Miller *et al.* 2008; Cardamone and Kirczenow 2009). However, so far convincing comparisons between these theories and the experiments have not been possible since not only have the atomic-scale geometries of the experimental three-terminal molecular devices been inaccessible to measurement but even the approximate location(s) of the molecule(s) within them have been unknown. Three-terminal scattering in a molecular nanowire context has been considered in some theoretical work, electron transmission between the molecule and third electrode being a source of decoherence in the molecular device (Cardamone *et al.* 2006). Whether it is appropriate to treat electrical conduction in the case of *electrochemically* gated molecular nanowires (see Section 4.3) as a true three-terminal *scattering* problem is unclear at present.

Following the observations of quantized two-terminal conductance described by eqn (4.4) in semiconductor devices, this phenomenon has also been observed in atomic contacts between electrodes of gold (Agraït *et al.* 1993; Pascual *et al.* 1993) and other metals; for a detailed discussion of conductance quantization mechanisms in metallic atomic contacts see Ludoph and van Ruitenbeek (2000). For a pair of gold electrodes bridged by a single gold atom or a chain of gold atoms conductances G near $G_0 = 2e^2/h$ are observed, the factor 2 arising from the two degenerate spin orientations. Thus, when a molecular nanowire is made by driving a gold STM tip into a gold substrate immersed in solution of the molecules and then retracting the tip, as was discussed in Section 4.3, observation of a conductance plateau with G near G_0 as the tip is retracted signals that the gold bridge between the tip and substrate has thinned to a single atomic diameter so that observation of another conductance plateau at still lower G is interpreted as meaning that the metal bridge has parted and conduction between the two electrodes is occurring via a bridging molecule or molecules (Xiao *et al.* 2004a).

The two-terminal theory of electrical conduction outlined above applies to electrical conduction at zero temperature for infinitesimal bias voltages V applied between the electrodes. However, the generalization of the above derivation of eqn (4.1) to non-zero temperatures and larger bias voltages (beyond the linear response regime) at which many of the experiments on molecular nanowires are carried out is again straightforward and yields

$$I = \frac{e}{h} \int_{-\infty}^{\infty} T(E, V)(f(E, \mu_S) - f(E, \mu_D)) dE. \tag{4.6}$$

Here, the transmission probability $T(E, V)$ is still of the form given in eqn (4.2) but depends on the applied bias voltage V as well as the electron energy E. $f(E, \mu_i) = 1/(\exp[(E - \mu_i)/k_B\Theta] + 1)$, μ_i is the electrochemical potential of the source ($i = S$) or drain ($i = D$) electrode with $V = (\mu_S - \mu_D)/e$ as before, and Θ is the temperature. The physical meaning of eqn (4.6) is simple: $T(E, V)f(E, \mu_S)$ represents electrons flowing from the source electrode through the molecule to the drain electrode, $T(E, V)f(E, \mu_D)$ represents electron flow in the opposite direction. The difference $f(E, \mu_S) - f(E, \mu_D)$ of the Fermi functions associated with the source and drain that appears in the integrand is largest (at low and moderate temperatures) in the range of energies E of width eV between the electrochemical potentials of the source and drain electrodes so that the main contribution to the *net* current through the molecule can be regarded as arising from that energy range.

Provided that the electron scattering being considered is elastic, many different formalisms that are used to calculate electric currents through molecular nanowires yield expressions for the electric current having the same form as or equivalent to eqn (4.1) or eqn (4.6). For discussions of some of these formalisms and their implementations the interested reader is referred to Datta (1995), Emberly and Kirczenow (1998a,b, 1999, 2001a, 2002a,b, 2003), Damle *et al.* (2001), Taylor *et al.* (2001b), Brandbyge *et al.* (2002), Di Ventra *et al.* (2002), Xue *et al.* (2002), Seideman and Guo (2003), Datta (2004), Kosov (2004), Tomfohr and Sankey (2004), Di Carlo *et al.* (2005), Dalgleish and Kirczenow (2005b), Kirczenow *et al.* (2005, 2009), Rocha *et al.* (2005, 2006), Stokbro *et al.* (2005), Novaes *et al.* (2006), Sanvito (2006), Ernzerhof (2007), Galperin *et al.* (2007a), Cardamone and Kirczenow (2008a,b), Solomon *et al.* (2008) and Datta (2009), among others. At the present time the main obstacle to gaining a better theoretical understanding of electrical conduction through molecular nanowires in the elastic scattering regime is not an inadequacy of the relevant transport formalisms but rather the prevailing uncertainty as to how to model the electronic structure of the molecular wires realistically, especially in the presence of a significant electric current. This issue is discussed further in Section 4.7. Here, it is worth mentioning, however, that although the derivation of the Landauer formula eqn (4.6) outlined above was in the context of a non-interacting (or mean-field) electron model, it has been shown that an expression having the same form applies to interacting, correlated electron systems provided that electron–electron interactions within the leads may be neglected and that a simplifying assumption regarding coupling to the leads is made (Meir and Wingreen 1992).

Inclusion of inelastic processes such as the emission and absorption of phonons in the theory requires the Landauer formula eqn (4.6) to be modified since the energy of an electron when it leaves the scattering region is no longer necessarily the same as when it enters, and furthermore the specific inelastic scattering processes that can occur and the probabilities with which they occur are constrained by the Pauli principle in a non-trivial way (Emberly and Kirczenow 2000a). A variety of formalisms have been developed to treat this situation. Some references to this literature may be found in Section 4.4, and an extensive review has recently been provided by Galperin and colleagues (2007a). This topic will be revisited in Section 4.9.

4.7 Model building: Principles and caveats

In order to evaluate the electric current flowing through a molecular nanowire theoretically whether from eqn (4.6) or otherwise, it is necessary to calculate the quantum-mechanical electron transmission probability $T(E, V)$ through the nanowire, or its equivalent. In order to do this an appropriate effective Hamiltonian H (or its equivalent) for the system including the molecule itself, and the metal or semiconductor contacts that pass electric current to and from the molecule is needed. Constructing a *realistic* model Hamiltonian of this kind that quantitatively captures the physics and chemistry of such systems that is relevant to electrical conduction, a non-equilibrium process that in the most interesting cases is *far* from equilibrium, is a challenging task. Finding a way to do this that is generally applicable and reliable is a difficult unsolved theoretical problem that is central to much of molecular nanoelectronics.

4.7.1   *Ab-initio* modelling

The most popular way to construct approximate model Hamiltonians for molecular nanowires at present is the first-principles approach based on time-independent density-functional theory. The standard way in which this approach is applied to molecular nanowire transport calculations (with differences in the details of the particular implementations) has been discussed by Damle *et al.* (2001), Brandbyge *et al.* (2002), Di Ventra *et al.* (2002), Xue *et al.* (2002), Seideman and Guo (2003), Ke *et al.* (2004), Tomfohr (2004), Di Carlo *et al.* (2005), Rocha *et al.* (2005, 2006), Stokbro *et al.* (2005), Novaes *et al.* (2006), Sanvito (2006), Ortiz and Seminario (2007), Varga (2007), Toher *et al.* (2008) and many others. Its advantages are that no empirical or *ad-hoc* input into the calculations is called for by this methodology (although this is often not adhered to in practice) and that it is sufficiently tractable numerically for it to be possible to carry out calculations of current–voltage characteristics for many of the molecular nanowires that are being studied experimentally today, incorporating modest numbers of atoms of the electrodes as well as those of the molecule explicitly in the calculation. However, as will be discussed below, the application of this *ab-initio* approach to electrical *conduction* in molecular nanowires is increasingly being questioned in the molecular nanowire literature today (Delaney 2004; Sai *et al.* 2005; Toher *et al.* 2005; Quek *et al.* 2006, 2007a; Darancet *et al.* 2007; Evers 2007; Ke *et al.* 2007; Pemmaraju *et al.* 2007; Prodan 2007; Thygesen 2008; Toher and Sanvito 2007, 2008; Koentopp *et al.* 2008; Wang *et al.* 2008, among others).

A fundamental concern is that the application of the *ab-initio* approach to transport calculations (in the way that this is routinely done today) does *not* have a rigorous justification: It is based on the work of Kohn and Sham (1965) who devised a way to calculate the ground-state energies and charge-density distributions of inhomogeneous electron systems by constructing a *fictitious* non-interacting electron system designed to reproduce the total electronic density distribution of the true interacting electron system in its ground state. The single-electron eigenstates (called Kohn–Sham orbitals) of this fictitious non-interacting electron system and their energy eigenvalues are a somewhat artificial construct and need *not* in general be good approximations to the wavefunctions and energies of the *physical* electronic quasi-particles of the real system whose quantum transmission probabilities appear in eqn (4.6) and whose wavefunctions and energies therefore determine its transport properties. Indeed, only the energy of the highest-occupied fictitious Kohn–Sham orbital is known to have a physical meaning: It has been shown by Perdew and colleagues (1982) that it would equal the negative of the ionization potential of the system, if it were calculated from the exact exchange-correlation energy functional (which is not known). However, in the first-principles calculations of electrical conduction in molecular nanowires that are standard today all of the single-electron eigenfunctions and energy eigenvalues of the *fictitious* non-interacting electron system are simply *substituted* for the corresponding properties of the true quasi-particles of the molecular nanowire.

How this substitution may impact the results of the calculations of the electric current can be understood by considering eqn (4.6) for the electric current together with the relevant features of the electronic structure of typical of molecular nanowires with metal electrodes depicted in Fig. 4.3. At low bias voltages applied between the source and drain electrodes, their electrochemical potentials μ_S and μ_D normally lie at energies between the highest-occupied energy level of the molecule broadened by hybridization with the electrodes (or the highest occupied molecule-electrode interface state; see Dalgleish 2006a,b) denoted H and the lowest-unoccupied level of this kind denoted L. Under these conditions the electron transmission probability $T(E, V)$ through the molecular nanowire in the energy window between μ_S and μ_D that dominates the integral in eqn (4.6) is small because there are no strongly transmitting states in that energy range. However, as the bias voltage increases and μ_S begins to approach L or μ_D begins to approach H, resonant transmission mediated by L or H begins to contribute to the electric current that can then increase greatly. Thus, electrical conduction through the molecular nanowire is sensitive to the energy *differences* between the strongly transmitting states H and L of the system and the Fermi level of the electrodes at zero applied bias, to the behavior of the energies of these states with increasing bias *in relation* to the electrochemical potentials of the leads, and to the details of the wavefunctions of both the resonantly and non-resonantly transmitting states of the nanowire. Unfortunately, the theorem of Perdew and colleagues (1982) that has been

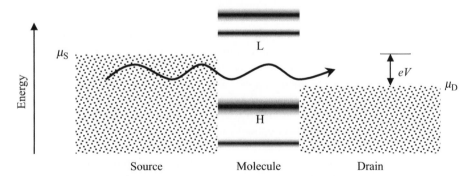

Fig. 4.3 Schematic drawing of the electronic structure of a molecular nanowire with a bias voltage V applied between metal source and drain electrodes. The shaded areas indicate the occupied electron states in the electrodes. The energies of the molecular orbitals broadened by hybridization with the electrodes and of resonant states associated with the molecule electrode interfaces are indicated by the broadened horizontal lines shown in the molecular region.

mentioned in the preceding paragraph shows *only* that the energy of the highest occupied Kohn–Sham orbital at zero bias has a physical meaning. However, the electric current is sensitive to the *differences* between the energies of the transmitting states and the electrochemical potentials (or Fermi levels) of the leads and the theorem says nothing about such energy differences. Indeed in the situation considered in Fig. 4.3 the highest-occupied Kohn–Sham orbital to which the theorem of Perdew *et al.* (1982) applies is simply the Fermi level of the *macroscopic* metal electrodes at zero applied bias (= $E_F = \mu_S = \mu_D$ at $V = 0$) and thus the theorem offers *no exact information* about the molecular nanowire itself or the energies at which the resonantly transmitting states (H and L in Fig. 4.3) occur. The theorem of Perdew and colleagues (1982) can also be applied to the ground state of the same system but with an added electron (so that now the system is negatively charged overall) but all that the theorem states in this case is that the energy of the highest-occupied Kohn–Sham orbital of the *charged* system (again simply equal to the Fermi level of the *macroscopic* metal electrodes at zero applied bias) has a physical meaning related to the electron affinity of the uncharged system. Therefore, arguments that often appear in the literature attempting to use the theorem of Perdew *et al.* (1982) to locate the energies of the transmitting *molecular* states of molecular nanowires are not rigorous and should be interpreted with caution.

To summarize the preceding discussion: Although the Kohn–Sham orbital eigenfunctions and energy eigenvalues are calculated self-consistently from first principles, the density-functional-based calculations of the current are not expected to be exact, and how realistic the results should be is not at all clear a priori. In this sense, the use of the fictitious Kohn–Sham orbitals in current calculations constitutes an *uncontrolled* approximation and therefore how good (or bad) it may be can only be assessed reliably by comparing the calculated currents with experiments. However, as was discussed in Section 4.4 the atomic-scale geometries of most molecular nanowires (especially for the most studied gold–thiol systems) are unknown, which makes such comparisons uncertain.

The mainstream first-principles calculations of the currents in molecular nanowires today have other weaknesses as well: One of these is that the theorems on which the Kohn–Sham theory is ultimately based (Hohenberg and Kohn 1964; Kohn and Sham 1965) apply to the (equilibrium) many-body *ground state*, whereas the electric current is a *non-equilibrium* (often far from equilibrium) property. Another is that the exchange-correlation energy functional that enters the Kohn–Sham theory is not known exactly and needs to be approximated. The simplest approximation in use is the local density approximation (LDA) in which the total electron density is assumed to be slowly varying spatially. This approximation can be improved by introducing gradient corrections (the GGA), with or without empirical parameters, into the theory at the expense of increased computational cost (Perdew *et al.* 2005), and this is in fact done in many of the density-functional-based molecular nanowire transport calculations that are being published today. However, while these improvements result in more accurate *total* ground-state energies for molecular and solid-state systems (these are the quantities that density-functional theory was in fact designed to calculate) they do not address the above *fundamental* shortcomings of the density-functional theory as applied to *transport* calculations. Thus, approximations to the exchange-correlation energy functional that yield more accurate ground-state energies do not necessarily yield more accurate calculated current–voltage characteristics for molecular nanowires.

Although these fundamental deficiencies of the mainstream density-functional-based *ab-initio* current calculations are well known, work directed at assessing them quantitatively and/or correcting for them in the molecular nanowire context has begun to appear in the literature only recently and is still at a very early stage. Different possible approaches for accomplishing this are being explored including those based on time-dependent density functional theory (Sai *et al.* 2005; Evers 2007; Koentopp *et al.* 2007; Prodan 2007), the GW approximation (Quek *et al.* 2006, 2007a; Darancet *et al.* 2007; Thygesen 2007, 2008; Wang *et al.* 2008) configuration interaction methods (Delaney 2004) and other many-body techniques (Hettler *et al.* 2003; Muralidharan 2006; Kletsov and Dahnovsky 2007).

Recent theoretical work (Toher *et al.* 2005; Ke *et al.* 2007; Pemmaraju *et al.* 2007; Toher and Sanvito 2007, 2008) has also questioned whether *within* today's mainstream time-independent density-functional-based *ab-initio* approach to molecular nanowire conduction the normally used approximations (LDA and GGA) to the exchange-correlation energy functional are appropriate for use in calculations of molecular nanowire current–voltage characteristics. It has been argued (see, for example, Toher and Sanvito 2008) that the LDA and GGA place the molecular HOMO levels of organic molecules too close to the Fermi level of the metal electrodes resulting in overestimates of the electric current by amounts that can be large, and that the molecular LUMO levels are similarly misplaced by the LDA and GGA. The argument of Toher and colleagues is based in part on the application of the theorem of Perdew *et al.* (1982). However, this theorem only provides rigorous information about the HOMO or LUMO levels of *isolated* molecules because, as has been discussed above, for a molecule *in the presence of macroscopic electrodes* it only provides rigorous information about the Fermi level of the macroscopic electrodes and *not* about the energy differences between that Fermi level and the *molecular* HOMO and LUMO levels. Thus, although it is plausible that the LDA and GGA do misplace the molecular HOMO and LUMO levels relative to the Fermi level of the electrodes significantly, the theorem of Perdew *et al.* (1982) leaves it uncertain as to how large these errors are. It may be relevant that recent work has found density-functional theory-based calculations to predict the Kohn–Sham HOMO orbital of a benzenedithiolate molecule in the vicinity of gold electrodes to move *upwards* in energy and locate very close to the Fermi level of the electrodes due to fractional charging of the molecule as the separation between the molecule and electrodes is *increased* (Andrews *et al.* 2006 and refs. therein), a counterintuitive prediction. Interestingly, spectroscopic experiments on pentacene films on gold have found the pentacene HOMO to move *downwards* in energy as the thickness of the pentacene film is increased (Amy *et al.* 2005) which may be indicative of a trend in the *opposite* direction for this system. However, pentacene lacks the sulfur atoms of benzenedithiolate.

One possible way to address these issues (Toher *et al.* 2005; Ke *et al.* 2007; Pemmaraju *et al.* 2007; Toher and Sanvito 2007, 2008; García and Sancho-García 2008) is to note that both the energies and wavefunctions of the Kohn–Sham orbitals are sensitive to self-interaction errors that cancel at the Hartree–Fock level of approximation but not in the LDA and the GGA, and to include corrections for this such as that proposed by Perdew and Zunger (1981) or those based on optimized effective potentials (Ke *et al.* 2007) in the transport calculations. Such calculations do indicate that the self-interaction errors of LDA and the GGA can result in large overestimates of the calculated current not only in cases where the coupling of the molecule to the electrodes is weak (Ke *et al.* 2007) but also in cases where the molecule–electrode coupling is strong (Toher and Sanvito 2008). However, these are *heuristic* corrections whose accuracy in the context of transport is not known, i.e. they themselves constitute *uncontrolled* approximations to the solution of a difficult many-body problem.

Perhaps not surprisingly in view of the above and given that, as has been discussed in Section 4.4, the interpretation of the experimental data that is available is subject to large uncertainties, very different opinions coexist even in the very recent literature as to what needs to be included in accurate *ab-initio* calculations of electric currents in molecular nanowires. For example, for gold–benzenedithiolate–gold molecular nanowires, Varga and Pantelides (2007) recently argued that the LDA provides a good description of electrical conduction at low bias while Toher and Sanvito (2007) argued that the LDA is inadequate for this and self-interaction corrections are necessary. Contributing to the confusion is that, in addition to the choice of approximate exchange-correlation energy functional, in practice the *ab-initio* calculations involve other approximations and idealizations that can strongly affect the results. These include both the type and size of the basis set used, how the atomic-scale geometry is estimated, how the Fermi energy of the electrodes at zero bias is estimated and the specifics of any pseudopotentials that are used. These important details vary widely in the literature and are usually not described fully so that meaningful comparison between theoretical studies published by different groups often cannot be made. On the other hand, if calculations of the conductance are carried out using the *same* set of assumptions and approximations for a variety of molecular nanowires thiol-bonded to gold contacts at zero bias (*in the tunnelling regime*), and compared with experiments also carried out consistently in the *same* way from system to system, then similar trends from system to system in the experimental and theoretical results can sometimes be discerned (Lindsay and Ratner 2007). Why this can work to the extent that it does despite the inadequacy of density-functional theory is pointed out in Section 4.8.3.

It is instructive to compare the present situation with that which existed in the past in solid-state physics when *ab-initio* calculations of the electronic band structures of crystals were being developed. Then, as now, the electronic single-particle wavefunctions and energy eigenvalues were being modelled with Kohn–Sham orbitals of time-independent density-functional theory, also without rigorous justification. The crucial difference between this and the present situation was that the atomic-scale geometries were known and unambiguous quantitative information regarding key aspects of the crystal electronic structures was available from experiments. This *empirical* input made it possible for *ab-initio* schemes to be found that, although based on *uncontrolled* approximations, matched the experimental data reasonably well or could be made to do so with the help of such corrections as empirical psuedopotentials. As a result, we now have accurate theories of the band structures of a wide variety of crystalline solids. Whether our understanding of transport in molecular nanowires will in the future follow a similar path is not clear, partly because it is not clear whether the difficulties being experienced in properly characterizing the experimental systems will be overcome. Another complication is that molecules may experience significant charging and large structural changes in response to application of a potential bias (see Section 4.9), phenomena that do not play a major role in typical solid-state systems.

4.7.2 Extended Hückel theory and other semi-empirical tight-binding models

While it is desirable to be able to calculate the current–voltage characteristics of molecular nanowires accurately from first principles, such *ab-initio* calculations (in addition to not having a rigorous fundamental justification and involving uncontrolled approximations) require *large* computing resources, which limits them in practice to molecular structures of quite modest size in dry environments. Furthermore, the complexity of the *ab-initio* calculations makes their results difficult to interpret physically and, in cases where comparison with experiment or theoretical analysis shows them to be clearly inadequate, their very nature as *first-principles* calculations makes correcting their deficiencies in any transparent way problematic. Thus, it is also highly desirable to develop simpler alternatives to the *ab-initio* calculations that are more transparent and can be applied to larger and more complex systems. Semi-empirical tight-binding models offer such an approach that is complementary to the *ab-initio* calculations. It is important, however, to distinguish between *semi-empirical* tight-binding models that incorporate the atomic-scale physics of molecular nanowires, and purely phenomenological modelling of experimental data with simple analytic expressions such as the Simmons (1963) formula that was derived to describe tunnelling through generic potential barriers but whose application to tunnelling through molecules is entirely phenomenological. The latter can be very effective in uncovering intriguing phenomena (Beebe *et al.* 2006) but it does not connect these to the atomic-scale physics and chemistry of molecular nanowires.

The semi-empirical tight-binding models most commonly used in studies of molecular nanowire transport are based on extended Hückel theory of molecular electronic structure developed by Wolfsberg and Helmholz (1952), Hoffmann (1963), Ammeter *et al.* (1978), Landrum and Glassey (2001), and others. Extended Hückel theory describes the electronic structures of molecular systems in terms of a *small* set (between 1 and 9 per atom) of Slater-type atomic orbitals $\{|\varphi_i\rangle\}$, their overlaps $S_{ij} = \langle\varphi_i|\varphi_j\rangle$ and a Hamiltonian matrix

$H_{ij} = \langle \varphi_i | H | \varphi_j \rangle$. The diagonal Hamiltonian elements $H_{ii} = \varepsilon_i$ are chosen to be the negatives of the respective *experimental* atomic orbital ionization energies. The non-diagonal matrix elements H_{ij} are assumed to be proportional to the respective overlaps S_{ij}; for example in the simple Wolfsberg–Helmholz form of the model, the non-diagonal elements are approximated by $H_{ij} = K S_{ij}(H_{ii} + H_{jj})/2$, where K is an empirical parameter typically chosen to be 1.75 for consistency with experimental molecular electronic structure data. A somewhat more elaborate (and more accurate) form for the non-diagonal Hamiltonian matrix elements H_{ij} was developed by Ammeter *et al.* (1978).

An important feature of extended Hückel models is that the correct (experimental) values of the atomic-orbital ionization energies are built into them from the outset by construction as diagonal matrix elements of the Hamiltonian $H_{ii} = \varepsilon_i$. To better appreciate the significance of this, it is instructive to compare this aspect of the semi-empirical tight-binding models with its counterpart in the *ab-initio* approach to molecular nanowire transport calculations: Recall that (as was discussed in Section 4.7.1) the only Kohn–Sham orbital energy of the *ab-initio* theories that is guaranteed to be physically meaningful by the theorem of Perdew *et al.* (1982) is the energy of the highest-occupied Kohn–Sham orbital that should equal the negative of the ionization potential, but in practice (as has been discussed by Toher and Sanvito 2008) the LDA and GGA approximations used in the *mainstream ab-initio* calculations of the currents in molecular nanowires today violate this theorem (often severely) for isolated atoms and molecules. For example, these *ab-initio* calculations of the energy of the highest-occupied Kohn–Sham orbital for an isolated carbon atom yield values near $-5\,\text{eV}$ for both the LDA and GGA and near $-6\,\text{eV}$ for the more realistic B3LYP and B3PW91 approximate exchange-correlation energy functionals. All of these numbers differ greatly from the correct value of $-11.3\,\text{eV}$, i.e. the negative of the ionization potential of the carbon atom that (according to the theorem of Perdew *et al.* (1982)) they should equal if the (unknown) exact exchange-correlation energy functional were used to calculate them. As discussed in Section 4.7.1 the self-interaction corrections to the exchange-correlation energy that have very recently begun to be introduced into the *ab-initio* calculations of molecular nanowire currents by Toher *et al.* (2005), Ke *et al.* (2007), Pemmaraju *et al.* (2007) and Toher and Sanvito (2007, 2008) are intended to correct this deficiency of the standard *ab-initio* calculations. To summarize: In the *ab-initio* approach finding a way to satisfy the theorem of Perdew *et al.* (1982) and to be consistent with such basic facts of quantum chemistry as the known values of atomic ionization potentials, is a many-body problem that has to be addressed in practice by a computer-intensive numerical procedure *and* heuristic approximations. On the other hand, an empirical solution to this problem (that is *exact* for isolated atoms) is built into extended Hückel theory at the outset. Whether the basic facts of quantum chemistry that are built into the *other* diagonal Hamiltonian matrix elements (corresponding to lower-energy atomic orbitals than the highest for each atom) and into the *non*-diagonal Hamiltonian matrix elements H_{ij} in extended Hückel theory are captured reasonably well by any of today's *ab-initio* molecular nanowire transport calculations is far from clear.

The discussion in the preceding paragraph offers some insight into why transport calculations based on the seemingly very simple extended Hückel model of electronic structure have been able to account for the results of some key experiments on electrical conduction through molecular nanowires thiol-bonded to gold electrodes and to shed light on the underlying physics (Datta *et al.* 1997; Emberly and Kirczenow 2001b,c; Kushmerick *et al.* 2002a,b; Cardamone and Kirczenow 2008a,b; see also Section 4.8).

While the extended Hückel model is semi-empirical in that the expressions of Ammeter *et al.* (1978) for the Hamiltonian matix H_{ij} involve experimentally determined parameters, a *standard* set of the values of these parameters based on experiments not involving electrical conduction is readily available (Landrum and Glassey 2001). Experience has shown that use of this standard parameter set without modification in calculations of the electrical conduction in molecular nanowires thiol-bonded to gold contacts can yield results in remarkably good agreement with experiments. One reason for this is that much of the conventional wisdom of quantum chemistry (that it is difficult to reproduce within *ab-initio* time-independent density-functional transport calculations, as has been discussed above) is built into this parameter set. Another is that the density of states of bulk *gold* metal near the Fermi energy is very simple (i.e. almost featureless; see for example Papaconstantopoulos 1986, p. 202) so that even an extended Hückel model fitted to chemical rather than solid-state data provides an acceptable first approximation to it for use in calculations of electrical conduction through a molecular nanowire whose resistance is determined mainly by the electronic structure of the molecule and its chemical bonding to the metal rather than that of the much more conductive metal electrodes themselves.

Gold electrodes are unusual in this regard because of the simplicity of their electronic structure: Standard extended Hückel parameter sets *chosen to match the quantum chemistry of molecules* are inadequate for describing the electronic structures of many other crystalline electrode materials in modelling molecular nanowire properties. These include semiconductors such as silicon that have bandgaps at the Fermi level, transition d-metals whose densities of states are sharply peaked in the vicinity of the Fermi level and ferromagnetic electrodes that in addition to being transition d-metals also have spin-split electronic structures. Some of these electrode materials have been modelled in studies of molecular nanowire transport (Emberly and Kirczenow 2002a; Dalgleish and Kirczenow 2005a, 2006a–c) using tight-binding models of the electrodes that are not of the extended Hückel form but still employ small basis sets and are parameterized so as to match known solid-state electronic band structures (Papaconstantopoulos 1986). Alternatively, parameterizations of extended Hückel theory have been developed that are good matches to solid-state electronic band structures (Cerda and Soria 2000; Kirczenow *et al.* 2005; Kienle *et al.* 2006a,b). However, in the solid-state-based parameterizations the atomic orbital energies are determined only up to an *arbitrary* constant. Thus, combining such alternate parameterizations of the electrodes with the standard quantum-chemical extended Hückel parameter set for the molecule can yield more accurate electronic structures for the molecular nanowire system as a whole, at the expense of the location of the Fermi level of the electrodes relative to the energies of the molecular orbitals at zero bias no longer being determined within the model and needing to be estimated in other ways, as discussed by Emberly and Kirczenow (2002a), Kirczenow *et al.* (2005), Dalgleish and Kirczenow (2005a, 2006a–c), Kirczenow (2007), and Raza *et al.* (2008a). It is also possible to modify extended Hückel theory and related tight-binding models in a variety of ways so as to combine them with self-consistent calculations of electrostatic potentials at the expense of added computational complexity (Emberly and Kirczenow 2000b,c, 2001, 2002b; Blomquist and Kirczenow 2004, 2005, 2006a,b; Zahid *et al.* 2005; Piva *et al.* 2008; Kirczenow *et al.* 2009) or with the help of appropriate analytic approximations (Cardamone and Kirczenow 2008a,b, 2009).

4.8   Theory confronts experiment: Some case studies

4.8.1   Oligopeptide molecular nanowires in electrolyte solution thiol-bonded to gold contacts: Combining semi-empirical and *ab-initio* modelling *without* adjustable parameters

As was discussed in Section 4.7.2, the extended Hückel theory provides a simple way to model electronic structure that incorporates much fundamental quantum chemistry that is not captured by the *ab-initio* methods as they are commonly applied today in molecular nanowire transport calculations. However, extended Hückel theory does not provide a way to calculate the electrostatic potential profiles along a molecular nanowires when a potential bias is applied between the electrodes bridged by the molecule. Potential profiles have been estimated in the *ab-initio* calculations discussed in Section 4.7.1, and it has been found that relatively simple model potential profiles can yield very similar results (Ke *et al.* 2004). The insights obtained in this way have been useful in model building (Dalgleish and Kirczenow 2005a, 2006a,b,c). However, much of the experimental work on electrical conduction in single-molecule nanowires being done today with the sophisticated statistical STM break junction technique (see Sections 4.3 and 4.4) is carried out with the molecule immersed in an organic solvent, or in water or ionic solutions. *Ab-initio* calculations of electrical conduction in molecular wires in these media have not yet been done and present significant challenges in part because of the large numbers of atoms that would have to be included. However, the potential profiles along molecular nanowires should be influenced by the dielectric medium surrounding the molecule, especially by the presence of such polar molecules as water and by screening due to dissolved ions. Examples of such systems that are also potentially interesting from the perspective of bionanotechnology are molecular nanowires consisting of oligopeptide molecules (i.e. protein fragments) thiol-bonded to gold electrodes in electrolyte solutions. Experimental STM break junction studies of electrical conduction in a number of such molecules whose structural formulae are shown in Fig. 4.4 have been carried out by Xiao *et al.* (2004b,c). By combining

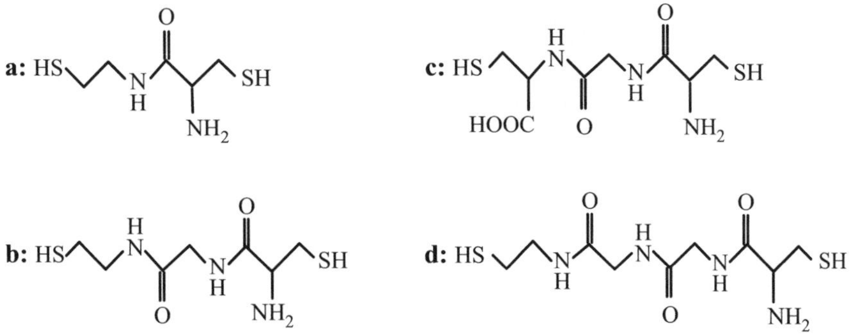

Fig. 4.4 Four oligopeptide molecules studied experimentally by Xiao *et al.* (2004b,c) and theoretically by Cardamone and Kirczenow (2008a,b). The hydrogens detach from the sulfur atoms when the latter bond to the gold contacts.

ab-initio and semi-empirical techniques, we have constructed a theoretical model of electron transport through these oligopeptide molecular nanowires with gold electrodes in the presence of the electrolyte solution (Cardamone and Kirczenow 2008a,b) that *without the use of any adjustable parameters* accounts very well for all of the experimental data (Xiao *et al.* 2004b,c). The molecular nanowire atomic geometries in our model were calculated by means of *ab-initio* density-functional theory-based total-energy minimizations. The potential profiles along the molecular nanowires under bias were calculated using the standard linearized Debye–Hückel model of screening in electrolytes (Landau and Lifshitz 1958) whose only parameters are the dielectric constant of water, the temperature and concentration of the electrolyte that are known experimentally, and the separation between the gold electrodes that was calculated from first principles as part of our geometry optimizations.

For the electronic structures of the molecules and electrodes we used the extended Hückel model of Ammeter and colleagues (1978) with its *standard parameter values* as implemented by Landrum and Glassey (2001) for *both* the molecules and the gold electrodes. The atomic orbital energies were shifted according to the potential profile that was calculated as above, making the standard corresponding corrections (Emberly and Kirczenow 2002a) to the non-diagonal Hamiltonian matrix elements that are required due to the non-orthogonality of the basis set of extended Hückel theory. *Thus, our model has no free parameters and no attempt was made to fit our results to any experimental molecular nanowire conductance data.*

The calculated conductance values (Cardamone and Kirczenow 2008a,b) at zero bias for the four molecules shown in Fig. 4.4 for different bonding geometries of the sulfur atoms at the two gold electrodes are compared with the experimental values in Table 4.1. The theoretically studied bonding arrangements in Table 4.1 are those usually considered to be the most likely ones for molecules thiol-bonded to gold in STM break junction experiments (Li *et al.* 2006a). Only the theoretical values for molecules bonded to the gold electrodes with *both* sulfur atoms located over hollow sites between gold atoms at the electrode surfaces (similar to the sulfur atom on the right in Fig. 4.1) are consistent with the experiment and for these bonding sites the agreement between theory and experiment is unusually good by the standards

Table 4.1 Conductances of the oligopeptide molecules in Fig. 4.4 thiol-bonded to gold electrodes in an electrolyte solution, in units of $2e^2/h$. Experimental values are from Xiao *et al.* (2004b,c). Calculated values for different bonding geometries of the sulfur atoms at the two gold electrodes are from Cardamone and Kirczenow (2008a,b)

Oligopeptide	Experiment	Hollow–hollow	Hollow–top	Top–hollow	Top–top
a	1.8×10^{-4}	5.4×10^{-5}	1.5×10^{-6}	9.6×10^{-7}	5.8×10^{-9}
b	4.2×10^{-6}	4.3×10^{-6}	5.3×10^{-8}	3.7×10^{-8}	1.5×10^{-9}
c	5.3×10^{-6}	4.2×10^{-6}	9.3×10^{-8}	1.0×10^{-7}	1.7×10^{-9}
d	5.0×10^{-7}	4.1×10^{-8}	7.3×10^{-10}	1.9×10^{-10}	2.2×10^{-11}

of molecular nanoelectronics research. The calculated conductances for the other geometries where one or more of the sulfur atoms bonds to a gold atom on top of the electrode surface are so much smaller than the experimental values that these bonding geometries appear to be ruled out as candidates for the geometries being probed experimentally. Furthermore, the same model accounts well (Cardamone and Kirczenow 2008a,b) for the rectification observed in the experimentally measured current–voltage characteristics of these molecules (Xiao *et al.* 2004c) due to their asymmetric structures, again without any adjustable parameters. In addition, the same theoretical methodology applied to alkanedithiolates bonded between gold electrodes also yields calculated conductances in similarly good agreement with experimental STM break junction data, for example, a calculated conductance value of $1.3 \times 10^{-4} \times 2e^2/h$ for a hollow–hollow gold–octanedithiol–gold junction (Cardamone and Kirczenow 2008a) compared with the experimental value of $2.5 \times 10^{-4} \times 2e^2/h$ (Li *et al.* 2006a). It also predicts that oligopeptides thiol-bonded to gold nanocontacts should exhibit negative differential resistance at higher bias voltages and act as transistors under electrochemical gating (Cardamone and Kirczenow 2008,a,b, 2009) raising the possibility of protein-based nanoelectronic technology.

These findings are quite encouraging: They confirm that the advent of the STM break junction technique with statistical analysis of the data represents a significant step forward in experimental methodology in that (although it still does not measure atomic-scale geometries) it does produce data that can be quite meaningfully compared with theory. On the theoretical side the results are also significant: They demonstrate that remarkably simple models with *no* adjustable parameters can be very successful in calculating the electrical transport properties even of molecular nanowires such as those in Fig. 4.4 that involve much more than just simple hydrocarbon chains and rings, even when the molecules are immersed in water with dissolved ions. This offers a much simpler and evidently quite viable alternative to the highly computer-intensive *ab-intio* transport calculations whose fundamental soundness is currently being questioned (see Section 4.7.1). It also opens the door to reliable modelling of electrical conduction in larger molecular systems in complex environments for which *ab-intio* transport calculations would require impracticably large computing resources. Accurate theoretical estimates of atomic-scale geometries are still necessary, however, practical alternatives to full *ab-initio* calculations of structures exist today (see, for example, Krüger *et al.* 2002; Kirczenow *et al.*

2005) and are continually being refined. Recently, density-functional theory-based transport calculations (although with an *ad-hoc* assumption regarding the location of the Fermi level of the electrodes) have been reported by Bautista *et al.* (2007) for several oligopeptides (namely, oligoglycines) thiol-bonded between gold electrodes in vacuum. However, experimental conductance data is not available at this time for those molecules for comparison.

4.8.2 Gold–benzenedithiolate–gold molecular nanowires

An early success of the use of the extended Hückel model with the *standard* parameter set in molecular nanowire transport calculations (Emberly and Kirczenow 1998a,b) was that it predicted (in retrospect, correctly) that for the gold–benzenedithiolate–gold molecular nanowires the Fermi level of gold (calculated within the extended Hückel model) should lie closer in energy to the first strongly transmitting molecular energy level below it (labelled H in Fig. 4.3 and often referred to loosely as the molecular HOMO resonance) than to that above it (labeled L in Fig. 4.3 and often referred to as the molecular LUMO resonance). This implies that conduction through the molecule is hole-like, being governed primarily by transmission through states having a HOMO-like character. However, this early theoretical work (Emberly and Kirczenow 1998a,b), in common with similar (unpublished) early calculations by Datta and collaborators, also yielded calculated differential conductances and currents for the gold–benzenedithiolate–gold molecular nanowires 2–3 orders of magnitude larger than those that had been observed in the pioneering experiment of Reed *et al.* (1997), the only experimental data available on gold–benzenedithiolate–gold molecular nanowires at that time. This *very* large discrepancy provided much of the initial motivation for the development of the *ab-initio* approach to transport calculations in molecular wires discussed in Section 4.7.1. However, the *ab-initio* calculations (Di Ventra *et al.* 2000a; Damle *et al.* 2001), like the earlier theoretical work based on the extended Hückel model (Emberly and Kirczenow 1998a,b) yielded calculated conductances and currents for gold–benzenedithiolate–gold molecular nanowires *orders of magnitude larger* than those observed by Reed *et al.* (1997). Furthermore the early *ab-initio* calculations (Di Ventra *et al.* 2000a; Damle *et al.* 2001) disagreed among themselves as to whether the gold Fermi level lies closer to the HOMO or LUMO of the molecular nanowire. Paulsson and Datta (2003) proposed that measurement of the sign of the thermoelectric voltage induced by a temperature difference between two electrodes bridged by the molecule would determine which of these possibilities is correct. Recently, this experiment was carried out (Reddy *et al.* 2007; Baheti *et al.* 2008). These measurements indicated that conduction through the molecule is hole-like, consistent with the result of the extended Hückel calculations (Emberly and Kirczenow 1998a,b). Meanwhile, most of the plethora of more recently published *ab-initio* calculations on the gold–benzenedithiolate–gold system have been yielding results consistent with this finding as well.

The fact that instead of resolving the very large discrepancy between the magnitudes of the conductances (and currents) observed by Reed *et al.* (1997)

and those found in the early extended Hückel theory-based calculations of Emberly and Kirczenow (1998a,b), the *ab-initio* calculations (Di Ventra *et al.* 2000a; Damle *et al.* 2001) yielded currents and conductances broadly consistent in magnitude with those obtained from the extended Hückel theory, spawned two lines of thought: One (Delaney and Greer 2004; Toher *et al.* 2005) was that a more sophisticated many-body theory was needed to bridge the gap between theory and the experiment of Reed *et al.* (1997). The other (Emberly and Kirczenow (2001b,c) was that the extended Hückel-based transport theory may not be a terrible first approximation after all and that the large discrepancy between theory and experiment may be an indication that the *interpretation of the experiment* should be reconsidered. In fact, the experimental work of Reed *et al.* (1997) provided no *direct* evidence that a single molecule bridged the two gold electrodes (bonding chemically to both of them as in Fig. 4.1 or in another similar bridging geometry) as had been simply *assumed* in all of the early theoretical work. Furthermore, the experiment had been performed by coating the two gold nanoelectrodes *densely* with benzenedithiol molecules and then bringing these coated electrodes together *gently* until an electric current flowing between them in response to applied bias could be measured. Thus, it seems quite plausible that in the experiment of Reed *et al.* (1997) the self-assembled molecular monlayers on the two gold electrodes were in contact with each other but that no individual molecule contacted *both* electrodes directly, i.e. that the conducting bridge between the two gold electrodes in the experiment of Reed *et al.* (1997) included at least *two molecules in series* (one bonded chemically to each electrode) with a weak electrical link between these two molecules, as is depicted schematically in Fig. 4.5. Our extended Hückel theory-based conductance calculations for this alternate model (Emberly and Kirczenow 2001b,c, 2002a) yielded the result shown as the solid curve in Fig. 4.5, which is within a factor of roughly 2 (at both low and high bias) of the experimental data of Reed *et al.* (1997) shown by the dashed curve in Fig. 4.5. Thus, a simple and very reasonable way to reconcile theory with the experimental data of Reed *et al.* (1997) had been found. If this line of reasoning is correct then it should be possible to observe experimentally

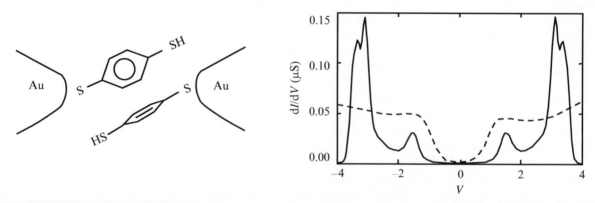

Fig. 4.5 Left: Two gold contacts bridged by two molecules *in series*. Right: Solid curve is the differential conductance $G = dI/dV$ of the junction on the left calculated using the standard extended Hückel electronic structure model averaging over 50 molecular geometrical configurations. The experimental result of Reed *et al.* (1997) is the dashed curve. Reprinted from Emberly and Kirczenow (2002a) with permission from Elsevier.

conductances $\sim$ two orders of magnitude larger than those measured by Reed *et al.* (1997) in gold–benzenedithiolate–gold junctions designed specifically to avoid the possibility of electrical conduction between the two gold contacts occurring via two molecular films in series and to make it more likely that a single benzeneditiolate molecule really does bridge the two gold nanocontacts. Some years after our explanation of the data of Reed *et al.* (1997) was proposed, just such experiments were carried out in different ways: Xiao *et al.* (2004a) used an STM break junction technique in which conductance measurements were taken at the same time as the nanoscale metal-to-metal contact between a gold tip and substrate was being broken in the presence of BDT molecules so that a dense double self-assembled molecular monolayer was not present between the metal contacts during the measurement. On the other hand, Tsutsui *et al.* (2006) brought together a pair of contacts in a way more closely resembling that of Reed *et al.* (1997) but the concentration of molecules on the gold surfaces in the system studied by Tsutsui *et al.* (2006) was low, so that again the possibility of conduction through two molecular layers in series could be avoided. Both of these experiments yielded molecular conductances up to two orders of magnitude larger than those found by Reed *et al.* (1997), consistent with our explanation (Emberly and Kirczenow 2001b,c, 2002a) of the experiment of Reed *et al.* (1997). These experiments rendered largely moot the search for exotic theories that might yield very low conductances (such as those observed by Reed *et al.* 1997) for *single* benzenedithiolate molecules bridging gold nanocontacts; recent work (such as that of Varga and Pantelides 2007 and Toher and Sanvito 2007, 2008) directed at improving the agreement between single-molecule *ab-initio* conductance calculations and experiment has focused instead on comparisons with the (two orders of magnitude larger) conductances observed at low bias by Xiao *et al.* (2004a).

However, even today, seemingly closely related experiments can still yield inconsistent measured conductance values for gold–benzenedithiolate–gold molecular nanowires, as a study reported recently by Lörtscher *et al.* (2007) has shown: Like Tsutsui *et al.* (2006), Lörtscher *et al.* (2007) brought together a pair of gold contacts with a very low concentration of benzenedithiol molecules on the gold surface and measured the conductances of molecules between the contacts as the contacts were brought together. However, the conductances that Lörtscher *et al.* (2007) found were *much* smaller than those of Tsutsui *et al.* (2006) (and also than those of Xiao *et al.* 2004a and of Tsutsui *et al.* 2008) but similar to those observed by Reed *et al.* (1997). Some differences between these earlier experiments and those of Lörtscher *et al.* (2007) are that the latter were carried out in ultrahigh vacuum and mainly at lower temperatures. At present there is no definitive understanding of why the conductances reported by Lörtscher *et al.* (2007) were very much smaller than those found in the single-molecule experiments of Xiao *et al.* (2004a) and Tsutsui *et al.* (2006, 2008). However, as is usual in this field, the *unknown* atomic-scale geometries most likely hold the answers. One possibility is that in the Lörtscher (2007) experiment the molecule bonds properly to only one of the two gold electrodes. That is, in the apparatus of Lörtscher *et al.* (2007), cleavage of the molecular H–S bond and formation of a strong S–gold chemical bond (a process only weakly favored thermodynamically, see Gronbeck *et al.* 2000) between the molecule and the bare gold electrode that approaches it (as the gap between the two gold electrodes is reduced) may be slow compared to the process of forming a direct metal-to-metal bridge between the electrodes that shorts out the molecular nanowire. Indeed, Lörtscher and colleagues (2007) reported that direct metal-to-metal bridging prevented their measuring molecular current–voltage characteristics at room temperature, unlike Xiao *et al.* (2004a) and Tsutsui *et al.* (2006, 2008). The difference here may be that in the ultrahigh-vacuum experiments of Lörtscher *et al.* (2007) there is no adsorbate on the gold surfaces that would impede direct gold-to-gold bridging. If indeed the molecule in the experiments of Lörtscher *et al.* (2007) was bonded strongly to only one of the two electrodes then the weak link between the molecule and the other electrode would result in an anomalously low conductance, consistent with their data. The strong asymmetry of the individual differential conductance traces under reversal of the sign of the bias observed by Lörtscher *et al.* (2007) suggests that the molecule did in fact couple differently to the two electrodes. However, it is evident that further experimental work is necessary before firm conclusions can be drawn regarding the meaning of the data of Lörtscher *et al.* (2007) and why it differs so much from that of Xiao *et al.* (2004a) and Tsutsui *et al.* (2006, 2008). In particular, it would be of interest to establish whether introducing insulating adsorbates in a controlled way onto the gold surfaces in the apparatus of Lörtscher *et al.* (2007) (so as to suppress premature direct metal-to-metal bridging between the gold electrodes) would allow them to recreate the much more conductive gold–benzenedithiolate–gold molecular nanowires realized by Xiao *et al.* (2004a) and Tsutsui *et al.* (2006, 2008). A recent experimental study (Martin *et al.* 2008) indeed concluded that the absence of a surrounding solvent is detrimental to the measurement of molecular-conductance histograms.

4.8.3   Tunnelling in alkanedithiolates and other molecules with gold electrodes

The transport properties of more strongly insulating molecular nanowires such as alkane-dithiolates bridging gold electrodes have been less controversial than gold–benzenedithiolate–gold and the experimental results for these systems have been more consistent. One reason for this is that the alkane chain itself is strongly insulating and for longer chains makes a large contribution to the resistance of the molecular nanowire so that the unknown details of the bonding between the molecule and the contacts are less crucial (Tomfohr and Sankey 2004). Another is that in both experimental and theoretical studies of conduction in these systems it has been possible to some extent to separate the effects of the unknown bonding geometry between the molecule and metal contacts from those of the alkane chain itself. Experimentally this is achieved by measuring the conductances of alkanedithiolates (or alkane thiolates) bridging gold electrodes for different numbers N of carbon atoms in the alkane chains of these molecules. Many such experimental studies have been carried out (for example, Wang *et al.* 2005; Chen *et al.* 2006a; Li *et al.* 2006a; Jang *et al.* 2006 and references therein; Song *et al.* 2007; Li *et al.* 2008) and it has been found that the measured conductances G_N fit well to the simple analytic form $G_N = Ae^{-\beta N}$ with $\beta \approx 1$. The use of the statistical STM break junction technique has even made it possible to distinguish experimentally between different values of the constant A that may correspond to different bonding geometries between the molecules and electrodes (Li *et al.* 2006a) and/or molecular chain conformations (Li *et al.* 2008). Theoretical conductance calculations based on both *ab-initio* and extended Hückel-based tight-binding models have also found this exponential dependence of the conductances of alkanedithiolates and alkanethiolates bridging metal contacts and have consistently yielded similar numerical exponents $\beta \approx 1$ for electrodes not only of gold but also of nickel and other transition d-metals (Kaun and Guo 2003; Tomfohr and Sankey 2004; Rocha *et al.* 2005; Dalgleish and Kirczenow 2006).

This agreement regarding the value of β between different theories (*even for different metal contacts*) and the *quantitative* agreement between theory and experiment may seem quite surprising at first, given that it relates to electrical conduction in molecular nanowires! However, an underlying physical reason for it can be found by considering the nature of the complex electronic band structures of infinite alkane chains: The eigenstates of infinite periodic alkane chains are Bloch states having the form $\psi_{kl} = e^{ikz}u_{kl}(\mathbf{r})$ where z is the co-ordinate along the chain and $u_{kl}(\mathbf{r})$ is a function that is periodic along the chain. At energies in the HOMO–LUMO gap of this alkane chain only eigenstates that decay or grow exponentially along the chain are possible so that for eigenstates in that energy range k must have a non-zero imaginary part. At each energy E in the HOMO–LUMO gap the eigenstate ψ_{kl} having the smallest positive value $\text{Im}(k)_{\min}$ of the imaginary part of k is the one that decays most slowly along the chain and therefore makes the dominant contribution to tunnelling along the chain at that energy. Thus, the decay constant β for electron transmission through the alkane chain at that energy is $\beta(E) = 2a\text{Im}(k)_{\min}$ where a is the length of the chain per carbon atom. The important point is that as the energy E approaches that of *either* the HOMO *or* the LUMO of the alkane chain, $\text{Im}(k)_{\min}$ approaches zero because within the molecular HOMO and LUMO bands electrons do not tunnel but travel freely as Bloch waves having real values of k. Thus, $\beta(E)$ has a maximum within the HOMO–LUMO gap of the alkane chain. The calculations of Tomfohr and Sankey (2002) carried out using density-functional theory showed this maximum value of $\beta(E)$ to be close to 1, the maximum to be very flat and broad and the Fermi energy of gold to fall within the range of energies around the maximum of $\beta(E)$ where β is not sensitive to the energy. Our calculations using an extended Hückel tight-binding model with the *standard* parameter set yield similar results, although in this model the Fermi level of gold lies considerably closer to the molecular HOMO band than in the model of Tomfohr and Sankey (2002). The *length* dependence of tunnelling through a *long* alkane chain thiol-bonded to metal contacts should be governed by the same physics and therefore for such systems β should be close to 1 and not very sensitive to the exact position the Fermi level of the metal contacts relative to the energies of the alkane HOMO and LUMO bands. This is why different theories consistently produce β close to 1 even for contacts made of different metals that have different values of the Fermi energy and despite the fact that, as has been discussed by Toher and Sanvito (2008), density-functional theory-based calculations within the LDA and GGA need not place the energies of molecular levels accurately relative to the Fermi energy of the electrodes. Thus, while theory and experiment agree well on the value of β, this agreement says little about how accurately (or inaccurately) the calculations place the molecular levels relative to the Fermi level of the electrodes.

The above physics is *not* readily apparent if (as is usually done in theoretical work on molecular nanowire transport) one considers *only* the calculated transmission probability $T(E, V)$ (defined in Section 4.6) of an electron through an alkanedithiolate or alkanethiolate molecule bridging two gold electrodes: At zero bias the transmission near the Fermi level of the gold leads is found to be in the tail of transmission resonances located between the Fermi level of gold and the *alkane* HOMO band and associated with states involving the *sulfur* atoms of the molecule and nearby gold atoms. This is not surprising since the HOMO states of *isolated* alkane*thiol* and alkane*dithiol* molecules have strong sulfur content. Thus, while β and its energy dependence are governed by the electronic structure of the alkane chain that makes up the central part of the molecular nanowire, the energy dependence of the transmission $T(E, V)$ near the Fermi level is strongly affected by the electronic structure of the molecular wire at its ends (where the sulfur and gold atoms are located) and tails of transmission resonances associated with this structure.

Experimental data on conduction through alkanedithiolates bridging gold electrodes is often analyzed phenomenologically using the Simmons (1963) formula that describes tunnelling through generic tunnel barriers. This can be misleading, however, because in such models β is a monotonically varying function of energy, whereas for alkanethiol and alkanedithiol molecules β has a broad maximum instead. Furthermore, as has been mentioned above, for these molecules the physics governing the energy dependence of $T(E, V)$ is different from the physics governing the energy dependence of β, which is not the case in the Simmons model.

It has been predicted that similar tunnelling exponentially decaying with the length of the molecule should occur through molecules involving hydrocarbon chains with unsaturated bonds or chains of aromatic rings (Samanta *et al.* 1996; Magoga and Joachim 1997; Tomfohr and Sankey 2002, 2004). However, for these systems the relevant HOMO–LUMO gaps are smaller, and the values of β are smaller accordingly, so that these systems are predicted to be less insulating than alkane chains, in agreement with experiment (He *et al.* 2005a).

The insensitivity of β to the energy of the tunnelling electron within much of the HOMO–LUMO gap may also be one reason why in the experiment of Venkataraman *et al.* (2006a) the conductance of a pair of phenyl rings in series was found, at first sight quite surprisingly, to be more sensitive to the *angle* between the rings than to changes in electronic structure due to the replacement of H atoms on the rings with Cl or F atoms. These substitutions, by withdrawing electrons from the rings, may be expected to shift the local electrostatic potentials on the rings through which the electrons tunnel and thus, in an overly simple potential barrier model of tunnelling such as the Simmons (1963) model, have a substantial direct effect on tunnelling through the molecules.

4.8.4   Self-assembled linear chains of molecules on silicon

Experiments on electrical conduction through molecules bridging gold electrodes are difficult to model theoretically in part because the atomic-scale geometries are unknown. On the other hand, the electronic density of states of gold is featureless in the vicinity of the Fermi energy, which makes molecular nanowires with gold electrodes somewhat forgiving of models that represent the electronic structure of the electrodes with limited accuracy. In the case of silicon electrodes the situation is reversed: As has already been mentioned in Section 4.5.3, the atomic-scale geometries of the bonding between many molecules and silicon are relatively simple and are understood much better. On the other hand, silicon has a bandgap located at the Fermi level so that it is essential to treat the electronic structure of silicon electrodes and the alignment of molecular nanowire electronic levels relative to this bandgap accurately in theoretical modelling. This presents additional challenges for theories of conduction in molecular nanowires on silicon whether *ab-intio* or semiempirical modelling or both are employed. On the positive side, inadequacies of a theory are more likely to manifest as clear *qualitative* differences with experiment for molecules on silicon than for molecules with gold electrodes, which together with the better knowledge of molecular-bonding geometries on silicon makes discriminating between more and less realistic theories subject to less uncertainty, greatly facilitating the benchmarking and improvement of models.

Thus, it is well established that *ab-initio* density-functional calculations utilizing the LDA and GGA approximations to the exchange-correlation energy functional underestimate significantly the size of the bandgap of silicon (which is accurately known experimentally), by ∼50% for the LDA and by ∼25% for the GGA. For molecules adsorbed on silicon the energy offset between the molecular HOMO (or the relevant frontier orbital of the adsorbate) and the silicon valence-band edge is also not predicted accurately by the LDA and GGA. The errors in these energy offsets obtained from the density-functional calculations have recently been estimated for a few molecules to range from 0.6 to 1.4 eV (Quek *et al.* 2006, 2007a). Because of these and other (Rakshit *et al.* 2004) deficiencies, transport calculations based on *ab-initio* density-functional modelling of the electronic structure are unreliable for molecules on silicon; they are able to capture some observable phenomena (Bevan *et al.* 2007) but are qualitatively incorrect for others (Kirczenow *et al.* 2005; Quek *et al.* 2006, 2007a).

The *standard quantum-chemical* parameterization of the extended Hückel model (Landrum and Glassey 2001) is also inadequate for modelling the silicon band structure since it predicts a direct bandgap instead of the correct indirect one for silicon and also overestimates the size of the bandgap. These deficiencies can be corrected and an accurate band structure for silicon obtained with more appropriate choices (Cerda and Soria 2000; Kirczenow *et al.* 2005) of the extended Hückel model's parameters for silicon. Within such models the energy offset between the molecular HOMO for adsorbed molecules and the silicon valence-band edge becomes an adjustable parameter. However, by carrying out calculations of electrical conduction through molecular nanowires on silicon for different values of this parameter and comparing the results with experiment much can be learned (Kirczenow *et al.* 2009).

A good example is the case of linear chains of styrene and methylstyrene molecules self-assembled on hydrogen-terminated (001) surfaces of silicon (Kirczenow *et al.* 2005). For these systems at low negative bias voltages applied to the silicon substrate relative to the STM tip (filled-state imaging), *ab-initio* calculations predict *minima* in the STM height profiles of the molecular chains over the centers of the molecules where *maxima* are observed experimentally (Kirczenow *et al.* 2005). However, transport calculations based on the modified extended Hückel model outlined above reveal the reason for this puzzling (from the *ab-initio* perspective) discrepancy (Kirczenow *et al.* 2005): The *ab-initio* calculations place the molecular HOMO levels states associated with the aromatic phenyl rings of the styrene and methylstyrene molecules too high in energy, very close to the silicon valence-band edge. Because of this *resonant* transmission between the silicon substrate and the STM tip via the molecular HOMO levels is predicted by the *ab-initio* models to begin already at very low bias. However, what occurs in reality at low bias is *tunnelling* via the molecule between tip and substrate, which gives rise to the qualitatively different STM height profiles of the styrene and methylstyrene molecules that are observed. The modified extended Hückel model with a more realistic offset between the molecular energy levels and the silicon valence-band edge not only yields qualitatively correct STM profiles of the molecules on silicon at low bias but also explains several counterintuitive features of the experimental data, including the experimentally observed reversal in the contrast between the styrene and methyl–styrene molecular chains with increasing STM tip bias, the observed increase in the apparent height of the molecules at the ends of the molecular chains relative to those far from the ends with increasing bias and the observed disappearance of the corrugation of the STM height profile along the molecular chains with increasing bias (Kirczenow *et al.* 2005). This example illustrates the advantages of a *well-chosen*, physically transparent and computationally simple model when the objective is to understand the essential physics governing the observed behavior of a specific experimental system, and also its usefulness in clarifying why the inherently less transparent and more rigid *ab-initio* approach is failing to make contact with the experiment. Given that this model has been able to account for several otherwise difficult to explain experimental facts, it may be of interest that it also predicts that an electron injected into the molecular LUMO level or a hole injected into the molecular HOMO level of a styrene molecule belonging to a self-assembled chain of styrene molecules on silicon should have a higher probability of travelling along the chain (at least by for short distances) by being transmitted directly from molecule to molecule than by taking the parallel path through the silicon substrate (Kirczenow *et al.* 2005).

4.9 Summary and outlook

The recent rapid progress in the study of molecular nanowires and their properties as electrical conductors has been driven primarily by the development of better experimental techniques. However, theoretical work has played an essential role in efforts to understand both what the systems being realized experimentally really are and the mechanisms underlying the phenomena being observed. This will most likely continue.

Over the last decade many different experimental methodologies have been tried for making molecular nanowires with a single molecule bridging a pair of metal contacts. The leading contender that has emerged to date is the STM break junction technique with statistical analysis of the data discussed in Sections 4.3, 4.4 and 4.8. Its importance is principally due to the fact that when this technique is employed in the way described by Tao and collaborators (Xiao *et al.* 2004a) or Venkataraman *et al.* (2006) one can be reasonably confident that conduction through a *single* molecule bridging the metal contacts really is being observed, that the molecule is *well bonded chemically to both contacts* and that there are no other weak electrical links in series with the molecular nanowire affecting conduction through the system. The ability of this technique to distinguish between different molecular nanowire geometries (Li *et al.* 2006a) is also very important even though it does not determine what these geometries are. In this methodology the junction is *stretched*, beginning with direct metal-to-metal contact and observed conductances quantized in units of $2e^2/h$, followed by conductance quantization in much smaller quanta signalling bridging of the electrodes by successively fewer molecules and ultimately by just a single one. As has been discussed in Section 4.8.2 this technique has made a decisive contribution to our understanding of the gold–benzenedithiolate–gold molecular nanowire system that has been an important focus of both theoretical and experimental work since the inception of the modern era in this field. It has also been found to be broadly applicable, yielding convincing results for different molecules and linker groups. There is every reason to expect it to continue to make definitive contributions to our understanding of the traditionally highly controversial topic of electrical conduction through biologically relevant molecules (Xiao *et al.* 2004b; Xu *et al.* 2004). It should also be able to shed much more light on the thorny but very important topics of electrical conduction through molecules with redox centers (Xiao *et al.* 2005) and through molecular nanowires exhibiting switching, hysteresis and negative differential resistance discussed in Section 4.3. It should also be possible to apply it to molecular nanowires with transition d-metal electrodes and thus to the realization of such potentially technologically important systems as single-molecule spin valves and spin-current rectifiers and new types of molecular nanowires exhibiting negative differential resistance, as discussed in Section 4.5.4. There is an ongoing debate as to how the statistics of the measured electrical conductances should be gathered using this technique: Some have argued that *all* measured conductances should be included in the statistics (Venkataraman *et al.* 2006) while others have found the technique to be most effective if the gathered statistics are restricted to conductance plateaus where the number of molecules bridging the junction (and possibly the bonding geometry) is temporarily stable as the junction is stretched (Chen *et al.* 2006a). Experience has shown that both are successful methodologies and both yield convincing results but the more selective approach reduces the level of statistical noise and is therefore applicable to a wider range of systems and can provide more detailed information, although it involves human interpretation of the data. As an alternative to human assessment, automatic data-filtering procedures designed to distinguish between experimental current traces in which a molecule does and does not bond to both of the metal contacts are being introduced at the present time (Xia *et al.* 2008). STM break junction techniques in which currents are measured as the junction is *compressed* starting from large separations between the electrodes are also being used, however, as discussed in Section 4.8.2, in order to perform single-molecule measurements in this way low concentrations of molecules on the electrode surfaces are needed and even then seemingly innocent changes in the apparatus can lead to very different results.

On the theoretical side, both *ab-initio* and semi-empirical models have been useful in interpreting the experimental data. The *ab-initio* approach is by far the most popular theoretical methodology in this field but it has serious fundamental deficiencies that have not received significant attention in the molecular-nanowire literature until quite recently. The absence of measurements of the atomic-scale geometries of molecular nanowires contacted with metal electrodes (Section 4.4) has made it impossible to make rigorous comparisons between the results of the *ab-initio* transport calculations and experiment in these systems. This, combined with the lack of a fundamental justification of the application of the *ab-initio* approach to transport calculations (see Section 4.7.1) in the way that is standard today (using the LDA and GGA approximations of time-independent density-functional theory) as well as a growing body of evidence of its inadequacy in specific cases, is unsettling to some of its practitioners while not being regarded as a concern by others. Molecular nanowires on silicon (Section 4.5.3 and 4.8.4) have considerable advantages as a testing ground for *ab-intio* and other theoretical methods because the bonding geometries between the molecules and the silicon are simpler and better understood and because the valence- and conduction-band edges of silicon provide clearer reference energies suitable for benchmarking molecular energy levels in transport experiments than are available for gold electrodes. Both experiments and the more sophisticated GW calculations indicate that the LDA and GGA place the energy levels of molecules adsorbed on silicon inaccurately relative to these reference levels, the estimated errors often being quite substantial, up to about 1.4 eV for the LDA (Section 4.8.4). If the errors made by the standard *ab-initio* transport calculations for molecules with gold electrodes are comparable to these then the appropriateness of this approach for calculating current–voltage characteristics of these systems should be seriously questioned. In this context it is worth noting that the *ab-initio* transport calculations have consistently yielded values of the bias voltage at which resonant conduction begins in the gold–benzenedithiol–gold system larger by factors of 2 or more than in the experiments, suggesting strongly that the energies of the resonantly transmitting states are not calculated at all accurately by these theories. Such criticism applies *a forteriori* to density-functional theory-based calculations of spintronic transport phenomena such as spin-valve effects since these are especially sensitive to the energies of resonantly transmitting states. On the other hand, as was discussed in Section 4.8.3, the values of the decay constants β associated with tunnelling through insulating molecules can be insensitive even to *large* errors in the calculated energies of the molecular states so that the good agreement that exists between *ab-initio* (and other) theories and experiments for the value of this parameter should *not* be interpreted as evidence of the theories placing the energies of the molecular levels in molecular nanowires correctly. The same is true for the values of the conductances at zero bias for strongly insulating molecules that, to a large extent, are determined just by the length of the molecule and the decay constant (Tomfohr and Sankey 2004). Thus, clearly, much research still needs to be done to properly test and improve on today's standard *ab-initio* methods for molecular nanowire transport calculations. As was discussed in Section 4.7.1 such efforts are being undertaken but at this stage it is too early to assess what the impact of this on the field will be; at present this work is still in a preliminary phase. Another important issue is that *ab-initio* transport calculations are a brute-force approach requiring large computer resources even for small systems. Thus, the *ab-initio* approach, even if it can be made more reliable, will become impractical as the molecules studied become more complex and molecular environments other than vacuum need to be taken into consideration, for example, in bionanoelectronics, or for real devices involving more than just one smallish molecule on a chip in *non*-periodic geometries.

However, the *ab-initio* approach is not the only route to a better understanding of experimentally observed molecular nanowire transport phenomena and of the underlying mechanisms: Semi-empirical tight-binding models discussed in Section 4.7.2 (possibly supplemented with *ab-initio* calculations of the molecular nanowire structure and *ground-state* electrostatic potential distributions (Piva *et al.* 2008; Kirczenow *et al.* 2009) for which the use of density-functional theory is justified, at least in principle) offer a practical alternative. As was discussed in Section 4.8, the semi-empirical approach has had success in shedding light on difficult to understand experiments, in some cases (somewhat paradoxically) without the use of *any* parameters fitted to the experiment under consideration. A disadvantage of the semi-empirical approach is that to use it effectively requires sophisticated human assessment of the system being studied. Thus, it does not lend itself well to implementation as commercial software. However, as the field progresses to studies of more complex molecules, and as nanoelectronic devices are developed that involve more than just single molecules, non-periodic molecular arrangements and/or molecules in complex environments, the advantage of the much smaller computing resources required by this approach will be increasingly important.

This chapter has focused mainly on electrical conduction mediated by elastic transmission of electrons through molecular nanowires since this is where the greatest progress in this field has been made. However, during the last few years inelastic scattering of electrons through molecular nanowires (with the emission and/or absorption of phonons) has begun to attract increased attention. As has been discussed in Section 4.4, inelastic spectroscopy has provided the first conclusive experimental evidence that particular molecular species are involved in electrical conduction through some metal–molecule–metal junctions. Within the Born–Oppenheimer approximation the vibrational modes of molecular nanowires and their frequencies are determined by the dependence on the nuclear co-ordinates of the *total* energy of the system in its *electronic* ground state. Since the use of density-functional theory for calculations of electronic *total ground-state energies* is well justified (at least in principle), *ab-initio* calculations of molecular vibrational frequencies and eigenmodes can also be justified. Thus, *ab-initio* calculations have been employed with some success in modelling the experimental inelastic spectroscopy data at low bias (references to some relevant papers can be found in Section 4.4) even though the criticism of the *ab-initio* approach to transport calculations discussed in Section 4.7.1 applies to *ab-initio* calculations of inelastic transport as well.

Inelastic scattering implies energy transfer to the molecule and therefore it can result in heating of the molecular nanowire. According to eqn (4.4) even *perfectly* transmitting nanostructures should have non-zero resistances $R = 1/G$. Since the resistance is not zero, a current passing through the nanostructure must result in Joule heating. Within the Landauer theory of *elastic* transport this heat is dissipated *not* in the nanostructure (molecule) itself but in the drain electrode where the electrons that carry the current through the nanostructure must eventually thermalize. As was mentioned in Section 4.4, it is clear from experimental inelastic spectroscopy data that for most of the molecules studied only a small fraction of the electrons passing through the molecular nanowire are involved in inelastic processes within the molecule. Thus, *most* of the Joule heat due to an electric current passing through such molecular nanowires is dumped in the drain electrode in accordance with Landauer theory. However, whether significant heating of the molecule occurs depends also on how quickly the Joule heat that the molecule receives through the occasional *inelastic* scattering events is transferred from the molecule to its environment. This is difficult to estimate theoretically with any confidence (Segal and Nitzan 2002; Chen *et al.* 2003, 2005a; D'Agosta *et al.* 2006; Galperin *et al.* 2007c; Pecchia *et al.* 2007) or to measure directly in the transport experiments. However, recent experimental work showed that the thermal equilibration of alkanethiolates on gold substrates heated by a femtosecond laser pulse occurs in a few tens of picoseconds (Wang *et al.* 2007). This is roughly equal to the average time between successive electrons passing through an octanedithiolate molecule bonded between gold electrodes under an applied bias of 1 V. Since only a very small fraction of these electrons is scattered inelastically within the molecule these numbers would seem to suggest that this molecule should be near *thermal* equilibrium with its environment under these conditions. However, in a recent experimental study the temperature of this molecular nanowire in toluene under a 1 V bias was estimated (albeit indirectly) to rise by up to a few tens of degrees above its room-temperature environment due to Joule heating (Huang *et al.* 2006, 2007). Clearly the topic of Joule heating of molecular nanowires and the related issue of the effects of dissipation and quantum dephasing (Raza *et al.* 2008a,b) within the molecule on electrical conduction needs further study.

The relationship between molecular vibrations and structural rearrangement of molecular nanowires that may occur in direct response to the application of bias (as discussed in Section 4.4), or in conjunction with redox reactions due to the application of bias and/or electrostatic or electrochemical gating (Section 4.3), or due to dynamic polarization of the molecule by the transiting electrons (Ness and Fisher 2002), is another interesting but poorly understood topic. Recent theoretical studies have focused on simplified model Hamiltonians that attempt to incorporate vibrational (phonon) modes, differently charged electronic states and electron–phonon coupling treating them using techniques from many-body physics, rate-equation approaches, semi-empirical modelling and/or reaction co-ordinate phenomenologies (Kuznetsov and Ulstrup 2004; Galperin *et al.* 2005, 2007a,b, 2008; May and Kühn 2006; Bratkovsky 2007; La Magna and Deretzis 2007; Yeganeh *et al.* 2007; Schultz *et al.* 2008, and others). This work has generally yielded results qualitatively resembling some of the intriguing experimentally observed phenomena such as negative differential resistance, hysteresis and switching. However, formulating the problem of structural transitions between different low-energy molecular geometries separated by energy barriers in terms of the vibrational eigenmodes corresponding to *one* of these low-energy geometries is problematic. Detailed comparisons with experiments on specific molecular systems have not as yet been possible partly because of the idealized nature of the theoretical models that have been studied to date. It is evident that a great deal of theoretical and experimental work still needs to be done in order to reach a definitive of understanding of this important topic.

Inelastic transmission of electrons through molecular nanowires under applied bias with the emission of photons (as distinct from phonons) can also occur. As has been outlined in Section 4.5.5, during the last few years experimental detection of photons emitted by molecular nanowires in this way has been reported and the evidence that these photons were indeed being emitted by the molecule itself is persuasive. Experimental spectra of the emitted photons together with the measured differential conductance characteristics and appropriate theoretical modelling (Buker and Kirczenow 2008) can provide definitive quantitative information regarding the evolution of the electronic states of molecular nanowires under applied bias that is needed to properly understand electrical conduction through molecular nanowires but has been lacking to date. Recently, correlations between electrical conduction in molecular nanowires and Raman scattering of photons have been reported (Ward *et al.* 2008), suggesting that another optical approach to probing molecular nanowire conduction experimentally may be feasible. However, the experimental and theoretical exploration of these topics still presents considerable challenges and has only just begun.

Acknowledgments

This work was supported by a Fellowship of the Canadian Institute for Advanced Research Nanoelectronics Program and by NSERC.

References

Agraït, N., Rodrigo, J.G., Vieira, S. *Phys. Rev. B* **47**, 12345 (1993).

Albrecht, T., Guckian, A., Ulstrup, J., Vos, J.G. *Nano Lett.* **5**, 1451 (2005a).

Albrecht, T., Guckian, A., Ulstrup, J., Vos, J.G. *IEEE Trans. Nanotech.* **4**, 430 (2005b).

Albrecht, T., Moth-Poulsen, K., Christensen, J.B., Hjelm, J., Bjørnholm, T., Ulstrup, J. *J. Am. Chem. Soc.* **128**, 6574 (2006a).

Albrecht, T., Guckian, A., Kuznetsov, A.M., Vos, J.G., Ulstrup, J. *J. Am. Chem. Soc.* **128**, 17132 (2006b).

Albrecht, T., Moth-Poulsen, K., Christensen, J.B., Guckian, A., Bjornholm, T., Vos, J.G., Ulstrup, J. *Faraday Discuss* **131**, 265 (2006c).

Albrecht, T., Mertens, S.F.L., Ulstrup, J. *J. Am. Chem. Soc.* **129**, 9162 (2007).

Alessandrini, A., Salerno, M., Frabboni, S., Facci, P. *Appl. Phys. Lett.* **86**, 133902 (2005).

Ammeter, J.H., Bürgi, H.-B., Thibeault, J.C., Hoffmann, R. *J. Am. Chem. Soc.* **100**, 3686 (1978).

Amy, F., Chan, C., Kahn, A. *Org. Electron.* **6**, 85 (2005).

Andrews, D.Q., Cohen, R., Van, Duyne, R.P., Ratner, M.A. *J. Chem. Phys.* **125**, 174718 (2006).

Andrews, D.Q., Van Duyne, R.P., Ratner, M.A. *Nano Lett.* **8**, 1120 (2008).

Aviram, A., Ratner, M.A. *Chem. Phys. Lett.* **29**, 277 (1974).

Baheti, K., Malen, J.A., Doak, P., Reddy, P., Jang, S.-Y., Tilley, T.D., Majumdar, A., Segalman, R. *Nano Lett.* **8**, 715 (2008).

Bautista, E.J., Yan, L., Seminario, J.M. *J. Phys. Chem. C* **111**, 14552 (2007).

Beebe, J.M., Engelkes, V.B., Miller, L.L., Frisbie, C.D. *J. Am. Chem. Soc.* **124**, 11268 (2002).

Beebe, J.M., Kim, B., Gadzuk, J.W., Frisbie, C.D., Kushmerick, J.G. *Phys. Rev. Lett.* **97**, 026801 (2006).

Beebe, J.M., Kushmerick, J.G. *Appl. Phys. Lett.* **90**, 083117 (2007).

Beenakker, C.W.J., van Houten, H. *Solid State Phys.* **44**, 1 (1991).

Benesch, C., Čížek, M., Thoss, M., Domcke, W. *Chem. Phys. Lett.* **430**, 355 (2006)

Berndt, R., Gaisch, R., Gimzewski, J., Reihl, B., Schlittler, R., Schneider, W., Tschudy, M. *Science* **262**, 1425 (1993).

Berndt, R., Gimzewski, J.K., Johansson, P. *Phys. Rev. Lett.* **67**, 3796 (1991).

Bevan, K.H., Zahid, F., Kienle, D., Guo, H. *Phys. Rev. B* **76**, 045325 (2007).

Blomquist, T., Kirczenow, G. *Nano Lett.* **4**, 2251 (2004).

Blomquist, T., Kirczenow, G. *Phys. Rev. B* **71**, 045301 (2005).

Blomquist, T., Kirczenow, G. *Nano Lett.* **6**, 61 (2006a).

Blomquist, T., Kirczenow, G. *Phys. Rev. B* **73**, 195303 (2006b).

Blum, A.S., Kushmerick, J.G., Long, D.P., Patterson, C.H., Yang, J.C., Henderson, J.C., Yao, Y.X., Tour, J.M., Shashidhar, R., Ratna, B.R. *Nature Mater* **4**, 167 (2005).

Bonča, J., Trugman, S.A. *Phys. Rev. Lett.* **75**, 2566 (1995).

Bonča, J., Trugman, S.A. *Phys. Rev. Lett.* **79**, 4874 (1997).

Brandbyge, M., Mozos, J.L., Ordejón, P., Taylor, J., Stokbro, K. *Phys. Rev. B* **65**, 165401 (2002).

Bratkovsky, A.M., Kornilovitch, P.E. *Phys. Rev. B* **67**, 115307 (2003).

Bratkovsky, A.M. *Polarons in Advanced Materials*. ed. Alexandrov A.S.) (Springer, Dordrecht, The Netherlands, 2007).

Buker, J., Kirczenow, G. *Phys. Rev. B* **66**, 245306 (2002).

Buker, J., Kirczenow, G. *Phys. Rev. B* **72**, 205338 (2005).

Buker, J., Kirczenow, G. *Phys. Rev. B* **78**, 125107 (2008).

Bumm, L.A., Arnold, J.J., Cygan, M.T., Dunbar, T.D., Burgin, T.P., Jones, L., Allara, D.L., Tour, J.M., Weiss, P.S. *Science* **271**, 1705 (1996).

Büttiker, M. *Phys. Rev. Lett.* **57**, 1761 (1986).

Cai, L.T., Cabassi, M.A., Yoon, H., Cabarcos, O.M., McGuiness, C.L., Flatt, A.K., Allara, D.L., Tour, J.M., Mayer, T.S. *Nano Lett.* **5**, 2365 (2005).

Cai, L.T., Skulason, H., Kushmerick, J.G., Pollack, S.K., Naciri, J., Shashidhar, R., Allara, D.L., Mallouk, T.E., Mayer, T.S. *J. Phys. Chem. B* **108**, 2827 (2004).

Cardamone, D.M., Stafford, C.A., Mazumdar, S. *Nano Lett.* **6**, 2422 (2006).

Cardamone, D.M., Kirczenow, G. *Phys. Rev. B* **77**, 165403 (2008a).

Cardamone, D.M., Kirczenow, G. *AIP Conf. Proc.* **995**, 135 (2008b).

Cardamone, D.M., Kirczenow, G. In press (2009).

Cerda, J., Soria, F. *Phys. Rev. B* **61**, 7965 (2000).

Chen, J., Reed, M.A., Rawlett, A.M., Tour, J.M. *Science* **286**, 1550 (1999).

Chen, J., Wang, W., Reed, M.A., Rawlett, A.M., Price, D.W., Tour, J.M. *Appl. Phys. Lett.* **77**, 1224 (2000).

Chen, Y.C., Zwolak, M., Di Ventra, M. *Nano Lett.* **3**, 1691 (2003).

Chen, Y.C., Zwolak, M., Di Ventra, M. *Nano Lett.* **5**, 621 (2005a).

Chen, Z.Z., Lu, R., Zhu, B.F. *Phys. Rev. B* **71**, 165324 (2005b).

Chen, F., Li, X., Hihath, J., Huang, Z., Tao, N. *J. Am. Chem. Soc.* **128**, 15874 (2006a).

Chen, F., Nuckolls, C., Lindsay, S. *Chem. Phys.* **324**, 236 (2006b).

Chen, L., Hu, Z., Zhao, A., Wang, B., Luo, Y., Yang, J., Hou, J.G. *Phys. Rev. Lett.* **99**, 146803 (2007).

Čížek, M., Thoss, M., Domcke, W. *Phys. Rev. B* **70**, 125406 (2004).

Collier, C.P., Mattersteig, G., Wong, E.W., Luo, Y., Beverly, K., Sampaio, J., Raymo, F.M., Stoddart, J.F., Heath, J.R. *Science* **289**, 1172 (2000).

Cornaglia, P.S., Grempel, D.R., Ness, H. *Phys. Rev. B* **71**, 075320 (2005).

Cornaglia, P.S., Ness, H., Grempel, D.R. *Phys. Rev. Lett.* **93**, 147201 (2004)

Cornil, J., Karzazi, Y., Brédas, J.L. *J. Am. Chem. Soc.* **124**, 3516 (2002).

Cui, X.D., Primak, A., Zarate, X., Tomfohr, J., Sankey, O.F., Moore, A.L., Moore, T.A., Gust, D., Harris, G., Lindsay, S.M. *Science* **294**, 571 (2001).

Dadosh, T., Gordin, Y., Krahne, R., Khivrich, I., Mahalu, D., Frydman, V., Sperling, J., Yacoby, A., Bar-Joseph, I. *Nature* **436**, 677 (2005).

de Leon, N.P., Liang, W., Gu, Q., Park, H. *Nano Lett.* **8**, 2963 (2008).

D'Agosta, R., Sai, N., Di Ventra, M. *Nano Lett.* **6**, 2935 (2006).

Dalgleish, H., Kirczenow, G. *Phys. Rev. B* **72**, 184407 (2005a).

Dalgleish, H., Kirczenow, G. *Phys. Rev. B* **72**, 155429 (2005b).

Dalgleish, H., Kirczenow, G. *Nano Lett.* **6**, 1274 (2006a).

Dalgleish, H., Kirczenow, G. *Phys. Rev. B* **73**, 245431 (2006b).

Dalgleish, H., Kirczenow, G. *Phys. Rev. B* **73**, 235436 (2006c).

Damle, P.S., Ghosh, A.W., Datta, S. *Phys. Rev. B* **64**, 201403 (2001).

Danilov, A.V., Kubatkin, S.E., Kafanov, S.G., Flensberg, K., Bjørnholm, T. *Nano Lett.* **6**, 2184 (2006).

Danilov, A., Kubatkin, S., Kafanov, S., Per Hedegård, P., Stuhr-Hansen, N., Moth-Poulsen, K., Thomas Bjørnholm, T. *Nano Lett.* **8**, 1 (2008).

Darancet, P., Ferretti, A., Mayou, D., Olevano, V. *Phys. Rev. B* **75**, 075102 (2007).

Datta, S. *Electronic Transport in Mesoscopic Systems* (Cambridge University Press, New York, 1995).

Datta, S., Tian, W.D., Hong, S.H., Reifenberger, R., Henderson, J.I., Kubiak, C.P. *Phys. Rev. Lett.* **79**, 2530 (1997).

Datta, S. *Nanotechnology* **15**, S433 (2004).

Datta, S. *The Oxford Handbook of Nanoscience and Technology: Frontiers and Advances* Vol. I ed. Narlikar, A.V., and Fu, Y.Y., (Oxford University Press, 2010), Ch. 1.

Delaney, P., Greer, J.C. *Phys. Rev. Lett.* **93**, 036805 (2004).

de Picciotto, A., Klare, J.E., Nuckolls, C., Baldwin, K., Erbe, A., Willett, R. *Nanotechnology* **16**, 3110 (2005).

de Picciotto, R., Stormer, H.L., Pfeiffer, L.N., Baldwin, K.W., West, K.W. *Nature* **411**, 51 (2001).

Di Carlo, A., Pecchia, A., Latessa, L., Frauenheim, T., Seifert, G. *Introducing Molecular Electronics*. Lect. Notes Phys. 680 ed. Cuniberti, G., Fagas, G. and Richter, K., (Springer, Berlin, 2005), p.153.

Di Ventra, M., Pantelides, S.T., Lang, N.D. *Phys. Rev. Lett.* **84**, 979 (2000a).

Di Ventra, M., Pantelides, S.T., Lang, N.D. *Appl. Phys. Lett.* **76**, 3488 (2000b).

Di Ventra, M., Kim, S.G., Pantelides, S.T., Lang, N.D. *Phys. Rev. Lett.* **86**, 288 (2001).

Di Ventra, M., Lang, N.D., Pantelides, S.T. *Chem. Phys.* **281**, 189 (2002).

Djukic, D., Thygesen, K.S., Untiedt, C., Smit, R. HM., Jacobsen, K.W., van Ruitenbeek, J.M. *Phys. Rev. B* **71**, 161402 (2005).

Dong, Z.C., Guo, X.L., Trifonov, A.S., Dorozhkin, P.S., Miki, K., Kimura, K., Yokoyama, S., Mashiko, S. *Phys. Rev. Lett.* **92**, 086801 (2004).

Dong, Z.C., Guo, X.L., Wakayama, Y., Hou, J.G. *Surf. Rev. and Lett.* **13**, 143 (2006).

Donhauser, Z.J., Mantooth, B.A., Kelly, K.F., Bumm, L.A., Monnell, J.D., Stapleton, J.J., Price, D.W., Rawlett, A.M., Allara, D.L., Tour, J.M., Weiss, P.S. *Science* **292**, 2303 (2001).

Donarini, A., Grifoni, M., Richter, K. *Phys. Rev. Lett.* **97**, 166801 (2006).

Dulic, D., van der Molen, S.J., Kudernac, T., Jonkman, H.T., de Jong, J.J.D., Bowden, T.N., van Esch, J., Feringa, B.L., van Wees, B.J. *Phys. Rev. Lett.* **91**, 207402 (2003).

Economou, E.N., Soukoulis, C.M. *Phys. Rev. Lett.* **46**, 618 (1981).

Emberly, E.G., Kirczenow, G. *Phys. Rev.* B58 10911; Ann. New York Acad. Sci **852**, 54 (1998a).

Emberly, E.G., Kirczenow, G. *Phys. Rev. Lett.* **81**, 5205 (1998b).

Emberly, E.G., Kirczenow, G.J. *Phys. Condens. Matter* **11**, 6911 (1999).

Emberly, E.G., Kirczenow, G. *Phys. Rev. B* **61**, 5740 (2000a).

Emberly, E.G., Kirczenow, G. *J. Appl. Phys.* **88**, 5280 (2000b).

Emberly, E.G., Kirczenow, G. *Phys. Rev. B* **62**, 10451 (2000c).

Emberly, E.G., Kirczenow, G. *Phys. Rev. B* **64**, 125318 (2001a).

Emberly, E.G., Kirczenow, G. *Phys. Rev. B* **64**, 235412 (2001b).

Emberly, E.G., Kirczenow, G. *Phys. Rev. Lett.* **87**, 269701 (2001c).

Emberly, E.G., Kirczenow, G. *Chem. Phys.* **281**, 311 (2002a).

Emberly, E.G., Kirczenow, G. *Ann. New York Acad. Sci.* **960**, 131 (2002b).

Emberly, E.G., Kirczenow, G. *Phys. Rev. Lett.* **91**, 188301 (2003).

Ernzerhof, M. *J. Chem. Phys.* **127**, 204709 (2007).

Evers, F., Burke, K. *Nano and Molecular Electronics Handbook.* ed. Lyshevski, S.E. (Taylor & Francis, Boca Raton, 2007) p. 24-1; cond-mat/0610413.

Fan, F.R.F., Lai, R.Y., Cornil, J., Karzazi, Y., Bredas, J.L., Cai, L.T., Cheng, L., Yao, Y.X., Price, D.W., Dirk, S.M., Tour, J.M., Bard, A.J. *J. Am. Chem. Soc.* **126**, 2568 (2004).

Fan, X.J., Rogow, D.L., Swanson, C.H., Tripathi, A., Oliver, S.R. *J. Appl. Phys. Lett.* **90**, 163114 (2007).

Fan, F.-R.F., Yang, J., Cai, L., Price, Jr., D.W., Dirk, S.M., Kosynkin, D.V., Yao, Y., Rawlett, A.M., James, M. Tour, J.M., Bard, A.J. *J. Am. Chem. Soc.* **124**, 5551 (2002).

Farajian, A.A., Belosludov, R.V., Mizuseki, H., Kawazoe, Y., Hashizume, T., Yakobson, B.I. *J. Chem. Phys.* **127**, 024901 (2007).

Fisher, D.S., Lee, P.A. *Phys. Rev. B* **23**, 6851 (1981).

Frederiksen, T., Paulsson, M., Brandbyge, M., Jauho, A. *Phys. Rev. B* **75**, 205413 (2007).

Galperin, M., Ratner, M.A., Nitzan, A. *J. Chem. Phys.* **121**, 11965 (2004).

Galperin, M., Ratner, M.A., Nitzan, A. *Nano Lett.* **5**, 125 (2005).

Galperin, M., Nitzan, A. *Phys. Rev. Lett.* **95**, 206802 (2005).

Galperin, M., Nitzan, A. *J. Chem. Phys.* **124**, 234709 (2006).

Galperin, M., Ratner, M.A., Nitzan, A. *J. Phys. Condens. Matter* **19**, 103201 (2007a).

Galperin, M., Ratner, M.A., Nitzan, A. *Phys. Rev. B* **76**, 035301 (2007b).

Galperin, M., Nitzan, A., Ratner, M.A. *Phys. Rev. B* **75**, 155312 (2007c).

Galperin, M., Nitzan, A., Ratner, M.A. *J. Phys.: Condens. Matter* **20**, 374107 (2008).

García, Y., Sancho-García, J.C. arXiv:0809.1031 (2008).

Gaudioso, J., Lauhon, L.J., Ho, W. *Phys. Rev. Lett.* **85**, 1918 (2000).

Gergel-Hackett, N., Majumdar, N., Martin, Z., Swami, N., Harriott, L.R., Bean, J.C., Pattanaik, G., Zangari, G., Zhu, Y., Pu, I., Yao, Y., Tour, J.M. *J. Vac. Sci. Technol. A* **24**, 1243 (2006).

Ghosh, A.W., Rakshit, T., Datta, S. *Nano Lett.* **4**, 565 (2004).

Goto, T., Degawa, K., Inokawa, H., Furukawa, K., Nakashima, H., Sumitomo, K., Aoki, T., Torimitsu, K. *Jpn. J. Appl. Phys.* **45**, 4285 (2006).

Goto, T., Inokawa, H., Nagase, M., Ono, Y., Sumitomo, K., Torimitsu, K. *Jpn. J. Appl. Phys.* **46**, 1731 (2007).

Grill, L., Moresco, F. *J. Phys. Condens. Matter* **18**, S1887 (2006).

Gronbeck, H., Curioni, A., Andreoni, W. *J. Am. Chem. Soc.* **122**, 3839 (2000).

Guisinger, N.P., Greene, M.E., Basu, R., Baluch, A.S., Hersam, M.C. *Nano Lett.* **4**, 55 (2004).

Guo, X.L., Dong, Z.C., Trifonov, A.S., Yokoyama, S., Mashiko, S., Okamoto, T. *Appl. Phys. Lett.* **84**, 969 (2004).

Guo, X.L., Dong, Z.C., Trifonov, A.S., Miki, K., Kimura, K., Mashiko, S. *Appl. Phys. A* **81**, 367 (2005).

Guo, X.F., Small, J.P., Klare, J.E., Wang, Y.L., Purewal, M.S., Tam, I.W., Hong, B.H., Caldwell, R., Huang, L.M., O'Brien, S., Yan, J.M., Breslow, R., Wind, S.J., Hone, J., Kim, P., Nuckolls, C. *Science* **311**, 356 (2006).

Haiss, W., Albrecht, T., van Zalinge, H., Higgins, S.J., Bethell, D., Hobenreich, H., Schiffrin, D.J., Nichols, R.J., Kuznetsov, A.M., Zhang, J., Chi, Q., Ulstrup, J. *J. Phys. Chem. B* **111**, 6703 (2007).

Harbola, U., Maddox, J.B., Mukamel, S. *Phys. Rev. B* **73**, 075211 (2006).

Harikumar, K.R., Polanyi, J.C., Sloan, P.A., Ayissi, S., Hofer, W.A. *J. Am. Chem. Soc.* **128**, 16791 (2006).

Härtle, R., Benesch, C., Thoss, M. *Phys. Rev. B* **77**, 205314 (2008).

He, H.Y., Pandey, R., Pati, R., Karna, S.P. *Phys. Rev. B* **73**, 195311 (2006).

He, J., Chen, F., Li, J., Sankey, O.F., Terazono, Y., Herrero, C., Gust, D., Moore, T.A., Moore, A.L., Lindsay, S.M. *J. Am. Chem. Soc.* **127**, 1384 (2005a).

He, J., Chen, F., Liddell, P.A., Andréasson, J., Straight, S.D., Gust, D., Moore, T.A., Moore, A.L., Li, J., Sankey, O.F., Lindsay, S.M. *Nanotechnology* **16**, 695 (2005b).

He, J., Fu, Q., Lindsay, S., Ciszek, J.W., Tour, J.M. *J. Am. Chem. Soc.* **128**, 14828 (2006).

Heersche, H.B., de Groot, Z., Folk, J.A., van der Zant, H.S.J., Romeike, C., Wegewijs, M.R., Zobbi, L., Barreca, D., Tondello, E., Cornia, A. *Phys. Rev. Lett.* **96**, 206801 (2006).

Henderson, J.J., Ramsey, C.M., del Barco, E., Mishra, A., Christou, G. *J. Appl. Phys.* **101**, 09E102 (2007).

Hettler, M.H., Wenzel, W., Wegewijs, M.R., Schoeller, H. *Phys. Rev. Lett.* **90**, 076805 (2003).

Hihath, J., Arroyo, C.R., Rubio-Bollinger, G., Tao, N.J., Agraït, N. *Nano Lett.* **8**, 1673 (2008).

Hoffmann, G., Libioulle, L., Berndt, R. *Phys. Rev. B* **65**, 212107 (2002).

Hoffmann, R. *J. Chem. Phys.* **39**, 1397 (1963).

Hohenberg, P., Kohn, W. *Phys. Rev.* **136**, B864 (1964).

Hu, Y., Zhu, Y., Gao, H., Guo, H. *Phys. Rev. Lett.* **95**, 156803 (2005).

Huang, Z.F., Xu, B.Q., Chen, Y.C., Di Ventra, M., Tao, N. *J. Nano Lett.* **6**, 1240 (2006).

Huang, Z.F., Chen, F., D'Agosta, R., Bennett, P.A., Di, Ventra, M., Tao, N. *J. Nature Nanotech.* **2**, 698 (2007)

Hush, N.S. *Ann. N.Y. Acad. Sci.* **1006**, 1 (2003).

Hybertsen, M.S., Venkataraman, L., Klare, J.E., Whalley, A.C., Steigerwald, M.L., Nuckolls, C. *J. Phys.: Condens. Matter* **20**, 374115 (2008).

Imry, Y. *Directions in Condensed Matter Physics*. ed. Grinstein, G., Masenko, G. (World Scientific, Singapore, 1986) Vol.1, p. 101 .

Jäckel, F., Watson, M.D., Müllen, K., Rabe, J.P. *Phys. Rev. Lett* **92**, 188303 (2004).

James, D.K., Tour, J.M. *Top. Curr. Chem.* **257**, 33 (2005).

Jang, S.Y., Reddy, P., Majumdar, A., Segalman, R.A. *Nano Lett*. **6**, 2362 (2006).

Jean, N., Sanvito S. *Phys. Rev. B* **73**, 094433 (2006).

Jiang, F., Zhou, Y.X., Chen, H., Note, R., Mizuseki, H., Kawazoe, Y. *J. Chem. Phys.* **125**, 084710 (2006).

Jo, M.H., Grose, J.E., Baheti, K., Deshmukh, M.M., Sokol, J.J., Rumberger, E.M., Hendrickson, D.N., Long, J.R., Park, H., Ralph, D.C. *Nano Lett*. **6**, 2014 (2006).

Joachim, C., Gimzewski, J.K. *Chem. Phys. Lett*. **265**, 353 (1997).

Jung, K.H., Hase, E., Yasutake, Y., Shin, H.K., Kwon, Y.S., Majima, Y. *Jpn. J. Appl. Phys*. Pt. 2 **45**, L840 (2006).

Katzenmeyer, A., Logeeswaran V.J., Tekin, B., Islam, M.S. arXiv:0801.0476v1 (2008)

Kaun, C.-C., Guo, H. *Nano Lett*. **3**, 1521 (2003).

Kaun, C.C., Seideman, T.J. *Comp. Theor. Nanosci.* (2006) **3**, 951 (2006).

Ke, S.H., Baranger, H.U., Yang, W.T. *Phys. Rev. B* **70**, 085410 (2004).

Ke, S.H., Baranger, H.U., Yang, W.T. *Phys. Rev. B* **71**, 113401 (2005).

Ke, S.H., Baranger, H.U., Yang, W.T. *J. Chem. Phys.* **126**, 201102 (2007).

Keane, Z.K., Ciszek, J.W., Tour, J.M., Natelson, D. *Nano Lett*. **6**, 1518 (2006).

Kergueris, C., Bourgoin, J.-P., Palacin, S., Esteve, D., Urbina, C., Magoga, M., Joachim, C. *Phys. Rev. B* **59**, 12505 (1999).

Kienle, D., Cerda, J.I., Ghosh, A.W. *J. Appl. Phys.* **100**, 043714 (2006a).

Kienle, D., Bevan, K.H., Liang, G.-C., Siddiqui, L., Cerda, J.I., Ghosh, A.W. *J. Appl. Phys.* **100**, 043715 (2006b).

Kiguchi, M., Miura, S., Hara, K., Sawamura, M., Murakoshi, K. *Appl. Phys. Lett.* **89**, 213104 (2006).

Kiguchi, M., Tal, O., Wohlthat, S., Pauly, F., Krieger, M., Djukic, D., Cuevas, J.C., van Ruitenbeek, J.M. *Phys. Rev. Lett.* **101**, 046801 (2008).

Kim, B., Beebe, J.M., Jun, Y., Zhu, X.Y., Frisbie, C.D. *J. Am. Chem. Soc.* **128**, 4970 (2006).

Kim, B., Beebe, J.M., Olivier, C., Rigaut, S., Touchard, D., Kushmerick, J.G., Zhu, X.Y., Frisbie, C.D. *J. Phys. Chem.* **111**, 7521 (2007).

Kirczenow, G. *Phys. Rev. B* **75**, 045428 (2007).

Kirczenow, G., Piva, P.G., Wolkow, R.A. *Phys. Rev. B* **72**, 245306 (2005).

Kirczenow, G., Piva, P.G., Wolkow, R.A. *Phys. Rev. B* **80**, 035309 (2009).

Kletsov, A., Dahnovsky, Y. *J. Chem. Phys.* **127**, 144716 (2007).

Koentopp, M., Chang, C., Burke, K., Car, R. *J. Phys.: Condens. Matter* **20**, 083203 (2008).

Kohn, W., Sham, L. *J. Phys. Rev.* **140**, A1133 (1965).

Kornilovitch, P.E., Bratkovsky, A.M. *Phys. Rev. B* **64**, 195413 (2001).

Kosov, D.S. *J. Chem. Phys.* **120**, 7165 (2004).

Krüger, D., Fuchs, H., Rosseau, R., Marx, D., Parrinello, M. *Phys. Rev. Lett.* **89**, 186402 (2002).

Kubatkin, S., Danilov, A., Hjort, M., Cornil, J., Brédas, J.L., Stuhr-Hansen, N., Hedegård, P., Bjørnholm, T. *Nature* **425**, 698 (2003).

Kula, M., Luo, Y. *J. Chem. Phys.* **128**, 064705 (2008).

Kushmerick, J.G., Blum, A.S., Long, D.P. *Analytica Chimica Acta* **568**, 20 (2006).

Kushmerick, J.G., Holt, D.B., Pollack, S.K., Ratner, M.A., Yang, J.C., Schull, T.L., Naciri, J., Moore, M.H., Shashidhar, R. *J. Am. Chem. Soc.* **124**, 10654 (2002a).

Kushmerick, J.G., Holt, D.B., Yang, J.C., Naciri, J., Moore, M.H., Shashidhar, R. *Phys. Rev. Lett.* **89**, 086802 (2002b).

Kushmerick, J.G., Lazorcik, J., Patterson, C.H., Shashidhar, R., Seferos, D.S., Bazan, G.C. *Nano Lett.* **4**, 639 (2004).

Kuznetsov, A.M., Ulstrup, J. *J. Electroanal. Chem.* **564**, 209 (2004).

La Magna, A., Deretzis, I. *Phys. Rev. Lett.* **99**, 136404 (2007).

Landau, L.D., Lifshitz, E.M. *Statistical Physics*. vol. **5** of *Course of Theoretical Physics*. (Pergamon, Bristol, 1958).

Landauer, R. *Philos. Mag.* **21**, 863 (1970).

Landrum, G.A., Glassey, W.V. *Yet Another Extended Hückel Molecular Orbital Package*. Freely available at http://sourceforge.net/projects/yaehmop (2001).

Lang, N.D. *Phys. Rev. B* **64**, 235121 (2001).

Lang, N.D., Solomon, P.M. *Nano Lett.* **5**, 921 (2005).

Lau, C.N., Stewart, D.R., Williams, R.S., Bockrath, M. *Nano Lett.* **4**, 569 (2004).

Leary, E., Higgins, S.J., van Zalinge, H., Haiss, W., Nichols, R.J., Nygaard, S., Jeppesen, J.O., Ulstrup, J. *J. Am. Chem. Soc.* **130**, 12204 (2008).

Lee, J.O., Lientschnig, G., Wiertz, F.GH., Struijk, M., Janssen, R.AJ., Egberink, R., Reinhoudt, D.N., Grimsdale, A.C., Müllen, K., Hadley, P., Dekker, C. *Ann. N. Y. Acad. Sci.* **1006**, 122 (2003).

Li, C., Pobelov, I., Wandlowski, T., Bagrets, A., Arnold, A., Evers, F. *J. Am. Chem. Soc.* **130**, 318 (2008).

Li, W., Kavanagh, K.L., Matzke, C.M., Talin, A.A., Léonard, F., Faleev, S., Hsu, J.W.P. *J. Phys. Chem. B* **109**, 6252 (2005).

Li, X., He, J., Hihath, J., Xu, B., Lindsay, S.M., Tao, N.J. *J. Am. Chem. Soc.* **128**, 2135 (2006a).

Li, X., Xu, B.Q., Xiao, X.Y., Yang, X.M., Zang, L., Tao, N.J. *Faraday Discuss* **131**, 111 (2006b).

Li, X., Hihath, J., Chen, F., Masuda, T., Zang, L., Tao, N.J. *J. Am. Chem. Soc.* **129**, 11535 (2007).

Liang, W., Shores, M.P., Bockrath, M., Long, J.R., Hongkun Park, H. *Nature* **417**, 725 (2002).

Lindsay, S.M., Ratner, M.A. *Adv. Mater.* **19**, 23 (2007).

Long, D.P., Lazorcik, J.L., Mantooth, B.A., Moore, M.H., Ratner, M.A., Troisi, A., Yao, Y., Ciszek, J.W., Tour, J.M., Shashidhar, R. *Nature Mater.* **5**, 901 (2006).

Long, M.-Q., Chen, K.-Q., Wang, L., Zou, B.S., Shuai, Z. *Appl. Phys. Lett.* **91**, 233512 (2007).

Loo, Y.L., Hsu, J.W.P., Willett, R.L., Baldwin, K.W., West, K.W., Rogers, J. A *J. Vac. Sci. Technol. B* **20**, 2853 (2002).

Lopinski, G.P., Wayner, D.D.M., Wolkow, R.A. *Nature* **406**, 48 (2000).

Lörtscher, E., Ciszek, J.W., Tour, J., Riel, H. *Small* **2**, 973 (2006).

Lörtscher, E., Weber, H.B., Riel, H. *Phys. Rev. Lett.* **98**, 176807 (2007).

Lucier, A.S., Mortensen, H., Sun, Y., Grütter P. *Phys. Rev. B* **72**, 235420 (2005).

Ludoph, B., van Ruitenbeek, J.M. *Phys. Rev. B* **61**, 2273 (2000).

Lüffe, M.C., Koch, J., von Oppen, F. *Phys. Rev. B* **77**, 125306 (2008).

Magoga, M., Joachim, C. *Phys. Rev. B* **56**, 4722 (1997).

Mahapatro, A.K., Ying, J., Ren, T., Janes, D.B. *Nano Lett.* **8**, 2131 (2008).

Majumdar, N., Gergel-Hackett, N., Bean, J.C., Harriott, L.R., Pattanaik, G., Zangari, G., Yao, Y., Tour, J.M. *J. Electron. Mater.* **35**, 140 (2006).

Malyshev, A.V. *Phys. Rev. Lett.* **98**, 096801 (2007).

Martin, C.A., Ding, D., van der Zant, H.S.J., van Ruitenbeek, J.M. *New J. Phys.* **10**, 065008 (2008).

Mativetsky, J.M., Pace, G., Elbing, M., Rampi, M.A., Mayor, M., Samori, P. *J. Am. Chem. Soc.* **130**, 9192 (2008).

May, V. *Phys. Rev. B* **66**, 245411 (2002).

May, V., Kühn, O. *Chem. Phys. Lett.* **420**, 192 (2006).

Meir, Y., Wingreen, N.S. *Phys. Rev. Lett.* **68**, 2512 (1992).

Mendes, P.M., Flood, A.H., Stoddart, J.F. *Appl. Phys. A* **80**, 1197 (2005).

Metzger, R.M. *Chem. Rev.* **103**, 3803 (2003).

Miao, F., Ohlberg, D., Stewart, D.R., Williams, R.S., Lau, C.N. *Phys. Rev. Lett.* **101**, 016802 (2008).

Miller, O.D., Muralidharan, B., Kapur, N., Ghosh, A.W. *Phys. Rev. B* **77**, 125427 2008.

Moore, A.M., Dameron, A.A., Mantooth, B.A., Smith, R.K., Fuchs, D.J., Ciszek, J.W., Maya, F., Yao, Y.X., Tour, J.M., Weiss, P.S. *J. Am. Chem. Soc.* **128**, 1959 (2006).

Mujica, V., Nitzan, A., Datta, S., Ratner, M.A., Kubiak, C.P. *J. Phys. Chem. B* **107**, 91 (2003).

Muralidharan, B., Ghosh, A.W., Datta, S. *Phys. Rev. B* **73**, 155410 (2006).

Natelson, D., Yu, L.H., Ciszek, J.W., Keane, Z.K., Tour, J.M. *Chem. Phys.* **324**, 267 (2006).

Naydenov, B., Teague, L.C., Ryan, P., Boland, J. *J. Nano Lett.* **6**, 1752 (2006).

Ness, N., Fisher, A. *J. Chem. Phys.* **281**, 279 (2002).

Ning, Z., Yu Zhu, Y., Wang, J., Guo, H. *Phys. Rev. Lett.* **100**, 056803 (2008).

Nishikawa, A., Tobita, J., Kato, Y., Fujii, S., Suzuki, M., Fujihira, M. *Nanotechnology* **18**, 424005 (2007).

Novaes, F.D., da, Silva, A. JR., Fazzio, A. *Brazil. J. Phys.* **36**, 799 (2006).

Ohnishi, H., Kondo, Y., Takayanagi, K. *Nature* **395**, 780 (1998).

Ortiz, D.O., Seminario, J.M. *J. Chem. Phys.* **127**, 111106 (2007)

Osorio, E.A., O'Neill, K., Stuhr-Hansen, N., Nielsen, O.F., Bjørnholm, T., van der Zant, H.S. *J. Adv. Mater.* **19**, 281 (2007a).

Osorio, E.A., O'Neill, K., Wegewijs, M., Stuhr-Hansen, N., Paaske, J., Bjørnholm, T., van der Zant, H.S. *J. Nano Lett.* **7**, 3336 (2007b).

Paaske, J., Flensberg, K. *Phys. Rev. Lett.* **94**, 176801 (2005).

Papaconstantopoulos, D.A. *Handbook of the Band Structure of Elemental Solids* (Plenum Press, New York, 1986).

Park, H., Park, J., Lim, A.K.L., Anderson, E.H., Alivisatos, A.P., McEuen P.L. *Nature* **407**, 57 (2000).

Park, J., Pasupathy, A.N., Goldsmith, J.I., Chang, C., Yaish, Y., Petta, J.R., Rinkoski, M., Sethna, J.P., Abruna, H.D., McEuen, P.L., Ralph, D.C. *Nature* **417**, 722 (2002).

Park, Y.S., Whalley, A.C., Kamenetska, M., Steigerwald, M.L., Hybertsen, M.S., Nuckolls, C., Venkataraman, L. *J. Am. Chem. Soc.* **129**, 15768 (2007).

Parks, J.J., Champagne, A.R., Hutchison, G.R., Flores-Torres S., Abruna H.D., Ralph, D.C. *Phys. Rev. Lett.* **99**, 026601 (2007).

Pascual, J.I., Méndez, J., Gómez-Herrero, J., Baró, A.M., García, N., Binh, V.T. *Phys. Rev. Lett.* **71**, 1852 (1993).

Pasupathy, A.N., Bialczak, R.C., Martinek, J., Grose, J.E., Donev, L.A.K., McEuen, P.L., Ralph, D.C. *Science* **306**, 86 (2004).

Pati, R., Senapati, L., Ajayan, P.M., Nayak, S.K. *Phys. Rev. B* **68**, 100407 (2003).

Patrone, L., Palacin, S., Bourgoina, J.P., Lagouteb, J., Zambellib, T., Gauthier, S. *Chem. Phys.* **281**, 325 (2002).

Paulsson, M., Datta, S. *Phys. Rev. B* **67**, 241403R (2003).

Paulsson, M., Frederiksen, T., Brandbyge, M. *Nano Lett.* **6**, 258 (2006).

Paulsson, M., Frederiksen, T., Ueba, H., Lorente, N., Brandbyge, M. *Phys. Rev. Lett.* **100**, 226604 (2008).

Pecchia, A., Di Carlo, A. *Nano Lett.* **4**, 2109 (2004).

Pecchia, A., Romano, G., Di Carlo, A. *Phys. Rev. B* **75**, 035401 (2007).

Petta, J.R., Slater, S.K., Ralph, D.C. *Phys. Rev. Lett.* **93**, 136601 (2004).

Peierls, R.E. *Quantum Theory of Solids* (Oxford University Press, London, 1956) p. 77.

Pemmaraju, C.D., Archer, T., Sánchez-Portal, D., Sanvito S. *Phys. Rev. B* **75**, 045101 (2007).

Peng, J., Chen, Z.G. *Phys. Lett. A* **365**, 505 (2007).

Perdew, J.P., Zunger, A. *Phys. Rev. B* **23**, 5048 (1981).

Perdew, J.P., Parr, R.G., Levy, M., Balduz, J.L. *Phys. Rev. Lett.* **49**, 1691 (1982).

Perdew, J.P., Ruzsinszky, A., Tao, J., Staroverov, V.N., Scuseria, G.E., Csonka, G.I. *J. Chem. Phys.* **123**, 062201 (2005).

Pitters, J.L., Wolkow, R.A. *Nano Lett.* **6**, 390 (2006).

Piva, P.G., DiLabio, G.A., Pitters, J.L., Zikovsky, J., Rezeq, M., Dogel, S., Hofer, W.A., Wolkow, R.A. *Nature* **435**, 658 (2005).

Piva, P.G., Wolkow, R.A., Kirczenow, G. *Phys. Rev. Lett.* **101**, 106801 (2008).

Poirier, G.E. *Phys. Rev. Lett.* **86**, 83 (2001).

Prodan, E., Car, R. *Phys. Rev. B* **76**, 115102 (2007).

Quek, S.Y., Neaton, J.B., Hybertsen, M.S., Kaxiras, E., Louie, S.G. *Phys. Status Solidi B* **243**, 2048 (2006).

Quek, S.Y., Neaton, J.B., Hybertsen, M.S., Kaxiras, E., Louie, S.G. *Phys. Rev. Lett.* **98**, 066807 (2007a).

Quek, S.Y., Venkataraman, L., Choi, H.J., Louie, S.G., Hybertsen, M.S., Neaton, J.B. *Nano Lett.* **7**, 3477 (2007b).

Qiu, X.H., Nazin, G.V., Ho, W. *Science* **299**, 542 (2003).

Qiu, X.H., Nazin, G.V., Ho, W. *Phys. Rev. Lett.* **92**, 206102 (2004).

Quinn, J.R., Foss, F.W., Venkataraman, L., Hybertsen, M.S., Breslow, R. *J. Am. Chem. Soc.* **129**, 6714 (2007a).

Quinn, J.R., Foss, F.W., Venkataraman, L., Breslow, R. *J. Am. Chem. Soc.* **129**, 12376 (2007b).

Rakshit, T., Liang, G.C., Ghosh, A.W., Datta, S. *Nano Lett.* **4**, 1803 (2004).

Rakshit, T., Liang, G.C., Ghosh, A.W., Hersam, M.C., Datta, S. *Phys. Rev. B* **72**, 125305 (2005).

Ramachandran, G.K., Hopson, T.J., Rawlett, A.M., Nagahara, L.A., Primak, A., Lindsay, S.M. *Science* **300**, 1413 (2003).

Raza, H., Bevan, K.H., Kienle, G. *Phys. Rev. B* **77**, 035432 (2008a).

Raza, H. arXiv:cond-mat/0703236v2 (2008b)

Reddy, P., Jang, S.-Y., Segalman, R., Majumdar, A. *Science* **315**, 1568 (2007).

Reed, M.A., Zhou, C., Muller, C.J., Burgin, T.P., Tour, J.M. *Science* **278**, 252 (1997).

Reichert, J., Ochs, R., Beckmann, D., Weber, H.B., Mayor, M., Löhneysen, H.V. *Phys. Rev. Lett.* **88**, 176804 (2002).

Rocha, A.R., García-Suárez, V.M., Bailey, S.W., Lambert, C.J., Ferrer, J., Sanvito, S. *Nature Mater.* **4**, 335 (2005).

Rocha, A.R., García-Suárez, V.M., Bailey, S., Lambert, C., Ferrer, J., Sanvito, S. *Phys. Rev. B* **73**, 085414 (2006).

Rosei, F., Schunack, M., Naitoh, Y., Jiang, P., Gourdon, A., Laegsgaard, E., Stensgaard, I., Joachim, C., Besenbacher, F. *Prog. Surf. Sci.* **71**, 95 (2003).

Saha, S., Stoddart, J.F. *Chem. Soc. Rev.* **36**, 77 (2007).

Sai, N., Zwolak, M., Vignale, G., Di Ventra, M. *Phys. Rev. Lett.* **94**, 186810 (2005).

Salomon, A., Cahen, D., Lindsay, S., Tomfohr, J., Engelkes, V.B., Frisbie, C.D., *Adv. Mater.* **15**, 1881 (2003).

Samanta, M.P., Tian, W., Datta, S., Henderson, J.I., Kubiak, C.P. *Phys. Rev. B* **53**, R7626 (1996).

Sanvito, S., Rocha, A.R. *J. Comp. Theor. Nanosci.* **3**, 624 (2006).

Schultz, M.G., Nunner, T.S., von Oppen F. *Phys. Rev. B* **77**, 075323 (2008).

Segal, D., Nitzan, A. *J. Chem. Phys.* **117**, 3915 (2002).

Shamai, T., Ophir, A., Selzer, Y. *Appl. Phys. Lett.* **91**, 102108 (2007).

Seideman, T., Guo, H. *J. Th. Comp. Chem.* **2**, 439 (2003).

Selzer, Y., Allara, D.L. *Annu. Rev. Phys. Chem.* **57**, 593 (2006).

Selzer, Y., Cai, L.T., Cabassi, M.A., Yao, Y.X., Tour, J.M., Mayer, T.S., Allara, D.L. *Nano Lett.* **5**, 61 (2005).

Seminario, J.M., Zacarias, A.G., Tour, J.M. *J. Am. Chem. Soc.* **122**, 3015 (2000).

Sergueev, N., Roubtsov, D., Guo, H. *Phys. Rev. Lett.* **95**, 146803 (2005).

Simmons, J.G. *J. Appl. Phys.* **34**, 1793 (1963).

Smit, R.H.M., Noat, Y., Untiedt, C., Lang, N.D., van, Hemert, M.C., van Ruitenbeek, J.M. *Nature* **906**, 658 (2002).

Smolyaninov, I.I., Khaikin, M.S. *Phys. Lett. A* **149**, 410 (1990).

Solomon, G.C., Andrews, D.Q., Hansen, T. Goldsmith, R.H., Wasielewski, M.R., Van Duyne, R.P., Ratner, M.A. *J. Chem. Phys.* **129**, 054701 (2008).

Song, H., Lee, T., Choi, N.-J. Lee, H. *Appl. Phys. Lett.* **91**, 253116 (2007).

Stewart, D.R., Ohlberg, D.AA., Beck, P.A., Chen, Y., Williams, R.S., Jeppesen, J.O., Nielsen, K.A., Stoddart, J.F. *Nano Lett.* **4**, 133 (2004).

Stokbro, K., Taylor, J., Brandbyge, M., Guo, H. *Introducing Molecular Electronics.* Lect. Notes Phys. 680 ed. Cuniberti, G., Fagas, G. and Richter, K., (Springer, Berlin, 2005).

Su, W., Jiang, J., Lu, W., Luo, Y. *Nano Lett.* **6**, 2091 (2006).

Tang, J.Y., Wang, Y.L., Klare, J.E., Tulevski, G.S., Wind, S.J., Nuckolls, C. *Angew. Chem. Int. Ed.* **46**, 3892 (2007).

Tao, N. *Phys. Rev. Lett.* **76**, 4066 (1996).

Tao, N.J. *Nature Nanotechnol.* **1**, 173 (2006).

Taylor, J., Guo, H., Wang, J. *Phys. Rev. B* **63**, 121104 (2001a).

Taylor, J., Guo, H., Wang, J. *Phys. Rev. B* **63**, 245407 (2001b).

Taylor, J., Brandbyge, M., Stokbro, K. *Phys. Rev. B* **68**, 121101 (2003).

Thouless, D.J. *Phys. Rev. Lett.* **47**, 972 (1981).

Thygesen, K.S., Rubio, A. *J. Chem. Phys.* **126**, 091101 (2007).

Thygesen, K.S., Rubio, A. *Phys. Rev. B* **77**, 115333 (2008).

Thygesen, K.S. *Phys. Rev. Lett.* **100**, 166804 (2008).

Toher, C., Filippetti, A., Sanvito, S., Burke, K. *Phys. Rev. Lett.* **95**, 146402 (2005).

Toher, C., Sanvito, S. *Phys. Rev. Lett.* **99**, 056801 (2007).

Toher, C., Sanvito, S. *Phys. Rev. B* **77**, 155402 (2008).

Tomfohr, J., Sankey, O.F. *Phys. Rev. B* **65**, 245105 (2002).

Tomfohr, J., Sankey, O.F. *J. Chem. Phys.* **120**, 1542 (2004).

Troisi, A., Ratner, M.A., Nitzan, A. *J. Chem. Phys.* **118**, 6072 (2003).

Troisi, A., Ratner, M.A. *Phys. Rev. B* **72**, 033408 (2005).

Troisi, A., Ratner, M.A. *Small* **2**, 172 (2006a).

Troisi, A., Ratner, M.A. *Nano Lett.* **6**, 1784 (2006b).

Tseng, H.-R., Wu, D.M., Fang, N.X.L., Zhang, X., Stoddart, J.F. *ChemPhys Chem* **5**, 111 (2004).

Tsutsui, M., Teramae, Y., Kurokawa, S., Sakai, A. *Appl. Phys. Lett.* **89**, 163111 (2006).

Tsutsui, M., Shoji, K., Morimoto, K., Taniguchi, M., Kawai, T. *Appl. Phys. Lett.* **92**, 223110 (2008).

van der Molen, S.J., van der Vegte, H., Kudernac, T., Amin, I., Feringa, B.L., van Wees, B.J. *Nanotechnology* **17**, 310 (2006).

Ulloa, S.E., MacKinnon, A., Casta no, E., Kirczenow, G. *Handbook of Semiconductors* Vol. 1 2nd edn ed. Moss., T.S. and Landsberg, P.T., (Elsevier Science Publishers B. V., Amsterdam, 1992) p. 863.

van der Zant, H.S.J., Kervennic, Y.V., Poot, M., O'Neill, K., de Groot, Z., Thijssen, J.M., Heersche, H.B., Stuhr-Hansen, N., Bjornholm, T., Vanmaekelbergh, D., van Walree, C.A., Jenneskens, L.W. *Faraday Discuss* **131**, 347 (2006).

van Wees, B.J., van Houten, H., Beenakker, C.W.J., Williamson, J.G., Kouwenhoven, L.P., van der Marel, D., Foxon, C.T. *Phys. Rev. Lett.* **60**, 848 (1988).

Varga, K., Pantelides, S.T. *Phys. Rev. Lett.* **98**, 076804 (2007).

Venkataraman, L., Klare, J.E., Nuckolls, C., Hybertsen, M.S., Steigerwald, M.L. *Nature* **442**, 904 (2006a).

Venkataraman, L., Klare, J.E., Tam, I.W., Nuckolls, C., Hybertsen, M.S., Steigerwald, M.L. *Nano Lett.* **6**, 458 (2006b).

Venkataraman, L., Park, Y.S., Whalley, A.C., Nuckolls, C., Hybertsen, M.S., Steigerwald, M.L. *Nano Lett.* **7**, 502 (2007).

Waldron, D., Haney, P., Larade, B., MacDonald, A., Guo, H. *Phys. Rev. Lett.* **96**, 166804 (2006).

Walczak, K., Platero, G. *Central Eur. J. Phys.* **4**, 30 (2006).

Walczak, K. *Physica E* **33**, 110 (2006).

Walczak, K. *Chem. Phys.* **333**, 63 (2007).

Wang, W.Y., Lee, T., Kretzschmar, I., Reed, M.A. *Nano Lett.* **4**, 643 (2004).

Wang, W., Lee, T., Reed, M.A. *Rep. Prog. Phys.* **68**, 523 (2005).

Wang, W.Y., Richter, C.A. *Appl. Phys. Lett.* **89**, 153105 (2006).

Wang, W., Scott, A., Gergel-Hackett, N., Hacker, C.A., Janes, D.B., Richter, C.A. *Nano Lett.* **8**, 478 (2008).

Wang, X., Spataru, C.D., Hybertsen, M.S., Millis, A. *J. Phys. Rev. B* **77**, 045119 (2008).

Wang, Z., Carter, J.A., Lagutchev, A., Koh, Y.K., Seong, N.-H., Cahill, D.G., Dlott, D.D. *Science* **317**, 787 (2007).

Ward, D.R., Halas, N.J., Ciszek, J.W., Tour, J.M., Wu, Y., Nordlander, P., Natelson, D. *Nano Lett.* **8**, 919 (2008).

Wassel, R.A., Fuierer, R.R., Kim, N., Gorman, C.B. *Nano Lett.* **3**, 1617 (2003).

Wharam, D.A., Thornton, T.J., Newbury, R., Pepper, M., Ahmed, H., Frost, J.E.F., Hasko, D.G., Peacock, D.C., Ritchie, D.A., Jones, G.A.C. *J. Phys. C* **21**, L209 (1988).

Wolfsberg, M., Helmholz, L. *J. Chem. Phys.*, **20**, 837 (1952).

Wolkow, R.A. *Annu. Rev. Phys. Chem.* **50**, 413 (1999).

Xia, J.L., Diez-Perez, I., Tao, N.J. *Nano Lett.* **8**, 1960 (2008).

Xiao, X., Xu, B., Tao, N.J. *Nano Lett.* **4**, 267 (2004a).

Xiao, X., Xu, B., Tao, N.J. *J. Am. Chem. Soc.* **126**, 5370 (2004b).

Xiao, X., Xu, B., Tao, N.J. *Angew. Chem. Int. Ed.* **43**, 6148 (2004c).

Xiao, X., Nagahara, L.A., Rawlett, A.M., Tao, N.J. *J. Am. Chem. Soc.* **127**, 9235 (2005).

Xiao, X.Y., Brune, D., He, J., Lindsay, S., Gorman, C.B., Tao, N. *J. Chem. Phys.* **326**, 138 (2006).

Xu, B., Zhang, P., Li, X., Tao, N.J. *Nano Lett.* **4**, 1105 (2004).

Xu, B.Q., Xiao, X.Y., Yang, X.M., Zang, L., Tao, N.J. *J. Am. Chem. Soc.* **127**, 2386 (2005).

Xue, Y.Q., Datta, S., Ratner, M.A. *Chem. Phys.* **281**, 151 (2002).

Yamada, R., Kumazawa, H., Noutoshi, T., Tanaka, S., Tada, H. *Nano Lett.* **8**, 1237 (2008).

Yan, L.J. *Phys. Chem. A* **110**, 13249 (2006).

Yang, Z.Q., Lang, N.D., Di Ventra, M. *Appl. Phys. Lett.* **82**, 1938 (2003).

Yasuda, S., Nakamura, T., Matsumoto, M., Shigekawa, H. *J. Am. Chem. Soc.* **125**, 16430 (2003).

Yeganeh, S., Galperin, M., Ratner, M.A. *J. Am. Chem. Soc.* **129**, 13313 (2007).

Yu, L.H., Keane, Z.K., Ciszek, J.W., Cheng, L., Stewart, M.P., Tour, J.M., Natelson, D. *Phys. Rev. Lett.* **93**, 266802 (2004).

Yu, L.H., Keane, Z.K., Ciszek, J.W., Cheng, L., Tour, J.M., Baruah, T., Pederson, M.R., Natelson, D. *Phys. Rev. Lett.* **95**, 256803 (2005).

Yu, L.H., Zangmeister, C.D., Kushmerick, J.G. *Phys. Rev. Lett.* **98**, 206803 (2007).

Zahid, F., Paulsson, M., Polizzi, E., Ghosh, A.W., Siddiqui, L., Datta, S. *J. Chem. Phys* **123**, 064707 (2005).

Zhang, R.Q., Feng, Y.Q., Lee, S.T., Bai, C.L. *J. Phys. Chem. B* **108**, 16636 (2004).

Zhitenev, N.B., Meng, H., Bao, Z. *Phys. Rev. Lett.* **88**, 226801 (2002).

Quasi-ballistic electron transport in atomic wires

Jan M. van Ruitenbeek

5.1   Introduction

Single-atom contacts can be characterized in great detail, and to this purpose a diverse set of experimental techniques has been developed in recent years (for a more extensive review, see Agraït *et al.* 2003a). It turns out to be possible to describe their transport properties in terms of a finite number of conductance channels and it is possible to experimentally determine the transmission probabilities for each of these channels. The binding of the atom to the surrounding lead atoms can be probed by measuring the vibration modes, where one exploits point-contact spectroscopy. Below, these developments are briefly reviewed, followed by a discussion of the application of these techniques for the study of conducting chains of individual metal atoms and for metal–molecule–metal junctions. For the latter we will focus on simple molecules, including H_2, C_{60}, and benzene, that can serve as benchmark systems for model calculations.

The nanoscale systems to be discussed fall into the class of quasi-ballistic conductors. For a truly ballistic conductor the electrons travel from source to drain without experiencing any scattering. The atomic and molecular bridges we will consider can be selected to have a single conductance channel having a transmission probability close to unity. The single transmission channel reduces the problem essentially to a one-dimensional form. The small deviations from unit transmission, i.e. the quasi-ballistic aspects of the problem, contain important experimental information on the nanosystem.

We will start by reviewing experiments on the conduction properties for single metal atoms. Nearly all the information on the properties of such nanocontacts should be extracted from the current and voltage only. Nevertheless, a wide range of techniques has been developed to obtain detailed information, e.g. by exploiting the intrinsic noise in the current (shot noise), the details of the differential conductance for superconducting leads (subgap structure), the inelastic scattering signals due to atomic vibrations (point-contact spectroscopy), and more. A recent review gives more details on these techniques and on the results for various metals (Agraït *et al.* 2003a). Some of these techniques can also be exploited to characterize single-molecule junctions, as will be discussed in the last section.

5.2   Experimental techniques

Many of the experiments on atomic-sized metallic contacts have been performed using scanning tunnelling microscopes (STM) or, alternatively, by conducting-tip atomic force microscopes (cAFM) (Agraït *et al.* 1993a,b; Pascual *et al.* 1993; Olesen *et al.* 1994; Brandbyge *et al.* 1995; Limot *et al.* 2005; Gai *et al.* 1996). The scanning probe methods have the advantage that the sample surface can be imaged and characterized before and after making contact. This imaging capability is largely sacrificed with the mechanically controllable break junction (MCBJ) technique (Muller *et al.* 1992; van Ruitenbeek *et al.* 1996), which is compensated by several other advantages. First, in order to avoid contaminations the contacting metal surfaces need to be clean. For scanning probe techniques this often implies that an ultrahigh-vacuum (UHV) surface preparation and characterization chamber is required, which increases the complication and cost of the system significantly. In the MCBJ technique a clean surface comes almost without effort. Second, the MCBJ technique allows for fast cycling between measuring and modifying the sample setup, which simplifies testing of new samples and new approaches. The STM technique is widely employed also for many other experiments and will not be discussed here in detail, since it is presented in many textbooks. Other methods for obtaining atomic-sized contacts are break junctions by electromigration (Park *et al.* 2000), and electrochemically fabricated junctions (Li and Tao 1998; Kervennic *et al.* 2003; Xie *et al.* 2004; Kiguchi *et al.* 2005). The former are very stable and have the advantage that one can add a gate electrode with reasonably strong coupling to the nanojunction. Many groups have explored electromigration break junctions for single-molecule transport studies (Liang *et al.* 2002; Park *et al.* 2002; Yu and Natelson 2004; Osorio *et al.* 2007), but this work is outside the scope of this review because in all cases the transmission of the junctions is low. It is also possible to form metallic atomic-sized contacts by these methods. Junctions formed electrochemically were even demonstrated to switch reproducibly between quantized conductance values controlled by the reference electrode (Xie *et al.* 2004).

The MCBJ technique is ideally suited for experiments at low temperatures. One of the main advantages compared to other low-temperature techniques (Park *et al.* 2000; Kubatkin *et al.* 2003) is the possibility to manipulate the atomic and molecular junctions and to modify the measuring configuration. This is particularly useful when measuring the vibration modes of the molecule, where the observation of stretching dependence in the vibration-mode energy allows identification of the type of vibration mode and gives further confirmation on the origin of the signal.

The principle of the MCBJ technique is illustrated in Fig. 5.1. By breaking the metal, two clean fracture surfaces are exposed, which remain clean due

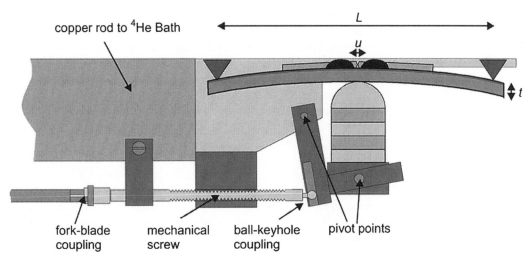

copper rod to ^{4}He Bath

fork-blade coupling

mechanical screw

ball-keyhole coupling

pivot points

Fig. 5.1 Schematic view of the mounting of a mechanically controllable break junction, where the metal to be studied has the form of a notched wire, which is fixed onto an insulated elastic substrate with two drops of epoxy adhesive very close to either side of the notch. The substrate is mounted in a three-point bending configuration between the top of a stacked piezo-element and two fixed counter supports. This setup is mounted inside a vacuum can and cooled down to liquid-helium temperatures. Then, the substrate is bent by moving the piezo-element forward. The bending causes the top surface of the substrate to expand and the wire to break at the notch. Typical sizes are $L \simeq 20\,\mathrm{mm}$, $t = 0.8\,\mathrm{mm}$, and $u \simeq 0.1\,\mathrm{mm}$.

to the cryo-pumping action of the low-temperature vacuum can. This method circumvents the problem of surface contamination of the tip and sample in STM experiments, where a UHV chamber with surface preparation and analysis facilities are required to obtain similar conditions. The fracture surfaces can be brought back into contact by relaxing the force on the elastic substrate, while a piezo-electric element is used for fine control. The roughness of the fracture surfaces results in a first contact at one point, and experiments usually give no evidence of multiple contacts. In addition to a clean surface, a further advantage of the method is the stability of the two electrodes with respect to each other. From the noise in the current in the tunnelling regime one obtains an estimate of the vibration amplitude of the vacuum distance, which is typically less than $10^{-3}\,\text{Å}$. The stability results from the reduction of the mechanical loop that connects one contact side to the other, from centimeters, in the case of an STM scanner, to $\sim$0.1 mm in the MCBJ.

Further improvements in stability and control can be achieved by replacing the sample wire with a microfabricated wire obtained by electron-beam lithography in a thin metal film on an insulating substrate (Zhou *et al.* 1995; van Ruitenbeek *et al.* 1996). Note that for such microfabricated MCBJ devices the ratio between the movement of the piezo-element and the displacement of the two wire ends with respect to each other is so small (of order 10^{-4}) that the range covered by the full piezo-extension is only a few tenths of a nanometer. The control over the displacement in such experiments is fully mechanical, often employing a stepper motor and a reduction gear near the MCBJ device.

For atomic-sized metallic contacts that have a resistance of order $1-10\,\text{k}\Omega$ one can either choose to source the current and measure the voltage in a four-point configuration, or source the voltage and measure the resulting current in a two-lead arrangement. For low-conductance molecules and for measurements in the tunnelling range the latter option is preferred. When measuring small variations of the differential conductance as described in Sections 5.3.3 and 5.3.5 it is advantageous to measure directly the differential conductance, dI/dV, by a lock-in technique. The absolute accuracy of the determination of the linear resistance is in the order of 1%, while relative changes down to 10^{-5} can be detected. Such precision is only achieved at low temperatures since higher temperatures lead to instabilities of the junctions. In particular, at high bias voltage ($V \sim 1$ V) and at room temperature, the junction may be altered during the measurement.

When measurements of the conductance as a function of the interelectrode distance are performed, e.g. for the recording of conductance histograms or the study of individual so-called opening or closing curves, a constant dc bias voltage is applied, the current is recorded and the position of the electrodes is controlled by ramping the piezo-voltage or running the motor linearly with a certain turning speed.

Introducing the molecules to the junction is often one of the least controlled and least understood aspects of the problem. When thiol anchoring groups are employed the bond with the metal needs to be formed at room temperature, otherwise there is not enough activation energy available to split off the protection groups that are usually present, and then bind to the metal. Once the molecule is in the junction one cannot simply cool down since thermal contraction of the various parts of the setup will change the interelectrode distance too much. Reichert *et al.* (2003) have devised a solution where they break the junction at room temperature after having introduced the molecules between the electrodes, relying on the strong Au–thiol bond to pull out a gold cluster from one of the electrodes. Maintaining a large electrode separation they cool down and remake contact with the gold cluster at low temperatures. They have shown sharp current–voltage characteristics obtained in this way, but the procedure needed to prevent contaminations remains a little delicate.

For the simple molecules that will be discussed below one relies on a different type of chemistry. Clean metal surfaces, in particular for nanosized metals, are highly reactive. Many common gaseous substances are believed to be able to chemically bind to those metal surfaces, even at cryogenic temperatures. Prominent examples are H_2 and CO binding to Pt, but the list of combinations is much longer, including nitrogen bonds as in pyridene, carbon double or triple bonds as in acetylene, etc. The advantage is that the metal surface can be exposed to cryogenic vacuum only and contamination with other substances can be minimized. Once the metal junction is characterized and has been verified to be clean, the molecules are introduced through a capillary that can be heated to avoid freezing of the substance. The dose of deposited material cannot be easily quantified, but one monitors the conductance properties of the junction while gently admitting the gases until a characteristic change in conductance is observed. Details depend on the type of gas being introduced, and often it is useful to continuously record conductance histograms (see Section 5.3.2) and monitor changes in those.

5.3 Atomic-sized metallic contacts

Figure 5.2 shows some examples of the conductance measured during breaking of a gold contact at low temperatures, using an MCBJ device. The conductance decreases by sudden jumps, separated by plateaux, which have a negative slope, the higher the conductance the steeper are the slopes. Some of the plateaux are remarkably close to multiples of the conductance quantum, $G_0 = 2e^2/h$; in particular the last plateau before losing contact is nearly flat and very close to $1G_0$. Closer inspection, however, shows that many plateaux cannot be identified with integer multiples of the quantum unit, and the structure of the steps is different for each new recording. Also, the height of the steps is of the order of the quantum unit, but can vary by more than a factor of 2, where both smaller and larger steps are found. Drawing a figure such as Fig. 5.2, with grid lines at multiples of G_0, guides the eye to the coincidences and may convey that the origin of the steps is in quantization of the conductance. However, in evaluating the graphs, one should be aware that a plateau cannot be farther away than one half from an integer value, and that a more objective analysis is required. Still, it is clear that we can use these fairly simple techniques to produce and study atomic-scale conductors, for which the conductance is dominated by quantum effects. The interpretation of graphs as in Fig. 5.2 will be the subject of this section.

5.3.1 Landauer theory of conductance

The metallic point contacts and molecular junctions that we will consider are all of atomic size, much smaller than all characteristic scattering lengths of the system. In particular, the electron mean-free path for elastic scattering on defects and impurities near the contact is assumed to be much longer than the contact size. The only elastic scattering considered is the scattering by the walls forming the boundary of the system. Also, it is assumed that the probability for scattering events that change the spin and phase of the electron wavefunction is negligible.[1] The atomic-sized junctions of interest here are strongly coupling the electronic states in left and right leads. In this limit, following the standard

[1] Note that we will introduce scattering on localized vibration modes shortly, but this occurs at fairly large bias, while we consider the zero-bias conductance at this point. In fact, even above the energy for vibration-mode excitations the approximation remains valid.

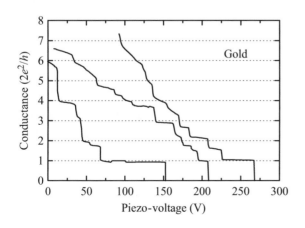

Fig. 5.2 Three typical recordings of the conductance G measured in atomic-sized contacts for gold at helium temperatures, using the MCBJ technique. The electrodes are pulled apart by increasing the piezo-voltage. The corresponding displacement is about 0.1 nm per 25 V. After each recording the electrodes are pushed firmly together, and each trace has new structure. From Krans (1996).

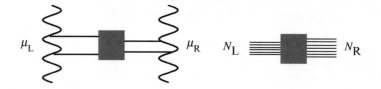

Fig. 5.3 Schematic representation for a ballistic two-terminal conductance problem. The gray box represents the atomic-sized contact, or the scattering area. The reservoirs (or electrodes) on the left and right of the wiggly lines emit electrons onto the sample with an energy distribution corresponding to the electrochemical potentials μ_L and μ_R, respectively, with $\mu_L - \mu_R = eV$. Electrons reflected from the sample are perfectly absorbed by the reservoirs. The straight sections connecting the reservoirs to the sample represent perfect leads, where the number of modes in the left and right lead is N_L and N_R, respectively, and these numbers are not required to be equal.

approach (for reviews, see Agraït *et al.* 2003a, Beenakker and van Houten 1991, Datta 1995, Imry 1997, and Chapter 1 of this volume) we will assume that the system can be schematically represented as in Fig. 5.3. The connection between the ballistic system and the outside world is represented by electron reservoirs on each side of the contact, which are held at a potential difference eV by an external voltage source. When the leads connecting the reservoirs to the contact are straight wires of constant width, there is a well-defined number of conducting modes in each of these wires, say N_L and N_R for the left and right lead, respectively. In a free-electron gas model the modes are simply plane waves, which can propagate in the current direction (to the left and right) and are standing waves in the perpendicular directions. The modes can be labelled by an index corresponding to the number of nodes in the perpendicular direction. The numbers N_L and N_R are not limited but all the modes that have an energy much more than $k_B T$ above the Fermi energy can be ignored.

The conductance of the system can now be simply expressed as

$$G = \frac{2e^2}{h}\mathrm{Tr}(\mathbf{t}^\dagger \mathbf{t}), \tag{5.1}$$

where e is the electron charge, h is Planck's constant, and $\mathbf{t}$ is an $N_L \times N_R$ matrix with matrix element t_{mn} giving the probability amplitude for an electron wave in mode n on the left to be transmitted into mode m on the right of the contact. It can be shown that the product matrix $\mathbf{t}^\dagger \mathbf{t}$ can always be diagonalized by going over to a new basis, consisting of linear combinations of the original modes in the leads. Further, the number, N_c, of non-zero diagonal elements is only determined by the number of modes at the narrowest cross-section of the conductor (Brandbyge *et al.* 1997; Cuevas *et al.* 1998). Equation (5.1) thus simplifies to

$$G = \frac{2e^2}{h}\sum_{n=1}^{N_c} T_n, \tag{5.2}$$

where $T_n = |t_{nn}|^2$ and the index refers to the new basis.

Under favorable circumstances, all transmission probabilities, T_n, can be close to unity. As an example, for a smooth ("adiabatic") and long wire the modes at the narrowest cross-section couple exclusively to a single mode at either side of the contact, and the expression for the conductance further simplifies to

$$G = N_c G_0, \tag{5.3}$$

where $G_0 = 2e^2/h$ is the conductance quantum. For increasing diameter of the contact the number N_c increases each time a new mode fits into the narrowest cross-section. This quantization of the conductance was first observed in experiments on 2D electron-gas devices by van Wees *et al.* (1988), where the Fermi wavelength $\lambda_F \simeq 400\,\text{Å}$ is much larger than the atomic scale. At zero temperature the number N_c is limited by the requirement that the kinetic energy for motion in the perpendicular direction is smaller than the Fermi energy. For a two-dimensional system this can be expressed as $(\hbar^2/2m)(\pi N_c/W)^2 < E_F$, with W the width of the contact, which leads to $N_c = \text{Int}(2W/\lambda_F)$, with λ_F the Fermi wavelength. For a three-dimensional metallic contact we have $N_c \approx (\pi R/\lambda_F)^2$, with R the contact radius.

In metals, the quantization effects are somewhat obscured by the discreteness of the atomic structure of the contacts because λ_F is of the order of the size of an atom. The lowest conductance plateau near $1G_0$ in Fig. 5.2 is due to a bridge of a single atomic bond, that has a single conductance channel with near-perfect transmission. Below, we will discuss how this statement can be verified experimentally. Only monovalent metals (the alkali metals and the noble metals) show this single-quantum mode character for a single atomic bond, and weaker quantization effects can be identified for somewhat larger contact sizes. For s-p metals and transition metals there are contributions from various orbitals and the picture is slightly more complicated. However, as we shall see, the number of channels for a given contact configuration is always well defined and can often be determined experimentally.

5.3.2 Conductance histograms

The atomic configuration of a contact adjusts itself in response to the externally applied stress and evolves according to the starting configuration of the contact at larger size. The fact that each conductance curve differs in many details from previous curves reflects the fact that the atomic configuration of the contact for given conductance is different in each run. However, as observed in the previous section, there appears to be a certain preference for conductance values near integer multiples of the quantum unit. For gold near $1G_0$ this is immediately obvious from the examples in Fig. 5.2. A general and objective method of analysis was introduced by Olesen *et al.* (1995) and Krans *et al.* (1995), which consists in recording histograms of conductance values encountered in a large number of runs. Figure 5.4 shows a histogram for gold measured using a room temperature STM under UHV conditions (Brandbyge *et al.* 1995). Up to four peaks are found centered near the first four multiples of G_0.

Not all metals show such pronounced histogram peaks near integer conductance values. The most clear-cut results are obtained only for monovalent metals. The noble metals Cu, Ag and Au show histograms as in Fig. 5.4. The details, such as the shift from the ideal position, the width and relative height of the peaks, can be different depending on the experimental conditions (Brandbyge *et al.* 1995; Krans *et al.* 1995; Gai *et al.* 1996; Hansen *et al.* 1997; Costa-Krämer et al. 1997a,b; Ludoph and van Ruitenbeek 2000). The alkali

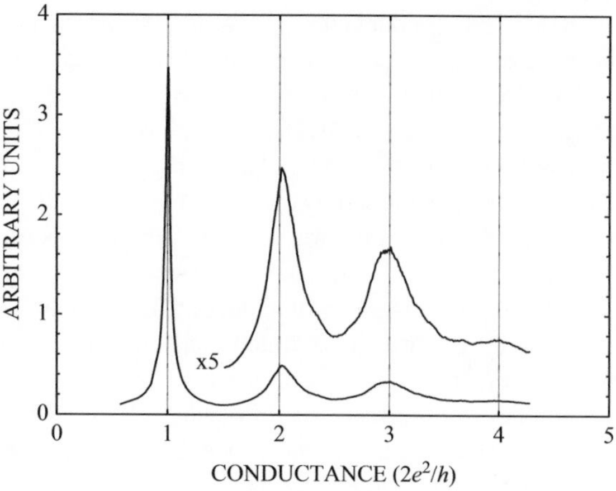

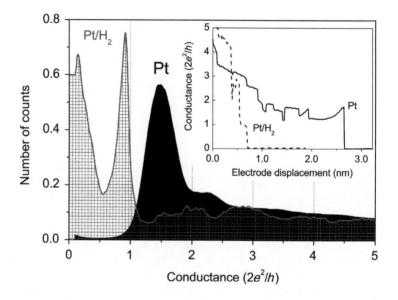

Fig. 5.4 Histogram representing the relative weight that each conductance value has in the experiments. The histogram is constructed from 227 conductance curves recorded while breaking Au contacts, using an STM under UHV at room temperature. From Brandbyge *et al.* (1995); copyright the American Physical Society.

Fig. 5.5 Conductance curves and histograms for clean Pt, and Pt in a H_2 atmosphere. The inset shows a conductance curve for clean Pt (solid line) at 4.2 K recorded with a bias voltage of 10 mV, before admitting H_2 gas into the system. About 10 000 similar curves are used to build the conductance histogram shown in the main panel (black), which has been normalized by the area under the curve. After introducing hydrogen gas the conductance curves change qualitatively as illustrated by the dashed curve in the inset, recorded at 100 mV. This is most clearly brought out by the conductance histogram (dashed; recorded with 140 mV bias). From Smit *et al.* (2002); copyright Nature Publishing Group.

metals demonstrate very clear peak structures at multiples of the conductance quantum (Krans *et al.* 1995; Yanson 2001). Most other metals only show a rather broad first peak, which cannot generally be identified with an integer value of the conductance; for example platinum shows a wide peak centered near 1.6 G_0 (Fig. 5.5, solid) and only weak features at larger contact size. Similar results have been obtained for many transition metals (Yanson 2001). The first peak in a histogram is usually associated with the typical conductance of a single atomic bond, or a chain of a few atoms. From the example in Fig. 5.5 it is evident that the conductance through this transition-metal atom is carried by more than a single channel, and that the conductance is not an integer multiple of the conductance quantum. We will next discuss how the actual number of channels can be extracted from experiment.

5.3.3 Conduction-channel composition

Instead of discussing the averaged properties of many contacts, we now concentrate on a single atomic bond bridging the electrodes at either side, and consider the question of what is the number of modes contributing to the conductance and the transmission probability, T_n, for each of these. The set of transmission probabilities $\{T_n\}$ fully determines the transport properties of the junction and we will refer to this set as the mesoscopic PIN code of the junction. From a measurement of the conductance alone we cannot obtain this information, since the conductance gives only the sum, $G = G_0 \sum_n T_n$. Scheer *et al.* (1997) have introduced a method that allows us to obtain this information from experiment. The method exploits the non-linearities in the current–voltage characteristic for contacts in the superconducting state. Other techniques, which give more limited information on the contribution of the various modes, include the measurement of shot noise (van den Brom and van Ruitenbeek 1999) or the measurement of conductance fluctuations (Ludoph *et al.* 1999).

5.3.3.1 *Superconducting subgap structure*

The principle of the method introduced by Scheer and colleagues can be illustrated by considering first a contact having a single mode with low transmission probability, $T \ll 1$. For $T \ll 1$ we have essentially a tunnel junction, and the current–voltage characteristic for a superconducting tunnel junction is known to directly reflect the gap, Δ, in the density of states for the superconductor (Wolf 1989). As illustrated in Fig. 5.6(a) no current flows until the applied voltage exceeds $2\Delta/e$ (where the factor 2 results from the fact that we have identical superconductors on both sides of the junction), after which the current jumps to approximately the normal-state resistance line. For $eV > 2\Delta$ single quasi-particles can be transferred from the occupied states at $E_F - \Delta$ on the low-voltage side of the junction to empty states at $E_F + \Delta$ at the other side. For $eV < 2\Delta$ this process is blocked, since there are no states available in the gap.

However, when we consider higher-order tunnel processes a small current can still be obtained. Figure 5.6(b) illustrates a process that is allowed for $eV > \Delta$ and consists of the simultaneous tunnelling of *two* quasi-particles from the low-bias side to form a *Cooper pair* on the other side of the junction. The onset of this process causes a step in the current at half the gap value, $V = 2\Delta/2e$. The height of the current step is smaller than the step at $2\Delta/e$ by a factor T, since the probability for two particles to tunnel is T^2. In general, one can construct similar processes of order n, involving the simultaneous transfer of n particles, which give rise to a current onset at $eV = 2\Delta/n$ with a step height proportional to T^n. An example for $n = 3$ is illustrated in Fig. 5.6(c). This mechanism is known as multiple-particle tunnelling and was first described by Schrieffer and Wilkins (1963). It is now understood that this is the weak-coupling limit of a more general mechanism known as

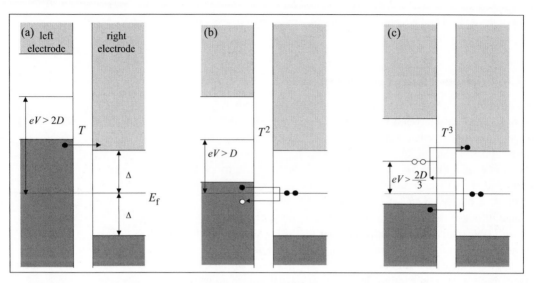

Fig. 5.6 Illustration of the multiple particle tunnelling (MPT), or multiple Andreev reflection (MAR) processes for the weak tunnelling limit. In each of the three diagrams the available quasi-particle states as a function of energy (vertical axis) at both sides of the tunnel barrier are given in the semiconductor representation. The occupied states are represented as dark gray, while the unoccupied states are light gray and the line in the middle of the gap represents the Cooper-pair energy, which is separated by an energy Δ from the occupied and the unoccupied quasi-particle states. Applying an external electrical potential V across the junction shifts the states on the left side of the junction up by an energy eV with respect to those in the right electrode. The ordinary superconducting tunnelling process is given in (a), which shows that a voltage $V > 2\Delta/e$ is required for single quasi-particles to cross the junction. The probability for tunnelling of a particle, determined by the transparency of the barrier, is T. This gives rise to the familiar jump in the current at $eV = 2\Delta$ in the current–voltage characteristic of a superconducting tunnel junction. When we consider higher-order processes, the next order is represented in (b). This can be described as two quasi-particles crossing simultaneously to form a Cooper pair in the right electrode (MPT). Alternatively, the process can be regarded as an electron-like quasi-particle falling onto the barrier, which is reflected as a hole-like quasi-particle, forming a Cooper pair on the right (MAR). The two descriptions are equivalent, and give rise to a current step in the current–voltage characteristic of the junction at $eV = 2\Delta/2$. The probability for the process is T^2, since it requires the crossing of two particles. (c) Shows the third-order process, which involves breaking up a Cooper pair on the left, combining it with a quasi-particle, to form a Cooper pair and a quasi-particle on the right. It is allowed for $eV > 2\Delta/3$ and has a probability T^3.

multiple Andreev reflection (Klapwijk *et al.* 1982; Arnold 1987; Averin and Bardas 1995; Bratus' *et al.* 1995, 1997; Cuevas *et al.* 1996). The theory could only recently be tested experimentally, since it requires the fabrication of a tunnel junction having a single tunnelling mode with a well-defined tunnelling probability T. For atomic-sized niobium tunnel junctions the theory was shown to give a very good agreement (van der Post *et al.* 1994), describing up to three current steps, including the curvature and the slopes, while the only adjustable parameter is the tunnel probability, which can be fixed by the measured normal state resistance.

Since the theory has now been developed to all orders in T (Arnold 1987; Averin and Bardas 1995; Bratus' *et al.* 1995, 1997; Cuevas *et al.* 1996), Scheer *et al.* (1997) realized that this mechanism offers the possibility of extracting the transmission probabilities for contacts with a finite number of channels contributing to the current, and that it is ideally suited for analyzing atomic-sized contacts. Roughly speaking, the current steps at $eV = 2\Delta/n$ are proportional to $\sum(T_m)^n$, with m the channel index, and when we can resolve sufficient details in the current–voltage characteristics, we can fit many independent sums of powers of T_ms. When the T_ms are not small compared

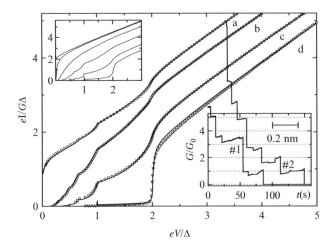

Fig. 5.7 Current–voltage characteristics for four atomic-sized contacts of aluminum using a lithographically fabricated mechanically controllable break junction at 30 mK (symbols). The right inset shows the typical variation of the conductance, or total transmission $T = G/G_0$, as a function of the displacement of the electrodes, while pulling. The bar indicates the approximate length scale. The data in the main panel have been recorded by stopping the elongation at the last stages of the contact (a–c) or just after the jump to the tunnelling regime (d) and then measuring the current while slowly sweeping the bias voltage. The current and voltage are plotted in reduced units, $eI/G\Delta$ and eV/Δ, where G is the normal-state conductance for each contact and Δ is the measured superconducting gap, $\Delta/e = (182.5 \pm 2.0)\,\mu$V. The left inset shows the current–voltage characteristics obtained from first-principles theory for a single-channel junction (Averin and Bardas 1995; Bratus' *et al.* 1995, 1997; Cuevas *et al.* 1996) with different values for the transmission probability T (from bottom to top: $T = 0.1, 0.4, 0.7, 0.9, 0.99, 1$). The full curves in the main panel have been obtained by adding several theoretical curves and optimizing the set of T values. The curves are obtained with: (a) three channels, $T_1 = 0.997$, $T_2 = 0.46$, $T_3 = 0.29$ with a total transmission $\sum T_n = 1.747$, (b) two channels, $T_1 = 0.74$, $T_2 = 0.11$, with a total transmission $\sum T_n = 0.85$, (c) three channels, $T_1 = 0.46$, $T_2 = 0.35$, $T_3 = 0.07$ with a total transmission $\sum T_n = 0.88$. (d) In the tunnelling range a single channel is sufficient, here $\sum T_n = T_1 = 0.025$. From Scheer *et al.* (1997); copyright the American Physical Society.

to 1, all processes to all orders need to be included for a description of the experimental curves. In practice, the full expression for the current–voltage characteristics for a single channel from theory (Averin and Bardas 1995; Bratus' *et al.* 1995, 1997; Cuevas *et al.* 1996) is numerically evaluated for a given transmission probability T_m (Fig. 5.7, inset), and a number of such curves are added independently, where the T_ms are used as fitting parameters. Scheer *et al.* (1997) tested this approach first for aluminum contacts. As shown in Fig. 5.7, all current–voltage curves for small contacts can be very well described by the theory. However, the most important finding was that at the last "plateau" in the conductance, just before the breaking of the contact, typically three channels with different Ts are required for a good description. Note that the total conductance for such contacts is of order $1G_0$, and would in principle require only a single conductance channel. Contacts on the verge of breaking are expected to consist of a single atomic bond, and this bond would then admit three conductance channels, but each of the three would only be partially open, adding up to a conductance close to $1G_0$. This very much contradicts a simple picture of quantized conductance in atomic-sized contacts. In fact, in a systematic study for a number of s, sp and sd-metals it has been shown that the number of conductance channels through an atom is associated with the number of valence orbitals (Scheer *et al.* 1998). A single channel is found for monovalent metals, three channels for sp metals and five for sd metals. These conclusions are consistent with further experiments on shot noise and conductance fluctuations to be discussed next.

5.3.3.2   *Shot noise*

Shot noise is the result of the discrete character of the current due to the passage of individual electrons. It was originally found in vacuum diodes, and first discussed by Schottky (1918). The passage of individual electrons can be regarded as delta functions of the current with time. The total current is the sum of a random distribution of such delta functions, giving a time averaged current I, and a frequency spectrum of fluctuations that is white (up to very high frequencies), with a noise power equal to $2eI$. This shot noise can be observed, for example, in tunnel junctions.

For a perfectly ballistic point contact in the absence of backscattering, i.e. a contact having all open-channel transmission probabilities equal to 1, shot noise is expected to vanish (Lesovik 1989; Büttiker 1990; Blanter and Büttiker 2000). This can be understood from the wave nature of the electrons, for which the wavefunction extends from the left bank to the right bank of the contact without interruption. When the state on the left is occupied for an incoming electron, it is occupied on the right as well and there are no fluctuations in this occupation number. In other words, in order to have noise the electron must be given the choice of being reflected at the contact. This will be the case when the transmission probability is smaller than 1 and larger than 0. In single-channel quantum point contacts, shot noise is predicted to be suppressed by a factor proportional to $T(1 - T)$, where T is the transmission probability of the conductance channel (Lesovik 1989; Büttiker 1990; Blanter and Büttiker 2000). This quantum suppression was first observed in point-contact devices in a two-dimensional electron gas (Reznikov *et al.* 1995; Kumar *et al.* 1996). For a general multichannel contact in the limit of very low temperatures the shot-noise power is predicted to be

$$P_I = 2eV G_0 \sum_n T_n(1 - T_n). \tag{5.4}$$

Since this depends on the sum over the second power of the transmission coefficients, this quantity is independent of the conductance, $G = G_0 \sum T_n$, and simultaneous measurement of these two quantities should give information about the mesoscopic PIN code of the contact. The relevant quantity is often conveniently expressed in terms of the Fano factor F, which is the ratio of the shot noise to the noise that the same current would produce in the classical Schottky limit,

$$F = \frac{P_I}{2eI} = \frac{\sum_n T_n(1 - T_n)}{\sum_n T_n}. \tag{5.5}$$

When measuring shot noise on atomic-sized point contacts (van den Brom and van Ruitenbeek 1999) it is necessary to work at low temperatures in order to reduce thermal noise, and to shield the contact carefully from external mechanical and acoustic vibrations. Using two sets of pre-amplifiers in parallel and measuring the cross-correlation of the noise for the two signals eliminates the noise of the preamplifier. Experimental results for shot noise have been reported in atomic-sized contacts for gold and aluminum (van den Brom and van Ruitenbeek 1999, 2000). From the two measured parameters, G and P_I, one can determine at most two independent transmission probabilities. For gold single-atom contacts with conductance below $1G_0$ the data are consistent with a single conductance channel having a transmission probability $T = G/G_0$. For larger contacts there is a tendency for the channels to open one-by-one, but admixture of additional channels grows rapidly. There is a very strong suppression, down to $F = 0.02$, for $G = 1\ G_0$, which unambiguously shows that the current is carried dominantly by a single channel. It needs to be stressed that this holds for gold contacts. There is a fundamental distinction between this monovalent metal and the multivalent metal aluminum, which shows no systematic suppression of the shot noise at multiples of the conductance quantum, and the Fano factors lie between about 0.3 and 0.6 for G close to G_0 (van den Brom and van Ruitenbeek 2000). This again agrees with the results from subgap structure analysis above. Every contact is different, but one can take, for example, curves b and c in Fig. 5.7 that have $G = 0.85$ and $G = 0.88 G_0$, respectively. From the mesoscopic PIN code given in the caption we obtain $F = 0.29$ and $F = 0.541$, respectively.

Cron *et al.* (2001) have investigated this connection more rigorously through an elegant experiment by measuring the subgap structure and shot noise for one and the same contact of aluminium. They first obtain the PIN code from the subgap structure and next measure the shot noise without modifying the contact. They obtain full quantitative agreement between the two measurements, illustrating that the PIN code suffices to predict all other low-energy transport properties for ballistic nanocontacts.

5.3.3.3 *Conductance fluctuations*

Interference between electron trajectories scattering on defects near atomic-sized metallic contacts gives rise to dominant contributions to the second derivative of the current with respect to bias voltage V, i.e. in dG/dV. This effect has the same origin as the well-known universal conductance fluctuations (UCF) in diffusive mesoscopic conductors (Lee and Stone 1985; Zyuzin and Spivak 1990). In experiments on gold contacts (Ludoph *et al.* 1999; Ludoph and van Ruitenbeek 2000) it was found that this voltage dependence is suppressed near multiples of the quantum value of conductance, $n(2e^2/h)$. By applying a constant modulation voltage at frequency ω and measuring the current with two lock-in amplifiers simultaneously, at ω and the second harmonic 2ω, the conductance and its derivative can be obtained during conductance scans. From the combined datasets of many such curves one can construct a conductance histogram together with the average properties of dG/dV. It was found that the standard deviation of the derivative of the conductance with bias voltage, $\sigma_{GV}^2 = \langle (dG/dV)^2 \rangle$, has a pronounced minimum near $G = 1 G_0$ for monovalent metals. Smaller minima can be found near $G = 2, 3$, and $4\, G_0$.

The explanation for this quantum suppression of the conductance fluctuations is illustrated in Fig. 5.8. The contact is modelled by a quasi-ballistic central part, which can be described by a set of transmission values for the conductance modes, sandwiched between diffusive banks, where electrons are scattered by defects characterized by an elastic scattering length l_e. An electron wave of a given mode falling onto the contact is transmitted with probability amplitude t and part of this wave is reflected back to the contact by the diffusive medium, into the same mode, with probability amplitude $a_n \ll 1$. This backscattered partial wave is then *reflected* again at the contact with probability amplitude r_n, where $T_n = |t_n|^2 = 1 - |r_n|^2$. The latter partial wave interferes with the original transmitted partial wave. This interference depends on the phase difference between the two waves, and this phase difference depends on the phase accumulated by the waves during the passage through the diffusive medium. The probability amplitude a_n is a sum over all trajectories of scattering, and the phase for such a trajectory of total length L is simply kL, where k is the wave vector of the electron. The wave vector can be influenced by increasing the voltage over the contact, thus launching the electrons into

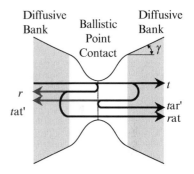

Fig. 5.8 Diagram showing the bare contact (light) sandwiched between diffusive regions (gray). The dark lines with arrows show the paths that interfere with each other and contribute to the conductance fluctuations in lowest order. From Ludoph and van Ruitenbeek (2000); copyright The American Physical Society.

the other electrode with a higher speed. The interference of the waves changes as we change the bias voltage, and therefore the total transmission probability, or the conductance, changes as a function of V. This describes the dominant contributions to the conductance fluctuations, and from this description it is clear that the fluctuations are expected to vanish either when $t_n = 0$, or when $r_n = 0$.

Elaborating this model Ludoph *et al.* (1999) obtained the following analytical expression for σ_{GV},

$$\sigma_{GV} = \frac{2.71\, e\, G_0}{\hbar k_F v_F \sqrt{1 - \cos \gamma}} \left(\frac{\hbar/\tau_e}{eV_m}\right)^{3/4} \sqrt{\sum_n T_n^2 (1 - T_n)}, \qquad (5.6)$$

where k_F and v_F are the Fermi wave vector and Fermi velocity, respectively, $\tau_e = l_e/v_F$ is the scattering time. The shape of the contact is taken into account in the form of the opening angle γ (see Fig. 5.8), and V_m is the applied voltage modulation amplitude.

The observed mean conductance fluctuation amplitude as a function of conductance for monovalent metals agrees with the assumption of a single conductance channel for single-atomic bond contacts. We will discuss conductance fluctuation data for Pt–H_2–Pt single-molecule junctions below.

5.3.4   Atomic chains

We may conclude that all evidence shows that for a single atomic bond for monovalent metals the current is carried by a single mode, with a transmission probability close to one. Guided by this knowledge in experiments on gold Yanson *et al.* (1998) discovered that during the contact-breaking process the atoms in the contact form stable chains of single atoms, up to 7 atoms long. Independently, Ohnishi *et al.* (1998) discovered the formation of chains of gold atoms at room temperature using an instrument that combines a STM with a transmission electron microscope, where an atomic strand could be directly seen in the images.

Some understanding of the underlying mechanism can be obtained from molecular dynamics simulations. Already before the experimental observations, several groups had observed the spontaneous formation of chains of atoms in computer simulations of contact breaking (Finbow *et al.* 1997; Sørensen *et al.* 1998). Figure 5.9 shows the results obtained by Sørensen *et al.* (1998) for gold. The authors caution that the interatomic potentials used in the simulation may not be reliable for this unusual configuration. However, the stability of these atomic wires has now been confirmed by various more advanced calculations (Häkkinen *et al.* 2000; Bahn and Jacobsen 2001; da Silva *et al.* 2001).

Only three metals are known to form purely metallic atomic chains, namely Au, Pt, and Ir (Smit *et al.* 2001). They are neighbors in the sixth row of the periodic table of the elements and they share another property: they make similar reconstructions of the surface atoms for clean [100], [110], and [111]

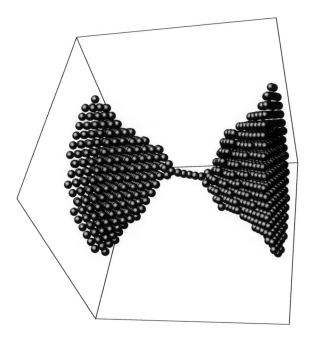

Fig. 5.9 Atomic configuration obtained during the last stages of breaking of a nanowire in a molecular dynamics simulation for gold, using a bath temperature of 12 K. The figure is produced from calculations by Sørensen *et al.* (1998).

surfaces. A common origin for these two properties has been suggested, in terms of a relativistic contribution to the linear bond strength (Smit *et al.* 2001).

5.3.5   Point-contact spectroscopy

Such chains constitute the ultimate one-dimensional metallic nanowires. For gold chains the current is carried by a single mode, with a transmission probability that is somewhat below 1 due to backscattering, but can be tuned to unity by adjusting the stress on the junction. A number of unusual properties for the atomic chains have been investigated, but here we want to focus on the scattering of electrons by vibration modes of the atomic chain. Atomic chains sustain enormous currents, up to 80 μA, due to the quasi-ballistic nature of the electron transport (Yanson *et al.* 1998). Nevertheless, electrons interact with the vibration modes as was observed in point-contact spectroscopy by Agraït *et al.* (2002), shown in Fig. 5.10.

For point-contact spectroscopy (PCS), in contrast to inelastic electron tunnelling spectroscopy (IETS), the conductance *decreases* at the voltage corresponding to the vibration mode energy, $eV = \hbar\omega$. This difference can be understood in a simple picture for the lowest-order process, as illustrated in Fig. 5.11. In this simplified scheme elastic processes of virtual emission and reabsorption that can modify the shape of the signal (Galperin *et al.* 2004) are ignored. IETS is becoming a very important tool in STM spectroscopy of individual molecules, and was first reported by Stipe *et al.* (1998). The crossover between PCS and IETS is expected to occur for a transmission probability of $T = 0.5$, as was recently observed (Tal *et al.* 2008).

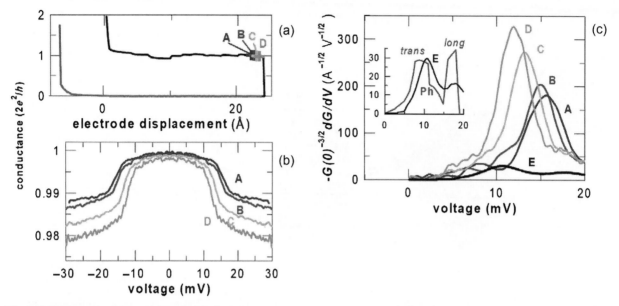

Fig. 5.10 Point-contact spectrum for a chain of gold atoms. Panel (a) shows the evolution of the conductance while the contact is being stretched (top curve). The long plateau at $1G_0$ indicates the formation of a monatomic chain. The return trace after breaking is also shown (lower curve). At each of the points marked A, B, C, and D the elongation was interrupted and a phonon spectrum, dI/dV vs. bias voltage V was recorded, as displayed in (b). The differential conductance was measured with a lock-in technique at a modulation of 0.5 mV. The numerical derivative of these curves is shown in (c). The experiments were performed at a bath temperature of 2 K. Curve E in (c) shows the point-contact spectrum for a larger gold contact, and the inset compares it to the bulk phonon density of states. Adapted from Agraït *et al.* (2002); copyright The American Physical Society.

PCS for relatively large contacts was pioneered by Yanson (Yanson and Shklyarevskii 1986). The spectrum for gold is shown as curve E in Fig. 5.10. It reflects the bulk phonon density of states, as illustrated in the inset, modified by the electron–phonon coupling function. The PCS signal for a chain of gold atoms in Fig. 5.10 is dominantly due to the zone-boundary longitudinal mode, with neighboring atoms moving in antiphase. As the atomic chain is stretched the energy of this mode is seen to decrease and the intensity of the signal increases. This is expected for this longitudinal mode, since stretching of the chain reduces the bond strength and therewith the effective spring constant for the harmonic oscillators. The signal strength is enhanced due to the larger corrugation of the atomic potential that the electrons experience. On further stretching the chain either breaks, or a new atom is inserted, which relaxes the strain and the vibration mode jumps back to the position of curve A.

Recently, Marchenkov *et al.* (2007) showed in an experiment on Nb junctions that vibration-mode information is possibly more clearly visible in the superconducting states of metallic leads. If confirmed for molecular junctions, superconducting leads may combine the advantages of providing information on the conductance channel composition with vibration spectra in one run. Spectroscopy for normal metal leads is discussed below for single-molecule junctions and this already is a powerful tool to obtain detailed information on the configuration and composition of the nanocontacts.

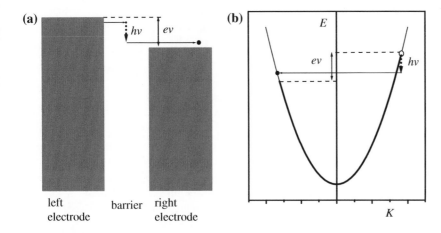

(a)

left
electrode
barrier
right
electrode

(b)

E

ev

hv

K

Fig. 5.11 Schematic illustration of the processes of (a) inelastic electron tunnelling spectroscopy (IETS) and (b) point-contact spectroscopy (PCS). The former (a) applies to weak coupling between the electrodes, with a nano-object (a molecule) inserted in the energy barrier. At low temperatures the electron states are filled up to the Fermi level, which is shifted by eV in the right electrode due to the application of an external potential. Electrons tunnel elastically from the left electrode to the right into the empty states. By exciting a vibration mode in the nano-object inside the tunnel barrier some of the electrons lose energy. The Pauli exclusion principle allows the scattered electrons to continue only into the right electrode. Since this scattering process is allowed above the vibration-mode energy, $eV > \hbar\omega$, and opens up a new forward current path, it is observed as step *up* in the conductance. In contrast, PCS (b) gives rise to a decrease of the conductance. We assume a single conductance mode with transmission probability equal to unity. The electrons are then completely delocalized over the two electrodes and the description naturally starts from considering momentum space. The right-moving electrons are occupied to a level eV higher than the left moving states (thick curve). When an electron suffers an inelastic scattering event and ends up at a lower energy, the Pauli exclusion principle dictates in this case that the electron should scatter backwards. Thus, we will see this as a step *down* in conductance.

5.4 Metal–molecule–metal junctions

Single organic molecules should be able to perform many functions that are presently realized in silicon semiconductor integrated circuits. Although there are still many roadblocks ahead it is very attractive to look into the possibilities of building molecular electronics circuits. A first step is to actually contact a single molecule.

A significant fraction of the experimental work on single-molecule transport was inspired by the paper by Reed *et al.* (1997). The experiment was performed using a mechanically controllable break junction device (Muller *et al.* 1992) working at room temperature, with the junction immersed in a solution of the organic compound of interest. The compound they selected was benzene-1,4-dithiol that has become the workhorse in this field of science. In the experiment the broken gold wire was allowed to interact with the molecules for a number of hours so that a self-assembled monolayer covered the surface. Next, the junction was closed and reopened a number of times and current–voltage ($I-V$) curves were recorded at the position just before contact was lost completely. The $I-V$ curves showed some degree of reproducibility with a fairly large energy gap feature of about $2\,\mathrm{eV}$, that was attributed to a metal–molecule–metal junction.

We will not attempt to give a full overview of the developments in this field here. More details on this research can be found in Chapters 4 and 23 of this volume and in a recent review by Natelson (2008). All of these experiments have been performed at room temperature, either in air, under inert gas atmosphere, or in solution. There is much to be gained by taking the junctions to helium temperatures, where the various tools described above become available for more detailed investigation of the metal–molecule–metal system. We will focus here on those experiments for which the coupling of the molecules to the leads is sufficiently strong that we remain in the limit of quasi-ballistic transport. One reason for specializing to this class of molecular junctions is the observation that for most of the experiments on single molecules in the regime of low conductance there is rather strong variation in the experimental results between various attempts, and the calculated conductance is generally between one and three orders of magnitude higher than the measured conductance. Therefore, it is useful to take a step back and first consider simple model systems that can be characterized in more detail by the methods outlined above.

5.4.1 Hydrogen

Smit *et al.* (2002) obtained molecular junctions of a hydrogen molecule between platinum leads using the mechanically controllable break junction technique. The inset to Fig. 5.5 shows a conductance curve for clean Pt (solid line) at 4.2 K, before admitting H_2 gas into the system. About 10 000 similar curves were used to build the conductance histogram shown in the main panel (black, normalized by the area). After introducing hydrogen gas the conductance curves were observed to change qualitatively as illustrated by the gray curve in the inset. The dramatic change is most clearly brought out by the conductance histogram (gray, hatched). Clean Pt contacts show a typical conductance of $1.5 \pm 0.2G_0$ for a single-atom contact, as can be inferred from the position and width of the first peak in the Pt conductance histogram. Below $1G_0$ very few data points are recorded, since Pt contacts tend to show an abrupt jump from the one-atom contact value into the tunnelling regime towards tunnel conductance values well below $0.1G_0$. In contrast, after admitting hydrogen gas a lot of structure is found in the entire range below $1.5G_0$, including a pronounced peak in the histogram near $1G_0$. We will now focus on the molecular arrangement responsible for this sharp peak. Clearly, many other junction configurations can be at the origin of the large density of data points a lower conductance, as discussed by Kiguchi *et al.* (2007).

Conductance histograms recorded using Fe, Co and Ni electrodes in the presence of hydrogen also show a pronounced peak near $1G_0$ (Untiedt *et al.* 2004), indicating that many transition metals may form similar single-molecule junctions. Also, Pd seemed a good candidate, but Csonka *et al.* (2004) find an additional peak at $0.5G_0$ in the conductance histogram, and it was argued that hydrogen is incorporated into the bulk of the Pd metal electrodes.

5.4.1.1   *Vibration modes*

The interpretation of the peak at $1G_0$ was obtained from a combination of measurements, including vibration spectroscopy, and density-functional theory (DFT) calculations. Experimentally, the vibration modes of the molecular structure were investigated by exploiting the principle of point-contact spectroscopy, for contacts adjusted to sit on a plateau in the conductance near $1G_0$.

Figure 5.12 shows examples for Pt–H_2 and Pt–D_2 junctions at a conductance near $1G_0$. The conductance is seen to drop by about 1 or 2%, symmetrically at positive and negative bias, as expected for electron–phonon scattering. The energies are in the range 50–60 meV, well above the Debye energy of $\sim$20 meV for Pt metal. A high energy for a vibration mode implies that a light element is involved, since the frequency is given by $\omega = \sqrt{\kappa/M}$ with κ an effective spring constant and M the mass of the vibrating object. The proof that the spectral features are indeed associated with hydrogen vibration modes comes from further experiments where H_2 was substituted by the heavier isotopes D_2

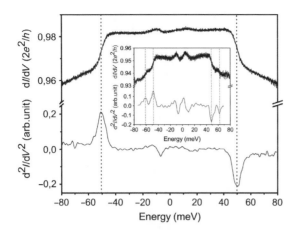

Fig. 5.12 Differential conductance curve for a molecule of D_2 contacted by Pt leads. The dI/dV curve (top) was recorded using a standard lock-in technique with a bias voltage modulation of 1 meV at a frequency of 7.7 kHz. The lower curve shows the numerically obtained derivative. The spectrum for H_2 in the inset shows two phonon energies, at 48 and 62 meV. All spectra show some, usually weak, anomalies near zero bias that can be partly due to excitation of modes in the Pt leads, partly due to two-level systems near the contact. From Djukic *et al.* (2005); copyright The American Physical Society.

and HD. The positions of the peaks in the spectra of d^2I/dV^2 vary within some range between measurements on different junctions, which can be attributed to variations in the atomic geometry of the leads to which the molecules bind. Figure 5.13 shows histograms for the vibration modes observed in a large number of spectra for each of the three isotopes.

Two pronounced peaks are observed in each of the distributions, that scale approximately as the square root of the mass of the molecules, as expected. The two modes can often be observed together, as in the inset to Fig. 5.12. For D_2 an additional mode appears near 90 meV. This mode cannot easily be observed for the other two isotopes, since the lighter HD and H_2 mass shifts the mode above 100 meV where the junctions become very unstable.

For a given junction with spectra as in Fig. 5.12 it is often possible to stretch the contact and follow the evolution of the vibration modes. The frequencies for the two lower modes were seen to increase with stretching, while the high mode for D_2 is seen to shift downward. This unambiguously identifies the lower two modes as transverse modes and the higher one as a longitudinal mode for the molecule. This interpretation agrees nearly quantitatively with DFT calculations for a configuration of a Pt–H–H–Pt bridge in between Pt pyramidally shaped leads (Djukic *et al.* 2005; Thygesen and Jacobsen 2005). The conductance obtained in the DFT calculations (Smit *et al.* 2002; Djukic *et al.* 2005; Thygesen and Jacobsen 2005) also reproduces the value of nearly $1G_0$ for this configuration. The fact that we observe vibration modes for HD that are intermediate between those for H_2 and D_2 confirms that the junction is formed by a molecule, not an atom. The number of conduction channels found in the calculations is one, which also agrees with experiment, as will be discussed next.

5.4.1.2 *Conductance fluctuations and shot noise*

As discussed above, the number of conduction channels can be obtained experimentally from a number of independent measurements. The mean amplitude of the conductance fluctuations can be exploited to verify whether the transport near multiples of G_0 is indeed due to full transmission of an integer number of conductance channels, as explained in Section 5.3.3.3 above. This has been done for Pt–H_2 (Smit *et al.* 2002) and Pd–H_2 (Csonka *et al.* 2004).

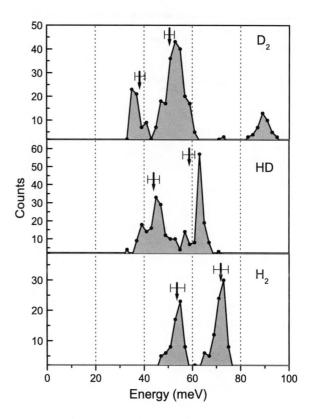

Fig. 5.13 Distribution of vibration mode energies observed for H_2, HD, and D_2 between Pt electrodes, with a bin size of 2 meV. The peaks in the distribution for H_2 are marked by arrows and their widths by error margins. These positions and widths were scaled by the expected isotope shifts, $\sqrt{2/3}$ for HD and $\sqrt{1/2}$ for D_2, from which the arrows and margins in the upper two panels have been obtained. From Djukic *et al.* (2005); copyright The American Physical Society.

For the former a clear suppression of the fluctuation amplitude was observed at the position of the peak near $1G_0$, although the minimum was not as deep as for gold atomic contacts. This can be attributed to the strong sensitivity of the fluctuation amplitude near the minimum to small deviations from perfect transmission. The transmission of the dominant channel was estimated to be 0.97, confirming that essentially a single channel carries the transport. This also implies that the junction is made up of just a single molecule. Although Pd with hydrogen may show a peak in the conductance histogram near $1G_0$, it is not as pronounced as for Pt, and Csonka *et al.* (2004) did not observe a clear minimum in the conductance fluctuations.

The conductance fluctuation experiment relies on averaging over many atomic configurations. A more direct test is provided by a measurement of shot noise, which was recently done for the hydrogen molecule bridge (Djukic and van Ruitenbeek 2006). A Pt–H_2 junction was adjusted that has a clear vibration-mode signal, and the shot noise signal was measured for the same junction. An example of this measurement is shown in Fig. 5.14. Although shot noise generally does not allow determining the full set of transmission values, we obtain information from the property that the noise increases the more channels are partially transmitted. We could redistribute the conductance over more than just the two channels considered in Fig. 5.14. When we break up the transmission $T_1 = 1.000$ into more channels this immediately increases the Fano factor. The only freedom we have is to redistribute the transmission

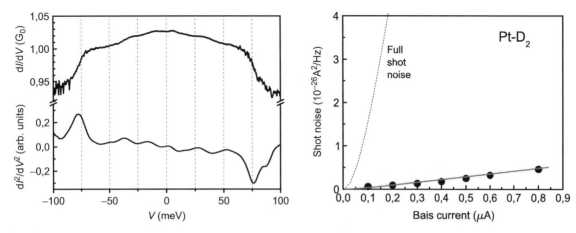

Fig. 5.14 Shot-noise measurements for a Pt–D_2 junction. The left panel shows the PCS signal for this junction, with a clear vibration mode at 76 meV. The right panel shows the excess noise (the white noise in the current above the thermal noise) as a function of the current. The noise is clearly strongly suppressed below the full Schottky noise for a tunnel junction. After each measurement of noise at a given current, the PCS was measured again to verify that the contact had not changed. The total conductance for this junction is $G = 1.021 G_0$, and the shot noise can be fitted with two channels, $T_1 = 1.000$, and $T_2 = 0.021$, giving a Fano factor of $F = 0.020$. Adapted from Djukic and van Ruitenbeek (2006); copyright American Chemical Society.

$T_2 = 0.021$ over two or more channels, that will all have a very small contribution. The dominant transmission by a single conductance channel with nearly perfect transmission is a very robust result of these measurements.

5.4.1.3   *Discussion of the results for hydrogen*

Several DFT calculations other than the ones mentioned above have been performed (see, e.g. García-Suárez *et al.* (2005), García *et al.* (2004), and Cuevas *et al.* (2003)). Using a slightly different approach García *et al.* (2004) obtain a conductance well below $1 G_0$. They propose an alternative atomic arrangement to explain the high conductance for the Pt–H bridge, consisting of a Pt–Pt bridge with two H atoms bonded to the sides in a perpendicular arrangement. However, this configuration gives rise to three conductance channels, which is excluded based on the analysis of shot noise and conductance fluctuations, as discussed above. The other calculations are in better agreement, and the origin of the discrepancy lies possibly in the choice of representing the leads by a Bethe lattice (García *et al.* 2004). This example illustrates the need for a reliable set of experimental data against which the various methods can be calibrated. The metal–hydrogen–metal bridge may provide a good starting point since it is the simplest and it can be compared in detail by virtue of the many parameters that have been obtained experimentally.

5.4.2   Organic molecules

Beyond the simplest molecule, H_2, many compounds have been tested. However, only a few will be discussed here, for we want to remain with simple systems that have a high transmission probability and for which the junctions have been characterized by the low-temperature techniques outlined above.

By its high degree of symmetry C_{60} is a good candidate for further study and it has been studied by several groups at low temperatures by STM, MCBJ, and other techniques (Joachim *et al.* 1995; Park *et al.* 2000; Champagne *et al.* 2005; Danilov *et al.* 2006; Böhler *et al.* 2007; Néel *et al.* 2007; Parks *et al.* 2007; Yoshida *et al.* 2007; Kiguchi and Murakoshi 2008). It appears that the conductance across a single C_{60} molecule is sensitive to the method of deposition and, to a lesser extent, to the type of metal electrodes used. When depositing the molecules from solution at room temperature the metal–molecule coupling varies from one device to the next and is typically well below $0.1G_0$. The weak coupling gives rise to a Kondo anomaly that can be tuned by the electrode spacing (Parks *et al.* 2007). However, depositing on clean metal electrodes by local sublimation gives rise to strong electronic coupling, which removes the Kondo effect. Danilov *et al.* (2006) use a special technique of *in-situ* fabrication of the Au metal electrodes by evaporation in a liquid-helium cooled chamber, followed by deposition of the molecules. They observe a conductance that is still low because the molecule is only strongly bound to one electrode and the technique employed does not permit adjusting the distance, but they observe nanomechanical oscillations of the molecule between the leads as in Park *et al.* (2000). MCBJ experiments that allow adjusting the gap size show a conductance for C_{60} between gold electrodes in the range of 0.1 to 0.2 G_0 (Böhler *et al.* 2007; Kiguchi and Murakoshi 2008). In the latter work it is demonstrated that the conductance is even considerably higher, $G = 0.5G_0$, when employing Ag electrodes.

The most detailed information on the conductance of a C_{60} molecular junction comes from STM experiments performed at 8 K under UHV (Néel *et al.* 2007). The molecules were deposited by sublimation onto a clean Cu(100) surface and were probed by a Cu-covered tip. The orientation of the molecules on top of the Cu(100) surface could be resolved, and only those molecules were selected that exposed a C–C bond between a hexagon and a pentagon at the top. When approaching the tip towards the molecule they observed a jump into contact from about $G = 2.5 \times 10^{-2}$ G_0 to $G = 0.25$ G_0 (see Fig. 5.15).

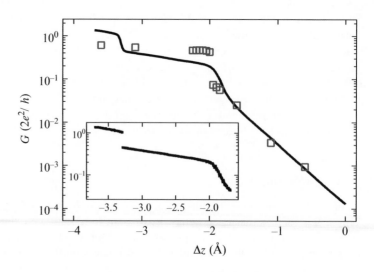

Fig. 5.15 Conductance G in units of G_0 vs. tip displacement Δz. Data are an average of 500 measurements. Zero displacement corresponds to the tip position before freezing the feedback loop at $V = 300$ mV and $I = 3$ nA. Experimental data appear as a line due to the high data point density, calculated data are depicted as squares. Inset: a single conductance curve revealing a discontinuity at $\Delta z \simeq -3.3$ Å. From Néel *et al.* (2007); copyright The American Physical Society.

The advantage of these low-temperature STM experiments is obviously the fact that one obtains "visual" confirmation of the presence of the molecule, the orientation and structure of at least one of the two metal electrodes is known, and one can even observe the orientation of the molecule on top of the surface. There is satisfactory agreement of the observed conductance with calculations based on DFT (Fig. 5.15).

The number of such STM studies is surprisingly scarce but beautiful results have recently been reported by Temirov *et al.* (2008) on a complex system, PTCDA (4,9,10-perylene-tetracarboxylic-dianhydrid), on a Ag(111) surface. They demonstrate that one can controllably contact the molecule to the STM tip at one of the four oxygen corner groups and peel the molecule gradually from the surface. The conductance clearly varies in the process of peeling, but when pulled to an upright position the conductance is approximately $0.15G_0$. During the process an interesting Kondo-like resonance develops that can be tuned by the electrode position.

Further results for high-transmission molecular junctions have been obtained for CO, C_2H_2, H_2O, and benzene between Pt electrodes by the MCBJ technique (Untiedt *et al.* 2004; Djukic 2006; Kiguchi *et al.* 2008; Tal *et al.* 2008). In order to preserve the advantages of the low-temperature break junction approach the molecules must be introduced at low temperatures. This has the disadvantage that the chemistry involved is largely unexplored, but it also opens many new avenues for exploring different metal electrodes and other types of chemical bonding. It turns out to be possible to deposit molecules while the atomic contact is held cold at helium temperatures. This was done by evaporating the substance from a source at room temperature through a capillary that is temporarily heated. Despite the fact that the mobility of molecules over a metal surface, once they are deposited at helium temperatures, should be negligible the metal–molecule–metal junctions can be formed repeatedly during the breaking/indentation cycles of the break-junction device. Likely, the mechanical deformation of the contact guides the molecules to neck, which is the most reactive spot of the exposed metal surface.

The most complete study of this kind was done for benzene (Kiguchi *et al.* 2008). The unsaturated carbon bonds provide a set of orbitals that are available for chemical bonding to clean transition-metal surfaces. A bonding configuration was found having the benzene molecule suspended between the Pt tips stretched to the point that the bond is confined to one carbon atom on each side. In this state the conductance is between 0.1 and 0.4 G_0, and good agreement was obtained between DFT calculations and experiment. Specifically, a vibration mode was identified near 42 meV that is insensitive to stretching of the junction over a wide range. Upon isotope substitution of all carbon atoms by ^{13}C a shift is observed that confirms that the signal is due to a vibration mode in benzene and that the vibration mode involves a large fraction of the carbons in the molecule (Fig. 5.16). Furthermore, by shot-noise measurements it could be demonstrated that the current in this configuration is carried by a single conductance mode. This substantiates the view that only a single molecule is involved and it agrees with the number of conductance channels for that molecule as obtained from the DFT calculations. These findings for benzene may proove relevant for further studies on more complicated organic molecules

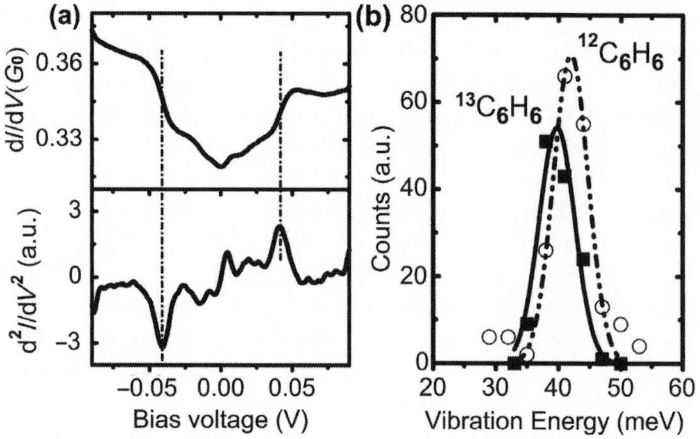

Fig. 5.16 (a) Differential conductance (top) and its derivative (bottom) for Pt contact after introduction of benzene taken at a zero bias conductance of $0.3G_0$. (b) Distribution of observed vibration energies. From Kiguchi *et al.* (2008); copyright The American Physical Society.

because they show that the unsaturated carbon bonds can be exploited for anchoring the molecule to the metal electrodes, which provides a more direct coupling of the metal electrodes to the carbon backbone of a molecule.

5.5   Conclusion

With this brief review we hope to have illustrated that clean atomic and molecular junctions at low temperatures offer many possibilities for obtaining detailed information, that reaches beyond the conductance or the current–voltage characteristics. The transport of electrons can be quasi-ballistic and the deviations from perfect transmission can be quantified and interpreted. Having this information allows us to distinguish between various approaches in computational modelling, and helps in refining the theory. The most active research is now towards coupling molecules to metal electrodes. The methods described here permit us to characterize such junctions in detail and, moreover, to explore different types of chemical bonding, with other metal electrodes than the ubiquitous gold electrodes, and with other anchoring groups than the commonly exploited thiols.

References

Agraït, N., Levy Yeyati, A., van Ruitenbeek J.M. *Phys. Rep.* **377**, 81–279 (2003).

Agraït, N, Rodrigo, J.G., Sirvent, C., Vieira, S. *Phys. Rev. B* **48**, 8499–8501 (1993a).

Agraït, N., Rodrigo, J.G., Vieira, S. *Phys. Rev. B* **47**, 12345–12348 (1993b).

Agraït, N., Untiedt, C., Rubio-Bollinger, G., Vieira, S. *Phys. Rev. Lett.* **88**, 216803 (2002).

Arnold, G.B. *J. Low Temp. Phys.* **68**, 1–27 (1987).

Averin, D., Bardas, A. *Phys. Rev. Lett.* **75**, 1831–1834 (1995).

Bahn, S.R., Jacobsen, K.W. *Phys. Rev. Lett.* **87**, 266101 (2001).

Beenakker, C.W.J., van Houten, H. *Solid State Phys.* **44**, 1–228 (1991).

Blanter, Y.M., Büttiker, M. *Phys. Rep.* **336**, 2–166 (2000).

Böhler, T., Edtbauer, A., Scheer, E. *Phys. Rev. B* **76**, 125432 (2007).

Brandbyge, M., Schiøtz, J., Sørensen, M.R., Stoltze, P., Jacobsen, K.W., Nørskov, J.K., Olesen, L., Lægsgaard, E., Stensgaard, I., Besenbacher, F. *Phys. Rev. B* **52**, 8499–8514 (1995).

Brandbyge, M., Sørensen, M.R., Jacobsen, K.W. *Phys. Rev. B* **56**, 14956–14959 (1997).

Bratus,' E.N., Shumeiko, V.S., Bezuglyi, E.V., Wendin, G. *Phys. Rev. B* **55**, 12 12666–12677 (1997).

Bratus,' E.N., Shumeiko, V.S., Wendin, G. *Phys. Rev. Lett.* **74**, 2110–2113 (1995).

Büttiker, M., *Phys. Rev. Lett.* **65**, 2901–2904 (1990).

Champagne, A.R., Pasupathy, A.N., Ralph, D.C. *Nano Lett.* **5**, 305–308 (2005).

Costa-Krämer, J.L., García, N., García-Mochales, P., Serena, P.A., Marqués, M.I., Correia, A. *Phys. Rev. B* **55**, 5416–5424 (1997a).

Costa-Krämer, J.L., García, N., Olin, H. *Phys. Rev. B* **55**, 12910–12913 (1997b).

Cron, R., Goffman, M.F., Esteve, D., Urbina, C. *Phys. Rev. Lett.* **86**, 4104–4107 (2001).

Csonka, S., Halbritter, A., Mihály, G., Shklyarevskii, O.I., Speller, S., van Kempen, H. *Phys. Rev. Lett.* **93**, 016802 (2004).

Cuevas, J.C., Heurich, J., Pauly, F., Wenzel, W., Schön, G. *Nanotechnology* **14**, R29–R38 (2003).

Cuevas, J.C., Levy Yeyati, A., Martín-Rodero, A., Rubio Bollinger, G., Untiedt, C., Agraït, N. *Phys. Rev. Lett.* **81**, 2990–2993 (1998).

Cuevas, J.C., Martín-Rodero, A., Levy Yeyati, A. *Phys. Rev. B* **54**, 7366–7379 (1996).

da Silva, E.Z., da Silva, A.J.R., Fazzio, A. *Phys. Rev. Lett.* **87**, 256102 (2001).

Danilov, A.V., Kubatkin, S.E., Kafanov, S.G., Bjørnholm, T. *Faraday Discuss* **131**, 337–345 (2006).

Datta, S., *Electronic Transport in Mesoscopic Systems* (Cambridge University Press, Cambridge, 1995).

Djukic, D. *Simple Molecules as Benchmark Systems for Molecular Electronics*. PhD thesis, Universiteit Leiden, The Netherlands (2006).

Djukic, D., Thygesen, K.S., Untiedt, C., Smit, R.H.M., Jacobsen, K.W., van Ruitenbeek, J.M. *Phys. Rev. B* **71**, 161402(R) (2005).

Djukic, D., van Ruitenbeek, J.M. *Nano Lett.* **6**, 789–793 (2006).

Finbow, G.M., Lynden-Bell, R.M., McDonald, I.R. *Molec. Phys.* **92**, 705–714 (1997).

Gai, Z., He, Y., Yu, H., Yang, W.S. *Phys. Rev. B* **53**, 1042–1045 (1996).

Galperin, M., Ratner, M.A., Nitzan, A. *Nano Lett.* **4**, 1605–1611 (2004).

García-Suárez, V.M., Rocha, A.R., Bailey, S.W., Lambert, C., Sanvito, S., Ferrer, J. *Phys. Rev. B* **72**, 045437 (2005).

García, Y., Palacios, J.J., SanFabián, E., Vergés, J.A., Pérez-Jiménez, A.J., Louis, E. *Phys. Rev. B* **69**, 041402 (2004).

Häkkinen, H., Barnett, R.N., Scherbakov, A.G., Landman, U. *J. Phys. Chem. B* **104**, 9063–9066 (2000).

Hansen, K., Laegsgaard, E., Stensgaard, I., Besenbacher, F. *Phys. Rev. B* **56**, 2208–2220 (1997).

Imry, Y. *Introduction to Mesoscopic Physics* (Oxford University Press, Oxford, 1997).

Joachim, C., Gimzewski, J., Schlittler, R., Chavy, C. *Phys. Rev. Lett.* **74**, 2102–2105 (1995).

Kervennic, Y.V., Vanmaekelbergh, D., Kouwenhoven, L.P., van der Zant, H.S. *J. Appl. Phys. Lett.* **83**, 3782–3784 (2003).

Kiguchi, M., Konishi, T., Miura, S., Murakoshi, K. *Physica E* **29**, 530–533 (2005).

Kiguchi, M., Murakoshi, K. *J. Phys. Chem. C, Lett.* **112**, 8140–8143 (2008).

Kiguchi, M., Stadler, R., Kristensen, I., Djukic, D., van Ruitenbeek, J. *Phys. Rev. Lett.* **98**, 146802 (2007).

Kiguchi, M., Tal, O., Wohlthat, S., Pauli, F., Krieger, M., Djukic, D., Cuevas, J., van Ruitenbeek, J. *Phys. Rev. Lett.* **101**, 046801 (2008).

Klapwijk, T.M., Blonder, Tinkham, M. *Physica B* **109 & 110**, 1657–1664 (1982).

Krans, J.M. *Size Effects in Atomic-scale Point Contacts*. PhD thesis, Universiteit Leiden, The Netherlands (1996).

Krans, J.M., van Ruitenbeek, J.M., Fisun, V.V., Yanson, I.K., de Jongh, L.J. *Nature* **375**, 767–769 (1995).

Kubatkin, S., Danilov, A., Hjort, M., Cornil, J., Brédas, J.L., Stuhr-Hansen, N., Hedegård, P., Bjørnholm, T. *Nature* **425**, 698–701 (2003).

Kumar, A., Saminadayar, L., Glattli, D.C., Jin, Y., Etienne, B. *Phys. Rev. Lett.* **76**, 2778 (1996).

Lee, P.A., Stone, A.D. *Phys. Rev. Lett.* **55**, 1622–1625 (1985).

Lesovik, G.B. *Sov. Phys. JETP Lett.* **49**, 592–594 (1989). [Pis'ma Zh. Eksp. Teor. Fiz. **49** (1989) 513].

Li, C.Z., Tao, N. *J Appl. Phys. Lett.* **72**, 894–896 (1998).

Liang, W., Shores, M.P., Bockrath, M., Long, J.R., Park, H. *Nature* **417**, 725–729 (2002).

Limot, L., Kröger, J., Berndt, R., Garcia-Lekue, A., Hofer, W. *Phys. Rev. Lett.* **94**, 126102 (2005).

Ludoph, B., Devoret, M.H., Esteve, D., Urbina, C., van Ruitenbeek, J.M. *Phys. Rev. Lett.* **82**, 1530–1533 (1999).

Ludoph, B., van Ruitenbeek, J.M. *Phys. Rev. B* **61**, 2273–2285 (2000).

Marchenkov, A., Dai, Z., Donehoo, B., Barnett, R., Landman, U. *Nature Nanotechnol.* **2**, 481–485 (2007).

Muller, C.J., van Ruitenbeek, J.M., de Jongh, L. *J. Phys. Rev. Lett.* **69**, 140–143 (1992).

Natelson, D. in H. S Nalwa, ed., *Handbook of Organic Electronics and Photonics*, Vol. 3 (American Scientific Publishers, Dordrecht, The Netherlands, 2008).

Néel, N., Kröger, J., Limot, L., Frederiksen, T., Brandbyge, M., Berndt, R. *Phys. Rev. Lett.* **98**, 065502 (2007).

Ohnishi, H., Kondo, Y., Takayanagi, K. *Nature* **395**, 780–785 (1998).

Olesen, L., Lægsgaard, E., Stensgaard, I., Besenbacher, F., Schiøtz, J., Stoltze, P., Jacobsen, K.W., Nørskov, J.K. *Phys. Rev. Lett.* **72**, 2251–2254 (1994).

Olesen, L., Lægsgaard, E., Stensgaard, I., Besenbacher, F., Schiøtz, J., Stoltze, P., Jacobsen, K.W. *Phys. Rev. Lett.* **74**, 2147 (1995).

Osorio, E., ONeill, K., Wegewijs, M., Stuhr-Hansen, N., Paaske, J., Bjørnholm, T., van der Zant, H. *Nano Lett.* **7**(11), 3336–3342 (2007).

Park, H., Park, J., Kim, A.K.L., Anderson, E.H., Alivisatos, A.P., McEuen, P.L. *Nature* **407**, 57 (2000).

Park, J., Pasupathy, A.N., Goldsmith, J.I., Chang, C., Yaish, Y., Petta, J.R., Rinkoski, M., Sethna, J.P., Abruña, H.D., McEuen, P.L., Ralph, D. *Nature* **417**, 722–725 (2002).

Parks, J.J., Champagne, A.R., Hutchison, G.R., Flores-Torres, S., Abruna, H.D., Ralph, D.C. *Phys. Rev. Lett.* **99**, 026601 (2007).

Pascual, J.I., Méndez, J., Gómez-Herrero, J., Baró, A.M., García, N., Binh, V.T. *Phys. Rev. Lett.* **71**, 1852–1855 (1993).

Reed, M.A., Zhou, C., Muller, C.J., Burgin, T.P., Tour, J.M. *Science* **278**, 252–254 (1997).

Reichert, J., Weber, H.B., Mayor, M., von Löhneysen, H. *Appl. Phys. Lett.* **82**, 4137–4139 (2003).

Reznikov, M., Heiblum, M., Shtrikman, H., Mahalu, D. *Phys. Rev. Lett.* **75**, 3340–3343 (1995).

Scheer, E., Agraït, N., Cuevas, J.C., Levy Yeyati, A., Ludoph, B., Martín-Rodero, A., Rubio Bollinger, G., van Ruitenbeek, J.M., Urbina, C. *Nature* **394**, 154–157 (1998).

Scheer, E., Joyez, P., Esteve, D., Urbina, C., Devoret, M.H. *Phys. Rev. Lett.* **78**, 3535–3538 (1997).

Schottky, W. *Ann. Phys. (Leipzig)* **57**, 541–567 (1918).

Schrieffer, J.R., Wilkins, J.W. *Phys. Rev. Lett.* **10**, 17–20 (1963).

Smit, R.H.M., Noat, Y., Untiedt, C., Lang, N.D., van Hemert, M.C., van Ruitenbeek, J.M. *Nature* **419**, 906–909 (2002).

Smit, R.H.M., Untiedt, C., Yanson, A.I., van Ruitenbeek, J.M. *Phys. Rev. Lett.* **87**, 266102 (2001).

Sørensen, M.R., Brandbyge, M., Jacobsen, K.W. *Phys. Rev. B* **57**, 3283–3294 (1998).

Stipe, B.C., Rezaei, M.A., Ho, W. *Science* **280**, 1732–1735 (1998).

Tal, O., Krieger, M., Leerink, B., van Ruitenbeek, J. *Phys. Rev. Lett.* **100**, 196804 (2008).

Temirov, R., Lassise, A., Anders, F., Tautz, F. *Nanotechnology* **19**, 065401 (2008).

Thygesen, K.S., Jacobsen, K.W. *Phys. Rev. Lett.* **94**, 036807 (2005).

Untiedt, C., Dekker, D.M.T., Djukic, D., van Ruitenbeek, J.M. *Phys. Rev. B* **69**, 081401(R) (2004).

van den Brom, H.E., van Ruitenbeek, J.M. *Phys. Rev. Lett.* **82**, 1526–1529 (1999).

van den Brom, H.E., van Ruitenbeek, J.M. in D. Reguera, G. Platero, L.L. Bonilla, Rubí, J.M. (eds) *Statistical and Dynamical Aspects of Mesoscopic Systems* (Springer-Verlag, Berlin, Heidelberg, 2000) pp. 114–122.

van der Post, N., Peters, E.T., Yanson, I.K., van Ruitenbeek, J.M. *Phys. Rev. Lett.* **73**, 2611–2613 (1994).

van Ruitenbeek, J.M., Alvarez, A., Piñeyro, I., Grahmann, C., Joyez, P., Devoret, M.H., Esteve, D., Urbina, C. *Rev. Sci. Instrum.* **67**, 108–111 (1996).

van Wees, B.J., van Houten, H.H., Beenakker, C.W.J., Williamson, J.G., Kouwenhoven, L.P., van der Marel, D., Foxon, C.T. *Phys. Rev. Lett.* **60**, 848–850 (1988).

Wolf, E.L. *Principles of Electron Tunneling Spectroscopy* (Oxford University Press, Oxford, 1989).

Xie, F.Q., Nittler, L., Ch. Obermair, Schimmel, T. *Phys. Rev. Lett.* **93**, 128303 (2004).

Yanson, A.I. *Atomic Chains and Electronic Shells: Quantum Mechanisms for the Formation of Nanowires.* PhD thesis, Universiteit Leiden, The Netherlands (2001).

Yanson, A.I., Rubio Bollinger, G., van den Brom, H.E., Agraït, N., van Ruitenbeek, J.M. *Nature* **395**, 783–785 (1998).

Yanson, I.K., Shklyarevskii, O.I. *Sov. J. Low Temp. Phys.* **12**, 509–528 (1986).

Yoshida, M., Kurui, Y., Oshima, Y., Takayanagi, K. *Jpn. J. Appl. Phys.* **46**, L67 (2007).

Yu, L.H., Natelson, D. *Nanotechnology* **15**, S517–S524 (2004).

Zhou, C., Muller, C.J., Deshpande, M.R., Sleight, J.W., Reed, M.A. *Appl. Phys. Lett* **67**, 1160–1162 (1995).

Zyuzin, A.Y., Spivak, B.Z. *Sov. Phys. JETP* **71**, 563–566 (1990). (Zh. Eksp. Teor. Fiz. **98**, 1011–1017 (1990)).

Thermal transport of small systems

Takahiro Yamamoto, Kazuyuki Watanabe, and Satoshi Watanabe

6

6.1 Introduction

According to Moore's law, the number of transistors in an integrated circuit on a chip roughly doubles every two years (Moore 1965), integration will enter the below 10 nm scale in the near future. As the device size shrinks down to the nanoscale, one needs to analyze the electronic transport properties of devices on the basis of fully quantum-mechanical approaches. In fact, the electronic transport in a mesoscopic or nanoscale system, whose dimension is much smaller than the mean-free path of electron, does not obey the well-known Ohm's law: the electrical conductance G is directly proportional to the cross-sectional area S of a sample and inversely proportional to its length L, namely $G = \sigma(S/L)$. Here, the proportionality coefficient σ is the electrical conductivity indicating an ability to conduct electric current through materials and is independent of sample dimensions. As a manifestation of quantum effects on the transport in such small systems, the electrical conductance is quantized in multiples of the universal value, $G_0 = e^2/h$, which can be understood within the framework of Landauer's theory. We refer further details of nanoscale electronic transport to Chapter 1 by Datta in this book.

Similar to electronic transport, phonon transport or thermal transport by phonons exhibits a remarkable behavior, which is different from that in bulk materials if a sample dimension is reduced to the nanoscale. For a normal bulk material, not only the electrical conductance G but also the thermal conductance κ is directly proportional to S and inversely proportional to L:

$$\kappa = \frac{S}{L}\lambda, \tag{6.1}$$

where the thermal conductivity λ indicates the ability to conduct heat in materials and is independent of its dimension. The thermal conductivity λ is defined as the ratio of the thermal current, J, to the temperature gradient dT/dx via Fourier's law:

$$J = -\lambda \frac{dT}{dx}. \tag{6.2}$$

The Fourier law well describes the thermal transport phenomena in normal bulk materials. However, it is no longer valid when the sample dimension reduces down to below the mean-free path of phonons. This problem has been a long-standing theoretical interest that goes back to Peierls' early work in the 1920s (Peierls 1929).

In such a small system, the phonons propagate coherently without interference with other phonons. We will learn that the thermal current by coherent phonons is not proportional to the temperature gradient dT/dx but to the temperature difference ΔT between the hot and cold heat reservoirs. We will also see that the thermal conductance κ is quantized in multiples of the universal value

$$\kappa_0 = \frac{\pi^2 k_{\mathrm{B}}^2 T}{3h} \equiv g_0 T, \tag{6.3}$$

where $g_0 = 9.4 \times 10^{-13}\,\mathrm{W/K^2}$ (Angelescu *et al.* 1998; Rego *et al.* 1998; Blencowe 1999). κ_0 gives the upper limit of the thermal conductance that a single acoustic phonon mode can carry. This fundamental upper limit, coming from quantum mechanics, can be regarded as a quantum of thermal conductance. This chapter gives an account of the theory of quantum thermal transport in small objects, which is a cornerstone of thermal management for the future nanodevices and of the field of phononics in general.

6.2   Boltzmann–Peierls formula of diffusive phonon transport

Before discussing the coherent phonon transport, we review the conventional theory of thermal transport by diffusive phonons undergoing frequent scattering events. Theoretical analyses of thermal transport began from the pioneering work by Fourier in the 1820s. According to Fourier's law (6.2), the thermal current J is proportional to the temperature gradient dT/dx. In this section, we derive the phenomenological Fourier's law (6.2) based on the familiar Boltzmann's kinetic theory and give a concrete expression of the thermal conductivity λ of quasi-1D materials with a length L and with a small cross-sectional area $S(\ll L^2)$. The length is assumed to be longer than the mean-free path of phonons.

The thermal current is caused by a deviation from equilibrium:

$$J = \frac{1}{V} \sum_{k>0,\nu} \hbar \omega_{k,\nu} \left(f_{k,\nu} - f_{\mathrm{B}}(\omega_{k,\nu}, T) \right) v_{k,\nu}, \tag{6.4}$$

where $V(= LS)$ is the volume of the system, $\hbar\omega_{k,\nu}$ is the energy of a phonon with the wave number k on the dispersion branch $\nu = 1, 2, \cdots, 3N$ (N = total number of atoms in a unit cell), $v_{k,\nu} = \partial \omega_{k,\nu}/\partial k$ is the group velocity of phonons labelled by $\{k, \nu\}$, $f_{\mathrm{B}}(\omega_{k,\nu}, T)$ is the Bose–Einstein distribution function that is given by

$$f_{\mathrm{B}}(\omega_{k,\nu}, T) = \frac{1}{\exp\left(\hbar\omega_{k,\nu}/k_{\mathrm{B}}T\right) - 1}, \tag{6.5}$$

and $f_{k,\nu}$ is the distribution function of non-equilibrium phonons. As seen in eqn (6.4), the thermal current vanishes in thermal equilibrium satisfying $f_{k,\nu} = f_B(\omega_{k,\nu}, T)$ as we expected. The time evolution of the phonon distribution function $f_{k,\nu}$ obeys the Boltzmann equation:

$$\frac{\mathrm{d}f_{k,\nu}}{\mathrm{d}t} + v_{k,\nu}\frac{\partial f_{k,\nu}}{\partial x} = \left[\frac{\partial f_{k,\nu}}{\partial t}\right]_{\mathrm{coll}}, \qquad (6.6)$$

where the first term in the left-hand side disappears for the steady-state thermal transport of interest here (i.e. $\mathrm{d}f_{k,\nu}/\mathrm{d}t = 0$), the second term in the left-hand side represents the drift dynamics of a single phonon, and the right-hand side denotes the effect of collisions among phonons. Since the collision term plays the role of shifting the non-equilibrium phonon state toward the equilibrium one, we assume a simple form to describe such a relaxation kinetics toward equilibrium,

$$\left[\frac{\partial f_{k,\nu}}{\partial t}\right]_{\mathrm{coll}} = -\frac{f_{k,\nu} - f_B(\omega_{k,\nu}, T)}{\tau_{k,\nu}}, \qquad (6.7)$$

where $\tau_{k,\nu}$ is the relaxation time for a phonon state with $\{k, \nu\}$.

Let us consider the case in which the spatial variation of temperature is very smooth over a wide area including many atoms. In this *local thermal equilibrium* situation, the distribution function can be described by a Bose–Einstein distribution function, $f_{k,\nu} \approx f_B(\omega_{k,\nu}, T(x))$, depending on position x through a spatially varying temperature $T(x)$. In the local thermal equilibrium situation, the drift term in eqn (6.6) is rewritten as

$$v_{k,\nu}\frac{\partial f_{k,\nu}}{\partial x} = v_{k,\nu}\frac{\partial f_B(\omega_{k,\nu}, T)}{\partial T}\frac{\mathrm{d}T}{\mathrm{d}x}. \qquad (6.8)$$

Substituting eqns (6.7) and (6.8) into eqn (6.6) under the steady-state condition $\mathrm{d}f_{k,\nu}/\mathrm{d}t = 0$, the deviation of the distribution function from the equilibrium is given by

$$f_{k,\nu} - f_B(\omega_{k,\nu}, T) = -\tau_{k,\nu}v_{k,\nu}\frac{\partial f_B(\omega_{k,\nu}, T)}{\partial T}\frac{\mathrm{d}T}{\mathrm{d}x}. \qquad (6.9)$$

Substituting eqn (6.9) into eqn (6.4), we obtain Fourier's law (6.2) and the Boltzmann–Peierls formula of thermal conductivity:

$$\lambda = \frac{1}{V}\sum_{k>0,\nu}\hbar\omega_{k,\nu}|v_{k,\nu}|\Lambda_{k,\nu}\frac{\partial f_B(\omega_{k,\nu}, T)}{\partial T}, \qquad (6.10)$$

where $\Lambda_{k,\nu} = \tau_{k,\nu}|v_{k,\nu}|$ is the mean-free path of phonons with $\{k, \nu\}$. Moreover, eqn (6.10) can be rewritten as

$$\lambda = \frac{1}{S}\sum_{\nu}\int_{\omega_\nu^{\mathrm{min}}}^{\omega_\nu^{\mathrm{max}}}\hbar\omega\frac{D_\nu(\omega)}{2}|v_\nu(\omega)|\Lambda_\nu(\omega)\frac{\partial f_B(\omega, T)}{\partial T}\mathrm{d}\omega$$

$$= \frac{1}{2\pi S}\sum_{\nu}\int_{\omega_\nu^{\mathrm{min}}}^{\omega_\nu^{\mathrm{max}}}\hbar\omega\left[\frac{\partial f(\omega, T)}{\partial T}\right]\Lambda_\nu(\omega)\mathrm{d}\omega, \qquad (6.11)$$

where ω_ν^{max} and ω_ν^{min} are the maximum and minimum values of the phonon dispersion curve for the mode ν, respectively. At the last line in eqn (6.11), we dropped the suffix B standing for "Bose" from $f_B(\omega, T)$ for simplicity, and we will use the simple form $f(\omega, T)$ as the Bose–Einstein distribution function hereafter. The expression of thermal conductivity in eqn (6.11) does not include the group velocity $v_\nu(\omega)$ and the density of states for each mode ν:

$$D_\nu(\omega) = \frac{1}{L} \sum_k \delta(\omega_\nu - \omega_{k,\nu}) = \frac{1}{\pi |v_\nu(\omega)|} \qquad (6.12)$$

because they cancel each other. The cancellation is peculiar to the 1D and quasi-1D systems. Consequently, the thermal conductivity behavior of the 1D and quasi-1D systems is essentially determined by the mean-free path Λ_ν.

As can be understood from the above derivation of Fourier's law (6.2) based on Boltzmann's equation (6.6), the thermal current described by eqn (6.2) is caused by randomly travelling phonons suffering frequent collisions. Even if the phonons travel coherently through the system without interference with other phonons, does the thermal transport obey the Fourier's law (6.2)? The answer is "No"! In the coherent case, there is no temperature gradient dT/dx and eventually the thermal current is no longer proportional to the temperature gradient but to the temperature difference $\Delta T = T_H - T_C$ between the hot and cold heat reservoir at both ends, as will be shown in the next section. In other words, the thermal conductivity λ cannot be well defined for coherent phonon systems satisfying $L \ll \Lambda_\nu$, because it diverges to infinity due to the lack of a temperature gradient. As a response function for the coherent phonon transport, we use the thermal conductance, defined by

$$\kappa = \lim_{T_H, T_C \to T} \frac{I}{T_H - T_C} \qquad (6.13)$$

instead of the thermal conductivity λ. Here, T is an averaged temperature defined as $T = (T_H + T_C)/2$, and $I = JS$ is the total thermal current passing through the cross-sectional area S of a quasi-1D system.

Moreover, there is another advantage to the use of the thermal conductance κ for quasi-1D systems. As seen in eqn (6.11), the thermal conductivity λ of a quasi-1D system depends on the cross-sectional area S, which is not a well-defined quantity for nanoscale quasi-1D systems like carbon nanotubes. In contrast to the thermal conductivity λ, the thermal conductance κ is a well-defined physical quantity even for quasi-1D systems as it does not depend on S.

In the case of diffusive phonon transport, the thermal conductance κ is connected to the thermal conductivity λ via the relation $\kappa = (S/L)\lambda$ in eqn (6.1). This relation can be derived from eqns (6.2) and (6.13). Substituting eqn (6.11) into eqn (6.1), the thermal conductance is expressed as

$$\kappa = \sum_\nu \int_{\omega_\nu^{\text{min}}}^{\omega_\nu^{\text{max}}} \frac{d\omega}{2\pi} \hbar\omega \left[\frac{\partial f(\omega, T)}{\partial T} \right] \frac{\Lambda_\nu(\omega)}{L} \qquad (6.14)$$

$$= \int_0^\infty \frac{d\omega}{2\pi} \hbar\omega \left[\frac{\partial f(\omega, T)}{\partial T} \right] \frac{\Lambda(\omega)}{L}, \qquad (6.15)$$

where $\Lambda(\omega) \equiv \sum_\nu \Lambda_\nu(\omega)$ is the total mean-free path. We refer to eqns (6.14) and (6.15) as the Boltzmann–Peierls formula for the thermal conductance in a quasi-1D system. We conclude that the thermal resistance defined by the reciprocal of the thermal conductance ($R_{th} \equiv 1/\kappa$) increases proportionally to the system length L, that is $R_{th} \propto L$. It should be noted that this conclusion is valid for diffusive phonon transport, but is not for the coherent case.

This section provided a brief introduction to the thermal transport theory based on Boltzmann's kinetic theory. For further details, see the excellent books by Peierls (1955) and Zimann (1960), for example.

6.3   Coherent phonon transport

In the previous section, we discussed the diffusive thermal transport in a quasi-1D material with a length longer than the mean-free path of phonons. As the length reduces to below the mean-free path, the phonons propagate coherently through a material, as interference among phonons can be neglected. In this section, we describe the thermal transport by coherent phonons in a quasi-1D system. The kinetic approach based on the Boltzmann equation (6.6) is no longer applicable to the coherent phonon transport. We thus introduce the Landauer approach for describing the coherent phonon transport in which the thermal current through a nanoscale conductor is expressed in terms of the transmission probability of each phonon through the conductor.

6.3.1   Landauer formulation of phonon transport

Let us consider the situation depicted in Fig. 6.1. A coherent phonon conductor in the absence of many-body interactions, such as phonon–phonon and electron–phonon scatterings, is connected to ballistic quasi-1D leads without any scattering, which are in turn connected to heat reservoirs. The left and right heat reservoirs are assumed to be in thermal equilibrium with well-defined temperatures T_H and $T_C(< T_H)$, respectively. In this situation, the relaxation of phonons occurs only in the heat reservoirs. Thus, a phonon injected from the left and the right heat reservoirs follows the Bose–Einstein distribution function $f(\omega, T_H)$ and $f(\omega, T_C)$, respectively. Here, ω is the frequency of the injected phonon.

Let us derive a general expression of thermal current I and thermal conductance κ for the above situation. For simplicity, we consider a symmetric system having the same leads for the left and right regions. The thermal current carried by phonons with mode ν and frequency ω injected from the left heat reservoir toward the central conductor is given by

$$i_\nu^L(\omega) = \hbar\omega |v_\nu(\omega)| D_\nu^+(\omega) f(\omega, T_H)$$

$$= \frac{1}{2\pi} \hbar\omega f(\omega, T_H), \qquad (6.16)$$

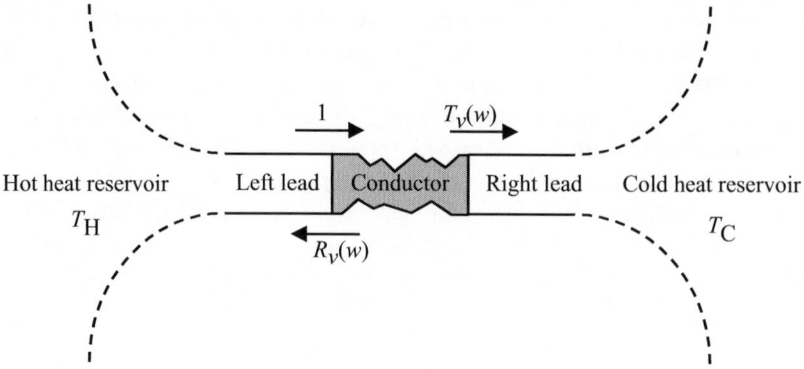

Fig. 6.1 Schematic view of quasi-1D conductor connected to the left and right leads, which are in turn connected to the hot and cold heat reservoirs at both ends. $\mathcal{T}_\nu(\omega)$ $(\mathcal{R}_\nu(\omega))$ is a transmission (reflection) function for injected phonons with mode ν and frequency ω.

while that injected from the right heat reservoir toward the central conductor is given by

$$i_\nu^R(\omega) = \hbar\omega|v_\nu(\omega)|D_\nu^-(\omega)f(\omega, T_C)$$

$$= \frac{1}{2\pi}\hbar\omega f(\omega, T_C). \qquad (6.17)$$

In the first line in eqns (6.16) and (6.17), $D_\nu^\pm(\omega)$ denotes the phonon density of states for a unit length in the left and the right leads and the superscript $+(-)$ stands for the density of states for phonons with positive (negative) group velocity $v_\nu(\omega)$. Since $D_\nu^\pm(\omega)$ is given by $D_\nu^\pm(\omega) = D_\nu(\omega)/2$ with $D_\nu(\omega) = 1/(\pi|v_\nu(\omega)|)$ for quasi-1D systems, it is cancelled by the group velocity $v_\nu(\omega)$ in the second line in eqns (6.16) and (6.17).

As shown in Fig. 6.1, $\mathcal{T}_\nu(\omega)$ $(\mathcal{R}_\nu(\omega))$ is the transmission (reflection) function for injected phonons with mode ν and frequency ω. The net thermal current at the left lead, which is carried by phonons with ν and ω, is thus given by

$$i_\nu(\omega) = i_\nu^L(\omega)\{1 - \mathcal{R}_\nu(\omega)\} - i_\nu^R(\omega)\mathcal{T}_\nu(\omega)$$

$$= \mathcal{T}_\nu(\omega)\{i_\nu^L(\omega) - i_\nu^R(\omega)\}$$

$$= \frac{1}{2\pi}\hbar\omega\mathcal{T}_\nu(\omega)\{f(\omega, T_H) - f(\omega, T_C)\}. \qquad (6.18)$$

Similarly, we can obtain the same result as eqn (6.18) for the net thermal current at the right lead. This can be interpreted as Kirchhoff's law for phonon transport. Using eqn (6.18), the total thermal current, referred to as Landauer's energy flux, can be expressed as

$$I = \sum_\nu \int_{\omega_\nu^{\min}}^{\omega_\nu^{\max}} i_\nu(\omega)d\omega$$

$$= \frac{1}{2\pi}\sum_\nu \int_{\omega_\nu^{\min}}^{\omega_\nu^{\max}} \hbar\omega\{f(\omega, T_H) - f(\omega, T_C)\}\mathcal{T}_\nu(\omega)d\omega, \qquad (6.19)$$

where $\omega_\nu^{\max}$ and $\omega_\nu^{\min}$ are the maximum and minimum values of the phonon-dispersion curve for the mode ν, respectively.

Let us consider the linear response situation with respect to the temperature difference $\Delta T \equiv T_H - T_L$ between hot and cold heat reservoirs. For a small temperature difference, $\Delta T \ll T (\equiv (T_H + T_C)/2)$, the thermal current is given by

$$I = \frac{\Delta T}{2\pi} \sum_{\nu} \int_{\omega_{\nu}^{\mathrm{min}}}^{\omega_{\nu}^{\mathrm{max}}} \hbar\omega \left[\frac{\partial f(\omega, T)}{\partial T} \right] \mathcal{T}_{\nu}(\omega) \mathrm{d}\omega. \tag{6.20}$$

Notice that this expression (6.20) is not proportional to the temperature gradient $\mathrm{d}T/\mathrm{d}x$ like Fourier's law (6.2) for diffusive phonons, but is in fact proportional to the temperature difference ΔT. Using eqns (6.13) and (6.20), we obtain the linear response form of thermal conductance as the following expression.

$$\kappa = \frac{1}{2\pi} \sum_{\nu} \int_{\omega_{\nu}^{\mathrm{min}}}^{\omega_{\nu}^{\mathrm{max}}} \hbar\omega \left[\frac{\partial f(\omega, T)}{\partial T} \right] \mathcal{T}_{\nu}(\omega) \mathrm{d}\omega \tag{6.21}$$

$$= \frac{1}{2\pi} \int_{0}^{\infty} \hbar\omega \left[\frac{\partial f(\omega, T)}{\partial T} \right] \mathcal{T}(\omega) \mathrm{d}\omega, \tag{6.22}$$

where $\mathcal{T}(\omega) = \sum_{\nu} \mathcal{T}_{\nu}(\omega)$ is the total transmission function (Rego *et al.* 1998). As seen in eqn (6.21), the thermal conductance for coherent phonons is determined by the phonon transmission function $\mathcal{T}_{\nu}(\omega)$, while eqn (6.14) for diffusive phonons is determined by the ratio of the phonon mean-free path to the system length $\Lambda_{\nu}(\omega)/L$. Since the transmission $\mathcal{T}(\omega)$ is independent of the system length L for coherent phonon transport, the thermal resistance $R_{\mathrm{th}} = 1/\kappa$ is constant with respect to L in this case. This is clearly different from the result derived by Fourier's law: the thermal resistance is proportional to L. In Section 6.4, we will discuss the cross-over from coherent to incoherent phonon transport when the system length increases at constant temperature.

6.3.2   Ballistic phonon transport and quantized thermal conductance

According to the Landauer formula in eqn (6.22), the thermal conductance is expressed by the phonon-transmission function $\mathcal{T}(\omega)$. For the moment, we consider an ideal situation where all phonons are transmitted ballistically through a conductor without any scattering, namely $\mathcal{T}_{\nu}(\omega) = 1$ for all phonon modes. In this ballistic transport case, the total transmission function is given by

$$\mathcal{T}(\omega) \equiv \sum_{\nu} \mathcal{T}_{\nu}(\omega)$$

$$= \text{Number of phonon modes at frequency } \omega. \tag{6.23}$$

As an example of ballistic quasi-1D phonon conductor, Figs. 6.2(a) and (b) show the low-energy phonon-dispersion relations and the corresponding phonon-transmission function of the (10,10) carbon nanotube, respectively. In

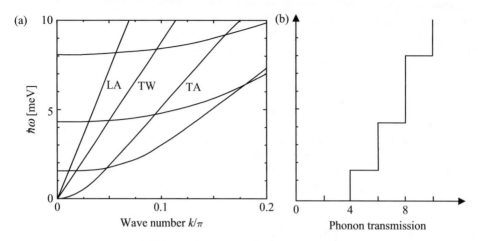

Fig. 6.2 (a) Phonon-dispersion relation and (b) phonon-transmission function of (10,10) carbon nanotube in low-frequency region. LA, TW, and TA represent a longitudinal, twisting, and two transverse (or flexural) acoustic modes.

general, quasi-1D systems like carbon nanotubes have four acoustic modes, namely a longitudinal, a twisting, and two transverse (or flexural) acoustic modes.

In the low-temperature region where the optical phonons are not excited, the thermal current is carried only by these 4 acoustic modes. Substituting $T(\omega) = 4$ into eqn (6.22), the low-temperature thermal conductance is given by

$$\kappa = 4 \times \frac{k_B^2 T}{h} \int_0^\infty \frac{x^2 e^x}{(e^x - 1)^2} dx = 4\kappa_0, \qquad (6.24)$$

where $\kappa_0 = \pi^2 k_B^2 T / 3h$ is the quantum of thermal conductance introduced by eqn (6.3), and the κ_0 gives the maximum value of thermal conductance that an acoustic phonon mode can take (Angelescu *et al.* 1998; Rego *et al.* 1998; Blencowe 1999). The quantization of thermal conductance is very similar to the quantization of electrical conductance $G_0 = e^2/h$ in quantum wires. It is noted that the thermal conductance quantum κ_0 contains the thermal energy "$k_B T$" instead of the electrical charge "e" in the electrical conductance quantum. This is because the thermal current carries the thermal energy "$k_B T$" while the electrical current transports the charge "e".

6.3.2.1   *Experimental observation of thermal conductance quantization*

It is very challenging to observe the quantized thermal conductance κ_0 experimentally, because its value is extremely small, as given in eqn (6.3). In 2000, Schwab and his colleagues succeeded in detecting it in a nanowire consisting of silicon nitride below 0.6 K using a sophisticated fabrication technique (Schwab *et al.* 2000). Figure 6.3(a) shows the experimental setup used to measure the quantum κ_0. The small square at the center is a silicon nitride (SiN) membrane. The membrane is heated by two C-shaped heaters made by a gold thin film and acts as a heat reservoir. Since this membrane is suspended by four SiN bridges and is in vacuum, the heat reservoir and the four bridges are thermally

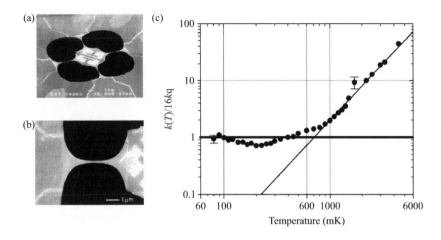

Fig. 6.3 (a) Experimental setup used to measure the quantum of thermal conductance κ_0. (b) The enlarged view of a nanowire consisting of silicon nitride in Fig. 6.3(a). (c) The temperature dependence of thermal conductance, normalized to a universal value of $16\kappa_0$, of the nanowire. From Schwab *et al.* (2000), copyright: Nature.

isolated from the environment and the heat flows from the membrane into 4 SiN nanowires without heat dissipation. The perfect phonon transmission was realized by constructing smooth connections between the membrane and the SiN nanowires, as shown in Fig. 6.3(b).

Figure 6.3(c) shows the temperature dependence of the phonon-derived thermal conductance, normalized to a universal value of $16\kappa_0$, of a SiN nanowire. The total thermal conductance of the device in Fig. 6.3(a) is expected to be $16\kappa_0$ because each of the four acoustic modes in each of the four bridges carries the quantum κ_0. At extremely low temperature, below a threshold temperature ~ 0.6 K, the normalized thermal conductance is one or lower, as we expected. Above the threshold temperature, the thermal conductance curve increases because the optical phonon modes begin to contribute to the thermal transport.

As well as in SiN nanowires, quantization of the thermal conductance has also been observed in carbon nanotubes (Chiu *et al.* 2005) just after its theoretical prediction (Yamamoto *et al.* 2004).

6.3.2.2   *Quantization of electron-derived thermal conductance*

Before closing this section, let us mention the thermal transport by ballistic electrons. As well known, the electrical conductance G is related to the electron-derived thermal conductance κ_{el} via the Wiedemann–Franz law for the degenerate free-electron gas:

$$\frac{\kappa_{\mathrm{el}}}{G} = L_0 T, \tag{6.25}$$

where T is the temperature and the factor $L_0 \equiv \pi^2 k_{\mathrm{B}}^2/3e^2$ is the Lorentz number consisting only of universal constants. Substituting $G = G_0 \equiv e^2/h$ for ballistic 1D conductors into eqn (6.25), the electron-derived thermal conductance κ_{el} is given by the same universal value of $\kappa_0 = \pi^2 k_{\mathrm{B}}^2 T/3h$ per single conduction channel. The quantum of thermal conductance by electrons has been observed by a quantum point contact (Molenkamp *et al.* 1992). This result implies that the electrons yield the same universal thermal conductance κ_0 as phonons in eqn (6.3), even though electrons obey different statistics. In general, the quantum of thermal conductance should be a universal value, independent of particle statistics (Rego *et al.* 1999). This is a unique property of quantum transport in one dimension.

6.3.3   Numerical calculation of the phonon-transmission function

The Landauer formula in eqn (6.22) tells us that the thermal conductance is determined by the phonon-transmission function $T(\omega)$. Except for the ideal case of perfect transmission for all modes, it is difficult to calculate analytically the transmission $T(\omega)$ for nanoscale objects with complex atomic structures. There are several alternative numerical techniques for calculating $T(\omega)$ for nanoscale objects, for example, the Green's function method (Mingo 2006; Wang *et al.* 2006; Yamamoto *et al.* 2006; Galperin *et al.* 2007) and the phonon wavepacket scattering method (Schelling *et al.* 2002; Kondo *et al.* 2006). We briefly review these two methods and their applications.

6.3.3.1   *Green's function method*

According to the scattering theory using the Green's function (GF) technique, the total phonon transmission function is expressed as

$$T(\omega) = 4\mathrm{Tr}\left[(\mathrm{Im}\boldsymbol{\Sigma}_L(\omega))\boldsymbol{G}(\omega)(\mathrm{Im}\boldsymbol{\Sigma}_R(\omega))\boldsymbol{G}^\dagger(\omega)\right] \tag{6.26}$$

in terms of the Green's function $\boldsymbol{G}(\omega)$ of the central conductor connected to the left and right leads. Here, the Green's function is given by

$$\boldsymbol{G}(\omega) = \left[\omega^2 \boldsymbol{M} - \boldsymbol{D} - (\boldsymbol{\Sigma}_\mathrm{L}(\omega) + \boldsymbol{\Sigma}_\mathrm{R}(\omega))\right], \tag{6.27}$$

where the boldface quantities represent matrices with a basis in the central region, $\boldsymbol{D}$ is the dynamical matrix of the central region, which is derived from the second derivative of the total energy with respect to the atom co-ordinates in the central region, $\boldsymbol{M}$ is a diagonal matrix with elements corresponding to the masses of the constituent atoms, and $\boldsymbol{\Sigma}_{\mathrm{L/R}}(\omega)$ is the self-energy due to the left/right lead that can be numerically calculated by, for example, the recursion algorithm (Guinea *et al.* 1983) or the mode-matching method (Ando 1991) modified for phonon transport. We refer for further details of the GF method to the literature (Mingo 2006; Wang *et al.* 2006; Yamamoto *et al.* 2006; Galperin *et al.* 2007), and we hereafter discuss the phonon transmissions of carbon nanotubes obtained by the GF method.

Figure 6.4(a) shows the phonon-transmission function of the (8,8) carbon nanotube (CNT) with no defects (dashed curve) and with a single-atom vacancy defect (solid curve) calculated using eqn (6.26). The dashed curve displays a clear stepwise structure that gives the number of phonon modes. In the low-energy region below 2.4 meV that is the excitation gap of the lowest optical phonons, the dashed curve shows $T(\omega) = 4$, indicating the number of acoustic branches corresponding to longitudinal, twisting, and doubly degenerated flexural modes. Reflecting the perfect transmission for all

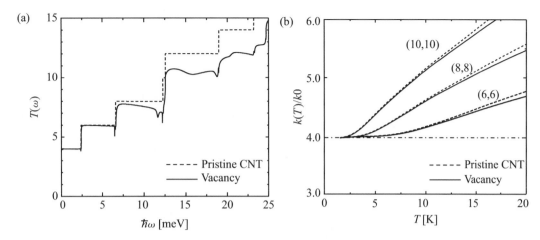

Fig. 6.4 (a) Phonon-transmission function of the (8,8) CNT with no defects (dashed curve) and with a single-atom vacancy defect (solid curve). (b) Low-temperature thermal conductances in (6,6), (8,8), and (10,10) CNTs without defects (dashed curve) and with a single-atom vacancy defect (solid curve). (Yamamoto *et al.* 2006), copyright: APS.

acoustic modes, the thermal conductance $\kappa(T)$ shows four universal quanta, $4\kappa_0 = 4(\pi^2 k_B^2 T/3h)$, in the low-temperature limit (see Fig. 6.4(b)).

The solid curve of $\mathcal{T}(\omega)$ for the (8,8)-CNT with a single-atom vacancy defect is dramatically deformed from the dashed curve for the pristine (8,8) CNT owing to defect scattering, particularly at high energies. However, it remains unchanged in the low-energy region. This is because the long-wavelength acoustic phonons with low energies in the CNTs are not scattered by vacancies. This leads to the important conclusion that the thermal conductance of CNTs exhibits $4\kappa_0$ at cryostatic temperatures even when the CNTs include a small number of defects (see Fig. 6.4(b)). In contrast to the low-temperature region, the vacancy defects decrease the room-temperature thermal conductance dramatically because phonons with short wavelength comparable to the defect size are strongly scattered by vacancies (Yamamoto *et al.* 2006). Recent molecular-dynamics simulation showed that the thermal conductance of an (8,0)-CNT is reduced by half with a vacancy concentration of only 1% (Kondo *et al.* 2006). Since the vacancy in CNTs is energetically metastable, a rebonding of two dangling-bond atoms occurs and only one dangling-bond atom remains (i.e. rearrangement of the single-atom vacancy) as a result of thermal treatments. Thus, the thermal conductance recovers somewhat (Yamamoto *et al.* 2007). However, the room-temperature thermal conductance is not dramatically increased by the structural rearrangement because the stable structure after rearrangement also disrupts the hexagonal network. In view of this, defect-repair techniques other than thermal annealing are required to restore the thermal conductivity by any appreciable amount. High-quality CNTs are desirable to retain the intrinsic high thermal conductance of CNTs.

6.3.3.2 *Phonon wave-packet scattering simulation*

In contrast to the GF method describing steady-state scattering of phonons, the phonon wave-packet scattering (PWS) simulation describes the time evolution

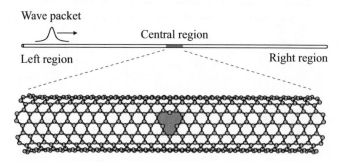

Fig. 6.5 Schematic of the (8,8) carbon nanotube with a single vacancy. The phonon wave packet is shot from the left region far away from the central region. Copyright: JJAP.

of the scattering process of a phonon wave packet (Schelling *et al.* 2002; Kondo *et al.* 2006). The main advantage of PWS simulations is that they allow us to visualize how a wave packet is scattered (i.e. reflected and transmitted) by impurities, defects, and so on.

Here, we demonstrate the PWS simulations by taking a CNT with an atom-vacancy defect as an example (see Fig. 6.5). We first set up a system to investigate phonon wave-packet propagation through defective CNTs. A vacancy defect is placed at the central region of the CNT, as shown in Fig. 6.5. To the left and right of this region, the CNT has perfect periodicity along the tube axis (x-direction). The unit cell of the left/right region is a *cis*-type ring with a primitive length along the tube axis. In the present PWS simulations, the phonon wave packet is injected from the left region rightward.

We next describe how to construct an incident phonon wave packet. In the defect-free left and right regions, the phonon-energy dispersions and the corresponding atomic displacements can be obtained by diagonalizing the dynamical matrix, given by

$$D_{i\alpha}^{j\beta}(k) = \frac{1}{M} \sum_m K_{ni\alpha}^{mj\beta} \exp\left(\mathrm{i}k\left(x_m - x_n\right)\right). \qquad (6.28)$$

Here, the summation index m runs over all unit cells, k is the wave number along the x-direction, x_m is the spatial co-ordinate specifying the mth unit cell, and M is the mass of carbon. $D_{i\alpha}^{j\beta}(k)$ is independent of n owing to the translational symmetry of the left region. The force constant $K_{ni\alpha}^{mj\beta}$ is defined as the second derivative of the potential energy V with respect to atomic co-ordinates:

$$K_{ni\alpha}^{mj\beta} = \frac{\partial^2 V}{\partial r_{ni\alpha} \partial r_{mj\beta}}, \qquad (6.29)$$

where $r_{ni\alpha}$ is the α component ($\alpha = x, y, z$) of the position vector of the ith atom in the unit cell n. The atomic displacement of a wave packet centered at a wave number k_0 in a phonon dispersion ν is given by

$$s_{ni\alpha}^{\nu} = A \frac{u_{i\alpha}^{\nu}(k_0)}{\sqrt{M}} \sum_k \exp\left[-\frac{(k - k_0)^2}{2\sigma^2}\right] \mathrm{e}^{\mathrm{i}(kx_n - \omega_\nu t)}, \qquad (6.30)$$

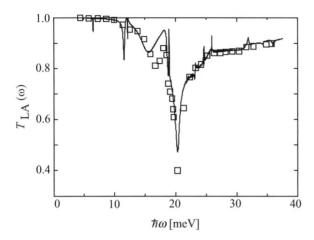

where $u_{i\alpha}^{\nu}(k)$ is a vibrational eigenstate, which is obtained by diagonalizing the dynamical matrix in eqn (6.28), σ is the broadening parameter of a Gaussian function and A is the amplitude of the wave. It is noted that the amplitude A should be small enough in order not to excite anharmonic components of force fields used in simulations.

Figure 6.6 plots the transmission function $\mathcal{T}_{LA}(\omega)$ for the longitudinal acoustic (LA) phonon wave packet in an (8,8) CNT with a single-atom vacancy. The $\mathcal{T}_{LA}(\omega)$ can be obtained by calculating the total energy for atoms in the left and right region after a long-time simulation (see the bottom panel in Figs. 6.7(a) and (b)). Here, the parameters in eqn (6.30) are chosen to be $\sigma\xi/\pi = 0.01$ (ξ is the length of the unit cell in the left/right region) and $A = 0.1$ for the normalized $u_{i\alpha}^{\nu}(k)$. The length of the CNT is $L \sim 160\,\text{nm}$. The rectangles and the solid curve in Fig. 6.6 are the transmission functions obtained by the PWS method and GF method, respectively. Since the PWS results are in

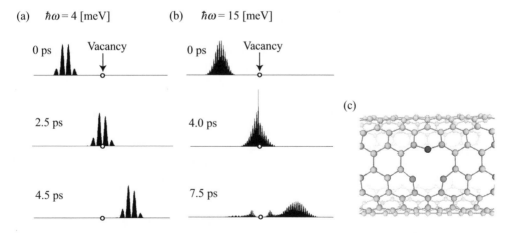

Fig. 6.7 Snapshots of the time evolution of kinetic energy of a wave packet with (a) $\hbar\omega = 4\,\text{meV}$ and (b) $\hbar\omega = 15\,\text{meV}$ in the (8,8) CNT (Kondo *et al.* 2006). (c) The magnitude of atomic vibration for each atom (or the local phonon number) for $\hbar\omega \approx 15\,\text{meV}$ calculated by the GF method (Yamamoto *et al.* 2006). The shading on the atom spheres indicates the magnitude of atomic vibration. Copyright: JJAP.

good agreement with the GF results, we can conclude that the PWS method is adequate for the phonon transport study of nanoscale materials. However, the PWS method has a shortcoming: it cannot capture the sharp peaks and dips in the transmission function that show up in the GF results. This is because the energy resolution of the PWS method is lower than that of the GF method owing to the Gaussian broadening in eqn (6.30). The broadening parameter σ cannot be made arbitrarily small to increase the energy resolution because the length of CNTs used in the simulations is finite.

We now focus on the perfect transmission in the low-energy region below ~ 10 meV in Fig. 6.6. Figure 6.7(a) shows snapshots of the time evolution of the wave packet with $\hbar\omega \sim 4.0$ meV ($k_0 T/\pi = 0.03$). The vertical axis of Fig. 6.7 is the kinetic energy averaged over N carbon atoms in each CNT unit cell, which is given by

$$K_n^{\mathrm{LA}}(t) = \frac{1}{N} \sum_{i=1}^{N} \sum_{\alpha=xyz} \frac{1}{2} M \left(\frac{\mathrm{d}s_{ni\alpha}^{\mathrm{LA}}}{\mathrm{d}t} \right)^2. \tag{6.31}$$

We can see that the wave packet propagates through a vacancy in the CNT without reflection. The transmission is reflectionless because the incident wave packet is composed of phonons with a wavelength much longer than the defect size. For the same reason, the other three acoustic modes (two transverse modes and a twisting mode) are not scattered by the vacancy in the low-energy region. According to the Landauer formula (6.22), the perfect transmission $\zeta(\omega) = 4$ of four acoustic phonons gives rise to the universal quantization of thermal conductance at low temperature. This result coincides with the GF result in Fig. 6.4.

In contrast to the low-energy region below ~ 10 meV in Fig. 6.6, the transmission function in the high-energy region above ~ 10 meV is critically influenced by the defect scattering and has two characteristic dips around ~ 15 and ~ 20 meV. Figure 6.7(b) shows snapshots of $K_n^{\mathrm{LA}}(t)$ with $\hbar\omega \sim 15$ meV ($k_0 \xi/\pi = 0.12$). In contrast to Fig. 6.7(a), the incident wave packet is definitely scattered by the vacancy. Observation of the atomic vibrations in real time reveals that the carbon atoms around the vacancy vibrate more strongly than other atoms as the wave packet collides with the vacancy (Fig. 6.7(c)). The localized phonon state at ~ 15 meV acts as a scatterer of incident phonons. By contrast, for an incident phonon with ~ 20 meV, no atomic vibration localized around the vacancy has been observed in both PWS or GF calculations. The strong reflection is attributed to the fact that the sound velocity of LA phonons with ~ 20 meV is very low due to the flat dispersion in this range of energies.

For further details of thermal transport in carbon nanotubes, see a comprehensive review book on carbon nanotubes (Yamamoto *et al.* 2008).

6.4 Quasi-ballistic phonon transport

If the system length is short in comparison with the mean-free path of phonons, the thermal current and the thermal conductance are expressed in terms of the transmission function $\mathcal{T}_\nu(\omega)$ of phonons through the conductor, as shown in eqn (6.21). Since the transmission function $\mathcal{T}_\nu(\omega)$ for the coherent phonon transport is independent of the system length L, the thermal conductance κ is also constant with respect to L. On increasing temperature, a phonon begins to be scattered by other phonons and electrons, and eventually the mean-free path Λ_ν becomes smaller than L. For this diffusive phonon transport, the thermal conductance is no longer described by the L-independent transmission function $\mathcal{T}_\nu(\omega)$; it is instead given by the well-known Boltzmann–Peierls formula eqn (6.14), which includes the ratio of the mean-free path Λ_ν to the length L. Thus, the thermal conductance varies inversely with L (i.e. $\kappa \propto 1/L$) in the diffusive region. In this section, we address the thermal transport in an intermediate region called the *quasi-ballistic region* between the ballistic and diffusive regions, and give a general expression of the transmission function, $\mathcal{T}_\nu(\omega) = \Lambda_\nu(\omega)/(L + \Lambda_\nu(\omega))$, which interpolates between the ballistic and diffusive cases.

6.4.1 Büttiker's fictitious probe

We provide a simple phenomenological method for including the effects of phonon–phonon scattering in the calculation of thermal transport. The original idea of this method was introduced first by Büttiker to treat the incoherent electronic transport in mesoscopic systems (Bütikker 1988). Here, we attempt to apply his idea to the thermal transport by interacting phonons (Yamamoto *et al.* 2009).

The phonon–phonon scattering processes drive the system toward equilibrium, and eventually the system reaches local thermal equilibrium having a spatially varying temperature. In order to describe such an effect phenomenologically, we introduce *Büttiker's fictitious probe* connected to heat reservoir with temperature T_B, as shown in Fig. 6.8. Büttiker's fictitious probe extracts a fraction of the phonons from the conductor and reinjects them after thermalization, thus effectively it acts as phonon–phonon scattering. The phonons travelling from the left to the right leads via the probe are the incoherent current component. In contrast to incoherent phonons, the remaining coherent phonons propagate from the left to right leads, passing through without

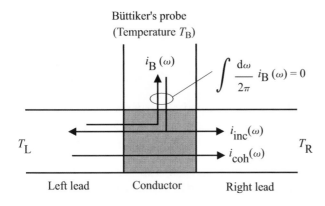

Fig. 6.8 Büttiker's fictitious probe to describe the phonon–phonon scattering effect. The net thermal current through Büttiker's probe should be zero.

entering the probe. Thus, the thermal current flowing through the left lead, $i_L(\omega) = i_L^{coh}(\omega) + i_L^{inc}(\omega)$, consists of the coherent and incoherent components, where the coherent component is given by

$$i_L^{coh}(\omega) = \hbar\omega \left[f(\omega, T_L) - f(\omega, T_R) \right] \mathcal{T}_{LR}(\omega) \qquad (6.32)$$

in terms of the transmission function $\mathcal{T}_{LR}(\omega)$ between the left and right leads, and the incoherent component is expressed by

$$i_L^{inc}(\omega) = \hbar\omega \left[f(\omega, T_L) - f(\omega, T_B) \right] \mathcal{T}_{LB}(\omega) \qquad (6.33)$$

in terms of the transmission function $\mathcal{T}_{LB}(\omega)$ between the left lead and the probe. The thermal current density $i_B(\omega)$ in the probe is also described by

$$i_B(\omega) = \hbar\omega \sum_{q=L,R} \left[f(\omega, T_B) - f(\omega, T_q) \right] \mathcal{T}_{Bq}(\omega). \qquad (6.34)$$

The current density $i_B(\omega)$ describes the energy flow referred to as *vertical heat flow* from one phonon channel to another due to the phonon–phonon scattering. Since Büttiker's probe is a conceptual probe that is introduced to express the phonon–phonon scattering effectively, the net thermal current integrated over all frequencies ω should be zero:

$$\int \frac{d\omega}{2\pi} i_B(\omega) = 0. \qquad (6.35)$$

In other words, the temperature T_B at the heat reservoir attached to Büttiker's probe is determined by the condition (6.35) that the net thermal current does not flow into the probe. Substituting eqn (6.34) into eqn (6.33), we can eliminate the Bose–Einstein distribution function $f(\omega, T_B)$ of the heat reservoir connected to Büttiker's probe from the expression of the incoherent current $i_{inc}(\omega)$ in eqn (6.33). Consequently, the thermal current density $i_L(\omega)$ is expressed as

$$i_L(\omega) = i_L^{coh}(\omega) + i_L^{inc}(\omega)$$

$$= \hbar\omega \left[f(\omega, T_L) - f(\omega, T_R) \right] \tilde{\mathcal{T}}_{LR}(\omega) - \frac{i_B(\omega)\mathcal{T}_{LB}(\omega)}{\mathcal{T}_{BL}(\omega) + \mathcal{T}_{BR}(\omega)}, \qquad (6.36)$$

where $\tilde{\mathcal{T}}_{LR}(\omega)$ is an effective transmission function of a phonon through a conductor including the phonon–phonon scattering, which is given by

$$\tilde{\mathcal{T}}_{LR}(\omega) = \mathcal{T}_{LR}(\omega) + \frac{\mathcal{T}_{LB}(\omega)\mathcal{T}_{BR}(\omega)}{\mathcal{T}_{BL}(\omega) + \mathcal{T}_{BR}(\omega)}. \qquad (6.37)$$

In the limit of coherent transport without phonon–phonon scattering, i.e. $\mathcal{T}_{BL}(\omega) = \mathcal{T}_{BR}(\omega) = 0$, the effective transmission reduces to the transmission function of coherent phonons $\tilde{\mathcal{T}}_{LR}(\omega) = \mathcal{T}_{LR}(\omega)$.

If we assume that the vertical flow is zero for each frequency ω, i.e. $i_B(\omega) = 0$, the second term of the right-hand side in eqn (6.36) can be dropped, and the net thermal current is given by exactly the same form as the expression of thermal current by coherent phonons,

$$I_L = \int_0^\infty \frac{d\omega}{2\pi} i_L(\omega)$$

$$= \int_0^\infty \frac{d\omega}{2\pi} \hbar\omega \left[f(\omega, T_L) - f(\omega, T_R) \right] \tilde{\mathcal{T}}_{LR}(\omega). \qquad (6.38)$$

Similarly, we can derive the thermal current I_R at the right lead by replacing the suffix L by R in eqn (6.38), to obtain $\tilde{\mathcal{T}}_{LR}(\omega) = \tilde{\mathcal{T}}_{RL}(\omega)$, according to Kirchhoff's law $I_L = -I_R$ for thermal current.

Even though the assumption of vanishing vertical current for each frequency seems to be rough, it actually performs quite well. In the next subsection, we will show an example of the successful application of Buttiker's probe idea under the present assumption.

6.4.2 Length dependence of the thermal conductance

As a typical application of Büttiker's fictitious probe, we discuss the system length dependence of thermal conductance arising from the phonon–phonon scatterings (Yamamoto *et al.* 2009). The thermal conductance for the purely ballistic phonon systems is constant with respect to the system length L, while that for diffusive phonon systems varies inversely with L. In this subsection, we address the cross-over of the thermal-transport behavior from the ballistic to the diffusive regime as the system length increases. The bottom line in this section is that the thermal conductance for the quasi-ballistic regions exhibits non-linear system-length dependence in the low-frequency region. In particular, the contribution of acoustic phonons to the thermal conductance displays a power-law behavior with respect to the system length.

As a first step to show this, let us consider a system where two conductors are connected in series as depicted in Fig. 6.9. Two conductors are attached to Büttiker's probes having effective transmission functions $\tilde{\mathcal{T}}_{v,1}(\omega)$ and $\tilde{\mathcal{T}}_{v,2}(\omega)$ as given by eqn (6.37). For the sake of simplicity, we drop the suffix v and the argument ω from $\tilde{\mathcal{T}}_{v,i}(\omega)$ $(i = 1, 2)$ hereafter. The total transmission function $\tilde{\mathcal{T}}_{12}$ is obtained by summing the transmission functions of all the multiple-scattering processes as shown in Fig. 6.9:

$$\tilde{\mathcal{T}}_{12} = \tilde{\mathcal{T}}_1\tilde{\mathcal{T}}_2 + \tilde{\mathcal{T}}_1\tilde{\mathcal{T}}_2\tilde{\mathcal{R}}_1\tilde{\mathcal{R}}_2 + \tilde{\mathcal{T}}_1\tilde{\mathcal{T}}_2\tilde{\mathcal{R}}_1^2\tilde{\mathcal{R}}_2^2 + \cdots = \tilde{\mathcal{T}}_1\tilde{\mathcal{T}}_2 \sum_{n=0}^\infty (\tilde{\mathcal{R}}_1\tilde{\mathcal{R}}_2)^n$$

$$= \frac{\tilde{\mathcal{T}}_1\tilde{\mathcal{T}}_2}{1 - \tilde{\mathcal{R}}_1\tilde{\mathcal{R}}_2} = \frac{\tilde{\mathcal{T}}_1\tilde{\mathcal{T}}_2}{\tilde{\mathcal{T}}_1 + \tilde{\mathcal{T}}_2 - \tilde{\mathcal{T}}_1\tilde{\mathcal{T}}_2}, \qquad (6.39)$$

so that

$$\frac{1}{\tilde{\mathcal{T}}_{12}} = \frac{1}{\tilde{\mathcal{T}}_1} + \frac{1}{\tilde{\mathcal{T}}_2} - 1 \quad \Rightarrow \quad \frac{1 - \tilde{\mathcal{T}}_{12}}{\tilde{\mathcal{T}}_{12}} = \frac{1 - \tilde{\mathcal{T}}_1}{\tilde{\mathcal{T}}_1} + \frac{1 - \tilde{\mathcal{T}}_2}{\tilde{\mathcal{T}}_2}, \qquad (6.40)$$

where $\tilde{\mathcal{R}}_i = 1 - \tilde{\mathcal{T}}_i$ $(i = 1, 2)$ is the reflection function of a phonon through the ith conductor. From eqn (6.40), we can understand that the transmission

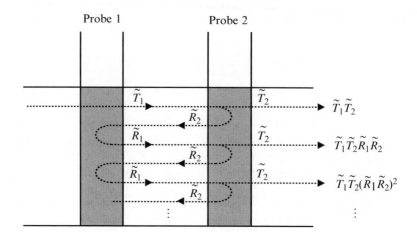

Fig. 6.9 Multiple-scattering processes in between two Büttiker fictitious probes connected in series, having phonon transmission (reflection) $\tilde{\mathcal{T}}_1$ $(\tilde{\mathcal{R}}_1)$ and $\tilde{\mathcal{T}}_2$ $(\tilde{\mathcal{R}}_2)$, respectively.

itself is not an additive quantity when two conductors are placed in series, while that $(1 - \tilde{\mathcal{T}}_i)/\tilde{\mathcal{T}}_i$ is additive. A similar discussion for electronic transport can be found in a book by Datta (1995).

Next, let us consider the total transmission function $\tilde{\mathcal{T}}^{\text{tot}}$ for the N conductors with Büttiker probes arranged in series. Using the additive property of eqn (6.40), the transmission function $\tilde{\mathcal{T}}^{\text{tot}}$ is given by

$$\frac{1 - \tilde{\mathcal{T}}^{\text{tot}}}{\tilde{\mathcal{T}}^{\text{tot}}} = N \frac{1 - \tilde{\mathcal{T}}}{\tilde{\mathcal{T}}}, \tag{6.41}$$

if each conductor has the same transmission function $\mathcal{T}$. As a result, the transmission function $\tilde{\mathcal{T}}^{\text{tot}}(\omega)$ for a conductor of length L can be written as

$$\tilde{\mathcal{T}}_\nu^{\text{tot}}(\omega) = \frac{\mathcal{L}_\nu(\omega)}{L + \mathcal{L}_\nu(\omega)} \sim \frac{\Lambda_\nu(\omega)}{L + \Lambda_\nu(\omega)}, \tag{6.42}$$

where $\mathcal{L}_\nu(\omega) \equiv \tilde{\mathcal{T}}_\nu(\omega)/\rho(1 - \tilde{\mathcal{T}}_\nu(\omega))$ is a characteristic length, which is of the order of a mean free path $\Lambda_\nu(\omega)$ (see Appendix), and $\rho = N/L$ is the density of scatterers in a conductor of length L. Thus, the thermal conductance is given by

$$\kappa = \sum_\nu \int_{\omega_\nu^{\text{min}}}^{\omega_\nu^{\text{max}}} \frac{d\omega}{2\pi} \hbar\omega \left[\frac{\partial f(\omega, T)}{\partial T} \right] \frac{\Lambda_\nu(\omega)}{L + \Lambda_\nu(\omega)}. \tag{6.43}$$

For a short system $L \ll \Lambda_\nu(\omega)$, eqn (6.43) reproduces the Landauer formula in eqn (6.22) with the perfect transmission $\mathcal{T}_\nu(\omega) = 1$. For a long system $L \gg \Lambda_\nu(\omega)$, it reduces to the Boltzmann–Peierls formula in eqn (6.14). Thus, eqn (6.43) interpolates the thermal conductance behaviors in the intermediate region between the ballistic and the diffusive regime (Wang *et al.* 2006; Yamamoto *et al.* 2009).

Let us now address the length dependence of the thermal conductance in the quasi-ballistic regime. In order to calculate thermal conductance from eqn (6.43), we need to know the mean-free path $\Lambda_\nu(\omega)$ as a function of frequency. In many cases, the frequency dependence of the mean-free path is given by $\Lambda_\nu(\omega) = a_\nu \omega^{-r_\nu}$. At the low frequencies of interest here, we can take $\hbar\omega[\partial f(\omega, T)/\partial T] \approx k_B$ out of the integral in eqn (6.43). Thus, the low-frequency thermal conductance is given by

$$\kappa \approx \frac{k_B}{2\pi} \sum_\nu \left(\frac{a_\nu}{L}\right)^{1/r_\nu} \int_{x_\nu^{min}(L)}^{x_\nu^{max}(L)} \frac{1}{x^{r_\nu} + 1} dx, \qquad (6.44)$$

where the lower/upper bound of the integral $x_\nu^{min/max}(L) = \omega_\nu^{min/max}(L/a_\nu)^{1/r_\nu}$ depends on the system length L. The contribution of optical phonons to κ is much smaller than that of acoustic phonons for a quasi-one-dimensional system with a small diameter. For acoustic phonon modes especially, the integral in eqn (6.44) becomes independent of L because $x_\nu^{min} = 0$ and $x_\nu^{max} \to \infty$, and eventually the thermal conductance decreases with L as a power law:

$$\kappa \approx \sum_{\nu=acoustic} C_\nu L^{-1/r_\nu}, \qquad (6.45)$$

and the thermal resistance increases non-linearly as $R_{th} \equiv 1/\kappa \approx 1/\sum_\nu C_\nu L^{-1/r_\nu}$, which is different from the Boltzmann–Peierls formula $R_{th} \propto L$. This result also demonstrates that the thermal conductivity $\lambda = (L/S)\kappa$ diverges as $\lambda \approx \sum_\nu C_\nu L^{(r_\nu-1)/r_\nu} \to \infty$ in the thermodynamic limit $L \to \infty$. This is closely related to the old problem pointed out by Pomeranchuk in the 1940s that the thermal conductivity diverges at low frequency for acoustic phonons, unless one considers higher-order interactions (Pomeranchuk 1941, 1942).

In fact, the power-law behavior of thermal conductivity in the quasi-ballistic regime has been reported by molecular-dynamics simulations on carbon nanotubes (Maruyama 2002; Shiomi *et al.* 2008), and has also been observed in recent experiments (Wang *et al.* 2007).

6.5 Conclusions

The thermal transport of small systems, focusing on quasi-one-dimensional systems like carbon nanotubes, has been reviewed from the standpoint of the phonon-conduction mechanisms on the basis of the thermal conductance. For a small system whose dimension is smaller than the mean-free path of phonons, the thermal-transport properties exhibit remarkable quantum-mechanical features and strong non-linear behaviors. They cannot be explained by the famous Fourier's law.

In this chapter, we discussed novel theoretical techniques to treat quantum thermal transport in small systems. For coherent phonon transport, we introduced the Landauer formulation combined with the Green's function method and the phonon-wave-packet scattering method. On the other hand, for the incoherent (or diffusive) phonon transport, we give a simple phenomenological technique to deal effectively with phonon–phonon scattering, referred to as Büttiker's fictitious probe method. As interesting applications of these theoretical techniques, we reviewed the quantization of thermal conductance by ballistic phonons and the non-linear length dependence of thermal conductance in quasi-ballistic phonon transport, respectively.

At extremely low temperatures where the optical phonons are not excited, the thermal current is carried ballistically by acoustic phonons without any scattering, and the thermal conductance is quantized in multiples of the universal value, $\pi^2 k_{\mathrm{B}}^2 T/3h$, which can be understood within Landauer's formula of coherent phonon transport. The quantization of thermal conductance by ballistic phonons has been experimentally observed in SiN nanowires and carbon nanotubes. The quantization phenomenon is a universal phenomenon independent of particle statistics such as the Fermion or Boson statistics. Therefore, the electrons' contribution to thermal conductance is quantized in the same universal way as that of phonons.

At intermediate temperatures where ballistic and diffusive phonons coexist, the thermal resistance is not proportional to the system length L, but increases non-linearly with the length. In particular, at low frequency where the acoustic phonons give the major contribution to the thermal transport, the thermal conductance exhibits a power-law behavior with respect to L.

The thermal transport at the nanoscale is a new field of materials science and technology. The recent progress in the field makes it possible to develop novel thermal devices such as a thermal rectifier that lets the thermal current flow more easily in one direction than in the other (Chang *et al.* 2006). However, many characteristic features remain to be elucidated. To design practical devices like a thermal rectifier, it is essential to gain an understanding of the underlying mechanisms. The authors hope that this chapter will serve as a useful guideline for students and researchers who enter into this new and challenging field.

Acknowledgments

The authors acknowledge partial financial support from the Japan Science and Technology Agency (CREST-JST). T.Y. and K.W. also acknowledge support from the Ministry of Education, Culture, Sports, Science and Technology of Japan through Grants-in-Aids (No. 19710083, No. 20048008 and No. 19540411) and through Holistic Computational Science (HOLCS) of the Tokyo University of Science. The authors would like to thank Eduardo R. Hernández and Satoru Konabe for reading the manuscript and making a number of helpful suggestions and comments.

Appendix: Derivation of eqn (6.42)

In this Appendix, we explain the characteristic length $\mathcal{L}_\nu(\omega) = \tilde{T}_\nu(\omega)/\rho(1 - \tilde{T}_\nu(\omega))$ in eqn (6.42) is of the order of the phonon mean-free path $\Lambda_\nu(\omega) = \tau_\nu(\omega)|v_\nu(\omega)|$, where $1/\tau_\nu(\omega)$ and $v_\nu(\omega)$ are a backscattering rate and a group velocity of a phonon labelled by mode ν and frequency ω, respectively. The effect of phonon–phonon scattering on the thermal transport can be described phenomenologically by Büttiker's fictitious probes, as discussed in Section 6.4. Let us consider the N probes arranged in series. The distance between the neighboring probes is given by $dL \equiv L/N = 1/\rho$, where ρ is the density of probes in the system. For propagation over a small distance dL, the reflection probability $\tilde{\mathcal{R}}_\nu(\omega) = 1 - \tilde{T}_\nu(\omega)$ is much smaller than unity (i.e. $T_\nu(\omega) \approx 1$) and is given by

$$1 - \tilde{T}_\nu(\omega) = \frac{1}{\tau_\nu(\omega)} \frac{dL}{|v_\nu(\omega)|} = \frac{1}{\rho \Lambda_\nu(\omega)}.$$

Thus, the phonon mean-free path $\Lambda_\nu(\omega)$ is expressed as

$$\Lambda_\nu(\omega) = \frac{1}{\rho(1 - \tilde{T}_\nu(\omega))},$$

and we found $\mathcal{L}_\nu(\omega) \sim \Lambda_\nu(\omega)$ if $\tilde{T}_\nu(\omega) \approx 1$ for a small distance dL between the neighboring probes.

References

Ando, T. *Phys. Rev. B* **44**, 8017 (1991).

Angelescu, D.E., Cross, M.C., Roukes, M.L. *Superlatt. Microstruct.* **23**, 673 (1998).

Blencowe, M.P. *Phys. Rev. B* **59**, 4992 (1999).

Bütikker, M. *IBM J. Res. Dev.* **32**, 63 (1988).

Chang, C.W., Okawa, D., Majumdar, A., Zettl, A. *Science* **314**, 1121 (2006).

Chiu, H.-Y., Deshpande, V.V., Postma, H.W., Lau, C.N., Miko, C., Forro, L., Bockrath, M. *Phys. Rev. Lett.* **95**, 226101 (2005).

Datta, S. *Electronic Transport in Mesoscopic Systems* (Cambridge University Press, Cambridge, 1995).

Galperin, M., Nitzan, A., Ratner, M.A. *Phys. Rev. B* **75**, 155312 (2007).

Guinea, F., Tejedor, C., Flores, F., Louis, E. *Phys. Rev. B* **28**, 4397 (1983).

Kondo, N., Yamamoto, T., Watanabe, K. *Jpn. J. Appl. Phys.* **45**, L963 (2006).

Kondo, N. e-J. *Surf. Sci. Nanotech.* **4**, 239 (2006).

Maruyama, S. *Physica B* **323**, 193 (2002).

Mingo, N. *Phys. Rev. B* **74**, 125402 (2006).

Molenkamp, L.W., Gravier, Th., van Houten, H., Buijk, O.J.A., Mabesoone, M.A.A. *Phys. Rev. Lett.* **68**, 3765 (1992).

Moore, G.E. *Electronics* **38**, 114 (1965).

Peierls, R.E. *Ann. Physik* **3**, 1055 (1929).

Peierls, R.E. *Quantum Theory of Solids* (Oxford University Press, New York, 1955).

Pomeranchuk, I. *J. Phys. (USSR)* **4**, 259 (1941).

Pomeranchuk, I. *J. Phys. (USSR)* **6**, 237 (1942).

Rego, L.G.C., Kirczenow, G. *Phys. Rev. Lett.* **81**, 232 (1998).

Rego, L.G.C. *Phys. Rev. B* **59**, 13080 (1999).

Schelling, P.K., Phillpot, S.R., Keblinski, P. *Appl. Phys. Lett.* **80**, 2484 (2002).

Schwab, K., Henriksen, E.A., Worlock, J.M., Roukes, M.L. *Nature* **404**, 974 (2000).

Shiomi, J., Maruyama, S. *Jpn. J. Appl. Phys.* **47**, 2005 (2008).

Yamamoto, T., Watanabe, S., Watanabe, K. *Phys. Rev. Lett.* **92**, 075502 (2004).

Yamamoto, T., Watanabe, K. *Phys. Rev. Lett.* **96**, 255503 (2006).

Yamamoto, T., Nakazawa, Y., Watanabe, K. *New J. Phys.* **9**, 245 (2007).

Yamamoto, T., Watanabe, K., Hernández, E.R. *Carbon Nanotubes: Advanced Topics in the Synthesis, Structure, Properties and Applications*, (eds) Jorio, A., Dresselhaus, M.S. and Dresselhaus, G., Topics in Applied Physics 111 (Springer-Verlag, Berlin, 2008) p. 165.

Yamamoto, T., Konabe, S., Shiomi, J., Maruyama, S. *Appl. Phys. Exp.* **2**, 095003 (2009).

Wang, J.-S., Wang, J., Zeng, N. *Phys. Rev. B* **74**, 033408 (2006).

Wang, J., Wang, J.-S. *Appl. Phys. Lett.* **88**, 111909 (2006).

Wang, Z.L., Tang, D.W., Li, X.B., Zheng, X.H., Zhang, W.G., Zheng, L.X., Zhu, Y.T. *Appl. Phys. Lett.* **91**, 123119 (2007).

Zimann, J.M. *Electron and Phonons* (Clarendon Press, Oxford, UK, 1960).

Atomistic spin-dynamics

M. Stamenova and S. Sanvito

7.1 Introduction

The spectacularly rapid development of the data-storage industry, followed from the discovery of the giant magnetoresistance effect (Baibich *et al.* 1988; Binasch *et al.* 1989), has no precedents in the history of technology. In less than ten years what was a laboratory curiosity has impacted over the mass production in such a dramatic way so as to increase the slope of the curve, describing the magnetic storage density as a function of time, and to exceed that of the Moore's law, established in the semiconductor industry. Intrinsic to this revolution is the full scalability of magnetism, whose ultimate boundary for applications is currently set by the paramagnetic limit, i.e. essentially by the magnetic anisotropy that one can engineer in a nanomagnet.

However, for many applications away from magnetic data recording, and in particular for non-volatile magnetic random access memories (MRAM), a second and important limitation appears. This is connected to the ability of achieving accurate and fast magnetic switching. In the most conventional MRAM design, switching is obtained with strong stray fields produced in a special "writing" circuit by large currents (see, for instance, Engel *et al.* 2002). Since the stray-field intensity depends on the total current and not on the current density, such a technology is not fully scalable. Dissipation increases with the current intensity and ultimately the electron-migration threshold establishes the final limit for the mechanical stability of the device. There is therefore a great demand for a novel, fully scalable, magnetic switching technology.

In the mid-1990s Slonczewski (1996) and Berger (1996) predicted the possibility of inducing spin-dynamics by means of spin-polarized current. This effectively was the prediction of the "inverse" GMR effect, i.e. the prediction that a spin-polarized current could change the magnetic state of a device. The essential idea is that a spin-polarized current can transfer angular momentum to the magnetization of a magnetic system, thus generating a torque. In the right conditions this current-induced torque can balance or even surpass the magnetic Gilbert damping, thus creating new dynamical solutions to the equations of motions for the magnetization. Magnetic switching, magnetic resonator and enhanced Gilbert damping can all be generated by a spin-polarized current. Importantly, the current-induced torque does depend on the current density and not on the total current. This clearly opens new prospects for switching MRAM, and largely justifies the growing interest in this area of research.

Theory and modelling of current-induced magnetic phenomena is largely based on solving the Landau–Lifshitz–Gilbert equations with additional terms describing the current-induced torques. At a more advanced level, transport theory for diffusive transport is also introduced in the description, effectively creating a sort of Kirchkoff magneto-circuitry theory (Brataas *et al.* 2006). A much less explored area is that of atomistic simulations of magnetodynamics. These, however, are expected to occupy an increasingly important place in theoretical magnetism, since nanoscale and even monoatomic one-dimensional (Gambardella 2003) magnetic devices are already available and examples of atomic-scale magnetic phenomena, such as transport in magnetic point contacts (Viret *et al.* 2002) and ultrathin domain walls (Pratzer *et al.* 2001), have been already demonstrated.

To date there is a notable and important attempt towards the atomistic calculation of quantities related to spin-dynamics. This is represented by the extension of density-functional theory (DFT) to time-dependent phenomena. Within this framework the theoretical foundation for time-dependent DFT for spin-dynamics was laid several years ago (Liu *et al.* 1989; Capelle *et al.* 2001) and practical calculations based on the adiabatic approximation have been rather successful in describing dynamical properties of magnetic transition metals, although in absence of an electron current (Antropov *et al.* 1995, 1996). Currents were introduced only recently and simulations for open systems, carried out by using non-equilibrium transport theory combined with DFT, are now available (Haney *et al.* 2007).

Importantly, all the calculations to date based on the atomistic evaluation of the spin-torque do not perform real molecular dynamics in the presence of an evolving current. Haney *et al.* (2007) for instance calculate the torques acting on the free layer of a spin-valve, but the magnetization itself is not relaxed. Since the conducting state of a device can be seriously affected by its magnetic configuration, particularly at the nanoscale, this is an important limitation for realistic magneto-device simulations. Note that there are no fundamental obstacles to molecular dynamics involving spins. Ultrafast spin-switching in the picosecond range has been demonstrated, indicating that the fastest timescale of atomic spin-dynamics is indeed in (or below) the ps range (Gerrits *et al.* 2002). Since the typical timescale for electronic processes is in the femtosecond range, one needs between 10^3 to 10^5 time steps to evolve the electronic structure to times relevant for the spin-dynamics. This is well in reach of state-of-the-art time-integration techniques.

This chapter aims at presenting recent advances towards the development of a truly atomistic time-dependent theory for spin-dynamics. The *s*-*d* tight-binding model including electrostatic corrections at the Hartree level will be our underlying electronic structure theory and whenever available we will make contact with other theoretical approaches. In particular, we will be focusing on introducing the main theoretical concepts behind our approach

and providing a range of examples where such a scheme offers insights well
beyond what is achievable by standard static theory. These include the investigation of the spin-wave dispersion in nanoscale magnets, spin-spin correlation between magnetic impurities in non-magnetic nanowires, current-induced
domain-wall motion, and the generation of an electromotive force as a result
of domain-wall precession.

7.2   Model spin Hamiltonian

Throughout this chapter we will always consider one-dimensional (1D) magnetic atomic wires and we will describe their electronic structure by means of
the *s-d* model (Yosida 1996), where conduction electrons (*s*) are exchange-
coupled to a number of classical spins $\mathbf{S}_i$ (*d*). When written in a tight-binding
form, the electronic Hamiltonian reads

$$\hat{H}_{\mathrm{e}}(t) = \sum_{i,j,\alpha} H_{ij}^{\mathrm{TB}} c_i^{\alpha\dagger} c_j^{\alpha} - \sum_{i,\alpha,\beta} c_i^{\alpha\dagger} \hat{\sigma}_{\alpha\beta} c_i^{\beta} \cdot \mathbf{\Phi}_i(t), \qquad (7.1)$$

where $c_i^{\alpha\dagger}$ $\left(c_i^{\alpha}\right)$ is the creation (annihilation) operator for an electron with spin
$\pm 1/2(\alpha = 1, 2)$ at the atomic site *i* and $\hat{\boldsymbol{\sigma}} = 1/2(\tilde{\sigma}^x, \tilde{\sigma}^y, \tilde{\sigma}^z)$ is the electron
spin operator, $\{\tilde{\sigma}^n\}$ being the set of Pauli matrices.[1]

The first term in eqn (7.1) is the spin- and time-independent tight-binding
(TB) part, while the second describes the spin interaction with a time-
dependent effective local field $\mathbf{\Phi}_i(t)$,

$$H_{ij}^{\mathrm{TB}} = \left(\mathcal{E}_0 + \sum_n \frac{\kappa \Delta q_n}{\sqrt{R_{in}^2 + (\kappa/U)^2}} \right) \delta_{ij} + \chi \delta_{i,j\pm 1} \qquad (7.2)$$

$$\mathbf{\Phi}_i(t) = J\mathbf{S}_i(t) + g_{\mathrm{e}}\mu_{\mathrm{B}}\mathbf{B}(t). \qquad (7.3)$$

Here, $\mathcal{E}_0$ is the on-site energy (identical for all sites), $\kappa = e^2/4\pi\varepsilon_0 =$
$14.4\,\mathrm{eV\,\AA}$ (or $\kappa = 0$ for describing non-interacting electrons), χ is the hopping
parameter, g_{e} is the electron *g*-factor and $\mathbf{B}(t)$ is the external magnetic field
(in general, time-dependent). The second term in brackets in eqn (7.2) is
a mean-field repulsive electrostatic potential with on-site strength *U* and a
Coulombic decay at large intersite distances R_{ij}. Finally, $\Delta q_i = q_i - q_i^{(0)}$ is
the excess number of electrons on site *i*, $q_i^{(0)}$ being the free-atom valence
electron-number. The total number of electrons N_{e} in the system is a model
parameter

$$N_{\mathrm{e}} = \sum_i q_i^{(0)} = N\rho_0, \qquad (7.4)$$

where *N* is the total number of sites and ρ_0 the band filling, i.e. the average
number of electrons per site.

In eqn (7.3) $\mathbf{S}_i(t)$ is the unit vector in the direction of the local spin at site *i*
at time *t*. $\{\mathbf{S}_i\}$ are treated as classical variables, nonetheless exchange-coupled
with strength $J > 0$ to the instantaneous conduction electron spin density and

[1]For a right-handed co-ordinate system (for
which $\mathrm{e}^x \times \mathrm{e}^y = \mathrm{e}^z$ is fulfilled for the basis
vectors), the basis of the Pauli matrix repre-
sentation reads

$$\tilde{\sigma}^x = \begin{pmatrix} 0 & 1 \\ 1 & 0 \end{pmatrix}, \quad \tilde{\sigma}^y = \begin{pmatrix} 0 & -i \\ i & 0 \end{pmatrix},$$

$$\tilde{\sigma}^z = \begin{pmatrix} 1 & 0 \\ 0 & -1 \end{pmatrix}.$$

to $\mathbf{B}(t)$ according to a classical Hamiltonian

$$H_S(t) = -\sum_i \mathbf{S}_i(t) \cdot \Theta_i(t) - \frac{1}{2} J_S \sum_{i,j \in \mathrm{nn}[i]} \mathbf{S}_i(t) \cdot \mathbf{S}_j(t) - J_z \sum_i \left(\mathbf{S}_i(t) \cdot \hat{\mathbf{z}} \right)^2,$$

(7.5)

where

$$\Theta_i(t) \equiv J \langle \hat{\boldsymbol{\sigma}} \rangle_i(t) + \frac{g_S \mu_B}{\hbar} S \mathbf{B}(t) \qquad (7.6)$$

is the effective field for the classical spins, analogous to Φ_i. In eqn (7.5) g_S is the g-factor of $\mathbf{S}_i$, which could be of a mixed spin and orbital origin, J_S is the direct intersite exchange coupling parameter (only first nearest neighbors, $\mathrm{nn}[i]$, are considered), $\hat{\mathbf{z}}(|\hat{\mathbf{z}}| = 1)$ is the unit vector and J_z the anisotropy constant along the easy z-axis. Note that the actual norm S of the classical spins is incorporated in the definition of the exchange parameters. Unless stated otherwise, we consider $S = \hbar$. Appearing in eqn (7.6) is also the expectation value of the electron spin at site i, $\langle \hat{\boldsymbol{\sigma}} \rangle_i(t) = \mathrm{Tr}[\rho_{ii}(t) \hat{\boldsymbol{\sigma}}]$, where $\rho_{ii} = \langle i | \hat{\rho} | i \rangle$ is the ith diagonal element of the density matrix $\hat{\rho}$ in the real-space orthonormal basis $\{|i\rangle\}$ (in a tight-binding fashion, $|i\rangle$ represents one electron s-orbital at site i) and the trace is over the spin coordinates. At $t = 0$ the density matrix is constructed as

$$\hat{\rho}(0) = \sum_n f_F(\epsilon_n - E_F) |n\rangle \langle n|, \qquad (7.7)$$

where $\{\epsilon_n, |n\rangle\}$ is the set of eigenvalues and eigenvectors of $\hat{H}_e(t = 0)$, f_F is the Fermi distribution and the Fermi level E_F is determined, so that $\sum_i \rho_{ii} = \sum_i \langle i | \hat{\rho} | i \rangle = N_e$. In the case of interacting electrons [$\kappa \neq 0$ in eqn (7.3)] $\hat{H}_e(t = 0, \hat{\rho})$ is determined self-consistently.

The corresponding quantum and classical Liouville equations of motion for the two subsystems are

$$\frac{d\hat{\rho}}{dt} = \frac{i}{\hbar} \left[\hat{\rho}, \hat{H}_e \right], \quad \frac{d\mathbf{S}_i}{dt} = \{\mathbf{S}_i, H_S\}, \qquad (7.8)$$

where $\{,\}$ represents the classical Poisson bracket and $[\ ,\ ]$ the quantum-mechanical commutator. In order to calculate the right-hand side of the classical Liouville equation we need the expression for the Poisson bracket of the classical spins. As the classical spins are essentially angular momenta they obey the same relation[2] as do the classical angular momenta (Yang *et al.* 1980), i.e.

$$\{S_i^n, S_i^m\} = \frac{1}{S} \varepsilon^{nmk} S_i^k, \qquad (7.9)$$

where $S_i^n = \mathbf{S}_i \cdot \hat{\mathbf{e}}^n$ is the nth Cartesian projection $(\{\hat{\mathbf{e}}^n\}_{n=1}^3 \equiv \{\hat{\mathbf{x}}, \hat{\mathbf{y}}, \hat{\mathbf{z}}\})$ of the spin on site i, ε^{nmk} is the Levi–Civita tensor and we assume summation over repeated indices (Einstein notation). Note that in eqn (7.9) we have factorized the spin magnitude S in order to obtain an expression involving only components of the dimensionless spin unit-vector $\mathbf{S}_i$. We thus obtain the following

[2] The Poisson bracket of two components of an angular momentum is $\{L^i, L^j\} = \{\varepsilon^{iab} x^a p^b, \varepsilon^{jcd} x^c p^d\} = \varepsilon^{iab} \varepsilon^{jca} (x^c p^b - x^b p^c) = x^i p^j - x^j p^i = \varepsilon^{ijk} L^k$, where we have used the Einstein notation for repeated indices, the Levi–Civita tensor contraction identity $\varepsilon^{iab} \varepsilon^{icd} = \delta^{ac} \delta^{bd} - \delta^{ad} \delta^{bc}$ and the canonical variables Poisson brackets $\{x^i, p^j\} = \delta^{ij}$. Note that upper indices for Cartesian components (here and throughout this chapter) are only used for aesthetic reasons as lower indices have already been used for designating site positions.

terms in the classical equation of motion

$$\{S_i, \mathbf{S}_j \cdot \Theta_j\}^n = \{S_i^n, S_i^m\} \, \Theta_i^m = \frac{1}{S} \varepsilon^{nmk} S_i^k \Theta_i^m$$

$$= -\frac{1}{S}(\mathbf{S}_i \times \Theta_i)^n \tag{7.10}$$

$$\sum_j \left\{ \mathbf{S}_i, \mathbf{S}_j \cdot \sum_{k \in \mathrm{nn}[j]} \mathbf{S}_k \right\}^n = \frac{2}{S} \sum_{k \in \mathrm{nn}[i]} \{S_i^n, S_i^m\} \, S_k^m$$

$$= -\frac{2}{S} \sum_{k \in \mathrm{nn}[i]} (\mathbf{S}_i \times \mathbf{S}_k)^n \tag{7.11}$$

$$\sum_j \{\mathbf{S}_i, (\mathbf{S}_j \cdot \hat{\mathbf{z}})^2\}^n = \frac{2}{S} S_i^m z^m z^k \left\{ S_i^n, S_i^k \right\}$$

$$= \frac{2}{S} S_i^m z^m z^k \varepsilon^{nkl} S_i^l = -\frac{2}{S} (\mathbf{S}_i \cdot \hat{\mathbf{z}})(\mathbf{S}_i \times \hat{\mathbf{z}})^n. \tag{7.12}$$

Hence, we finally obtain the following equation of motion for the classical spin at site i

$$\frac{d\mathbf{S}_i}{dt} = \frac{1}{S} \left(\mathbf{S}_i \times \Theta_i + J_S \sum_{j \in \mathrm{nn}[i]} \mathbf{S}_i \times \mathbf{S}_j + 2J_z(\mathbf{S}_i \cdot \hat{\mathbf{z}})(\mathbf{S}_i \times \hat{\mathbf{z}}) \right). \tag{7.13}$$

The above equations for all classical spins in the system are integrated together with the quantum Liouville equation (7.8) for the conduction electrons to give the combined quantum-classical time evolution of the *s-d* system.

7.3   Test simulations

7.3.1   Spin-wave dispersion in a *s-d* monoatomic chain

Molecular-dynamics (MD) simulations, based on an empirical tight-binding model and realistic interatomic potentials, provide a valuable tool for accessing the vibrational properties of condensed-matter systems, especially in situations where temperature- and pressure-related effects are investigated (Wang *et al.* 1990; Kim *et al.* 1994). Phonon dispersion and density of states are calculated from the Fourier transform of the velocity–velocity correlation function, resulting from MD simulation (Car *et al.* 1988). The phonon spectral intensity (Wang *et al.* 1990) is simply

$$\mathcal{I}(\mathbf{k}, \omega) \sim \int dt e^{i\omega t} \sum_n e^{-i\mathbf{k} \cdot \mathbf{R}_n(t)} \left| \frac{\mathbf{v}_n(t) \cdot \mathbf{v}_0(0)}{\mathbf{v}_n(0) \cdot \mathbf{v}_0(0)} \right|, \tag{7.14}$$

where $\mathbf{k}$ and ω are the wave vector and the frequency of the phonon, $\mathbf{R}_n(t)$ and $\mathbf{v}_n(t)$ are the position and velocity of the nth atom at time t. Similar schemes are often used in the framework of time-dependent density-functional theory (Car *et al.* 1988) for calculating optical excitation of molecules.

[3]Magnons are the bosonic quasi-particles corresponding to the spin waves.

We extend this idea to calculating the spin wave (*magnon*[3]) spectra in our *s-d* spin systems, which, within the nearest-neighbor direct-exchange approximation, are the spin analog to the discrete harmonic chain. Similarly, our approach is based on the time evolution of the local spins as described in the previous section. Let us briefly introduce the spin-wave approximation in a Heisenberg exchange-coupled ferromagnet by following Ashcroft and Mermin (1976). If $|0\rangle$ is the ground state of the spin system, where all spins are aligned along the same direction (say z, Fig. 7.1), then the most elementary spin excitation corresponds to the state $|n\rangle \equiv \left|S_n^z - 1\right\rangle = \frac{1}{\sqrt{2S}}\mathbf{S}_n^-|0\rangle$ (normalized to unity, i.e. $\langle n|n\rangle = 1$). Here, $\mathbf{S}_n^- \equiv \mathbf{S}_n^x - i\mathbf{S}_n^y$ is the lowering of the spin ladder operators acting on the nth site (Ashcroft and Mermin 1976)

$$\mathbf{S}_n^\pm \left|S_n^z\right\rangle = \left(\mathbf{S}_n^x \pm i\mathbf{S}_n^y\right)\left|S_n^z\right\rangle = \sqrt{\left(S \mp S_n^z\right)\left(S + 1 \pm S_n^z\right)}\left|S_n^z \pm 1\right\rangle, \quad (7.15)$$

which reduces the S^z spin component at site n from S to $S - 1$. This state is, however, not an eigenstate of the Heisenberg Hamiltonian

$$\mathcal{H}_H = \sum_{n,m} J_S(\mathbf{R}_n - \mathbf{R}_m)\mathbf{S}_n \cdot \mathbf{S}_m. \quad (7.16)$$

With the help of the following relation for the spin operators acting on the elementary excitation $|n\rangle$

$$\mathbf{S}_m^- \mathbf{S}_n^+ |n\rangle = 2S|m\rangle \quad \text{and} \quad \mathbf{S}_m^z |n\rangle = \begin{cases} S|n\rangle, & n \neq m \\ (S-1)|n\rangle, & n = m \end{cases}, \quad (7.17)$$

and the fact that $\mathbf{S}_n \cdot \mathbf{S}_m = \frac{1}{2}\left(\mathbf{S}_n^+ \mathbf{S}_m^- + \mathbf{S}_n^- \mathbf{S}_m^+\right) + \mathbf{S}_n^z \mathbf{S}_m^z$ we can express the action of the Heisenberg Hamiltonian (7.16) on the $|n\rangle$ state as

$$\mathcal{H}_H|n\rangle = E_0|n\rangle + 2S\sum_m J_S(\mathbf{R}_n - \mathbf{R}_m)[|n\rangle - |m\rangle], \quad (7.18)$$

where $E_0 = -\sum_{n,m} J_S(\mathbf{R}_n - \mathbf{R}_m)S^2$ is the ground-state energy (no spin-excitations) and the terms of the order of S^0 are neglected as the spin-wave approximation assumes $S \gg 1$ (in the classical analog that would correspond to small misalignments). Thus, $\mathcal{H}_H|n\rangle$ is a linear combination of states $|n\rangle$.

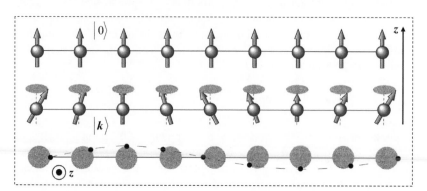

Fig. 7.1 Schematic representation of the ground state, $|0\rangle$, and an excited spin-wave state, $|\mathbf{k}\rangle$, of a Heisenberg exchange-coupled monoatomic spin chain.

Because of the translational invariance of J a Bloch wave

$$|\mathbf{k}\rangle = \frac{1}{\sqrt{N}} \sum_n e^{i\mathbf{k}\cdot\mathbf{R}_n} |n\rangle, \qquad (7.19)$$

where N is the number of sites in the system, is an eigenstate of $\mathcal{H}_H$ with an eigenvalue $E_k = \langle\mathbf{k}|\mathcal{H}_H|\mathbf{k}\rangle$ corresponding to an excitation energy

$$\mathcal{E}(\mathbf{k}) = E_k - E_0 = 4S \sum_n J_S(R_n) \sin^2\left(\frac{1}{2}\mathbf{k}\cdot\mathbf{R}_n\right). \qquad (7.20)$$

This represents the magnon dispersion relation for a finite Heisenberg spin chain. If only nearest-neighbor exchange interaction is considered, i.e. $J_S(\mathbf{R}) = J_S \neq 0$ only for $|\mathbf{R}| = a$, then eqn (7.20) reduces to the familiar expression

$$\mathcal{E}(k) = 4SJ_S \sin^2\left(\frac{ka}{2}\right), \qquad (7.21)$$

where $k = |\mathbf{k}|$. As a test of our time-dependent scheme we will reproduce this result with time-dependent dynamics by using the s-d model and setting the exchange coupling $J = 0$.

We consider a uniformly polarized (ferromagnetic) monoatomic chain with N sites, in which at $t = 0$ the spin at site i_0 is tilted to a finite angle (e.g. $\phi_0 = \pi/6$) with respect to the rest of the spins, i.e. at $t = 0$ the ferromagnetic chain is subjected to a spin excitation. We then integrate numerically the time evolution of this system, described by eqn (7.8), to obtain a two-dimensional array of values for each local spin Cartesian component as a function of position and time (see Fig. 7.2) for a sequence of N_T discrete timesteps $\{t_j\}_{j=1}^{N_T} : \mathbf{S}_i(t_j) \to \mathbf{S}_{ij}$. These arrays undergo a two-dimensional discrete Fourier transformation (dFT), resulting in an approximation of the system response spectrum in the (k, E)-domain. Out of the three orthogonal local spin components, S^x and/or S^y are more useful when transformed, since in the common limit of small excitations (small angles like in the conventional ferromagnetic resonance (FMR)) the major spin component $S^z \approx const$ and the variation is strictly transverse. Only one of the transverse components is normally sufficient, as the dominant precessional motion of the local spins implies a definite phase relation between the two, while the major one is insensitive to the precession phase.

As an example, the dFT (as implemented in the mathematical programming environment Maple^{TM} 10) of S_{ij}^y yields the reciprocal image

$$\text{dFT}[S^y]_{l,m} = \frac{1}{\sqrt{N N_T}} \sum_{i=1}^N \sum_{j=1}^{N_T} S_{ij}^y e^{-\frac{2\pi i}{N}(i-1)(l-1)} e^{-\frac{2\pi i}{N_T}(j-1)(m-1)} S_{ij}^y, \qquad (7.22)$$

where $l \in [1, N]$ and $m \in [1, N_T]$. This array is then mapped into the (k, E)-space by the change of variables $l \to k = \frac{(l-1)\pi}{(N_{\frac{1}{2}}-1)a}$ for $1 \leq l \leq N_{\frac{1}{2}}$ and $l \to k = -\frac{(l-N_{\frac{1}{2}}+1)\pi}{(N_{\frac{1}{2}}-1)a}$ for $l > N_{\frac{1}{2}}$ where $N_{\frac{1}{2}} \equiv \lceil N/2 \rceil$ is the closest integer greater

Fig. 7.2 Example of how to extract the spin-wave dispersion from a time-dependent spin-dynamics simulation. Here: $J = 0$, $J_z = 0$, $N = 101$. On the left-hand side we present the time and space evolution of S_i^y along the chain ($i \equiv z/a$), with the gray-scale shade representing the magnitude of the $S_i^y(t)$. Note that the spin-excitation is emitted from one initially tilted spin at $i_0 = 51$ and then propagates from it and is reflected from the boundaries, forming a complex interference pattern. On the right-hand side panel we present the natural logarithm (for better contrast) of the absolute value of $\mathrm{dFT}[S^y](k, E)$ as described by eqn (7.22) and mapped into the (k, E)-space. The brighter the region, the higher is the intensity.

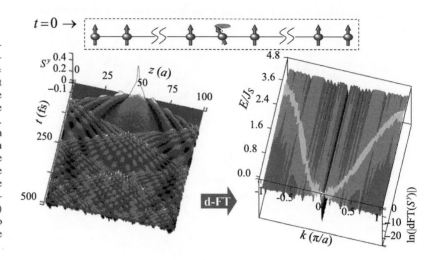

than $N/2$, and similarly $m \to E = \frac{(m-1)\pi\hbar}{T_{\max}}$ for $E > 0$, where $T_{\max} = t_{N_T}$ is the total time of the simulation and $N_T \sim 10^5 \gg N$ (so that we have neglected corrections of the order of 1 in the denominator of the right-hand side). Thus obtained, the (k, E)-space portrait of $S_i^y(t)$ (Fig. 7.2) exhibits a preferential and almost uniform population of modes obeying the correct dispersion relation, given by eqn (7.21).

This calculation of a Heisenberg spin-chain serves as an example of the applicability of the time-dependent spin-dynamics simulation, defined by eqns (7.8) and (7.13), in extracting the magnon spectrum for atomistic structures. We note that the form of the initial spin excitation is of crucial importance for this method to work and to produce a full portrait of the magnon dispersions in the reciprocal space. The excitation should contain enough energy to populate all the available spin-modes and should not carry restricting spatial symmetries. We now demonstrate how an initial excitation with a certain spatial symmetry produces incomplete reciprocal-space portraits, i.e. it is unable to populate some regions of the spin-wave spectrum. For the following spin-dynamics simulations we use parameter values $J_S = 0.2\,\mathrm{eV}$, $J_z = 0$, $0 \leq J \leq 3\,\mathrm{eV}$. The hopping parameter is $\chi = 1\,\mathrm{eV}$ and the band filling is $\rho_0 = 1.75$.[4]

Similarly to vibrational modes, the natural spin-wave modes of a one-dimensional system can be visualized as standing waves in a suspended wire. For a N-atom wire of length $L = (N - 1)a$ the natural modes have wavelengths λ_n with an integer n such that $n\lambda_n/2 = L$. Hence their wave vectors are

$$k_n = \frac{2\pi}{\lambda_n} = \frac{n\pi}{L} = \frac{n\pi}{(N-1)a} \quad \text{for} \quad n = 0, 1, \ldots, N - 1. \quad (7.23)$$

This accounts for $N - 2$ non-trivial modes as the $n = 0$ mode represents the fully aligned chain (not necessarily along z) and $n = N - 1$ is the trivial spin-wave with one node at each site. Note that this mode labelling is valid for both

---

[4]These values mimic a half-metallic electronic structure (for the chains with more than 20 atoms and $J \gtrsim 1\,\mathrm{eV}$) with a Fermi level $\gtrsim 0.5\,\mathrm{eV}$ above the fully occupied spin-up band and a spin polarization $|P(E_F)| = 1$. This promotes the highest attainable spin-flux signal and it has been especially chosen for our simulations of current-driven DW motion (see Section 7.4.2) where the pressure produced by the current scales linearly with P.

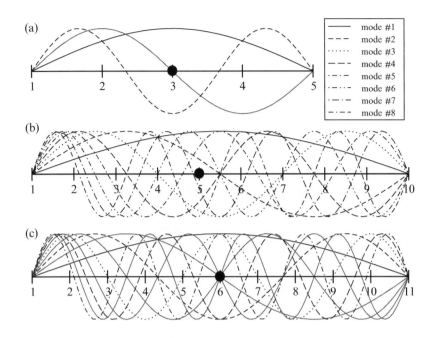

Fig. 7.3 Standing-wave modes in chains of 5, 10 and 11 sites. The black circle represents the site where the excitation is applied. The modes plotted in gray (solid lines) cannot be excited because they have a node at the site where excitation is applied.

fixed–fixed and loose–loose boundary conditions for the wire (corresponding to nodes or antinodes at both ends), the latter being the case of our spin-chain simulations.

If the initial excitation is in the form of a tilted spin at one or more atomic sites, there are two reasons for which certain modes cannot be excited. First, they may have nodes at those particular sites, where the excitation is applied. Secondly, they may not comply with the spatial symmetry inherent to the excitation. In particular, if the $t = 0$ spin excitation is applied only to the middle site of a chain with an odd number of sites then nearly half the spin modes cannot be excited as they have a node at that site. In this situation only $N_M = \lfloor N/2 \rfloor$ modes are excited (see Fig. 7.3, where $N_M = 2$ for $N = 5$ in (a) and $N_M = 5$ for $N = 11$ in (c)). If there are no nodes at the site of the excitation, all $N_M = N - 2$ modes can in principle be excited (Fig. 7.3(b)).

These considerations are verified by the calculations presented in Fig. 7.4. Here, the number of excited spin-modes (bright spots) for $k > 0$ (and identically for $k < 0$ due to the time-reversal symmetry), resulting from the applied excitation, agrees with the scheme pictured in Fig. 7.3. The top panels of Fig. 7.4 represent classical Heisenberg chains non-interacting with the conducting electrons ($J = 0$), while in the bottom panels the exchange interaction with the conduction electrons is switched on ($J = 1 \, \text{eV}$) to yield our full s-d model. As a result of the s-d interaction, both the energy and the intensity of the excited magnons are somewhat altered. We observe a shift in the energies and some level splitting but the overall spectral pattern remains the same (especially for the longer chains). One eye-catching difference is the less pronounced dark stripe at $k = 0$, corresponding to zero signal in the logarithmic scale. Despite the appearance, the offset that electrons produce at $k = 0$ is actually very small ($\approx 10^{-5}$ of the bright spots intensity).

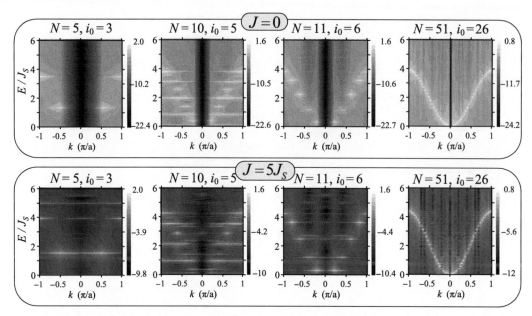

Fig. 7.4 Calculated magnon spectra for several short Heisenberg spin-chains either decoupled ($J = 0$, upper panels) or exchange-coupled to the electron gas ($J = 1\,\text{eV}$, lower panels). The gray-scale bar to the right of each contour plot corresponds to the value of $\ln|\text{dFT}[S_i^y(t)]|$. From left to right panels $N = 5, 10, 11, 51$ and the location of the $t = 0$ excitation is $i_0 = 3, 5, 6, 26$, respectively.

The most significant effect of the *s-d* exchange, which is independent of the particular type of excitation, is an energy blue-shift by as much as 10% in the short-wavelength range. From simulations for a 51-atom chain, where the dispersion is close to continuous, we can clearly extract a magnon dispersion relation $\propto \sin^2(ka/2)$. This maps onto a standard Heisenberg chain with an effective interatomic exchange coupling that is about 10% greater than the actual J_S. Hence, for this particular choice of parameters, the interaction with the conduction electrons effectively adds an extra positive contribution to the ferromagnetic spin-spin interaction, making it stronger.

In order to further investigate the interplay of the coupled dynamics of the electronic and local-spin subsystems we look at the time evolution of a longer spin-chain ($N = 101$) after a transverse spin excitation at the $i_0 = 51$ site for different values of the *s-d* exchange constant J. The gray-scale plots of the $\ln|\text{dFT}[]|$ of both $S_i^y(t)$ and $\sigma_i^y(t)$, the latter defined as

$$\sigma_i^y(t) \equiv \langle \hat{\sigma}^y \rangle_i(t) = \sum_{\alpha,\beta} \frac{1}{2} \{\tilde{\sigma}^y\}^{\alpha\beta} \rho_{ii}^{\beta\alpha}(t) \tag{7.24}$$

(with $\tilde{\sigma}^y$ designating the second Pauli matrix), for 500 fs simulations are presented in Fig. 7.5 for three particular values of J. For the given band filling $\rho_0 = 1.75$, these correspond to Fermi levels lying (a) inside both y spin-polarized bands (for $J = 0.05\,\text{eV}$); (b) at the edge of the y spin-up band (for $J = 0.5\,\text{eV}$) and (c) well above the edge of the fully filled up-band and inside the y spin-down band (for $J = 10\,\text{eV}$).

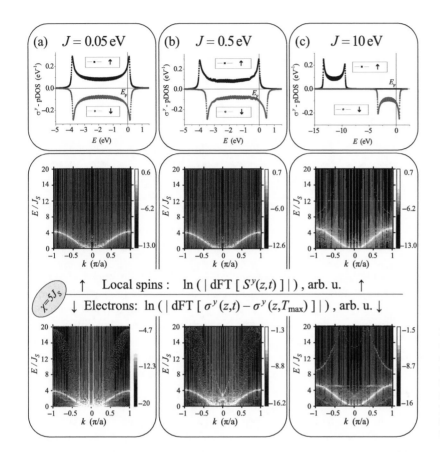

Fig. 7.5 Spin-wave spectra in a 101-atom spin-chain for three values of the *s-d* exchange strength $J = 0.05, 0.5, 10$ eV (columns from left to right). Top panels illustrate the y spin-polarized DOS at $t = 0$ when the spin on site $i_0 = 51$ is tilted to $\pi/6$ in the yz-plane. Middle panels represent the classical spin-wave spectra and the bottom ones are for the electron spin density. $J_S = 0.2$ eV.

For all three regimes the Fourier transforms of the local spins are relatively featureless—they all show the typical $4J_S^{\text{eff}} \sin^2(ka/2)$ band with an effective direct exchange J_S^{eff} that increases when J gets bigger (Fig. 7.5, second row of panels). Interestingly, only the (b) case (where the Fermi level is very close to the spin-up band edge) shows some deviations from the classical dispersion (see Figs. 7.6 (a) and (b)), while the case of very big J is again very close to the classical. The latter is attributed to the increased spin-up localization for large J, in which case the electrons from the fully filled band are so deeply bound that they behave as a classical complement to the local spin.

It is also interesting to look at the evolution of the electron (*s*) spins. The three spectra of the electronic transverse-spin density $\sigma_i^y(t)(\equiv \sigma^y(z_i, t))$ also show the very same magnon band from their corresponding local-spin spectra. Here, however, this band shows an increasing intensity with J. The presence of the classical spin-wave band in the electronic spectra is interpreted as a parametric excitation—as the electrons tend to align to the local spins, they reproduce the local transverse-spin spatial distribution at any instant of time. In the adiabatic limit (large J) the electrons are nearly in their ground state for each local spin arrangement and indeed for these cases we observe an increase of the intensity of the local-spin band. Another recognizable (though with very low intensity) feature in the electronic spectra is the silhouette

Fig. 7.6 Calculated magnon dispersion curves in a 101-atom chain with $J_S = 0.2\,\text{eV}$ (a,b) in each of the three different regimes of exchange parameters considered in the text. These are fitted to the analytic expression of eqn (7.20) to extract an effective Heisenberg exchange J_S^{eff}. In panel (c) is the dependence of J_S^{eff} on J and the dashed line marks the transition between metallic and half-metallic system.

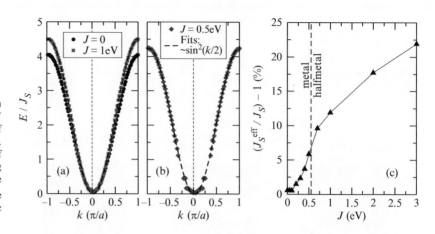

[5] The modification of the direct intersite exchange, which we find in the presence of the conduction electrons, actually has an oscillatory signature similar to that of the conventional Ruderman–Kittel–Kasuya–Yosida (RKKY) interaction (see Litvinov and Dugaev 1998, for the 1D result). We have observed that a different value of the band filling (defining a different Fermi wave vector) could result instead in an antiferromagnetic indirect exchange, reducing J_S^{eff}. For instance, at half-filling (1e/atom) we obtain a much lower $J_S^{\text{eff}} \approx \frac{1}{4} J_S$.

[6] Note that the middle region of this band appears folded down because of the sampling rate we are using (our sampling period $\Delta t = 0.5$ fs corresponds to a maximum energy of about $\pi\hbar/\Delta t = 4.1\,\text{eV} \approx 20 J_S$).

of a 4χ ($= 20J_S$ in our parameterization) wide band in the two lower J cases. This is apparently the signature of the nearest-neighbor tight-binding dispersion $\mathcal{E}(k) \propto 2\chi \cos(ka)$, with $\chi = 1\,\text{eV}$ being the hopping parameter. This structure in the electron spin-density spectrum is very diffuse and it looks like a superposition of bands for the different local-spin configurations. This is supported by the fact that its actual spread increases as J becomes larger. However, for J in the half-metallic regime (fully filled up-band) the purely electronic excitations are suppressed due to the enhanced localization of the spin-up electrons. The actual magnon spectrum is still very similar in shape to the Heisenberg one ($J = 0$) and fits rather well to the expression given in eqn (7.20) with an enhanced effective direct exchange.[5] In this regime we also observe the contour[6] of the massively spin-split s-band.

It is worth noting here that whatever signal is seen in the electron reciprocal-space portrait, this is due to excitations originating from the local spin subsystem as it is the only one excited at $t = 0$. This means that the local spins can exchange energy with the electron bath. In the next section we will prove that the energy can also be exchanged in the other direction, i.e. that our classical–quantum mixed Ehrenfest spin-dynamics is also able of describing energy transfer from the quantum electrons to the classical local spins. We note that this aspect may appear to be in contrast with the well-established fact that the Ehrenfest approximation suppresses thermal equilibration (Theilhaber 1992; Horsfield *et al.* 2004). However, although microscopic thermal noise cannot be transferred from quantum to classical degrees of freedom in the Ehrenfest dynamics, because of the mean-field description of the interaction between the two, temporal and spatial excitations at the level of the mean electron density can be captured and transferred between the two subsystems. This is indeed the case here where the energy exchange is driven by the short-wavelength spin excitations.

Finally, by fitting our calculated local-spin dispersions for different values of J (for $J_S = \text{const}$) to eqn (7.20), we are able to determine the dependence of the effective intersite exchange J_S^{eff} on the *s-d* exchange (Fig. 7.6(c)). We observe a monotonic increase of J_S^{eff} with J. For small J (when neither of the two spin bands is fully occupied) the dependence is non-linear (seemingly parabolic). As the system passes into the half-metallic regime, the dependence $J_S^{\text{eff}}(J)$ slows down for large J and tends to saturation (not shown on this graph, for instance $J_S^{\text{eff}}/J_S - 1 \approx 32\%$ for $J = 10\,\text{eV}$). In simple terms we could interpret the increase and the tendency to saturation of $J_S^{\text{eff}}(J)$ with electron localization for large J. That is, the local spins "dressed" with the localized spin-polarized electron cloud are effectively larger in magnitude, hence their direct coupling becomes stronger. The precise interpretation of this effect has to be along the lines of the 1D RKKY interaction. Our simple examples clearly indicate that the screening properties of the conduction electrons are important to the dynamics of magnetization in metallic media.

7.3.2 Spin impurities in a non-magnetic chain

This section is devoted to illustrate another aspect that can be addressed by time-dependent simulations. We shall investigate the indirect (electron-mediated) temporal spin correlations between two well-separated localized spin impurities $\mathbf{S}_{j_1}$ and $\mathbf{S}_{j_2}$ immersed in a finite metallic wire. The understanding of the indirect exchange coupling of magnetic impurities implanted in low-dimensional metallic structures could be important for the engineering of their response to magnetic fields in molecular spintronics applications. For instance, a sizable change in the conductance of a nanotube with the applied magnetic field is a strongly desired effect (Hueso *et al.* 2007). In a recent theoretical work (Kirwan *et al.* 2008) the static indirect exchange of magnetic moments sitting on metallic nanotubes has been found to strongly depend on their actual positions, due to interference effects.

Here, we consider the following dynamical situation. An excitation at $t = 0$ is produced only upon one of the two spins, say $\mathbf{S}_{j_1}$, but it is then mediated through the electron cloud and detected in the dynamical response of the second spin $\mathbf{S}_{j_2}$. We anticipate that the latter response bears the signature of the electronic structure and depends on the size of the system, the location of the two spins and the local exchange coupling parameter J. Without aiming for thoroughness we describe a few cases that capture some of the specifics of this problem. The calculation is fully in the spirit of the quantum-classical spin dynamics described in Section 7.2. The electron Hamiltonian is simplified by removing the electrostatic term from eqn (7.2) ($\kappa = 0$) and the external magnetic field from eqn (7.3). The actual system of equations now reads (in a matrix form)

$$\frac{d\rho_{ij}^{\alpha\beta}}{dt} = \frac{i}{\hbar}\left[\hat{\rho}, \hat{H}_e\right]_{ij}^{\alpha\beta}, \tag{7.25}$$

$$\frac{d\mathbf{S}_i}{dt} = \frac{1}{\hbar}\{\mathbf{S}_i, -J\mathbf{S}_i \cdot \langle\hat{\boldsymbol{\sigma}}\rangle_i - g_S\mu_B B_i\hat{\mathbf{z}}\}, \quad \text{for} \quad i = j_1, j_2, \tag{7.26}$$

where $\langle \hat{\boldsymbol{\sigma}} \rangle_i$ is the expectation value of the spin density on the ith site (see eqn (7.24)), $g_S = 2$, B_i is a magnetic field applied locally only to $\mathbf{S}_i$ and

$$\left\{ \hat{H}_e \right\}_{ij}^{\alpha\beta} = -|\chi| \delta_{\alpha\beta} (\delta_{i,j+1} + \delta_{i,j-1}) - J \{ \mathbf{S}_i \cdot \hat{\boldsymbol{\sigma}} \}^{\alpha\beta} (\delta_{ij_1} + \delta_{ij_2}) \qquad (7.27)$$

is the electronic Hamiltonian matrix, coupling the conduction electron spin to the local one only at the two spin-impurity sites, j_1 and j_2.

As an example, we consider a monoatomic chain with $N = 31$, intersite hopping $\chi = 1\,\text{eV}$ and a band filling of $\rho_0 = 1$ (i.e. 1 e/atom). The system is initialized with the spin at j_1 tilted to a small angle in the xz-plane, e.g. $\angle(\mathbf{S}_{j_1}, \hat{\mathbf{z}})_{t=0} = \pi/30$. A magnetic field is then applied $B_{j_1} \neq 0$ in order to achieve a steady precession of $\mathbf{S}_{j_1}$, while $B_{j_2} = 0$. We thus call $\mathbf{S}_{j_1}$ the "driven" spin, as opposed to the "free" spin $\mathbf{S}_{j_2}$, which is not coupled to the magnetic field. The exchange interaction to the conduction electrons is tuned through the $J > 0$ parameter. In our first set of simulations, we set $J = \chi = 1\,\text{eV}$ and place the two spins at the end sites of the chain, i.e. $j_1 = 1$ and $j_2 = N = 31$, as we are interested mainly in the transportation of the spin excitation between the two spins.

Typical time evolutions of the various spin quantities is shown in Fig. 7.7. We start our analysis by comparing the field-driven dynamics of $\mathbf{S}_1$ in the presence of the electron gas only (gray dashed curves) to that obtained when also the second spin is present. When the driven spin is decoupled from the electrons ($J = 0$) it simply performs a uniform steady precession at the Larmor frequency $\omega_L = (2\mu_B/\hbar) B_1$. In these simulations we use $B_1 = 1000\,\text{T}$, which corresponds to $\omega_L = 0.028\,\text{fs}^{-1}$ in order to obtain a large enough number of oscillations for the time of the simulation in order to provide for a reasonable resolution in our spectral analysis. In this regime the local spin-precession period ($\approx 35\,\text{fs}$) is comparable to the times for the electron passage along the chain.

When $J \neq 0$ and the free spin $\mathbf{S}_N$ is present in the system, the latter starts to oscillate after a small time delay ($\approx 10\,\text{fs}$). The small drift in S_N^x after the start of the simulation is due to the evolution of the small transverse electron spin-density present on the site since the moment $t = 0$ when $\mathbf{S}_1$ is tilted. At around $t = 10\,\text{fs}$ we find a step in σ_N^z (see Fig. 7.7(d)) and that is when the excited by $\mathbf{S}_1$ spin wave-packet arrives at the Nth site. Because of the finite group velocity of the conduction electrons, the spin-excitation takes a time $T \sim L/v_F$ (v_F is the Fermi velocity) to propagate from $\mathbf{S}_1$ to $\mathbf{S}_N$. The free spin also starts precessing about the z-axis but with a smaller transverse amplitude S_N^x (dashed curves). Of course, neither of the two spins performs a pure precession. The z-components S_1^z and S_N^z oscillate too, with similar rates and opposite phases. Such a pattern is present also in the evolution of the electron spin densities on the two magnetic sites. This is a manifestation of two restrictions acting on the system: (i) the conservation of the z-component (along the external field) of the total spin and (ii) the spin adiabaticity, i.e. the tendency of the electron spin to follow the direction of the local spins. The latter property is provided by the high strength of the s-d exchange, which in this simulation is $J = 1\,\text{eV}$.[7]

We investigate how the electron gas mediates the interaction between the two spins by varying the s-d exchange J. In order to quantify the temporal

[7] This value of J is typically large enough to provide for good spin-transport adiabaticity in such systems, as will become evident in the next section on current-induced domain-wall motion.

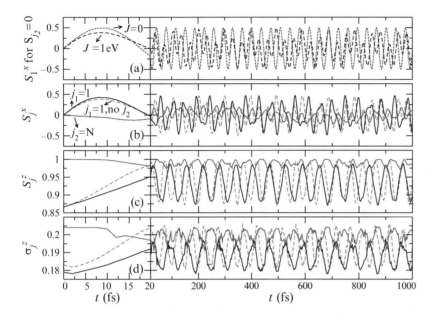

Fig. 7.7 Time evolution of spin components: (a) x-component of the driven spin with (dashed curve) and without (dotted curve) interaction to the electron gas when the free spin is absent (no spin at j_2); (b,c) x- and z-components of the two spins for $J = 1\,\mathrm{eV}$; (d) the electron spin polarization (with respect to the quantization axis z, i.e. $\sigma_i^z \equiv \left(\rho_{ii}^{11} - \rho_{ii}^{22}\right)/2$) at the spin-impurity sites. Black and gray solid curves represent site 1 and site $N = 31$, respectively, while the dashed-gray underlying curves represent site $j_1 = 1$ when there is no spin-impurity at j_2.

correlations between the transverse components of the two spins, we use the following normalized spin-spin correlation function

$$C\left(S_1^x, S_N^x, \Delta t\right) = \frac{1}{\int \left(S_1^x(t)\right)^2 dt \int \left(S_N^x(t)\right)^2 dt} \int S_1^x(t) S_N^x(t + \Delta t) dt.$$

(7.28)

The right-hand side panels of Fig. 7.8 represent the square of the above quantity as a function of the time delay Δt for a wide range of values of J. All results show a similar high-rate oscillatory decay pattern as Δt increases. The maxima of all $C^2\left(S_1^x, S_N^x, \Delta t\right)$ are reached in the very early stages (compared to the whole length) of the time simulation. This demonstrates the rapid ($\sim 10\,\mathrm{fs}$ for 30 atoms separation) transfer of angular momentum carried out by the itinerant electrons. As the electron ticking rates (the inverse of the time to propagate forth and back between the two spins) are comparable to the precession rate of the driven spin, we find a very rapidly oscillating correlation function. In order to roughly estimate the maximum correlation amplitude (which is very close to the origin) we fit these results to a decaying exponent $C_0 \exp(-\Delta t/\tau)$ and the results for $C_0(J)$ are presented in Fig. 7.9. The constant $C_0 \sim C^2(t \approx 0)$ is a measure of the level of dynamical correlation and reaches its maximum value for J in the range between $1.5\,\mathrm{eV}$ and $4\,\mathrm{eV}$. For smaller J there is little energy transfer between the local spin and the electron gas and for very large J the majority (minority) electrons are too localized (delocalized)[8] to convey the spin information, hence the correlation is suppressed in both cases.

In order to analyze the dynamical behavior of the coupled system we look at the discrete Fourier transforms of the transverse x-components of the two spin-impurities. We compare the case when they are both present at each of the two ends of the chain with those obtained when $\mathbf{S}_1$ is the only localized

[8] In connection with this last point one should note that the larger J, the deeper (shallower) the potential associated with the on-site spin interaction for majority (minority) conduction electrons (see eqn (7.3)).

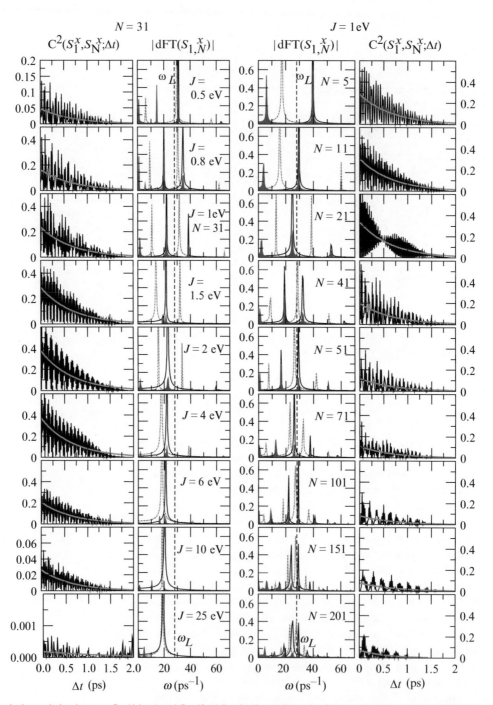

Fig. 7.8 Dynamical correlation between $\mathbf{S}_1$ (driven) and $\mathbf{S}_N$ (free) local spins at the ends of a metallic chain as a function of the *s-d* exchange strength J for a $N = 31$-atom chain (pairs of panels to the left) and as function of N for $J = 1\,\mathrm{eV}$ (pairs of panels to the right). Each case is represented by a pair of plots: one for the time-correlation function C^2 (see eqn (7.28)) and one for the spectra of $S_1^x(t)$ (black solid line), $S_N^x(t)$ (gray hatched curve) and again $S_1^x(t)$ but for the case when $\mathbf{S}_N$ is absent from the chain (gray dotted curve). The thick gray solid curves in the external panels represent the exponential-decay fits used to determined C_0 (see Fig. 7.9).

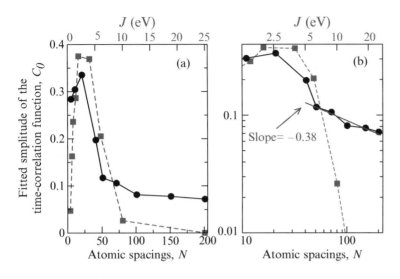

Fig. 7.9 The amplitude of the computed time-correlation functions (see Fig. 7.8) as a function of J (top axis) or number of atoms N separating the two spins (bottom axis, the distance is $L = Na$) in linear (a) and logarithmic scale (b). The tail of the dependence on distance is fitted to a power law decay $\sim L^{\alpha}$, where $\alpha = -0.38$.

spin in the chain (Fig. 7.8, left-hand side). We observe that as J increases the spectrum of S_1^x evolves from having one low-frequency peak and a peak with a higher amplitude slightly above the Larmor frequency ($\omega_L = 0.028\,\text{fs}^{-1}$) for $J = 0.5\,\text{eV}$, through two equal-amplitude peaks on each side of ω_L for $J < 1.5\,\text{eV}$, and then saturates to a single peak somewhat below ω_L for very large J. The other spin S_N^x shares the same modes for $J < 10\,\text{eV}$, although the amplitudes are different.

Qualitatively similar is also the evolution of the spectrum of S_1^x when the second spin is absent (underlying dotted-gray curves) and this becomes identical to that of the two spins for extremely large values of the *s-d* couplings J. In general, there is a red-shift tendency of the natural modes of the field-driven spin system as J is increased. Finally, for extremely large J the two spins are completely decoupled, because of the delocalization of the conducting minority electrons, which are expelled from the spin sites. As a result $\mathbf{S}_1$, "dressed" with its localized spin-up electrons (hence larger in magnitude), precesses at a rate below ω_L, while $\mathbf{S}_N$ is practically still (note that in our model dynamical simulation there is no Zeeman split in the electron gas).

Without aiming at analyzing all the natural modes of the combined system of itinerant electrons and local spins (an impractical task given the high number of degrees of freedom and the non-linearity of the dynamical response), we investigate numerically the dependence of the frequencies of the major modes on some model parameters. For instance, on the right-hand-side of Fig. 7.8 we illustrate the effect of the separation on the dynamical correlation between the spins situated at the two ends of the wire by varying its length (Na) for a fixed $J = 1\,\text{eV}$. In general, as N increases the precessional spectrum of S_1^x becomes richer. All natural modes show a tendency to red shift as the chain becomes longer. This correlates to the increased electron ticking time (the time for an electron round trip). The modes also gain amplitude when they get close to ω_L. As a result, the spectrum of the field-driven spin condenses about ω_L. The "free" spin spectrum shows the same modes although they appear much more evenly populated. Evidently, the temporal correlations between the spins start off being rather high at very small separations, peak at around 40 atoms and decrease in amplitude as the chain length increases (see also Fig. 7.9). Above $N = 50$ we find a slower decrease of the correlation amplitude. We have fitted with a power law $\propto L^\alpha$ (linear on the log-log scale in Fig. 7.9(b)) and obtained $\alpha \approx -0.38$. Despite the crudeness of the way α is extracted from the time correlations, it clearly suggests a rather long-range indirect exchange coupling between the spins in the wire. This observation comes in agreement with recent theoretical findings (Costa *et al.* 2008) of an enhanced range of the dynamical indirect exchange coupling between adsorbed magnetic moments in metallic nanotubes, compared to the static version of it, suggesting that $|\alpha|$ could be smaller than 1.

Since the finite atomic chain acts as an electronic wave resonator, the correlations between the two spin impurities can be amplified or suppressed depending on the total length of the chain and the position of the site at which they are located. We have determined numerically that in an even-sited chain the field-driven spin can transfer spin excitation to the electron gas mainly locally, as illustrated in Figs. 7.10(a) and (b). In this case the standard deviation of the transverse spin distribution, which is a measure of the spin-density oscillation in the chain, peaks only at the site containing the spin impurity. The spectrum of $S_j^x(t)$ of the driven spin shows a single peak at the Larmor frequency for both even or odd j (Fig. 7.10(c)). In order for the precessing spin to be able to transfer effectively spin excitation to the electron gas in the rest of the chain, this has to have an odd number of sites and the spin has to be at an odd-site location as illustrated in Fig. 7.10(e). The difference can be seen clearly from the spectrum (Fig. 7.10(f)), where two peaks, corresponding to combined modes of the coupled spin system, replace the peak at ω_{L}.

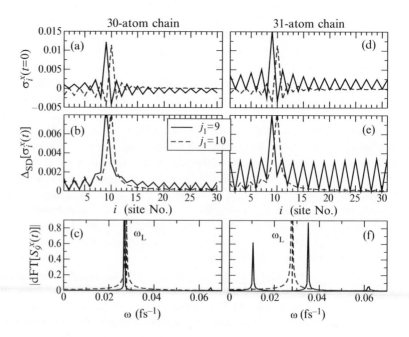

Fig. 7.10 Simulations of a single magnetic-field-driven spin $\mathbf{S}_{j_1}$ in 30 and 31-atom chains (left- and right-hand side panels, respectively). (a,d) The x-component of the on-site electron-spin density at $t = 0$ and (b,e) its standard deviation for the time of the simulation vs. the site position. (c,f) The absolute value of the Fourier transform of $S_{j_1}^x$ for $j_1 = 9$ (black) and $j_1 = 10$ (gray dashed curves). The position of ω_{L} is marked by a vertical dashed line.

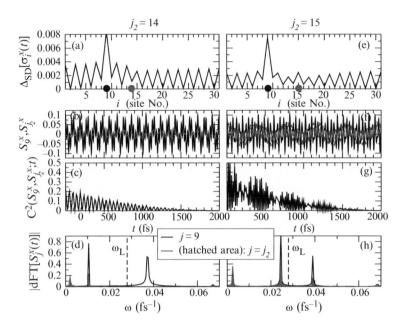

Fig. 7.11 Simulations of two spins $\mathbf{S}_{j_1=9}$ ("driven"), $\mathbf{S}_{j_2=14,15}$ ("free") in a 31-atom chain. (a,e) Standard deviations $\Delta_{\mathrm{SD}}[\langle \sigma^x \rangle_i(t)]$ vs. position i, (b,f) x-components of the two local spins and (c,g) time-correlation functions (see eqn (7.28)) between those vs. time, (d,h) absolute value of the discrete Fourier transforms of the data in panels (b,f).

Similar parity rules apply for the position of the second (free) spin-impurity. In the case of odd chains (apparently more susceptible to excitations) the temporal correlation and the mere amplitude of the excitation transferred between the two spins are substantially higher if the spins are both in odd positions (see Fig. 7.11). This evidently can be seen from the spectra of $S^x_{j_1}$ and $S^x_{j_2}$ too. The position of ω_{L} is marked by a vertical dashed line.

Further calculations (not presented here) have shown that typically the spin correlations do not depend as much on the location of the two-spin complex in the chain as they depend on the distance between them and the parity of the position j_1 of the driven spin. These seemingly peculiar odd-even effects originate from the fact that our electronic band structure is that of half-filling (1 e/atom) and the electronic temperature is low. We note that the notion of electronic temperature actually enters our model with the choice of eigenstate occupation used for constructing the ground-state density matrix. In all calculations presented so far we have used some low electronic temperature $T = 50\,\mathrm{K}$. The spatial distribution of a spatially abrupt spin (or charge) excitation is rather corrugated at half-band filling because of the absence of modes with small enough wavelengths in the occupied spectrum. The Friedel-like fringes of spin density around the driven spin can be seen, for example, in Fig. 7.10(d).

The rough spin-density texture becomes smoother as the temperature is increased or a band filling different from $\rho_0 = 1$ is considered. As an illustration of this effect, the case of a single driven spin in the 31-atom chain is repeated for different T and ρ_0, and the results are presented in Fig. 7.12. As the temperature is increased to introduce a Fermi level smearing $k_{\mathrm{B}}T \approx 0.43\,\mathrm{eV}$ comparable to the bandwidth (4 eV), the interference pattern occurring in the spatial distribution of the transverse electron spin is no longer present. As more modes are allowed to contribute at higher T, the electron localization

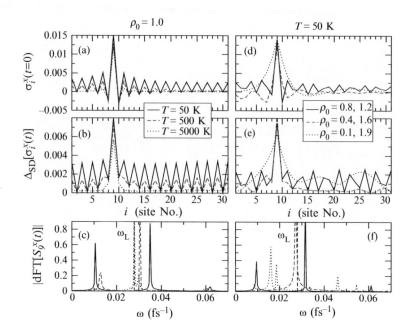

Fig. 7.12 Effects on the spin dynamics of a single field-driven spin placed at position $j_1 = 9$ in a 31-atom chain as a function of the band filling, ρ_0, and the electron temperature (left- and right-hand side panels, respectively). The quantities plotted are the same as those represented in Fig. 7.10.

is better defined spatially and the local spin spectra reduce to a single peak at $\omega_L = 0.028\,\text{fs}^{-1}$.

The effect of changing the band filling (e.g. by charging or discharging system) is in many ways similar to that of the temperature. Values of ρ_0 that are symmetric with respect to 1 e/atom produce the same spin-density pattern because of the electron–hole symmetry. We observe that intermediate band fillings $1.3 < \rho_0 < 1.7$ quite resemble the high-T case, i.e. electron-spin excitation is localized and precession of the local spin is almost unaffected. For higher band fillings, because of the big wavelengths at the Fermi level, a variety of resonant phenomena can occur. Figure 7.13 depicts some curious results of simulations of a driven and a free spin in an electron gas with Fermi level close to the band edge ($\rho_0 = 1.9$). The real-space and time evolution of the transverse spin density reminds us of a "tsunami" effect. The long-wavelength spin-density excitations build up on the free spin, causing it to shake with an enormous (compared to the initial tilt of the driven spin) amplitude at some instances of time. As there is no dissipation in our system these are regularly recurring events.

7.4 Current-induced domain-wall motion

The prospect of manipulating the magnetization texture of a device with a spin-polarized current is at the heart of a whole area in the field of spin-electronics, which takes the name of *spin-transfer torque*. This unifying notion originates from the work of Slonczewski (1996), who made a pioneering theoretical prediction of current-induced magnetization dynamics in single-domain magnetic multilayers for the current perpendicular to the plane (CPP) geometry. Having

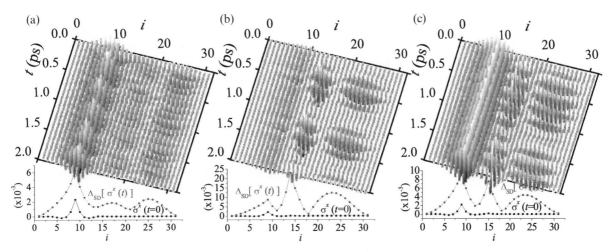

Fig. 7.13 Trajectories of the expectation value of the transverse on-site spin polarizations $\sigma_i^x(t)$ in a 31-atom chain with 1.9 e/atom and two spin-impurities $\mathbf{S}_{j_1}$ (magnetic field-driven) and $\mathbf{S}_{j_2}$ ("free"), for $j_1 = 9$ and (a) no second spin; (b) $j_2 = 14$; (c) $j_2 = 15$. The bottom plots represent the initial spatial distribution $\sigma_i^x(t = 0)$ (black lower curve) and its temporal standard deviation (gray curve).

a spin-valve in mind, the main idea is that the transversely spin-polarized electron flux generated by the first magnetic layer (polarizer) produces a torque upon the magnetization of the second layer (analyzer), effectively transferring spin between the two. A steady precession of the analyzer magnetization is predicted at a constant current, strong enough to dominate the Larmor precession about the Öersted field, and a repetitive magnetization switching is expected under a pulsed current. These phenomena can be detected through the magnetoresistance effect and both switching (Katine *et al.* 2000; Grollier *et al.* 2001) and precession (Rippard *et al.* 2004; Krivorotov *et al.* 2005) have been observed experimentally in spin-valve structures. More recently time-resolved ultrafast X-ray images (Acremann *et al.* 2006) have been able to peek into the very process of magnetization reversal due to STT. The huge interest in the spin-torque phenomena is constantly fuelled by the industry as the miniaturization of magnetic storage devices progresses and new spin-torque-based magnetic random access memory (MRAM) devices are about to become commercially available.

Another founding contribution to spin-transfer torque was made even earlier. In the 1970s Berger (1978), by looking at the low-field magnetoresistance of some ferromagnets, noticed that "electrons crossing a (domain) wall apply a torque to it, which tends to cant the wall spins". This is again a spin-torque phenomenon similar to Slonczewski's one for single-domain dynamics. In fact, the spin torque occurs in any non-uniform magnetization landscape that is traversed by a spin-polarized current. In the case of the smoothly changing direction pattern of a magnetic domain wall (DW), the spin-transfer torque is converted to pressure, which pushes the wall in the direction of the electron flow (we shall explain this in detail later in this section) and this effect has been confirmed experimentally (for recent works see Grollier *et al.* (2003), Klaui *et al.* (2003), Tsoi *et al.* (2003) and Krivorotov *et al.* (2005)). The current-induced domain-wall (DW) drag is particularly interesting from the information storage point of view because the adjacent domains could represent binary bits. Hence, proposals for magnetic memories where the magnetic domains are driven across a device by pulsed spin-polarized currents, e.g. the so-called magnetic race track (Parkin *et al.* 2008), have emerged.

Theoretical work is typically based on either semi-classical theory (Tserkovnyak *et al.* 2005) or on the *s-d* model, where the local magnetization $\mathbf{M}(t, z)$ is continuously varied only along the direction (say, along z) of the current flow. The problem is usually addressed in a micromagnetic fashion by solving the Landau–Lifshitz–Gilbert (LLG) equation for the magnetization with additional *spin-torque* terms describing the effect of the electron flux. For one-dimensional DWs Zhang and Li (2004) have derived two current-induced spin-torque contributions, by starting from the *s-d* model and integrating out the electronic degrees of freedom. Their working assumption is that of a slowly varying magnetization direction. The length scale is set by their intrinsic spin-relaxation mechanism so that in the adiabatic limit, which they consider, the width of the wall is much bigger than the spin-relaxation length. The two current-induced torques, derived by Zhang and Li (ZL), are

$$\mathbf{T}_{\mathrm{adiab}} = -\frac{b_J}{M^2}\mathbf{M} \times \left(\mathbf{M} \times \frac{\partial \mathbf{M}}{\partial z}\right), \tag{7.29}$$

$$\mathbf{T}_{\mathrm{nonad}} = -\frac{c_J}{M}\mathbf{M} \times \frac{\partial M}{\partial z}. \tag{7.30}$$

The so-called *adiabatic* torque, $\mathbf{T}_{\mathrm{adiab}}$, is analogous to the Slonczewski torque in spin-valves. It is due to the fact that the electron spin tends to align to the local magnetic field as it passes through regions of spatially varying magnetization. For $|\mathbf{M}(z)| = M$ (only the direction of the magnetization varies) this can be rewritten (Zhang and Li 2004) as $-b_J\frac{\partial \mathbf{M}}{\partial z}$, where by definition $b_J = Pj_e\mu_B/eM$, j_e being the electric current density, P the spin polarization. The other torque $\mathbf{T}_{\mathrm{nonad}}$ (eqn (7.30)) is perpendicular to $\mathbf{T}_{\mathrm{adiab}}$ and is described as a *non-adiabatic* torque. In other theoretical works this torque, derived from different microscopic viewpoints, is instead known as the β-torque (Tatara *et al.* 2008) and the term non-adiabatic is reserved for all other additional terms (Xiao *et al.* 2006).

A simple mechanism for understanding the origin of the adiabatic torque in a 1D Néel wall is illustrated in Fig. 7.14 (see, for example, Stiles *et al.* 2007). An electron entering from the left-hand side with spin-up and following the wall magnetization adiabatically, eventually flips its spin on exit, i.e. $\Delta S_e = 2s = \hbar$. If all non-conservative spin-relaxation mechanisms are neglected (for example electron scattering with impurities having a strong spin-orbit coupling) (Vanhaverbeke and Viret 2007) then all of the angular-momentum variation must be absorbed by the local magnetization. In the case envisaged in Fig. 7.14, the angular momentum gained by the DW for a single adiabatic electron crossing is $\Delta S_{\mathrm{DW}} = -\Delta S_e > 0$, which corresponds to the DW shifting to the right (i.e. in the direction of the electron flow).

In contrast, the origin and the effect of the non-adiabatic torque in eqn (7.30) is not as easy to cartoon. As introduced by Zhang and Li (2004), this torque is due to the non-conservative spin-flip processes during the electron's crossing

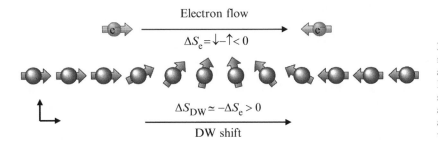

Fig. 7.14 Illustration of the angular-momentum transfer mechanism behind the adiabatic contribution to the current-induced DW motion. An incoming electron flips its spin direction by passing across a DW. Such a spin rotation is compensated by an equal and opposite change of the total local spin, which generates a DW motion.

of the DW. These are modelled by a phenomenological spin-relaxation term in the equation of motion of the electron spin density. Although for transition-metal ferromagnets $c_J/b_J \approx 10^{-2}$ (Zhang and Li 2004), the role of the non-adiabatic term is often claimed to be crucial for sustaining the motion of the wall. According to Zhang and Li, the non-adiabatic term acts as a non-uniform time-dependent magnetic field with just the right distribution to sustain the DW motion. More recently, Zhang and Li's picture was questioned by Stiles *et al.* (2006) and calculations based on the Stoner model demonstrated that indeed the pre-factor c_J associated to the non-adiabatic torque is actually non-local and it is not necessarily associated with an intrinsic mechanism for spin-relaxation, as also pointed out by Berger (2007).

The other extreme limit (contrary to the adiabatic one) that can also be addressed analytically is that of the very thin wall with a width w much smaller than the Fermi wavelength, i.e. $k_F w \ll 1$. In this case the STT effect vanishes, the wall can be described as a quasi-particle and the current-induced pressure is due to the linear-momentum transfer of the backscattering conduction electrons (Tatara and Kohno 2005; Dugaev *et al.* 2006).

In order to investigate the spin-transfer torques in the range of DW widths comparable with the Fermi wavelength (a few atomic spacings wide), we have developed an open-boundary spin-dynamics simulation, which is described in the following section.

7.4.1 Computational method and static properties

Within our atomistic *s-d* model simulations of the spin dynamics we treat the electrons quantum-mechanically and integrate the set of Liouville equations for the classical spins. This reads (as in eqn (7.8) but without the external magnetic field)

$$\frac{d\mathbf{S}_i}{dt} = \frac{1}{\hbar}(\mathbf{S}_i \times \mathbf{B}_i), \qquad (7.31)$$

where

$$\mathbf{B}_i = J\langle\boldsymbol{\sigma}\rangle_i + J_S \sum_{j=i\pm1} \mathbf{S}_j + 2J_z(\mathbf{S}_i \cdot \hat{\mathbf{z}})\hat{\mathbf{z}} \qquad (7.32)$$

is the effective local magnetic field at site i. Thus, the classical spins are locally exchange-coupled to the instantaneous expectation value of the electronic spin $\langle\boldsymbol{\sigma}\rangle_i$, which produces all the contributions to the STT. In other words, there is no need for additional empirical current-induced torques in our LLG-like equation as the effective-field term accounts for both the adiabatic and non-adiabatic torques originating from the electron flow as mean-field (Ehrenfest) forces.

In order to simulate the DW dynamics under bias at the atomic level, we have used a semi-empirical method of sustaining an electron flux in a finite system designed for Ehrenfest-type dynamical simulations and described by Sanchez *et al.* (2006). In this method the atomistic system is partitioned into three subregions: two external regions, acting as a source (S) and a drain (D), coupled to a central region, where the atomistic device to be studied under bias is located (see Fig. 7.15). A modified Liouville equation is used to describe the electronic dynamics

$$\frac{\partial \hat{\rho}}{\partial t} = \frac{i}{\hbar} \left[\hat{\rho}, \hat{H}_V \right] - \Gamma(\hat{\rho} - \hat{\rho}_0), \tag{7.33}$$

where Γ is a real parameter and $\hat{\rho}_0$ is a modified density matrix such that

$$\{\hat{\rho}_0\}_{ij}(t) = \begin{cases} \rho_{ij}(t=0) & \text{for } i, j \in \text{S, D} \\ \rho_{ij}(t) & \text{otherwise} \end{cases}. \tag{7.34}$$

Within this definition the source/drain term proportional to Γ is applied only to the S and D regions. If the initial density matrix $\rho_{ij}(t=0)$ is set in such a way that the source and the drain carry an electron imbalance, this will be maintained during the time evolution of the electron density with a relaxation time of $1/\Gamma$, and it will produce a current flow through the system. Such a source/drain initial density in our simulations is constructed from the eigenstates of the Hamiltonian with an applied external potential of $2\Delta V$

$$\{\hat{H}_V\}_{ij}(\hat{\rho}, t) = H_{ij}(\hat{\rho}, t) + \begin{cases} -e\Delta V f_V(t)\delta_{ij} & \text{for } i, j \in \text{S} \\ e\Delta V f_V(t)\delta_{ij} & \text{for } i, j \in \text{D} \end{cases}, \tag{7.35}$$

where $H_{ij}(\hat{\rho}, t)$ is the *s-d* exchange tight-binding Hamiltonian as in eqn (7.2) and $f_V(t)$ is a smoothly decaying to 0 polynomial ramp. This is designed to vanish in a few tens of time-steps and it has been introduced for damping the charge oscillations that might originate from an abrupt removal of the initial bias. There is no external magnetic field and the hopping integral is $\chi = 1\,\text{eV}$.

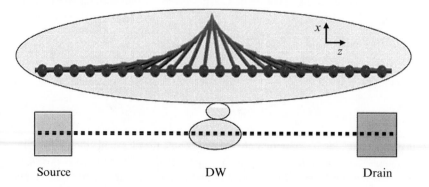

Fig. 7.15 Scheme of the electron-transport calculations through a monoatomic wire containing a DW. The source and drain are introduced in the equation of motion as a phenomenological term that maintains charge imbalance at the edges of the wire (see eqn (7.33)).

Source   DW   Drain

In order to mimic a half-metallic electronic structure, we have chosen a band filling of 1.75, for which the Fermi level lies about 0.5 eV above the band edge of the fully occupied spin-up band.

Ideally, the source/drain term in eqn (7.33) produces a constant carrier imbalance between the two ends of the wire therefore simulating an open system. Moreover, since ρ_0 is purely real it also provides for a phase-breaking mechanism in the source. This phenomenological method has been proven to be able to produce, under certain conditions, a current-carrying state equivalent to the Landauer steady-state (Sanchez *et al.* 2006; McEniry *et al.* 2007; Todorov 2007). A major problem with the method remains the correct definition of the applied bias as the system is finite and there are no real electron reservoirs (as thermodynamically defined in the Landauer picture). However, if one is not interested in the precise $I–V$ characteristics of the system but only in the effect of the current, and if the current does not alter significantly the resistance of the device, then tuning the value of Γ is sufficient to dynamically achieve a steady-current state. In the case of the current-induced DW motion studied here a steady current is achieved for $\Gamma = 2\,\text{fs}^{-1}$.

We thus have all the ingredients for atomistic simulations of DW motion in such an open-boundary one-dimensional atomic wire under a bias. The wire is 200 atoms long and the DW is set in the middle. In the initial state (before the bias is applied) the DW and the electrons are relaxed self-consistently so that all torques in the system vanish within a certain tolerance. For an infinite Heisenberg spin chain with a longitudinal anisotropy $J_z \neq 0$ (but $J = 0$, i.e. no coupling to the electron gas) the DW ground state is known. This is a planar (Nèel) DW with a longitudinal spin distribution $S_i^z = S \cos \theta_{z_i}$, where

$$\theta_{z_i} = 2 \, \arctan \left[\exp \left(\frac{z_i - z_0}{z_{\text{w}}} \right) \right]. \tag{7.36}$$

Here, z_0 is the center of the DW and z_{w} is the DW width at approximately 89% of its transverse-profile height (see Fig. 7.16(a)). In the anisotropic Heisenberg model this is related to the direct exchange and the anisotropy parameters (Barnes and Maekawa 2005)

$$z_{\text{w}} = a\sqrt{J_S/2J_z}. \tag{7.37}$$

First, we reproduce numerically this result for our finite spin-chain and investigate the effect of the exchange coupling to the itinerant electrons J. After initializing the DW in the xz-plane, we perform a damped dynamical spin relaxation (by adding a Landau–Lifshitz damping term $-\alpha \mathbf{S}_i \times (\mathbf{S}_i \times \mathbf{B}_i)$ to eqn (7.31)) until all torques in the system are below a given threshold. The relaxed DW is still planar as can be seen in Fig. 7.16(a), where in the DW centre $S_0^x \approx 1$. By fitting the $S_i^x(i)$ profile to the angular distribution in eqn (7.36) we calculate z_{w} for different values of J_z and $J \neq 0$. By comparing with eqn (7.37) we extract an effective direct exchange parameter J_S^{eff} and find that, for our choice of parameters, the exchange with the electrons acts towards increasing J_S^{eff} (see Fig. 7.16(c)). For $J = 1\,\text{eV}$ this contribution is nearly as big as our model direct exchange $J_S = 20\,\text{meV}$. A similar result was obtained also in the previous section by studying the magnon spectra (see, for example, Fig. 7.6).

If one keeps the anisotropy constant J_z and number of electrons (band filling $\rho_0 = 1.75\,e/$atom) fixed, as J is increased the DW width z_w tends to saturate (see Fig. 7.16(b)). The dip near the beginning of the $z_w(J)$ curve corresponds to a J value for which the Fermi level is at the spin-up band edge. Above this value the system is a half-metal with only minority-spin electrons at the Fermi level. For our finite chain with a DW in the middle, the local spin along z is fully compensated and thus the σ^z-projected DOS (pDOS) are identical for the two spin species (see Fig. 7.17). However, the transition to half-metallicity at $J = 0.5\,$eV can be seen in the σ^x-pDOS as the wall lies in the xz-plane. As the spin-split increases the transverse spin-up presence at the Fermi level goes through a peak and drops. For the critical value $J = 0.5\,$eV for which the Fermi level is closest to the edge of the spin-up band for a uniformly polarized 1D spin-chain (not shown in this figure), we find that the planar DW is no longer possible and the relaxed DW develops a small twist out of the xz-plane, which manifests itself in the non-negligible and asymmetric σ^y-pDOS.

In summary, for the half-metallic regime the DW thickness varies very little with J, for the range of values between 1 and 3 eV and for $J_z = 5\,$meV. In this case $z_w \approx 2$–$3a$ and the wall profile starts to deviate from the one of the classical Heisenberg model eqn (7.37) but it is still not too far from it. These wall thicknesses are indeed comparable to the Fermi wavelength in our simulations. A rough estimation of the latter can be made from the σ^x-pDOS (see top panel of Fig. 7.17, $J = 1\,$eV), where we clearly distinguish two 1D nearest-neighbor TB bands and the Fermi level splits the spin-down band as 7:1 (approximately) from the bottom up. This determines that $k_F a = \arccos((E_c - E_F)/2\chi) \approx \arccos(-3/4) \approx 2.42$, which corresponds to $\lambda_F = 2\pi/k_F \approx 2.6a$. Thus, our model parameters indeed provide walls in the intermediate thickness range that are not directly accessible to both the possible analytical descriptions.

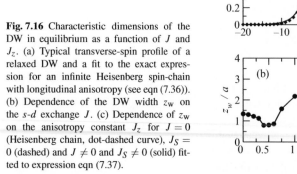

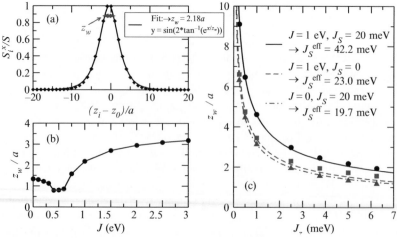

Fig. 7.16 Characteristic dimensions of the DW in equilibrium as a function of J and J_z. (a) Typical transverse-spin profile of a relaxed DW and a fit to the exact expression for an infinite Heisenberg spin-chain with longitudinal anisotropy (see eqn (7.36)). (b) Dependence of the DW width z_w on the s-d exchange J. (c) Dependence of z_w on the anisotropy constant J_z for $J = 0$ (Heisenberg chain, dot-dashed curve), $J_S = 0$ (dashed) and $J \neq 0$ and $J_S \neq 0$ (solid) fitted to expression eqn (7.37).

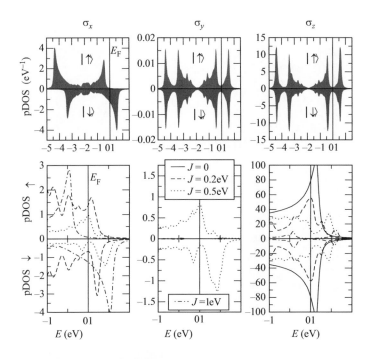

Fig. 7.17 Spin-projected density of states for the conduction electrons in a 200-atom wire, containing a DW in xz-plane. Top panels: the three spin components of the DOS for $J = 1\,\text{eV}$ and a relaxed DW (no voltage is applied). The Fermi level is at $0\,\text{eV}$. Bottom panels: a zoom around the Fermi level of the spin-projected DOS for J ranging from 0 to $1\,\text{eV}$.

7.4.2 Domain-wall motion

After the DW is relaxed self-consistently in the presence of the electron gas, a finite bias voltage is applied to two 10-atom-long segments at each wire end (S and D) by shifting their on-site energies according to eqn (7.35). Then, the quantum-classical system of equations of motion (eqns (7.31) and (7.33)) is integrated numerically. Typical real-space and real-time contour plots of the dynamical observables are presented in Fig. 7.18. Note that the DW in our model is moving opposite to the electron flux, which does not contradict the illustration of Fig. 7.14 since the carriers in our half-metallic system are down-spin polarized, i.e. they are spin-polarized in the opposite direction to that of the local spin.

Currents in a TB description can be expressed in terms of the bond current expectation value I_{nm} (Todorov 2002). This is the electron flow between two sites n and m, i.e. $\sum_{m \neq n} I_{nm} = \{\dot{\rho}\}_{nn}$ with the right-hand side representing the rate of change of the on-site occupation. Following Todorov (2002), we define two partial bond currents for the two spin species with respect to the z-axis

$$I_{nm}^{\sigma} = \frac{2e}{\hbar} \text{Im} \left[\rho_{nm}^{\sigma\sigma} H_{nm}^{\sigma\sigma} \right], \tag{7.38}$$

where $\sigma = \uparrow, \downarrow$ and there is no summation over repeating indices. Because our Hamiltonian extends only to the nearest neighbors, the bond currents are non-zero only for $m = n \pm 1$. We denote the total particle current and the z-polarized spin current through the bond $(n, n+1)$ for $n = 1, \ldots, N-1$ as

$$I_n = I_{n,n+1}^{\uparrow} + I_{n,n+1}^{\downarrow} \qquad \text{and} \qquad I_n^{\sigma_z} = I_{n,n+1}^{\uparrow} - I_{n,n+1}^{\downarrow}. \tag{7.39}$$

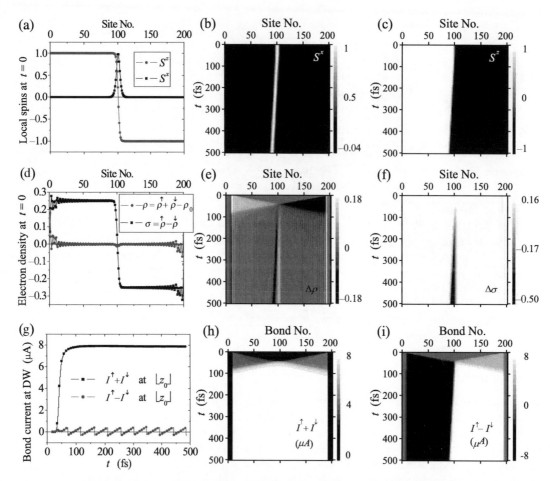

Fig. 7.18 Results from a typical simulation of DW motion in a 200-atom chain with initial bias of $\Delta V = 1$ V. Self-consistently relaxed $t = 0$ configuration of (a) the local spins and (b) the electron charge excess and spin density (quantization axis is z). Space and time evolutions of: (b, c) local spins $S_i^{x,z}$; (h, i) charge and spin bond currents (see eqns (7.38) and (7.39)); the relative variation with respect to $t = 0$ of (e) the on-site density $\Delta\rho \equiv \rho(t) - \rho(0)$, and (f) the spin density $\Delta\sigma \equiv \sigma(t) - \sigma(0)$. In panel (g) is the time evolution of the charge and spin currents at the bond where the center of the wall is (discontinuities correspond to moments when the wall center goes through a site).

[9]Here, and throughout, by charge density we actually mean number of electrons and ρ is always a dimensionless quantity.

These two quantities as functions of time and the atomic position are shown in Fig. 7.18 for an initially applied bias voltage of $\Delta V = 1$ V and $J = 1$ eV. The contour plots illustrate how the steady state is established when the charge imbalances from the two ends propagate to the center of the chain (see also the charge-[9] and spin-density plots in panels (c, d) and their variations in panels (e, f)). Since the steady-state current flow is not established instantaneously it takes a finite time before the DW starts to move (see panels (a, b) for the evolution of the local spin components). However, once the charge current is steady, it is not affected by the DW motion (see panel (g)). Because of the half-metallic electronic structure with a minority band at the Fermi level, we have $P = -100\,\%$ spin polarization of the current. Hence, $I_n^\uparrow = 0$ on the left-hand side of the DW, where the local spin is up, and $I_n^\downarrow = 0$ to the right-hand side. In other words, the spin and the charge currents are the same on the right-hand side of the wall $\left(I_n = I_n^{\uparrow} = I_n^{\sigma_z} \right)$ and have opposite signs on the other side $\left(I_n = I_n^{\downarrow} = -I_n^{\sigma_z} \right)$.

We first investigate the evolution of the spin in our open-boundary system. The mere fact that the DW is moving means that our fictitious battery also acts as a spin sink in the system since a spin-dissipation mechanism can only be provided by the itinerant electrons through the open boundaries. As shown in Fig. 7.19, the variation of the electron-spin density in the steady-state regime (with respect to the distribution at $t = 0$) is significant only inside the region swept by the moving wall (see Fig. 7.18(f)). In that region the on-site spin polarization changes sign following the new local spin direction and if non-adiabatic torques are present a transverse spin could be dynamically accumulated. We calculate the total spin along the z-direction as a function of time

$$S_{\text{tot}}^z = \sum_{i=1}^{N} \langle \sigma^z \rangle_i + \sum_{i=1}^{N} S_i^z. \tag{7.40}$$

This consists of an electronic and a local-spin contribution (Fig. 7.19). Initially only the electron spin decreases in time until the DW is reached by the electron flux (the classical spin is constant while the wall is static). When steady-state transport is established, the wall moves uniformly and $\sum_i S_i^z$ decreases linearly with nearly eight times (8.067) the slope of $\sum_i \langle \sigma^z \rangle_i (t)$. The ratio of the spin loss rates of electrons and local spins for adiabatic rotation is fixed by the band filling. Since $\rho_0 = 1.75$, there is an average on-site spin density of $0.25 S_i^z / 2$ (in the uniform regions away from the wall). A local spin flip costs $2 S_i^z$ and it is accompanied by an electron spin loss of $0.25 S_i^z$. Hence, this produces the 8:1 ratio. The ratio that we find in this simulation is 8.067, which differs from the expected 8 by only 0.8%. This gives an estimation of the magnitude of electron transverse-spin accumulation due to non-adiabatic effects.

If the DW is pinned (the local spins are affixed and not included in the dynamics), however, we find no net spin variation for steady-current flow (dotted curve in Fig. 7.19). This demonstrates that we have a spin-flipping battery and no longitudinal-spin accumulation—an electron entering the chain from the left-hand side as a spin-down flips at the wall and leaves the wire as a spin-up at the right-hand side boundary but then another electron with a spin-down enters the wire from the other side.

We define the current at the wall I_{DW} as the spatially averaged bond current in a region of $\mathcal{Z}_{\text{DW}} = \{ z_0 \pm 20 z_w \}$ around the center of the wall (which contains all, say N_{DW}, sites with substantial transverse spins) and then averaged over the steady-state time of the simulation and the corresponding DW velocity, V_{DW}, for that period as

$$I_{\text{DW}} = \frac{1}{N_{\text{DW}}(N_T - N_0 + 1)} \sum_{n=1}^{N_{\text{DW}}} \sum_{m=N_0}^{N_T} I_n(t_m),$$

$$V_{\text{DW}} = \frac{1}{N_T - N_0 + 1} \sum_{m=N_0}^{N_T} v_{\text{DW}}(t_m), \tag{7.41}$$

[10]The fact, pointed out by Li and Zhang (2004), that the factor b_J in the adiabatic torque in eqn (7.29) represents the DW velocity V_{DW} can be understood from the following argument. Let us consider our 1D DW made of discrete classical spins from the moving (classical) reference frame of an electron from the flux. There, the resting electron spin s is experiencing a passing-by at a constant velocity $\widetilde{V}_{DW}$ planar spin-wall. If the spin is aligning instantaneously (we are not interested in the actual mechanism of alignment for this argument) to the spin of the wall then it will be rotating with an angular velocity $\omega_s = \dot{\theta}_z = \widetilde{V}_{DW}\nabla_z\theta_z(z_s + \widetilde{V}_{DW}t)$, where z_s is the position of the spin, θ_z is the wall angular distribution with respect to $\hat{\mathbf{z}}$ (e.g. eqn (7.36)). Such a rotation could effectively be produced by a torque $|T_e| = s\omega_s = s\widetilde{V}_{DW}\nabla_z\theta_z(z_s + \widetilde{V}_{DW}t)$ and by the second Newton law such a torque, but in the opposite direction, must be acting on the wall. In the frame of the wall the torque on the wall at location z_i is similarly $T_{DW}(z_i) = s\widetilde{V}_e\nabla_z\theta_z(z_i)$. We go back to the laboratory reference frame. As the spin conservation holds (see Fig. 7.14), if the spin s passes the DW of discrete spins $\{S_i\}$ for time Δt then $\Delta s = -2sV_e\Delta t/z_w = -\Delta S = -2SV_{DW}\Delta t/a$, where z_w is the whole length of spatial variation of spin in the wall (not the same quantity as in eqn (7.36)), and therefore the velocities of the spin and the wall are related as $V_e/V_{DW} = Sz_w/sa$ so that in the adiabatic limit $z_w \gg a$ and, moreover, if $S \gg s$ (classical local spins) then $V_e \gg V_{DW}$ and $\widetilde{V}_e \approx V_e$. We obtain for the torque due to one electron $T_{DW}(z_i) = V_{DW}S(z_w/a)\nabla_z\theta_z(z_i)$, which is indeed proportional to the wall velocity.

Fig. 7.19 Time evolution of the S^z_{tot} in a 200-atom chain and its electronic (dashed curve) and the local-spin (dot-dashed curve) contributions. The applied bias at $t = 0$ is 0.1 V and the DW is initially in the center of the chain ($z_0 = 100.5a$). The dotted curve represents a test system, where an identical DW is initially prepared at $z_0 = 150.5a$ and is kept fixed for the whole simulation. This curve has actually been rigidly shifted vertically by $S^z_{tot}(t = 0)$ to allow for a comparison.

where N_0 is some time step after the steady-state transport is established and $v_{DW}(t)$ is the momentary DW velocity, obtained by a numerical differentiation of the wall position $z_0(t)$, which is defined through the equation $S^z(z_0) = 0$ and $S^z(z)$ is an interpolation of the set $\left\{\left[z_i, S^z_i\right]\right\}_{z_i \in \mathcal{Z}_{DW}}$. We then examine how I_{DW} and V_{DW} depend on the choice of model parameters ΔV, J, J_z (see Fig. 7.20). Our reference set of parameters is

$$\Delta V = 0.1\,\text{V}, \quad J = 1\,\text{eV}, \quad J_z = 5\,\text{meV}, \quad J_S = 20\,\text{meV}. \quad (7.42)$$

As the only notion of distance in our model is the lattice spacing a, we define our current density at the DW as $j_{DW} = I_{DW}/a^2$. For the reference case above and a typical value $a = 2\,\text{Å}$ our current density is about $8\,\mu A/4\,\text{Å}^2 = 20 \times 10^9\,\text{A/cm}^2$ and DW velocity is about 4.3 km/s.

By varying the initializing bias voltage $0 < \Delta V < 0.15\,\text{V}$ we find a perfectly linear correlation between V_{DW} and I_{DW} (Fig. 7.20(b)) with a slope of around $2.725\,a/(\text{ps}\,\mu A)$ or $2.725 \times 10^{18}\,a/\text{C}$. We compare this result to the expected maximal (when there is spin damping) DW velocity $V_{ad} = -b_J = -Pj\mu_B/eM_s$ in the adiabatic limit[10] (Li and Zhang 2004; Stiles *et al.* 2007). In the expression for V_{ad}, P is the spin polarization of the current, j is its density and M_s is the saturation magnetization in the system. These quantities translate to our one-dimensional system as $j \to I_{DW}/a^2$, $P = -1$ (half-metal) and $M_s \to 2\mu_B S_s/a^3$, where $S_s = 1 + (2 - \rho_0)/2 = 1.125$ (in units of $\hbar$ and $\rho_0 = 1.75$) is the total on-site spin (localized plus conduction band). Hence, we obtain approximately $V_{ad}/I_{DW} \approx a/eS_s = 2.743 \times 10^{18}\,a/\text{C}$. This value is different from our calculated slope (Fig. 7.20(b)) by less than 1%. This spectacular agreement suggests that the main diving torque for the DW motion we simulate is indeed adiabatic. Because of the different dimensionality of the systems we are comparing, we cannot say accurately what is the contribution

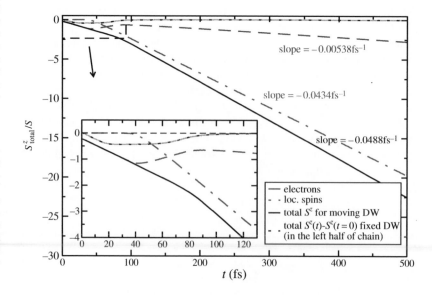

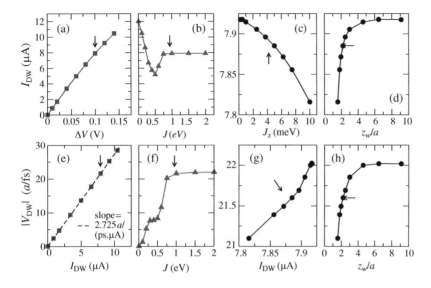

Fig. 7.20 Dependence of the average steady-state current I_{DW} and the corresponding average DW velocity on bias ΔV ($J = 1$ eV, $J_z = 5$ meV) or on J, J_z and z_w for fixed $\Delta V = 0.1$ V. Same symbols correspond to the same calculation. In particular, the variation of the arguments z_w (DW width) in panels (d,h) and I_{DW} in (g) are generated by changing the anisotropy J_z in (c). Arrows are pointing to calculations with the reference set of parameters (7.42).

of the non-adiabatic torques in our simulations but apparently it is small compared to the adiabatic torque.

Next, we examine the dependence of the current and the DW velocity on the parameters in our model. We vary J (or J_z) with respect to the reference state (7.42). Dependence of both current and velocity on J (Figs. 7.20(b,f)) show some peculiarity around the critical J of the transition to half-metallicity—a dip and a plateau, respectively. In the half-metallic regime, however, they both saturate. That is where our reference state is. The change of the anisotropy J_z with respect to the reference state affects the width of the wall z_w (see Fig. 7.16(c)). Both I_{DW} and V_{DW} appear rather insensitive to the variations of z_w. We observe an overall change in V_{DW} of less than 5% and less than 2% in I_{DW} for a nearly fivefold increase in the DW thickness (Figs. 7.20(d,h)). The fact that $V_{DW}(I_{DW})$, obtained by the variation of the anisotropy, is almost linear (panel (g)) suggests that the main effect of the different J_z is on the conductance of the chain.

We find that thinner walls are slightly more resistive and because of the decreased current move slower. Though the overall variation in the conductance is small, the drop for thin walls is relatively steep and suggests non-adiabaticity, i.e. the inability of the conduction electron spins to follow the local spin direction while traversing the wall (van Hoof *et al.* 1999). The adiabaticity is controlled by the *s-d* exchange parameter J. In our reference case ($J = 1$ eV) corresponds to a half-metallic state and we find that I_{DW} and V_{DW} are quite insensitive to a further increasing J (enhancing the adiabaticity) for a fixed anisotropy (i.e. DW width) (panels (b,f)), meaning that our reference state is already close to the adiabatic. Again, what we see is a very small difference in the conductance for the range of wall widths we are working with (varying by a factor of 5). Both V_{DW} and I_{DW} show a tendency to saturation with the DW thickness and our reference state is on the verge of the saturation thickness, where the transport is adiabatic.

We now take a closer look at the details of the time evolution of walls with different thicknesses. In particular, we compare our reference case (eqn (7.42), relaxed width is $z_w = 2.2a$) to a more than 4 times thicker wall with $z_w = 9.1a$ (obtained for $J_z = 0.25$ meV). Here, the overall length of the chain is increased (from 200 to 300 atoms) to allow for longer-lasting simulations and all remaining parameters (apart from anisotropy and voltage) are as in the reference case. The real-space evolutions of the local spins in the two cases are presented in Fig. 7.21, where the initially applied bias is $\Delta V = 0.2$ V. The length of our simulations is limited by the accumulation of numerical error during the time integration and the approach of the DW to the boundaries but we can safely reach times of about 2 ps (8×10^5 time steps) during which the DWs appear to be moving rather uniformly. In both cases we see a very pronounced spin-wave pattern developing in time and the onset of deformations in the shape of the thicker wall (see Fig. 7.21). Interestingly, spin-waves develop at a much earlier stage in time for the thin wall than for the thick one. This indicates that the current-induced torques on the wall clearly depend on its width. There are qualitative differences even for the small range of widths investigated here. The thicker wall firstly deforms (see Fig. 7.21(f)) by wriggling out of plane until some critical deformation point, which is the onset of the spin-waves.

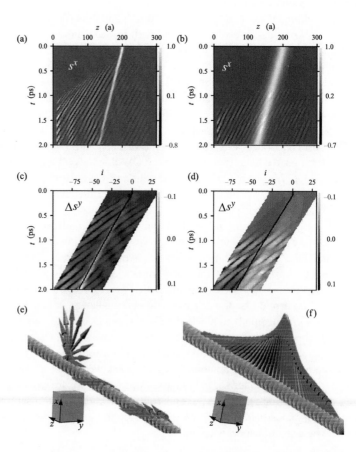

Fig. 7.21 Real-space and -time evolution of two DWs of different width. The left-hand side panels describe a thin DW ($z_w = 2.2a$ obtained with $J_z = 5$ meV), while the right-hand side panels are for a thick one ($z_w = 9.1a$ obtained with $J_z = 0.25$ meV). Panels (a,b) represent $S_i^x(t)$ and (c,d) represent the temporal variation of S_i^y in the vicinity of the wall (we define $\Delta S_i^y \equiv S_{z_0(t)+i}^y - S_{z_0(0)+i}^y$ for $i \in [-30, 30]$). The magnitude of these quantities is specified by individual grayscale bars. The bottom panels (e,f) depict the DWs at 0.75 ps. The initially applied bias is $\Delta V = 0.2$ V and the solid black lines in the panels (c) and (d) mark the exact DW center $z_0(t)$.

This suggests that indeed non-adiabatic effects (like the spin waves) arise from the increased gradients in the local-spin texture (Berger 2007).

Another non-adiabatic effect we find is that thin walls ($z_w < 3a$) show a tendency to precess rigidly about the z-axis. This is illustrated in Fig. 7.22(e) where the time evolution of the S^x of the classical spins in the middle of the domain wall is fitted to $\cos(\omega t)$. We find that the precession rate ω scales almost linearly with the applied bias voltage.

In contrast, thick walls develop a butterfly-like out-of-plane deformation with the spins in the center remaining in the xz-plane (Fig. 7.22(f)). Apart from that, both types of walls move rather uniformly with very similar velocities. One could speculate that the thin wall decelerates slightly although the timeframe of our simulations does not allow us to conclude whether or not the actual average DW velocity changes in time in the long-term limit.

Evidently, both the DW velocity and the angular velocity of precession are characterized by large fluctuations. These are clearly correlated as the oscillations in the velocity match in time those of the transverse spin in the

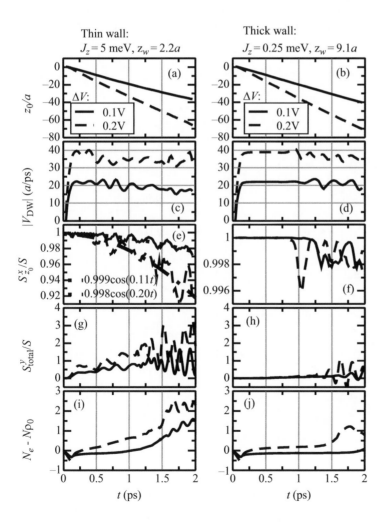

Fig. 7.22 Time evolution of (a,b) the position of the DW's center z_0 along the wire, (c,d) the average velocity V_{DW}, (e,f) the transverse local-spin component S^x at z_0, (g,h) the total transverse spin in the system $S_{total}^y = \sum_i \left(\langle \sigma^y \rangle_i + S_i^y \right)$ and (i,h) the variation of the number of itinerant electrons N_e with respect to the equilibrium. All the quantities are plotted for two different bias voltages, 0.1 V and 0.2 V. The left- and right-hand side panels represent thin and thick DW, respectively. Note that z_0 is defined such that $S_{z_0}^z = 0$, where $S^z(z)$ is an interpolation of $\{S_i^z\}$ with the classical DW-profile function (eqn (7.36)). $S_{z_0}^x$ is also the interpolation of $\{S_i^x\}$. The fluctuations in (c), (d), (e), (f), (g) and (h) correlate to the spin-wave patterns in Fig. 7.21.

DW center (see Fig. 7.22). They also both correlate with the spin-wave pattern illustrated in Fig. 7.21. Hence, we attribute variations in both the DW velocity and precession angular velocity to spin-wave emission. In our computational scheme the spin-waves are not fully absorbed at the boundaries, nevertheless, they can provide an actual spin-relaxation mechanism.

We also find a small electron-number and transverse spin-density accumulation that scales with the bias and is more pronounced for the thinner wall (Figs. 7.22 (g–j)). In order to isolate this effect we calculate the trajectory of the non-equilibrium spin in the system that is defined as

$$\mathbf{S}_{ne} = \sum_i \mathbf{S}_i - 8 \sum_i \langle \sigma \rangle_i . \tag{7.43}$$

In Fig. 7.23 we plot the time variation of this quantity $\Delta \mathbf{S}_{ne}(t) = \mathbf{S}_{ne}(t) - \mathbf{S}_{ne}(0)$. The factor 8 before the total electronic spin in eqn (7.43) is due to the equilibrium ratio of the on-site spin magnitudes – local spin $\hbar$ and itinerant spin-density $\hbar/8$ for a band filling of 1.75e/atom. Thus, $\Delta \mathbf{S}_{ne}(t)$ represents the accumulated non-equilibrium itinerant spin during the simulation. We find that after the initial, mainly longitudinal (along z), spin disruption preceding the steady state, $\Delta \mathbf{S}_{ne}(t)$ falls into a precession very much confined to the xy-plane. Hence, the accumulated spin during the current-induced DW motion is primarily transverse and it scales with the bias. With respect to this observation

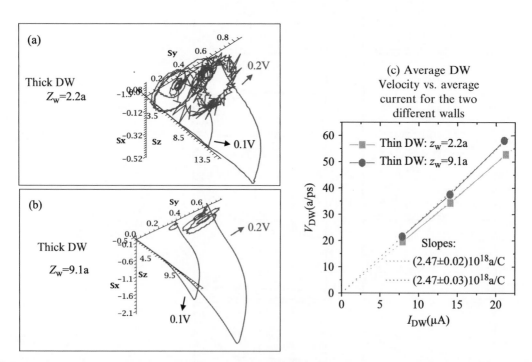

Fig. 7.23 Trajectories in spin-space of the non-equilibrium itinerant spin accumulation defined as $\Delta \mathbf{S}_{ne}(t) = \mathbf{S}_{ne}(t) - \mathbf{S}_{ne}(0)$, where $\mathbf{S}_{ne}(t)$ (eqn (7.43)) is the total non-equilibrium itinerant spin at time t for the thin (a) and for the thick wall (b) at two values of the initial bias voltage $\Delta V = 0.1$ V and 0.2 V. These are the same simulations as those in Fig. 7.22. The origin corresponds to the tip of the spin vector at $t = 0$. In panel (c) are the extracted plots of the averaged over the whole simulation DW velocity versus the averaged current at the wall.

the two walls are very much alike. The difference is mainly in the amplitude of precession of $\Delta \mathbf{S}_{ne}(t)$ about the y-axis, which is perpendicular to the plane of the wall and we find higher amplitude oscillations for the thinner (and expectedly more non-adiabatic) wall.

In Fig. 7.23(c) are the plots of the averaged DW velocity V_{DW} against the average current I_{DW} at the wall during the whole simulation for each of the two walls. We again find a signature for the higher level of spin-transport non-adiabaticity for the thin wall as it is slower for nearly the same value of the current, which suggests a reduced efficiency of the STT mechanism. The relative difference in the slopes is about 8%.

7.4.3 Comparison with the LLG equations

In this last section we compare our combined quantum-classical simulations with the analytical predictions of Zhang and Li (2004). We recall that the Zhang and Li (ZL) theory is derived from the s-d model by integrating out the electronic degrees of freedom in the linear response limit. As a result, the dynamical evolution of the magnetization is then driven by two current-induced torques, one originating from the adiabatic dynamics and one taking into account non-adiabatic effects. Their effective Landau–Lifshitz–Gilbert (LLG) equation of motion, adapted to our discrete local-spin realization, reads

$$\frac{d\mathbf{S}_i}{dt} = \frac{1}{\hbar} \left(J_S \sum_{j \in nn[i]} \mathbf{S}_i \times \mathbf{S}_j + 2J_z(\mathbf{S}_i \cdot \hat{\mathbf{z}})(\mathbf{S}_i \times \hat{\mathbf{z}}) \right)$$
$$+ b_J \frac{\partial \mathbf{S}_i}{\partial z} + c_J \mathbf{S}_i \times \frac{\partial \mathbf{S}_i}{\partial z} + \alpha \mathbf{S}_i \times \frac{\partial \mathbf{S}_i}{\partial t}, \qquad (7.44)$$

where the terms in the parentheses are the local effective field-like torques and the last three torques are, respectively, the adiabatic term, the non-adiabatic one and a phenomenological spin-damping term. We have then carried out atomistic simulations of eqn (7.44) and in this sense, our simulations are similar to those carried out by Schieback *et al.* (2007). Our results, investigating the various empirical parameters over an order-of-magnitude range, are summarized in Fig. 7.24. As reference values we consider $b_J = 5$ km/s [11], $c_j = 10^{-2} b_j$ (Li and Zhang 2004) and $\alpha = 0.1$.

Bearing in mind that our simulations effectively inspect only the short-timescale region (up to 4 ps) and that a thorough analysis is difficult, we can still conclude that the adiabatic torque is indeed the one driving the DW motion, at least at the beginning of its development. This is demonstrated by the fact that whenever $b_j = 0$ the DW does not move. Such a dynamics is in agreement with the quantum-classical simulations and with our previous observation that even for small DW thicknesses the electron spin-evolution is to a great extent adiabatic. The role of the non-adiabatic torque of ZL form is instead more complicated. In general, it drives the DW precession (see $S_{z_0}^y(t)$ in Fig. 7.24) and has very little effect on the DW velocity. A similar effect is produced by the spin-damping term and indeed we observe situations where

[11] The value of b_J is actually chosen to reproduce similar velocities as the ones from our reference quantum-classical simulation (eqn (7.42)) where $V_{DW} \approx 20a/\text{ps} \approx 5$ km/s for $a = 2.5$Å.

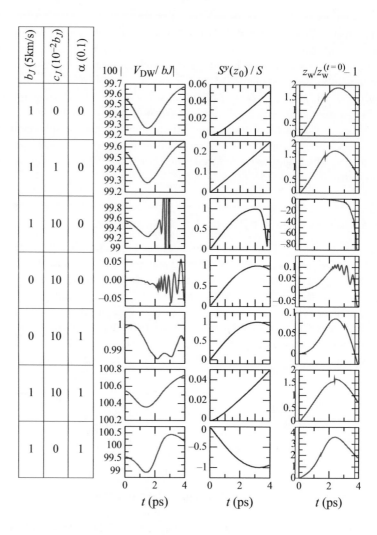

b_J (5km/s)	c_J ($10^{-2}b_J$)	α (0.1)
1	0	0
1	1	0
1	10	0
0	10	0
0	10	1
1	10	1
1	0	1

Fig. 7.24 Results from simulations of a DW in a 400-atom spin-chain modelled by eqn (7.44) for different values of b_J, c_J and α, listed in the table. Plots from left to right represent the time evolution of the DW velocity (factorized by b_J), out-of-plane (y-component) spin at the DW center and relative DW width variation.

the damping and the non-adiabatic torque produce opposite dynamical evolution (see the $(1, 10, 1)$ case, where the DW precession is massively reduced, compared to the $(1, 10, 0)$ and $(1, 0, 1)$ cases where it evolves in different directions). For large values of c_J, however, it can also trigger DW deformation and spin-wave emission.

In conclusion, in our quantum-classical simulations of current-induced DW motion we find similar qualitative features in the DW dynamics as those related to the ZL's non-adiabatic torque, i.e. the tendency to precession, deformation and spin-wave excitation. We, however, do not explicitly include any transverse-spin relaxation mechanism intrinsic to their model. The only source of spin-transport non-adiabaticity in our systems stems from the dynamically evolving texture of the spin-DW structure. As predicted by a few authors (Waintal and Viret 2004; Xiao *et al.* 2006; Berger 2007) non-adiabatic torques can arise from the mere "sharpness" of the wall, determined by non-negligible second gradients in the angular distribution (Xiao *et al.* 2006). Obviously, these effects decay with the width of the wall but the different models predict different exponents in the power-law decay ranging from –1 to –5 (Berger 2007). One other analysis indicates an exponential decay (Xiao *et al.* 2006).

The advantage of our simulations is that they are based on fully quantum ballistic description of the itinerant electrons, there are no assumptions for the spin-density distribution, no linear-response approximation and the texture of the wall, starting from its equilibrium state, is governed by Ehrenfest dynamics. The disadvantage remains in the computational cost and for this reason we restrict ourselves to systems of reduced dimensions with at most ~500 atoms and simulation times of a few picoseconds. For these times it is hard to extract information on the long-time behavior but we can semi-quantitatively analyze the processes that occur at the beginning of the current-induced dynamics. The spatial scale we cover allows us to look at the medium-width walls that are too narrow to be treated in the adiabatic (diffusive) regime, but still not exactly point-like objects to be treated by scattering theory. Though we only observe the non-adiabatic effects related to the abruptness of the wall, we find that, at least in the short-timescale range (~4 ps), the non-adiabatic torques upon the wall on the verge of the atomic-scale abruptness have a minor effect on its average velocity. They, however, cause DW precession, deformation and spin-wave generation.

7.5 Spin-motive force

7.5.1 Background and concepts

In the previous sections we have investigated how an electric current can affect the magnetization landscape of a magnetic nano-object. Here, we explore the opposite effect, namely whether or not a driven magnetization dynamics can generate an electrical signal. Such an effect has been recently predicted by Barnes and Maekawa (2007), who have proposed a generalization of Faraday's law to account for a non-conservative force of spin origin. This arises in systems with time-dependent order parameters as a result of Berry phase (BP) accumulation (Berry 1984). As an example they have considered a DW, formed in a finite ferromagnetic wire and precessing about a static coaxial external magnetic field, i.e. a magnetic system similar to those investigated in the previous sections. They demonstrated that in the adiabatic limit a constant potential shift $\Delta\phi$ is generated between the two ends of the wire. This is directly proportional to the angular frequency of precession of the wall ω,

$$\Delta\phi = \frac{\hbar}{e}\omega. \tag{7.45}$$

Moreover, if the ferromagnet is described by the Stoner mechanism $\Delta\phi$ exactly cancels out the potential produced by the Zeeman interaction. Such a potential, described as a spin-motive force (SMF), has been recognized previously in the context of the Aharonov–Casher (Oh *et al.* 1994; Ryu *et al.* 1996) and Stern's (1992) effects. These are all manifestations of BP-related phenomena, where holonomies arise as a result of a parallel transport of some kind

(Anandan 1992). The latter does not need to be a quantum effect, another example being the classical Foucault pendulum (Wilczek and Shapere 1996).

In what follows we first demonstrate computationally the result of eqn (7.45) through time-dependent quantum-classical simulations of a finite atomic wire incorporating a precessing DW. Then, we also present an analytical classical argument for the driving mechanism of the SMF in this system (Stamenova *et al.* 2008). Our approach has the benefit of being "Berry-phase-free" in the sense that it does not need to call for a Berry-phase argument to explain the SMF and demonstrates the Newtonian nature of the conversion of the magnetic response of electronic spins into an electrostatic voltage drop. This is further illustrated with classical dynamical simulations for a system of classical magnetic dipoles in a rotating magnetic field mimicking the DW. In addition, we show that if one abandons the Stoner model and accounts for a non-spin component of the magnetic moments forming the DW, the cancellation between the SMF and the Zeeman potential predicted by Barnes and Maekawa is incomplete. This leaves behind a non-zero net SMF, which can be experimentally measured.

In order to investigate the SMF generation, we implement the quantum-classical (Ehrenfest) dynamics scheme presented in Section 7.2 with the Hamiltonian from eqn (7.1) and (7.5). The parameters used for the simulations here are $\chi = -1\,\text{eV}$, $\rho_0 = 1.75$, $U = 7\,\text{eV}$, $J = 1\,\text{eV}$, $J_z = 0.5\,\text{meV}$, $g_S = 2$, $a \equiv R_{i,i+1} = 2.5\,\text{Å}$. Note that here we consider interacting electrons and $U \neq 0$, which facilitates the definition of the intrinsic Coulombic potential in the system. $N = 400$ atoms so that the chain is much longer than the typical width of the relaxed DW (about 10 atomic spacings). The values of χ, J and ρ_0 are the same as before and provide for half-metallicity.

The time-dependent simulations proceed as the following. Initially, the set of classical spins $\{\mathbf{S}_i\}^{(0)}$ is prepared in a DW arrangement (see Fig. 7.25) and relaxed self-consistently in the electronic environment. At time $t = 0$ the external magnetic field $\mathbf{B} = B\hat{\mathbf{z}}$ is switched on along the wire and a new initial electronic state is self-consistently determined for $\{\mathbf{S}_i\}^{(0)}$. The system is then propagated according to eqn (7.8).

The electrostatic potentials $V_{\text{L(R)}}$, developing away from the DW on the left(right)-hand-side of the chain, are computed as the spatial (over two identical segments $L(R)$ of N_V atoms at each wire end) and temporal (over the evolution time T) averages of the onsite potential, i.e.

$$V_{\text{L(R)}}(T) = 1/(T N_V) \sum_{i \in L(R)} \int_0^T dt \sum_{n=1}^N \frac{\kappa \, \Delta q_n(t)}{\sqrt{R_{in}^2 + (\kappa/U)^2}}. \tag{7.46}$$

We investigate the stationary voltage drop $\Delta V_{\text{calc}} = \lim_{T \to T_{\text{tot}}} [V_{\text{L}}(T) - V_{\text{R}}(T)]$ building up across the system for the whole duration of the simulation T_{tot}. In the limit of local charge neutrality ($U \to \infty$) this build-up corresponds to the negative of the actual potential-energy landscape in those segments. We anticipate two contributions to the potential

$$\Delta V = \Delta \phi - g_e \mu_B B / e, \tag{7.47}$$

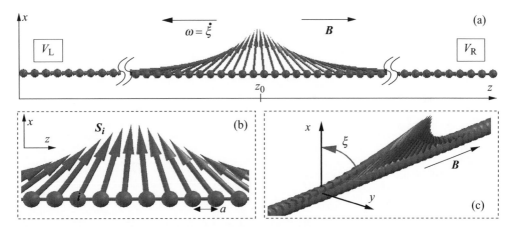

Fig. 7.25 Different prospect views of the DW formed by the local spin $\{\mathbf{S}_i\}$ in the middle of a monoatomic chain. An external magnetic field, applied along the wire, induces a clockwise rotation of the DW about the z-axis. (Reprinted from Stamenova, M., Todorov, T. N., Sanvito, S. (2008) *Phys. Rev. B* **77**, 054439, © 2008 with permission from American Physical Society).

where the first term is due to the proposed non-conservative SMF from eqn (7.45), while the second is due to the Zeeman split.

7.5.2 Quantum-classical simulations

In order to extract the effect of the SMF itself in the first set of simulations we have set $g_e = 0$, so that only the first term in eqn (7.47) remains. In Fig. 7.26 are the calculated time evolutions of some representative dynamical quantities. The DW in these simulations, driven by some large magnetic field, undergoes a steady-rate clockwise rotation (precession) with an angular frequency ω about the direction of the field (see Fig. 7.26(a)) and oscillates gently about a center z_0, slightly displaced to the left (see Fig. 7.26(b)). Note that the wall does not propagate, since there is no net current or dissipation in the system. The steady rotation generates a SMF manifested in a potential drop with small oscillations that correlate with those of the DW center (since the projection of the total spin in the system on the direction of the field is conserved) and that has an asymptotic time-averaged value ΔV_{calc} (see Fig. 7.26(c)). The dependence $\Delta V_{\text{calc}}(\omega)$, obtained by sweeping the external field between 20 T and 500 T, is linear (see Fig. 7.27(a)) with a slope $\hbar_{\text{calc}} = 0.606\,\text{eV fs} \approx 0.92\,\hbar$. Note that the extremely large magnetic fields employed in our simulation are only instrumental and facilitate a fast DW precession and therefore a higher SMF generation. This is necessary to guarantee that our time-dependent simulations do not run over long times, and therefore become numerically unstable. Importantly, our system remains adiabatic even at such large rotation frequencies (see also Fig. 7.30).

The deviation of $\hbar_{\text{calc}}$ from the exact $\hbar$ (from eqn (7.45)) is studied with respect to the two main assumptions in our model: (i) the adiabaticity, which is governed by the strength of the exchange coupling J, and (ii) the local charge neutrality, which allows us to identify $|\Delta V|$ with the SMF and is exact only for $U \to \infty$. The first criterion is found to be well satisfied for $J = 1\,\text{eV}$.

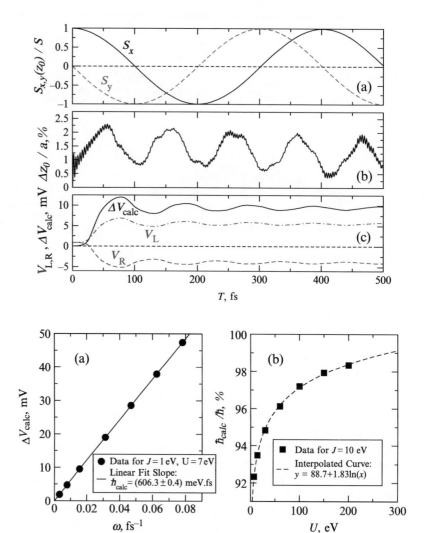

Fig. 7.26 Time evolution of some dynamical variables at $B = 100\,\mathrm{T}$ and for $g_e = 0$: (a) $\mathbf{S}_x$ and $\mathbf{S}_y$ local-spin components at the DW center z_0, showing the clockwise rotation of the DW about the z-axis. The angular frequency ω of the DW precession is extracted by fitting $S_x(T)$ to $\cos(\omega T)$; (b) longitudinal displacement of the DW center z_0; (c) averaged potentials V_L, V_R and ΔV_calc (see text). (Reprinted from Stamenova, M., Todorov, T. N., Sanvito, S. (2008) *Phys. Rev. B* **77**, 054439, © 2008 with permission from American Physical Society).

Fig. 7.27 Here, $g_e = 0$ and $g_S = 2$. (a) The calculated stationary ΔV depends linearly on ω with a slope $\hbar_\mathrm{calc} \approx 0.92\hbar$ for realistic values of the parameters J and U; (b) $\hbar_\mathrm{calc}$ tends to saturate at the exact value of $\hbar$ with increasing U. (Reprinted from Stamenova, M., Todorov, T. N., Sanvito, S. (2008) *Phys. Rev. B* **77**, 054439, © 2008 with permission from American Physical Society).

Increasing J ten times results in less than 1% improvement in $\hbar_\mathrm{calc}$. The ratio $\hbar_\mathrm{calc}/\hbar$, however, is found to be sensitive to U and it asymptotically tends to 1 as U is increased and it is demonstrated in Fig. 7.27(b). This result confirms the validity of eqn (7.45) and indeed demonstrates that a SMF originates from the precession of the DW.

In reality, however, the effect of the applied magnetic field on the electrons cannot be switched off. We, therefore, return to eqn (7.47) and rewrite it in the form

$$\Delta V = \frac{\hbar}{e}\omega - \frac{g_e}{g_S}\frac{\hbar}{e}\omega_S = \left(1 - \frac{g_\mathrm{e}}{g_S^*}\right)\frac{\hbar}{e}\omega. \qquad (7.48)$$

Here, $\omega_S = g_S\mu_\mathrm{B}B/\hbar$ is the Larmor frequency of the local spins. The actual angular frequency of precession of the DW ω differs slightly from ω_S due to the exchange interaction with the conduction electrons. In order to account for

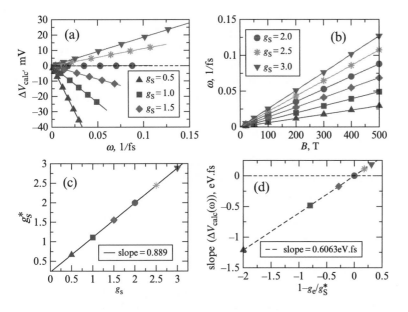

Fig. 7.28 Here, $g_e = 2$ and $g_S = 0.5\text{--}3$. Panel (a) shows the linear dependence of the stationary potential build-up ΔV_{calc} on ω; (b) is used to determine the effective g-factors g_S^* and they are compared to the input values g_S in (c). Note that $g_S^* = g_S$ for $g_S = g_e = 2$. Panel (d) illustrates the agreement with eqn (7.48). (Reprinted from Stamenova, M., Todorov, T. N., Sanvito, S. (2008) *Phys. Rev. B* **77**, 054439, © 2008 with permission from American Physical Society).

this effect, we have introduced an effective g_S^* such that $\omega = g_S^* \mu_B B / \hbar$. We have verified eqn (7.48) numerically by varying the value of g_S (see Fig. 7.28). The effective value g_S^* is determined by the calculated precession frequency of the wall (Fig. 7.28(c)). It is equivalent to $g_S = g_S^* = 2$ only when $g_e = g_S = 2$. In any other case the conduction electrons act either as a friction ($g_e < g_S$) or as a driving force ($g_e > g_S$) to the DW rotation. The first manifestation of such effect was observed in Section 7.3.1, where we have shown an increase of the spin-wave dispersion bandwidth as a function of the exchange parameter J (see Fig. 7.6). Finally, we have again obtained a value of $\hbar_{\text{calc}} \approx 0.92\,\hbar$, identical to the previous finding in the case $g_e = 0$ for this choice of exchange parameter and charging strength.

Apparently, the voltage drop across the system fully disappears when $g_S = g_S^* = g_e$, as derived by Barnes and Maekawa (2007) for the Stoner model. However, in s-d systems, where g_S^* has a partially orbital origin, this is not the case and the SMF would manifest itself as a measurable quantity. This could, in principle, be used to determine the effective g-factor of the localized spins. In particular, if the DW precession is blocked, the measured drop would be just equal to the Zeeman split, i.e. a measurement could determine if the wall is precessing or not. In the next section we discuss the mechanism for the SMF from a classical viewpoint.

7.5.3 Purely classical simulations

Instead of quantum electrons (as in the previous section) we now consider non-interacting classical particles with an intrinsic angular momentum $\mathbf{s}(|\mathbf{s}| = s = \hbar/2)$ and with the electron mass m_e. These are trapped in a one-dimensional box, immersed in a magnetic field of the form

$$\mathbf{b}(z, t) = b(\cos(ft)\sin(\theta_z), \sin(ft)\sin(\theta_z), \cos(\theta_z)). \qquad (7.49)$$

[12]The energy-minimizing angular distribution of a planar DW in the one-dimensional anisotropic Heisenberg spin model is $\theta_z \equiv \theta(z) = 2\arctan(\exp((z-z_0)/z_w))$, where z_0 is the center of the wall and z_w is its width. Note that such a magnetic field is irrotational, i.e. it cannot be derived as a curl of a vector potential. This means that, strictly speaking, our predictions are not experimentally verifiable. In this case **b** is simply instrumental, used to map our quantum simulations onto classical ones.

Here, $\theta_z = \theta(z)$ is chosen such as to mimic a continuous DW structure,[12] rotating rigidly with an angular frequency f. In this classical problem f is analogous to ω from the quantum simulation, although here $f > 0$ corresponds to an anticlockwise rotation about the longitudinal axis. The classical Hamiltonian of the spin-particles (or classical magnetic dipoles) in the field **b** is analogous to that of the quantum electrons interacting with local spins $\{S_i\}$,

$$\mathcal{H}_{\text{class}} = \frac{p^2}{2m_{\text{e}}} - \gamma \mathbf{s} \cdot \mathbf{b}(z,t), \tag{7.50}$$

where γ is the coupling strength (replacing J of the quantum case) and p is the canonical momentum of the particles. Then, Hamilton's equations of motion (Aharonov and Stern 1992) are

$$m_{\text{e}}\ddot{z} = \gamma \mathbf{s} \cdot \nabla_z \mathbf{b}(z,t), \quad \dot{\mathbf{s}} = \gamma \mathbf{s} \times \mathbf{b}(z,t), \tag{7.51}$$

where $\nabla_z \equiv (\partial/\partial z)_{\mathbf{s},t}$.

We consider the limit of large γb, in which the dynamics of the spin-particle becomes adiabatic, in the sense that **s** remains closely aligned with **b** and its precession about **b** is by far the fastest motion in the system. However, for **s** to follow $\mathbf{b}(z,t)$, there must always be *some* residual misalignment (Berger 1986) between the two. This is necessary in order to generate those torques, which, when averaged over the quick precession of **s**, enable **s** to keep up with $\mathbf{b}(z,t)$. This small misalignment, marked by the angle φ in Fig. 7.29, is also the origin of the effective Newtonian force on the spin-particle that manifests itself as SMF.

The cartoon in Fig. 7.29 depicts **s** and **b** at some instance of the particle's migration and helps us to calculate the forces acting on the classical dipoles. This is achieved by differentiating the relation between the angles at the bottom vertex of the tetrahedron $\cos \alpha = \cos \varphi \cos \theta - \sin \varphi \sin \theta \cos \beta$ under the condition $s_{\varphi 1} = s \sin \varphi \sin \beta = \text{const}$ (which corresponds to keeping **s** and t fixed). In the adiabatic limit $\varphi \rightarrow 0$, we obtain

$$\nabla_z \varphi = -\cos(\beta)\nabla_z \theta. \tag{7.52}$$

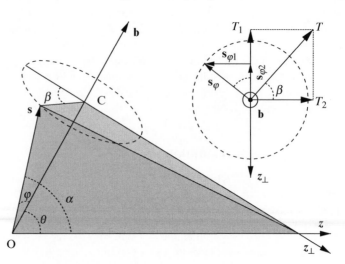

Fig. 7.29 A snapshot of a spin-particle's passage through the DW-like region of $\mathbf{b}(z,t)$. (Reprinted from Stamenova, M., Todorov, T. N., Sanvito, S. (2008) *Phys. Rev. B* **77**, 054439, © 2008 with permission from American Physical Society).

From eqn (7.51) the longitudinal force F_z and the torque $T = |\mathbf{T}| = \gamma s_\varphi b$ are related by

$$F_z = -\gamma|\mathbf{s}||\mathbf{b}|\sin(\varphi)\nabla_z\varphi = -T\nabla_z\varphi = T_2\nabla_z\theta, \qquad (7.53)$$

where eqn (7.52) has been applied and the full torque T decomposed into two orthogonal torques with magnitudes $T_1 = T\sin\beta$ and $T_2 = T\cos\beta$ (see the inset of Fig. 7.29).

In the adiabatic regime ($s_\varphi \ll s$), we average the two components of the torque over the fast precession of $\mathbf{s}$ about $\mathbf{b}$. These averaged torques $\overline{T_1}$ and $\overline{T_2}$ must be driving the two separate motions of the spin as the particle crosses the region of the rotating magnetic field. A rotation in the bz-plane enables $\mathbf{s}$ to keep up with the spatial variation of $\mathbf{b}$, and another rotation in a plane perpendicular to the z-axis, which makes $\mathbf{s}$ follow the anticlockwise precession of $\mathbf{b}$, and thus

$$\overline{T_1} \approx s\dot{z}\nabla_z\theta, \quad \overline{T_2} \approx -|\mathbf{s} \times \mathbf{f}| = -sf\sin\theta. \qquad (7.54)$$

From eqns (7.53) and (7.54), we obtain for the averaged linear force upon the spin-particle in $\mathbf{b}(z,t)$ $\overline{F_z} = \overline{T_2}\nabla_z\theta = sf\sin(\theta)\nabla_z\theta$ and therefore the work done by the rotating DW-like field (or the SMF) on the spin-particle for one passage (left to right) is

$$W_{L\to R} = \int_{z_L}^{z_R} \overline{F_z}\mathrm{d}z = -s\ f \int_0^\pi \sin(\theta)\mathrm{d}\theta = -2sf, \qquad (7.55)$$

where $z_{L,R}$ are the leftmost and the rightmost positions of the spin-particle on the wire far from the region of spatial variation of the field. This result has been derived with the single assumption of adiabaticity, i.e. $s_\varphi \ll s$, which translates into

$$\overline{s_{\varphi 1}} = \overline{T_1}/\gamma b = sv\nabla_z\theta/\gamma b \ll s, \quad \overline{s_{\varphi 2}} = \overline{T_2}/\gamma b = sf\cos(\theta)/\gamma b \ll s. \qquad (7.56)$$

Thus, considering the maximum attainable values of the right-hand sides and using the fact that $\max(\nabla_z\theta) = 1/z_w$, the necessary conditions for adiabaticity become

$$1/t_w \ll f_L, \quad f \ll f_L, \qquad (7.57)$$

where $t_w = vz_w$ and $f_L = \gamma b$ is the Larmor precession frequency of the itinerant spin about $\mathbf{b}$.

Simulations show that the adiabatic conditions in eqn (7.57) apply also to the quantum case. As illustrated in Fig. 7.30, the SMF effect dies out completely above the Larmor precession frequency $\omega_L = J/\hbar$ of the exchange-coupled spins for any choice of band filling ρ_0 (corresponding to a different Fermi level spin-polarization η). This is another demonstration of both the relevance of the foregoing classical interpretation and the good local spin-adiabaticity in such model one-dimensional metallic systems with atomic-scale DWs and strong $s - d$ exchange (~ 1 ev).

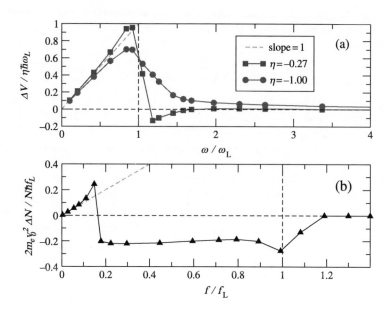

Fig. 7.30 The normalized SMF as a function of the DW rotation frequency in units of the Larmor precession frequency of the itinerant spins about the local field. We present the results from simulations away from the adiabatic regime eqn (7.57). Here, $\eta = (DOS_\uparrow(E_F) - DOS_\downarrow(E_F))/(DOS_\uparrow(E_F) + DOS_\downarrow(E_F))$ is the Fermi-level spin-polarization. The dashed pale line corresponds to a slope of 1. (Reprinted from Stamenova, M., Todorov, T. N., Sanvito, S. (2008) *Phys. Rev. B* **77**, 054439, © 2008 with permission from American Physical Society).

Classical simulations based on eqn (7.51), reported in Stamenova *et al.* (2008), further support the classical interpretation that the $\hbar$-factor in eqn (7.45) represents the magnitude of the electron spin $2s = \hbar$.

7.6   Conclusions

In this chapter we have presented and discussed the theoretical foundations of atomistic time-dependent simulations for both open and closed systems. In particular, we have discussed a range of phenomena including domain-wall motion and spin-wave emission, where truly atomistic spin-dynamics can provide unique insights into the device behavior, well beyond what is possible with steady-state quasi-static approaches. Although our calculations have been carried out in the framework of the self-consistent tight-binding method, these are easily extendable to more general first-principles approaches such as DFT. We envision that these methods will soon become an invaluable tool for investigating spin-dynamics in nano- and atomic devices.

Acknowledgments

This work is supported by Science Foundation of Ireland under the grant 07/IN.1/I945. Computational resources have been provided by the HEA IITAC project managed by the Trinity Center for High Performance Computing and by the Irish Center for High End Computing (ICHEC). The authors wish to warmly thank Tchavdar Todorov for inspiring discussions and for his critical review of the manuscript.

References

Acremann, Y., Strachan, J.P., Chembrolu, V., Andrews, S.D., Tyliszczak, T., Katine, J.A., Carey, M.J., Clemens, B.M., Siegmann. H.C., Stöhr, J. *Phys. Rev. Lett.* **96**, 217202 (2006).

Aharonov, Y., Stern, A. *Phys. Rev. Lett.* **69**, 3593 (1992).

Anandan, J. *Nature* **360**, 307 (1992).

Antropov, V.P., Katsnelson, M.I., van Schilfgaarde, M., Harmon, B.N. *Phys. Rev. Lett.* **75**, 729 (1995).

Antropov, V.P., Katsnelson, M.I., Harmon, B.N., van Schilfgaarde, M., Kusnezov, D. *Phys. Rev. B* **54**, 1019 (1996).

Ashcroft, N.W., Mermin, N.D. *Solid State Physics* (Saunders College Publishing, Philadelphia, PA, 1976).

Baibich, M.N., Broto, J.M., Fert, A., Van Dau, F.N., Petroff, F., Eitenne, P., Creuzet, G., Friederich, A., Chazelas, J. *Phys. Rev. Lett.* **61**, 2472 (1988).

Barnes, S.E., Maekawa, S. *Phys. Rev. Lett.* **95**, 107204 (2005).

Barnes, S.E., Maekawa, S. *Phys. Rev. Lett.* **98**, 246601 (2007).

Berger, L. *J. Appl. Phys.* **49**, 2156 (1978).

Berger, L. *Phys. Rev. B* **33**, 1572 (1986).

Berger, L. *Phys. Rev. B* **54**, 9353 (1996).

Berger, L. *Phys. Rev. B* **75**, 174401 (2007).

Berry, M.V. *Proc. R. Soc. Lond.* **45**, 392 (1984).

Binasch, G., Grnberg, P., Saurenbach, F., Zinn, W. *Phys. Rev. B* **39**, 4828 (1989).

Brataas, A., Bauer, G.E.W., Kelly, P.J. *Phys. Rep.* **427**, 157 (2006).

Capelle, K., Vignale, G., Györffy, B.L. *Phys. Rev. Lett.* **87**, 206403 (2001).

Car, R., Parrinello, M. *Phys. Rev. Lett.* **60**, 204 (1988).

Costa, A.T., Muniz, R.B., Ferreira, M.S. *New J. Phys.* **10**, 063008 (2008).

Dugaev, V.K., Vieira, V.R., Sacramento, P.D., Barna, J., Arajo, A.N., Berekdar, J. *Phys. Rev. B* **74**, 054403 (2006).

Engel, B.N., Rizzo, N.D., Janesky, J., Slaughter, J.M., Dave, R., DeHerrera, M., Durlam, M., Tehrani, S. *IEEE Trans. Nanotech.* **1**, 32 (2002).

Gambardella, P.J. *Phys.: Condens. Matter* **15**, S2533 (2003).

Gerrits, Th., van den Berg, H.A.M., Hohlfeld, J., Bär, L., Rasing, Th. *Nature* **418**, 509 (2002).

Grollier, J., Cros, V., Hamzic, A., George, J.M., Jaffrs, H., Fert, A., Faini, G., Ben Youssef, J., Legall, H. *Appl. Phys. Lett.* **78**, 3663 (2001).

Grollier, J., Boulenc, P., Cros, V., Hamzic, A., Vaurs, A., Fert, A., Faini, G. *Appl. Phys. Lett.* **83**, 509 (2003).

Haney, P.M., Waldron, D., Duine, R.A., Nzqez, A.S., Guo, H., MacDonald, A.H. *Phys. Rev. B* **76**, 024404 (2007).

van Hoof, J.B.A.N., Schep, K.M., Brataas, A., Bauer, G.E.W., Kelly, P.J. *Phys. Rev. B* **59**, 138 (1999).

Horsfield, A.P., Bowler, D.R., Fisher, A.J., Todorov, T.N., Montgomery, M.J. *J. Phys.: Condens. Matter* **16**, 3609 (2004).

Hueso, L.E., Pruneda, J.M., Ferrari, V., Burnell, G., Valdes-Herrera, J.P., Simons, B.D., Littlewood, P.B., Artacho, E., Fert, A., Mathur, N.D. *Nature* **445**, 410 (2007).

Katine, J.A., Albert, F.J., Buhrman, R.A., Myers, E.B., Ralph, D.C. *Phys. Rev. Lett.* **84**, 3149 (2000).

Kim, E., Lee, T.H. *Phys. Rev. B* **49**, 1743 (1994).

Kirwan, D.F., Rocha, C.G., Costa, A.T., Ferreira, M.S. *Phys. Rev. B* **77**, 085432 (2008).

Kläui, M., Vaz, C.A.F., Bland, J.A.C., Wernsdorfer, W., Faini, G., Cambril, E., Heyderman, L.J. *Appl. Phys. Lett.* **83**, 105 (2003).

Krivorotov, I.N., Emley, N.C., Sankey, J.C., Kiselev, S.I., Ralph, D.C., Buhrman, R.A. *Science* **307**, 228 (2005).

Li, Z., Zhang, S. *Phys. Rev. Lett.* **92**, 207203 (2004).

Litvinov, V.I., Dugaev, V.K. *Phys. Rev. B* **58**, 3584 (1998).

Liu, K.L., Vosko, S.H. *Can. J. Phys.* **67**, 1015 (1989).

Maple™ 10, Waterloo Maple Inc.

McEniry, E.J., Bowler, D.R., Dundas, D., Horsfield, A.P., Sanchez, C.G., Todorov, T.N. *J. Phys.: Condens. Matter* **19**, 196201 (2007).

Oh, S., Ryu, C.M., Suck Salk, S.H. *Phys. Rev. A* **50**, 5320 (1994).

Parkin, S.S.P., Hayashi, M., Thomas, L. *Science* **320**, 190 (2008).

Pratzer, M., Elmers, H.J., Bode, M., Pietzsch, O., Kubetzka, A., Wiesendanger, R. *Phys. Rev. Lett.* **87**, 127201 (2001).

Rippard, W.H., Pufall, M.R., Kaka, S., Russek, S.E., Silva, T.J. *Phys. Rev. Lett.* **92**, 027201 (2004).

Ryu, C.M. *Phys. Rev. Lett.* **76**, 968 (1996).

Sanchez, C.G., Stamenova, M., Sanvito, S., Bowler, D.R., Horsfield, A.P., Todorov, T.N. *J. Chem. Phys.* **124**, 214708 (2006).

Schieback, C., Kläui, M., Nowak, U., Rüdiger, U., Nielaba, P. *Eur. Phys. J. B* **59**, 429 (2007).

Slonczewski, J. *J. Magn. Magn. Mater.* **159**, L1 (1996).

Stamenova, M., Todorov, T.N., Sanvito, S. *Phys. Rev. B* **77**, 054439 (2008).

Stern, A. *Phys. Rev. Lett.* **68**, 1022 (1992).

Stiles, M.D., Saslow, W.M., Donahue, M.J., Zangwill, A. *Phys. Rev. B* **75**, 214423 (2007).

Stiles, M.D. *Theory of Current Induced DW Motion*, http://cnst.nist.gov/epg/Projects/Theory/theory_cidwm.html (2007).

Tatara, G., Kohno, H.E. *Electron Microc.* **54**, 169 (2005).

Tatara, G., Kohno, H., Shibata, J.J. *Phys. Soc. Jpn.* **77**, 031003 (2008).

Theilhaber, J. *Phys. Rev. B* **46**, 12990 (1992).

Todorov, T.N. *J. Phys.: Condens. Matter* **14**, 3049 (2002).

Todorov, T.N. private correspondence (2007).

Tserkovnyak, Y., Brataas, A., Bauer, G.E.W., Halperin, B.I. *Rev. Mod. Phys.* **77**, 1375 (2005).

Tsoi, M., Fontana, R.E., Parkin, S.S. P. *Appl. Phys. Lett.* **83**, 2617 (2003).

Vanhaverbeke, A., Viret, M. *Phys. Rev. B* **75**, 024411 (2007).

Viret, M., Berger, S., Gabureac, M., Ott, F., Olligs, D., Petej, I., Gregg, J.F., Fermon, C., Francinet, G., Le Goff, G. *Phys. Rev. B* **66**, 220401 (2002).

Waintal X., Viret, M. *Europhys. Lett.* **65**, 427 (2004).

Wang, C.Z., Chan, C.T., Ho, K.M. *Phys. Rev. B* **42**, 11276 (1990).

Wilczek, F., Shapere, A. *Geometrical Phases in Physics* (World Scientific, Singapore, 1996).

Xiao, J., Zangwill, A., Stiles, M. D. *Phys. Rev. B* **73**, 054428 (2006).

Yang, K.H., Hirschfelder, J.O. *Phys. Rev. A* **22**, 1814 (1980).

Yosida, K. *Theory of Magnetism* (Spinger-Verlag, Berlin, 1996).

Zhang, S., Li, Z. *Phys. Rev. Lett.* **93**, 127204 (2004).

8 Patterns and pathways in nanoparticle self-organization

MO Blunt, A. Stannard, E. Pauliac-Vaujour, CP Martin, Ioan Vancea, Milovan Šuvakov, Uwe Thiele, Bosiljka Tadić, and P. Moriarty

8.1 Introduction

The assembly of 2D and 3D "supracrystals" from colloidal nanoparticles is a well-established and, in some respects, maturing subfield of twenty-first century nanoscience. Nevertheless, there remain many unresolved questions with regard to the control of the assembly process and, in particular, the degree to which the dynamics of solvent dewetting can be coerced to produce self-organized structures with predictable and tunable morphologies. Key advances in understanding and controlling the self-assembly and self-organization of nanoparticles on solid substrates have, however, recently been made and in this chapter we review the work of a number of groups, including our own, who have contributed in this area over the past decade. We focus on the assembly and properties of *2D* nanoparticle arrays. Moreover, we draw a strong distinction between the terms *self-assembly* and *self-organisation* where the former is used to describe a close-to-equilibrium process that produces a nanoparticle array whose correlation length/periodicity is dictated by interparticle interactions; the latter term describes a far-from-equilibrium process involving matter/energy flow and where correlations over distances much greater than the interparticle spacing can develop.

A number of important reviews of nanoparticle synthesis and self-assembly/organization have been published over the past decade by leaders in the field including Murray *et al.* (2000), Pileni *et al.* (2001, 2006, 2007), Mirkin *et al.* (Storhoff and Mirkin 1999), and Heath *et al.* (Collier *et al.* 1998). Our aim in this chapter is to complement, rather than revisit, the material covered in those reviews and thus we largely focus on relatively recent forms of self-assembly and self-organization that have demonstrated particular potential for the assembly of nanostructured matter, namely biorecognition and solvent-mediated dynamics. Having discussed the key elements of these processes, we then in Section 8.2 move on to describe the mechanisms and pathways for charge transport in nanoparticle superlattices and, in particular, the relationship between the current–voltage characteristics and the topology of the lattice.

8.2 Self-assembled and self-organized nanoparticle arrays

8.2.1 Self-assembly of nanoparticle superlattices

Self-assembly and self-organization of nanoparticle superlattices can lead to a vast variety of complex micro-, meso-, and macroscopic structures. Indeed, the appearance of emergent properties on large length scales is the basis for new functional nanomaterials, thus shifting the horizons of nanotechnology well beyond the limits of classical crystalline solids. There are numerous examples of structure–function interdependencies in assembled nanosystems. Here, we list only a few:

- Inorganic nanoparticles assembled via reverse micelles have been organized into two- and three-dimensional supracrystals (Pileni 2001, 2006). It was shown that the collective dynamical effects induced by the regular arrangements of the nanoparticles cause new optical, magnetic and charge-transport properties (Pileni 2006, 2007), compared to isolated nanoparticles and to the crystalline bulk form of the material;

- Gold nanoparticles may form different three-dimensional structures if they are linked via programable biorecognition between attached compatible parts of DNA strands (Mirkin *et al.* 1996, 2000; Winfree *et al.* 1998; Nykypanchuk *et al.* 2008; Park *et al.* 2008). In recent experiments (Nykypanchuk *et al.* 2008; Park *et al.* 2008) regular fcc structures were observed;

- Antigen–antibody recognition with complementary molecules attached to nanoparticles in a solvent, often exploited in biosensing applications (Liron *et al.* 2001), can be used to tune nanoparticle aggregation to form particular types of structure;

- Chemical functionalization of gold nanoparticles allows interactions between particles to take place along pre-specified directions (Archer *et al.* 2007). Attaching such nanoparticles to a branched polymer structure produces an assembly with a novel magnetic response. Theoretical investigations (Tadić 2007) of the field-driven magnetization reversal processes on strongly inhomogeneous spin networks suggest entirely different hysteresis-loop properties, compared to the classical memory materials with moving domain walls. In particular, the coercive field increases with topological disorder. Hysteresis curves have been measured for homogeneous nanoparticle arrays (Duruöz *et al.* 1995; Pileni 2006);

- Arrays of metallic nanoparticles on substrates are examples of systems that conduct current via single-electron charging, a property of crucial importance in nanoelectronics (Ferry *et al.* 1997; Guéron *et al.* 1999). A striking feature of conducting nanoparticle superlattices is the emergent non-linearity of the current–voltage curve, which, as discussed below, strongly varies with the structure of the array (Parthasarathy *et al.* 2001; Blunt *et al.* 2007). In Section 8.2 we discuss the processes that underpin these non-linear effects.

In the following, we expand on a number of these topics in detail. Starting with a discussion of the control of self-assembly via biorecognition, we subsequently review solvent-mediated self-organization and pattern formation, before, in Section 8.2, discussing the interplay of morphology/topology and charge transport in 2D nanoparticle supracrystals.

8.2.2 Exploiting biorecognition for nanoparticle assembly

A number of methods have been developed for the assembly of nanoparticles ligated with complementary chemical entities (Mirkin 1996, 2000; Nykypanchuk *et al.* 2008; Park *et al.* 2008). Consider two types of colloidal particles, A and B, which are ligated by biopolymers, as shown schematically in Fig. 8.1(a), such that the complementary parts of the molecule are attached to different particle types. Some well-known examples are proteins with specific interactions, e.g., in antigen–antibody recognition (Huber 1986), and in complementary DNA strands (Nykypanchuk *et al.* 2008; Park *et al.* 2008), where the strength of bonding is directly related to the number of complementary base pairs. It was demonstrated recently (Nykypanchuk *et al.* 2008; Park *et al.* 2008) that by varying the lengths of the free and the bonding parts of the ligands, different three-dimensional structures of nanoparticles may emerge, including regular bcc-like supracrystals.

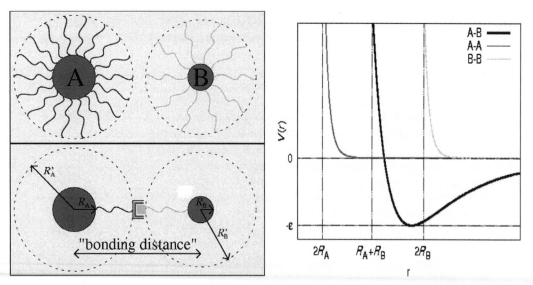

Fig. 8.1 (a) Schematic view of biorecognition bonding between pairs of ligated colloidal particles A and B, and (b) the associated potentials in the binary colloidal mixture with particle radii R_A and R_B (Šuvakov and Tadić 2008).

The attractive interactions between such particles arise due to both van der Waals interactions and hydrogen bonding and are readily described in the literature by the attractive part of the α-Lennard-Jones potential (Mossa *et al.* 2004; Zaccarelli 2007). The attraction occurs at distances above the "bonding distance", d_B, which depends on the particle sizes and the lengths of the attached biopolymers (see Fig. 8.1). At the "bonding distance" particle pair A–B binds together via hydrogen bonding (Nykypanchuk *et al.* 2008) or click-chemistry mechanisms (Brennan *et al.* 2006). At smaller distances $d < d_B$, however, a repulsive interaction starts building up, first as a soft interaction between attached polymer chains, and eventually reaching hard-core repulsion at the "touching distance" $d = R_A + R_B$, as shown in Fig. 8.1(b). We also assume that a large number of polymer chains of the same kind are symmetrically distributed over the particle surface, so that the radial symmetry of the potential is preserved. Hence, in the absence of long-range electrostatic repulsion, the potential for the effective interaction between pairs of particles of different kind is given by:

$$V_{A-B}(r) = 4\epsilon \left(\frac{\sigma^{12}}{(r - R_A - R_B)^{12}} - \frac{\sigma^6}{(r - R_A - R_B)^6} \right), \qquad (8.1)$$

where r denotes the distance between particles, R_A and R_B are their radii and σ and ϵ are the usual Lennard-Jones parameters. Here, we use $\alpha = 6$, the classical version of the Lennard-Jones potential, given in eqn (8.1), which corresponds to a wide potential well and, thus, allows particle attraction across a wider area. (The potential well becomes increasingly narrow for larger α values (Mossa 2004).) The depth ϵ measures directly the binding strength of the biorecognition pair (Jelesarov 1996). Similarly, the optimal binding distance, corresponding to the potential minimum in Fig. 8.1(b), is related to the parameter σ in view of eqn (8.1). Thus, σ should be appropriately selected for a given particle size. For pairs of equivalent particles, A–A and B–B, we assume repulsive interactions described by the first term in the Lennard-Jones potential:

$$V_{X-X}(r) = 4\epsilon' \frac{\sigma'^{12}}{(r - 2R_X)^{12}}, \qquad (8.2)$$

where R_X stands for R_A or R_B, and the parameters σ' and ϵ' may, in principle, be different from the those in the A–B interaction.

The kinetics of the binary colloidal mixture can be described by the Langevin equation with the interaction between pairs of nanoparticles given by eqns (8.1) and (8.2). The Langevin equation of motion for particles in high-viscosity media is given by:

$$\nu_i \dot{\mathbf{r}}_i = -\nabla_{\mathbf{r}_i} \sum_j V_{i-j}(|\mathbf{r}_i - \mathbf{r}_j|) + F_{i,T}, \qquad (8.3)$$

where r_i is the position of ith particle and ν_i represents $\nu_A \equiv 6\pi\eta R_A$ or $\nu_B \equiv 6\pi\eta R_B$, when the ith particle is of the type A or type B, respectively. η is the fluid viscosity coefficient. The stochastic Langevin force $F_{i,T}$ can be seen as originating from the integration over the fluid degrees of freedom. In the simplest case the fluctuations of the Langevin term are uncorrelated, and described by the moments of a normal distribution with

$$\langle F_{i,T}(t) \rangle = 0, \tag{8.4}$$

$$\langle F_{i,T}(t) F_{j,T}(t') \rangle = 6k_B T v_i \delta_{i,j} \delta_{t,t'}. \tag{8.5}$$

The standard meaning of these terms, with respect to the Einstein–Smoluchowski relation, is that the diffusion coefficient D_i of the ith particle is given by $D_i = k_B T / v_i$, which is generally different for the particles of A and B type, and $k_B T$ is the thermal energy.

We used the fourth-order Runge–Kutta method $(RK4)$ to solve the differential eqn (8.3) for the trajectories of colloids of both A and B type. In the simulations the following set of parameters was specified: the size R_A, R_B and number N_A, N_B of the particles of type A and B, the respective viscosity (or diffusion) coefficients v_A and v_B, the temperature T, and the parameters of the potential σ and $\epsilon/k_B T$. We start with a random distribution of $N = N_A + N_B$ particles in a box of size $L \times L$ with an infinite potential at the boundaries. At each time step $k \times \Delta t$ the positions of the particles are updated in parallel according to eqn (8.3). A small time step Δt is chosen so that the updated positions of the particles result in a smooth change of the potential. The random Langevin force $F_{i,T}$ is distributed once at each time step according to the symmetric normal distribution with the second moment given by eqn (8.5). Details of the numerical implementation with increased efficiency algorithm are given in Šuvakov and Tadić (2008).

Note that, due to the ligands attached to each particle, the larger radii marked with R'_A and R'_B in Fig. 8.1(a), are relevant for the definition of the *coverage* of the box with particles. In the simulations, when two particles bind with compatible chemistry, they stay at the binding distance (which may be larger than the particle radii). A free particle, however, cannot move across the bond. Such constraints on particle motion induce additional effective interactions, making the system condense at non-equilibrium global energy states. These effects depend on the overall particle density and the fraction of the attractive particle pairs compared to the total number, N_A/N. In principle, at any finite temperature a bond may break. The process is, however, less probable at low temperatures. Here, we keep $k_B T \epsilon \ll 1$. The system can be additionally "equilibrated" by a sudden increase and subsequent drop of temperature.

For a given ratio of the particle sizes R_A/R_B, emergent structures of the nanoparticle aggregates depend on the coverage, $\Phi = A/L^2$, where A is the total area covered by the particles, and on the relative concentrations of two particle types N_B/N_A. In Figs. 8.2(a) and (b) we show assemblies obtained with the same total number of particles but with different fractions of the large (A) and small (B) particles. Note that the ratio of the particle sizes R_A/R_B determines the maximum number of small particles that can attach to a single large particle. Local structures with large particles linked via small particles occur, which are joined in an irregular pattern, with voids of different sizes. The walls of such voids are built of large particles in the case of

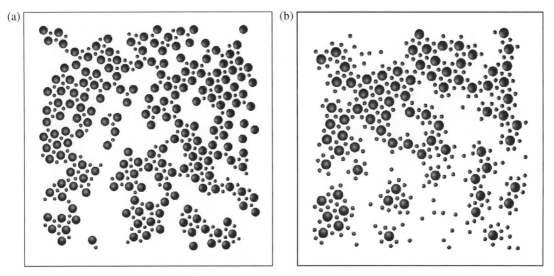

Fig. 8.2 Three-dimensional rendering of simulated nanoparticle assemblies obtained by diffusion and aggregation with biorecognition bonding: two types of particles with size ratio $R_A/R_B = 2.5$ and fixed total number $N_A + N_B$. Concentration ratio, $N_B/N_A = $ (a) 1 and (b) 4.

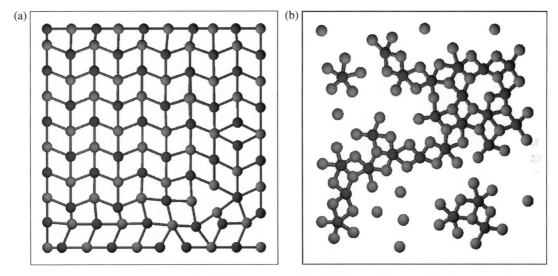

Fig. 8.3 Simulated nanoparticle structures obtained by diffusion and aggregation with biorecognition potentials: Occurrence of periodic structure for equal particle sizes with (a) full coverage with equal concentrations of A and B particles and (b) 50% coverage with concentration ratio 1:3 (Šuvakov and Tadić 2008).

equal concentrations. However, for an abundance of the small particles, as in Fig. 8.2(b), aggregates of different sizes are often isolated by small particles.

An entirely different situation is found when the particle sizes are equal, i.e. the particles differ only by the ligand type. In Fig. 8.3(a) we show the situation corresponding to equal concentrations of particles, when a periodic structure arises. A regular pattern with alternating A-B-A-B particles spans almost the whole system, with only a few topological defects. In the case when one particle type is abundant compared to the other, such structural regularity

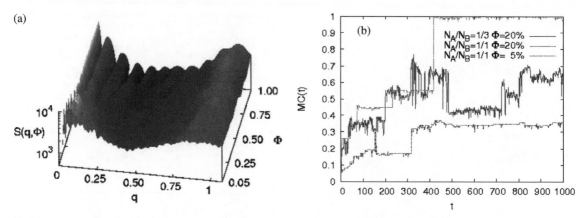

Fig. 8.4 (a) Structure factor $S(q)$ for nanoparticle aggregates obtained by biorecognition bonding for equal size particles and varied coverage Φ. (b) Growth of the largest cluster of connected nanoparticles against time for various values of coverage Φ and concentration ratios N_A/N_B.

remains only at the local level. Large as well as isolated clusters may be found, as in Fig. 8.3(b). In both figures the bonding links between the compatibly ligated nanoparticles are shown.

The regularity of the nanopatterns in our 2-dimensional simulations can be quantitatively measured by computing the structure factor $S(q)$ defined by

$$S(\mathbf{q}) = \left| \sum_{i=1}^{N} e^{-i\mathbf{q}\mathbf{r}_i} \right|, \qquad (8.6)$$

where r_i are positions of nanoparticles in the 2-dimensional box. In Fig. 8.4(a) we show $S(q)$ obtained after angular averaging of $S(\mathbf{q})$ and different coverages Φ. The figure shows how the regular structure of nanoparticle assemblies arises, leading to the characteristic oscillatory pattern of the surface $S(q, \Phi)$ (seen in the upper left part), with increased coverage and equal particle size (Šuvakov and Tadić 2008).

The time evolution of the largest cluster, shown in Fig. 8.4(b), reveals the aggregation process in more detail. Depending on the coverage Φ and relative concentrations of the A and B particle types, one can observe that either large clusters join together (marked by large jumps in the upward curves), or small A–B complexes are being attached to a large cluster. In addition, the process is affected by the initial growth of the large cluster, which increasingly slows down cluster diffusion.

8.2.3   Beyond self-assembly: Self-organising nanoparticles

Although liquid-phase self-assembly is an essential component of nanoparticle science, of paramount importance is the intriguing and complex dynamics associated with the transfer of the particles from a colloidal solution to a solid substrate. This involves the evaporative and, in some cases, convective dewetting of the solvent in which the nanoparticles are dissolved. The coupling of the solvent dewetting dynamics with the motion of the nanoparticles leads to a rich parameter space and, thus, a wide variety of complex and correlated nano- and microstructured patterns result from the drying process. In the following sections we review recent developments in the understanding and control of dewetting-mediated pattern formation in nanoparticle assemblies, focusing in particular on the "archetypal" nanoparticle system: thiol-capped Au particles in organic solvents.

8.2.3.1 *Drying-mediated pattern formation*

Figure 8.5 shows a representative subset of self-organized patterns formed by thiol-passivated Au nanoparticles ($\sim$2 nm diameter) via dewetting of a droplet of colloidal solution on native oxide-terminated Si(111) substrates, taken from the Nottingham group's work on this system. The patterns span a range of morphologies observed not only for Au nanoparticles but a considerable number of other colloidal particle types (including Ag, Co, ferrite, and CdS): isolated droplets, dendritic aggregates, branching and fingering structures, labyrinthine assemblies, cellular networks (on a number of different scales), and isolated rings. While the precise formation mechanisms are in some cases not fully understood, much of the key physics driving the self-organisation process has been elucidated by a number of groups.

Key insights into the physics of the self-organization of colloidal nanoparticles have arisen from the work of Ge and Brus (2000) and Rabani *et al.* (2003). Ge and Brus (2000) proposed that self-organization of colloidal nanoparti-

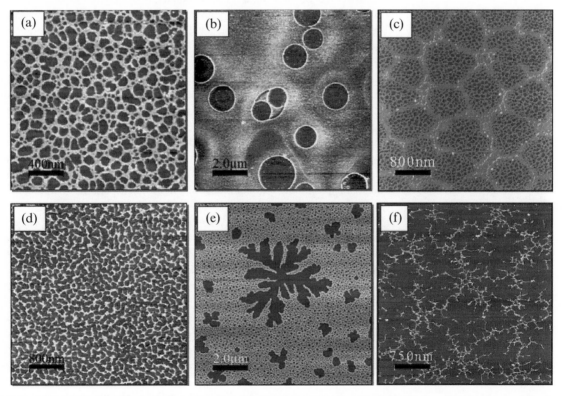

Fig. 8.5 A selection of tapping-mode atomic force microscope (AFM) images of the patterns formed by thiol-passivated Au nanoparticles deposited from an organic solvent onto silicon substrates. The scale bars in the images are: (a) 400 nm, (b) 2 μm, (c) 800 nm, (d) 800 nm, (e) 2 μm, (f) 750 nm.

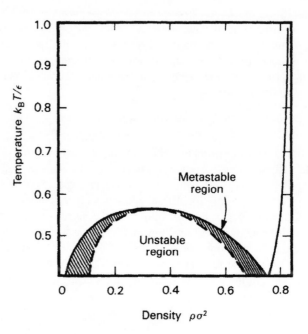

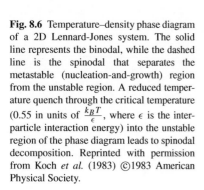

Fig. 8.6 Temperature–density phase diagram of a 2D Lennard-Jones system. The solid line represents the binodal, while the dashed line is the spinodal that separates the metastable (nucleation-and-growth) region from the unstable region. A reduced temperature quench through the critical temperature (0.55 in units of $\frac{k_B T}{\epsilon}$, where ϵ is the interparticle interaction energy) into the unstable region of the phase diagram leads to spinodal decomposition. Reprinted with permission from Koch *et al.* (1983) ©1983 American Physical Society.

cles occured through fluid–fluid spinodal phase separation of a 2D system of particles interacting via a Lennard-Jones-type potential. Figure 8.6 is the universal 2D phase diagram (taken from Koch *et al.* (1983)) around which Ge and Brus based their arguments. Perhaps the most important point with regard to colloidal nanoparticle systems is that evaporation of the solvent is equivalent to a reduced-temperature quench, driving the system from the gas–liquid coexistence region (i.e. the upper region of Fig. 8.6) to phase separation, either in the unstable region of the phase diagram (spinodal decomposition) or in the metastable (nucleation and growth) regime. The critical temperature, in units of $k_B T/\epsilon$ is 0.55, at a coverage of 0.325 monolayer (ML). Removal of the solvent results in less effective screening of the van der Waals interactions between the nanoparticles, driving the phase transition, and producing intricate, spatially correlated patterns of nanoparticles, highly reminiscent of those observed in other systems that phase separate via a spinodal process (binary fluids, polymer blends)—see, for example, Fig. 8.5(d).

Ge and Brus' model captured essential elements of the self-organization of colloidal nanoparticles but, as pointed out by Rabani *et al.* (2003), it neglected the role of solvent fluctuations. Rabani and colleagues put forward an important Monte Carlo model of colloidal nanoparticle (increasingly called "nanofluid") systems that took into account the dynamics of solvent evaporation. In their lattice gas model, the 2D cells are of the size of the liquid coherence length. A dynamic variable ℓ_i, representing fluid density at the ith cell may take two states $\ell_i = 1, 0$, corresponding to the presence or the absence of the liquid. Nanoparticles occupy a given fraction of cells with a density $n_i = 1$ if the nanoparticle occupies the cell i excluding the fluid at that cell, and $n_i = 0$, otherwise. Generally, a single nanoparticle occupies several cells.

Typically, in the simulations the linear size of a nanoparticle is taken to be three times the liquid cell size (Rabani *et al.* 2003; Martin *et al.* 2004).

Attractive interactions occur between pairs of nanoparticles (interaction strengths ϵ_n), liquid–liquid (ϵ_ℓ) and nanoparticle–liquid ($\epsilon_{n\ell}$) according to the Hamiltonian

$$\mathcal{H} = -\epsilon_\ell \sum_{<ij>} \ell_i \ell_j - \epsilon_n \sum_{<ij>} n_i n_j - \epsilon_{n\ell} \sum_{<ij>} n_i \ell_j - \mu \sum_i \ell_i, \qquad (8.7)$$

where $<ij>$ indicates interaction pairs, either nearest neighbors (Rabani *et al.* 2003) or next-nearest neighbors (Martin *et al.* 2004; Yosef and Rabani 2006; Martin *et al.* 2007). The chemical potential μ and the temperature T of the heat bath regulate the equilibrium liquid–vapor concentration.

The kinetics of the model comprise two processes: nanoparticle diffusion and liquid evaporation/condensation. The particle mobility rate (MR) is an important control parameter and determines the relative frequency of events due to these processes. A key element of the model is that nanoparticles can diffuse only in the presence of solvent in cells neighboring the particle. The Monte Carlo dynamics for each of the two processes is performed with the acceptance probability:

$$p_{\text{accept}} = \min(1, \exp[-\Delta H_\kappa / k_B T]), \qquad (8.8)$$

where ΔH_{evap} and ΔH_{diff} are the changes in the total energy associated with the attempted process, computed from the Hamiltonian (8.7). (Modifications to the Hamiltonian have been recently introduced (Martin *et al.* 2007; Pauliac-Vaujour *et al.* 2008) for reasons to be discussed below.) Figure 8.7 shows a region of the "phase space" for the Rabani model where some of the primary types of nanoparticle assembly observed experimentally—networks, worm-like domains and labyrinths, and droplets—are reproduced.

8.2.4 Regimes of organization: Near-to- and far-from-equilibrium

An impressive aspect of the Rabani *et al.* (2003) model is that it accurately reproduces patterns formed in both the metastable (nucleation-and-growth) and unstable (spinodal) regions of the nanoparticle–solvent phase diagram. Rabani and colleagues refer to these as the heterogeneous and homogeneous limits of solvent evaporation, respectively. Experimentally, these regimes can be accessed via control of solvent volatility–higher vapor pressure solvents drives the system towards the spinodal limit of homogeneous solvent evaporation, whereas slower solvent evaporation moves the system into the heterogeneous limit of solvent evaporation. It is also possible to move between these limits simply by controlling solvent-evaporation time. This has the benefit that a variety of nanoparticle patterns, spanning the homogeneous (spinodal) to heterogeneous (metastable) evaporation limits, can be formed on a single sample.

To systematically and controllably vary solvent-evaporation times, we adopted a deposition approach borrowed from previous work on latex spheres

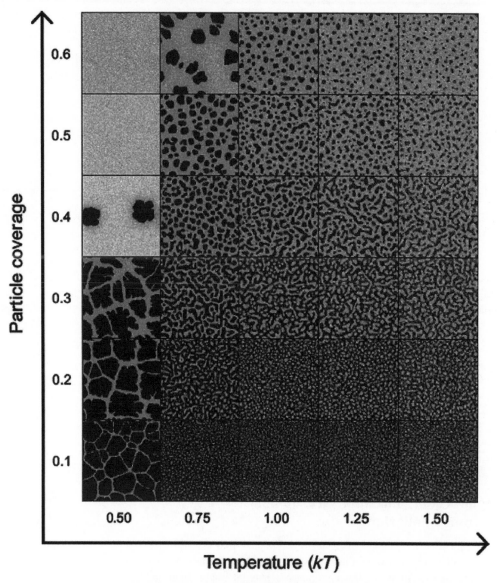

Fig. 8.7 A representative set of nanoparticle assemblies generated using the Rabani *et al.* (2003) Monte Carlo model discussed in the text. A coverage vs. temperature "phase space" is shown where each image is the result of running the simulation for 1000 Monte Carlo steps in a 1024×1024 solvent cell system with $\epsilon_l = 2$, $\epsilon_n = 2\epsilon_l$, $\epsilon_{nl} = 1.5\epsilon_l$, and a mobility ratio of 20.

(Denkov *et al.* 1992; Gigault *et al.* 2001), where a PTFE (Teflon) ring is used to form a meniscus (rather than droplet) geometry for the nanoparticle solution (Pauliac-Vaujour and Moriarty 2007). Figure 8.8 shows a series of atomic force microscope images taken from different areas within the Teflon ring following complete evaporation of the solvent. There is a clear progression from worm-like domains at the center of the ring, arising from rapid (~ a few tens of seconds) solvent evaporation to a close-packed and relatively void-free monolayer at the edge of the ring, where solvent evaporation can take a

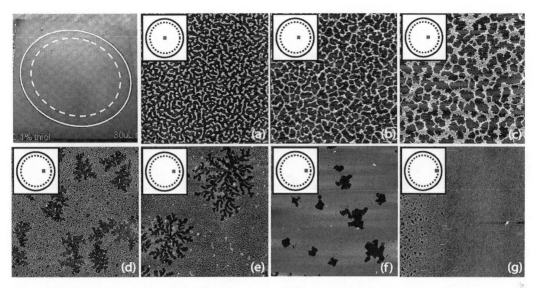

Fig. 8.8 Tapping-mode AFM images of dodecanethiol-passivated Au nanoparticle assemblies on a native-oxide-terminated silicon sample prepared with the meniscus technique described in the text. A photograph of the substrate following evaporation of the Au nanoparticle solution within a teflon ring is shown in the top left. The outer and inner diameters of the ring are shown. The nanoparticle solution used in this case contained 0.1% excess dodecanethiol by volume. In each of (a)–(f), the insert shows schematically the macroscopic region on the sample where the scan is taken. Scans (a)–(c) and (f) are 5 μm × 5 μm in size;(d),(e) and (g) are 20 μm × 20 μm. Reprinted with permission from Pauliac-Vaujour and Moriarty (2007), ©2007 American Chemical Society.

few hours. A variety of patterns are observed moving from the center to the edge of the ring. Importantly, in many cases there is a coexistence of patterns with radically different signature length scales. This is particularly clear in Fig. 8.8(f) where fingered structures reminiscent of those formed in Hele–Shaw cells or during solidification from a melt are surrounded by a high density of small voids forming a nanoparticle network. This observation of pattern formation on distinctly different length scales provides important clues as to the mechanisms of solvent dewetting for nanofluids on solid surfaces and we return to this point below.

In Fig. 8.9, we compare Fig. 8.8(b) with an AFM image taken from the work of Ge and Brus (2000) on CdSe nanoparticle assembly on graphite substrates. Despite the use of entirely different nanoparticle types (Au vs. CdSe) and substrates (SiO$_2$/Si(111) and graphite), the similarity between the images is striking. In each case, solvent evaporation has been rapid, driving the system into the spinodal regime of phase separation. We also include in Fig. 8.9 a nanoparticle distribution generated by the Monte Carlo code described above. Again, there is extremely good qualitative agreement between the simulated pattern and the experimental data. However, and as we have described at length elsewhere (Martin *et al.* 2004, 2007b), a variety of morphological metrics (Fourier analysis; Voronoi tesselations; Minkowski functionals) also show that the images are *quantitatively* virtually indistinguishable.

Rabani *et al.* (2003) have shown that their Monte Carlo code not only captures the final state of the nanoparticle–solvent system (i.e. in the limit where all, or the vast majority, of solvent has evaporated) but the *dynamics*

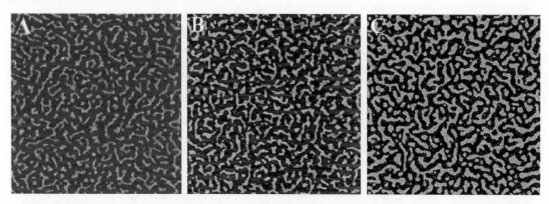

Fig. 8.9 Spinodal nanoparticle assemblies. (A) AFM image of a spinodal pattern formed by CdSe nanoparticles on a highly oriented pyrolytic graphite (HOPG) surface. (Reprinted with permission from Ge and Brus (2000) ©2000 American Chemical Society. The image size is ∼3.5 × 3.5 μm². (B) 5 μm × 5 μm AFM image of dodecanethiol-passivated Au nanoparticles (∼2 nm diameter) on a native-oxide-terminated Si(111) sample. (C) Nanoparticle distribution generated by Rabani *et al.* Monte Carlo algorithm in the spinodal (homogeneous) limit of solvent evaporation. (See Fig. 8.7).

of the system's approach to equilibrium from a far-from-equilibrium state. This evolution towards thermodynamic equilibrium proceeds via self-similar morphologies, is known as *coarsening* or *ripening*, and has been studied, in the context of Ostwald ripening, for example, for a very wide range of systems for over a century. For colloidal nanoparticles (CdSe on graphite), it has been shown that the island distribution scales such that the average size of the domains, r, grows with time, t, as $t^{1/4}$, consistent with a cluster-diffusion-and-coalescence mechanism (Rabani *et al.* 2003). In addition to coarsening via thermally driven nanoparticle diffusion, the AFM probe can be used to *mechanically* drive the transport of nanoparticles and thus locally coarsen a given area. This transport process is akin to a detachment-limited Ostwald ripening process, whereby the tip moves nanoparticles from poorly coordinated sites at the edges of small nanoparticle islands to more highly coordinated, and thus energetically favorable, "bonding" positions at larger islands. The result is again an evolution towards equilibrium via self-similar morphologies but with a scaling exponent of $\frac{1}{2}$, rather than $\frac{1}{4}$, consistent with that expected for detachment-limited Ostwald ripening (Blunt *et al.* 2007a).

Although, as compared to a close-to-equilibrium process, far-from-equilibrium organization generally gives rise to significantly lower levels of long-range order, in a fascinating recent paper Bigioni *et al.* (2006) have shown that a highly non-equilibrium growth mechanism can produce exceptionally ordered Au nanoparticle superlattices (spanning areas of ∼10^8 particles). The key difference between the arrays produced by Bigioni *et al.* (2006) and the nanoparticle assemblies described thus far (and below) is that organization occurs at the solvent *air*, rather than solvent substrate, interface. By ensuring rapid evaporation, so that the solvent air interface moves more rapidly than the diffusing particles, the nanoparticles are trapped at the interface and, as shown by the real-time video microscopy of Bigioni and colleagues, form 2D islands that eventually coalesce into a large-area, and highly ordered, monolayer. In the final stage of drying, the monolayer is deposited on the substrate. Bigioni *et al.*'s work built on earlier important experiments by Narayanan *et al.* (2004) that elucidated, through X-ray scattering measurements, the "interface crushing" effect due to rapid solvent evaporation. With slower evaporation rates, Narayanan *et al.* (2004) found that nanoparticles could diffuse away from the solvent air interface and form 3D supracrystals in the interior of the solvent droplet.

8.2.4.1 *Marangoni convection and network formation*

Although the Rabani *et al.* (2003) Monte Carlo model very successfully predicts and accounts for the observation of a wide range of 2D nanoparticle patterns, the algorithm ignores the contribution of hydrodynamics to the self-organization process (as pointed out by Rabani *et al.* (2003). Prior to the publication of the Rabani *et al.* (2003) model, the important role that Marangoni convection can play in controlling pattern formation in nanoparticle assemblies had been elucidated by Pileni's group at Université Pierre et Marie Curie, Paris. A Bénard–Marangoni instablity—usually referred to as the Marangoni effect—arises when inhomogeneities in surface/interfacial tension, due to either variations in temperature or solute (e.g. nanoparticle) concentration, are minimized by the flow of fluid across the surface from warmer to cooler regions. The net result is a surface-tension-driven convective flow whose stationary state is a hexagonal or polygonal pattern. This pattern has a characteristic wavelength, λ, given by:

$$\lambda = \frac{2\pi h}{\alpha},\tag{8.9}$$

where h is the thickness of the liquid film and α is a dimensionless constant related to the Marangoni number, M_a (also dimensionless) as $M_a = 8\alpha^2$. The Marangoni number was put forward by Pearson (1958) in the late 1950s and provides a measure of the balance of the destabilizing effects of the surface tension gradient ($B = \frac{d\gamma}{dT}$) and the suppression of fluctuations due to the thermal diffusivity, κ, and dynamic viscosity, v, of the liquid:

$$M_a = \frac{B\Delta T h}{\rho v \kappa},\tag{8.10}$$

where ΔT is the temperature gradient across the liquid film and ρ is the liquid density. Far above a critical value of the Marangoni constant, $M_c = 80$, the expression for λ given above (eqn (8.9)) holds and there is thus a well-established relationship between the wavelength of the instability, the temperature gradient across the film, and the variation of surface tension with temperature.

For the volatile solvents used in the nanoparticle experiments described throughout this chapter, the evaporation of the solvent sets up a temperature gradient across the nanofluid film. The more volatile the solvent, the larger the temperature gradient, and thus, from eqn (8.10), the higher the Maranagoni number. Maillard *et al.* (2000) also proposed that higher concentrations of nanoparticles lead to an increase in M_a. Figure 8.10(a), taken from the work of the Pileni group (Maillard *et al.* 2000), shows a TEM image of a network of

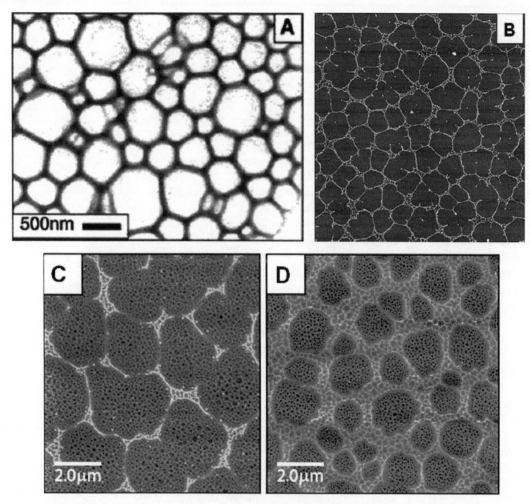

Fig. 8.10 Examples of network formation in nanoparticle self-organization. (A) Transmission electron microscope image of polygonal pattern arising from Marangoni convection in drying Ag nanoparticle solutions. Image reprinted with permission from Maillard *et al.* (2000) ©2000 American Chemical Society. (B) Dual-scale network formation in a submonolayer coverage of pentanethiol-passivated Au nanoparticles (~2 nm diameter) on a native-oxide-terminated Si(111) sample. Image reprinted with permission from Moriarty, Taylor, and Brust (2002), ©2002 American Physical Society. (C), (D) Multiple-scale network formation in assemblies of dodecanethiol-passivated Au nanoparticle films.

silver nanoparticles formed via the Marangoni effect. The polygonal nature of the pattern is clear and the spacing of the cells is broadly in line with estimates of λ based on the experimental parameters. Following the work published by Maillard *et al.* (2000), a detailed study of the formation of honeycomb networks of Au nanoparticles via the Marangoni effect was carried out by Stowell and Korgel (2001). Importantly, they found that a combination of changes in thermal conductivity and *B* as a function of nanoparticle concentration made *"the most dilute solutions more prone to Marangoni convection"*. However, counterbalancing this effect is the necessity to have a sufficiently large number of particles to produce deposits at the edges of the convection cells. Thus, the observation of Marangoni-driven polygonal networks occurs only within a relatively narrow "window" of nanoparticle concentrations.

However, Marangoni convection is not the only mechanism by which nanoparticle networks can be generated. Holes in a dewetting nanofluid open up and grow in either the homogeneous (spinodal) or heterogeneous (nucleation and growth) regimes of solvent evaporation described in the preceding section. For very dilute concentrations of nanofluid solution, rings of nanoparticles can form (see below). With increasing nanoparticle concentration, evaporative dewetting alone—i.e. *without any form of convective fluid motion*—can produce polygonal patterns arising from the collision of expanding holes in the solvent film. Indeed, and as shown in Fig. 8.7, these types of network are readily produced by the Rabani *et al.* Monte Carlo code. Moreover, networks spanning a number of length scales can coexist within the same area on the surface of a sample. Figure 8.10 includes a number of striking images of this effect. In an AFM and statistical crystallography analysis of cellular/polygonal structures formed by nanoparticle assemblies (Moriarty *et al.* 2002) it was proposed that both Maranagoni convection and a spinodal mechanism could give rise to "nested" networks of the type shown in Fig. 8.10.

As described by Martin *et al.* (2004), although a spinodal dewetting/decomposition mechanism is associated with a well-defined preferred length scale that one might not initially expect for networks formed by a nucleation-and-growth process, distinguishing between the morphologies generated by the two types of process is not always straightforward. Although for nucleation-and-growth one would expect the initial dewetting sites to be spatially uncorrelated (i.e. follow a Poisson distribution), the expansion and coalescence of the nucleated holes "washes out" the Poisson distribution, producing a correlation function reminiscent of that for a 2D ensemble of hard discs. That is, there is a reasonably well-defined mean hole separation that appears as a peak in a radially averaged Fourier transform or is evident from Voronoi tesselation and Minkowski functional analyses of AFM images (Martin *et al.* 2004). Similar difficulties in distinguishing spinodal- and nucleation-mediated morphologies are described in the literature on polymer and thin-film dewetting phenomena (see Jacobs *et al.* (1998) and references therein).

8.2.4.2 *Rings and fingering instabilities*

In their studies of the role of the Marangoni effect in driving the self-organization of nanoparticles, Maillard *et al.* (2000) observed the formation of rings of particles whose diameters ranged from a few hundred nm to a few micrometers. Rings of this type had first been reported by Ohara *et al.* (1997) and a typical example, taken from Ohara *et al.*'s work, is shown in Fig. 8.11(a). Many groups, including our own (see Fig. 8.11(d)), have observed ring formation as a result of drying nanofluids on solid substrates: FePt (Zhou *et al.* 2003), polypyrrole (Jang *et al.* 2004), Ni (Cheng *et al.* 2006), barium hexaferrite (Shafi *et al.* 1999), Au, Cu, Co, ferrite, and CdS (Maillard *et al.* 2000) nanoparticles all form well-defined sub-micrometer or micrometer-scale rings.

Ohara and Gelbart (1998) put forward a general theory to explain the organization of nanoparticles into rings. They argued that ring formation is caused by the nucleation of holes in a nanofluid film due to evaporation (as discussed in previous sections) or disjoining pressure. As a hole expands, nanoparticles

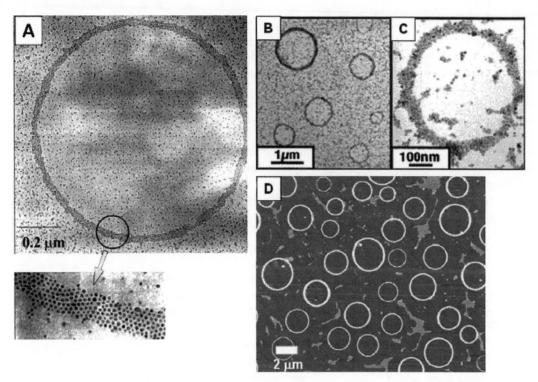

Fig. 8.11 Examples of ring formation in nanoparticle self-organization. (A) TEM image of a ring formed by dodecanethiol-passivated 2.5-nm diameter Ag nanoparticles. Reprinted with permission from Ohara, Heath, and Gelbart (1997), ©1997 American Chemical Society. (B), (C) TEM images of CdS nanoparticle rings. Reprinted with permission from Maillard *et al.* (2000), ©2000 American Chemical Society. (D) AFM image of rings formed by octanethiol-passivated ~2-nm Au nanoparticles deposited from dichlormethane via spin casting onto a native-oxide-terminated Si(111) surface. The solution contained an excess of 0.1% by volume of octanethiol.

"track" its rim (due to solvation forces). In Ohara and Gelbart's model, the hole stops expanding when a sufficiently high density of nanoparticles has been reached. That is, the ring diameter is fundamentally controlled by the frictional force between the nanoparticles and the substrate. A modified version of the Rabani *et al.* Monte Carlo model described in previous sections, which models a 3D rather than 2D nanofluid, has also been used to study ring formation (Yosef and Rabani 2006). Use of a 3D model enabled the rate of solvent evaporation, rate of hole growth, and rate of hole nucleation to be effectively decoupled. Yosef and Rabani found both similarities and key differences with the theoretical approach put forward by Ohara and Gelbart (1998). In particular, their results indicated that the growth of the ring halts not when a critical nanoparticle density is reached but when the thin solvent layer evaporates. Moreover, Yosef and Rabani found that there was a negligible dependence of ring size on nanoparticle density.

In the course of their simulations of ring formation, Yosef and Rabani observed the generation of well-developed fingering instabilities in certain regions of parameter space, as shown in Fig. 8.12. We have recently focused our experiments on the generation of fingering instabilities (Pauliac-Vaujour *et al.* 2008) and find that the nanoparticle diffusion rate (or the effective

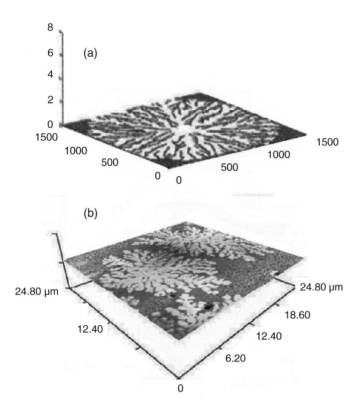

Fig. 8.12 A comparison of fingering structures produced in (a) a 3D Monte Carlo simulation (reprinted with permission from Yosef and Rabani (2006), ©2006 American Chemical Society), and (b) in experiment. AFM image of structure formed in a dewetting nanofluid comprising dodecanethiol-passivated Au nanoparticles ($\sim$2 nm diameter) in toluene with a 0.1% by volume excess of dodecanethiol.

viscosity of the nanoparticle solution) must be tuned appropriately to produce well-developed fingering patterns of the type simulated by Yosef and Rabani. In Fig. 8.12 we show a direct comparison between the patterns generated by Yosef and Rabani and an AFM image taken from the Nottingham group's work on fingering instabilities in colloid Au nanoparticle solutions. Figure 8.13 shows a series of experimental AFM images of self-organized Au nanoparticle monolayers where we have systematically varied the thiol chain length so as to modify the nanoparticle diffusion rate. Addition of a small amount of excess thiol (Figs. 8.13(f)–(j)) increases the viscosity of the nanofluid and dramatically enhances the development of fingered structures.

8.2.4.3 *Combined nucleation-driven and spinodal dewetting*

A striking characteristic of many of the patterns formed by dewetting nanofluids is the presence of self-organization on two or more distinct length scales. Although we touched on this point in the section on network formation above, it is important that we revisit the question of dual-scale (or multiscale) organization via dewetting for two reasons: (i) the organization of matter across a hierarchy of length scales has particular potential in the generation of self-organized materials and devices; and (ii) the presence of multiple length scales in self-organized colloidal nanoparicle arrays provides key insights into the solvent dewetting processes.

The combined effects of nucleation-mediated and spinodal dewetting are responsible for pattern formation on two distinct length scales in the AFM

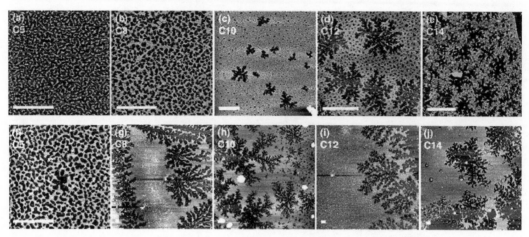

Fig. 8.13 AFM scans of (a,f) C_5-, (b,g) C_8-, (c,h) C_{10}-, (d,i) C_{12}- and (e,j) C_{14}-thiol-passivated gold nanoparticles deposited from toluene onto native-oxide-terminated Si(111). Samples (a–e) were prepared without excess thiol, samples (f–j) with solutions containing 0.1% excess thiol by volume. Figure and caption reproduced from Pauliac-Vaujour *et al.* (2008) with permission ©2008 American Physical Society.

images of fingering structures shown in Fig. 8.13. In Fig. 8.14 we show a high-resolution image where a high density of holes surrounding the branched structure is clearly visible. Importantly, the small holes have a rather narrow size distribution, suggesting that they appeared within a relatively narrow time window. In order to reproduce "dual-scale" pattern formation of this type, we have modified the Hamiltonian of the Rabani *et al.* (2003) model (eqn (8.11)) so that the chemical-potential term is no longer constant but is a function of solvent coverage, ν:

$$E = -\epsilon_l \sum_{<ij>} l_i l_j - \epsilon_n \sum_{<ij>} n_i n_j - \epsilon_{nl} \sum_{<ij>} n_i l_j - \mu(\nu) \sum_i l_i. \qquad (8.11)$$

An important consequence of mapping the chemical potential onto the solvent coverage in this fashion is that we effectively build in some degree of 3D character to the simulated film of solvent (although the Monte Carlo algorithm is still carried out on a 2D grid). Alternatively, and equivalently, if we consider the solvent coverage as representative of the mean thickness of the solvent film, by introducing a solvent-coverage-dependent chemical potential we introduce a disjoining pressure contribution to the Hamiltonian (Martin *et al.* 2007a; Pauliac-Vaujour *et al.* 2008).

If we now make the chemical potential a step (sigmoidal) function of the solvent coverage (Pauliac-Vaujour *et al.* 2008) we can drive the system from a nucleation regime to a spinodal regime of dewetting (or, in the language of Rabani *et al.* (2003), from heterogeneous to homogeneous evaporation) at a critical solvent coverage. Figure 8.14 shows a number of frames from a simulation of fingering instabilities in a nanofluid where we have carried out this switch in chemical potential. Initially, a hole at the center of the simulation grid nucleates and grows until the solvent coverage reaches a value that "triggers" rapid spinodal dewetting and to a large extent freezes further development of the fingers.

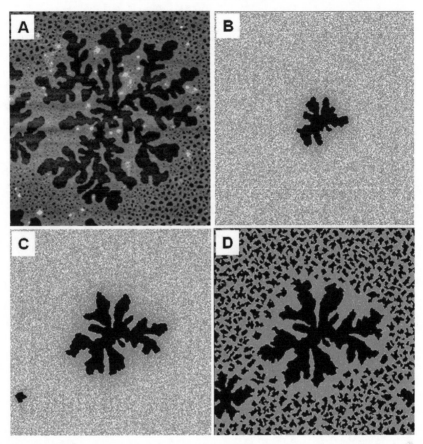

Fig. 8.14 (A) Fingering pattern formed in a submonolayer coverage of dodecanethiol-passivated Au nanoparticles deposited from toluene with an excess of dodecanethiol (0.1% by volume) using the meniscus technique described above. Note the high areal density of holes that surround the fingered structure. (B)–(D) Frames from a Monte Carlo simulation, using the coupled chemical potential–solvent coverage approach described in the text, which show the evolution of a fingered structure and, at late stages of solvent dewetting, the formation of a high areal density of holes due to a spinodal process.

Returning to the subject of dual-scale networks, it is possible to use the coverage-dependent chemical-potential approach to reproduce a number of other morphologies observed experimentally. This is shown in Fig. 8.15 for a power-law (rather than sigmoidal) dependence of the chemical potential on solvent coverage. At values of the exponent A between $\sim$0.3 and 4.0, a very clear *bimodal* distribution of cell sizes is present, in good agreement with experiment. Thus, although Marangoni convection coupled with a spinodal dewetting mechanism certainly can give rise to dual-scale networks (Moriarty *et al.* 2002) (as discussed in Section 8.1.4), the simulation results shown in Fig. 8.15 illustrate that the presence of hydrodynamically driven solvent flow is not a pre-requisite for the generation of multiscale networks. A combination of nucleation-driven and spinodal dewetting alone is sufficient. Similar arguments regarding the joint contribution of nucleation- and spinodal-driven mechanisms have been put forward for dewetting of suspensions of macromolecules.

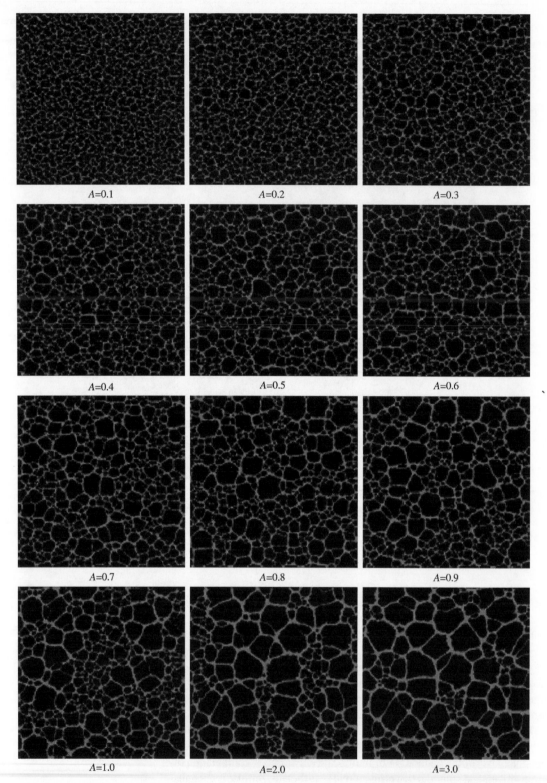

Fig. 8.15 Generation of dual-scale nanoparticle networks via a "dynamic" chemical potential of the form $\mu(v) = \mu_0 + (0.1\mu_0 \times v^A)$, where v is the fraction of solvent that has become vapor, and μ_0 is the value of μ (the chemical potential) at the start of the simulation. All images are 2048×2048 lattice sites with $k_B T = \epsilon_l/2$, $\epsilon_n = 2\epsilon_l$, $\mu_0 = -2.25\epsilon_l$, coverage = 30% and mobility ratio = 30. The value of A is given under each image.

The competition of nucleation and spinodal processes has been discussed by Thiele *et al.* (1998) and Thiele (2003) for systems without evaporation.

8.2.4.4 *Controlling pattern formation via solvent dewetting*

There have been a number of very impressive examples of coercing the solvent dewetting process so as to form highly ordered 1D or quasi-1D structures over very large length scales (up to millimeters square). In particular, elegant control of the motion of the three-phase contact line through the use of a sphere-on-flat geometry (a spherical lens on a silicon wafer) has been demonstrated by Xu *et al.* (2007). They have formed concentric, submicrometer-wide rings of 5.5-nm CdSe/ZnS nanoparticles and, with slightly smaller (4.4 nm) particles, have found that a fingering instability produces "spokes" that propagate transverse to the dewetting front direction. Similar stripe patterns (albeit for somewhat larger particles) have been produced by Huang *et al.* (2005) using a Langmuir–Blodgett approach.

Thus far throughout this chapter we have focused on pattern formation on homogeneous, unpatterned surfaces. An intriguing question, and an issue that has been explored in some depth by the polymer community (see Geoghegan and Kraush 2003 for a review), relates to the influence that a topographically or chemically heterogeneous substrate has on the dewetting process. We have explored this by exploiting a scanning probe technique pioneered by Dagata *et al.* (1991), Lyding *et al.* (Shen *et al.* 1995), and Avouris *et al.* (1997) in the 1990s: local oxidation of hydrogen-passivated silicon surfaces. In our case, a metallized AFM probe is biased with respect to a HF-treated Si(111) sample and used to write patterns of silicon oxide with linewidths down to approximately 20 nm (and feature heights ∼2 nm). The HF treatment renders the surface exceptionally hydrophobic, in contrast to the polar oxide regions created by the tip. Thus, the AFM local oxidation procedure produces a topographically and chemically patterned substrate with almost arbitrary control over the topography of the pattern. Some simple patterns are shown in Fig. 8.16.

Our "directed dewetting" experiments were motivated not only by the analogous experiments carried out by the polymer community but by three earlier key papers related to control of spatial nanoparticle distributions via changes in solvent behaviour. Korgel and Fitzmaurice (1998), in a seminal study, showed that the polarity of the solvent directly controlled the thickness of a nanoparticle film deposited from that solvent onto a substrate. The question we initially sought to address was whether *substrate* polarity could be used to have a similar—but locally controllable—effect on nanoparticle organization. The first steps in this direction had been made by researchers at the Nanoscale Physics Laboratory at Birmingham in the late 1990s (Parker *et al.* 2001) who used lithographically patterned resists (at the micrometer scale) to influence the flow of solvent and thus influence nanoparticle distributions. Lu *et al.* (2004) subsequently showed that an anisotropically patterned L-R-dipalmitoyl-phosphati-dycholine (DPPC) film could be used to control solvent dewetting and thus form linear stripes of CdSe or CdSe/ZnS nanocrystals. An important objective of our experiments was to ascertain the extent to which pattern formation in dewetting nanofluids could be controlled at the

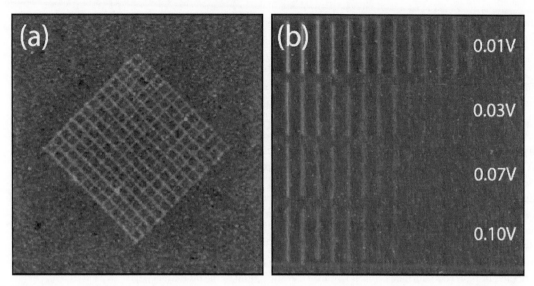

Fig. 8.16 (a) Oxide grid pattern produced on a hydrogen-passivated Si(111) surface using scanning probe oxidation. The grid is 1 μm × 1 μm in area and comprises lines that are ∼2 nm in height and ∼50 nm wide. (b) Effect of tapping-mode voltage set point and tip bias on the width and height of scanning-probe-generated oxide lines on H:Si(111). Set of lines written at set points of 0.01 V, 0.03 V, 0.07 V, and 0.1 V are shown. For each set of lines the tip bias voltage was varied between 10 V and 3V in 0.5 V steps from left to right. The scan speed and relatively humidity were kept constant at 0.5 μm s^{-1} and 70%, respectively.

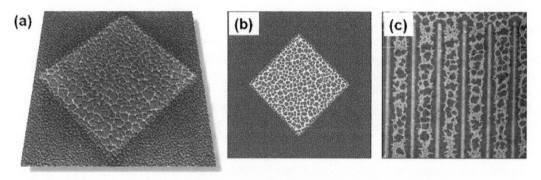

Fig. 8.17 (a) 3D representation of an AFM image showing the effect of an AFM tip-generated oxide square of 4 μm × 4 μm area on the organization of octanethiol-passivated Au nanoparticles on a H:Si(111) surface. (b) Result of a simulation using the original Rabani *et al.* (2003) Hamiltonian with |μ|6% smaller on the simulated oxide region. (Reprinted with permission from Martin *et al.* (2007a), ©2007 American Physical Society.) To reproduce the sharp transition between nucleation-dominated and spinodal-dominated dewetting oberved in the experiment, a 100-nm wide region at the edge of the oxide feature was included over which the chemical potential varies linearly from its value on the H:Si(111) surface to that on the oxide. (c) Narrower oxide features lead to the rupture of the solvent film, producing a region denuded of nanoparticles in the vicinity of the surface heterogeneity.

submicrometer or nanometer length scale via topographic and chemical surface heterogeneities.

The degree to which spatial control of pattern formation in dewetting nanofluids can be achieved is best exemplified by Fig. 8.17(a) where the oxide square at the center of the image has clearly dramatically affected the solvent evaporation dynamics. A Monte Carlo simulation using the original Hamiltonian of the Rabani model, but with a reduction of the absolute value of the chemical potential in a square region whose area was chosen to match that formed in the experiment, reproduces the experimental data extremely well (Fig. 8.17). There is a remarkably seamless transition from a wormlike pattern, characteristic of a spinodal dewetting mechanism, on the hydrophobic H:Si(111) surface, to a network structure arising from nucleation-driven dewetting on the oxide square. What is particularly intriguing about the images shown in Fig. 8.17 is the remarkably sharp switch from one pattern type to another–there is no evidence of a rupture of the solvent film that would give rise to a region denuded of particles close to the edges of the oxide square. To replicate this in the simulations required a linear variation of μ, over an approximate 100 nm length scale, at the edge of the oxide region. Narrower (higher aspect ratio) oxide structures (as in Fig. 8.17(c)), however, exhibit a region free of nanoparticles in their vicinity.

The scanning probe oxidation technique has the advantage that arbitrary oxide patterns can be written with relative ease on a H:Si(111) surface. We are currently pursuing the use of oxide features to induce solvent dewetting in highly localized surface areas and, in particular, exploring the extent to which arrays of oxide sites can be used to direct self-organization of a variety of nanoparticle (and nanorod) types.

8.3 Pathways for charge transport in nanoparticle assemblies

In terms of its electronic and charge-transport properties, a nanoparticle array can be described as a set of coupled Coulomb-blockade devices (Ferry and Goodnick 1997). Coulomb blockade occurs when the charging energy of a metal particle, $e^2/2C$, is significantly larger than the thermal energy, $k_B T$, such that transfer of an electron onto the particle blocks the addition of further electrons until the voltage is increased to offset the charging energy. In a seminal paper in the early 1990s, Middleton and Wingreen (MW) (1993) established and studied a theoretical model of non-linear charge transport in arrays of capacitively coupled metal dots, of sufficiently small capacitance to exhibit single-electron charging (Coulomb blockade) effects, where electrons tunnel from dot to dot. Their key result was that, in the presence of random offset charges (which locally gate the metal particles), the current through the array, above a voltage threshold, V_T, exhibits a power-law dependence on voltage:

$$I \sim (V/V_T - 1)^\zeta. \tag{8.12}$$

MW highlighted that there is an elegant analogy that can be drawn between charge transport in metal particle arrays of the type they studied and interface growth problems, including, in particular, those involving the Kardar–Parisi–Zhang (1986) (KPZ) model. From the KPZ model MW determined critical exponents for current and correlation lengths in the metal-particle array system, predicting $\zeta = \frac{5}{3}$ for 2D arrays.

Almost a decade later, Parthasarathy *et al.* (2001) carried out a pioneering study of the charge-transport properties of a highly ordered 2D monolayer of thiol-passivated Au nanoparticles (diameter ~ 2 nm), reproducing experimentally the $I \sim (V/V_{\mathrm{T}} - 1)^\zeta$ dependence predicted by MW but measuring a substantially higher value of the scaling exponent, $\zeta = 2.25 \pm 0.1$. Parthasarathy *et al.* (2001) found that topological disorder in nanoparticle arrays (the presence of voids) gives rise to deviations from simple single power-law behavior, where values of ζ varying between ~ 1 and >2.5 could be observed. They argued that bottlenecks in charge transport due to the presence of voids led to the absence of single scaling behavior, a proposal that was confirmed by the subsequent molecular-dynamics simulations of Reichhardt and Olson Reichhardt (2003a,b).

While a number of groups have explored the charge-transport properties of ordered or relatively weakly disordered lattices of nanoparticles (Geigenmuller and Schon 1989; Bakhalov *et al.* 1991; Middleton and Wingreen 1993; Parthasarathy *et al.* 2001, 2004; Reichhardt and Olson Reichhardt 2003a,b; Elteto *et al.* 2005), we have focused on combined experimental and numerical simulation work on topologically complex self-organized assemblies of the type discussed at length in the preceding sections. In the following we discuss how the dewetting-induced complex topology of nanoparticle networks affects the single-electron charging-mediated conduction paths focusing, in particular, on networks of the types shown in Fig. 8.18.

In order to quantify the effects of network structure and topology on conduction, we construct a network (graph) of nanoparticles, by inserting a link between each pair of the nanoparticles that are spaced from one another within a tunnelling radius for the electrons. Thus, each link between the adjacent nanoparticles, which is given by a non-zero element $A_{ij} = 1$ of the respective adjacency matrix, represents a possible junction for tunnelling. The tunnelling radius r, which defines the maximum length of the junction, depends on the physical properties of the nanoparticles. It is clear that the morphology of holes in the evaporated nanoparticle films will affect the network of junctions. Some topological features of these networks that are relevant to the conduction are

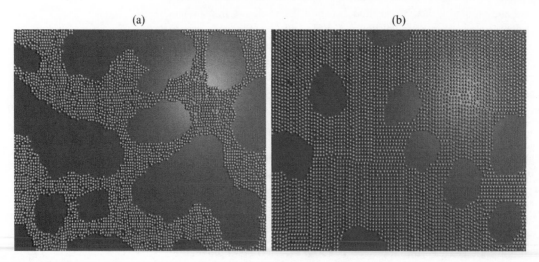

(a)   (b)

Fig. 8.18 Three-dimensional renderings of typical nanoparticle assemblies used for our charge-transport studies, simulated using the Monte Carlo model discussed in Section 8.2. (a) 40% coverage (NNET1) and (b) 80% coverage (NNET2).

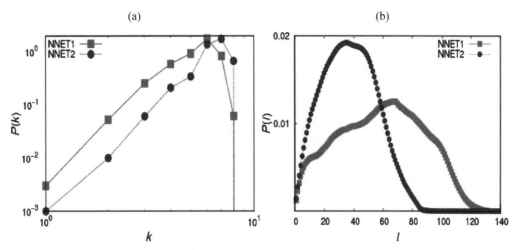

Fig. 8.19 Topological properties of two networks representing tunnelling junctions between nanoparticles in Fig. 8.18: 40% coverage (NNET1) and 80% coverage (NNET2): (a) Histograms of the number of junctions per nanoparticle and (b) Histograms of the lengths of all the shortest paths on the graph.

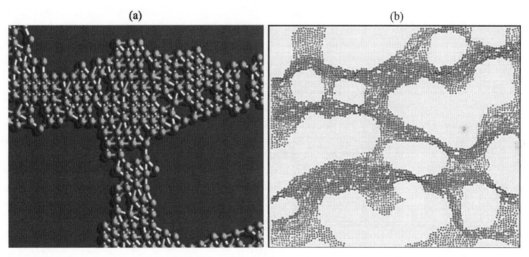

Fig. 8.20 (a) Enlarged part of nanoparticle network NNET1; (b) Simulated single-electron paths for NNET1 (panel (b) reproduced with permission from Blunt *et al.* (2007a), ©2007 American Chemical Society).

given in Fig. 8.19a. Specifically, in the inhomogeneous array the number of junctions attached to a nanoparticle varies within a larger range compared to the homogeneous film (see enlarged part of the array in Fig. 8.20(a)). Other topological features that strongly influence charge transport are the distribution of the lengths of the shortest paths and the number of such paths through each node (topological centrality). The distribution of the lengths of shortest paths between all pairs of nodes on the networks shown in Fig. 8.18 are given in Fig. 8.19(b), showing further quantitative differences between the homogeneous and inhomogeneous nanoparticle arrays.

8.3.1   Modelling single-electron conduction in nanoparticle films

As discussed above, an array of metallic nanoparticles on a substrate and under an applied voltage represents a *capacitively coupled system*, in which current transport occurs under Coulomb-blockade conditions. Following a single electron tunnelling through the junction from one nanoparticle to another, the charging energy of the nanoparticle increases inversely proportional to its capacitance. This increases the barrier for further tunnelling events through the junction. A subsequent increase of the voltage will reduce the barrier and facilitate further tunnelling processes, pushing the moving front of the electrons towards the zero-voltage electrode.

For a voltage above a certain threshold V_T (which depends on the sample size), the profile of the voltage across the sample is sufficient to push the front of charge to reach the opposite electrode. Thus, the current through the sample can be measured for voltages $V > V_T$. Theoretical models of single-electron charge transport in nanoparticle arrays (Geigenmuller and Schon 1989; Bakhalov *et al.* 1991; Middleton and Wingreen 1993; Parthasarathy *et al.* 2001, 2004; Reichhardt and Olson Reichhardt 2003a,b; Elteto *et al.* 2005) have mostly considered regular (or slightly disordered) 2-dimensional and 1D geometries. Recently, we generalized the model for any 2D structure formed by the Rabani *et al.* (2003) Monte Carlo approach and variants thereof (Martin *et al.* 2004, 2007a; Pauliac-Vaujour *et al.* 2008). Each nanoparticle array generated by the Monte Carlo self-organization code can be represented by a network of junctions, as described above, and given by the network adjacency matrix **A**. Note that the elements of the adjacency matrix $A_{ij} = 1$ if the particles (i, j) are within a tunnelling distance, and $A_{ij} = 0$ otherwise. Here, we will describe the main features of single-electron conduction within the generalized model. A more detailed description and the numerical implementation of the model can be found in Šuvakov and Tadić (2009).

Following Geigenmuller and Schon (1989), Bakhalov *et al.* (1991), and Middleton and Wingreen (1993), the tunnelling process through arrays of nanoparticles separated by RC junctions can be described by considering the electrostatic energy of the assemble. We use the generalized expression for charge, Q_i, on the ith nanoparticle within the array, which has a given topology of connections A_{ij} (Blunt *et al.* 2007b; Šuvakov and Tadić 2009):

$$Q_i = \sum_j C_{ij}(\Phi_i - \Phi_j) + \sum_\mu C_{i,\mu}(\Phi_i - \Phi_\mu), \qquad (8.13)$$

where Φ_i is the potential of the ith nanoparticle, $C_{ij} \equiv C A_{ij}$ is the capacitance between the ith and the jth nanoparticle, $C_{i,\mu}$ is the capacitance between the ith nanoparticle and electrode $\mu \in \{+, -, gate\}$, Φ_μ is the chemical potential, and Q_μ is the charge on the electrode μ. This system of equations can be written in matrix form $\mathbf{Q} = M\Phi - \mathbf{C}_\mu \Phi^\mu$ and solved for Φ:

$$\Phi = M^{-1}\mathbf{Q} + M^{-1}\mathbf{C}_\mu \Phi^\mu, \qquad (8.14)$$

where M is the capacitance matrix of the whole system (nanoparticle array and electrodes):

$$M_{ij} = \delta_{i,j} \left(\sum_k C_{ik} + \sum_\mu C_{i,\mu} \right) - C_{ij}. \tag{8.15}$$

With these definitions the electrostatic energy of the entire system can be written in matrix form as (Geigenmuller and Schon 1989; Bakhalov *et al.* 1991; Middleton and Wingreen 1993; Šuvakov and Tadić 2009):

$$E = \frac{1}{2}\mathbf{Q}^\dagger M^{-1}\mathbf{Q} + \mathbf{Q}\cdot V^{\text{ext}} + Q_\mu \Phi^\mu, \quad V^{\text{ext}} = M^{-1}\mathbf{C}_\mu \Phi^\mu. \tag{8.16}$$

The system is driven by increasing the external voltage V^{ext} at one of the electrodes, which causes electrons to first tunnel to nearest-neighbor particles and then forward through the junctions between the nanoparticles. Due to the overall voltage profile, tunnelling is predominantly towards the zero-voltage electrode as $+V \to a \to b \to \dots \to -V$. The gate voltage is kept fixed $V_g = 0$.

Following a single electron tunnelling from nanoparticle $a \to b$, which causes a change of the local charge as $Q'_i = Q_i + \delta_{ib} - \delta_{ia}$, the energy change $\Delta E(a \to b)$ of the array can be computed from eqn (8.16) for a given voltage V^{ext}. Introducing the variable

$$V_c \equiv \sum_i Q_i M_{ic}^{-1}; \tag{8.17}$$

one can express the energy change $\Delta E(a \to b)$ as follows

$$\Delta E(a \to b) = V_b - V_a + \frac{1}{2}(M_{aa}^{-1} + M_{bb}^{-1} - M_{ab}^{-1} - M_{ba}^{-1}), \tag{8.18}$$

where V_a and V_b are computed via eqn (8.17) and an additional contribution $V_b^{\text{ext}} - V_a^{\text{ext}}$ comes from the second term in expression (8.16). Note also that the energy changes due to tunnellings between the electrodes and a nanoparticle can be expressed as $\delta E(a \leftrightarrow \pm) = \pm V_a + \frac{1}{2}M_{aa}^{-1}$. These forms are suitable for numerical implementation, where the updated value V'_c at a node c after the tunnelling $a \to b$ can be calculated recursively according to:

$$a \to b: \quad V'_c = V_c + M_{bc}^{-1} - M_{ac}^{-1}, \quad a \leftrightarrow \pm. \quad V'_c = V_c \pm M_{ac}^{-1}. \tag{8.19}$$

All details regarding the numerical implementation are discussed in Šuvakov and Tadić (2009). We describe here the main features of the simulations that are necessary to understand the results. For a given voltage V at one electrode, the tunnelling processes are simulated as follows. At each time step, following a tunnelling event $a \to b$, the energy changes $\Delta E_{i \to j}$ at each junction $i \to j$ are calculated, including the junctions between nanoparticles and the electrodes $\pm$. The *tunnelling rate* $\Gamma_{i \to j}$ of an electron tunnelling from $i \to j$ is then determined as:

$$\Gamma_{i \to j}(V) = \frac{1}{eR_{i \to j}} \frac{\Delta E_{i \to j}/e}{1 - e^{-\Delta E_{i \to j}/k_B T}}. \tag{8.20}$$

Here, $R_{i \to j}$ is the so-called *quantum resistance* of the junction, which appears as a parameter in the macroscopic theory. It determines the characteristic timescale of the tunnellings as $\tau \sim RC_g$. Here, we assume $R_{i \to j} = R$ for all junctions. (For a junction between metallic nanoparticles the tunnelling resistance is obtained within the transfer Hamiltonian method as $R = \hbar/2\pi e^2 |T_0|^2 D_0^2$, where T_0 and D_0 are the transmission coefficient and the density of states for each nanoparticle (Ferry and Goodnick 1997). Generally, for a junction between small nanoparticles the quantum resistance (conductance) is readily computed, for instance within density-functional theory (Reimann and Manninen 2002), exhibiting a characteristic dependence on the gate voltage and the electronic states of the dots.)

In our continuous-time implementation (Šuvakov and Tadić 2009), we determine from the tunnelling rates $\Gamma_{i \to j}$ the time t_{ij} for each tunnelling event. Then, the processes whose tunnelling times are in a given small time window are performed first, and the energy is updated at all junctions. In the limit when the capacitance of the gate $C_g \gg C$, the diagonal elements of the capacitance matrix are dominant and one can reduce the number of terms to be included in the calculations (Šuvakov and Tadić 2008). The net current through a junction $i \leftrightarrow j$ is then given by

$$I_{i \leftrightarrow j}(V) = e \left(\Gamma_{i \to j}(V) - \Gamma_{j \to i}(V) \right). \tag{8.21}$$

The current through the system is measured at the zero-voltage electrode as the total current at the junctions between the last layer of nanoparticles and that electrode. In these stochastic tunnelling processes, the number of electrons through each junction is fluctuating in time. For large enough voltage above the threshold voltage $V > V_T$ there is a finite average number of electrons arriving at the electrode at each time step, which gives a constant current $I(V)$. In the simulations we keep track of the number of tunnel events (flow) at each junction in the network. In Fig. 8.20(b) we show the conduction paths throughout the network (the electrodes are along the left and the right boundary): Darker color represents larger flow on the junction. Clearly, some junctions (and their adjacent nanoparticles) are used more often than others, depending on their position between the electrodes and the overall nanoparticle array structure. A complete statistical analysis of the flow (dynamical centrality) of junctions is given by Šuvakov and Tadić (2009).

In conducting nanoparticle arrays a non-linear current–voltage dependence of the following form was measured (Parthasarathy *et al.* 2001, 2004; Blunt *et al.* 2007a)

$$I(V) = B \left(\frac{V - V_T}{V_T} \right)^\zeta, \tag{8.22}$$

where B is a constant. The degree of non-linearity measured by the exponent ζ varies with the structure of the array that as described throughout Section 8.1, is strongly dependent on the solvent dewetting dynamics. Experimentally measured values for 2D nanoparticle arrays vary from $\zeta = 2.25$ (Pathasarathy *et al.* 2001) to $\zeta = 4.3$ (Blunt *et al.* 2007a). In Fig. 8.21(a) we show simulation results for several representative regular arrays, and, in Fig. 8.21(b), for the

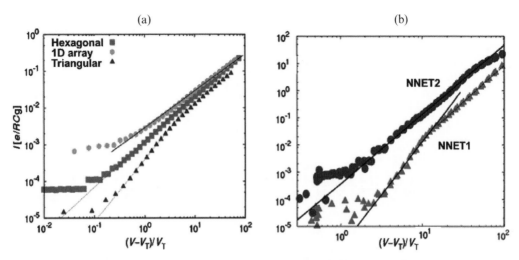

Fig. 8.21 $I(V)$ characteristics for (a) hexagonal, triangular, and 1D nanoparticle arrays in the presence of quenched charge disorder, and (b) for the topologically disordered structures shown in Fig. 8.18.

two nanoparticle arrays simulated by the Rabani *et al.* (2003) code (NNET1 and NNET2; see Figs. 8.18(a,b)). In the case of a linear chain of nanoparticles, the $I(V)$ dependence is linear, despite the stochastic nature of the tunnelling processes. This suggests that the non-linearity in eqn (8.22) is a topological effect due to multiple conducting paths in 2D arrays. Furthermore, the geometry of the shortest paths between the electrodes, which contribute to the onset of conduction at the threshold voltage, depend strongly on the topology of the nanoparticle network. These effects are combined with the next-shortest paths, etc., enhancing non-linear effects when the voltage is increased. For very large voltages the barriers for tunnelling across the junctions along the paths are too low and the $I(V)$ curve bends towards the classical linear behavior.

Non-linear current–voltage characteristics are found for a wide range of voltages above threshold, both in simulations and experiments. In a regular hexagonal array of nanoparticles, $\zeta \approx 3$ was found (Šuvakov and Tadić 2009) in the absence of charge disorder. In Fig. 8.21(a) we show the results for regular arrays in the presence of *quenched charge disorder*: at each nanoparticle the charge Q_i has a random non-integer offset q_i, taken from a uniform distribution in the range $0 \le q_i \le e$. The presence of the charge disorder affects the local relation between the charge Q_i of moving electrons and the voltage. This blocks the tunnelling process along some junctions, thus reducing the number of topologically available shortest paths. The total current through the sample is reduced compared with the non-disordered case for the same voltage. Consequently, we find lower values of the exponent ζ in the presence of quenched charge disorder. Again, the effects depend on the geometry of the regular array: the exponent reduction is stronger in the case of an hexagonal than in a triangular array. The exponent in the case of the linear chain of nanoparticles is not affected by charge disorder.

In contrast to the quenched charge disorder discussed above, *topological disorder*, as observed in the self-assembled nanoparticle arrays in Fig. 8.18,

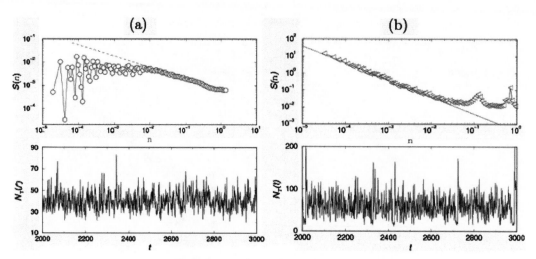

Fig. 8.22 Time series of the total number of tunnellings (bottom panels) and their power spectra (top panels) for: (a) regular triangular nanoparticle array with charge disorder and (b) nanoparticle network NNET1 (see Fig. 8.18). Reproduced with permission from Šuvakov and Tadić (2009), ©2009 IOP Publishing Ltd and SISSA.

enhances the $I(V)$ non-linearity. The curves in Fig. 8.21(b) are the simulation results for the networks of Figs. 8.18(a,b). Network NNET1, with 40% coverage and large topological inhomogeneity (cf. Fig. 8.19), exhibits a larger non-linearity of the current with the exponent $\zeta \approx 3.9$. For the more homogeneous network, NNET2, $\zeta \approx 2.6$, comparable with that observed for regular arrays without charge disorder.

In stochastic tunnelling processes the charge at each nanoparticle fluctuates in time. We thus observe a fluctuating time series $\{Q_i(t)\}, i = 1, 2, \cdots N$ for each nanoparticle in the network. The number of tunnel events occurring at the same instance of time in the whole system, $N_T(t)$, provides an important measure of time correlation in the conduction process. $N_T(t)$ fluctuates over a well-defined average value when current flows through the array, i.e. for $V > V_T$, making a *stationary time series*. Two examples of such time series are shown in Fig. 8.22 for the regular array with charge disorder and for the topologically inhomogeneous network NNET1. The power spectra of these two time series, shown in the top panels in Fig. 8.22, appear to be entirely different. In particular, the total number of tunnel events in NNET1 exhibits long-range correlations in time with the power spectrum

$$S(\nu) \sim \nu^{-\beta}, \tag{8.23}$$

and the exponent $\beta \approx 1$ ($1/\nu$ noise) in a wide range of frequencies. On the contrary, the spectrum of the tunnellings in the regular triangular array with quenched charge disorder shows only weak correlations in the high-frequency region, and white-noise behavior for a range of low frequencies. Somewhat similar long-range correlations in the power spectra of current fluctuations have recently been measured for conducting nanowires (Kohno and Takeda 2007).

The long-range correlations in the power spectrum in the case of self-assembled nanoparticle arrays, together with other statistical properties of the conduction process (correlations between successive events at each nanoparticle, distribution of flow along conduction paths, dispersion of the time series at each node, and current fluctuations) studied in detail by Šuvakov and Tadić (2009), reveal that the non-linearity in the current–voltage curves arises as a collective dynamical effect in these systems.

8.4 Conclusion

In this chapter we have given a brief overview of a number of key themes in state-of-the-art nanoscience: biomimetic strategies for nanoparticle self-assembly, pattern formation and self-organization in dewetting nanofluids, and charge transport in extended arrays of metal nanoparticles. Our aim has not been to present an exhaustive review of the (extensive) literature in these areas but to provide the reader with some insight into recent advances in these exciting subfields of nanoscience. There remain many challenges with regard to extending our ability to direct the organization and assembly of matter across a hierarchy of length scales (nanometers–micrometers–millimeters) but it is clear that very simple experiments, bearing much in common with work on the physics of coffee stains (Deegan *et al.* 1997), can yield remarkable and facile new methods of controlling nanoparticle organization. While the experiments are straightforward, their interpretation, modelling, and analysis are, however, far from simple. The physics of nanoparticle ensembles is a rich and complex area of modern condensed-matter science that will continue to provide key insights into the dynamics of self-assembly and self-organization in nanostructured matter.

Acknowledgments

We are grateful for the financial support of the EU Framework Programme 6 Marie Curie scheme (under grant MRTN-CT-2004005728 (PATTERNS)). We also acknowledge funding from the UK Engineering and Physical Sciences Research Council (in the form of a doctoral training account (DTA) PhD studentships). BT acknowledges program P1-0044 (Slovenia) for funding. We thank the members of the PATTERNS network for helpful discussions regarding pattern formation in nanoparticle assemblies including, in particular, Uwe Thiele, Ulli Steiner, James Sharp, and Mathias Brust. In addition, we thank Bryan Gallagher and Laurence Eaves for providing access to the cryostats used in this work, and Eran Rabani and Juan P. Garrahan for very helpful discussion in relation to Monte Carlo simulations.

References

Archer, P.I., Santangelo, S.A., Gamelin, D.R. *Nano Lett.* **7**, 1037 (2007).

Avouris, P., Hertel, T., Martel, R. *Appl. Phys. Lett.* **71**, 285 (1997).

Bakhvalov, N., Kazacha, G., Likharev, K., Serdyukova, S.T. *Physica B* **173**, 319 (1991).

Bigioni, T.P., Lin, X.M., Nguyen, T.T., Corwin, E.I., Witten, T.A., Jaeger, H.M. *Nature Materials* **5**, 265 (2006).

Blunt, M.O., Martin, C.P., Ahola-Tuomi, M., Pauliac-Vaujour, E., Sharp, P., Nativo, P., Brust, M., Moriarty, P.J. *Nature Nanotech.* **2**, 167 (2007a).

Blunt, M.O., Šuvakov, M., Pulizzi, F., Martin, C.P., Pauliac-Vaujour, E., Stannard, A., Rushforth, A.W., Tadić, B., Moriarty, P. *Nano Lett.* **7**, 855 (2007b).

Brennan, J.L., Hatzakis, N.S., Tshikhudo, T.R., Dirvianskyite, N., Razumas, V., Patkar, S., Vind, J., Svendsen, A., Nolte, R.J.M, Rowan, A.E., Brust, M. *Bioconjug. Chem.* **17**, 1373 (2006).

Cheng, G.J., Puntes, V.F., Guo, T.J. *Colloid Interf. Sci.* **293**, 430 (2006)

Collier, C.P., Vossmeyer, T., Heath, J.R. *Ann. Rev. Phys. Chem.* **49**, 371 (1998).

Dagata, J.A., Schneir, J., Harary, H.H., Bennett, J., Tseng, W.J. *Vac. Sci. Technol. B* **9**, 1384 (1991)

Deegan, R.D., Bakajin, O., Dupont, T.F., Huber, G., Nagel, S.R., Witten, T.A. *Nature* **389**, 827 (1997).

Denkov, N.D., Velev, O.D., Kralchevsky, P.A., Ivanov, I.B., Yoshimura, H., Nagayama, K. *Langmuir* **8**, 3183 (1992).

Duruöz, C.I., Clarke, R.M., Marcus, C.M., Harris, Jr., J.S. *Phys. Rev. Lett.* **74**, 3237 (1995).

Elteto, K., Lin, X.-M., Jaeger, H. *Phys. Rev. B* **71**, 205412 (2005).

Ferry, D.K., Goodnick, S.M. *Transport in Nanostructures* (Cambridge University Press, Cambridge, UK, 1997).

Ge, G., Brus, L.J. *Phys. Chem. B* **104**, 9573 (2000).

Geigenmuller, U., Schon, G. *Europhys. Lett.* **10**, 765 (1989).

Gigault, C., Dalnoki-Veress, K., Dutcher, J.R. *J. Colloid Interf. Sci.* **2001**, **243**, 143 (2001).

Guéron, S., Deshmukh, M.M., Myers, E.B., Ralph, D.C. *Phys. Rev. Lett.* **83**, 4148 (1999).

Huang, J., Kim, F., Tao, A.R., Connor, S., Yang, P. *Nature Mater* **4**, 896 (2005).

Huber, R. *Science* **233**, 702 (1986).

Jacobs, K., Herminghaus, S., Mecke, K.R. *Langmuir*, **14**, 965 (1998).

Jang, J., Oh, J.H. *Langmuir* **20**, 8419 (2004)

Jelesarov, I., Leder, L., Bosshard, H.R. *Methods: A Companion to Methods in Enzymology* **9**, 533 (1996).

Kardar, M., Parisi, G., Zhang, Y.-C. *Phys. Rev. Lett.* **56**, 889 (1986).

Koch, S.W., Desai, R.C., Abraham, F.F. *Phys. Rev. A* **27**, 2152 (1983).

Kohno, H., Takeda, S. *Nanotechnology* **18**, 395706 (2007).

Korgel, B.A., Fitzmaurice, D. *Phys. Rev. Lett.* **80**, 3531 (1998).

Liron, Z., Bromberg, A., Fisher, M. *Novel Approaches in Biosensors and Rapid Diagnostic Assays* (Kluwer Academic/Plenum Publishers, Eilat, Israel, 2001).

Lu, N., Chen, X.D., Molenda, D., Naber, A., Fuchs, H., Talapin, D.V., Weller, H., Muller, J., Lupton, J.M., Feldmann, J., Rogach, A.L., Chi, L.F. *Nano Lett.* **4**, 885 (2004).

Maillard, M., Motte, L., Ngo, A.T., Pileni, M.P. *J. Phys. Chem. B* **104**, 11871 (2000).

Martin, C.P., Blunt, M.O., Moriarty, P. *Nano Lett.* **4**, 2389 (2004).

Martin, C.P., Blunt, M.O., Pauliac-Vaujour, E., Stannard, A., Moriarty, P., Vancea, I., Thiele, U. *Phys. Rev. Lett.* **99**, 116103 (2007a).

Martin, C.P., Blunt, M.O., Vaujour, E., Fahmi, A., D'Aleo, A., De Cola, L., Vögtle, F., Moriarty, P. *"Self-organised Nanoparticle Assemblies; A Panoply of Patterns"*, in *Systems Self-assembly: Interdisciplinary Snapshots*, (eds) Krasnogor, N., Gustafson, S., Pelta, D., Verdegay, J.L. (Elsevier, Amsterdam, 2007b).

Middleton, A., Wingreen, N. *Phys. Rev. Lett.* **71**, 3198 (1993).

Mirkin, C.A., *Inorg. Chem.* **39**, 2258 (2000).

Mirkin, C.A., Letsinger, R.L., Mucic, R.C., Storhoff, J.J. *Nature* **382**, 607 (1996).

Moriarty, P., Taylor, M.D.R., Brust, M. *Phys. Rev. Lett.* **89**, 248303 (2002).

Mossa, S., Sciortino, F., Tartaglia, P., Zaccarelli, E. *Langmuir* **20**, 10756 (2004).

Murray, C.B., Kagan, C.R., Bawendi, M.G. *Ann. Rev. Mater. Sci.* **30**, 545 (2000).

Narayanan, S., Wang, J., Lin, X.-M. *Phys. Rev. Lett.* **93**, 135503 (2004).

Nykypanchuk, D., Maye, M.M., van der Lelie, D., Gang, O. *Nature* **451**, 549 (2008).

Ohara, P.C., Gelbart, W.M. *Langmuir* **14**, 3418 (1998).

Ohara, P.C., Heath, J.R., Gelbart W.M. *Angew. Chem. Int. Ed.* **36**, 1078 (1997).

Park, S.Y., Lytton-Jean, A.K.R., Lee, B., Weigand, S., Schatz, G.C., Mirkin, C.A. *Nature* **451**, 553 (2008).

Parker, A.J., Childs, P.A., Palmer, R.E., Brust, M. *Nanotechnology* **12**, 6 (2001).

Parthasarathy, R., Lin, X.-M, Elteto, K., Rosenbaum, T., Jaeger, H. *Phys. Rev. Lett.* **92**, 076801 (2004).

Parthasarathy, R., Lin, X.-M., Jaeger, H. *Phys. Rev. Lett.* **87**, 186807 (2001).

Pauliac-Vaujour, E., Moriarty, P. *J. Phys. Chem. C* **111**, 16255 (2007).

Pauliac-Vaujour, E., Stannard, A., Martin, C.P., Blunt, M.O., Moriarty, P., Vancea, I., Thiele, U. *Phys. Rev. Lett.* **100**, 176102 (2008).

Pearson, J.R.A. *J. Fluid. Mech.* **4**, 489 (1958).

Pileni, M.P. *J. Phys. Chem. B* **105**, 3358 (2001).

Pileni, M.P. *J. Phys.: Condens. Matter* **18**, S67 (2006).

Pileni, M.P. *Acc. Chem. Res.* **40**, 685 (2007).

Rabani, E., Reichman, D.R., Gleissler, P.L., Brus, L.E., *Nature* **426**, 271 (2003).

Reichhardt, C., Olson Reichhardt, C. *J. Phys. Rev. B* **68**, 165305 (2003a).

Reichhardt, C., Olson Reichhardt, C. *J. Phys. Rev. Lett.* **90**, 046802 (2003b).

Reimann, S.M., Manninen, M. *Rev. Mod. Phys.* **74**, 1283 (2002).

Shafi, K.V.P.M., Felner, I., Mastai, Y., Gedanken, A. *J. Phys. Chem. B* **103**, 3358 (1999).

Shen, T.C., Wang, C., Abeln, G.C., Tucker, J.R., Lyding, J.W., Avouris, Ph., Walkup, R.E. *Science* **268**, 1590 (1995).

Storhoff, J.J., Mirkin, C.A. *Chem. Rev.* **99**, 1849 (1999).

Stowell, C., Korgel, B.A. *Nano Lett.* **1**, 595 (2001).

Šuvakov, M., Tadić B. *Structure of Colloidal Aggregates with Bio-recognition Bonding*, unpublished (2008).

Šuvakov, M., Tadić, B. *J. Stat. Mech.: Theory and Exp.* 1742–4568/09/P02015 (2009).

Tadić, B. *From Microscopic Rules to Emergent Cooperativity in Large-scale Patterns*. in "Systems Self-Assembly: Multidisciplinary Snapshots", (ed.) N. Krasnogor *et al.* (Elsevier, Amsterdam, 2007).

Thiele, U., Mertig, M., Pompe W. *Phys. Rev. Lett.* **80**, 2869 (1998).

Thiele, U. *Eur. Phys. J. E.* **12**, 409 (2003).

Winfree, E., Liu, F., Wenzler, L.A., Seeman, N. C. *Nature* **394**, 539 (1998).

Xu, J., Xia, J., Lin, Z. *Angew. Chem. Int. Ed.* **46**, 1860 (2007).

Yosef, G., Rabani, E. *J. Phys. Chem. B* **110**, 20965 (2006).

Zaccarelli, E. *J. Phys.: Cond. Matter* **19**, 323101 (2007).

Zhou, W.L., He, J.B., Fang, J.Y., Huynh, T.A., Kennedy, T.J., Stokes, K.L., O'Connor, C.J. *J. Appl. Phys.* **93**, 7340 (2003).

Self-organizing atom chains

Arie van Houselt and Harold J.W. Zandvliet

9

9.1 Introduction

9.1.1 Towards one-dimensional physics

Nanowires play an important role in the current downscaling of electronic components. Besides the fact that nanowires can act as interconnects between nanoscale devices they can also be active components themselves. It is natural to assume that the nanowire-based quasi-one-dimensional materials will be the focus of the next decade of nanomaterials research.

One-dimensional (1D) electron systems are of particular fundamental interest too since they can exhibit a wealth of exotic phenomena, such as quantization of conductance, Peierls instability, spin-and charge-density waves and Luttinger liquid behavior. Compared to two-dimensional (2D) and three-dimensional (3D) electron systems, the 1D electron systems are relatively easy to handle from a theoretical point of view. For instance, the incorporation of electron–electron and electron–phonon interactions is less cumbersome in the 1D case as compared to the 2D and 3D cases. The 1D electron system is often described in terms of the Luttinger liquid theory. In a Luttinger liquid the electron loses its identity and splits up into two quasi-particles: a quasi-particle that only carries the spin (spinon) and a quasi-particle that only carries the charge (holon) (Luttinger 1963; Haldane 1981; Voit 1994).

From an experimental point of view the situation is, however, reversed: 1D electron systems are much harder to realize than their 2D and 3D counterparts. Among the most elegant practical realizations of 1D or quasi-1D electron systems belong without any doubt to the mechanically controlled break junction experiments. Using the mechanically controlled break junction technique one can create constrictions of only a few or just a single atom in diameter by controllably breaking a metallic wire. Some metals, such as for instance Au and Pt, can form freely suspended one-atom thick chains with lengths up to 6 or 7 atoms. The conductance of these one-atom thick chains is predicted to be quantized in units of $2e^2/h$. For single-atom noble-metal contacts, such as Au and Ag, the conductance is indeed quantized in units of $2e^2/h$ (Scheer *et al.* 1997, 1998). However, if more subbands are involved and transmission probabilities (T) of less than unity occur, the total conductance

G is given by $\sum_i T_i \left(2e^2/h\right)$ (Landauer 1957). Besides this mechanically controlled break junction technique there are several other techniques that allow the formation of nanowires, such as scanning tunnelling microscopy and various lithographic techniques. In contrast to the chains produced by the mechanically controlled break junction technique these chains or nanowires are not freely suspended, but positioned on a substrate. Furthermore, since the lithographic techniques do not have the ability yet to create single-atom chains we will not discuss them here. The scanning tunnelling microscope (STM) on the other hand can be used to produce an atom chain on a surface by simply picking up, dragging, pushing or pulling atoms in a one-by-one fashion to a pre-defined location (Nilius *et al.* 2002). Although the latter approach is very time consuming it allows one to study in a very systematic way:

(1) the physical properties of chains as a function of their length (Wallis *et al.* 2002);
(2) the influence of defects on the properties of the chains (Nilius *et al.* 2003).

An attractive route to produce nanowires is to decorate pre-existing steps of semiconductor surfaces with metal atoms via the deposition of submonolayer amounts of material. This method works for a variety of metals, e.g. Pt, Au, Ag, Co and In, on Si, Ge and Pt surfaces with different orientations and vicinalities (Gambardella *et al.* 2002; Ahn *et al.* 2003; Crain *et al.* 2003, 2006).

There is yet another attractive technique that can produce nearly perfect atom chains on surfaces, namely self-organization (Evans *et al.* 1999; Gurlu *et al.* 2003). The self-organization method can be roughly divided in two branches, a kinetic branch and a thermodynamic branch. In the former case a system is forced to follow a certain pathway, so that reaching the global free-energy minimum is kinetically hindered. The latter method simply relies on thermodynamics and the realization time depends on the heights of the kinetic barriers that are involved in order to reach the energetic minimum. In either of these approaches nature does the job. The technique often requires not less than a few seconds and leaves the researchers with an almost unimaginable number density of nanostructures. Though the self-organization approach also suffers from a number of drawbacks, namely the freedom to create any desired nanostructure is very limited and the produced nanostructures exhibit a natural density of thermodynamically induced deviations form the ideal ordering.

In the first part of this chapter we first briefly address a number of interesting phenomena of 1D electron systems. First, we introduce the 1D free-electron model. We derive an expression for the 1D density of states, which exhibits a singularity at the bottom of the band (Van Hove singularity, Section 9.1.2). Second, we extend the free-electron model and include a weak periodic potential that is induced by the lattice. No matter how weak this potential is, it will always result in the opening of an energy gap at the edges of the Brillouin zones (Section 9.1.3). The next step is to include the electrostatic interactions between the electrons as well. The interacting electron models (Fermi liquid,

Tomogana model and Luttinger liquid model) are only briefly discussed (Section 9.1.4). Subsequently, we address two interesting features of 1D systems:

(1) the quantization of conductance (Section 9.1.5);
(2) the Peierls instability (Section 9.1.6).

In the second part of this chapter we report on the experimental results of a nearly ideal one-dimensional system, namely self-organizing Pt atom chains on a Ge(001) surface. We discuss their formation (Section 9.2), quantum confinement between the Pt chains (Section 9.3) and the occurrence of a Peierls transition within the chains (Section 9.4). The last Section (Section 9.5) contains the conclusions and a short outlook.

9.1.2 1D free-electron model

In a free-electron gas the electrons are assumed to be free. This means that electron–electron and electron–phonon interactions are completely ignored. Despite this very crude approach the free-electron model is quite successful in the description of a number of electronic phenomena (Kittel 2005). Let us assume that the free electrons are confined to a line segment with length L. The eigenfunctions, $\psi_n(x)$, and eigenvalues (E_n) can be extracted from the time-independent Schrödinger equation:

$$\frac{-\hbar^2}{2m}\frac{\partial^2}{\partial x^2}(\psi_n(x)) = E_n\psi_n(x). \tag{9.1}$$

If we assume periodic boundary conditions,[1] i.e. $\psi_n(x = 0) = \psi_n(x = L)$, we find

$$\psi_n(x) = Ce^{ikx} \tag{9.2a}$$

$$E_n = \frac{\hbar^2 k^2}{2m} = \frac{\hbar^2}{2m}\left(\frac{2\pi n}{L}\right)^2 \text{ with } n = 0, \pm1, \pm2, \dots \pm\infty. \tag{9.2b}$$

The energy dispersion curve, i.e. $E(k)$, is parabolic in k and the density of states in k-space, $D(k)$, is $L/2\pi$. The total number of free electrons, N, allows us to determine the Fermi wave vector

$$N = \int_{k=-k_F}^{k_F} 2D(k)\mathrm{d}k = \frac{2L}{\pi}k_F, \tag{9.3}$$

where we have assumed that both N and L are very large. The factor 2 as a prefix of the $D(k)$ term arises due to the spin degeneracy. The density of states in energy space can easily be derived by setting $2D(k)\mathrm{d}k = D(E)\mathrm{d}E$:

$$D(E) = \frac{2D(k)}{\mathrm{d}E/\mathrm{d}k} = \frac{L/\pi}{\hbar^2 k/m} = \frac{L}{\pi\hbar}\sqrt{m/2E}. \tag{9.4}$$

For an energy band with its minimum at $E = E_0$ the $1/\sqrt{E}$ term should be replaced by $1/\sqrt{(E - E_0)}$. Since both positive and negative k-values lead to the same energy the total density of states should be multiplied by an additional

[1] An alternative approach is to require that the wavefunction is zero at both boundaries, i.e. $\psi_n(x = 0) = \psi_n(x = L) = 0$. In this case, the eigenfunctions and eigenvalues are: $\psi_n(x) \propto \sin(kx)$ and $E_n = \frac{\hbar^2}{2m}\left(\frac{\pi n}{L}\right)^2$ with $n = 1, 2, \dots \infty$, respectively.

factor of 2, i.e.

$$D(E) = \frac{2L}{\pi \hbar} \sqrt{m/2(E - E_0)}. \tag{9.5}$$

Sometimes, the density of states is given per unit length, $D(E) = \frac{2}{\pi \hbar} \sqrt{m/2(E - E_0)}$.

9.1.3   1D nearly free-electron model

The free-electron model is particularly useful to describe a number of proper-ties such as the heat capacity, the magnetic susceptibility and the electrical and thermal conductivity. However, it fails for instance to describe the difference between metals, semiconductors and insulators, and it also does not describe all the transport properties in an adequate way.

In order to address the difference between metals and insulators we need to understand why and how an energy gap in a solid material can emerge. Therefore, we must extend the free-electron model to take account of the peri-odic lattice of the chain. The one-dimensional time-independent Schrödinger equation for a single electron in a periodic potential $V(x)$ is,

$$\left(-\frac{\hbar^2}{2m} \frac{\partial^2}{\partial x^2} + V(x) \right) \psi_n(x) = E \psi_n(x), \tag{9.6}$$

where the potential is periodic with the lattice constant, a, of the chain,

$$V(x + a) = V(x). \tag{9.7}$$

Since the potential $V(x)$ has the same periodicity as the chain, it can be expanded in the following Fourier series:

$$V(x) = \sum_G V_G e^{iGx}, \tag{9.8}$$

where G is a reciprocal lattice vector ($G = 2\pi m/a, m \in Z$).

Bloch showed that the solutions of the Schrödinger equation for a periodic potential must have a particular form (Bloch 1928):

$$\psi_k(x) = u_k(x) e^{ikx}, \tag{9.9}$$

where the function $u_k(x)$ has the periodicity of the crystal lattice, i.e. $u_k(x + a) = u_k(x)$. Equation (9.9) is known as the Bloch theorem.

In the free-electron model the dispersion relation is a simple parabola, i.e. $E(k) = \frac{\hbar^2 k^2}{2m}$. The incorporation of a periodic potential gives rise to forbidden and allowed energy bands due to the opening of bandgaps at the edges of the Brillouin zone, $k = \pm \frac{\pi n}{a}$:

$$E_{\pm} = \frac{1}{2} \left(E_{k-G}^0 + E_k^0 \right) \pm \sqrt{ \left(\frac{1}{4} \left(E_{k-G}^0 - E_k^0 \right)^2 + |V_G|^2 \right)}, \tag{9.10}$$

where

$$E_{k-G}^0 = \frac{\hbar^2\,(k-G)^2}{2m} = \frac{\hbar^2\left(k-\frac{2\pi m}{a}\right)^2}{2m}, \tag{9.11}$$

and $G = \pm\frac{2\pi m}{a}$ is a reciprocal lattice vector. At the edges of the Brillouin zones $(G/2)$ one finds $E_{k-G}^0 = E_k^0$ (and thus $E^{\pm} = E_k^0 \pm |V_G|$). The energy gap has a value of $2\,|V_G|$, i.e. twice the Gth Fourier component of the potential. The main conclusion of this section is that no matter how weak the periodic potential is, it always leads to the opening of an energy gap at the edges of the Brillouin zones.

9.1.4   Interacting electrons: Fermi liquid, Tomogana model and Luttinger liquid

So far we have ignored the electrostatic interaction between electrons and have assumed that the electrons are non-interacting (this system is often referred to as a Fermi gas). If interactions between electrons are incorporated we denote the electron system as a Fermi liquid. Moving electrons will cause an inertial reaction in the surrounding electron gas, resulting in an increase of the effective mass of the electrons. In two and three dimensions the Landau Fermi liquid describes the low-lying single-particle excitations rather well, but fails to explain the strong-coupling regime adequately. In 1D metals, where electrons are forced to move on a line, the single-particle description completely breaks down by even a very small Coulomb repulsion. In 1950 Tomogana put forward a model (Tomogana 1950) for a 1D interacting electron system that does not suffer from this problem. In addition, the Tomogana model is also applicable for stronger interactions. The important step forward is the recognition that the excitations of the electron gas are approximate bosons, despite the fact that the elementary particles are electrons (i.e. fermions). Within the framework of the Tomogana model it is assumed that these excitations are ideal bosons. About a decade later Luttinger (1960) proposed a model that only deviates slightly from the Tomogana model. The advantages of the Luttinger model are: (1) it is exactly solvable and (2) it contains fewer assumptions than the Tomogana model. In contrast to the Tomogana model, there are two types of fermions in the Luttinger model. One has an energy spectrum given by $E(k) = kv_{\mathrm{F}}$, while the other has an energy spectrum represented by $E(k) = -kv_{\mathrm{F}}$ (v_{F} is the Fermi velocity). There are several characteristic features of a Luttinger liquid:

(1) the charge and spin degrees of freedom of the electrons are separated;
(2) the spin velocity is comparable to the Fermi velocity, whereas the charge velocity is larger than the Fermi velocity;
(3) near the Fermi level the density of states exhibits a power-law suppression.

Up to now there have only been a handful of experimental claims of Luttinger liquid behavior (Bockrath *et al.* 1999; Auslaender *et al.* 2002; Ishii *et al.* 2003; Lee *et al.* 2004). It should be pointed out here that most of these observations of Luttinger liquid behavior are still under debate. Here, we will briefly mention them and we will not touch upon the still ongoing discussion regarding the interpretation of these experiments. Most of the experimental studies that were aiming to verify Luttinger liquid behavior were performed on carbon nanotubes and focused in particular on the power-law behavior of the density of states near the Fermi edge. Bockrath and coworkers (1999) reported temperature-and voltage-dependent measurements of the conductance of bundles of single-wall carbon nanotubes that agree with the predictions of the Luttinger liquid theory. Their experiments revealed that the conductance and the differential conductance scale as power laws with respect to the temperature and the bias voltage, respectively. In addition, the functional forms and the exponents are in agreements with the theoretical predictions. Auslaender and coworkers (2002) measured the collective excitation spectrum of electrons in one dimension by varying the energy and momentum of electrons tunnelling between two parallel quantum wires in a GaAs/AlGaAs heterostructure. They observed a significant enhancement of the excitation velocity of this system of interacting electrons as compared to a non-interacting electron system. Ishii *et al.* (2003) reported angle-integrated photoemission measurements on single-wall carbon nanotubes. Their experiments revealed that the spectral function and intensities at the Fermi edge exhibit power-law behavior. More recently, Lee *et al.* (2004) probed the electronic standing waves near one end of a single-wall carbon nanotube with a scanning tunnelling microscope. They observed near the end of the nanotube two standing waves, which they claim are caused by separate spin and charge bosonic excitations. The enhanced group velocity of the charge excitation, the power-law decay of their amplitudes away from the end of the nanotube and the suppression of the density of states near the Fermi edge all seem to point in the direction of Luttinger liquid behavior.

9.1.5   1D transport and quantization of conductance

A 1D chain of atoms can, for a given voltage drop applied across both its ends, only carry a finite amount of current. Thus, the conductance is finite even if there is no scattering in the chain. This transport is termed ballistic transport. Let us consider the following system: a thin wire is connected between two electrodes that can be considered as large reservoirs. A voltage difference V is applied to the two reservoirs. The wire has one partly occupied band that is responsible for the transport. The electronic states of the right (left) electrode are populated up to an electrochemical potential $\mu_1(\mu_2)$, where $\mu_1 - \mu_2 = -eV$ and $\mu_1 - \mu_2 = +eV$ for electrons and holes, respectively. The net current, I, that flows from the right to the left due to the excess carrier density Δn is,

$$I = -\Delta nev = (D(E)eV)\,ev = \frac{2e^2}{h}V, \qquad (9.12)$$

where $D(E) = \frac{1}{\pi\hbar}\sqrt{\frac{m}{2E}}$ is expressed per unit length. Since only positive k-values are considered here, the additional factor of 2 in the density of states should be omitted. The two-terminal conductance $G = I/V$ is then

$$G = \frac{I}{V} = \frac{2e^2}{h} \approx 7.7483 \times 10^{-5} \quad \Omega^{-1}. \qquad (9.13)$$

Equation (9.13) is known as the Landauer formula. A perfectly transmitting 1D channel has a conductance of $2e^2/h$. When more subbands are involved in the transport process and transmission probabilities (T) of less than unity occur, the total conductance G is given by $\sum_i T_i \left(2e^2/h\right)$.

9.1.6   Peierls instability

A Peierls instability can best be understood by considering a chain consisting of atoms with a nearest-neighbor distance a. Furthermore, we assume that each atom has one valence electron. Peierls suggested that such a linear chain is unstable with respect to a static lattice deformation of wavevector $2k_F$ (Peierls 1955, 1991). The energy band is exactly half-filled and the Fermi wave vector is located half-way in the Brillouin zone, i.e. at $k = \pm\pi/2a$. Under these conditions the atom chain is metallic. Doubling of the periodicity will lead to a modulation of the electronic density, often referred as a charge-density wave (CDW), and a folding of the Brillouin zone in reciprocal space. The edges of the new Brillouin zone coincide with the Fermi wave vector. Since the presence of a periodic potential leads to the development of an energy gap at the edges of the Brillouin zone, the chain undergoes a metal-to-insulator transition. The opening of the energy gap at the edges of the Brillouin zone lowers the energy of the electrons below the energy gap. The deformation proceeds until the increase in elastic energy is balanced by the above mentioned gain in electronic energy (see Fig. 9.1).

   For an ideal 1D chain the Peierls transition will occur at 0 K. When the chain is coupled, for instance, to an underlying substrate or neighboring chains, the systems loses its ideal 1D character and becomes quasi-one-dimensional. The Peierls transition temperature of such a quasi-1D system lies above 0 K.

   So far, we have restricted ourselves to monovalent atoms. This is not a strict requirement for the occurrence of a Peierls transition. Fractional band fillings deviating from 1 electron per band provide an interesting subset of 1D systems. For a band filling of $1/n$ electrons the chain will adopt a real-space unit cell with a size $2na$. At the edges of the new Brillouin zone, i.e. $k = \pm\pi/2na$, an energy gap develops.

9.2   Formation of monoatomic Pt chains on Ge(001)

9.2.1   Fabrication of atom chains

In 2003 the first measurements of large arrays of Pt chains on Ge(001) were reported (Gurlu *et al.* 2003). The self-organizing Pt chains only grow on specific Pt-modified, so-called β-terraces. They have a cross-section of only one atom, are perfectly straight (i.e. kink-free), very long and literally

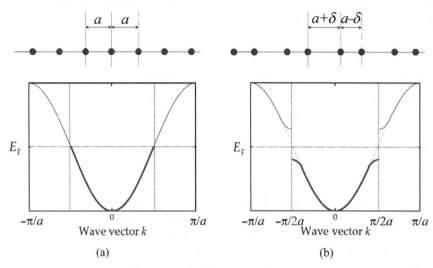

Fig. 9.1 (a) Dispersion for a one-dimensional equidistant atom chain with nearest-neighbor distance a (the chain is shown schematically above the graph). With one valence electron per atom the band is half-filled (thick part of the solid line) and the chain is metallic. (b) Dispersion for a Peierls distorted atom chain. A cartoon of the distortion is shown above the graph. The doubling of the periodicity leads to a modulation of the electronic density and folding of the Brillouin zone in reciprocal space. The edges of the new Brillouin zone coincide with the Fermi wave vector. The presence of a periodic potential opens an energy gap at the edges of the new Brillouin zone and the chain undergoes a metal-to-insulator transition.

defect free. For these reasons these Pt chains can be considered as the ultimate nanowires. Using similar procedures the formation of wires is reported for the Au/Ge(001) (Wang *et al.* 2004, 2005) and Au/Si(001) (Kageshima *et al.* 2001) systems, while deposition of Pd or Ag on Ge(001) favors the formation of three-dimensional clusters (Chan and Altman 2002; Wang *et al.* 2006).

Since the discovery of these Pt chains several studies on the Pt/Ge(001) system were reported (Schäfer *et al.* 2006; Schwingenschlögl *et al.* 2008a,b). However, the actual formation process of these atomic Pt chains has remained unclarified. Here, we focus on the formation process of these Pt nanowires in a detailed STM study. Although high-temperature (i.e. ~1100 K) measurements with STM are in principle possible, the study of the formation of the Pt atomic chains near the melting temperature is complicated by the low diffusion barrier of Ge ad-atoms and ad-dimers. At elevated temperatures the diffusing Ge dimers will hamper atomic resolution. However, an a posteriori inspection following a quick freeze in, at different stages of the formation process, can provide relevant insight into the formation process too. Our measurements reveal that Pt atoms deposited at room temperature, initially intermix with the bulk, but upon a high-temperature annealing treatment they pop-up from the bulk as Pt dimers. Initially, these Pt dimers are located in between the substrate dimer rows and are oriented parallel to the substrate dimers. Upon reaching a critical coverage the Pt dimers spontaneously rotate by 90° and form an atomic Pt chain in between the substrate dimer rows.

Experiments were performed in ultrahigh-vacuum (UHV) systems, with base pressures of 3×10^{-11} mbar. Measurements are performed with a commercially available Omicron STM-1, which is a room-temperature STM, and with a commercially available Omicron LT-STM, which has a working regime from 4.7 K to room temperature.

Ge(001) substrates were cut from nominally flat 3 in. by 0.3 mm, about 25 Ωcm resistance, single-side-polished n-type wafers. The samples were mounted on Mo holders and contact of the samples to any other metal during preparation and experiment was carefully avoided. The Ge(001) samples were cleaned by alternating cycles of ion bombardment (800-eV Ar^+ ions, angle of incidence 45°, 2 μA/cm^2 sputter yield) and annealing via resistive heating at 1100 (±25) K for a few minutes. The temperature is measured with a pyrometer. After several cleaning cycles the Ge(001) samples were atomically clean and exhibited a well-ordered $(2 \times 1)/c(4 \times 2)$ domain pattern (Zandvliet *et al.* 1998; Zandvliet 2003). An example of this striped pattern of alternating (2×1) and $c(4 \times 2)$ domains on clean Ge(001) is presented in Fig. 9.2. The dimer rows in the $c(4 \times 2)$ domains have a zigzag appearance. The dimer rows in the (2×1) domains have a symmetric appearance and consist of dimers that rapidly flip-flop between their two buckled configurations (van Houselt *et al.* 2006b).

Subsequently, an equivalent of 0.20–0.30 monolayers of platinum was deposited onto the clean Ge(001) surface. The Pt deposition was performed at room temperature. Platinum was evaporated by resistively heating a W wire wrapped with high-purity Pt (99.995 %). After Pt deposition the sample was annealed at 1050 (±25) K and then cooled down to room temperature by

Fig. 9.2 STM image ($V = -1.5$ V, $I = 0.4$ nA, $T = 293$ K, 20×20 nm^2) of a clean Ge(001) substrate. A striped pattern of alternating (2×1) and $c(4 \times 2)$ domains is visible.

radiation quenching for at least half an hour. Subsequently, the sample was placed into the STM for observation.

9.2.2   The α- and β-terraces

After deposition of Pt on Ge(001) and subsequent annealing at 1050 K, two different types of terraces are formed on the Ge(001) surface, denoted as α- and β-terraces (Gurlu *et al.* 2003).

An STM image of an α-terrace is displayed in Fig. 9.3(a). The α-terraces are comprised of symmetric and asymmetric Ge dimers and resemble the normal dimer reconstructed Ge(001) terraces rather well (Gurlu *et al.* 2004). The amount of missing dimer defects and ad-atoms is, however, much higher as compared to the bare Ge(001) surface. Most of these missing dimer defects can be identified as so-called $2 + 1$ missing dimer defects (two missing dimer defects followed by a normal dimer and a missing dimer defect). It is believed that these defects are induced by the presence of a metal atom, such as for instance Ni, Ag, Cu, Co or, in our case, Pt atom, sitting in a subsurface position (Zandvliet *et al.* 1995). Figure 9.3(b) shows an STM image of a β-terrace. The surface is again dimer terminated, but in this case the termination consists of an ordered array of dimers, that clearly deviate from normal Ge–Ge dimers, as they are apparently Pt modified (Gurlu *et al.* 2003). A remarkable feature of these β-terraces is the presence of dimer vacancy lines, which are always aligned along the [310] and [110] directions. These vacancy lines have never been observed on the bare Ge(001) surfaces and the Pt-contaminated α-terraces. Interestingly, patches of monoatomic Pt chains are exclusively observed on the β-terraces.

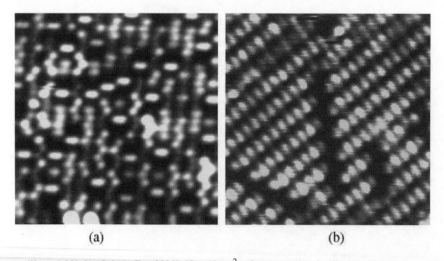

(a)   (b)

Fig. 9.3 (a) STM image ($V = -1.0$ V, $I = 0.4$ nA, $T = 293$ K, 10×10 nm^2) of an α–terrace after Pt deposition and subsequent annealing steps. Note the high concentration of defects. (b) STM image ($V = -0.3$ V, $I = 0.5$ nA, $T = 293$ K, 10×10 nm^2) of a β-terrace after Pt deposition and subsequent annealing steps. Note the vacancy clusters aligned in the [310] and [110] directions.

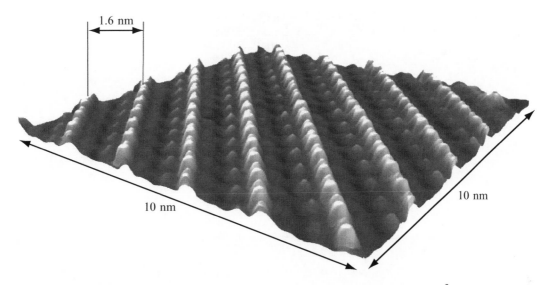

Fig. 9.4 Three-dimensional representation of an STM image ($V = -1.4\,\text{V}$, $I = 0.5\,\text{nA}$, $T = 293\,\text{K}$, $10 \times 10\,\text{nm}^2$) of an array of self-organizing Pt chains. The Pt atoms within a chain dimerize. The spacing between adjacent Pt chains is only 1.6 nm.

9.2.3 The birth of Pt chains

An STM image of such a patch of monoatomic Pt chains is shown in Fig. 9.4. The chains have a cross-section of one atom, are kinkless and defect free. The monoatomic chains are located in troughs between the substrate dimer rows. The Pt atoms within a chain dimerize. The interchain distance is found to be mostly 1.6 nm, although spacings of 2.4 nm and even sometimes 3.2 nm, 4.0 nm, etc., are found as well. The majority of the monoatomic chains is found in patches. Occasionally isolated Pt chains are also observed.

The high-temperature (i.e. annealing at $\pm 1050\,\text{K}$) treatment turns out to be essential for the formation of the monoatomic Pt chains. In order to obtain more detailed insight into the nanowire formation mechanism, we used an anneal time of only a second, in contrast to our previous work, where we used an anneal time of about 10 min (Gurlu *et al.* 2003). In fact, independent of the actual duration of annealing at 1050 K we have always observed the formation of atomic Pt chains.

Figure 9.5(a) shows an STM image of the Pt chains covered Ge(001) surface after high-temperature annealing at 1050 K for 1 s. Large fractions of the surface are covered with Pt chains running along the substrate dimer row directions.

Around the center of Fig. 9.5(a) a narrow Pt-chain-free area is observed. Within this chain-free region the trough between two substrate dimer rows widens. The widening starts from two ordinary substrate dimer rows in the upper right corner of Fig. 9.5(a). This widening process resembles the behavior of a zipper. In the widened trough two isolated and one cluster of two elongated features are visible. The elongation direction is along the substrate–dimer bond direction, perpendicular to the Pt-chain direction. The widened troughs are often found near the edges of a patch of Pt chains (Figs. 9.5(c) and (d)).

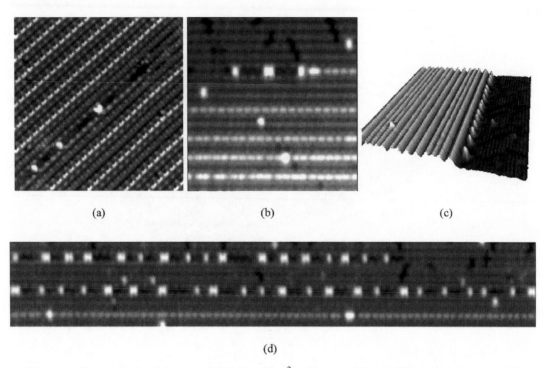

(a)   (b)   (c)

(d)

Fig. 9.5 (a) STM image ($V = -0.5$ V, $I = 0.5$ nA, $T = 293$ K, 15×15 nm^2) of the Pt-modified Ge(001) surface, after a 1-s high-temperature anneal at 1050 K. Note the widening of the β-terrace rows near the center of the image. (b) STM image ($V = -1.2$ V, $I = 0.4$ nA, $T = 77$ K, 14×14 nm^2) of the Pt-modified Ge(001) surface, after a 1-s high-temperature anneal. (c) Three-dimensional STM image ($V = 1.5$ V, $I = 0.2$ nA, $T = 77$ K, 37.5×37.5 nm^2) of the Pt-modified Ge(001) surface after a high-temperature anneal. (d) STM image ($V = 1.5$ V, $I = 0.4$ nA, $T = 77$ K, 10×50 nm^2) of the Pt-modified Ge(001) surface after a high-temperature anneal (images from Fischer *et al.* 2007).

We suggest that the protrusions in this widened trough are comprised of Pt dimers that are aligned along the substrate–dimer bond direction. The Pt chain eventually forms because these Pt dimers rotate by 90°, leading to a Pt chain in the trough between the substrate rows. This rotation process is quite similar to the rotation of a Si dimer on top of the substrate–dimer rows of a Si(001) surface (Zhang *et al.* 1995): The ad-dimer axis is initially perpendicular to the substrate dimer row direction and it rotates to a parallel configuration. However, the Pt rotation process takes place in between the substrate dimer rows rather than on top of the substrate–dimer rows.

In Fig. 9.5(b) a broadened trough that contains a fragment of a well-ordered Pt chain in the trough as well as several Pt dimers in the "wrong" orientation can be seen. The "wrongly" orientated Pt dimers in the widened trough are always found as single ones or in pairs (see Fig. 9.5(b)). This suggests that the Pt dimers rotate by 90 degrees and form a chain as soon as the Pt dimer cluster becomes larger than two dimers. The amount of Pt dimers we have observed varies from troughs that contain only a few Pt dimers (Fig. 9.5(a)) up to troughs that contain about half the number of misaligned Pt dimers needed for a complete Pt chain (see Figs. 9.5(c) and (d)).

Using the STM tip as a nanotweezer we were able to pick up Pt dimers from the Pt chain. Figure 9.6(a) shows an STM image of region covered with

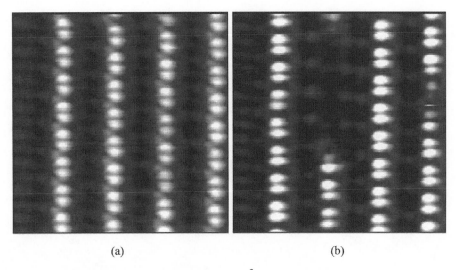

(a)   (b)

Fig. 9.6 STM images ($V = -0.020$ V, $I = 0.5$ nA, $T = 293$ K, 7.0×7.0 nm^2) of a region covered with Pt chains, before a manipulation experiment (a) and after the removal of some Pt dimers from the second Pt chain (b). Note the structure of the underlying (widened) trough (images from Fischer *et al.* 2007).

several Pt chains. In Fig. 9.6(b) several Pt dimers have been removed from the second Pt chain by the STM tip (Gurlu *et al.* 2007). Beneath the Pt chain we see again the widened trough. Thus, the widened trough remains intact after the rotation of the Pt dimers from the "wrong" orientation to the "right" orientation.

An STM image of a prolonged (10 min) annealed Pt-modified Ge(001) surface is shown in Fig. 9.7. A variety of interesting features can be observed. Firstly, it reveals that, once the chains have nucleated, the chains expand easily along the dimer-row direction. As a consequence, the Pt chains are as long as possible, i.e. their length is only limited by defects in the terrace, pre-existing step edges or out-of-registry neighboring chains. The latter act as an antiphase boundary (denoted by number 1 in Fig. 9.7). Antiphase boundaries also occur in the direction perpendicular to the Pt chains, as a result of the coalescence of neighboring chain patches. In this case the antiphase boundary consists of a 2.4-nm wide trough between the neighboring Pt chain (denoted by number 2 in Fig. 9.7). In Fig. 9.4(c) it can be seen that nucleation of a new Pt chain occurs preferably at the edge of a patch with an interchain distance of 1.6 nm. This explains the formation of large patches of Pt chains (see Fig. 9.7). Besides these patches quite a number of isolated atomic Pt chains are also formed. The majority of these isolated Pt chains consists of Pt dimers in a "wrong" orientation, i.e. these Pt dimers have their dimer bond aligned in a direction perpendicular to the trough. Obviously these unrotated Pt dimer line segments are relatively stable. In addition, a more careful analysis of these line segments reveals that the number of unrotated Pt dimes that is required to form a well-oriented Pt chain in the trough is too high. It remains, however, unclear why this occurs mainly for the isolated atomic Pt chains. We believe that the rotation process of the Pt dimers in the unrotated dimer line segments is sterically hindered. The latter leads to Pt chains with thick "tails" of Pt dimers in the

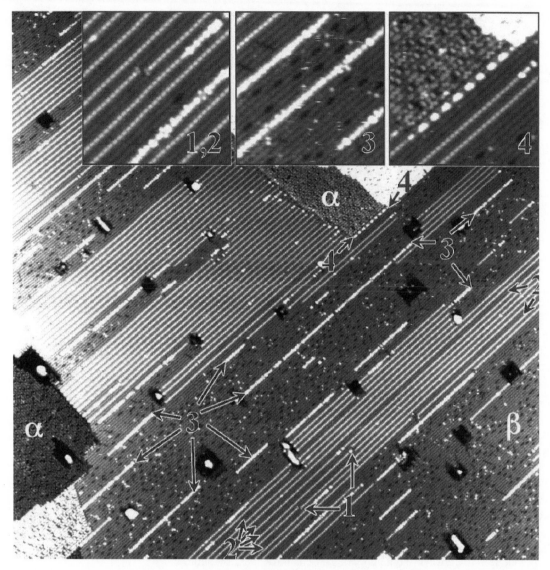

Fig. 9.7 STM image ($V = -1.5$ V, $I = 0.4$ nA, $T = 293$ K, 150×150 nm^2) of a prolonged annealed Pt-modified Ge(001) surface. α- and β-terraces are indicated. On the β-terraces atomic Pt chains with varying lengths are visible. Antiphase boundaries along and perpendicular to the Pt-chain direction are, respectively, indicated with number 1 and 2. Pt chains containing in the "wrong" rotation, which are sterically hindered in their rotational motion, are pointed out by number 3. The numbers 4 mark a double step on an α-terrace decorated with Pt atoms. The insets are 15×15 nm^2 enlargements of the indicated features.

"wrong" orientation (denoted by number 3 in Fig. 9.7). Additionally, double step edges on α-terraces are found occasionally to be decorated with a row of Pt atoms (denoted by number 4 in Fig. 9.7).

In summary, in this section the formation of well-defined, defect- and kink-free Pt atomic chains on a Pt-modified Ge(001) surface has been studied with STM. Initially, the Pt atoms deposited at room temperature submerge subsurface. After annealing at elevated temperatures the surface opens locally in a zipper-like manner and the Pt atoms emerge as pairs within the widened through between the two substrate dimers rows on the β-terraces. Finally, this is followed by a 90° rotation of the Pt dimers, giving rise to well-defined atomic Pt chains.

9.3 Quantum confinement between monoatomic Pt chains

9.3.1 Electron confinement on surfaces

Much progress in the fabrication of nanostuctures enabled the study of low-dimensional physics. Nice experiments regarding the scattering and confinement of electrons in low-dimensional structures have been described (Crommie *et al.* 1993; Bürgi *et al.* 1998; Nilius *et al.* 2002; Meyer *et al.* 2003). The monoatomic Pt chains described in the prevouis section have a cross-section of only one atom, are literally defect and kink free, and have lengths up to hundreds of nanometers. They thus promise to satisfy a pre-requisite for the further confinement of surface states in an additional dimension, leading to a truly one-dimensional electron gas. The fact that the Pt atoms are ordered so nicely on the substrate enables us to image the quantum-mechanical interference of surface-state eletrons within these self-organized Pt chains. The spectroscopic capabilities of the STM allow the observation of the spatial distribution of a pre-selected electronic state. We have been able to unveil and explore novel electronic states, which are, surprisingly, confined between the Pt chains, rather than inside them, as we will show below (Oncel *et al.* 2005; van Houselt *et al.* 2006).

9.3.2 Novel electronic states and their spatial origin

Besides the majority of the Pt chains, that have a spacing of 1.6 nm (see Fig. 9.8(a)), Pt chains with a spacing of 2.4 nm are also found. The latter actually form light domain walls between the (4×2) domains established by the Pt chains that have a spacing of 1.6 nm, see Fig. 9.8(b).

We have performed scanning tunnelling spectroscopy on the arrays with Pt chains at both 77 and 300 K. Figure 9.9(a) shows the (spatially averaged) derivative dI/dV recorded, at 77 K and at 300 K, on an array of Pt chains spaced 1.6 nm apart. Figure 9.9(b) shows the local density of states (LDOS, i.e. the normalized derivative $(dI/dV)/(I/V)$) of the bare underlying terrace (lower curve) and of an array of Pt chains that are 1.6 nm (middle curve) or 2.4 nm (upper curve) spaced apart. All spectra are recorded at 77 K. It is commonly accepted that the $(dI/dV)/(I/V)$ curve provides a better measure of the local density of states (LDOS) than the direct derivative dI/dV (Lang 1986; Stroscio *et al.* 1986; Feenstra *et al.* 1987). In the middle $(dI/dV)/(I/V)$ curve in Fig. 9.9(b) two peaks are resolved. Most noticeable is the appearance of a peak at 77 K at $\sim$0.1 eV above the Fermi level in the $(dI/dV)/(I/V)$ curve, which is not present in the room-temperature data (note that this peak shifts to $\sim$0.15 V in the dI/dV curve). Another peak positioned just below

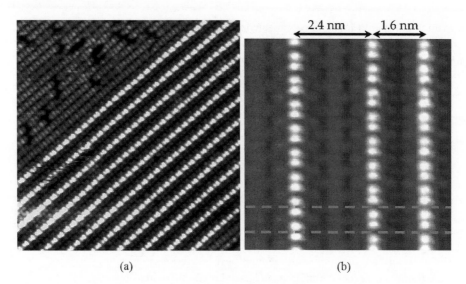

(a)   (b)

Fig. 9.8 (a) STM image ($V = -0.25$ V, $I = 0.5$ nA, $T = 293$ K, 20×20 nm^2) of a large array of Pt chains that are spaced 1.6 nm apart from each other. (b) STM image ($V = -1.4$ V, $I = 0.5$ nA, $T = 293$ K, 6×6 nm^2) of Pt chains with different interchain distances. The 2.4-nm wide trough acts as a domain wall between the (4×2) domains established by the Pt chains that are 1.6 nm separated. Dotted lines are added as a guide to the eye.

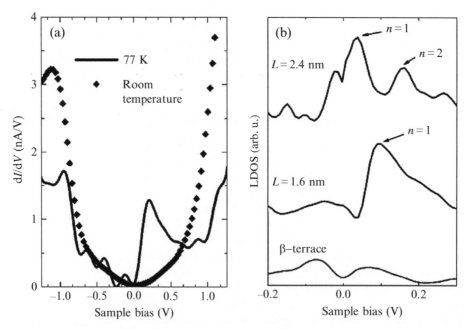

Fig. 9.9 (a) Differential conductivity ($\mathrm{d}I/\mathrm{d}V$) of a Pt chain array recorded at 300 K (diamonds) and 77 K (solid line). The tunnelling current setpoints at -1.5 V are 2.5 nA (300 K) and 1.5 nA (77 K). The Pt chains are 1.6 nm spaced apart. (b) Local density of states ($\mathrm{d}I/\mathrm{d}V)/(I/V)$ of the bare β-terrace (lower curve), troughs between Pt chains with a spacing of 1.6 nm (middle curve) and troughs with a spacing of 2.4 nm (top curve). All spectra are recorded at 77 K. For $L = 1.6$ nm only one novel peak at 0.1 eV above the Fermi level is observed, whereas for $L = 2.4$ nm two novel peaks are found at 0.04 and 0.16 eV above the Fermi level.

the Fermi level is present in the local density of states at 77 K. Scanning tunnelling spectroscopy experiments (Fig. 9.9(b), lower curve) reveal that the present novel metallic peak actually originates from the underlying β-terrace on which the Pt chains can form. Note that the "bare underlying terrace", the so-called β-terrace, is not a clean Ge(001) surface, but is actually "platinum-modified", as discussed in Section 9.2.2

The novel electronic state located just above the Fermi level can be ascribed to either the Pt chains or to the troughs between the Pt chains. In order to identify the exact origin of this one-dimensional peak we have recorded spatial maps of the derivative dI/dV at 0.15 V, simultaneously with the topography. dI/dV maps were measured through lock-in detection of the ac tunnel current driven by a 797 Hz, 10 mV (rms) signal added to the junction bias. In Fig. 9.10 we present the simultaneously acquired topography and dI/dV spatial map.

From this image it is immediately clear that the novel electronic state originates from the troughs and *not* from the Pt chains. We attribute the novel one-dimensional electronic state to confinement of the electronic surface state near the Fermi level. One should keep in mind that the apparent barrier height above a local topographic protrusion is larger, i.e. the effective decay length is smaller than above a local topographic depression. This spatial variation in the transmission coefficient will show up in spatially resolved measurements of dI/dV as a "background" that is essentially an inverted constant-current topography. The observation that missing atoms in the troughs result in a local decrease of the dI/dV signal implies that the latter effect is of minor importance (see the four dotted circles and ellipses in Fig. 9.10). The shift in energy of the one-dimensional state with respect to the two-dimensional surface state is convincingly attributed to the confinement of the electrons between neighboring Pt chains. Near defects in a Pt chain the peak position is either shifted towards higher energy (when the defect narrows the trough) or lower energy (when the defect widens the trough). Typically, the peak shifts and decays over a distance of a few nms measured along the trough direction.

The shift in energy of the one-dimensional state with respect to the two-dimensional surface state is attributed to the confinement of the electrons between neighboring Pt chains. The Pt chains thus act as a potential barrier for the electrons in the troughs. Using the particle in an infinite well[2] description (see Section 9.1.2) the energy positions are given by:

$$E_n = \frac{\hbar^2}{2m}\left(\frac{\pi n}{L}\right)^2 \text{ with } n = 1, 2, \ldots \infty, \tag{9.14}$$

where $\hbar$ is Planck's constant, m is the electron mass and L the width of the well. For a 1.6-nm spacing between adjacent Pt chains one finds an energy shift of 0.147 eV, which is only slightly higher than in our observations. Additional subbands (e.g. $n = 2$ at 0.4 eV and $n = 3$ at 0.9 eV) are, however, not observed for the 1.6 nm spacing between the Pt chains. This is most probably related to a finite barrier height and the influence of the band edge of the conduction band of the underlying substrate. This will also result in a renormalization of the energy position of all subbands. It can be seen immediately that the energy positions of the novel electronic states in Fig. 9.9 (0.10 eV for

[2]The use of a well of finite depth alters the picture somewhat: the energy levels are renormalized and the wavefunctions slightly penetrate into the walls. Using this model, we are unable to determine the exact value of the effective electron mass, but we are able to determine the product $\nu\sqrt{\lambda}$ (which is around 0.6), where $\nu = L_{\text{eff}}/L_0$ and $\lambda = m_{\text{eff}}/m_0$. Inserting realistic values for the effective well width L_{eff} ($\nu = 0.6$–0.8) yields an effective mass in the range from 0.6 to 1 m_0.

(a) Topography   (b) Derivative (d*I*/d*V*)

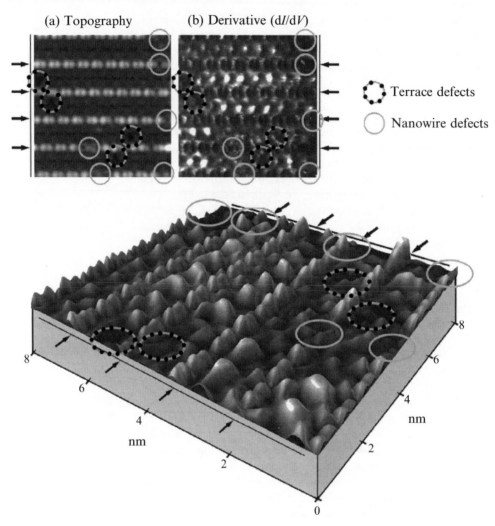

Terrace defects

Nanowire defects

Fig. 9.10 Topography (a) and spatial map of the differential conductivity (b) of a $8 \times 8 \, \text{nm}^2$ area with several Pt chains that are spaced 1.6 nm apart recorded at 77 K. The sample bias is 0.15 V and the tunnel current is 0.437 nA. The d*I*/d*V* map is recorded with a modulation voltage of 10 mV and an oscillation frequency of 797 Hz. In the lower image a 3-dimensional representation of the topography and the d*I*/d*V* map is shown. The one-dimensional electronic state is exclusively located in the troughs of the Pt chains (black arrows refer to the position of the Pt chains). The circles (ellipses) refer to defects in the Pt chain (gray ones) or the underlying substrate (dotted ones). The confinement of the electronic state disappears near these defects. (Images in Figs. 9.9 and 9.10 are taken from Oncel *et al.* 2005.)

$L = 1.6 \, \text{nm}$ and $n = 1$; 0.04 and 0.016 eV for $L = 2.4 \, \text{nm}$ and $n = 1, 2$) obey the $(n/L)^2$ scaling law accurately.

The differential conductivity in Fig. 9.10 also shows a strong corrugation along the chain direction. The periodicity of this corrugation is equal to the periodicity of the Pt chains and the periodicity of the β-terrace dimer row, namely 8 Å. The Bloch theorem requires that a wavefunction in a periodic potential will always adapt the periodicity of the potential (Bloch 1928; Kittel 2005). The corrugation along the chain direction is considered as a Bloch wave, since it follows exactly the periodicity of the chains and the substrate.

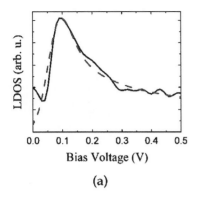

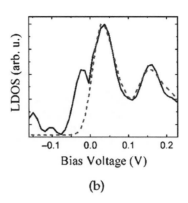

Fig. 9.11 Spatially averaged LDOS between Pt chains that are spaced 1.6 nm apart (a) and 2.4 nm apart (b) at $T = 77$ K. In (a) the dashed line shows a Van Hove singularity for a one-dimensional system using $E_1 = 100$ meV, convoluted with a thermal broadening using a Gaussian of $\sigma = 35$ meV. For the Van Hove singularities in (b) $E_1 = 40$ meV, $E_2 = 160$ meV and the same Gaussian is used for the thermal broadening (images in Figs. 9.11–9.13 are taken from van Houselt *et al.* 2006a).

9.3.3   Spatial maps of the confined states

For a one-dimensional system, the DOS exhibits a Van Hove singularity (eqn (9.5)): $E \propto 1/\sqrt{E - E_n}$ with E_n the energy position of the nth state.

Figure 9.11(a) shows the LDOS between Pt chains with a spacing of 1.6 nm. The dashed line shows a Van Hove singularity using $E_1 = 100$ meV convoluted with a thermal broadening using a Gaussian of $\sigma = 35$ meV. Figure 9.11(b) shows the LDOS between Pt chains with a spacing of 2.4 nm. The dashed line shows two Van Hove singularities using $E_1 = 40$ meV and $E_2 = 160$ meV and the same Gaussian is used for the thermal broadening. Most noticeable is the relatively large contribution (∼40%) of the tail of the $n = 1$ state to the total DOS at the onset of the $n = 2$ peak at E_2.

Figure 9.12 shows the spatially averaged cross-section of the differential conductivity between the Pt chains with an interwire distance of 2.4 nm recorded at the $n = 1$ and $n = 2$ peak positions, 40 and 160 mV. The $n = 1$ state has a maximum in the middle of the trough and gradually fades away near the edges of the trough. The $n = 2$ state appears as a rather broad peak with a small minimum in the middle of the through. The lateral distribution of the LDOS is proportional to $|\Psi_n(x)|^2$. Therefore, one expects a lateral distribution that behaves as $\sin(n\,\pi\,x/L)^2$, that is, a maximum at $x = L/2$ for the $n = 1$ state and a minimum at $x = L/2$ for the $n = 2$ state. The cross-section of the spatial map of the $n = 1$ state agrees well with the expected distribution $|\Psi_1(x)|^2 = \sin(\pi\,x/L)^2$ (dashed line in Fig. 9.11(a)). At first sight, the measured distribution of the $n = 2$ state seems to disagree with the expectations. However, one should take into account that at 160 meV both states contribute, as can be observed in the LDOS curves (Figs. 9.9(b) and 9.11(b)). To resolve the spatial variation of the $n = 2$ state, the contribution of the $n = 1$ state at 160 mV should be subtracted from the measured intensity at 160 mV. The corrected measured spatial variation of the $n = 2$ state dashed line in Fig. 9.11(b) now resembles the expected lateral distribution of $|\Psi_2(x)|^2 = \sin(2\,\pi\,x/L)^2$, with a minimum in the middle of the trough.

In addition, spatial mapping at energies in the energy range from 170 to 250 meV, shown in Fig. 9.13, clearly shows that with increasing energy the dip in the LDOS distribution at $x = L/2$ deepens. The latter can be understood

Fig. 9.12 Spatially averaged cross-section of the differential conductivity between the Pt chains with an interchain distance of 2.4 nm recorded with a sample bias of 40 mV (a) and 160 mV (b), which corresponds to the $n = 1$ state and $n = 2$ state, respectively (solid lines). The dashed line in (a) shows a representation of the expected squared wavefunction $|\Psi_1(x)|^2 = \sin(\pi\,x/L)^2$. In the STM images, the topography and spatial map of the differential onductivity (dI/dV) of the same area taken at a sample bias of 40 mV (a) and 160 mv (b) are shown. The dashed line in (b) shows a corrected distribution, in which the contribution of the $n = 1$ state is subtracted from the measured distribution. The dashed-dotted line shows a representation of the expected squared wavefunction $|\Psi_2(x)|^2 = \sin(2\,\pi\,x/L)^2$.

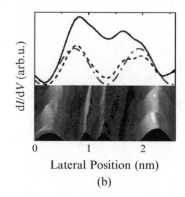

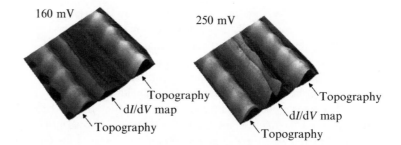

Fig. 9.13 Topography and spatial map of the differential conductivity (dI/dV) of the same area, taken at sample biases of 160 mV (left) and 250 mV (right) at $T = 77$ K.

by the fact that the influence of the $n = 1$ tail becomes less and less with increasing energy (sample bias).

In summary, one-dimensional states confined between self-organizing Pt chains with a mutual distance of 1.6 and 2.4 nm has been investigated by means of STM and STS. The existence of a set of novel electronic states at 77 K, which are absent at room temperature has been discovered. The energy of the novel electronic states are 0.10 eV ($n = 1$ and $L = 1.6$ nm), 0.04 and 0.16 eV ($n = 1, 2$ and $L = 2.4$ nm) higher than that of a two-dimensional surface state on the Pt-modified β-terraces, which is located just below the Fermi level. A careful analysis of the differential conductivity data reveals that these states are actually located within the troughs between the Pt chains. The increase in energy is a result of the confinement of the nearly free surface-state electrons between the Pt chains and its values agree well with basic theoretical estimates. The fact that the density of states at the center of the troughs gradually fades away near defects in either the Pt chains or the underlying terrace provides strong additional proof in favor of the assignment of the electronic state to quantum confinement of electrons in between Pt chains. The spatial maps of the $n = 1$ and $n = 2$ states in the 2.4-nm wide trough agree nicely with the lowest-energy bands of a 1D quantum-mechanical particle in a box.

9.4 Peierls instability in monoatomic Pt chains

9.4.1 Introduction

In this section the structural and electronic properties of the atomic Pt chains are described. The Pt chains on a Pt-modified Ge(001) surface dimerize at room temperature. This dimerization is not related to a Peierls distortion. Below, we will show that this dimerization is a direct consequence of the difference in Pt–Pt binding length (~ 3 Å) and the periodicity of the underlying terrace (4 Å). Furthermore, we show that the Pt chains undergo an intriguing and unanticipated structural phase transition from a $2\times$ periodicity at 293 K to a $4\times$ periodicity at 4.7 K. Combined $I(V)$ and $I(z)$ tunnelling spectroscopy experiments recorded on top of the Pt chains reveal that the $2\times$ to $4\times$ transition is accompanied by a substantial reduction of the metallicity of the Pt chains. The coupling between the atomic chains is of essential importance, since isolated Pt chains and chains located at the edge of an array of chains maintain their $2\times$ periodicity at temperatures as low as 4.7 K. The $2\times$ to $4\times$ transition is interpreted as a Peierls instability.

9.4.2 Pt chains on β-terraces: Frenkel–Kontorova pairing

The Pt chains on a Pt-modified Ge(001) surface dimerize at room temperature. In this section we discuss the origin of this $2\times$ periodicity in terms of Frenkel–Kontorova pairing. The lattice mismatch between the Ge nearest-neighbor distance in the (001) plane (4.0 Å) and the Pt nearest-neighbor distances (2.77 Å) is so large that dimerization is a logical consequence. In order to supplement these experimental results with some theoretical support, *ab-initio* total-energy and electronic-structure calculations within the framework of the density-functional theory (DFT) have been carried out. The calculations have been performed with the VASP package (Kresse *et al.* 1993, 1996) using the local density approximation (LDA, Hohenberg *et al.* 1964) and the projector-augmented wave method (PAW, Blöchl 1994; Kresse *et al.* 1999). The Ge(001) surface is set up in a repeated slab geometry. Each slab contains eight layers of Ge atoms and a passivating H layer at the bottom. The two bottom Ge layers have been fixed at their theoretical bulk positions, as determined in a prior calculation. All other atoms are relaxed into the equilibrium by the conjugate-gradient algorithm. In all cases the vacuum region amounts to approximately 14 Å. The calculations are performed with eight k points in the irreducible Brillouin zone. For the model of the underlying β-terraces the $c(4 \times 2)$-reconstruction of the pure Ge(001) surface forms the starting point. Twenty-five per cent of the Ge atoms in the top layer are replaced by Pt atoms. Since the exact positions of these Pt atoms are not known, several a priori reasonable combinations have been investigated. On top of each model of the β-terraces the Pt chain has been placed in such a manner, that the Pt chain shows no dimerization at the beginning of each calculation.

Regardless of the model used for the β-terraces the results are all very similar. In all cases the Pt chain dimerizes after the relaxation of the atomic positions. For the majority of the calculations the distances between the Pt atoms in the chain are in excellent agreement with the experiment. While the experimental values are 5 Å and 3 Å the calculated values lie between 5.2–5.3 Å and 2.7–2.8 Å. For comparison, an additional calculation of a Pt chain on a pure Ge(001)-$c(4 \times 2)$ surface has been performed. Also, for this system the Pt chain dimerizes and the distances between the atoms are again 5.3 Å and 2.7 Å. These findings corroborate the experimental observation that the dimerization within the atomic Pt chains is driven by Frenkel–Kontorova pairing.

9.4.3 Coupling between the Pt chains and periodicity doubling

A Peierls distortion converts a one-dimensional chain from a metal into a semiconductor, while doubling its periodicity. Figure 9.14 displays 3D STM images of an array of self-organized atomic Pt chains (left image) and a single atomic Pt chain (right image) on a β-terrace recorded at 4.7 K. The Pt chains labeled a–c show a $4\times$ periodicity. In these chains the dimers of adjacent chains buckle always in the same direction (i.e. up or down, see the line profiles in Fig. 9.14). The other Pt chains (d and e), respectively positioned near the edge of a chain array and an isolated wire, exhibit a $2\times$ periodicity. In these chains (d and e) there is hardly any buckling of the Pt dimers observable. The strong chain-to-chain coupling within the chain array, which is either substrate mediated or occurs via a direct interaction between the chains, points towards a Peierls instability: The isolated chains and the chains located at the edge of an array ($2\times$ periodicity) are not Peierls distorted as a consequence of their weaker neighbor coupling, which lowers their transition temperature

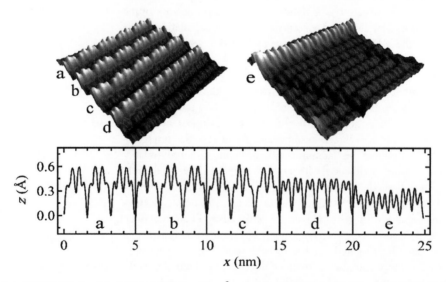

Fig. 9.14 STM images ($V = -1.5$ V, $I = 0.5$ nA, $T = 4.7$ K, 10×10 nm^2 or both images) of an array of Pt chains (left image) and an isolated Pt chain (right image) recorded at 4.7 K. Line scans along the Pt chains labelled a–e are depicted below the STM images. Isolated Pt chains (chain e) of Pt chains located near the edge of an array (chain d) exhibit a $2\times$ periodicity, whereas the Pt chains within the patch (chains a–c) show a $4\times$ periodicity (images in Figs. 9.14, 9.15, 9.18, and 9.19 are taken from van Houselt *et al.* 2008).

to below 4.7 K. The chains within a chain array (4× periodicity) have in this view already undergone a Peierls distortion, since their transition temperature has significantly increased due to the strong chain-to-chain coupling. This interpretation in terms of a Peierls transition would require that, upon heating the substrate to above the Peierls transition temperature for these strongly coupled chains, the 4× periodicity changes into a 2× periodicity, accompanied by an insulator-to-metal transition.

In Fig. 9.15 we compare STM images (2D and 3D) of an array of atomic Pt chains recorded at 4.7 K and 293 K. In the images on the left-hand side (4.7 K) the chains near the edge of an array exhibit a 2× periodicity, while the chains within the array show a 4× periodicity (the patch expands to the right outside the measurement area). The room-temperature images on the right-hand side show, however, a 2× periodicity for each Pt chain, irrespective of a position near the edge or within a patch. The transition from a 4× periodicity at 4.7 K to a 2× periodicity at 293 K confirms our interpretation in terms of a Peierls transition.

Occasionally one finds phase slips in the domain pattern established by the Peierls distorted Pt chains at 4.7 K. Figure 9.16 shows a series of

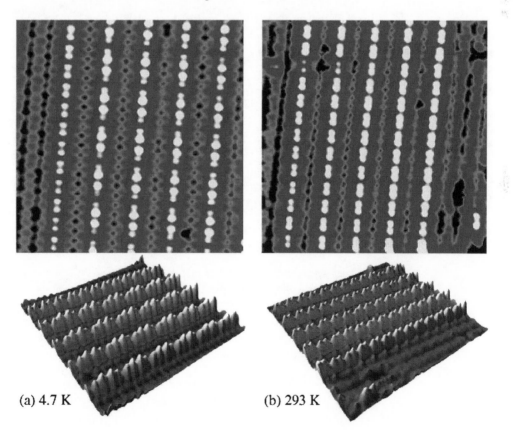

(a) 4.7 K   (b) 293 K

Fig. 9.15 2D and 3D STM images ($V = -1.5$ V, $I = 0.5$ nA, $T = 4.7$ K, 10×10 nm^2 for both images) of an array of Pt chains recorded at 4.7 K (a) and 293 K (b). At 4.7 K the outermost left chain (near the edge of an array) exhibits a 2× periodicity, while the dimers of Pt chain within an array buckle alternating up and down, leading to a 4× periodicity. At room temperature every chain shows a 2× periodicity.

Fig. 9.16 Series of STM images ($I = 0.5\,\text{nA}$, $T = 4.7\,\text{K}$, $10 \times 10\,\text{nm}^2$ for all images) of an array of Pt chains at bias voltage range from $-1.5\,\text{V}$ (image 1) to $-0.1\,\text{V}$ (image 15). The step size is $0.1\,\text{V}$. The areas highlighted with a gray line are out of phase with their neighboring chains.

bias-dependent STM images of the Pt chains at $4.7\,\text{K}$. In the images recored at a bias voltage range from -1.1 to $-0.6\,\text{V}$ areas with Pt chains that are out of phase with their neighboring chains are marked with a gray line. Even in one single Pt chain the two periodicities can show up, as shown in Fig. 9.17.

9.4.4 Probing the metal-to-insulator transition

Regarding the accompanying insulator-to-metal transition, one should realize that the Pt chains are exclusively found on the β-terraces, which are metallic in nature (see Section 9.3.2). This metallicity complicates the measurement of the insulating behavior of the Pt chains at low temperatures, since the contribution of the metallic terrace to the tunnelling current will hide the much lower contribution of the insulating chains to the tunnelling current. dI/dV curves recorded on top of the Pt chains of Fig. 9.14 are displayed in Fig. 9.18. The chain that exhibits a $2\times$ periodicity (chain d) has a more distinct metallic character ((dI/dV) at zero bias) than the chains that exhibit a $4\times$ periodicity (chains a–c from Fig. 9.14). Since the underlying β-terrace is metallic too, we need to figure out a way to separate the contribution of the Pt chain from the contribution of the underlying terrace to the tunnelling current. In an attempt to separate these contributions we have determined the inverse decay length, κ, for both types of Pt chains. The inverse decay length can provide information on the spatial character of the tunnelling electrons. Using the Wannier–Kramers–Brillouin (WKB) approximation, the tunnelling current can be written as (Bonnel *et al.* 1993):

$$I(V) = \int_0^V \rho(E)T(E, eV)\mathrm{d}E, \tag{9.15}$$

where $\rho(E)$ is the density of states of the surface (assuming a constant density of states of the tip) and $T(E, eV)$ refers to the transmission probability. The latter is defined as: $T(E, eV) = \exp(-2\kappa z)$. κ is the inverse decay length of the wavefunction, i.e.

$$\kappa = \sqrt{\frac{2m}{\hbar^2}\left(\frac{\Phi_t + \Phi_s}{2} - E + \frac{eV}{2}\right) + k_{//}^2}, \tag{9.16}$$

where $\Phi_{t(s)}$ is the work function of the tip (substrate), V the applied voltage between tip and substrate, E the energy of the state relative to the Fermi level and m the mass of an electron. It is evident from eqn (9.16) that a wavefunction decays slowly into the vacuum, if the corresponding parallel wave vector, $k_{//}$, is zero. The electronic states near the Γ point of the surface Brillouin zone usually have the largest contribution to the tunnelling current because of their small inverse decay length. In most cases it is even justified to consider only the contribution of the electronic states near the Γ point. However, if no electronic state is present at the Γ point, other electronic states with non-zero parallel momentum and thus larger inverse decay lengths will dominate the tunnelling current. This is, for instance, quite common for semiconductors

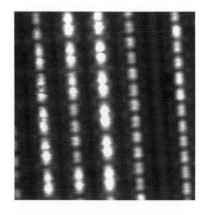

Fig. 9.17 STM image ($V = -0.7\,\mathrm{V}$, $I = 0.5\,\mathrm{nA}$, $T = 4.7\,\mathrm{K}$, $10 \times 10\,\mathrm{nm}^2$) of an array of Pt chains. In the second chain on the left both the $2\times$ periodicity and the $4\times$ periodicity are observable.

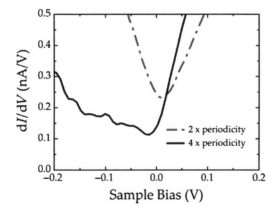

Fig. 9.18 dI/dV curves recorded on top of the Pt chains in Fig. 9.14 at 4.7 K ($2\times$ curve: averaged over wire d and $4\times$ curve: averaged over wire a).

where often either an occupied state lying close to the Fermi level disperses upward in energy from the Γ point or a low-lying unoccupied state disperses downward in energy from the Γ point (Stroscio *et al.* 1986; Feenstra *et al.* 1987; Bonnel *et al.* 1993). Conservation of energy requires that under these conditions tunnelling occurs via electrons with non-zero parallel momentum. For small sample biases, we can extract the inverse decay length, provided that we record I–V curves at two different heights, z_1 and z_2:

$$I_n(V) = e^{-2\kappa z_n} \int_0^V \rho(E)\,\mathrm{d}E. \qquad (9.17)$$

Hence:

$$\kappa = \frac{\ln\left(\frac{I_1}{I_2}\right)}{2\,(z_2 - z_1)}. \qquad (9.18)$$

Thus, the natural logarithm of the ratio of the current values measured on the same spot at two different tip–sample separations, is directly proportional to the inverse decay length of the electronic state under study (at that particular bias voltage V). The inverse decay length is a slowly varying function of the sample bias (Feenstra *et al.* 1987). Therefore, large variations in the measured inverse decay length should be attributed to electrons that tunnel from other electronic states (slightly further away from the Γ point) with non-zero parallel momentum vectors.

Figure 9.19 shows the inverse decay length, κ, versus sample bias as extracted from I–V curves recorded at different tip–sample separations on top of Pt chains that exhibit a $4\times$ periodicity (upper trace) and a $2\times$ periodicity (lower trace). The inverse decay length of the Pt chains with a $2\times$ periodicity exhibits a rather broad minimum near the Fermi level (zero bias), which is indicative of a metallic character of the Pt chains. For the chains with a $4\times$ periodicity the inverse decay length, κ, is significantly larger. In addition, the

Fig. 9.19 Inverse decay length κ versus sample bias recorded on top of Pt chains with a $4\times$ periodicity (upper trace, $\triangle$) and a $2\times$ periodicity (lower trace, $\bullet$). The distance between tip and sample was set by the tunnelling conditions ($V = -1.0$ V, $I = 0.5$ nA). The upper cartoon is an artistic impression of the tunnelling process on an insulating chain and the lower cartoon of the tunnelling process on a metallic chain.

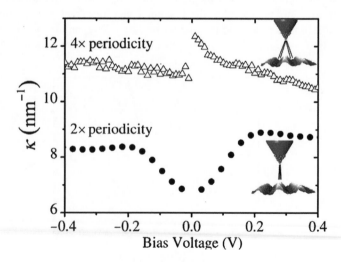

4× curve does not display a minimum near the Fermi level. We interpret the absence of this minimum near the Fermi level as a strong indication that the metallicity of the 4× chains has been reduced significantly as compared to the metallicity of the 2× chains. The sudden jump in the inverse decay length just above the Fermi level of the 4× Pt chains might be related to the development of novel electronic states between the Pt chains [4]. The cartoons in Fig. 9.4 give an artist's impression of the tunnelling on an insulating chain (upper cartoon) and metallic chain (lower cartoon).

We conclude that the 4× reconstructed chains are less metallic than the 2× reconstructed ones. In a one-band model obviously a Peierls transition is leading to an insulator. However, in a multiband case metallic behavior can still survive if more than one band is crossing the Fermi level. In this case the Peierls transition will show up only as reduced metallicity. Due to the fact that the Peierls transition depends critically on the topology of the Fermi surface *ab-initio* calculations require a detailed structural model for the underlying β-terrace. Unfortunately, this is not known. Different approximate models that have been explored only confirmed the expected high sensitivity of the Fermi topology on the structural model used and are not conclusive.

In Section 9.3.2 the observation of novel electronic states is reported. These electronic states, with a distinct one-dimensional character, are confined between the Pt chains. It is emphasized here that confinement of the electrons only occurs at low temperature in regions where the Pt dimers exhibit a 4× periodicity. Therefore, it is tempting to believe that the Peierls instability with its corresponding change in electronic properties of the chains is also responsible for the observed electron confinement.

In summary, at low temperatures self-organizing Pt chains on Pt-modified Ge(001) undergo an unanticipated structural transition from a 2× to 4× periodicity. Isolated Pt chains and chains located at the edge of an array of Pt chains maintain the 2× periodicity down to temperatures as low as 4.7 K. Spectroscopic experiments reveal that the 2× to 4× phase transition is accompanied by a substantial reduction of the metallicity of the Pt chains. The 2× to 4× transition is driven by a Peierls instability.

9.5   Conclusions and outlook

In this chapter we have given a brief overview of the intriguing physical properties, such as quantization of conductance, Peierls instability and Luttinger liquid behaviour, that nanowires can exhibit. In the second part of this chapter we focused our attention on one particular example, namely self-organizing Pt chains on a germanium (001) surface. These chains can be considered at the ultimate nanowires since they have a cross-section of only one atom, are perfectly straight and virtually defect free. In the preceding sections we touch upon a few interesting properties, such as the development of novel one-dimensional electronic states and a Peierls instability, that these chains exhibit at low temperatures. Despite the fact that the above-mentioned properties are rather elementary we believe that they nicely illustrate that the experimental discovery of (quasi-) one-dimensional physical properties has just begun.

References

Ahn, J.R., Yeom, H.W., Yoon, H.S., Lyo, I.-W. *Phys. Rev. Lett.* **91**, 196403 (2003).

Auslaender, O.M, Yacoby, A., de Picciotto, R., Baldwin, K.W., Pfeiffer, L.N., West, K. W. *Science* **295**, 825 (2002).

Bloch, F.Z. *Physik* **52**, 555 (1928).

Blöchl, P.E. *Phys. Rev. B* **50**, 17953 (1994).

Bockrath, M., Cobden, D.H., Lu, J., Rinzler, A.G., Smalley, R.E., Balents, L., McEuen, P.L. *Nature* **397**, 598 (1999).

Bonnel, D.A. (ed.), *Scanning Tunneling Microscopy and Spectroscopy* (VCH Publishers, New York, 1993).

Bürgi, L., Jeandupeux, O., Hirstein, A., Brune, H., Kern, K. *Phys. Rev. Lett.* **81**, 5370 (1998).

Chan, L.H., Altman, E.I. *Phys. Rev. B* **66**, 155339 (2002).

Crain, J.N., Kirakosian, A., Altmann, K.N., Bromberger, C., Erwin, S.C., McChesney, J.L., Lin, J.-L., Himpsel, F. *J. Phys. Rev. Lett.* **90**, 176805 (2003).

Crain, J.N., Himpsel, F. *J. Appl. Phys. A* **66**, 431 (2006).

Crommie, M.F., Lutz, C.P., Eigler, D.M. *Nature* **363**, 524 (1993).

Crommie, M.F., Lutz, C.P., Eigler, D.M. *Science* **262**, 218 (1993).

Evans, M.M.R., Nogami, J. *Phys. Rev. B* **59**, 7644 (1999).

Feenstra, R.M., Stroscio, J.A., Fein, A.P. *Surf. Sci.* **181**, 295 (1987).

Fischer, M., van Houselt, A., Kockmann, D., Poelsema, B., Zandvliet, H.J.W. *Phys. Rev. B* **76**, 245429 (2007).

Gambardella, P., Dallmeyer, A., Maiti, K.M.C., Malagoli, W., Eberhardt, K., Kern, C. *Carbone, Nature* **416**, 301 (2002).

Gurlu, O., Adam, O.A.O., Zandvliet, H.J.W., Poelsema, B. *Appl. Phys. Lett.* **83**, 4610 (2003).

Gurlu, O., Zandvliet, H.J.W., Poelsema, B., Dag, S., Ciraci, S. *Phys. Rev. B* **70**, 085312 (2004).

Gurlu, O., van Houselt, A., Thijssen, W.H.A., van Ruitenbeek, J.M., Poelsema, B., Zandvliet, H.J.W. *Nanotechnology* **18**, 365305 (2007).

Haldane, F.D.M. *J. Phys. C: Solid State Phys.* **14**, 2585 (1981).

Hohenberg, P., Kohn, W. *Phys. Rev.* **136**, 864 (1964).

van Houselt, A., Oncel, N., Poelsema, B., Zandvliet, H.J.W. *Nano Lett.* **6**, 1439 (2006a).

van Houselt, A., van Gastel, R., Poelsema, B., Zandvliet, H.J.W. *Phys. Rev. Lett.* **97**, 266104 (2006b).

van Houselt, A., Gnielka, T., Aan de Brugh, J.M.J., Oncel, N., Kockmann, D., Heid, R., Bohnen, K.P., Poelsema, B., Zandvliet, H.J.W. *Sur. Sci.* **602**, 1731 (2008).

Ishii, H., Kataura, H., Shiozawa, H., Yoshioka, H., Otsubo, H., Takayama, Y., Miyahara, T., Suzuki, S., Achiba, Y., Shimada, K., Namatame, H., Taniguchi, M. *Nature* **426**, 540 (2003).

Kageshima, M., Torii, Y., Tano, Y., Takeuchi, O., Kawazu, A. *Surf. Sci.* **472**, 51 (2001).

Kittel, C, *Introduction to Solid State Physics* (John Wiley & Sons Inc, New York, Chichester, 2005).

Kresse, G., Hafner, J. *Phys. Rev. B* **47**, 558 (1993).

Kresse, G., Furthmüller, J. *Phys. Rev. B* **54**, 11169 (1996).

Kresse, G., Joubert, D. *Phys. Rev. B* **59**, 1758 (1999).

Landauer, R. *IBM J. Res. Dev.* **1**, 223 (1957).

Lang, N.D. *Phys. Rev. B* **34**, 5947 (1986).

Lee, J., Eggert, S., Kim, H., Kahng, S.-J., Shinohara, H., Kuk, Y. *Phys. Rev. Lett.* **93**, 166403 (2004).

Luttinger, J.M. *Phys. Rev.* **119**, 1153 (1960).

Luttinger, J.M. *J. Math. Phys.* **4**, 1154 (1963).

Meyer, Chr., Klijn, J., Morgenstern, M., Wiesendanger, R. *Phys. Rev. Lett.* **91**, 076803 (2003).

Nilius, N., Wallis, T.M., Ho, W. *Science* **297**, 1853 (2002).

Nilius, N., Wallis, T.M., Ho, W. *Phys. Rev. Lett.* **90**, 186102 (2003).

Oncel, N., van Houselt, A., Huijben, J., Hallbäck, A.-S, Gurlu, O., Zandvliet, H.J.W., Poelsema, B. *Phys. Rev. Lett.* **95**, 116801 (2005).

Peierls, R.E. *Quantum Theory of Solids* (Clarendon Press, Oxford, 1955).

Peierls, R.E. *More Surprises in Theoretical Physics* (Princeton University Press, Princeton, NJ, 1991).

Schäfer, J., Schrupp, D., Preisinger, M., Claessen, R. *Phys. Rev. B* **74**, *R*041404 (2006).

Scheer, E., Joyez, P., Esteve, D., Urbina, C., Devoret, M.H. *Phys. Rev. Lett.* **78**, 3535 (1997).

Scheer, E., Agraït, N., Cuevas, J.C., Levy Yeyati, A., Ludoph, B., Martin-Rodero, A., Rubio Bollinger, G., van Ruitenbeek, J.M., Urbina, C. *Nature* **394**, 154 (1998).

Schwingenschlögl, U., Schuster, C. *Europhys. Lett.* **81**, 26001 (2008a).

Schwingenschlögl, U., Schuster, C. *Eur. Phys. J. B* **60**, 409 (2008b).

Stroscio, J.A., Feenstra, R.M., Fein, A.P. *Phys. Rev. Lett.* **57**, 2579 (1986).

Tomogana, S. *Prog. Theor. Phys.* **5**, 544 (1950).

Voit, J. *Prog. Phys.* **57**, 977 (1994).

Wallis, T.M., Nilius, N., Ho, W. *Phys. Rev. Lett.* **89**, 236802 (2002).

Wang, J., Li, M., Altman, E.I. *Phys. Rev. B* **70**, 233312 (2004).

Wang, J., Li, M., Altman, E.I. *Surf. Sci.* **596**, 126 (2005).

Wang, J., Li, M., Altman, E.I. *J. Appl. Phys.* **100**, 113501 (2006).

Zandvliet, H.J.W., Louwsma, H.K., Hegeman, P.E., Poelsema, B. *Phys. Rev. Lett.* **75**, 3890 (1995).

Zandvliet, H.J.W., Swartzentruber, B.S., Wulfhekel, W., Hattink, B.J., Poelsema, B. *Phys. Rev. B* **57**, *R*6803 (1998).

Zandvliet, H.J.W., Elswijk, H.B. *Phys. Rev. B* **48**, 14269 (1993).

Zandvliet, H.J.W. *Phys. Rep.* **388**, 1 (2003).

Zhang, Z., Wu. F., Zandvliet, H.J.W., Poelsema, B., Metiu, H., Lagally, M.G. *Phys. Rev. Lett.* **74**, 3644 (1995).

10 Designing low-dimensional nanostructures at surfaces by supramolecular chemistry

Nian Lin and Sebastian Stepanow

10.1 Introduction

Two strategies—top-down and bottom-up—are employed to fabricate nanoscale materials and devices. In top-down approaches nanoscale materials or devices are generated by means of microfabrication methods where externally controlled tools are used to nanostructure bulk materials. Top-down approaches, which play a major role in nanofabrication, are of great advantage to produce structures with long-range order and for making macroscopic connections. The drawbacks are (1) the fabricated structures are not perfect at molecular/atomic scales; (2) the minimum feature size that can be produced by these approaches is limited, for example, in photolithography, by the wavelength of the light used to define the patterns and (3) the fabrication processes are slow and expensive. In contrast, the bottom-up approaches start with smaller (1–100 nanometer-scale) components, which arrange into more complex nanostructures automatically at appropriate external conditions. In particular, molecular self-assembly, which uses the specific intermolecular binding to organize single-molecule components spontaneously into some forms of nanostructures, represents a powerful bottom-up approach. The bottom-up approaches are advantageous to assemble and establish short-range order at nanoscale dimensions, with molecular-level accuracy, able to produce devices in parallel processes and much cheaper and faster than top-down methods.

One widely used bottom-up approach is supramolecular self-assembly, a process based on supramolecular chemistry. While traditional chemistry examines the covalent bonding interactions that construct molecules (1–100 Å length scale) from atoms, supramolecular chemistry focuses on the non-covalent interactions between molecules (Lehn 2002). Inter-molecular forces, including hydrogen bonding, metal–ligand co-ordination, hydrophobic forces, van der Waals forces, pi–pi interactions and electrostatic effects, are reversible–an essential feature for self-assembly processes. Among these forces, hydrogen bonds and metal–ligand co-ordination are superior because these two types of bonds possess specific bond direction and are highly selective. In the past three decades supramolecular self-assembly based on hydrogen bonds and metal–ligand co-ordination has been used to synthesize a wide range of novel organic materials, inorganic materials and biomaterials not only with complex structures, but new functions as well.

Recently, many groups have been working on employing supramolecular chemistry concepts at well-defined planar substrates with the aim of fabricating low-dimensional molecular architectures (De Feyter *et al.* 2003; Lin *et al.* 2005, 2006). As illustrated in Fig. 10.1, the assembly of supramolecular architectures can be directly conducted at surfaces following the deposition of the components, i.e. organic linkers or metal atoms. Due to 2D confinement provided by surfaces, the molecular/atomic components deposited on solid surfaces will organize through non-covalent intermolecular binding as zero-, one- or two-dimensional nanostructures of high structural ordering. The formation of a given supramolecular shape is driven by the inherent information stored in the molecular/atomic building blocks, including size, shape and functionality. Therefore, a synergic selection of the building components can create predetermined structures, i.e. synthesis by rational design.

In general, the fabrication of 2D supramolecular nanostructures on surfaces can be divided into two categories according to how molecules/atoms are brought onto surfaces: (1) bring molecular building blocks to liquid solid interfaces in solution phases and (2) bring molecular building blocks to solid surfaces in vacuum conditions by means of molecular beam deposition techniques. The first method is technically easy and cheap, and can be applied to a wide range of molecules. In contrast, the second method is more challenging technically and requires sophisticated vacuum equipment. In addition, this method is not applicable for molecules with high sublimation temperatures due to thermal decomposition problems. However, the advantage of the second method is that the vacuum condition provides an environment to employ advanced surface-science tools, e.g. high-resolution scanning tunnelling microscopy, photoemission spectroscopy, low-energy electron diffraction, etc., to characterize the supramoleuclar systems with atomic/molecular-level resolution, which offers unprecedented insight into these systems.

In contrast to the conventional solution-based supramolecular chemistry in 3D space, several issues have to be taken into account for the 2D supramolecular assembly. First, surfaces impose two-dimensional confinement towards

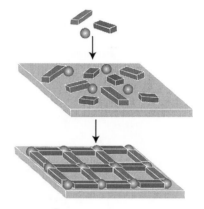

Fig. 10.1 Supramolecular assembly on a surface.

adsorbed molecular components as well as the formed supramolecular structures. As a result, the freedom of motion of molecular components is restricted to two-dimensional surface diffusion and rotation and the final supramolecular structures are two-dimensional in nature. Second, interaction between adsorbed molecular/atomic components and surfaces plays a role of comparable importance as that of non-covalent bonds that organize supramolecular structures. The balance between intermolecular interaction and surface interactions assumes therefore a critical role in determining the structures of supramolecular nanostructures. For instance, molecules or atoms usually prefer certain adsorption sites and orientations with respect to surface atomic lattices, hence in many cases, to minimize free energy, supramolecular structures are deviated from the "ideal" intrinsic structures regardless of a surface's influence. Third, in the case of metal–ligand co-ordination, the well-established pictures of conventional 3D co-ordination chemistry have to be adapted to 2D, for example, tetrahedral or octahedral co-ordination must be replaced by two-fold linear, three-fold triangular or four-fold square co-ordination. Fourth, the surface plays an important role regarding the functional properties of the low-dimensional supramolecular systems it supports. In particular, the properties of transition-metal ions embedded in the co-ordination nanostructures can be affected by the intricate interplay between the present interactions, e.g. the hybridization of the metal d-states with substrate electrons can effectively screen their magnetic moments. The interplay of metal–ligand interaction, intermolecular interaction and surface interaction assumes therefore a critical role in determining the chemical or electronic properties of self-assembled supramolecular nanostructures.

In this chapter we will discuss the design strategies of two types of low-dimensional supramolecular nanostructures: structures stabilized by hydrogen bonds and structures stabilized by metal–ligand co-ordination interactions. A hydrogen bond is the attractive force that exists between an electronegative atom of one molecule and a hydrogen atom bonded to another electronegative atom of a different molecule (intermolecular hydrogen bond) or the same molecule (intramolecular hydrogen bond). Usually the electronegative atom is oxygen, nitrogen, or fluorine, which has a partial negative charge. The hydrogen then has the partial positive charge. Hydrogen bonds, which play an important role in determining the three-dimensional structures of biological entities such as proteins and nucleic bases, are of fundamental importance in biological systems. For example, the double-helical structure of DNA is stabilized by hydrogen bonding between the base pairs. Hydrogen bonding has been widely used to design 3D supramolecular architectures, including 0D discrete aggregates or 1D chains, 2D sheets, or 3D infinite frameworks. On surfaces various nanostructures stabilized by hydrogen bonds have been reported in recent years. It has been realized that similar to what happens in 3D space, molecular building units containing end groups as complementary donors or acceptors may form hydrogen bonds on surfaces. The nanostructures formed so far include 0D, 1D and 2D structures. 0D structures are discrete clusters of dimers, trimers, tetramers, etc. 1D structures are linear chains. 2D structures consist of close-packed 2D crystals as well as porous open networks.

Metal–ligand co-ordination interactions allow for the self-assembly of supramolecular architectures as diverse as polygonal clusters, polyhedra, cages, and grid structures in 3D space. Co-ordination chemistry has gathered a vast database of metal–ligand pairs with a huge variety of specific binding modes (Bersuker 1996). Therefore, the use of transition-metal centers, or in general secondary building blocks, and co-ordination chemistry for directing the formation of complex structures has evolved into one of the most successful strategies for organizing molecular building blocks into 2D supramolecular nanostructures. Similar to hydrogen-bond systems, surface-supported co-ordination systems are 0D, 1D and 2D structures, but with much high thermal stability. The formation of a given supramolecular shape is driven by the inherent symmetry of the available metal orbitals and the spatial organization of the donor atoms in the organic ligand system. Therefore, careful consideration must be given to the preferred co-ordination environment of the metal to be used as well as the binding mode of organic linker molecules. Given a co-ordination environment around the metal centers, the symmetric and rigid extension of the ligand system from mono- to multitopicity will automatically lead to infinite 0D, 1D and 2D co-ordination nanostructures. The supramolecular structure is encoded in the electronic structure of both the metal ions and in the organic ligands and the interpretation of this information during the self-assembly process leads to a mutually acceptable structure.

10.2 Hydrogen-bond systems

At surfaces, hydrogen bonds can organize molecules as 0D discrete clusters, 1D chains, and 2D open networks and close-packed arrays through judicious choice of molecular building components with defined shape, size and functionality.

10.2.1 0D clusters

Discrete clusters consisting of a finite number of components can be obtained in two ways, as illustrated in Fig. 10.2: (1) using molecules that contain monotopic moieties that can join together through hydrogen binding and (2) using molecules with complementary multitopic moieties forming structures of closed-loop features.

10.2.1.1 *Macrocycle clusters*

A macrocycle molecule, **mt-33** (Fig. 10.3, left), has been used to form molecular clusters of dimer, trimer and tetramer (Ruben *et al.* 2006). Figure 10.3(a) is a STM topograph of macrocycle molecules **mt-33** deposited on the pristine Ag(111) surface at low molecule coverage. The isolated molecules appear as a doughnut-shape object, which reflects the ring structure of **mt-33**. The observation of the doughtnut-shape points to a flat absorption of the isolated molecules with their aromatic ring planes largely parallel to the surface.

The high-resolution data in Figs. 10.3(b) and (c) elucidate an asymmetrically ellipsoidal shape of isolated molecules of **mt-33** consisting of a broader 2, 2′, 6′, 2″-terpyridine (denoted as head) part and a sharper 1,10-phenanthroline part (denoted as tail). Within the aggregates, the molecules form head-to-head arrangements for both the dimer and the tetramer. The intermolecular interactions are attributed to hydrogen bonds between the nitrogen atoms and the protons of the aromatic rings of *all-transoid* conformers of neighboring 2, 2′, 6′, 2″-terpyridine units, as proposed by the dashed lines in Figs. 10.3(b) and (c). Although C–H $\cdots$ N hydrogen bonds are considered to be weaker than O–H $\cdots$ O or O–H $\cdots$ N hydrogen bonds, DFT calculations indicate that two pyridine nuclei can interact via C–H $\cdots$ N bonds with

Fig. 10.2 Dark tips: hydrogen-binding moieties; Gray blocks: molecular backbones.

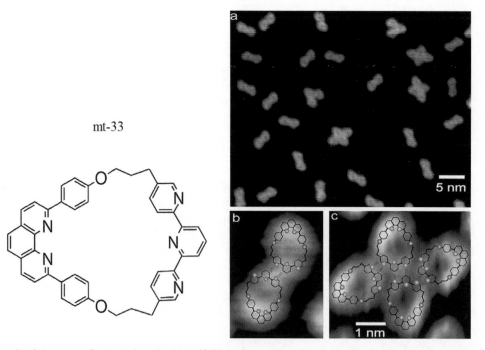

mt-33

Fig. 10.3 Left: molecular structure of macrocycle molecule **mt-33.** Right: Spontaneous clustering of macrocycle molecule **mt-33** at a clean Ag(111) surface. (a) STM image showing the aggregation of **mt-33** at a low coverage. (b) and (c) (in same scale) High-resolution STM image showing the dimeric and tetrameric aggregates with inscribed models and intermolecular hydrogen bonds. (Reprinted with permission from J. Am. Chem. Soc. Copyright (2006) American Chemical Society.)

a binding energy of 3.7 kcal/mol and the formation of pyridine dimers on surfaces was also confirmed experimentally. Such surface hydrogen-bonding behavior is in strong contrast to the observed situation in the crystal of **mt-33**: There, the $2, 2', 6', 2''$-terpyridine units (also in the all-*transoid* conformation) form exclusively intramolecular hydrogen bonds and no intermolecular contacts could be observed. The different behavior in bulk can rely on the predominance of π–π stacking interactions of the 1,10-phenanthroline moieties within the crystal lattice, which are hampered when the molecules are adsorbed flat on the metallic substrate. In addition, in the solid-state crystal, intercalated aromatic solvent molecules efficiently block the intermolecular terpyridine–terpyridine interactions by formation of C–H $\cdots$ N contacts.

10.2.1.2 *Carboxyphenyl porphyrin clusters*

In contrast to the random clustering of **mt-33**, a precise control of the cluster size can be realized by using specific synthesized molecules that carry defined number of binding moieties at defined sites (Yokoyama *et al.* 2004). An example is to use a family of carboxyphenyl-substituted porphyrins to design well-defined discrete structures. The carboxyl function is known to be able to form dimerized hydrogen bonds, as shown in Fig. 10.4, where double OH $\cdots$ O hydrogen bonds are formed between carboxyl groups, providing good directionality and high stability. In this optimized structure, the coplanar orientations are formed between two phenyl rings, resulting from further stabilization by the π conjugation between the phenyl rings.

It was demonstrated that through carboxylic substantiation of 5, 10, 15, 20-tetrakis-(3,5-di-t-butyl-phenyl porphyrin (H2-TBPP, shown in Fig. 10.5(a)), well-defined supramolecular nanostructures could be assembled. Three derivatives were used in this study: CaTBPP (Fig. 10.5(d)) has a carboxyphenyl group replacing a tBP substituent of H$_2$–TBPP. *trans*-BCaTBPP and *cis*-BCaTBPP (Figs. 10.5(g) and (j), respectively) have two carboxyphenyl groups that are substituted at the *trans* and *cis* positions, respectively.

CaTBPP molecules assemble as dimers at an Au(111) surface, as shown by the STM image of Fig. 10.6(a). The dimers nucleate at the elbows of the herringbone reconstruction patterns of Au surface. The distance between two central porphyrins of CaTBPP, close to 2.24 nm, suggests that the CaTBPP dimer is stabilized by the optimal hydrogen bonding between the carboxyl groups, as illustrated in Fig. 10.4. The authors observed an exclusive formation of dimer structures that obviously is determined by the saturation of carboxylic functions in dimers, as illustrated by the model in Fig. 10.6a. As a sharp contrast, *cis*-BCaTBPP molecules assemble as supra-molecular tetramers on the surface at low molecule coverage. As shown in Fig. 10.6(b), each *cis*-

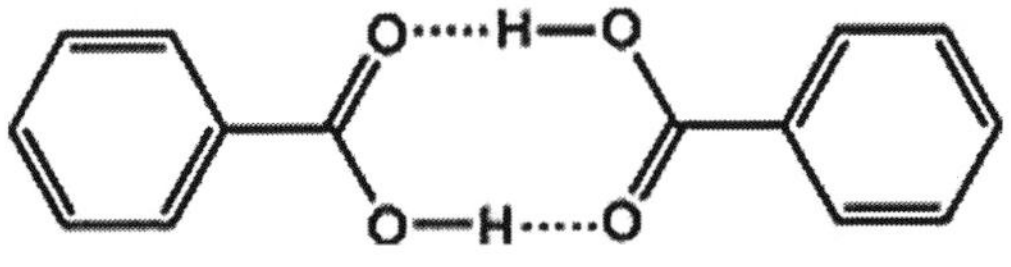

Fig. 10.4 Dimerized hydrogen bond formed by carboxyl groups.

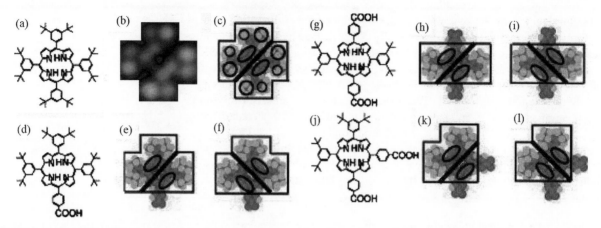

Fig. 10.5 (a) H$_2$–TBPP, which is composed of a free-base porphyrin and four tBP substituents. (b) and (c) STM image of an adsorbed H$_2$–TBPP molecule and its conformation on the Au (111) surface. The molecular orientation of the central non-planar porphyrin can be distinguished from the direction of the dark line in which the two bright oblong protrusions in the central porphyrin correspond to the tilt-up pyrrole rings. (d–f) CaTBPP. (g-i) *trans*-BCaTBPP and (j–l) *cis*-BCaTBPP. (Reused with permission from Takashi Yokoyama, Toshiya Kamikado, Shiyoshi Yokoyama, and Shinro Mashiko, *Journal of Chemical Physics*, **121**, 11993 (2008). Copyright 2008, American Institute of Physics.)

BCaTBPP molecule connect to two *cis*-BCaTBPP molecules through the two carboxylic functions at its *cis* positions. In total, eight hydrogen bonds between the carboxyphenyl groups are formed from the four involved molecules, forming the closed-loop arrangement of a tetramer, as illustrated by the model in Fig. 10.6(b).

10.2.2   1D structures

1D structures can be obtained by using molecules that contain complementary ditopic moieties, either by a homotopic molecular assembly or by heterotopic molecular assembly of binary molecular systems, as illustrated in Fig. 10.7.

10.2.2.1   *Single chains*

A diimide derivative (NTCDI) of naphthalene tetracarboxylic dianhydride (NTCDA) on the Ag/Si(111)-$\sqrt{3} \times \sqrt{3}$R30° surface form 1D rows stabilized by homotopic hydrogen bonds involving two imide groups (Keeling *et al.* 2003). As shown in Fig. 10.8 rows of molecules up to ~20 nm in length with an intermolecular separation of 1.5 lattice constants were formed. The incorporation of the double hydrogen bonds of NH $\cdots$ O between two neighboring imide groups in NTCDI controls the supra-molecular ordering. It is interesting to note that a closely related molecule, perylene-3,4,9,10-tetra-carboxylic-di-imide (PTCDI), in which similar H-bonding moieties are present, grow into 2-dimensional islands instead of 1D rows. The 1D growth of NTCDI was suggested due to an interplay of inter-molecular and molecular–substrate interactions, whereas in the case of PTCDI this interplay leads to the 2D organization.

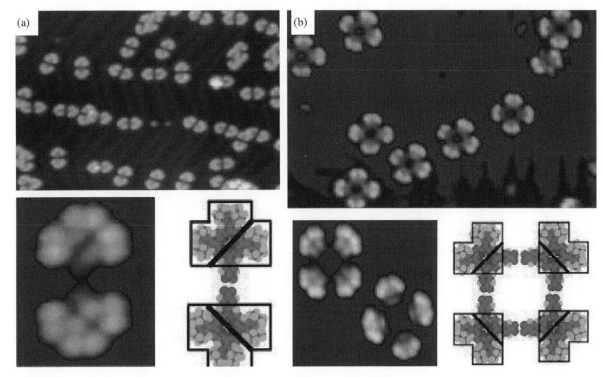

Fig. 10.6 (a) STM image (40 nm × 60 nm) of CaTBPP on the Au (111) surface and an enlarged STM image of the single CaTBPP dimer with its model. (b) STM image of *cis*-BCaTBPP on the Au (111) surface with an enlarged STM image of two *cis*-BCaTBPP tetramers and its model. (Reused with permission from Takashi Yokoyama, Toshiya Kamikado, Shiyoshi Yokoyama, and Shinro Mashiko, *Journal of Chemical Physics*, **121**, 11993 (2008). Copyright 2008, American Institute of Physics.)

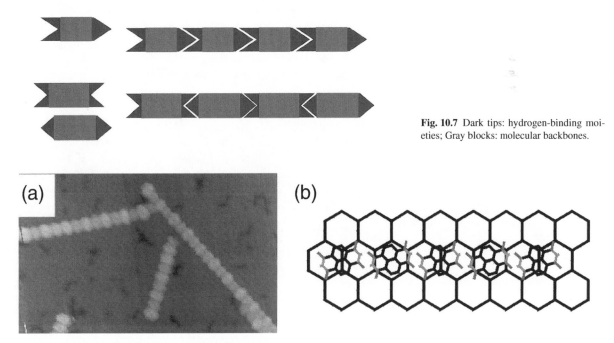

Fig. 10.7 Dark tips: hydrogen-binding moieties; Gray blocks: molecular backbones.

Fig. 10.8 (a) STM image showing three molecular chains of NTCDI. (b) Schematic diagram showing adsorption sites of molecules within a chain on the Ag/Si(111)-√3 × √3R30° surface. (Reprinted with permission from NanoLett. Copyright (2003) American Chemical Society.)

10.2.2.2   *Double chains*

A spectacular example of one-dimensional supra-molecular nanostructure at surfaces is the double-chain structures formed by the co-operative self-assembly of 4-[trans-2-(pyrid-4-yl-vinyl)]benzoic acid (PVBA) (Barth *et al.* 2000). The overview picture (Fig. 10.9(a)) reveals the formation of highly regular, one-dimensional supramolecular arrangements in a domain that extends over two neighboring terraces that are separated by an atomic step. A close-up view of some molecular stripes (Fig. 10.9(b)) reveals that the one-dimensional superstructure actually consists of two chains of PVBA. The molecular axis is oriented along the chain direction, in agreement with the expected formation of hydrogen bonds between the PVBA end groups. The observed features of the supramolecular structure are rationalized by the model proposed in Fig. 10.9(c), in which the OH group of the benzoic acid function forms a hydrogen bond with the nitrogen of the pyridine. When an unrelaxed molecular configuration is assumed, the length of the OH $\cdots$ N hydrogen bond is 2.5 Å. Within the twin chains, weak OH $\cdots$ N and possible weak CH $\cdots$ OC hydrogen bonds are formed. The orientation of the molecular chains reflects a good match between the PVBA subunits and the high-symmetry lattice positions. This result also accounts for the chain periodicity being a multiple of Ag lattice units. In agreement with the threefold symmetry of Ag(111) three rotational domains of this structure exist, which usually extend in the mm range on the surface.

10.2.2.3   *Zigzag chains*

The *cis*-BCaTBPP molecules used previously to form closed-loop tetramers also form zigzag-shaped chains if the molecule coverage on the surface is higher (Yokoyama *et al.* 2004). As shown in the STM images of Fig. 10.10(a), in contrast to the loop-like assembly of the *cis*-BCaTBPP tetramer, at increasing coverage the carboxylic functions at the *cis* positions form sequential hydrogen bonding, resulting in the formation of the zigzag arrangement, as

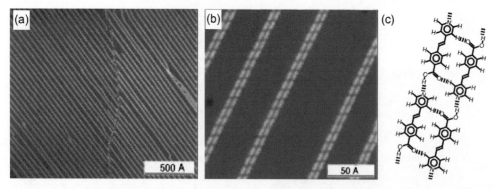

Fig. 10.9 Formation of a one-dimensional supramolecular PVBA twin-chain structure on an Ag(111) surface at 300 K (measured at 77 K). (a) An STM topograph of a single domain extending over two terraces demonstrates ordering at the mm scale. (b) A close-up image of the self-assembled twin chains reveals that they consist of coupled rows of PVBA molecules. (Copyright Wiley-VCH Verlag GmbH&Co. KGaA.) (c) Repeat motif of the PVBA supramolecular twin chains with weak OH $\cdots$ N and possible weak CH $\cdots$ OC hydrogen bonds indicated.

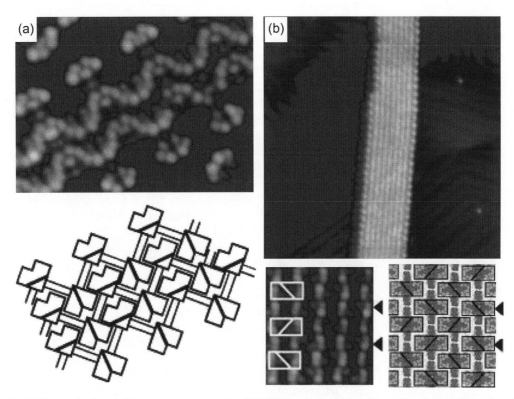

Fig. 10.10 (a) STM images for about half-monolayer coverage of *cis*-BCaTBPP on the Au(111) surface and its schematic illustration of aggregated zigzag *cis*-BCaTBPP. (b) STM image of the bundled *trans*-BCaTBPP wires and an enlarged view and its model. (Reused with permission from Takashi Yokoyama, Toshiya Kamikado, Shiyoshi Yokoyama, and Shinro Mashiko, *Journal of Chemical Physics*, **121**, 11993 (2008). Copyright 2008, American Institute of Physics.)

illustrated by the schematic model in Fig. 10.10(a). Energetically, the zigzag structure should be equivalent to the tetrameric structure since both have the same number of H bonds. But due to the efficient packing on the surface, the zigzag structure is preferred at high molecule coverages.

10.2.2.4   *Chain bundles*

The *trans*-BCaTBPP molecules are expected to form straight linear supramolecular chains through the two opposite carboxylic functions (Yokoyama *et al.* 2004). As shown in Fig. 10.10(b), *trans*-BCaTBPP does form very long and straight supra-molecular wires by sequential hydrogen bonding between carboxyphenyl groups. In addition, bundled supra-molecular wires are observed for *trans*-BCaTBPP on the surface. The high-resolution STM image and the model in Fig. 10.10(b) shows the straight supramolecular wires are densely aggregated into the bundled structure. The formation of bundles instead of single wires indicates there exist attractive interactions between neighboring wires.

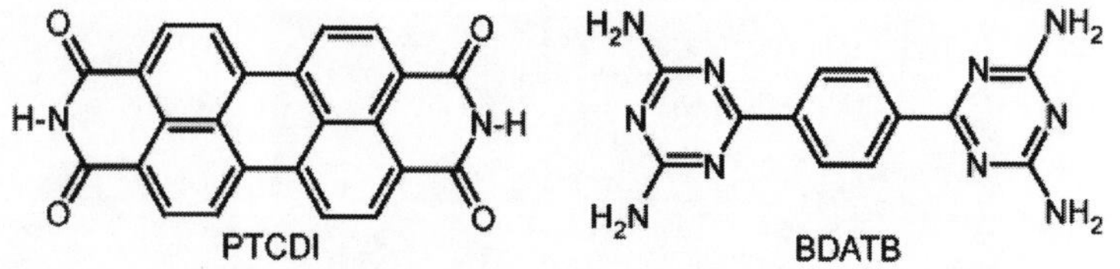

Fig. 10.11 Molecular structures of 3,4,9,10-perylenetetracarboxylic diimide (PTCDI) and 1,4-bis-(2,4-diamino-1,3,5,-triazine)-benzene (BDATB).

10.2.2.5   *Bimolecular chains*

Binary molecular chains may possess enhanced functions compared with the single-component chains. Cañas-Ventura *et al.* (2007) chose molecular species of 1,4-bis-(2,4-diamino-1,3,5,-triazine)-benzene (BDATB) and 3,4,9,10-perylenetetracarboxylic diimide (PTCDI), as shown in Fig. 10.11, to assemble a regular superlattice of 1D heteromolecular wires consisting of one or two molecular rows, as well as 2D supramolecular ribbons depending on coverage. PTCDI exhibits a -CO-NH-CO- (imide) sequence on the two opposite sides with a NH hydrogen-bond donor (D) and two CO hydrogen-bond acceptors (A), thus giving rise to an A-D-A sequence.

Figure 10.12 shows a STM image of a highly ordered superlattice consisting of heteromolecular chains. The brighter spots are the fingerprints of isolated PTCDI molecules, which alternate with dimmer spots of the BDATB molecules. The hydrogen-bonding distances are $(2.8 \pm 0.2)\text{Å}$ for the N–H $\cdots$ O and N $\cdots$ H–N bonds, which are similar to hydrogen-bond lengths reported in other studies. The structure was prepared by the codeposition of PTCDI and BDATB in a 1 to 1 ratio. The deposition sequence of the two components is unimportant for the formation of the mixed network—both molecules are mobile enough to intermix homogeneously whenever the stoichiometry is close enough to 1:1. It was found that annealing further enhances the molecular mobility and increases the length scale on which the order of the mixed layer prevails; the optimum annealing temperature is 390 K.

10.2.3   2D structures

10.2.3.1   *Open square networks*

The first strategy has been demonstrated by supramolecular networks formed from tetrakis(4-carboxy-phenyl)-porphyrin cobalt(II) (CoTCPP) (Yoshimoto *et al.* 2006). A highly ordered square packing arrangement of the two-dimensional network array was formed by CoTCPP molecule on Au(111) (Fig. 10.14(a)). Each isolated CoTCPP molecule can be identified based on the bright spot for the Co(II) ion surrounded by less intense spots at the corners associated with the tetrakis(4-carboxy-phenyl) porphyrin ligand (Fig. 10.14(b)). The intermolecular distance between the isolated CoTCPP complexes is 2.45 ± 0.08 nm. A model for the hydrogen bonding of this twodimensional network structure is shown in Fig. 10.9(c). The orientation of each CoTCPP molecule appears to be determined by the formation of hydrogen bonds between four adjacent carboxyl groups.

Figure 10.15 shows an example of the second approach (Weigelt *et al.* 2006)—square networks formed by linear molecules. The molecule consists of

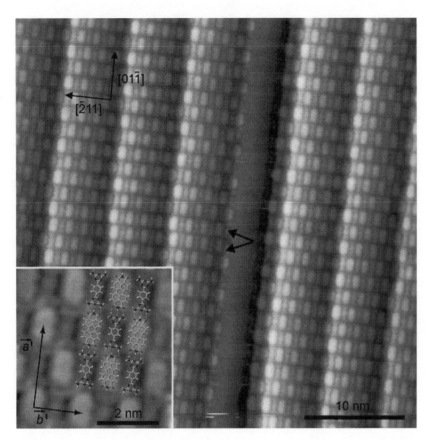

Fig. 10.12 BDATB and PTCDI bicomponent supramolecular organization on Au (11,12,12) at 390 K. STM image of seven terraces fully covered with the heteromolecular superlattice. Inset: High-resolution STM image and corresponding structural model of the binary supramolecular lattice with three hydrogen bonds between the heteromolecular pairs. (Copyright Wiley-VCH Verlag GmbH&Co. KGaA.)

a linear backbone formed from three benzene rings connected by ethynylene spokes, and is functionalized at both ends with an aldehyde, a hydroxyl and a bulky *t*-butyl group. Four molecules are linked by totally four hydrogen bonds between the aldehyde and the hydroxyl groups in a windmill-like node. Each molecule connects two nodes at its two terminals, producing an extended, ordered square-shape network with openings boarded by four molecules.

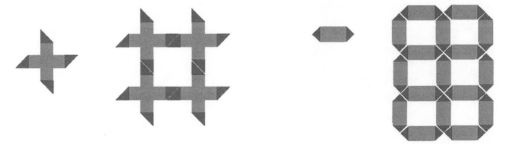

Fig. 10.13 Dark tips: hydrogen-binding moieties; Gray blocks: molecular backbones. Two strategies are used to generate square networks : (1) bridging the molecules with orthogonal branched arms by linear hydrogen bonds or (2) joining linear-shaped molecules through quadruple hydrogen bonds involving four moieties.

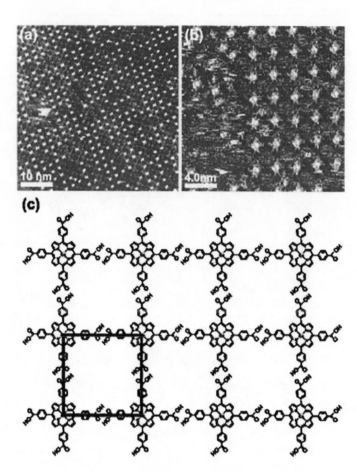

Fig. 10.14 (a) Large-scale $(50 \times 50\,nm^2)$ and (b) high-resolution $(20 \times 20\,nm^2)$ STM images of CoTCPP array formed on Au(111) acquired at (a) 0.1 V and (b) 0.85 V (stepped from 0.2 V) vs. RHE in 0.1 M $HClO_4$. Tip potentials and tunnelling currents were 0.45 V and 0.8 nA for (a) and 0.47 V and 0.83 nA for (b), respectively. (c) Structural model of highly ordered square packing array formed by potential modulation of the CoTCPP adlayer on Au(111). (Copyright Royal Society of Chemistry.)

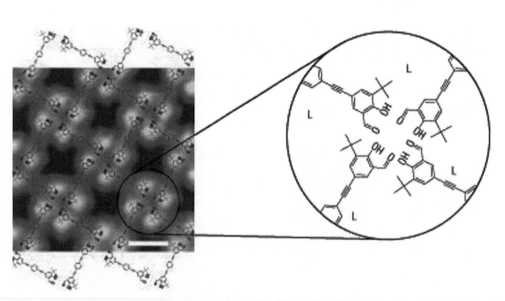

Fig. 10.15 STM images and schematic models of the square network structure produced by the windmill-like hydrogen bonds (scale bar 2 nm). (Copyright Nature Publishing Group.)

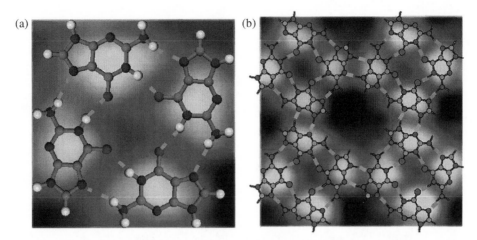

Fig. 10.16 (a) Comparison of a high-resolution STM image of the G-quartet unit cell with the Hoogsteen-bonded G-quartet structure determined by X-ray crystallography. (b) Comparison of an STM image of several G-quartet unit cells with the relaxed structure obtained by DFT calculations. The lateral interaction between G quartets occurs by eight new hydrogen bonds between the peripheral N3 and N9 atoms of neighboring guanine molecules. (Copyright Wiley-VCH Verlag GmbH&Co. KGaA.)

In another example, a square network system may contain a superstructure of large squares and small squares where guanine is used as building components (Otero *et al.* 2005). Guanine can assemble as G quartets mediated by H-bonds in quadruplex DNA. The authors found that on a Au(111) substrate guanine molecules self-assembles into a square network structure. High-resolution STM data reveal that the networks are made up by a similar structure as the G quartets. As shown in Fig. 10.16, square networks are formed in which each unit cell is composed of four molecules. They found that by superimposing the G-quartet structure determined by X-ray crystallography on G-quadruplex DNA crystals and the STM images, a good correspondence is observed between the former and the unit cell of the G network reported here. Neighboring unit cells consisting of G quartets are connected by so-called interquartet hydrogen bonds, resulting in a superstructure consisting of the G quartets that contains large squares.

10.2.3.2 *Hexagonal networks*

Similar to the square networks, the formation of hexagonal networks can be categorized into two approaches: (1) molecules of threefold arms are linked by linear hydrogen bonds or (2) linear molecules are connected by a threefold hydrogen-bond junction, as shown in Fig. 10.17.

A prototype molecular components of hexagonal networks is trimesic acid (TMA). As shown in Fig. 10.18(a) TMA has three carboxyl groups (forming an angle of 120° with respect to each other) in the molecular plane of the benzyl core. The carboxyl groups can form intermolecular hydrogen bonds in either dimeric (Fig. 10.18(b)) or trimeric (Fig. 10.18(c)) form. TMA can assemble at various substrates such as graphite, Au, Cu, and Au/Si, in liquid or ultrahigh-vacuum (UHV) environments (Dmitriev *et al.* 2002; Griessl *et al.* 2002; Su *et al.* 2004; Lin *et al.* 2005).

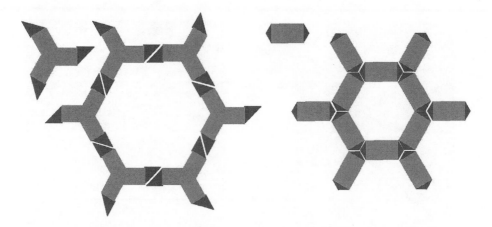

Fig. 10.17 Dark tips: hydrogen-binding moieties; Gray blocks: molecular backbones.

Fig. 10.18 (a) TMA molecular structure, (b) hydrogen bond in dimeric form, and (c) hydrogen bond in trimeric form. The dashed circles highlight the zones of intermolecular hydrogen bonds in dimeric and trimeric forms. (Reprinted with permission from J. Phys. Chem. C. Copyright (2007) American Chemical Society.)

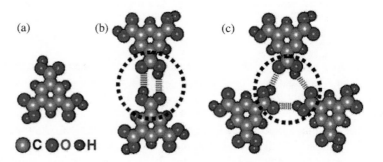

(a)   (b)   (c)

●C ◐O ◑H

Ye *et al.* have conducted a systematic study (Ye *et al.* 2007) of TMA self-assembly as a function of molecular coverage on a gold surface. At low coverage TMA forms an hexagonal open honeycomb network, as shown in Fig. 10.19(A). With increasing TMA coverage, various phases appear, including a flower phase (Fig. 10.19(B)), hexagonal phases consisting of pores with gradually increased pore-to-pore spacing (Figs. 10.19C–H) and finally a close-packed phase (Fig. 10.19(I)). They suggested that the various phases could be well described by a unified model in which the TMA molecules inside the half-unit cells (equilateral triangles) were bound via trimeric hydrogen bonds and all half-unit cells were connected to each other via dimeric hydrogen bonds, as illustrated by the models in Figs. 10.19(a)–(i). In all assemblies, there are only two types of hydrogen bonds, dimeric and/or trimeric ones. The bond length in a dimer is slightly shorter than that in a trimer. Energetic analysis unveils that the assembling structures less than one molecular layer are optimally driven by maximization of the dimeric hydrogen bonds.

Various derivatives of TMA have been used as building components to assemble open hexagonal networks. We discuss two examples—1,3,5-Tris(carboxymethoxy)benzene (**TCMB**) and 4, 4′, 4″-benzene-1,3,5-triyl-tri-benzoic acid (**BTA**), as shown in Fig. 10.20.

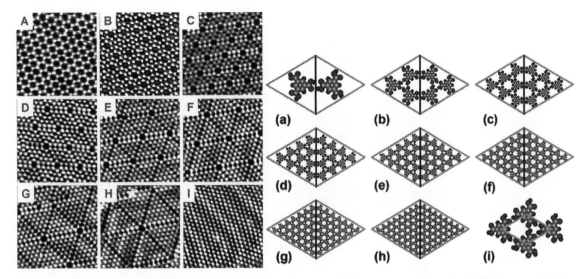

Fig. 10.19 Coverage-induced evolution of TMA self-assembling structures on Au(111) at room temperature. (A–I) STM images of the H_{TMA-n} ($n = 1$–8 and DC) structures. Each STM image size: $16.5 \times 16.5\,\text{nm}^2$. (a–i) Corresponding unit-cell models for the experimentally observed structures in A-I. The shorter diagonal separates each unit cell into two halves. (Reprinted with permission from J. Phys. Chem. C. Copyright (2007) American Chemical Society.)

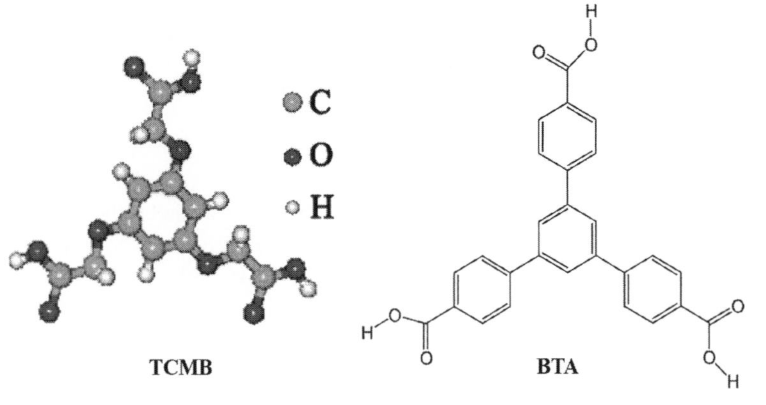

Fig. 10.20 Derivatives of TMA used as building components to assemble open hexagonal networks. (Reprinted with permission from J. Phys. Chem. B. Copyright (2004) and J. Am. Chem. Soc. Copyright (2006) American Chemical Society.)

●C
●O
○H

TCMB

BTA

Figure 10.21(a) is a large-scale STM image showing well-ordered 2D honeycomb networks ad-layer of **TCMB** molecules formed on Au(111) surface (Yan *et al.* 2004), in which each bright spot corresponds to one **TCMB** molecule. More details of isolated **TCMB** molecules in the honeycomb networks are revealed in the higher-resolution image of Fig. 10.21(b). The feature of a propeller with three blades is consistent with the chemical structure of isolated **TCMB** molecules. The three blades of each propeller should be attributed to the three carboxymethoxyl groups of **TCMB** molecules. The neighboring molecules with the same orientation are separated with a distance of $1.7 \pm 0.1\,\text{nm}$, about 6 times the lattice parameter of Au(111). A 6×6 structure for the molecular adlayer can be concluded on the basis of intermolecular distances and orientation of molecular rows. A unit cell is superimposed in Fig. 10.21(b). A tentative model is proposed in Fig. 10.21(c),

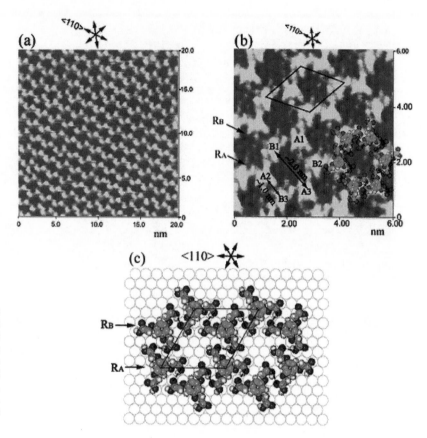

Fig. 10.21 (a) Large-scale STM image of TCMB molecules on an Au(111) surface. E_{Au} (Au(111) substratepotential) = 0.63 V, I = 1 nA. (b) Higher-resolution STM image of TCMB molecules. E_{Au} = 0.52 V, I = 900 pA. (c) Tentative model for the monolayer of TCMB molecules. (Reprinted with permission from J. Phys. Chem. B. Copyright (2004) American Chemical Society.)

in which the phenyl rings of **TCMB** molecules are proposed to be located on the threefold sites of the Au(111) substrate. The hydrogen atom in carboxyl groups can revolve toward the oxygen atom of another carboxyl group to form H-bonding, as illustrated by the dashed lines in Fig. 10.21(c). This model is in good agreement with the results of STM images.

The **BTA** molecules form a regular 2D honeycomb network on a Ag(111) surface at room temperature (Fig. 10.22(a)), which incorporates a regular array of cavities of 2.95 nm inner diameter and 2.84 nm periodicity (Fig. 10.22(b)) (Ruben *et al.*, 2006). The supramolecular interconnection of the **BTA** molecules is achieved by symmetric hydrogen bonds of carboxylic acid dimers, as shown in Fig. 10.22(c), whereby the oxygen–oxygen distance is estimated at a value of 3.1 Å, as expected for this type of hydrogen bond.

As an example of networks formed through threefold hydrogen bonds, Stöhr *et al.* reported on the formation of hexagonal networks from a perylene derivative, 4,9-diaminoperylene-quinone-3,10-diimine (DPDI, see Fig. 10.23(b)) (Stöhr *et al.* 2005). The authors reported that at room temperature, only a mobile phase and no ordered arrangement is found, as the nitrogen functions are exclusively hydrogen-bond donors and there are no appropriate acceptor functionalities. After an annealing process at 300 °C, the deposited molecules are chemically modified by dehydrogenation of the DPDI, as shown

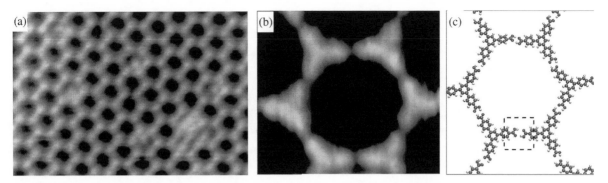

Fig. 10.22 Open 2D honeycomb network formed by **BTA** molecules on a Ag(111) surface. (a) An overview STM image, (b) High-resolution STM image, (c) Proposed atomistic model. (Reprinted with permission from J. Am. Chem. Soc. Copyright (2006) American Chemical Society.)

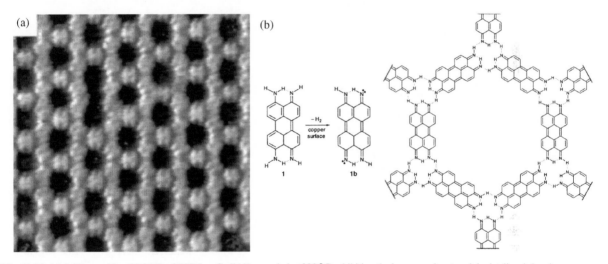

Fig. 10.23 (a) Self-assembly of 0.3 ML of DPDI on Cu(111) annealed at 300 °C exhibiting the honeycomb network in detail and showing a vacancy defect. (b) Schematic model of the network showing the proposed threefold hydrogen-binding motif. (Copyright Wiley-VCH Verlag GmbH&Co. KGaA.)

in Fig. 10.23(b). As a result, DPDI molecules form a well-ordered 2D honeycomb network (Fig. 10.23(a)), in which the now-modified nitrogen functions may act as both hydrogen-bond acceptors and donors. The molecules are connected through hydrogen-bonding interactions with each DPDI monomer binding to a total of four neighboring monomers (Fig. 10.23(b)). The author found that hydrogen-bonding interactions between the nitrogen functions is consistent with the structural data from the STM and LEED experiments, from which a N · · · N distance between adjacent molecules of about 3.1 Å is derived. Based on the STM data and the information obtained from the LEED pattern, a commensurate arrangement with regard to the Cu substrate of the DPDI monomers in a p(10 × 10) superlattice with a lattice constant of 2.55 nm is determined.

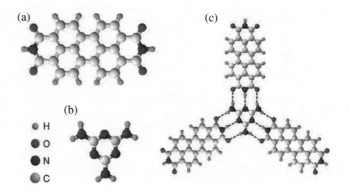

Fig. 10.24 Chemical structure of PTCDI (a) and melamine (b). (c) Schematic diagram of a PTCDI–melamine junction. Dotted lines represent the stabilizing hydrogen bonds between the molecules. (Copyright Nature Publishing Group.)

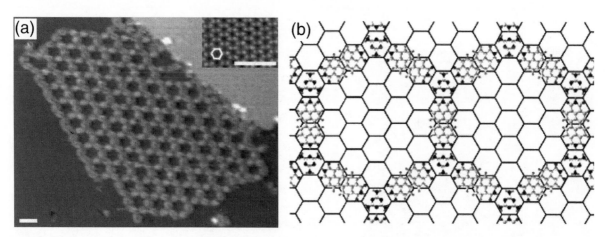

Fig. 10.25 (a) STM image showing the hexagonal PTCDI-melamine network in which PTCDI resolved as brighter rectangular features forming network edges and melamine is the triangular lower-contrast network vertex. Inset, high-resolution view of the Ag/Si(111)-$\sqrt{3} \times \sqrt{3}$R30° substrate surface; the vertices and centers of hexagons correspond, respectively, to the bright (Ag trimers) and dark (Si trimers) topographic features in the STM image. Scale bar, 20 nm. (b) Schematic diagram showing the registry of the network with the surface. (Copyright Nature Publishing Group.)

10.2.3.3   Bimolecular hexagonal networks

Beton and his collaborators have studied bimolecular self-assembly extensively. A particular interesting system is erylene tetra-carboxylic di-imide (PTCDI; Fig. 10.17(a)) with melamine (1,3,5-triazine-2,4,6-triamine; Fig. 10.24(b)) (Theobald *et al.* 2003). This bimolecular system reveals very rich self-assembled structures through the triple hydrogen bonds.

Sequential deposition of PTCDI and melamine onto a Ag/Si(111)-$\sqrt{3} \times \sqrt{3}$R30° surface leads to the formation of an intermixed hexagonal networks. (see Fig. 10.25(a)). The network has principal axes at 30° to those of the Ag/Si(111)-$\sqrt{3} \times \sqrt{3}$R30° surface, and a lattice constant $3\sqrt{3}a_0 = 34.6$ Å. with melamine and PTCDI forming, respectively, the vertices and edges of the network, which is stabilized by the hydrogen bonding illustrated in Fig. 10.24(c). Figure 10.25(b) shows the PTCDI/melamine molecular structure overlaid on the surface. Numerical calculations give the center-to-center spacing of a single PTCDI–melamine pair as 10.2 Å, close to the observed

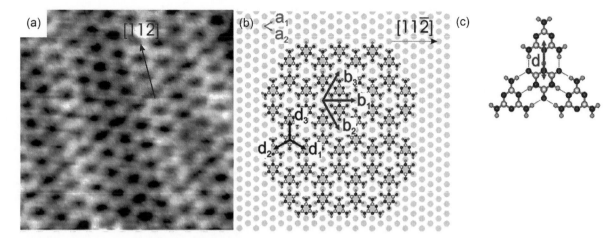

Fig. 10.26 (a) STM image of the CA · M honeycomb network domain on Au(111), 80 Å × 80 Å, the [11$\bar{2}$] direction of the Au(111) is labelled. (b) Schematic of the CA · M honeycomb network on the reconstructed Au(111) surface; the unit-cell vectors along with their linear combination are shown $\mathbf{b}_1$ and $\mathbf{b}_2$, $\mathbf{b}_3$; the cyanuric acid–melamine center–center molecular separation in the three directions are shown by d_1, d_2 and d_3; the reconstructed Au(111) lattice vectors are shown by $\mathbf{a}_1$ and $\mathbf{a}_2$. (c) The arrangement of three melamine molecules bonded to a single central cyanuric acid molecule. (Reprinted with permission from J. Phys. Chem. C. Copyright (2007) American Chemical Society.)

separation of $1.5a_0 = 9.98$ Å. Thus, the calculated melamine–melamine separation has a near-commensurability with the surface lattice.

Another bimolecular system is cyanuric acid-melamine mixture (CA · M) (Staniec *et al.* 2007). When cyanuric acid and melamine are combined on the surface, the pure cyanuric or melamine phases do not appear, but distinct intermixed phases of cyanuric acid and melamine (CA · M) evolve. Figure 10.26(a) shows an STM image of the first structure, a honeycomb structure similar to the rosette structures reported in earlier studies of bulk CA · M. In these structures, cyanuric acid–melamine pairs are stabilized by triple hydrogen bonds (Figs. 10.26(b) and (c)), forming hexagons containing three cyanuric acid and three melamine molecules.

10.3   Metal-co-ordination systems

Co-ordination metal–ligand bonding constitutes an appealing class of interactions for the formation of highly robust and functional architectures. The directional and selective character of this bonding mechanism enables the construction of complex architectures and together with other interaction, e.g. hydrogen bonding and electrostatic interaction, even more intricate structures can be obtained through hierarchical assembly protocols. In general, a co-ordination system consists of a co-ordinating atom (co-ordination center) ligated to other atoms or groups of atoms (ligands) by co-ordination bonds that are delocalized over all or several ligands. That delocalization of the one-electron bonding orbitals is realized for the transition-metal compounds, by the spatial distribution of the d orbitals of the central atom. Thus, the formation of a given supramolecular shape is driven by the inherent symmetry of the co-ordination sites available on the metal center depending on the

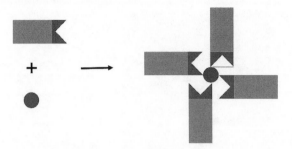

Fig. 10.27 Dark tips: co-ordination ligands; Gray blocks: molecular backbones; Spheres: metal centers.

element and its oxidation state. Therefore, careful consideration must be given to the preferred co-ordination environment of the metal to be used and the binding mode of the linkers. Starting from such a co-ordination environment at the metal centers, the symmetric and rigid extension of the ligand system from mono- to multitopicity will automatically lead to a grid-like one-, two-, or three-dimensional co-ordination network with regularly arrayed metal ions. The supramolecular shape is programmed in both the metal ions and in the organic ligands and the interpretation of this information during the self-assembly process leads to a mutually acceptable structure. The point of interest for their potential functional properties in the two-dimensional metal-organic architectures is the common feature of (co-ordinatively unsaturated) metal centers. As a consequence of the components' surface confinement, the reduction to two dimensions is frequently accompanied by unsaturated co-ordination sites, which opens the way to realize novel compounds exhibiting unusual co-ordination sites. Thus, the choice of donor atoms, bridging groups, paramagnetic metal ions, and systematic synthetic design strategies might render these systems ideal for designing receptor sites with tailorable molecular recognition properties and catalysts with tunable reactivities. In particular, a series of systematic investigations demonstrate the construction of mono- and dinuclear metal–carboxylate clusters, polymeric co-ordination chains, and fully reticulated networks based on polyfunctional exodentate aromatic species on surfaces under ultrahigh-vacuum conditions. The functional endgroups include carboxylate, pyridine, hydroxyl and carbonitrile. In particular, carboxylates represent a versatile class of building blocks to engineer robust 3D metal-organic frameworks or functional co-ordination polymers.

10.3.1 0D clusters

10.3.1.1 *Isolated clusters*

Figure 10.27 shows the concepts of forming isolated co-ordination compounds–molecules carrying a specific ligand are joined through a well-defined metal–ligand interaction.

Iron atoms react with TMA molecules at a Cu(100) surface represents an example of 0D co-ordination clusters (Messina *et al.* 2002). The iron was codeposited at low temperatures in order to inhibit intermixing reactions with the surface. The necessary deprotonation of the carboxylic acid groups of TMA molecules is thermally activated on the Cu surface (Lin

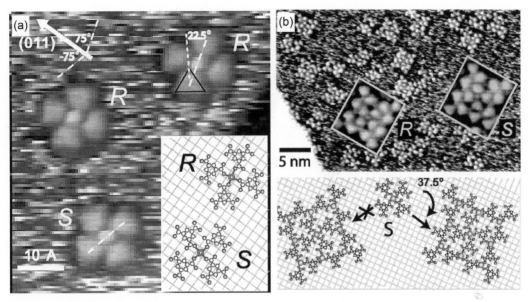

Fig. 10.28 (a) STM image showing the two Fe(TMA)$_4$ stereoisomers on the Cu(100) surface, labelled R and S, representing mirror-symmetric species with respect to the [011] substrate direction. The corresponding model depicts a unidentate co-ordination of the carboxylate ligands to the central Fe atom (placed on the hollow site) with a bond length of about 2 Å. The corresponding rotation of the carbon backbone is strictly correlated for all TMA molecules in a given complex. The resulting symmetry break accounts for the chirality of the complexes. (b) Assembly of square-shaped polynuclear nanogrids evolve upon annealing at 350 K. The insets and model below reveal that the respective core units of the dissymmetric metal-organic motifs are related to the chiral secondary Fe$_9$TMA$_{16}$ compounds. (Reprinted with permission from J. Am. Chem. Soc. Copyright (2002) American Chemical Society.)

et al. 2002). The resulting complexes appear exclusively in the presence of Fe on the surface. The STM observations at room temperatures reveal two mirror-symmetric Fe(TMA)$_4$ complexes consisting of fourfold square-planar co-ordination, where the correlated attachment of the ligands defines the handedness of the entity (Fig. 10.28(a)). Upon annealing at 350 K the Fe(TMA)$_4$ complexes form larger entities consisting of 16 TMA and 9 Fe, aggregating as a (4×4)-grid pattern (Spillmann *et al.* 2003). These grid-like structures inherit the chiral nature of the central Fe(TMA)$_4$ complexes and are randomly distributed at the surface (Fig. 10.28(b)).

10.3.1.2 *Arrays of clusters*

In subsequent systematic investigations it was shown that by employing the symmetric linker 1,4-benzoic dicarboxylic acid (terephthalic acid, TPA), the linear analog of TMA, one can achieve distinct regular two-dimensional arrays consisting of co-ordination complexes interconnected by hydrogen bonds on a Cu(100) surface in the low Fe concentration regime (Lingenfelder *et al.* 2004). The molecules form mononuclear iron complexes Fe(TPA)$_4$ where four molecules coordinate each with one carboxylate oxygen to the Fe center. The isolated Fe centers span a (6×6)-superstructure commensurate with the Cu(100) lattice (Fig. 10.29(a)), and this square array extends over entire substrate terraces. The high degree of long-range organization is presumably mediated by secondary intercomplex carboxylate-phenyl hydrogen bonds (see

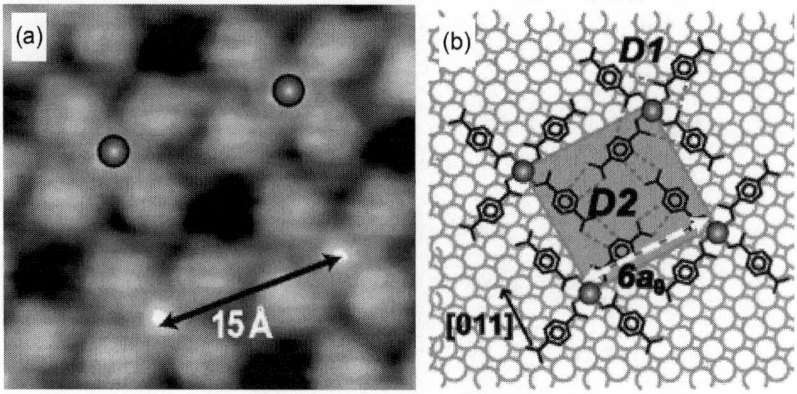

Fig. 10.29 (a) High-resolution STM image of the Fe(TPA)$_4$ cloverleaf phase on Cu(100). (b) Geometrical model of the co-ordination structure shown in (a). Each Fe atom (gray spheres) co-ordinates four carboxylate ligands unidentately in a square-planar configuration. (Copyright Wiley-VCH Verlag GmbH&Co. KGaA.)

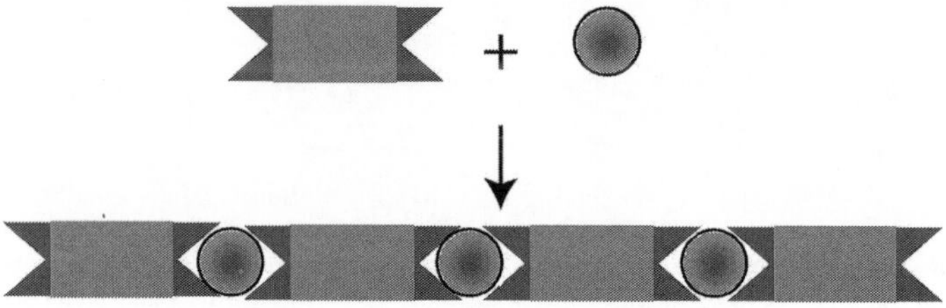

Fig. 10.30 Dark tips: co-ordination ligands; Gray blocks: molecular backbones; Spheres: metal centers.

model in Fig. 10.29(b)). This rather unusual hydrogen bond represents a particular member of the class of ionic hydrogen bonds (Payer *et al.* 2007).

10.3.2   1D chains

The formation of 1D structures can be deliberately steered through two strategies: (1) applying anisotropic surfaces and (2) utilizing linear molecules carrying dipotic co-ordination moieties that participate linear co-ordination modes. We focus only on the second approach here, which is illustrated in Fig. 10.30.

10.3.2.1   *Single chains*

A recent study in which linear aromatic dipyridyl linkers were investigated on the isotropic Cu(100) surface demonstrated 1D co-ordination chains (Tait *et al.* 2007). Upon deposition on the substrate held at room temperature linear chains evolve where the molecules are linked by a linear co-ordination motif of pyridine-Cu-pyridine (Fig. 10.31(a)). The Cu centers are not imaged, presumably due to an electronic effect. This unconventional twofold co-ordination of Cu centers has not been observed in bulk co-ordination compounds. However, it is suggested that the 2D environment enforces this rather special co-ordination mode.

10.3.2.2   *Chain bundles*

An example demonstrating bundles of 1D co-ordination chains is metal-organic co-ordination polymer monolayers consisting of copper ions and 2,5-dihydroxybenzoquinone (DHBQ) ligands that were directly synthesized on the

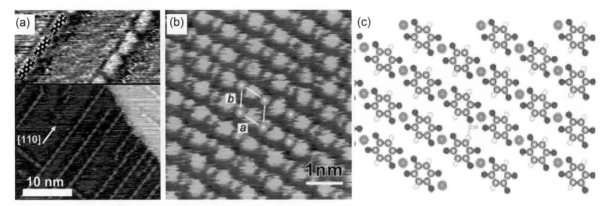

Fig. 10.31 (a) STM images of 1,4-bis(4-pyridyl)benzene adsorbed on Cu(100) at 300 K. The structural model overlaid on the image illustrates the N–Cu–N co-ordination bonding. The lower STM topograph shows an overview of the chains attached to the lower side of the terrace step or running parallel on the upper side of the step. (b) STM image of DHBQ-Cu monolayers showing the close-packed chains with repeating parts of big and small bumps. (c) Schematic model of the DHBQ-Cu monolayers. (Reprinted with permission from J. Phys. Chem. C. Copyright (2007) American Chemical Society.)

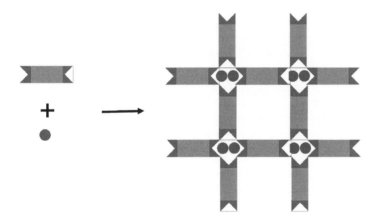

Fig. 10.32 Dark tips: co-ordination ligands; Gray blocks: molecular backbones; Spheres: metal centers.

pre-heated Au(111) surfaces from aqueous solutions (Zhang *et al.* 2007) (see Figs. 10.31(b) and (c)). The molecules coordinate with the copper ions to form well-ordered close-packed chain-like –[Cu-DHBQ]- structures.

10.3.3 2D open networks

10.3.3.1 *Square networks*

Open co-ordination networks were realized by the direct reticulation of linker molecules in two dimensions. As illustrated in Fig. 10.32, linker molecules carry ditopic co-ordination functional groups. The functional groups co-ordinate to proper metal center(s) forming extended 2D networks.

With TPA linkers regular 2D network structures can be realized by complexation with appreciable amounts of Fe. One achieves a fully reticulated structure comprising arrays of diiron co-ordination centers (Figs. 10.33(a) and (b))

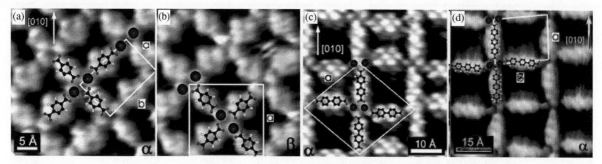

Fig. 10.33 (a,b) STM topographs of isomeric FeTPA network phases: (a) identical and (b) alternating Fe dimer arrangement. (c) STM image of the FeBDA network. (d) STM image of the FeTDA network phase. Tentative models are superimposed on the STM images. (Reprinted with permission from J. Phys. Chem. B. Copyright (2006) American Chemical Society.)

(Lingenfelder *et al.* 2007). The Fe–Fe spacing within a dimer amounts to about 4.7 Å, slightly smaller than two times the substrate lattice constant (2.55 Å). The co-ordination geometry for each Fe ion assumes a distorted square-planar geometry. There exist two equivalent isomeric structures that differ in the orientation of the neighboring Fe pairs in the network nodes, i.e. they are either equally oriented or alternate. Both isomeric networks reside commensurate on Cu(100) with a (6 × 4) and (5 × 5) unit cell, respectively. These structures possess cavities of well-defined size and shape, exposing the underlying Cu surface. Two longer analogs of TPA, 4, 4′-biphenyl dicarboxylic acid (BDA) and 4, 1′, 4′, 1″-terphenyl-1, 4″-dicarboxylic acid (TDA) having two and three phenyl groups in the molecular backbone, respectively, form networks with increasing size similarly a containing diiron center as the essential coupling motif of the carboxylate groups (Figs. 10.33(c) and (d)) (Stepanow *et al.* 2006).

10.3.3.2 *Rectangular networks*

Rectangular networks with controlled aspect ratios can be realized through co-ordination assembly involving multicomponent ligands. A recent systematic study demonstrated that the equatorial linker, represented by carboxylic acids bridging the Fe pairs, and the axial linkers, represented by pyridyl linkers binding the Fe pairs laterally, can be chosen independently, leaving the overall rectangular structure preserved, as illustrated in Fig. 10.34(a) (Langner *et al.* 2007). The rigid binding of the carboxylate linkers ensures the stability of the networks, whereas the axially engaged pyridyl linkers account for the flexibility of the structures accounting for incommensurability with the substrate and misengaged bonding. This ability of error correction involving the efficient reversibility of the Fe co-ordination bonds and the co-operative binding of the ligand groups is of prime importance to achieve self-selection of the components. Hence it was shown that the mixtures of the two homofunctional molecules form well-ordered arrays of rectangular multicomponent networks, whose size and shape can be deliberately tuned by selecting ligands of desired length from complementary ligand families (Fig. 10.34(b)).

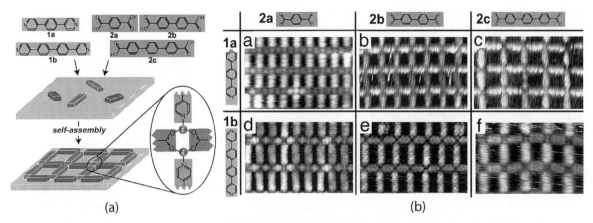

Fig. 10.34 (a) Concept of multicomponent co-ordination assembly. (b) STM images showing six rectangular networks formed by multicomponent ligands of bipyridine (**1a**, **1b**) and *bis*-carboxylic acid (**2a–c**) ligands. All images are 9.4 nm × 6.0 nm. Structure periodicity is (*a*) 1.1 nm × 1.8 nm, (*b*) 1.5 nm × 1.8 nm, (*c*) 1.8 nm × 1.8 nm, (*d*) 1.1 nm × 2.3 nm, (*e*) 1.5 nm × 2.3 nm, and (*f*) 1.8 nm × 2.3 nm. (Copyright The National Academy of Science of the USA.)

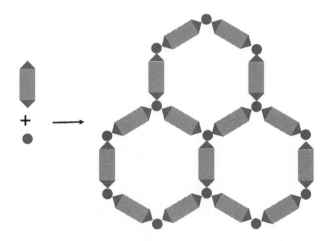

Fig. 10.35 Dark tips: co-ordination ligands; Gray blocks: molecular backbones; Spheres: metal centers.

10.3.3.3   *Hexagonal networks*

Hexagonal networks can be realized through threefold co-ordination motifs, as illustrated in Fig. 10.35. Complex structures such as honeycomb or kagomé lattices are scarce in 3D compounds since threefold co-ordination modes are less frequent. At surfaces, the strict two-dimensional confinement of the ligands and metal ions imposed by the substrate substantially influences the metal-to-ligand binding modes. On surfaces, networks comprising trigonal mononuclear co-ordination nodes have been achieved by the Co and Fe directed assembly of ditopic dicarbonitrile- and hydroxyl-terminated, respectively, polybenzene linkers (Stepanow *et al.* 2007), shown in Fig. 10.36. The occurrence of threefold co-ordination motifs on substrates with different symmetries signifies that the binding motif is an intrinsic characteristic of the metal co-ordination and not due to the geometrical coupling to the supporting surface. The two binding modes of the different functional groups differ with respect of the orientation of the ligand termination. The cyano-terminated ones

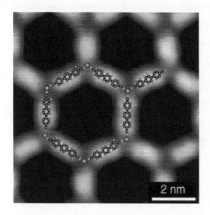

Fig. 10.36 High-resolution STM image of the Co-dicarbonitrile honeycomb network assembled on Ag(111). The tentative models are superimposed over the data. (Copyright Wiley-VCH Verlag GmbH & Co. KGaA.)

point directly towards the metal ion, whereas the hydroxy ligands are directed slightly off center, which accounts for the chirality of the binding motif. These features are intrinsic properties of the ligand system and have to be taken into account when designing co-ordination architectures. The results demonstrate that surface-assisted assembly can lead to unusual co-ordination motifs that are generally not found in conventional 3D bulk phases. These findings are attributed to the presence of the surface, where hybridization of the metal orbitals with the metal states of the substrate causes unusual redox states. In addition, the preferred flat configuration of the aromatic system can stabilize such binding modes.

Furthermore, a series of hexagonal networks can be assembled from a series of simple ditopic dicarbonitrile molecular bricks and Co atoms on Ag(111) (Schlickum *et al.* 2007). This approach enabled fabrication of size- and shape-controlled open hexagonal pores with dimensions up to 5.7 nm. For the investigations, a series of C_2-symmetric dicarbonitrile-polyphenyl molecular linkers (abbreviated NC-Ph$_n$-CN, whereby n can be 3, 4 or 5) was synthesized. All ditopic molecular linkers have the same functional carbonitrile endgroups, while their lengths increase with n from 1.66 via 2.09 up to 2.53 nm. The area of the enclosed hexagons increases stepwise with the number of phenyl rings incorporated in the molecular linkers' backbone (Fig. 10.37). Accordingly, the cell size expands stepwise from ≈ 10 via 15 up to $20\,\text{nm}^2$ for $n = 3, 4, 5$, respectively. The $20\,\text{nm}^2$ nanopores achieved with NC-Ph$_5$-CN linkers represented a record for the most open surface-confined nanomesh realized by self-assembly.

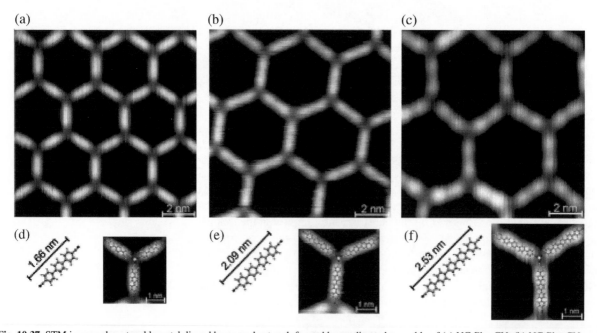

Fig. 10.37 STM images show tunable metal–ligand hexagonal network formed by co-directed assembly of (a) NC-Ph$_3$-CN, (b) NC-Ph$_4$-CN, and (c) NC-Ph$_5$-CN. (d–f) Molecular structure and length along with models of the threefold Co-carbonitrile co-ordination motif resolved in (a–c). (Reprinted with permission from NanoLett. Copyright (2007) American Chemical Society.)

10.4 Conclusions

The presented examples of design of low-dimensional nanostructures at surfaces through supramolecular assembly constitute part of the future tool kit needed to put into reality bottom-up fabrication of functional interfaces and surfaces. The described results reveal that both hydrogen bonds and metal co-ordination offer protocols to achieve unique nanostructured systems on 2D surfaces or interfaces. As the conventional 3D supramoelcular self-assembly has generated a vast number of nanostructures revealing high complexity and functionality, the 2D approaches are conceivable for a great variety of molecular systems and can be applied to substrates with different symmetries, as well as physical and chemical properties.

Many applications have been envisioned and explored: 2D nanopores or nanocavities provide by the open networks can be used to arrange guest species in well-defined nanoscale environments, both for patterning purposes or investigations of surface chemical reactions in controlled surroundings. They bear potential to control large biomolecules in tunable spaces and their molecular motions. Furthermore, they may serve as templates for the organization of separated, regularly distributed functional entities, such as magnetic nanoclusters. A particular appealing property of metal–ligand systems is the co-ordination metal centers embedded in the co-ordination nanostructures. The intrinsic physical properties of these metal centers (e.g. redox-, spin-, magnetic, electronic states) are affected by the intricate interplay between the hybridization of the metal d-states with substrate electrons. The interplay of metal–ligand interaction, intermolecular interaction and surface interaction assumes therefore a critical role in determining the chemical or electronic properties of self-assembled supramolecular nanostructures. Thus, it presents an attractive topic to tailor the metal center's properties via judicious choice of ligands and surfaces, which needs to be addressed in more detail in future.

Porous metal-organic co-ordination frameworks have been extensively studied as a new class of porous materials because of their potential industrial applications, such as separation, heterogeneous catalysis, and gas storage. They are completely regular, have high porosity, and highly designable frameworks (Yaghi *et al.* 2003; Kitagawa *et al.* 2004). However, even when they have nanosized channels or cavities, the crystals of the compounds themselves are of micrometer size, and insoluble in any solvents. Therefore, it is desirable to develop thin-layer compounds that attached at substrates. It is of great interest to use 2D metal–ligand networks formed on surfaces as templates to grow 3D supramolecular metal-organic frameworks with high quality and large size.

References

Barth, J.V., Weckesser, J., Cai, C., Günter, P., Bürgi, L., Jeandupeux, O., Kern, K. *Angew. Chem. Int. Ed.* **39**, 1230 (2000).

Bersuker, I.B. *Electronic Structure and Properties of Transition Metal Compounds: Introduction to the Theory* (Wiley, New York, 1996).

Cañas-Ventura, M.E., Xiao, W., Wasserfallen, D., Müllen, K., Brune, H., Barth, J.V., Fasel, R. *Angew. Chem. Int. Ed.* **46**, 1814 (2007).

De Feyter, S., De Schryver, F.C. *Chem. Soc. Rev.* **32**, 139 (2003).

Dmitriev, A., Lin, N., Weckesser, J., Barth, J.V., Kern, K. *J. Phys. Chem. B* **106**, 6907 (2002).

Griessl, S., Lackinger, M., Edelwirth, M., Hietschold, M., Heckl, W.M. *Single Mol.* **3**, 25 (2002).

Keeling, D.L., Oxtoby, N.S., Wilson, C., Humphry, M.J., Champness, N.R., Beton, P.H. *Nano Lett.* **3**, 9 (2003).

Kitagawa, S., Kitaura, R., Noro, S. *Angew. Chem. Int. Ed.* **43**, 2334 (2004).

Langner, A., Tait, S.L., Lin, N., Rajadurai, C., Ruben, M., Kern, K. *Proc. Nat. Acad. Sci. USA* **104**, 17927 (2007).

Lehn, J.M. *Proc. Natl. Acad. Sci. USA* **99**, 4763 (2002).

Lin, N., Payer, D., Dmitriev, A., Strunskus, T., Woll, C., Barth, J.V., Kern, K. *Angew. Chem. Int. Ed.* **44**, 1488 (2005).

Lin, N., Stepanow, S., Vidal, F., Kern, K., Alam, S., Strömsdörfer, S., Dremov, S., Müller, P., Landa, A., Ruben, M. *Dalton Trans.* 2794 (2006).

Lingenfelder, M., Spillmann, H., Dmitriev, A., Lin, N., Stepanow, S., Barth, J.V., Kern, K. *Chem. Eur. J.* **10**, 1913 (2004).

Messina, P., Dmitriev, A., Lin, N., Spillmann, H., Abel, M., Barth, J.V., Kern, K. *J. Am. Chem. Soc.* **124**, 14000 (2002).

Otero, R., Schöck, M., Molina, L.M., Lægsgaard, E., Stensgaard, I., Hammer, B., Besenbacher, F. *Angew. Chem. Int. Ed.* **44**, 2270 (2005).

Payer, D., Comisso, A., Dmitriev, A., Strunskus, T., Lin, N., Wöll, C., De Vita, A., Barth, J.V., Kern, K. *Chem. Eur. J.* **13**, 3900 (2007).

Ruben, M., Payer, D., Landa, A., Comisso, A., Gattinoni, C., Lin, N., Collin, J.-P., Sauvage, J.-P., De Vita, A., Kern, K. *J. Am. Chem. Soc.* **128**, 15644 (2006).

Schlickum, U., Decker, R., Klappenberger, F., Zoppellaro, G., Klyatskaya, S., Ruben, M., Silanes, I., Arnau, A., Kern, K., Brune, H., Barth, J.V. *Nano Lett.* **7**, 3813 (2007).

Spillmann, H., Dmitriev, A., Lin, N., Messina, P., Barth, J.V., Kern, K. *J. Am. Chem. Soc.* **125**, 10725 (2003).

Staniec, P.A, Perdigao, L.M.A., Rogers, B.L., Champness, N.R., Beton, P.H. *J. Phys. Chem. C* **111**, 886 (2007).

Stepanow, S., Lin, N., Payer, D., Schlickum, U., Klappenberger, F., Zoppellaro, G., Ruben, M., Brune, H., Barth, J.V., Kern, K. *Angew. Chem. Int. Ed.* **46**, 710 (2007).

Stepanow, S., Lin, N., Barth, J.V., Kern, K. *J. Phys. Chem. B* **110**, 23472 (2006).

Stöhr, M., Wahl, M., Galka, C.H., Riehm, T., Jung, T.A., Gade, L.H. *Angew. Chem. Int. Ed.* **44**, 7394 (2005).

Su, G.J., Zhang, H.M., Wan, L.J., Bai, C.L., Wandlowski, T. *J. Phys. Chem. B* **108**, 1931 (2004).

Tait, S.L., Langner, A., Lin, N., Stepanow, S., Rajadurai, C., Ruben, M., Kern, K. *J. Phys. Chem. C* **111**, 10982 (2007).

Theobald, J.A., Oxtoby, N.S., Phillips, M.A., Champness, N.R., Beton, H. *Nature* **424**, 1029 (2003).

Weigelt, S., Busse, C., Petersen, L., Rauls, E., Hammer, B., Gothelf, K,V., Besenbacher, F., Linderoth, T. *Nature Mater.* **5**, 112 (2006).

Yaghi, O.M., O'Keeffe, M., Ockwig, N.W., Chae, H.K., Eddaoudi, M., Kim, J. *Nature* **423**, 705 (2003).

Yan, H.J., Lu, J., Wan, L.J., Bai, C.L. *J. Phys. Chem. B* **108**, 11251 (2004).

Ye, Y., Sun, W., Wang, Y., Shao, X., Xu, X., Cheng, F., Li, J., Wu K. *J. Phys. Chem. C* **111**, 10138 (2007).

Yokoyama, T., Kamikado, T., Yokoyama, S., Mashiko, S. *J. Chem. Phys.* **121**, 11993 (2004).

Yoshimoto, S., Yokoo, N., Fukud, T., Kobayashi, N., Itaya, K. *Chem. Commun.* **500**, (2006).

Zhang, H.M., Zhao, W., Xie, Z.-X., Long, L.-S., Mao, B.-W., Xu, X., Zheng, L.-S. *J. Phys. Chem. C* **111**, 7570 (2007).

11 Nanostructured surfaces: Dimensionally constrained electrons and correlation

E. Bertel and A. Menzel

11.1 Introduction

Electronic correlation is a topic of pre-eminent relevance for electrons confined in nanostructures. In order to understand the relation between spatial confinement and correlation it is useful to start with one of the simplest electronic systems, in which correlation effects have been studied in detail, namely the He atom. In the Hartree–Fock approximation, the two-electron state vector is represented by Slater determinants of one-electron wavefunctions. The one-electron wavefunctions correspond to radially symmetric charge distributions. Thus, correlation of the movement of the two electrons caused by the Coulomb repulsion is not taken into account in the single-particle wavefunctions. However, antisymmetrization of the total (spatial and spin) electron wavefunction by forming the Slater determinants results in a different spatial charge distribution for electrons with parallel and antiparallel spin, respectively. Thus, the Hartree–Fock approximation does take into account correlated electron movements to some extent, although the solution is constructed from single-electron wavefunctions, which individually do not reflect this correlation. Electronic correlation in a more specific sense, as being used in the literature to delineate so-called "(strongly) correlated systems" from, e.g. Fermi-liquid metals, refers to effects of electron–electron interaction, which go beyond the Hartree–Fock approximation and are not just a consequence of the Pauli principle.

In many cases, considering the movement of a particular electron in the time-averaged charge distribution of all the other electrons in the system is a good approximation. The time-averaged charge distribution has to reflect the total symmetry of the system. Hence, in an atom it has to be radially symmetric, in a periodic solid it has to be translationally invariant with respect to multiples of the lattice constant, etc. The single-electron wavefunctions in such a system therefore have to obey the Bloch theorem and are eigenfunctions of the momentum operator. As a consequence, they are completely delocalized. The approximation works well for metallic systems with a high density of mobile electrons, where any local perturbation of the charge distribution is screened out on a scale shorter than the Fermi wavelength $\lambda_F = 2\pi/|k_F|$. Here, k_F is the Fermi wave vector. In systems with restricted electron mobility, e.g. most oxides, heavy-Fermion systems, and—particularly relevant in the present context—systems with electrons confined in nanostructures, this approximation breaks down. The localized character of the electron wave function has to be taken into account explicitly and leads to much stronger effects of electron correlation. Unfortunately, there is no *ab-initio* procedure to determine in a self-consistent way the exchange and correlation interaction of localized electrons and the resulting correlated electron dynamics. Typically, parameterized Hamiltonians are used for the description of such strongly correlated electronic systems. The Mott–Hubbard Hamiltonian is a paradigmatic case. Parameters are the transfer matrix element determining the kinetic energy of the electrons and the on-site Coulomb repulsion of electronic charges. Strong correlation in this model arises, if the electron–electron interaction becomes comparable to the kinetic energy. Electrons in this case are described as localized charges, which is complementary to the Bloch description, where the electron wavefunction extends over the whole solid. Loosely speaking, correlated systems are therefore those with more or less localized electrons. The localization of electronic wavefunctions results in strongly directional interactions and correspondingly strongly correlated electron dynamics.

From this—somewhat intuitive—description of electronic correlation it is clear that correlation effects are important to consider, as soon as electrons are spatially confined by nanostructures. The effect of correlation will become increasingly dominant as the dimensionality of the electron wavefunction is reduced. In the following, the discussion will focus on quasi-one-dimensional (quasi-1D) confinement, i.e. more or less strongly coupled one-dimensional nanostructures, with occasional reference to 2D and 0D systems. Of course, 0D nanostructures—quantum dots—on surfaces are highly interesting for applications. We here just mention the Coulomb blockade in transport through quantum dots or the problem of magnetism in quantum-dot structures, which is relevant for the construction of magnetic memory devices on the nanoscale. Perpendicular magnetization, magnetic anisotropy, and super-paramagnetism are just a few of the questions that have to be addressed when dealing with magnetism in quantum dots. However, these subjects go beyond the scope of the present chapter. Furthermore, we will concentrate on nanostructures on surfaces as opposed to 3D nanoarchitectures. Nanostructured surfaces are particularly apt for studying fundamental interactions in correlated systems: First, they allow tuning the system parameters by a variety of methods not available in other systems. For example, the geometry can be modified within a wide range by using template growth of nanostructures. Taking advantage of pseudomorphic growth, phases can be stabilized, which would not naturally be found in bulk material. Strained-overlayer growth can be achieved, which puts the material under both compressive and tensile stress depending on the substrate. Secondly, the use of surface systems allows access with a variety of surface-sensitive probes, in particular direct-space imaging methods. This is a severe problem in bulk materials, where structural information has to be derived mostly from reciprocal space (i.e. diffraction) methods with limited spatial resolution.

11.2   Motivation

The understanding of interactions in correlated systems is of fundamental importance, because it turns out that such materials tend to exhibit a variety of electronically driven phase transitions. More importantly, the phases occurring in the generic phase diagram of correlated materials are extremely interesting, because on the one hand they challenge our understanding of basic physical phenomena, on the other hand they lend themselves to extremely attractive technical applications from high-temperature superconductivity to magnetoresistant devices. Figure 11.1, for example, depicts a generic phase diagram of quasi-1D materials. It is similar to a version published by Bourbonnais (2000) for organic superconductors, but it bears several features also found in the phase diagram of high-T_c superconductors and transition-metal oxides in general. Various phases are observed as a function of temperature and of transverse coupling $t_\perp$ between the 1D subunits. $t_\perp$ designates the transverse hopping matrix element and is proportional to the wavefunction overlap of adjacent chains in the structure. It is immediately clear from looking at this phase diagram that it encompasses phases that are mutually exclusive in their properties, for instance magnetically ordered phases and superconducting phases, yet seem to be closely related, because they always appear together and even share common phase boundaries. It will be one of the goals of the following discussion to provide some insight into why this is the case.

A few words about the phases depicted in Fig. 11.1 are in place here. If we start at the upper right, i.e. moderately high temperatures and large transverse

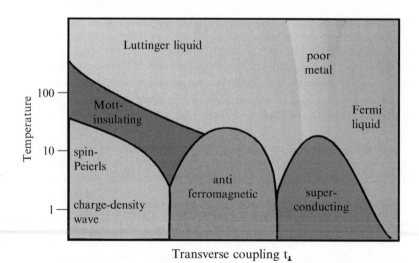

Fig. 11.1 Generic phase diagram of quasi-one-dimensional materials. Depending on the transverse coupling a variety of technically relevant phases occur, which are only partly understood. After Bourbonnais (2000).

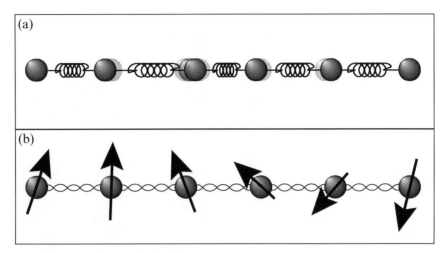

Fig. 11.2 Collective excitations in a one-dimensional fermionic system: (a) charge excitation (holon), (b) spin excitation (spinon). The two excitations propagate each with a different group velocity.

mobility, the system can be described as a Fermi liquid and behaves as a normal metal. At $T = 0$ K the system is in the ground state, i.e. the electrons occupy all states up to the Fermi energy (and nothing happens). Any physical process requires an excitation above the ground state. This implies adding a particle into an unoccupied state. Although the electrons in the system are strongly interacting, such an excitation can still be described in the independent particle picture. The effect of the interactions can be captured essentially by ascribing to the particle an effective mass different from the free-electron mass and a finite lifetime. The quasi-particles can be envisioned as electrons dragging with them a cloud of electron–hole excitations, yet the wavefunction describing the "quasi-particle" possesses a well-defined momentum and energy. Hence the wavefunction is delocalized.

As the transverse mobility $t_\perp$ is reduced, the concept of a particle propagating in that direction eventually breaks down. The coherence length of the wavefunction shrinks below the Fermi wavelength λ_F of the particle. In this range the material behaves as a "poor metal" (Emery and Kivelson 1995) with anomalous transport properties. With further reduction of $t_\perp$ the wavefunction eventually becomes localized to a single chain. This is the truly one-dimensional case, where the Fermi-liquid paradigm has to be replaced by the Tomonaga–Luttinger liquid model (Voit 1995). It is intuitively clear that electrons confined to one dimension cannot do a very good job in screening. Furthermore, like cars on an overcrowded highway, they are no longer able to move as anything like independent particles. Any excitation is collective.

As illustrated in Fig. 11.2, there are two possible types of excitations, one in the charge and one in the spin system. In 1D, the group velocity generally is different in the charge and the spin system and hence the two excitations are able to separate. Each of the excitations carries an energy and an associated momentum. This has a fundamental effect on one of the most useful experimental techniques for the exploration of the electronic structure: Angle-resolved UV-photoemission spectroscopy (ARPES). This technique uses UV

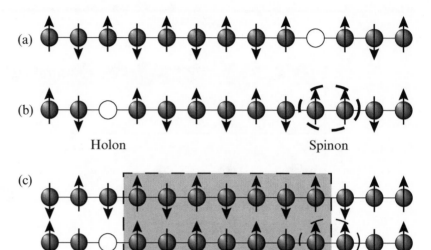

Fig. 11.3 Spinon–Holon separation in a photoemission process: (a) Photoemission removes an electron from an antiferromagnetically ordered chain. (b) Electrons hopping into the hole from the left create a spin excitation (spinon) and a charge excitation (holon), which in the one-dimensional chain are able to separate. (c) In a two-dimensional antiferromagnetically ordered system, the separation of the holon from the spinon creates a string of misaligned spins, which costs additional energy. Hence, spinon and holon are kept in a bound state here due to the string interaction.

photons (usually in the energy range between 15 and 150 eV from an electron storage ring; in the laboratory typically He I radiation with $h\nu = 21.22$ eV) to excite the electrons in a solid to energies above the vacuum level. The electrons are then able to leave the solid and to propagate towards an electron energy analyzer. After passing the analyzer, electrons are detected and counted. From the measured final-state energy, the photon energy and the emission direction the initial-state energy and the initial momentum parallel to the surface can be retrieved. For a normal Fermi liquid the electron energy band dispersion can be reconstructed in this way (Hüfner 1995).

In a Luttinger liquid, however, photoemission of an electron leaves two excitations behind, one in the charge and one in the spin system, as shown in Figs. 11.3(a) and (b). As energy and momentum are now shared between the photoelectron and two independent excitations, it is no longer possible to reconstruct the band dispersion from the photoemission spectrum. The very concept of an electron energy band loses its meaning. The spectral function of such a system, however, can be calculated (Voit 2001; see also Fig. 11.4) and shows two edge singularities corresponding to the spin and charge channel, respectively, and a finite intensity in between, as could be expected on the basis of the argument given above for a continuous distribution of energy and momentum between the excitations. A rather conclusive experimental observation of spin-charge separation in angle-resolved photoemission was published recently (Kim *et al.* 2006) for the 1D compound $SrCuO_2$, which features a CuO chain structure. At this point we shall not delve into the detailed properties of the Luttinger liquid phase, but address the next phase in the phase diagram, i.e. the Mott insulating phase.

Conceptually, a Mott insulating phase occurs in systems with a half-filled band, whenever the Coulomb interaction U of electrons on a lattice site ("on-site Coulomb interaction") becomes larger than the kinetic energy

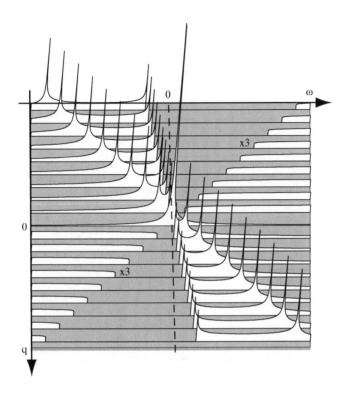

Fig. 11.4 Spectral function of a Luttinger liquid. ω is the energy with respect to the Fermi level and q is the wave vector relative to the Fermi wave vector k_{F}. After Voit (2001).

measured by the band width $W \propto t$ (t is the hopping matrix element). In this case it is always favorable to localize each electron on a lattice site, because the penalty in the kinetic energy is lower than the energy gain obtained from the minimization of the Coulomb repulsion. If the band filling deviates from 1/2, for instance due to doping, the particles or holes can propagate without experiencing an additional repulsion and the system is metallic again. In a one-dimensional system close inspection shows that the system becomes Mott insulating for any repulsive interaction, no matter how weak it is. Furthermore, a metal–insulator transition can occur also for other commensurate filling factors, provided the repulsive interaction is strong enough and has a finite range (Giamarchi 2004). For instance at a quarter filling, next-nearest-neighbor repulsion has to be included into the Hamiltonian in order to get a Mott insulating phase. By the way, this discussion illustrates that the phases in the schematic phase diagram of Fig. 11.1 of course do not only depend on T and $t_{\perp}$, but also on parameters like doping, repulsive (or attractive) interactions between particles, etc.

Very closely related to the Mott insulating phase is the antiferromagnetic phase. It has been mentioned above that the localization of electrons onto individual lattice sites costs a kinetic energy penalty. This penalty is reduced if one admits virtual hopping processes between nearest-neighbor sites. While such hopping is classically not allowed due to the on-site repulsion $U \gg t$, it can be perturbatively admixed to the ground state in a quantum-mechanical calculation. The resulting energy correction is of the order of t^2/U (Feng and Jin 2005). Obviously, the Pauli principle allows such hopping to occur only

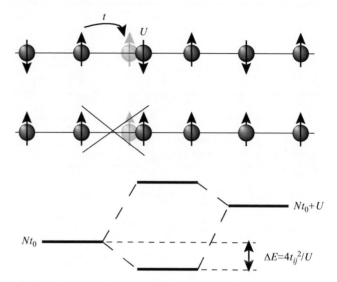

Fig. 11.5 Virtual hopping processes in the one-dimensional Hubbard model. Upper part: Hopping processes are only allowed for antiparallel spin configuration. Lower part: The virtual hopping processes yield an energy lowering of the ground state that scales as t^2/U.

for particles with opposite spin, which favors antiferromagnetic ordering as illustrated in Fig. 11.5.

We now turn to the phase designated as spin-Peierls/charge-density wave phase. In this phase the charge (and/or the spin) density as well as the lattice constant is periodically modulated. In order to get an idea about the origin of this phase we concentrate on the charge-density wave (CDW), setting aside a more detailed discussion of the relevant interactions to further below. In our discussion of quasi-1D phases we have considered so far only electron–electron interactions. The CDW phase involves a lattice distortion and therefore we now have to include the electron–phonon interaction into our considerations. For simplicity, we restrict ourselves here to the purely 1D case (Tosatti 1975; Grüner 1994). Figure 11.6(a) shows a nearly free-electron band in 1D. Electron–hole excitations in such a system can be classified according to the excitation energy and the momentum transfer. Figure 11.6(b) depicts the result: Zero-energy excitations are possible at two locations in momentum space, namely at $q = 0$ and at $q = 2k_F$.

Note that the energy scale in Fig. 11.6(b) is set by the Fermi energy, which is typically of the order of a few eV. The electron–hole excitation continuum of Fig. 11.6(b) can be compared to the phonon dispersion, which typically spans an energy range of some tens of meV (Fig. 11.6(c)). Close to $q = 2k_F$ the curves cross, i.e. the two types of excitation agree with respect to energy and momentum. Therefore, they are able to hybridize. The result is shown in Fig. 11.6(d): At the crossing point a gap opens up. The phonon branch develops a cusp at $2k_F$ eventually reaching down to zero energy, hence zero phonon frequency, which corresponds to a static lattice distortion. The electron–hole branch in contrast stays at finite energy indicating the opening of a bandgap, the so-called Peierls gap. Simply speaking, one could say that due to the resonance at $2k_F$ the electrons respond so strongly to the phonon-induced lattice distortion that the restoring force is screened out by the charge rearrangement

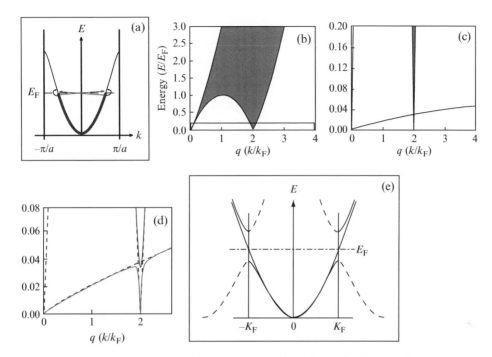

Fig. 11.6 (a) One-dimensional nearly free-electron energy band with an occupied part (below E_F; thick solid line) and an unoccupied part (above E_F; thin solid line). Possible zero-energy excitations are illustrated by arrows. (b) The electron–hole excitation continuum in the band shown in (a). (c) Degeneracy of the electron–hole and the phonon excitations at $2k_F$. (d) Hybridization of the electron and lattice degrees of freedom at $2k_F$. (e) New Brillouin zone boundaries at $\pm k_F$ resulting from the frozen phonon with $q = 2k_F$. The occupied states are lowered in energy around $\pm k_F$, thus providing an electronic energy stabilization for the lattice-distorted state.

(overscreening). The energy gain that drives the system into the Peierls phase can be envisioned as follows: The periodic lattice distortion introduces a new periodicity $d = 2\pi/2k_F = \lambda_F/2$ in direct space. In reciprocal space this corresponds to the introduction of a new Brillouin zone (BZ) boundary at $\pm k_F$. At the BZ, the electronic band structure is backfolded and a gap opens up (compare with Fig. 11.6(e)). Consequently, the occupied electronic states are lowered in energy. Of course, the lattice distortion in a real system costs elastic energy and it remains to be investigated under which conditions the electronic energy gain dominates the energy balance. This point will be addressed later.

The last phase to be mentioned is the superconducting phase. For this phase to be stable an attractive interaction between the electrons is needed. Such an attractive interaction can be provided by phonons, as in the BCS theory, but there are other possibilities as well. For instance, a tendency to antiferromagnetic ordering in the system may favor the formation of local singlet pairs. The 2D model sketched in Fig. 11.7 gives a remote idea how singlet pairing in a system with antiferromagnetic fluctuations could be associated with d-wave symmetry of the pair wavefunction (Monthoux *et al.* 2007).

At this point it should have become clear that electronic correlation is indeed an essential aspect of the interactions responsible for the phase diagram of Fig. 11.1. We have, for instance, explicitly referred to the on-site Coulomb correlation energy U in the above discussion. On the other hand, in the brief

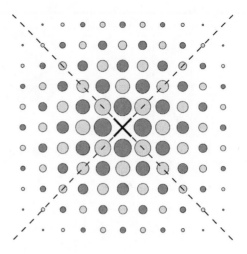

Fig. 11.7 Schematic representation of the interaction potential of quasi-particles with opposite spin (singlet pairing) in a 2D system with antiferromagnetic fluctuations. One of the particles is assumed to reside at the center (cross). The effective interaction potential arises from kinetic exchange as illustrated in Fig. 11.5. On the nearest-neighbor positions of the central site the second quasi-particle experiences an attractive potential, while the potential is repulsive for the diagonally positioned next-nearest-neighbor sites. Hence, the pair wavefunction exhibits d-wave symmetry. After Monthoux *et al.* (2007).

account of the Peierls transition, the role of electronic correlation may not be obvious. In fact, the instability is obtained in principle in the absence of electron–electron interactions as well. However, if interactions are switched on, the energy scale is renormalized and the phase transition becomes considerably more complex, as will be seen in a discussion of real systems below.

Another issue has yet to be raised: As we are dealing with 1D systems here, the phase diagram is actually not so neat and tidy as suggested in Fig. 11.1. The correlation length in a 1D system with nearest-neighbor interactions diverges only as $T \rightarrow 0\,\mathrm{K}$ (Mermin and Wagner 1966). The formation of domain walls is favorable because of entropic reasons (Kagoshima *et al.* 1988). The domain walls are mobile and therefore 1D systems are subject to fluctuations in a wide range of parameter space. Fluctuations in space and time are a severe problem in characterizing the materials properties, because most probes average over space or time or both and are therefore incapable of revealing detailed information about the fluctuation pattern itself. The latter, however, contains essential physics, as already mentioned above in the context of unconventional superconductivity. One could even say that fluctuations by themselves constitute a new type of material with unconventional properties. First, fluctuations can mediate different interactions between quasi-particles. Secondly, by virtue of the fluctuation-dissipation theorem, fluctuations are also equivalent to a high susceptibility of the material, which in turn provides the possibility to change the materials properties drastically by small changes in external (e.g. pressure) or internal (e.g. chemical composition) parameters.

Therefore, tuning a material into a state with critical fluctuations appears to be a promising strategy for the construction of sensors, memory devices, switches and so on. As we have seen in the discussion of the quasi-1D phase diagram, one and the same material can be used in principle to switch between conducting, non-conducting or magnetic/non-magnetic, etc., by just minor variations of, for instance, the overlap between neighboring chains or the doping. This applies for bulk material, but if one were able to arrange 1D systems on surfaces several additional advantages would result. Such systems would be accessible to the whole toolbox of surface-science techniques, in particular to real-space imaging methods. In this way, much more could be learned about the details of microscopic interactions. Furthermore, several additional pathways for external parameter variation would open up. Gate electrodes can be used to shift potentials, appropriate choices of the substrate allow variation of the through-substrate coupling between the 1D subsystems and last, but not least, a host of different nanostructuring methods apart from changes in chemical composition can be used to shape the materials in an optimal way for the designated application. This is why nanostructured surfaces are no less attractive from the academic point of view than they are from the standpoint of the device engineer.

11.3 Interactions in low-dimensional systems

In this section we shall make ourselves a little more familiar with the previously mentioned electron–electron and electron–phonon interactions. An extremely instructive and comprehensive discussion is given in (Giamarchi 2004). The material presented below is mainly taken from this book. As a starting point we choose the generalized response function $\chi_0(q, \omega)$:

$$\langle \rho(q, \omega) \rangle = \chi_0(q, \omega) \, \hat{h}_{\text{tot}}(q, \omega). \tag{11.1}$$

Here, $\rho(q, \omega)$ is some response of the system to the (time-dependent) perturbation $\hat{h}_{\text{tot}}(q, \omega)$. For instance, $\rho(q, \omega)$ might be the charge density induced by an electric field, a spin density induced by a magnetic field or a pair density induced by a pairing potential. Note that we are dealing in general with fluctuating systems, so the left side is the expectation value of the response. The perturbation $\hat{h}_{\text{tot}}(q, \omega)$ designates the local field seen by the particles. In the following, we concentrate on the static susceptibility $\chi_0(q, \omega = 0) = \chi_0(q)$ and first consider the charge density response in a non-interacting Fermi gas. Application of perturbation theory yields the following expression for $\chi_0(q)$:

$$\chi_0(q) = \frac{1}{\Omega} \sum_k \frac{f_{k+q} - f_k}{\varepsilon_{k+q} - \varepsilon_k}, \tag{11.2}$$

with Ω being the sample volume, f_k the Fermi function and ε_k the energy of the kth state. Note that, since f_k can be considered as a monotonously decreasing function of ε_k, $\chi_0(q)$ is negative. The response function is a measure of the charge redistribution under the influence of an external potential, which involves the transfer of electrons from states k to states $k + q$ (or vice versa). Such a transfer is only possible if the source states are occupied and the target states are empty, which is expressed by the Fermi function difference in the numerator. Obviously, the main contribution to the response function arises from processes where the denominator vanishes, i.e. source and target states are at the same energy. This implies that both lie at the Fermi surface. The possible processes are pictorially shown in Fig. 11.8. In 1D these processes can be grouped into two classes, either $q = 0$ or $q = 2k_F$, see Fig. 11.8(a). This is not so in two and higher dimensions, where $2k_F$ is not a singular point (Fig. 11.8(b)). Assuming a linearized dispersion around E_F the resulting

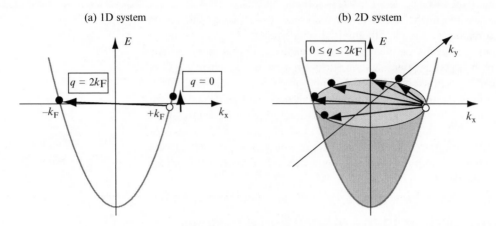

Fig. 11.8 Excitation processes yielding the dominant contribution to the charge-density response function in 1D (a) and 2D (b).

response function is easily calculated for different dimensionalities and shown in Fig. 11.9 for $T = 0$ K. We notice that in 1D $\chi_0(q)$ has a singularity at $q = 2k_F$, indicating a strong response of the charge density to an arbitrarily small perturbation. This signals an instability in the system, giving rise to a phase transition.

Several qualifying remarks have to be made, though. First, the singularity appears only in the 1D response function in our model. However, some traits of the 1D physics are recovered, if real band structures in higher dimensions are considered. A maximum in the response function of higher-dimensional systems may reappear, if there are extended parallel sections of the Fermi surface, because in this case there is a finite volume in phase space, where the so-called Fermi-surface nesting condition $\varepsilon_k = \varepsilon_{k+q} = \varepsilon_F$ is fulfilled (see Fig. 11.10(a)). Actually, there is not only a nesting condition in k space, but also in the dispersion relation: If $\xi(k)$ measures the energy relative to E_F

$$\xi(k) = \varepsilon(k) - E_F \tag{11.3}$$

Fig. 11.9 The normalized charge-density response function for the one to three-dimensional non-interacting Fermi gas.

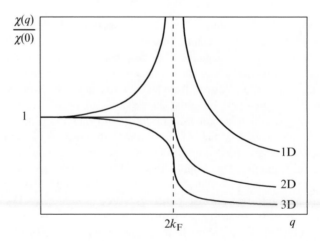

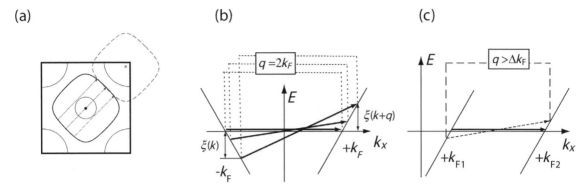

Fig. 11.10 (a) Fermi surface nesting in a 2D system. The black square is the surface Brillouin zone. Thin gray lines are constant-energy surfaces. Shifting of the Fermi surface (black line) by $2k_F$ results in a common line in k space shared by the original and the shifted Fermi surface (dashed gray). (b) Illustration of the nesting condition (eqn (11.4)). There is a finite range of k vectors for which $\xi(k) = -\xi(k + 2k_F)$ is fulfilled (black arrows). (c) Absence of nesting for two parallel bands crossing the Fermi level at k_{F1} and k_{F2}, respectively. There is only one point in k space where eqn (11.4) is fulfilled.

then nesting requires

$$\xi(k + q) = -\xi(k). \tag{11.4}$$

This nesting condition is illustrated in Fig. 11.10(b). Although it is automatically fulfilled in single-band systems for the Fermi surface points at $\pm k_F$ (provided we have inversion symmetry), it is not necessarily so for multiband systems, where the Fermi surfaces of two different bands could nest in k space, but at the same time may not comply with the nesting condition (11.4) (Fig. 11.10(c)). Another possibility to get a diverging response apart from Fermi surface nesting is to have a diverging density of states (DOS) at E_F. In 2D, a saddle point in the band structure is associated with a diverging DOS (Bertel and Lehmann 1998). Hence, if the saddle point coincides with E_F, this may also give rise to a CDW instability (Rice and Scott 1975).

Thus, a singular response function at $2k_F$ and $T = 0\,\text{K}$ may occur not only in 1D systems but also in higher dimensions, which is satisfying, because there is hardly a real system that is truly 1D. On the other hand, we have so far discussed the phase transition at $T = 0\,\text{K}$, but real physics happens at finite temperatures and at $T > 0\,\text{K}$ the singularity in $\chi_0(q)$ is rapidly suppressed:

$$\chi_0(q, T) \propto -\text{DOS}(E_F) \ln(E/T), \tag{11.5}$$

where E is an energy cut-off of the order of the energy range where the nesting condition (11.4) is fulfilled (Giamarchi 2004). Function (11.5) is qualitatively shown in Fig. 11.11. So far we have considered the non-interacting Fermi gas. What happens, if we switch on electron–electron interactions? To show this, we go back to eqn (11.1).

In eqn (11.1) $\hat{h}_{\text{tot}}$ is the effective perturbation seen by an electron. However, $\hat{h}_{\text{tot}}$ consists of two parts: First, the external perturbation $\hat{h}_{\text{ext}}$ (here external implies that the perturbation is external to the electronic system, not necessarily to the system as a hole; for instance, the potential set up by a phonon, i.e. the displacement of the ion cores, is such an external perturbation); second, the effect of the external perturbation on the other electrons. Due to

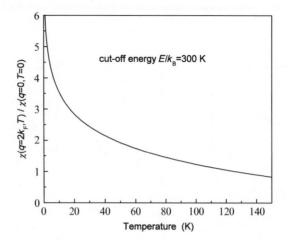

Fig. 11.11 Temperature dependence of the response function $\chi\,(2k_{\text{F}},\,T)$ normalized to $\chi\,(0,0)$.

electron–electron interaction this adds to the external perturbation:

$$\hat{h}_{\text{tot}} = \hat{h}_{\text{ext}} + \hat{h}_{\text{mol}}. \qquad (11.6)$$

The "internal" contribution $\hat{h}_{\text{mol}}$ is termed the "molecular field" for historic reasons. Obviously, $\hat{h}_{\text{mol}}$ is related to the charge response:

$$\hat{h}_{\text{mol}}\,(q) = -g\,\langle \rho\,(q)\rangle. \qquad (11.7)$$

From eqns (11.1), (11.6), and (11.7) we can define an effective response function $\chi_{\text{eff}}\,(q)$ in the following way:

$$\chi_{\text{eff}}\,(q) = \frac{\chi_0\,(q)}{1 + g\,\chi_0\,(q)}. \qquad (11.8)$$

As mentioned before, $\chi_0\,(q) < 0$. Hence, for positive g, $\chi_{\text{eff}}\,(q)$ might become singular even if $\chi_0\,(q)$ is not! Thus, g is an important quantity that influences the response in a fundamental way and we have to discuss it for various cases in order to appreciate its physical significance.

The most obvious way to relate the molecular field to the charge response is to calculate the Coulomb field by means of the Poisson equation:

$$\Delta V = -\frac{e\,\langle \rho \rangle}{\varepsilon_0}, \qquad (11.9)$$

where ε is the dielectric constant of the material. Fourier transformation yields

$$\hat{h}_{\text{mol}}\,(q) = \frac{e^2\,\langle \rho\,(q)\rangle}{\varepsilon_0 q^2}. \qquad (11.10)$$

And, consequently,

$$g = \frac{-e^2}{\varepsilon_0 q^2}. \qquad (11.11)$$

Inserting into eqn (11.8) finally yields

$$\chi_{\text{eff}}(q) = \frac{\chi_0(q)}{1 - \frac{e^2}{\varepsilon_0 q^2}\chi_0(q)}. \tag{11.12}$$

From this result—note that $\chi_0(q) < 0$ – it is evident that the Coulomb interaction reduces χ_{eff} and therefore tends to suppress the formation of a CDW in a Fermi gas (at this point the influence of the lattice is not taken into account).

Next, we consider the Hubbard model, because this is one of the standard models for correlated systems and allows discussion of even spin-dependent electron–electron interactions in a rather elementary way. The model Hamiltonian includes the ingredients of the tight-binding approximation together with a simplified form of (local) Coulomb interaction:

$$\hat{H} = \sum_{i,j,\sigma} t_{ij} c_{i\sigma}^\dagger c_{j\sigma} + U \sum_i n_{i\sigma} n_{i\bar{\sigma}}. \tag{11.13}$$

Here, i and j denote lattice sites and σ ($\bar{\sigma}$) denotes the spin $\uparrow$ ($\downarrow$) of the electrons. U is the so-called on-site Coulomb interaction energy. $c_{j\sigma}$ $\left(c_{j\sigma}^\dagger\right)$ is the annihilation (creation) operator for electrons on site j with spin σ. In the simplest version, t_{ij} is zero unless $i = j$ or $i = j \pm 1$. Then, $t_{ii} = t_0$ is the mean energy of the band and $t_{ii\pm1}$ is the nearest-neighbor hopping integral. $n_{i\sigma}$ can take the values zero or one. If there are two electrons at the same site the potential energy goes up by U due to the on-site repulsion and of course this is only possible if the electrons have opposite spin, because the presence of two electrons with identical spin at the same site is excluded by the Pauli principle. For $U = 0$ one recovers the tight-binding approximation.

Let us now consider the case of a half-filled band with $U \gg t_{ij}$. Electron mobility is restricted by the fact that displacing an electron to the next-neighbor site costs an energy U (Fig. 11.5). Thus, we can consider the state with each lattice site being occupied by one electron as the unperturbed ground state. However, the kinetic energy t acts as a perturbation. It mixes excited states into the ground state. The lowest excited states are those with one double occupancy (Fig. 11.5). The perturbation results in a lowering of the ground state. The calculation yields an energy lowering

$$\Delta E = -\frac{a t_{ij}^2}{U}, \tag{11.14}$$

where a depends on the dimensionality of the system. Clearly, this lower ground state can only be realized for antiferromagnetic ordering. This introduces an effective exchange interaction J into the system, because the model Hamiltonian (11.13) can now be expressed equivalently as

$$\hat{H} = \sum_{i,j,\sigma} t_{ij} c_{i\sigma}^\dagger c_{j\sigma} - J \sum_{i<j} S_i \cdot S_j, \tag{11.15}$$

S_i being the spin operator at site i with eigenvalue ± 1. $J = -a t_{ij}^2 / U$ is the so-called kinetic exchange. In this model it is easy to show that the exchange energy can result in an attractive interaction between quasi-particles. In order

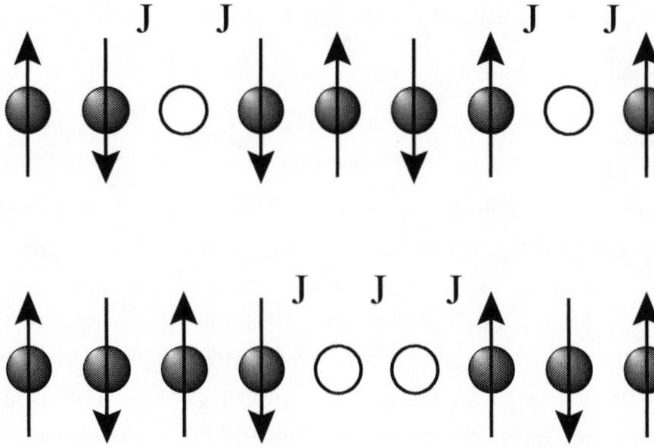

Fig. 11.12 The doping of holes into an anti-ferromagnetic chain structure costs twice the exchange energy J per hole. Hole bunching reduces this energy cost to $(n+1)J$, where n is the number of holes. There is a penalty for hole bunching in the kinetic energy of order t_{ij} per hole, but for large enough J the energy balance results in an effective attraction leading to "phase separation". After Giamarchi (2004).

to see this, we consider hole doping of a half-filled band in an atomic chain. If the holes are distributed randomly, each isolated hole costs an exchange energy $2J$. For two isolated holes we have therefore an energy cost of $4J$, while pairing the holes reduces the energy cost to $3J$, as illustrated in Fig. 11.12.

Taking the Hubbard model as the starting point, the corresponding value of g can be recalculated. As the interaction is now spin dependent, one has to consider the spin-up and spin-down charge density separately. The total potential energy can be written as (Giamarchi 2004):

$$\hat{H}_{\text{pot}} = U \int \rho_\uparrow(x)\rho_\downarrow(x)\, d^D x \qquad (11.16)$$

because only the on-site interaction is taken into account. $d^D x$ designates the volume element in D dimensions. As we need to relate the energy with the expectation value of the charge density we express the density in terms of the fluctuation $\delta\rho_\sigma = \rho_\sigma - \langle\rho_\sigma\rangle$. Expansion up to first-order terms in the fluctuation yields:

$$\hat{H}_{\text{pot}} = U \int \left[\langle\rho_\uparrow(x)\rangle\langle\rho_\downarrow(x)\rangle + \langle\rho_\uparrow(x)\rangle\delta\rho_\downarrow(x) + \langle\rho_\downarrow(x)\rangle\delta\rho_\uparrow(x)\right] d^D x$$

$$= U \int \left[\langle\rho_\uparrow(x)\rangle\rho_\downarrow(x) + \langle\rho_\downarrow(x)\rangle\rho_\uparrow(x) - \langle\rho_\uparrow(x)\rangle\langle\rho_\downarrow(x)\rangle\right] d^D x. \qquad (11.17)$$

The last term amounts to a renormalization of the energy, while the first two terms can be considered as the potential created by the electrons with up (down) spin, which is experienced by the electrons of the opposite spin. Thus, the potential energy shift experienced by the up (down) electrons is given as

$$\hat{h}^{\uparrow(\downarrow)}(x) = U\langle\rho_{\downarrow(\uparrow)}(x)\rangle. \qquad (11.18)$$

Using this expression we can rewrite eqn (11.1) in two parts:

$$\langle\rho_\uparrow(q)\rangle = \chi_0(q)\,\hat{h}^\uparrow_{\text{tot}}(q) = \chi_0(q)\left(\hat{h}_{\text{ext}}(q) + U\langle\rho_\downarrow(q)\rangle\right) \qquad (11.19)$$

$$\langle\rho_\downarrow(q)\rangle = \chi_0(q)\,\hat{h}^\downarrow_{\text{tot}}(q) = \chi_0(q)\left(\hat{h}_{\text{ext}}(q) + U\langle\rho_\uparrow(q)\rangle\right).$$

Solving for $\langle \rho_{\uparrow(\downarrow)}(q) \rangle$ yields

$$\langle \rho_{\uparrow(\downarrow)}(q) \rangle = \frac{\chi_0(q)}{1 - U\chi_0(q)} \hat{h}_{\text{ext}}(q). \tag{11.20}$$

Thus, we recover eqn (11.12) showing that the Coulomb repulsion tends to suppress a CDW also for the Hubbard model. Equation (11.19), however, has further implications. Consider for instance an external magnetic field. In that case the effect of $\hat{h}_{\text{ext}}$ is opposite for up and down spins, respectively:

$$h_{\text{ext}}^{\uparrow} = -h_{\text{ext}}^{\downarrow}. \tag{11.21}$$

Inserting this in eqn (11.19) yields

$$\langle \rho_{\uparrow(\downarrow)}(q) \rangle = \frac{\chi_0(q)}{1 + U\chi_0(q)} \hat{h}_{\text{ext}}^{\uparrow(\downarrow)}(q). \tag{11.22}$$

Here, the denominator can become zero producing a pole in the effective spin susceptibility $\chi_{\text{eff}}(q)$. As one of the consequences, ferromagnetic order in an external magnetic field is promoted by the repulsive Coulomb interaction between electrons (or holes). If we replace the Coulomb field in eqn (11.9) by an attractive interaction, a pole in eqn (11.12) is obtained. Thus, it is plausible that the exchange interaction, which results in an attractive interaction between quasi-particles, as shown above, is able to enhance the density response function, thereby promoting the formation of a CDW state.

So far, we have only considered the response of a Fermi gas without taking into account the interaction with the lattice. The positive ions do not just form a rigid background, but are able to adjust according to density fluctuations in the Fermi gas. This reduces the potential energy and may contribute significantly to the stabilization of a Peierls state, i.e. a state with a CDW in the Fermi gas and a concomitant lattice distortion. Inclusion of the interaction with the lattice into the Hamiltonian of a 1D system leads to the so-called Fröhlich Hamiltonian (Grüner 1994):

$$H = \sum_k \varepsilon_k c_k^{\dagger} c_k + \sum_q \hbar\omega_q b_q^{\dagger} b_q + \sum_{k,q} \gamma_q c_{k+q}^{\dagger} c_k (b_{-q}^{\dagger} + b_q). \tag{11.23}$$

The spin variable has been suppressed here, because it is not relevant for the following consideration. $c_k^{\dagger}$ and c_k are the creation and annihilation operators, respectively, for electrons in state k with unperturbed energy $\varepsilon_k = \hbar^2 k^2/2m$, $b_q^{\dagger}$ and b_q are creation and annihilation operators for phonons with wave vector q and unperturbed energy $\hbar\omega_q$. γ is an electron–phonon coupling constant. The lattice displacement is expressed through the phonon operators as

$$u(x) = \sum_q \left(\frac{\hbar}{2NM\omega_q} \right)^{1/2} \left(b_q + b_{-q}^{\dagger} \right) e^{iqx}, \tag{11.24}$$

with N being the number of lattice sites per unit length and M the ion mass. Solving the Fröhlich Hamiltonian for the phonon frequency yields

$$\omega_{\text{ren},q}^2 = \omega_q^2 + \frac{2\gamma^2\omega_q}{\hbar} \chi(q, T), \tag{11.25}$$

where γ is assumed to be independent of q. Obviously, for large negative $\chi(q, T)$ a frozen phonon can be obtained corresponding to a Peierls transition. Switching on the quasi-particle interaction g renormalizes $\chi(q, T)$ as discussed before, but it will also affect the electron–phonon coupling γ. The phase diagram is then modified in a non-trivial way. For short-range electron–lattice interaction and a Hubbard-type e–e interaction one obtains the Hubbard–Holstein Hamiltonian, which has been discussed, for instance, by Berger *et al.* (1995).

Returning to eqn (11.25) we observe that the bare response function reaches its maximum at $2k_F$. Inserting for $q = 2k_F$ from eqn (11.5) one obtains

$$\omega_{\text{ren}, 2k_F}^2 = \omega_{2k_F}^2 - \frac{2\gamma^2 \omega_{2k_F} \text{DOS}(E_F)}{\hbar} \ln\left(\frac{E}{k_B T}\right). \tag{11.26}$$

From this, the mean-field Peierls transition temperature can be calculated as

$$k_B T_{\text{MF}} = E e^{-1/\lambda}, \tag{11.27}$$

with

$$\lambda = \frac{\gamma^2 \text{DOS}(E_F)}{\hbar \omega_{2k_F}} \tag{11.28}$$

defining a dimensionless electron–phonon coupling constant. Thus, the transition temperature depends critically on the DOS at the Fermi level and on the sign and value of g in eqn (11.7), which governs the details of the phase diagram in correlated systems.

For the transition a complex order parameter can be defined:

$$|\Delta| e^{i\Phi} = \gamma \left(\langle b_{2k_F} \rangle + \langle b_{-2k_F}^\dagger \rangle\right). \tag{11.29}$$

Inserting this definition into eqn (11.24) shows that the expectation value for the lattice displacement $\langle u(x) \rangle$ and the order parameter are linearly correlated, as it should be. Using $\langle u(x) \rangle$ one can calculate the electronic energy gap opening up at the new Brillouin zone boundary. The energy gap turns out to be 2Δ. The total energy gain resulting from the Peierls transition is

$$E_{\text{tot}} = \frac{\text{DOS}(E_F)}{2} \Delta^2. \tag{11.30}$$

From eqns (11.26), (11.27) and (11.28) one finally obtains a relation between the energy gap and the mean-field transition temperature:

$$2\Delta = 3.52 k_B T_{\text{MF}}. \tag{11.31}$$

Before concluding this section it is worthwhile to note another aspect of eqn (11.17). The latter establishes a connection between the generalized susceptibility and the fluctuation of the induced charge density, which may serve as order parameter in describing, for instance, the Peierls transition. In fact, this is a very general result. The fluctuation-dissipation theorem relates the susceptibility of a system to the fluctuations of the order parameter. While this is a familiar result from the theory of phase transitions, it should be emphasized again in the present context. In low-D materials we have seen that the susceptibility can be strongly enhanced and therefore fluctuations persist over a wide temperature range. Although they are of paramount importance for the material properties, they may sometimes go unnoticed, as most experimental probes are integrating over time and space.

11.4   Self-assembled nanostructures on surfaces

The study of correlation on nanostructured surfaces offers valuable insights into the peculiarities of correlated systems, because the easy access to surface structures provides additional possibilities for the control of system parameters and the response of the system can be studied directly by a variety of methods, for which a bulk material is not accessible, at least not directly. Low-energy electron diffraction (LEED), He-atom scattering, and scanning tunnelling microscopy (STM) are among those techniques, to name just a few. In this section we give a brief account of techniques to obtain low-dimensional nanostructures on surfaces. We focus on the bottom-up approach, which is based on self-assembly, because at the moment this is the only method that is able to provide long-range ordered arrays of atomic-scale structures.

Nanostructures on surfaces can either be metastable systems, self-assembled due to kinetic limitations governing the growth of low-dimensional structures, or equilibrium systems, i.e. energetically optimized structures in thermodynamic equilibrium. Self-organization of nanostructures can be achieved far from equilibrium in an open system. An example is the nucleation of islands during deposition of material onto a substrate surface. Depending on the temperature and the flux from the gas phase the nucleation density and hence the average island size can be tuned. The decisive parameter here is the ratio between incoming flux and surface diffusion, which of course depends on the substrate temperature. As there are different diffusion barriers for diffusion on free surfaces, at differently oriented edges of the islands and across island corners, relative diffusion speeds can be varied by appropriate temperature selection. In this way, island shapes can be tuned within wide margins between dendritic morphologies for diffusion-limited aggregation (DLA) and thermodynamic equilibrium shapes dictated by the optimization of the perimeter energies in analogy to the Wulff construction of 3D crystal equilibrium shapes. The nucleation can either be homogeneous or one can use pre-structured templates to achieve the growth of ordered arrays. A splendid account of such methods is provided in the book of Michely and Krug (2004). Instead of promoting self-organization of nanostructures during deposition of thin films one can, for instance, also use ion sputtering to provide the energy flux for non-equilibrium growth of nanostructures (Frost *et al.* 2000; van Dijken *et al.* 2001).

An alternative to kinetically determined structures are thermodynamic equilibrium structures. In principle, such structures are preferable, because they tend to be more stable if exposed to high temperatures or a reactive environment. A possible approach is to create long-range structures exploiting the competition of two similar periodicities as a spatial analog to the beat of similar frequencies in acoustics. One could use, for instance, the lattice mismatch between an ad-layer and the substrate in order to realize such a

Fig. 11.13 STM image of Ag clusters deposited on a R(15 × 12)-C/W(110) template (100 × 85 nm²). 97% of the clusters have a uniform size of seven Ag atoms per cluster (**?**).

structure. If the interaction between the lattices is weak in comparison to the elastic energy required for lattice distortion, one obtains in good approximation a superposition of the undistorted lattices, i.e. a Moiré pattern. Prominent examples attracting interest recently are the BN nanogrid (Corso *et al.* 2004) and the graphene layer on Ir(111) (N'Diaye *et al.* 2006). Of course, choosing lattices with unidirectional mismatch could allow producing 1D Moiré patterns as templates for nanowire growth. To our knowledge, such structures have not been realized yet. A system that perhaps comes close to this idea is the (15 × 12)-C/W(110) surface. Presently, it is not clear, however, whether the long-range superstructure is actually due to a Moiré pattern (Bode *et al.* 1995). It does form an anisotropic template, which is excellently suited for uniform cluster growth and perhaps for the deposition of ordered nanowires, as illustrated in Fig. 11.13 (Bachmann *et al.* **?**; Varykhalov *et al.* 2008).

In most cases, heteroepitaxy with lattice mismatch does not produce Moiré patterns in the proper sense. Rather, the ad-layer relaxes into commensurate domains and the resulting stress is relieved in periodic domain boundaries. These domain boundaries, which are associated with a phase slip in the superstructure, are mathematically described by a sine-Gordon differential equation and therefore also called solitons. The Frenkel–Kontorova model offers a conceptually simple framework for the semi-quantitative evaluation of such structures (Bak 1982). A well-studied example is the hexagonally reconstructed Au(111) surface, where the top layer has a 4% higher packing density than the bulk (111) planes. The surface exhibits three equivalent types of stress domains rotated 120° with respect to each other. In every one of these domains the stress is uniaxially released, thereby producing a 1D pattern of soliton walls (Harten *et al.* 1985) as shown in Fig. 11.14. One should note here that STM studies led to the conclusion that the atoms are uniformly compressed

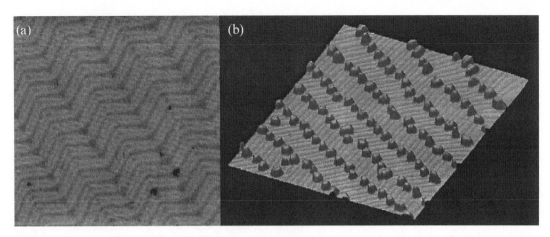

Fig. 11.14 (a) STM image of soliton walls on the reconstructed Au(111) surface ($115 \times 115\,nm^2$). The narrow stripes between the soliton walls correspond to hcp, the wider stripes to fcc stacking. (b) Co clusters deposited on a Au(111) template. Reprinted from Chado *et al.* (2000) with permission from Elsevier.

(Barth *et al.* 1990). In that case the walls are completely delocalized and the pattern is a true Moiré structure. In any case, the resulting nanostructure can also be used as a template for metal-cluster deposition (Chado *et al.* 2000). In addition, the Au(111) surface features a surface state in the center of the surface Brillouin zone (SBZ) and the soliton walls confine the surface state to the 1D domains in between. This was one of the early systems, where properties of 1D confined electrons have been studied (Chen *et al.* 1998). This concept for producing 1D nanopatterns is definitely appealing, but one should note that it is not universally applicable. Although the formation of regular soliton arrays is the thermodynamically most stable solution to the stress-release problem, the domain-wall motion is in many cases strongly activated and this may prevent the evolution of a periodic arrangement. Often, the solitons are caught in a metastable structure lacking long-range order. This is a phenomenon of general relevance. It is, for example, one of the reasons why, starting from a high-temperature long-range ordered parent state, electronically driven phase transitions sometimes produce a low-temperature glassy state with "phase separation" and randomly distributed "phase boundaries". The proper low-T phase in such a case could be an incommensurate Peierls phase. It decays into commensurate domains of, for example, two- and threefold periodicity. The equilibrium structure would be a periodic succession of such domains. However, particularly at low temperature, reaching equilibrium is kinetically hindered and consequently a disordered domain array is the experimentally observed result (Le Daeron and Aubry 1983).

Another self-assembly strategy of nanostructures is closely related to the stress-domain principle outlined above: If the substrate is strongly strained by the stress imposed from a pseudomorphically growing ad-layer, the stress may not just be relieved by domain walls, but the ad-layer may be disrupted into individual islands, each allowing for local elastic relaxation along the island boundary (Tersoff and Tromp 1993). In the case of a 1D strain source, for instance a 1D island boundary, the strain field decays as $1/r$, where r is the

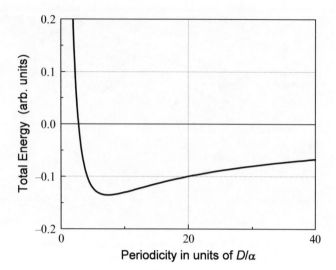

Fig. 11.15 Total energy of a stress domain structure as a function of periodicity according to eqn (11.31).

distance from the boundary. Neighboring domain boundaries repel each other due to the associated stress field and therefore tend to form a periodic array. Calculation of the associated elastic deformation energy ε yields

$$\varepsilon = \frac{C_1}{D} \ln \frac{D}{\alpha}. \tag{11.32}$$

Here, C_1 contains the elastic constants of the material, D is the periodicity of the step array and α is a cut-off parameter of the order of the atomic lattice constant. The pre-factor accounts for the fact that the elastic relaxation energy increases linearly with the number of unit cells of the 1D periodic structure, which is proportional to $1/D$. On the other hand, formation of steps or domain boundaries costs some energy η:

$$\eta = \frac{C_2}{D}, \tag{11.33}$$

where C_2 is the domain-wall formation energy per unit length. Balancing the two energy contributions yields:

$$E = \frac{C_2}{D} - \frac{C_1}{D} \ln \frac{D}{\alpha}. \tag{11.34}$$

The total energy E has a minimum at a certain periodicity as shown in Fig. 11.15. At closer distances the domain-wall formation energy dominates the energy balance, while at larger periodicities both the domain-wall formation energy per unit surface area and the elastic relaxation energy tend to zero. A famous example of this type of nanostructure self-assembly is provided by the growth of oxide stripes on the Cu(110) surface.

Figure 11.16 shows the structure and an STM image of the surface. The adsorbed oxygen forms Cu/O chains, which grow perpendicular to the close-packed atom rows of the clean Cu(110) surface. The Cu/O chains attach to each other and form 1D islands, but the island size perpendicular to the chain direction is limited by the elastic energy accumulated due to a slight

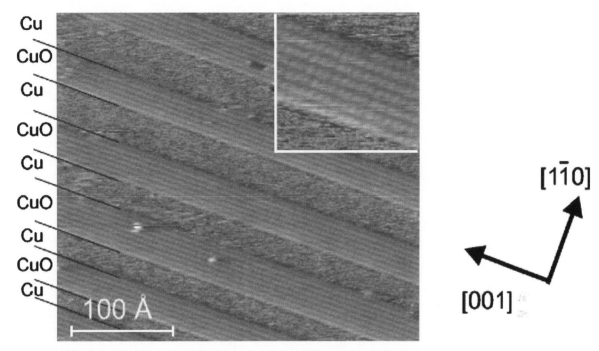

Fig. 11.16 STM image of the striped O/Cu(110) surface with an oxygen coverage of $\Theta_O = 1/4\,ML$. (Reprinted from Sun *et al.* (2008) with permission from Elsevier).

mismatch between the oxide layer and the substrate. Preferred periodicities are of the order of 6–7 nm (Kern *et al.* 1991). Under continued exposure to oxygen the island width increases, while the periodicity tends to stay constant. Hence, the clean copper stripes become narrower until at 0.5 ML they finally disappear and the oxide layer closes.

As in the case of Au(111) there exists an sp-derived surface state on Cu(110), albeit not in the SBZ center, but around the $\bar{Y}$ point of the SBZ. On the striped surface this surface state is confined to the clean Cu channels. Photoemission spectroscopy reveals the typical hallmarks of a quasi-1D electron system with coupled 1D electron states forming minibands in a strongly anisotropic SBZ. Saddle points in the quasi-1D band structure give rise to pronounced peaks in the DOS (Bertel and Lehmann 1998). The physics is very similar to the 1D confined surface state on Au(111) mentioned before (Chen *et al.* 1998). It should be mentioned that the 1D confinement can also be achieved by deposition of molecules, as shown for the Ag(111) surface by Barth and coworkers (Pennec *et al.* 2007). Similarly, Shockley surface states can be confined by steps and in all these cases the fingerprint of a quasi-1D miniband structure has been found (Baumberger *et al.* 2004a). A brilliant review on surface-state confinement on stepped surfaces is given by (Mugarza and Ortega, 2003).

Stepped surfaces are in fact the most common template for the investigation of 1D electronic states, either by directly looking at step-edge states, by confining surface states between steps or by step decoration with metal chains. In order to investigate coupled 1D chains an ordered array of steps is required. If the step–step distance is not sufficiently well controlled, each terrace forms a quantum-well with different width. Accordingly, the quantum well state energy (or coupling strength, respectively) varies from terrace to terrace. If this variation is of the order of the hopping matrix element between the wells or larger, Anderson localization will take place (Anderson 1958). An ensemble of incoherent quantum-well states with a certain energy distribution is observed in this case. Only if the step–step distance is kept constant, is coherence between the well states possible and a quasi-1D system can be realized. Of course, these considerations apply to the coupling of all types of 1D nanostructures (Baumberger *et al.* 2004b; Berge *et al.* 2004). In the case of stepped surfaces, periodic step arrangements require the existence of long-ranged step–step repulsion. There are essentially two mechanisms that provide the required long-range interaction. One is related to the omnipresent stress in surfaces, which arises from the absence of binding partners in the half-space above the surface. In most cases, this leads to an inward relaxation of the top layer and to a tensile stress within the top layer. The absence of bonding towards one side tends to strengthen the remaining bonds of the surface atoms. As a consequence, bond lengths tend to decrease and a tensile stress is established. This is also the origin of the hexagonal reconstruction of Au(111) mentioned above. Here, the tendency to increase the packing density is so large that additional atoms are actually incorporated into the top layer.

Surface reconstruction is in fact the rule on semiconductor surfaces. On many metal surfaces the atomic packing is essentially maintained at the cost of a considerable surface stress. For more details on surface stress the review by Ibach should be consulted (Ibach 1997). Around steps, the stress field becomes asymmetric, which results in a long-range strain field. In the presence of such a strain field steps repel each other and a periodic step array is the result. However, on metal surfaces featuring surface states or resonances an additional interaction of electronic origin seems to stabilize preferred step–step distances. Such "magic" step distances occur, if the terrace width (nearly) coincides with a multiple integer of one half of the Fermi wavelength of a surface state confined on the terrace (Baumberger *et al.* 2004b). This can be considered as a special case of the more general RKKY interaction. In the presence of partially filled electronic states any defect induces charge-density oscillations due to scattering. Neighboring defects interact through their induced charge-density oscillations. The charge-density oscillations are described by

$$\rho(x) \propto \frac{1}{r^n} \cos\left(2k_{\mathrm{F}}x + \delta\right). \tag{11.35}$$

Here, δ is a phase shift depending on the scattering potential of the defect. If surface states are involved, the interaction between two point defects falls off as $1/r^2$, for line defects, such as steps, as $1/r$ (Lau und Kohn 1978).

In the following we discuss quasi-1D systems on the Si(111) surface, because here a variety of 1D systems has been realized in recent years. Early work has focused on alkali metal chains on Si(100) and Si(111). For a brief review see, for instance, Yakovkin (2004). Here, we address more recent work involving Au and transition-metal chains on Si(111) and vicinal Si(111) surfaces (Crain and Himpsel 2006). The vicinal Si(111) surface is a preferred substrate for 1D step lattices, because the unit cell is so large. Hence, spontaneous kink formation at steps is suppressed, because the kink size and hence the formation energy scales with the number of atoms in the unit cell. For higher miscut angles the step spacing is more regular since the higher the step density the stronger the step–step interaction. This is in keeping with eqn (11.34), where the minimum in the total energy curve is much better defined for small periodicity D. Therefore, out of the variety of vicinal surfaces metal chains on the Si(557) and the Si(553) are particularly well investigated. Unexpectedly, the metals do not decorate the steps. Instead, a honeycomb chain, i.e. a flat, graphene-like arrangement of Si atoms runs along the steps in [1$\bar{1}$0] direction (Riikonen and Sánchez-Portal 2005a). The metal atoms are embedded in the middle of the terraces occupying substitutional positions in the top layer. On the flat Si(111) surface, 1D and 2D structures are found, depending on coverage. Most notably, Au forms a 1D Si(111)5 × 2-Au structure with Au double chains. For a detailed discussion of the structures see Riikonen and Sánchez-Portal (2005a,b, 2007).

11.5 The phase diagram of real quasi-1D systems

11.5.1 Case study 1: Metal chains on Si surfaces

Au adsorption on the Si(557) vicinal surface results in Au atom chains adsorbed in the middle of the terraces (Crain and Himpsel 2006). ARPES reveals the existence of 2 closely spaced bands dispersing up to the Fermi energy and crossing it in the middle of the SBZ, resulting in half-filling (Altmann *et al.* 2001). These bands were originally interpreted as the spinon and holon band of a Luttinger liquid (Segovia *et al.* 1999), but soon afterwards this explanation was discarded (Losio *et al.* 2001). There have been different explanations put forward to account for the split bands, but the available evidence seems to concur in the spin-orbit splitting of a Si–Au hybrid band (Barke *et al.* 2006; Nagao *et al.* 2006).

Given two half-filled bands in a quasi-1D system one could eventually expect a Peierls instability. Indeed, a metal–insulator transition was found in ARPES with a critical temperature of $T_c = 270$ K (Ahn *et al.* 2003). In parallel to the ARPES results, STM images showed a doubling of the periodicity along the step direction as expected for a Peierls transition.

In a theoretical analysis of the transition Sánchez-Portal and coworkers arrived at the following conclusions: The Peierls-type symmetry breaking is not primarily caused by a $2k_F$ interaction in the Au chains. Rather, the period doubling arises from the buckling of the step-edge atoms. Since each of these atoms has a dangling bond, one should expect a flat, half-filled band at E_F. This is an extremely unstable situation, the instability being removed by the introduction of a buckling with alternating "up" and "down" atoms. This leads to a period doubling and splits the flat band into an occupied and an unoccupied part with a gap in between. The buckling mechanism is thus similar to the asymmetric dimer reconstruction of the Si(100) surface. The period doubling is very similar to what is expected from a Peierls mechanism, except that the stabilization of the occupied band is not restricted to a small area in k space around the original Fermi level crossing. The DFT calculations indicate an almost constant binding energy increase throughout the SBZ (Riikonen and Sánchez-Portal 2007). According to the usual classification this could also be termed a strong-coupling CDW (Tosatti 1995; Aruga 2006). As the energy gain is delocalized in k space, the relevant interaction is localized in real space and may be considered as a local bond distortion rather than a Peierls distortion driven by a $2k_F$ interaction. In any case, following Riikonen *et al.* the metal–insulator transition in the Au-derived bands is not the driving mechanism, but a consequence of the step-edge buckling, which periodically distorts the Si–Au bond angle. The interpretation was challenged by Yeom *et al.* (2005), but a detailed theoretical analysis suggests that most of the experimental data are in fact consistent with the model (Riikonen and Sánchez-Portal 2007).

At room temperature, the DFT study in combination with a molecular-dynamics simulation predicts strong fluctuations of the step-edge atoms (Riikonen and Sánchez-Portal 2007). A period doubling seen even at room temperature (Krawiec *et al.* 2006) is attributed to the pinning of the fluctuations by defects. These results led to a central point of controversy often raised in connection with the observation of spontaneous symmetry breaking (Uhrberg and Balasubramanian 1998; Avila *et al.* 1999): Is the transition from the distorted low-temperature state to the seemingly undistorted high-T state really a Peierls transition, that is to say a displacive transition or just an order–disorder transition as for instance realized in the Ising model? There is a smooth cross-over between these types of phase transition (Rubtsov *et al.* 2000), therefore it is difficult to discriminate between the two mechanisms. Fluctuations are predicted in both cases, for the order–disorder transition around T_c, for a displacive phase transition in a temperature range below the mean-field temperature T_{MF}, with the correlation length diverging substantially below T_{MF} (Lee *et al.* 1973). There might actually be a way for unambiguously discriminating between the two alternatives by comparing the fluctuations parallel and perpendicular to the coupled chains. According to our knowledge, however, this has not yet been done for the quantum wires on Si, but for the system Si(111)-In, to be discussed further below, there are STM observations available that allow a comparison with fluctuations on the metal systems and shed some light on this controversy.

The Si(553)–Au system is even more difficult to understand. Experimental results show a spontaneous symmetry breaking with the appearance of a three-fold periodicity and a less well-defined appearance of a twofold periodicity as seen in Fig. 11.17 (Ahn *et al.* 2005; Snijders *et al.* 2006). The period doubling appears at higher temperature than the 3-fold modulation. The threefold periodicity seems to be associated with step-edge atom chains and the ×2 periodicity with terrace chains (Ahn *et al.* 2005). ARPES data reveal three quasi-1D bands, where two are nearly half-filled, while one has a filling between 1/4 and 1/3 (Crain *et al.* 2003; Crain and Himpsel 2006), see Fig. 11.18.

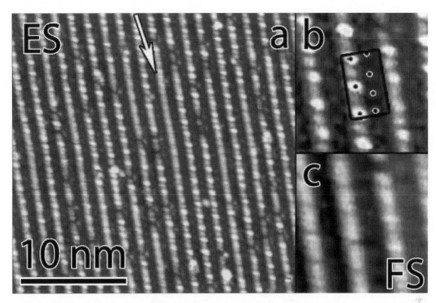

Fig. 11.17 STM image of the Si(553)-Au atomic chain surface. (a) Empty state STM image (40 K) showing distinct chain structures. (b) Close up with a 6 × 1 unit cell highlighted. Both the threefold and the double periodicity are clearly visible. (c) Filled state (FS) image of the same area. Reprinted with permission from Snijders *et al.* (2006) © APS.

It is natural to associate the period doubling in the terrace chains with a gap opening in the half-filled bands, while the 3-fold periodicity of the step-edge chains could be associated with the remaining band. The warping of the Fermi surface is larger for the 1/4–1/3-filled band consistent with the lower transition temperature for the 3-fold periodicity. A DFT study similar to the ones carried out for the Si(557) surface showed no tendency to form

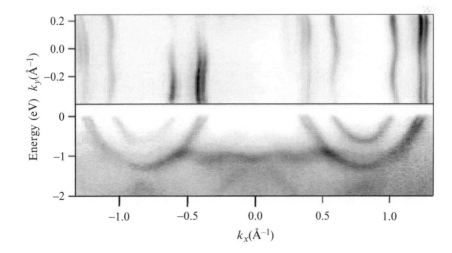

Fig. 11.18 ARPES results for the Si(553)-Au atomic chain surface. Upper panel: Fermi surfaces of three bands. Two of them are approximately half-filled, closely spaced and almost purely one-dimensional, while the third band is between 1/4 and 1/3 filled and produces a somewhat warped Fermi surface indicating a stronger perpendicular coupling. Lower panel: Dispersion of the corresponding bands. Reprinted with permission from Crain *et al.* (2003) © APS.

tilted dimers, nor was it able to reproduce the band structure seen in ARPES (Riikonen and Sánchez-Portal 2005a). From the experimental results it appears as if the new periodicities developed independently of each other in different structural subunits. This implies a weak, if any, coupling between them. Some STM images show phase slips (solitons) occurring close to each other, but with a slight shift in neighboring 3-fold modulated chains (Snijders *et al.* 2006). This too implies a rather weak coupling even between equivalent chains. The modulation is apparently somehow correlated, but not really coherent from chain to chain. This points to a rather localized interaction, again suggesting a strong-coupling-type CDW. As we shall see further below, one should also take into account the possibility of tip-induced contrast in a fluctuating system. Fluctuations are associated with a high susceptibility of the system to external perturbations, hence details of the contrast could depend in part on the tip-induced perturbation. For a better understanding of the mechanisms determining the Si(553)-Au system more information about possible fluctuations and the coherence length parallel and perpendicular to the chains would be helpful. Such information could be derived from diffraction measurements (Aruga 2006; Dona *et al.* 2007).

Next, we address the system Si(111)–In, which has some similarity to the Si(553)–Au system. There are three metallic bands seen in the room-temperature spectra, with filling degrees of 0.11, 0.38, and 0.50, respectively (1 corresponds to complete filling). A symmetry-breaking transition $(4 \times 1) \rightarrow (8 \times 2)$ is observed at around 120 K, which results in a period doubling both along and perpendicular to the 1D subunits. From ARPES results it is concluded that during the electronic metal–insulator (or semi-metal as argued by Jiandong Guo *et al.* 2005), transition a charge transfer takes place from the least-filled band to the intermediate band thus resulting in two half-filled bands that then open bandgaps of 340 and 40 meV below E_F, respectively (Ahn *et al.* 2004). On the basis of DFT calculations the phase transition was classified as an order–disorder transition from the ordered low-temperature state to a dynamically fluctuating high-T state (González *et al.* 2006). The apparent (4×1) symmetry of the high-T state is attributed to an averaging over different, degenerate (8×2) configurations, between which the system randomly fluctuates at room temperature. This interpretation is similar to the one promoted by Riikonen and Sánchez-Portal (2007) for the Si(557)–Au system. It is refuted, however, on account of ARPES and core-level XPS data (Ahn *et al.* 2007), which seem to exclude dynamical fluctuations in the metallic state at room temperature.

In a detailed STM study fluctuations in the Si(111)-In system were analyzed above and below the metal–insulator transition (Geunseop Lee *et al.* 2005). At 140 K fluctuating 1D stripes of the low-T phase were found embedded in the (4×1) phase. As T decreases, these stripes develop in elongated islands, still strongly fluctuating. Below $T = 115$ K the (8×2) phase dominates, but now variable 2D clusters of the (4×1) structure are embedded in the low-T structure. From their study the authors derive three characteristic temperatures: First, the mean-field temperature T_{MF}, which, based on eqn 11.31, is estimated as 430 K. Below this temperature, CDW fluctuations occur in single chains; second, a "cross-over" temperature T^*, where fluctuations spread across neighboring chains and thus 2D (8×2) domains are formed. Finally, a critical temperature $T_c \approx 120$ K, where the long-range ordered (8×2) structure becomes the dominant phase with only few embedded (4×1) islands. This scenario is strikingly similar to the one found in the Br/Pt(110) system discussed below, except that the order of T_{MF} and T^* is apparently reversed in the latter case. The value of $T_{MF} \approx 430$ K for Si(111)–In is a mystery, however, since it is estimated from a CDW gap of 150 meV (Geunseop Lee *et al.* 2005), while later ARPES data give a lower limit for one of the gaps of 340 meV below E_F (Ahn *et al.* 2004). This should result in a much higher T_{MF} of around 1100 K. In this case it would be difficult to understand that fluctuations prevail down to 120 K. Furthermore, it is somewhat disquieting that the *ab-initio* calculations do not converge with the experimental results. This could be interpreted as an indication of strong correlation in the nanostructured Si surfaces, which is not properly taken into account by the DFT calculations. On the other hand, the ARPES spectra exhibit no sign of anomalous many-body effects, in contrast to the results on metal surfaces discussed below, where one would expect to see a reduced amount of correlation.

A possible way out can be found along the lines proposed in the paper by (Riikonen and Sánchez-Portal 2007). They point out that the bands seen in photoemission are very likely not the bands that dominate the energy balance. In many cases the latter bands seem to be purely Si derived. They often have little dispersion and show significant shifts during the phase transition throughout the SBZ. This has two consequences: First, the phase transitions tend to be of the localized strong-coupling CDW type, rather than the typical weak-coupling Peierls transition. Second, correlation may be strong in these electronic states. In fact, the many-body response could be the reason that their intensity in photoemission is hidden in the background, while the metal–Si hybrid bands retain their quasi-particle character and therefore dominate in the spectra. The coupling between the two different electronic systems is mediated by the strong lattice deformation that also affects the Si–metal chain geometry. Due to the strong-coupling character, the phase transition may be driven by lattice entropy rather than electronic entropy (Aruga 2006). In this scenario the metal chains are not providing the 1D states that drive a Peierls transition, but act more or less as catalysts for a Si reconstruction in which Si-derived orbitals, e.g. dangling-bond orbitals, provide the high density of states at the Fermi level, which is responsible for the instability (Crain and Himpsel 2006).

Interestingly, Yakovkin has pointed out that metal-chain structures on Si may not provide the ideal model systems to investigate quasi-1D behavior, because the metal orbitals are involved in strong covalent bonding with the Si substrate (Yakovkin 2004). As a consequence, the metal-related states are strongly localized and if they participate at all in a Peierls transition, then it is of the strong-coupling CDW type. In contrast, metal surface states or surface resonances can be dimensionally constrained on appropriately nanostructured surfaces and then also function as 1D electronic systems. Correlation may still play a prominent role despite the metallic substrate, if d-derived (Tamm) surface states are involved. We therefore turn in the following to case studies on metal surfaces.

11.5.2 Case study 2: Quasi-1D structures on metallic surfaces

Metal surfaces may seem unlikely candidates in order to search for correlation effects. However, electronic surface states are already dimensionally constrained and further restriction to one or zero dimensions is possible, as discussed in the previous section. Shockley surface states occurring on the low-index surfaces of the noble metals (Memmel 1998) can be laterally confined (Bertel and Lehmann 1998; Pennec *et al.* 2007), but so far none of the characteristic phase transitions discussed above has been observed in this context, perhaps due to the large dispersion of these states and the ensuing small contribution to the DOS at E_F. Exceptions are, for instance, Be and Bi, where surface states, albeit strongly dispersing, are almost solely responsible for the DOS at E_F. To our knowledge, a 1D confinement has not yet been tried on these surfaces. In contrast, d-derived (Tamm) surface states are associated in general with a much larger DOS. Tamm states are split off from d bands, so the transition metals with nearly filled d shell (e.g. Ni, Pd and Pt) are obvious candidates to look for surface states or resonances, which contribute significantly to the DOS at E_F and can eventually be confined within 1D nanostructures.

A surface resonance crossing E_F was found previously on the Pt(111) surface (Bertel 1995; Roos *et al.* 1995) and shown to have a significant effect on the surface chemistry of Pt(111). In contrast to the (111) surfaces, the (110) surfaces of the fcc metals are strongly anisotropic. They consist of close-packed atom chains, separated from each other by $\sqrt{2}$ times the nearest-neighbor distance. The close-packed chains may be considered as intrinsic 1D nanostructures, but the question is of course, whether the coupling through the substrate is not too strong to prevent 1D confinement. Pt(110) is a particular case, because here the surface spontaneously reorganizes, ejecting every second close-packed row in a so-called missing-row reconstruction (see Fig. 11.19(a)). This surface is therefore a natural starting point to look for 1D surface states and associated correlation effects. ARPES investigations indeed reveal the existence of quasi-1D surface resonances at E_F and a variety of anomalies indicating the presence of correlation effects even on this metallic substrate. However, we will not immediately discuss the details of photoemission from Pt(110), because this might be of interest primarily for the more specialized reader. Instead, we first address the phase diagram of Pt(110) and associated adsorbate structures. The clean Pt(110) surface is spontaneously reconstructed in a (1×2) missing-row reconstruction as mentioned before, but there are no other anomalies in the phase diagram nor in the response of the clean surface, which would indicate the presence of an electronic instability.

After adsorption of halogens, for example Br or Cl, the reconstruction of the surface is lifted. The required mass transport is thermally activated and can be suppressed at low temperature. In the following, we concentrate on the bromine adsorbate structures, because Cl and F are chemically too aggressive, causing erosion of the surface already at room temperature. A Br coverage of half a monolayer (one monolayer here corresponds to the atom density of the unreconstructed Pt(110)-(1×1) surface) leads to the formation of a

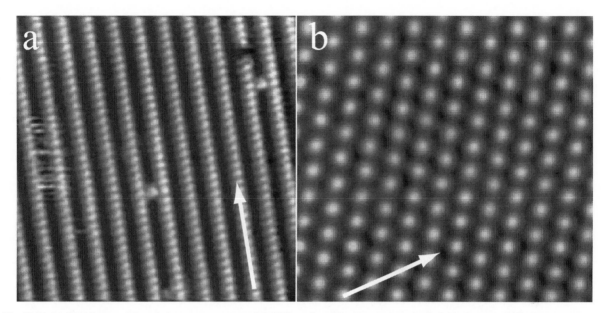

Fig. 11.19 (a) STM image of the missing-row reconstructed Pt(110) surface. The distance between individual rows is 0.78 nm, the nn distance along the rows is 0.28 nm. (b) STM image of the c(2 × 2)-Br/Pt(110) structure. Here, the missing-row structure is lifted. Arrows mark the direction of the close-packed Pt rows.

c(2 × 2)-Br/Pt(110) adsorbate structure as depicted in Fig. 11.19(b). This seems to be an entirely conventional adsorbate structure, where the quasi-hexagonal c(2 × 2) arrangement of the Br atoms results from a nearest-neighbor repulsion. The geometry does not suggest the existence of a strong anisotropy in the system. However, this impression turns out to be fallacious, as soon as the response of the system to a local perturbation is probed. Figure 11.20(a) shows the response of the nearly free-electron gas in a Shockley surface state on Ag(111) to local defects (ad-atom or missing atom) at the Ag(111) surface (image by courtesy of Omicron GmbH). The defect causes a concentric, circular standing-wave pattern in the surface-state charge density, which is reflected in the STM image. The system's response is entirely isotropic. Figure 11.20(b), in contrast, shows the response of the c(2 × 2)-Br/Pt(110) surface to local defects. Here, a plane-wave pattern is observed around the defects with a wave vector $k = 2\pi/3a$ along the $[1\bar{1}0]$ direction, which is the direction of the close-packed atom rows of the substrate surface (x-direction). The appearance of a plane wave with just one unique wave vector is the hallmark of a 1D response. It is not strictly 1D, as the finite width of the wave pattern indicates a finite interchain coupling. In the absence of such a coupling one would expect just the particular row to respond to the perturbation, in which the perturbing defect is located. Thus, the pattern shown in Fig. 11.20b reveals a quasi-1D character of the system.

One should note that the contrast mechanism in Fig. 11.20(b) is different from that in Fig. 11.20(a). Figure 11.20(b) is a topographical image. The protrusions imaged as bright spots correspond to Br atoms. The response to the

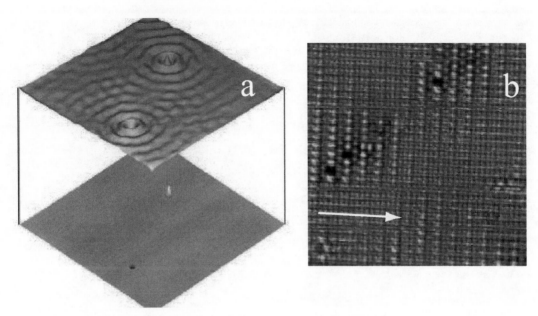

Fig. 11.20 (a) Bottom: Topographical STM image of a Ag(111) surface with one ad-atom and one missing atom. Top: Standing waves produced by the scattering of the Shockley surface state at the defects. The perturbation propagates isotropically (courtesy of Omicron GmbH). (b) Topographical STM image of defects on a c(2 × 2)-Br/Pt(110) surface. The perturbation propagates as a plane wave rather than a circular wave (Swamy *et al.* 2001).

defect shown in Fig. 11.20(b) therefore consists in a local rearrangement of the Br atoms into an adsorbate structure with a threefold periodicity. The domains appear to have a (3 × 1) structure with two Br atoms per unit cell (a corner atom and, with different brightness, a central atom). In between the perturbed domains the c(2 × 2) order is partially replaced by a (1 × 1) pattern. The local coverage derived from a (3 × 1) structure is $\Theta_{Br} = 0.67$ and for (1 × 1) it is $\Theta_{Br} = 1$. According to DFT calculations the energy for these higher packing densities is considerably less favorable than the c(2 × 2) structure. Actually, for a (1 × 1) packing, repulsive energies are prohibitively large. Therefore, the observations raise two questions: Why do the Br atoms appear with such a different brightness in the STM pattern and how is it possible that a single defect induces an energetically much less favorable structure in a finite area? The answer is: It doesn't. The Br atoms fluctuate between different adsorbate sites and the structures observed are just time averages of the Br distribution. Sites with high Br occupation probability appear bright, others with smaller probability less bright and only sites that are never occupied remain dark. This explains both the continuously varying contrast of the Br atoms, and the seemingly high local coverages. Of course, STM images such as the one shown in Fig. 11.20(b) suggest the presence of fluctuations, but do not prove them. Further experiments to be discussed below, however, provide unequivocal evidence for the dynamic character of the Br layer. This leads to the next question: Why and how does a defect induce fluctuations in the Br layer next to it?

An indirect answer is provided by studying the structure with threefold periodicity in more detail. This can be achieved by enlarging the Br coverage to $\Theta_{Br} = 0.67$ and carrying out a dynamical LEED structure analysis (Deisl *et al.* 2004; see Fig. 11.21). It turns out that the Br corner atoms sit in short-bridge sites on the close-packed rows, while the centered atoms sit in long-bridge sites between the rows. Remarkably, this structure is associated with a strong buckling (0.22 Å) of the Pt substrate atoms: Every third Pt atom along the rows is outward relaxed, while the other two move inward. Thus, the threefold-periodic structure arises either because the Br rearranges into a (3×1) short-bridge/long-bridge structure, which in turn leads to a substrate buckling, or because the substrate develops a periodic buckling, which in turn pushes the Br ad-layer into the threefold pattern. The fact that the threefold pattern around defects is associated with a fluctuating Br structure is just one of several clues indicating that the substrate buckling is the primary cause and destabilizes the $c(2 \times 2)$ Br structure. Several other experimental findings, partly reported below, substantiate this view and yield the following scenario: The $c(2 \times 2)$-Br/Pt(110) surface is very close to a Peierls transition. In fact, the $c(2 \times 2)$ structure is not static, but fluctuates throughout a wide temperature range. In the presence of a defect, the fluctuations are pinned and therefore become visible in the STM image. The buckling with threefold periodicity is actually a soft or frozen phonon, respectively (Rayleigh phonon). Lowering the temperature pushes the system towards the Peierls phase. Anderson and coworkers (Lee *et al.* 1973) have shown that in a quasi-1D system the transition into the CDW phase is delayed to much below the mean-field temperature, but in the present case an additional smearing seems to take place due to the incommensurability of the CDW: The natural periodicity in the present system is between two and three. Hence, as the $2k_F$ interaction increases with falling temperature, the system tends to switch into an incommensurate CDW, but this is energetically less favorable than to rearrange into alternating commensurate domains with twofold periodicity, i.e. a (2×1) structure, and threefold periodicity, i.e. a (3×2) structure. Both correspond to a local coverage of $\Theta_{Br} = 0.5$. Of course this "lattice pinning" of commensurate domains requires the proliferation of domain walls or "solitons". The walls are rather mobile and therefore support the prevalence of fluctuations over a large temperature range.

In the following we report experimental observations supporting this scenario. Figure 11.19(b) shows an STM image of the $c(2 \times 2)$-Br/Pt(110) structure at room temperature. The same structure at 50 K completely loses its long-range order and decays into a glassy state with local $c(2 \times 2)$, (2×1), and (3×2) structural elements, as shown in Fig. 11.22. DFT calculations indicate that at 0 K the three structures differ by only 14 meV/atom in energy (Deisl *et al.* 2004). One cannot interpret, however, the 50-K structure as a coexistence of three different phases, because this would violate Gibbs's phase rule: The number of degrees of freedom equals the number of components (here a single component: Br) minus the number of phases plus 2. Three phases would result in zero degrees of freedom, therefore the coexistence could be observed just at one single temperature. This is not the case. The conflict is resolved, if the pattern shown in Fig. 11.22 is interpreted as the coexistence of two "phases", namely the $c(2 \times 2)$ structure and an incommensurate CDW, which in turn readjusts into a commensurate domain structure with domain walls.

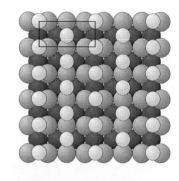

Fig. 11.21 Ball model of the (3×1)-Br/Pt(110) structure derived from quantitative low-energy electron diffraction (IV-LEED) and *ab-initio* calculations. Note the strong buckling of the top Pt layer ($b_1 = 0.22$ Å). The buckling in the second Pt layer is about 0.1 Å, in the third 0.01 Å (Deisl *et al.* 2004).

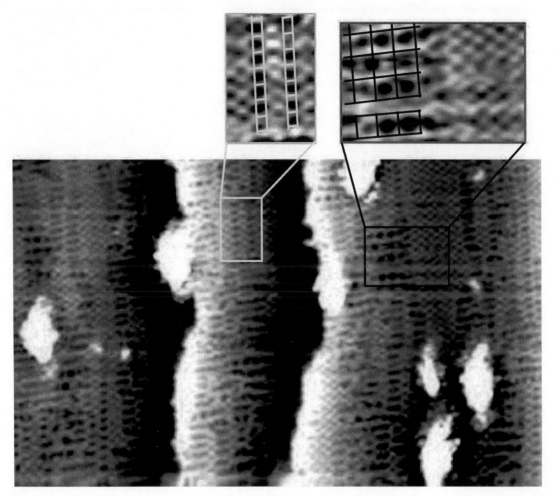

Fig. 11.22 STM image of a Br/Pt(110) surface taken at 50 K. The global c(2 × 2) structure observed at room temperature decays reversibly into a glassy phase with local c(2 × 2), (2 × 1) (upper left panel) and (3 × 2) structure (upper right panel) (Doná *et al.* 2007).

This interpretation is actually consistent with the phase separation predicted to occur in correlated systems with competing phases (Le Daeron and Aubry 1983; Nussinov *et al.* 1999).

In Section 11.2 we discussed the generalized susceptibility, which diverges close to a phase transition. As a consequence, the system is extremely sensitive to external perturbations. This effect is illustrated in Fig. 11.23. The starting point here was again a c(2 × 2) preparation at room temperature. At 150 K the STM topographical contrast can be reversibly switched between a c(2 × 2) and a pattern with threefold periodicity as the tunnelling parameters are changed. Thus, the STM observations are all consistent with the scenario outlined above. Additional support stems from low-energy electron diffraction (LEED) spot-profile analysis.

The theoretical background of diffraction spot profile analysis is as follows: Close to a phase transition the structure function $S(q, T)$ consists of three

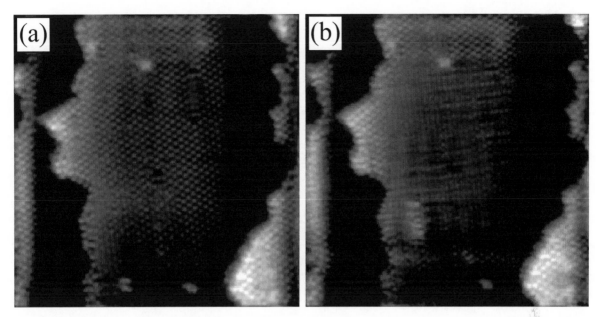

Fig. 11.23 (a) STM image of a c(2 × 2)-Br/Pt(110) preparation recorded at 150 K with a bias voltage of 1.2 V. (b) The same area was scanned again, but scan parameters were switched after the first quarter of the scan to low bias voltage, thus reducing the tip–sample distance, and reset to the initial values shortly before completing the scan. In the areas recorded with low bias voltage, a threefold periodicity is clearly observable, while the position of individual Br atoms can no longer be discerned. This demonstrates the susceptibility of the surface to tip-induced perturbations.

contributions:

$$S(q, T) = I_0(T)\,\delta(q - q_0) + \frac{\chi(T)}{1 + \xi^2(T) \cdot (q - q_0)^2} + S_{\text{bg}}, \qquad (11.36)$$

where q_0 is the position of the diffraction spots in reciprocal space, $\delta(q - q_0)$ is a delta function arising from the coherent scattering on the long-range ordered array of scattering centers. In practice, the delta function is broadened into a Gaussian due to the instrument function, i.e. the finite coherence length of the primary electron beam (the coherence length ranges from ~ 10 nm in a conventional LEED instrument to ~ 1000 nm in a special spot-profile analysis LEED (SPA-LEED) system). The intensity of the first term is thus a measure of the long-range order parameter m. The second term yields a Lorentzian contribution to the spot profile and stems from the fluctuating short-range order. Its intensity measures the magnitude of the fluctuations $\chi(T) = \left(\langle m^2(r) \rangle - \langle m(r) \rangle^2\right)_T$, while $\xi(T)$ measures the correlation length as a function of temperature. S_{bg} is an asymmetric background contribution.

Figure 11.24 shows the LEED pattern of the c(2 × 2)-Br/Pt(110) surface. Taking the cross-section through a half-order spot, whose intensity derives essentially from the adsorbate, in the direction perpendicular to the close-packed rows ([001] or y-direction) and decomposition of the profile into a Gaussian and a Lorentzian contribution at various temperatures yields the temperature dependence of the long-range order parameter, the fluctuations and the coherence length shown in Fig. 11.25. At about 370 K the long-range

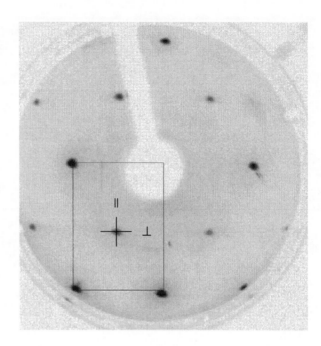

Fig. 11.24 Low-energy electron diffraction (LEED) pattern of the c(2×2)-Br/Pt(110) surface at room temperature. The rectangle delineates the (1×1) unit cell in reciprocal space, the cross marks a half-order spot and the directions parallel and perpendicular to the close-packed rows.

order parameter drops, the system starts to fluctuate and, as the temperature is further increased, the ad-layer disorders totally. The long-range order parameter does not fall completely to zero, because steps in the system act as pinning centers preserving traces of the c(2×2) structure at even higher temperatures. This is the normal behavior expected for an order–disorder transition in an adsorbate system. Note that the integer order spots, which arise mainly from the Pt substrate, do not show any change as the system goes through the transition.

A different pattern emerges, as the spot profiles are examined in the direction parallel to the close-packed rows ([1$\bar{1}$0] or x-direction). Although both Gaussian and Lorentzian contribution rise in a similar way as before, when the temperature falls below 370 K, the Lorentzian contribution does not die out, as the temperature is lowered further. It increases down to 270 K and below this temperature the system starts to lose again both the long-range and the short-range order. This signals the onset of the phase separation described above.

From the spot-profile analysis one can therefore conclude that in the temperature range from $\sim$ 270 K to 370 K the system exhibits a c(2×2) long-range order, albeit with fluctuations in the close-packed row direction. Perpendicular to the rows there appears perfect long-range order, hence the fluctuations are coherent from row to row. Only above 370 K is this coherence destroyed and the system totally disorders. Most significantly, integer-order spots also show a Lorentzian contribution in x-direction spot profiles. This clearly indicates that the substrate participates in the fluctuations.

Summarizing the experimental observations we recall that the c(2×2)-Br/Pt(110) surface exhibits critical fluctuations over a wide temperature range.

Towards the high-temperature side the system goes through an order–disorder transition leaving it totally disordered. Towards the low-temperature side the system phase separates into the c(2 × 2) phase and the incommensurate CDW phase with local commensurate domains of two- and threefold periodicity, respectively. These sequence of transitions fits perfectly into part of a phase diagram of quasi-1D systems discussed previously in the literature (Imada *et al.* 1998) and shown in Fig. 11.26. The mean-field temperature T_{MF} is in this case significantly below the coherence temperature T_{coh}, which is defined by

$$T_{coh} = \frac{t_\perp}{k_B}. \qquad (11.37)$$

Here, $t_\perp$ is the interchain coupling and k_B the Boltzmann constant. In contrast, the Si(111)-In system discussed in the previous case study shows the reversed order of transitions. Upon cooling, individual 1D chains appear with broken symmetry. Only if the temperature is lowered further, broken-symmetry islands with finite size in the lateral direction start to develop. Hence, in this system, T_{MF} is larger than T_{coh} (which is to be identified with T^*, the "cross-over" temperature discussed by (Geunseop Lee *et al.* 2005)).

So far, we have discussed the phase diagram of the Br/Pt(110) system in terms of a transition into an incommensurate CDW, but we have not yet examined whether the photoemission results support a CDW mechanism at all. It has to be noted right away that we are dealing with a fluctuating system, as pointed out above. Hence, the interpretation of the photoemission spectra is not straightforward. Nevertheless, they yield important information on the system and the presence of correlation effects. Figures 11.27(a) and (b) show photoemission spectra of the c(2 × 2)-Br/Pt(110) phase measured along the lines indicated in the schematic diagram of the surface Brillouin zone (Fig. 11.27(c)). At the $\bar{S}$ point an intense photoemission feature is observed close to E_F, which can be identified as a surface resonance for a variety of

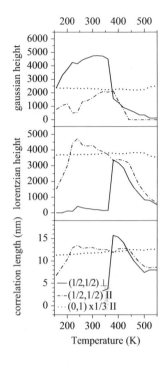

Fig. 11.25 Temperature dependence of the LEED spot profiles for the c(2 × 2)-Br/Pt(110) structure. The Gaussian height is a measure of the long-range order parameter, while the Lorentzian height characterizes the intensity of the fluctuations. The correlation length is derived from the width of the Lorentzian contribution. The profile of the (0,1) spot is dominated by scattering from substrate atoms.

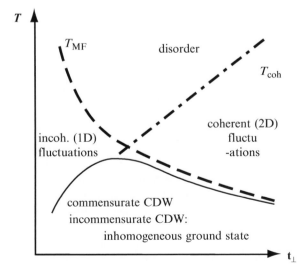

Fig. 11.26 Schematic phase diagram for CDW systems: Above both the mean-field transition temperature T_{MF} and the coherence temperature T_{coh} the system is disordered. For small transverse coupling $t_\perp$, T_{MF} is crossed first as T is lowered. Charge-density fluctuations develop within the chains, but there is no interchain phase correlation. For larger $t_\perp$, T_{coh} is crossed first and an ordered state appears. As T_{MF} is approached, CDW fluctuations develop here as well, but in this case the correlation length across the rows is finite. Below both T_{MF} and T_{coh} the Peierls state is entered. In the case of a commensurate CDW, a homogeneous Peierls phase is obtained, while an incommensurate CDW results in a "phase-separated," glassy state.

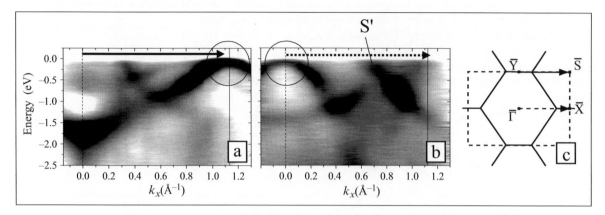

Fig. 11.27 Photoemission from c(2 × 2)-Br/Pt(110). (a) ARPES intensities measured along $\bar{Y}\bar{S}$. The circle denotes the surface resonance at $\bar{S}$. (b) ARPES intensities measured along $\bar{\Gamma}\bar{X}$. The circle denotes the surface resonance, which in the c(2 × 2) SBZ is backfolded onto $\bar{\Gamma}$. Note the band S' crossing the Fermi level at ∼ 0.7 Å^{-1}. (c) Solid lines: The SBZ of the c(2 × 2) structure; broken line: The (1 × 1) SBZ.

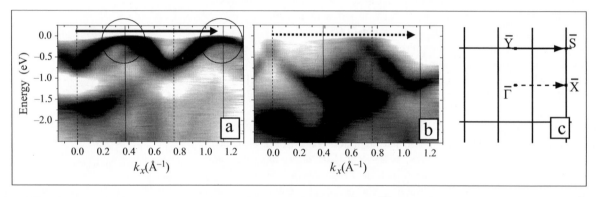

Fig. 11.28 Photoemission from (3 × 1)-Br/Pt(110). (a) ARPES intensities measured along $\bar{Y}\bar{S}$. The circle denotes the surface resonance at $\bar{S}$ and its backfolded replica at the first SBZ boundary. (b) ARPES intensities measured along $\bar{\Gamma}\bar{X}$. Solid lines are SBZ boundaries, broken lines mark the zone centres. (c) SBZ of the (3 × 1) structure.

reasons discussed elsewhere (Minca *et al.* 2007; Doná *et al.* 2008). Due to the symmetry of the SBZ, the $\bar{\Gamma}$ and the $\bar{S}$ point are equivalent, hence the feature is backfolded to $\bar{\Gamma}$ (Fig. 11.27(b)). A second band starting at about 1.5 eV below E_F at $\bar{S}$ is seen to disperse upwards as $\bar{Y}$ is approached. It crosses the Fermi level at $k_\parallel \approx 0.3$ Å^{-1}. The band is backfolded, too, along the $\bar{\Gamma}\bar{X}$ line, yielding a band that starts at 1.5 eV below E_F at $\bar{\Gamma}$ and would cross E_F at ∼ 0.82 Å^{-1}. This crossing, however, is obscured by another structure labelled S' that comes down through E_F at ∼ 0.7 Å^{-1} and drops to –1.5 eV at $\bar{X}$. The latter structure has no correspondence on the $\bar{Y}\bar{S}$ line. On the other hand, it is not readily explained as a bulk transition. A very similar, only sharper, structure is found in the photoemission spectra of the (3 × 1)-Br/Pt(110) surface (Fig. 11.28). In our first studies on this system we assumed that the driving force for the CDW formation was the removal of the Fermi surface of feature S' (Swamy *et al.* 2001), i.e. the backfolding of this band and the opening of a bandgap (Peierls gap) at 0.7 Å^{-1}. This is in between $1/3G_\parallel$(0.74 Å^{-1}) and $1/4G_\parallel$(0.55 Å^{-1}),

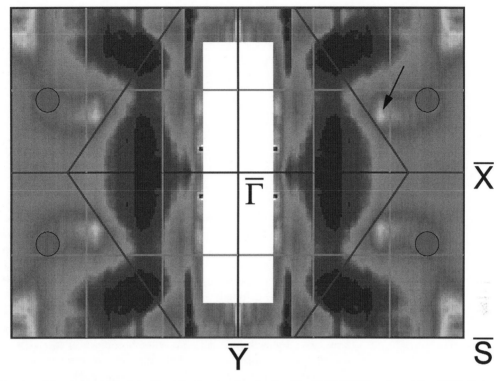

Fig. 11.29 Mapping of the photoemission intensity at the Fermi surface for the c(2 × 2)-Br-Pt(110) surface at $T = 300\,$K. Dark corresponds to low, white to high intensity (photon energy $h\nu = 21.22\,$eV). Gray circles mask an artefact (reflection of the capillary outlet into the analyzer). The black arrow highlights an area of extremely high photoemission intensity at E_F. Note that this "hot spot" occurs very close to the point $(2\pi/3, \pi/2)$ which is the crossing point of the gray lines.

where $G_\parallel$ is a reciprocal lattice vector of the SBZ of the clean Pt(110) surface in the x-direction. The Fermi-level crossing is consistent with the appearance of Br structures with 3-fold and 2-fold periodicity at low temperatures, as described above. Meanwhile, an ARPES survey revealed the existence of a saddle point at E_F in the interior of the SBZ close to $(2\pi/3, 0.5\pi)$ (in this notation $\bar{S}$ is the (π, π) point), as shown in Fig. 11.29. One of the connecting vectors, namely $q_\perp = (0, \pi)$, is very close to the reciprocal lattice vector $G'_\perp$ of the Pt(110)-(1 × 2) surface, while the orthogonal one, $q_\parallel = (1.42\pi, 0)$ is less than 10% off from $2/3G_\parallel$. As mentioned in Section 11.3, a saddle point yields a diverging DOS in 2D and therefore may induce a CDW as well (Rice and Scott 1975). On the basis of the experimental data accumulated so far, we therefore favor such a saddle-point mechanism as the origin of the Peierls instability. In summary, the electronic structure does not conform to the naïve expectation of a Fermi surface nesting with $q = 2k_F = 1/3G_\parallel$. Instead, a saddle-point mechanism seems to apply.

The ARPES spectra also show a very unusual temperature dependence. For both the $\bar{S}$ point and the saddle point at $(0.71\pi, 0.5\pi)$ we observe a quasi-particle peak at E_F, which is rapidly quenched with rising temperature. The T dependence is much stronger than expected on the basis of the Debye–Waller model and the surface Debye temperature (110 K) of Pt(110) (Menzel *et al.*

2002, 2005). Quantitative examination suggests that the quasi-particle intensity loss is linked to the order–disorder transition observed in LEED, i.e. the peak is destroyed as the interchain coherence is lost (see also the phase diagram in Fig. 11.26). The loss of coherence between the rows can be viewed as a dimensional transition from 2D to 1D. Thus, the present observation can be put in the more general context of quasi-particle peak quenching upon dimensional cross-over, which is a feature generally observed in strongly correlated compounds. Examples are given in Fig. 11.30.

A further indication of correlation effects is a distinct renormalization of the effective mass observed in the dispersion of some bands (Menzel *et al.* 2005). The effect is shown schematically in Fig. 11.31(a) and results from the scattering of electrons with bosonic excitations, for instance phonons. For strongly bound electrons electron–phonon scattering is negligible, because the phonon energies are not sufficient to promote the scattered electrons into unoccupied states. However, electrons close to E_F have final states available within a narrow energy range and therefore can exchange energy and momentum with phonons. One may visualize these electrons as being associated with a surrounding phonon cloud. Both together constitute a quasi-particle, whose effective mass is distinctly larger than that of a bare electron. The onset of the mass renormalization takes place within about a typical phonon energy (usually of the order of 50 meV or less) from E_F. However, scattering may occur not only with phonons, but other bosonic excitations as well. In the present case we observe a threshold for mass renormalization of the order of 200 meV, which practically excludes electron–phonon scattering as the origin (see Fig. 11.31(b)). It appears more likely that interaction with plasmon excitations, as recently postulated for similar ARPES structures on graphene (Bostwick *et al.* 2007), is the cause.

In summary, the Peierls transition is masked by strong fluctuations extending over a large temperature range. In addition, the CDW seems to be incommensurate, but decays into commensurate domains. As the domain walls do not order (at least under the conditions applied so far) a glassy state without long-range order appears at low temperature. Photoemission results exhibit a number of features peculiar to correlated systems, which indicates that correlation becomes important in quasi-1D nanostructures even on metallic substrates.

Another case study has been carried out by Aruga and coworkers on Cu(001) covered with submonolayers of In and Sb (Aruga 2006). Although the undistorted phase is not quasi-1D in this case, Fermi-surface nesting is observed in ARPES and depending on coverage there are displacive transitions that bear the hallmarks of a Peierls mechanism. In the present discussion we focus on the In/Cu(001) system. At $\Theta_{In} = 0.63(5/8)$ a transition between a low-temperature (LT) $\left(2\sqrt{2} \times 2\sqrt{2}\right) R45°$ structure and a high-temperature (HT) (2×2) structure is observed. The LT structure contains 5 In atoms per unit cell, while the observed HT structure is inconsistent with a coverage of 5/8. It is therefore concluded that the (2×2) structure is just a temporal average of a fluctuating structure. In a surface X-ray diffraction study (XRD) a transition temperature of 345 K was observed (Hatta *et al.* 2005). Around the transition

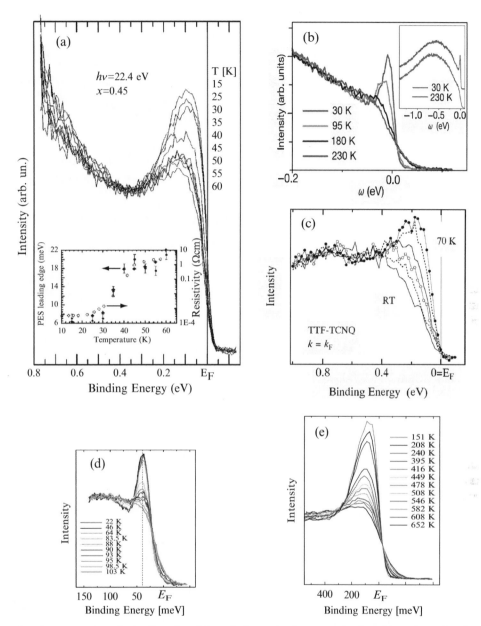

Fig. 11.30 Anomalous temperature dependence of quasi-particle peaks in correlated systems: Quasi-particle peak as a function of temperature (a) at the Mott–Hubbard metal–insulator transition of $NbS_{2-x}Se_x$ (Matsuura *et al.* 1998, copyright APS). (b) At a dimensional cross-over in the layered metallic system $(Bi_{0.5}Pb_{0.5})_2Ba_3Co_2O_y$ (Valla *et al.* 2002, copyright Macmillan Publishers Ltd.); (c) in a fluctuating CDW system, i.e. TTF-TCNQ (Zwick *et al.* 1998, copyright APS); (d) at the transition temperature of the optimally doped high-T_c superconductor $Bi_2Sr_2CaCu_2O_8$ (Fedorov *et al.* 1999, copyright APS); (e) close to an order–disorder transition in the surface system c(2 × 2)-Br/Pt(110) (Menzel *et al.* 2002, copyright APS).

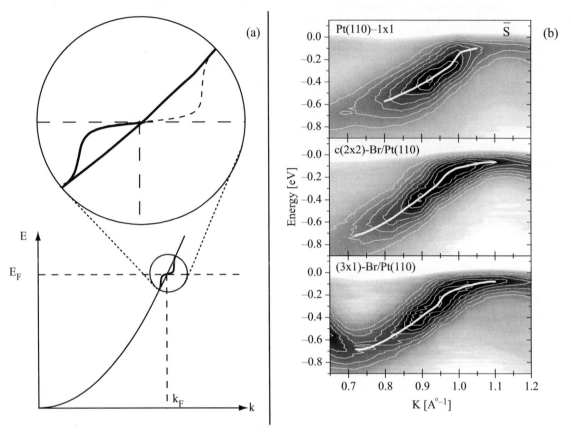

Fig. 11.31 (a) Renormalization of the effective mass (i.e. slope of the band) close to the Fermi energy due to electron–phonon (boson) interactions. A kink appears in the dispersion, but the Fermi crossing point is not changed. (b) Experimentally observed kinks in the dispersion of the surface resonance close to the $\bar{S}$ point on various Pt(110) and Br/Pt(110) surface structures.

temperature the peak profiles showed the typical characteristics of an order–disorder transition as discussed above for the LEED profiles in Br/Pt(110). An ARPES study of a surface resonance band showed the backfolding and opening of a bandgap expected for a Peierls mechanism, albeit with a critical temperature some 60 K above the order–disorder transition associated with the structural change. Extrapolation of the electronic energy gap to 0 K yielded a value of 860 meV. Furthermore, ARPES data show that the CDW phase is incommensurate just as in the Br/Pt case.

It is interesting to note the similarities and differences between the Br/Pt(110) and the In/Cu(001) system. In both cases the CDW phase is incommensurate, but for In/Cu(001) extended domains are formed. There are two possible reasons: First, the deviation from commensurability might by significantly smaller, thus allowing the formation of more extended domains. Second, perhaps more importantly, the system is 2D rather than quasi-1D, which enhances the commensurability pinning significantly. Thus, the "phase separation" observed in Br/Pt(110) is suppressed in In/Cu(001). Also, the bandgap in In/Cu(001) is apparently much larger, which certainly stabilizes the commensurate LT phase. The width of the relevant bandgap in Br/Pt(110) is not yet known due to the complex band topology, but Aruga has pointed out that the CDW stability with respect to thermal electron excitation in metallic surface systems is determined by the difference between the lower band and the Fermi energy rather than the total gap width, because the electrons can be thermally excited to unoccupied bulk states (Aruga 2006). This "effective gap" is certainly less than 100 meV in Br/Pt(110), while it amounts to ~ 600 meV for In/Cu(001).

A further interesting difference between the two systems refers to the position of the order–disorder transition with respect to the Peierls temperature. While the latter in Br/Pt(110) is obviously much lower – note that the phase separation occurs below 100 K – the reverse seems to apply to the In/Cu(001) system: Upon heating, the LT phase is not converted first to an undistorted phase and then disordered, but it directly disorders into a fluctuating phase without long-range order. This can be easily understood on the basis of the large effective energy gap, which causes the LT structure to be destroyed on account of the lattice entropy rather than the electronic entropy (Aruga, 2006). In essence, one can state that the Peierls transitions tend to be masked in real systems by fluctuations on the one hand and order–disorder transitions on the other hand. Particularly in low-D systems fluctuations dominate the phase diagram over a wide temperature range. This has to be borne in mind when interpreting data, which are often obtained from space- and time-integrating measurements. Finally, we mention that the ARPES data show no sign of mass renormalization close to E_F in In/Cu(110) indicating that correlation effects play a minor role in the latter system. This is not surprising in view of the fact that the relevant surface resonance bands are strongly dispersing delocalized sp-derived bands and that the system is 2D rather than quasi-1D in character.

There are several other examples of quasi-1D nanostructures on metal surfaces. A very interesting class of examples are metal, in particular alkali-metal and alkaline-earth metal, chains on the furrowed Mo(112) and W(112) surfaces. These substrates exhibit a rather low DOS at E_F, thus facilitating the study of the adsorbate-induced bands near the Fermi level. A review of these systems has been presented by Yakovkin (2004). Interestingly, at low coverage ad-atom chains form perpendicular to the furrows (Katrich *et al.* 1994). It is believed that the chain–chain distance correlates with k_F of the substrates according to an RKKY-type interaction as mentioned in the section on nanostructure formation. If this idea is correct, then the position of the chains is associated with the charge-density modulation in the substrate surface resonances. Consequently, a periodic chain spacing implies a periodic charge-density modulation, i.e. a CDW. Although it is likely that this CDW is associated with a periodic lattice distortion in the Mo surface layer and hence conforms to the definition of a Peierls phase, detailed structural information is still missing for this system. A "non-metal to metal" transition is observed for instance in the Mg/Mo(112) system by ARPES (Zhang *et al.* 1994, 1995) as the Mg–Mg distance is reduced with increasing coverage. The transition can be understood on the basis of a metallic band formation within the ad-layer. The transition seems to be associated with the commensurate–incommensurate transition of the compressed ad-layer. It would be especially interesting to see whether this transition is associated with a change in the Mo surface geometry. The role of electron–electron correlation in these systems is not yet clear, although the relationship between screening parameter and correlation length derived from ARPES and resonant photoemission experiments agrees quite well with predictions derived from the Mott–Hubbard model (Zhang *et al.* 1995). The non-metal–metal transition can also be explained solely on the basis of distance-dependent band dispersion and the decoupling from the substrate as commensurability is lost. However, it has to be emphasized here, that more structural information is needed and particular attention has to be paid to fluctuations in such systems, because the latter tend to obscure the structural changes in conventional methods. Both, W and Mo have a propensity for Peierls-type interactions (Plummer *et al.* 2003) and further investigations of these systems are certainly worthwhile.

11.6   Conclusions

In summary, quasi-1D systems on both semiconductor and metal surfaces exhibit a rather complex phase diagram with apparently electron-driven phase transitions as expected for correlated systems. In particular, spontaneous symmetry breaking is observed for many of these systems at low temperature. A rather fundamental difficulty arises from the fact that the corresponding high-temperature phase is rarely a simple, long-range ordered undistorted phase as naively expected for a textbook Peierls transition. Instead, fluctuations are observed over a wide temperature range. In many cases it is therefore inferred that the observed transition is of the order–disorder rather than the Peierls type. A clear distinction is sometimes difficult, as there is a smooth cross-over between the two types of transition, but there are cases where the critical temperature for symmetry breaking is clearly separated from the order–disorder temperature, as in the case of Si(111)-In (Geunseop Lee *et al.* 2005) and Br/Pt(111) (Doná *et al.* 2007). A further problem with the quasi-1D systems on semiconductors is the discrepancy between experimental and theoretical studies. While the former yield, for instance, ARPES spectra, which support a weak-coupling CDW mechanism for the symmetry-breaking phase transition, the theoretical results point rather to a local, strong-coupling-type mechanism. The discrepancy may be attributable to the larger correlation in semiconductor as opposed to metal substrates. On the one hand, correlation leads to a strong many-body response, which suppresses quasi-particle peaks and therefore renders some electronic bands almost unobservable in ARPES. On the other hand, DFT calculations may miss essential details of the electronic structure due to the inadequate treatment of correlation. One has to note, however, that the bands observed in ARPES do not show clear signs of correlation effects, such as effective-mass renormalization or anomalous temperature dependence of quasi-particle peaks, in contrast to the ARPES results on Br/Pt(110).

While ARPES results on 1D structured metal surfaces clearly show the fingerprints of electronic correlation, the details are not yet fully understood.

Nevertheless, the application of local probes (STM and STS) in combination with reciprocal space probes (ARPES, LEED and XRD) has already provided a wealth of interesting insights into the details of the phase transitions and in particular the character of the dynamic fluctuations governing the behavior in the temperature range at and above the critical temperature. However, a quantitative description allowing, for instance, to extract the g values that characterize the quasi-particle interactions in the system is not yet in sight. Furthermore, what is missing almost completely at present is information on the spin structure of these systems. Magnetic-ordering phenomena are very likely, but more difficult to measure than the charge ordering. Considering the possibility of tuning magnetic properties and the fascinating role of fluctuations in shaping materials properties from unconventional superconductivity to magnetism (Monthoux *et al.* 2007) it is clear that the study of 1D nanostructured systems is a topic of foremost interest in modern materials science.

Acknowledgments

Financial support by the Austrian Science Fund (FWF) is gratefully acknowledged. Furthermore, we thank Mikhail Baranov and Erich Vass for useful discussions.

References

Ahn, J.R., Yeom, H.W., Yoon, H.S., Lyo, I.-W. *Phys. Rev. Lett.* **91**, 196403 (2003).

Ahn, J.R., Byun, J.H., Koh, H., Rotenberg, E., Kevan, S.D., Yeom, H.W., *Phys. Rev. Lett.* **93**, 106401 (2004).

Ahn, J.R., Kang, P.G., Ryang, K.D., Yeom, H.W., *Phys. Rev. Lett.* **95**, 196402 (2005).

Ahn, J.R., Byun, J.H., Kim, J.K., Yeom, H.W., *Phys. Rev. B* **75**, 033313 (2007).

Altmann, K.N., Crain, J.N., Kirakosian, A., Lin, J.-L., Petrovykh, D.Y., Himpsel, F.J., Losio, R., *Phys. Rev. B* **64**, 035406 (2001).

Anderson, P.W. *Phys. Rev.* **109**, 1492 (1958).

Aruga, T. *Surf. Sci. Rep.* **61**, 283 (2006).

Avila, J., Mascaraque, A., Michel, E.G., Asensio, M.C., LeLay, G., Ortega, J., Pérez, R., Flores, F. *Phys. Rev. Lett.* **82**, 442 (1999).

Bachmann, M., Gabl, M., Deisl, C., Memmel, N., Bertel, E. *Phys. Rev. B* **78**, 235410 (2008).

Bak, P. *Rep. Prog. Phys.* **45**, 587 (1982).

Barke, I., Fan Zheng, Rügheimer, T.K., Himpsel, F.J. *Phys. Rev. Lett.* **97**, 226405 (2006).

Barth, J.V., Brune, H., Ertl, G., Behm, R.J. *Phys. Rev. B* **42**, 9307 (1990).

Baumberger, F., Hengsberger, M., Muntwiler, M., Shi, M., Krempasky, J., Patthey, L., Osterwalder, J., Greber, T. *Phys. Rev. Lett.* **92**, 016803 (2004a).

Baumberger, F., Hengsberger, M., Muntwiler, M., Shi, M., Krempasky, J., Patthey, L., Osterwalder, J., Greber, T. *Phys. Rev. Lett.* **92**, 196805 (2004b).

Berge, K., Gerlach, A., Meister, G., Goldmann, A., Bertel, E. *Phys. Rev. B* **70**, 155303 (2004).

Berger, E., Valásek, P., von der Linden, W. *Phys. Rev. B* **52**, 4806 (1995).

Bertel, E. *Surf. Sci.* **331–333**, 1136 (1995).

Bertel, E., Lehmann, J. *Phys. Rev. Lett.* **80**, 1497 (1998).

Bode, M., Pascal, R., Wiesendanger, R. *Surf. Sci.* **344**, 185 (1995).

Bostwick, A., Ohta, T., Seyller, Th., Horn, K., Rotenberg, E. *Nature Physics* **3**, 36 (2007).

Bourbonnais, C. *J. Phys. IV France* **10**, 81 (2000).

Chado, I., Padovani, S., Scheurer, F., Bucher, J.P. *Appl. Surf. Sci.* **164**, 42 (2000).

Chen, W., Madhavan, V., Jamneala, T., Crommie, M.F. *Phys. Rev. Lett.* **80**, 1469 (1998).

Corso, M., Auwärter, W., Muntwiler, M., Tamai, A., Greber, Th., Osterwalder, J. *Science* **303**, 217 (2004).

Crain, J.N., Himpsel, F. *J. Appl. Phys. A* **82**, 431 (2006).

Crain, J.N., Kirakosian, A., Altmann, K.N., Bromberger, C., Erwin, S.C., McChesney, J.L., Lin, J.-L., Himpsel, F. *J. Phys. Rev. Lett.* **90**, 176805 (2003).

Deisl, C., Swamy, K., Memmel, N., Bertel, E., Franchini, C., Schneider, G., Redinger, J., Walter, S., Hammer, L., Heinz, K. *Phys. Rev. B* **69**, 195405 (2004).

Doná, E., Loerting, Th., Penner, S., Minca, M., Menzel, A., Bertel, E., Schoiswohl, J., Berkebile, St., Netzer, F.P., Zucca, R., Redinger, J. *Phys. Rev. Lett.* **98**, 186101 (2007).

Doná, E., Amann, P., Cordin, M., Menzel, A., Bertel, E. *Appl. Surf. Sci.* **254**, 4230 (2008).

Emery, V.J., Kivelson, S.A. *Phys. Rev. Lett.* **74**, 3253 (1995).

Fedorov, A.V., Valla, T., Johnson, P.D., Li, Q., Gu, G.D., Koshizuka, N. *Phys. Rev. Lett.* **82**, 2179 (1999).

Feng, D., Jin, G. *Introduction to Condensed Matter Physics* (World Scientific, Singapore, 2005).

Frost, F., Schindler, A., Bigl, F. *Phys. Rev. Lett.* **85**, 4116 (2000).

Geunseop Lee, Jiandong Guo, Plummer, E.W. *Phys. Rev. Lett.* **95**, 116103 (2005).

Giamarchi, Th. *Quantum Physics in One Dimension* (Oxford University Press, New York, 2004).

Gonzalez, C., Flores, F., Ortega, J. *Phys. Rev. Lett.* **96**, 136101 (2006).

Grüner, G. *Density Waves in Solids* (Perseus Publishing, Cambridge MA, 1994).

Harten, U., Lahee, A.M., Toennis, J.P., Wöll, Ch. *Phys. Rev. Lett.* **54**, 2619 (1985).

Hatta, S., Okuyama, H., Aruga, T., Sakata, O. *Phys. Rev. B* **72**, 081406(R) (2005).

Hüfner, St. *Photoelectron Spectroscopy – Principles and Applications* (Springer-Verlag, Berlin, 1995).

Ibach, H. *Surf. Sci. Rep.* **29**, 193 (1997).

Imada, M., Fujimori, A., Tokura, Y. *Rev. Mod. Phys.* **70**, 1039 (1998).

Jiangdong Guo, Geunseop Lee, Plummer, E.W. *Phys. Rev. Lett.* **95**, 046102 (2005).

Kagoshima, S., Nagasawa, H., Sambongi, T. *One-dimensional Conductors* (Springer-Verlag, Berlin, 1988).

Katrich, G.A., Klimov, V.V., Yakovkin, I.N. *J. Electron Spectrosc. Relat. Phenom.* **68**, 369 (1994).

Kern, K., Niehus, H., Schatz, A., Zeppenfeld, P., Goerge, J., Comsa, G. *Phys. Rev. Lett.* **67**, 855 (1991).

Kim, B.J., Koh, H., Rothenberg, E., Oh, S.-J., Eisaki, H., Motoyama, N., Uchida, S., Tohyama, T., Maekawa, S., Shen, Z.-X., Kim, C. *Nature Phys.* **2**, 397 (2006).

Krawiec, M., Kwapinski, T., Jalochowski, M. *Phys. Rev. B* **73**, 075415 (2006).

Lau, K.H., Kohn, W. *Surf. Sci.* **75**, 69 (1978).

Le Daeron, P.Y., Aubry, S. *J. Phys. C: Solid State Phys.* **16**, 4827 (1983).

Lee, P.A., Rice, T.M., Anderson, P.W. *Phys. Rev. Lett.* **31**, 462 (1973).

Losio, R., Altmann, K.N., Kiakosian, A., Lin, J.-L., Petrovykh, D.Y., Himpsel, F. *J. Phys. Rev. Lett.* **86**, 4632 (2001).

Malterre, D., Grioni, M., Baer, Y. *Adv. Phys.* **45**, 299 (1996).

Matsuura, A.Y., Watanabe, H., Kim, C., Doniach, S., Shen, Z.-X., Thio, T., Bennett, J.W. *Phys. Rev. B* **58**, 3690 (1998).

Memmel, N. *Surf. Sci. Rep.* **32**, 91 (1998).

Menzel, A., Beer, R., Bertel, E. *Phys. Rev. Lett.* **89**, 076803 (2002).

Menzel, A., Zhang, Zh., Minca, M., Loerting, Th., Deisl, C., Bertel, E. *New J. Phys.* **7**, 102 (2005).

Mermin, N.D., Wagner, H. *Phys. Rev. Lett.* **17**, 1133 (1966).

Michely, T., Krug, J. *Islands, Mounds and Atoms* (Springer-Verlag, Berlin, 2004).

Minca, M., Penner, S., Doná, E., Menzel, A., Bertel, E., Brouet, V., Redinger, J. *New J. Phys.* **9**, 386 (2007).

Monthoux, P., Pines, D., Lonzarich, G.G. *Nature* **450**, 1177 (2007).

Mugarza, A., Ortega, J.E. *J. Phys.: Condens. Matter* **15**, S3281 (2003).

Nagao, T., Yaginuma, S., Inaoka, T., Sakurai, T. *Phys. Rev. Lett.* **97**, 116802 (2006).

N'Diaye, A., Bleikamp, S., Feibelman, P.J., Michely, Th. *Phys. Rev. Lett.* **97**, 215501 (2006).

Nussinov, Z., Rudnick, J., Kivelson, S.A., Chayes, L. N. *Phys. Rev. Lett.* **83**, 472 (1999).

Pennec, Y., Auwärter, W., Schiffrin, A., Weber-Bargioni, A., Riemann, A., Barth, J.V. *Nature Nanotechnol.* **2**, 99 (2007).

Plummer, E.W., Junren Shi, Tang, S.-J., Rotenberg, E., Kevan, S.D. *Prog. Surf. Sci.* **74**, 251 (2003).

Rice, T.M., Scott, G.K. *Phys. Rev. Lett.* **35**, 120 (1975).

Riikonen, S., Sánchez-Portal, D. *Nanotechnology* **16**, S218 (2005a).

Riikonen, S., Sánchez-Portal, D. *Phys. Rev. B* **71**, 235423 (2005b).

Riikonen, S., Sánchez-Portal, D. *Phys. Rev. B* **76**, 035410 (2007).

Roos, P., Bertel, E., Rendulic, K.D. *Chem. Phys. Lett.* **232**, 537 (1995).

Rubtsov, A.N., Hlinka, J., Janssen, T. *Phys. Rev. E* **61**, 126 (2000).

Segovia, P., Purdie, D., Hengsberger, M., Baer, Y. *Nature* **402**, 504 (1999).

Snijders, P.C., Rogge, S., Weitering, H.H. *Phys. Rev. Lett.* **96**, 076801 (2006).

Sun, L.D., Denk, R., Hohage, M., Zeppenfeld, P. *Surf. Sci.* **602**, L1 (2008).

Swamy, K., Menzel, A., Beer, R., Bertel, E. *Phys. Rev. Lett.* **86**, 1299 (2001).

Tersoff, J., Tromp, R.M. *Phys. Rev. Lett.* **70**, 2782 (1993).

Tosatti, E. *Festkörperprobleme* **15**, 113 (1975).

Tosatti, E. in: *Electronic Surface and Interface States on Metallic Systems*, (eds) Bertel, E., and Donath, M. (World Scientific, Singapore, 1995) p. 67.

Uhrberg, R.I.G., Balasubramanian, T. *Phys. Rev. Lett.* **81**, 2108 (1998).

Valla, T., Johnson, P.D., Yusof, Z., Wells, B., Li, Q., Loureiro, S.M., Cava, R.J., Mikami, M., Mori, Y., Yoshimura, M., Sasaki, T. *Nature* **417**, 627 (2002).

van Dijken, S., de Bruin, D., Poelsema, B. *Phys. Rev. Lett.* **86**, 4608 (2001).

Varykhalov, A., Rader, O., Gudat, W. *Phys. Rev. B* **77**, 035412 (2008).

Voit, J. *Rep. Prog. Phys.* **58**, 977 (1995).

Voit, J. *J. Electron Spectrosc. Relat. Phenom.* **117–118**, 469 (2001).

Yakovkin, I.N. *Atomic Wires.* in: *Encyclopedia of Nanoscience and Nanotechnology*, Vol. 1, (ed.) Nalwa, H.S. (American Scientific Publishers, Valencia CA, 2004) p. 169.

Yeom, H.W., Ahn, J.R., Yoon, H.S., Lyo, I.-W., Hojin Jeong, Sukmin Jeong *Phys. Rev. B* **72**, 035323 (2005).

Zhang, J., McIlroy, D.N., Dowben, P.A. *Phys. Rev. B* **49**, 13780 (1994).

Zhang, J., McIlroy, D.N., Dowben, P.A. *Phys. Rev. B* **52**, 11380 (1995).

Zwick, F., Jérome, D., Margaritondo, G., Onellion, M., Voit, J., Grioni, M. *Phys. Rev. Lett.* **81**, 2974 (1998).

Reaction studies on nanostructured surfaces

12

Adolf Winkler

12.1   Introduction

For many decades heterogeneous catalysis has played a key role in the chemical industry. Indeed, more than 80% of the industrial chemical processes in use nowadays rely on catalytic reactions. Most of these powerful catalysts have been designed by exhaustive trial and error studies. It is known that subtle changes of the structure and chemical composition of catalyst surfaces can tremendously influence the activity and selectivity of catalysts. With the advent of surface science in the second half of the twentieth century a more basic understanding of the surface chemistry underlying heterogeneous catalysis has been developed. It became more and more possible to delineate the correlations between the chemical and physical properties of catalyst surfaces and their effectiveness for special reactions. However, a complete understanding of the basis behind catalysis is still quite far off.

The systematic design of catalysts and their investigation with respect to activity and selectivity is a classical example of a research area that is now known as nanoscience and nanotechnology. Nanostructured surfaces can be produced, either by top-down processes (e.g. nanolithography) or bottom-up procedures (e.g. self-assembly), and these structures can be characterized on the nanometer and subnanometer level with techniques like scanning tunnelling microscopy (STM), scanning force microscopy (AFM) or with other scanning probe techniques. A large number of further surface-analytical tools is nowadays available to characterize catalyst surfaces with respect to the chemical composition, as well as regarding their geometric and electronic structure on a nanometer scale. Studies of this type have to be carried out under well-defined conditions, meaning that in most cases the experiments have to be carried out under ultrahigh-vacuum conditions.

In this chapter I will discuss the properties of some self-organized nanostructured surfaces with respect to specific model reactions, from a surface-science point of view. First, I will give an overview of the most important types of nanostructured surfaces, their preparation and characterization. Then, I will deal with the fundamentals of reaction processes, where I will primarily concentrate on the adsorption and desorption kinetics, as well as on the dynamics of adsorption and desorption. After that I will briefly describe the experimental techniques used in the context of reaction studies under ultrahigh-vacuum conditions. Finally, I will present some experimental results of model reactions as investigated in our laboratory in recent years: Hydrogen adsorption and desorption on stepped nickel surfaces, methanol reaction on self-assembled copper–copper oxide surfaces and hydrogen desorption and water formation on vanadium-oxide nanostructures on palladium surfaces.

12.2   Nanostructured surfaces

Material structures assembled from atoms or molecules with a size in the range of 1 to 100 nanometers are called nanostructures. Depending on the number of dimensions that are *not* in the nanoscale, one can distinguish between zero-dimensional objects (single atoms, clusters or other 2-dimensionally constrained arrangements), one-dimensional objects (nanowires, stepped surfaces, facetted surfaces) and two-dimensional objects (surface layers, ultrathin films). Since we are restricting our considerations to heterogeneous reactions, i.e. reactions of gases on solid surfaces, in principle all such reactions take place on 2-dimensional nanostructured objects. However, in the context of reactions one typically denotes only surfaces with *lateral* structures in the nanometer scale as nanostructured surfaces, i.e. 0D and 1D objects.

12.2.1   One-dimensional nano-objects

The simplest way to obtain 1D objects is to form regular steps on a surface. If a single crystal is cut slightly off parallel to a dense-packed plane one obtains a stepped or vicinal surface (Fig. 12.1). The cutting angle α defines the distance Λ between the individual step rows, which is simply given by $\Lambda = d / \tan \alpha$, where d is the distance between the net planes. Typically, cutting angles between $0.5°$ and $15°$ are applied, yielding step distances between about 230 Å and 8 Å. Since the lateral distance between the atoms is similar to the distances between the atomic net planes (about 2 Å), this yields stepped surfaces with step densities between 1% and 25%.

It is clear that by using higher cutting angles one would just approach another dense-packed plane and thus the number of steps would decrease

Fig. 12.1 Schematic drawing of a stepped surface by cutting a crystal surface by an angle α with respect to a dense-packed surface plane (hkl). $(h^*k^*l^*)$: Miller indices of the vicinal surface, $(h'k'l')$: Miller indices of the step plane.

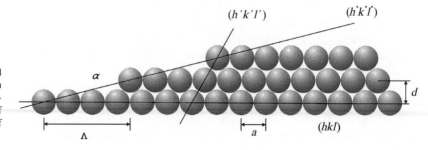

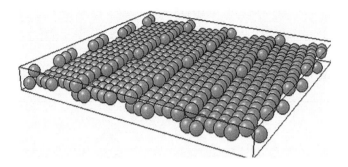

Fig. 12.2 Schematic drawing of a kinked and stepped surface that is described by Miller indices (11,12,16). Such a surface can be produced by cutting the sample with a polar angle of 9.8° relative to the (111) plane. Azimuthal cut angle with respect to the [110] direction: 13°. Somorjai notation: 7(111) × (710). The figure was produced with the help of the BALSAC software (copyright K. Hermann, Fritz-Haber-Institut Berlin (Germany)).

again. In addition to the cutting angle (polar angle) the azimuthal angle of the tilt axis is also important. Only when the tilt axis is parallel to a dense-packed atomic row will the steps be "infinitely" long. If the tilt axis is off a dense-packed row the step line will be intermitted by so-called kinks (Fig. 12.2). Such a kinked surface is by definition already a 0D nanostructure. Furthermore, the sign of the polar angle will define the local geometry at the steps. Consider the (111) plane of a face-centered cubic crystal with a tilt axis along the [110] direction. If the surface normal ([111] direction) is tilted into the direction of the [11$\underline{2}$] normal, the step plane is a (111) plane. If it is tilted into the opposite direction (towards the [$\underline{11}$2] plane normal) the step plane is a (100) plane. Thus, by proper cutting of single crystal surfaces one can prepare nanostructured surfaces with different geometric features.

The notation of a stepped surface can be done by using Miller indices ($h^*k^*l^*$). However, a more meaningful notation is the step-terrace notation (Somorjai notation) after Lang *et al.* 1972: E(S)-[$m(hkl) \times n(h'k'l')$] E: element, m: number of atom rows on the terrace, n: number of atom rows in the step, h, k, l: Miller indices of the terrace plane, $h,' k,' l'$: Miller indices of the step plane. Most often the step height at metal surfaces is monoatomic. In this case the number n is omitted in the notation. As an example (Fig. 12.3), a nickel single crystal cut by 6.2° with respect to the (111) plane and with the [110] direction as tilt axis yields a stepped surface that can be described according to the Somorjai notation by Ni(S)-[9(111) × (100)]. This is equivalent to a Ni(445) plane according to the Miller-indices notation.

As mentioned above, stepped metal surfaces prevalently exhibit monoatomic steps. However, so-called step bunching can appear, which means that several terraces with monoatomic steps transform into a single terrace with step heights of several atoms. Strong step bunching finally leads to a

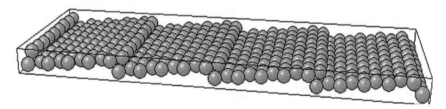

Fig. 12.3 Schematic representation of a stepped fcc surface produced by cutting the sample by 6.2° with respect to the (111) plane, with the [110] direction as tilt axis. The figure was produced with the help of the BALSAC software (copyright K. Hermann, Fritz-Haber-Institut Berlin (Germany)).

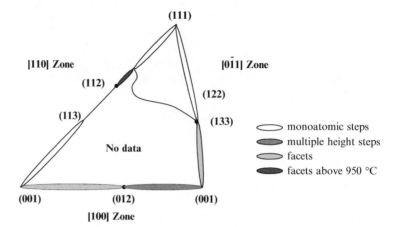

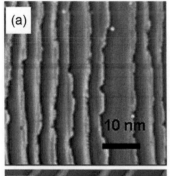

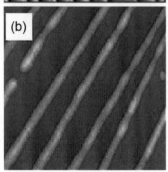

faceted surface, i.e. both the terrace width as well as the step-plane width are now of similar size. All these surface changes may happen on the clean surface by annealing the sample. But, in addition, adsorbates on the surface can induce step bunching and faceting as well. These processes may be reversible but can also be irreversible. The driving force for step bunching and faceting is the minimization of the surface free energy γ. Consider a vicinal surface plane with a surface free energy γ_0 and a total surface area A_0. This surface will transform into a faceted surface with facets of orientation 1 and orientation 2, characterized by surface free energies γ_1 and γ_2 and areas A_1 and A_2, if the condition: $A_0 \cdot \gamma_0 > A_1 \cdot \gamma_1 + A_2 \cdot \gamma_2$ is fulfilled. This may happen even if $A_1 + A_2 > A_0$.

Depending on the individual surface free energies, individual vicinal surfaces will be more or less prone to step bunching and faceting. In Fig. 12.4 a stereographic projection unit triangle for the fcc platinum crystal is shown, indicating regions of stability of monoatomic height steps, multiple height steps (step bunching) and faceting, according to Blakely and Somorjai 1977.

Stepped surfaces can also be used to produce a specific group of one-dimensional nanostructures, namely 1D nanowires, by step decoration. By taking advantage of the fact that the sticking probability at step sites is usually much higher than on terrace sites, small amounts of material A (in the submonolayer range) will preferentially line up along the steps of substrate material B. There are now many examples in the literature of nanowires formation on vicinal surfaces. To exemplify such structures nickel nanowires on a vicinal rhodium surface (Fig. 12.5(a)) and rubrene nanowires on a vicinal sapphire surface (Fig. 12.5(b)) are shown. In particular, by step decoration of an insulating substrate with a metal or a conducting organic material one can obtain nanowires with unique electronic features. However, nanowires on surfaces also play an important role for reaction processes, as will be shown in more detail below.

A further method to prepare 1D nanostructured surfaces is to take advantage of self-organization due to surface strain. Strain can generally be built up on a surface if a monolayer or an ultrathin film of a material is evaporated onto a substrate surface that possesses a different lattice constant (Teichert 2002).

But strain-induced surface reconstructions can also be accomplished by gas adsorption. The relaxation energy and the long-range interactions mediated by the strain field are then responsible for the periodic ordering of domains on the nanoscale. Since such *self-assembled* nanostructures are equilibrium structures they are generally quite stable. While stress-induced nanostructuring can also be used to produce 0D nanostructures on surfaces (see below), we will here briefly describe a now famous example of a 1D structure. When a Cu(110) surface is exposed to oxygen up to saturation at $T > 450$ K, it is observed that a well-ordered (2×1) superstructure forms, which corresponds to a coverage of 0.5 monolayers. However, if smaller amounts of oxygen are adsorbed alternating stripes of CuO and bare Cu are formed, generating a periodic supergrating (Coulman *et al.* 1990; Jensen *et al.* 1990; Kern *et al.* 1991) (Fig. 12.6(a)). According to Kern *et al.* (1991) there are three interaction regimes present in this unique chemisorption system. Along the [001] direction there is a strongly attractive short-range interaction between Cu and O atoms, giving rise to the formation of long and stable Cu–O strings, already at very low oxygen coverage. This is facilitated by the rather large mobility not only of the oxygen atoms but also of the copper atoms at $T > 450$ K. In the CuO island formation these strings are held together by medium-range attractive forces along the [1$\bar{1}$0] direction, forming long Cu–O stripes (Fig. 12.6(b)). The aggregation of the single strings to stripes saturates at about 8–12 strings, and each stripe repels its neighboring stripe by long-range repulsive interaction, so that they become more or less equidistant. It is assumed that the aggregation of the strings as well as the repulsion of the stripes is due to substrate-mediated strain forces. It is important to note that the CuO stripe width and the distance between the stripes can be varied over a wide range. At half of the oxygen saturation layer, at 0.25 monolayers, the width of the CuO and bare Cu areas are nearly equal and the grating distance is about 6.5 nm (Kern *et al.* 1991)

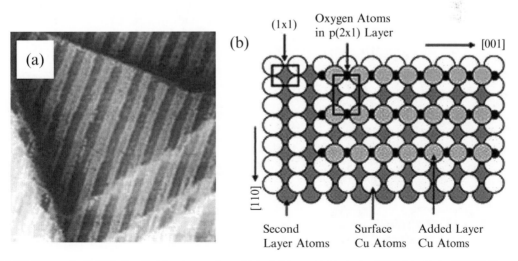

Fig. 12.6 (a) STM image of a Cu(110)-(2×1) stripe phase surface, obtained at an oxygen coverage of 0.26 monolayers at 550 K (Kern *et al.* 1991, copyright by the American Physical Society). (b) Model of the added row (2×1) oxygen structure on Cu(110), leading to the Cu–CuO stripe phase.

(Fig. 12.6(a)). This offers the intriguing opportunity to study the influence of the size and the separation of the stripes on the reaction properties of the Cu(110)-(2 × 1) stripe phase.

12.2.2   Zero-dimensional nano-objects

Most of the 0D nano-objects are based on surface strain. Contrary to 2-fold symmetric surfaces (fcc(110)), which tend to form 1D nano-objects, 3-fold (fcc(111)) and 4-fold (fcc(100)) symmetric surfaces have a tendency to form 0D objects, i.e. they are laterally nanostructured in two dimensions. In particular, non-metallic surfaces tend to laterally reconstruct, even without an additional strain-inducing surface layer. A typical example is the famous (7 × 7) reconstruction of Si(111), as shown in Fig. 12.7 (Takayanagi *et al.* 1985; Stipe *et al.* 1997). This special structure is the result of forming dimers, ad-atoms and stacking faults on the surface in a special arrangement. Regular arrays of 0D nanostructures can also be easily obtained by covering a substrate with a material of different lattice constants. As an example in Fig. 12.8 a layer of Ga atoms on a Ge(111) surface forms a discommensurate structure, yielding a (8 × 8) superstructure, due to the misfit of about 10% in the lattice constants (Zegenhagen *et al.* 1997).

The STM patterns obtained for such discommensurate structures stem from the Moire effect of the two layers with different lattice constants. Structures of this type can be used to produce 0D entities (nanodots) by evaporation of another material on such a surface. In this case, the evaporated material will start to nucleate at distinct sites of the nanostructured surface, typically at the threefold hollow sites of the Moire pattern. As an example in Fig. 12.9 the formation of well-ordered iron clusters on the dislocation network of a Cu bilayer on Pt(111) is shown (Brune *et al.* 1998). This type of magnetic nanostructure is relevant for data storage, which in this case would allow a storage density of 10 Terabit/inch^2. But of course, such well-defined structures are also desired objects for the study of surface reactions on nano-objects.

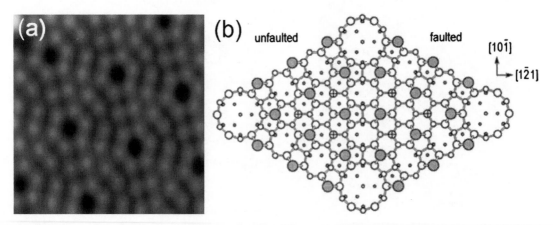

Fig. 12.7 (a) STM image of the Si(111) (7 × 7) reconstructed surface (Stipe *et al.* 1997, copyright by the American Physical Society). (b) Schematic diagram showing the top view of the (7 × 7) dimer-ad-atom-stacking fault model for Si(111) (Takayanagi *et al.* 1985, with permission from Elsevier, copyright 1985).

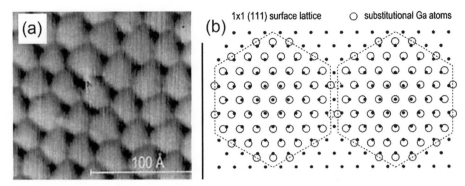

1x1 (111) surface lattice   ○ substitutional Ga atoms

Fig. 12.8 (a) STM image of a superstructure of a Ga film (0.7 monolayer) on a Ge(111) surface due to a discommensurate layer formation. The periodicity of this superstructure is about 32 Å (Zegenhagen *et al.* 1997, copyright Wiley-VCH Verlag GmbH&Co. KGaA. Reproduced with permission). (b) Schematic model of the atomic structure of the Ga/Ge(111) surface, which leads to a (8×8) superstructure. (Zegenhagen *et al.* 1997, copyright Wiley-VCH Verlag GmbH&Co. KGaA. Reproduced with permission.)

Finally, regular nanostructures can also be produced by preparing oxides or nitrides on metal surfaces. Such structures are in particular relevant for heterogeneous catalysis, because they represent a model system for an inverse catalyst (Schoiswohl *et al.* 2007). If vanadium is evaporated on a Pd(111) surface under an oxygen background pressure in the 10^{-7} mbar range, various surface oxides can be formed, depending on the oxygen partial pressure and the surface temperature. These oxides form well-defined nanostructures. The evaporation of 0.3 monolayers of vanadium on Pd(111) at 523 K in an oxygen atmosphere of 2×10^{-7} mbar yields a surface oxide with a stoichiometry of V_5O_{14} (Fig. 12.10(a)). This is a rather open network structure, forming a honeycomb-like arrangement with a (4×4) periodicity with respect to the surface unit cell of Pd(111). However, this structure reacts readily with hydrogen and is therefore quite unstable under reducing conditions. The stable structure that forms in this case is a V_2O_3 surface oxide, with a (2×2) superstructure (Fig. 12.10(b)) (Surnev *et al.* 2002; Kratzer *et al.* 2006). This leads us to one of the fundamental questions in the context of reactions on nanostructures surfaces. What is the stability of such structures under specific reaction conditions and how do structural changes influence the reaction rates? This will be discussed in Section 12.5.3.

12.3 Fundamentals of reaction processes

For the understanding of reactions on nanostructured surfaces one has first to consider the individual reaction processes on simple surfaces. Therefore, I will first briefly introduce the main aspects of molecular interaction with single-crystal surfaces and then extend the discussion to nanostructured surfaces. The reaction process of a molecule on a surface can be subdivided into the fundamental reaction steps of adsorption, dissociation, surface diffusion, reaction and desorption. In this chapter I will mainly focus on the adsorption and desorption processes. Most frequently the *kinetics* of adsorption and desorption is investigated in the experiments, i.e. the rates of adsorption and desorption as a function of coverage and temperature are measured. To obtain a more

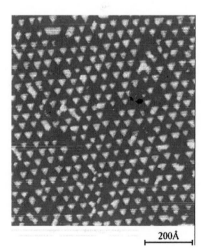

Fig. 12.9 STM image of a periodic array of Fe islands nucleated on the dislocation network of a Cu bilayer on Pt(111) at 250 K (Brune *et al.* 1998, reprinted with permission from Macmillan Publishers Ltd., copyright 1998).

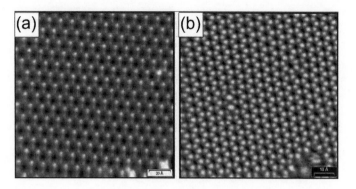

Fig. 12.10 (a) STM image of a V_5O_{14} (4×4) superstructure on Pd(111) (Kratzer *et al.* 2006, reprinted with permission from the American Institute of Physics, copyright 2006). (b) STM image of a V_2O_3 (2×2) superstructure on Pd(111) (Kratzer *et al.* 2006, reprinted with permission from the American Institute of Physics, copyright 2006).

detailed understanding of the physics behind this one has to determine the *dynamics* of adsorption and desorption. This means the study of differential adsorption and desorption rates as a function of the adsorption/desorption angle, the translational energy and the internal energies.

12.3.1   Adsorption and desorption kinetics

The characteristic value for the adsorption probability is the sticking coefficient S. It is defined as the number of adsorbed particles versus the number of impinging particles per time and square unit. This quantity will depend on the amount of already adsorbed particles, the coverage Θ. The adsorption probability at the very beginning of adsorption (on the clean surface) is called the initial sticking coefficient S_0. The change of the sticking coefficient with coverage depends on the special type of adsorption. In this chapter we will restrict our considerations to chemisorption, in contrast to physisorption. In the case of physisorption the particles condense and can form multilayers. In that case the sticking coefficient is typically unity and does not decrease with increasing coverage. In the case of chemisorption typically only a maximum coverage can be reached, which is of the order of one monolayer. A monolayer exists if on each surface atom one particle (atom or molecule) is adsorbed. For solid surfaces 1 monolayer corresponds to about 10^{15} particles/cm^2.

For a 1st-order adsorption (Fig. 12.11(a), curve 1), where the particles do not dissociate upon impact, the coverage dependence of the sticking coefficient $S(\Theta)$ is simply determined by the fraction of the uncovered surface area:

$$S(\Theta) = S_0(1 - \Theta). \tag{12.1}$$

For the dissociative adsorption of a diatomic molecule (2nd-order adsorption) it follows (Fig. 12.11(a), curve 2):

$$S(\Theta) = S_0(1 - \Theta)^2. \tag{12.2}$$

In fact, only in very few cases can this coverage dependence, which is called Langmuir adsorption, be verified experimentally in this simple form. The main reason for the discrepancy is usually that not all adsorption sites are equivalent and that the adsorption is not direct, but involves a transient state, a

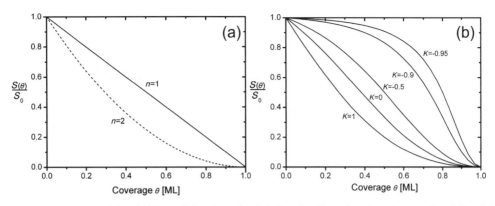

Fig. 12.11 (a) Coverage dependence of the sticking coefficient for molecular adsorption (1st-order adsorption, $n = 1$) and dissociative adsorption (2nd-order adsorption, $n = 2$). (b) Coverage dependence of the sticking coefficient for dissociative adsorption via a precursor state, for different K.

so-called precursor state. The precursor state is a weakly bound state, in which the molecule can be trapped and can diffuse along the surface until it finds a preferable site for adsorption.

There exist several models for molecular and dissociative adsorption via precursor states (Gorte and Schmidt 1978; Cassuto and King 1981). One of the most frequently used description of precursor-assisted adsorption stems from Kisliuk 1958. In this model it is assumed that the molecule in the precursor state, if located above an empty adsorption site, can either chemisorb (P_a), desorb (P_b) or migrate to another site in the precursor state (P_c). If it is located above an already occupied adsorption site no chemisorption can take place ($P_a^* = 0$), but the probabilities for desorption are P_b^* and for migration P_c^*, respectively. By an iteration procedure one can arrive at the total sticking coefficient, which is the sum of the chemisorption probabilities on the first, second, third, ... etc. site, visited by the molecule in the precursor. With the additional conditions of $P_a + P_b + P_c = 1$ and $P_b^* + P_c^* = 1$, for non-dissociative adsorption the following expression results for the coverage-dependent sticking coefficient (Kisliuk 1958):

$$S(\Theta) = \frac{S_0(1 - \Theta)}{1 - \Theta + K\Theta}, \text{ with } S_0 = \frac{P_a}{P_a + P_b} \text{ and } K = \frac{P_b^*}{P_a + P_b}. \quad (12.3)$$

A similar expression can be derived for dissociative adsorption via a precursor state:

$$S(\Theta) = \frac{S_0(1 - \Theta)^2}{1 - \Theta(1 - K) + \Theta^2 S_0}, \text{ with } S_0 = \frac{P_a}{P_a + P_b} \text{ and } K = \frac{P_b^* - P_a}{P_a + P_b}. \quad (12.4)$$

The main feature of precursor-assisted adsorption is the nearly coverage-independent sticking coefficient over a wide coverage range. In Fig. 12.11(b) the coverage dependence of the sticking coefficient for precursor-assisted, dissociative adsorption is plotted, with different K as parameter. The larger the probability to migrate away from an occupied site, P_c^*, the smaller is the probability of desorption, P_b^*. This leads to increasingly negative values K, with the maximum value $K = -1$, when the adsorption probability is 1 at an empty site.

The above-described Kisliuk model assumes that all chemisorption sites are equivalent, i.e. the chemisorption probability P_a is the same all over the surface. This is of course not true for the adsorption on nanostructured surfaces, where inherently non-equivalent adsorption sites exist. Such non-equivalent adsorption sites can be defects on the surface (ad-atoms, vacancies, steps, grain boundaries), but also well-defined nanostructures on surfaces. Generally, at such specific sites the adsorption probability may be larger than at regular surface sites. Another frequently observed phenomenon is that the sticking coefficient increases rather than decreases with increasing coverage of a particular adsorbate. This is particularly true on flat surfaces, which exhibit quite small sticking coefficients. In such a case adsorbed particles act as active sites for further adsorption, in particular for dissociative adsorption. Such behavior is called an autocatalytic adsorption process.

One can extend the Kisliuk model to be able to describe such a physical situation, by introducing two types of precursor states (Winkler and Rendulic 1982). To actually obtain the adsorption kinetics one best applies a formalism developed by Schönhammer (1979). In our model we used the following assumptions and definitions:

(1) Two types of precursor states exist on the surface: Sites that are located above an empty chemisorption site (intrinsic precursor) and sites that are located above an occupied chemisorption site (extrinsic precursor).

(2) A molecule in the precursor state above an empty chemisorption site can dissociate and the fragments are transferred into the chemisorption sites with probability P_a, if all adjacent chemisorption sites are empty. The molecule can also move away in the precursor state (probability for the site change is P_c), or it can desorb (probability P_b).

(3) Similarly, the corresponding transition probabilities above an occupied chemisorption site are denoted by P_a^*, P_b^*, P_c^*.

(4) The coverage dependence of the adsorption probabilities above an empty site, $P_a(\Theta)$, and an occupied chemisorption site, $P_a^*(\Theta)$, can be described as:

$$P_a(\Theta) = P_a(1 - \Theta) \quad \text{and} \quad P_a^*(\Theta) = P_a^*(1 - \Theta)^2.$$

(5) The change of the adsorption probability with coverage should not influence the desorption probability significantly. This assumption is also implicitly made in the Kisliuk formalism. But the sum of the three probabilities for each transition event has to be unity.

From the evaluation procedure as outlined in detail by Winkler and Rendulic (1982) one derives the sticking coefficient as a function of coverage on a surface with twofold adsorption sites:

$$S(\Theta) = \frac{(1 - \Theta)^2(1 + B\Theta)}{C + (D - 1)\Theta + (1 - 2B + B\Theta)\Theta^2}, \tag{12.5}$$

with $B = P_a^*/P_a$, $C = (1 - P_c)/P_a$ and $D = (P_c - P_c^*)/P_a$.

This equation changes to the Kisliuk formula for dissociative adsorption (eqn (12.4)) for the special case $P_a^* = 0$.

The desorption kinetics is typically described by the Polanyi–Wigner equation. The coverage dependence of the desorption rate $R_d(\Theta)$ is given by:

$$R_d(\Theta) = -\nu \cdot f(\Theta)\exp(-E/kT), \tag{12.6}$$

with ν: pre-exponential factor, $f(\Theta)$: coverage dependence, E: activation energy of desorption, k: Boltzmann's constant, T: surface temperature.

Comparable to the coverage dependence of the adsorption rate (sticking coefficient) the coverage dependence of direct desorption is:

$f(\Theta) = const.$ for 0th-order desorption (evaporation from multilayers)
$f(\Theta) = \Theta$ for 1st-order desorption (atomic or molecular desorption)
$f(\Theta) = \Theta^2$ for 2nd-order desorption (associative desorption)

For desorption via a precursor state the functional dependence of the desorption rate is more complicated. Alnot and Cassuto (1981) have developed rate equations with the following form of the coverage dependence:

$$f(\Theta) = \frac{P_b(1 - P_a)\Theta}{P_a P_c(1 - \Theta) + P_b} \quad \text{(1st-order desorption)} \tag{12.7}$$

$$f(\Theta) = \frac{P_b(1 - P_a)\Theta^2}{P_a P_c(1 - \Theta)^2 + P_b} \quad \text{(2nd-order desorption).} \tag{12.8}$$

Further complications will show up in the description of the desorption kinetics when intrinsic and extrinsic precursor states have to be taken into account. To the best of my knowledge no analytical formulation of the desorption rate exists for such cases.

12.3.2 Adsorption and desorption dynamics

In the preceding section we have seen that at different surface sites different adsorption and desorption probabilities may exist. The reason for this behavior is that different activation barriers for adsorption exist. Studies of the *adsorption/desorption dynamics* aim at the determination of such activation barriers and their dependence on various parameters. In this section I will summarize the most important aspects of adsorption and desorption dynamics from the experimental point of view. Actually, this topic has also gained enormous interest from the theoretical point of view, because modern computer calculations are capable of describing dynamical processes at a classical as well as the quantum-mechanical level. Several review articles are available on this subject, e.g. by Darling and Holloway (1995), Gross (1998), and Kroes (1999). The experimental approach to obtain information on the dynamics of adsorption is to measure the sticking coefficient (usually the initial sticking coefficient) as a function of the kinetic energy, the rotational and vibrational energy of the impinging particles, as well as a function of the impingement angle. Similarly, one can perform desorption experiments where the angular distribution, the translational and the internal state distribution of the desorbing molecules are measured. This will allow delineating the microscopic processes that govern the adsorption/desorption processes.

12.3.3   Influence of the kinetic energy on sticking

The simplest characterization of the activation barrier and its influence on the adsorption probability can be obtained from the 1-dimensional Lennard-Jones potential (Fig. 12.12). It represents the total energy of the surface–molecule system as a function of the distance between the surface and the molecule. Starting from a large distance, with the reference total energy equal to zero, the molecule first enters a weak attractive potential due to dispersion (Van der Waals) forces. Upon further approach the total energy increases again due to the interaction of the electrons of the molecule with the surface electrons (Pauli repulsion) (Fig. 12.12, curve $M + 1/2H_2$), physisorption potential). On the other hand, this interaction weakens the intermolecular forces, which finally can lead to a dissociation of the molecule. The potential that is seen by the individual atoms is now described by the atom–surface interaction (Fig. 12.12, curve $M + H$, chemisorption potential). The point where the two potential curves cross each other defines the activation barrier for adsorption, E_a. The height of this activation barrier can be determined from the kinetic energy dependence of the sticking coefficient, $S(E)$.

Qualitatively, the sticking coefficient increases with kinetic energy E if an activation barrier exists. In the simplest approximation the sticking coefficient is zero if the kinetic energy is smaller than the activation barrier height and unity if it is larger than the barrier height. This leads to a step function of $S(E)$. Actually, there may be some tunnelling through the barrier, in particular for hydrogen, and also some reflection from the repulsive part of the physisorption potential, which will round off the step function (Harris 1989). Furthermore,

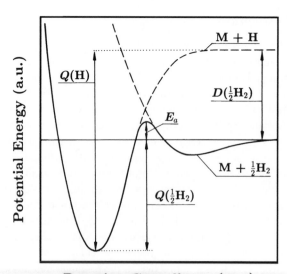

Fig. 12.12 Lennard-Jones potential for dissociative adsorption.

Reaction Co-ordinate (a.u.)

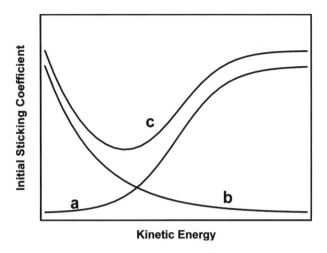

Fig. 12.13 Qualitative behavior of the sticking coefficient as a function of the kinetic energy of the impinging molecules for direct activated adsorption (a), precursor-assisted adsorption (b) and a mixture of both processes (c).

the activation barrier is not equal at all sites of impact within the surface unit cell. The activation barrier-height distribution results in a further smearing out of the step function (Fig. 12.13, curve a). This S-shaped function is frequently approximated by the following expression:

$$S(E) \propto \frac{1}{2}\left[1 + \tanh\left(\frac{E - E_a}{W}\right)\right], \qquad (12.9)$$

where E is the kinetic energy of the impinging molecule, E_a the activation barrier height and W describes the broadening of the curve, i.e. the activation barrier height distribution. The deflection point of eqn 12.9 defines the mean activation barrier height. In the case of a nanostructured surface, e.g. on a bimetallic surface, the overall sticking coefficient can be determined by summing up functions of the type in eqn 12.9 with different activation barriers E_a.

The above-described kinetic-energy dependence is only valid for direct adsorption. When precursor states are involved in the adsorption process, the lifetime in the precursor state becomes the most important parameter. During the sojourn time in the precursor state the temporarily trapped molecule can visit several sites on the surface, so that it can eventually adsorb dissociatively at a site with the smallest activation barrier. Surface defects typically represent such "active" sites where no activation barrier exists. The lifetime of the molecule in the physisorption state, and hence the probability to find an active site for dissociative adsorption, is inversely proportional to the kinetic energy. Thus, adsorption via a precursor state is characterized by a sticking coefficient decreasing with kinetic energy (Fig. 12.13, curve b). Such a behavior has been frequently found experimentally, e.g. for the adsorption of oxygen on Pt(111) and Pt(112) (Winkler *et al.* 1988) or the adsorption of hydrogen on Ni(110) (Rendulic *et al.* 1989). In many cases, but in particular on nanostructured surfaces, both activated and precursor-assisted adsorption takes place at the same time. Precursor-assisted adsorption dominates at low kinetic energy, whereas activated adsorption prevails at higher kinetic energy (Fig. 12.13, curve c).

An additional mechanism has been proposed to explain sticking coefficients decreasing with kinetic energy, in particular for hydrogen adsorption on metal surfaces, by Gross *et al.* (1995), Kay *et al.* (1995) and Darling *et al.* (1998), by introducing the concept of dynamical steering. In this case the impinging molecule is assumed to be steered upon impact, due to the distribution of the activation barrier heights, directly into the region of smallest barrier heights (translational steering). Thus, molecules with lower kinetic energy will be more easily steered to these favorable adsorption sites, and hence a sticking coefficient increasing with decreasing translational energy will appear. Actually, this behavior cannot be explained via the simple 1-dimensional potential-energy diagrams. When it comes to take into account dimensions other than the molecule–surface distance, multidimensional potential-energy surfaces (PES) have to be calculated and the sticking coefficients have to be determined by trajectory calculations in these PES. This allows not only taking into account the exact point of impingement within the surface unit cell, but also the vibrational and rotational degrees of freedom.

12.3.4   Influence of the vibrational energy on sticking

Dissociative adsorption involves breaking of the intermolecular bonds. This is facilitated if the molecule is already in a vibrationally excited state. Therefore, vibrationally excited molecules generally possess a higher sticking probability. Since the intramolecular distance now becomes an important co-ordinate, the total energy of the molecule–surface system is now a function of at least two co-ordinates. This can be visualized by a 2D potential-energy diagram (elbow potential), as shown in Fig. 12.14 (Hand and Holloway 1989). Here, it is assumed that the molecule axis is parallel to the surface (x-axis). When the molecule approaches the surface (in the z-direction) and enters the weakly bound physisorption state (region I), the interatomic distance is not yet changed significantly. However, on further approach the occupation

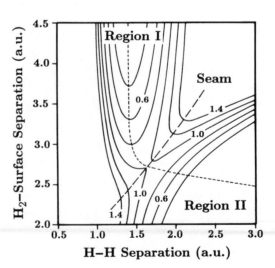

Fig. 12.14 Two-dimensional potential-energy contour plot for a hydrogen molecule approaching a surface with its axis parallel to the surface. Region I: Physisorption potential, Region II: Chemisorption potential (Hand and Holloway 1989, reprinted with permission from the American Institute of Physics, copyright 1989).

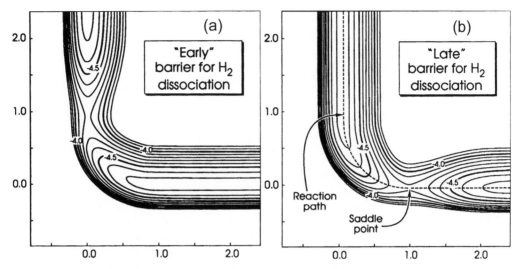

Fig. 12.15 (a) Two-dimensional PES with a barrier in the entrance channel of adsorption. (b) Two-dimensional PES with a barrier in the entrance channel of adsorption (Halstead and Holloway 1990, reprinted with permission from the American Institute of Physics, copyright 1990).

of anti-bonding electronic levels leads to a weakening of the inter-atomic bonds and therefore to an increase of the bond length, until the molecule eventually fully dissociates and enters the chemisorption state (region II). Now, the potential energy of the individual atoms on the surface is relevant. The cross-over between the molecular potential and the atomic potential (seam) is a saddle point in the potential-energy surface (PES) and characterizes the activation barrier for dissociative adsorption. Depending on the location of this saddle point in the PES, either in the entrance channel (Fig. 12.15(a)) or in the exit channel for adsorption (Fig. 12.15(b)), one speaks of an early or late activation barrier (Halstead and Holloway 1990). In the case of an early barrier the vibrational energy does not play an important role, but in the case of a late barrier vibrationally excited molecules possess a much higher sticking coefficient at a given translational energy.

In the latter case vibrational energy is partially transferred into translational energy that helps to overcome the activation barrier (bootstrapping mechanism, Harris 1991). The influence of the vibrational energy in dissociative adsorption has been frequently verified experimentally (Hayden and Lamont 1989; Berger *et al.* 1990; Michelsen *et al.* 1993). It is particularly pronounced in those adsorption systems where the sticking coefficient is very small.

12.3.5 Influence of the rotational energy on sticking

In a more precise description of the adsorption process the rotational degree of freedom also has to be taken into account. This means that the orientation of the molecule in the interaction region is not necessarily parallel to the surface, as assumed in the preceding section. The orientation can be described by two angles, the polar angle and the azimuthal angle of the molecule axis with respect to the surface normal. Contrary to the influence of the vibrational

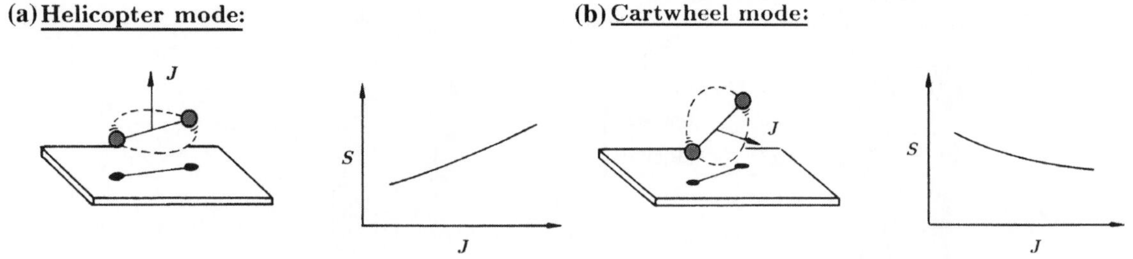

(a) Helicopter mode: **(b) Cartwheel mode:**

Fig. 12.16 (a) For a molecule with the rotational axis parallel to the surface normal (helicopter mode) the sticking coefficient is increasing with rotational energy, due to rot–trans energy conversion. (b) For a molecule with the rotational axis parallel to the surface plane (cartwheel mode) the sticking coefficient decreases with rotational energy, due to decreased orientational steering.

energy, which is always increasing the sticking coefficient, the influence of the rotational energy can be both favorable as well as unfavorable. There exists an energetic effect as well as a geometric effect. On the one hand, the rotational energy can be partially converted into translational energy by the bootstrapping effect, thus increasing the dissociative sticking. On the other hand, an unfavorable orientation of the molecule, i.e. an orientation off-parallel to the surface, will decrease the sticking probability (Darling and Holloway 1995). There exist two extreme orientations of the molecule rotational axis with respect to the surface. If the rotational axis is normal to the surface this is called a helicopter mode (Fig. 12.16(a)). If the rotational axis is parallel to the surface it is called a cartwheel mode (Fig. 12.16(b)).

In the helicopter mode the molecule is always in a favorable position with respect to an electron transfer into anti-bonding electronic levels, therefore rotational to translational energy transfer will dominate and the sticking coefficient increases with rotational energy. For the cartwheel mode the molecule spends a considerable time in an orientation with components normal to the surface, thus hindering dissociative adsorption. Since some time is needed to break the bonds this hindering effect is more pronounced when the molecules are rotating faster. Therefore, this contribution leads to a decrease of the sticking coefficient with increasing rotational energy. In this context steering also plays an important role again. Molecules in a low rotational state can be more easily steered into a favorable position (and can stay longer in this orientation) and thus the sticking coefficient increases with decreasing rotational energy. Since all these effects play together it is obviously difficult to make a specific forecast for the influence of the rotational energy. For the systems H_2–Pt(110) (Beutl *et al.* 1996) and H_2-Pd(111) (Beutl *et al.* 1995) a decrease of the sticking coefficient with increasing rotational energy has been observed.

There are now theoretical models available to describe the influence of all degrees of freedom on the sticking probability (trajectory calculations in a 4D potential-energy surface). When the exact impact within the surface unit cell is also taken into account, 6D PES have to be used for the calculations. This is now state-of-the-art, at least for the adsorption of small molecules. However, there are still several possibilities to calculate the movement of the molecule in the multidimensional PES. One can either use a classical trajectory calculation or a full quantum-mechanical approach. The pros and cons of these methods have been discussed in several papers by Kinnersley *et al.* (1996), Gross (1998) and Kroes (1999).

12.3.6 Influence of the angle of incidence on sticking

As outlined above, the effective activation barrier is actually a function of several parameters. Nevertheless, the translational energy still plays in most cases the dominant role in the adsorption process. Two different extreme scenarios can be considered: If the potential energy surface is very smooth all over the geometric surface the assumption of a 1D barrier height is sufficient to describe the adsorption process. If, in addition, no significant precursor state exists (direct adsorption), only the normal component of the translational energy is relevant to surmount the activation barrier (normal energy scaling, NES). On the other hand, when the molecule is heavily scrambled in a precursor state prior to dissociation the total energy is relevant for the adsorption probability (total-energy scaling, TES). The effectiveness of these two scenarios will have a significant influence on the angular dependence of the sticking coefficient. In the case of total-energy scaling the sticking coefficient is independent of the angle of incidence. In this context we have to differentiate between the sticking coefficient S (defined as the ratio between adsorbed and impinged molecule) and the adsorption probability A (defined as the adsorption rate per square unit). Since the impingement rate decreases according to $\cos \Phi$, with Φ being the angle between the surface normal and the molecule trajectory, an angle-independent sticking coefficient ($S(\Phi) = $ constant) corresponds to a cosine dependence of the adsorption probability:

$$A(\Phi) = S(\Phi) \cos \Phi. \qquad (12.10)$$

In fact, for many adsorption systems a cosine dependence of the adsorption probability can be observed, in particular for dissociative adsorption on unactivated surfaces (H_2 on Ni(110), Steinrück *et al.* 1985a) and in general for non-dissociative adsorption (CO on nickel, Steinrück *et al.* 1984). For direct, activated adsorption and also for precursor-assisted adsorption frequently strong deviations from a cosine distribution are observed (Winkler and Rendulic 1992). In the case of normal energy scaling a clear correlation between the translational energy dependence and the angular dependence of the adsorption probability should exist. Sticking coefficients increasing with translational energy exhibit a strongly forward focused angular distribution of adsorption, whereas sticking coefficients decreasing with translational energy show a broad angular dependence. It is common to describe such angular distributions in the form of a $\cos^n \Phi$ distribution (Comsa and David 1985; Rendulic 1991):

$$A(\Phi) = A(0°) \cos^n \Phi \qquad (12.11)$$

$$S(\Phi) = S(0°) \cos^{n-1} \Phi, \qquad (12.12)$$

with $n > 1$ for activated adsorption, $n < 1$ for precursor-assisted or steering-assisted adsorption, and $n = 1$ for unactivated adsorption.

As the angular distribution of the sticking coefficient is a probe for the local activation barrier, its determination allows information to be obtained on the contribution of individual surface sites to the integral adsorption behavior on nanostructured surfaces.

12.3.7   Angle dependence, translational and internal energy dependence of the desorption flux and the principle of detailed balancing

Adsorption and desorption processes are intimately connected with each other, therefore the determination of the adsorption behavior should allow information to be obtained on the desorption characteristics of a particular system. It is obvious that for a gas–surface system under equilibrium conditions each concentration in the gas phase (gas pressure p) is correlated with a particular surface coverage Θ. This correlation is described by the adsorption isotherm, the simplest one for non-interaction monoatomic particles being the Langmuir isotherm:

$$\Theta(p) = \frac{Kp}{1 + Kp} \qquad \text{with } K = \frac{\exp(E/kT)}{\nu\sqrt{2\pi mkT}}, \qquad (12.13)$$

with E: adsorption energy, m: molecule mass, T: temperature, k: Boltzmann's constant, ν: pre-exponential factor.

Whereas this relationship allows the adsorption energy to be obtained (more precisely the isosteric heat of adsorption) from equilibrium measurements, it does not yield information on the microscopic processes involved in adsorption and desorption. However, one can make the statement that each microscopic process leading to adsorption has to be compensated by a correlated desorption process. This is due to the microscopic reversibility of equilibrium systems, also called the principle of detailed balancing. That means that on the average for each adsorbing particle, described by its angle of incidence Φ, translational energy E, vibrational state v and rotational state j, a desorbing particle with the same angle of desorption, and the same translational, vibrational and rotational energy has to exist, i.e.:

$$A(\Phi, E, j, v) = D(\Phi, E, j, v) = S(\Phi, E, j, v)\cos\Phi. \qquad (12.14)$$

Whereas this is obvious for a true equilibrium situation between gas phase and a surface, one can put the question if this condition is also fulfilled if the system is not in equilibrium, but in quasi-equilibrium. According to Comsa (1977) quasi-equilibrium means that the distribution of the molecules adsorbed on the different kinds of adsorption sites continues to be an equilibrium-like distribution, even if the supply of the molecules from the gas phase is interrupted. In other words, the desorption characteristic is only defined by the momentary surface coverage and surface temperature and it does not play a role if there exists a compensation by adsorbing molecules or not.

The applicability of detailed balancing under quasi-equilibrium conditions can only be experimentally verified by independent adsorption and

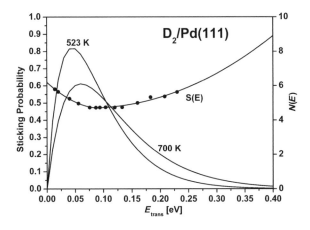

Fig. 12.17 Initial sticking coefficient for deuterium on Pd(111) as a function of the kinetic beam energy. The solid line through the data points stems from a polynomial best fit function. The kinetic energy distribution of Maxwellian beams with 523 K and 700 K are included (Kratzer *et al.* 2007, with permission from Elsevier, copyright 2007).

desorption experiments. But generally one cannot perform adsorption and desorption experiments at the same temperature and coverage. Nevertheless, there now exists substantial evidence that for many adsorption systems the principle of detailed balancing is fulfilled with respect to all degrees of freedom even if the experiments have been done only close to quasi-equilibrium conditions. This has been shown for the angular distribution of hydrogen on several metal surfaces (Cardillo *et al.* 1975; Steinrück *et al.* 1985c), for the kinetic energy distribution (Dabiri *et al.* 1971; Comsa and David 1982; Rendulic *et al.* 1989) as well as for the internal state distributions (Michelsen *et al.* 1992; Rettner *et al.* 1993).

Very recently, we re-examined the system D_2-Pd(111) and have again observed very good agreement between the adsorption and desorption dynamics, by applying the principle of detailed balancing and normal energy scaling. The angular distribution of desorbing deuterium was measured with the help of a permeation source (Kratzer *et al.* 2007). The obtained distribution at a given sample temperature was compared with the calculated distribution for adsorption at the same temperature. For this purpose we used the energy-dependent sticking coefficient as recently measured with a quasi-monoenergetic nozzle beam by Riedler (1996) (Fig. 12.17) and assumed normal energy scaling. The sticking coefficient S of a Maxwellian gas under normal incidence at this temperature is given by

$$S_T(0°) = \frac{\int N_T(E) \cdot S(E) dE}{\int N_T(E) dE},$$
(12.15)

The applicability of normal energy scaling can be described as:

$$S(E, \Phi) = S(E_\perp, 0°) = S(E \cdot \cos^2 \Phi, 0°),$$
(12.16)

where $E \cdot \cos^2 \Phi$ means the energy of the molecules impinging at an angle Φ to the surface normal, associated with the momentum normal to the surface. The number of particles $N(E)$ of a Maxwellian beam impinging at an angle Φ should then be equal to the number of particles with energy $E_\perp / \cos^2 \Phi$ of a Maxwellian beam impinging normal to the surface. Hence, the sticking coefficient $S_T(\Phi)$ for a Maxwellian gas flux at temperature T impinging under

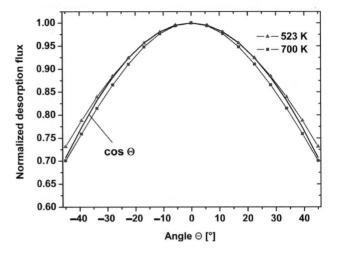

Fig. 12.18 Angular desorption distribution as obtained from the beam data in Fig. 12.17 by applying normal energy scaling and detailed balance, for two different temperatures, 523 K and 700 K. The $\cos \Theta$ distribution (full line) is shown for comparison. (Kratzer *et al.* 2007, with permission from Elsevier, copyright 2007.)

an angle Φ can be calculated:

$$S_T(\Phi) = \frac{\int N_T(E_\perp / \cos^2 \Phi) \cdot S(E_\perp, 0°) dE_\perp}{\int N_T(E_\perp / \cos^2 \Phi) dE_\perp}. \tag{12.17}$$

From the principle of detailed balancing between the angular distribution of the adsorption flux $A(\Phi)$ and desorption flux $D(\Phi)$ given by

$$A(\Phi) = S(\Phi) \cdot \cos \Phi = D(\Phi) \tag{12.18}$$

we can finally calculate the angular distribution for desorption. Quite good agreement between the calculated and measured angular distribution for two different sample temperatures has been obtained (Fig. 12.18).

12.4   Experimental techniques

Reaction studies on nanostructured surfaces require both the characterization of the particular surface involved, as well as the determination of the reaction kinetics and dynamics. With respect to the surface characterization scanning tunnelling microscopy (STM) is the most useful technique, since it allows surface characterization with atomic resolution. Atomic force microscopy (AFM) is also a possible method, albeit atomic resolution is usually not achieved. There are now already several examples available in the literature where indeed surface reactions have been studied on the atomic level by using STM (Ertl 2002; Hla and Rieder 2003; Schoiswohl *et al.* 2005). However, it is in general extremely difficult to combine the requirements for proper STM studies with reaction measurements on a sophisticated level. Therefore, in most cases the structural characterization of the surface and the reaction studies are performed in separate apparatus. Since the experimental techniques for STM and AFM have been reviewed many times (e.g. Bray *et al.* 1995; Wiesendanger and Güntherodt 1997; Chen 2007); I will concentrate in this chapter on the experimental techniques to study reaction kinetics and in particular reaction dynamics.

12.4.1 Determination of the sticking coefficient

The determination of the sticking coefficient requires both the determination of the adsorbed amount as well as the number of the impinged molecules per time unit and surface area. The integral of the adsorption rate over time is the exposure, which is usually given in Langmuir (L), where 1 L corresponds to 10^{-6} Torr s. To obtain quantitative values of the impingement rate one has to obtain the gas pressure absolutely. Sticking coefficients obtained by dosing the sample from the isotropic gas phase are called *integral sticking coefficients*. The impingement rate per surface unit area A is given by:

$$\frac{\mathrm{d}N_{imp}}{A\mathrm{d}t} = \frac{p}{\sqrt{2\pi mkT}}, \tag{12.19}$$

with p: gas pressure, m: mass of the molecule, T: gas temperature, k: Boltzmann's factor.

Usually, ion gauges that are most frequently used for pressure measurements in the ultrahigh-vacuum range have to be calibrated. This can be done with the help of a spinning rotor gauge (Fremerey 1985; Winkler 1987). The same is true for the mass spectrometers, most frequently quadrupole mass spectrometers (QMS), equipped with secondary electron multipliers (SEM), when it comes to determining partial pressures absolutely. For the quantitative determination of the adsorbed amount several methods can be applied. If the adsorbate shows a clear saturation coverage and is also well ordered, then low-energy electron diffraction (LEED) can be used to obtain the coverage from the superstructure of the LEED pattern. Although this method is not always unambiguous it is frequently used. Of course, STM can also be used to obtain the adsorbate coverage when a well-defined structure is observed. However, one has to make sure that this structure is indeed representative of the whole surface.

Thermal desorption spectroscopy (TDS) is also a powerful method to determine the amount of adsorbed molecules. In this case the adsorbed particles are desorbed by heating the sample. This yields a pressure increase Δp in the vacuum chamber from which the amount of adsorbed particles can be calculated:

$$N_{\mathrm{ads}} = N_{\mathrm{des}} = KS\int \Delta p\mathrm{d}t, \tag{12.20}$$

with S: pumping speed [l/s], K: 3.3×10^{19} [molecules/Torr l].

However, this again requires the calibration of the pressure gauge or the mass spectrometer, as well as the pumping speed in the vacuum chamber. This is, in fact, not an easy task. A relatively simple method in this context is to compare the desorption spectra of interest with a standard desorption spectrum. This can be the desorption spectrum of a well-known sample or the spectrum from a sample that has been previously calibrated. We have recently described such a method in detail for hydrogen desorption (Winkler 1987).

In short, a tungsten filament, which can easily be cleaned by heating to high temperature, shows a pronounced saturation desorption spectrum for hydrogen. This spectrum has been calibrated by a so-called "synthetic desorption spectrum" simulated by the effusion of a known amount of hydrogen from a glass vessel into the vacuum chamber via a valve. The amount of gas in the glass vessel with a known volume has been determined by measuring the gas pressure absolutely via a spinning rotor gauge (Fremerey 1985). The so-calibrated tungsten filament is then used as a secondary standard.

12.4.2 Determination of the angle dependence of adsorption and desorption rates

For the experimental determination of the angular dependence of the sticking coefficient unidirectional gas dosing and a tiltable sample have to be used. For unidirectional dosing a capillary array may be used, but also the effusion of the gas from a simple tube or a Knudsen cell may be sufficient if the distance between the gas source and the sample is not too small. Typically, capillary arrays of 1–2 cm in diameter are used with capillary diameters of about $10 \, \mu$m and 0.5 mm length. The high aspect ratio of these capillaries yields a strongly forward focused angular distribution and thus a nearly parallel gas beam. Such a capillary array can also be used as a unidirectional detector to determine the angular dependence of the desorption probability. In this case the capillary array forms the entrance port of a differentially pumped detector chamber (Winkler and Yates 1988). However, there also exist other possibilities to collimate the molecule flux that enters the detector chamber. Steinrück *et al.* (1985b) have described a collimator consisting of a tube with several apertures where the inner walls can be covered with titanium, which strongly pump those molecules that are scattered at the aperture walls. There are several possibilities to determine the angular distribution of the desorption probability: a) The sample can be tilted in front of the detector. This is the most accurate way and most frequently applied (Steinrück *et al.* 1985b). b) The sample–detector distance can be changed. From the distance dependence of the measured signal the angular distribution can also be calculated (Winkler and Yates 1988). Another possibility is to laterally shift the sample in front of the differentially pumped detector. Again, the change of the signal as a function of the lateral displacement can be determined and then compared with a Monte Carlo calculation, by assuming a particular angular dependence in the form of a $\cos^n \Phi$ function (Kratzer *et al.* 2007). In all of these setups one has to verify that the gas molecules entering the detector chamber do not directly strike the ionization region of the mass spectrometer. Since the mass spectrometer is a density detector rather than a flux detector, the measured signal depends also on the kinetic energy of the molecules. To suppress this influence the mass spectrometer has to be placed out of line-of-sight.

12.4.3 Determination of the kinetic-energy dependence of adsorption and desorption rates

The simplest way to influence the kinetic energy of the gas flux is to heat the gas in a Knudsen cell. If the pressure in the cell is small enough the effusing gas has a kinetic energy distribution according to a Maxwell distribution. With this method at least the mean energy of the gas can be varied by varying the temperature of the gas source. It should be noted that the energy distribution and hence the mean energy of a *gas flux* is different from that of an *isotropic gas* (Comsa and David 1985). The energy distribution $N(E)$ and the velocity distribution $N(v)$ of a Maxwellian flux is given by:

$$\frac{dN(E)}{dE} \propto E \cdot \exp\left(-\frac{E}{kT}\right) \tag{12.21}$$

$$\frac{dN(v)}{dv} \propto v^3 \exp\left(-\frac{mv^2}{2kT}\right). \tag{12.22}$$

As a consequence, the mean kinetic energy of a gas flux is $\langle E \rangle = 2kT$, instead of $\langle E \rangle = 3/2kT$ for an isotropic gas.

A much better defined gas beam can be produced by high-pressure expansion from a nozzle source. Due to the expansion, adiabatic cooling leads to a very narrow energy and velocity distribution of the molecules in the jet beam (Scoles 1988). The velocity distribution can be described by a Maxwellian flux corresponding to a very small temperature, $T_{\parallel}$, superimposed on the drift velocity v_s:

$$\frac{dN(v)}{dv} \propto v^3 \exp\left(-\frac{m(v - v_s)^2}{2kT_{\parallel}}\right). \tag{12.23}$$

The quality of the quasi-monoenergetic beam can be characterized by the speed ratio $S = v_s/u$, with v_s being the drift velocity and u the most probable velocity of a gas with temperature $T_{\parallel}$, but other criteria can also be used for this characterization. Since the velocity of sound depends on the temperature of the gas such a nozzle beam is also called a supersonic beam.

Applying thermodynamic considerations, namely that the enthalpy during expansion should be conserved, one can determine the maximum kinetic energy of the beam. For a monoatomic gas the mean energy of a jet beam for optimum expansion conditions amounts to $\langle E \rangle = (5/2)kT$, where T is the nozzle temperature. For a two-atomic gas the rotational and vibrational energy can also be partially converted into translational energy, thus $\langle E \rangle \geq (5/2)kT$ can be possible (Gallagher and Fenn 1974). The actual velocity or energy distribution, and hence the (mean) kinetic energy has to be measured by determining the time-of-flight distribution. For this purpose the beam has to be chopped and the arrival time on a mass spectrometer can be measured by a multichannel analyzer. The beam energy cannot only be adjusted by the nozzle temperature, but also by mixing the gas with another gas, typically an inert gas, with different mass. This so-called seeding technique allows us to expand the attainable energy range. Such quasi-monoenergetic molecular beams have frequently been used to study the energy dependence of the sticking coefficient, this subject has been reviewed by Rendulic and Winkler (1994).

The energy distribution of particles *desorbing* from a surface can also be measured by time-of-flight spectroscopy, similar to that used for the determination of the velocity distribution of molecular beams (Comsa and David 1985; Matsushima 2003; Winkler *et al.* 2007). Here, the desorbing particles first enter a differentially pumped first stage, via a small aperture, which contains the chopper. Typically, two-slit chopper blades are used with rotational frequencies of several hundred Hz. The molecule packages with an initially close to delta function in the space regime travels to a quadrupole mass spectrometer, located in a separate differentially pumped chamber, where the arrival times of the molecules of the dispersed package are measured with a multichannel analyzer, which has been triggered by a chopper signal. The obtained time-of-flight (TOF) signal can be converted into the velocity or energy regime and a mean energy for the desorbing molecules can be calculated.

12.4.4   Determination of the internal-energy dependence of adsorption and desorption rates

The most demanding task in the context of determining differential adsorption and desorption rates concerns the influence of the rotational and vibrational state distribution. The simplest approach of course would be to use Maxwellian beams from a Knudsen cell, where not only the translational energy changes as a function of the cell temperature, but also the rotational and vibrational state distribution, which can be described by the corresponding Boltzmann distribution. However, this change is in most cases not significant enough to be useful for this purpose. In particular, the vibrational states are typically separated by several hundred meV, so the higher vibrational states cannot be populated at reasonable cell temperatures. The most severe problem is, however, that the three degrees of freedom cannot be influenced individually. Therefore, it is not possible to figure out the contribution of the individual degrees of freedom to sticking. A similar problem arises when using supersonic molecular beams. However, here the seeding technique allows at least to partially disentangle the influence of translational and internal energies (Miller 1988). By mixing the gas of interest (minority species) into an inert seeding gas (majority species) the translational energy of the minority species accommodates to that of the majority species. By this method, e.g. hydrogen can be translationally cooled by seeding in argon, or oxygen can be translationally heated by seeding in helium. The internal state distribution, however, is still mainly determined by the nozzle temperature. By continuously changing the mixing ratio and the nozzle temperature adequately, one can change the internal energies within some range, while the translational energy can be held constant, or vice versa. With this method the influence of the rotational energy on the sticking of hydrogen on Pd(111) has been investigated (Beutl *et al.* 1995).

Another method takes advantage of the ortho–para modifications of symmetric molecules, e.g. for hydrogen or deuterium. Due to the nuclear spin degeneracy hydrogen with even j-states are onefold degenerated, whereas odd j-states are threefold degenerated. The population of the individual states is given by:

$$N(j) = g_n(2j + 1) \exp\left(-\frac{E(j)}{kT}\right), \qquad (12.24)$$

with g_n: nuclear spin degeneracy, j: rotational quantum number, $E(j)$: rotational energy of state j.

Between the two branches of the ortho and para modifications only little interconversion takes place. Therefore, for normal hydrogen (n-H_2) at low temperature ($< 70\,K$) nearly all para-H_2 will end up in the $j = 0$ state and all ortho-H_2 will be in the $j = 1$ state. The population of these two states is 1:3. On the other hand, quite pure para-hydrogen can be produced by a catalytic conversion method and purchased commercially, which after cooling will result in a 100% population of the $j = 0$ state. Thus, one can produce hydrogen beams with significantly different populations of the two lowest rotational states. This technique has been used by Beutl *et al.* (1996) to study the influence of the rotational energy on the sticking behavior of hydrogen on Pt(110).

The most direct way to influence the rotational-state distribution in a gas beam would be to selectively populate individual states by direct optical excitation. Another possibility would be to select special states by an electrostatic filter. The experimental realization of both techniques has recently been reviewed by Sitz (2002).

For the determination of the rotational and vibrational state distribution of desorbing molecules also spectroscopic techniques can be applied. Laser-induced fluorescence (LIF) and resonance-enhanced multiphoton ionization (REMPI) are the most frequently used successful techniques (Zacharias 1988). In this case the desorbing molecules are resonantly excited, with the help of an intense tunable laser radiation, from the rovibrational states of the electronic ground state into the rovibrational levels of an electronically excited state, obeying the proper selection rules. One can then either detect the fluorescence light as a consequence of the de-excitation of the molecule with a photo-multiplier (LIF) or one can further ionize the already-excited molecule and detect the ions with an electron multiplier (REMPI). In particular, the latter method has turned out to be quite sensitive because ions can be collected very effectively (Pozgainer *et al.* 1994). This technique has been used by several groups, especially to study the rotational and vibrational (rovibrational) state population of desorbing hydrogen molecules from various surfaces, e.g. by Kubiak *et al.* (1985), Schröter *et al.* (1991), Michelsen *et al.* (1993), Winker *et al.* (1994), and others.

In a further improved version of the REMPI technique the kinetic energy of the state-selected ions can be measured by a TOF spectrometer, thus allowing the simultaneous determination of all degrees of freedom (Michelsen *et al.* 1993; Schröter *et al.* 1993; Murphy and Hodgson 1997; Gleispach and Winkler 2003). If such an experiment is performed even for different desorption angles the most comprehensive information on a desorption system can be gained.

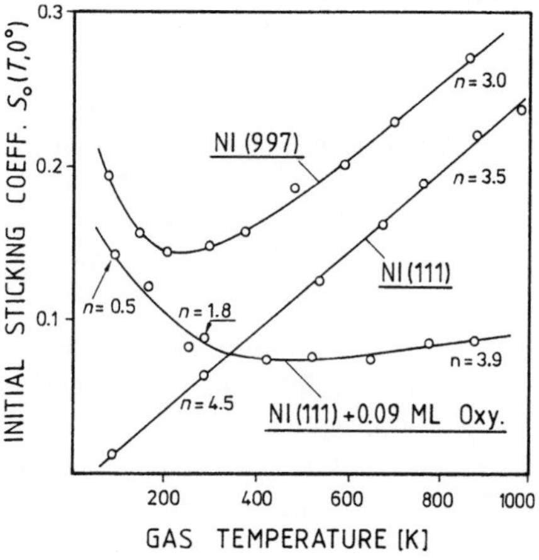

Fig. 12.19 Comparison of the initial sticking coefficient of hydrogen as a function of the mean translational energy (gas temperature) on a flat Ni(111), a stepped Ni(997) surface and a partially oxygen covered Ni(111) surface (Rendulic *et al.* 1987, reprinted with the permission of the American Institute of Physics).

12.5   Selected experimental results

12.5.1   Hydrogen adsorption/desorption on stepped surfaces

Stepped surfaces are ideal nanostructured templates to study the specific adsorption properties of active sites. We have studied the adsorption and desorption kinetics of hydrogen on stepped nickel single-crystal surfaces, Ni(997) and Ni(445), and compared the results with the flat Ni(111) surface. The Ni(997) surface consists of terraces of (111) orientation 9-atom rows wide and of monoatomic steps with (111) orientation. The Ni(445) surface consists again of terraces with (111) orientation 8-atom rows wide with monoatomic steps of (100) orientation. Figure 12.19 shows the strong difference of the sticking coefficient for hydrogen on the flat (111) and stepped (997) surface, as a function of the mean kinetic energy of the impinging molecules. In this case a Knudsen cell was used to vary the mean translational energy (Rendulic *et al.* 1987). Whereas on the flat surface the sticking probability decreases continuously with decreasing translational energy, indicating activated adsorption, on the stepped surface the sticking coefficient starts to increase again at very low kinetic energy. This is a clear indication of a precursor-assisted or steering-induced adsorption. Apparently, at the step sites the activation barrier for adsorption is strongly reduced or even totally eliminated. Comparing both curves at high translational energy one can see a constant offset that corresponds to a difference in the sticking coefficient of about 0.07. Since the step density is about 10% one can conclude that the local sticking coefficient at a step site is close to unity, i.e. that no activation barrier exists.

In Fig. 12.19 the influence of oxygen on the dissociative adsorption of hydrogen on the flat Ni(111) surface is also shown. Whereas at higher beam energies the initial sticking coefficient is smaller than on the clean surface, at low kinetic energy again an increase of S_0 is observed. There are apparently two different reasons responsible for this behavior. At a coverage of

0.09 monolayers of oxygen there exist oxygen islands forming a (2×2) superstructure. On these islands no hydrogen adsorption can take place (Winkler and Rendulic 1982). Therefore, the decreased sticking at high kinetic energy just reflects the reduced free surface areas. On the other hand, at the oxygen island borders as well as on single adsorbed oxygen atoms (forming a so-called lattice gas between the condensed oxygen islands) the activation barrier for dissociative hydrogen adsorption is reduced. At these active sites adsorption can take place with high sticking probability via a precursor state. Since the accommodation into this precursor is facilitated at lower beam energy, the sticking coefficient increases with decreasing translational energy of the impinging hydrogen molecules.

Following the described scenario one should also expect an influence on the angular distribution of the sticking coefficient. According to eqn (12.12) the angular distribution can be characterized by the exponent n of the $\cos^{n-1} \Phi$ function. On the clean Ni(111) surface n is between 4.5 and 3.5 over the whole energy range, reflecting the activated adsorption behavior. At the oxygen-covered surface at high translational energy $n = 3.9$, again reflecting the activated adsorption on the clean parts of the surface. With decreasing kinetic energy the value n also decreases to 1.8 for the room-temperature gas and even further to 0.5 for the lowest gas temperature of 90 K. This fits nicely to the assumption of precursor-assisted adsorption where in fact the exponent n should be smaller than unity. The oxygen islands and the oxygen monomers in the lattice gas are evenly distributed on the surface. Therefore, the measured angular distributions are symmetric with respect to the surface normal and do not show any azimuthal dependence. However, on the stepped Ni(997) surface one could expect an asymmetric azimuthal-dependent angular distribution. This is actually observed in the experiment.

Figure 12.20 shows the angular distribution of the adsorption probability $A(\Phi) = S(\Phi) \cos \Phi$ as a function of the tilt angle of the sample with the [01̄1] direction as tilt axis (Steinrück *et al.* 1985a). This is the direction parallel to

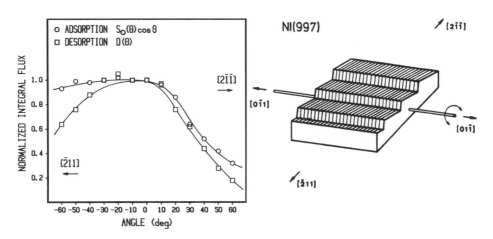

Fig. 12.20 Angular distribution of the adsorption probability of hydrogen on the stepped Ni(997) surface, as a function of incidence angle. Tilt angle is parallel to the [01̄1] direction, as sketched on the right-hand side. Negative angles are "step-up". For comparison the angular distribution of the desorption flux is also shown (Steinrück *et al.* 1985a, reprinted with the permission of the American Physical Society).

the steps. Asymmetric behavior is clearly observed. In the case of negative angles the molecules approach the surface "step-up", leading to a significantly increased sticking coefficient. In the case of "step-down" impingement no strong difference to the flat Ni(111) surface is observed. This shows that step sites act as adsorption sites where dissociation can take place without an (or with a small) activation barrier if the molecule has a velocity component "into" the steps. The angular distribution if measured with the tilt axis normal to the steps shows a symmetric behavior, as one would expect (Steinrück *et al.* 1985a).

Quite interesting is the effect of oxygen on the sticking behavior of hydrogen on the stepped surface. Whereas on the flat Ni(111) surface oxygen has a rather activating effect, on the stepped surface it has a deactivating effect. The integral sticking coefficient of hydrogen on the clean Ni(111) surface is about 0.02. This value increases to 0.16 on the stepped surface, but decreases again to 0.05 on the partially oxygen-covered surface. In this case an oxygen coverage of only 5% has been used. This is sufficient to fully decorate the steps that leads to a deactivation of the steps with respect to hydrogen dissociation. This is also nicely seen in the changed angular distribution (Fig. 12.21). Now, the angular distribution is again close to symmetric, with the symmetry axis 6° off the surface normal. This corresponds to the cutting angle between the macroscopic surface plane and the (111) net plane. The exponent n of the angular distribution is 3.6, similar to that on the flat (111) plane. The implication of this experiment is straightforward: Oxygen adsorbed on the step sites deactivates these sites and only activated adsorption on the clean terrace sites can take place. The important conclusion that can be drawn from these experiments is also that impurity atoms (like oxygen) can act both as a promoter (on a flat surface) as well as an inhibitor (on a stepped or defective surface).

The described adsorption and desorption angular distribution experiments on the clean Ni(997) and the partially oxygen covered Ni(997) surface (Fig. 12.20), compared with the equivalent experiments on the flat Ni(111) surface, allow us to draw some important conclusions with respect to the

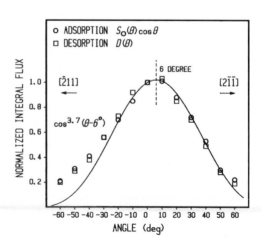

Fig. 12.21 Comparison of the adsorption and desorption flux angular distribution of hydrogen on a partially oxygen (0.05 ML) covered Ni(997) surface. The steps are blocked and the distribution function is symmetric to the normal of the (111) terraces (Steinrück *et al.* 1985a, reprinted with the permission of the American Physical Society).

application of detailed balancing on nanostructured surfaces. On the flat surface both the adsorption as well as the desorption angular distribution follow a $\cos^{4.5} \Phi$ function (Steinrück *et al.* 1985c). This means that detailed balancing is fulfilled with high accuracy even for the non-equilibrium situation. Adsorption was performed at a surface temperature of 220 K, a gas temperature of 300 K and the coverage for the determination of the initial sticking coefficient was 2–5% of a monolayer (Steinrück *et al.* 1984). Desorption of hydrogen takes place at a peak temperature of about 390 K and the coverage was about 85% of the β_2 state. Similar experimental conditions were applied also for the stepped surface. But in this case a clear deviation, in particular for the step-up situation, can be observed between the adsorption and desorption experiment (Fig. 12.20). For the initial sticking coefficient measurement the steps play the dominant role, where unactivated adsorption can take place. Therefore, the distribution function is very asymmetric. For the desorption measurement the β_2 adsorption state is nearly saturated, i.e. the terraces are already to a large extent occupied. Only those hydrogen molecules that recombine in the vicinity of the steps will show the strong asymmetric desorption behavior, whereas those from the terraces will desorb with a distribution function symmetric to the terrace normal. That this is indeed the case is clearly seen on the oxygen-deactivated surface (Fig. 12.21). In this case the adsorption and desorption angular distribution functions are very similar, because only on the terraces does adsorption and desorption take place. The main message of these experiments is that in the case of inhomogeneous surfaces, as typically represented by nanostructured surfaces, the principle of detailed balancing is generally not applicable for non-equilibrium measurements.

12.5.2   Methanol reaction on the Cu-CuO stripe phase

Two-fold symmetric metal surfaces in general and the Cu(110) surface in particular tend to form one-dimensional nanostructures due to surface reconstructions (added row or missing row superstructures). This can be enhanced by the adsorption of other material on the surface. Oxygen induces the formation of long Cu–O strings, which finally add up to stripes. At a nominal coverage of 0.25 monolayers of oxygen a well-defined nanostructured surface of alternating stripes of bare copper and copper oxide appears with a mean width of 6.5 nm (see Fig. 12.6(a)). Such a surface is well suited to act as a template for the growth of other nanostructures or to serve as a well-defined nanostructure for catalytic model reactions. In the group of Zeppenfeld the fabrication of such Cu–CO stripe phase surfaces (Kern *et al.* 1991; Zeppenfeld *et al.* 1994, 1998) and their application has been intensively studied in recent years. For example, the formation of regular arrays of cobalt on the Cu–CuO stripe phase was studied, because of its relevance for ultrathin magnetic films (Tölkes *et al.* 1998). The adsorption and film growth of large organic molecules, like para-hexaphenyl, was also investigated on this surface (Oehzelt *et al.* 2007). With respect to surface reactions the adsorption of carbon monoxide (Sun *et al.* 2003) and nitrogen (Zeppenfeld *et al.* 2002) has been studied in detail.

(a)   (b)   (c)

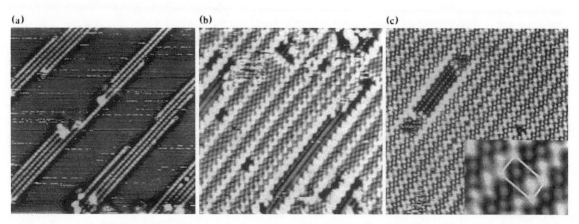

Fig. 12.22 Scanning tunnelling microscopy (STM) images of an oxygen-covered Cu(110) surface. (a) After dosing 5 L oxygen leading to the stripe phase, (b) after dosing 10 L methanol and (c) after waiting for 40 min. The STM image areas are 200 Å × 200 Å. (Leibsle *et al.* 1994, reprinted with permission, copyright 1994 by the American Physical Society.)

The interaction of methanol on the oxygen-covered Cu(110) surface in general and on the Cu–CuO stripe phase in particular has attracted considerable interest due to its importance in the oxidation and synthesis of methanol. Whereas the general adsorption and reaction kinetics was already investigated in the pioneering works by Madix (Wachs and Madix 1978; Bowker and Madix 1980) and Sexton *et al.* (1985), additional details concerning the microscopic processes have been unearthed by STM studies in the groups of Leibsle and Bowker (Francis *et al.* 1994; Leibsle *et al.* 1994; Poulston *et al.* 1996; Jones *et al.* 1997). In Fig. 12.22 STM images of a Cu(110) surface covered with some CuO stripes (a), following dosing of 10 L methanol (b) and after waiting 40 min at room temperature (c) are shown (Leibsle *et al.* 1994). One can clearly see that methanol adsorbs on the clean copper areas only. Furthermore, it is known from spectroscopic experiments that actually methoxy species are adsorbed instead of methanol. The methoxy species form a regular (5×2) superstructure. The important result is, however, that during dosing the surface with methanol at room temperature the CuO stripes shrink along the [001] direction. This is due to the reaction of stripped off hydrogen from methanol with the oxygen atoms at the end of the stripes. Recombination of the formed OH with a further hydrogen atom leads to the formation of water, which immediately desorbs at this temperature. It was also shown that along the long side of the stripes a more stable zigzag-like structure exists.

We have studied this system by using thermal desorption spectroscopy (TDS) and angle-resolved TDS to get a comprehensive understanding of the desorption kinetics of the reaction products. Figure 12.23 shows a multiplexed thermal desorption spectrum after dosing a 0.25-ML oxygen-dosed Cu(110) surface to 5 L methanol at 180 K. Between 200 K and 300 K methanol and water desorption can be observed. Around 350 K nearly concomitantly methanol, formaldehyde and hydrogen desorption takes place. Finally, at around 450 K carbon dioxide desorbs, accompanied by small amounts of hydrogen and water.

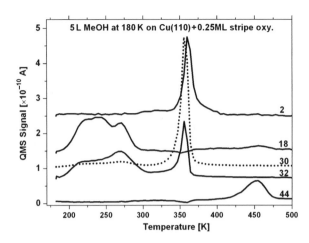

Fig. 12.23 Multiplexed TD spectrum after dosing a Cu(110)-(2 × 1)O stripe phase with 5 L methanol at 180 K. Heating rate 2 K/s. Mass 2: hydrogen, 18: water, 30: formaldehyde, 32: methanol, 44: carbon dioxide.

The desorption peaks around 350 K exhibit an extremely narrow FWHM. This indicates a very specific desorption kinetics. Detailed desorption experiments with increasing methanol exposure reveal a shift of these narrow peaks to higher temperature with increasing methanol coverage. On the other hand, the methanol desorption peak around 250 K shows a decrease of the peak maxima with increasing coverage. This is what one would expect for recombinative desorption (recombination between methoxy and hydrogen to methanol) (Demirci *et al.* 2007). The reason for the peculiar desorption at 350 K is that all desorbing species are reaction limited. In this temperature range hydrogen atoms are detached from the methyl group of methoxy leading to formaldehyde. These nascent species desorb immediately because adsorbed formaldehyde would already desorb around 220 K (Sexton *et al.* 1985). On the other hand, some of the detached hydrogen atoms can recombine with methoxy to methanol. The nascent methanol species also desorbs immediately because molecularly adsorbed methanol would already desorb below 300 K. Finally, some of the detached hydrogen atoms can recombine to H_2. The desorption temperature of hydrogen from Cu(110) is around 330 K (Anger *et al.* 1989), thus this species is at least partially desorption limited. It can be seen in Fig. 12.23 that the desorption peak of hydrogen is indeed delayed by about 5 K compared to methanol and formaldehyde.

To get more information on the specific desorption kinetics of the reaction products for methanol on Cu(110) we have determined the angular distribution of the desorption products. For this purpose we have applied the method of lateral sample displacement with respect to the flux-sensitive detector, as described elsewhere (Kratzer *et al.* 2007). In Fig. 12.24 the experimental result is shown for hydrogen desorption at 350 K, for the sample displacement in X- and Y-directions. In our experimental setup the Y-direction is parallel to the [001] direction of Cu(110), i.e. parallel to the stripes. One can see that the angular distribution is quite different in both directions. Whereas the displacement in the Y-direction shows a smooth change of the signal, which can be approximated by an angular distribution of $\cos^{11} \Phi$, the displacement along the X-direction leads to a bimodal desorption behavior. The borders between the CuO stripes and the bare Cu stripes apparently lead to a desorption distribution

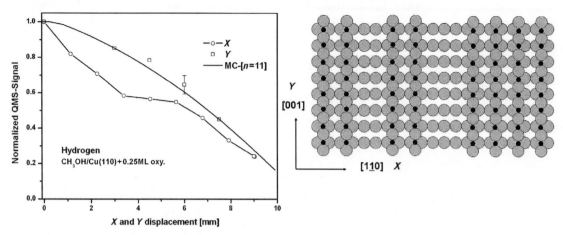

Fig. 12.24 Angular desorption distribution of hydrogen after exposing the Cu–CuO stripe phase surface to 3 L methanol at 190 K, as measured by sample displacement in the X-direction (parallel to the Cu[1$\bar{1}$0] direction) and the Y-direction (parallel to the Cu[001] direction), as shown in the right-hand sketch. The smooth line for the Y-direction is obtained from Monte Carlo simulations using the distribution function in the form of $\cos^{11} \Phi$. The data points for the X-direction are connected to guide the eye (Demirci *et al.* 2007, reprinted with permission from the American Institute of Physics, copyright 2007).

that is focused off normal by about 20° (Demirci *et al.* 2007). A similar behavior was also observed for formaldehyde and methanol desorbing in the same temperature range. The strongly forward-focused desorption distribution indicates strongly activated desorption. Indeed, recent DFT calculations by Sakong and Gross (2005) have shown that a high activation barrier of 1.6 eV exists in order to break the C–H bond of the methoxy species next to an adsorbed hydrogen atom, which then leads to desorption of formaldehyde, leaving two hydrogen atoms on the surface. The energy of the produced formaldehyde molecule is still 1.2 eV above the zero-energy level. This excess energy leads inherently to a hyperthermal desorption flux and to a forward-focused angular desorption distribution. We assume that part of this excess energy is also channelled into the other reaction products, namely methanol and hydrogen, which therefore also leads to a forward-focused angular desorption distribution. These investigations clearly show the specific effect of well-defined nanostructures on the desorption kinetics of reaction products.

The *adsorption* kinetics of methanol on Cu(110) is also strongly influenced by adsorbed oxygen on the surface. We have determined the uptake of methanol at different oxygen pre-coverage with the help of reflectance difference spectroscopy (RDS) (Sun *et al.* 2009). This method is sensitive to electronic changes on the surface as induced, e.g by oxygen or methanol. In this case the strong absorption at 2.1 eV is quenched. Details of this method can be found elsewhere (Hohage *et al.* 2005). The RDS signal change at 2.1 eV is over some coverage range proportional to the adsorbed amount and can be measured continuously. In Fig. 12.25 a series of RD signals as a function of methanol exposure for different amounts of pre-adsorbed oxygen is shown. The oxygen-covered substrate was always annealed at 600 K in order to yield a well-defined striped surface. The most interesting result is the strong difference of the RD signal change for $\Theta_O = 0$ and $\Theta_O = 0.05$ ML oxygen. The initial

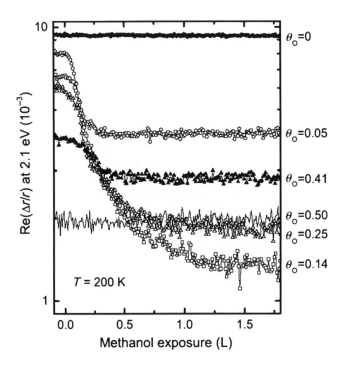

Fig. 12.25 RD signal at 2.1 eV recorded during methanol adsorption at 200 K on Cu(110) as a function of methanol exposure for different initial oxygen coverage. From Sun *et al.* (2009).

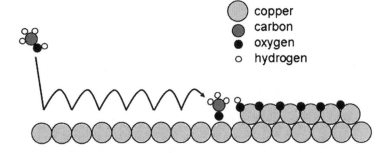

Fig. 12.26 Schematic model of a precursor adsorption of methanol on the edge of a Cu-CuO stripe, where hydrogen is detached, forming OH and methoxy.

slope can be correlated with the initial sticking coefficient, which is larger by more than three orders of magnitude on the 0.05-ML oxygen-covered surface than on the clean surface. This behavior can be explained by an adsorption process involving a strong precursor. The methanol molecule first enters a weakly bound physisorption state where it can freely move along the surface. The barrier that has to be overcome to break the O–H bond of methanol is apparently very high on the clean Cu(110) surface. Only when the molecule approaches an adsorbed oxygen atom, either in the form of a single atom or in the form of a boundary atom of the CuO stripes, can the H atom be efficiently attracted by the oxygen atom to form OH and the nascent methoxy species can chemisorb in the neighborhood (see Fig. 12.26).

12.5.3   Hydrogen desorption and water formation on vanadium-oxide nanostructures

Surface vanadium oxides, which exhibit a large variety of different stoichiometries and form different structures on Pd(111), are excellent model systems for reaction studies on nanostructured surfaces. In Fig. 12.10 STM images of two oxide structures, a $(4 \times 4)V_5O_{14}$ and a (2×2) s-V_2O_3 structure, are shown. The most important question in this context is of course the stability of such surface oxides under reaction conditions, e.g. for the reaction of hydrogen or deuterium with oxygen to form water. For the supply of hydrogen we have applied a special experimental technique, namely the permeation technique. In this case, the Pd(111) crystal is part of a permeation source, as described in detail by Pauer *et al.* (2005b). Briefly, the palladium single crystal is soldered onto a nickel cylinder that can be filled with hydrogen or deuterium up to about 1000 mbar. When the whole device is heated, hydrogen permeates through the 1-mm thick palladium sample and recombines on the surface, to desorb as H_2. If at the same time oxygen is supplied from the gas phase, water can be produced that again desorbs at sufficient high surface temperature. The advantage of the permeation technique is that a continuous hydrogen flux can be maintained, which can be analyzed in detail, e.g. with respect to the angular and energy distribution of the desorption and reaction products. In our case we have measured the time-of-flight (TOF) spectra of desorbing deuterium and water.

On the clean sample deuterium desorption is close to being thermalized, both at 523 K and 700 K (Fig. 12.27(a)). This shows that associative desorption of deuterium from Pd(111) is largely un-activated, as already described for hydrogen desorption in previous work (Comsa and David.1985; Pauer *et al.* 2005a). The TOF spectrum for deuterium from a Pd(111) surface covered with 0.3 MLE of vanadium oxide is shown in Fig. 12.27(b) (curve 1). By MLE we mean monolayer equivalents: 0.3 monolayers of vanadium were evaporated under an oxygen atmosphere forming the oxide on the Pd surface (Kratzer *et al.* 2006). We have checked the surface structure during the reaction with LEED and have compared the LEED patterns with those obtained by Surnev *et al.* (2003) in combined LEED-STM investigations. The surface vanadium-oxide film, which is stable in our particular case, is characterized by the s-$V_2O_3(2 \times 2)$ structure. A least squares fit of a Maxwellian TOF distribution to the experimental data points leads to $T_{\text{fit}} = 481$ K. Although this value is not far off the surface temperature of 523 K, this decrease in the mean translational energy was reproducibly measured within the margin of error.

The TOF measurement of D_2 during molecular oxygen exposure at 1×10^{-6} mbar leads to a TOF spectrum as depicted in Fig. 12.27(b) (curve 2). In addition to water formation associative desorption of deuterium molecules still takes place, due to the much higher deuterium permeation rate compared to the oxygen impingement rate. The best fit temperature is now $T_{\text{fit}} = 476$ K. Apparently, the TOF spectrum is not changed significantly and again exhibits a "cooled" energy distribution. It would be tempting to ascribe this feature of a cooled deuterium flux to particular changes of the potential-energy surface

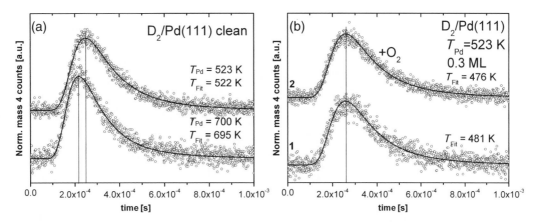

Fig. 12.27 (a) Time-of-flight spectra for deuterium from clean Pd(111) at 523 K and 700 K, respectively. The best fit temperatures for a Maxwellian distribution are also given, indicating a thermalized desorption (Kratzer *et al.* 2006, reprinted with permission from the American Institute of Physics, copyright 2006). (b) Time-of-flight spectra for pure deuterium desorption (curve 1) and deuterium desorption during concomitant oxygen exposure (curve 2) from 0.3 MLE VO_x on Pd(111) at 523 K (Kratzer *et al.* 2006, reprinted with permission from the American Institute of Physics, copyright 2006).

(PES) for the desorption process due to the vanadium-oxide film. However, careful examination of the experiment leads to a different and quite simple explanation of this result. One has to consider that the permeating deuterium leads to an increase of the partial deuterium pressure in the main chamber, which results in quite a high impingement rate of isotropic deuterium from the gas phase (thermalized to 300 K). By fitting the sum of two Maxwellians with $T = 523$ K and $T = 300$ K to the experimental data we have calculated that about 15% of the total deuterium flux, which leaves the surface, originates from reflected or adsorbed/desorbed deuterium from the gas phase. On the bare Pd(111) surface the sticking coefficient for hydrogen (deuterium) is quite high ($S_0 = 0.45$, Beutl *et al.* 1995). In this case the permeated/recombined deuterium molecules and the adsorbed/desorbed deuterium molecules are both closely thermalized to the sample temperature (523 K), as observed experimentally. In the case of the 0.3 MLE vanadium oxide about 60% of the surface is covered by the oxide. We assume that deuterium molecules impinging on the oxide areas do not adsorb but are reflected without a notable change of their kinetic energy, corresponding to 300 K. This contribution to the total desorption flux yields on average a mean translational energy that is equivalent to a smaller temperature than the corresponding sample temperature.

We have also measured the angular distribution of desorbing deuterium from a 0.3 and 0.5 MLE VO_x covered surface (Fig. 12.28). This has been done by lateral displacement of the sample in front of a directional detector and comparing the result with Monte Carlo simulations assuming a $\cos^n \Phi$ distribution, as described in section 4.2. One can see that the obtained angular distributions are well within $\cos \Phi$ and $\cos^2 \Phi$. Apparently, no significant increase of the activation barrier for the adsorption/desorption process has been induced by the vanadium oxide. How can we explain that the desorption behavior of deuterium on the vanadium-oxide-covered surface is nearly the same as on the

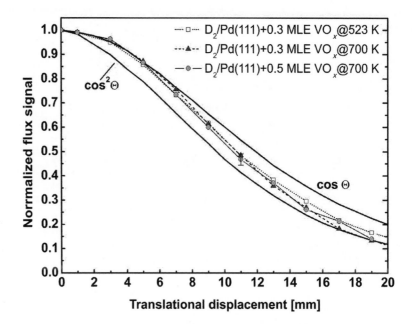

Fig. 12.28 Experimentally obtained change of the desorption flux signal for deuterium from the Pd(111) surface, pre-covered with 0.3 and 0.5 MLE VO_x. The full lines stem from Monte Carlo calculations for $\cos^n \Phi$ functions with $n = 1$ and 2, respectively (Kratzer *et al.* 2007, with permission from Elsevier, copyright 2007).

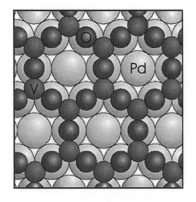

Fig. 12.29 Structure model of the s-V_2O_3 surface oxide on Pd(111) (after Schoiswohl *et al.* 2005).

clean Pd(111) surface? We believe that the particular structure of the vanadium surface oxide, as shown in Fig. 12.29, is responsible for this behavior. This structure, which is called the "pocket structure", is quite porous and allows for recombinative desorption in these pockets. The electronic influence of the vanadium and oxygen atoms is apparently of very short range character. Indeed, high-resolution XPS and HREELS experiments for CO adsorption on this s-V_2O_3 surface, in combination with DFT calculations, led to the conclusion that CO adsorbs in these "pockets" similarly as on the clean Pd(111) surface (Surnev *et al.* 2003).

During molecular-oxygen exposure we have also measured the TOF distribution of the water reaction product (Fig. 12.30). A best-fit temperature $T_{fit} = 544$ K is obtained, indicating that water desorption is close to being thermalized, but slightly hyperthermal. The surface vanadium oxide was not changed in the case of concomitant oxygen exposure. This is most probably due to the fact that even at 1×10^{-6} mbar oxygen pressure the number of impinging oxygen molecules (2.7×10^{14} O_2/cm^2 s) is much smaller than the number of permeating deuterium atoms (1.2×10^{16} H/cm^2 s), i.e. there still exist reducing conditions. The fact that desorbing water is slightly hyperthermal suggests that those deuterium atoms with higher kinetic energy in the energy distribution react with higher probability with the intermediate hydroxide species on the surface to form water. Taking into account that the above considerations for the influence of isotropic gas impingement on the measured TOF distribution also holds for water desorption, indicates an even "hotter" energy distribution for D_2O than determined from the best fit.

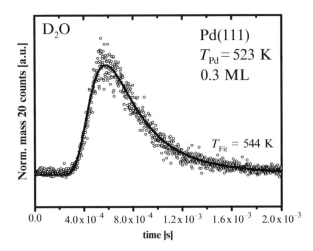

Fig. 12.30 Time-of-flight spectrum of D_2O resulting from the reaction of permeating D and impinging O_2 on 0.3 MLE VO_x on Pd(111) at 523 K (Kratzer *et al.* 2006, reprinted with permission from the American Institute of Physics, copyright 2006).

12.6 Summary

Surface-analytical techniques have been used to study model reactions on some selected nanostructured surfaces. The focus of these studies was put on the adsorption and desorption kinetics, as well as on the dynamics of adsorption and desorption. The latter comprises the energy and angle dependence of adsorption and desorption.

a) Hydrogen adsorption and desorption studies on a stepped Ni(997) surface revealed the strong influence of the steps. Whereas on the flat Ni(111) surface the sticking coefficient is very small, it increases significantly on the stepped surface. The steps create sites with negligible activation barrier for dissociative adsorption. This has a strong influence on the energy dependence of the sticking coefficient. Hydrogen molecules with low translational energy can adsorb at the steps via a precursor state, but also by translational steering of the impinging molecules to the favorable adsorption sites at the steps. This also leads to a characteristic angle dependence of the sticking coefficient, with a higher sticking if the hydrogen molecules are aiming into the steps. The active behavior of the steps can be poisoned by oxygen, which occupies the step sites preferentially. A partially oxygen covered stepped Ni(997) surface behaves similar to a flat, clean Ni(111) surface.

b) Methanol adsorption on the so-called Cu–CuO stripe phase again revealed the importance of island borders on the reactivity. Both on the clean Cu(110) surface as well as on the fully oxygen covered Cu(110) surface, where a $(2 \times 1)O$ superstructure is formed, the methanol adsorption probability is exceedingly small. On a partially oxygen covered surface CuO stripes are formed and at the borders of these stripes the sticking of methanol is increased by three orders of magnitude. This can again be explained by a pronounced precursor state for methanol over the clean and the oxygen-covered areas of the copper surface. Angular distribution measurements on the desorption flux of the individual reaction products (hydrogen, methanol, formaldehyde) revealed that the reaction and desorption also takes place predominantly at the rim of the CuO islands.

c) The dynamics of hydrogen desorption and water formation on a vanadium-oxide-modified Pd(111) surface turned out to be not very different from that on a clean Pd(111) surface. This could be explained by the special honeycomb-like structure of the palladium surface oxide (s-V_2O_3). Recombinative desorption of hydrogen and water takes place in the pockets of this structure, without a significant increase of the activation barrier. Therefore, the energy distribution of the desorption flux from the s-V_2O_3 covered Pd(111) surface is close to thermalized and the angular distribution is close to cosine, similar as for the clean surface. However, the rate of desorption is significantly reduced in the case of the oxide-covered surface.

Acknowledgments

I would like to acknowledge all the coworkers who have contributed to this work over the years. Significant contributions to this work came particularly from Klaus Rendulic, Hans-Peter Steinrück, Gernot Pauer, Markus Kratzer and Erkan Demirci. The financial support throughout the years by the Austrian Science Fund (FWF) is gladly acknowledged.

References

Alnot, M., Cassuto, A. *Surface Sci.* **112**, 325 (1981).

Anger, G., Winkler, A., Rendulic, K.D. *Surface Sci.* **220**, 1 (1989).

Berger, H.F., Leisch, M., Winkler, A., Rendulic, K.D. *Chem. Phys. Lett.* **175**, 425 (1990).

Beutl, M., Riedler, M., Rendulic, K.D. *Chem. Phys. Lett.* **247**, 249 (1995).

Beutl, M., Riedler, M., Rendulic, K.D. *Chem. Phys. Lett.* **256**, 33 (1996).

Blakely, D.W., Somorjai, G.A. *Surface Sci.* **65**, 419 (1977).

Bowker, M., Madix, R. *J. Surface Sci.* **95**, 190 (1980).

Bray, M.T., Cohen, S.H., Lightbody, M.L. *Atomic Force Microscopy and Scanning Tunneling Microscopy* (Springer Verlag, Berlin, 1995).

Brune, H., Giovanni, M., Bromann, K., Kern, K. *Nature* **394**, 451 (1998).

Cardillo, M.J., Balooch, M., Stickney, R.E. *Surf. Sci.* **50**, 263 (1975).

Cassuto, A., King, D.A. *Surf. Sci.* **102**, 88 (1981).

Chen, C.J. *Introduction to Scanning Tunneling Microscopy* (Oxford University Press, New York, Oxford, 2007).

Comsa, G., David, R. *Surf. Sci.* **117**, 77 (1982).

Comsa, G., David, R. *Surf. Sci. Rep.* **5**, 145 (1985).

Comsa, G. in *Proc. 7th Intern. Vac. Congr. and 3rd Intern. Conf. Solid Surf.* Vienna, p. 1317 (1977).

Coulman, D.J., Winntterlin, J., Behm, R.J., Ertl, G. *Phys. Rev. Lett.* **64**, 1761 (1990).

Dabiri, A.E., Lee, T.J., Stickney, R.E. *Surf. Sci.* **26**, 522 (1971).

Darling, G.R., Holloway, S. *Rep. Prog. Phys.* **58**, 1595 (1995).

Darling, G.R., Kay, M., Holloway, S. *Surf. Sci.* **400**, 314 (1998).

Demirci, E., Stettner, J., Kratzer, M., Schennach, R., Winkler, A. *J. Chem. Phys.* **126**, 164710 (2007).

Ertl, G. *Faraday Disc.* **1121**, 1 (2002).

Francis, S.M., Leibsle, F.M., Haq, S., Xiang, N., Bowker, M. *Surf. Sci.* **315**, 284 (1994).

Fremerey, J.K. *J. Vac. Sci. Technol. A* **3**, 1715 (1985).

Gallagher, R.J., Fenn, J.B. *J. Chem. Phys.* **60**, 3487 (1974).

Gleispach, D., Winkler, A. *Surf. Sci.* **537**, L435 (2003).

Gorte, R., Schmidt, L.D. *Surf. Sci.* **76**, 559 (1978).

Gross, A. *Surf. Sci. Rep.* **32**, 291 (1998).

Gross, A., Wilke, S., Scheffler, M. *Phys. Rev. Lett.* **75**, 2718 (1995).

Halstead, D., Holloway, S. *J. Chem. Phys.* **93**, 2859 (1990).

Hand, M.R., Holloway, S. *J. Chem. Phys.* **91**, 7209 (1989).

Harris, J. *Surf. Sci.* **221**, 335 (1989).

Harris, J. *Langmuir* **7**, 2528 (1991).

Hayden, B.E., Lamont, C. A. *Phys. Rev. Lett.* **63**, 1823 (1989).

Hermann, K. http://w3.rz-berlin.mpg.de/~rammer/surfexp_prod/SXinput.html

Hla, S.W., Rieder, K.H. *Ann. Rev. Phys. Chem.* **54**, 307 (2003).

Hohage, M., Sun, L.D., Zeppenfeld, P. *Appl. Phys. A* **80**, 1005 (2005).

Jensen, F., Besenbacher, F., Laensgaard, E., Steensgard, I. *Phys. Rev. B* **41**, 10233 (1990).

Jones, A.H., Poulston, S., Bennett, R.A., Bowker, M. *Surf. Sci.* **380**, 31 (1997).

Kay, M., Darling, G.R., Holloway, S., White, J.A., Bird, D.M. *Chem. Phys. Lett.* **245**, 311 (1995).

Kern, K., Niehus, H., Schatz, A., Zeppenfeld, P., Goerge, J., Comsa, G. *Phys. Rev. Lett.* **67**, 855 (1991).

Kinnersley, A.D., Darling, G.R., Holloway, S., Hammer, B. *Surf. Sci.* **364**, 219 (1996).

Kisliuk, P. *J. Phys. Chem. Solids* **3**, 78 (1958).

Kratzer, M., Surnev, S., Netzer, F.P., Winkler, A. *J. Chem. Phys.* **125**, 074703 (2006).

Kratzer, M., Stettner, J., Winkler, A. *Surf. Sci.* **601**, 3456 (2007).

Kroes, G. *J. Progr. Surf. Sci.* **60**, 1 (1999).

Kubiak, G.D., Sitz, G.O., Zare, R.N. *J. Chem. Phys.* **83**, 2583 (1985).

Lang, B., Joyner, R., Somorjai, G. A. *Surf. Sci.* **30**, 440 (1972).

Leibsle, F.M., Francis, S.M., Davis, R., Xiang, N., Haq, S., Bowker, M. *Phys. Rev. Lett.* **72**, 2569 (1994).

Matsushima, T. *Surf. Sci. Rep.* **52**, 1 (2003).

Michelsen, H.A., Rettner, C.T., Auerbach, D. *J. Phys. Rev. Lett.* **69**, 2678 (1992).

Michelsen, H.A., Rettner, C.T., Auerbach, D.J., Zare, R.N. *J. Chem. Phys.* **98**, 8294 (1993).

Miller, D.R., in: *Atomic and Molecular Beams I*, (ed.) Scoles, G. (Oxford University Press, New York, Oxford, 1988) p. 14.

Murphy, M.J., Hodgson, A. *Surf. Sci.* **390**, 29 (1997).

Oehzelt, M., Grill, L., Berkebile, S., Koller, G., Netzer, F.P., Ramsey, M. *G. Chem. Phys. Chem.* **8**, 1707 (2007).

Pauer, G., Kratzer, M., Winkler, A. *J. Chem. Phys.* **123**, 204702 (2005a).

Pauer, G., Kratzer, M., Winkler, A. *Vacuum* **80**, 81 (2005b).

Poulston, S., Jones, A.H., Bennett, R.A., Bowker, M. *J. Physics-Cond. Mat.* **8**, L765 (1996).

Pozgainer, G., Windholz, L., Winkler, A. *Meas. Sci. Technol.* **5**, 947 (1994).

Rendulic, K.D. *Surf. Sci.* **272**, 34 (1991).

Rendulic, K.D., Anger, G., Winkler, A. *Surf. Sci.* **208**, 404 (1989).

Rendulic, K.D., Winkler, A. *Surf. Sci.* **299/300**, 261 (1994).

Rendulic, K.D., Winkler, A., Karner, H. *J. Vac. Sci. Technol.* A **5**, 488 (1987).

Rettner, C.T., Michelsen, H.A., Auerbach, D.J. *J. Vac. Sci. Technol.* A **11**, 1901 (1993).

Ribic, P.R., Bratina, G. *Surf. Sci.* **601**, L25 (2007).

Riedler, M. Master thesis, Graz University of Technology (1996).

Sakong, S., Gross, A. *J. Catal.* **231**, 420 (2005).

Schoiswohl, J., Surnev, S., Netzer, F. *P. Topics Catal.* **36**, 91 (2005)

Schoiswohl, J., Mittendorfer, F., Surnev, S., Ramsey, M.G., Andersen, J.N., Netzer, F.P. *Surf. Sci.* **600**, L274 (2006).

Schoiswohl, J., Sock, M., Chen, Q., Thornton, G., Kresse, G., Ramsey, M.G., Surnev, S., Netzer, F.P. *Topics Catal.* **46**, 137 (2007).

Schönhammer, D. *Surf. Sci.* **83**, L633 (1979).

Schröter, L., David, R., Zacharias, H. *J. Vac. Sci. Technol.* A **9**, 1712 (1991).

Schröter, L., Trame, C., Gauer, J., Zacharais, R., Brenig, W. *Faraday Disc. Chem. Soc.* **96**, 55 (1993).

Scoles, G. (ed.) *Atomic and Molecular Beams, Vols. I and II* (Oxford University Press, New York and Oxford, 1988).

Sexton, B.A., Hughes, A.E., Avery, N. *R. Surf. Sci.* **155**, 366 (1985).

Sitz, G.O. *Rep. Prog. Phys.* **65**, 1165 (2002).

Steinrück, H.P., Luger, M., Winkler, A., Rendulic, K.D. *Phys. Rev. B* **32**, 5032 (1985a).

Steinrück, H.P., Winkler, A., Rendulic, K.D. *J. Phys. C* **17**, L311 (1984).

Steinrück, H.P., Winkler, A., Rendulic, K.D. *Surf. Sci.* **152/153**, 323 (1985b).

Steinrück, H.P., Rendulic, K.D., Winkler, A. *Surf. Sci.* **154**, 99 (1985c).

Stipe, B.C, Rezaei, M.A., Ho, W. *Phys. Rev. Lett.* **79**, 4397 (1997).

Sun, L.D., Hohage, M., Zeppenfeld, P., Balderas-Navarro, R.E., Hingerl, K. *Phys. Rev. Lett.* **90**, 106104 (2003).

Sun, L.D., Demirci, E., Balderas-Navarro, R.E., Winkler, A., Hohage, M., Zeppenfeld, P. *Surf. Sci.* submitted (2009).

Surnev, S., Schoiswohl, J., Kresse, G., Ramsey, M.G., Netzer, F.P. *Phys. Rev. Lett.* **89**, 246101 (2002).

Surnev, S., Sock, M., Kresse, G., Andersen, J.N., Ramsey, M.G., Netzer, F.P. *J. Phys. Chem. B* **107**, 4777 (2003).

Takayanagi, K., Tanishiro, Y., Takahashi, S., Takahashi, M. *Surf. Sci.* **164**, 367 (1985).

Teichert, C. *Phys. Rep.* **365**, 335 (2002).

Tölkes, C., Struck, R., David, R., Zeppenfeld, P., Comsa, G. *Phys. Rev. Lett.* **80**, 2877 (1998).

Wachs, I.E., Madix, R.J. *J. Catal.* **53**, 208 (1978).

Wiesendanger, R., Güntherodt, H.J. *Scanning Tunneling Microscopy III*. Springer Series in Surface Science (Springer, Berlin, Heidelberg, New York, 1997).

Winkler, A. *J. Vac. Sci. Technol. A* **5**, 2430 (1987).

Winkler, A., Eilmsteiner, G., Novak, R. *Chem. Phys. Lett.* **226**, 589 (1994).

Winkler, A., Kratzer, M., Pauer, G., Eibl, C., Gleispach, D. *Topics Catal.* **46**, 189 (2007).

Winkler, A., Rendulic, K.D. *Surf. Sci.* **118**, 19 (1982).

Winkler, A., Rendulic, K.D. *Int. Rev. Phys. Chem.* **11**, 101 (1992).

Winkler, A., Yates, Jr. J.T. *J. Vac. Sci. Technol. A* **6**, 2929 (1988).

Winkler, A., Guo, X., Siddiqui, H.R., Hagans, P.L., Yates, J.T. Jr. *Surf. Sci.* **201**, 419 (1988).

Zacharias, H. *Appl. Phys. A* **47**, 37 (1988).

Zegenhagen, J., Lyman, P.F., Böhringer, M.J., Bedzyk, M. *J. Phys. Status Solidi. B* **204**, 587 (1997).

Zeppenfeld, P., Krzyzowski, M., Romainczyk, C., Comsa, G., Lagally, M.G. *Phys. Rev. Lett.* **72**, 2737 (1994).

Zeppenfeld, P., Diercks, V., Tölkes, C., David, R., Krzyzowski, M.A. *Surf. Sci.* **130–132**, 484 (1998).

Zeppenfeld, P., Diercks, V., David, R., Picaud, F., Ramseyer, C., Girardet, C. *Phys. Rev. B* **66**, 085414 (2002).

Nanotribology

13

S.K. Biswas

13.1 Introduction

Is the "nano" prefix to tribology anything more than a fundraiser's gimmick? To answer that we first consider that tribology is a discipline encompassing friction, wear and lubrication where two bodies are in proximity and are moving relative to each other, transmitting power. At large practical scales of engines, machineries or human joints these processes are well described by continuum models that provide reproducible formulations that can be used for design. The Reynolds equation describes hydrodynamic lubrication well, assuming the solid shaft and bearings to be rigid and the liquid to be Newtonian. Continuum descriptions are still valid at small gaps except now the solid is considered to be deformable yielding elasto—or plasto—hydrodynamic lubrication. The Laplace biharmonic equation can still be used to derive the appropriate stress functions to describe deformations. Similarly changes in viscosity with pressure, frequency/velocity as well as non-Newtonian fluid dynamics may be incorporated even if the relations are non-linear into the continuum equations of fluid mechanics to describe lubrication. In a similar vein the Coulomb's law of friction, Tabor's two-term model of friction and Archard's wear equations are very useful descriptions of the tribological processes. The general statement we can make at this stage is that tribology is well served by continuum solid and fluid mechanics as long as the processes are within the framework of reversible thermodynamics and the properties of matter change uniformly with changing intensive parameters such as temperature and pressure.

In modelling sliding friction in ultrahigh vacuum the large-scale (non-scalable) semi-empirical models (Fig. 13.1) may provide the right answers but they may do so by averaging information over a multitude of asperity contacts each one occurring at the nanoscale. Taking the simple case of single crystals sliding on each other, if the sliding is incommensurate the friction is nil. If the contact is commensurate the friction is high. In three-dimensional contact between single crystals or between polycrystals the friction is, however, finite. The aggregate result we get provides little information about commensurabilty that within nanodomains control friction and is obviously important phenomenologically to understand friction. Similarly conducting a sliding experiment in air may yield aggregate laws of friction that are useful but provide little information about the fact that solid surfaces generally carry

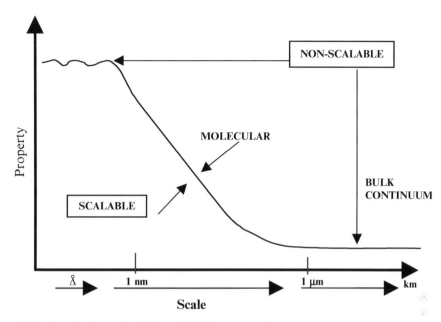

Fig. 13.1 Properties and behaviors of matter scale with length in the nm to μm regime. At length scales less than nm the generally non-scalable behavior is dominated by quantum effects. At lengths more than 1 μm continuum mechanics well predict the behavior, independent of length.

different types of oxides, adsorbed fragments of hydrocarbon and the contacts are either hydrated or form, capillary junctions using water condensed from the environment. Experiments and modelling therefore need to be done on these isolated ideal phenomena in their nanodomains to get an adequate explanation, for why there are specific numbers associated with friction and wear in large-scale experiments.

There are many other phenomena at large length scales that have defied comprehensive explanations. The boundary lubrication regime encountered at piston reversals in engines is one such. In the ultramild wear boundary regime, within tens of nanometers of the surface the grain size is found dramatically reduced by an order. This defies continuum explanations. In the same regime aliphatic hydrocarbons have been found to lubricate steel creating a 2–3-nm layer where the long-chain molecules are entropically chilled and behave in a non-random cooperative manner. The performance of the additives, which are generally of 1–2 nm length, is complex and not fully understood. These additives principally made of carbon backbones have been found to have Young's moduli of 2–10 GPa and large relaxation times. Such properties clearly enable large-load support macroscopically but the genesis of the property needs research on the small scale. Additives and their interaction with metals, water and base oil are matters of intense investigation as efficiency of practical engineering systems such as metal cutting/working emulsions, and engines are critically dependent on their interactions. One cannot help citing one more interesting example where large submarine engines have been immobilized because the detergent molecule used had the wrong pH value. What is important is to appreciate at this point that the regime and nature of forces at the nanoscale of interaction are no longer cohesive, in bulk systems one deals principally with cohesive forces. At this scale secondary intermolecular forces dominate; hydrophobic/hydrophilic interactions, Van der Waals forces, electrostatic and ionic forces operate close to the solid surfaces often involving liquid films.

Considering that different authors and learned meetings define nanotribology diversely it is relevant to state the scope of this chapter. We limit ourselves to the scalable regime where contact dimensions, topographical perturbations, confinement scale and molecular dimensions are of the same order. In this regime changing scale yields a qualitative shift in the physics and mechanics of matter such that the forces and interactions that are absent, averaged out or smoothed at larger scales (large-scale non-scalable, Fig. 13.1) become explicit and important to influence the constitutive relations. Our primary motivation to understand nanotribology and therefore partake research in the scalable regime stems from boundary lubrication. Here, the phenomena that controls the dissipation and attrition of large-scale machineries and joints occur genuinely at the nano-scale where the constitutive relations and mechanisms are not mere derivatives of larger-scale relations. Much work has been done to characterize solid surfaces, interfacial matter and friction at this scale. To do this required the use of instruments specially developed to record very small scale signatures and phenomena. Even this approach leaves mysteries, which presently are addressed by simulation using theoretical techniques such as quantum chemistry, density-functional theory, molecular dynamics and thermodynamics. The second general motivation for work in nanotribology comes from miniaturization of machinery. These pose problems similar to that of boundary lubrication but not exclusively so.

Our inability to decipher nanotribological phenomena in their entire complexities has often led to simplifications at the experimental and simulation levels. For example, model surfaces such as mica, gold and silica have been used in experiments to get around the issue of roughness and also to allow the use of ordered/crystalline self-assembly of model linear molecules such as thiols and silanes. A study of the latter is considered to elucidate the action of additives that diffuse or desorb from the oil/solvent in actual tribology to protect the sliding surfaces. Ultrahigh-vacuum studies have been undertaken to deconvolute dissipation mechanisms in the absence of contaminants. In a similar fashion much simulation has been done using diamond/DLC surfaces, single crystals of copper and nickel, substrates on which simple alkane molecules have been chemisorbed. Nanotribology may thus be considered as a discipline or an interdiscipline in its infancy. Addressing complex real surfaces, complex additive packages, commercial solvents of complex structures and the resulting interactions is the task of the future, for which good groundwork has been done.

The chapter is written in the following sections.

1. Introduction—defining nanotribology
2. Nanotribological tools
3. The interfacial phenomena and interaction forces
4. The microscopic origins of friction
5. Oil in confinement—boundary lubrication

6. Self-assembled additives in cofinement—boundary lubrication
7. Summary

13.2 Nanotribological tools

One of the main reasons why nanotribology has come of age is that researchers have access to instruments that can probe distances and forces at very small scales with high resolutions and repeatability. We still do not have direct visual access to the microscopic contact region, unless transparent probes are used, but are able to assess the physics and chemistry of materials in the contact region using instruments such as the atomic force microscope (AFM), surface force apparatus (SFA) and quartz crystal microbalance (QCM). The experimental force and displacement data and quality factor in the case of QCM are used as signatures of processes that happen at the atomic and molecular scale. This knowledge is ably supplemented by *ex situ* information obtained using various spectroscopic, electrical, diffraction, microscopy and interferometric techniques to build up a realistic understanding of processes. The present trend is to combine the two approaches to yield *in situ* chemical, physical and mechanical information. An example of the latter is to use Raman spectroscopy that allows high spatial resolution and micrometric spot sizes, a laser beam is focused into the tribological contact to collect information *in situ* in the normal and attenuated total reflection (ATR) modes. We will briefly describe the workings of AFM, SFA and QCM. More detailed information of these techniques are available in published books and papers (Israelachvili 1992; Bhushan 1994; Carpick and Salmeron 1997; Mate 2008).

13.2.1 Atomic force microscope

In AFM, the force acting on the last few atoms of a sharp tip is measured as the tip moves over a surface in contact, non-contact or intermittent modes. The tip is attached to a compliant cantilever that deflects when the surface is brought into close proximity of the tip. Knowing the stiffness of the cantilever (typically in the 0.01 to 1000 N/m range) the interaction force is estimated (Fig. 13.2). In the lateral mode the cantilever twists and the torque gives an estimate of the friction force. Experiments can be done in ambient, in liquids or in ultrahigh-vacuum environments.

At its early stage of development the AFM was used for topographic measurements and imaging. It is now used extensively to generate force curves from which adhesion forces and surface energy can be obtained. Hydrophilic, hydrophobic, electrostatic and capillary interactions can be deduced from such measurements. Modulating the lever-surface position vertically (force-modulation mode) at low frequencies it is possible to obtain viscoelastic properties such as the relaxation time of polymers and liquids. The force-modulation mode is also particularly useful in its contact mode to obtain mechanical properties of small volumes situated near the surface. This is a powerful method to map local variations in elasticity of a multiple-phase

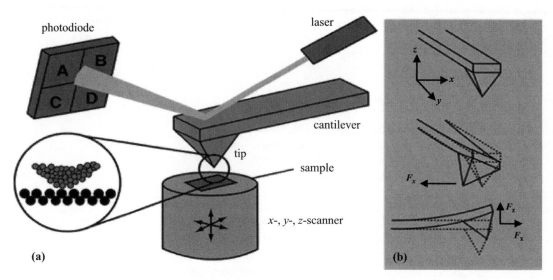

Fig. 13.2 (a) Principle of the scanning force microscope. Bending and torsion of the cantilever are measured simultaneously by measuring the lateral and vertical deflection of a laser beam while the sample is scanned in the xy-plane. The laser-beam deflection is determined using a four-quadrant photodiode: $(A + B) - (C - D)$ is a measure of the bending and $(A + C) - (B + D)$ a measure of the torsion of the cantilever, if A, B, C, and D are proportional to the intensity of the incident light of the corresponding quadrant. (b) The torsion of the cantilever (middle is solely due to lateral force acting in the x-direction, whereas both forces acting normal to the surface (F_z) as well as acting inplane in the y-direction (F_y) cause a bending of the cantilever (bottom). Reprinted with permission from Schwarz and Hölscher (2001). Copyright 2008, CRC Press.

sample. The efficiency of such measurements is increased if the contact is intermittent (the tapping mode). An extension of this mode to operations in the ultrasonic frequency range allows imaging of subsurface features. Mechanical measurements involving plasticity can be made in the force-controlled mode where the force and displacement are known simultaneously, here the tip is subjected to known electromagnetic forcing by applying a field to a magnetically coated lever. Another variation of this technique is the replacement of the sharp tip by micrometer-size silica balls to probe colloidal interaction by making use of the Derjaguin approximation. This mode is also useful for investigating interaction between two molecular species, the probe is functionalized with the other molecule. The Kelvin probe principle may be incorporated into the AFM to measure the potential difference between the tip and the surface. A dc bias potential or an oscillating potential is applied to a conducting tip to make the current measurement over a grounded substrate. This is a useful tool to investigate *in situ* film development in microscopic tribology and also the effect of surface charge distribution developed *in situ* or imposed *ex situ*, on friction.

13.2.2   Surface force apparatus

The SFA contains two curved molecularly smooth surfaces of mica (radius is millimetric) between which the interaction forces are measured using force-measuring springs. The separation between the surfaces can be measured optically from micrometer to molecular scales using multiple-beam interferometry. This is a powerful technique in spite of its gross lateral resolution, as force and distance, unlike in the case of the AFM, are measured independent of each other. The technique has been successfully used to measure interaction between surfaces in aqueous and non-aqueous media to track Van der Waals forces, electrostatic double-layer repulsion, oscillatory solvation forces, hydrophobic forces and capillary and adhesive interactions. SFA and its variations in complex modulation modes have been used to track interfacial viscosity and phase transitions in nanotribological contacts.

13.2.3 Quartz crystal microbalance

Phononic and electronic contributions to kinetic friction are investigated using the QCM. As atoms, molecules or films adsorbed on a substrate slip on a transversely oscillating (5–10 MHz) substrate the quality factor of the oscillator changes as the resonance, which is sharp in the pre-oscillatory stage, broadens. Tracking the change provides a measure of the slip time (inversely proportional to viscosity), the characteristic time over which the friction acts to reduce the relative motion. A long slip time indicates a low viscous resistance and low kinetic friction. QCM consists of a thin single crystal of quartz with metal electrodes deposited on its top and bottom surfaces. Molecules are adsorbed on the crystal surface and the coverage is detected accurately by the corresponding shift in resonance frequency.

13.3 Interfacial phenomena and interaction forces

We have noted that adhesion plays an important role in pegging the level of frictional forces between two surfaces. The work done to separate two surfaces 1 and 2 in a medium 3 is (Israelachvili 1992),

$$W_{132} = \gamma_{13} + \gamma_{23} - \gamma_{12},$$

where γ_{13}, γ_{23}, and γ_{12} are interaction energies of materials 1, and 3 in vacuum.

When surfaces are brought together Van der Waals forces build up to increase the interaction energy. Strong attraction forces act between the two surfaces when they are in contact so much so that a finite tensile loading is needed to separate the two surface. This is the force of adhesion. For a sphere (radius R) and a flat in contact, the force of adhesion $L_{ad} = \frac{3}{2}\Pi W_{1,2}R$. This force acting in conjunction with applied compression L results in a contact area (Maugis 2000) A.

$$A = \Pi \left[\frac{3R}{4E*} \left(L + 3\Pi R W_{1,2} + \sqrt{6\Pi R W_{1,2}L + (3\Pi R W_{1,2})^2} \right) \right]^{2/3},$$

where E^* is the reduced Young's modulus.

The work of adhesion W principally influenced by the dispersion (Van der Waals) forces are modulated when the interaction takes place in a medium. When there is liquid in the gap and the gap is small the Van der Waals forces may be superposed upon by other forces. If the two surfaces are slid against each other the friction force may be expressed in the following ways.

(1) $F = \mu L + F_0 = \mu L + \mu L_{ad}$, L_{ad} is the force of adhesion

     independent of L (13.1)

(2) $F = \tau A$ where τ is a shear stress and A is a function of

     L and $W_{1,2}$ (13.2)

It is generally agreed that adhesion plays a significant role in determining the F_0 intercept. Chandross *et al.* (2008) have in fact shown F_0 to scale with tip radius R and F to scale linearly with L. The implication is that L_{ad} influences F_0. The precise nature of this influence is unfortunately not clear yet as the implied contact-area dependence is found only at very small loads and for very smooth surfaces under dry adhesive conditions.

13.3.1   Capillary forces

When a contact takes place in the presence of a vapor the vapor at the contact gives rise to an attractive force between the two surfaces. For water vapor, practically, this happens at relative humidities of 15–20% and above. For a

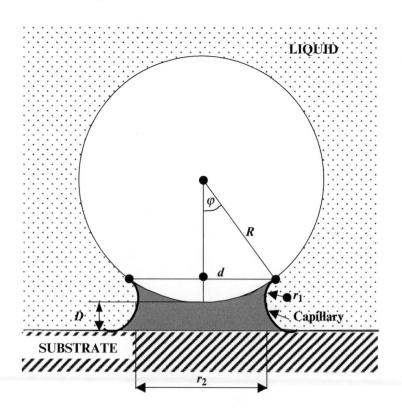

Fig. 13.3 Geometry of capillary condensation at sphere (radius, R) on a flat contact. r_1 and r_2 are the principal radii of curvature and D is the gap distance between the sphere and flat.

sphere on a flat configuration (Fig. 13.3),

$$\left(\frac{1}{r_1} + \frac{1}{r_2}\right)^{-1} = r_K = \frac{\gamma_L \, V}{RT \log\left(P/p_s\right)},$$

where the relative vapor pressure is P/p_s, r_1 and r_2 are the principal radii of curvature of the meniscus, R is the radius of the sphere, T is the temperature in Kelvin, θ is the angle of contact of the liquid with the sphere, V is the molecular volume and γ_L is the surface tension of the liquid. The total adhesion force

$$L_{ad} = 4\Pi R \left(\gamma_{LV} \cos\theta + \gamma_{SL}\right),$$

where the first term within the bracket is due to the Laplace pressure of the meniscus and the second term is due to the direct attraction between solid surfaces in the liquid. If the liquid wets the solid, $\theta \to 0$ and the adhesion is maximum. If, on the other hand, $\theta \to \Pi$, as would happen for example with water on a methyl-terminated monolayer, the adhesion force is small.

The Laplace pressure P_{cap} is given by (Israelachvili 1992),

$$P_{cap} = \gamma_L \left(\frac{1}{r_1} + \frac{1}{r_2}\right) \approx \frac{\gamma_L}{r_1} \; (\text{since } r_1 > r_2).$$

When the liquid is confined in a very narrow gap ($d \to 0$), between two flat plates ($r_2 = R \to \alpha$) and the liquid wets the plates, $r_1 < 0 \approx \frac{2\cos\theta}{d}$ (Mate 2008),

$$P_{cap} = \gamma \left(\frac{1}{R} - \frac{\cos\theta}{d/2}\right) \approx -\frac{2\gamma \cos\theta}{d}.$$

The attractive (−ve) P_{cap} continues to increase with decreasing d till solid–solid contact is achieved. If, on the other hand, $\theta > 90$ (($r_1 > 0$) the plates repel each other and the gap tends to increase, reducing the capillary pressure.

13.3.2 Structural forces

When the liquid is very well confined and the gap is less than 10 molecular diameters high the above continuum approach ceases to apply, especially when the liquid molecules are spherical in shape or are unbranched long hydrocarbons. The confinement forces the liquid molecules to order into quasi-discrete layers that are energetically or entropically favored. This results in an additional solvation force that oscillates varying between attraction and repulsion, with a periodicity equal to some mean dimension (σ) of the liquid molecule. The mean force that decays with distance (D) is repulsive if the surfaces are hydrophilic and is attractive if the surfaces are hydrophobic (Israelachvili 1992). The oscillatory force may be described by an exponentially decaying cosine function of the form,

$$F = F_0 \cos\left(\frac{2\Pi D}{\sigma}\right) e^{D/\sigma}.$$

Solvation forces do not arise because liquid molecules become structured into ordered layers (Israelachvili 2004). They arise because of the disruption or change of this ordering during the approach of the second surface. The phenomenon has been observed in the SFA (Gee and Israelachvili 1990; Zhu and Granick 2004) and in the AFM (Lim and O'Shea 2002; Lim *et al.* 2002) experiments as well as simulated using molecular dynamics (Gao *et al.* 2000).

13.3.3   Adhesion hysteresis

Adhesion hysteresis occurs (Mate 2008) because the work expended to separate two surfaces is greater than the work originally gained when the two surfaces were brought together. Adhesion hysteresis may be observed even when the two surfaces are perfectly smooth, chemically homogenous and perfectly elastic. When two surfaces approach and separate, the process does not go through (Israelachvili 2001) a continuous series of thermodynamic equilibrium states. A series of spontaneous—thermodynamically irreversible—instabilities and transitions occur. Energy is liberated during these events and lost, causing the hysteresis in frictional systems. The hysteresis is observed in contact-angle measurements—advancing and receding—due to surface roughness, chemical heterogeneity, molecular rearrangement such as that due to diffusion of species and contamination. Figure 13.4 shows a schematic of a capillary layer being dragged across a self-assembled organic monolayer. The difference between the surface forces experienced ahead and to the rear of the slider may therefore be considered to be equal to friction, $F = 2r\gamma_{LV}(\cos\theta_a - \cos\theta_r)$. Yoshizawa *et al.* (1993) have observed for surfactants deposited on mica that it is adhesion hysteresis, and not adhesion, which should be correlated to friction. The authors correlate the irreversible energy component of a contact–separation experiment in a SFA to frictional dissipation.

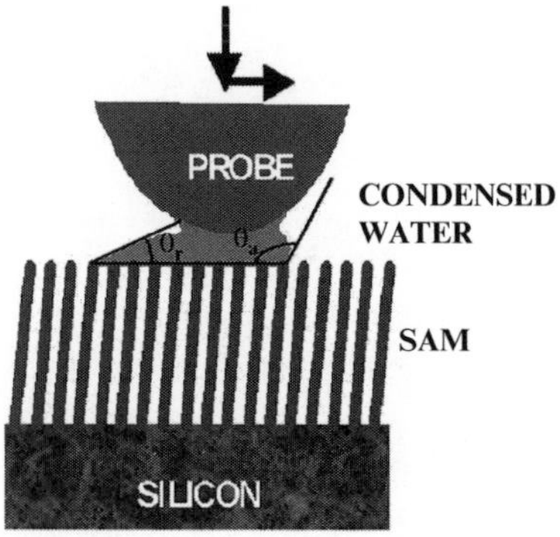

Fig. 13.4 A meniscus of water on a self-assembled organic monolayer being dragged by a probe under load. Note the advancing (θ_a) and receding (θ_r) contact angles.

13.3.4   Forces in confined aqueous medium

Unless the two surfaces are brought together in a dry environment or in a UHV (ultrahigh vacuum) the environmental water vapor intervenes at contact. Capillary forces may be one outcome. If the approach and retraction happens immersed in a water medium other forces specific to an aqueous medium, such as that due to the electric double layer, hydration repulsion and hydrophobic attraction modulate the interaction energy of surfaces.

13.3.4.1   *Electrostatic double layer*

Solid surfaces become ionized and acquire charge in the presence of water. Counterions of opposite charge that exist close to the surface and in a diffused region (double layer) in the water balance the surface charge. There is electrostatic attraction between the opposite charges but this is often much smaller than a repulsive entropic force that becomes active when two similarly charged surfaces are brought towards each other, increasing the counterion concentration near the surfaces. This lowers the entropy. The repulsive force, the magnitude of which is dependent (Mate 2008) on the pH value of the medium, decays over a length called the Debye length, which may be tens of nanometers this phenomenon is of particular relevance to colloid stability and aquaeous lubrication.

13.3.4.2 *Entropic interactions in water*

Non-polar particles or surfaces in contact with water disturb the random orientation of the four hydrogen bonds and make them reorient to accommodate the solid. This lowers the entropy and as this disruption happens leading to more ordered local structures, repulsive structural forces come into existence. The lack of solubility of hydrocarbon molecules in water can be largely accounted for by this effect (Israelachvili 1992).

A similar effect (Mate 2008), which also extends over one or two molecular diameters (1–2 nm), is the repulsion observed between two hydrophilic surfaces interacting in water. Strong hydrogen bonds form between the surfaces and water to disrupt the random orientation of the water molecules. This is also an entropic effect that results in generating enough energy to squeeze water out from between the surfaces. A somewhat longer-range interaction is observed when two hydrophobic surfaces approach each other and attract each other in a water medium. The origin of this force is not well understood but it is thought (Israelachvili 1992) to be entropic in origin and related to the rearrangement of hydrogen-bond configurations in the overlapping solvation zones.

13.4 Microscopic origin of friction

When a surface is moved tangentially relative to another, external work is done on the system to overcome resistance to shear. The work done is dissipated as heat and the more possible channels of dissipation there are the greater is the frictional resistance. Experimental investigation into the origin of friction between two materials in their solid state is hampered by the environment where the experiment is done. Experiments in the ambient air introduces third bodies of contaminants such as water and hydrocarbon fragments into the contact. Experimental work reported is therefore often done in ultrahigh vacuum or in dry ultra violated environments.

13.4.1 Amonton's Law at the single asperity

One of the earliest statements of what constitutes friction was made by Amonton (1663–1706). Accordingly, friction force (F) is linearly proportional to normal load (L) and independent of velocity and contact area. An alternative approach is to write $F = \tau^* A$ where the area A, of course, varies non-linearly with the normal load for non-adhesive and adhesive junctions. The latter gives a non-linear variation of F with the normal load. The law is now debated at the single asperity level, at the nanometer length scale. Pristine single-asperity (AFM) and large-area (SFA) experiments were done do verify the law yielding a plethora of mutually contradictory results. A number of authors (Schwarz and Holscher 2001) have found a 2/3 exponent of load, while others (Gao *et al*. 2004) have found the exponent to be 1 when the interface is smooth and non-adhesive but 2/3 when the interface is adhesive. MD simulations of these interfaces have found the adhesion-induced non-linearity only at very small normal loads, the friction characteristics at higher loads is linear. The latter work finds simple separation of load-dependent and adhesion-dependent terms such as $F = \mu L + sA$ (Berman and Israelachvili 1999), where s is an aggregate adhesion-induced stress and μ is the coefficient of friction as unrealistic. The authors (Gao *et al.* 2004) suggest that the contact area that anyway does not feature in Amonton's Law is a red herring, for the whole friction process is determined by intermolecular potential between the two surfaces and is a derivative of energy per unit area. Figure 13.5 shows the

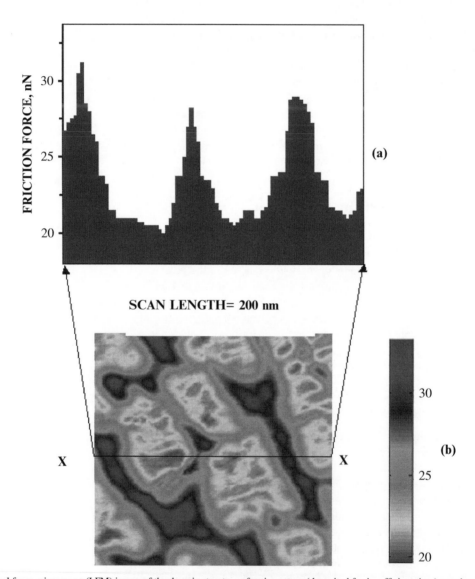

Fig. 13.5 Lateral force microscope (LFM) image of the domain structure of an immature (deposited for insufficient time) octadecyltrichlorosilane monolayer self-assembled on silicon. The shades of gray indicate levels of friction force scanned by the tip. (a) Shows the variation of friction force along the line XX in (b).

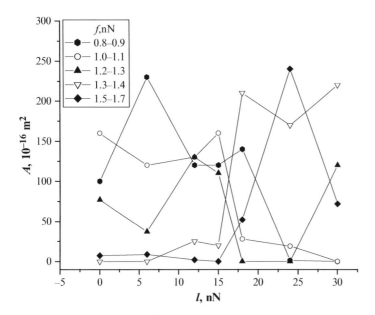

Fig. 13.6 Area of domains showing characteristics frictional (f) response (LFM) in a $200 \times 200 \, nm^2$ scan at L, the load on a Si_3N_4 tip, of tip radius 30 nm. Octadecyltrichlorosilane monolyer self-assembled on silicon.

domain structure of an immature self-assembled silane monolayer on silicon. The following shows (Khatri and Biswas 2008) the consequence of a single asperity scan (AFM tip) of a $200 \times 200 \, nm^2$ area at different tip loads. Each subdomain (i) has a characteristic friction force f_i value associated with it. Figure 13.6 shows the area covered by each characteristic subdomain to change with the normal load.

$$F(L) = \sum_{i=1}^{n} [f_i] A_i(L)$$

$$F = F_0 + \mu L,$$

$A_i(L)$ is the subdomain scan area fraction,
f_i is the characteristic friction force at subdomain i,
L is the total normal load on the AFM tip,
$F(L)$ is the area-averaged friction force at a contact load of L.

Figure 13.7 shows the plot of friction with load to be linear with an offset at zero load. These results would tend to support the contention that the heterogeneity in any contact averages out a linear dependence of friction on load. Strong support for this statement also comes from the fact that the AFM and SFA experiments at different contact scales give identical results.

13.4.2 Atomistic modelling of adhesion and friction

Figure 13.7 shows a finite friction intercept at zero load. The genesis of this finite intercept may be sought by exploring the contribution of adhesion between two molecularly smooth surfaces to the friction force. Figure 13.8 (Israelachvili 2001) shows that the top siding surface needs to be lifted by ΔD to slide by a distance Δ. The input energy to cause lateral translation is

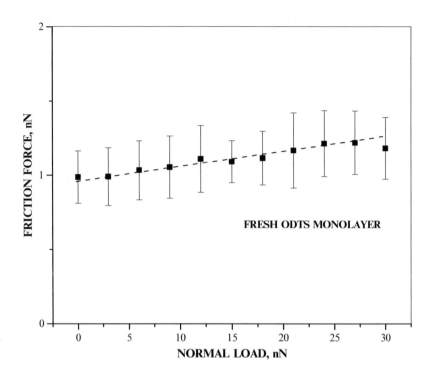

Fig. 13.7 Friction force per unit area F estimated by summing friction in domains as a function of normal tip load in L, a 200 × 200 nm^2 scan area.

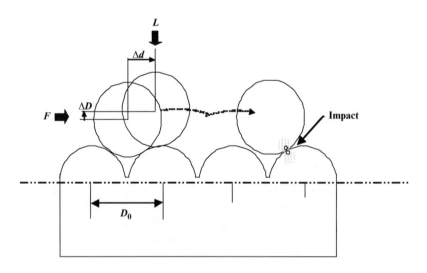

Fig. 13.8 Schematic illustration of how one molecularly smooth surface moves over another when a lateral force F is applied. As the upper surface moves laterally by some fraction of lattice dimension Δd, it must also move up by some fraction of an atomic or molecular dimension ΔD, before it can slide across the lower surface. On impact, some fraction ε of the kinetic energy is "transmitted" to the lower surface, the rest being "reflected" back to the colliding molecules (upper surface). Reprinted with permission from Israelachvili (2001). Copyright 2008, CRC Press.

$F\Delta d$, this energy may be spent in lifting the top surface against the surface forces, such as the Van der Waals forces. The latter may be expressed in terms of surface energy as $\gamma \propto \frac{1}{D^2}$. The change in surface energy from $D = D_0$ to $D = (D_0 + \Delta D)$, is

$$\Delta E_s = 4\gamma A(\Delta D/D_0).$$

Equating this with the input energy, $F\Delta d$ gives a frictional stress τ as $\tau_a = \frac{F}{A} = \frac{4\gamma\varepsilon}{D_0}\frac{\Delta D}{\Delta d}$, where ε is the fraction of energy loss due to collision.

This is independent of normal load, a normal-load term can be added to this by considering the work done by the normal load L, as $L\Delta d$. Expressing this in terms of contact pressure $p_m = L/A$ and adding it to the adhesion stress gives,

$$\tau = \frac{4\gamma\varepsilon}{D_0}\frac{\Delta D}{\Delta d} + \frac{p_m\varepsilon\Delta D}{\Delta d}$$
$$= C_1 + C_2 p_m,$$

so

$$F = C_1 A + C_2 p_m A = C_1 A + C_2 L.$$

The equation is structurally the same as that proposed by Berman and Israelachvili (1999).

When $C_1 \to 0$, $F = C_2 L \to \mu L$, recovers Amonton's Law. The same cobblestone model (Tomlinson 1929) can be used to investigate the genesis of stick slip in friction.

This treatment lends itself to direct experimental verification by the AFM (Schwarz and Holscher 2001). When one considers an AFM tip to be connected by a spring of stiffnesses C_x, C_y to a rigid support (Fig. 13.9; Schwarz and Holscher 2001), the elastic energy of the spring changes as the tip moves interacting through a periodic potential $V_{(x,y)}$ representing the physical interaction between a tip and a substrate. The tip movement $x(t)$ can be obtained by solving the differential equation,

$$m_x\ddot{x} = C_x(v_m t - x) - \frac{\partial V_{(x,y)}}{\partial x} - \gamma_x\dot{x},$$

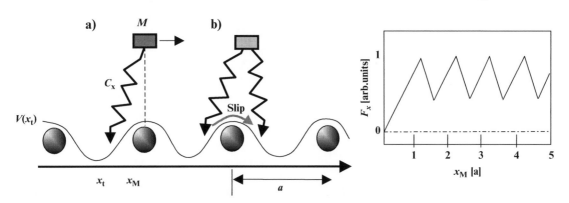

Fig. 13.9 (a) A simple model for a tip sliding on an atomically flat surface based on the Tomlinson model. A point-like tip is coupled elastically to the body M by a spring constant c_x in the direction; x_t represents its position within an external potential $V(x_t)$ with periodicity a. If $x_t = x_M$, the spring is in its equilibrium position. For sliding, the body M is moved with the velocity v_M in the x-direction. (b) A schematic view of the tip movement in a sinusoidal interaction potential. The tip shows the typical stick-slip-type movement, i.e. it jumps from one potential minimum to another. (c) If the tip moves with stick-slip over the sample surface, the lateral force F_x manifests as a sawtooth-like function. Reprinted with permission from Schwarz and Hölscher (2001). Copyright 2008, CRC Press.

where V may take the form for a single crystal of lattice constants a_x and a_y, as

$$V_{(x,y)} = V_0 \cos\left(\frac{2\Pi}{a_x}x\right) \cos\left(\frac{2\Pi}{a_y}y\right),$$

where v_m is the sliding velocity, t is the time, $\gamma_x \dot{x}$ is a velocity-dependent damping term and V_0 is a constant that depends on load.

The friction force $F = C_x(x_m - x)$.

The friction force at any instant is the gradient of the periodic potential V and the stick-slip phenomena is controlled by the inequality of spring stiffness and the gradient $\frac{\partial V}{\partial x}$. As the tip moves laterally in a low-stiffness configuration, $C_x < -\left[\frac{\partial^2 V}{\partial x^2}\right]$ the elastic energy builds up. When $C_x(x - x_m) = \left[\frac{\partial V}{\partial x}\right]$, there is slip and the tip jumps to a new potential-energy minimum, generating discontinuous motion. According to this model when the spring is very stiff it is always in a stable equilibrium position, the motion is continuous and the average friction force is zero. As V_0 is proportional to the normal load L, decreasing the load decreases the amplitude of the periodic potential. When the load reduces below a threshold $C_x(x - x_m) > \left[\frac{\partial V}{\partial x}\right]$, the slip disappears (Mate 2008). Since no slip occurs, there is no hysteresis, no dissipation and the average friction force is zero.

In the stick-slip situation if the direction of scan is reversed, the spatial position that marked slip in the forward scan registers a stick. The resultant is a hysteresis. The area of the friction loop equals the energy dissipated by friction during the cyclic scan. This energy stored during the 'stick' or the 'positive friction—distance gradient' part of the traverse is released as heat during slip. If there is no slip the forward and reverse scan signatures overlap, no energy is dissipated and the friction is zero. SFA experiments (Yoshizawa *et al.* 1993) with surfactants on mica showed a strong correlation between friction and adhesion hysteresis for solid-like, amorphous and liquid-like surfactants. The above phenomenological model of friction has been verified by molecular-dynamics simulation (Landman *et al.* 1989; Harrison *et al.* 1993a,b; Shimizu *et al.* 1998), which captured the stick-slip motion, it has been validated by AFM experiments on NaCl crystal (Socoliuc *et al.* 2004), graphite, MoS_2, NaF (Holscher 1997, 1998) and MoS_2 (Fuzisawa *et al.* 1995).

While Tomlinson's model captures the lateral force microscope data in a UHV environment, as an independent oscillator model, Frenkel and Kontorova (1938, 1939) (FK) considered the slides as made up of atoms interconnected with each other by overdamped springs, moving as a chain in a periodic potential imposed by the substrates (Mate 2008). Friction is greatest when the periodicity of slider atoms is the same as that of the substrate potential. This is the commensurate situation. If the atomic density of the slider is increased or a misfit angle is introduced between the rows of the slider and the substrate atoms, local defects analogous to dislocations are generated. When there is relative motion the energies associated with these defects lowers the activation-energy barrier and the defects glide easily, lowering the friction force. For such incommensurate systems it is possible to achieve zero friction, as has been shown experimentally for graphite (Dienwiebel *et al.* 2004).

In practical systems the slider and the substrate are incommensurate but the friction is always finite. Simulation by Muser *et al.* (2001) and Robins and Muser (2001) showed that in a non-UHV environment third-body contaminants diffuse into the interface to find local energy minima between the opposing surface atoms. These locked bodies provide resistance to sliding and stresses generated are sufficient to rearrange the atoms and molecules towards commensurability. Similar rearrangement of atoms occurs at metallic junctions, Bowden and Tabor (1951) had proposed that when asperities of two metallic surfaces are normally pressed against each other they deform plastically and cold weld. Large tangential forces are needed to shear and fracture these junctions. A more recent molecular-dynamic simulation of sliding of a nickel tip pressed on a gold surface (Landman *et al.* 1990) showed that junctions are formed on pressing where the gold atoms diffuse and wet the nickel. When the junction is sheared the shear plane is in the bulk gold underneath the zone where the gold adheres to the nickel. The sliding nickel tip, out of the junction now, carries gold atoms. The Frenkel and Kontorova model explains the phenomena as one where the high surface energies of metals in the UHV, compel diffusion of atoms at the interface to maximize the metal–metal contact (Mate 2008). The diffusion rearranges atoms at the junction to promote commensurability and high friction. Bowden and Tabor (1951) showed that friction is lowered significantly in the presences of contaminants at the interface. In the light of the FK model this would indicate promotion of incommensurability.

The motion of atoms or molecules as they slide over each other, are also restricted by the viscosity of the interface. The sliding motion gives rise to atomic vibration and consequent phonon excitation. Phonon excitation contributes to interfacial viscosity and is dissipative. The slip of adsorbed gas atoms on a laterally excited quartz crystal has been recorded (Krim and Widom 1988) in a quartz crystal microbalance. The slip broadens the resonance of the system. The change provides a measure of the viscous loss at this interface. While these experiments demonstrate the importance of the viscous and kinetic components of friction there is as yet no suitable technique available to estimate interfacial viscosity isolated from other dissipation mechanisms buried in the overall friction associated with practical systems.

13.4.3 Eyring equation and coherence in molecular lubrication

A physically explicit way to approach the problem of friction is to use the Eyring equations to analyze friction data. Eyring discussed a pure liquid at rest in terms of a thermal activation model (Eyring 1935). The individual liquid molecules experience a cage-like barrier that hinders molecular free motion because of packing in liquids. To escape from the cage an activation barrier needs to be surmounted. The barrier is enhanced by the applied pressure through a pressure-activation volume (Fig. 13.10) and is lowered by the shear stress acting on a shear-activation volume. The larger the shear-activation volume the greater is the cooperative movement of molecules and the lower

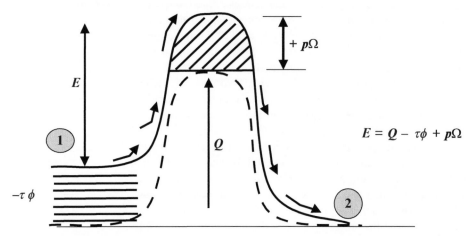

$$E = Q - \tau\phi + p\Omega$$

Fig. 13.10 Energy barrier encountered by a particle moving from molecular position (1) to molecular position (2) Q, activation energy (electrostatic); τ, shear stress; p, mean pressure; ϕ, shear activation volume; Ω, pressure-activation volume.

is the barrier height to be overcome. The Eyring model has been successfully used (He *et al.* 2002) to distinguish between the frictional mechanisms of two liquids in confinement; n-hexadecane, long-chain alkane, and octamethylcyclotetrasiloxane (OMCTS), spherical. The authors (He *et al.* 2002) used shear modulated scanning force microscopy to detect a 2-nm layer of the liquid, next to the substrate where the hexadecane molecules were found to be aligned with respect to each other in the direction of shear. The authors described the layer as entropically chilled. No such coherent layer was found when the OMCTS was slid. The authors used the Eyring equation to determine the coherence length (shear activation volume/contact area) which was found to be in the order; hexadecane>OMCTS> water. The friction of hexadecane was found to be smaller than that of OMCTS molecules where each molecule is deformed "plastically" by shear instead of aligning to form a coherent layer. The Eyring equations have been used by Briscoe and Evans (1982) to physically distinguish the frictional responses of aliphatic carboxylic acid and their soaps self-assembled on mica substrates, by McDermott *et al.* (1997) to investigate the effect of alkane chain length on friction, by Buenviage *et al.* (1999) to study the effect of normal load on the friction of polyethylene copropylene polymers and by Drummond and Israelachvili (2000) who showed that complex branched polymers are less likely to exhibit coherent response due to internal constraints and poor mixing within the contact area. Gnecco *et al.* (2000) showed in a ultrahigh-vacuum study of sodium chloride that the Eyring model is also useful to analyze dry friction studies.

We have recently used the Eyring equation to distinguish the frictional responses of octadecyltrichlorosilane (ODTS) and perflurooctyltrichlorosilane (FOTS) self-assembled on a silicon wafer. The work was prompted by the observed anomalous response of these molecules, the friction of FOTS is higher than that of ODTS, although the surface energies of FOTS is significantly lower than that of ODTS. Assuming the average time for single-molecular barrier hopping is given by the Boltzmann distribution in a process where there are a regular series of barriers that are overcome repeatedly, the Eyring equation gives (Briscoe and Evans 1982; He *et al.* 2002; Overney *et al.* 2004) the shear stress as

$$\tau = \frac{K_B T}{\varphi} \ln\left(\frac{v}{v_o}\right) + \frac{Q + p\Omega}{\varphi},$$

where the stress limit is high ($\tau\phi K_B T > 1$, here $\tau\phi K_B T \approx 10$), T is temperature, K_B, Q is activation energy, J, p is mean contact pressure, Pa, Q is shear activation volume, m^3, Ω is pressure activation volume of the junction, m^3, v is sliding velocity, m/s, φ is process coherence volume or the size of the molecular segment that moves collectively in the tangential direction acted upon by the shear stress.

The potential-energy barrier that may account for the overall frictional work is given by

$$E = Q + p\Omega - \tau\varphi.$$

The Eyring equation thus provides a means to relate some properties of a monolayer such as Q and v_0 as well as some structural parameters such as stress ϕ and pressure (Ω) activation volumes of the assembly to friction. Going back to the motivation for the present work we take the cue from Bouhacina *et al.* (1997). Knowing that sliding friction is a thermally activated process, we attempt to relate the difference between the barrier heights (E) of the two test molecules to their frictional difference and then differentiate between the structural parameters to understand the cause for this difference.

Figure 13.11 shows that the friction–velocity relationship is logarithmic for both the test molecules at room temperature at two different normal loads. A logarithmic friction–velocity relationship corresponds (He *et al.* 2002) to

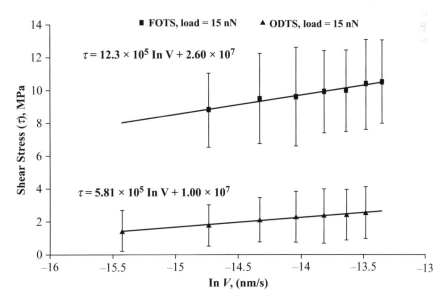

Fig. 13.11 Shear stress vs. sliding velocity, V. FOTS, perfluroalkylsilane; ODTS, octadecyltrichlorosilane. Room temperature, lateral force microscope, normal load 15 nN.

a discontinuous sliding process. For a given set of molecular assembly such discontinuities are observed in appropriate velocity ranges. They are caused by activation barriers that are repeatedly overcome during the sliding process. The process is thermally activated and exhibits a declining monotonic trend with temperature (Briscoe and Evans 1982) above a critical temperature (Glosli and McClelland 1993). Briscoe and Evans (1982) described an experimental method to obtain the potential energy barrier. They use the surface force apparatus (SFA) to measure contact area (A) and therefore the shear stress τ as function of (1) p (keeping v and T as constants), (2) v (keeping p and T as constants) and (3) T (keeping p and v as constants) to yield the following equations,

$$\tau = \tau_0 + \alpha p; \text{ at constant } v \text{ and } T; \text{ where, } \tau_0$$

$$= \frac{1}{\varphi}\left(K_B T \ln\left(\frac{v}{v_o}\right) + Q\right) \text{ and, } \alpha = \frac{\Omega}{\varphi}$$

$$\tau = \tau_1 - \beta T; \text{ at constant } p \text{ and } V; \text{ where, } \tau_1$$

$$= \frac{1}{\varphi}(Q + p\Omega) \text{ and, } \beta = -\frac{K_B}{\varphi}\ln\left(\frac{v}{v_o}\right)$$

$$\tau = \tau_2 + \theta \ln V; \text{ at constant } p \text{ and } T; \text{ where, } \tau_2$$

$$= \frac{1}{\varphi}(Q + p\Omega - K_B T \ln V_o) \text{ and, } \theta = \frac{K_B T}{\varphi}$$

$\tau_0, \tau_1, \tau_2, \alpha, \beta$ and θ are measured from the sliding friction experiments and we need to estimate the four unknowns Ω, φ, Q and v_0 to obtain E. The three sets of experiments should ideally yield mutually consistent values of these four unknowns. Lateral force microscopy was used (Subhalaksmi *et al.* 2008) to generate the friction data for the two test molecules (FOTS and ODTS) under controlled conditions; (1) (friction vs. load) v,t (2) (friction vs. velocity) $_{L,T}$ and (3) (friction vs. temperature)$_{L,v}$. Table 13.1 shows the estimated Eyring parameters. For perflurooctyltrichlorosilane the shear coordination was found to be much less and the change in local volume to accommodate unit shear was found to be greater than those for octadecyltrichlorosilane. These differences in structural responses between these molecules were found to modulate the differences between their barrier energies to molecular motion in sliding. The dominant factor that differentiates their barrier energies and therefore friction was, however, found to be the intermolecular electrostatic repulsion energy.

Table 13.1 Eyring equation parameters for FOTS and ODTS.

SAM	α	ϕ, nm^3	ϕ/A, nm	Ω, nm^3	$\tau\phi$, kJ/mol	$p\Omega$, kJ/mol	$(p\Omega - \tau\phi)$, kJ/mol	Q, kJ/ mol	E, kJ/ mol
FOTS $P = 42.8\,$nN $T = 295\,$K	0.13	2.08	0.0046	0.2704	21.8219	15.5899	−6.2320	50.75	44.51
ODTS $P = 46.5\,$nN $T = 295\,$K	0.042	4.69	0.0089	0.1972	14.5395	10.5824	−3.9571	45.05	41.09

The latter is primarily determined by the size of the terminal group and the separation distance between adjacent molecules. The possibility of such a rationale has been discussed (Kim *et al.* 1997).

13.5   Oil in confinement and boundary lubrication

One of the most significant contributions of nanotribology has been in the domain where surfaces slide or roll over each other maintaining a thin lubricant film in the narrow gap between the surfaces. Both MD simulation (Gao *et al.* 2000, Gao *et al.* 2004, Robbins and Muser 2000) and experimental work using SFA and tribometers (Gee *et al.* 1990; Berman and Israelachvili 1999; Drummond and Israelachvili 2000; Israelachvili 2001; Krupka *et al.* 2006) have elucidated physical mechanisms that control liquid-film behavior in boundary and EHD regimes. The knowledge is of practical importance as the film thicknesses at the TDC (top dead center) of an engine or a heavy-duty gear assembly are very thin indeed.

The first condition to be met in these situations is that oil, water or polymeric liquids must wet the solid surfaces. The disjoining pressures due to Van der Waals, bonding and structural forces (Mate 2008) between the solid and the liquid needs to be high to prevent a squeeze out at the contact pressures, which can go up to 3–4 GPa. There is much debate (Alba-Simionesco *et al.* 2001; Krupka *et al.* 2006) as to how thin the film can be. Krupka *et al.* (2006) show complex as well as linear oils to reach thicknesses of the order of 1–2 nm at pressures up to 3 GPa. Depending on the structure of the liquid it is also shown that the Reynolds equation and linear elasticity (Hamrock and Dowson 1977) based equations break down at film thicknesses less than 10 nm although it has been shown by Alba-Simionesco *et al.* (2001) that incorporating the pressure dependence of viscosity and solvation forces in the EHL equations improves the predictions of the older theory.

Moving to the boundary-lubrication regime marks a significant departure from Newtonian and Hookean mechanics. The first issue is that of the pressure dependence of viscosity. At very low sliding speeds the viscosity can increase by several orders, the matter in confinement may still be a liquid that exhibits a gradual transition with speed to the bulk viscosity. This is certainly true at moderate loads. At very high loads and low shear rates there is a phase transition and the matter at the interface becomes solid-like, often exhibiting (Israelachvili 2001) strong stick-slip behavior. Geometric confinement aids in enhancing viscosity (Hu *et al.* 1997), confinement of liquid molecules in a narrow gap between two stiff surfaces restricts molecular transport by diffusion. Attractive interactions between the liquid molecules and the solid surface further slow (Alba-Simionesco *et al.* 2006) down the motion and lowers the entropy. As in the case of high pressure and low shear rates, confinement in the 1–10 nm gap can bring about dramatic increase in viscosity and the resulting friction (Gao *et al.* 2004).

One of the signatures of high friction is the discontinuous movement of the slider. This is stick-slip. Conventionally, stick-slip has been associated with the

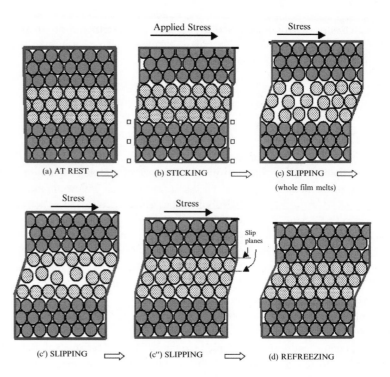

Fig. 13.12 Idealized schematic illustration of molecular rearrangements occurring in a molecularity thin film of spherical or simple chain molecules between two surfaces during shear. Note that, depending on the system, a number of different molecular configurations within the film are possible during slipping and sliding, shown here as stage (c) total disorder as whole film melts, (c′) partial disorder, and (c″) order persists even during sliding with slip occurring at a single slip plane either within the film or at the walls. Reprinted with permission from Israelachvili (2001). Copyright 2008, CRC Press.

dynamics of the mechanical system that constitutes tribology, especially for contacts that exhibit a strong velocity dependence of friction. A more current explanation of stick-slip is the structurization of the liquid in the gap (Robbins and Thompson 1991; Gao *et al*. 2004). The molecules develop inplane and through thickness order when the thickness is of a few molecular diameters. When such matter is slid there is alternating melting and freezing at lattice periodicity resulting in stick-slip. The crystalline part corresponds to stick where the friction is high, while slip occurs when the film melts periodically (Fig. 13.12) to give rise to low friction. Stick-slip, whether of mechanical or physical origin, is associated with declining average friction with increasing velocity. With increasing velocity it is postulated that there is more time spent in the slip region than in the stick region and the friction decreases with increasing velocity (Thompson and Robbins 1991).

It is considered (Israelachvili 2005) that structurization in confinements is best suited to spherical molecules such as OMCTS. Characteristically, these molecules can organize themselves into an ordered state easily to institute periodic phase transitions and corresponding stick-slip on sliding. Long-chain alkanes molecules, on the other hand, exist in a random state in the bulk, align themselves cooperatively in narrow confinement to lower friction. We have discussed this in the previous section in the context of coherence in molecular lubrication. One of the most important conclusions of practical importance to emerge from these studies is an elucidation of relationship of oil structure with friction in the boundary regime. When the oil structure becomes branched, asymmetric and complex (Gao *et al*. 2004), the molecules become entangled (Gee *et al*. 1990; Alba-Simionesco *et al*. 2001) and are unable to form ordered structures in confinement. Compared to linear alkane oils the napthenic oils, for example, exhibit smooth sliding (no stick-slip) and low friction. Curiously, this means that as the oil becomes more viscous by making its structure more complex it also exhibits low friction. A corollary to this is the effects of substrate roughness and that of mixing hydrocarbons of different molecular lengths (Landman 2008) that on sliding would break up ordered molecular structures in the gap, again thereby lowering the friction.

13.6   Additives in confinement-boundary lubrication

In micromachine or information-storage devices long-chain hydrocarbons consisting of reactive or polar head groups and generally a hydrophobic terminal group are assembled on a surface to institute load bearing, to protect a substrate from contact with the opposing surface and to generate a low shear interfacial plane. In large-scale machine such molecules are dispersed in oil. At low sliding speed, high-pressure situations where the liquid oil thickness is drastically reduced the dispersed molecules self-assemble on the surface to serve functions as they do in micromachines. The performance of these molecules to yield low friction under conditions of high normal load and low sliding velocity is then crucial to the performance and life of the machine and, as realized now, of human joints. We present below our understanding of the response of self-assembled organic monolayers (SAM) to normal loading, heating and tangential loading.

13.6.1   Amonton's Law

To verify Amonton's Law for the functionalized surfaces, we present below some lateral force measurements done on alkane silane monolayers self-assembled on an aluminum surface (Devaprakasam *et al.* 2005; Khatri *et al.* 2005a,b, 2006). We first explore the effect of molecular order on friction. We heat the monolayer to different temperatures and generate (Khatri *et al.* 2004) residual molecular disorder that scales with the peak heat-treatment temperature (Fig. 13.13). Figure 13.14(a) shows the lateral force as obtained by sliding a silicon nitride tip on the monolayer. The behavior is very similar to that reported by others (Gao *et al.* 2004) where smooth mica surfaces were coated with octodecylphosphonic acid. Figure 13.14(A) shows that for both the test molecues (OTS-octadecyltricholorosilane, FOTS-perflurooctyltrichlorosilane) the coeffecient of friction μ, and the intercept F_0 eqn (1) increase with increasing a priori molecular disorder. Further, μ and F_0 are both functions of molecular structure and composition and the friction of FOTS is consistently higher than that of OTS. The finite intercept F_0 and linearity of F with L suggests that the level of friction is controlled by the parameter F_0, which is most probably related to a load-independent adhesion parameter (L_{ad}) and is not controlled by the contact area, the latter would yield a non-linear F—L characteristic. The data presented shows the linear load dependence of friction

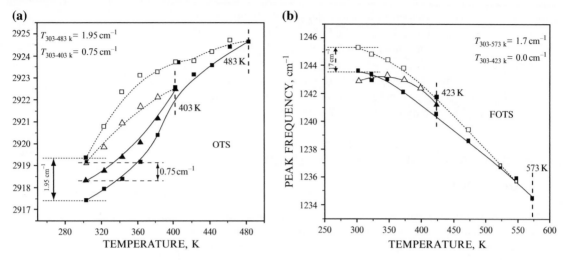

Fig. 13.13 Temperature dependence of peak frequency for different thermal cycles. (a) CH_2 antisymmetric stretch for OTS SAM on aluminum. Peak temperatures in the cycles are (■) 483 K and (▲) 403 K. (b) CF_2 antisymmetric stretch for FOTS SAM on aluminum. (■) 573 K and (▲) 423 K. Filled symbol is for heating and open symbol is for cooling. Reprinted with permission from Khatri *et al.* (2005). Copyright 2008, Elsevier Ltd.

force but also points to the state of molecular order, which governs avenues of energy dissipation, as an important parameter that determines frictional force as well as the coefficient of friction.

Variation of friction force F with load L may be summarized as follows.

$$F = \mu L + F_0,$$

where $\mu = \mu(O_p)$,

$$F_0 = F_0(O_p),$$

(O_p) is an order and structural parameter.

A validation of Amonton's Law, however, necessitates a validation of $\mu = Constant$ and independent of load at all loads (contact pressure) and contact areas.

13.6.2 The scaling effect; the relevance of nanotribology

For nanotribology to be of practical use it is important first to establish the phenomenological validity across a wide range (3–6 orders) of length and pressure scales. The contact dimension/pressure vary as nm/GPa, 10–100 μ m/100 MPa, 10–100 μ m/100–1000 MPa and mm/1–100 MPa in AFM, SFA, nanotribometer and (namely, pin-on-disc) macrotribometer, respectively. If an identity can be established between the friction characteristics observed over such a wide scale, it may be possible to extrapolate the molecular-level tribophysics and tribochemistry to an industrial application where large-scale objects are involved. This, if possible, of course opens up the scope for new approaches in industrial design for optimization of energy and materials.

Two approaches are undertaken to establish this identity. The first one involves simply scaling up an experiment from the AFM to the pin-on-disc

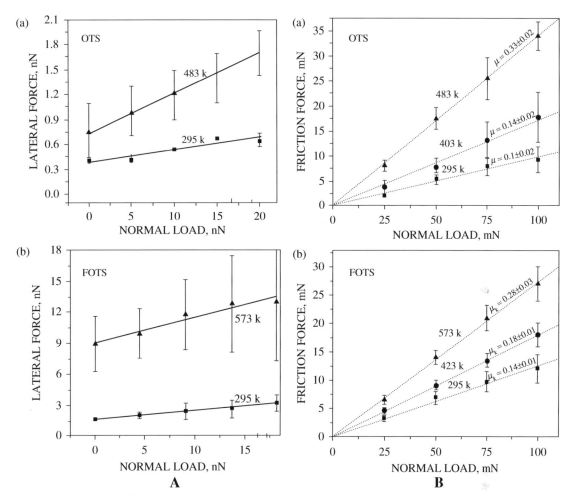

Fig. 13.14 (A) Influence of adhesion on lateral force (in dry air, 0% RH); (a) Total lateral force due to OTS SAM after heat treatment, temperatures of (■) 295 K and (▲) 573 K and residual friction force indicated by open symbol at same temperatures. Lateral force increases with heat treatment temperature; lateral force due to adhesion also increases with heat treatment temperature. (B) Friction forces measured by using the nanotribometer at different loads for (a) OTS SAM after heat treatment, peak temperatures of (■) 295K, (●) 403 K and (▲) 483 K. (b) FOTS SAM after heat treatment, peak temperatures of (■) 295 K, (●) 423 K and (▲) 573 K. Reprinted with permission from Khatri *et al.* (2005). Copyright 2008, Elsevier Ltd.

level. This is fraught with problems as artifactual machine characteristics such as inertia and resolution interfere with the results. For example, periodic fluctuations in friction force may be due to a phase transition in trapped material or related to machine inertia. It may be difficult to distinguish between these two factors. The second approach is to track the differential in performance between two systems over a wide range of length and pressure scales. The latter, while being more immune than the first approach to machine artifact, also provides a bench-top methodology for industrial screening of materials and lubricants.

Gao *et al.* (2004) observed the identity of friction coefficient and F vs. *L* characteristics obtained using AFM and SFA over a six-order scaling span in contact radius, pressure, load and friction force. This identity was observed for smooth and damaged mica and alumina surfaces under dry, lubricated, adhesive and non-adhesive contact. Similar observations have been made by Choi *et al.* (2002) and Ruths *et al.* (2003). In changing from the nN to mN load range, others have, however, observed the disappearance of the zero-load intercept F_0 (Devaprakasam *et al.* 2005; Ishida *et al.* 2005; Khatri *et al.* 2005) and an enhancement of the friction coefficient (Ahn *et al.* 2003; Devaprakasam 2005). Figures 13.14 (A) and (B) show the effects of scaling for two silane monolayers self-assembled on aluminum.

Moving on to the second approach to explore the scaling effect Ahn *et al.* (2003) have observed a change in the friction coefficient (0.07–$0.08 \rightarrow 0.15$) brought about by replacing a hydrophobic monomolecular layer (octade-canethiol) by a hydrophilic layer (11-mercaptondecanoic acid). This difference recorded in an AFM experiment is reproduced in a microtribological experiment (contact radius 14 μm, contact pressure 660 MPa). Choi *et al.* (2002) have similarly recorded that the difference between the friction of perfluropolyether and perflurodecyltriethoxysilane self-assembled monolayers (SAM) is maintained in changing from an AFM experiment (nN) to a mN-level tribological experiment.

We have explored (Devaprakasam *et al.* 2005; Khatri *et al.* 2005) the perpetuation of an apparent contradiction between the relative magnitude of surface energy and friction of two silane monolayers, over a ten-order length scale. Figure 13.15 shows that a fluorinated silane molecule (FOTS) is consistently less surface energetic than a hydrogenated silane (OTS) monolayer. Figure 13.14 shows that at the nN and mN levels of load the FOTS SAM, however, yields a higher level of friction than what is observed for the OTS SAM. This reversal has been observed by others (Burnham *et al.* 1990;

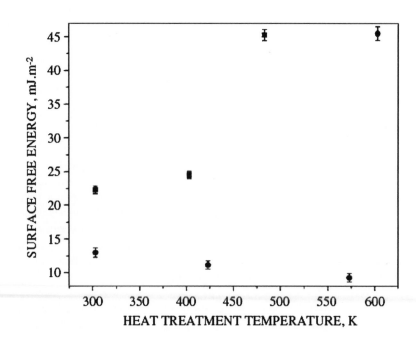

Fig. 13.15 Surface free energy of SAMs after heat treatment at different peak temperatures for (■) OTS and (●) FOTS, obtained from contact-angle measurements. Reprinted with permission from Devaprakasam *et al.* (2005). Copyright 2008, Elsevier Ltd.

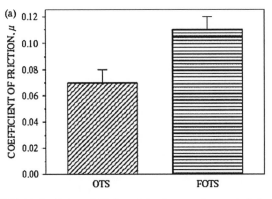 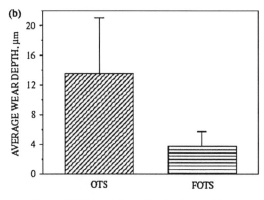

Fig. 13.16 (a) Coefficient of friction of aluminum against steel pin in presence of OTS and FOTS additives (5% v/v) dispersed in n-hexadecane. (b) Wear measurement of aluminum against steel pin in the presence of OTS and FOTS additives for 250 m sliding distance. Load: 60 N, speed: 0.2 m/s. Reprinted with permission from Devaprakasam *et al.* (2005). Copyright 2008, Elsevier Ltd.

Graupe *et al.* 1999; Kojio *et al.* 2000; Takahara *et al.* 2002; Sung *et al.* 2003). To take this one step further, Fig. 13.16 shows that when these molecules are dispersed in n-hexadecane and used as a lubricant in a pin-on-disc machine to lubricate a steel (pin 3 mm diameter) on an aluminum contact at a normal load of 60 N and sliding speed of 0.2 m/s, the relative coefficient friction between the two molecules is maintained at this load, which is about 10 orders greater than that used in the AFM. Interestingly, the wear of aluminum is lower when the FOTS molecules are dispersed than when the OTS molecules are dispersed in hexadecane. A similar exercise was undertaken to investigate the effect of a double bond or saturation on the friction of fatty acids self-assembled on steel. This effect has been under some debate in the past (St Pierre *et al.* 1966; Kondo 1997; Adhvaryu *et al.* 2006; Smith *et al.* 2006). Figure 13.17 shows the two molecules of the same chemistry and conformation, the stearic acid saturated and linoleic acid unsaturated. Figure 13.18 shows the comparison of friction recorded in (a) AFM and (b) nanotribometer of both the acids self-assembled on a smooth steel surface. Figure 13.19(c) shows the comparison when the acids are dispersed in hexadecane and used to lubricate in a pin-on-disc machine. Unsaturation or double bonding is clearly seen in Fig. 13.18 to reduce friction at all test length (contact) scales and pressure regimes.

Amonton's Law appears to be valid even when the molecules are not a priori self-assembled but dispersed in a hydrocarbon oil. This implies that the performance of an additive in an industrial machine is ultimately related to its self-assembly state on the surface for however brief a time interval. The latter observation must be taken with some caution as there is always a possibility of thermal- and pressure-activated chemical and physical change specific to a molecule that can alter its relative merit with respect to a more passive molecule–substrate interaction. Figure 13.19 shows the results of an AFM experiment on a sodium monolayer self-assembled on steel. There is a critical load above which the formation of an oleate soap generates a permanent low-friction regime (Kumar *et al.* 2008).

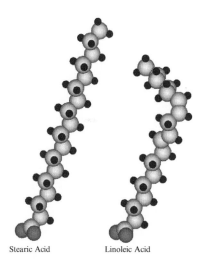

Stearic Acid   Linoleic Acid

Fig. 13.17 A schematic of the stearic and linoleic acid structures showing the presence of a double bond in the latter.

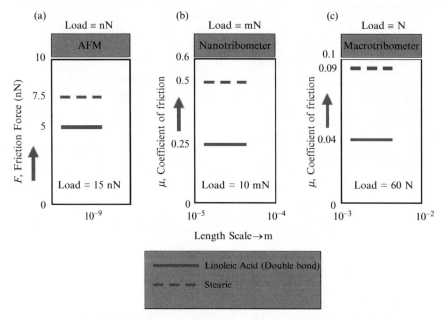

Coefficient of friction for stearic and linoleic acid SAM deposited and
dispersed in hexadecane on steel from nano to macro scale

Fig. 13.18 Comparison of friction of stearic and linoleic acids self-assembled on steel. (a) AFM, (b) Nanotribometer, (c) Pin-on-disc, additives
dispersed (5% v/v) in hexadecane, load 60 N, speed 0.2 m/s.

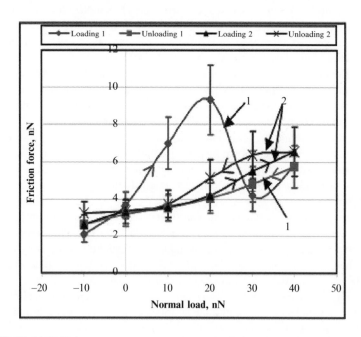

Fig. 13.19 Friction force vs. normal load.
Lateral force microscopy using a Si_3N_4 tip
on a sodium oleate monolayer self-assembled
on silicon. The figure shows a chemical transformation of the monolayer at 17 nN load in
the first loading cycle.

13.6.3 Some properties of relevance to friction of self-assembled additives

13.6.3.1 *Thermal stability of the SAM*

When self-assembled organic monolayers are heated or mechanical work is done on them the monolayers are viscoelastically deformed. Heating the monolayer tilts the molecule with respect to an axis normal to the substrate and there is rotation of and about the chemical bonds of the molecules. These deformations are well mapped using vibrational spectroscopy (Porter *et al.* 1987; Bryant *et al.* 1991; Dubois *et al.* 1993), X-ray diffraction (Fenter *et al.* 1993, 1996) and electrochemical methods (Wizrig *et al.* 1991) for thiol molecules self-assembled on gold and silane molecules (Devaprakasam *et al.* 2003; Khatri *et al.* 2006). Molecular-dynamic simulation (Hautman *et al.* 1989, 1991; Prathima *et al.* 2005) has corroborated these findings and supported them with subtle explanations. The rotational movements are generally grouped as creating a generic type of defect called the gauche defect. The term arises out of a deviation from the all-*trans* configuration. These defects as they change molecular conformation also change and disturb crystallographic symmetry and cause steric strains. In general, these defects cause molecular disorder. Some of the deformations generated are reversible, while others are not. After heating the molecule to a peak temperature if it is cooled back some of the disorder generated during heating is annealed out but there is residual disorder, which often scales with the peak temperature to which the molecules was heated (Bhatia and Garrison 1997; Benesebaa *et al.* 1998; Devprakasham *et al.* 2003; Khatri *et al.* 2005, 2006).

13.6.3.2 *Compression of the SAM*

Load-bearing capacity is an important property of the self-assembled monolayers that has a major bearing (Garcia-Parajo *et al.* 1993; Takahara *et al.* 2002; Ahn *et al.* 2003) on the boundary lubrication tribology of these molecules used as additives. Experiments have been done using AFM and SFA to record the response of these monolayers under compression (Joyce *et al.* 1992; Garcia-Parajo *et al.* 1993; Jeffery *et al.* 2004; Devaprakasam and Biswas 2005). The initial contact is with the terminal group that enjoys a large degree of freedom of movement and yields a "liquid-like" flat response with pressure. The sum frequency generation (SFG) spectroscopic experiments of Salmeron (2001) done in tandem with a SFA has clearly shown the bending of the terminal CH_3 stretch mode (alkane silane monolayer) that becomes nearly parallel to the surface, there is an accompanying $120°$ rotation of this methyl group around the next C–C bond axis. This generates gauche defects. As penetration increases, the anchored chains cannot relax and all the adjacent chains pinned in the gap between the indentor and the substrate continue to become compressed against each other, thus reducing the interchain distance with increasing penetration. We note at this stage that the interchain repulsion supports the applied indentation load (Joyce *et al.* 1992). A large increase in force is needed to make a very small difference in the interchain distance as well as the confined volume (Fig. 13.20). Consequently, there is a large increase in stiffness with load (Devaprakasam and Biswas 2005) as can be

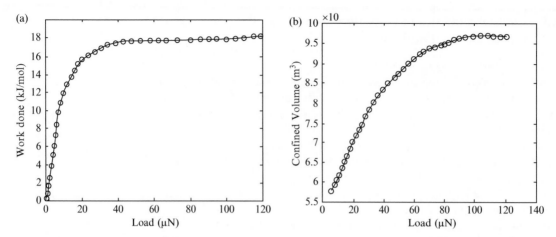

Fig. 13.20 (a) Work done on the system per mole by the probe, estimated knowing the load–displacement characteristics and the volume under real contact area. (b) Estimated confined volume under the real area of contact as a function of applied load. Reprinted with permission from Devaprakasam and Biswas (2005). Copyright 2008, American Physical Society.

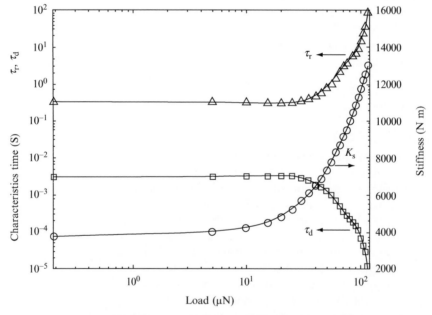

Fig. 13.21 The relaxation time τ_r, the retardation time τ_d, and the stiffness of FOTS SAM as a function of applied load. Reprinted with permission from Devaprakasam and Biswas (2005). Copyright 2008, American Physical Society.

seen in Fig. 13.21. As the molecules are collectively pressed together, they are likely to constrain each other's movement by interdigitation and entanglement exhibiting a large increase in the relaxation time (Fig. 13.21) (Salmeron 2001; Jeffrey *et al.* 2004). Experiments by Salmeron and coworkers (Lio *et al.* 1997; Barrena *et al.* 1999; Salmeron 2001) suggest that at high loads the load is borne by collective tilt and slipping of the molecules past each other. They argue that such a behavior is possible when the molecules are tightly packed. When the molecules are less densely packed, space is available between the molecules for gauche defects to be generated down the backbone without causing too much stearic hindrance. Thus, compression should generate significantly larger disorder in loosely packed molecules than in the more closely packed one, perhaps accounting for the larger friction generally observed in the former than in the latter.

When two surfaces slide against each other, the same process is allowed to happen as they are kept pressed together by the normal load. To understand dissipation in the sliding process it is therefore important for us to briefly supplement our experimentally obtained understanding of the consequence of normal load application by some elegant insights that have emerged from molecular-dynamic simulation of the same process by Harrison and coworkers (Tutein *et al.* 1999, 2000; Mikulski and Harrison 2001a,b). In molecular-dynamic simulation, the atoms in confinement are allowed to move freely according to classical dynamics. The equation of motion for all non-rigid atoms are integrated using a velocity algorithm. The force on each atom is derived using a chosen potential.

Tutein *et al.* (2000) considered linear hydrocarbon chains anchored to a semi-infinite diamond plate being compressed by a semi-infinite hydrogen-terminated diamond plate. Gauche defects in this case are immediately generated on load application at the terminal-group level of the hydrocarbon, as was in fact observed experimentally (Salmeron 2001). The confined molecules also tilt, and the tilt angle increases with the increasing load. On release of compression, the induced tilt disappears and the molecular axis springs back to its original position. It is important to note that the energy released when the molecules spring back on unloading is used to generate further gauche defects, the population of which reaches a maximum only when unloading is complete. There is thus always a phase lag. This is an important finding for tribology, as will be discussed later. This phase lag generates a hysteresis observed in the difference in damping ability between the loading and unloading paths as seen in Fig. 13.22 (Devaprakasam and Biswas 2005). The situation does not change qualitatively when the flat compressor is replaced by a sharp indentor (Tutein *et al.* 1999), there is penetration to great depths along the backbone because of much higher levels of local energy. There are some bond 0. Molecules tilt in a less synchronized manner than in the case of a flat compressor.

The nature of the increase in gauche-defect count remains the same as in the case of a flat compressor. The count increases with penetration and is independent of chain length in the C_{13}–C_{22} range. Comparing these results with the experimental finding (Salmeron's tips are of large radius as compared to those of Harrison) it is clear that the deformation becomes more synchronized and coherent as the tip geometry becomes increasingly flat.

13.6.3.3 *Sliding friction of SAM*

To continue with process simulation, Tutein *et al.* (1999, 2000) show that as the terminal end of a SAM molecule is pushed by the counterface in the sliding direction the friction force increases linearly. In this process the molecular length increases as it is stretched (Mikulski and Harrison 2001a,b). As the molecule is mechanically constrained, no gauche defects are generated during

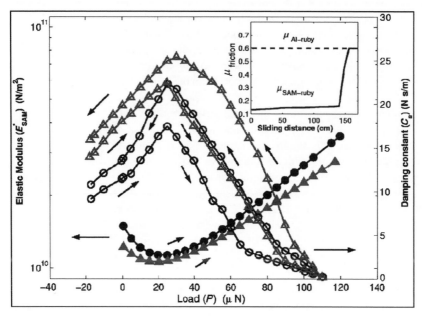

Fig. 13.22 Young's modulus deconvoluted from stiffness data (Fig. 13.21) using eqn (5) and damping constant to show their opposite trends with load. The inset shows the friction coefficient measured for FOTS/Al by a ruby ball under 100 mN normal load. Filled symbols, elastic modulus; unfilled symbols, damping constant. O, FOTS/Al; ▲, FOTS/Si. Arrows show loading direction. Reprinted with permission from Devaprakasam and Biswas (2005). Copyright 2008, American Physical Society.

this phase; in fact, the defect count decreases. Once the molecule is released, it relaxes the stretching energy, which is utilized to generate gauche defects so much so that when the stretching force is minimum the gauche defect count is maximum (Tutein *et al.* 2000). This is the avenue of energy dissipation. Combining this with the effect of pure normal compression (Tutein *et al.* 1999) one would expect the friction force to increase with normal load. The relationship Mikulski and Harrison (2001a,b) demonstrate is almost linear at high loads, though some anomaly related to packing density exists at low loads. They demonstrate quite clearly that due to the low stearic constraint in a loosely packed monolayer the gauche-defect generation is high and this results in higher friction. In a similar vein in a short-chain monolayer the low interchain Van der Waals force results in higher a priori defect population than is the case in a longer-chain monolayer. The higher friction, as Tutein *et al.* (1999, 2000) estimate, is therefore for the shorter chain monolayer due to the higher level of a priori defect population in the shorter chain. Mikulski and Harrison (2001a,b) conclude that gauche defects may not be the principal course of friction as torsional energy does plateau at a critical level of work input and they opine that the stretching mode is a good source of major energy dissipation. To explore the above, we performed tribology (Khatri and Biswas 2004, 2006; Devaprakasam *et al.* 2005; Khatri *et al.* 2005a,b) on three self-assembled monolayers of the same chemistry but of different packing densities. We changed the packing density and disorder further by subjecting each species to heat treatment to different peak temperatures: Alkylsilane monolayers octadecyltrichlorosilane (ODTS-C_{18}), dodecyltrichlorosilane (DDTS-C_{12}) and octyltrichlorosilane (OTS-C_8) self assembled on pure (97.25%) aluminum surface were subject to heat-treatment cycles to peak temperatures of 297 K, 253 K, 413 K, and 473 K. Figure 13.23 shows the representative vibrational spectra acquired by infrared reflection absorption spectroscopy (IRRAS). The figure shows the temperature-dependent shifts of methylene antisymmteric (d^-) and symmetric (d^+) stretch modes. Decreasing the chain length and increasing the heat-treatment peak temperature bring about a pronounced reduction in the packing density of in these monolayers and crystallographic order in them. Figure 13.24 shows that there is a linear increase in the coefficient of friction with a reduction in packing density. The contention that emerged from the experimental work of Salmeron and coworkers and the molecular dynamic simulation work of Harrison and coworkers appears to be valid. Creation of greater free volume by a reduction of the packing density

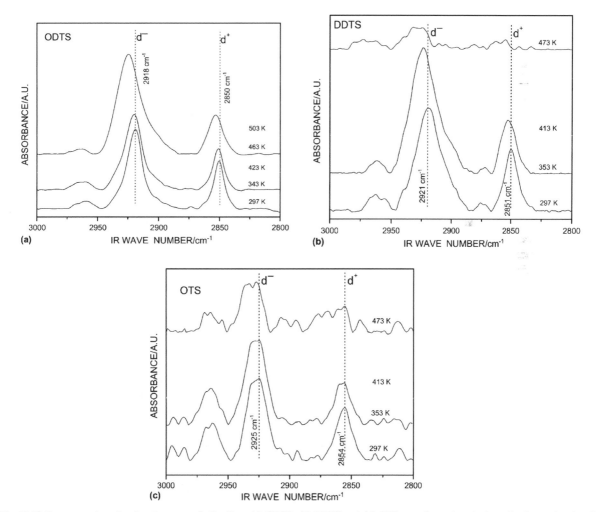

Fig. 13.23 Representative vibrational spectra of alkysilane (a) ODTS, (b) DDTS and (c) OTS monolayer deposited on aluminum showing the temperature-dependent shifts of methylene antisymmetric (d^-) and symmetric (d^+) stretch modes. Reprinted with permission from Khatri and Biswas (2006). Copyright 2008, Elsevier Ltd.

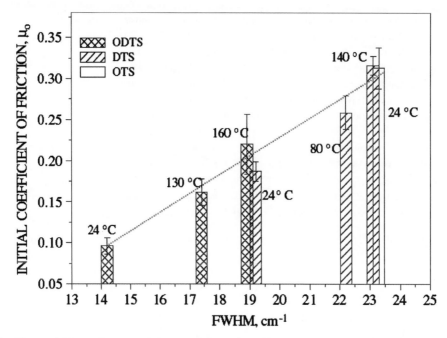

Fig. 13.24 Plot of FWHM vs. initial coefficient of friction (μ_0) for alkylsilane (ODS, DDTS and OTS) monolayers at different heat-treatment temperatures. Values of FWHM were obtained from methylene antisymmetric (d^-) stretch mode. Here, μ_0 shows a rough linear correlation with FWHM. Reprinted with permission from Khatri and Biswas (2007). Copyright 2008, American Chemical Society.

creates more avenues for rotation of molecular bonds and more frictional energy can thus be dissipated.

The MD simulation work quoted above clearly shows that normal loading and tangential loading of confined organic molecules, both generate gauche defects and cause stretching and bending of the chemical bonds. These deformations are reversible but have their characteristic life cycles that are not necessarily in phase with each other. In a real tribological situation microscopic events at the interface are more likely to be random than strictly periodic in time and space. For example, even if the SAM is strictly ordered the load on a molecule would vary and the time interval between collisions would vary from collision to collision depending on the spatial distribution of the asperities on the counterface and the substrate. The randomness is further enhanced if the molecules are disordered. Thus, depending on the sliding velocity and load (we will demonstrate this effect later) the life of energy-dissipating modes generated by normal and tangential loadings may not have the uniform wave-like morphology in time predicted by the MD simulation (Mikulski and Harrison 2001a,b). It is of importance to note at this stage the simulation work of Gao *et al.* (2004) that predicts a similar space-time variation at the microscopic scale.

The above argument suggests that it is possible for a 0–0 load life cycle of a dissipation mode to be interrupted by another collision event, the response to which would be modulated by a priori population of defects or a partially stretched bond. What this implies is that a time element, related to the periodicity of the collision events and the characteristic relaxation times of the dissipation modes, may play important roles in the tribology of organic SAMs. We will experimentally demonstrate below that this randomness does not get fully time averaged out into a time-independent steady-state average but manifests as an average friction that varies with time on a scale that is much larger than that for individual collisions. That friction varies with sliding velocity in certain regimes of sliding velocity (Yoshizawa *et al.* 1993; Liu *et al.* 2001; Ruths *et al.* 2003; Ishida *et al.* 2005) with driving frequency and relaxation time (Garcia-Parajo *et al.* 1997; Duwez *et al.* 2003) and with sliding time (Yoshizawa *et al.* 1993; Nakano *et al.* 2003) have been observed. Yoshizawa *et al.* in fact show friction increases with a reducing slope with time to finally achieve a steady-state value invariant with time.

13.6.3.4   *Friction in intermittent dry tribological contact*

In the experiments reported below a molecule is subjected to two types of intermittent contacts; t_r the time in between two contacts of a molecule with an asperity tip and the relaxation time τ_s. When it is under the probe it is slid over the contact diameter. A line (2 μm) scan of AFM topographical image of the steel ball ($2 \times 2\ \mu m^2$ scan area) showed five asperities of height more than 1 nm above a mean line. Taking these asperities as the contacting asperities, at the lower test velocity of 0.02 cm/s, a molecule is loaded by an asperity every 2×10^{-3} s. Taking l as the stroke length, D as the contact diameter and v as an average sliding velocity the time period $\left(t_r = \frac{(l-D)}{v} \right)$ over which a molecule is not under the probe or the time that elapsed between its leaving a probe contact and the next probe contact, t_r is less than 1 s at the lowest sliding velocity and 0.01 s at the highest test sliding velocity. Of the two intermittent contacts the one that gives the longer out-of-contact time is therefore the important one from the point of view of molecular relaxation. For the four different experimental conditions used here (Khatri and Biswas 2007) τ_s is greater than t_r.

Figure 13.25 shows the change in the coefficient of friction with sliding time measured in a 0% RH environment at 0.02 cm/s velocity and 363 MPa mean contact pressure. Table 13.2 shows that under these experimental conditions, the value of τ_s is significantly greater than t_r. The friction is seen to increase with time. At the commencement of sliding, the friction force jumps to a certain value, which we designate as static friction, almost instantaneously. The static friction force needs to be reached before a dynamic friction regime is initiated where the friction increases gradually with time. The rate of rise of friction decreases with time till at large times when the friction becomes insensitive to sliding time. If the dynamics friction is assumed to be related to defect generation in sliding, it has been shown (Garcia-Parajo *et al.* 1997) that as defects accumulate it becomes more difficult to generate more defects.

When a test is interrupted, unloaded, stopped for a prolonged period of time and recommenced (loaded); the level of static friction of an unslid ($t = 0$) monolayer is recovered initially; with increasing number of interruptions the recommenced static friction decreases slightly. The gradient of friction with time at recommencement, however, remains more or less the same as that at

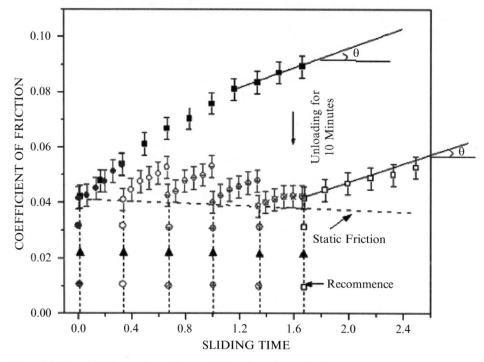

Fig. 13.25 Coefficient of friction of ODTS monolayer self-assembled on silicon: load 100 mN, sliding velocity 0.02 cm/s, and 0% relative humidity. The monolayer is unloaded after different sliding times and rested for 10 min before being loaded again and the test recommenced. The figure shows the existence of a static friction obtained immediately on recommencement and a kinetic friction that changes perceptively with sliding time. It is to be noted that the friction–time slope (θ) at interruption and recommencement of a test after prolonged stoppage (unloaded) are the same. Reprinted with permission from Khatri and Biswas (2007). Copyright 2008, American Chemical Society.

Table 13.2 Relaxation time τ_s and time allowed for relaxation t_r as a function of experimental load and velocity.[a] Reprinted with permission from Khatri and Biswas (2007). Copyright 2008, American Chemical Society.

load, N	velocity = 0.02 cm/s					velocity = 1.05 cm/s				
	D, 10^{-6} m	p_m, MPa	t_r, s	τ_s, s	$(D/1)$ $[1-(t_r/t_s)]$	D, 10^{-6} m	p_m, MPa	t_r, s	τ_s, s	$(D/1)$ $[1-(t_r/t_s)]$
0.1	18.7	363	0.9	7.91	0.082	18.7	363	0.018	7.91	0.093
0.8	37.5	726	0.8	15.97	0.17	37.5	726	0.016	15.97	0.1875

[a] τ_s is measured using a contact-force apparatus as a function of mean contact pressure p_m. p_m and D are estimated from a contact mechanical analysis of bilayer systems.

interruption. This is an important finding as it implies that it is possible to maintain a low level of friction with frequent interruptions.

We may now combine the above observations to construct a simple model. Figure 13.26(a) shows an idealistic schematic where the population of defects or dissipation modes start to increase when the force experienced by the counterface asperity reaches its maximum and where the work done on the system is maximum. Figure 13.26(b) shows that if the time available (as dictated by

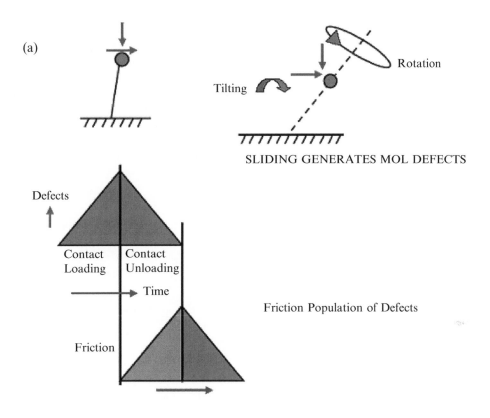

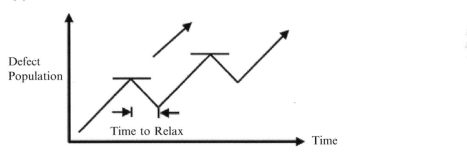

Fig. 13.26 (a) Schematic showing a cycle of defect generation in a molecule as a function of sliding. (b) Schematic showing accumulation of defects in a molecule with sliding time.

a ratio of $\frac{t_r}{\tau_s}$) to the defects to relax is not sufficient for the defect population to be reduced to nil, there is an accumulation of defects with each collision. If g is the defect population, the rate of defect generation is given by Biswas (2007).

$$\frac{dg}{dt} = \frac{x\,D}{l\,g^m}\left[1 - \frac{l - D}{\tau_s\,v}\right], \qquad (13.3)$$

where g^m is a function of the current defect population, $1 > m > 0$.

If we now assume that $\frac{\mathrm{d}f}{\mathrm{d}t} \approx \frac{\mathrm{d}g}{\mathrm{d}t}$, where f is a frictional stress, the above model for friction implies that if the sliding velocity is reduced, t_r increases and there is a corresponding reduction in $\frac{\mathrm{d}g}{\mathrm{d}t}$ and $\frac{\mathrm{d}f}{\mathrm{d}t}$. If the normal load is increased, the relaxation time τ_s (Khatri and Biswas 2007) and D are increased. According to eqn (13.3) any such increase would increase $\frac{\mathrm{d}g}{\mathrm{d}t}$ and therefore the overall level of friction as well. Figure 13.27 shows the gradient of friction

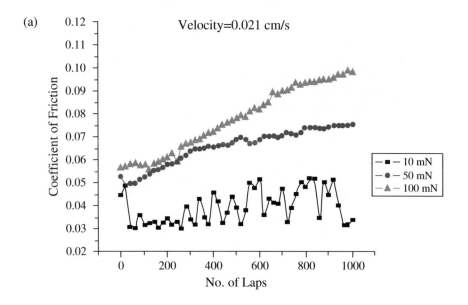

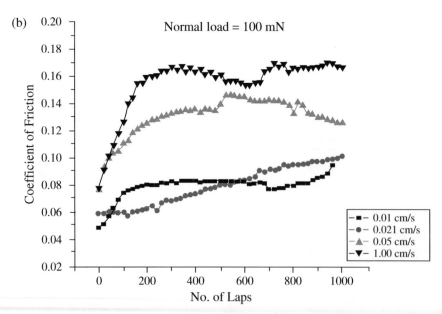

Fig. 13.27 (a) The effect of normal load on coefficient of friction on ODTS monolayer, sliding velocity = 0.21 cm/s. (b) The effect of sliding velocity on coefficient of friction on ODTS monolayer, normal load = 100 mN.

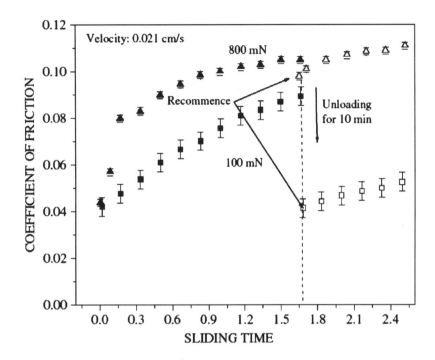

Fig. 13.28 Effect of load on coefficient of friction (static) obtained on recommencement after a 10-min unloaded rest. Sliding velocity: 0.02 m/s. The solid symbols represent continuous loading, while the open symbols represent recommenced loading after interruption.

with time to increase with increasing velocity and normal load. Figure 13.25 indicates a complex response of organic self-assembled molecules to sliding traction. We present below a possible simple model to account for the experimental observations. The frictional response of the monolayer is a composite of responses of two structurally distinct parts of a molecule. The stiff backbone is anchored chemically to the substrate and tightly bound to adjacent backbones by Van der Waals forces, the response of the backbone may be visualized as "solid like" and elastic. The terminal end of the molecule has a large degree of freedom of movement and may be visualized as "liquid like" and "plastic". When the monolayer is slid, the backbone bends and stretches elastically until the terminal end "yields" marking a level of static friction. Once it yields, the terminal end becomes active and gauche defects are generated. In a cycle when the molecule is out of contact, the backbone is restored instantaneously to its original state and the terminal end recovers partially, no doubt determined by the relative magnitude of τ_s and t_r. This is repeated with each cycle and the terminal end defects accumulate to yield a rising friction with time. When the test is recommenced after prolonged stoppage the SAM, by virtue of retention of some permanent defects at its terminal end, has a memory. This is reflected in the recommenced friction gradient after "yield" being the same as that at the time of interruption. Other effects of the sliding time on the SAM structure are (1) backbone fatigue that lowers the static friction with time (Fig. 13.2) and (2) permanent deformation of the terminal end. At high loads the difference between the static and dynamic friction is greatly reduced with increasing sliding time (Fig. 13.28). It is important to note that the magnitude of the relaxation time that allows a positive friction—time gradient (which is experimentally observed) is several orders greater than what is expected

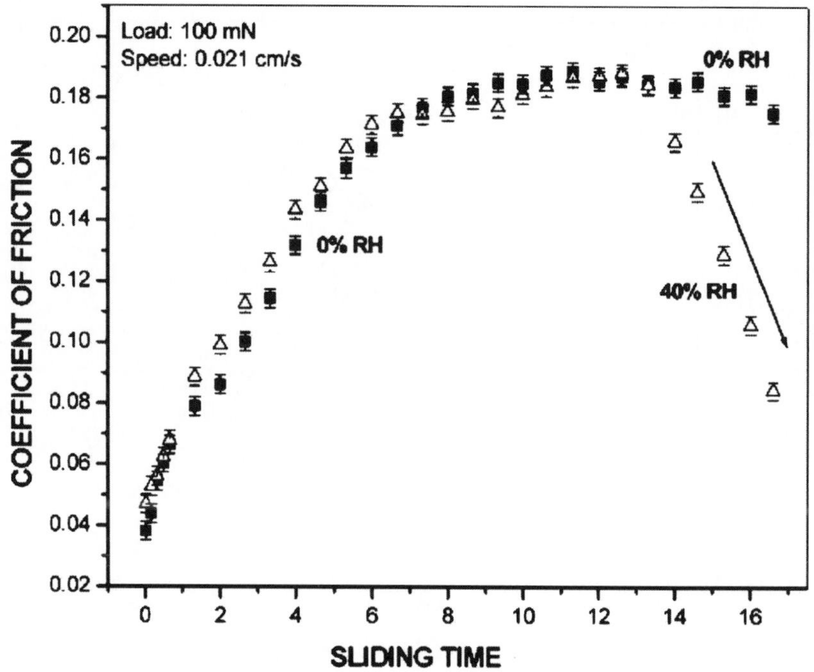

Fig. 13.29 Effect of introduction of water vapor into a sealed enclosure housing the nanotribometer after 13 h of sliding at a nominal 0% RH, a steel ball sliding against an ODTS monolayer. The figure records the variation in friction due to sliding between 13 and 17 h because of the introduction of water vapor into the chamber. Load 100 mN, mean contact pressure 370 MPa. Reprinted with permission from Khatri *et al.* (2007). Copyright 2008, American Chemical Society.

for a single molecule (Barrena *et al.* 1999). As a matter of fact, if τ_m is the relaxation time of a single molecule and all the molecules in contact relax fully in the relaxation time of a single molecule and all the molecules in contact relax fully in time τ_m, when $\tau_m <<< t_r$, this gives $\frac{df}{dt} = 0$, which is of course contrary to experimental findings (Khatri and Biswas 2007). Overney (2004) has shown that in efficient boundary lubrication, the molecules do not respond to traction individually but as constituents of a coherent volume. This cooperative behavior lowers the activation-energy barrier that needs to be overcome for sliding motion. The large relaxation times, which appear to be relevant to the tribological performance, suggest that the material response to traction should be tracked not on the basis of individual molecular behavior but by recognizing an aggregate phenomenon, which involves a large number of interacting molecules. This we believe is a phenomenon that may be unique for self-assembled organic molecules that when collectively pressed are likely to interdigitize to minimize free volume. Such high values of relaxation time in comparison to that of single molecules, have been observed in previous experimental works (Joyce *et al.* 1992; Demirel and Granick 1996; Devaprakasam and Biswas 2005) on compression and shear of molecules in confinement.

13.6.3.5 *Friction in humid environment*

Continuing their nanotribological experiments with silane monolayer Khatri *et al.* (2007) showed the friction to drop (Fig. 13.29) to a very low value when water vapor was bled into a dry chamber where the SAM was slid against a steel ball. The effect of water, confined between hydrophobic surfaces, on friction has been much debated. Doshi *et al.* (2005) have discussed the genesis of low friction at such interfaces as due to the formation of nanobubbles and nanobridges. This happens due to the lowering of water viscosity in the 1-nm gap between the monolayer and the water. Zhang *et al.* (2002) argue the case for dewetting of the surface, while the experiments of Scherge *et al.* (1999) show the presence of solvated water at the hydrophobic interface even in high vacuum. In the Khatri *et al.* (2007) experiment the tip is hydrophilic, which would tend to collect some water. Capillary condensation as a controlling mechanism was ruled out on the grounds that a hysteresis-induced friction force was found to be several orders smaller than the measured friction. Elastohydrodynamic calculations again yielded unrealistic confinements and forces. The authors explained their finding on the basis of a water-induced softening of the monolayer, which they argued reduces the molecular relaxation time dramatically to that of the monolayer slid in a dry environment. They argue that the hydrophilic probe carries some water, which has a very low relaxation time, to the interface. Given this, the friction is controlled by the periodicity of intermittent contact where the timescales are large, of the order of seconds. Molecules perturbed by sliding have sufficient time to relax and anneal, lowering the friction.

13.7 Summary

Whenever two surfaces are brought into close proximity of each other contact occurs. Contact is always discrete and except in the case of atomically smooth (or bumpy) surfaces, it is heterogeneous over an apparent area. The area involved in a single contact between two opposing protrusions is generally of the order of nm^2 and the corresponding forces are of the order of nN. The contact pressures can therefore be large, generally of the order of 10^9 Pa. Friction under such large pressures can be highly dissipative and destructive, leading to large-scale deformation, fracture and wear. This situation prevails where large surfaces come together to transmit power, the interfacial response there is an aggregate of a multitude of elemental responses. The same situation prevails when the bodies in contact are small, of micrometric dimensions, as in the case of micro-electro-mechanical systems. In the latter case the discrete elemental responses hold sway and govern the machinery response. To ameliorate the severity of interfacial conditions that dictate the overall efficiency of power transmission, the interfaces are packed with soft matter that by promoting a low-shear regime reduce dissipation. Such soft matters are chosen as per their inherent normal load-bearing capacity, their weak interaction potentials and their ability to undergo favorable physical and/or chemical changes and transformations in confinement. The introduction of the third body at the interface is generally called lubrication.

In this chapter we have described our prevailing understanding of frictional response at the elemental level. Having briefly described the novel tools that have made possible study of contact and shear at this level we first discuss the state of the contact entrapping layers of lubricant. Adhesion, capillary and structural/entropic forces operate in the gap carrying a fluid, when the separation is at the level of a few nanometers or less. If the confined fluid is water the characteristic hydrogen bonding and polarity give rise to additional normal forces due to the presence of an electrostatic double layer and the hydrophobic effect. These forces in addition to the applied pressure define the contact area and modulate normal traction. The friction, according to Amonton's Law or otherwise, scale with normal traction and/or contact area.

Given the complexity of the normal force regime that may exist in the confined space questions arise as to what governs friction. We first examine a model of the atomistic origion of friction where the bumpy geometry, surface energy and a periodic potential between a slider and the substrate atoms determine the dynamics of motion and yield the friction force. The model known popularly as the cobblestone model recovers Amonton's Law when the loads are large and predicts the stick-slip, a phenomenon characteristic of most sliding processes. The phenomena of stick-slip is also observed when organic matter is trapped in confinement. Here, the sliding process may be regarded as thermally activated, causing molecular jumps to facilitate sliding. The energy required to cause molecular jumps may be estimated using the Eyring thermodynamic model of the process. The model provides information on frictional energy, stick-slip and coherence in molecular sliding motion. The model is sensitive to initial microstructure and its entropic state and distinguishes the response of different molecular species being slid in narrow confinement. In this section we also examine the changes in the dissipative capacity of a morphologically heterogeneous microstructure, caused by a change in normal traction and record the linear variation of such dissipation with a normal load. Does this provide a mechanistic explanation of the empirically observed Amonton's Law, the validity of which at the microscopic scale remains unresolved? The section ends with an alternative explanation of stick-slip when organic matters are trapped in a 3 to 5 molecular diameter gap. The explanation relates stick-slip to structurization and phase change in the confined matter, brought about by the sliding motion and provides a rationale for the velocity dependence of friction and the associated stick-slip motion.

The final section of this chapter is devoted to the problem of boundary lubrication where the wisdom of molecular-level mechanisms are examined in the context of large-scale experiments and practice. The effect of initial conformational order, intermittency of contact, viscoelastic properties (relaxation times) and environmental humidity on friction are critically examined over several orders of forces and contact scales.

While much insight has emerged in the last decade or so due to concerted efforts from many research groups to simulate microscopic contact and to measure parameters of relevance to dissipation, many questions remain unanswered. The main issue that has troubled tribologists is the load and contact-area dependence of friction, the final mechanistic/physical explanation remains elusive. The other issue of great practical importance is the role of environmental water vapor on friction. The presence of such vapor in practical applications is inevitable but the role it plays in friction is still unclear and the problem needs serious attention by nanotribologists in the near future.

Acknowledgments

The author is grateful to Mr S.R. Binu for his help with the figures and formatting. The financial support for some of the author's own work reported here came from Indian Oil Corp. (Ltd), Bharat Petroleum Corp (Ltd), Hindustan Petroleum Corp. (Ltd), General Motors (R&D), Michigan, US and the Centre For High Technology, Min. of Petroleum, Govt of India.

References

Adhvaryu, A., Biresaw, G., Sharma, B.K., Erham, S.Z. *Ind. Eng. Chem. Res.* **45**, 3735 (2006).

Ahn, H., Cuongo, P.D., Park, S., Kim, Y., Lim, J. *Wear* **255**, 819 (2003).

Alba-Simionesco, Coasne, B., Dosseh, G., Dudziak, G., Gubbins, K.E., Radhakrishnan, R., Al-Samieh, M., Rahnejat, H. *J. Phys D: Appl. Phys.* **34**, 2610 (2001).

Barrena, E., Kopta, S., Ogletree, D.F., Charych, D.H., Salmeron, M. *Phys. Rev. Lett.* **82**, 14 2880 (1999).

Benesebaa, F., Ellis, T.M., Badia, M., Lennox, R.B. *Langmuir* **14**, 2361 (1998).

Berman, A., Israelachvili, J. *N. Handbook of Micro/Nanotribology*, (ed.) Bhushan, B. (CRC Press, Boca Raton, 1999) p. 371.

Bhatia, R., Garrison, B.J. *Langmuir* **13**, 765 (1997).

Bhushan, B. *Modern Tribology Handbook*, (ed.) Bhushan, B. (CRC Press, Boca Raton, 2001) Ch.19.

Biswas, S.K. *J. Ind. Inst. Sci.* **87**(1), 15 (2007).

Bouchina, T., Aime, J.P., Gauthier, S., Michel, D. *Phys. Rev. B* **56**(12), 7694 (1997).

Bowden, F.P., Tabor, D. *The Friction and Lubrication of Solids* (Clarendon Press, Oxford, 1951) Ch. 5.

Briscoe, B.J., Evans, D.C. B. *Proc. Roy. Soc, London, A* **380**, 389 (1982).

Bryant, M.A., Pemberton, J.E., *J. Am. Chem. Soc.* 1138284 (1991).

Buenviage, C., Ge, S., Rafaelovich, M., Sokolov, J., Drake, J.M., Overney, R.M. *Langmuir* **19**, 6446 (1999).

Burnham, N.A., Domingez, D.D., Mowrey, R.L., Colton, R. *J. Phys. Rev. Lett.* **64**, 1931 (1990).

Carpick, R.W., Salmeron, M. *Chem. Rev.* **97**, 1163 (1997).

Chandross, M., Lorenz, Christian. D., Slevens, M.J., Grest, G.S. *Langmuir* **24**, 1420 (2008).

Choi, J., Kawaguchi, M., Kato, T. *J. Appl. Phys.* **91**(10), 2574 (2002).

Demirel, A.L., Granick, S. *Phys. Rev. Lett.* **77**, 3261 (1996).

Devaprakasam, D., Biswas, S. *K. Phys. Rev. B* **72**, 125434 (2005).

Devaprakasam, D., Sampath. S., Biswas, S.K. *Langmuir* **20**, 1329 (2003).

Devaprakasam, D., Khatri, O.P., Shankar, N., Biswas, S.K. *Tribol. Int.* **38**, 1022 (2005).

Derjaguin, B.V. *Z. Physics* **88**, 661 (1934).

Dienwiebel, M., Verhoeven, G.S., Pradeep, N., Frenken, J.W.M., Heimberg, J.A., Zandbergen, H.W. *Phys. Rev. Lett.* **92**(12), 126101 (2004).

Doshi, D.A., Watkins, E.B., Israelachvili, J.N., Majjewski, J. *PNAS* **102**, 9458 (2005).

Drummond, C., Israelachvili, J.N. *Macromolecules* **33**(13), 4910 (2000).

Dubois, L.H., Zegarski, B.R., Nuzzo, R.G. *J. Chem. Phys.* **98**, 678 (1993).

Duwez, A.S., Jonas, U., Klein, H., *Chem. Phys. Chem.* **4**, 1101 (2003).

Eyring, H. *J. Chem. Phys.* **3**, 107 (1935).

Fenter, P., Eisenberger, P., Liang, K.S. *Phys. Rev. Lett.* **70**, 2447 (1993).

Fenter, P., Eisenberger, P., Burrows, P., Forrest, Liang, K.S. *Physica B* **221**, 145 (1996).

Frenkel, Y.I., Konotorova, T. *Phys. Z Sowietunion* **13**, 1 (1938).

Fuzisawa, S., Kishi, E., Sugawara, Y., Morita, S. *Phys. Rev. B* **51**, 7849 (1995).

Gao, J., Luedtke, W.D., Gourdon, D., Ruths, M., Israelachvili, J.N., Landman, U. *J. Phys. Chem. B* **97**, 4128 (2004).

Gao, J.P., Luedtke, W.D., Landman, U. *Tribol. Lett.* **9**(1–2), 3 (2000).

Garcia-Parajo, M., Longo, C., Servat, J., Gorostriza, P., Sanz, F. *Langmuir* **13**, 2333 (1997).

Gee, M.L., Israelachvili, J.N. *J. Chem. Faraday Trans.* **86**(24), 4049 (1990).

Gee, M.L., Mcguiggan, P.M., Israelachvili, J.N., Hamola, A.M. *J. Chem. Phys.* **93**(3), 1895 (1990).

Glosli, J.N., McClleland, G.M. *Phys. Rev. Lett.* **70**(13), 1960 (1993).

Gnecco, E., Bennewitz, R., Gyalog, T., Loppacher, C., Bammerlin, M., Meyer, E., Guntherodt, H. *J. Phys. Rev. Lett.* **84**(6), 1172 (2000).

Graupe, M., Koini, T., Kim, H.I., Garg, N., Miura, Y.F., Takanga, M., Perry, S.S., Lee, T.R. *Mater. Res. Bull.* **34**(3), 447 (1999).

Hamrock, B.J., Dowson, D. *Trans. ASME, J. Tribol.* **99**, 264 (1977).

Harrison, J.A., Colton, R.J., White, C.T., Brennar, D.W. *Wear* **168**, 127 (1993).

Harrison, J.A., White, C.T., Colton, R.J., Brenner, D.W. *J. Phys. Chem.* **97**, 6573 (1993).

Hautman, J., Klein, M.L. *J. Chem. Phys.* **91**, 4994 (1989).

Hautman, J., Bareman, J.P., Mar, W., Klein, M.L. *J. Chem. Soc. Faraday Trans.* **87**, 2031 (1991).

He, M., Schumacher, Blum. A., Overney, G., Overney, R.M. *Phys. Rev. Lett.* **88**(15), 154302 (2002).

Holscher, H., Schwarz, U.D., Zworner, O., Wiesendanger, R. *Phys. Rev. B* **57**, 2437 (1998).

Holscher, H., Schwarz, U.D., Wiesendanger, R. *Surf. Sci.* **375**, 395 (1997).

Hu, H.W., Carson, G.A., Granick, S. *Phys. Rev. Lett.* **66**(21), 2758 (1991).

Ishida, H., Koga, T., Morita, M., Ostuka, H., Takahara, A. *Tribol. Lett.* **19**(1), 1 (2005).

Israelachvili, J.N. *Intermolecular and Surface Forces* (Academic Press, London, 1992).

Israelachvili, J.N. *Modern Tribology Handbook*, (ed.) Bhushan, B. (CRC Press, Boca Raton, 2001) Ch.16.

Israelachvili, J.N. *MRS Bull* **30**(7), 533 (2005).

Jeffery, S., Hoffman, P.M., Pethica, J.B., Ramanujan, C., Ozger, H.O., Oral, A. *Phys. Rev. B* **70**, 054114 (2004).

Joyce, S.A., Thomas, R.C., Homston, J.E., Michalske, T.A., Crooks, R.M. *Phys. Rev. Lett.* **68**, 2790 (1992).

Khatri, O.P., Biswas, S.K. *J. Phys. Chem. C* **111**, 2696 (2007).

Khatri, O.P., Bain, C.D., Biswas, S.K. *J. Phys. Chem. B* **109**, 23405 (2005a).

Khatri, O.P., Devaprakasam, D., Biswas, S.K. *Tribol. Lett.* **20**(3), 243 (2005b).

Khatri, O.P., Math, S., Bain, C.D., Biswas, S.K. *J. Phys. Chem.* **111**, 16339 (2007).

Khatri, O.P., Biswas, S.K. *Surf. Sci.* **572**, 228 (2004).

Khatri, O.P., Biswas, S.K. *Surf. Sci.* **600**, 4399 (2006).

Khatri, O.P., Biswas, S.K. Unpublished (2008).

Kim, H.I., Koini, T., Lee, T.R., Perry, S.S. *Langmuir* **13**, 7192 (1997).

Kojio, K., Takahara, A., Kajiyama, T. *Langmuir* **16**, 9314 (2000).

Kondo, H. *Wear* **202**, 149 (1997).

Konotorova, T.A., Frenkel, Y.I. *Zh. Eksp. Teor. Fiz* **8**, 89 (1939).

Krim, J., Widom, A. *Phys. Rev. B* **38**(17), 12184 (1988).

Krupka, I., Hartl, M., Liska, M. *Tribol. Int.* **39**, 1726 (2006).

Kumar, D., Biswas, S.K. *Trib. Lett.* **30**(2), 159–60 (2008).

Landman, U. (Private communication, 2008).

Landman, U., Luedtke, W.D., Burnham, N.A., Colton, R.J. *Science* **248**, 454 (1990).

Landman, U., Luedtke, W.D., Nitzan, A. *Surf. Sci.* **210**, L177 (1989).

Lim, R., O'Shea, S. *J. Phys. Rev. Lett.* **88**(24), 246101 (2002).

Lim, R., Li, S.P.Y., O'Shea, S.J. *Langmuir* **18**(16), 6116 (2002).

Lio, A., Morant, C., Ogletree, D.F., Salmeron, M. *Phys. Chem. B* **101**, 4767 (1997).

Liu, H., Ahmed, S.I., Scherge, M. *Thin Solid Films* **381**, 135 (2001).

Mate, M.C. *Tribology on the Small Scale* (Oxford University Press, Oxford, 2008).

Maugis, D. *Contact, Adhesion, and Rupture of Elastic Solids* (Springer-Verlag, Berlin, 2000).

McDermott, M.T., Green, J.-B.D., Porter, M.D. *Langmuir* **13**, 2504 (1997).

Mikulski, P.T., Harrison, J.A. *Tribol. Lett.* **10**(1–2), 29 (2001a).

Mikulski, P.T., Harrison, J.A. *J. Am. Chem. Soc.* **123**, 6873 (2001b).

Muser, M., A., Wenning, L., Robbins, M.O. *Phys. Rev. Lett.* **86**(7), 12 (2001).

Nakano, M., Ishida, T., Numata, T. Numata, T., Ando, Y., Sasaki, S. *Jpn. J. Appl. Phys.* **42**, 4734 (2003).

Overney, R.M., Tyndall, G., Frommer, J. *Nanotechnology Handbook*, (ed.) Bhushan, B. (Springer-Verlag, Heidelberg, Germany, 2004) Ch. 29.

Porter, M.D., Bright, T.B., Allora, D.L., Chidsey, E.E.D. *J. Am. Chem. Soc.* **109**, 3559 (1987).

Prathima, N., Harini, M., Rai, Neeraj., Chandrashekara, R.H., Ayappa, K.G., Sampath, S., Biswas, S.K. *Langmuir* **21**(6), 2364 (2005).

Robbins, M.O., Muser, M.H. *Handbook of Modern Tribology*, (ed.) Bhushan. B. (CRC Press, Boca Raton, 2001) Ch. 20.

Robbins, M.O., Thompson, P.A. *Science* **1253**, 916 (1991).

Ruths, M., Alcanter, N.A., Israelachvili, J.N. *J. Phys. Chem.* **107**, 11149 (2003).

Salmeron, M. *Tribol. Lett.* **10**(1–2), 69 (2001).

Scherge, M.L., Li, X., Schaefer, J.A. *Tribol. Letts.* **6**, 215 (1999).

Schwarz, U.D., Holscher, H. *Modern Tribology Handbook*, (ed.) Bhushan, B. (CRC Press, Boca Rotan, 2001) Ch. 18.

Shimizu, J., Eda, H., Yoritsune, M., Ohmura, E. *Nanotechnology* **9**, 118 (1998).

Smith, O., Priest, M., Taylor, R.I., Price, R., Cantlay, A., McCoy, R.C. *Proc. I. Mech. E., Part J. Eng. Tribol.* **220**, 181 (2006).

Socoliuc, A., Bennewitz, R., Gnecco, E., Meyer, E. *Phys. Rev. Lett.* **92**(13), 134301 (2004).

St. Pierre, L.E., Owens, R.S., Klint, R.V. *Wear* **9**, 160 (1966).

Subhalakshmi, K., Devaprakasam, D., Math, S., Biswas, S.K. *Tribol. Lett.* **32**(1), 1–11 (2008).

Sung, I.H., Yang, J.C., Kim, D., Shin, B. *Wear* **255**, 808 (2003).

Takahara, A., Kojio, K., Kajiyama, T. *Ultramicroscopy* **91**, 203 (2002).

Tomlinson, G.A. *Phil. Mag.* **7**, 905 (1929).

Tutein, A.B., Stuart, S.J., Harrison, J.A. *J. Phys. Chem. B* **103**, 11357 (1999).

Tutein, A.B., Stuart, S.J., Harrison, J.A. *Langmuir* **16**, 291 (2000).

Wizrig, C.A., Chung, C., Porter, M.D., *J. Electroanal. Chem.* **310**, 335 (1991).

Yoshizawa, H., Chen, You-Lung Israelachvili, J.N. *J. Phys. Chem.* **97**(16), 4128 (1993).

Zhang, X., Zhu, Y., Granick, S. *Science* **295**, 663 (2002).

Zhu, Y., Granick, S. *Phys. Rev. Lett.* **93**(9), 096101 (2004).

The electronic structure of epitaxial graphene—A view from angle-resolved photoemission spectroscopy

14

S. Y. Zhou and A. Lanzara

14.1 Introduction

Graphene, the fundamental building block for all graphitic materials, is a two-dimensional sheet of carbon atoms arranged in a honeycomb lattice with sp^2 bonding. The sp^2 hybridization between one s-orbital and two p-orbitals leads to the formation of σ bonding between carbon atoms, while the π orbitals perpendicular to the plane lead to the formation of the half-filled bands, the π-bands. These half-filled bands are the basis for most of the fascinating properties of graphene, and its full potential is still not reached. Although the theoretical study of graphene started in the 1950s (McClure 1957), the experimental study of graphene had not been realized until the recent discovery and characterization of exfoliated graphene by Novoselov *et al.* (2004) and epitaxial graphene by Berger *et al.* (2004). Immediately after that, graphene was found to exhibit various intriguing properties unexpected from conventional materials. For example, the charge carriers in graphene are massless Dirac fermions with mobility higher than that of silicon and the doping can be tuned from electrons to holes through a gate voltage (Novoselov *et al.* 2004). Also, the quantum Hall effect shows half-integer numbers (Novoselov *et al.* 2005; Zhang *et al.* 2005). Because of its fundamental importance in physics as a realization of a relativistic condensed-matter system, as well as its application potentials in next-generation electronics, the research interest in graphene has been rising rapidly (Geim *et al.* 2007; Castro Neto *et al.* 2008).

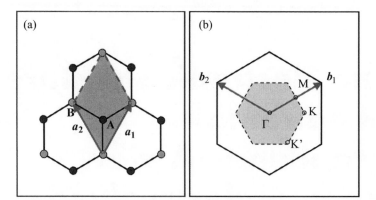

Fig. 14.1 Unit vectors of graphene in the real space (a) and the reciprocal space (b).

14.2   Electronic structure of graphene

The carbon atoms in graphene are arranged in a honeycomb lattice as shown in Fig. 14.1. The unit cell of graphene contains two geometrically different carbon sublattices A and B. The lattice vectors can be written as
$$\vec{a}_1 = a\left(\tfrac{1}{2}, \tfrac{\sqrt{3}}{2}\right)$$
$$\vec{a}_2 = a\left(-\tfrac{1}{2}, \tfrac{\sqrt{3}}{2}\right),$$
where a is the carbon–carbon lattice constant $a = 1.42$ Å. The corresponding reciprocal lattice vectors are
$$\vec{b}_1 = \tfrac{4\pi}{\sqrt{3}a}\left(\tfrac{\sqrt{3}}{2}, \tfrac{1}{2}\right)$$
$$\vec{b}_2 = \tfrac{4\pi}{\sqrt{3}a}\left(-\tfrac{\sqrt{3}}{2}, \tfrac{1}{2}\right)$$
and the Brillouin zone also shows a hexagonal shape.

In graphene, the p electrons form the sp^2 bonding. The intraplanar interaction between the 2s, $2p_x$, $2p_y$ atomic orbitals form the strongly covalent σ orbitals, which give rise to three bonding σ bands (Fig. 14.2) and three antibonding σ* bands. The out-of-plane p_z atomic wavefunction forms weakly van

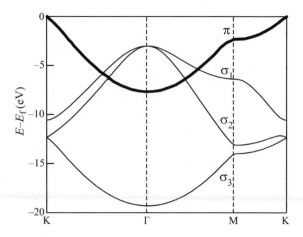

Fig. 14.2 The band structure of graphene.

der Waals π bonds, which give rise to the π and π^* bands that cross the Fermi energy at the six corners of the hexagonal Brillouin zone.

The dispersion of the π and π^* bands is

$$E(\vec{k}) = \pm\gamma \sqrt{1 + 4\cos\left(\frac{k_x a}{2}\right)\cos\left(\frac{\sqrt{3}k_y a}{2}\right) + 4\cos^2\left(\frac{k_x a}{2}\right)},$$

where $\gamma \sim 3\,\text{eV}$ (Dresselhaus *et al.* 1981) is the nearest-neighbor hopping integral, $a = \sqrt{3}a_0$ with the carbon–carbon lattice constant $a_0 = 1.42$ Å. The valence and conduction bands touch at the six corners of the Brillouin zone. Since each carbon atom contributes one electron, the valence band is completely filled up to the Fermi level, while the conduction band is empty. Because of the zero separation between the bottom of the conduction band and the top of the valence band, graphene is also known as a semi-metal or zero-gap semiconductor.

The most peculiar property of graphene is in the region near the Brillouin zone corners where the valence and conduction bands merge. Expanding the dispersions near the Brillouin zone corner K point, $\vec{k} = \vec{K} + \vec{\kappa}$, the dispersion can be written as $E(k) = \pm\frac{\sqrt{3}a}{2}\gamma|\vec{\kappa}| = \pm\hbar v_F \kappa$, where $v_F = \frac{\sqrt{3}a\gamma}{2\hbar}$. This dispersion relation is analogous to that of relativistic particles $E = \pm\sqrt{m^2 c^4 + c^2 p^2}$ with zero effective mass $m = 0$, and the speed of light c is replaced by the Fermi velocity $\hbar v_F$, which is approximately 300 times smaller. Therefore, electrons in graphene are governed by a two-dimensional version of the relativistic theory introduced by Dirac. Because of this similarity, the low-energy electrons in graphene are also described as massless "Dirac fermions" and the points where the valence and conduction bands merge are termed "Dirac points".

Figure 14.3 shows a schematic drawing of the low-energy dispersions near E_F. The peculiar linear dispersion near E_F results in many intriguing

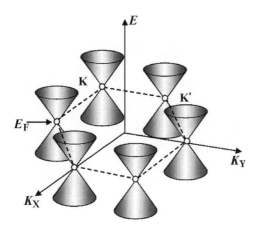

Fig. 14.3 Schematic drawing of the low-energy dispersions near E_F.

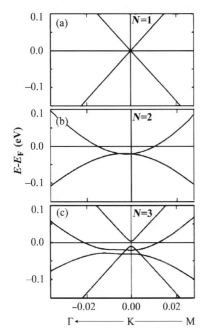

Fig. 14.4 (a–c) Dispersions near the K point for single layer, bilayer and trilayer graphene. From Partoens and Peeters (2007), copyright American Physical Society.

properties. First, the "Fermi surface" in graphene contains only six points, different from a real *surface* common in metals. Because of the finite number of points at the Fermi surface, the density of states vanishes at E_F. Moreover, electrons in graphene travel with a constant velocity, in contrast to electrons in a semiconductor, where the dispersion shows a quadratic behavior near the top and bottom of the bands. An electron in the latter case is modelled as a quasi-particle with a finite ("massive") effective mass $m^* = \hbar^2 \left(\dfrac{d^2 E}{dk^2} \right)^{-1}$ and can be described by non-relativistic theory formulated in Schrödinger's equation. The effective mass m^* is usually different from the non-interacting electron mass and this mass renormalization is used to take into account the effects of electron–electron and electron–phonon interactions.

The completely linear dispersion near E_F and the massless Dirac fermions are expected in single-layer graphene only in the case of a perfect crystal. In the presence of additional interactions, e.g. breaking of the carbon-sublattice symmetry, the valence and conduction bands will hybridize; causing a finite gap between the valence and conduction bands. One extreme case is BN, which has a similar hexagonal structure and two completely different sublattices, which results in a gap of up to a few eV (Blase *et al.* 1995). Similar gap opening can be expected if the potentials on the two carbon sublattices are inequivalent. When a gap exists between the valence and conduction bands, the Dirac fermions will acquire a finite mass. However, despite a small deviation near E_F from the linear dispersion, the overall dispersion will still show a linear behavior when moving far enough away from the Dirac point.

In the case of bilayer and trilayer graphene, the number of π bands will increase with the number of graphene layers. The overall linear dispersion is still preserved, along with some additional parabolic perturbations near E_F (Fig.14.4; Partoens and Peeters 2007). When the number of graphene layers exceeds 10, the electronic structure of multilayer graphene is basically that of graphite.

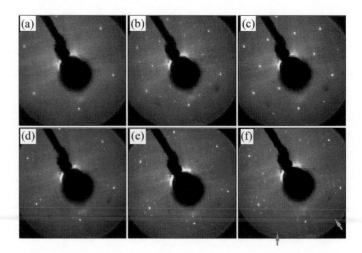

Fig. 14.5 LEED patterns for different stages of the graphene growth at 180 eV: (a) 1×1 of SiC, (b) 3×3, (c) $(\sqrt{3} \times \sqrt{3})R30°$, (d) $(6\sqrt{3} \times 6\sqrt{3})R30°$, (e) sharper $6\sqrt{3} \times 6\sqrt{3}R30°$. Panel (f) is taken at the same stage as panel (e) but with a lower energy of 130 eV, where the graphene spots (pointed to by the arrows) can be observed.

14.3   Sample growth and characterization

Graphene research has focused mainly on two types of samples, exfoliated graphene and epitaxial graphene. The exfoliated graphene sample is produced by mechanical exfoliation of graphite flakes followed by deposition on a Si wafer coated with 300 nm of SiO_2 (Novoselov *et al.* 2004, 2005). Exfoliated graphene samples have been widely used in transport measurements as they have very high mobility and its structure makes it easy to apply a gate voltage to tune the charge carriers. However, exfoliated graphene samples have low yield and they are typically small (of the order of 10 μm, much smaller than the size of the synchrotron beam ~100 μm used in ARPES) and therefore it is difficult to study the electronic structure of exfoliated graphene using ARPES. So far, the electronic structure of graphene has been mostly reported only in the other type of graphene sample, epitaxial graphene. It is well known that thick graphite samples can be grown by cracking of hydrocarbon gas on various metals (McConville *et al.* 1986; Land *et al.* 1992) or by thermal decomposition of carbide wafers (Van Bommel *et al.* 1975; Nagashima *et al.* 1993). However, it was not until recent years that the single-layer epitaxial graphene sample on SiC was fully characterized (Berger *et al.* 2004). The advantage of epitaxial graphene is that the samples can be made much larger (mm scale) and therefore they are suitable for ARPES study to extract the electronic structure directly. Technologically, epitaxial graphene is important since the existing Si technology can be incorporated to mass produce epitaxial graphene samples (Berger *et al.* 2006; de Heer *et al.* 2007; Kedzierski *et al.* 2008).

Graphene samples on SiC wafer can be grown by thermal decomposition of the SiC wafer. The choice of SiC substrate has the advantage that SiC has a bandgap of ~2.6 eV and therefore its electronic structure does not interfere with that of graphene near E_F. There are two terminations for the SiC wafers, Si-terminated and C-terminated. The graphene on the C-terminated 4H-SiC shows more inplane orientational disorder (Hass *et al.* 2006) than graphene on Si-terminated 6H-SiC. Therefore, graphene on Si-terminated face is more favorable for ARPES studies. The 6H-SiC is doped with nitrogen to have a resistivity of ~0.2Ω cm^{-1} unless mentioned otherwise. The growth process is monitored with low-energy electron diffraction (LEED) or low-energy electron microscopy (LEEM). The wafer was cut into pieces of a few millimeters in size by using a diamond saw, and cleaned with acetone and isopropyl alcohol in an ultrasonic bath. The clean wafers were then mounted to a Ta sample holder and loaded into an ultrahigh-vacuum preparation chamber with base pressure 1×10^{-10} torr. The wafer was then annealed at 850 °C under silicon flux with deposition rate about 3 Å per minute to remove the native surface oxides by the formation of volatile SiO. Figure 14.5 shows the LEED pattern at different stages during the growth process. After the initial cleaning under Si flux, the LEED pattern shows a 3×3 reconstruction with respect to the SiC substrate, as has been well documented in the literature (Kaplan 1989; Bermudez *et al.* 1995; Forbeaux *et al.* 1998; Starke 1998). A subsequent annealing at 1000 °C in the absence of Si flux shows a 1×1 pattern. More annealing at 1100 °C shows the $(\sqrt{3} \times \sqrt{3})R30°$ reconstruction.

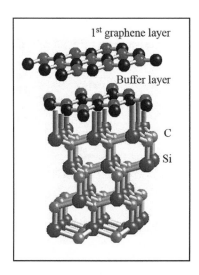

Fig. 14.6 Structure of epitaxial graphene. The buffer layer grows on top of SiC before a graphene layer is formed.

This corresponds to a 1/3 layer of Si ad-atoms in threefold symmetric sites on top of the outermost SiC layer (Northrup 1995; Owman 1995). Further annealing at 1250 °C results in the $(6\sqrt{3} \times 6\sqrt{3})R30°$ pattern, which indicates the formation of the carbon-rich layer. This carbon-rich layer is formed by carbon atoms arranged in the same structure as graphene. However, this layer forms only the σ bands but no π bands characteristic of single-layer graphene (Emtsev *et al.* 2007). Therefore, this carbon-rich layer is called a buffer layer (Emtsev *et al.* 2007; Mattausch *et al.* 2007; Varchon *et al.* 2007). Finally, annealing at 1400 °C forms the graphene layer on top of the buffer layer (Fig. 14.6).

Even though LEED is a good indicator for the different stages of the growth, the distinction between graphene and the carbon-rich (buffer layer) is not obvious, and LEED is not sensitive to the graphene thickness. Therefore, the sample thickness needs to be characterized using other methods.

Low-energy electron microscopy is a powerful technique to study the surface topography and the dynamics of the growth process (Bauer 1994; Phaneuf *et al.* 2003), as well as characterizing the sample thickness. LEEM is a surface-sensitive imaging tool that collects the backscattered electrons ranging from 1 to 100 eV. The low energy of the electrons used distinguishes LEEM from other electron microscopies, e.g. transmission electron microscopy (TEM) where the electron energy is typically 100 000 eV. The use of low-energy electrons has the advantage that a large fraction of electrons will be backscattered elastically and that the backscattered electrons in this low-energy range are very sensitive to the physical and chemical properties of the surface. Thus, depending on the energy of the incident electrons, subtle differences in the local atomic structure or composition can result in dramatic reflectivity contrast in LEEM. Another advantage of LEEM over other microscopy is that the images can be taken instantly, allowing real-time study of the dynamics. For example, Fig. 14.7 shows the LEEM image before cleaning the SiC wafer at 850 °C under Si flux. The LEEM image shows that after the initial cleaning, the sample surface becomes smoother. The direct visualization of the growth process allows a fine tuning of the growth parameters to achieve high-quality samples, and to obtain graphene samples with various characteristics, e.g. different graphene terrace size.

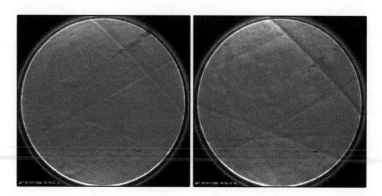

Fig. 14.7 LEEM images taken before and after annealing the SiC substrate under Si flux with a field of view of 5 μm.

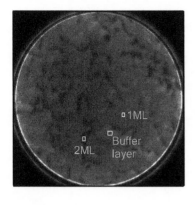

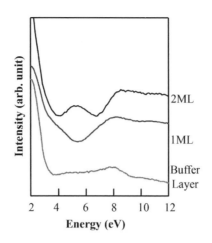

Fig. 14.8 LEEM images taken at electron energy of 6.6 eV with a 3 μm field of view and the energy scans for the buffer layer, single layer and bilayer graphene.

In addition to studying the surface topography, LEEM can also yield direct information about the graphene sample thickness by studying the quantum-well states on a substrate (Hibino *et al.* 2007; Ohta *et al.* 2008). The coherent reflected electrons from the surface and the interface between the graphene and the substrate will interfere with each other and form interference patterns in the reflectivity as a function of electron energy. The number of quantum oscillations directly reflects the number of states, or the thickness of the graphene sample, and can therefore determine the sample thickness. Figure 14.8 shows a LEEM image with intensity ranging from white, gray and black. The energy scan shows zero, one and two minima in the energy scan, allowing us to determine these regions to be buffer layer, single layer and bilayer graphene, respectively. Therefore, the LEEM image shows that the epitaxial graphene has regions of single and bilayer graphene, as well as the buffer layer. The single-layer graphene is the majority and the terrace size of the single-layer graphene can be controlled by tuning the annealing temperature and annealing time.

Additional ways to characterize the sample thickness are: Auger electron spectroscopy (Berger *et al.* 2004), X-ray photoelectron spectroscopy (Rolling *et al.* 2006) and angle-resolved photoemission spectroscopy (Ohta *et al.* 2006).

14.4 Electronic structure of epitaxial graphene

Angle-resolved photoelectron spectroscopy (ARPES) is a direct and powerful tool to probe the electronic structure of a material (Damascelli *et al.* 2003). The basic principle behind this technique is the photoelectric effect. When a beam of light with photon energy $h\nu$ shines on a clean sample surface, photoelectrons will be emitted from the sample, which are collected by an electron analyzer as a function of kinetic energy E_k and angle θ (Fig. 14.9). The energy is conserved in the photoemission process, therefore from the kinetic energy E_k of photoelectrons and the work function ϕ, the binding energy of electrons inside the solid can be determined as $E_B = h\nu - \phi - E_k$. For single-crystal samples, because of the translational symmetry in the crystal plane and the negligible wave vector of the photons, the inplane momentum $p_{//}$ is preserved. Therefore, the electron binding energy and inplane momentum

Fig. 14.9 Schematic drawing of ARPES measurements. A monochromatic light shines on the sample and the photoelectrons are collected by an electron analyzer.

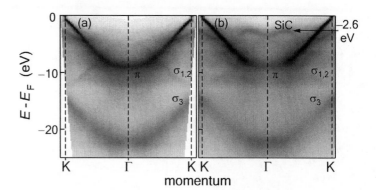

Fig. 14.10 ARPES data taken along the ΓK direction at a photon energy of (a) 80 eV and (b) 100 eV.

can be determined as $\begin{aligned} E_B &= h\upsilon - \phi - E_k \\ p_{//} &= \hbar k_{//} = \sqrt{2mE_k}\sin\theta \end{aligned}$. Therefore, ARPES can directly measure the dispersion $E(k)$ and map out the Fermi surface. Moreover, ARPES directly measures the single-particle spectral function $A(k,\omega)$, which not only contains the band dispersion, but also the self-energy due to different interactions.

Figure 14.10 shows ARPES data taken in single-layer epitaxial graphene along the high-symmetry direction ΓK in a large energy range. The π bands at low binding energy and the σ bands at high binding energy can be clearly observed, similar to the calculated dispersion shown in Fig. 14.2. In addition to the π and σ bands, the bands from the SiC substrate (pointed to by the arrow in panel (b)) can also be observed and the intensity of these bands increases with higher photon energy. The advantage of using semiconducting SiC as substrate is that the SiC bands are at much higher binding energy, and therefore the low-energy π band near E_F, which is the main focus of this study, can be easily separated from the substrate.

Figure 14.11 shows the intensity map measured at E_F, −0.4 and −1.2 eV taken on single-layer epitaxial graphene near the Brillouin zone corner K. At

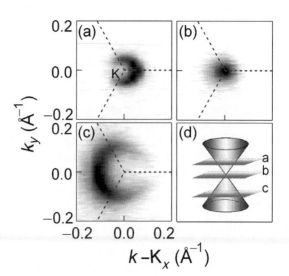

Fig. 14.11 (a–c) Intensity maps taken at E_F, −0.4 eV and −1.2 eV. (d) Schematic drawing of the conical dispersion and the relative positions for data shown in panels a–c.

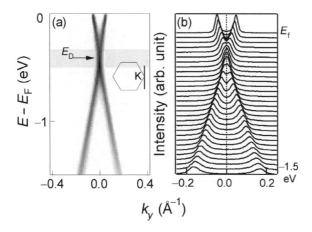

Fig. 14.12 (a) Intensity map as a function of energy and momentum taken along a cut through the K point (black line in the inset). (b) MDCs at energies from E_F to -1.5 eV.

the E_F (panel (a)), the intensity map shows a small circular pocket instead of a single point as expected, suggesting that the sample is doped. This pocket decreases in size when going down in energy and it appears to be a point near -0.4 eV. Beyond -0.4 eV, the size of the pocket increases again and it shows an overall circular behavior. The modulation of the intensity inside and outside the first Brillouin zone is known to be caused by the dipole matrix element in ARPES (Shirley *et al.* 1995). These constant energy-intensity maps are in overall agreement with the conical dispersion. However, the sample is electron doped and the Dirac point energy E_D lies below the Fermi energy E_F.

The conical dispersion and the electron doping can also be studied from the dispersion. Figure 14.12 shows the dispersion along a line through the K point. Two cones dispersing in opposite directions, one upward and another downward, can be clearly observed. The center point between these two cones, which is defined as the Dirac point energy E_D, lies at -0.4 eV. Figure 14.12(b) shows the momentum distribution curves (MDCs), intensity as a function of momentum, for energies between E_F and -1.5 eV. The dispersion can be extracted by following the peak positions in the MDCs, which shows an almost linear behavior for both the upper cone and lower cone. Overall, the data presented so far on single-layer epitaxial graphene show that the dispersion is in agreement with the conical dispersion and that the sample is electron doped.

Figure 14.13 shows the evolution of the dispersion with graphene sample thickness. From single-layer to bilayer graphene, the π band splits into two bands as a result of interlayer interaction. In addition, the bilayer graphene sample is also electron doped and there is a finite electron pocket at E_F. For trilayer graphene, E_D moves even closer to E_F, and there is still a finite electron pocket at E_F (Ohta *et al.* 2007; Zhou *et al.* 2007). Eventually, when the sample thickness is infinite, E_D lies almost at E_F (Zhou *et al.* 2006).

The thickness dependence in Fig. 14.13 suggests that the electron doping comes from the interface. To test if the doping is related to the doping of the SiC substrate or the Si atoms at the interface, we measured two bilayer graphene samples on a different substrate 4H-SiC with resistivity of $10^5 \Omega \text{cm}^{-1}$ compared to $0.2 \Omega \text{cm}^{-1}$ in the 6H-SiC substrate studied previously.

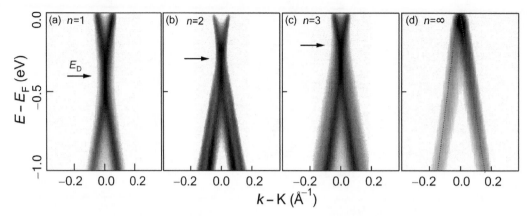

Fig. 14.13 Intensity map measured along a line through the K point for (a) single layer, (b) bilayer (c) trilayer epitaxial graphene, and (d) graphite. The arrows mark the position of the Dirac point energy, which moves closer to E_F with increasing sample thickness.

Figure 14.14 shows that even though the resistivity of the substrate is different by a factor of 10^5, E_D is still at the same energy of -0.29 eV within an error bar of 20 meV. Figure 14.15 summarized the data taken on graphene samples on both 6H-SiC and 4H-SiC. It is clear tat even though E_D changes with graphene sample thickness, it is independent of the doping of the SiC substrate. This comparison shows that the doping of the graphene sample is related to the charge transfer at the interface from the Si atoms (Zhou *et al.* 2007), and it *not* determined by the doping of the substrate. This is in agreement with the thickness dependence, since for the thicker graphene sample, it is farther away from the interface, and therefore the change in E_D is smaller and E_D is closer to E_F.

Data presented so far show that epitaxial graphene samples are electron doped. The electron doping is most likely associated with the charge transfer from the Si atoms at the interface and not related to the doping of the substrate. The dispersions show an overall conical behavior. However, near the Dirac point energy (Fig. 14.13(a)), there is an anomalous vertical region. To understand this vertical region near E_D, a more detailed analysis of the dispersion near E_D is required.

Fig. 14.14 Data taken through the K point for bilayer graphene (a) on more conductive 6H-SiC substrate with resistivity of $0.2 \Omega \text{cm}^{-1}$ (b) on more insulating 4H-SiC substrate with resistivity of $10^5 \Omega \text{cm}^{-1}$.

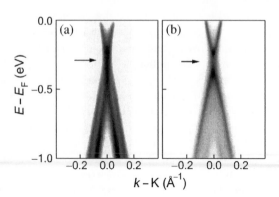

14.5   Gap opening in single-layer epitaxial graphene

Figure 14.16 shows a more detailed analysis of the data near E_D at the Brillouin zone corner K. The dispersions can be extracted from the energy-distribution curves (EDCs), intensity scans as a function of energy for fixed momentum. In the EDCs, there are two peaks that correspond to the valence and conduction bands. It is interesting to note that even at the K point, where these two bands are expected to meet for a perfect conical dispersion, there are still two peaks in the EDCs. This presence of two peaks at the K point suggests that the valence and conduction bands do not merge at the K point. Instead, there is a finite separation between these two bands. To extract the dispersion for both the valence and conduction bands, we fit EDCs with two Lorentzians multiplied by the Fermi–Dirac function. The extracted dispersions are shown as white lines in panel (a), which clearly show a finite separation of ~0.26 eV between the valence and conduction bands. This anomalous region between the valence and conduction bands can also be observed in the MDCs shown in panel (c). Away from E_D, two peaks are clearly observed in the MDCs, while in an extended region between -0.3 eV and -0.5 eV, only one peak is observed in the MDC and the peak position does not change with energy. The non-dispersive MDC peak is a typical ARPES feature of the gap (Shen *et al.* 2005). Panel (d) shows the angle-integrated intensity curve, which is proportional to the density of states. At high binding energy, the angle-integrated intensity curve shows a linear behavior, while near the Dirac point energy, there is a dip suggesting suppression of the density of states below the expected linear behavior. This is in agreement with the gap opening near E_D. In summary, Fig. 14.16 shows that the dispersion near E_D deviates from the linear behavior and there is a finite gap opening between the valence and conduction bands. One interesting observation is that there is still finite intensity inside the gap region.

Figure 14.17 shows how the dispersions evolve across the K point. Panels (a,b) show the intensity maps at E_F (above E_D) and -1 eV (below E_D),

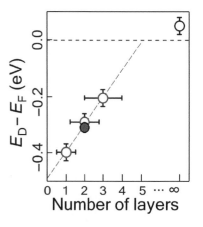

Fig. 14.15 Summary of the Dirac point energy E_D as a function of graphene sample thickness. The open symbols are taken on graphene on 6H-SiC and the filled symbol is taken on a bilayer graphene sample on 4H-SiC.

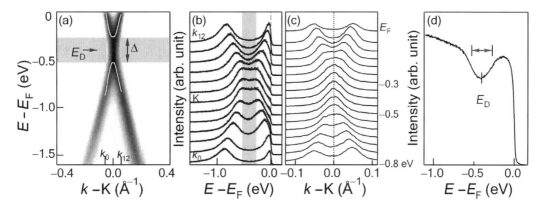

Fig. 14.16 (a) Data taken through the K point. The white lines show the dispersions extracted from panel (b). (b) EDCs taken from k_0 to k_{12} as labelled at the bottom of panel (a). The dots are the raw data and the black curves are the fits using two Lorentzians multiplied by the Fermi–Dirac function. (c) MDCs from E_F to -0.8 eV. (d) Angle-integrated intensity as a function of energy for data shown in panel (a).

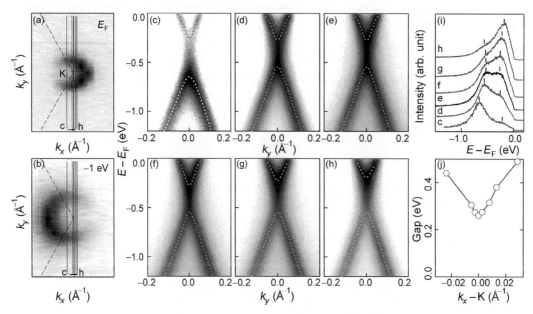

Fig. 14.17 (a, b) Intensity maps taken at E_F (a) and -1 eV (b) near the K point. The vertical lines show the position of the cuts for data shown in panels (c–h). (c–h) Data taken near (c,d,f,g,h) and through (panel e) the K point. The dotted lines are guides for the eye for the dispersions of the valence and conduction bands. (i) EDCs taken at $k_y = 0$. The tick marks above the EDCs label the EDC peak positions. (j) Plot of the gap as a function of k_x when approaching the K point.

respectively. Above E_D, the intensity is stronger outside the first Brillouin zone, while below E_D the intensity is stronger inside the first Brillouin zone. This intensity modulation is related to the dipole matrix element in graphene (Shirley 1995) and can be utilized to identify the exact K point. Panels (c–h) show the data taken before and after crossing the K point, with a very fine step of 0.05 deg close to the K point, which corresponds to a step of 0.0026 Å^{-1} to make sure to catch the K point as accurately as possible. Before crossing the K point (panel (c)), the data shows a stronger intensity for the valence band, while after crossing the K point (panels (g,h)), the data show a stronger intensity for the conduction bands, in agreement with the intensity modulation in panels (a,b). In all the panels, the valence and conduction bands can be followed from the EDC peak positions. The most important information comes from the EDC at $k_y = 0$, where the separation between these two bands is the smallest in each panel. Panel (i) shows the EDCs at $k_y = 0$. Two peaks can be identified in the EDCs, and the intensity modulation between these two peaks switches before and after crossing the K point. This intensity modulation enables us to identify that the K point is for cut (e), since for this cut the EDC shows an almost identical intensity for the two peaks and the gap is smallest. The gaps for the various cuts taken can be plotted a function of the momentum k_x, as shown in panel (j). The consistent trend of the gap with the minimum value of $\sim$0.26 eV at the K point shows that the gap is *not* an artifact due to sample misalignment, therefore the upper cone and lower cones are separated by a finite amount and there is a bandgap of $\sim$0.26 eV in single-layer graphene. In addition, a finite intensity is always observed inside the gap region for all the cuts.

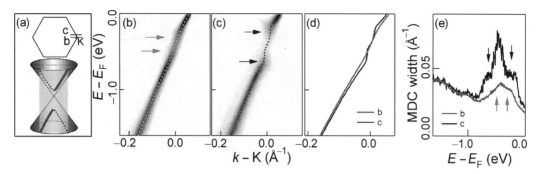

Fig. 14.18 (a) Schematic drawing of the cuts for data shown in panels (b,c). The cartoon in the inset shows schematically the gap opening for the cut away from the K point even when there is no gap at the K point. (b,c) Data taken through and off the k point. The dotted lines are the extracted dispersions from the MDCs. The arrows mark the position of the top valence band and the bottom of the conduction band. (d) Comparison of the dispersions for the cut through and off the K point. (e) Extracted MDC width as a function of energy for data shown in panels (b,c).

The dispersions near the K point can also be studied by measuring along another geometry (see Fig. 14.18(a)), which shows asymmetric intensity for the dispersions on the two sides of the K point. This geometry has the advantage of showing only one branch of the dispersion, thereby enabling an easier fit of the MDCs to extract the dispersion and the MDC width, with the additional disadvantage that the Dirac point cannot be envisioned directly. In addition, one caveat of the MDC analysis is that the presence of a peak does not necessarily mean the presence of a quasi-particle, therefore one also needs to be careful about the interpretation of the MDC dispersions (Zhou *et al.* 2008b). Panels (b) and (c) show two cuts, one through the K point and another one off the K point. In panel (b), the top of the valence band and the bottom of the conduction band for the cut through the K point (panel (b)) are determined from the symmetric geometry discussed previously and are labelled by the two gray arrows in panel (b). The black arrows in panel (c) label the top and bottom of the bands determined for the cut off the K point. The data show an overall similarity between these two cuts with a suppression of intensity in the region near the Dirac point energy. The extracted MDC dispersions show an even larger deviation from the linear dispersion for the cut off the K point. In addition, the MDC width shows an anomalous region near E_D, where multiple peak structures appear. This anomalous region has been discussed previously by Bostwick *et al.* (2006) and was attributed to electron–plasmon interaction. However, it is interesting to note that energies for the top of the valence band and bottom of the conduction band coincide with the two peak positions in the anomalous region for both cuts (b) and (c), and the amplitude of the peaks is larger for the cut off the K point. This suggests that the anomalous region near E_D has a similar origin for the cuts through and off the K point (Zhou *et al.* 2008b). In the latter case, it is known from the conical dispersion that a gap is definitely expected. This is in agreement with previous EDC analysis of Fig. 14.16 and Fig. 14.17, where a gap is present.

Figure 14.19 shows that the anomalous peaks in the MDC width near E_D can be explained by the gap (Zhou *et al.* 2008b). Panels (a) and (b) show the simulated data with the dipole matrix element that enhances one branch of the dispersion, similar to the case in graphene. Applying a similar MDC fit

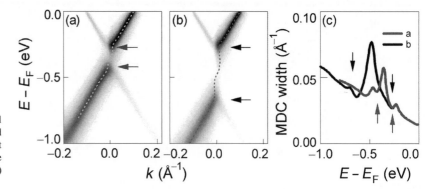

Fig. 14.19 (a,b) Simulation of the conical dispersions with a gap of 150 meV and 400 meV with added dipole matrix element to reproduce similar intensity pattern as the experimental data on epitaxial graphene. (c) Extracted MDC width from the dispersions.

and extracting the MDC width, the MDC width shows two peaks in the gap region and the peak is larger for the larger gap. This similarity with respect to the experimental data discussed in Fig. 14.18 shows that the gap opening is the most likely origin for explaining the deviation near E_D and the anomalous peaks in the MDC width, and not due to electron–plasmon interaction.

14.6   Possible mechanisms for the gap opening

There are a few possible mechanisms that can open up a gap in graphene: quantum confinement (Nakada *et al.* 1996; Brey *et al.* 2006; Son *et al.* 2006; Han *et al.* 2007; Nils *et al.* 2007; Nilsson *et al.* 2007), mixing of the states between the K and K' points induced by scattering, and hybridization of the valence and conduction bands caused by breaking of carbon sublattice symmetry (Manes *et al.* 2007). We will examine all of these possibilities and we think that the most likely scenario is the breaking of the carbon-sublattice symmetry.

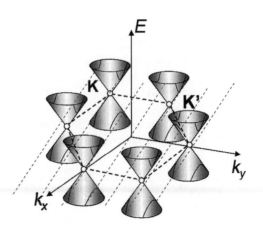

Fig. 14.20 Schematic diagram showing the electronic states for carbon nanotubes.

14.6.1 Quantum confinement

The gap induced by quantum confinement is well known in the case of carbon nanotubes (Saito *et al.* 1992) and has been recently extended to graphene nonoribbons (Nakada *et al.* 1996; Son *et al.* 2006; Han *et al.* 2007; Trauzettel *et al.* 2007). Carbon nanotube is a sheet of rolled graphene, therefore additional boundary conditions apply to carbon nanotube. When the graphene sheet is rolled along a certain direction to form carbon nanotubes, the electronic states will be quantized along this direction (Fig. 14.20). For certain directions where the bands do not cross the Dirac point, the carbon nanotube will be a semiconductor with a finite size gap (Saito *et al.* 1992). The electronic states near E_F are determined by the intersection of the allowed K points with the dispersion cones at the K point. The size of the gap increases as the carbon nanotube becomes smaller, and scales inversely with the width of the ribbons. The dependence of the gap size on the graphene ribbon width has been recently shown by Han *et al.* in exfoliated graphene (Han *et al.* 2007). It is important to note that for ribbons as small as 10 nm, the gap size can be as big as 200 meV, while for ribbons larger than 30 nm, the gap would be smaller than 10 meV. Therefore, it is important to measure the gap in our graphene samples with various graphene ribbon sizes to test whether the scenario of quantum confinement can account for the large gap observed here.

Single-layer graphene samples with various terrace size can be obtained by controlling annealing temperature and annealing time. The real-time observation of the growth process is a unique advantage to control the ideal growth conditions and obtain graphene samples with various sizes. As discussed previously in Section 14.2, LEEM is a powerful tool to study the surface topography and identify the single-layer graphene regions. Therefore, the graphene terrace size can be directly measured from the LEEM image. Figure 14.21 shows a LEEM image with a majority of single-layer graphene with smaller percentage of bilayer graphene and buffer layer. The graphene terrace size and the error bar is quantified by taking the average and the standard deviation of the graphene terrace sizes crossed by the two lines along the diagonal direction.

Figure 14.22 shows the LEEM images for a few graphene samples with various single-layer graphene terrace sizes. The samples contain a majority of single-layer graphene with different terrace sizes, along with a much smaller fraction of bilayer graphene and buffer layer. The single-layer graphene regions are the focus here. The single-layer graphene terrace size in each panel is characterized using the method described above. In these samples, with increasing annealing, the graphene terrace size can be controlled from 50 nm (panel (e)) to 180 nm (panel (a)). After the LEEM measurements, the same samples were transported to the ARPES chamber. After annealing at 700 °C to remove the gas adsorbed on the graphene surface, the samples were measured with ARPES to yield direct information about the electronic structure. Figure 14.23 shows the corresponding ARPES data for the same samples measured in the same geometry. Following the same procedure, the gap can be determined by measuring the distance between the valence band and conduction band extracted from the EDC peak positions. When the

Fig. 14.21 LEEM image with a 2-μm field of view taken on a graphene sample to show how the graphene terrace size is quantified. The gray, black and white regions are the single-layer, bilayer graphene and the buffer layer. The tick marks along the diagonal direction label the edges of the single-layer graphene (gray area), which were used to quantify the single-layer graphene terrace size.

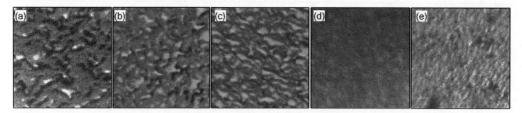

Fig. 14.22 LEEM image for various graphene samples width different graphene terrace sizes of (a) 57 nm, (b) 85 nm, (c) 95 nm, (d) 121 nm, (e) 179 nm. The circle in each panel marks the characteristic width of the graphene terrace.

graphene terrace size decreases from panel a to panel e, the gap increases slightly.

Figure 14.24 shows the direct correlation of the gap size measured from ARPES with the graphene terrace size obtained from the LEEM measurements (Zhou *et al.* 2008a). It is clear that the gap slightly increases for the smallest graphene terrace size. However, the gap does not change within the experimental uncertainty even when the graphene terrace size exceeds 150 nm. This is in contrast to the gap induced by quantum confinement in the exfoliated graphene nanoribbons where the gap is below 10 meV when the graphene terrace size is larger than 30 nm. This direct comparison shows that even though quantum confinement can contribute to enhance the gap size, quantum confinement cannot explain the large gap of 180 meV still observed in largest graphene terrace size. Therefore, quantum confinement is not the main mechanism for the gap opening in epitaxial graphene (Zhou *et al.* 2008a).

14.6.2   Inter-Dirac-point scattering

Another possible way to open up a gap is to hybridize the electronic states at K and K′ (Manes *et al.* 2007) as schematically shown in Fig. 14.25. This requires breaking of the translational symmetry. It is known that in graphene there are reconstructions of 6 × 6 and $(6\sqrt{3} \times 6\sqrt{3})R30°$ (Tsai *et al.* 1992; Forbeaux *et al.* 1998). However, the scattering vectors related with these two constructions are much smaller than the K–K′ distance, which is required to mix the states at K and K′. Higher-order scattering process involving consecutive small

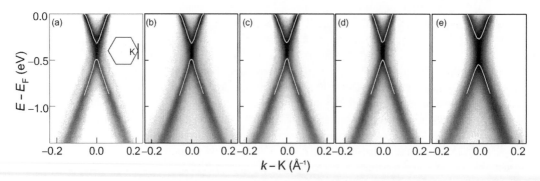

Fig. 14.23 Corresponding ARPES data taken along a line through the K point for samples shown in Fig. 14.22 with various graphene terrace sizes. From panel (a) to panel (e), the single-layer graphene terrace size decreases. The white lines show the dispersions extracted by fitting the EDCs.

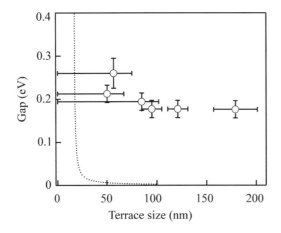

Fig. 14.24 Gap size measured from ARPES as a function of single-layer graphene terrace size measured from LEEM. The dotted line is the gap measured in exfoliated graphene reported by Han *et al.* (2007), copyright American Physical Society, where the gap is induced by quantum confinement.

scattering vectors is weak in general and is an unlikely source for the gap opening. Impurity scattering can also mix the states at K and K′ (Rutter *et al.* 2007). However, this would give rise to a gap that strongly depends on impurity concentration. This is in contrast to our findings, where the gap is similar for all the samples studied.

14.6.3 Breaking of the carbon-sublattice symmetry

The most likely scenario to explain the gap that we observed in epitaxial graphene is the breaking of the carbon sublattices (see Fig. 14.26). We note that a perfectly conical dispersion is expected only in the case of perfect single-layer epitaxial graphene, where the potentials on the two carbon sublattices are the same. When this symmetry breaks down, the hybridization between the valence and conduction bands can open up a gap. This has been well known in the extreme case of BN, which has a similar honeycomb structure and completely different sublattices in the unit cell. In the case of BN, the bandgap can be as big as 5.8 eV (Blase *et al.* 1995).

In the case of single-layer epitaxial graphene, since the sample is grown on top of the SiC substrate and there is a buffer layer between graphene and the SiC substrate, the buffer layer can break the symmetry between the two carbon sublattices. A prediction of this scenario is that breaking of the sixfold rotational symmetry of graphene near the Dirac point energy. Figure 14.27(e) shows the calculated intensity map at E_D using a tight-binding model in the

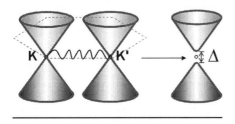

Fig. 14.25 Schematic drawing of inter-Dirac-point scattering between the K and K′ points to induce a gap at the Dirac point.

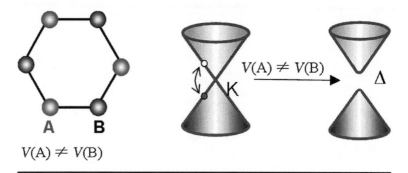

Fig. 14.26 Schematic drawing to show that when the two carbon sublattices have different potentials, the valence and conduction bands can hybridize to open a gap at the Dirac point.

$V(A) \neq V(B)$

extreme case when only one of the two carbon sublattices have non-zero potential. The potential modulation imposed by the $(6\sqrt{3} \times 6\sqrt{3})R30°$ reconstruction has been added as a perturbation to the Hamiltonian (Zhou *et al.* 2007). For energy well above or below E_D, the symmetry is restored. Figures 14.27(a–d) show the constant energy-intensity maps measured at E_F, E_D and below E_D. The dominant feature in all these intensity maps is the small pockets centered at the K points. In addition, there are six weaker replicas surrounding each K point. The intensity of these replicas is only ~4% that of the main peak at the K point. These replicas are associated with the $(6\sqrt{3} \times 6\sqrt{3})R30°$ observed in LEED. At E_F (panel (a)), the six replicas show similar intensity, while near E_D (panel (b)), three of the six replicas (pointed to by the arrows) are enhanced, suggesting that the sixfold symmetry is broken near E_D. At higher binding energy far away from the Dirac energy (panel (d)), the six replicas show similar intensity again, showing the restoration of the sixfold symmetry. Our experimental data shown in panels (a–d) are in good agreement with theoretical predictions that show the breaking of the sixfold symmetry near E_D. We note that scanning tunnelling microscopy STM measurements (Brar *et al.* 2007) on epitaxial single-layer graphene did not report the evidence

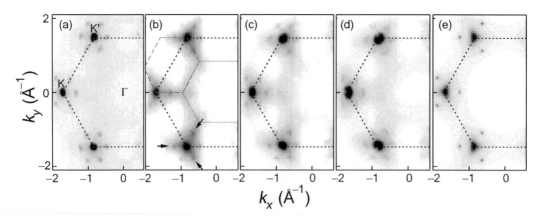

Fig. 14.27 Constant energy intensity maps taken at (a) E_F, (b) $-0.4\,\text{eV}$, (c) $-0.8\,\text{eV}$, (d) $-1.0\,\text{eV}$. (e) Calculated ARPES intensity map at E_D in the presence of symmetry breaking on the two carbon sublattices. The arrows in panel (b) point to the three replicas that are enhanced compared to the other three. Note that the intensity for the main spots at the K points is saturated to enhance the weaker replicas surrounding the K points.

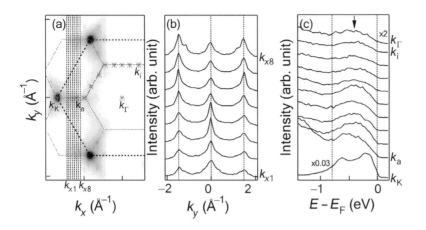

Fig. 14.28 (a) Intensity map at E_D. The dotted vertical lines mark the positions where the MDCs in panel (b) are taken and the markers (*) label the position of the EDCs show in panel (c). (b) MDCs taken from k_{x1} to k_{x8}. (c) EDCs taken from k_K, k_Γ, k_a to k_i. Note that the EDCs at K and Γ are scaled by 0.03 and 2, respectively.

of sixfold symmetry breaking. However, the STM data are taken far away from the Dirac point, where it is known that the effect of the symmetry breaking is much weaker and difficult to observe. Therefore, the lack of evidence of sixfold symmetry breaking in STM studies does not contradict our ARPES results.

In addition, there are some interesting structures in the intensity maps near E_D. In particular, we observed hexagonal patterns where the intensity is enhanced (see Fig. 14.27(b)). There are two observations associated with these additional patterns. First, the center mid-sized hexagon around Γ (gray dotted line in panel (b)) almost overlaps with the first Brillouin zone of SiC. Second, all other hexagons (gray broken lines in panel (b)) are not regular, i.e. the six sides that form the hexagon do not have the same length. However, they all pass through the K points. These two observations suggest that the states associated with the hexagonal pattern are related to the buffer layer between the SiC substrate and the graphene.

Figure 14.28 shows a more careful characterization of the in-gap states. Figure 14.28(b) shows the MDCs at E_D for momenta from k_{x1} to k_{x8}. Three non-dispersive peaks are observed in the MDCs, which correspond to the three horizontal segments of the additional hexagonal patterns. This shows that the mid-gap states have maxima along the hexagonal patterns. Panel (c) shows the EDCs taken at various points in the Brillouin zone. At the K point, the

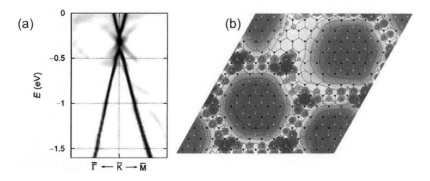

Fig. 14.29 (a) Dispersion near the K point from *ab-initio* calculation. (b) Contour of the potential on the graphene layer generated by the buffer layer. From Kim *et al.* (2008).

EDC shows two peaks separated by a finite amount, which correspond to the top of the valence band and the bottom of the conduction band, in agreement with the gap. Along the hexagonal pattern, the EDCs show a very broad peak from E_F to -0.8 eV, while away from the hexagonal pattern (see, e.g. EDC at the Γ point), the broad peak in the EDC can still be observed. However, the intensity is only half of that along the additional hexagons. The analysis of Fig. 14.28 shows that there are additional states that are peaked at the Dirac-point energy, and these mid-gap states form interesting hexagonal pattern in the momentum space. The exact origin of these mid-gap states still requires further studies.

The proposed mechanism of breaking carbon-sublattice symmetry to explain the gap opening in epitaxial graphene is also supported by recent *ab-initio* calculation (Kim *et al.* 2008), which shows a gap near E_D with a gap value similar to ARPES measurements. The calculation shows that the graphene layer follows the atomic structure of the buffer layer underneath and there is a corrugation of the height of the graphene layer, which results in a 140-meV average potential difference in the two carbon sublattices. Moreover, mid-gap states are also predicted, which are caused by the interlayer coupling between the π states in graphene and the localized π states in the buffer layer. This is in agreement with the observation of anomalous intensity in the gap region. However, further studies are still needed to understand why the mid-gap states form such hexagonal patterns in the momentum space as discussed above.

Our proposed mechanism of graphene–substrate interaction as the origin for the opening of the bandgap has important implications, since engineering the bandgap in graphene is an important topic for its applications. It was predicted that when graphene is grown on BN substrate, a gap can also be expected and the gap depends on the orientation of graphene relative to the BN substrate as well as the distance between graphene and the substrate (Giovannetti *et al.* 2007). If one can tailor graphene–substrate interaction by growing graphene on different substrates, it is possible to engineer the bandgap of graphene in a wider range. Further study to move the Fermi energy inside the gap region with hole dopants to eventually make graphene a semiconductor is also an important topic.

14.7   Conclusions

In summary, we have studied the electronic structure of epitaxial graphene. Clear deviations from the conical dispersions are observed near the Dirac-point energy, and interpreted as a gap opening due to graphene–substrate interaction. This points to graphene–substrate interaction as a promising route to engineer the bandgap in graphene. Even though it might take a long time before graphene's full application potentials can be fully realized, graphene is definitely a very intriguing system and there is still a lot more to be explored.

Acknowledgments

We thank A.V. Fedorov, D.A. Siegel, G.-H. Gweon, F. El Gabaly, A.K. Schmid, K.F. McCarty and J. Graf for experimental assistance, P.N. First, W.A. De Heer for helping with the graphene samples in the initial stage of this project, D.-H. Lee, F. Guinea, and A.H. Castro Neto for useful discussions.

References

Bauer, E. *Rep. Prog. Phys.* **57**, 895 (1994).

Blase, X., Rubio, A., Louie, S.G., Cohen, M.L. *Phys. Rev. B* **51**, 6868 (1995).

Berger, C., Song, Z.M., Li, T.B., Li, X.B., Ogbazghi, A.Y., Feng, R., Dai, Z.T., Marchenkov, A.N., Conrad, E.H., First, P.N., de Heer, W.A. *J. Phys. Chem. B* **108**, 19912–19916 (2004).

Berger, C., Song, Z., Li, X., Wu, X., Brown, N., Naud, C., Mayou, D., Li, T., Hass, J., Marchenkov, A.N., Conrad, E.H., First, P.N., de Heer, W.A. *Science* **312**, 1191 (2006).

Bermudez, V.M. *Appl. Surf. Sci.* **84**, 45 (1995).

Blase, X., Rubio, A., Louie, S.G., Cohen, M.L. *Phys. Rev. B* **51**, 6868 (1995).

Bostwick, A., Ohta, T., Seyller, Th., Horn, K., Rotenberg, E. *Nature Phys.* **3**, 36 (2006).

Brar, V.W., Zhang, Y., Yayon, Y., Bostwick, A., Ohta, T., McChesney, J.L., Horn, K., Rotenberg, E., Crommie, M.F. *Appl. Phys. Lett.* **91**, 122101 (2007).

Brey, L., Fertig, H.A. *Phys. Rev. B* **73**, 235411 (2006).

Castro Neto, A.H., Guinea, F., Peres, N.M.R., Novoselov, K.S., Geim, A.K. arXiv:Cond-mat/0709.1163 (2008).

Damascelli, A., Hussain, Z., Shen, Z.-X. *Rev. Mod. Phys.* **75**, 473 (2003).

De Heer, W.A., Berger, C., Wu, X.S., First, P.N., Conrad, E.H., Li, X.B., Li, T.B., Sprinkle, M., Hass, J., Sadowski, M.L., Potemski, M., Martinez, G. *Solid State Commun.* **143**, 92 (2007).

Dresselhaus, M.S., Dresselhaus, G. *Adv. Phys.* **30**, 139 (1981).

Emtsev, K.V., Seyller, Th., Speck, F., Ley, L., Stojanov, P., Riley, J.D., Leckey, R.G.C. *Mater. Sci. Forum* **556–557**, 525 (2007).

Forbeaux, I., Themlin, J.-M., Debbever, J.-M. *Phys. Rev. B* **58**, 16396 (1998).

Geim, A.K., Novoselov, K.S. *Nature Mat.* **6**, 183 (2007).

Giovannetti, G., Khomyakov, P.A., Brocks, G., Kelly, P.J., van den Brink, *J. Phys. Rev. B* **76**, 073103 (2007).

Hass, J., Feng, R., Li, T., Li, X., Zong, Z., de Heer, W.A., First, P.N., Conrad, E.H., Jeffrey, C.A., Berger, C. *Appl. Phys. Lett.* **89**, 143106 (2006).

Han, M.Y., Özyilmaz, B., Zhang, Y., Kim, P. *Phys. Rev. Lett.* **98**, 206805 (2007).

Hibino, H., Kageshima, H., Maeda, F., Nagase, M., Kobayashi, Y., Yamaguchi, H. arXiv:cond-mat/0710.0469 (2007).

Kaplan, R. *Surf. Sci.* **215**, 111 (1989).

Kedzierski, J., Hsu, P.-L., Healey, P., Wyatt, P., Keast, C., Sprinkle, M., Berger, C., de Heer, W.A. arXiv:cond-mat/0801.2744 (2008).

Kim, S., Ihm, J., Choi, J.J., Son, Y.-W. *Phys. Rev. Lett.* **100**, 176802 (2008).

Land, T.A., Michely, T., Behm, R.J., Hemminger, J.C., Comsa, G. *Surf. Sci.* **264**, 261 (1992).

Mattausch, A., Pankratov, O. *Phys. Rev. Lett.* **99**, 076802 (2007).

McClure, J.W. *Phys. Rev.* **108**, 612 (1957).

McConville, C.F., Woodruff, D.P., Kevan, S.D. *Surf. Sci.* **171**, L447 (1986).

Nagashima, A., Nuka, K., Itoh, H., Ichinokawa, T., Oshima, C. *Surf. Sci.* **291**, 93 (1993).

Nakada, K., Fujita, K., Dresselhaus, M., Dresselhaus, G. *Phys. Rev. B* **54**, 17954 (1996).

Nilsson, J., Castro Neto, A.H., Guinea, G., Peres, N.M.R. *Phys. Rev. B* **76**, 165416 (2007).

Northrup, J.E., Neugebauer, J. *Phys. Rev. B* **52**, 17001 (1995).

Novoselov, K.S., Geim, A.K., Morozov, S.V., Jiang, D., Zhang, Y., Dubonos, S.V., Grigorieva, I.V., Firsov, A.A. *Science* **306**, 666 (2004).

Novoselov, K.S., Jiang, D., Schedin, F., Booth, T.J., Khotkevich, V.V., Morozov, S.V., Geim, A.K. *Proc. Natl. Acad. Sci. USA* **102**, 10451 (2005).

Novoselov, K.S., Geim, A.K., Morozov, S.V., Jiang, D., Katsnelson, M.I., Grigorieva, I.V., Dubonos, S.V., Firsov, A.A. *Nature* **438**, 197 (2005).

Ohta, T., Bostwick, A., Seyller, Th., Horn, K., Rotenberg, E. *Science* **313**, 951 (2006).

Ohta, T., Bostwick, A., McChesney, J.L., Seyller, Th., Horn, K., Rotenberg, E. *Phys. Rev. Lett.* **98**, 206802 (2007).

Ohta, T., Gabaly, F. El, Bostwick, A., McChesney, J., Emtsev, K.V., Schmid, A.K., Seyller, Th., Horn, K., Rotenberg, E. arXiv:cond-mat/0710.0877 (2008).

Owman, F., Martensson, P. *Surf. Sci.* **330**, L639 (1995).

Partoens, B., Peeters, F.M. *Phys. Rev. B* **75**, 193402 (2007).

Phaneuf, R.J., Schmid, A.K. *Phys. Today* **56**, 50 (2003).

Rollings, E., Gweon, G.-H., Zhou, S.Y., Mun, B.S., McChesney, J.L., Hussain, B.S., Fedorov, A.V., First, P.N., de Heer, W.A., Lanzara, A. *J. Phys. Chem. Solids* **67**, 2172 (2006).

Rutter, G.M., Crain, J.N., Guisinger, N.P., Li, T., First, P.N., Stroscio, J.A. *Science* **13**, 219 (2007).

Saito, R., Fujita, M., Dresselhaus, G., Dresselhaus, M.S. *Appl. Phys. Lett.* **60**, 2202 (1992).

Shen, K.M., Ronning, F., Armitage, N.P., Damasceilli, A., Lu, D.H., Ingel, N.J.C., Lee, W.S., Meevasana, W., Baumberger, F., Kohsaka, T., Azuma, M., Takano, M., Takagi, H., Shen, Z.-X. *Science* **307**, 901 (2005).

Shirley, E., Terminello, L.J., Santoni, A., Himpsel, F. *J. Phys. Rev. B* **51**, 13614 (1995).

Son, Y.W., Cohen, M.L., Louie, S.G. *Phys. Rev. Lett.* **97**, 216803 (2006).

Starke, U. *Phys. Rev. Lett.* **80**, 758 (1998).

Trauzettel, B., Bulaev, D.V., Loss, D. Burkard, G. *Nature Phys.* **3**, 192 (2007).

Van Bommel, A.J., Crombeen, J.E., Van Tooren, A. *Surf. Sci.* **48**, 463 (1975).

Varchon, F., Feng, R., Hass, H., Li, X., Ngoc Nguyen, B., Naud, C., Mallet, P., Veuillen, J.-Y., Berger, C., Conrad, E.H., Magaud, L. *Phys. Rev. Lett.* **99**, 126805 (2007).

Zhang, Y.B., Tan, Y.-W., Stormer, H.L., Kim, P. *Nature* **438**, 201 (2005).

Zhou, S.Y., Gweon, G.-H., Graf, J., Fedorov, A.V., Spataru, C.D., Diehl, R.D., Kopelevich, Y., Lee, D.-H., Louie, S.G., Lanzara, A. *Nature Phys.* **2**, 595 (2006).

Zhou, S.Y., Gweon, G.-H., Fedorov, A.V., First, P.N., de Heer, W.A., Lee, D.-H., Guinea, G., Castro Neto, A.H., Lanzara, A. *Nature Mater* **6**, 770 (2007).

Zhou, S.Y., Siegel, D.A., Fedorov, A.V., Gabaly, El Gabaly, F., Schmid, A.K., Castro Neto, A.H., Lee, D.-H., Lanzara, A. *Nature Mater* **7**, 259 (2008a).

Zhou, S.Y., Siegel, D.A., Fedorov, A.V., Lanzara, A. *Physica E* **40**, 2642 (2008b).

15

Theoretical simulations of scanning tunnelling microscope images and spectra of nanostructures

Jinlong Yang and Qunxiang Li

15.1 Introduction

The invention of the scanning tunnelling microscope (STM) (Binnig *et al.* 1982, 1983) introduced a new and revolutionary tool in surface science, and started a development that is leading worldwide towards the new field of nanotechnology. Figure 15.1 shows the setup of an STM. The main components include a sample holder, on which the sample/surface under study is mounted; a piezo-tube, which connects the STM tip or sample; an electronic feedback loop; and a computer to monitor and record the operation. In most cases the STM is built into an ultrahigh-vacuum (UHV) chamber. The basic principle of STM is now well established. A metal tip, of which the terminal is sharp at the nanoscale, is brought close to a sample with a vacuum separation equal to a few angstroms (about 5–7 Å), and a bias voltage is applied between them, and the tunnelling current due to the quantum tunnelling effect is measured. The tunnelling current varies approximately exponentially with the sample–tip separation, so this tunnelling current can be used as a parameter to adjust the sample–tip separation by a feedback loop. In the standard constant-current mode, the tip scans over the sample surface keeping the tunnelling current at the given value (generally measured in nanoamps, nA). Movement of the tip in the other two directions, x and y, is controlled so as to scan the sample surface, giving rise to a two-dimensional map of sample–tip separation (or the tip height over the sample). If the feedback loop is closed, a current map is obtained as a constant-height image. They are two typical imaging models in normal STM experiments.

Now, STM is widely used to study the atomic-scale configuration of nanostructures especially for ad-atoms and molecule-adsorption systems because it

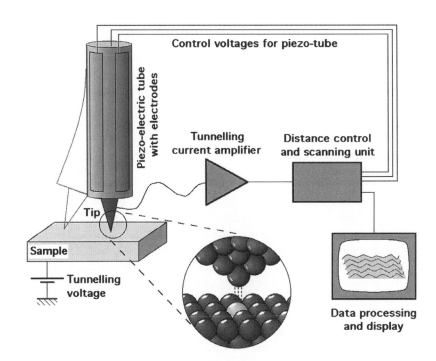

Fig. 15.1 Setup of a STM. The piezo-tube deformation by applied electric fields translates into lateral and vertical manipulation of the tip. The height of the tip is adjusted according to the tunnelling current using an electronic feedback loop. Then, a two-dimensional current contour is recorded. Courtesy of M. Schmid (Schmid 1998).

possesses the highest spatial resolution (lateral resolution 0.1 nm, and vertical resolution 0.01 nm). However, the tunnelling current and the internal patterns in STM images are not directly related to the atomic structure and topography of the sample, but to the local density of states (LDOS) distribution near the Femi energy. Actually, the electronic structures and LDOS distribution of molecules (even a single molecule) on various substrates or nanostructures are usually more complicated than metal and semiconductor surfaces with two-dimensional spatial periodicities. Moreover, the existence of various intrinsic and extrinsic factors, such as tip effect, bias polarity, substrate–molecule interaction, that can influence STM imaging process. Clearly, it is still a challenge to directly image molecules by STM with high resolution and unveil their internal structures.

In addition to capturing a sample image, the STM is also used as a tool for spectroscopy. The STM can precisely vary the bias voltage and tip-to-sample distance for measurement at any lateral position. It can, therefore, perform different types of spectroscopic measurement at a given location: I–V (current versus voltage), dI/dV and d^2I/dV^2 versus bias voltage, such as at constant tip–sample separation and at constant tunnelling resistance V/I. Therefore, scanning the sample surface allows one to record both the topographic image and the local spatially resolved $I(V)$, $dI(V)/dV$ and $d^2I(V)/dV^2$ characteristics (which are named as scanning tunnelling spectroscopy, STS). It is clear that STS will be very useful for studying samples with pronounced local effects in their electronic structure. However, to achieve reproducible and reliable STS is very difficult due to the uncertainties in the geometrical and chemical nature of the tip. These I–V curves depend on the electronic properties of both tip and sample, and confident interpretation demands knowledge of the tip, which is in general simply not available. Interestingly, based on inelastic electron tunnelling spectroscopy (IETS) obtained by measuring the second derivative of the tunnelling current with respect to bias voltage, $d^2 I(V)/dV^2$, some of the vibrational modes of single adsorbed molecules on substrate surfaces can also be identified.

During STM imaging, the location of STM tip at the proximity of the surface often causes perturbations due to tip–sample interactions. In a normal imaging mode, these perturbations are not desirable. However, these undesired perturbations became one of the most fascinating subjects to be pursued since 1990. One can manipulate atoms and molecules on surfaces based on this kind of tip–sample interaction. Using STM manipulation techniques, quantum structures can be constructed on an atom-by-atom basis, single molecules and artificial clusters (i.e. quantum corrals) can be synthesized, and detailed physical/chemical properties of atoms/molecules can be accessed at an atomic level. A STM manipulation procedure to relocate single atoms/molecules across a surface is known as lateral manipulation. Vertical manipulation is another useful manipulation technique and it involves the transfer of atoms/molecules between the tip and surface.

As for STM and STS theory, techniques, and applications, many excellent review articles and books have been published in the past years, such as those by Tromp (1989); Sautet (1997), Briggs and Fisher (Briggs *et al.* 1999), Bonnell (2001), Drakova (2001), Ho (2002), Hofer (2003), Hofer *et al.* (2003), Pascual (2005), and Hla (2005). They gave comprehensive reviews of theories and applications of STM. We recommend readers to these publications. This chapter will focus on the current understanding of the observed STM images and the measured STS of nanostructures. The rest of this chapter is organized as follows. The theories of STM and STS are introduced briefly in Section 15.2. Based on the combination of high-resolution STM measurements with theoretical simulations, the conventional STM and STS investigations of various systems including clean surfaces, ad-atoms, single molecules, self-assembled monolayers, and nanostructures are presented in Section 15.3. We address these beyond conventional STM investigations, such as the functionalized STM tip, dI/dV mapping technique, inelastic tunnelling spectroscopy, manipulation, molecular electronics and molecular machines in Section 15.4. Finally, we present a short concluding remark.

15.2   Theories of STM and STS

In principle, one could directly calculate the transmission coefficient for an electron incident on the tunnelling vacuum barrier between sample and tip. However, such a calculation is not feasible for a realistic model of the three-dimensional STM problem. One has to obtain a correct description of the tunnelling vacuum barrier between the tip and sample surfaces, a detailed description of the sample and tip's electronic states, and a determination of the wavefunction tail of the tunnel electrons in the tunnel gap. Moreover, the exact geometric structure of the tip is commonly unknown except for some outstanding STM measurements, where the tip structure was determined before and after a scan by field-ion microscopy (Cross *et al.* 1998).

In what follows, we briefly introduce four main approaches including the perturbation or Bardeen approach (Bardeen 1961), the Tersoff–Hamann approach (Tersoff and Hamann 1985) and its extension, the scattering theory or Landauer–Bütticker approach (Bütticker *et al.* 1986; Sautet 1986), and the non-equilibrium Green's function or Keldysh approach (Keldysh 1965; Levy Yeyati *et al.* 1993; Mingo *et al.* 1996) as well as the theory of STS.

15.2.1   Perturbation approach

Fortunately, for typical tip–sample distances (of order 6 Å nucleus-to-nucleus) the coupling between the tip and sample is weak, and the tunnelling can be treated with first-order perturbation theory. That is to say, if the interaction between the two electrodes is neglected, the tunnel current can be evaluated from the eigenfunctions ψ_μ and ψ_ν (the corresponding eigenenergies are E_μ and E_ν) of the isolated tip and sample electrodes, respectively. Over the last decades, several tunnelling theories have been developed based on the "perturbative-transfer Hamiltonian" formalism introduced by Bardeen (1961). In this formalism, the tunnel current I_t is calculated from the overlap of the tails of ψ_μ and ψ_ν, in the region of the tunnelling vacuum barrier, and the following is obtained on the basis of Fermi's golden rule:

$$I_t = \frac{2\pi e}{\hbar} \sum_{\mu,\nu} f(E_\nu)[1 - f(E_\nu + eV_t)] \left| M_{\mu\nu} \right|^2 \delta(E_\mu - E_\nu), \qquad (15.1)$$

where the tunnel matrix element is given by:

$$M_{\mu\nu} = \frac{\hbar^2}{2m} \int\limits_{S_0} \mathrm{d}S \cdot (\psi_\mu^*(\vec{r})\nabla\psi_\nu(\vec{r}) - \psi_\nu(\vec{r})\nabla\psi_\mu^*(\vec{r})). \qquad (15.2)$$

Here, the integral extends over the separation surface S_0 between sample and tip, the summation includes all eigenfunctions within a given interval from the Fermi level. This interval is determined by experimental conditions, e.g. the temperature of STM. The delta function in eqn (15.1) implies that the electron does not lose energy during tunnelling (elastic tunnelling), whereas $f(E)$, the Fermi–Dirac function, takes into account that tunnelling occurs, for instance, from an empty tip state into a filled sample state. The energy shift eV_t is the result of the tunnel bias voltage V_t.

15.2.2   Tersoff–Hamann approach

Tersoff and Hamann (TH) (1983, 1985) were the first to apply the transfer Hamiltonian approach to STM. Clearly, to quantitatively calculate the current, the main practical difficulty in evaluating the matrix element $M_{\mu\nu}$ in eqn (15.2) from density-functional theory (DFT) is to obtain sufficiently accurate wavefunctions of both tip and sample at the separation surface. This difficulty is due to two main reasons. First, the exponentially decaying wavefunctions require a good convergence in the vacuum region due to their negligible contribution to the total energy. Second, the description of the wavefunctions in the vacuum region is limited by the incompleteness of the basis set. Moreover, these problems get worse with increasing tip–sample separation, DFT simulations are frequently performed at unrealistically close separations, thus allowing only for qualitative comparisons. To simplify the calculations, Tersoff and Hamann considered the limit of small V_t and made the simplified assumption that the tip is spherical with s-wavefunctions only. With these assumptions, the tunnelling current is proportional to the local density of states (LDOS) at the position of the STM tip

$$I(r_t) \propto \rho_S(\vec{r}, E_F)$$
$$\rho_S(\vec{r}, E_F) = \sum_\nu \left|\psi_\nu(\vec{r})\right|^2 \delta(E_\nu - E_F), \tag{15.3}$$

where $\rho_S(r_t, E_F)$ is the LDOS of the sample surface at the Fermi level evaluated at the center position r_i of the tip.

Now, the Tersoff–Hamann method is incorporated in many state-of-the-art DFT codes. Despite an extension of existing simulation methods, especially with respect to quantitative comparisons between experiments and theory, it continues to be the most popular theoretical model of STM because it is a very convenient approach and gives a simple picture of the operation in STM. In many situations, for example, adsorbates on surfaces and surface reconstructions, this model provides a reliable qualitative picture of the surface topography. However, there are situations in which one needs to include more details of the electronic structure of the tip to understand the experiments. In general, these situations arise because the tip electronic structure has significant modulation (rather than being an s-wavefunction as assumed by Tersoff and Hamann) or because the tip density of states shows an obvious dependence on energy with the tunnelling energy range.

Lang (1985) demonstrated the importance of adopting realistic atomic potentials for the surface adsorbates using the so-called jellium model. Actually, more elaborate and detailed models may be required for a more detailed and even quantitative simulation of the experimental STM images. Chen (1990a,b) has generalized the perturbation approach to p and d states as tip orbitals, by calculating the corresponding tunnelling matrix elements. This case is very relevant in practice, since many of the materials commonly used for tips are transition metals with valence d-electrons. Tsukada *et al.* (1990, 1991) have extended Bardeen perturbation formalism and formulated a method for theoretical simulations of STM images based on DFT calculations. The electronic structures for the tip and the surface are calculated separately, using a cluster model for the tip and a slab model for the surface. The wavefunctions are represented by a linear combination of atomic-orbital numerical basis. They have explored the dependence of the STM image on the tip structure and found that about 80% of the tunnel current is concentrated at the apex atom of the tip.

Recently, Paz *et al.* (2005) proposed a very efficient and accurate method to simulate STM images and spectra from DFT calculations. The tip and sample were treated as weakly interacting systems with no corrections to the wavefunctions of two isolated systems due to their interaction. They found that the tip wavefunction ψ_μ is replaced by the Green function G, which obeys the following equation

$$\nabla^2 G(\vec{r} - \vec{R}) - \kappa^2 G(\vec{r} - \vec{R}) = \delta(\vec{r} - \vec{R}),\qquad (15.4)$$

where $\vec{R}$ denotes the tip position, $\kappa^2 = \frac{2m}{\hbar^2}(\phi - E)$, and ϕ is the work function. Then, the transfer matrix element $M_{\mu\nu}$ can be written as

$$M_{\mu\nu}(\vec{R}) \propto \int_{S_0} dS[G^*(\vec{r} - \vec{R})\nabla\psi_\nu(\vec{r}) - \psi_\nu(\vec{r})\nabla G^*(\vec{r} - \vec{R})] = \psi_\nu(\vec{R}).$$

$$(15.5)$$

Note that in the Tersoff–Hamann approach the above equation is used to replace $M_{\mu\nu}$ by ψ_ν in eqn (15.1), eliminating the uncertainties from the tip composition and structure and greatly simplifying the simulations. Paz *et al.* (2005) represented the separation surface S_0 by the constraint function $S(\vec{r}) \equiv \ln(\rho(\vec{r})/\rho_0)$, so that surface integrals in eqn (15.2) can be efficiently transformed into volume integrals via $\int_{S_0} f(\vec{r})dS = \int f(\vec{r}) \cdot C(\vec{r})d\tau$, here, $C(\vec{r}) = \delta(S(\vec{r}))\frac{\nabla\rho(\vec{r})}{\rho(\vec{r})}$. This kind of three-dimensional integral can be performed conveniently in a regular grid after smoothing the delta function. The convolution theorem can be applied in eqn (15.5) to express the wavefunction as an inverse Fourier transform. Then, the tunnelling matrix elements can be evaluated for all tip positions and bias voltages in a single computation using fast Fourier transforms for the convolutions. The input values of $C(\vec{r})$, $\psi_\mu(\vec{r})$, and $\psi_\nu(\vec{r})$ can be stored in the points of a uniform grid within the regions of the broadened surfaces S_{tip} and S_{sample}.

15.2.3 Scattering theory

A quite different type of approach is necessary when perturbation theory itself can no longer be applied. The other class of methods goes beyond perturbation theory by a proper description of the interacting sample and tip with scattering theory formalism. The tunnel event in STM is viewed as a scattering process. For example, the incoming electrons from the sample scatter from the tunnel junction and have a small probability to penetrate into the tip, and a large one to be reflected toward the bulk. It has been pointed out that, at small tip–sample distances, the strong tip–sample interaction may cause the interstitial regions between the atoms of a clean metal surface to appear as maxima in the variation of the tunnel current, an effect referred to as "image inversion". Sautet *et al.* (Sautet and Bocquet 1986; Sautet 1997) have shown that their electron scattering quantum chemistry (ESQC) approach can predict surface contrasts in a semi-quantitative way to allow comparison with experimentally recorded images. In this way, quantitative structural information may be extracted from the STM images. However, these more elaborate STM calculations are fairly complicated and time consuming. Therefore, only very few selected metal–surface structures have been simulated so far.

Within a scattering approach the transition matrix is usually calculated based on the Landauer–Bütticker formula (Landauer 1957; Bütticker 1986). In this formulation the conductance G across a tunnelling junction is given by the ratio of transmission probability T to reflection probability R as

$$G = \frac{I}{V_t} = \frac{e^2 T}{\pi \hbar R}. \tag{15.6}$$

The tunnelling current in this approach is usually written

$$I(V_t) = \frac{2e}{\hbar} \int_0^{eV_t} T(E)dE. \tag{15.7}$$

The transition probability matrix $T(E)$, for multiple channels of the tunnelling current, could in principle be evaluated in any basis set. This approach has mathematical rigor. It should thus yield a more accurate description of the tunnelling condition. In addition, the treatment includes interference effects between conductance channels. However, the limitation of this method is that the lateral position of the STM tip needs to be continuously varied to simulate STM images. Within a DFT treatment of the problem, such a lateral variation is difficult for a reasonable system size because of periodic boundary conditions in the z-direction.

15.2.4   Non-equilibrium Green's function approach

During previous years the tunnelling process has been treated on the basis of non-equilibrium Green's function formalism (Keldysh 1965). The most complete treatment of the problem considers the Hamiltonian of a system (comprising two electrodes and a central scattering region) including the Hamiltonians of the left and right leads, the Hamiltonian of the scattering region, and the interactions between the scattering region and left/right electrode. The tunnelling current in this case is given by the following expression:

$$I = \frac{2e}{\hbar} \int dE[f_L(E) - f_R(E)] \times \text{Im}[\text{Tr}\left(\frac{\Gamma^L \Gamma^R}{\Gamma^L + \Gamma^R}\right) G^R]. \tag{15.8}$$

In the above equation, it is assumed that the coupling to the left and right leads are given by Γ^L and Γ^R, respectively. The full retarded Greens function G^R of Keldysh's approach includes all scattering events. In general, Keldysh's formalism allows one to use perturbation theory and the diagram technique to calculate the non-equilibrium Green functions of a system of interacting particles in terms of the Green functions for a non-interacting electron gas.

This approach has been used to compute the tunnelling current in STM by several groups in previous years, regarding the system as in a stationary non-equilibrium state. In the theories of Levy Yeyati *et al.* (1993) and Mingo *et al.* (1996) the Green functions for the two decoupled tip and sample electrodes with different chemical potentials (μ_t and μ_s) are defined as the unperturbed Green functions. The non-equilibrium situation, as created by applying an external bias voltage V, is modelled by shifting the chemical potentials on the two electrodes relative to each other by $eV = \mu_t - \mu_s$. The unperturbed

Green functions are then used to obtain the total Green function for the two electrodes coupled via the scattering region by solving Dyson's equation. However, since the interaction of electrons within the vacuum barrier in STM is negligible, tunnelling currents are not usually calculated within this theoretical framework. The Landauer–Bütticker formalism becomes more convenient, and the tunnelling current is calculated by

$$I = \frac{2e}{\hbar} \int dE [f_L(E) - f_R(E)] \text{Tr}[t^+(E)t(E)]. \qquad (15.9)$$

Due to the wide range of interactions included in the formalism, Keldysh's method is the most accurate. Its main problem is the computational cost, which either has to be made up for by approximations in the description of subsystems, or by limiting the number of atoms in the interaction range. In recent years, both scattering theory and the non-equilibrium Greeen's function technique have been widely used in molecular electronics to obtain the current–voltage characteristics of various molecular junctions.

15.2.5 Scanning tunnelling spectroscopy (STS)

As mentioned above, the first STM theory was put forward by Tersoff and Hamann and only the case of small bias voltage was considered. Their formalism did not consider finite bias voltage. If one has a tip and a sample with a bias voltage (V) applied between tip and sample, then the tunnelling current flowing between these two electrodes is given by

$$I = \int_0^{eV} \rho_S(r, E)\rho_T(r, \pm eV \mp E)T(r, E, eV)dE, \qquad (15.10)$$

here, ρ_S and ρ_T are the density of states of sample and tip at the energy E relative to the Fermi level (E_F) and the position r. The upper signs are for positive sample bias voltage, and lower signs for negative. The tunnelling transmission probability density T is given by

$$T(E, eV) = \exp\left[-\frac{2z}{\hbar} \sqrt{2m \left(\frac{\phi_S + \phi_T + eV}{2} - E \right)} \right]. \qquad (15.11)$$

Clearly, $T(E, eV)$ depends exponentially on the sample–tip separation z and the square root of the sum of the sample and tip work functions (ϕ_S and ϕ_T). Then, the effect of the finite tunnel voltage in STS can be understood. Lang (1986) has proved that it is very difficult to probe low-lying occupied sample states, a fact that has important consequences for STS measurements of adsorbate-covered metal surfaces as the main adsorbate-induced states are situated well below E_F for common adsorbates, such as oxygen.

In order to divide out a large part of the $T(E, eV)$ influence on the density of state features in tunnelling current, Feenstra *et al.* (1987a) showed that the normalized differential conductivity $(V/I)dI/dV = d\ln I/d\ln V$ versus tunnelling bias voltage has a close resemblance to the LDOS, at least for the empty states, whereas the occupied states are reduced by a factor increasing with energy below E_F. In general, it is difficult to image low-lying occupied surface states. Moreover, since STS depends crucially on the extension of the relevant states, it should be pointed that in STS only the states that protrude into the vacuum and overlap with the tip wavefunctions are probed, as opposed to photoemission and inverse photoemission spectroscopy measurements where information on filled and empty states is integrated over several surface layers.

A magnetically sensitive imaging technique based on STM is named spin-polarized STM. The tunnelling current depends on the relative orientation of the quantization axes of both magnetic sample and/or tip. In a simplified Stoner model, the spin-up band is energetically shifted with respect to the spin-down band, which results in a spin polarization (P) at the Fermi level

$$P_{T,S}(E_F) = \frac{\rho_{T,S}^{\uparrow}(E_F) - \rho_{T,S}^{\downarrow}(E_F)}{\rho_{T,S}^{\uparrow}(E_F) + \rho_{T,S}^{\downarrow}(E_F)}. \tag{15.12}$$

In general, the tunnelling of spin-up and spin-down electrons are regarded as independent processes. This is true in the limit of a low bias voltage since inelastic tunnelling scattering including spin-flip has only a small contribution to the total tunnelling current in a first-order approximation. The tunnelling current is dominated by elastic processes and the electron spin is preserved during the tunnelling process. Considering the electron spin, the conductivity at zero bias can be written as:

$$\frac{dI}{dV} \propto \rho_T^{\uparrow}(E_F)\rho_S^{\uparrow}(E_F) + \rho_T^{\downarrow}(E_F)\rho_S^{\downarrow}(E_F). \tag{15.13}$$

In spin-polarized STM, the tunnel magnetoresistance (TMR) is defined by

$$R_{TMR} = \frac{R_{\uparrow\downarrow} - R_{\uparrow\uparrow}}{R_{\uparrow\uparrow}} = \frac{2P_T P_S}{1 - P_T P_S}. \tag{15.14}$$

If the tip and sample have non-collinear quantization axes with an angle θ, the above equation is redefined as:

$$\frac{R_{TMR}}{\cos\theta} = \frac{2P_T P_S}{1 - P_T P_S}. \tag{15.15}$$

15.3   Conventional STM and STS investigations

In conventional STM experimental investigations, the topographic and LDOS of the sample are often characterized by using the high-resolution images and STS, respectively. In this section, we focus on these conventional STM and STS investigations on various clean surfaces, ad-atoms, single molecules, self-assembled monolayers as well as nanostructures (i.e. nanoislands).

15.3.1   Clean surfaces

Since the topographic pictures of $CaIrSn_4(110)$ surface on an atomic scale were obtained for the first time by Binnig *et al.* (1982), STM is now widely used to characterize various metallic and semiconducting clean surfaces with atomic resolution. Metal surfaces have been extensively studied by STM. The main investigations focus on reconstructions, relaxations and surface states, which include surfaces with a different arrangement of surface atoms than in the bulk, the trend for surface layers to possess a different interlayer spacing than bulk crystals, and electron states trapped in the surface region due to the potential boundary, respectively. Actually, the surface chemical composition plays an important role in STM imaging of metal surfaces. For example, in STM experiments, sulfur, oxygen or carbon atoms on a surface change the image locally in a subtle way.

For simple metals, there is typically no strong variation of the LDOS or wavefunctions with energy near the Fermi level. This is true for noble and even transition metals since the d shell apparently does not contribute significantly to the tunnelling current. It is therefore convenient in the case of metals to ignore the bias voltage dependence, and consider the limit of small bias voltage. It implies that the observed STM image reveals the LDOS of the bare surface. For example, Tersoff and Hamann (1985) have calculated the LDOS of Au(110) 2×1 and 3×1 surfaces and compared them with experiments.

Semiconductor surfaces have an enormous range of technical applications and have therefore attracted many experimental and theoretical studies. In general, unlike metals, semiconductors show a very strong variation of LDOS with bias voltage. In particular, the LDOS changes discontinuously at the band edges. With negative sample bias voltage, the current tunnels out of the valence band, while for positive bias voltage, current tunnels into the conduction band. The corresponding images reflect the spatial distribution of valence- and conduction-band wavefunctions, respectively. They can be qualitatively different. A particularly simple and illustrative example, which has been studied in great detail, is the GaAs (110) surface. The valence states are preferentially localized on the As atoms, and the conduction states on the Ga atoms. The negative and positive sample bias voltages should reveal the As and Ga atoms of GaAs(110) surface, respectively. Such an atom-selective imaging has been confirmed by the theoretical calculations of Tersoff and Hamann (1985), and subsequently observed experimentally by Feenstra *et al.* (1987b).

The principal silicon surface under investigation is Si(100) surface. The Si(111) surface, for example the Si(111)-7×7 surface, even though widely imaged by STM, is substantially less important technologically. In fact, it is today mostly imaged to calibrate the instrument and to study adsorption of various species. Actually, a detailed structural analysis of Si(111)-7×7 surface, directly from the experiments, is still a difficult task, because structural and electronic properties are intermixed in the STM images. At the same time, the composition and structure of the tip is not easily determined. This is particularly crucial in the case of STS, where tip states can entirely modify the spectra. In practice, the atomic-resolution STM images of the Si(111)-7×7 surface, however, are usually obtained after intentional slight tip–sample contacts, which may lead to the sample-atom termination of the original STM tips. As a result of these uncertainties, a careful comparison with theoretical simulations is generally needed to interpret safely the experimental information. For example, Paz *et al.* (2005) have simulated the STM images of the Si(111)-7×7 surface and compared them with the experimental data, as

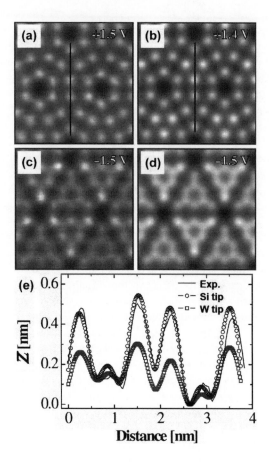

Fig. 15.2 (a) and (b) Empty state (1.5 V and 1.4 V), (c) and (d) Occupied state images (−1.5 V) of the Si(111)-7 × 7 surface at 0.2 nA. The left panels show experimental data, while the right panels represent simulations with a Si tip, using the same gray scale. (e) Corrugation profile comparison between experimental and theoretical images for positive sample bias voltage case, along the solid lines in (a) and (b). (Paz *et al.* 2005) Copyright American Physical Society.

shown in Fig. 15.2. They found that the agreement between the theoretical and experimental images is excellent for the Si tip and considerably worse for the W tip.

Recently, other surfaces are also intensively investigated. For example, Ni_2P, one of transition-metal phosphides, is a new class of catalyst active for hydrodesulfurization and hydrodenitrogenation for petroleum fuels. Along the (0001) direction in bulk Ni_2P, there is an alternation between two non-equivalent atomic layers with stoichiometry of Ni_3P_2 and Ni_3P. Two planes arranging alternately along (0001) give the full Ni_2P stoichiometry of the bulk. So, a simple (0001) termination of the bulk structure allows for a surface layer with either a Ni_3P_2 or Ni_3P plane as shown in Figs. 15.3(a) and (b), respectively. Using the Tersoff–Hamann model, the STM images for the Ni_2P surfaces at the positive bias voltage (in the range of 1.0–2.0 V) have been simulated by Li and Hu (2006), which are shown in Figs. 15.3(c) and (d). The direct comparison with the atomic STM images observed by Golam Moula *et al.* (2006) and the simulated results have suggested that the bright STM spots correspond well to the P positions on both Ni_3P- and Ni_3P_2-terminated surface. Clearly, this example indicates that these STM images using the Tersoff–Hamann model reproduce the main feature of the observed experimental STM results and its interpretations is rather simple and reliable.

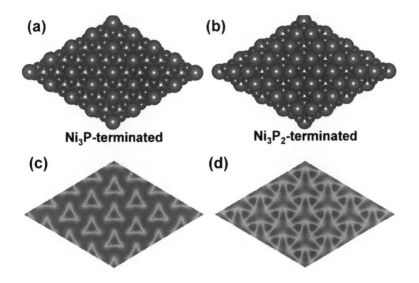

Ni₃P-terminated   **Ni₃P₂-terminated**

Fig. 15.3 (a) and (b) Top-view illustration of Ni_3P- and Ni_3P_2-terminated Ni_2P (0001) surfaces with the size of a 4×4 superlattice, where the large and small spheres stand for P and Ni sites, respectively. (c) and (d) Simulated STM images for Ni_3P- and Ni_3P_2-terminated surfaces under the sample positive bias voltage of 1.7 and 1.2 V, respectively. Here, the tip–sample distance is 4.5 Å (Li and Hu 2006). Copyright American Physical Society.

Besides characterizing the clear perfect surface, STM actually is often used to examine various surfaces with defects, such as vacancies, impurities, steps, and facet boundaries, by observing with atomic-resolution STM images. Defects at transition-metal (TM) and rare-earth oxide surfaces (Ganduglia-Pirovano *et al.* 2007; Bonnell and Garra 2008), in particular oxygen vacancies, play a major role in a variety of technological applications. Titanium dioxide (TiO_2) is the model system in the surface science of TM oxides (Diebold 2003). TiO_2 has three main phases, rutile, anatase, and brookite. However, only rutile and anatase play a role in the applications of TiO_2. The oxygen-terminated (110) and (101) oriented (1×1) surfaces are the most stable rutile and anatase surfaces, respectively. Both surfaces have sixfold and fivefold coordinated Ti atoms ($Ti^{(6)}$ and $Ti^{(5)}$), as well as threefold and twofold coordinated O atoms ($O^{(3)}$ and $O^{(2)}$), respectively. The removal of the latter, so-called bridging oxygen (BO) atoms is a subject of much interest. Moreover, the BO vacancy sites that appear as defects in the BO rows can play an important role in the catalytic activity of rutile TiO_2(110). Wendt *et al.* (2006) have studied the interaction of water with the reduced TiO_2(110) surface through a combination of high-resolution STM images and DFT simulations. The paired hydroxyl groups are formed as the direct product of water dissociation in BO vacancies. Zhang *et al.* (2007) have reported the first measurements and calculations of the *intrinsic* mobility of BO vacancies on a rutile TiO_2(110). The STM images show that BBO vacancies migrate along BO rows.

Since Novoselov *et al.* (2005) fabricated ultrathin monolayer graphite devices, the electronic properties of graphite monolayer (graphene) have attracted a great deal of research interest due to its Dirac-type spectrum of charge carriers in this gapless semiconductor material. Many interesting properties of single-layer graphene, such as Landau quantization, defect-induced localization, and spin-current states, have been studied experimentally and theoretically by many groups (Geim and Novoselov 2007). STM images of a perfect graphite surface show a substantial asymmetry between *A* atoms (over another carbon atom) and B atoms (over the center of an hexagon). This asymmetry is a consequence of the inter-layer interactions on the basis of electronic considerations. For graphite surface with point defects, Ruffieux *et al.* (2005) have observed the standing waves in the LDOS, which are due to backscattering of electron wavefunctions at individual point defects. Ferro and Allouche (2007) have interpreted STM images of a single atomic vacancy on single and double graphene sheets based on DFT calculations. They also found that the interlayer interaction plays a crucial role in the interpretation of STM images. On an α vacancy, this causes the $\sqrt{3} \times \sqrt{3}R30°$ modulation based on A atoms to disappear, while the three-branch star feature comes closer to the Fermi level. On the contrary, on a β vacancy, the LDOS is dominated by the $\sqrt{3} \times \sqrt{3}R30°$ modulation based on B-type atoms, while the unit-cell network based on A atoms is moved away from the Fermi level. These calculations are in good agreement with experimental observations of Ruffieux *et al.* (2005) and allow differentiation between α and β atomic vacancies on a graphite surface.

15.3.2   Ad-atoms

Adsorbates (even a single ad-atom) on metal surfaces generally induce pronounced changes in the LDOS near the Fermi level (E_F) by electronic resonances derived from the interaction of the adsorbate states with the surface electron states. Lang (1986) has investigated the degrees to which STM images of chemically different atoms vary, such as Na, S, C, O atoms. He modelled the STM tunnel gap region as two planar metal electrodes on which adsorbates are chemisorbed. For simplicity, the metal electrodes were described by the so-called jellium model in which the ionic lattice of each metal is smeared out into a uniform, positive background. Lang found that the Na 3s resonance and the S 2p resonances generate an increase in the LDOS at the E_F, whereas for the electronegative C and O atoms, the 2p resonances lead to a depletion of the LDOS at the E_F. In the constant-current STM mode, the adsorbates will therefore appear either as protrusions (Na, S) or as holes (C, O), depending on whether the adsorbates add to or deplete the LDOS at the E_F by depicting the calculated vertical tip displacement obtained by moving the tip laterally over the atom in question. These results confirm the Tersoff–Hamann prediction for low bias voltage, that is, the STM images reflect the LDOS at E_F of the sample at the tip position rather than reflecting the geometrical position of the adsorbates.

It is well known that the tip and the surface may markedly influence the appearance of ad-atoms in STM images. Tilinin *et al.* (1998) have shown that atomic size and electronegativity are the dominant factors that determine the image contrast. In their low-temperature STM experiments, isolated carbon and sulfur atoms on the Pd(111) surface scanned by a platinum tip appear as bumps with a height of about 0.3–0.4 Å and 0.8 Å, respectively, whereas oxygen atoms are characterized by a depression with a negative corrugation of -0.35 Å. The shape and size of the experimental image corrugations are in agreement with their simulations.

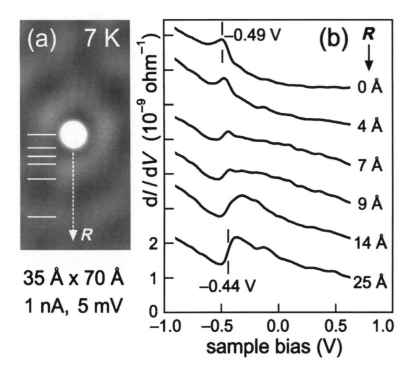

Fig. 15.4 Experimental STM image and STS spectra. (a) Constant-current STM image of a single Cu ad-atom on Cu(111) surface (35 Å × 70 Å, 5 mV, 1 nA, 7 K), the gray-scale is enhanced to display the standing-wave pattern of the surface state; horizontal white bars mark lateral distances, R, relative to the ad-atom where the differential tunnelling conductance was measured at constant tip height. (b) Corresponding spectra measured with the tip positioned on top of the ad-atom (topmost curve) and at different lateral distances R (Olsson *et al.* 2004). Copyright American Physical Society.

From single ad-atoms to assemble nanostructures (such as quantum corrals) by using an STM technique provides a means to tailor the propagation of surface states in confined geometries with potential use in nanotechnology. Clearly, there is a need to characterize the scattering properties of surface states from single ad-atoms. Fortunately, STM and STS can provide detailed information about ad-atoms interacting with surface states, for example, standing-wave patterns around ad-atoms and Kondo resonances induced by magnetic ad-atoms. Olsson *et al.* (2004) have reported an ad-atom-induced localization of the surface state for a single Cu ad-atom on Cu(111) surface. The STM image in Fig. 15.4(a) shows a single Cu ad-atom characterized by a bright spot and the faint wave pattern around the ad-atom arises from standing waves of the surface state. As shown in Fig. 15.4(b), the topmost dI/dV spectra recorded over the Cu ad-atom exhibits a clear peak at the sample bias voltage of -0.49 V. With increasing lateral tip displacement (R), this peak decays rapidly in intensity and the Cu(111) surface-state band edge evolves. The spectra measured at $R = 25$ Å shows the surface-state band edge located at the sample bias voltage of -0.44 V. To gain insight into the physics underlying these resonances, Olsson *et al.* (2004) have simulated the LDOS, which is based on a time-dependent wave-packet-propagation technique and confirmed that the Cu ad-atom-induced peaks at -0.7 eV in the LDOS originate from the surface-state localization by the ad-atom.

Remarkable developments in spin-polarized STM (Bode 2003) are expected to capture the spatial variation in the magnetic density, which might provide an understanding of spin-polarization behavior of a single magnetic ad-atom

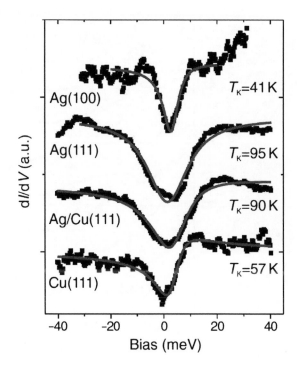

Fig. 15.5 dI/dV spectra taken on a cobalt ad-atom on Ag(100) as well as on cobalt ad-atoms on Ag(111), one monolayer of Ag on Cu(111), and on Cu(111). The solid lines depict the fit of a Fano line shape to the data (Wahl *et al.* 2004). Copyright American Physical Society.

on a surface. Based on their *ab-initio* calculations of quantum mirages and the magnetic interactions in quantum corrals, Stepanyuk *et al.* (2005) have shown that the spin polarization of surface electrons caused by magnetic ad-atoms can be projected to a remote location by quantum states of corrals. The exchange interaction between magnetic atoms can be manipulated at large distances. Lazarovits *et al.* (2006) have reported a theoretical study of a surface state close to 3d transition-metal ad-atoms (Cr, Mn, Fe, CO, Ni, and Cu) on a Cu(111) surface in terms of an embedding technique using the fully relativistic Korringa–Kokn–Rostoker (KKR) method. They found the resonance in the s-like states to be attributed to a localization of the surface states in the presence of an impurity. A magnetic impurity causes spin polarization of the surface states.

Recently, the interest has been revived through the investigation of Kondo phenomena in a single magnetic impurity on various metallic substrates using low-temperature STM and STS since the understanding of the physics of a single spin supported on a metal host is at the basis of a bottom-up approach to the design of high-density magnetic recording. The Kondo resonance shows up as a sharp peak in the local density of states, which is pinned to the Fermi level and has a width proportional to the Kondo temperature (T_K).

Wahl *et al.* (2004) have studied the Kondo behavior of cobalt ad-atoms at the (111) and (100) surfaces of Cu, Ag, and Au. The experimental dI/dV curves are shown in Fig. 15.5. The typical STS were measured with a tip placed above a single cobalt atom. It is clear that, for Co on Ag(100), the spectrum shows a distinct dip. The line shape is the same as for Co on Ag(111), but the value of the Kondo temperature for Co/Ag(100) is lower than that for Co/Ag(111). This indicates that the scaling of the Kondo temperature is solely based on the number of nearest neighbors of the Co impurity. The line shape and Kondo temperature of Co on one monolayer of Ag on Cu(111) is almost the same as on a bulk Ag sample, although the Kondo temperatures of Ag and Cu differ by almost a factor of 2. This provides evidence that the Kondo behavior is governed by the interaction with the nearest neighbors of the impurity. These experimental observations can be understood using the Anderson model. Wahl *et al.* (2004) concluded that the Kondo temperature depends on the occupation of the d level determined by the hybridization between the ad-atom and the substrate. After that, Wahl *et al.* (2005) have further demonstrated that the strength of the coupling between the spin of individual cobalt adatoms with their surroundings is determined via the Kondo resonance by the low-temperature STS measurements and this coupling can be tuned by controlling the number of CO ligands.

Direct observation of the spin-polarization state of isolated ad-atoms is very difficult since isolated atoms have low magnetic anisotropy energy (of the order of meV) which causes their spin to fluctuate in time due to environmental interactions. Yayon *et al.* (2007) have measured the spin polarization of individual Fe and Cr ad-atoms on a metal surface. In their experiment, spin-polarized tips were created by coating etched tungsten tips with a thin film of Cr to produce an out-of-plane magnetization. To fix the ad-atom spin in time, the Fe and Cr atoms were deposited onto the ferromagnetic Co nanoislands, thereby coupling the ad-atom spin to the island magnetization through the direct exchange interaction. Then, they characterized the spin-polarized electronic structures of individual Fe and Cr ad-atoms on the cobalt nanoislands via dI/dV spectra. The spin-up and spin-down Co islands were distinguished spectroscopically via contrast arising from a spin-polarized surface state centered 0.28 eV below the Fermi energy, as seen in Fig. 15.6(a) with dashed lines. This resonance comes from a spin-polarized minority surface state of d_z^2 symmetry.

To clearly describe the spectroscopic differences in ad-atom behavior, the dI/dV spectra are normalized and shown in Figs. 15.6(a)–(c) with solid lines. It is clear that spin-polarized spectra measured for individual Fe and Co ad-atoms (Figs. 15.6(a) and (b), solid lines) show strong spin-polarization contrast depending on whether they lie on spin-up or spin-down Co islands. Fe atoms on spin-down islands have a stronger dI/dV signal than Fe atoms on spin-up islands. Cr atoms, on the other hand, show opposite spin polarization in this energy range compared to the Co island spin polarization: Cr atoms on spin-down islands exhibit a lower dI/dV signal than Cr atoms on spin-up islands. Clearly, the magnetic contrast for Cr atoms, however, qualitatively differs from the contrast seen for Fe atoms. Figure 15.6(d) presents the non-spin-polarized spectra with a tungsten STM tip. No discernible contrast for Fe and Cr ad-atoms on different Co nanoislands can be observed.

15.3.3 Single molecules

Compared to a single ad-atom, the interpretation of the contrast of chemisorbed molecules becomes a little more complex. It is well known that the STM image

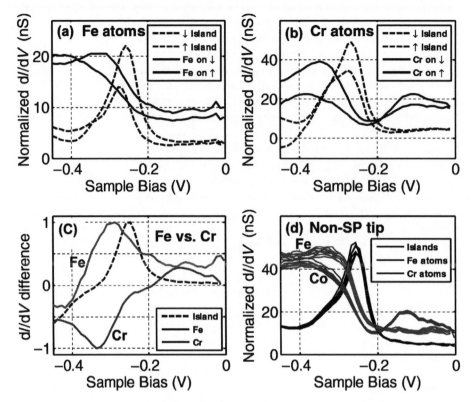

Fig. 15.6 (a) and (b) Normalized spin-polarized dI/dV spectra of two Co islands with opposite spin orientation (dashed lines), as well as spin-polarized dI/dV spectra of Fe and Co ad-atoms on these two islands (solid lines), respectively. (c) Difference between spin-up and spin-down spectra for Co islands, Fe ad-atoms, and Cr ad-atoms. (d) Normalized dI/dV spectra measured with a non-spin-polarized tip held over 8 different Co islands as well as corresponding Fe ad-atoms and Cr ad-atoms on these islands. Initial tunnelling parameters for (a): $V = -25\,\text{meV}$, $I = 3\,\text{pA}$, and for (b): $V = -50\,\text{meV}$, $I = 2\,\text{pA}$. Fe and Cr ad-atom spectra in (a), (b) and (f) multiplied by a constant factor of 3 for better clarity in plots (Yayon *et al.* 2007). Copyright American Physical Society.

depends on the surface, adsorption site, metal Fermi level, and tunnelling conditions. The image dependence for a single benzene molecule on Pt(111) has been studied by Sautet and Bocquet (1996) with the ESQC approach. These simulated STM images strongly depend on the chemisorption site and they allow the assignment of each experimental image (Weiss and Eigler 1993) of benzene to a given site and orientation of the molecule. The hollow, atop, and bridge sites give an image with three lobes on a triangle, a sixfold volcano shape, and a simple bump shape with only a weak twofold aspect, respectively.

Note that in characterizing the adsorption behavior, one of central issues is the adsorption orientation of molecules on a substrate since it is the basis for designing new catalysts with functionalized molecules or for fabricating thin films of desirable orientational orders. For molecules, especially those with complicated orientation degrees, of which a typical example is a C_{60} molecule, there exist a multitude of possibilities. There are five different C_{60} high-symmetry rotational orientations including a hexagon, a 6–6 bond, a 5–6 bond, a pentagon, and an edge atom facing towards the substrate, as shown in the bottom panel of Fig. 15.7. If the substrate effect can be neglected, it

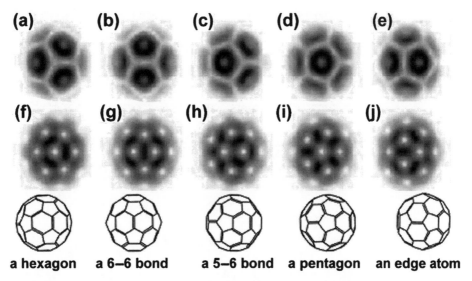

Fig. 15.7 The simulated STM images of a free C_{60} molecule with five different high-symmetry rotational orientations. (a)–(e) stand for the positive bias voltage STM images, and (f)–(j) for the negative bias voltage. Here, the height of tip is about 5 Å.

is interesting to compare the feature of simulated STM images of C_{60} with different orientations. The positive and negative bias voltage simulated STM images with different orientations are shown in Figs. 15.7(a)–(e) and (f)–(j), respectively. It is clear that the internal patterns of the STM image of C_{60} depend on its orientation. Of course, these results give useful information to determine the adsorption orientation of C_{60} on various surfaces.

Hou *et al.* (1999) have proved that the molecular orientation of the adsorbed C_{60} molecules with respect to a Si(111)-(7 × 7) surface can be determined unambiguously by combining low-temperature STM experiments with density-functional theory (DFT) calculations. After depositing submonolayer C_{60} molecules on a Si(111)-(7 × 7) surface, the STM images were obtained using ultrahigh-vacuum and low-temperature STM with an electrochemically etched W tip at 78 K. The observed image shows that the internal pattern of the C_{60} depends strongly on the bias voltage and the tip–sample distance. There are four possible adsorption sites. At site A (faulted half), the feature of the large positive bias image is one bright pentagon ring plus two curved strokes. At adsorption site B (corner hole), the internal pattern of C_{60} at positive sample bias is a bright pentagon ring plus three curved strokes to its right.

To identify the C_{60} molecular orientations on a Si(111)-(7 × 7) surface, the Tersoff–Hamann method was adopted by Hou *et al.* to simulate STM images. Cluster models were used to mimic an individual C_{60} adsorbed on Si(111)-(7 × 7), and in cluster models the Si dangling bonds, irrelevant to C_{60} adsorption, are saturated by hydrogen atoms. For C_{60} adsorbed on site A, the cluster model is $C_{60}Si_{57}H_{42}$, while it is $C_{60}Si_{67}H_{45}$ for the B site, respectively. The simulated STM images show that the positive-bias images depend strongly on the orientation of C_{60} on the substrate and less on the adsorption sites. On the other hand, the simulated negative-bias images have a weak dependence on either the adsorption site or the orientation, and all display four bright stripes.

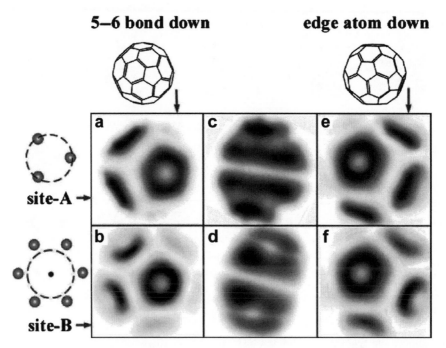

Fig. 15.8 Simulated STM images: (a) 5–6 bond adsorption on A site (2.5 V); (b) 5–6 bond adsorption on B site (2.5 V, the sample bias voltage); (c) 5–6 bond adsorption on A site (−1.8 V); (d) edge atom adsorption on B site (−1.8 V); (e) edge-atom adsorption on A site (2.5 V); and (f) edge-atom adsorption on B site (2.5 V). For all cases, the separation between tip and sample is about 5 Å (Hou *et al.* 1999). Copyright American Physical Society.

Comparing these simulations with the experimental observations, Hou *et al.* (1999) found that the experimental image of C_{60} on the A site best matches Fig. 15.8(a) (a bold pentagon ring plus two bright curved strokes to its left); while the image of C_{60} on the B site best matches with Fig. 15.8(f) (one bright pentagon ring plus three bright curved strokes to its right). Therefore, they conclude that that C_{60} is adsorbed on site B with one of the edge atoms facing towards the Si substrate, and has one of the 5–6 bonds facing towards the surface on site A. Moreover, Hou *et al.* have revealed that the positive-bias images mainly result from C_{60} single bonds bordering a hexagon and a pentagon, and hence the five-membered rings are highlighted. The negative-bias images, however, have contributions from both C_{60} double bonds and Si bonds. The difference results from the C_{60}–substrate interaction. Because of this strong interaction, the electronic structure of C_{60} molecule is modulated and the stripe-like internal structure appears in the negative-bias image.

Since an individual C_{60} adsorbs on a Si(111)-(7 × 7) surface with different orientations, this leads to the different local electronic structures. Wang *et al.* (1999) have proved that the local density of states of the adsorbed C_{60} molecules was site dependent based on their STS experimental measurements and the corresponding DFT calculations.

As STM relies on the non-zero conductance of its tunnelling junction to produce an image, almost all STM studies of individual molecules have been limited to molecules on metals or semiconductors. In these cases the electronic

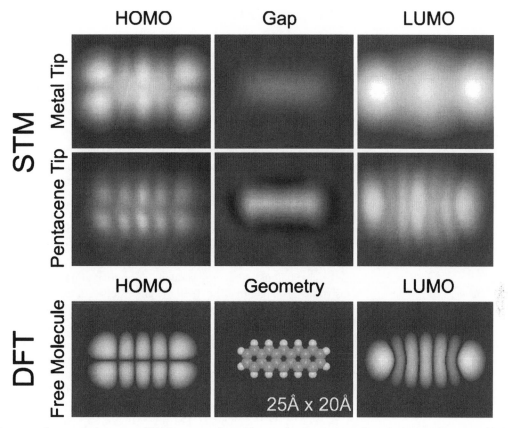

Fig. 15.9 From top to bottom, the images are STM images acquired with a metal and a pentacene tip, and contours of constant orbital probability distribution of the free molecule. Whereas the STM images for bias voltages in the HOMO–LUMO bandgap are relatively featureless (center), the images at bias voltages exceeding the HOMO (left) or LUMO (right) exhibit very pronounced features, resembling the electron density of the HOMO (left) and LUMO (right) of the free molecule. The geometry of the free pentacene molecule is displayed in the lower center image (Repp and Meyer 2005). Copyright American Physical Society.

structure of the molecules is perturbed by the presence of the substrate electrons. To understand the electronic properties of an individual molecule in mesoscale devices and for monomolecular electronics, an electronic decoupling of the molecules from the supporting substrate is therefore desirable. Repp and Meyer (2005) have reported that ultrathin insulating NaCl films can be used to decouple individual pentacene molecules from the metallic Cu(111) substrate. In their experiment Cu(111) single-crystal samples were cleaned by several sputtering and annealing cycles. NaCl was evaporated thermally, while the sample temperature was kept at 220 to 300 K, so that defect-free, (100)-terminated NaCl islands of up to three atomic layers were formed. Individual pentacene molecules were adsorbed on the upmost layer of NaCl at a sample temperature of $T = 5$ K. The corresponding STM images at voltages below -2.4 and above 1.7 V very closely resemble the native highest-occupied molecular orbital (HOMO) and the lowest-unoccupied molecular orbital (LUMO) of the free molecule in Fig. 15.9.

15.3.4   Self-assembled monolayers

Self-assembled monolayers (SAMs) have shown advanced applications in chemical sensing, biosensing, biomimetics, molecular electronics, and significantly contribute to the general understanding of the fundamental physics and chemistry of complex surfaces and interfaces. But little is known about the electronic properties of SAMs system, especially at the molecular level, even for the most typical example: alkanethiol SAMs on Au(111), which is of great importance to fully understand the SAMs–substrate–interface interactions. Actually, it is a challenge to fully understand the electronic properties of the insulating alkanethiol SAMs on the Au(111) surface, and to interpret the internal patterns in STM images due to the long insulating alkyl chain and almost standing molecular configuration.

Zeng *et al.* (2003) have carried out STM in a high vacuum at 78 K. The observed STM images show c(4 × 2) superlattices of a $\sqrt{3} \times \sqrt{3}$R30° hexagonal lattice. When the sample bias voltage ranges from 0.5 to 1.5 V, only bright intensity modulation of the spots in the hexagonal lattice can be resolved with one bright spot and three gray spots in a primitive unit cell. When the sample bias voltage is larger than 1.5 V, all the gray spots turn to stripe shapes, and the bright spots mainly keep their original shapes. When the sample bias voltage is changed to negative, the gray spots observed at low positive sample bias voltage turn to dumbbell shapes, and the bright spots are almost unchanged. No obvious variation is observed when the sample bias voltages vary from −0.5 to −2.5 V. The variations of the intramolecular patterns can only be explained by the twists (angle θ) of the all-*trans* hydrocarbon backbone along the molecular axis. These experimental observations were interpreted self-consistently by Li *et al.* (2003). The simulated STM images showed that for the sp mode the molecular pattern is always a bright spot at the sample bias voltage ranged from −2.5 to 2.5 V. For the sp^3 mode, when the sample bias voltage is in the range of 0 to 1.5 V, the intramolecular pattern is one spot with less bright intensity than that for the sp mode, which turns to a stripe shape when the sample bias voltage is larger than 1.5 V, and is always a dumbbell shape for the negative sample bias voltage. These experimental and theoretical studies show that all patterns in STM images reflect information of several groups in the alkyl terminal.

The chiral chemistry in two-dimensional (2D) molecules/substrate systems is of both technological and fundamental importance and has attracted great interest in the past decade. With the help of STM, various chiral phenomena such as chiral resolution, chirality amplification, chiral phase transition, and loss of chirality have been directly observed at the submolecular scale in 2D molecules/substrate systems. A 6-nitrospiropyran (SP6) molecule consists of two halves of near-planar heterocyclic fused rings (substituted indoline and chromene) connected at a sp^3 hybrid carbon atom. In solid-state SP6, due to the very small thermal barrier, left- (S) and right-handed (R) SP6 molecules can interconvert to each other spontaneously at room temperature, thereby forming a natural racemic mixture. The adsorption and chiral expression of SP6 molecules on a Au(111) surface have been investigated by Huang *et al.* (2007) with STM measurements and DFT calculations. An experimental STM image recorded at a sample bias of $+1.8\,V$ reveals the empty-state intramolecular patterns of adsorbed SP6 molecules. Each molecule within the SP6 rows is resolved into triangular geometry with fine structures, which can be classified into two types designated as A- and B-type, respectively. Each triangle appears as three lobes with distinct nodes in the A-type row, while each triangle is composed of an asymmetric *dumbbell* side and a very weak apex in the B-type row. Thus, four different triangular patterns in the 2D SP6 domain designated as A_1, A_2, B_1, and B_2 were found. Interestingly, A1 and A2 are mirror images of each other and so are B1 and B2. These experimental observed features were reproduced by DFT calculations with the Tersoff–Hamann model. Due to the remarkable conformity between simulated and experimental STM images, the chirality and the adsorption orientation of each adsorbed SP6 molecule are then determined. Simultaneously, the packing models of A- and B-type rows are obtained. Within each row SP6 molecules with opposite chirality pack alternately, arrange their long axis perpendicular to the row, attach to the Au(111) surface with the same function groups, and expose the same parts to be imaged. Each two neighboring molecules within an SP6 row display enantiomorphous shapes in the STM images. The chromene moieties of adjacent rows (A- or B-type) interdigitate into the clefts of each other to form the close-packed 2D structure.

The C_{60} native cage structure was seen in 2001 by Hou *et al*. In their STM experiments, to reduce the substrate effect, a self-assembled alkylthiol monolayer was inserted between the C_{60} molecules and gold substrate. The observed STM image is shown in Fig. 15.10(a). The C_{60} molecules form close-packed hexagonal arrays, with a nearest-neighbor distance of 10 Å. At room temperature, molecules at the edge of an array can detach readily and diffuse to another part of the same array or to other nearby arrays. Each C_{60} displays a smooth hemispherical protrusion, suggesting that the C_{60} molecules are rotating freely at this temperature. This contrasts with the image of C_{60} adsorbed on metal or semiconductor surfaces, when the molecular rotation is frozen even at room temperature owing to strong C_{60}–substrate binding. When the sample is cooled to 5 K, all C_{60} molecules in the STM image start to reveal an identical internal fine structure that closely matches the well-known cage structure.

The inset in Fig. 15.10(a) shows a simulated STM image of a C_{60} structure obtained by integrating the electron density of states on a C_{60} from the Fermi level to the bias voltage. The C_{60} orientation is tuned for the best agreement between the simulation and the experiment. The final molecule orientation has an edge atom facing towards the substrate and can be specified by the azimuthal ($\theta = 1.5°$) and polar ($\phi = 0.6°$) rotation, as defined in Fig. 15.10(b). This remarkable conformity between simulation and experiment allows us confidently to identify C_{60} domains with different molecular orientations.

Hou *et al*. (2001) have also observed a single array consisting of domains of two different orientations, in which the boundary separates two distinguishable domains by their internal features. Through model computations, Yuan *et al*. (2003) have shown that orientational ordering in a 2D C_{60} is drastically different from that in a C_{60} solid. The reduced dimensionality allows C_{60} molecules

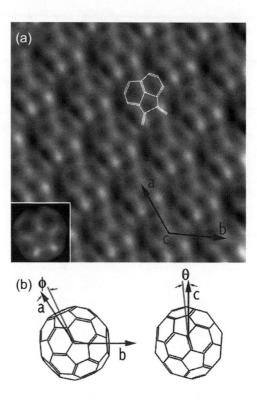

Fig. 15.10 (a) STM image (35 × 35 Å) of a C_{60} lattice taken at 5 K with −2.0 V sample bias. Detailed internal features of the C_{60} molecule are evident that closely resemble the C_{60} cage structure and match the theoretical simulation shown in the inset. (b) Top view (left) and side view (right) of a stick model outlining the C_{60} orientations obtained by simulation (Hou *et al.* 2001).

a greater degree of freedom in adjusting their mutual orientations. Although the interface orientations have lower symmetry than those in the bulk case, they better minimize the system energy and the domain boundary energies and lead to a deliberate uniorientational molecular order for a 2D C_{60} and a new topological order for the orientational domains.

15.3.5   Nanostructures

In order to gain new insights into the properties of novel types of nanostructures such as nanoparticles, nanoislands and ultrathin films, well-defined nanostructures have to be grown or deposited on equally well-defined substrates. It was already shown earlier by STM and STS that metallic nanoparticles exhibit fascinating size-dependent physical properties. Nanoparticles that are isolated from the metallic substrate by an artificial tunnel barrier, e.g. exhibit pronounced single-electron tunnelling phenomena with Coulomb charging and discretization of the electron energy levels. Nanoparitcles can be designed to be both amorphous and crystalline forms. In general, the features of STM images are very similar for these two types of configurations with the same size, but their electronic structures and transport properties are different. Since single-electron tunnelling (SET) has been proven to be a powerful technique, Hou *et al.* (2003) have conducted a comparative study of thiol-stabilized crystalline Pd (c-Pd) and amorphous Pd (a-Pd) nanoparticles by measuring the SET spectra of two types of Pd nanoparticles with a STM set up. For the *c*-Pd

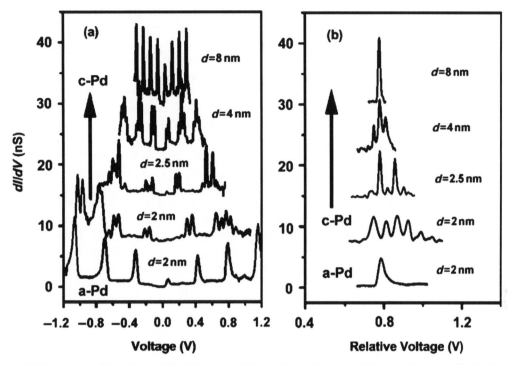

Fig. 15.11 (a) dI/dV spectra of c-Pd particles and an a-Pd particle. (b) Comparison of fine spectral features of the second Coulomb-blockade steps for various particle sizes. Here, the curves in (a) are shifted vertically, and peaks in (b) are shifted in the voltage coordinate for clarity (Hou *et al.* 2003). Copyright American Physical Society.

particles, it is evident from the top four curves in Fig. 15.11(a) that there is an increasing complexity in the spectral features as the particle size reduces. For the *a*-Pd particle of comparable sizes, however, the discrete energy level effect is completely suppressed. It is interesting to see in Fig. 15.11(b) that the shape of the SET peak of a 2-nm *a*-Pd particle appears to be closer to that of a *c*-Pd particle that is 4 times larger (8 nm in diameter). Moreover, one can see that the peak width increases with deceasing particle size. These observations suggest that the quantum effects associated with the quantized states are suppressed significantly in the *a*-Pd particle with a small size (about 2 nm).

An alternative route for studying the intrinsic properties of nanostructures is to grow or deposit them directly on a metallic substrate. Such configurations exhibit intriguing electronic phenomena such as surface-state scattering and effects due to the lateral confinement of electrons within nanosized islands, vacancy islands and corrals. The Au(111) surface reconstruction is an ideal candidate to use as a template for the controlled growth of nanostructures by atomic deposition. It grows in large atomically flat islands of over 1000 nm^2 and exhibits a remarkable type of surface reconstruction, known as the "herringbone" reconstruction. Till now, various materials, including Fe, Co, Ni and Cr, have shown island formation with mono-, bi- or multilayers on Au(111) surfaces. For example, Schouteden *et al.* (2008) have observed both strong shape and energy dependence of the electron-density variation with nanoscale bilayer Co islands on Au(111) surface by means of STM and STS measurements. They have also performed particle-in-a-box calculations, which reproduce the main features of their experimental results. The surface-state electrons are strongly confined laterally inside the Co islands, with their wavefunctions reflecting the symmetry of the islands. These observed standing-wave patterns are identified either as individual eigenstates or as a mixture of two or more energetically close-lying eigenstates of the Co islands.

Magnetism in one-dimensional (1D) systems has been the subject of continuous theoretical and experimental research. Progress in atomic engineering makes it possible today to build 1D arrays of transition-metal chains by self-assembly epitaxial techniques on suitable substrates. For example, Elmers *et al.* (1994) have designed Fe chains locating at the step of W(110) and Cu(111) surfaces. Gambardella *et al.* (2002) have demonstrated that both short- and long-range ferromagnetic orders for one-dimensional monatomic chains of Co formed on a Pt substrate exist. The Co monatomic chains consisting of thermally fluctuating segments of ferromagnetically coupled atoms (below a threshold temperature) evolve into a ferromagnetic long-range ordered state owing to the presence of anisotropy barriers. Using spin-polarized STS, Pietzsch *et al.* (2006) have revealed how the standing-wave patterns of confined surface-state electrons on top of nanometer-scale ferromagnetic Co islands on Cu(111) are affected by the spin character of the pertinent state. Their experimental results confirmed the theoretical predictions of Niebergall *et al.* (2006). Moreover, Pietzsch *et al.* (2006) found that at the rim of the islands a spin-polarized state enhances the zero-bias conductance and this polarization is opposite to that of the islands.

The property of ultrathin magnetic films and nanostructures on non-magnetic metal surfaces has also been a hot research topic. This is due to their unique properties such as enhanced magnetic moments, modified magnetic anisotropy arising from the interface, coupling phenomena, etc. The understanding of how the magnetism is governed by the electronic structure may help to control the magnetic properties of nanostructures, which is important to advance magnetic-storage technology and other magnetoelectronics applications. Diekhöner *et al.* (2003) have investigated the surface states of cobalt nanoislands on Cu(111). A strong localized peak located at 0.31 eV below the Fermi level and a mainly unoccupied dispersive state were found in their experiments. Two observed surface states originate from $3d_{Z^2}$-minority and spin-polarized majority bands, respectively.

It is well known that the properties of a material at the nanometer scale will be dominated by quantum effects. In the past decade, many studies have been focused on the correlations between the properties and the size, shape, and composition of nanostructures. Guo *et al.* (2004) have reported the first definitive and quantitative demonstration of quantum size effects (QSE) on superconductivity in the Pb/Si(111) system with atomically flat films. By using a low-temperature growth method, uniform Pb films with precisely controlled thicknesses in terms of atomic layers, were prepared on Si(111)-7×7 substrates. As a result, the quantum oscillations in superconducting transition temperature were observed, which correlate perfectly with the confined electronic structure. Other novel properties induced by QSE, such as bilayer growth, selective strip-flow growth and growth-rate modulation, oscillating perpendicular upper critical field, thermal expansion, adhesion force, local work function, and surface diffusion barrier, have also been observed. These activities of QSE in metal thin films were reviewed critically by Milun *et al.* (2002) and Jia *et al.* (2007).

15.4 Beyond conventional STM investigations

In this section, we turn to the STM activities using advanced capabilities, which go beyond the conventional STM images and STS. For example, the recent progresses on the functionalized STM tip, dI/dV mapping, inelastic spectroscopy identification, manipulation, molecular electronics and single-molecule machine, are briefly introduced in the following subsections.

15.4.1 Functionalized STM tip

In STM experiments, the status of the tip is crucial and determines the contrast of pattern as well as the quality of the obtained images. Different STM images are accidentally and often observed on the same area with the same parameters of measurement but different scans, which should result from occasional changing of the tip terminal structure when the tip scans. Many studies have been done on the effects of different tips on STM images. Very recently, Chen *et al.* (2007a) have measured I–V curves of cobalt phthalocyanine (CoPc) molecules on a Au(111) surface with W and Ni tips. For two different metal tips used, Ni and W, negative differential resistance (NDR) occurs only with Ni tips and shows no dependence on the geometrical shape of the tip. A new mechanism for originating from local orbital symmetry matching between the electrode and the adsorption molecule in a single-molecular electronic device has been proposed and demonstrated by a joint experimental and theoretical STM study.

Theoretical studies showed that the electronic structure of an idealistic tip should be of $m = 0$ (m is the magnetic quantum number) type orbitals, and using this kind of tip one can obtain images that mainly reflect information of the sample. But various occasional events, such as the atomic displacements in the tip terminal, may occur during the tip scanning and thus result in a change of electronic structure of the tip. Another origin of tip effects is transfer of sample components onto the tip, which makes the electronic states of sample components adsorbed onto tip become active tunnelling states of the new "tip" and influences the internal pattern of the STM image. Of course, the tip approaching the sample may induce changing STM images when there is a strong tip–sample interaction.

In general, theoretical work, including electronic structure analyses for every subsystem in STM and simulation for STM images, are important and even necessary to understand STM images and related imaging mechanisms. For an $O(2 \times 2)/Ru(0001)$ surface, in order to quantify the influence of the tip electronic structure in the images, Calleja *et al.* (2004) have simulated the effect of an O atom at the tip apex. An image recorded in standard conditions with such an oxygen-terminated tip reveals the existence of vacancies

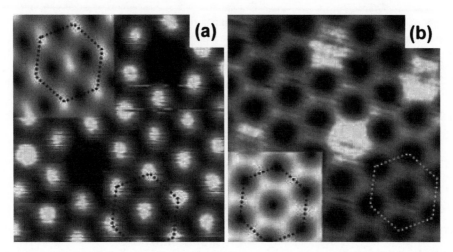

Fig. 15.12 (a) STM image recorded with an oxygen atom at the tip apex. The sample voltage was −0.35 V. (b) STM image taken with a W tip at −0.1 V. The size of STM images is about 4 nm × 5.2 nm. The insets in (a) and (b) show the images simulated with an O-terminated tip displaying the inversion of contrast (Calleja *et al.* 2004). Copyright: American Physical Society.

(missing the maxima) in a 2 × 2 hexagonal pattern of bright bumps as shown in Fig. 15.12(a). The inset in Fig. 15.12(a) shows the simulated image obtained with an O-terminated tip, which confirms that the O atoms are now imaged as maxima. This indicates that the bumps now reflect the position of oxygen. The contrast has thus been inverted with respect to images obtained with a W-terminated tip, such as shown in Fig. 15.12(b). This reversal STM originates from the different tip electronic structures. The LDOS of the W tip is much sharper than that of the O tip. Moreover, it is an order of magnitude larger at the same distance. As a result, the O tip is located around 1 Å closer to the surface than the W tip. In addition, there is a relative minimum in the LDOS at the apex of the O tip. Because of this, when the O tip is located right above an O atom of the surface, the tunnelling current from the neighboring Ru atoms is the maximum and the simulations predict a complete change of contrast, i.e. the oxygen depressions would turn into protrusions.

In recent years, several research groups have performed STM experiments in which the chemically modified tip was used to identify molecule adsorbed on substrate. For example, C_{60}-functionalized metal tips (Kelly *et al.* 1996), and carbon-nanotube tips (Dai *et al.* 1996), and heteropolyacid-functionalized Pt–Ir tip (Song *et al.* 2002) have demonstrated the potential of controlling the chemical identity and geometric structure of tip atoms in STM. Such a molecule-terminated tip yields enhanced resolution in topographical imaging of substrate atoms and molecular orbitals of molecules adsorbed on the surface. Bartels *et al.* (1998) found that electrons tunnelling from a STM tip to individual CO molecules on a Cu(111) surface can cause their hopping from the surface to the tip if the bias exceeds a threshold of 2.4 V. There is a depressed region at the position of CO molecule in the STM image when using an ordinary tip, but when using a chemically modified CO tip, a protrusion appears at the center of the original depressed region in image. Ho's group (Hahn and Ho 2000; Hahn *et al.* 2001) have also observed some small molecules and atoms by using this kind of tip, and obtained a series of STM images from which the positions and orientations of adsorbates can be inferred. Of course, the STM experiment using these chemically modified tips, which possess special electronic states, can bring new phenomena and functions that do not appear in conventional STM experiments. For example, Zeng *et al.* (2000) have shown that an NDR molecular device can be realized involving two C_{60} molecules, one is adsorbed on the STM tip and the other is on the surface of the hexanethiol SAM. The observed NDR effect comes from the narrow local density of states features near the Fermi energy of the C_{60} molecules.

15.4.2 dI/dV mapping technique

Since a normal STM topological image reflects the integrated LDOS of the sample from the Fermi level to the sample bias voltage V, a dI/dV spatial map is approximately proportional to the LDOS at the energy of eV. The dI/dV mapping, a very powerful technique, has been used to determine the local spatial and energy-resolved electronic structures of various systems. For example, the spatial variations in C_{60} spectral density have been mapped by Lu *et al.* (2003) using energy-resolved STM mapping (dI/dV) technique for the first time combined with DFT calculations. This kind of combination of experiment and theory has general implications for the interpretation of STM spectral maps of electronically inhomogeneous molecular systems. Almost at the same time, the localized pentagon–heptagon pair defect states in carbon nanotubes were spatially observed by Kim *et al.* (2003) using the atomically resolved LDOS (STS).

Electronic states of surfaces are a major subject in surface science because of their importance in fundamental science and device applications. As a classical example of metal-induced silicon reconstruction, Si(111)-$\sqrt{3} \times \sqrt{3}$-Ag has been investigated intensively. With the help of the dI/dV mapping technique, Chen *et al.* (2004) have detected its surface states. Besides three well-known surface bands of S_1, S_2 and S_3, two surface resonance (SR) bands were discovered. It was found that one of the SR states lies in the energy gap between the S_1 and S_2 bands and the other one lies deep into the valence bands. The existence of these new SR bands was confirmed by their first-principles calculations. It should be pointed out from the Chen *et al.* (2004) studies that the correct information about surface states must be acquired via adopting a slab model with sufficient Si double layers.

Of course, it is very interesting to explore the vacancy-induced surface state by using the powerful dI/dV mapping technique. For example, Chen *et al.* (2007b) have employed a low-temperature STM to investigate the energy-resolved electronic structure of the Si(111)-7×7 surface containing ad-atom vacancies. A typical electronic state associated with the single ad-atom vacancy in the surface is experimentally found to be at about 0.55 eV below the Fermi energy level, which is attributed to the two upward backbone atoms around the ad-atom vacancy in the dimer-ad-atom-stacking-fault reconstruction after comparing the measured dI/dV images and curves with their calculated local electronic structures. More interestingly, they also found that ad-atom vacancies can induce the images of some rest atoms to be invisible in the dI/dV maps.

Endohedral metallofullerenes have been a subject of intensive investigation in recent years, not only because of their structural and electronic novelties but also because of their promising electronic, optical, and biomedical applications. The location of metal atoms inside the fullerene cage and the metal–cage interaction are two central issues. Various diffraction, spectroscopy, and microscopy techniques have been used to characterize metallofullerenes. Most of them require macroscopic quantities of metallofullerenes and give either ensemble-averaged or spatially averaged results. It has been a challenging task to characterize the local properties of isolated metallofullerene molecules. Wang *et al.* (2003) have shown that STM can be used to detect the encapsulated metal atom inside a fullerene cage. To systematically characterize the structural and electronic properties of individual $Dy@C_{82}$ molecules, they took one adsorbed $Dy@C_{82}$ molecule as an example to measure its STM images and dI/dV spatial maps with different bias voltages. The internal pattern of the STM images depends strongly on the bias voltages. At the negative voltage, the molecule appears as several slightly curved bright stripes, while the large positive-bias STM images show some bright pentagon and hexagon rings. Since the $Dy@C_{82}$ molecule has many non-equivalent sites due to its low symmetry, the orientation of $Dy@C_{82}$ cannot be deduced with these STM images alone. The experimental dI/dV spatial maps of $Dy@C_{82}$ at most bias voltages show either a fully dark hollow or a bright mesh structure. Only at certain positive bias voltages (i.e. 2.0 and 2.1 V), Wang *et al.* (2003) observed a locally bright ring or dot. These dI/dV maps reveal details in the electronic structures that cannot be observed in topographs.

In order to understand qualitatively the above experimental observations, Wang *et al.* (2003) have studied the electronic structures of an isolated $Dy@C_{82}$ molecule using DFT calculations. The orbital population analysis shows that the molecular orbitals of $Dy@C_{82}$ can be divided into three types: cage-dominated (type I), metal-dominated (type II), and metal–cage hybrid (type III) orbitals. In the occupied valence-band region, only types I and II orbitals exist. In the unoccupied conduction-band region, however, there are two metal–cage hybrid orbitals at around 2 eV above the HOMO. In these orbitals, there is a strong hybridization between the Dy 6s orbital and the 2s and 2p orbitals of some specific C atoms. This kind of hybridization can result in a very local distribution of the LDOS on the fullerene cage. Since type-I orbitals are irrelevant to the encapsulated metal atom, the LDOS of these orbitals show no direct information on the metal atom position in the fullerene cage. Type-II orbitals belong to the metal atom, but they are localized inside the fullerene cage, and the LDOS of these orbitals cannot be detected by the STS. In order to detect the metal atom in the fullerene cage, mapping the metal–cage hybrid orbitals becomes the crux of the matter. In principle, one can determine where the metal atom is in the cage, hence, the molecular orientation on the surface by matching the simulated LDOS distribution of the hybrid orbitals and STM images with the experimental dI/dV maps and STM images. According to their DFT calculations, Wang *et al.* (2003) found that the existence of the metal–cage hybrid states in metallofullerenes is a universal phenomenon. The experimental and theoretical observations of the metal–cage hybrid states support the view that the orbital hybridization and charge transfer constitute a complete picture for the interaction between the cage and metal atom. Detecting the energy-resolved metal–cage hybrid states of a single endohedral metallofullerene provides not only the information for the metal–cage interaction, but also the geometrical and orientational information about the metallofullerene. This technique may find interesting applications in the fields of *in-situ* characterization and diagnostics of metallofullerene-based nanodevices.

Grobis *et al.* (2005) have presented a unified experimental and theoretical inelastic electron tunnelling spectroscopy (IETS) study of isolated Gd@C82 molecules on a Ag(001) surface. In their experiments, several elastic and inelastic conductance channels were well resolved. They found that spatial dI/dV mapping of the dominant inelastic channel is highly localized to a small area of the molecule's surface and the inelastic channel arises from a vibrational cage mode. This study suggests that spatial sampling significantly influences the number of channels detected via inelastic tunnelling.

15.4.3 Inelastic electron tunnelling spectroscopy identification

IETS was first developed by Jacklevic and Lambe in 1966. They observed that tunnelling electrons were able to excite vibrational modes of a thin molecular layer buried between two metallic electrodes and an oxide layer (i.e. a tunnelling barrier). The vibrational excitation occurs when the tunnelling electron energy matches with that of a vibrational eigenmode. Upon activation of these inelastic scattering processes an additional transport channel is opened, inducing a slight increase of the conductivity of the tunnel junction. Within the past decade, STM has been proposed as a strategy to obtain the vibrational structure of a single molecule based on a technique called IETS.

In general, molecules adsorb on an atomically clean metal surface with a well-defined geometry. The adsorption geometry and orientation of the adsorbate can be well known. The energy of the tunnelling electron is donated to the adsorbate through the scattering with molecular states. In a molecule/metal chemisorption system, the surface and molecular states are strongly coupled, causing tunnelling transitions typically faster than nuclear motions. Fast ionic transition states do not favor the energy-redistribution processes. Thus, it is possible to find that one specific vibration can be excited simply by adjusting the sample bias voltage above to its frequency. Note that there are important differences between inelastic electron tunnelling spectroscopy with STM (STM-IETS) and traditional IETS, related to the nature of the excitation mechanism. In the STM configuration, the majority of the electron current tunnels though the adsorbate itself, therefore having a strong component of molecular-derived resonances. This results in local inelastic scattering mechanisms strongly mediated by the electronic configuration of the molecule–surface complex. In IETS, however, the tunnelling current is less focused on

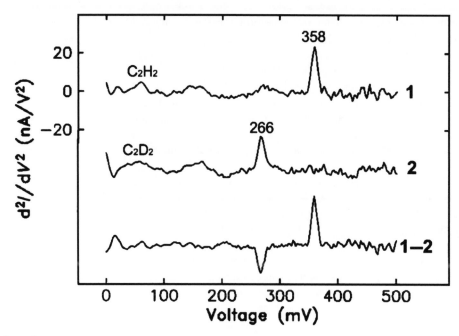

Fig. 15.13 The d^2I/dV^2 spectra for C_2H_2 (1) and C_2D_2 (2). The difference spectrum (1–2) yields a more complete background subtraction. The peaks at 358 mV and 266 mV correspond to the C–H and C–D stretch vibrational modes, respectively (Stipe *et al.* 1998b). Reprinted with permission from AAAS.

the adsorbate and the excitation follows predominantly a non-local mechanism as, for example, dipolar interaction between electron and molecule.

The resolution of inelastic effects depends on two fundamental mechanisms: excitation and detection. The excitation is based on inelastic scattering processes, thus connecting initial and final states with different energy. It is very sensitive to the nature of the interaction between adsorbate and surface. The detection relies on the effect of the new (inelastic) channel on experimentally observable magnitudes, i.e. the junction differential conductivity. In practice, the change of the total conductance due to inelastic processes is less than 10%. To help with the detection of such a small signal, the second derivative of the tunnelling current (d^2I/dV^2) is usually measured. This magnitude shows peaked features at energies corresponding to the opening of an inelastic channel. To detect the weak d^2I/dV^2 signal lock-in techniques are usually employed.

Stipe *et al.* (1998b) have shown the first successful detection of a vibrational mode using a STM. The C–H and C–D stretch modes of acetylene adsorbed on a Cu(100) surface were measured by the IETS technique to 358 and 266 meV, respectively, as shown in Fig. 15.13. Actually, the C–H stretch has been detected in several hydrocarbons by Ho *et al.* (2002) and Kim *et al.* (2002). By combining the STM-IETS with the atomic-scale spatial resolution of the STM image, the sample atomic structure may be determined in real space with chemical specificity. Lee and Ho (1999) have demonstrated that this kind of technique can be used to characterize a single bond. Their measured vibrational frequencies agree well with the calculated results by Yuan *et al.* (2002). Now, the STM-IETS technique is widely used to probe the vibrational properties of single molecules. The energies and intensities of the spectral features are characteristic of the chemical bonds and their interactions with the surrounding medium. Chemical sensitivity can thus be obtained. In addition, the molecules are not perturbed in STM-IETS since low-energy tunnelling electrons are involved.

Bocquet *et al.* (2006) have performed the first-principles IETS simulations on the identification of the dehydrogenation products of benzene chemisorbed on Cu(100). Their results show that the product of STM-induced dehydrogenation of a benzene molecule is a phenyl (C_6H_5) molecule rather than the initially assigned benzyne (C_6H_4) molecule in the experiments conducted by Lauhon *et al.* (2000). Recently, Katano *et al.* (2007) have successfully broken the N–H bond of a methylaminocarbyne ($CNHCH_3$) molecule on a Pt(111) surface to form an adsorbed methylisocyanide molecule ($CNCH_3$). The $CNCH_3$ product was identified using the STM-IETS measured vibrational spectrum and the DFT calculations. These studies imply that a thorough DFT analysis permits one to trace back the origin of the IETS spectra to the underlying vibrational modes, molecule–substrate electronic states and electron–vibration couplings. The quality of the experiment–theory agreement is such that this type of simulation can be developed to routinely analyze and identify unknown products in STM-IETS experiments. Clearly, a strong combination of experimental measurements and theoretical simulations must be used in order to understand the fundamentals of STM-IETS. It should be pointed out that a major drawback of STM-IETS is the lack of sensitivity for many of the molecular modes of a molecular–surface system, which remain either inactive or undetected. There is no relation with excitation selection rules from other vibrational spectra like IR, Raman, or EELS. The initial results of STM-IETS point to the fact that active modes depend on the symmetry and configuration of the adsorption geometry. Although the exact origin of selection rules for excitation and detection is unknown, there is general agreement that it must be related to the resonant character of tunnelling transport.

15.4.4 Modification and manipulation

The central aim of nanotechnology is to controllably modify material properties on a nanometer scale. Now, STM is the powerful and unique technique to manipulate single atoms and molecules on various surfaces as well as the surface atomic structures. Stipe *et al.* (1998) have studied the reversible rotations of O_2 molecules among three equivalent orientations on a Pt(111) surface by applying bias voltage pulses via a STM tip. This kind of controlled motion of a single molecule enhances our understanding of the basic chemical processes involved.

Kliewer *et al.* (2001) have constructed artificial structures via controlled relocation of individual Mn atoms, an array of 28 Mn atoms forming a rectangle on Au(111) surface. The Mn adatoms appear as bright protrusions in all cases. Inside this structure a weak corrugation is found, originating from the confinement of the surface-state electrons. At negative bias voltages (i.e. when

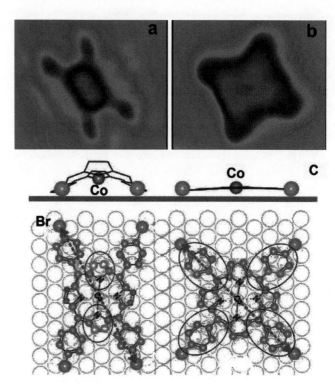

Fig. 15.14 Molecular conformations. (a) STM images of saddle conformation (the width is about 11 Å, while the length is about 18 Å) and (b) planar conformation (the length is about 15.5 Å) of TBrPP-Co on Cu(111). (c) The corresponding models of saddle (left) and planar (right) conformation. The encircled regions of the molecule indicate the area providing higher current in the STM images (Iancu *et al.* 2006a). Copyright American Chemical Society.

tunnelling from occupied levels in the sample) the constant-current topographs exhibit a rectangular pattern of four maxima within the confining array of Mn atoms. At positive voltages, imaging unoccupied states, the pattern becomes more complex and varies with bias voltage. These experimental features are reproduced well by their model calculations. It should be pointed out that the experimental dI/dV maps are not directly comparable to LDOS maps. The reason is that these experimental dI/dV data recorded by the methods used are affected by the variations in tip–sample separation, which give rise to the corrugation in the topographs.

The conformational structure of molecules adsorbed on surfaces is dictated by the subtle balance between the molecule–substrate interactions and the internal mechanics of the molecules. Understanding this complex interplay and being able to control the conformational structures of adsorbed molecules would provide a deeper understanding of chemical dynamics on surfaces and enable molecule-based applications in functional devices. Recently, by applying certain bias voltage pulses through the STM tip, the molecular conformation has been manipulated via an electrically induced mechanism by several groups (Qiu *et al.* 2004; Iancu *et al.* 2006a). The 5,10,15,20-tetrakis(4-bromophenyl) porphyrin (TBrPP-Co) molecules anchor on the Cu(111) surface via their four bromine atoms positioning at the 3-fold hollow sites of the copper surface and form two molecular conformations: the saddle and the planar, as shown in Fig. 15.14. They found that on the Cu(111) substrate 25% of the molecules were adsorbed in the saddle conformation. In the saddle conformation, the central part of the molecule is bent by the elevation of two pyrrole units of the porphyrin macrocycle. The STM images acquired at a 4.6 K substrate temperature in a UHV environment show two protrusions for the saddle originated from the two lifted pyrrole units, as shown in Fig. 15.14(a). If the STM tip is positioned at a fixed height above the center of a saddled molecule, and a fixed voltage pulse of +2.2 V is applied for a few seconds, the TBrPP-Co molecule can be switched from the saddle to the planar conformation. The STM images of planar TBrPP-Co reveal a set of four lobes, as shown in Fig. 15.14(b). This observation suggests that the planar molecule has an approximately square shape, and the porphyrin plane is positioned parallel to the Cu(111) surface and four bromophenyl groups of the molecule can interact strongly with the surface via π-interactions.

Iancu *et al.* (2006a) further investigated the localized spin–electron interaction between isolated TBrPP-Co molecules and the free electrons of Cu(111) surface by measuring the energy dependence of the LDOS around the Fermi level via differential tunnelling spectroscopy (dI/dV) over both molecular conformations. For the shape and width of the measured resonant peaks near the Fermi level, the Kondo temperature of saddle and planar TBrPP-Co at 4.6 K is about 130 K and 170 K, respectively. Clearly, the Kondo temperature of TBrPP-Co with the saddle configuration is lower than that of the planar one. Both molecule conformations bind to Cu(111) via bromine atoms, indicating that the spin–electron coupling through molecular binding is permitted in two cases. However, for the planar conformation, there is another spin–electron coupling path: a direct coupling of Co d_{z^2} orbital to the substrate, which results in the higher Kondo temperature of TBrPP-Co with the planar conformation.

It is clear that the magnetism of a single-atom or single-molecule regime is an especially interesting research topic, because the magnetism arises in this case from very few unpaired electronic spins and is thus quantum mechanical in nature. This property opens up new opportunities that range from basic quantum impurity studies to quantum information and spintronics applications. Moreover, molecular systems provide a useful means for "packaging" quantum spin centers. Zhao *et al.* (2005) have successfully manipulated the magnetism of CoPc molecules on a Au(111) surface by tailing the molecular structures. In their experiments, single CoPc molecules adsorbed on the terraces of an Au(111) surface exhibit a protruding four-lobed structure, as shown in Fig. 15.15(b).

Dehydrogenation of a CoPc molecule was realized with a given bias voltage pulse from the STM tip. A typical current trace simultaneously measured during the application of a 3.6-V pulse on one of the four lobes of a CoPc molecule shows two sudden drops in the current signal, indicating the sequential dissociation of the two H atoms from the benzene ring. The dehydrogenation threshold voltage is in the range of 3.3 to 3.5 V, depending on the structure of the tip apex. Topographic images of the dehydrogenation product show that the bright lobes disappear sequentially. When all four lobes (these eight end hydrogen atoms) are cut, the dehydrogenated CoPc (d-CoPc) molecule is obtained and the STM image is shown in Fig. 15.15(d) with a bright spot. The height of the central Co atom in d-CoPc case increases by about 0.8 Å. These experimental observations indicate either a strong conformational change of

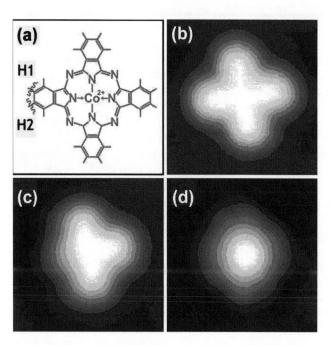

Fig. 15.15 (a) Structure formula of the CoPc. The end hydrogen atoms, such as H1 and H2, can be dissociated by a local bias voltage pulse via a STM tip. (b) and (c) STM images, (b) a single CoPc, (c) for a CoPc with cutting two end hydrogen atoms, (d) stands for d-CoPc (Zhao A.D. *et al.* 2005). Reprinted with permission from AAAS.

the molecular structure or a redistribution of the local density of states of the molecule.

To understand the experimental results, Zhao A.D. *et al.* (2005) have carried out DFT calculations on the structural and electronic properties of CoPc and d-CoPc molecules adsorbed on Au(111). The spin-polarized partial density of states (PDOS) of the Co atom in a free CoPc molecule, in the CoPc and d-CoPc adsorption systems is shown in Figs. 15.16(a), (b), and (c), respectively. The calculated DOS reveals that the spin-down states are filled more than the spin-up states for the free CoPc molecule as shown Fig. 15.16(a). It is easy to understand this since the Co atom has unpaired d electrons. The calculated magnetic moment of the Co atom is 1.09 Bohr magnetons. In the CoPc adsorption system, they found that a CoPc molecule adsorbs on Au(111) surface with a nearly planar configuration and the distance between the molecule and the gold substrate is 3.0 Å. This molecule–substrate interaction clearly changes the electronic structure and magnetic property of the CoPc molecule. The magnet moment is completely quenched since the filling difference disappears for the spin-down and spin-up states in CoPc adsorbed on Au(111) system, as shown in Fig. 15.16(b).

Dehydrogenation results in a marked change of the adsorption structure. The d-CoPc molecule on a Au(111) surface is no longer planar. The distance between the edge carbon atom and the Au substrate is about 1.9 Å and the central Co atom shifts upwards obviously (3.8 Å). The filling difference between the spin-down and spin-up states (Fig. 15.16(c)) indicates that the magnetic moment is recovered for the d-CoPc adsorption system. As seen from Fig. 15.16(c), the spin-polarized PDOS of the Co atom in the d-CoPc adsorption system near the Fermi level has an empty minority spin peak that comes

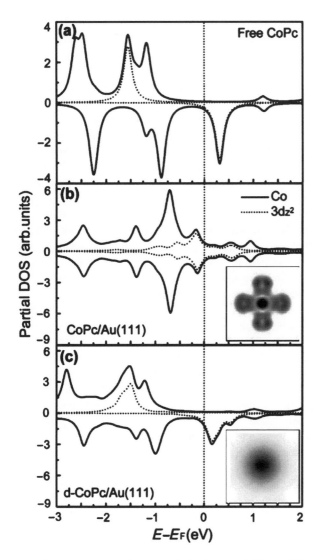

Fig. 15.16 The calculated partial DOS. (a) Free CoPc molecule, (b) CoPc molecule on Au(111) surface, (c) for d-CoPc on Au(111). Here, the solid and short dotted lines stand for the partial DOS of $3d$ and $3d_z{}^2$, respectively. The insets in (b) and (c) are simulated STM images for CoPc and d-CoPc molecules on Au(111) surface, respectively (Zhao A.D. *et al.* 2005). Reprinted with permission from AAAS.

from the magnetic quantum number $m = 0(d_{z^2})$ states. This peak is consistent with their experimental spectra measured at different temperatures, in which an observable peak appears near 135 meV (Zhao A.D. *et al.* 2005). The magnetic moment of the d-CoPc molecule is now 1.03 Bohr magnetons, very close to the value of a free CoPc molecule. As seen in insets in Figs. 15.16(b) and (c), the simulated STM image with the Tersoff–Hamann formula displayed with four bright lobes and a large bright spot for the CoPc and d-CoPc molecules on Au(111) surface, respectively. It is clear that the simulations agree quite well with the observed images (Figs. 15.15(b) and (d)).

Since Zhao A.D. *et al.* (2005) have elevated the Kondo temperature of CoPc on Au(111) based on the dehydrogenation processes, now, various experimental schemes have been explored to control the Kondo effect at a single atomic or molecular level. For example, Wahl *et al.* (2005) have demonstrated the ability to tune the coupling between the spin of individual cobalt ad-atoms with their surroundings by controlled attachment of CO molecular ligands. Their results showed that the coupling of the spin to the substrate conduction electrons can be enhanced through increasing the number of ligands. The Kondo temperature increases from 165 to 283 K when the number of CO molecules changes from 2 to 4. Iancu *et al.* (2006a) have reported that the Kondo temperature of TBrPP-Co (5, 10, 15, 20-Tetrakis-(4-bromophenyl)-porphyrin-Co, a porphyrin unit with a cobalt (Co) atom caged at its center and four bromophenyl groups at the end parts) molecules on Au(111) surface increase from 105 to 170 K on decreasing the number of nearest-neighbor molecules from six to zero. This observation indicates that the spin–electron coupling of the central molecule can be tuned by manipulating nearest-neighbor molecules with a STM tip in a controlled manner. Iancu *et al.* (2006a) have also found that TBrPP-Co molecules on a Cu(111) surface can be switched from the saddle to a planar molecular conformation by a bias voltage pulse with a STM tip, which enhances the spin–electron coupling and results in increasing the associated Kondo temperature from 130 to 170 K. This result implies that the Kondo temperature can be manipulated just by changing the molecular conformation without altering the chemical composition of the molecule.

Gao *et al.* (2007) have indicated that for an iron phthalocyanine (FePc) molecule a Kondo temperature is well above room temperature and the resonant signal depends on the adsorption site of the molecule on a Au(111) surface. Fu *et al.* (2007) have shown a more technologically feasible way to modulate the Kondo resonance by the quantum confinement of a nanostructure whose physical size can now be precisely and conveniently tuned with current nanotechnology. The Kondo resonance (Kondo temperatures) of individual manganese phthalocyanine (MnPc) molecules adsorbed on the top of Pb islands spectroscopy oscillated as a function of Pd film thickness, which was attributed to the formation of the thickness-dependent quantum-well states in the host Pb islands. Since the quantum size effect in precisely tailored nanostructures provides a general platform to manipulate the spin degree of freedom at a single-molecule level, this controlled approach can open a new avenue for the study along this direction.

15.4.5   Molecular electronics and single-molecule machine

Stimulated by its potential applications in the field of nanoscale electronic devices, the charge transport through a single molecule is a rapidly developed field (Joachim *et al.* 2000; Ratner 2002). With the help of the various experimental techniques such as STM and the mechanical break junction, the transport behaviors of the molecular wires based on small conjugated molecules, carbon nanotubes, fullerene, and DNA molecules have been investigated. In a STM setup, the typical distance between tip and sample is about 6 Å. This implies that these junctions always have the built-in asymmetry (leading to the asymmetric *I–V* characteristics). DFT is not good at dealing with this kind of weak tip–sample coupling. Therefore, the developed non-equilibrium Green's function techniques (Taylor *et al.* 2001; Brandbyge *et al.* 2002) based on DFT are not suitable to address the STM-measured transport properties.

Various molecular systems exhibit some amazing current–voltage (*I–V*) characteristics, including molecular switching and rectification, negative differential resistance, and a molecular transistor. Among them, as the simplest unit, a molecular rectifier should play a key role in the development of future molecular electronic components. Since the first molecular diode was proposed by Aviram and Ratner (1974), various molecular rectifiers or diodes have been synthesized and their rectifying properties have been measured in the past decades. Recently, a new class of relatively simple diode molecules has been synthesized based on conjugated diblock oligomer molecules by Yu's group, such as the molecules containing two thiophene (C_4S) and thiazole (C_3NS) or pyrimidinyl (C_4N_2) and phenyl (C_6) units at both ends (Ng and Yu 2002; Ng *et al.* 2002; Jiang *et al.* 2004; Morales *et al.* 2005). The molecules chemisorb on metal substrate via the S–Au bonds and metallic nanoparticles are deposited on the above molecules. Subsequently tunnelling is probed over the nanoparticles by an STM tip under external bias voltage. Importantly, it is claimed that the experimental observed rectification effect does not come from the built-in chemical asymmetry, but it is an intrinsic property of the molecule.

Wang *et al.* (2006) have built a donor-barrier-acceptor (D-σ-A) architecture based on pyridyl aza[60]fulleroid oligomers, abbreviated as $C_{60}NPy$ using STM. They found that the HOMO and LUMO are well localized either on the Py moiety (donor) or on the C_{60} moiety (acceptor), indicating the σ-bridge decouples the LUMO and the HOMO of the donor and the acceptor, respectively. This structure accords well with the unimolecular rectifying Aviram–Ratner model. By directly comparing the experimental conductance peaks and the calculated density of states of the $C_{60}NPy$, they revealed that the observed rectification is attributed to the asymmetric positioning of the LUMOs and the HOMOs of both sides of the acceptor and the donor of the $C_{60}NPy$ molecules with respect to the Fermi level of the electrodes. Zhao J. *et al.* (2005) have also realized a new type of single-molecular rectifier device based on $C_{59}N$. In their experiments, the substrate was prepared by self-assembling high-quality monolayer alkanethiols (SAM) on Au(111) surface. The experiments were conducted on a double-barrier tunnel junction formed by positioning a STM Pt-Ir tip above a $C_{59}N$ that resides on top of the SAM on Au(111). The positive onset voltage is about 0.5–0.7 V, while the negative onset voltage is about 1.6–1.8 V. A typical measured *I–V* curve is shown in Fig. 15.17. The orthodox theoretical analyses combining with DFT calculations show that the half-occupied Fermi level of the neutral $C_{59}N$ molecule and the asymmetric shift of the Fermi level when the molecule is charged are responsible for the molecular rectification.

It should be pointed out that, in molecular electronics based on STM, one has to carefully deal with several fundamental questions, such as, the molecular structure within the applied electrical field, the vibrational coupling, temperature effect, the band lineup, electronic correlation, dissipation mechanism, cooperative behaviors, etc.

Molecular machines, and in particular molecular motors with synthetic molecular structures and fuelled by external light, bias voltage or chemical conversions, have recently been reported by Hernandez *et al.* (2004), Keeling *et al.* (2005), Pijper *et al.* (2005), Fletcher *et al.* (2006),

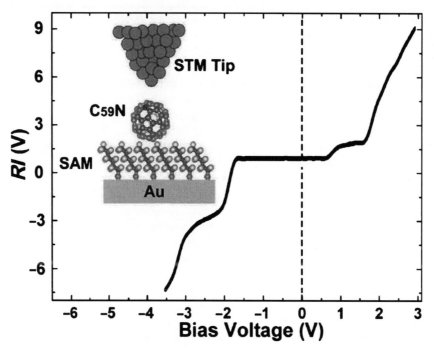

Fig. 15.17 A typical *I–V* curve for an individual $C_{59}N$ molecule residing on top of the SAM on a Au(111) surface measured at 5 K (Zhao J. *et al.* 2005). Copyright American Physical Society.

and Chiaravalloti *et al.* (2007). The design of a single-molecule machine consisting of functional components requires a detailed understanding of its mechanical motion. The STM is the only available tool for driving and imaging such a nanoscale machine on a surface. Both lateral hopping motions and conformational changes of single molecules can be induced using the STM tip (Jung *et al.* 1996). Keeling *et al.* (2005) have shown that C_{60} can be manipulated along the troughs of Si(100)-(2 × 1) surface through either an attractive or a repulsive model of manipulation. They recorded the trajectories of the STM tip, which are characteristic of molecular rolling. This motion induced by the STM tip for covalently bound molecules is controlled by bond breaking.

Grill *et al.* (2007) have demonstrated the rolling of a single molecule equipped with two wheels (0.8 nm in diameter) induced by the STM tip. The characteristics of the rolling were recorded in the STM feedback loop manipulation signal and in real time. They captured unambiguous signatures of the conformational change happening during the rolling. Chiaravalloti *et al.* (2007) have presented a molecular rack-and-pinion device for which an STM tip drives a single pinion molecule at low temperature. The pinion is a 1.8-nm diameter molecule (hexa-*t*-butyl-hexaphenylbenzene, $C_{66}H_{78}$) functioning as a six-toothed wheel interlocked at the edge of a self-assembled molecular island acting as a rack. They monitored the rotation of the pinion molecule tooth by tooth along the rack by a chemical tag attached to one of its cogs. Clearly, this kind of approach of controlling the intra-molecular mechanics provides a path towards the bottom-up assembly of more complex molecular machines.

15.5 Concluding remarks

In summary, STM is an instrument not only used to *see* individual atoms by imaging, but also used to *touch* and *take* the atoms or to *hear* their vibration by means of manipulation. In this chapter, we have presented the current understanding of the conventional STM and STS investigations on nanostructures, including clean surfaces, single ad-atoms and molecules, SAMs, and nanostructures. These STM and related theoretical investigations using advanced capabilities, for example, the functionalized STM tip, dI/dV mapping, inelastic spectroscopy identification, manipulation, molecular electronics and single-molecule machine, are also briefly introduced. It is clear that to explore STM imaging and manipulating mechanisms, the related theoretical calculations/simulations are needed. Now, the cutting edge in theory is an exact description of current, electronic structure, magnetism, inelastic effect, and so on.

Combining its imaging, manipulation, spectroscopic characterization, and chemical modification capabilities, the STM has enabled direct visualization of chemistry by revealing the fundamental properties of atoms, molecules, surfaces and their interactions with each other as well as various nanostructures. It is believed that further progress in this field should eventually allow us to study not only surface topography, but also the local electronic structures and magnetism, surface dynamics, excitations, and chemical processes. To achieve these goals, we must combine theory and experiment more closely to allow STM to make all the more unique and important contributions to our understanding of the microscopic world at the atomic level than it has done so far.

References

Aviram, A., Ratner, M.A. *Chem. Phys. Lett.* **29**, 277 (1974).

Bardeen, J. *Phys. Rev. Lett.* **6**, 57 (1961).

Bartels, L., Meyer, G., Rieder, K.H., Velic, D., Knoesel, E., Hotzel, A., Wolf, M., Ertl, G. *Phys. Rev. Lett.* **80**, 2004 (1998).

Binnig, G., Rohrer, H., Gerber, C., Weibel, E. *Phys. Rev. Lett.* **49**, 57 (1982).

Binnig, G., Rohrer, H., Gerber, C., Weibel, E. *Phys. Rev. Lett.* **50**, 120 (1983).

Bocquet, M.L., Lesnard, H., Lorente N. *Phys. Rev. Lett.* **96**, 096101 (2006).

Bode, M. *Rep. Prog. Phys.* **65**, 523 (2003).

Bonnell, D.A. Ed. *Scanning Probe Microscopy and Spectroscopy: Theory, Techniques, and Applications* (Wiley-VCH, New York, 2001).

Bonnell, D.A., Garra, J. *Rep. Prog. Phys.* **71**, 044501 (2008).

Brandbyge, M., Mozos, J.L., Ordejón, P., Taylor, J., Stokbro, K. *Phys. Rev. B* **65**, 165401 (2002).

Briggs, G.A.D., Fisher, A. *J. Surf. Sci. Rep.* **33**, 1 (1999).

Bütticker, M. *Phys. Rev. Lett.* **57**, 1761 (1986).

Calleja, F., Arnau, A., Hinarejos, J.J., Vázquez de Parga, A.L., Hofer, W.A., Echenique, P.M., Miranda, R. *Phys. Rev. Lett.* **92**, 216101 (2004).

Chen, J.C. *Phys. Rev. Lett.* **65**, 448 (1990a).

Chen, J.C. *Phys. Rev. B* **42**, 8841 (1990b).

Chen, L., Xiang, H.J., Li, B., Zhao, A.D., Xiao, X.D., Yang, J.L., Hou, J.G., Zhu, Q.S. *Phys. Rev. B* **70**, 245431 (2004).

Chen, L., Hu, Z.P., Zhao, A.D., Wang, B., Luo, Y., Yang, J.L., Hou, J.G. *Phys. Rev. Lett.* **99**, 146803 (2007a).

Chen, L., Pan, B.C., Xiang, H.J., Wang, B., Yang, J.L., Hou, J.G., Zhu, Q.S. *Phys. Rev. B* **75**, 085329 (2007b).

Chiaravalloti, F., Gross, L., Rieder, K.H., Stojkovic, S.M., Gourdon, A., Joachim, C., Moresco, F. *Nature Mater.* **6**, 30 (2007).

Cross, G., Schirmeisen, A., Stalder, A., Grütter, P., Tschudy, M., Dürig, U. *Phys. Rev. Lett.* **80**, 4685 (1998).

Dai, H., Hafner, J.H., Rinzler, A.G., Colbert, D.T., Smalley, R.E. *Nature* **384**, 147 (1996).

Diebold, U. *Surf. Sci. Rep.* **48**, 53 (2003).

Diekhöner, L., Schneider, M.A., Baranov, A.N., Stepanyuk, V.S., Bruno, P., Kern, K. *Phys. Rev. Lett.* **90**, 236801 (2003).

Drakova, D. *Rep. Prog. Phys.* **64**, 205 (2001).

Elmers, H.J., Hauschild, J., Höche, H., Gradmann U., Bethge, H., Heuer, D., Köhler, U. *Phys. Rev. Lett.* **73**, 898 (1994).

Feenstra, R.M., Stroscio, J.A., Fein, A.P. *Surf. Sci.* **181**, 295 (1987a).

Feenstra, R.M., Stroscio, J.A., Tersoff, J., Fein, A.P. *Phys. Rev. Lett.* **58**, 1192 (1987b).

Ferro, Y., Allouche, A. *Phys. Rev. B* **75**, 155438 (2007).

Fletcher, S.P., Dumur, F., Pollard, M.M., Feringa, B.L. *Science* **310**, 80 (2006).

Fu, Y.S., Ji, S.H., Chen, X., Ma, X.C., Wu, R., Wang, C.C., Duan, W.H., Qiu, X.H., Sun, B., Zhang, P., Jia, J.F., Xue, Q.K. *Phys. Rev. Lett.* **99**, 256601 (2007).

Ganduglia-Pirovano, V.M., Hofmann, A., Sauer, J. *Surf. Sci. Rep.* **62**, 219 (2007).

Gambardella, P., Dallmeyer, A., Maiti, K., Malagoli, M.C., Eberhardt, W., Kern, K., Carbone, C. *Nature* **416**, 301 (2002).

Gao, L., Ji, W., Hu, Y., Cheng, Z., Deng, Z., Liu, Q., Jiang, N., Lin, X., Guo, W., Du, S., Hofer, W., Xie, X., Gao, H. *Phys. Rev. Lett.* **99**, 106402 (2007).

Geim, A.K., Novoselov, K.S. *Nature Mater.* **6**, 183 (2007).

Golam Moula, M., Suzuki, S., Chun, W.J., Otani, S., Oyama, S.T., Asakura, K. *Chem. Lett.* **35**, 90 (2006).

Grill, L., Rieder, K.H., Moresco, F., Rapenne, G., Stojkovic, S., Bouju, X., Joachim, C. *Nature Nanotechnol.* **2**, 95 (2007).

Grobis, M., Khoo, K.H., Yamachika, R., Lu, X.H., Nagaoka, K., Louie, S.G., Crommie, M.F., Kato, H., Shinohara, H. *Phys. Rev. Lett.* **94**, 136802 (2005).

Guo, Y., Zhang, Y.F., Bao, X.Y., Han, T.Z., Tang, Z., Zhang, L.X., Zhu, W.G., Wang, E.G., Niu, Q., Qiu, Z.Q., Jia, J.F., Zhao, Z.X., Xue, Q.K. *Science* **306**, 1915 (2004).

Hahn, J.R., Lee, H.J., Ho, W. *Phys. Rev. Lett.* **85**, 1914 (2000).

Hahn, J.R., Ho, W. *Phys. Rev. Lett.* **87**, 196102 (2001).

Hernandez, J.V., Kay, E.R., Leigh, D.A. *Science* **306**, 1532 (2004).

Hla, S.W. *J. Vac. Sci. Technol. B* **23**, 1351 (2005).

Hofer, W.A. *Prog. Surf. Sci.* **71**, 147 (2003).

Hofer, W.A., Foster, A.S., Shluger, A.L. *Rev. Mod. Phys.* **75**, 1287 (2003).

Hou, J.G., Wang, B., Yang, J.L., Wang, K.D., Lu, W., Li, Z.Y., Wang, H.Q., Chen, D.M., Zhu, Q.S. *Phys. Rev. Lett.* **90**, 246803 (2003).

Hou, J.G., Yang, J.L., Wang, H.Q., Li, Q.X., Zeng, C.G., Lin, H., Wang, B., Chen, D.M., Zhu, Q.S. *Phys. Rev. Lett.* **83**, 3001 (1999).

Hou, J.G., Yang, J.L., Wang, H.Q., Li, Q.X., Zeng, C.G., Yuan, L.F., Wang, B., Chen, D.M., Zhu, Q.S. *Nature* **409**, 304 (2001).

Huang, T., Hu, Z.P., Zhao, A.D., Wang, H.Q., Wang, B., Yang, J.L., Hou, J.G. *J. Am. Chem. Soc.* **129**, 3857 (2007).

Iancu, V., Deshpande, A., Hla, S.W. *Nano Lett.* **6**, 820 (2006a).

Iancu, V., Deshpande, A., Hla, S.W. *Phys. Rev. Lett.* **97**, 266603 (2006b).

Jia, J.F., Li, S.C., Zhang, Y.F., Xue, Q.K. *J. Phys. Soc. Jpn.* **76**, 082001 (2007).

Jiang, P., Morales, G.M., You, W., Yu, L.P. *Angew. Chem. Int. Ed.* **43**, 4471 (2004).

Joachim, C., Gimzewski, J.K., Aviram, A. *Nature* **408**, 541 (2000).

Jung, T.A., Schlittler, R.R., Gimzewski, J.K., Tang, H., Joachim, C. *Science* **271**, 181 (1996).

Katano, S., Kim, Y., Hori, M., Trenary, M., Kawai, M. *Science* **316**, 1883 (2007).

Keldysh, L.V. *Sov. Phys. JETP* **20**, 1018 (1965).

Kelly, K.F., Sarkar, D., Hale, G.D., Oldenburg, S.J., Halas, N.J. *Science* **273**, 1371 (1996).

Kim, H., Lee, J., Kahng, S.J., Son, Y.W., Lee, S.B., Lee, C.K., Ihm, J., Kuk, Y. *Phys. Rev. Lett.* **90**, 216107 (2003).

Kim, Y., Komeda, T., Kawai, M. *Phys. Rev. Lett.* **89**, 126104 (2002).

Kliewer, J., Berndt, R., Crampin, S. *New J. Phys.* **9**, 22 (2007).

Landauer, R. *IBM J. Res. Dev.* **1**, 233 (1957).

Lang, N.D. *Phys. Rev. Lett.* **55**, 230 (1985).

Lang, N.D. *Phys. Rev. B* **34**, 5947 (1986).

Lauhon, L.J., Ho, W. *J. Phys. Chem. A* **104**, 2463 (2000).

Lazarovits, B., Szunyogh, L., Weinberger, P. *Phys. Rev. B* **73**, 045430 (2006).

Lee, H.J., Ho, W. *Science* **286**, 1719 (1999).

Levy Yeyati, A., Martin-Rodero, A., Flores, F. *Phys. Rev. Lett.* **71**, 2991 (1993).

Li, B., Zeng, C.G., Li, Q.X., Wang, B., Yuan, L.F., Wang, H.Q., Yang, J.L., Hou, J.G., Zhu, Q.S. *J. Phys. Chem. B* **107**, 972 (2003).

Li, Q.X., Hu, X. *Phys. Rev. B* **74**, 035414 (2006).

Lu, X.H., Grobis, M., Khoo, K.H., Louie, S.G., Crommie, M.F., *Phys. Rev. Lett.* **90**, 096802 (2003).

Milun, M., Pervan, P., Woodruff, D.P. *Rep. Prog. Phys.* **65**, 99 (2002).

Mingo, N., Jurczyszyn, L., García-Vidal, F-J., Saiz-Pardo, R., de Andres P.L., Flores, F., Wu, S.Y., More, W. *Phys. Rev. B* **54**, 2225 (1996).

Morales, G.M., Jiang, P., Yuan, S.W., Lee, Y., Sanchez, A., You, W., Yu, L.P. *J. Am. Chem. Soc.* **127**, 10456 (2005).

Ng, M.K., Lee, D.C., Yu, L.P. *J. Am. Chem. Soc.* **124**, 11862 (2002).

Ng, M.K., Yu, L.P. *Angew. Chem. Int. Ed.* **41**, 3598 (2002).

Niebergall, L., Stepanyuk, V.S., Berakdar, J., Bruno, P. *Phys. Rev. Lett.* **96**, 127204 (2006).

Novoselov, K.S., Geim, A.K., Morozov, S.V., Jiang, D., Katsnelson, M.I., Grigorieva, I.V., Dubonos, S.V., Firsov, A.A. *Nature* **438**, 197 (2005).

Olsson, F.E., Persson, M., Borisov, A.G., Gauyacq, J.P., Lagoute, J., Folsch, S. *Phys. Rev. Lett.* **93**, 206803 (2004).

Pascual, J.I. *Eur. Phys. J. D* **35**, 327 (2005).

Paz, Ó., Brihuega, I., Gómez-Rodríuez, J.M., Soler, J.M. *Phys. Rev. Lett.* **94**, 056103 (2005).

Pietzsch, O., Okatov, S., Kubetzka, A., Bode, M., Heinze, S., Lichtenstein, A., Wiesendanger, R. *Phys. Rev. Lett.* **96**, 237203 (2006).

Pijper, D., van Delden, R.A., Meetsma, A., Feringa, B.L. *J. Am. Chem. Soc.* **127**, 17612 (2005).

Qiu, X.H., Nazin, G.V., Ho, W. *Phys. Rev. Lett.* **93**, 196806 (2004).

Ratner, M.A. *Mater. Today* **5**, 20 (2002).

Repp, J., Meyer, G. *Phys. Rev. Lett.* **94**, 026803 (2005).

Ruffieux, P., Melle-Franco, M., Gröning, O., Bielmann, M., Zerbetto, F., Gröning, P. *Phys. Rev. B* **71**, 153403 (2005).

Sautet, P., Bocquet M.L. *Phys. Rev. B* **53**, 4910 (1986).

Sautet, P. *Chem. Rev.* **97**, 1097 (1997).

Schmid, M., 1998, http://www.iap.tuwien.ac.at/www/surface/STM-Gallery/stm-schematic.html

Schouteden, K., Lijnen, E., Janssens, E., Ceulemans, A., Chibotaru, L.F., Lievens, P., Van Haesendonck, C. *New J. Phys.* **10**, 043016 (2008).

Song, In K., Kitchin, J.R., Barteau, M.A. *PNAS* **99**, 6471 (2002).

Stepanyuk, V.S., Niebergall, L., Hergert, W., Bruno, P. *Phys. Rev. Lett.* **94**, 187201.

Stipe, B.C., Rezaei, M.A., Ho, W. *Science* **279**, 1907 (1998a).

Stipe, B.C., Rezaei, M.A., Ho, W. *Science* **280**, 1732 (1998b).

Taylor, J., Guo, H., Wang, J. *Phys. Rev. B* **63**, 245407 (2001).

Tersoff, J., Hamann, D.R. *Phys. Rev. Lett.* **50**, 1998 (1983).

Tersoff, J., Hamann, D.R. *Phys. Rev. B* **31**, 805 (1985).

Tilinin, I.S., Rose, M.K., Dunphy, J.C., Salmeron, M., Van Hove, M.A. *Surf. Sci.* **418**, 511 (1998).

Tromp, R.M. *J. Phys.: Condens. Matter* **1**, 10211 (1989).

Tsukada, M., Kobayashi, K., Ohnishi, S. *J. Vac. Sci. Technol. A* **8**, 160 (1990).

Tsukada, M., Kobayashi, K., Shima, N., Ohnishi, S. *J. Vac. Sci. Technol. B* **9**, 492 (1991).

Wahl, P., Diekhöner, L., Schneider, M.A., Vitali, L., Wittich, G., Kern, K. *Phys. Rev. Lett.* **93**, 176603 (2004).

Wahl, P., Diekhöner, Wittich, G., Vitali, L., Schneider, M.A., Kern, K. *Phys. Rev. Lett.* **95**, 166601 (2005).

Wang, B., Zhou, Y.S., Ding, X.L., Wang, K.D., Wang, X.P., Yang, J.L., Hou, J.G. *J. Phys. Chem. B* **110**, 24505 (2006).

Wang, H.Q., Zeng, C.G., Li, Q.X., Wang, B., Yang, J.L., Hou, J.G., Zhu, Q.S. *Surf. Sci.* **442**, L1024 (1999).

Wang, K.D., Zhao, J., Yang, S.F., Chen, L., Li, Q.X., Wang, B., Yang, S.H., Yang, J.L., Hou, J.G., Zhu, Q.S. *Phys. Rev. Lett.* **91**, 185504 (2003).

Weiss, P.S., Eigler, D.M. *Phys. Rev. Lett.* **71**, 3139 (1993).

Wendt, S., Matthiesen, J., Schaub, R., Vestergaard, E.K., Lægsgaard, E., Besenbacher, F., Hammer B. *Phys. Rev. Lett.* **96**, 066107 (2006).

Yayon, Y., Brar, V.W., Senapati, L., Erwin, S.C., Crommie, M.F. *Phys. Rev. Lett.* **99**, 067202 (2007).

Yuan, L.F., Yang, J.L., Wang, H.Q., Zeng, C.G., Li, Q.X., Wang, B., Hou, J.G., Zhu, Q.S., Chen, D.M. *J. Am. Chem. Soc.* **125**, 169 (2003).

Yuan, L.F., Yang, J.L., Li, Q.X., Zhu, Q.S. *Phys. Rev. B* **65**, 035415 (2002).

Zeng, C.G., Li, B., Wang, B., Wang, H.Q., Wang, K.D., Yang, J.L., Hou, J.G., Zhu, Q.S. *J. Chem. Phys.* **117**, 851 (2003).

Zeng, C.G., Wang, H.Q., Wang, B., Yang, J.L., Hou, J.G. *Appl. Phys. Lett.* **77**, 3595 (2000).

Zhang, Z.R., Ge, Q.F., Li, S.C., Kay, B.D., White, J.M., Dohnálek, Z. *Phys. Rev. Lett.* **99**, 126105 (2007).

Zhao, A.D., Li, Q.X., Chen, L., Xiang, H.J., Wang, W.H., Pan, S., Wang, B., Xiao, X.D., Yang, J.L., Hou, J.G., Zhu, Q.S. *Science* **309**, 1542 (2005).

Zhao, J., Zeng, C.G., Cheng, X., Wang, K.D., Wang, G.W., Yang, J.L. Hou, J.G., Zhu, Q.S. *Phys. Rev. Lett.* **95**, 045502 (2005).

16

Functionalization of single-walled carbon nanotubes: Chemistry and characterization

R. Graupner and F. Hauke

16.1   Introduction

Soon after single-walled carbon nanotubes (SWCNTs) became available in macroscopic amounts scientists started to exploit the remarkable potential of this extraordinary new carbon allotrope. Their unique one-dimensional structure and their outstanding electronic, thermal and mechanical properties are unrivalled by any other substance class. In order to pave the way for CNT applications in various molecular-based devices, a multitude of hurdles had to be overcome related to the intrinsic poor solubility of SWCNTs in organic and aqueous solvents. It was rapidly recognized that chemical functionalization is a keystone in resolving this task and it would also help to solve another problem based on the production of SWCNTs. Up to now, all the different production techniques for CNTs yield a mixture of carbon nanotubes in terms of length, diameter, chirality and, directly related to that, a mixture of metallic and semiconducting tubes that hampers their entry into nanoworld devices. Therefore, chemical functionalization and structural alteration of carbon nanotubes has flourished in the last half decade, which is nicely documented by a series of review articles related to this topic (Bahr and Tour 2002; Hirsch 2002; Niyogi *et al.* 2002; Sun *et al.* 2002; Lin *et al.* 2003; Tasis *et al.* 2003; Banerjee *et al.* 2005; Hirsch and Vostrowsky 2005; Tasis *et al.* 2006).

The chemical modification of the carbon framework of CNTs can improve their solubility, can modify and fine tune the physical properties of carbon nanotubes and their derivatives, can combine the extraordinary properties of CNTs with the properties of other substance classes and, as a consequence, can open the door for the entry of CNTs into CNT-based technological applications.

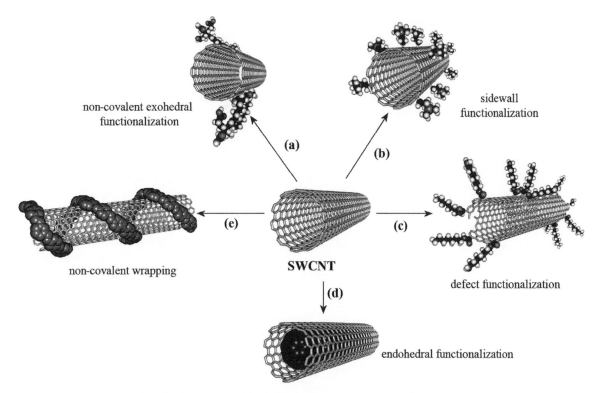

Fig. 16.1 Different possibilities of the functionalization of SWCNTs. (a) Non-covalent exohedral functionalization with molecules through $\pi - \pi$ stacking. (b) Covalent sidewall functionalization. (c) Defect-group functionalization. (d) Endohedral functionalization, in this case C_{60}@SWCNT. (e) Non-covalent exohedral wrapping with polymers.

16.2   Chemical functionalization of single-walled carbon nanotubes

When dealing with the functionalization of SWCNTs, different types of CNT modifications have to be distinguished (Fig. 16.1). The non-covalent and π-π-stacking functionalization of CNTs is based on a supramolecular complexation of various kinds of organic and inorganic moieties by the use of different interaction forces like van der Waals, charge transfer and $\pi - \pi$ interactions. On the other hand, endohedral functionalization deals with the filling of the hollow CNT cavities with a broad variety of small molecules.

In this chapter we deal with the covalent functionalization of the SWCNT framework that is the covalent attachment of functional entities onto the CNT scaffold. This structural alteration can take place at the termini of the tubes and/or at the sidewalls. The direct sidewall functionalization is associated with a rehybridization of one or more sp^2-carbon atoms of the carbon network into a sp^3-configuration and a simultaneous loss of conjugation. In contrast, the so-called defect functionalization of CNTs is based on the chemical transformation of defect sites introduced or already present. These defect sites can be the open ends and holes in the sidewalls, terminated by functional entities like carboxylic acid groups, or pentagon and heptagon irregularities in the hexagon graphene framework of the CNTs.

16.2.1   Reactivity of single-walled carbon nanotubes

The chemical modification and reactivity of SWCNTs needs to be discussed in the context of the reactivity of their closest carbon relatives—graphite and fullerenes. Graphite and similarly graphene are chemically inert due to their planar, aromatic sp^2-bonded carbon structure. Only few, very reactive compounds like fluorine are capable of derivatizing the planar conjugated π-surface. In contrast, the functionalization of the 3D curved π-surface of fullerenes can be carried out rather easily, which is nicely documented by the large variety of chemical derivatization sequences published up to now (Hirsch and Vostrowsky 2005). This tremendous difference in reactivity of the two carbon allotropes is based on the introduction of strain by the bending of the planar graphene sheet into a three-dimensional carbon framework. In non-planar conjugated carbon structures two basic factors are responsible for the chemical reactivity: (a) curvature-induced pyramidalization at the individual carbon atoms and (b) π-orbital misalignment between adjacent carbon atoms (Haddon 1988; Chen *et al.* 2003). Fullerenes and carbon nanotubes both represent spherical carbon structures. Fullerenes are curved in two dimensions, whereas carbon nanotubes are curved in one dimension only. As a consequence, the reactivity of fullerenes is primarily driven by the release of pyramidalization strain energy. In contrast to sp^2-carbon atoms with a pyramidalization angle of $\Theta_p = 0°$, sp^3-carbon atoms exhibit a pyramidalization angle of $\Theta_p = 19.5°$. In the spherical C_{60}, the smallest stable fullerene, the pyramidalization angle of the 60 equivalent carbon atoms is $\Theta_p = 11.6°$. Therefore, an addition reaction towards the convex surface of C_{60} and the release of strain energy based on the rehybridization of a sp^2-carbon atom into a sp^3-carbon atom is the basis for the wide variety of addition chemistry observed with fullerenes. A carbon nanotube with the same radius as C_{60} exhibits a less distorted π-framework due to its 1D curvature. As a consequence, CNTs are less reactive than fullerenes and the fullerene-like end-caps of CNTs are more susceptible towards addition reactions than their sidewalls.

As mentioned above, the other factor that governs reactivity in conjugated non-planar molecules is the π-orbital misalignment and for carbon nanotubes this has the greater influence on their reactivity (Niyogi *et al.* 2002; Hirsch and Vostrowsky 2005). The difference between the π-orbital alignment of a C_{60} fullerene and of an armchair (5,5)-SWCNT is depicted in Fig. 16.2. The π-orbital alignment in fullerenes is almost perfect with a π-orbital misalignment angle of $\Phi = 0°$. Although all sidewall atoms in the nanotube are equivalent as well, two sets of different bonds can be found in the (5,5)-SWCNT: one type runs parallel to the circumference (or perpendicular to the nanotube axis) with a π-orbital misalignment angle of $\Phi = 0°$ and the other set of bonds exhibit an

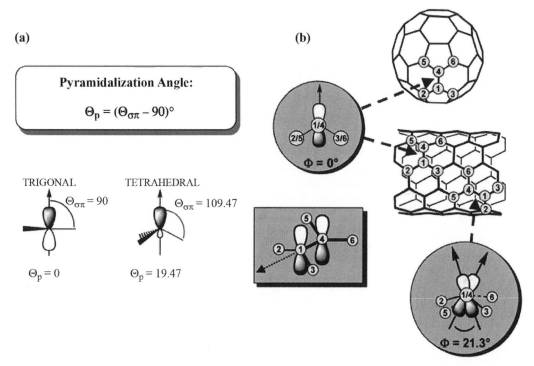

(a)

Pyramidalization Angle:

$$\Theta_p = (\Theta_{\sigma\pi} - 90)°$$

TRIGONAL

$\Theta_{\sigma\pi} = 90$

$\Theta_p = 0$

TETRAHEDRAL

$\Theta_{\sigma\pi} = 109.47$

$\Theta_p = 19.47$

(b)

$\Phi = 0°$

$\Phi = 21.3°$

Fig. 16.2 Representation of (a) the pyramidalization angle (Θ_p) and (b) the π-orbital misalignment angle (Φ) along the C1–C4 bond in a (5,5)-SWCNT and C_{60}. Reprinted from Hirsch and Vostrowsky (2005) with kind permission from Springer Science and Business Media.

angle to the circumference with a π-orbital misalignment angle of $\Phi = 21.3°$. This π-orbital misalignment is the origin of torsional strain in nanotubes and the relief of this strain energy controls the extent to which addition reactions occur with carbon nanotubes.

Based on the two topological-induced key factors for the reactivity of a conjugated carbon network, the chemical inertness of graphite and graphene can easily be understood. With a pyramidalization angle of $\Theta_p = 0°$ and a π-orbital misalignment angle of $\Phi = 0°$, no strain energy is stored within the carbon framework of graphite and graphene. Therefore, addition reactions will build up strain energy in contrast to the situation with the spherical carbon allotropes. Since the pyramidalization angle as well as the π-orbital misalignment angles scale inversely with the diameter of the nanotubes, smaller SWCNTs are expected to be more reactive than their larger counterparts.

In addition to these considerations on the localized electronic structure (steric effect), Joselevich (2004) discussed that the delocalized electronic structure (π-band structure) of the CNT carbon framework (resonance effect) has a similar influence on the reactivity of SWCNTs with different chiralities. By merging the solid-state physics band-structure model with the chemical molecular orbital description, the different reactivities of metallic and semiconducting SWCNTs can be interpreted and predicted.

16.2.2   Defect functionalization of single-walled carbon nanotubes

Single-walled carbon nanotubes can be produced by different production methods. All production techniques yield SWCNTs with varying amounts of impurities like amorphous carbon, fullerenes and catalyst particles. Therefore, the SWCNT material has to be purified. Very efficient purification sequences are based on an oxidative treatment of the starting material by liquid-phase or gas-phase oxidation. Harsh oxidative purification methods, like for instance boiling nitric acid can be seen as the primary chemical functionalization step of SWCNTs, as oxygen-bearing functionalities, mainly carboxylic acid functions, are introduced into the carbon framework of the nanotubes. By a systematic study of the evolution of gaseous CO and CO_2 in heat-treated oxidized SWCNT material, Mawhinney *et al.* (2000) determined a degree of functionalization of about 5%. If the high aspect ratio of carbon nanotubes is taken into account, this value implies that in addition to an oxidative opening of the SWCNT caps also an introduction of defects into the CNT sidewall takes place. In another study the expected diameter-dependent reactivity of SWCNTs was demonstrated (Zhou *et al.* 2001).

In a pioneering work, Liu *et al.* (1998) showed that a combination of liquid-phase oxidation in combination with ultrasonication leads to a shortening of the SWCNT material and that the introduced carboxylic acid functionalities can be used as a basic platform for a further chemical derivatization. Using thionyl chloride ($SOCl_2$), these carboxylic acid functionalities are converted into acid chlorides, which can subsequently be reacted with NH_2-$(CH_2)_{11}$-SH, yielding a covalently coupled carbon nanotube amide derivative (Fig. 16.3). The free thiol groups were attached to gold nanoparticles of 10 nm diameter and investigated by AFM. Chen *et al.* (1998a) expanded this defect functionalization concept by reacting the intermediately generated SWCNT acyl chloride functionalities with long-chain amines (e.g. octadecylamine (ODA)). This reaction protocol resulted in the first chemically modified SWCNT material that was soluble in organic media.

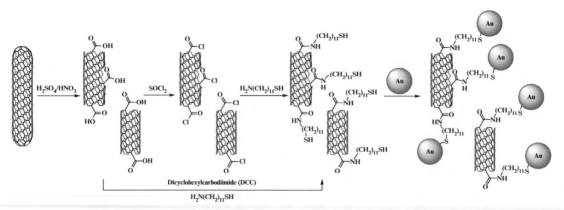

Fig. 16.3 Schematic representation of the oxidatively induced generation of carboxylic acid functionalities, their subsequent conversion into amide derivatives and the attachment of the functionalized moieties onto gold nanoparticles.

A subsequent quantitative IR spectroscopic study (Hamon *et al.* 2001) of the degree of functionalization of SWCNTs that had been amidated with octadecylamine gave a mass percentage of about 55% for the added groups, which can be translated to a functional defect density of 6%, in good agreement with the value obtained by Mawhinney *et al.* (2000). The activation of the SWCNT carboxylic acid functions can also be carried out under mild, neutral conditions by the use of cyclohexylcarbodiimide (DCC). Due to its versatility, the carboxylic-acid-based amidation sequence became a widely used tool for the construction of functional single-walled carbon-nanotube derivatives.

In their seminal work on the advantage of CNT-based applications, Wong *et al.* (1998) coupled the carboxylic functionalities that are present at the SWCNT ends with benzylamine and used the derivatized nanotube as an AFM tip (Fig. 16.4). By this method they were able to record chemically sensitive images with a nanometer-scale resolution of a patterned hydroxyl-terminated self-assembled monolayer surface.

The coupling of functional entities to the present carboxylic acid functions via amidation reactions was used for the synthesis of a broad variety of SWCNT derivatives with different physical properties. Li *et al.* (2001) were able to generate water-soluble SWCNT derivatives (Fig. 16.5(a)) by a solid-state reaction between oxidized nanotubes and 2-aminoethanesulfonic acid (taurine). Water solubilization of SWCNTs was also achieved by a couple of other groups by the coupling of a variety of sugar derivatives (Fig. 16.5(b)) to the oxidatively treated starting material (Tasis *et al.* 2006). The reaction of an appropriate bifunctional amine, like α,ω-diamines with acylated nanotubes

$$\xrightarrow[\text{amide bond formation}]{H_2N-R}$$

R = -CH$_2$CH$_2$NH$_2$; -H$_2$C—⟨⟩

COOH

Fig. 16.4 Schematic illustration of a SWCNT force microscope probe and modification of an oxidized SWCNT tip by coupling an amine RNH$_2$ to a terminal carboxylic acid function. The probe is able to sense specific interactions between the functional group R and surface OH groups.

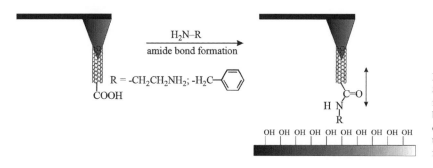

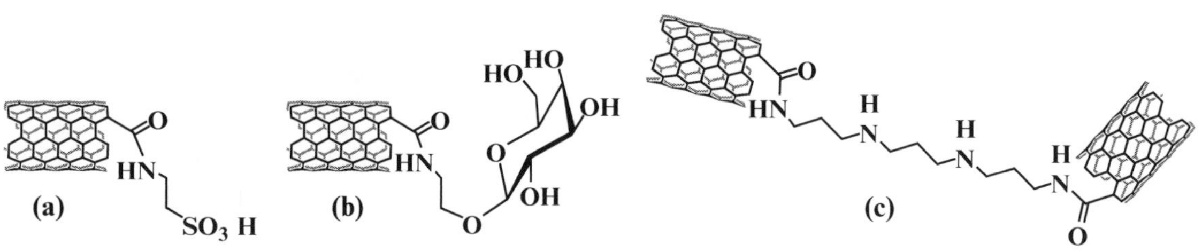

Fig. 16.5 Examples of the versatility of the defect-group-based amidation protocol. (a) Water-soluble CNT derivative by the coupling of taurine. (b) Water-soluble CNT material based on sugar attachment. (c) Interconnection of CNTs by a diamine spacer.

leads to an interconnection of the individual tubes (Fig. 16.5(c)) and the formation of CNT junctions (Chiu *et al.* 2002). By this reaction sequence end-to-end and end-to-side nanotube interconnections with interesting properties were formed and investigated by AFM and Raman spectroscopy.

The use of the carboxylic acid groups as an anchor for further transformations can easily be extended towards the reaction with functional entities equipped with alcohol functions, yielding SWCNT ester derivatives. Based on similar considerations Sun *et al.* (2002) attached a variety of lipophilic and hydrophilic dendrimers, photoactive moieties and polymers onto oxidized SWCNT via the amidation and an esterification sequence (Fig. 16.6). In order to provide evidence about the existence of ester linkages in the functionalized SWCNT derivatives, an acid- and base-catalyzed ester cleavage of the soluble material was carried out. This experiment led to the recovery of the starting CNT material, which was completely insoluble in any solvent.

Similar to the acylation-esterification approach, the carboxylic acid groups of oxidized CNTs were converted to carboxylate salts by the treatment with a base (Qin *et al.* 2003). Subsequently, the carboxylates reacted with alkyl halides in the presence of a phase-transfer agent to yield alkyl-modified nanotubes. The solubility of the adducts was found to be a function of the chain length of the alkyl group (Fig. 16.7).

The versatile utilization of the carboxylic-acid-based defect group functionalization is documented by the coupling of SWCNT to a large variety of substance classes including biological systems, like peptides, proteins and

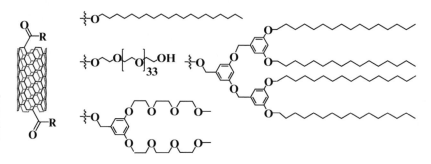

Fig. 16.6 Examples of functional CNT derivatives based on CNT esters. The solubility of the SWCNT material is a function of the solvent and of the functional group attached.

Fig. 16.7 Formation of CNT-esters by electrophilic alkylation of the CNT carboxylate ions; TOAB: tetra-*n*-octylammonium bromide. The solubility of the derivatized CNT material in organic media is a function of the alkyl chain length. The reaction was performed on MWCNTs.

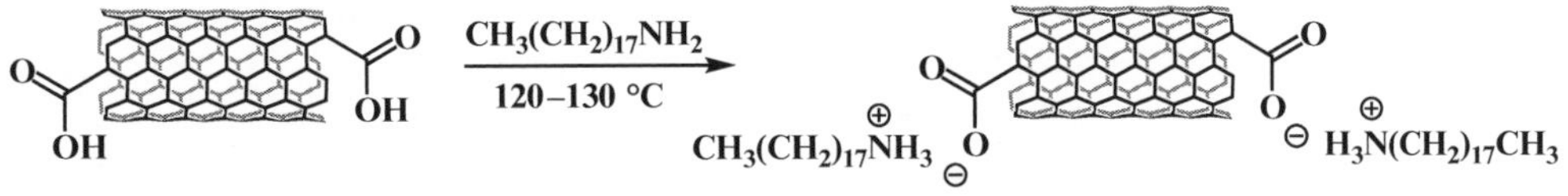

Fig. 16.8 Reaction of SWCNT-COOH with octadecylamine melt leads to zwitterionic-functionalized CNT derivatives with good solubility in organic solvents.

DNA, as well as for the construction of SWCNT-reinforced materials. Especially in the field of reinforced polymers, it has been shown that a covalent connection of the surrounding matrix to the embedded CNT material is vital for an improvement of the mechanical properties. Further interesting examples of the use of carboxylic acid decorated SWCNTs in a defect functionalization sequence are summarized in the review articles published on this topic (Coleman *et al.* 2006).

Another CNT derivatization sequence that is based on the carboxylic acid defects introduced into the carbon framework is the acid-base reaction with amines. Hamon *et al.* (1999) were able to show that the treatment of shortened SWCNT-COOH material with long-chain alkylamines yields zwitterionic SWCNT derivatives with good solubilities in organic solvents (Fig. 16.8). This ionic CNT functionalization represents a simple route for CNT solubilization and leads to an exfoliation of the majority of the large SWCNT bundles into smaller ropes and individual tubes, which was shown by AFM investigations (Chen *et al.* 2001). An intriguing example of the use of the zwitterionic SWCNT derivatization was given by Chattopadhyay *et al.* (2002), where THF-soluble zwitterionic-functionalized SWCNTs were separated by GPC chromatography according to their lengths.

In this chapter on defect-based CNT functionalization we were only able to show the underlying principles of this intriguing and versatile method, which is now widely used to combine the outstanding properties of carbon nanotubes with other substance classes. Therefore, defect functionalization of carbon nanotubes has become a cornerstone for the development of CNT–polymer composite materials (Coleman *et al.* 2006) and biological applications (Katz and Willner 2004).

16.2.3 Sidewall functionalization of single-walled carbon nanotubes

One unique characteristic of carbon nanotubes is their high aspect ratio. A defect-functionalization-based chemical attachment of functional moieties on CNTs will primarily proceed only at the nanotube tips. Due to their higher curvature, the end-caps of CNTs are more reactive and the introduction of carboxylic-acid-based defects will predominantly occur at these locations. As a consequence, the functionalization degree will be relatively low as the huge surface area of the SWCNT sidewalls is hardly involved in the transformation sequence. Direct sidewall functionalization therefore offers the opportunity to install a large amount of functional entities on the nanotube framework and should in principle lead to higher degrees of functionalization. On the other hand, the direct sidewall functionalization deals with two drawbacks. In contrast to defect-functionalization sequences, sidewall derivatization and the formation of a covalent bond between the attacking species and a carbon atom of the CNT framework is always associated with a rehybridization of a sp^2-carbon atom into a sp^3-configuration, leading to a disturbance of the π-system of the CNTs. This is directly correlated with a negative influence on the electronic and mechanical properties of the carbon nanotubes. Moreover, the direct chemical alteration of the CNT sidewall is a challenging task due to the close relationship of the CNT sp^2-carbon network with planar graphite and its low reactivity. On the other hand, this leads directly to the question: Which reaction sequences of a chemist's toolkit are in principle capable of functionalizing nanotube sidewalls? And further, is there a selectivity of these reactions in terms of the diameter of the tube or even between semiconducting and metallic SWCNTs?

The first point will be the topic of this section where a short and general overview of the different possibilities for a direct CNT sidewall functionalization will be given. The latter point is discussed in a different chapter in this handbook.

16.2.3.1 *Fluorination of SWCNTs*

In their pioneering work, Mickelson *et al.* (1998) reported the first successful sidewall derivatization. The treatment of SWCNTs with elemental fluorine, which is also capable of functionalizing graphite, in the temperature range between 150 °C and 600 °C leads to fluorinated carbon nanotubes (Fig. 16.9). In contrast to the high electrical conductivity of the pristine starting material, the fluorinated SWCNTs turned into insulators. This change in the electronic properties reflects the structural changes associated with the covalent attachment of functional entities onto the carbon sidewall.

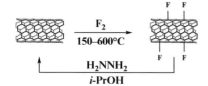

Fig. 16.9 Treatment of SWCNTs with elemental fluorine yields sidewall-fluorinated material. The functionalization degree depends on the reaction temperatures. Defluorination by hydrazine leads to a regeneration of the starting SWCNT material.

The fluorination of SWCNTs has been extensively investigated, which is described in detail in a review article of Khabashesku *et al.* (2002). The highest functionalization degree is estimated to be C_2F, which can be translated into a uniform coverage of the outer SWCNT surface. Fluorination at higher temperatures and the use of small-diameter SWCNTs leads to a shortening of the SWCNT material due to C–C bond cleavage. One common property of fluorinated SWCNT derivatives is their good solubility in alcoholic solvents, which opened the door for solution-based chemical transformations. By the treatment with hydrazine (Fig. 16.9) Mickelson *et al.* (1998) were able to detach the fluorine atoms, restoring the characteristic spectroscopic properties of intrinsic SWCNTs.

Furthermore, fluorinated SWCNTs can be used as a starting material for substitution reactions leading to a variety of SWCNT derivatives (Fig. 16.10). The reaction with Grignard reagents (Boul *et al.* 1999) and the treatment with organolithium compounds (Saini *et al.* 2003) leads to the formation of alkylated SWCNT derivatives that are well soluble in many organic solvents. These systems can easily be dealkylated upon heating in an oxygen atmosphere, restoring the pristine SWCNTs with their initial physical properties. The nucleophilic substitution of fluorine with diamines (Stevens *et al.* 2003)

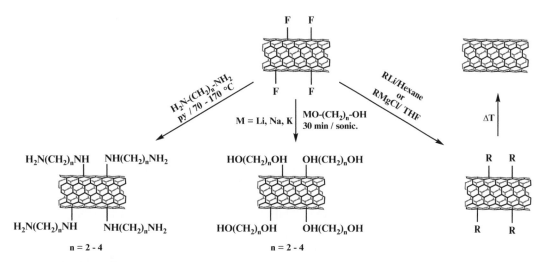

Fig. 16.10 Chemical transformation based on fluorinated single-walled carbon nanotubes. The treatment of the fluorinated SWCNTs with strong nucleophiles leads to a substitution of the fluorine atoms and the generation of heterofunctionalized SWCNT derivatives. Thermal treatment in an oxygen atmosphere regenerates the unfunctionalized starting material.

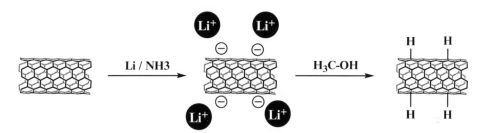

Fig. 16.11 The reduction of SWCNTs with lithium yields anionic SWCNT intermediates, which subsequently can be protonated by the reaction with methanol.

and diols (Zhang *et al.* 2004) leads to SWCNT heteroatom derivatives that can be used as building blocks for SWCNT-based molecular architectures.

16.2.3.2 *Hydrogenation of SWCNTs*

The hydrogenation of SWCNTs was the first reaction sequence that was based on the reduction of SWCNTs with alkali metals in liquid ammonia, the so-called Birch reduction. Pekker *et al.* (2001) showed that the protonation of Li-reduced SWCNTs intermediates with methanol leads to the formation of hydrogenated SWCNTs (Fig. 16.11). The chemical composition of the derivatized material was determined by thermogravimetry-mass spectrometry analysis to be $C_{11}H$. The obtained material is thermally stable up to 400 °C, while above this temperature a decomposition takes place.

16.2.3.3 *Radical additions onto SWCNTs*

16.2.3.3.1 Functionalization based on free-radical species

The reactive radical species can be generated from a variety of precursors either thermally or by photophysical routes. The experimental demonstration

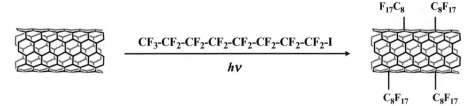

Fig. 16.12 Radical-based sidewall functionalization of SWCNTs. Addition of heptadecafluorooctyl radicals obtained from photo-induced cleavage of heptadecafluorooctyl iodide.

of a covalent sidewall functionalization of SWCNTs by a radical species was given one year after the theoretical prediction. The addition of perfluorinated alkyl radicals (Holzinger *et al.* 2001), obtained by photo-induced cleavage of the carbon–iodine bond from heptadecafluorooctyl iodide, onto SWCNTs yields perfluorooctyl-derivatized CNTs (Fig. 16.12). The covalent attachment of the perfluorated alkyl chain was detected by ^{19}F NMR spectroscopy of the soluble reaction product.

Alkyl- or aryl peroxides are commonly used radical initiators. It has been shown that they are suitable candidates for the sidewall functionalization of SWCNTs. The thermal cleavage of benzoyl and lauroyl peroxids (Umek *et al.* 2003; Peng *et al.* 2003a) leads to the generation of very reactive phenyl and alkyl radicals that attack the SWCNTs very efficiently, leading to sidewall-derivatized materials that have been characterized by a variety of spectroscopic methods (Fig. 16.13, left). The combination of the benzoyl peroxide as radical starter with alkyl iodides (Ying *et al.* 2003) allows the covalent attachment of a large variety of different functionalities to the sidewalls of single-walled carbon nanotubes (Fig. 16.13, right).

By the reaction of SWCNTs with succinic or glutaric acid acyl peroxides (Peng *et al.* 2003b) this concept of radical-based functionalization can be extended to the generation of highly functional SWCNT building blocks (Fig. 16.14, left). The accessible sidewall acid-functionalized SWCNT derivatives can easily be converted into amides by the reaction with various terminal amines. Liu *et al.* (2006, 2007) used for their sidewall functionalization

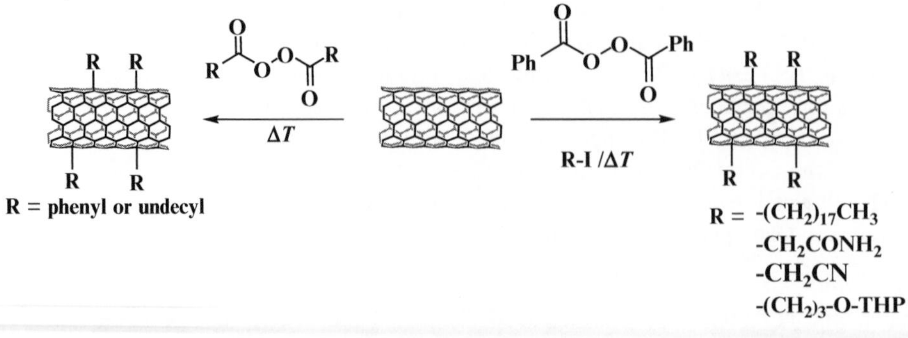

Fig. 16.13 Radical-based sidewall functionalization of SWCNTs. Addition of alkyl radicals generated from acylperoxides (left). The combination of benzoyl peroxide with alkyl iodides yields higher functional SWCNT derivatives (right).

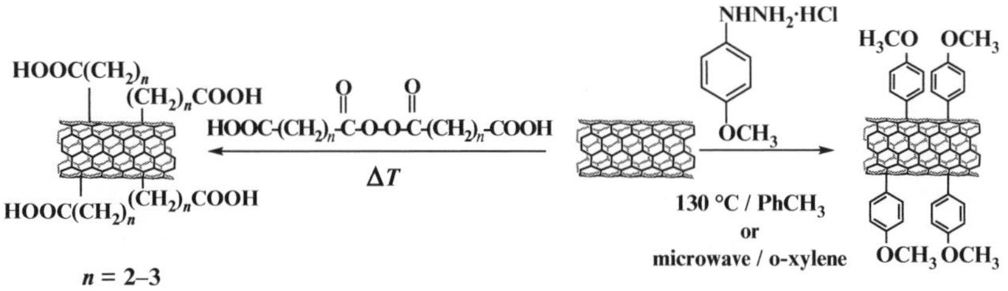

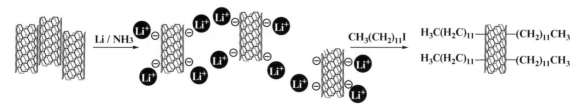

Fig. 16.14 Radical-based sidewall functionalization of SWCNTs. Addition of carboxyalkyl radicals (left). 4-Methoxyphenylhydrazine as radical precursor for the SWCNT derivatization (right).

sequence 4-methoxyphenylhydrazine as a radical precursor (Fig. 16.14, right). The hydrazine derivative is converted into a phenyl radical in the presence of oxygen either in a classical thermal reaction or a microwave-assisted reaction yielding SWCNT sidewall-derivatized species.

16.2.3.3.2 Functionalization of SWCNTs by reductive alkylation

Beside the SWCNT hydrogenation, described above, reductive alkylation developed by the group of Billups (Liang *et al.* 2004) is also based on negatively charged SWCNT derivatives generated through Birch reduction conditions. The reaction of SWCNTs with alkali metals like lithium or sodium in liquid ammonia leads to the reduction of the carbon nanotubes and the generation of negatively charged CNT intermediates (Fig. 16.15).

The lithium intercalation and the Coulomb repulsion leads to a very efficient debundling of the CNTs and therefore to an individualization of the tubes. These individualized CNT intermediates react with alkyl halides or aryl halides (Chattopadhyay *et al.* 2005) by an electron-transfer mechanism, generating alkyl radicals, which subsequently functionalize the SWCNT material. The alkylated/arylated nanotubes exhibit a high solubility in organic media, a SWCNT-functional group ratio of about 1/20 and AFM investigations showed that the SWCNTs remain individualized after functionalization. In a selectivity investigation of the covalent sidewall functionalization of SWCNTs by reductive alkylation Wunderlich *et al.* (2008a) was able to show that SWCNTs with smaller diameter are considerably more reactive than tubes with larger diameter.

16.2.3.3.3 Diazonium-based SWCNT functionalization

In 2001 Bahr *et al.* (2001) discovered a convenient route to functionalized carbon nanotubes by the reaction of aryl diazonium salts with SWCNTs in

Fig. 16.15 Reductive alkylation of SWCNTs. The negatively charged SWCNT intermediates react with alkyl or aryl halides by an electron-transfer reaction.

an electrochemical reaction, using a bucky paper as the working electrode (Fig. 16.16). Upon variation of the aryl diazonium salt, a broad variety of highly functionalized and well-soluble SWCNT adducts are accessible. The corresponding reactive aryl radicals are generated from the diazonium salt via one-electron reduction.

The authors (Bahr and Tour 2001) extended that concept onto a solvent-based thermally induced reaction with *in-situ* generation of the diazonium compound by reacting aniline derivatives with isoamyl nitrite (Fig. 16.16). Shortly thereafter, these basic studies of diazonium-based SWCNT derivatization led to a very efficient functionalization sequence of individualized carbon nanotubes (Dyke and Tour 2003b). The use of water-soluble diazonium compounds in combination with sodium dodecyl sulfate (SDS) coated individual SWCNTs yielded highly functionalized nanotubes, having up to 1 in 9 carbon atoms along their backbones bearing an organic moiety. The diazonium-based functionalization of SWCNTs is a very versatile reaction as it is even possible to conduct the reaction in the total absence of any solvent (Dyke and Tour 2003a). This solvent-free method yields heavily functionalized and very soluble material.

In a groundbreaking investigation Strano *et al.* (2003) showed that the chemical sidewall functionalization of SWCNTs can be selective in terms of metallic and semiconducting CNTs. By a very detailed UV/Vis and Raman study they were able to show that diazonium-based SWCNT functionalization reactions lead to a higher degree of functionalization with metallic carbon nanotubes compared to their semiconducting relatives. Based on these facts the diazonium-based functionalization has become a widely used tool for the sidewall derivatization of SWCNTs.

16.2.3.4 Cycloaddition reactions

16.2.3.4.1 Addition of carbenes

The first functionalization of the carbon nanotube sidewall with a carbon-based substituent was reported in 1998 by Chen *et al.* (1998b). Dichlorocarbene

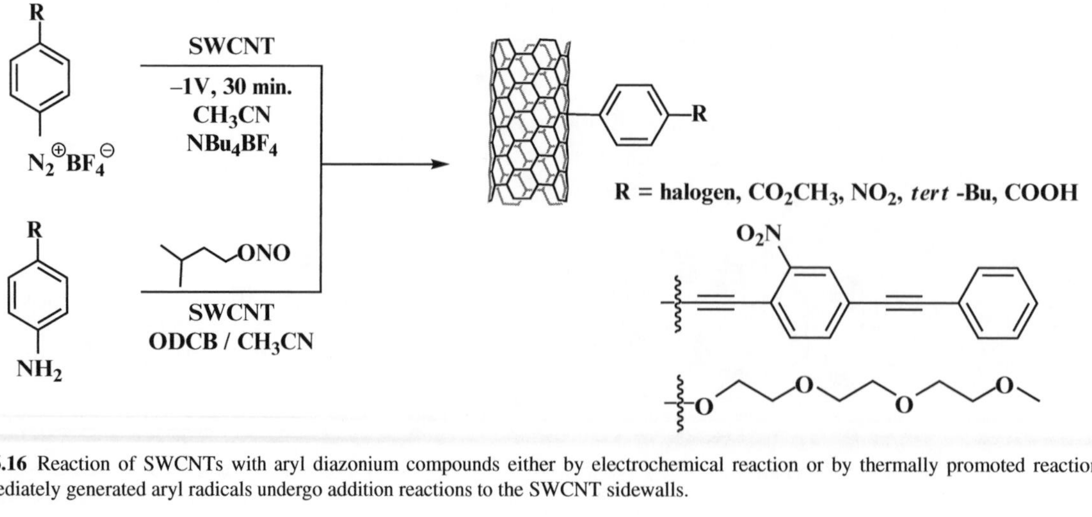

Fig. 16.16 Reaction of SWCNTs with aryl diazonium compounds either by electrochemical reaction or by thermally promoted reaction. The intermediately generated aryl radicals undergo addition reactions to the SWCNT sidewalls.

generated *in situ* from chloroform with potassium hydroxide or from phenyl(bromodichloromethyl)mercury (Chen *et al.* 1998a) was covalently attached to soluble SWCNTs (Fig. 16.17). The electrophilic carbene addition led to a comparatively low degree of functionalization of about 2%.

16.2.3.4.2   Addition of nitrenes

Alkoxycarbonyl nitrenes are capable of attacking SWCNT sidewalls in a [2+1] cycloaddition reaction leading to the formation of an aziridine ring (Holzinger *et al.* 2001). The nitrenes are generated thermally *in situ* out of alkoxycarbonyl azide precursors (Fig. 16.18, right). The driving force for the generation of the highly reactive nitrene intermediates is the thermally induced N_2 extrusion. As alkoxycarbonyl azides with different functional moieties are easily accessible, this SWCNT-sidewall-functionalization sequence is a versatile tool leading to a broad variety of SWCNT adducts with different functionalities, like alkyl chains, aromatic units, crown ethers and even dendritic structures (Holzinger *et al.* 2003). The solubility of the different SWCNT derivatives in organic media depends on the functional moiety attached and can be as high as 1.2 mg/ml.

In a similar reaction, Moghaddam *et al.* (2004) used photochemically generated nitrene intermediates from a azidothymidine precursor (Fig. 16.18, left) to functionalize the sidewalls of solid support-bound MWCNTs.

16.2.3.4.3   Nucleophilic cyclopropanation

One common reaction for the functionalization of fullerenes is the so-called Bingel reaction, a nucleophilic cyclopropanation of one double bond of the fullerene framework by the reaction of a bromomalonate in the presence

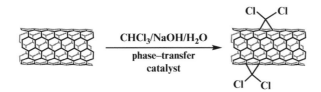

Fig. 16.17 Cycloaddition of *in-situ* generated dichlorocarbene to SWCNT sidewalls.

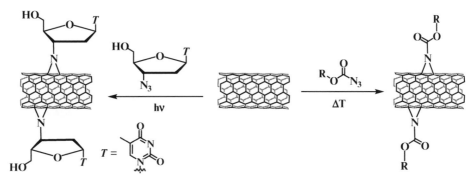

Fig. 16.18 Sidewall derivatization by nitrenes. Azidothymidine precursor for photochemical nitrene generation (left). Alkoxycarbonyl azides are versatile species for thermal-based SWCNT-sidewall functionalization (right).

of a base. This reaction sequence was transferred to SWCNTs by Coleman *et al.* (2003). The reaction of diethyl bromomalonate with SWCNT led to the sidewall-functionalized material that was subsequently transformed into a thiolated derivative (Fig. 16.19). By "tagging" the cyclopropane moiety with gold nanoparticles and visualizing the adduct by AFM, the authors were able to locate the covalent attachment of the malonate addend. To further confirm the derivatization of the SWCNTs, a perfluorinated marker was introduced by transesterification, making the adduct accessible for ^{19}F NMR and XPS. The degree of functionalization by the Bingel reaction was estimated to be about 2%.

16.2.3.4.4 Osymylation

The [3 + 2] cycloaddition of the strong oxidant osmium tetroxide to carbon–carbon double bonds of the SWCNT framework was theoretically predicted by Lu *et al.* (2002a). Shortly afterwards, Cui *et al.* (2003) were able to show that sidewall osmylation of SWCNTs can be achieved with OsO_4 vapor under UV photo-irridation. The covalent attachment of osmium tetroxide was found to increase the electrical resistance of the tubes by several orders of magnitude. The reaction is reversible and the original SWCNTs can easily be regained by irradiation with UV light in the presence of oxygen.

In an extensive subsequent solution-based study, Banerjee and Wong (2004b) proposed a mechanism for the photostimulated osmylation of the SWCNTs that involves a photoinduced charge transfer from the nanotubes to OsO_4 and a subsequent formation of the osmate ester adduct. As the reaction proceeds by the intermediate formation of an electron-rich charge-transfer complex, the sidewall osmylation favors metallic over semiconducting SWCNTs.

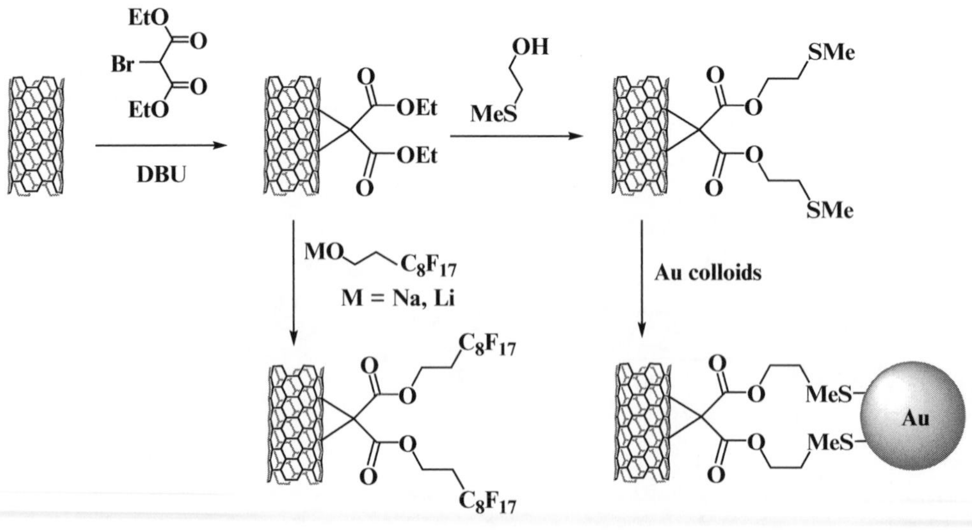

Fig. 16.19 Nucleophilic cyclopropanation of SWCNTs by the Bingel reaction with subsequent transformation of the ethylester unit.

16.2.3.4.5   Ozonolysis

The possibility for the [3 + 2] cycloaddition of ozone onto SWCNT sidewalls was confirmed in a detailed study by Banerjee and Wong (2002). They were able to show that the primary CNT-ozonide can be cleaved by the treatment with hydrogen peroxide, dimethyl sulfide and sodium borhydride, yielding carboxylic acids, aldehydes/ketones and alcohols, respectively. In a Raman-based study (Banerjee and Wong 2004a) they were also able to show that the addition reaction of ozone onto SWCNT sidewalls is diameter dependent, i.e. smaller-diameter tubes are more extensively functionalized than the larger ones.

16.2.3.4.6   Addition of azomethine ylides

Georgakilas *et al.* (2002a) adapted the 1,3-dipolar cycloadditions of azome-thine ylides, originally developed for the exohedral modification of C_{60}, for CNT-sidewall derivatization. The thermal treatment of SWCNTs in DMF in the presence of an aldehyde and α-amino acid derivatives results in the formation of substituted pyrrolidine moieties on the SWCNT surface and the formation of a well-soluble material in common organic solvents (Fig. 16.20).

In a subsequent report, Georgakilas *et al.* (2002b) used this functionaliza-tion sequence for the purification of HiPco-SWCNTs from metal particles and amorphous carbonaceous species by a chemical modification of the raw material. The subsequent separation of the soluble adducts, the reprecipitation of the derivatized material and finally the thermal removal of the functional groups by annealing of the SWCNT material at high temperatures yielded high-purity SWCNTs.

The azomethinylide-based functionalization sequence has become a valu-able tool for CNT-sidewall functionalizations due to its versatility. In principle, this methods allows the attachment of any moiety to the carbon framework of the CNTs by a simple variation of the chemical structure of the educts. Therefore, the use of CNTs derivatives for applications in the field of medicinal

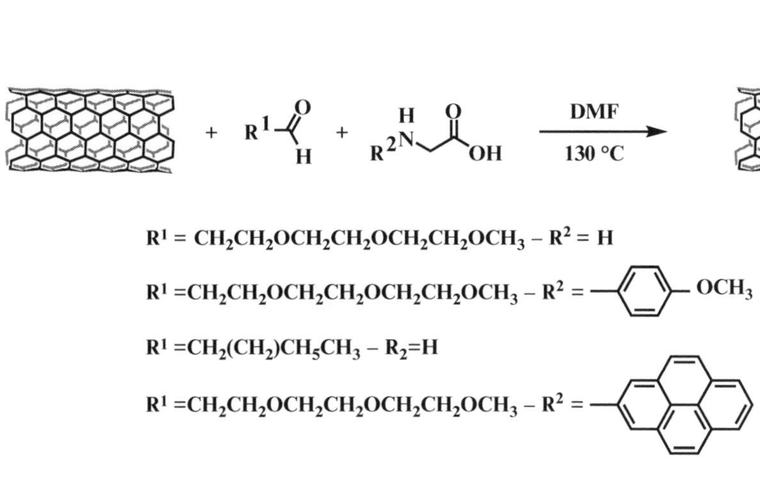

$R^1 = CH_2CH_2OCH_2CH_2OCH_2CH_2OCH_3 - R^2 = H$

$R^1 = CH_2CH_2OCH_2CH_2OCH_2CH_2OCH_3 - R^2 = $ ⟨benzene ring⟩— OCH_3

$R^1 = CH_2(CH_2)CH_5CH_3 - R_2 = H$

$R^1 = CH_2CH_2OCH_2CH_2OCH_2CH_2OCH_3 - R^2 = $ ⟨pyrene ring⟩

Fig. 16.20 1,3-Dipolar cycloadditions of azomethine ylides by the reaction of SWCNTs with aldehydes and α-amino acid derivatives.

chemistry, solar-energy conversion and recognition of chemical species is a current field of research.

16.2.3.4.7   Diels–Alder reaction

The Diels–Alder cycloadditions of *o*-quinodimethane onto the sidewalls of SWCNTs has been theoretically predicted (Lu *et al.* 2002b) to be viable due to the aromatic stabilization of the transition state and reaction product. In this reaction sequence SWCNTs should perform as dienophiles. This was experimentally verified by Delgado *et al.* (2004) by a microwave-assisted reaction of soluble SWCNTs (Fig. 16.21) and 4,5-benzo-1,2-oxathiin-2-oxide (a *o*-quinodi-methane precursor). The modified tubes were characterized by Raman spectroscopy and TGA.

As the rate of Diels–Alder addition reactions is enhanced by the use of electron-withdrawing substituents on the monoene (the "dienophile"), fluorinated SWCNTs were used as starting materials by Zhang *et al.* (2005a). These SWCNTs with activated sidewalls were reacted thermally with 2,3-dimethyl-1,3-butadiene, anthracene and 2-trimethylsiloxyl-1,3-butadiene yielding highly functionalized SWCNT derivatives. XPS and IR investigations showed that the addition of the butadiene unit is associated with a significant reduction of the fluorine content.

16.2.3.5   *Electrophilic addition*

The first report on the derivatization of SWCNTs by electrophilic addition of chloroform in the presence of $AlCl_3$ was given in 2002 (Tagmatarchis *et al.* 2002). The hydrolysis of the labile intermediate yielded hydroxyl-group-terminated SWCNTs. These functionalized SWCNTs were subsequently transformed into ester derivatives allowing their structural characterization based on their increased solubility in organic solvents.

The concept of electrophilic addition to the SWCNT sidewall was generalized by Xu *et al.* (2008) who developed a microwave-based functionalization

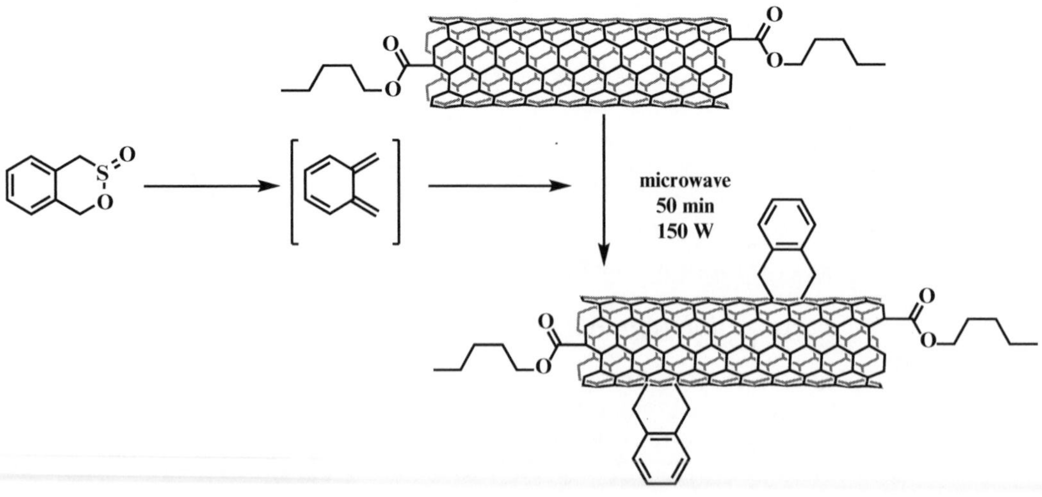

Fig. 16.21 Microwave-assisted [4 + 2] cycloadditions of *o*-quinodimethane to SWCNT sidewalls.

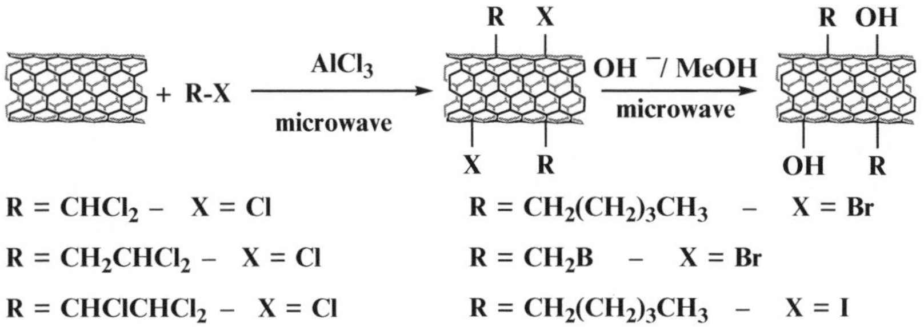
AlCl3
+ R-X
microwave
R X
X R
OH ⁻/ MeOH
microwave
R OH
OH R
R = CHCl2 – X = Cl
R = CH2CHCl2 – X = Cl
R = CHClCHCl2 – X = Cl
R = CH2(CH2)3CH3 – X = Br
R = CH2B – X = Br
R = CH2(CH2)3CH3 – X = I

Fig. 16.22 Microwave-assisted electrophilic addition of alkylhalides to SWCNTs and basic hydrolysis of the intermediate halogen derivatives.

sequence that allows the use of a variety of alkyl halides for the CNT derivatization (Fig. 16.22). By this sequence SWCNT derivatives are accessible that are functionalized simultaneously with alkyl and hydroxyl substituents.

16.2.3.6 *Nucleophilic addition*

16.2.3.6.1 Carbon-atom-based nucleophiles

Holzinger *et al.* (2001) showed that the SWCNT sidewalls could also be attacked by nucleophilic carbenes (Fig. 16.23). In contrast to the proposed cyclopropane structure in the case of dichlorocarbene, the addition of the nucleophilic dipyridyl imidazolidene leads to a zwitterionic reaction product due to the extraordinary stability of the resultant aromatic 14π perimeter of the imidazole addend.

In 2003 Viswanathan *et al.* (2003) grafted SWCNTs with polystyrene based on the anionic polymerization of the styrene monomer. They reacted SWCNTs with *sec*-butyllithium. This led to the nucleophilic addition of the alkyl chain and the intermediate generation of negatively charged SWCNT anions. The charge transfer led to an efficient exfoliation of the SWCNT bundles and the SWCNT anions initiated the polymerization upon styrene addition. An analogous concept was used by Blake *et al.* (2004) for the generation of MWCNT composites.

The SWCNT anions generated by the nucleophilic addition of organolithium compounds can be used as a versatile tool for the generation of

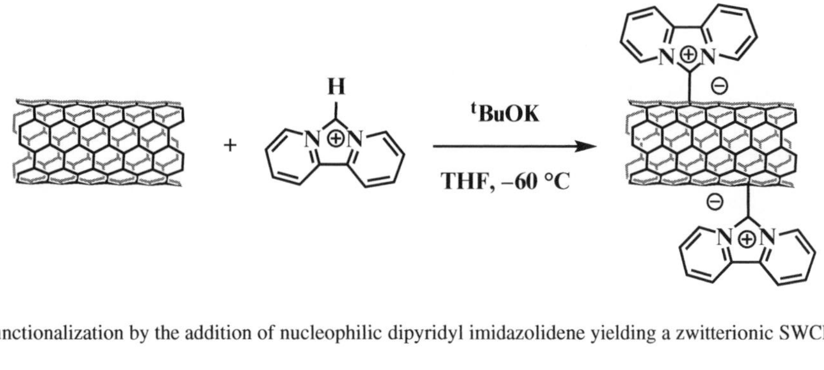
H
+
N⊕N
ᵗBuOK
THF, –60 °C
N⊕N
⊖
⊖
N⊕N

Fig. 16.23 Sidewall functionalization by the addition of nucleophilic dipyridyl imidazolidene yielding a zwitterionic SWCNT derivative.

higher-functionalized SWCNT assemblies. The reaction of SWCNTs with *sec*-butyllithium and the subsequent treatment with carbon dioxide followed by an aqueous workup leads to functionalized SWCNTs bearing simultaneously alkyl groups and carboxylic acid functionalities (Fig. 16.24, left), which can be used for further derivatizations (Chen *et al.* 2005).

In a detailed study Graupner *et al.* (2006) were able to show that the SWCNT carbanions generated by the nucleophilic addition of organolithium compounds could easily be reoxidized, yielding alkylated SWCNT derivatives (Fig. 16.24, right). This nucleophilic alkylation reoxidation cycle could be repeated and the degree of functionalization increases with each cycle. Another important finding was that the nucleophilic addition of organolithium compounds leads to preferred functionalization of metallic SWCNTs. In an extended study, Wunderlich *et al.* (2008b) was able to confirm this finding and, furthermore, showed that this kind of addition reaction is diameter dependent, namely smaller tubes exhibit a higher degree of functionalization than larger CNTs. In addition, the range of the nucleophilic compounds was extended onto magnesium-based organometallic system, the so-called Grignard reagents.

16.2.3.6.2   Nitrogen-atom-based nucleophiles

In a solvent-free approach Basiuk *et al.* (2004) used 1-octadecylamine as a nitrogen-based nucleophile. They proposed a direct sidewall addition of the nitrogen atom onto closed-cap MWCNTs and suggested that the addition takes place only on five-membered rings of the graphitic network of the nanotubes and that the benzene rings are inert to the direct amination.

One year later, Yokoi *et al.* (2005) reported the successful amino-functionalization of SWCNTs by a reaction with aromatic hydrazines. The functionalized material exhibits a high solubility in a variety of organic solvents. They also found a direct dependence of the degree of functionalization with the type of substituent attached in the *para* position to the hydrazine unit. Electron-donating substituents yield higher degrees of functionalization than electron-withdrawing entities.

Syrgiannis *et al.* (2008) showed that the nucleophilic sidewall addition to SWCNTs is also possible by using amine-based nucleophiles, namely *in-situ* generated lithium amides. Although they are less potent nucleophiles than their carbon-based relatives, they yield SWCNT derivatives with a degree

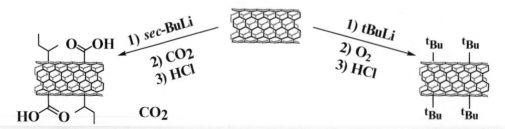

Fig. 16.24 Nucleophilic addition of carbon-based nucleophiles to SWCNTs. The electrophilic quenching of the intermediately generated SWCNT anions yields mixed functionalized materials (left). Oxidation of the anionic intermediates leads to alkylated SWCNT derivatives (right).

of functionalization in the order of 3%. The reaction of *n*-propylamine with *n*-buthyllithium yields the corresponding amide, which subsequently attacks the SWCNTs generating negatively charged SWCNT derivatives, which are subsequently reoxidized by air to the corresponding neutral amino-functionalized SWCNT derivatives (Fig. 16.25). The reaction product exhibits a drastically improved solubility in organic solvents and there is a first hint that the functionalization degree is directly related to the concentration of the amide used.

16.3 Characterization

Although the chemical strategies for the covalent functionalization are numerous, as discussed above, the characterization of the obtained products is a very challenging task on its own. As stated by Bahr and Tour (2002), a standard suite of techniques does not exist for an adequate characterization of chemically modified SWCNTs. One of the main problems is caused by the fact that often the starting materials are very inhomogeneous, i.e. depending on the preparation conditions they may differ in purity, diameter distribution of the SWCNTs and defect density. After chemical processing, an optimal characterization of the processed material should provide answers to the following questions: Was the functionalization successful? Did it functionalize the sidewalls of the SWCNTs, pre-existing defects or other graphitic or amorphous by-products? What is the degree of functionalization, i.e. the number of functional groups per carbon atom of the nanotube sidewall? And, finally: Is there any specific reactivity, either depending on the diameter of the SWCNTs, the electronic structure, or locally, for example at the nanotubes' ends?

It is evident that no single characterization technique is able to answer all the questions posed above. Instead, a complete picture may only be obtained by a combination of several techniques and we would like to introduce some of the most common methods below. With some exceptions, we concentrate on techniques that are not too common in organic chemistry, being aware that our list is far from complete.

Generally, a functionalization of SWCNTs is verified either by a direct detection of the functional group in the material itself or by a change of the intrinsic properties of the SWCNTs upon functionalization. The disadvantage of the first approach is that it is not clear whether the functional group is indeed bound to the nanotube sidewall, the latter does not guarantee that

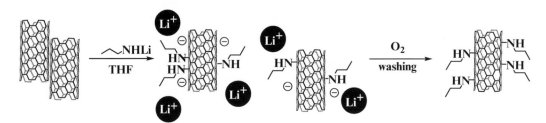

Fig. 16.25 SWCNT-sidewall functionalization sequence based on primary amines.

changes of the intrinsic properties of SWCNTs are caused by the attachment of the functional group and not by other processing steps like purification or ultrasonication.

Techniques that verify a functionalization by a change of the intrinsic properties of the SWCNTs include several spectroscopic techniques, like NIR-Vis-UV absorption or Raman spectroscopy. Here, in defect group functionalization these effects are observed upon the generation of the defects, and a subsequent derivatization of these defects should in principle not lead to further changes in the spectra. This is in contrast to a covalent functionalization technique where the attachment of the functional moieties is investigated.

To identify functional groups themselves, the atomic entities of the functional group (e.g. in photoelectron spectroscopy), the detection of characteristic vibrational modes (IR-absorption) or the weight loss of the material upon thermal detachment of the functional groups (thermogravimetric analysis, TGA) can be detected. In addition, microscopic techniques are used to reveal the presence of functional groups on the sidewalls of the SWCNTs.

In the following, some of the most common techniques are presented and examples will be given. In all cases, it has to be noted that all techniques have their own strengths and weaknesses and in many cases only a combination of different methods may doubtless confirm that the carbon nanotubes in a sample bear functional groups.

16.3.1   Thermogravimetric analysis (TGA)

In thermogravimetric analysis, a certain amount of functionalized SWCNT material is annealed in an inert atmosphere using a controlled temperature ramp and the mass change with respect to temperature is measured. The mass loss due to the thermal detachment of the functional groups is detected and the degree of functionalization, i.e. the number of functional groups per sidewall carbon atom can be calculated. In addition, using a mass spectrometer the fragments of the functional groups may be identified.

As an example, Fig. 16.26 shows the weight loss of SWCNTs, prepared by two different techniques, which were then functionalized by alkylation of fluorinated SWCNTs (Saini *et al.* 2003). The higher weight loss in the case of HiPco SWCNTs is attributed to the smaller average diameter compared to the case of SWCNTs prepared by laser ablation, leading to a higher reactivity and therefore a higher degree of functionalization.

However, there are several pitfalls to this technique. One corresponds to the fact that in addition to chemisorbed groups, species that are only adsorbed on the outside of the sidewalls or that were incorporated into the interior of the SWCNTs also lead to a weight loss upon desorption and are therefore detected as well. In principle, the desorption temperature of these moieties should be significantly lower, however, if these groups are located in between the individual tubes of a bundle or a rope, inside the tubes or from deeper regions of the material, a complete desorption of these groups may not have occurred while the detachment of the covalently functionalized moieties takes place. To minimize these effects, a measurement of a reference sample that

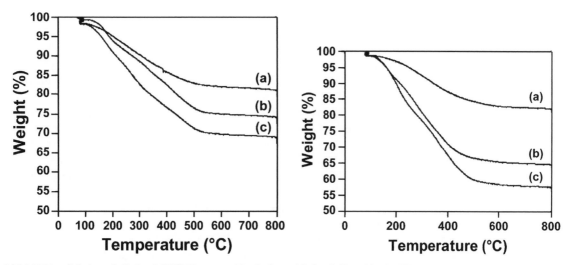

Fig. 16.26 TGA weight loss of alkylated SWCNTs prepared by the laser ablation (left) and by the HiPco process (right) in argon. (a) Methylated SWCNTs, (b) *n*-butylated SWCNTs, (c) *n*-hexylated SWCNTs. Reprinted with permission from Saini *et al.* (2003). Copyright 2003 American Chemical Society.

did undergo similar treatments, i.e. for purification or solubilization except the functionalization step is recommended and the weight loss with respect to this reference sample is taken as a measure of the degree of functionalization. Another severe limitation of this technique is due to the fact that TGA measurements *per se* do not differentiate between functionalized nanotubes and functionalized material that is present in the starting material besides the single-walled carbon nanotubes, like amorphous or graphitic carbon or fullerene molecules.

16.3.2 Spectroscopic techniques

16.3.2.1 *UV-Vis-NIR spectroscopy*
Electromagnetic radiation in the spectral range between ultraviolet and near-infrared light induces electronic transitions between the van-Hove singularities (vHs) of the SWCNTs. For solubilized SWCNTs these transitions manifest as series of peaks in absorption spectra. The vHs are a consequence of the regular, one-dimensional arrangement of carbon atoms in the SWCNTs. Dipole-selection rules and the depolarization or antenna effect (Duesberg *et al.* 2000; Fagan *et al.* 2007a) limits these transitions to those between the mirror-like vHs. Depending on the diameter distribution of the SWCNTs in the sample, these structures in the absorption spectra group together and may partially overlap.

For HiPco SWCNTs (Nikolaev *et al.* 1999; Bronikowski *et al.* 2001); which are reported to exhibit a diameter distribution between 0.7 nm and 1.6 nm (Nikolaev *et al.* 1999), the optical transitions that are lowest in energy are those of semiconducting SWCNTs between the first vHs that form the valence-band maximum and the conduction-band minimum (E_{11}^S). These transitions are observed between 1600 nm and 800 nm ($E = 0.78$ to 1.55 eV), followed by the transition between the second vHs of the valence and conduction band of semiconducting SWCNTs (E_{22}^S, $\lambda = 900$ nm to 550 nm, $E = 1.38$ eV to 2.25 eV). The lowest-energy optical transition for metallic SWCNTs (E_{11}^M) occur between $\lambda = 600$ nm and 400 nm ($E = 2.07$ eV to 3.10 eV, O'Connell *et al.* 2002; Nair *et al.* 2006, 2007). An example of absorption spectra measured on solubilized SWCNTs is given in Fig. 16.27.

In contrast to spectrofluorimetric measurements (Bachilo *et al.* 2002; Weisman and Bachilo 2003), or Raman measurements (see below) an assignment of chiral indices in the absorption spectra is not as straightforward, as several (n, m) usually contribute to a peak in the absorption spectra of SWCNTs. Nevertheless, in some cases an assignment to individual chiral indices is given (Lian *et al.* 2003; Nair *et al.* 2006). UV-Vis-NIR absorption spectra can be used to estimate the purity of SWCNT material, as shown by Itkis *et al.* (2003) and Lian *et al.* (2005). Here, the intensity of the characteristic transitions in the absorption spectra of SWCNTs relative to the unstructured absorption of amorphous or graphitic carbon is taken as a measure of the SWCNT content in the sample (this issue is discussed in detail by Haddon and Itkis (2008)). It has to be noted, however, that an individualization of the tubes is necessary to observe these features as a bundling of the tubes suppresses these features. This also results in the fact that the intensities of these vHs transitions in solution vary if different surfactants are used, reflecting the variations in the ability to solubilize individual SWCNTs (Blackburn *et al.* 2006). In addition, the position of the peaks in absorption spectra is found to vary in different

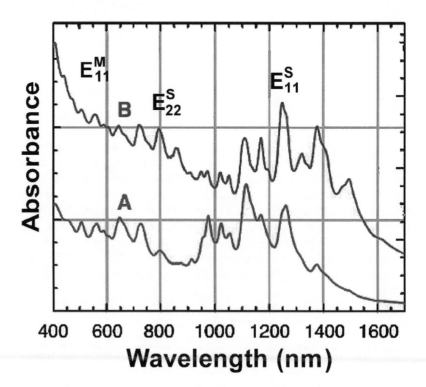

Fig. 16.27 UV-Vis-NIR absorption spectrum of SWCNTs, solubilized using SDS in D$_2$O. Curve A displays the spectrum of SWC-NTs with a smaller average diameter than those of spectrum B, which shifts the vHs to higher energies, i.e. lower wavelengths. From O'Connell *et al.* (2002). Reprinted with permission from AAAS.

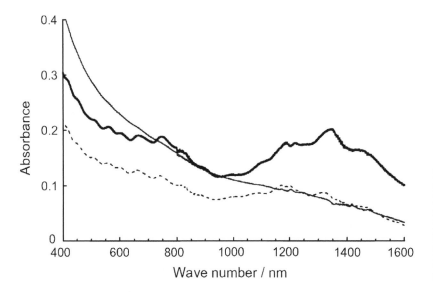

Fig. 16.28 UV-Vis-NIR absorption spectra of SWCNTs functionalized using organic hydrazines. The dotted line is taken from the unfunctionalized material, the full line from the functionalized SWCNTs, while the bold line is measured on the functionalized SWCNTs after thermal defunctionalization. Reprinted from Yokoi *et al.* (2005) with permission from Elsevier.

solvents (Blackburn *et al.* 2006), which reflects variations in the excitonic binding energies due to variations in dielectric screening (Perebeinos *et al.* 2004).

Both the generation of defects as well as the covalent sidewall functionalization disturbs the regular arrangement of the carbon atoms in the SWCNTs, which leads to an energetic broadening and reduction of the density of states in the vHs. Therefore, the spectral intensity in the characteristic maxima of the absorption spectra is reduced, which is demonstrated in Fig. 16.28, where absorption spectra of SWCNTs are shown that were functionalized using organic hydrazines. While the spectrum of the unfunctionalized raw material exhibits the characteristic peaks of the respective vHs, the one taken on the functionalized SWCNTs is almost completely featureless. After detachment of the functional groups by thermal annealing, the characteristic absorption maxima of intrinsic SWCNTs are restored.

The fact that the peaks in the absorption maxima of the SWCNTs are grouped together according to their respective electronic structure (i.e. in spectral regions where absorption of metallic or semiconducting SWCNTs prevails) makes this method an important tool for the characterization of sidewall-functionalization techniques that preferentially react with metallic or semiconducting SWCNTs. One of the most prominent examples is given in Fig. 16.29, where absorption spectra of SWCNTs functionalized by diazonium salts are shown (Strano *et al.* 2003). Diazonium salts react preferentially with metallic SWCNTs, leading to a gradual decrease of the characteristic absorption maxima of metallic SWCNTs while those of the semiconducting tubes remain largely unaffected.

There seems to be no consensus, however, up to which degree of functionalization the vHs in the absorption spectra are retained. While most of the authors report on a reduction or even a disappearance of the vHs upon sidewall functionalization, the vHs peaks are retained in a microwave-assisted Bingel

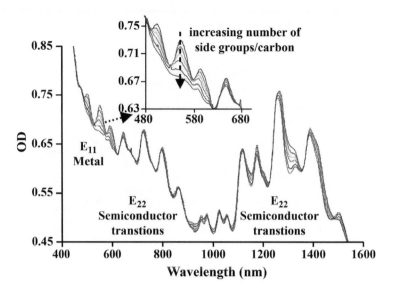

Fig. 16.29 UV-Vis-NIR absorption spectra of SWCNTs functionalized by successive addition of 4-chlorobenzenediazonium tetrafluoroborate. The electronic transitions of the metallic SWCNTs are gradually reduced while those of the semiconducting SWCNTs remain largely unaffected. The inset shows an expanded view of the region where the metallic transitions occur. Reprinted from Strano *et al.* (2003) with permission from AAAS.

reaction up to a degree of functionalization of one diester group per 75 carbon atoms of the SWCNTs sidewall (Umeyama *et al.* 2007).

In addition, it was shown by Fagan *et al.* (2007b) that the intensity of the absorption signal depends on the length of the SWCNTs. If a functionalization leads to a shortening of the SWCNTs, a reduction in the vHs absorption signals is expected as well. However, thermal defunctionalization then should not restore the spectrum of pristine SWCNTs, in contrast to the example depicted in Fig. 16.28.

16.3.2.2   *Raman spectroscopy*

Raman spectroscopy is one the most frequently used techniques for the characterization of SWCNTs in general (see, e.g. Dresselhaus *et al.* 2005). For Raman spectroscopy on SWCNTs, it has to be noted that the scattering process is resonant, i.e. for a mixture of SWCNTs of different chiralities in the form of a bucky paper or in solution, only those SWCNTs are probed where either the incoming or the scattered radiation matches an electronic transition (single resonant process).

The main structure in the Raman spectra of graphitic materials (see Fig. 16.30) is the G-band, located around a wave number of $1600\,\mathrm{cm}^{-1}$ and first observed by Tuinstra and Koenig (1970) in graphite. For isolated SWC-NTs, this band consists of several components of different vibrational symmetries. In semiconducting SWCNTs these components form two distinct and symmetric maxima at wave numbers of $1590\,\mathrm{cm}^{-1}$ (G^+) and around 1560–$1580\,\mathrm{cm}^{-1}$ (G^-). While the position of the G^+-component is independent of the diameter of the tubes, the measured Raman shift of the G^- component increases with increasing diameter (Jorio *et al.* 2002; Paillet *et al.* 2006). Due to this diameter dependence this peak is interpreted for semiconducting SWCNTs to transverse optical vibrations, while G^+ is attributed to the longitudinal optical vibrational modes. For metallic SWCNTs, the

G$^-$-component is found to be asymmetrically broadened and shifted to lower wave numbers. While this line shape was attributed to a Breit–Wigner–Fano mechanism in the past (Brown *et al.* 2001a), more recently the existence of a Kohn anomaly in SWCNTs seems more probable (Popov and Lambin 2006; Caudal *et al.* 2007; Piscanec *et al.* 2007). Measurements on individual metallic SWCNTs revealed that the occurrence of a shifted and broadened G$^-$-component depends on the chiral index (Wu *et al.* 2007) and therefore from the occurrence of this feature the presence of metallic SWCNTs can be deduced, the reverse, however, is not true, i.e. the absence of a broadened G$^-$-component does not necessarily imply the absence of metallic SWCNTs. In addition, the G-band line shape of metallic SWCNTs changes upon shifts of the Fermi level (Rafailov *et al.* 2005; Abdula *et al.* 2007; Nguyen *et al.* 2007; Wu *et al.* 2007), i.e. a charge transfer to or from the SWCNTs may lead to a removal of the characteristic metallic G-band line shape, as already observed by Kataura *et al.* (2000).

Structures in the Raman spectra that are unique to SWCNTs are the so-called radial breathing modes (RBMs) at wave numbers between 100 and 300 cm^{-1}. These modes correspond to inphase radial vibrations of all sidewall carbon atoms of SWCNTs. The RBM frequency is inversely related to the diameter d of the tube. This fact, together with the resonant nature of the Raman scattering process allows the identification of individual chiral indices in the RBM Raman spectra of SWCNTs (Strano 2003; Fantini *et al.* 2004; Telg *et al.* 2004, 2006; Jorio *et al.* 2005; Maultzsch *et al.* 2005; Araujo *et al.* 2007) using so-called "Kataura plots" (Kataura *et al.* 1999). This makes Raman spectroscopy of the RBMs an ideal tool for the characterization of chemical transformations that are sensitive to the diameter or electronic structure of the SWCNTs.

For the interpretation of the RBMs it is important to note that the transition energies reported above are commonly measured for SWCNTs that are solubilized using surfactants. In direct comparisons between SWCNTs in different morphologies, slightly different transition energies have been reported (Heller *et al.* 2004; Luo *et al.* 2004; O'Connell *et al.* 2004; Ericson and Pehrsson 2005). Upon bundling of SWCNTs, the resonant electronic transitions are red shifted and broadened as a consequence of the differences in dielectric screening of excitons, as discussed above for the optical absorption spectra. This has to be taken into account if RBM spectra of functionalized and pristine SWCNTs are compared, as one of the main intentions of functionalization is the individualization of the SWCNTs.

In addition, it was reported by Fantini *et al.* (2007) that a functionalization itself may induce shifts in the transition energies of SWCNTs. Here, red as well as blue shifts have been reported for different functional groups at the sidewalls of SWCNTs, which again might imply that different SWCNTs are probed before and after functionalization if the same excitation wavelength is used.

The most discussed feature in the Raman spectra of functionalized SWCNTs is the defect induced D-band at a Raman shift around 1300 cm^{-1}. This feature was already observed by Tuinstra and Koenig (1970) in disordered graphite and it was shown that its intensity depends linearly on the inverse crystallite

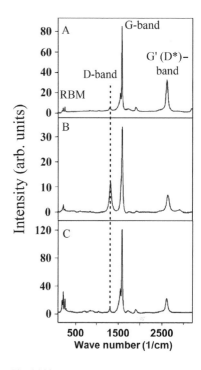

Fig. 16.30 Raman spectra ($\lambda_{\text{exc}} = 633$ nm) of pristine SWCNTs (A), after functionalization by aryl-groups (B) as well as after thermal detachment of the functional groups at 850 °C in argon (C). Notice the strong increase in D-band intensity upon functionalization and the decrease after defunctionalization. Reprinted with permission from Stephenson *et al.* (2006). Copyright 2006 American Chemical Society.

size, i.e. it is a measure for the amount of disorder in graphitic systems. The fact that the measured D-band frequency depends on the incident laser energy for disordered graphite (Vidano *et al.* 1981; Pócsik *et al.* 1998) as well as for SWCNTs (Brown *et al.* 2001b; Kürti *et al.* 2002) with a slope of $\approx 50\,\text{cm}^{-1}/\text{eV}$ led to the interpretation of the D-band structure as a double-resonance effect (Thomsen and Reich 2000; Saito *et al.* 2003). The fact that the presence of a defect is necessary for the D-band makes it an important tool for the study of functionalization of SWCNTs and the increase of the D-band intensity is taken as one of the crucial fingerprints for a successful functionalization.

To account for variations in absolute Raman scattering intensities, the D-band intensity (or area) has to be normalized to some intrinsic Raman feature of the SWCNTs that is independent of defect concentration. In most cases, the I_D/I_G ratio is given. As there have been concerns that the G-band itself is defect induced, a normalization to the overtone of the D-band, the G′-band (sometimes called the D^*-band) at a Raman shift of around $2600\,\text{cm}^{-1}$ was proposed by Maultzsch *et al.* (2002). This mode does not require the presence of a defect, therefore its intensity is expected to be independent of disorder introduced by functionalization. However, recently it was shown by Dossot *et al.* (2007), that for semiconducting SWCNTs both methods lead to the same results.

However, although the intensity of the D-band depends on the concentration of defects in the SWCNTs, a definite relationship between the D/G ratio and the degree of functionalization cannot be given. First, it is not clear whether a linear relationship between the number of defects and the D-band intensity exists (Graupner 2007, and references therein). In addition, it was found that the D/G ratio depends on the length of the tubes (Chou *et al.* 2007; Fagan *et al.* 2007b) which leads to an increase of the D-band intensity if the SWCNTs are shortened by the functionalization process. Again, this can be excluded if the samples are annealed and the functional groups are detached. Furthermore, the D-band intensity for the very same functionalized sample depends on the laser wavelength (Chou *et al.* 2007; Fantini *et al.* 2007), which implies that only I_D/I_G ratios can be compared if the same excitation wavelength is used. For low degrees of functionalization, a dependence on the state of aggregation has also been reported (Dossot *et al.* 2007).

Even if all these limitations are taken into account, the I_D/I_G ratio is surely one of the most important fingerprints for a qualitative investigation of a functionalization of SWCNTs. In Fig. 16.30 the Raman spectra for the pristine SWCNT material, SWCNTs functionalized by aryl groups as well as the functionalized SWCNTs material after thermal detachment of the functional groups at 850°C in argon are shown. While the D-band intensity is low for the pristine SWCNTs, its intensity considerably increases after functionalization, while after annealing a spectrum comparable to the pristine material is obtained.

16.3.2.3 *Photoelectron spectroscopy*

In photoelectron spectroscopy, a solid sample is illuminated by monochromatic light and the energy distribution of the photoemitted electrons is detected that allows the determination of the binding energy of the electrons prior to the photoexcitation process. If laboratory X-ray sources are used (X-ray-induced photoelectron spectroscopy, XPS), the atomic core levels are accessible and the surface elemental composition of the sample can be quantified via the core-level intensities using tabulated photoionization cross sections (Scofield 1976; Yeh 1993). If the functional groups contain atoms other than carbon and hydrogen, the degree of functionalization can therefore be estimated. In addition, the exact binding energy of an element is characteristic of its chemical environment, leading to so-called "chemical shifts". To a first approximation, these chemical shifts can be correlated with the charge on the atom that is photoexcited. With increasing positive charge, the binding energy of a core level increases, while a negative charge decreases the binding energy (Hüfner 1995; Bagus *et al.* 1999). A positive charge in turn is induced if a covalent bond to a different atom with higher electronegativity is formed. Many illustrative examples of experimentally determined chemical shifts in XPS are found in the book of Beamson and Briggs (1992), which compiles high-resolution core-level spectra measured on many different polymers. One of the strengths of this method is that the peak decomposition is quantitative, i.e. the area ratios between the individual components reflect the ratios of the carbon atoms in the different chemical environments.

For pristine single-walled carbon nanotubes, the XPS C 1s core-level line shape is asymmetric (Goldoni *et al.* 2002), characteristic of delocalized, sp^2-bonded systems like graphite (van Attekum and Wertheim 1979; Sette *et al.* 1990; Leiro *et al.* 2003). This intrinsic line shape complicates the discrimination of small, chemically shifted components induced by small degrees of functionalization, common for most techniques of sidewall or defect functionalization.

An example, where the intrinsic line shape of the C 1s core level plays a minor role due to high degrees of functionalization, is the case of fluorinated SWCNTs. In addition, the difference in electronegativities between carbon and fluorine is large. Figure 16.31 displays the C 1s core level of fluorinated SWCNTs (Alemany *et al.* 2007). The high degree of functionalization leading to a stoichiometry close to C_2F results in a high intensity of the chemically shifted component induced by the carbon atoms that are bound to fluorine, indicated by the component *C–F* in Figure 16.31, compared to the unperturbed carbon atoms (denoted by SWNT) and those C atoms next to a C atom bearing a functional group (*C–C–F*). Using quantum-chemical methods, Wang and Sherwood (2004) determined the charge on the carbon sidewall atoms of fluorinated carbon nanotubes for different arrangements of the fluorine atoms. Thereby, the expected chemical shifts can be calculated that were then compared to the measured spectra.

Sometimes, a distinction between sp^2- and sp^3-bonded carbon is made in the main C 1s core-level component, based on results of XPS measurements on amorphous carbon (Díaz *et al.* 1996; Haerle *et al.* 2001). Estrade-Szwarckopf (2004) reports on the observation of a defect-related C 1s component in disordered graphite. Although these results would offer an easy way for a purity estimation of a SWCNT sample, the small binding energy difference

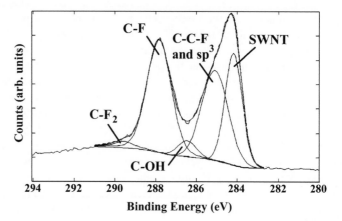

Fig. 16.31 C 1s core level of a fluorinated SWCNT sample. The stoichiometry of the sample, as determined by the intensities of the F 1s and C 1s core levels is close to C_2F. For the interpretation of the individual components, see text. Reprinted with permission from Alemany *et al.* (2007). Copyright 2007 American Chemical Society.

to the main C 1s component makes a quantitative determination not very reliable.

For all interpretations of XPS results, it has to be kept in mind that this technique is extremely surface sensitive due to the small inelastic mean-free path of the photo-electrons that is of the order of a few Ångströms (Hüfner 1995). This implies that only the topmost surface atom layers of a solid sample are probed, which may not necessarily reflect the bulk composition and may be influenced, e.g. by the presence of adsorbates.

16.3.3   Microscopic techniques

16.3.3.1   *Transmission electron microscopy (TEM)*

Surely the transmission electron microscopy (TEM) experiments by Iijima (1991), Iijima and Ichihashi (1993) and Bethune *et al.* (1993) are of the most cited publications in the field of carbon-nanotube research. Here, the tubular nature of the nanotubes was first demonstrated. Using electron diffraction, the chiral index of the SWCNTs can be determined (see Qin 2006, and references therein). The high resolution of advanced transmission electron microscopes allows the observation of individual defects in SWCNTs (Hashimoto *et al.* 2004). Nevertheless, the direct observation of functional groups on the sidewalls was only rarely reported. Dyke and Tour (2003b) and Price *et al.* (2005) report on the presence of functional groups on SWCNTs after sidewall functionalization by diazonium salts. Similar results have been obtained by Liang *et al.* (2004) on SWCNTs after reductive alkylation (Fig. 16.32, left). Small spots that are stable under TEM conditions on defect-functionalized SWCNTs were interpreted as iodine atoms by Coleman *et al.* (2007).

Often, TEM measurements are used as a tool for the investigation of the structural integrity of SWCNTs after purification and functionalization. Monthioux *et al.* (2001), for example performed a detailed TEM investigation of SWCNTs after acid purification techniques (Fig. 16.32, right). They concluded that severe structural changes in the sidewalls of single-walled carbon nanotubes may occur, depending on the temperature and acid used for the purification of the tubes.

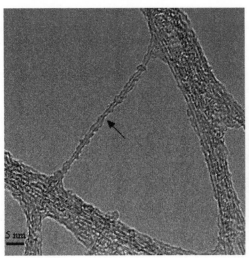
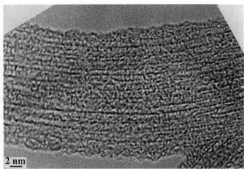

Fig. 16.32 Left: TEM image of a dodecylated SWCNT. The functionalized tubes exhibit a morphology that can be attributed to functionalization by the dodecyl groups (arrow). Reprinted with permission from Liang *et al.* (2004). Copyright 2004 American Chemical Society. Right: TEM image of a rope of SWCNTs prepared by laser ablation after a subsequent HNO_3 and $HNO_3 + H_2SO_4$ treatment. Short and distorted SWNCT walls are associated with amorphous-like material. Reprinted from Monthioux *et al.* (2001) with permission from Elsevier.

For TEM measurements the samples are usually deposited on a holey grid and the structures that protrude the bars into the holes are imagined. It has to be kept in mind that this deposition technique may selectively enrich elongated structures, i.e. long SWCNTs. Another potential problem is the possible modification of the SWCNTs under the electron beam. Kiang *et al.* (1996) noticed the formation of ripples and the deformation of SWCNTs until breakage as a result of electron-beam irradiation. In addition, the deposition of amorphous carbon around CNTs was observed. While in the beginning these effects were explained by e-beam heating, the removal of carbon atoms was later made responsible for these structural changes by Smith and Luzzi (2001). In a combined theoretical and experimental investigation, a minimum energy of 86 keV was determined for the removal of a carbon atom in a knock-on geometry. It was speculated that defects in the SWCNTs, which are primarily invisible in the TEM may lead to nucleation sites for impurity molecules in the atmosphere of the transmission electron microscope. Moreover, due to the presence of functional groups, functionalized SWCNTs may be more sensitive to electron-beam irradiation, as demonstrated by An *et al.* (2003), where structural transformations of fluorinated SWCNTs during TEM investigations were observed. Here, it was found that the functionalized single-walled carbon nanotubes undergo a structural change into a multiwalled structure at 300 keV electron energy, while the unfunctionalized samples were found to be stable under the same conditions.

16.3.3.2 *Atomic force microscopy (AFM)*

In AFM, a cantilever is used that is brought into close contact with the sample surface and the force between sample surface and cantilever is held constant while the surface is scanned (Meyer *et al.* 2004). To investigate SWCNTs, the tubes are commonly deposited from solution onto a substrate surface, for example by spin coating. As the interaction between SWCNTs and substrate material is weak, lateral forces between tip and substrate have to be minimized. Therefore the AFM measurements are performed in the dynamic mode, i.e. in *tapping* or *non-contact* mode (García and Pérez 2002), where the cantilever is oscillating close to its free eigenfrequency and variations of the amplitude or frequency of the cantilever are detected if the AFM tip is in close contact to the sample surface.

AFM images of chemically processed carbon nanotubes can be used for an investigation of the length distributions of the SWCNTs after purification (Furtado *et al.* 2004), oxidative cutting (Ziegler *et al.* 2005) or functionalization (Gu *et al.* 2002; Liu *et al.* 2005). As the topographic image is the result of a convolution of the shape of the tip and the substrate, the lateral dimension of the structures recorded in the AFM image is broadened by approximately the tip diameter. This does not hold for the height of the structures, provided that the amplitude of the tip oscillation is low enough to exclude a deformation of the substrate or the tube. Upon measuring line profiles across the SWCNT structures in the AFM images, it can be estimated whether the tubes are present as bundles or individual tubes (Furtado *et al.* 2004; Hudson *et al.* 2004; Umeyama *et al.* 2007). After exposure of SWCNTs to atomic hydrogen, an increase in the measured height of CVD-grown SWCNTs in AFM images has been observed by Zhang *et al.* (2006). This "swelling" of the tubes was attributed to a hydrogenation of the tube walls. Using harsher conditions, an etching of the SWCNTs was observed in the AFM images, where the smaller diameter tubes were found to be less stable. An increase in the diameter of SWCNTs has also been observed by Hemraj-Benny and Wong (2006) as a result of a silylation of SWCNTs.

Although the direct observation of functional groups in an AFM has not yet been reported, a decoration of functional groups by nanoparticles that are large enough to be observed by AFM has been reported by several groups. An example is shown in Fig. 16.33, where Au nanoparticles were attached to functional groups on SWCNTs, introduced by the Bingel reaction by Coleman *et al.* (2003).

16.3.3.3   *Scanning tunnelling microscopy (STM)*

Scanning tunnelling microscopy (STM) employs a sharp tip that tapers off in a single or only a few atoms. If this tip is brought into close contact within a few Å to a conductive substrate surface and a voltage is applied, a tunnelling current flows between tip and sample surface that depends exponentially on the distance z. In conventional STM images this tunnelling current is held constant while the tip is scanned in the xy-direction over the sample surface. The movement in the z-direction that is necessary to keep the tunnelling current constant is monitored as a function of the x- and y-position and gives the *topography image* of the sample (constant-current mode). Moreover, by measuring the tunnelling current as a function of the applied voltage, local information on the density of states can be obtained by taking the derivative of the tunnelling characteristic, dI/dV (scanning tunnelling spectroscopy, STS (Hamers 1989)). A major breakthrough was the simultaneous imaging and spectroscopy of an individual SWCNT by STM and STS by Wildöer *et al.*

(1998) and Odom *et al.* (1998) as well as the observation of curvature-induced energy gaps in metallic SWCNTs by Ouyang *et al.* (2001).

The ability to obtain atomically resolved images of carbon nanotubes makes STM an extremely exciting tool for the investigation of functionalized SWC-NTs. Early STM investigations on functionalized nanotubes were reported by Kelly *et al.* (1999) on fluorinated SWCNTs. As already discussed above, high degrees of functionalization up to C_2F can be obtained. In their images, bands across the circumference of the tubes were observed that were attributed to fluorinated regions of the SWCNTs, as shown in Fig. 16.34. In these regions no atomic resolution was visible, whereas in the sections in between these bands atomic resolution could be obtained.

In STM studies of fluorinated SWCNTs that were reacted with substituted amines, Zhang *et al.* (2005b) determined different substituent patterns for thiol- and thiophene-terminated moieties. While the thiol-bearing substituents are found to be grouped in bands, the thiophene-terminated groups are distributed uniformly along the sidewalls of the SWCNTs.

In functionalized SWCNTs using the Bingel reaction (Coleman *et al.* 2003), STM images were found to exhibit regular patterns that are attributed to regular arrangements of functional groups about 5 nm apart (Worsley *et al.* 2004). This regular arrangement of the functional groups is attributed to long-range electronic effects that occur upon addition of a functional group to the sidewall of the SWCNT.

STM images of SWCNTs that were covalently functionalized with phenol groups by a 1,3-dipolar cycloaddition reaction are presented by Cahill *et al.* (2004). Although no atomic resolution was obtained, an uneven distribution

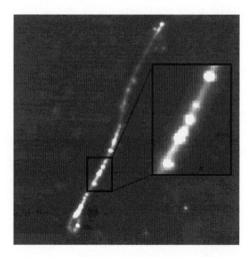

Fig. 16.33 AFM image of a SWCNT functionalized using the Bingel reaction (image size 900 nm × 900 nm). The bright dots are the Au nanoparticles connected to the functional groups by gold–sulfur bonds. Reprinted with permission from Coleman *et al.* (2003). Copyright 2003 American Chemical Society.

Fig. 16.34 STM image of fluorinated SWC-NTs. The bright regions correspond to areas on the tube that are covered by fluorine atoms. Reprinted from Kelly *et al.* (1999) with permission from Elsevier.

of functional groups was observed with densities of about one functional group per 20 nm. In STM images of SWCNTs functionalized by nucleophilic addition of *t*-butyllithium, *t*-butylgroups have been observed by Graupner *et al.* (2006), which exhibit the expected threefold symmetry.

Bonifazi *et al.* (2006) investigated defect-functionalized SWCNTs, where short, oxidized SWCNTs were functionalized with aliphatic chains via an amide reaction. In the pristine, although already HNO_3-treated SWCNTs about 25% of the tubes were found to exhibit irregularities either on the sidewalls or the nanotube ends, which were attributed to defects. After oxidative cutting of the SWCNTs the number of these irregularities increased, in accordance with an increased number of defects on the sidewalls. The STM images of the functionalized SWCNTs, shown in Fig. 16.35, were found to exhibit characteristic brush-like structures at the nanotubes' ends, which are attributed to the aliphatic chains.

Scanning tunnelling microscopy of SWCNTs functionalized by high molecular weight polymers was studied by Czerw *et al.* (2001). Here, regular structures are reported that are ascribed to a "nanotube-driven" crystallization process of the polymer where the structure of the polymer was determined by the chirality of the underlying SWCNT.

One big problem in STM, as in every microscopy technique, is the question whether the results that were obtained using this method are really representative of the whole sample. In most cases STM images are presented that, naturally, only represent a very small part of the sample. In the future, more quantitative investigations would be desirable, e.g. by a statistical analysis that tries to get information about degrees of functionalization and the role of defects and the ends of the nanotubes. One good example is the work by Bonifazi *et al.* (2006), which is a step in this direction. However, using STM, this is an extremely tedious task.

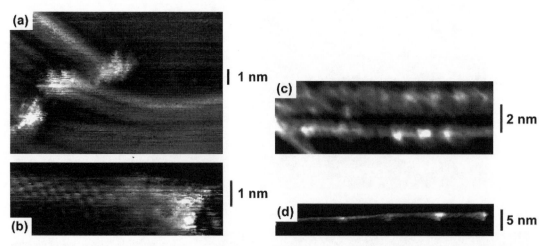

Fig. 16.35 STM images of SWCNTs that were functionalized with aliphatic chains. In a) the ends of the SWCNTs are shown that exhibit a characteristic, brush-like structure, which is attributed to the aliphatic chains. Besides the nanotube terminations (a and b), the sidewalls appear to be functionalized as well, as shown by the lumps on the sidewalls in the two right images (c and d). Reprinted with permission from Bonifazi *et al.* (2006). Copyright 2006 American Chemical Society.

16.4 Conclusion

Meanwhile, the ways to functionalize carbon nanotubes are numerous and we have presented the main reaction pathways above. The defect-group functionalization, which, due to its relative ease and due to the fact that no harsh reaction conditions are required, is a well-established technique and is already used, e.g. for the preparation of composite materials. For the direct sidewall functionalization of carbon nanotubes, the structural integrity of the backbone of the CNT structure is not disturbed, which is expected to have a beneficial effect on the mechanical properties of the functionalized nanotubes. Moreover, a higher degree of functionalization can be obtained with the disadvantage that rather reactive chemicals are necessary to ensure an addition to the sidewall of the SWCNT. For both methods, however, one big problem is the quality of the starting materials used. Different production methods of SWCNTs lead to materials that differ in their properties in terms of diameter, length distribution as well as defect density. In addition, in most cases additional purification techniques are necessary to remove catalyst particles as well as carbonaceous, non-SWCNT by-products. If an oxidizing environment is used, this also leads to an increase in the number of defects, an opening of the tube ends and a shortening of the SWCNTs. These variations in the homogeneity of the starting materials imply that a comparison of different strategies for SWCNT functionalization should be done with great care. Concerning the characterization of functionalized SWCNTs, the main information, the degree of functionalization, is usually given as an average value over a large number of SWCNTs. Together with the non-uniformity of the SWCNTs stated above, hardly anything is known, e.g. on the degree of functionalization for sidewall additions depending on the number of pre-existing defects or the role of the SWCNTs ends. This also holds for the local distribution of the functional groups, i.e. with very few exceptions, hardly anything is known on whether the functional groups are really distributed uniformly along the tubes. While all these issues might not affect the use of functionalized SWCNTs in various applications, from a scientific point of view, a lot of work still has to be done to obtain a complete picture of the chemistry of single-walled carbon nanotubes.

References

Abdula, D., Nguyen, K., Shim, M. *J. Phys. Chem.* C **111**, 17755 (2007).

Alemany, L.B., Zhang, L., Zeng, L.L., Edwards, C.L., Barron, A.R. *Chem. Mater.* **19**, 735 (2007).

An, K.H., Park, K.A., Heo, J.G., Lee, J.Y., Jeon, K.K., Lim, S.C., Yang, C.W., Lee, Y.S., Lee, Y.H. *J. Am. Chem. Soc.* **125**, 3057 (2003).

Araujo, P.T., Doorn, S.K., Kilina, S., Tretiak, S., Einarsson, E., Maruyama, S., Chacham, H., Pimenta, M.A., Jorio, A. *Phys. Rev. Lett.* **98**, 067401 (2007).

Bachilo, S.M., Strano, M.S., Kittrell, C., Hauge, R.H., Smalley, R.E., Weisman, R.B. *Science* **298**, 2361 (2002).

Bagus, P.S., Illas, F., Pacchioni, G., Parmigiani, F.J. *Electron Spectrosc.* **100**, 215 (1999).

Bahr, J.L., Yang, J.P., Kosynkin, D.V., Bronikowski, M.J., Smalley, R.E., Tour, J.M. *J. Am. Chem. Soc.* **123**, 6536 (2001).

Bahr, J.L., Tour, J.M. *Chem. Mater.* **13**, 3823 (2001).

Bahr, J.L., Tour, J.M. *J. Mater. Chem.* **12**, 1952 (2002).

Banerjee, S., Hemraj-Benny, T., Wong, S.S. *Adv. Mater.* **17**, 17 (2005).

Banerjee, S., Wong, S.S. *J. Phys. Chem. B* **106**, 12144 (2002).

Banerjee, S., Wong, S.S. *Nano Lett.* **4**, 1445 (2004a).

Banerjee, S., Wong, S.S. *J. Am. Chem. Soc.* **126**, 2073 (2004b).

Basiuk, E., Monroy-Pelaez, M., Puente-Lee, I., Basiuk, V. *Nano Lett.* **4**, 863 (2004).

Beamson, G., Briggs, D. *High Resolution XPS of Organic Polymers* (John Wiley & Sons Ltd., Chichester, 1992).

Bethune, D.S., Klang, C.H., de Vries, M.S., Gorman, G., Savoy, R., Beyers, J.V.R. *Nature* **363**, 605 (1993).

Blackburn, J., Engtrakul, C., McDonald, T., Dillon, A., Heben, M. *J. Phys. Chem. B* **110**, 25551 (2006).

Blake, R., Gun'ko, Y.K., Coleman, J., Cadek, M., Fonseca, A., Nagy, J.B., Blau, W.J. *J. Am. Chem. Soc.* **126**, 10226 (2004).

Bonifazi, D., Nacci, C., Marega, R., Campidelli, S., Ceballos, G., Modesti, S., Meneghetti, M., Prato, M. *Nano Lett.* **6**, 1408 (2006).

Boul, P.J., Liu, J., Mickelson, E.T., Huffman, C.B., Ericson, L.M., Chiang, I.W., Smith, K.A., Colbert, D.T., Hauge, R.H., Margrave, J.L., Smalley, R.E. *Chem. Phys. Lett.* **310**, 367 (1999).

Bronikowski, M.J., Willis, P.A., Colbert, D.T., Smith, K.A., Smalley, R.E. *J. Vac. Sci. Technol. A* **19**, 1800 (2001).

Brown, S.D.M., Jorio, A., Corio, P., Dresselhaus, M.S., Dresselhaus, G., Saito, R., Kneipp, K. *Phys. Rev. B* **63**, 155414 (2001a).

Brown, S.D.M., Jorio, A., Dresselhaus, M.S., Dresselhaus, G. *Phys. Rev. B* **64**, 073403 (2001b).

Cahill, L.S., Yao, Z., Adronov, A., Penner, J., Moonoosawmy, K.R., Kruse, P., Goward, G.R. *J. Phys. Chem. B* **108**, 11412 (2004).

Caudal, N., Saitta, A.M., Lazzeri, M., Mauri, F. *Phys. Rev. B* **75**, 115423 (2007).

Chattopadhyay, D., Lastella, S., Kim, S., Papadimitrakopoulos, F. *J. Am. Chem. Soc.* **124**, 728 (2002).

Chattopadhyay, J., Sadana, A., Liang, F., Beach, J., Xiao, Y., Hauge, R., Billups, W. *Org. Lett.* **7**, 4067 (2005).

Chen, J., Hamon, M.A., Hu, H., Chen, Y.S., Rao, A.M., Eklund, P.C., Haddon, R.C. *Science* **282**, 95 (1998a).

Chen, J., Rao, A.M., Lyuksyutov, S., Itkis, M.E., Hamon, M.A., Hu, H., Cohn, R.W., Eklund, P.C., Colbert, D.T., Smalley, R.E., Haddon, R.C. *J. Phys. Chem. B* **105**, 2525 (2001).

Chen, S., Shen, W., Wu, G., Chen, D., Jiang, M. *Chem. Phys. Lett.* **402**, 312 (2005).

Chen, Y., Haddon, R.C., Fang, S., Rao, A.M., Lee, W.H., Dickey, E.C., Grulke, E.A., Pendergrass, J.C., Chavan, A., Haley, B.E., Smalley, R.E. *J. Mater. Res.* **13**, 2423 (1998b).

Chen, Z., Thiel, W., Hirsch, A. *Chem. Phys. Chem.* **4**, 93 (2003).

Chiu, P.W., Duesberg, G.S., Dettlaff-Weglikowska, U., Roth, S. *Appl. Phys. Lett.* **80**, 3811 (2002).

Chou, S.G., Son, H., Kong, J., Jorio, A., Saito, R., Zheng, M., Dresselhaus, G., Dresselhaus, M.S. *Appl. Phys. Lett.* **90**, 131109 (2007).

Coleman, J.N., Khan, U., Blau, W.J., Gun'ko, Y.K. *Carbon* **44**, 1624 (2006).

Coleman, K.S., Chakraborty, A.K., Bailey, S.R., Sloan, J., Alexander, M. *Chem. Mater.* **19**, 1076 (2007).

Coleman, K.S., Bailey, S.R., Fogden, S., Green, M.L.H. *J. Am. Chem. Soc.* **125**, 8722 (2003).

Cui, J., Burghard, M., Kern, K. *Nano Lett.* **3**, 613 (2003).

Czerw, R., Guo, Z., Ajayan, P.M., Sun, Y.P., Carroll, D.L. *Nano Lett.* **1**, 423 (2001).

Delgado, J.L., de la Cruz, P., Langa, F., Urbina, A., Casado, J., López Navarrete, J.T. *Chem. Commun.* 1734 (2004).

Díaz, J., Paolicelli, G., Ferrer, S., Comin, F. *Phys. Rev. B* **54**, 8064 (1996).

Dossot, M., Gardien, F., Mamane, V., Fort, Y., Liu, J., Vigolo, B., Humbert, B., McRae, E. *J. Phys. Chem. C* **111**, 12199 (2007).

Dresselhaus, M., Dresselhaus, G., Saito, R., Jorio, A. *Phys. Rep.* **409**, 47 (2005).

Duesberg, G.S., Loa, I., Burghard, M., Syassen, K., Roth, S. *Phys. Rev. Lett.* **85**, 5436 (2000).

Dyke, C.A., Tour, J.M. *J. Am. Chem. Soc.* **125**, 1156 (2003a).

Dyke, C.A., Tour, J.M. *Nano Lett.* **3**, 1215 (2003b).

Ericson, L.M., Pehrsson, P.E. *J. Phys. Chem. B* **109**, 20276 (2005).

Estrade-Szwarckopf, H. *Carbon* **42**, 1713 (2004).

Fagan, J.A., Simpson, J.R., Landi, B.J., Richter, L.J., Mandelbaum, I., Bajpai, V., Ho, D.L., Raffaelle, R., Hight Walker, A.R., Bauer, B.J., Hobbie, E.K. *Phys. Rev. Lett.* **98**, 147402 (2007a).

Fagan, J.A., Simpson, J.R., Bauer, B.J., De Paoli Lacerda, S.H., Becker, M.L., Chun, J., Migler, R.B., Walker, A.R.H., Hobbie, E.K. *J. Am. Chem. Soc.* **129**, 10607 (2007b).

Fantini, C., Jorio, A., Souza, M., Strano, M.S., Dresselhaus, M.S., Pimenta, M.A. *Phys. Rev. Lett.* **93**, 147406 (2004).

Fantini, C., Usrey, M., Strano, M. *J. Phys. Chem. C* **111**, 17941 (2007).

Furtado, C.A., Kim, U.J., Gutierrez, H.R., Pan, L., Dickey, E.C., Eklund, P.C. *J. Am. Chem. Soc.* **126**, 6095 (2004).

García, R., Pérez, R. *Surf. Sci. Rep.* **47**, 197 (2002).

Georgakilas, V., Kordatos, K., Prato, M., Guldi, D., Holzinger, M., Hirsch, A. *J. Am. Chem. Soc.* **124**, 760 (2002a).

Georgakilas, V., Voulgaris, D., Vázquez, E., Prato, M., Guldi, D.M., Kukovecz, A., Kuzmany, H. *J. Am. Chem. Soc.* **124**, 14318 (2002b).

Goldoni, A., Larciprete, R., Gregoratti, L., Kaulich, B., Kiskinova, M., Zhang, Y., Dai, H., Sangaletti, L., Parmigiani, F. *Appl. Phys. Lett.* **80**, 2165 (2002).

Graupner, R. *J. Raman Spectrosc.* **38**, 673 (2007).

Graupner, R., Abraham, J., Wunderlich, D., Vencelova, A., Lauffer, P., Röhrl, J., Hundhausen, M., Ley, L., Hirsch, A. *J. Am. Chem. Soc.* **128**, 6683 (2006).

Gu, Z., Peng, H., Hauge, R.H., Smalley, R.E., Margrave, J.L. *Nano Lett.* **2**, 1009 (2002).

Haddon, R., Itkis, M. In S. Freiman, S. Hooker, K. Migler, S. Arepalli, (eds), *Measurement Issues in Single Wall Carbon Nanotubes*. NIST Recommended Practice Guide (2008).

Haddon, R.C. *Acc. Chem. Res.* **21**, 243 (1988).

Haerle, R., Riedo, E., Pasquarello, A., Baldereschi, A. *Phys. Rev. B* **65**, 45101 (2001).

Hamers, R.J. *Annu. Rev. Phys. Chem.* **40**, 531 (1989).

Hamon, M.A., Hu, H., Bhowmik, P., Niyogi, S., Zhao, B., Itkis, M.E., Haddon, R.C. *Chem. Phys. Lett.* **347**, 8 (2001).

Hamon, M.A., Chen, J., Hu, H., Chen, Y., Itkis, M.E., Rao, A.M., Eklund, P.C., Haddon, R.C. *Adv. Mater.* **11**, 834 (1999).

Hashimoto, A., Suenaga, K., Gloter, A., Urita, K., Iijima, S. *Nature* **430**, 870 (2004).

Heller, D.A., Barone, P.W., Swanson, J.P., Mayrhofer, R.M., Strano, M.S. *J. Phys. Chem. B* **108**, 6905 (2004).

Hemraj-Benny, T., Wong, S. *Chem. Mater.* **18**, 4827 (2006).

Hirsch, A. *Angew. Chem. Int. Ed.* **41**, 1853 (2002).

Hirsch, A., Brettreich, M. *Fullerenes—Chemistry and Reactions* (Wiley-VCH, Weinheim, 2005).

Hirsch, A., Vostrowsky, O. *Top. Curr. Chem.* **245**, 193 (2005).

Holzinger, M., Abraham, J., Whelan, P., Graupner, R., Ley, L., Hennrich, F., Kappes, M., Hirsch, A. *J. Am. Chem. Soc.* **125**, 8566 (2003).

Holzinger, M., Vostrowsky, O., Hirsch, A., Hennrich, F., Kappes, M., Weiss, R., Jellen, F. *Angew. Chem. Int. Ed.* **40**, 4002 (2001).

Hudson, J.L., Casavant, M.J., Tour, J.M. *J. Am. Chem. Soc.* **126**, 11158 (2004).

Hüfner, S. *Photoelectron-Spectroscopy*, 2nd edn. (Springer-Verlag, Berlin, 1995).

Iijima, S. *Nature* **354**, 56 (1991).

Iijima, S., Ichihashi, T. *Nature* **363**, 603 (1993).

Itkis, M., Perea, D., Niyogi, S., Rickard, S., Hamon, M., Hu, H., Zhao, B., Haddon, R. *Nano Lett.* **3**, 309 (2003).

Jorio, A., Fantini, C., Pimenta, M.A., Capaz, R.B., Samsonidze, G.G., Dresselhaus, G., Dresselhaus, M.S., Jiang, J., Kobayashi, N., Grüneis, A., Saito, R. *Phys. Rev. B* **71**, 075401 (2005).

Jorio, A., Souza Filho, A.G., Dresselhaus, G., Dresselhaus, M.S., Swan, A.K., Ünlü, M.S., Goldberg, B.B., Pimenta, M.A., Hafner, J.H., Lieber, C.M., Saito, R. *Phys. Rev. B* **65**, 155412 (2002).

Joselevich, E. *Chem. Phys. Chem.* **5**, 619 (2004).

Kataura, H., Kumazawa, Y., Kojima, N., Maniwa, Y., Umezu, I., Masubuchi, S., Kazama, S., Ohtsuka, Y., Suzuki, S., Achiba, Y. *Mol. Cryst. Liq. Cryst.* **340**, 757 (2000).

Kataura, H., Kumazawa, Y., Maniwa, Y., Umezu, I., Suzuki, S., Ohtsuka, Y., Achiba, Y. *Synth. Met.* **103**, 2555 (1999).

Katz, E., Willner, I. *Chem. Phys. Chem.* **5**, 1084 (2004).

Kelly, K.F., Chiang, I.W., Mickelson, E.T., Hauge, R.H., Margrave, J.L., Wang, X., Scuseria, G.E., Radloff, C., Halas, N. *Chem. Phys. Lett.* **313**, 445 (1999).

Khabashesku, V.N., Billups, W.E., Margrave, J.L. *Acc. Chem. Res.* **35**, 1087 (2002).

Kiang, C.H., Goddard, W., Beyers, R., Bethune, D. *J. Phys. Chem.* **100**, 3749 (1996).

Kürti, J., Zólyomi, V., Grüneis, A., Kuzmany, H. *Phys. Rev. B* **65**, 165433 (2002).

Leiro, J.A., Heinonen, M.H., Laiho, T., Batirev, I.G. *J. Electron Spectrosc.* **128**, 205 (2003).

Li, B., Shi, Z., Lian, Y., Gu, Z. *Chem. Lett.* **30**, 598 (2001).

Lian, Y.F., Maeda, Y., Wakahara, T., Akasaka, T., Kazaoui, S., Minami, N., Choi, N., Tokumoto, H. *J. Phys. Chem. B* **107**, 12082 (2003).

Lian, Y.F., Maeda, Y., Wakahara, T., Nakahodo, T., Akasaka, T., Kazaoui, S., Minami, N., Shimizu, T., Tokumoto, H. *Carbon* **43**, 2750 (2005).

Liang, F., Sadana, A.K., Peera, A., Chattopadhyay, J., Gu, Z., Hauge, R.H., Billups, W.E. *Nano Lett.* **4**, 1257 (2004).

Lin, T., Bajpai, V., Ji, T., Dai, L. *Aust. J. Chem.* **56**, 635 (2003).

Liu, J., Zubiri, M.R.I., Dossot, M., Vigolo, B., Hauge, R.H., Fort, Y., Ehrhardt, J.J., McRae, E. *Chem. Phys. Lett.* **430**, 93 (2006).

Liu, J., Rinzler, A.G., Dai, H., Hafner, J.H., Bradley, R.K., Boul, P.J., Lu, A., Iverson, T., Shelimov, K., Huffman, C.B., Rodriguez-Macias, F., Shon, Y.S., Lee, T.R., Colbert, D.T., Smalley, R.E. *Science* **280**, 1253 (1998).

Liu, J., Zubiri, M.R.i., Vigolo, B., Dossot, M., Fort, Y., Ehrhardt, J.J., McRae, E. *Carbon* **45**, 885 (2007).

Liu, M.H., Yang, Y.L., Zhu, T., Liu, Z.F. *Carbon* **43**, 1470 (2005).

Lu, X., Tian, F., Feng, Y., Xu, X., Wang, N., Zhang, Q. *Nano Lett.* **2**, 1325 (2002a).

Lu, X., Tian, F., Wang, N., Zhang, Q. *Org. Lett.* **4**, 4313 (2002b).

Luo, Z.T.F., Li, R., Kim, S.N., Papadimitrakopoulos, F. *Phys. Rev. B* **70** (2004).

Maultzsch, J., Reich, S., Thomsen, C., Webster, S., Czerw, R., Carroll, D.L., Vieira, S.M.C., Birkett, P.R., Rego, C.A. *Appl. Phys. Lett.* **81**, 2647 (2002).

Maultzsch, J., Telg, H., Reich, S., Thomsen, C. *Phys. Rev. B* **72**, 205438 (2005).

Mawhinney, D.B., Naumenko, V., Kuznetsova, A., Jr., J.T.Y., Liu, J., Smalley, R.E. *Chem. Phys. Lett.* **324**, 213 (2000).

Meyer, E., Hug, H.J., Bennewitz, R. *Scanning Probe Microscopy* (Springer Verlag, Berlin, 2004).

Mickelson, E.T., Huffman, C.B., Rinzler, A.G., Smalley, R.E., Hauge, R.H., Margrave, J.L. *Chem. Phys. Lett.* **296**, 188 (1998).

Moghaddam, M., Taylor, S., Gao, M., Huang, S., Dai, L., McCall, M. *Nano Lett.* **4**, 89 (2004).

Monthioux, M., Smith, B.W., Burteaux, B., Claye, A., Fischer, J.E., Luzzi, D.E. *Carbon* **39**, 1251 (2001).

Nair, N., Kim, W.J., Usrey, M.L., Strano, M.S. *J. Am. Chem. Soc.* **129**, 3946 (2007).

Nair, N., Usrey, M., Kim, W.J., Braatz, R., Strano, M. *Anal. Chem.* **78**, 7689 (2006).

Nguyen, K.T., Gaur, A., Shim, M. *Phys. Rev. Lett.* **98**, 145504 (2007).

Nikolaev, P., Bronikowski, M.J., Bradley, R.K., Rohmund, F., Colbert, D.T., Smith, K., Smalley, R.E. *Chem. Phys. Lett.* **313**, 91 (1999).

Niyogi, S., Hamon, M.A., Hu, H., Zhao, B., Bhowmik, P., Sen, R., Itkis, M.E., Haddon, R.C. *Acc. Chem. Res.* **35**, 1105 (2002).

O'Connell, M.J., Bachilo, S.M., Huffman, C.B., Moore, V.C., Strano, M.S., Haroz, E.H., Rialon, K.L., Boul, P.J., Noon, W.H., Kittrell, C., Ma, J., Hauge, R.H., Weisman, R.B., Smalley, R.E. *Science* **297**, 593 (2002).

O'Connell, M.J., Sivaram, S., Doorn, S.K. *Phys. Rev. B* **69**, 235415 (2004).

Odom, T.W., Huang, J.L., Kim, P., Lieber, C.M. *Nature* **391**, 62 (1998).

Ouyang, M., Huang, J.L., Cheung, C.L., Lieber, C.M. *Science* **292**, 702 (2001).

Paillet, M., Michel, T., Meyer, J.C., Popov, V.N., Henrard, L., Roth, S., Sauvajol, J.L. *Phys. Rev. Lett.* **96**, 257401 (2006).

Pekker, S., Salvetat, J.P., Jakab, E., Bonard, J.M., Forro, L. *J. Phys. Chem. B* **105**, 7938 (2001).

Peng, H.P., Reverdy, P., Khabashesku, V.N., Margrave, J.L. *Chem. Comm.* **362** (2003a).

Peng, H.Q., Alemany, L.B., Margrave, J.L., Khabashesku, V.N. *J. Am. Chem. Soc.* **125**, 15174 (2003b).

Perebeinos, V., Tersoff, J., Avouris, P. *Phys. Rev. Lett.* **92**, 257402 (2004).

Piscanec, S., Lazzeri, M., Robertson, J., Ferrari, A.C., Mauri, F. *Phys. Rev. B* **75**, 035427 (2007).

Pócsik, I., Hundhausen, M., Kos, M., Ley, L.J. *Non-Cryst. Solids* **227–230**, 1083 (1998).

Popov, V.N., Lambin, P. *Phys. Rev. B* **73**, 85407 (2006).

Price, B.K., Hudson, J.L., Tour, J.M. *J. Am. Chem. Soc.* **127**, 14867 (2005).

Qin, L.C. *Rep. Prog. Phys.* **69**, 2761 (2006).

Qin, Y., Shi, J., Wu, W., Li, X., Guo, Z.X., Zhu, D. *J. Phys. Chem. B* **107**, 12899 (2003).

Rafailov, P.M., Maultzsch, J., Thomsen, C., Kataura, H. *Phys. Rev. B* **72**, 045411 (2005).

Saini, R., Chiang, I., Peng, H., Smalley, R., Billups, W., Hauge, R., Margrave, J. *J. Am. Chem. Soc.* **125**, 3617 (2003).

Saito, R., Grneis, A., Samsonidze, G.G., Brar, V.W., Dresselhaus, G., Dresselhaus, M.S., Jorio, A., Cancado, L.G., Fantini, C., Pimenta, M.A., Filho, A.G.S. *New J. Phys.* **5**, 157.1 (2003).

Scofield, J.H. *J. Electron Spectrosc.* **8**, 129 (1976).

Sette, F., Wertheim, G.K., Ma, Y., Meigs, G., Modesti, S., Chen, C.T. *Phys. Rev. B* **41**, 9766 (1990).

Smith, B.W., Luzzi, D.E. *J. Appl. Phys.* **90**, 3509 (2001).

Stephenson, J.J., Hudson, J.L., Azad, S., Tour, J.M. *Chem. Mater.* **18**, 374 (2006).

Stevens, J.L., Huang, A.Y., Peng, H., Chiang, I.W., Khabashesku, V.N., Margrave, J.L. *Nano Lett.* **3**, 331 (2003).

Strano, M.S. *J. Am. Chem. Soc.* **125**, 16148 (2003).

Strano, M.S., Dyke, C.A., Usrey, M.L., Barone, P.W., Allen, M.J., Shan, H., Kittrell, C., Hauge, R.H., Tour, J.M., Smalley, R.E. *Science* **301**, 1519 (2003).

Sun, Y.P., Fu, K., Lin, Y., Huang, W. *Acc. Chem. Res.* **35**, 1096 (2002).

Syrgiannis, Z., Hauke, F., Röhrl, J., Hundhausen, M., Graupner, R., Elemes, Y., Hirsch, A. *Eur. J. Org. Chem.* 2544 (2008).

Tagmatarchis, N., Georgakilas, V., Prato, M., Shinohara, H. *Chem. Commun.* 2010 (2002).

Tasis, D., Tagmatarchis, N., Bianco, A., Prato, M. *Chem. Rev.* **106**, 1105 (2006).

Tasis, D., Tagmatarchis, N., Georgakilas, V., Prato, M. *Chem. Eur. J.* **9**, 4000 (2003).

Telg, H., Maultzsch, J., Reich, S., Hennrich, F., Thomsen, C. *Phys. Rev. Lett.* **93**, 177401 (2004).

Telg, H., Maultzsch, J., Reich, S., Thomsen, C. *Phys. Rev. B* **74**, 115415 (2006).

Thomsen, C., Reich, S. *Phys. Rev. Lett.* **85**, 5214 (2000).

Tuinstra, F., Koenig, J.L. *J. Chem. Phys.* **53**, 1126 (1970).

Umek, P., Seo, J., Hernadi, K., Mrzel, A., Pechy, P., Mihailovic, D., Forro, L. *Chem. Mater.* **15**, 4751 (2003).

Umeyama, T., Tezuka, N., Fujita, M., Matano, Y., Takeda, N., Murakoshi, K., Yoshida, K., Isoda, S., Imahori, H. *J. Phys. Chem. C* **111**, 9734 (2007).

van Attekum, P.M.T.M., Wertheim, G.K. *Phys. Rev. Lett.* **43**, 1896 (1979).

Vidano, R.P., Fischbach, D.B., Willis, L.J., Loehr, T.M. *Solid State Commun.* **39**, 341 (1981).

Viswanathan, G., Chakrapani, N., Yang, H., Wei, B., Chung, H., Cho, K., Ryu, C.Y., Ajayan, P.M. *J. Am. Chem. Soc.* **125**, 9258 (2003).

Wang, Y.Q., Sherwood, P.M.A. *Chem. Mater.* **16**, 5427 (2004).

Weisman, R., Bachilo, S. *Nano Lett.* **3**, 1235 (2003).

Wildöer, J.W., Venema, L.C., Rinzler, A.G., Smalley, R.E., Dekker, C. *Nature* **391**, 59 (1998).

Wong, S.S., Joselevich, E., Woolley, A.T., Cheung, C.L., Lieber, C.M. *Nature* **394**, 52 (1998).

Worsley, K.A., Moonoosawmy, K.R., Kruse, P. *Nano Lett.* **4**, 1541 (2004).

Wu, Y., Maultzsch, J., Knoesel, E., Chandra, B., Huang, M., Sfeir, M.Y., Brus, L.E., Hone, J., Heinz, T.F. *Phys. Rev. Lett.* **99**, 027402 (2007).

Wunderlich, D., Hauke, F., Hirsch, A. *J. Mater Chem.* **18**, 1493 (2008a).

Wunderlich, D., Hauke, F., Hirsch, A. *Chem. Eur. J.* **14**, 1607 (2008b).

Xu, Y., Wang, X., Tian, R., Li, S., Wan, L., Li, M., You, H., Li, Q., Wang, S. *Appl. Surf. Sci.* **254**, 2431 (2008).

Yeh, J.J. *Atomic Calculation of Photoionization Cross Sections and Asymmetry Parameters* (Gordon & Breach, Langhorne, PA, 1993).

Ying, Y., Saini, R.K., Liang, F., Sadana, A.K., Billups, W.E. *Org. Lett.* **5**, 1471 (2003).

Yokoi, T., Iwamatsu, S., Komai, S., Hattori, T., Murata, S. *Carbon* **43**, 2869 (2005).

Zhang, G., Qi, P., Wang, X., Lu, Y., Mann, D., Li, X., Dai, H. *J. Am. Chem. Soc.* **128**, 6026 (2006).

Zhang, L., Kiny, V.U., Peng, H.Q., Zhu, J., Lobo, R.F.M., Margrave, J.L., Khabashesku, V.N. *Chem. Mater.* **16**, 2055 (2004).

Zhang, L., Yang, J.Z., Edwards, C.L., Alemany, L.B., Khabashesku, V.N., Barron, A.R. *Chem. Commun.* 3265 (2005a).

Zhang, L., Zhang, J., Schmandt, N., Cratty, J., Khabashesku, V.N., Kelly, K.F., Barron, A.R. *Chem. Commun.* 5429 (2005b).

Zhou, W., Ooi, Y.H., Russo, R., Papanek, P., Luzzi, D.E., Fischer, J.E., Bronikowski, M.J., Willis, P.A., Smalley, R.E. *Chem. Phys. Lett.* **350**, 6 (2001).

Ziegler, K.J., Gu, Z., Peng, H., Flor, E.L., Hauge, R.H., Smalley, R.E. *J. Am. Chem. Soc.* **127**, 1541 (2005).

Quantum-theoretical approaches to proteins and nucleic acids

17

Mauro Boero and Masaru Tateno

17.1 Introduction

Despite remarkable advances in theoretical approaches, numerical algorithms, computer codes and computer performance, quantum calculations of biomolecules represent still one of the most challenging tasks. The obvious reason is the intrinsic complexity of quantum mechanics for a many-body system such as a protein or a nucleic acid. If conformational changes are the main issue of the problem, there is no need to use quantum mechanics, since many versatile codes, accurate classical model potentials (or *force fields*, as they are generally indicated, referring to the forces computed through their gradients) and algorithms exist. Instead, if electronic-structure modifications, chemical reactions, polarization effects and charge-transfer processes are involved, the use of a quantum approach is unavoidable, since these phenomena are inherently *quantum* and can be correctly described only if electrons are treated in terms of wavefunctions given by the solution of the corresponding Schrödinger or Dirac equations (or one of their many-body generalizations). As a matter of fact, it must be acknowledged that nanobiochemistry and nanobiology, although still at a pioneering stage, are now pushing the studies of proteins and nucleic acids towards the domain of quantum mechanics; typical examples are the design of RNA enzymes in cancer gene therapy (Pley *et al.* 1994), or the charge transport in DNA with related oxidative damage problems (Douki *et al.* 2004) and futuristic applications in nanoelectronic devices (Di Ventra and Zwolak 2004). The emerging field of nanoscience and the intrinsic complexity of proteins and nucleic acids brought to the attention of the theoretical community the problem of simulations of system on the nanometer scale. In the specific cases treated in this chapter, one has to deal with a system size ranging from hundreds of atoms, as in the 10-residues chignolin (Honda *et al.* 2004) or the 20-residues Trp-cage (Neidigh *et al.* 2002) (Fig. 17.1), to more than 200 000 atoms, as in the case of protein–nucleic acid interactions in the aminoacyl-tRNA synthetases (Nagel *et al.* 1991).

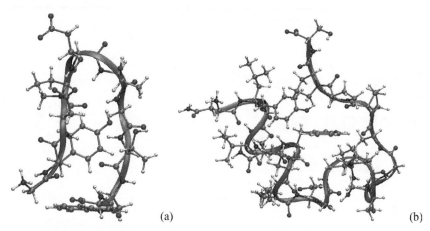

Fig. 17.1 Smallest known proteins: the 10-residues chignolin (a) and the 20-residues Trp-cage (b). The ribbon is the conventional schematic drawing joining all the sp^3 C atoms (C$_\alpha$) along the chain. Atoms are shown as stick and balls where light gray balls indicate C, dark gray O, black N and white H.

This is often accompanied by a growing chemical complexity and a huge variety of possible different conformations representing stable or metastable structures (*multiple minima problem*), that make much more challenging the calculation of even the "simple" electronic structure of a protein with respect, for instance, to a bulk piece of silicon of comparable size. On the other hand, the computational effort required by a full quantum treatment of the atoms makes even the optimization of the wavefunctions of the valence electrons sometimes prohibitively expensive for such large biomolecules.

An additional difficulty arises from the fact that the structure of proteins and nucleic acids obtained experimentally, and whose coordinates are available via the Protein Data Bank or are provided to theoreticians directly by the experimental groups, are generally obtained by X-ray scattering. Now, it is well known that X-rays are insensitive to hydrogen atoms, thus only the coordinates of heavier atoms can be supplied by these experimental probes. Hydrogens are then added at a later analysis stage and this leaves in general a certain freedom in the protonation state of the various residues or bases; to complement the missing information, a careful analysis a posteriori of the local hydrogen bond (H-bond) network and theoretical estimations of the pK are performed to recover the coordinates of H atoms. While the former operation is easily done with most of the available graphic software, the second one is more demanding and generally implies the solution of the Poisson–Boltzmann equation. The method has been well assessed over the years and makes use of the Poisson equation,

$$\nabla^2 V(\mathbf{r}) = \frac{4\pi}{\varepsilon} \cdot \rho(\mathbf{r}), \qquad (17.1)$$

where the total electrostatic potential $V(\mathbf{r})$ is generated by the charge distribution $\rho(\mathbf{r})$ due to all the positive and negative ions composing the system, thus including also the solvent, if present. The dielectric function ε is generally assumed to be a constant and typical values $\varepsilon \sim 80$ represent the standard assumption in the case of aqueous solutions (Andelman 1995). In the limit of a continuum density distribution, the chemical potential μ_i of the ith charge in the system is defined by

$$\mu_i(\mathbf{r}) = e Z_i V(\mathbf{r}) + k_B T \ln \rho_i(\mathbf{r}), \qquad (17.2)$$

where e is the electron charge, Z_i the valence of the ith ion in the system, V the electrostatic potential, k_B the Boltzmann constant, T the temperature and ρ_i the charge density of the ith ion. The first addend in the right-hand side of eqn (17.2) comes from the pure electrostatic contribution, while the second one is due to the entropy contribution of the ions under the hypothesis of weak solution. This leads to a Boltzmann distribution of the charges $\rho_i(\mathbf{r}) = \rho_i^0 \exp(-e Z_i V(\mathbf{r})/k_B T)$ and the explicit form of the Poisson–Boltzmann equation reads

$$\nabla^2 V(\mathbf{r}) = \frac{4\pi \cdot e}{\varepsilon} \left[Z_+ \rho_+^0 e^{-e Z_+ V(\mathbf{r})/k_B T} + Z_- \rho_-^0 e^{-e Z_- V(\mathbf{r})/k_B T} \right], \quad (17.3)$$

in which the contributions to the total charge density due to positive and negative ions have been written separately and indicated by a subscript for the sake of clarity. This equation is non-linear in the electrostatic potential $V(\mathbf{r})$ and can be solved numerically in a self-consistently way, as done by several computer codes (Baker *et al.* 2001) and web-based services (Miteva *et al.* 2005). The original works date back to the early 1900s after Gouy (1910) and Chapman (1913). Besides being a milestone in the field, they are still widely applicable whenever charged and flexible membranes are treated. The electrostatic free energy is computed by integrating over the volume Ω,

$$F_{el} = \frac{\varepsilon}{8\pi} \int_\Omega [\nabla V(\mathbf{r})]^2 \, d^3 r + k_B T \cdot$$

$$\int_\Omega \{\rho_+(\mathbf{r}) \ln [\rho_+(\mathbf{r})/\rho_0(\mathbf{r})] + \rho_-(\mathbf{r}) \ln [\rho_-(\mathbf{r})/\rho_0(\mathbf{r})] - [\rho_+(\mathbf{r})$$

$$+ \rho_-(\mathbf{r}) - 2\rho(\mathbf{r})]\} \, d^3 r, \qquad (17.4)$$

and then it is possible to estimate the theoretical pK by introducing the free-energy difference between two (or more) states, A and B, $\Delta F_{el} = F_{el}(B) - F_{el}(A)$ (Karplus and Bashford 1990; Yang *et al.* 1993).

$$pK = \frac{1}{\ln 10 k_B T} \Delta F_{el}. \qquad (17.5)$$

In principle, the charge-density distribution $\rho(\mathbf{r})$ could be obtained via a full quantum treatment of the electrons, without assuming any continuum limit or *ad hoc* hypothesis. However, in practice, this is a computationally demanding task that has so far been attempted only a few times, due to the size problems mentioned at the beginning of the present section, that limit full quantum approaches to small portions of the whole protein or nucleic acid. Furthermore, the same expensive calculation must be repeated on the same system differing only in the number of saturating H for a complete characterization of the protonated state. Quantum calculations of the structure and electronic states of biomolecules at different protonation states represent indeed one of the very early applications of quantum mechanics at the semi-empirical Austin Model 1 (AM1) level to proteins and nucleic acids (Lahti *et al.* 1997). Only recently, with the introduction, after Kohn and Sham (1965) of the computationally less demanding density-functional theory (DFT) pK and redox potentials became affordable quantum calculations (Ullmann *et al.* 2002).

Once the protonation state of the system has been addressed, the protein or nucleic acid system has to be solvated in water, since this would correspond to the natural environment of the biomolecule. Indeed, very few cases have been reported over the years about experiments conducted on completely dry proteins; in the particular case of the Trp-cage shown in Fig. 17.1, only recent fluorescence experiments have been attempted on the unsolvated protein (Iavarone and Parks 2005). In the general case, biomolecular systems form hydrogen bonds with the solvating H_2O with all the exposed hydrophilic parts and repel water with the hydrophobic moieties. To this aim, several tools in most of the available software packages can very easily add an (almost) arbitrarily large amount of solvent water around the crystallographic structure of the protein or nucleic acid at standard liquid water density ($1.0 \, \text{g/cm}^3$); the system constructed in this way can then be equilibrated via classical molecular dynamics. Of course, the addition of the solvent water implies a dramatic increase in the number of atoms constituting the whole system that even for the smallest proteins shown in Fig. 17.1 amounts to 1130 atoms in the case of chignolin and 2306 atoms in the case of Trp-cage.

The various steps described so far represent just the preparation stage of the system that one intends to study; the major part of the work has still to begin and the calculation of the electronic structure is the first obstacle to face. To this aim, the standard procedures adopted so far fall in one of these two categories: (i) full quantum-mechanical calculations on reduced models extracted from the real system, (ii) hybrid quantum-mechanics/molecular-mechanics (QM/MM) approaches in which a small (few tens, hundreds or, at very best, about one thousand of atoms) portion is treated via quantum mechanics and the largest part via classical force fields. Linearized algorithms that scale as $O(N)$, where N is the number of atoms, can still be regarded as belonging to the first category, since they can partly solve the problem of the size of the system up to few thousands of atoms (de Pablo *et al.* 2000). The same applies to partitioning methods in which the whole system is divided into smaller portions and each subsystem computed separately, thus transforming a true full computation into a chain of microcanonical ensembles. This is done, for instance, in divide-and-conquer methods (Zhao and Yang 1995), in the linear scaling density functional methods (Bowler *et al.* 2000) or in the fragment molecular orbitals (FMO) approach (Kitaura *et al.* 2001). However, it must be remarked that biomolecules are generally on the verge of the reach also of these approaches, in the sense that very rarely can one go beyond the calculation of the electronic structure, of the geometry optimization or a few ps dynamics (Gogonea *et al.* 2000). Long dynamical simulations are routinely conducted via classical force fields to sample the phase space, then uncorrelated configurations are extracted from these ns-long trajectories and static quantum calculations (Barnett *et al.* 2001) or shorter *ab-initio* molecular-dynamics runs (Gervasio *et al.* 2005) are performed subsequently.

This brings us to the second problem involved in proteins and nucleic acids: they are chemically active systems that undergo folding and unfolding, docking, cleavage and formation of chemical bonds, charge dislocation, etc. These are exactly the processes in which one is generally interested. Yet, most of the time, they are beyond the reach of any quantum and classical dynamical simulation. In fact, the typical timescale of a biochemical reaction falls in the range from microsecond (μs) to millisecond (ms) or even second, since rather large energy barriers must be overcome. In a schematic view, the system makes a transition from a certain reactant A to a generic product B by overcoming a free-energy barrier ΔF. Typical values of ΔF are some tens of kcal/mol and these values are considerably greater than the fluctuations of the system during a standard molecular-dynamics simulation and, on average, these oscillations are of the order of $k_B T = 0.6$ kcal/mol at room temperature. Larger deviations occur extreme rarely and would require unaffordable computer times, since the transition time τ scales as the exponential of the free-energy barrier, $\tau \sim \exp(\Delta F / k_B T)$. The purpose of the following sections is just to have a closer look at these two classes of problems.

17.2 Hartree–Fock and all-electron approaches

The set of coordinates $\{\mathbf{R}_I\}$ representing the classical Cartesian positions of all the atoms composing the protein or nucleic acid under investigation represents the first ingredient of quantum approaches. They are provided by X-ray and neutron scattering experiments on crystallized forms of the biomolecular system. Of course, these are not the only variables: the system has to be complemented with all the electrons concurring to form the structure and the chemical bonds; these must be treated as quantum objects, hence, described in terms of wavefunctions ψ_i. As a consequence, the Hamiltonian of the system must contain all the interactions between all the variables involved. The main quantum approaches that have been used to study systems composed of many electrons (and atoms) over the years rely all some many-body formulation of the fundamental Schrödinger equation and they are classified according to the basis set or functional form adopted: Hartree–Fock (HF), generalized valence bond (GVB), density-functional theory (DFT), configuration interaction (CI), complete active space self-consistent field (CASSCF), etc. (Marx and Hutter 2000).

Indeed, solving the Schrödinger equation for a many-body system is in general a rather demanding task. The first difficulty is how to write a many-body wavefunction $\Psi(\mathbf{q})$, where $\mathbf{q} = (\mathbf{x}_{1\sigma}, \mathbf{x}_{2\sigma}, \dots, \mathbf{x}_{N\sigma})$ is a multidimensional vector defining the position and spin state σ of each electron of the system. The associated eigenvalues problem is the steady-state Schrödinger equation

$$\hat{H}(\mathbf{q}, \mathbf{R}_I)\, \Psi(\mathbf{q}) = E\, \Psi(\mathbf{q}). \tag{17.6}$$

Among all the methods proposed over the years, one of the most popular and widely used is the HF approach. Without any claim of completeness, since many good textbooks are available, let us recall that this specific approach is based on the variational principle applied in a subspace of wavefunctions where the electronic ground state is assumed to be an antisymmetric combination of single-particle orbitals $\psi_{i\sigma}(\mathbf{x})$, called Slater determinants, with an orthonormal constraint imposed on the wavefunctions. In the single-particle wavefunction $\psi_{i\sigma}(\mathbf{x})$ the first index i identifies the ith electron, while the second one σ labels the up/down spin state,

$$\Psi(\mathbf{q}) = \frac{1}{\sqrt{N!}} \det\left[\psi_{i\sigma}(\mathbf{x}_j)\right]. \tag{17.7}$$

The second approximation involved in the HF approach concerns the analytic form of the Hamiltonian of the system. In fact, the HF version of the many-body Schrödinger equation, written in terms of the single-particle orbitals, in atomic units, i.e. assuming as a charge unit the proton (or absolute value of the electron) charge ($e = 1$) and as a mass unit the electron mass ($m_e = 1$), is written as

$$\left\{-\frac{1}{2}\nabla^2 + V_{\text{eI}}(\mathbf{x}) + V_{\text{H}}(\mathbf{x})\right\} \psi_{i\sigma}(\mathbf{x})$$

$$- \sum_{j\sigma'} \int d^3x' \frac{\psi_{j\sigma'}^*(\mathbf{x}')\psi_{i\sigma}(\mathbf{x}')}{|\mathbf{x}-\mathbf{x}'|} \psi_{j\sigma'}(\mathbf{x}) = \varepsilon_i \psi_{i\sigma}(\mathbf{x}), \tag{17.8}$$

and the terms inside the parentheses on the left-hand side of eqn (17.8) are the kinetic energy operator, the electron–ion and ion–ion interactions, often referred to as "external Coulomb potential", and the Coulomb (Hartree) potential acting between two electrons, respectively. These two latter interactions have the following explicit form

$$V_{\text{eI}}(\mathbf{x}) = -\sum_I \frac{Z_I}{|\mathbf{x}-\mathbf{R}_I|} + \frac{1}{2}\sum_{I,J} \frac{Z_I Z_J}{|\mathbf{R}_I-\mathbf{R}_J|} \tag{17.9}$$

$$V_{\text{H}}(\mathbf{x}) = \sum_{j\sigma'} \int d^3x' \frac{\psi_{j\sigma'}^*(\mathbf{x}')\psi_{j\sigma'}(\mathbf{x}')}{|\mathbf{x}-\mathbf{x}'|} = \int d^3x' \frac{\rho(\mathbf{x}')}{|\mathbf{x}-\mathbf{x}'|}, \tag{17.10}$$

where the single particle electron density is just the result of summing up on all the spin and particle indexes the $\psi_{j\sigma'}(\mathbf{x}')$ square modulus wavefunctions. In practice, this integral form is rarely used, since the Hartree potential V_{H} can be obtained more easily from the solution of the associate Poisson equation

$$\nabla^2 V_{\text{H}}(\mathbf{x}) = 4\pi\rho(\mathbf{x}). \tag{17.11}$$

The second term in eqn (17.8) is the exchange operator. The name clearly refers to the fact that this operator, acting on the orbital $\psi_{i\sigma}(\mathbf{x})$, exchanges the index i with j and the index σ with σ', and accounts for the fact that from the quantum point of view, electrons are indistinguishable particles. Although the exchange is correctly accounted for in HF approaches, it must be remarked that this term is exactly zero if the ground-state wavefunction is not assumed to be anti-symmetric. Another issue worthy of a comment is the fact that the Coulomb potential represented by eqn (17.10) acts only between electron pairs described in terms of single-particle wavefunctions. This means that three-body and higher-order terms are neglected and, thus, correlations are not included in HF approaches; these are generally added a posteriori in the so-called "post-HF" approaches. In practical applications, the geometry of the model system representing the protein or nucleic acid is optimized at the HF level, then higher-order corrections to the energy, coming from perturbation theories are applied. This is, for instance, the case of the popular Møller–Plesset (MP2) approach (Møller and Plesset 1934): the total energy of the system is corrected to the second order, but the wavefunctions used to compute the second-order perturbation are the ones obtained in the uncorrelated HF calculation. As a consequence, wavefunctions and energy functionals are not consistent and the first derivatives of the MP2 energy functional would not provide the correct forces needed, for example, to do a molecular-dynamics run. Instead, HF structure optimizations and successive MP2 calculations can be repeated iteratively within a Born–Oppenheimer approach to sample the potential-energy surface (Keshari and Ishikawa 1994). This methodology is in principle very powerful and can lead to rather precise results. However, it has the drawback that the scaling with the system size is very unfavorable. In fact, despite recent pioneering attempts at reducing the computational cost (Kobayashi *et al.* 2007), the typical scaling of an MP2 procedure is at least $O(N^4)$, whereas the scaling of HF calculations is $O(N^3)$. This implies that only systems with relatively few electrons can actually be treated.

The selection of a specific Hamiltonian and single-particle expression of the many-body wavefunction does not include all the ingredients needed to actually perform the calculations: one has also to select an appropriate (finite) basis set good enough to approximate the (infinite) Hilbert space spanned by the eigenfunctions of the adopted Hamiltonian and able to represent the single-particle orbitals $\psi_{i\sigma}(\mathbf{x})$ in terms of simple linear combinations of atom-centered analytic functions $\phi_k(\mathbf{x}; \{\mathbf{R}_I\})$,

$$\psi_{i\sigma}(\mathbf{x}) = \sum_{k=1}^{M} c_{i\sigma}^k \phi_k(\mathbf{x}; \{\mathbf{R}_I\}), \tag{17.12}$$

and the number of analytic functions used, M, is also an indicator of the computational cost of the quantum calculation, in the obvious sense that the larger the basis set, the higher the computational workload. In most of the HF applications, a popular, although minimal, basis set is represented by Slater-type orbitals (STO) or Gaussian-type orbitals (GTO). It must be remarked that orbitals expanded in a localized basis set, depend on the atomic positions $\mathbf{R}_I$. As a consequence, in any calculation in which the forces acting on the ions are required, the explicit derivatives of these wavefunctions with respect to $\mathbf{R}_I$ must be computed, leading to non-Hellmann–Feynman force components, known in the literature as Pulay forces. These forces are, instead, absent if a non-site-dependent basis set is adopted, such as, for instance, the plane waves that will be discussed in the next section. As a further word of warning, let us stress that the cubic scaling of the HF method with the system size and the additional computational cost of post-HF calculations pose severe limitations to the size of the systems.

Coming to specific applications to proteins and nucleic acids, the widespread use of HF approaches, since the early days of computer-based electronic structure calculations, makes it next to impossible to illustrate all the calculations that have been put forward over the years. We are then forced to focus on a few selected cases that are also representative of the kind of analyses that these techniques make possible. An illustrative example of general interest in nucleic acid research is represented by the enzyme ribonuclease (RNase) catalysis, responsible for the degradation of RNA.

This mechanism occurs in RNA enzymes (ribozymes) and is at the basis of three main processes: (i) it contributes to the transfer of the genetic information from DNA to proteins, (ii) it is an important cofactor in the genetic evolution of living organisms, (iii) it inhibits gene expression, such as oncogene, and thus has promising applications in cancer gene therapy (Perreault and Anslyn 1997; Zhou and Taira 1998). Even a single strand of the nucleic acid, such as the one sketched in Fig. 17.2(a), amounts to a number of atoms unaffordable by HF approaches. However, one can extract a smaller model by observing that not the whole RNA enzyme is involved in the RNase enzymatic reaction, known in the related literature as *transesterification*, but only a small portion of the system around a specific phosphorous site as sketched in panel (b) of Fig. 17.2. Namely, upon deprotonation of the $O^{2'}-H$ group (Fig. 17.5, panel (b), scheme (1)), the oxygen labelled as $O^{2'}$ forms a new bond with the P atom.

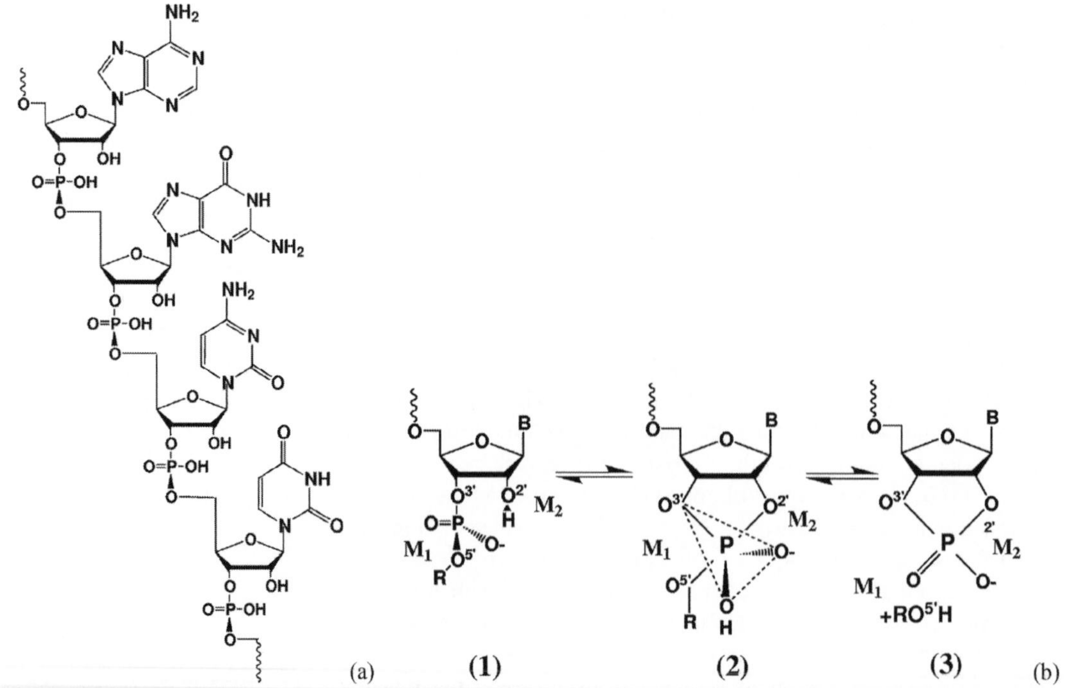

Fig. 17.2 (a) Scheme of a single strand of RNA with the four bases, adenine, guanine, cytosine and uracil, from the top to the bottom, respectively. (b) Schematic reaction pathway of the RNase enzymatic reaction catalyzed by metal ions M_1 and M_2.

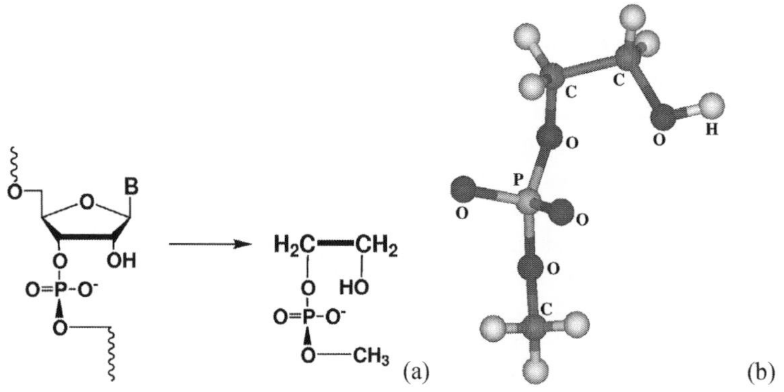

Fig. 17.3 The methyl $2'$-hydroethyl phosphate (MHEP) system used in HF calculations. In panel (a), the left-hand side shows the ribose ring and the phosphate group, while on the right the MHEP model is sketched. Panel (b) shows the atomic structure of the MHEP.

This step of the reaction, named *nucleophilic attack*, leads to the formation of a 5-fold phosphate, representing the transition state shown in scheme (2) of panel (b) of Fig. 17.2. The final product of this reaction is the cleavage of the $P-O^{5'}$ bond, illustrated in scheme (3) of panel (b) in the same figure. The superscripts $2'$, $3'$ and $5'$ labelling the various O atoms in the reaction scheme are a standard notation in biochemistry and indicate the position of the atoms with respect to the sugar ring. The counting is clockwise and starts from the C atom of the ribose carrying the base B, which is then site number 1, then position 2 becomes the one of $-$O-H, position 3 the phosphate group, etc., while M_1 and M_2 indicate two divalent metal ions that are expected to catalyze the reaction. It is evident, then, that formation and breaking of chemical bonds involve only the P atom and its closest neighbors.

This observation led several groups (Dejaegere *et al.* 1991; Storer *et al.* 1991; Uchimaru *et al.* 1991; Chang and Lim 1997) to consider as representative of the chemical environment around the phosphorous site smaller structures such as, for instance, the methyl $2'$-hydroethyl phosphate (MHEP). This is reproduced in Fig. 17.3 and in practice includes up to the second nearest-neighbor site of the P atom involved in the reaction mechanism. It must be stressed that whenever a smaller model is extracted from a full protein or nucleic acid, the nature of each chemical bond involved must be preserved. The same procedure holds also in the case of the $P-O-CH_3$ group, where the sp^3 character of the C atom linking the $P-O$ bond to the next residue is ensured by the CH_3 methyl group. As a general rule, special attention must be paid in the construction of model systems by cut and saturation of bonds; failing to do so is one of the major sources of errors, especially when simulation results on reduced models are extrapolated to the full system. Despite the reduced size of the MHEP system, a single geometry optimization cycle at the HF level using a localized Gaussian basis set nearly minimal required about two hours with the computer facilities of the 1990s (Lim and Tole 1992). In the example illustrated by Fig. 17.3 and used in the quoted literature, the 4-fold P group is chemically identical to the one in the RNA structure. Then, the ribose ring is reduced to only two carbon atoms and the dangling bonds of these carbon atoms are saturated by hydrogen atoms. In this way, the $C-C$ carbon is still a single bond and the two C atoms keep their sp^3-bonding character in the MHEP model system.

The outcome of typical HF calculations is a set of optimized configurations of the model-system adopted. For each configuration corresponding to a stable local minimum, the HF Hamiltonian, H^{HF}, is diagonalized and the related total energy minimized until the structure reaches an optimum configuration. Total energies are generally further corrected at the MP2 level, providing the relative stabilities of the various conformers. Transitions states (TS), instead, can be searched by computing the second derivative of the HF total energy (Hessian) with respect to the atomic positions $\mathbf{R}_I$, and by maximizing the energy in the direction of one eigenmode ω_α of the Hessian, while minimizing it in all the other directions, thus one of the eigenmodes ω_α is imaginary.

In the case of MHEP, the TS looks like the structure (2) in panel (b) of Fig. 17.2. The total energy difference between the MHPE reactant, according to the calculations of Storer and coworkers (Storer *et al.* 1991) and Lim and Tole (1992) amounts to about 38–40 kcal/mol. This is larger than the experimental activation energy in ribozymes for this specific reaction, estimated to range between 25 and 30 kcal/mol, depending on the specific enzyme, catalyst and environment considered. The agreement improves if MP2 corrections are included; in this case energy barriers of about 35 kcal/mol were obtained. Slightly larger model systems, including the ribose ring, the metal cations and the few water molecules belonging to the first solvation shell of the metals, could be afforded only recently (Leclerc and Karplus 2006), showing that the H_2O solvent is not an inert background but participates actively to this specific enzymatic reaction. This extended model has the merit of improving the energy barrier obtained by the former MEHP calculations and provides a reaction path at least qualitatively closer to the one expected in the full ribozyme. Yet, these are still gas-phase static optimization in which solvation effects are accounted for only in a subsequent refinement making use of a continuum polarizable model. Indeed, small model systems are unable to reproduce all the complexity represented by the whole nucleic acid or protein system, the non-trivial influence of the surrounding environment, in particular the solvating water, the fluctuations due to finite temperature, the entropy effects, etc. Nonetheless, stable species and activation barriers at a qualitative or semi-quantitative level provide useful hints to experimentalists in the rationalization of the outcome of the experiments and can also serve as general guidelines in the molecular design of enzymes and catalysts.

Another fertile domain in which HF approaches have been used is represented by the Watson–Crick DNA base-pair interaction. The typical problems that have been addressed so far are the H-bond interactions between guanine (G) and cytosine (C), GC, and between adenine (A) and thymine (T), AT. One of the basic questions that the theoretical research has attempted to address over the years is the planarity vs. non-planarity of the base pairs. This is a fundamental issue in addressing the problem of the stability of DNA. Yet, also in the case of DNA, the whole double-helix system is too large to be handled at a full quantum level. Thus, the model systems extracted from the double-strand DNA are reduced to single base pairs. By looking at Fig. 17.4, it is evident that the GC pairing is more stable than AT, because the former involves the formation of three H-bonds, whereas the AT pairing is realized by only two inter-molecular H-bonds. By remembering that HF approaches do not include

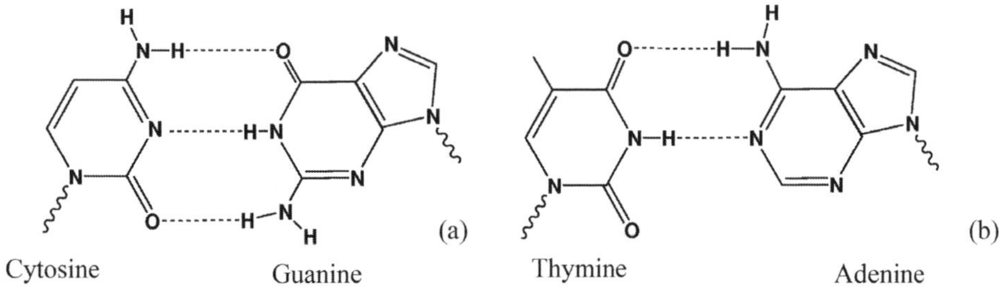

Fig. 17.4 Base pairs in DNA. Panel (a) shows the guanine–cytosine (GC) hydrogen bonding and panel (b) the adenine–thymine (AT).

electron correlations, the immediate consequence is that if the target of HF calculations is DNA or RNA, then the base stacking cannot be reproduced. In fact, base pairs piled up in a double-strand architecture interact among each other via the π orbitals of the aromatic rings present in the bases. Now, these interactions, mostly van der Waals, are exactly the ones not accounted for in the HF scheme. Conversely, it has been shown that the stability of the hydrogen bonding in Watson–Crick base pairs is mainly due to electrostatic interactions and post-HF corrections at the MP2 level have also evidenced that a certain degree of non-planarity can indeed be present (Danilov and Anisimov 2005). In the absence of MP2 corrections, the GC and AT pairs turn out to be perfectly planar upon HF geometry optimization; this seems to be due to the insufficient description of the amino nitrogen atoms present in the bases (Hobza and Šponer 2002).

Quantum-chemical calculations at the semi-empirical level, HF and post-HF levels were also used in an attempt at addressing the problem of the donor–acceptor coupling between neighboring Watson–Crick base pairs, related to the charge-transfer problem in DNA. These results, representing a considerable computational effort, underscored the fact that charge-transfer processes are considerably affected by the type of base pairs involved (Rösch and Voityuk 2004). Since HF approaches, especially in connection with post-HF corrections, are able to handle with appreciable accuracy this class of problems, they were used to study base–amino acid interactions. These are at the basis of the DNA–protein recognition processes, that play a crucial role in gene expression of living organisms. A wealth of X-ray and NMR experiments have provided accurate geometries (Ollis and White 1987) available for theoretical calculations. Proteins suitable to bind to DNA often make use of certain motifs and patterns to realize a specific fit to a given DNA sequence and these DNA–protein contacts turn out to be H-bond-type interactions.

For instance, an extensive investigation has been reported for the interaction of the asparagine amino acid (Asn), schematically shown in panel (a) of Fig. 17.5, with both GC and AT (Pichierri *et al.* 1999) with force fields parameterized at the HF level. This particular use of HF calculations was driven by the fact that a direct HF analysis was not possible because of the large amount of conformations that would make the number of HF stationary points to be computed unaffordable. The outcome of these HF-based investigations is nonetheless worthy of note, since this study provided a quite exhaustive energy map of where stable configurations could be located and

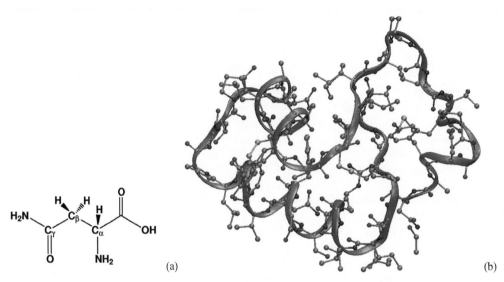

Fig. 17.5 Scheme of the asparagine (Asn) amino acid (a) and structure of the Crambin protein (b). In panel (b) only the heavy atoms of Crambin are shown as sticks and balls. Hydrogens are omitted for the sake of clarity, and the ribbon is the conventional representation joining all the sp^3 C sites.

the relative stability of the various conformers analyzed in detail. Specifically, for each possible position of the carbon atom of Asn labelled as C_α in the literature, the identification of the various stationary points allowed for the estimation of thermodynamic quantities, such as free energy, enthalpy and entropy. In particular, the global minimum has shown that the most stable Asn–A interaction is achieved by the formation of two specific H-bonds N-H...N- and $-C=O$...H-N- that represent a sort of peculiarity common to several DNA–protein docking structures. To date, the largest system that has been afforded at a full HF level is the crambin protein (Fig. 17.5(b)). This relatively small protein consists of 46 residues and is found in nature in the hydrophobic plant seeds. Its primary structure was identified both by X-ray crystallographic analysis (Hendrickson and Teeter 1981) and solid-phase sequencing techniques (Teeter *et al.* 1981). HF calculations were aimed at the geometrical determination of crambin, using a 4-21G Gaussian basis set and performing a plain geometry optimization (Van Alsenoy *et al.* 1998). In order to circumvent the heaviest part of the computations represented by the calculation of the gradients, necessary in the self-consistent optimization procedure, each product of two basis functions was expanded in a single set of auxiliary functions. This mathematical trick allows us to reduce the $O(N^4)$ scaling of the Fock matrix to $O(N^3)$, N being the number of basis functions. Nonetheless, this amounted to 3597 basis functions and each optimization cycle required about 80 hours of CPU time on a DEC/Alpha Station 600. The outcome of this milestone calculation was an accurate set of geometrical parameters, in particular disulfide bridges and peptide groups, that allowed a precise refinement of the X-ray diffraction data.

Despite remarkable progresses in the computational power, it is generally acknowledged that high-level *ab-initio* methods, such as HF and post-HF approaches, will not be a viable possibility to study dynamics and reactions of large proteins and nucleic acids. Instead, smaller systems or reduced basis sets in the so-called low-level HF calculations can be adopted to study, for instance, peptide bonds conformations. This has been done by Perczel and coworkers (Perczel *et al.* 2003) for the alanine (Ala) dipeptide, using as a reference molecule the compound HCO-L-Ala-NH$_2$, performing HF calculations at the restricted HF level with a nearly minimal Gaussian basis set (3–21G). In this work, first the various stationary points, minima and transition states of the different conformers were identified by varying the torsion angles of the reference molecule, then the results were refined at a higher level. This accurate analysis allowed locatation of several *cis–trans* isomers and a total of nine stable conformers of the alanine amino acid residue, in agreement with the results of X-ray scattering experiments.

Attempts at overcoming these difficulties were pioneered by partitioning large proteins into smaller subsystems, and doing separate calculations on each one of these subunits. Worthy of note are the lego method of Walker and Mezey (1994) and the fragment molecular orbital (FMO) scheme of the group of Kitaura (Kitaura *et al.* 2001). In these approaches, a large protein is divided into smaller units, then on each one of these units HF-type calculations are performed in parallel, as if they were isolated microcanonical ensembles; at the end of the calculations, the results properly merged and joined. Examples of applications of the FMO approach have been reported for the ubiquitin (Komeiji *et al.* 2007), a protein involved in ATP-dependent protein degradation (Hershko and Chiechanover 1982), consisting of 76 amino acid residues and whose inner part is shaped as an α-helix and a β-sheet. The FMO partitioning of hydrated structures of this protein amounted to 76 fragments (Model-P), 2238 fragments (Model-PS) and 1487 fragments (Model-S) according to the number of atoms composing the whole system. The larger number of fragments (2238) refers to a hydrated ubiquitin amounting to 7717 atoms. The outcome of this calculation can be summarized in a detailed electronic structure of the protein, its relaxed geometry and the local dipole moments, with special attention to the identification of hydrophilic and hydrophobic segments. The problems that cannot be solved with these partitioning methods are unfortunately some of the fundamental questions in biochemistry; namely, charge transfer along the biomacromolecule or with the solvent, dynamics and chemical reactions and all non-static processes. In fact, each subsystem or fragment, is treated as a microcanonical (N, V, E) or canonical (N, V, T) ensemble and, as such, electron or atom transfer among fragments is not possible.

17.3 Density-functional theory approaches

The density-functional theory (DFT) was originally proposed in the early 1960s by Hohenberg and Kohn (1964) and Kohn and Sham (1965) with important contributions also from the group of Pople (Pople *et al.* 1981, 1989). Its importance in the advancement of computational quantum chemistry and related fields was internationally acknowledged by the Nobel

Prize in Chemistry in 1998 awarded jointly to Walter Kohn and John A. Pople. The DFT is a formulation of the many-body quantum mechanics in terms of an electron-density distribution, $\rho(\mathbf{x})$, which describes the ground state of a general system composed of interacting electrons and classical nuclei at given positions $\{\mathbf{R}_I\}$. Several excellent books and review articles have been published on the fundamentals of DFT (Parr and Yang 1989; Marx and Hutter 2000). For this reason, we limit the discussion to the basic details necessary to the ongoing discussion and refer the reader to the rich literature on the subject, part of which is cited in the references listed at the end of this chapter. The first step in DFT consists in giving an explicit form for the electron-density distribution; in a way similar to the hypotheses introduced in the HF approach, also here single-particle wavefunctions $\psi_i(\mathbf{x})$ are used to express the many-body mathematical function $\rho(\mathbf{x})$. The major difference and, at the same time, dramatic simplification, is the fact that not even the specific analytic form of the complex function $\psi_i(\mathbf{x})$ matters, but only its square modulus, so that

$$\rho(\mathbf{x}) = \sum_{i=1}^{N^{\mathrm{occ}}} f_i \, |\psi_i(\mathbf{x})|^2. \tag{17.13}$$

This expression is a single Slater determinant constructed from the single-particle wavefunctions representing all the N^{occ} occupied orbitals. The coefficients f_i are the integer occupation numbers, and they are equal to 1 in the case in which the spin is explicitly considered (spin-unrestricted) or equal to 2 if the spin is neglected and energy levels are considered as doubly occupied (spin-restricted). Furthermore, the wavefunctions $\psi_i(\mathbf{x})$ are subject to the orthonormality constraint

$$\int \psi_i^*(\mathbf{x})\psi_j(\mathbf{x}) \mathrm{d}^3 x = \delta_{ij} \tag{17.14}$$

as in any quantum-mechanics approach. The Kohn–Sham (KS) DFT total energy of the system in its ground state is then written as

$$E^{\mathrm{KS}}[\{\psi_i\}] = E_{\mathrm{k}}[\{\psi_i\}] + E_{\mathrm{H}}[\rho] + E_{\mathrm{xc}}[\rho] + E_{\mathrm{eI}}[\rho] + E_{\mathrm{II}}, \tag{17.15}$$

where the first three terms on the right-hand side (E_{k}, E_{H}, E_{xc}) describe all the electron–electron interactions, the fourth term (E_{eI}) refers to the electron–nucleus interaction and the fifth one (E_{II}) the nucleus–nucleus interaction. Let us revise briefly the explicit form and the meaning of each one of these terms. E_{k} is the Schrödinger;-like kinetic energy expressed in terms of the single-particle wavefunctions $\psi_i(\mathbf{x})$ as

$$E_{\mathrm{k}}[\{\psi_i\}] = \sum_{i=1}^{N^{\mathrm{occ}}} f_i \int \mathrm{d}^3 x \, \psi_i^*(\mathbf{x}) \left(-\frac{1}{2}\nabla^2\right) \psi_i(\mathbf{x}), \tag{17.16}$$

and it is completely diagonal both in the index i and in the argument $\mathbf{x}$ of the wavefunctions, as in a non-interacting system of N^{occ} electrons. We remark, in passing, that this expression for the kinetic energy does not depend on the density $\rho(\mathbf{x})$ but directly on the wavefunctions. The second term, E_{H}, is the Hartree energy already encountered in the HF approach, it accounts for the

Coulomb electrostatic interaction between two charge distributions that reads

$$E_{\mathrm{H}}[\rho] = \iint \mathrm{d}^3x\,\mathrm{d}^3y \frac{\rho(\mathbf{x})\rho(\mathbf{y})}{|\mathbf{x}-\mathbf{y}|}. \tag{17.17}$$

This is generally computed via the associated Poisson equation. The exchange interaction and the electron correlations due to many-body effects are represented by the term $E_{\mathrm{xc}}[\rho]$, whose exact analytical expression is unfortunately unknown; this represents in a sense a limit of the DFT. There are good approximations derived from the homogeneous electron-gas limit for the exchange interaction (Becke 1988). Namely, the pure exchange part of the functional, $E_{\mathrm{x}}[\rho]$ can be written in the so-called local density approximation (LDA) as

$$E_{\mathrm{x}}[\rho] = -\frac{3}{2}\left(\frac{3}{4\pi}\right)^{1/3} \int \mathrm{d}^3x\,[\rho(\mathbf{x})]^{4/3}, \tag{17.18}$$

and the name comes from the fact that an interacting but homogeneous electron distribution is assumed, in which the (unknown) density is given by the local density $\rho(\mathbf{x})$ at a specific point $\mathbf{x}$ in the inhomogeneous system. Similarly, the LDA version of the correlation energy, originally proposed by Kohn and Sham (1965) and successively improved by Perdew and Zunger (1981), reads

$$E_{\mathrm{xc}}[\rho, \nabla\rho] = \int \mathrm{d}^3x\,\varepsilon_{\mathrm{xc}}(\rho(\mathbf{x}), \nabla\rho(\mathbf{x})), \tag{17.19}$$

where the explicit analytic form of the function ε_{c} comes from a parameterization of the results of random phase approximation calculations. Due to the insufficiency of a simple LDA approximation in the treatment of many real systems, such as proteins and nucleic acids, non-local approximations including the gradient of the density are often adopted.

In practical applications, however, the gradient enters only with its modulus, thus adding only a modest computational cost to DFT calculations. These generalized gradient corrections (GGA) are indeed a little arbitrary, in the sense that they do not represent a regular perturbation expansion as, for instance, in the case of MP2 corrections. Nonetheless, they are generally based on solid physical and mathematical argumentations and anyhow their accuracy can be assessed a posteriori by test calculations and comparisons with both exact results and experiments (Johnson *et al.* 1993). As far as the exchange term is concerned, the original work of Becke (1988) represents a milestone in the field and it is still regarded as a standard. It was subsequently refined and improved by the same author (Becke 1993) with remarkable success in a wealth of applications. Also, correlations have received special attention from several research groups, sometimes including also the exchange part into a single functional (Vosko *et al.* 1980; Lee *et al.* 1988; Perdew and Wang 1992; Perdew *et al.* 1996; Hamprecht *et al.* 1998). Among all the possible functional forms for the exchange and correlation functional, the ones that have been most widely used in the applications of DFT to biomolecules are the BLYP (Becke 1988; Lee *et al.* 1988) and the HCTH (Hamprecht *et al.* 1998) ones, because of the fact that they provide the best—or at least the more acceptable—performance in terms of geometrical parameters and relative energies for a wide variety of hydrocarbons and phosphates, two of the main ingredients in proteins and nucleic acids. In general, the explicit form adopted to describe the exchange and correlation interactions represents a delicate step in the setup of the simulation framework and must be carefully tested and benchmarked. As a word of warning, let us also stress the fact that none of the present versions of E_{xc} included van der Waals interactions (Langreth *et al.* 2005; Grimme 2006; Silvestrelli 2008).

The analytic form of the electrostatic interaction between the two sets of variables, electrons and nuclei, is simply given by the Coulomb attraction between a point-like charge at nuclear positions $\mathbf{R}_I$ and the electron-density distribution. However, for most of the applications that will be discussed in the following sections, this turns out to be computationally expensive. In fact, in a large protein or nucleic acid system there are two different length scales that come into play: a small one for the core electrons, characterized by rapidly varying wavefunctions, especially in the region very close to the nucleus, and a longer one for the valence electrons that form chemical bonds and vary more smoothly. Clearly, in an all-electron calculation, the first one would dominate and add a computational workload that would make impractical dynamical simulations, and often even static optimizations, of large biomolecules. Provided that the model system is small enough and long dynamical quantum simulations are not required, it is certainly possible to use an all-electron DFT approach. However, one can observe that core electrons are generally inert and do not participate to chemical bonds. This crucial observation led to the use of pseudopotentials (PPs), where core electrons are eliminated and the core–valence interaction is built by fitting to the all-electron solutions of the Schrödinger or Dirac equation for a single atom (Hamann *et al.* 1979). Alternatively, one can use frozen orbitals for the inner electrons described by angular momentum projectors (Vanderbilt 1990; Blöchl 1994). The electron–nucleus interaction is then

$$E_{\text{el}}[\rho] = \int \mathrm{d}^3x \, V_{\text{ps}}(\mathbf{x} - \mathbf{R}_I) \cdot \rho(\mathbf{x}), \qquad (17.20)$$

and the elimination of the core electrons allows one to eliminate the short length scale problem. The computational cost can be further reduced, provided that a separable form is chosen for the pseudopotential (Kleinman and Bylander 1982),

$$V_{\text{ps}}(\mathbf{r}) = V_{\text{loc}}(\mathbf{r}) + \sum_{l,m} \phi_{lm}(\mathbf{r}) V_{\text{NL}}^{lm} \phi_{lm}^*(\mathbf{r}), \qquad (17.21)$$

where $\mathbf{r} = \mathbf{x} - \mathbf{R}_I$ and on the right-hand side we have separated the PP into a local term $V_{\text{loc}}(\mathbf{r})$, depending only on the position $\mathbf{r}$, and a non-local (NL) part represented by a sum over all the orbital angular momenta l,m. The functions $\phi_{lm}(\mathbf{r}) = f_l(r)Y_{lm}(\Omega)$ are eigenfunctions of the atomic Hamiltonian in which the core–valence interaction has been replaced by the PP, $f_l(r)$ indicates the radial part of the solution, whereas the angular dependence is represented by the spherical harmonics $Y_{lm}(\Omega)$. Several analytical forms for the PPs have been reported over the years (Bachelet *et al.* 1982; Gonze *et al.* 1991; Troullier and Martins 1991; Goedecker *et al.* 1996) and widely used. In general, PPs are *norm-conserving*, meaning that the square norm of the wavefunction is preserved in the pseudopotential construction. This constraint can be released as shown by Vanderbilt (1990), leading to the so-called ultrasoft pseudopotentials, provided that an augmented charge is added to restore the total charge density. As a final observation, let us point out that in some cases the presence of a core electron-charge density can be accounted for via the introduction of a non-linear core correction (Louie *et al.* 1982).

Finally, the fifth and last term on the right-hand side of eqn (17.15) is simply the Coulomb interaction between two classical nuclei I and J and is written as

$$E_{II} = \sum_{I<J}^{M} \frac{Z_I Z_J}{|\mathbf{R}_I - \mathbf{R}_J|}, \tag{17.22}$$

where Z_I and Z_J must be intended as the net valence charge only in a PP approach.

The total energy E^{tot} of the ground state of such a system of interacting electrons and nuclei can then be obtained by minimizing the KS functional with respect to the single-particle orbitals $\psi_i(\mathbf{x})$, which, in practice, means solving the KS Schrödinger-like equations given by the variational derivative of the KS functional with respect to the single-particle wavefunctions $\psi_i(\mathbf{x})$,

$$\frac{\delta E^{\text{KS}}[\rho]}{\delta \psi_i^*} \equiv f_i H^{\text{KS}} \psi_i(\mathbf{x}) = f_i \varepsilon_i \psi_i(\mathbf{x}). \tag{17.23}$$

In a way similar to what has been done in the case of HF, also in DFT approaches we need to select a proper basis set on which orbitals can be expanded. One possible choice is to adopt a Slater or Gaussian set of functions, as successfully done in many quantum-chemistry approaches. An alternative choice is represented by plane waves (PW). In this expansion, the single-particle wavefunctions $\psi_i(\mathbf{x})$ become

$$\psi_i(\mathbf{x}) = \sum_{\mathbf{G}=0}^{\mathbf{G}^{\text{max}}} c_i(\mathbf{G}) e^{i\mathbf{G}\cdot\mathbf{x}}, \tag{17.24}$$

where the Hilbert space spanned by the plane waves $\exp(i\mathbf{G}\mathbf{x})$ is truncated at a suitable cut-off generally expressed as an energy and measured in Rydberg, $E_{\text{cut}} = (\mathbf{G}^{\text{max}})^2/2$. The advantage of plane waves is that they do not depend on the atomic positions $\mathbf{R}_I$, the accuracy can be systematically (variationally) improved by increasing the cut-off and they form a complete orthonormal basis set. The first property is useful in the calculation of the forces required for instance in dynamics, since Pulay forces are exactly zero, and the Hellmann–Feynman theorem, applies. Furthermore, the calculations of gradients or Laplace operators, such as the electronic kinetic energy, is reduced to simple products in the reciprocal space; the use of the fast Fourier transform (FFT) makes the calculation easier and distributable in parallel processing (Press *et al.* 1992).

DFT calculations have been extensively used across the years in a wealth of applications, with a success comparable to HF approaches, because of their reduced computational cost. For this reason, it is impossible to cover here

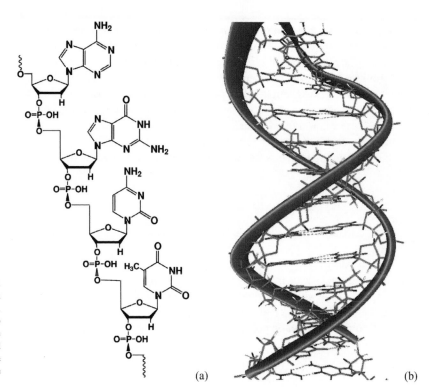

Fig. 17.6 Panel (a) shows the general scheme of a single strand of DNA with its four bases, adenine (A), guanine (G), cytosine (C) and thymine (T), from the top to the bottom, respectively. Panel (b) shows the double-helix structure of a DNA, identified by the two twisted ribbons representing the imaginary lines joining all the backbone P atoms.

all the biochemical systems studied by this approach. Thus, the discussion will be limited to the more recent results. A particularly active field in recent years concerns DNA. As mentioned in the analysis of the HF approaches, the deoxyribonucleic acid (DNA) is a long double-stranded structure of base pairs where the backbone is composed of sugar rings and phosphate groups (Franklin and Gosling 1953; Watson and Crick 1953). The length and size of DNA made quantum calculations prohibitively expensive, even PP-based DFT approaches, until the late 1990s. The major difference with respect to RNA stems from one base, thymine (T), that in DNA substitutes uracil (U) of RNA and the replacement of the −OH group in the sugar ring, in the $2'$ position, with a H as shown in Fig. 17.6(a). The base pairing results in a sequence of H-bonded GC and AT motifs and the greater chemical stability of DNA with respect to RNA is ascribed to the −H, replacing the $-O^{2'}H$, that hinders nucleophilic attacks and deprotonation processes responsible for the enzymatic reactions in RNA-based systems. This stability makes DNA suitable for the genetic data storage and its self-assembled double-helix structure (Fig. 17.6(b)) has inspired several recent applications, and a renovated interest for quantum calculations, in the emerging field of nanotechnology.

The main directions towards which quantum calculations of the DNA electronic structure properties have been pushed forward are the understanding of the oxidative and radiation damage of DNA (Odom and Barton 2001; Douki *et al.* 2004) and its potential applications in nanoelectronic devices (Forbes 2000; Dekker and Ratner 2001). In fact, DNA is a stable polymer with a strong one-dimensional character that can be rather easily engineered. Furthermore, specific DNA operations, such as the bonding of two strands to form the double helix, are about 10^9 times more efficient than any computing process on any computer. Also, DNA stores information at densities of one bit per nm^3, which amounts to more than 10^{11} times the typical densities of the best DVD on the market. Pioneering studies in the construction of DNA junctions and nanostructures have been reported by the group of Seeman (Seeman and Kallenbach 1983; Chen and Seeman 1991; Zhang and Seeman 1994). These two important research lines in nanoscience are, by their own nature, "quantum", since the length scale of the interactions involved in keeping together these nanoarchitectures and to transport electrons across them are describable only in terms of quantum mechanics. This emerging field inspired the first DFT calculations of full double-stranded DNA systems, either with localized basis sets (de Pablo *et al.* 2000) or plane waves (Gervasio *et al.* 2002) aimed at understanding both the ground-state electronic properties and different charge states of both native and synthetic DNA. The ultimate target is to unravel the conducting properties of DNA and its derivatives in order to understand whether or not this specific biopolymer can become an effective component in nanoelectronic devices. These exploratory calculations have shown, for instance, that a synthetic DNA made up of only GC bases has a gap of 1.28 eV, which is comparable to semiconductors, that separates a manifold of 12 occupied states, originating from the π orbitals of G, from a conduction band of empty states in which electrons are transferred from the DNA counterions Na^+ to the phosphate groups PO_4^- and the solvating water molecules, in wet DNA. Since electronic wavefunctions are directly provided by DFT calculations, the transition matrix elements of the momentum operator $\mathbf{p}$ can be directly computed, giving the optical conductivity via the Kubo–Greenwood formula

$$\sigma(\omega) = \frac{2\pi e^2}{9m_e^2 \Omega \omega} \sum_{v,c} |\langle \psi_v | \mathbf{p} | \psi_c \rangle|^2 \delta(E_c - E_v - \hbar\omega), \qquad (17.25)$$

where the subscripts v and c indicate the valence and conduction wavefunctions and energies, respectively, and Ω is the volume of the simulation cell adopted. This analysis has shown that at low frequency the conductivity is dominated by $\pi \rightarrow Na^+$ transitions and that the hole doping of DNA is possible at a relatively low energetic cost (Gervasio *et al.* 2002). This calculation was very demanding, because the 3960 valence electrons of the system, in a PW basis set with a cut-off $E_{cut} = 70$ Ry, amount to 408 238 PWs, thus requiring a large vector-parallel machine (Hitachi SR8000 model).

Calculations on smaller systems, reduced to one or two base pairs, have been done to study more specific problems in which the phenomena of interest are localized and, as such, they are relatively insensitive to the length of the DNA strand considered. This is, for instance, the case of protonation of the AT pair, where a set of geometry relaxation performed within the DFT approach with a localized Gaussian basis set were done on both the neutral and the protonated AT base pair. The results have shown that the addition of a proton to adenine induces a strengthening of about 4–5 kcal/mol on the base pairing and a double-proton transfer between the two bases can occur

(Noguera *et al.* 2007). We can anticipate here, that a double-proton-transfer mechanism holds also in the case of the GC base pair in a full double-helix DNA, and occur during the general charge-transfer process along the double helix, although a detailed discussion will be given later in this chapter. Not only native DNA, but also artificial DNA has received attention, because of its potential nanotechnological applications mentioned above. In this case, artificial base pairs can bind together not only via the formation of regular H-bonds, but also by forming a complex with a metal ion. Control and tuning of the conducting properties of these artificial DNA systems seem to be possible by using Cu^{2+} as intercalated metal cations, giving rise to DNA-like double helices consisting of stacked $-H-Cu^{2+}-H-$ units, where the $-H$ atoms are the terminal-capping hydrogens of the synthetic base pairs. DFT-based electronic structure calculations have shown that these systems behave as an insulating ferromagnet (Jishi and Bragin 2007).

Coming to proteins, one of the hot topics in this field is certainly represented by their interactions with nucleic acids. This is of fundamental importance in gene expression, regulation and a wealth of functions that cells exert. In this respect, radiation-induced effects can alter the nature of the interaction by changing the charge state of the system and, hence, leading to malfunctions and transcription errors that could degenerate DNA in strand breaking, mutations and various forms of cancer. A pioneering DFT study of the interaction between guanine in the ground state, anion and radical forms with lysine (Lys) in different charge states has shown that the G–Lys complex, when Lys is in a radical cation state, induces a remarkable charge flow from Lys to G with the net effect that G is converted from its standard neutral form into a radical cationic G that can even induce CO_2 release and proton-transfer processes from Lys (Jena and Mishra 2007). This constitutes a pioneering attempt at performing quantum calculations aimed at unravelling the nature and the biological consequences of the interaction between nucleobases and protein residues. Another important protein possessing a special role and on which theoreticians have spent efforts is represented by the so-called hOGG1. This is the protein involved in repair processes of human DNA and has several crucial roles that can be summarized in the promotion of the enzyme activation, cleavage of the glycosidic bond and expulsion of the DNA base that has been damaged. A series of static DFT calculations, with localized basis sets, on the catalytic cycle of the glycosylase activity in hOGG1 has been done on a rather large model system including all the protein residues involved (Calvaresi *et al.* 2007). These calculations, done on selected configurations representative of the whole reaction path, provided a first insight into the reaction mechanism, allowed to discriminate which are the residues actively involved in the catalysis and showed that the rate-limiting step of the whole process is the nucleophilic attack leading to expulsion of the damaged base.

A fundamental reaction in a wide variety of proteins is the conversion of adenosine triphosphate (ATP) into diphosphate (ADP). This reaction, operated by living organisms in several cell subunits and subsystems, is of fundamental importance in all organisms and has been defined as "the principal net chemical reaction occurring in the whole world" by P.D. Boyer in his Nobel Lecture (Boyer 2003). ATP consists of an adenosine bonded to three phosphate groups

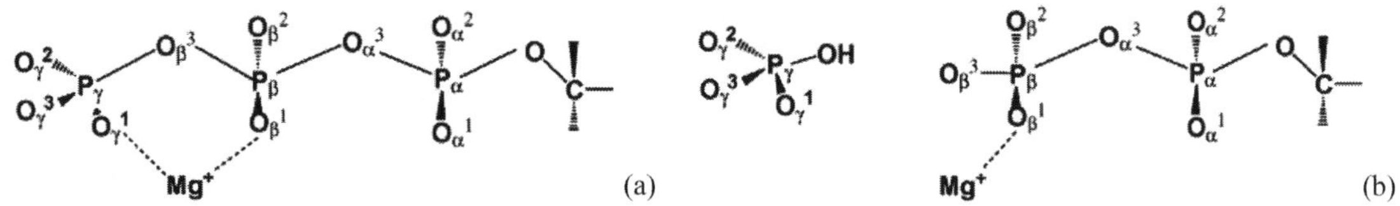

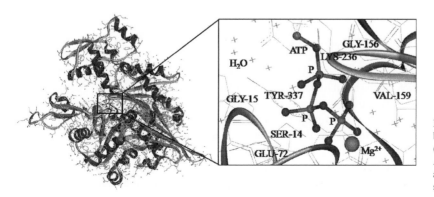

Fig. 17.7 Panel (a) shows the general scheme of a single strand of DNA with its four bases, adenine (A), guanine (G), cytosine (C) and thymine (T), from the top to the bottom, respectively. Panel (b) shows the double-helix structure of a DNA, identified by the two twisted ribbons representing the imaginary lines joining all the backbone P atoms.

(Fig. 17.7) and the removal of the last one of these groups, named γ-phosphate, provides, in living organisms, the energy required for a wealth of processes, ranging from cell movement (Wang and Oster 1998; Stock *et al.* 1999) to response to stress and adaptation to the external environment (Chapell *et al.* 1986; Liu *et al.* 2006).

The main phases of the ATP-to-ADP reaction consist in an attack of an H_2O molecule to ATP. In general, this water molecule undergoes dissociation into a proton H^+ and a hydroxyl anion OH^-. The proton is expected to attack the bridging oxygen indicated as $O_\beta{}^3$ in Fig. 17.7, and this induces the cleavage of the P_γ–O_β^3 bond that eventually leads to the release of the terminal phosphate group. In this way, ATP reverts to ADP with a general reaction mechanism summarized as $ATP + H_2O \rightarrow ADP + P_i$. The symbol P_i indicates the inorganic phosphate resulting from the released phosphate at the end of the process and the hydroxyl anion is supposed to bond to the departing group as sketched in the figure. The general reaction occurs in the presence of metal ions (Mg^{2+}, Ca^{2+}, K^+, etc.), experimentally identified, that are located in proximity of the phosphate chain. These cations are expected to act as a catalyst for the hydrolysis reaction, although their actual role is still under debate.

Pioneering DFT calculations have been done on a specific ATP system, actin, (Fig. 17.8), which is a protein involved in muscle contraction, cellular mobility and division and vesicle transport (Vorobiev *et al.* 2003). Because of the size of the system, full quantum DFT calculations of the whole protein

Fig. 17.8 Structure of actin (left) as provided by X-ray data and adenosine triphosphate (ATP) catalytic center (right) undergoing the ATP-to-ADP reaction. The main residues and atoms are labelled and small crosses refer to solvating water molecules.

were not possible and the model system extracted from the X-ray data included only on the triphosphate group, few residues and solvating water molecules crystallographically identified around the ATP site (Akola and Jones 2006) as shown in the left panel of Fig. 17.8. These calculations allowed for an atomic-level inspection of the reaction path and for the identification of the rate-limiting step. The latter turns out to be the cleavage of the $P_\gamma - O_\beta{}^3$ terminal bond that leads to the release of the inorganic phosphate. During the process, the water molecule that dissociates was found to be one of the H_2O molecules of the first solvation shell of Mg^{2+}, and the migration of the released group has been evidenced as a necessary step for a subsequent conformational change of actin. Different possible pathways were inspected, working out activation barriers ranging from 21 to 39 kcal/mol and providing the atomic-scale details of the process.

All the application of quantum mechanics to proteins and nucleic acids presented so far, either in the HF or in the DFT version, are static calculations of the electronic structure performed either on stable experimental config-urations or on stationary points obtained via geometry optimization. These are indeed instructive and rich in information. Nonetheless, finite temperature and entropy effects are two of the dominant features in soft matter and their role is often far from negligible. In this respect, the so-called first-principles molecular dynamics (FPMD) has represented a huge step forward in quantum simulations of condensed phases in general and, more recently, in biological systems. In practice, the interactions among atoms, instead of being described by an analytic function of the atomic coordinates $\mathbf{R}_I$, is directly computed from the total energy E^{tot}, which is simultaneously a function of the electron wavefunctions and of the atomic coordinates in the sense specified at the beginning of this section. The interactions are supposed to be electrostatic, i.e. dependent just on the charges and positions of the particles and not on their velocities; in particular, the Born–Oppenheimer (BO) approximation, at a given instant t in time, consists in an optimization of the electronic structure at the corresponding (fixed) nuclear positions $\mathbf{R}_I(t)$. Then, the nuclear forces are computed as gradients of the total energy with respect to the ionic position and the variables $\mathbf{R}_I(t)$ updated to $\mathbf{R}_I(t + \delta t) = \mathbf{R}'_I(t')$. The set of equations that one has to solve iteratively reads

$$M_I \ddot{\mathbf{R}}_I = -\nabla_{\mathbf{R}_I} \min_{\{\psi_i\}} E^{tot}[\{\psi_i\}, \{\mathbf{R}_I\}] \qquad (17.26)$$

$$\frac{\delta E^{tot}}{\delta \psi_i^*} \equiv H^{tot}\psi_i(\mathbf{x}) = \varepsilon_i \psi_i(\mathbf{x}), \qquad (17.27)$$

and imply the recalculation of the electronic ground state after each displace-ment of the atoms. The BO approximation assumes that electrons stay "frozen" on their ground state, whereas nuclei move and for this reason it is referred to as the "adiabatic approximation".

An alternative to this scheme has represented a real breakthrough in first-principles dynamical simulations: the Lagrangean-based method proposed by Car and Parrinello in 1985 (Car and Parrinello 1985). Two major problems arise in FPMD: On the one hand, one has to integrate the equations of motion for the nuclear positions, which represent the long-timescale part to the problem. On the other hand, one has to propagate dynamically the smooth time-evolving (ground state) electronic subsystem. The Car–Parrinello molecular dynamics (CPMD) is able to satisfy this second requirement in a numerically stable way and makes an acceptable compromise for the time step length of the nuclear motion. The formulation is an extension of a classical molecular-dynamics Lagrangean in which the electronic degrees of freedom are added to the system, along with any other dynamical variable $q_\alpha(t)$, such as a thermostat (Nosé 1984; Hoover 1985) or a barostat (Andersen 1980; Parrinello and Rahman 1980).

$$\mathcal{L}^{CP} = \frac{1}{2} \sum_I M_I \dot{\mathbf{R}}_I^2 + \sum_i \mu \int d^3x \left| \dot{\psi}_i(\mathbf{x}) \right|^2 + \frac{1}{2} \sum_\alpha \eta_\alpha \dot{q}_\alpha^2$$

$$- E^{tot}[\rho, \{\mathbf{R}_I\}, q_\alpha] + \sum_{ij} \lambda_{ij} \left(\int d^3x \psi_i^*(\mathbf{x}) \psi_j(\mathbf{x}) - \delta_{ij} \right). \quad (17.28)$$

The first three terms in the right-hand side of eqn (17.28) are the kinetic energies of the nuclei, of the electrons and of the additional dynamical variables, the fourth one is the total energy, in practice the DFT functional in the applications that will be discussed, and the last addendum is the orthonormality constraint for the wavefunctions. The kinetic energy for the electronic degrees of freedom is the novelty of the CPMD approach. The fictitious mass μ assigned to the orbitals $\psi_i(\mathbf{x})$ is the parameter that controls the speed of the updating of the wavefunctions with respect to the nuclear positions and, for this reason, it determines the degree of adiabaticity of the two subsystems, electrons and nuclei. A rigorous mathematical proof has been given (Bornemann and Schütte 1998), showing that the CPMD trajectory $\{\mathbf{R}^{CP}(t)\}$ stays close to the BO one $\{\mathbf{R}^{BO}(t)\}$ and the upper bound is given by $|\mathbf{R}^{CP}(t) - \mathbf{R}^{BO}(t)| < C \mu^{1/2}$, where C is a positive constant. This is simply a strategy to update on-the-fly the wavefunctions when ions undergo a displacement. The related Euler–Lagrange equations of motion read

$$\mu \ddot{\psi}_i(\mathbf{x}) = -\frac{\delta E^{tot}}{\delta \psi_i^*} + \sum_j \lambda_{ij} \psi_j(\mathbf{x}) \quad (17.29)$$

$$M_I \ddot{\mathbf{R}}_I = -\nabla_{\mathbf{R}_I} E^{tot} \quad (17.30)$$

$$\eta_\alpha \ddot{q}_\alpha = -\frac{\partial E^{tot}}{\partial q_\alpha}. \quad (17.31)$$

Let us observe that it is straightforward to give a Hamiltonian, instead of a Lagrangean, formulation of the CPMD method, via a simple Legendre transform. Despite the fact that the method dates back more than a quarter of century, applications to proteins and nucleic acids became possible only in the late 1990s, when large systems could be computationally afforded.

One of the earliest and more interesting CPMD simulations in this domain was performed on the protease of the human immunodeficiency virus type 1 (HIV-1), one of the major targets against AIDS (Piana and Carloni 2000; Piana *et al.* 2002). This particular enzyme catalyzes the hydrolysis reaction of a

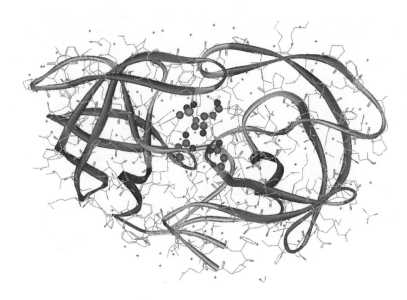

Fig. 17.9 Structure of the HIV-1 protease system with C_2 symmetry. The catalytic center (peptide group) undergoing a cleavage is located at the center of the figure, between the two subunits (shown as ribbons), and is shown with ball and sticks.

specific peptide bond and the cleavage site is located at the interface of the two subunits forming the system (Fig. 17.9).

First-principles simulations could be afforded due to the fact that the protein backbone surrounding the active site is relatively rigid and does not undergo conformational changes on the ps timescale typical of quantum-dynamical simulations. Hence, quantum calculations could focus on the active site only and a model system could be extracted from the whole structure, including only the closer residues (Asp dyad, Thr-26 and Gly-27) plus the solvating water molecules. This allowed for a detailed investigation of the chemical reactivity of the enzyme at different configurations and for the inspection of its evolution during the reaction at an atomistic level. The mechanism resulting from the simulation can be summarized in a specific action of the enzyme that brings the substrate into a configuration such that dynamical fluctuations of the protein frame favor the enzymatic catalysis. These findings represent the first direct inspection of the general pathway proposed by Lumry (1995) for which experiments fail in providing direct evidence. In particular, the use of dynamics demonstrated rather clearly that the transition state realized by the HIV-1 system in the first stage of the reaction is stabilized by mechanical fluctuations of the protein, a detail that is beyond the reach of any static calculation (Carloni 2002).

Another interesting application of the CPMD method was done for the low molecular weight GTO binding protein that, in a way similar to ATPase systems, hydrolyzes GTP into GDP (Cavalli and Carloni 2002). The overall reaction is $GTP + H_2O \rightarrow GDP + HPO_4^{2-}$ and is catalyzed by the divalent metal cation Mg^{2+}. The system used in the dynamical simulations was extracted from X-ray data of a specific member of this family of proteins, Cdc42/Cdc42GAP/GDP. This model was shown to account for the most relevant contributions of the protein electrostatic field on the reactants and its size is tractable within DFT-based approaches. These simulations have shown for the first time that a catalytic water molecule forms H-bonds with the residues Gln-61 and Thr-35 during the nucleophilic attack and that a proton transfer from the water to GTP represents the trigger of the reaction. This is consistent with the X-ray data and with the site-directed mutagenesis experiments reported for the Q61E mutant (Lerm *et al.* 1999).

Nucleic acids are computationally more demanding than proteins if one wants to go beyond a simple analysis of a base pair, because, in general, the double-helix sequenced structure of DNA and RNA does not make it easy to extract a specific catalytic center. In fact, in most of the cases of biochemical interest, long portions of the nucleic acid, often accompanied by the surrounding solvent water, are involved in the phenomena on which experiments focus, such as, for instance, charge localization and transport along DNA fibers. From experiments, charge localization and displacement along a double-stranded DNA are mostly deduced from indirect evidence. Nonetheless, three basic ideas have been put forward in recent years to explain the stabilization mechanism of a charge (generally a hole) on one or more bases in DNA: (i) changes in the tilt angle of the bases carrying the localized electronic charge (Bruinsma *et al.* 2000), (ii) fluctuations in the position of the metal counter ions or their partial desolvation (Barnett *et al.* 2001), and (iii) a change in the protonation state of the nucleobases (Stenkeen 1997; Giese and Wessely 2001; Shafirovic *et al.* 2001). Simulations within a classical molecular-dynamics approach of a seven base-pair duplex DNA, fully hydrated, at room temperature allowed sampling of the configuration space for a sufficiently long time to extract uncorrelated configurations. On these configurations, DFT calculations within the BO scheme were performed and allowed to inspect in detail the ion-gated mechanism, namely the localization process induced by the counterion fluctuations mentioned above (Barnett *et al.* 2001). The first set of quantum-dynamical calculations relative to the other two mechanisms were made possible only by the use of the NEC SX-6 Earth Simulator system, one of the largest resources in the world. The simulated system consisted of a solvated synthetic double-strand poly-GC Z-DNA in an oxidized state and composed of twelve G:C base pairs, the global system is described by the formula $C_{228}N_{96}O_{144}P_{24}Na_{24}H_{264}*138(H_2O)$ (Fig. 17.10). Such a GC polymer contains all the ingredient of an active DNA, apart from the chemical disorder that would result by the presence of A:T base pairs.

After classical molecular-dynamics simulations at $T = 300$ K, uncorrelated configurations were extracted and CPMD simulations were performed both at room temperature and by cooling the system to $T = 0$ K. In these calculations, because of the use of a PW basis set with an energy cut-off of 70 Ry, the resulting 3959 valence electrons were represented by 403 840 plane waves and the calculations required a time of 111.5 s/step on 32 nodes of the Earth Simulator. Since each node of such a machine consists of 8 vector CPUs, a total number of 256 CPUs was necessary. The first outcome of these simulations is that at very low temperature, such as $T = 0$ K, the charge injected in the system is symmetrically distributed and no localization arises. On the contrary, when the temperature increases and thermal fluctuations occur, the tilt of a specific base with respect to the major axis of the DNA polymer is sufficient to break the symmetry and to localize the charge; the larger localization was

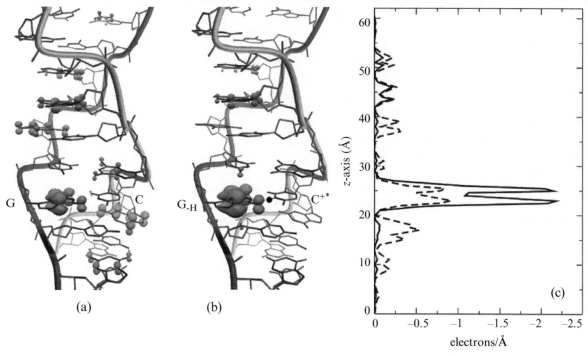

Fig. 17.10 Snapshots of the CPMD simulations on the oxidized Z-DNA before the proton transfer from G to C (a) and after the proton transfer (b). The transferred proton is the black ball and gray clouds show the spin density at isosurfaces of 10^{-4} $e/\text{Å}$. Panel (c) shows the projection of the spin density along the z-axis. The dashed and solid lines refer to the system before and after the G–C proton transfer, respectively.

observed on those nucleobases that showed the larger tilting, with a one-to-one correspondence. This is consistent with the mechanism proposed by Bruinsma *et al.* (2000) and it is an effect that dynamical simulations, as opposed to static calculations, are able to capture. The major problem, if DNA were to be used in nanotechnological applications, is the fact that such fluctuations are uncontrollable. However, these same simulations showed that the third mechanism proposed—the change of the protonation state of a G base— represents a possible controlled way to localize a charge on a specific base or group of bases, as shown in Fig. 17.10. By dislodging a proton from the N1 nitrogen site of the G where a high localization was found, and by transferring it to the adjacent H-bonded C base, these CPMD simulations have shown that the localization increases by more than a factor of two. This was quantified in terms of the projected spin density along the z-axis of the DNA, as shown in panel (c) of Fig. 17.10. Simulations repeated by deprotonating different G bases, also in the absence of any tilt, gave a similar scenario. These results represent the first atomic-level investigation of the strict relationship between proton transfer and charge localization (Giese and Wessely 2001).

Photochemical reactions are another interesting field whose development has been so far more experimental than theoretical. In fact, photoactive proteins are not only an important class of biological systems, but they are also the target of potential technological applications in molecular optoelectronics and photosensors (Kolodner *et al.* 1997). In this respect, the time-dependent formulation of DFT (TDDFT) has shown its potential in the treatment of electron excitations, although specific applications have so far focused mostly on inorganic molecules and solid-state physics. As far as biomolecules are concerned, an alternative DFT approach involving only low excitations, has been set up and applied with a certain success. This rather dramatic limitation in the number of the excited states included in the calculations makes the method computationally cheaper than TDDFT. Yet, there are good reasons to limit the analysis to the low excitations; in fact, the majority of the known organic photoreactions involve just the first excited singlet state (S_1) and the lowest triplet state (T_1). Other excited states have too short a lifetime to be of practical interest and can generally be neglected. Now, if a single valence electron is excited from the highest-occupied (ground-state) orbital **a** to the lowest-unoccupied orbital **b**, four different determinants can be obtained according to Pauli's principle. Instead of separate wavefunctions for the triplet t and singlet m states, it has been shown that it is possible to determine a single set of spin-restricted single-particle orbitals $\psi_i(\mathbf{x})$ for the states $i = 1, \ldots, N+1$ in such a way that the total electronic densities for both the t and the m states are identical, whereas their spin densities are different (Frank *et al.* 1998). In this way, a new DFT functional, the restricted open-shell Kohn–Sham (ROKS) functional, generalized from the ground-state Kohn–Sham total energy, can be written as

$$H^{\text{ROKS}}\left[\{\psi_i(\mathbf{x})\}\right] = 2E_m^{\text{KS}}[\rho] - E_t^{\text{KS}}[\rho]$$

$$- \sum_{i,j=1}^{N+1} \varepsilon_{ij} \left(\int d^3x \, \psi_i^*(\mathbf{x}) \psi_j(\mathbf{x}) - \delta_{ij} \right) \qquad (17.32)$$

in which the last term is ordinarily the orthonormality constraint and

$$\rho(\mathbf{x}) = \rho_\alpha^m(\mathbf{x}) + \rho_\beta^m(\mathbf{x}) = \rho_\alpha^t(\mathbf{x}) + \rho_\beta^t(\mathbf{x}). \qquad (17.33)$$

The functionals with the superscript KS are, instead, Kohn–Sham total-energy functionals with the difference reduced only to the exchange-correlation term. The minimization of the functional $H^{\text{ROKS}}[\psi_i(\mathbf{x})]$ with respect to the orbitals leads to two sets of Schrödinger-like equations, one for the doubly occupied orbitals and one for the singly occupied **a** and **b** states, which can be solved either by iterative diagonalization or by minimization. This approach has been used to study the isomerization of the rhodopsin chromophore (Fig. 17.11). This is the photosensitive protein in the rod cells of the retina of vertebrates and the process of vision, as we understand it, involves the photoisomerization of this specific protein as a response to the absorption of photons. This isomerization is remarkably fast (about 200 fs) and triggers a cascade of slower reactions that eventually produce a specific biological signal. Due to the very short timescale of this process, experimental probes fail in the direct observation of the reaction. This missing information can be recovered by quantum simulations within the ROKS scheme (Molteni *et al.* 1999).

The 11-*cis* isomer of the rhodopsin protein used in the simulations is shown in panel (a) of Fig. 17.11. This system consists in a protonated Schiff base containing 54 atoms and the simulations included also a glutamate amino

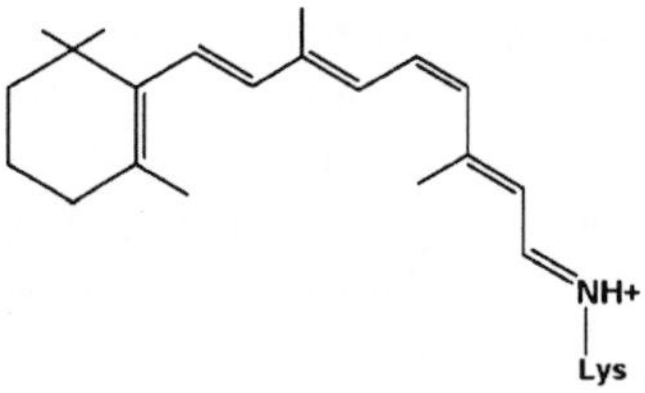

Fig. 17.11 Scheme of the 11-*cis* isomer of the rhodopsin protein (a) and its *trans* isomer (b) formed upon exposure to light. The link via a protonated Schiff base to the rest of the protein through the lysine (Lys) residue is explicitly labelled.

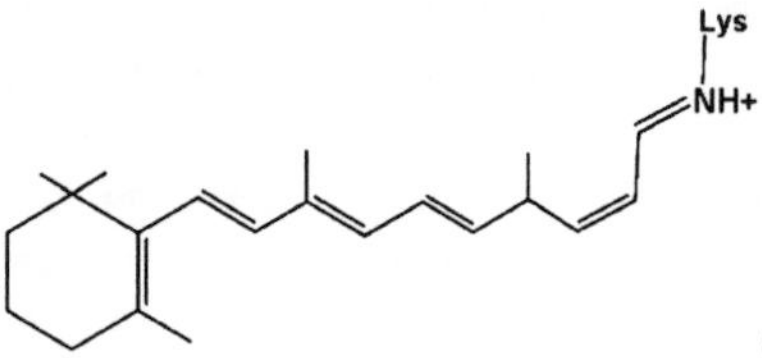

acid and one water molecule that is expected to be involved in the photoisomerization process. As the TDDFT, also the ROKS approach suffers from an underestimation of the excitation energies, a fact that is generally pathological in any DFT approach, nonetheless, the dynamics turns out to be correct and the reaction path could be simulated with appreciable accuracy, offering for the first time an insight at a molecular level into this important class of biological reactions. The structural changes from the *cis* to the *trans* isomer could be followed in detail, although the fact that a small model was extracted by the entire protein and, basically, a gas-phase reaction was simulated prevented accurate description of the energetic of the reaction.

17.4   Hybrid QM/MM approaches

The size is a common denominator to any computational approach dealing with nanoscale objects. A workaround, particularly successful in the field of macromolecules and biosystems, is represented by a combination of classical and quantum-mechanics methodologies, the so-called quantum-mechanics/molecular-mechanics (QM/MM) approach. Namely, extended systems can be split into two or more domains in which the level of accuracy decreases by going far from the portion undergoing important structural and/or electronic changes. This part of the system, generally involving chemically active sites, is described quantum mechanically, whereas the rest of the system, important in terms of potentials and electrostatic interactions but chemically inert, is described by classical force fields. Such a partitioning is somehow arbitrary and generally driven by "chemical intuition", nonetheless, it has been successfully applied to a variety of problems since its first introduction (Warshel and Levitt 1976; Momay 1978) until recent applications (Andreoni 1998; Andreoni *et al.* 2001; Lin and Truhlar 2007).

In QM/MM approaches, as can be easily guessed, two major problems must be faced. In the first instance, those atoms that will be treated quantum mechanically must be carefully selected and they must be representative of the chemically active center on which one wants to focus. On the second instance, the type of QM approach, the type of classical force field and the type of interaction between the QM and the MM worlds must be established. All their advantages, limitations and possible drawbacks must be carefully analyzed and kept in mind when these approaches are applied to a specific problem. In a general QM/MM scheme, the total energy of the whole system is written as the sum of three terms, $E^{tot}[FS] = E[QM] + E[MM] + E^{int}[QM/MM]$, where the first contribution on the right-hand side represents the total energy of the QM subsystem, the second one the total energy of the classical MM subsystem and the third one their mutual interaction. The acronym FS indicates the full system, $FS = QM + MM$. In general, the classical MM subsystem is described by an analytic potential that keeps into account all the fundamental inter-molecular and intra-moelcular interactions. All the various parameters of a force field are derived by fitting to experimental data or from QM calculations on a reduced model system. In the case of proteins and nucleic acids, several excellent force fields have been proposed and assessed over the years: ECEPP/2 (Momany *et al.* 1975), CHARMM (Brooks *et al.* 1983), AMBER (Cornell *et al.* 1995), GROMOS (Hermans *et al.* 1984), GROMACS (Van der Spoel *et al.* 2005) just to mention the most popular ones. These classical potentials are very powerful in modelling thermal effects and dynamical properties of proteins and nucleic acids.

Nonetheless, as can be easily guessed by looking, for instance, at the harmonic term for the bond stretching, they are unable to describe the formation or breaking of chemical bonds. This is one of the main reasons that led to the development of the hybrid QM/MM methods. A discussion about classical force fields is beyond the scope of the present chapter; instead, we shall focus on the interface between the QM subsystem and the MM part. The interactions between the QM and MM subsystems are generally rather strong and pose several problems. In the easiest case, QM atoms interact with the MM atoms via H-bonds and/or non-bonding interactions between QM and MM atoms chemically unbound. In these conditions, the selection of the QM/MM frontier does not pose particular difficulties. Representative examples are aqueous solutions in which a QM solute is surrounded by MM water molecules or peculiar cases of ligand–protein where the interaction between the two occurs via H-bonds only. An instructive example is the formation of a complex between the HIV-1 integrase and its inhibitor named S-1360 (Alves *et al.* 2007). This complex represents a rather intriguing case, since integrase is one of the three human immunodeficiency virus type 1 (HIV-1) enzymes crucial in virus-replication processes. The molecule indicated as S-1360 is a powerful and selective inhibitor of HIV-1 integrase. Hybrid QM/MM simulations have allowed computation of the protein–ligand interaction energy for S-1360 and two analogs, showing that the strongest protein–inhibitor interactions are established between Lys-159 residue and Mg^{2+} cation, a crucial detail that has a direct relationship with the experimentally observed anti-HIV activity and that can give important hints in anti-AIDS drug design.

However, apart from these particular cases, in the majority of the systems, the frontier between the QM and the MM ensembles passes across a (covalent) chemical bond as sketched in Fig. 17.12.

In this case, some suitable termination of the boundary   between the two subsystems is required in order not to create artificial dangling bonds. As a rule of thumb, in the case of proteins, the QM/MM frontier is placed preferentially at an aliphatic carbon, whereas in the case of nucleic acids it is assumed to pass across the sp^3 C atom connecting a ribose ring to the phosphate group, in order to have the smallest possible charge distribution at the frontier. To this aim, the

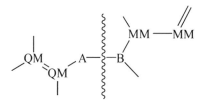

Fig. 17.12 Schematic view of the border between the QM (left side) and MM (right side) subsystems in which a chemical bond between atoms A (QM) and B (MM) is crossed.

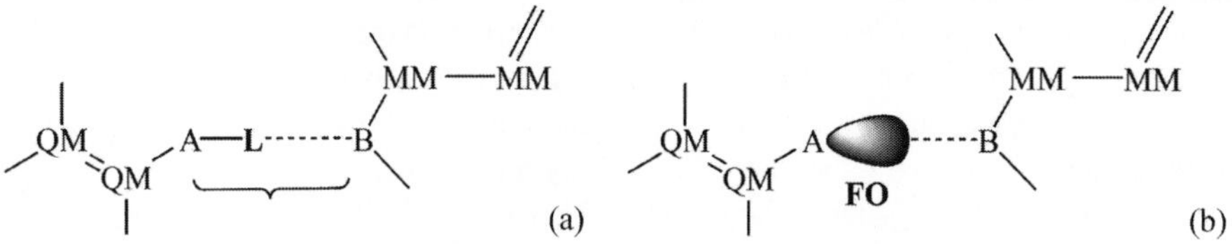

Fig. 17.13 Schematic representation of (a) a link atom (L) and (b) frontier orbital (FO) used to saturate dangling bonds generated at the QM/MM boundary whenever a covalent bond crosses the border between the two subsystems.

methods proposed in the literature over the years can be classified into three groups: Link atoms (1), frontier orbitals (2) and effective pseudopotentials (3).

Link (L) atoms are additional monovalent hydrogen-like atoms added to the QM subsystem to saturate the covalent bonds across which passes the QM/MM boundary (Singh and Kollman 1986; Field *et al.* 1990; Bakowies and Thiel 1996). They saturate the dangling bonds generated upon creation of a QM subsystem with some covalent bonds crossing the QM/MM border. Alternatively, the monovalent hydrogen-like atom can be replaced by a fluorine or a methyl group. L atoms are generally invisible to the MM atoms, which interact via the force field directly with the QM atoms at the border, as sketched in panel (a) of Fig. 17.13.

This is done to ensure that the QM–MM covalent bonds and related lengths, angles and dihedrals are not affected by the cut across them. For the sake of completeness, it must be mentioned that there are cases in which a L atom must be kept into account also from the MM side; this happens for C species with non-negligible polarization effects. In fact, it has been shown (Reuter *et al.* 2000) that omitting L atoms in the calculation of the MM interactions could give poor results because of the polarization of L atom–carbon bonds.

Each L atom should be placed far from any other L atom, otherwise this would lead to spurious Coulomb interactions between them. Furthermore, a L atom should be able to reproduce the local chemical environment. For instance, if the A–B bond in Fig. 17.13 is a sp^3 or sp^2 bond, then this character must be preserved if a capping L atom is used to saturate the dangling bond due to the missed B atom. If chemical differences between the hydrogen-like L atom and the original chemical group it replaces arise, empirical and semi-empirical adjustments are possible by an *ad hoc* parameterization of L and its charge (Antes and Thiel 1999; Zhang *et al.* 1999).

In the case of frontier orbitals (FO), the unsaturated covalent bond of a QM atom at the border, such as atom A in Fig. 17.13, is compensated by an additional localized orbital $\psi_{FO}(\mathbf{x} - \mathbf{R}_A)$ that is treated as frozen during the calculation (Assfeld and Rivail 1996; Gao *et al.* 1998). This scheme avoids the use of L atoms and works well in self-consistent field-optimization procedures. It can instead give some problems when dynamics is attempted, in particular when a variational basis set, such as plane waves, is adopted, since this orbital contributes to the total electron density $\rho(\mathbf{x})$, but it does not undergo dynamical variations as, for instance, the Car–Parrinello orbitals, and its contribution to the forces can result in spurious components that, in turn, can bias the dynamics. This has often been regarded as a drawback in terms of applications to dynamical simulations. Nonetheless, interesting applications making use of FO are not lacking. One of the most recent (and remarkable) ones is the study of hydrogen transfer via a tunnelling to the active site catalyzed by coenzyme B_{12}-dependent methylmalonyl-CoA mutase (Dybala-Defratyka *et al.* 2007). The system used by the authors consisted of 672 amino acid residues solvated in 1388 H_2O molecules, plus 299 atoms composing the substrate and coenzyme. The QM subsystem is formed by 45 atoms, including the ligand and a portion of the methylmalonyl-CH_2- substrate. The FO were placed at the carbon atoms C_2 of the β-mercaptoethylamine part of the CoA chain representing the QM/MM boundary. The QM subsystem was treated at the unrestricted HF level, while the MM force field adopted was CHARMM. The major outcome of these simulations was an accurate explanation of the kinetic isotope effect; a multidimensional tunnelling effect seems to be responsible for an increase of a factor of 3.6 in the computed intrinsic hydrogen kinetic isotope effect, a result in fairly good agreement with experiments. These calculations represent the first confirmation that tunnelling contributions can be large enough to account for a remarkable kinetic isotope effect, not as a consequence of a thin barrier but because of the fact that corner-cutting tunnelling decreases the distance over which the system tunnels without an appreciable increase in the effective potential barrier or in the effective mass.

A source of concern in saturating dangling bonds at the QM/MM frontier is the perturbation that an artificial capping such as FO or L atoms would induce on the QM subsystem. In the case of monovalent L atoms, this problem has been solved by the introduction of the optimized effective core potential (OECP) approach (DiLabio *et al.* 2002). This scheme makes use of pseudopotentials written in a fairly standard way,

$$V_I^{OECP}(\mathbf{r}, \mathbf{r}') = V^{loc}(\mathbf{r})\,\delta(\mathbf{r} - \mathbf{r}') + \sum_l V_l^{NL}(\mathbf{r}, \mathbf{r}'), \qquad (17.34)$$

where $\mathbf{r} = \mathbf{x} - \mathbf{R}_I$, and $\mathbf{R}_I$ the position of the capping atom at the QM/MM interface. The two terms in eqn (17.34) are written as an analytic function depending on a set of parameters,

$$V^{loc}(\mathbf{r}) = -\frac{Z_I}{r}\mathrm{erf}\left(\frac{r}{r_0\sqrt{2}}\right) + e^{-(r/r_0)^2/2}\left[c_1 + c_2\left(\frac{r}{r_0}\right)^2\right.$$
$$\left. + c_3\left(\frac{r}{r_0}\right)^4 + c_4\left(\frac{r}{r_0}\right)^6\right] \qquad (17.35)$$

$$V_l^{NL}(\mathbf{r}, \mathbf{r}') = \sum_{m=-l}^{l} Y_{lm}^*(\hat{\mathbf{r}})Y_{lm}(\hat{\mathbf{r}}) \sum_{i,j=1}^{3} p_{lj}(r)h_{lji}\,p_{li}(r'), \qquad (17.36)$$

where $p_{ij}(r) = \mathrm{const}\; r^{l+2(h-1)}\,\exp(-0.5\,r^2/r_l^2)$ and Y_{lm} are the spherical harmonics. All the parameters $\{r_0, c_1, c_2, c_3, c_4, h_{lji}, r_l\}$ are optimized by minimizing iteratively the differences in electron density between the QM subsystem and a full quantum reference configuration including atoms beyond the QM/MM boundary. It must be remarked, in passing, that the dimensionality of the parameter space is determined by the maximum angular momentum component in the non-local part of the OECP. In practical applications (von Lilienfeld *et al.* 2005) it has been shown that a maximum value $l = $ s or, rarely, $l = $ p is enough to achieve a good optimization for oxygen in water or carbon in acetic acid. Optimized potentials are particularly suitable in the cases in which the QM subsystem embedded in the MM environment is characterized by the presence of highly ionic species. In general, this is not the case for proteins and nucleic acids; however, it must be kept in mind that highly ionic atoms at the QM/MM boundary could lead to unphysical effects, from an overpolarization of the QM charge density to a leakage of electrons towards the MM subsystem. (Winter and Pitzer 1988).

Recent interesting applications of the QM/MM approach focused on cytochrome P450 (Zheng *et al.* 2007). This specific system belongs to the ubiquitous large family of mono-oxigenases proteins, fundamental in the regulation of drug metabolism and biosynthesis of metabolites. The cytochrome P450 enzyme is one of the best characterized structures in which the most relevant catalytic center is represented by a heme Fe(IV) oxoferryl site (Rovira and Parrinello 2000). In these calculations, the unsaturated C-bonds at the QM/MM border were terminated with L atoms. In the calculations of Freindorf *et al.* (2006), QM/MM simulations were performed by keeping fixed the MM subsystem of the protein, while the QM part underwent both dynamics and relaxation. The molecular oscillations of the heme active site evidenced that some amino acids of the active site pocket can be replaced by other amino acids. This kind of protein modification would be rather expensive to obtain experimentally; in this respect, QM/MM simulations offer a way to perform virtual experiments and a practical tool to predict new structures and mutations. Previous calculations on the same system (Guallar *et al.* 2003) focused on hydroxylation of camphor operated by cytochrome P450, using as QM driver a DFT approach with a localized basis set. These simulations evidenced for the first time a remarkably low activation barrier (11.7 kcal/mol) for the hydrogen-abstraction step in the enzyme catalysis. On these bases, experiments that would otherwise be difficult to interpret (Davydov *et al.* 2001) found a satisfactory rationalization. Specifically, any experimental attempt at trapping the catalytic high-valent heme Fe(IV) oxoferryl species responsible for this specific hydroxylation process failed.

Another interesting application of the L atom method approaches was the calculation of the one- and two-electron reduction potentials of Flavin adenine dinucleotide (FAD) cofactor (Fig. 17.14) in solution (Bhattacharyya *et al.* 2007). FAD is a common cofactor in redox proteins, and its reduction potentials are controlled by the protein environment. Calculations have been performed with the scope of determining the reduction potentials of FAD in aqueous solution, in medium-chain acyl-CoA dehydrogenase (MCAD) and cholesterol oxidase (CHOX). In this QM/MM approach, the flavin ring was treated as the QM subsystem at DFT level, while the rest of the enzyme–solvent system was the MM part. The computed two-electron/two-proton reduction potentials turned out to be in rather good agreement with experimental data and allowed us to infer that a bending of the flavin is responsible for a destabilization of about 5 kcal/mol of the oxidized state.

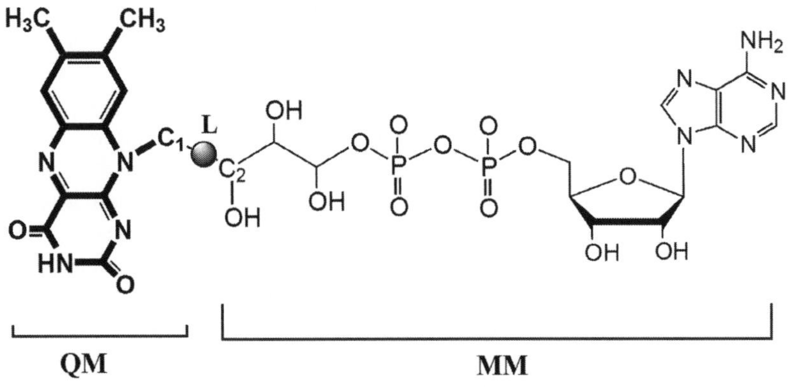

Fig. 17.14 Scheme of FAD. The QM/MM frontier crosses the covalent bond C_1–C_2. The 7,8-dimethyl isoalloxazine (flavin), highlighted by thick sticks, is the QM subsystem, including C_1, which is end capped by a L atoms, whereas C_2 is included in the MM part.

This showed that the first coupled electron–proton addition is stepwise for both the free and the two enzyme-bound flavins. On the contrary, the second coupled electron–proton addition, although also being stepwise for the free flavin, seems to be concerted when the flavin is bound to either the dehydrogenase or the oxidase enzyme at physiological pH.

The study of molecular oscillations, as in the two cases discussed above, could be partly biased by the presence of monovalent hydrogen-like L atoms capping the QM subsystem. In fact, these L atoms turn out to be chemically bonded to the QM atom that they saturate and are subject to dynamical fluctuations during the simulation, as any other atom present in the structure. However, they do not represent real chemical bonds and do not reproduce the correct bond length of the MM atom that they replace. In an attempt at overcoming this problem, two main prescriptions have been proposed. The first one (Echinger *et al.* 1999) targets C–C bonds. As mentioned, non-polar carbon single bonds joining CH_2 groups are ubiquitous in proteins and their cut represents one of the best choices to terminate a QM region. Instead of simply capping the C dangling bond with a hydrogen, the position of the monovalent saturating H-like L atoms is rescaled from the C–H bond length to that of the original C–C bond distance. This is the reason that gives the acronym SPLAM, scaled position link atom method, to this approach. The scaling algorithm is rather simple: given the C–C equilibrium distance $r_{CC}{}^0$, the actual bond length would be $r_{CC} = |\mathbf{r}_C{}^{QM} - \mathbf{r}_C{}^{MM}|$ and the H-like L atom would give a distance $r_{CH} = |\mathbf{r}_C{}^{QM} - \mathbf{r}_L|$; thus the scaled position becomes

$$r_{CL} = r_{CH} + \frac{k_{CC}}{k_{CH}}\left(r_{CC} - r_{CC}^0\right), \qquad (17.37)$$

where the force constants k_{CC} and k_{CH} are deduced from the corresponding bond stretching. Such a scaling is chosen in a way that the C–L stretching force approximates the C–C one. However, this somewhat artificial way of elongating a C–H chemical bond has the drawback of introducing spurious force components that in some cases can affect the dynamics of the system. Instead, it has been shown to be a versatile tool in structural optimizations as, for instance, in heme conformation in myoglobin (Rovira and Parrinello 2000; Rovira *et al.* 2001). These calculations have shown that the heme–CO structure (QM part) is quite rigid and not influenced by the distal pocket conformation

(MM part), thus excluding any relation between FeCO distortions and different CO absorptions observed in the infrared spectra. In any case, the SPLAM approach requires an energy correction written as a harmonic term. This approach has been benchmarked on several molecular systems, ranging from simple ethane to protonated Schiff bases and has been proposed as a viable possibility to enlarge the system size of the retinal chromophore in bacteriorhodopsin. As mentioned in the previous section, the photoisomerization of the rhodopsin chromophore, simulated on a model system extracted from the protein, displayed system-size problems that affected the isomerization barriers (Molteni *et al.* 1999). These QM/MM approaches allowed extension of the system size to the whole protein and the solvent water, showing that the flexibility of the rhodopsin chromophore, observed in vacuum, is highly reduced by the environment (Röhrig *et al.* 2005).

The termination of unsaturated QM bonds at the QM/MM interface does not exhaust all the problems involved in a hybrid approach. Another important issue concerns the electrostatic interaction between the QM and the MM subsystems. A rather general three-step scheme was proposed some years ago (Bakowies and Thiel 1996). The first step consists in a mechanical embedding, in which the electronic structure and electrostatic potential of the QM subsystem is computed in vacuum, disregarding the coupling to the surrounding MM environment. The electrostatic interaction between the two parts is then added a posteriori using a classical potential-derived point-charge model for the QM charge distribution. The second step is represented by an electrostatic embedding, in which the Coulomb potential due to the MM part is treated as an external charge distribution in the QM Hamiltonian, thus accounting for the polarization induced by the MM subsystem to the QM region. In first-principles schemes, such as HF or DFT, the electrostatic embedding is generally implemented, at least at long range, by adding the contribution of the MM point charges to the HF or DFT Hamiltonian (Maurer *et al.* 2007). The third step is in the inclusion of the polarization of the MM subsystem as a response to the QM charge-density distribution.

In the case of a Car–Parrinello approach, an elegant and fully Hamiltonian scheme has been proposed to reduce the computational cost due to the Coulomb interaction of the QM electron density with the MM atoms (Laio *et al.* 2002). Since this method has been extensively used in QM/MM simulations of proteins and nucleic acids (Biswas and Gogonea 2005), we summarize here the basic points. The coupling between the QM and the MM parts is achieved by extending the Car–Parrinello Hamiltonian H^{CP} as $H^{tot} = H^{CP} + H^{MM} + E^{int}[\rho, \{\mathbf{r}_I\}]$. The coupling functional,

$$E^{int}[\rho, \{\mathbf{r}_I\}] = \sum_{I=1}^{MM} q_I \int d^3x \frac{\rho(\mathbf{x})}{|\mathbf{x} - \mathbf{r}_I|} \qquad (17.38)$$

represents the major computational cost. This sum runs over all the MM classical atoms (MM > 10 000) and the integration has to be performed on the whole space. This particle-mesh calculation would then scale as $O(\mathrm{MMx}N_G)$ where MM is the number of classical atoms and N_G the number of mesh points on which the electronic density is mapped or, in a plane-wave basis set, the number of reciprocal-space vectors. Analogously, the calculation of the contribution to the electronic potential and of the forces acting on the MM charges would result in a hard computational task. The proposed workaround starts from the observation that for MM atoms very far from the QM region, the distance $|\mathbf{x} - \mathbf{r}_I|$ is large enough to make the integral of eqn (17.38) vanishing. It is then possible to distinguish three domains in which the electrostatic interactions can be treated at different levels of accuracy; a region $r < r_1^{cut}$ very close to the QM subsystem in which the interacting functional must be computed "as is". An intermediate region, $r_1^{cut} < r < r_2^{cut}$, whose extension depends on the system and the strength of the QM–MM interaction, where the exact expression (38) becomes

$$E_{RESP}^{int} = \sum_{I=1}^{MM'} q_I \sum_{J=1}^{QM} \frac{q^{RESP}(\rho, \mathbf{r}_I)}{|\mathbf{r}_J - \mathbf{r}_I|} \qquad (17.39)$$

that will be explicitly specified in the ongoing discussion, and in which $MM' < MM$ is a subset of the MM atoms. A multipolar expansion is then used for the region at $r > r_2^{cut}$. In eqn (17.39), q^{RESP} is an atomic point-like charge similar to the one introduced by Bayly *et al.* (1993) by fitting the electrostatic potential (ESP) due to the QM charge density $\rho(\mathbf{x})$ "seen" by the closer MM' atoms. A restrain penalty function is included, since unphysical charge fluctuations occur in unrestrained ESP charges during dynamics. In practice, to compute q^{RESP} the quantity that is minimized during the dynamics is

$$\chi = \sum_{I=1}^{MM''} \left(\sum_{J=1}^{QM} \frac{q_J^{RESP}}{|\mathbf{r}_I - \mathbf{r}_J|} - V_I \right)^2 + w_q \sum_{J=1}^{QM} \left(q_J^{RESP} - q_J^{H} \right)^2, \qquad (17.40)$$

where V_I is simply the Coulomb potential modified at short range to avoid unphysical overpolarization effects, w_q is a weight factor generally in the range 0.1–0.25 and q_J^{H} is the Hirshfeld charge (Hirshfeld 1977). In short, the RESP potential is requested to reproduce the correct Coulomb interaction and, at the same time, the RESP charge to be as close as possible to the corresponding Hirshfeld charge. As can be noticed, eqn (17.40) is just a least-square procedure and, as such, it reduces to a simple matrix inversion operation that does not add significant computational cost to the simulation. In many embedding procedures, the QM subsystem is polarized by the MM subsystem, but the MM subsystem is not polarized by the QM one (Zhang *et al.* 2007); this may result in an unbalanced treatment of the electrostatic interactions. To overcome this problem, a polarized-boundary redistributed charge (PBRC) procedure and a polarized-boundary redistributed charge and dipole (PBRCD) scheme have been adopted, consisting in an addition of a self-consistent mutual polarization of the boundary region of the MM subsystem to the existing embedding schemes.

These QM/MM methods have been widely used in computational biochemistry; for further details we call the attention of the reader to Carloni *et al.* 2002, Sulpizi *et al.* (2002, 2005), Spiegel and Magistrato (2006) and Dal Peraro *et al.* (2007). Just to mention a few of the most relevant applications, worthy of note is the study of the binding of Cu^{2+} to the murine prion protein, a prototype model for the general problem of transition metal–protein interactions. This specific case is representative of a wide range of practical problems, since prions are the infectious agents that play a key role in neurodegenerative diseases affecting animals and humans, the most relevant case being the cattle BSE disease. A controversy that QM/MM approaches helped to settle is the assumption, postulated a priori, that Cu^{2+} cations bind to the His-rich unstructured part. Experimentally, EPR and ENDOR measurements (Van Doorslaer *et al.* 2001) suggested that Cu^{2+} binds instead to sites located in the structured part of the protein. Nonetheless, the specific position of the metal-binding sites in the protein could not be identified. This would be of great interest, because divalent metal cations are expected to influence the structural stability of the protein and its transition to the infectious form (Price *et al.* 1999). Finite-temperature QM/MM dynamical simulations, accompanied by a statistical analysis of experimental crystal structures, allowed for the determination of the relative stability of various amino acids–copper ligation, providing an accurate ranking of possible sites classified according to their relative probability (Colombo *et al.* 2002).

Interesting QM/MM calculations within the same framework were performed on the protease of the human immunodeficiency virus type 1 (HIV-1) as an extension of the small model system discussed in the previous section (Piana *et al.* 2002).

The same research group took into account the whole system using a QM/MM approach in order to unravel the effects of the full protein. One of the novelties was the discovery of the triggering of the cleavage reaction at the active site, consisting in the formation of a low-barrier H-bond (Northrop 2001) connecting the two Asp groups (Fig. 17.15). Furthermore, it was shown that the electrostatic interaction between the aspartyl negative charge and the dipole moment of the rather rigid group Thr 26(26′)-Gly 27(27′) plays a non-negligible role in the stabilization of the coplanar orientation of the two carboxy groups, shown in the left panel of Fig. 17.15, and this is consistent with mutagenesis experiments. As far as activation barriers are concerned, QM/MM calculations show that at conformations in which the distance between substrate and enzyme is large, the free-energy barrier is several tens of kcal/mol,

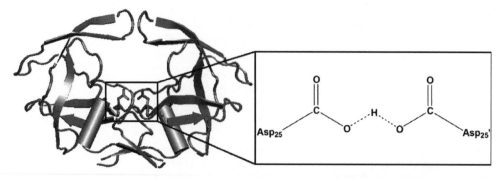

Fig. 17.15 Structure of HIV-1 protease enzyme that cleaves peptide segments at specific sites. The enzyme is composed of two identical subunits facing each other, conferring a C_2 symmetry to the system. The symmetry axis passes through the cleavage site, formed by the Asp catalytic dyad (thick sticks); its carboxy groups are shown in the left panel.

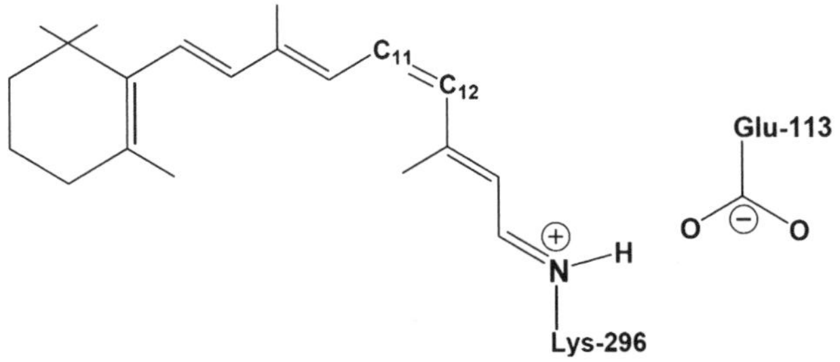

Glu-113

Fig. 17.16 Scheme of the 11-*cis* isomer of the rhodopsin protein with the main carbon atoms involved in the isomerization labelled as C_{11} and C_{12}.

in contrast with the experimental outcome. On the contrary, when the substrate/Asp dyad distances are short, the barrier reduces to 16–18 kcal, in good agreement with experiments. In this case, a concerted double-proton transfer among the Asp dyad, the catalytic water and the substrate can occur.

More recently, photochemical reactions have been pioneered by QM/MM approaches. In particular, the ROKS method has been used as a QM driver in simulations targeting the photoisomerization of the rhodopsin chromophore (Röhrig *et al.* 2003). The model system adopted was based on the crystal structure of bovine rhodopsin (Teller *et al.* 2001) embedded in a membrane mimetic environment. The results show that, in rhodopsin, the photoreaction process is characterized by three basic features. First, the protein-binding pocket selects and accelerates the isomerization exclusively around the carbon bond C_{11}–C_{12} (Fig. 17.16) via a twisted structure.

Secondly, 11-*cis* to all-*trans* isomerization occurs within the binding pocket without requiring large structural modifications. Thirdly, the photon energy in bovine rhodopsin is stored in internal strain of the retinal protonated Schiff base and in steric interaction energy with the protein. This offers a suggestive picture of the initial stage of the process of vision, namely, this first step can be rationalized in terms of a compression of a molecular spring that releases its strain by altering the protein environment in a specific manner.

Provided that the electronic excitations are well localized on a restricted portion of the system including a specific site of the solute and/or the solvent molecules directly interacting with it, TDDFT (Ghosh and Deb 1982) can be coupled to a MM force field. For the sake of clarity, we recall that TDDFT is generally applied to the time propagation of electrons and nuclei. Here, instead, we focus on applications dealing with weak electric fields, involved in photoabsorption, treated as a small perturbation within the linear response theory. According to the Runge–Gross theorem (Runge and Gross 1984), there is a one-to-one correspondence between a time-dependent external potential $v_{ext}(\mathbf{x}, t)$ and the time-dependent electron density $\rho(\mathbf{x}, t)$ written as a superposition of single-particle orbitals of a given initial state so that the many-body problem can be mapped into a non-interacting electrons system,

$$\left\{ -\frac{1}{2}\nabla^2 + \int \frac{\rho(\mathbf{y}, t)}{|\mathbf{x} - \mathbf{y}|} d^3 y + v_{xc}[\rho(\mathbf{x}, t)] + v_{ext}(\mathbf{x}, t) \right\} \psi_i(\mathbf{x}, t) = i\frac{\partial}{\partial t}\psi_i(\mathbf{x}, t),$$

$$(17.41)$$

where the exchange-correlation potential is assumed to be the derivative with respect to the electron density of the static ground-state exchange-correlation functional. If an electric field, $\mathbf{E}(\mathbf{x}, t) = -\nabla\Phi(\mathbf{x}, t)$ is applied as a linear

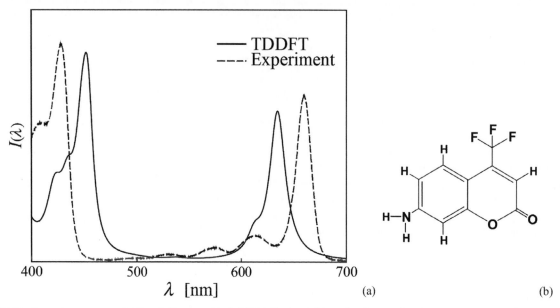

(a)   (b)

Fig. 17.17 Absorption spectrum of chlorophyll computed via TDDFT in the visible region for wavelengths between 400 and 700 nm compared with experiments (a) and scheme of the aminocoumarin C151 (b).

time-dependent perturbation, this perturbation is responsible for the electronic excitations and its simplest expression reads $\Phi(\mathbf{x}, t) = |\mathbf{E}|xe^{i\omega t}$. The accuracy of the approach is at least semi-quantitative. An interesting example, directly related to photoactive proteins, is the optical absorption spectrum of chlorophyll computed by the group of Baroni (Rocca *et al.* 2008), compared with the experimental data (Du *et al.* 1988) and shown in Fig. 17.17(a).

One of the earliest biochemical applications of this particular QM/MM approach focused on aminocoumarins (Sulpizi *et al.* 2003, 2005). Aminocoumarins (Fig. 17.17(b)) are organic compounds with promising applications as anticancer agents, antibiotics and anticoagulants that show a strong fluorescence in the visible region (350–500 nm). This aspect makes aminocoumarins interesting non-linear optical chromophores and, as a matter of fact, they have been used to monitor the enzymatic activities of various types of proteases. Experiments have shown that they are characterized by a red shift of the absorption spectrum when they are solvated in water (Nad and Pal 2001; Karmakar and Samanta 2002; Krolocki *et al.* 2002). The QM/MM approach adopted in this case did not make use of L atoms or FO, since the QM subsystem is represented by the solute, interacting with the solvent, treated with a classical force field, only via hydrogen bonds. Instead, the electrostatic coupling is computed within the RESP scheme. The ground state and the excited states of the QM solute were calculated by DFT and TDDFT, respectively. The calculations evidenced a strong solvent–solute electrostatic interaction, resulting in an enhanced dipole moment of C151 in water (15.3 Debye) with respect to the system in vacuum (6.5 Debye). The interaction of this enhanced dipole with the solvent increases the absorption red shift with respect to the vacuum, providing an atomistic explanation to experiments.

17.5   Beyond the local-minima exploration

Besides an analysis of the electronic structure, one of the main reasons that justify the use of quantum-mechanical approaches to simulate proteins and nucleic acids is the study of cleavage and formation of chemical bonds in biochemical reactions. This is still a computational challenge, because long timescales due to the presence of relatively high free-energy barriers separating reactants and products, and these times are beyond the reach of any classical or quantum molecular dynamics. Several methods have been proposed to overcome this difficulty. Just to mention the most important ones, we call the attention of the reader to enhanced temperature methods such as high-temperature MD (Sørensen and Voter 2000), parallel tempering (Geyger and Thompson 1995) and replica exchange (Frenkel and Smit 2002). These are mostly used in classical simulations to study, for instance, protein folding and generally do not allow recovery of direct information about rare events at room temperature. An alternative class of methods make use of the knowledge of the initial and final states and the reaction path joining these two states is sampled and/or optimized (Dellago *et al.* 1998; Henkelman *et al.* 2000; Passerone and Parrinello 2001; Bolhuis *et al.* 2002; Branduardi *et al.* 2007). These methods work excellently, provided that the final product is known. Unfortunately, this is sometimes one of the unknown variables in the field of biochemical reactions.

Alternative methods not relying on the knowledge a priori of the final products make use of biasing potentials to push the initial system away from its local minimum and to enhance the sampling of the free-energy landscape. These approaches are known as: elevated potential (Huber *et al.* 1994), flooding potential (Grubmüller 1995), umbrella sampling (Torrie and Valleau 1997), Blue Moon (Sprik and Ciccotti 1998) and metadynamics (Laio and Parrinello 2002; Iannuzzi *et al.* 2003). However, the biasing potential is generally a function of one or more order parameters that identify the important reaction coordinates driving the reaction and these crucially depend on the specific system. In this section, we shall focus on a couple of these biasing techniques applied with success to the biosystems that make up the subject of this chapter.

The first biasing technique, also in a chronological sense, applied to biosystems was the Blue Moon. The method starts from the identification of a reaction coordinate, or order parameter, $\xi = \xi(\mathbf{R}_I)$ of a given subset of atomic coordinates (Lagrangean variables) $\mathbf{R}_I$ able to track the chemical reaction on which one wants to focus. A simple example is the distance between two atoms that form or break a chemical bond. This analytical function is added to, e.g. a Car–Parrinello Lagrangean as a holonomic constraint, $\mathcal{L} = \mathcal{L}^{\mathrm{CP}} + \lambda_k(\xi(\mathbf{R}_I) - \xi_k)$, where λ_k is a Lagrange multiplier and ξ_k is a given value of $\xi(\mathbf{R}_I)$ to be sampled. For each value, a constrained dynamics is performed and the average constraint force f_ξ computed as $f_\xi = <\lambda_k>_t = \mathrm{d}F/\mathrm{d}\xi$, resulting in the gradient of the free energy F along ξ. The thermodynamic integration gives the free energy as

$$\Delta F = \int_{\xi_0}^{\xi_f} f_\xi \, \mathrm{d}\xi, \tag{17.42}$$

where ξ_0 is the initial value of the reaction coordinate and ξ_f the final one.

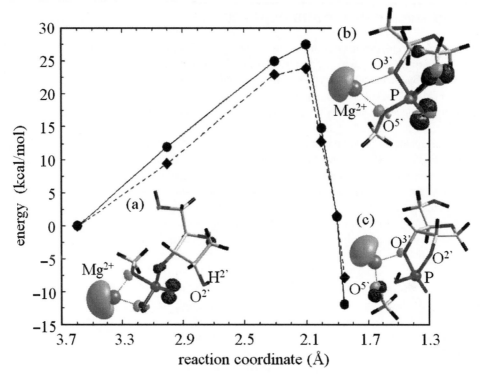

Fig. 17.18 Free (black diamonds, dashed line) and total (black circles, solid line) energies for the RNase reaction with one Mg^{2+} metal catalyst. The initial state (a) forms a transition state (b) and then to the final product (c).

In the field of nucleic acids, this approach was used to study the RNase enzymatic reaction catalyzed by divalent metal cations on a model system extracted from a hammerhead ribozyme. These calculations provided an insight into the catalytic role of Mg^{2+} cations, showing that they lower the activation barriers and make the reaction path highly selective (Boero *et al.* 2002). The sampled reaction coordinate was the distance between the phosphorus atom, labelled as P in Fig. 17.18, and the $O^{2'}$ oxygen; this specific choice was driven by the fact that this is the bond that has to be formed to realize the trigonal–bipyramidal transition state leading to the cleavage of the $P-O^{5'}$ bond, expected to be the rate-limiting step of the reaction. Another interesting application of the thermodynamic integration technique was the study of a novel metal Mg^{2+}-dependent phosphatase activity discovered in the N-terminal domain of the soluble epoxide hydrolase, opening a new branch of fatty-acid metabolism and drug targeting (De Vivo *et al.* 2007). On the basis of experimental results, the two-step mechanism investigated showed that the formation of a phosphoenzyme intermediate is followed by a hydrolysis of the intermediate, suggesting metaphosphate-like transition states for these phosphoryl transfers, and evidencing for the first time that the enzyme promotes water deprotonation and facilitates shuttling of protons via a metal–ligand connecting water bridge.

When a single reaction coordinate is insufficient, a recently proposed scheme, the so-called metadynamics, can be used. In this approach, the Lagrangean of the system, for instance the Car–Parrinello one, is extended to include as additional dynamical variables all the reaction coordinates s_α that describe the process under study, plus a history-dependent Gaussian potential $V(s_\alpha, t)$ that prevents the system from visiting a region of the phase space already explored. The Lagrangean reads

$$\mathcal{L} = \mathcal{L}^{CP} + \sum_\alpha \tfrac{1}{2} M_\alpha \dot{s}_\alpha^2 - \sum_\alpha \tfrac{1}{2} k_\alpha \left[s_\alpha(q) - s_\alpha \right]^2 - V(s_\alpha, t), \qquad (17.43)$$

where the argument q of $s_\alpha(q)$ can be any function of an arbitrary subset of atomic coordinates, electronic wavefunctions, etc. $V(s_\alpha, t)$ has the explicit Gaussian form

$$V(s_\alpha, t) = \int_0^t dt' \, |\dot{\mathbf{s}}(t')| \, \delta \left[\frac{\dot{\mathbf{s}}(t')}{|\dot{\mathbf{s}}(t')|} \left(\mathbf{s} - \mathbf{s}(t') \right) \right] A(t') \exp \left[-\frac{\left(\mathbf{s} - \mathbf{s}(t') \right)^2}{2(\Delta s)^2} \right],$$
$$(17.44)$$

where $\mathbf{s} = (s_1, \ldots, s_\alpha, \ldots)$ and the Gaussian amplitude $A(t')$ has the dimensions of an energy. In a dynamical simulation, a new Gaussian contribution is added to the potential $V(s_\alpha, t)$ at given time intervals Δt (generally a few hundreds of molecular dynamics steps). A careful tuning of the two input parameters $A(t')$ and Δs is required, since these two quantities determine how accurate is the free-energy landscape exploration and how fine is the reaction coordinate sampling, respectively. The free-energy hypersurface $F(s_\alpha)$, results as

$$F(s_\alpha) = -\lim_{t \to \infty} V(s_\alpha, t) + const. \qquad (17.45)$$

One of the first examples of application in nucleic acids research was the RNase enzymatic reaction (Boero *et al.* 2005) discussed previously. The simultaneous use of three reaction coordinates, i.e. the co-ordination number between P and $O^{2'}$, $s_1(t) = N_{coord}(P-O^{2'})$, the co-ordination number between P and $O^{5'}$, $s_2(t) = N_{coord}(P-O^{5'})$, allowed all the bond breaking and formation processes occurring during the reaction (see Fig. 17.18 for the labelling) to be followed. On the one hand, this provided a comparison of free-energy barriers for the same reaction obtained by two different methods, Blue Moon and metadynamics, making use of different reaction coordinates. On the other hand, the results showed that in RNA enzymes, at least *in vitro*, the concerted action of two metal cations can efficiently catalyze the reaction and that the presence of an OH^- anion is a triggering agent able to favor the initial nucleophilic attack.

As far as proteins are concerned, metadynamics has been used to study the proton-pump mechanism in bovine cytochrome c oxidase, fundamental in cell respiration. In particular, one of the proposed proton-transfer pathways, called the H-path, is characterized by two distinct H-bond networks separated by a chemical bond of a peptide group that could hinder the proton propagation (Tsukihara *et al.* 2003). A DFT-based metadynamics inspection, on a model system extracted from the X-ray data led to the discovery that despite such a peptide group, the pumping activity is possible via a double-proton exchange between the incoming H^+ and hydrogens bonded to N atoms belonging to the peptide group (Kamiya *et al.* 2007). A modest activation barrier (13 kcal/mol) characterizes the process, the transfer is one-directional and the process does not induce permanent conformational changes to the protein, thus being compatible with the crystallographic structure.

In combination with QM/MM methods, metadynamics was used to study the oxidative damage of DNA via radical cation formation (Gervasio *et al.* 2004). The mechanism that triggers the first stages of this process was inspected, finding that guanine (G), which has the lowest oxidation potential, and the phosphate backbone play a central role. The rate-limiting step of the reaction was found to be the water deprotonation, providing an insight into the role of the local environment in catalyzing the reaction and lowering the activation barrier. One of the most remarkable simulations of this kind, both for the scientific insight and in terms of large-scale calculations, was the study of charge-transfer processes in a fully hydrated double-strand B-DNA using the NEC SX-6 Earth Simulator, ranked No. 1 supercomputer in the world until 2004. The system amounted to 25 000 MM atoms and 242 QM atoms, and calculations were performed on 64 vector-parallel CPUs (Gervasio *et al.* 2006; Boero *et al.* 2007). The collective variables selected to sample the free-energy profile were the co-ordination number of the N atom of a C base with nearby H-bonded protons (Fig. 17.19) and the distance of this same N

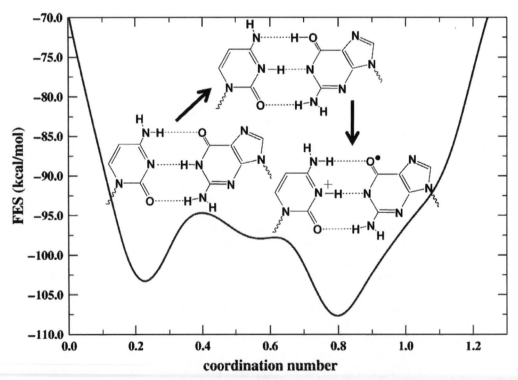

Fig. 17.19 Double-proton-exchange mechanism in a fully hydrated double-strand B-DNA system provided by QM/MM metadynamics simulations, assuming as a reaction coordinate the coordination number of the N atom of the C base with nearby H atoms.

atom from the nearest proton chemically bonded to the adjacent G base, used separately in two distinct simulations. This study inspected the mechanism of hole transfer in a solvated double-helical DNA system, providing direct evidence for a coherent single-step charge transfer, with a free-energy barrier of about 7 kcal/mol, and proved that a double-proton transfer is the trigger for charge hopping along the DNA. This supported and complemented for the first time the experimental evidence from H/D substitution and electron paramagnetic resonance (Giese *et al.* 2001; Shafirovich *et al.* 2001).

Coming to proteins, a first simulation performed making joint use of QM/MM techniques and metadynamics focused on the ATP hydrolysis in the heat shock cognate-70 (Hsc-70) protein. This protein has a wealth of important biological roles: protects cells from stress, maintains cellular homeostasis, influences metabolism and muscular adaptation, etc. (Liu *et al.* 2006). These simulations targeted the adenosine triphosphate (ATP) of the Hsc70 ATPase protein (Boero *et al.* 2006), showing which specific water molecule of the solvation shell of the Mg^{2+} metal cation, present in the protein, acts as a trigger in the ATP hydrolysis reaction. Different possible reaction coordinates were used in a series of simulations; in particular, the free-energy landscape shown in Fig. 17.20 refers to the sampling of the distance between the O atom of the catalytic water (O_{wat}) from the terminal phosphate ($P_\gamma - O_{wat}$)

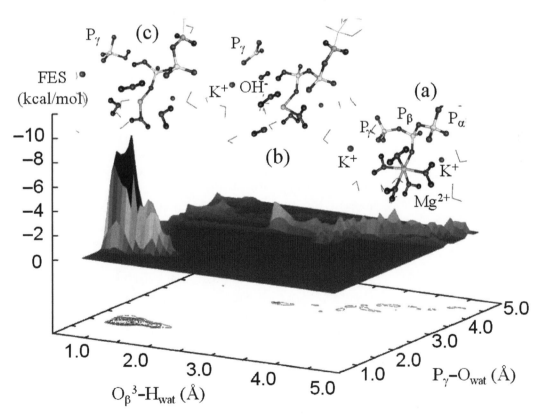

Fig. 17.20 ATPase (a) hydrolysis reaction in Hsc-70. The QM atoms are indicated as sticks and balls; thin sticks are MM atoms. The transition state is reached via deprotonation of a specific H_2O molecule leading to the cleavage of the P_γ-O bond (b) and to the ADP formation (c).

and the distance between the $P_\gamma - O - P_\beta$ bridging oxygen ($O_\beta{}^3$) and the proton H_{wat} of the catalytic H_2O molecule ($O_\beta{}^3 - H_{wat}$). The detailed picture of the reaction mechanism obtained, not accessible to experiments, allowed two important issues previously unravelled to be addressed: (i) The pathway followed by H^+ and OH^-, produced by the dissociation of this H_2O molecule. And (ii) the cooperative role of the metal cations K^+ and Mg^{2+} in promoting the release of the inorganic phosphate via an exchange of the OH^- hydroxyl anion between their respective solvation shells. This is deeply different from the proton wire mechanism evidenced, for instance, in actin (Akola and Jones 2006) and significantly lowers the free-energy barrier.

17.6   Final remarks

The overview presented here about quantum methods suitable to study proteins and nucleic acids is far from being complete and exhaustive of such a rapidly developing field. Within the limitations of the present chapter, we just aimed at offering to the reader a collection of basic computational techniques supported by selected examples that we consider as milestones or pioneering works in the field. Many more are available in the rich literature, part of which is reported in the included list of references, and that represent a more complete panorama to researchers interested or working in the field.

References

Alves, C.N., Martí, S., Castillo, R., Andrés, J., Moliner, V., Tuñón, I., Silla, E. *Bioorg. Med. Chem.* **15**, 3818 (2007).

Akola, J., Jones R.O. *J. Phys. Chem. B* **110**, 8121 (2006).

Andelman, D. Electrostatic properties of membranes: The Poisson-Boltzmann Theory in *Handbook of Biological Physics*, vol. 1, (eds) R. Lipowsky and E. Sackmann (Elsevier Science B.V., 1995) p. 603–638.

Andersen, H.C. *J. Chem. Phys.* **72**, 2384 (1980).

Andreoni, W. *Persp. Drug Disc. Des.* **9**, 161 (1998).

Andreoni, W., Curioni, A., Mordasini, T. *IBM J. Res. & Dev.* **45**, 397 (2001).

Antes, I., Thiel, W. *J. Phys. Chem. A* **103**, 9290 (1999).

Assfeld, X., Rivail, J.L. *Chem. Phys. Lett.* **263**, 100 (1996).

Bachelet, G.B., Hamann, D.R., Schluter, M. *Phys. Rev. B* **26**, 4199 (1982).

Baker, N.A., Sept, D., Joseph, S., Holst, M.J., McCammon, J.A. *Proc. Natl. Acad. Sci. USA* **98**, 10037 (http://apbs.sourceforge.net/) (2001).

Bakowies, D., Thiel, W. *J. Phys. Chem.* **100**, 10580 (1996).

Barnett, R.N., Cleveland, C.L., Joy, A., Landman, U., Schuster, G.B. *Science* **294**, 567 (2001).

Bayly, C.I., Cieplak, P., Cornell, W.D., Kollman, P.A. *J. Phys. Chem.* **97**, 10269 (1993).

Becke, A.D. *Phys. Rev.* **38**, 3098 (1988).

Becke, A.D. *J. Chem. Phys.* **98**, 5648 (1993).

Bhattacharyya, S., Stankovich, M.T., Truhlar, D.G., Gao, J. *J. Phys. Chem. A* **111**, 5729 (2007).

Biswas, P.K., Gogonea, J. *Chem. Phys.* **123**, 164114 (2005).

Blöchl, P.E. *Phys. Rev. B* **50**, 17953 (1994).

Boero, M., Terakura, K., Tateno, M. *J. Am. Chem. Soc.* **124**, 8949 (2002).

Boero, M., Tateno, M., Terakura, K., Oshiyama, J. *Chem. Theory Comput.* **1**, 925 (2005).

Boero, M., Ikeda, T., Ito, E., Terakura, K. *J. Am. Chem. Soc.* **128**, 16798 (2006).

Boero, M., Gervasio, F.L., Parrinello, M. *Mol. Simul.* **33**, 57 (2007).

Bolhuis, P.G., Chandler, D., Dellago, C., Geissler, P.L. *Annu. Rev. Phys. Chem.* **53**, 291 (2002).

Bornemann, F.A. *Schütte Numerische Matematik* **78**, 359 (1998).

Bowler, D.R., Bush, I.J., Gillan M.J. *Int. J. Quantum Chem.* **77**, 831 (2000).

Boyer, P.D. *Nobel Lectures, Chemistry 1996–2000*, ed. I. Grenthe (World Scientific Publishing, Singapore, 2003).

Branduardi, D., Gervasio, F.L., Parrinello, M. *J. Chem. Phys.* **126**, 054103 (2007).

Brooks, B.R., Bruccoleri, B.R., Olafson, B.D., States, D.J., Swaminathan, S., Karplus, M. *J. Comp. Chem.* **4**, 187 (1983).

Bruinsma, R., Grüner, G., D'Orsogna, M.R., Rudnik, J. *Phys. Rev. Lett.* **85**, 4393 (2000).

Calvaresi, M., Bottoni, A., Garavelli, M. *J. Phys. Chem. B* **111**, 6557 (2007).

Car, R., Parrinello, M. *Phys. Rev. Lett.* **55**, 2471 (1985).

Carloni, P. *Quant. Struct.-Act. Relat. (QSAR)* **21**, 166 (2002).

Carloni, P., Röthlisberger, U., Parrinello, M. *Acc. Chem. Res.* **35**, 455 (2002).

Cavalli, A., Carloni, P. *J. Am. Chem. Soc.* **124**, 3763 (2002).

Chang, N., Lim, C. *J. Phys. Chem. A* **101**, 8706 (1997).

Chapell, T.G., Welch, W.J., Schlossman, D.M., Palter, K.B., Schlesinger, M.J., Rothman, J.E. *Cell* **45**, 3 (1986).

Chapman, D.L. *Philos. Mag.* **25**, 475 (1913).

Chen, J., Seeman, N.C. *Nature* **350**, 631 (1991).

Colombo, M.C., Guidoni, L., Laio, A., Magistrato, A., Maurer, P., Piana, S., Roehrig, U., Spiegel, K., VandeVondele, J., Zumstein, M., Röthlisberger, U. *Chimia* **56**, 11 (2002).

Cornell, W.D., Cieplak, P., Bayly, C.I., Gould, I.R., Merz, K.R., Ferguson, D.M., Spellmeyer, D.C., Fox, T., Caldwell, J.W., Kollman, P.A. *J. Am. Chem. Soc.* **117**, 5179 (1995).

Dal Peraro, M., Ruggerone, P., Raugei, S., Gervasio, F.L., Carloni, P. *Curr. Op. Struct. Biol.* **17**, 149 (2007).

Danilov, V.I., Anisimov, V.M. *J. Biomol. Structure Dynamics* **22**, 471 (2005).

Davydov, R., Makris, T.M., Kofman, V., Werst, D.E., Sligar, S.G., Hoffman, B.M. *J. Am. Chem. Soc.* **123**, 1403 (2001).

de Pablo, P.J., Moreno-Herrero, F., Colchero, J., Gómez Herrero, J., Herrero, P., Baró, A.M., Ordejón, P., Soler, J.M., Artacho, E. *Phys. Rev. Lett.* **85**, 4992 (2000).

Dejaegere, A., Lim, C., Karplus, M. *J. Am. Chem. Soc.* **113**, 4353 (1991).

Dekker, C., Ratner, M.A. *Phys. World* **14**, 29 (2001).

Dellago, C., Bolhuis, P.G., Csajka, F.S., Chandler, D. *J. Chem. Phys.* **108**, 1964 (1998).

De Vivo, M., Ensing, B., Dal Peraro, M., Gomez, G.A.D., Christianson, W., Klein, M.L. *J. Am. Chem. Soc.* **129**, 387 (2007).

Di Labio, G.A., Hurley, M.M., Christiansen, P.A. *J. Chem. Phys.* **116**, 9578 (2002).

Di Ventra, M., Zwolak, M. *Encyclopedia of Nanoscience and Nanotechnology*, (ed.) H. Singh-Nalwa (American Scientific Publishers, New York, 2004).

Douki, T., Ravanat, J.L., Angelov, D., Wagner, J.R., Cadet, J. *Top. Curr. Chem.* **236**, 1 (2004).

Du, H., Fuh, R.C.A., Li, J., Corkan, L.A., Lindsey, J.S. *Photochem. Photobiol.* **68**, 141 (1988).

Dybala-Defratyka, A., Paneth, P., Benrjee, R., Truhlar, D.G. *Proc. Nat. Acad. Sci. USA* **104**, 10774 (2007).

Echinger, M., Tavan, P., Hutter, J., Parrinello, M. *J. Chem. Phys.* **110**, 10452 (1999).

Field, M.J., Bash, P.A., Karplus, M. *J. Comp. Chem.* **11**, 700 (1990).

Forbes, N. *Comp. Sci. Eng.* **2**, 83 (2000).

Frank, I., Hutter, J., Marx, D., Parrinello, M. *J. Chem. Phys.* **108**, 4060 (1998).

Franklin, R., Gosling, R.G. *Nature* **171**, 740 (1953).

Freindorf, M., Shao, Y., Kong, J., Furlani, T.R. *BIOCOMP* 391 (2006).

Frenkel, D., Smit, B. *Understanding Molecular Simulation*, 2nd edn. (Academic Press, San Diego, 2002).

Gao, J., Amaram, P., Alhambra, C., Field, M.J. *J. Phys. Chem. A* **102**, 4714 (1998).

Gervasio, F.L., Laio, A., Parrinello, M., Boero, M. *Phys. Rev. Lett.* **94**, 158103 (2005).

Gervasio, F.L., Carloni, P., Parrinello, M. *Phys. Rev. Lett.* **89**, 108102 (2002).

Gervasio, F., Laio, A., Iannuzzi, M., Parrinello, M. *Chem. Eur. J.* **10**, 4846 (2004).

Gervasio, F.L., Boero, M., Parrinello, M. *Angew. Chem. Int. Ed.* **45**, 5606 (2006).

Geyger, C.J., Thompson, A. *J. Am. Stat. Assoc.* **90**, 909 (1995).

Ghosh, S.K., Deb, B.M. *Chem. Phys.* **71**, 295 (1982).

Giese, B., Wessely, S. *Chem. Commun.* **20**, 2108 (2001).

Giese, B., Amaudrut, J., Köhler, A., Spormann, M., Wessely, S. *Nature* **412**, 318 (2001).

Goedecker, S., Teter, M., Hutter, J. *Phys. Rev. B* **54**, 1703 (1996).

Gogonea, V., Suárez, D., van der Vaart, A., Kenneth, M. *Curr. Op. Struct. Biol.* **11**, 217 (2000).

Gonze, X., Stumpf, R., Scheffler, M. *Phys. Rev. B* **44**, 8503 (1991).

Gouy, G. *J. Phys.* **9**, 457 (1910).

Grimme, S. *J. Comput. Chem.* **27**, 1787 (2006).

Grubmüller, H. *Phys. Rev. E* **52**, 2893 (1995).

Guallar, V., Baik, M.-H., Lippard, S.J., Friesner, R.A. *Proc. Nat. Acad. Sci. USA* **100**, 6998 (2003).

Hamann, D.R., Schluter, M., Chang, C. *Phys. Rev. Lett.* **43**, 1494 (1979).

Hamprecht, F.A., Cohen, A.J., Tozer, D.J., Handy, N.C. *J. Chem. Phys.* **109**, 6264 (1998).

Hendrickson, W.A., Teeter, M.M. *Nature* **290**, 107 (1981).

Henkelman, G., Uberuaga, B.P., Jonsson, H. *J. Chem. Phys.* **113**, 9901 (2000).

Hermans, J., Berendsen, H.J.C., van Gunsteren, W.F., Postma, J.P.M. *Biopolymers* **23**, 1 (1984).

Hershko, A., Chiechanover, A. *Annu. Rev. Biochem.* **51**, 335 (1982).

Hirshfeld, F.L. *Theo. Chim. Acta* **44**, 129 (1977).

Hobza, P., Šponer, J. *J. Am. Chem. Soc.* **124**, 11802 (2002).

Hohenberg, P., Kohn, W. *Phys. Rev.* **136**, B864 (1964).

Honda, S., Yamasaki, K., Sawada, Y., Morii, H. *Structure* **12**, 1507 (2004).

Hoover, W.G. *Phys. Rev. A* **31**, 1695 (1985).

Huber, T., Torda, A.E., van Gunsteren, W.F. *J. Comput. Mol. Design* **8**, 695 (1994).

Iannuzzi, M., Laio, A., Parrinello, M. *Phys. Rev. Lett.* **90**, 238302 (2003).

Iavarone, A.T., Parks, J.H. *J. Am. Chem. Soc.* **127**, 8606 (2005).

Jena, N.R., Mishra, P.C. *J. Phys. Chem. B* **111**, 5418 (2007).

Jishi, R.A., Bragin *J. Phys. Chem. B* **111**, 5357 (2007).

Johnson, B.G., Gill, P.M.W., Pople, J.A. *J. Chem. Phys.* **98**, 5612 (1993).

Kamiya, K., Boero, M., Tateno, M., Shiraishi, K., Oshiyama. A. *J. Am. Chem. Soc.* **129**, 9663 (2007).

Karmakar, R., Samanta, A. *J. Phys. Chem. A* **116**, 4447 (2002).

Karplus, M., Bashford, D. *Biochemistry* **29**, 10219 (1990).

Keshari, V., Ishikawa, Y. *Chem. Phys. Lett.* **218**, 406 (1994).

Kitaura, K., Sugiki, S., Nakano, T., Komeiji, Y., Uebayashi, M. *Chem. Phys. Lett.* **336**, 163 (2001).

Kleinman, L., Bylander, D.M. *Phys. Rev. Lett.* **48**, 1425 (1982).

Kobayashi, M., Imamura, Y., Nakai, H. *J. Chem. Phys.* **127**, 074103 (2007).

Kohn, W., Sham, L. *J. Phys. Rev.* **140**, A1133 (1965).

Kolodner, P., Lukashev, E.P., Ching, Y.-C., Druzhko, A.B. *Thin Solid Films* **302**, 206 (1997).

Komeiji, Y., Ishida, T., Fedorov, D.G., Kitaura, K. *J. Comput. Chem.* **28**, 1750 (2007).

Krolocki, R., Jarzeba, W., Mostafavi, M., Lampre, I. *J. Phys. Chem. A* **106**, 1708 (2002).

Lahti, A., Hotokka, M., Neuvonen, K., Äyräs, P. *Struct. Chem.* **8**, 331 (1997).

Laio, A., VandeVondele, J., Röthlisberger, U. *J. Chem. Phys.* **116**, 6941 (2002).

Laio, A., VandeVondele, J., Röthlisberger, U. *J. Phys. Chem. B* **106**, 7300 (2002).

Laio, A., Parrinello, M. *Proc. Nat. Acad. Sci. USA* **99**, 12562 (2002).

Langreth, D.C., Dion, M., Rydberg, H., Schröder, E., Hyldgaard, P., Lundqvist, B.I. *Int. J. Quantum Chem.* **101**, 599 (2005).

Leclerc, F., Karplus, M. *J. Phys. Chem. B* **110**, 3395 (2006).

Lee, C., Yang, W., Parr, R.G. *Phys. Rev. B* **37**, 785 (1988).

Lerm, M., Selzer, J., Hoffmeyer, A., Rapp, U.R., Aktories, K., Schmidt, G. *Infect. Immun.* **67**, 496 (1999).

Lim, C., Tole, P. *J. Am. Chem. Soc.* **114**, 7252 (1992).

Lin, H., Truhlar, D.G. *Theor. Chem. Acc.* **117**, 185 (2007).

Liu, Y., Gampert, L., Nething, K., Steinacker, J.M. *Frontiers Biosci.* **11**, 2802 (2006).

Louie, S.G., Froyen, F., Cohen, M.L. *Phys. Rev. B* **26**, 1738 (1982).

Lumry, R. *Methods Enzymol.* **259**, 628 (1995).

Marx, D., Hutter, J. Ab initio molecular dynamics: Theory and Implementation in *Modern Methods and Algorithms of Quantum Chemistry*, (ed.) J. Grotendorst (John von Neumann Institute for Computing, Jülich, 2000).

Maurer, P., Laio, A., Röthlisberger, U. *J. Chem. Theory Comput.* **3**, 628 (2007).

Miteva, M.A., Tufféry, P., Villoutreix, B.O. *Nucleic Acid Res.* **33**, W372 (http://bioserv.rpbs.jussieu.fr/PCE) (2005).

Møller, C., Plesset, M.S. *Phys. Rev.* **46**, 618 (1934).

Molteni, C., Frank, I., Parrinello, M. *J. Am. Chem. Soc.* **121**, 12177 (1999).

Momany, F.A. *J. Phys. Chem.* **82**, 592 (1978).

Momany, F.A., McGuire, R.F., Burgess, A.W., Scheraga, H.A. *J. Phys. Chem.* **79**, 2361 (1975).

Nad, S., Pal, H. *J. Phys. Chem. A* **105**, 1097 (2001).

Nagel, G.M., Doolittle, R.F. *Proc. Nat. Acad. Sci. USA* **88**, 8121 (1991).

Neidigh, J.W., Fesinmeyer, R.M., Andersen, N.H. *Nature Struct. Biol.* **9**, 425 (2002).

Noguera, M., Sodupe, M., Bertran, J. *Theor. Chem. Acc.* **118**, 113 (2007).

Northrop, D. *Acc. Chem. Res.* **34**, 790 (2001).

Nosé, S. *J. Chem. Phys.* **81**, 511 (1984).

Odom, D.T., Barton, J.K. *Biochemistry* **40**, 8727 (2001).

Ollis, D.L., White, S.W. *Chem. Rev.* **87**, 98 (1987).

Parr, R.G., Yang, W. *Density-functional Theory of Atoms and Molecules* (Oxford University Press, New York, 1989).

Parrinello, M., Rahman, A. *Phys. Rev. Lett.* **45**, 1196 (1980).

Passerone, D., Parrinello, M. *Phys. Rev. Lett.* **87**, 8302 (2001).

Perczel, A., Faraks, Ö., Jákli, I., Topol, I.A., Csizmadia, I.G. *J. Comp. Chem.* **24**, 1026 (2003).

Perdew, J.P., Zunger, A. *Phys. Rev. B* **23**, 5048 (1981).

Perdew, J.P., Wang, Y. *Phys. Rev. B* **45**, 13244 (1992).

Perdew, J.P., Burke, K., Ernzerhof, M. *Phys. Rev. Lett.* **77**, 3865 (1996).

Perreault, D.M., Anslyn, E.V. *Angew. Chem. Int. Ed. Engl.* **36**, 433 (1997).

Piana, S., Carloni, P. *Proteins: Struct. Funct. Genet.* **39**, 26 (2000).

Piana, S., Carloni, P., Parrinello, M. *J. Mol. Biol.* **319**, 567 (2002).

Pichierri, F., Aida, M., Gromiha, M.M., Sarai, A. *J. Am. Chem. Soc.* **121**, 6152 (1999).

Pley, H.W., Flaherty, K.M., McKay, D.B. *Nature* **372**, 68 (1994).

Pople, J.A., Schlegel, H.B., Krishnan, R., Defrees, D.J., Binkley, J.S., Frisch, M.J., Whiteside, R.A., Hout, R.F., Hehre, W.J. *Int. J. Quantum Chem. Symp.* **15**, 269 (1981).

Pople, J.A., Head-Gordon, M., Fox, D.J., Raghavachari, K., Curtiss, L.A. *J. Chem. Phys.* **90**, 5622 (1989).

Press, W.H., Teukolsky, S.A., Vetterling, W.T., Flannery, B.P. *Numerical Recipes in Fortran 77* (Cambridge University Press, New York, 1992).

Reuter, N., Dejaegere, A., Maigret, B., Karplus, M. *J. Phys. Chem. A* **104**, 1720 (2000).

Price, D.L., Sisodia, S.S., Borchelt, D.R. *Science* **282**, 1079 (1999).

Rocca, D., Gebauer, R., Saad, Y., Baroni, S. (2008).

Röhrig, U.F., Guidoni, L., Röthlisberger, U. *Chem. Phys. Chem.* **6**, 1836 (2005).

Röhrig, U.F., Frank, I., Hutter, J., Laio, A., VandeVondele, J., Röthlisberger, U. *Chem. Phys. Chem.* **4**, 1177 (2003).

Rösch, N., Voityuk, A.A. *Topics Curr. Chem.* **237**, 37 (2004).

Rovira, C., Parrinello, M. *Int. J. Quantum Chem.* **80**, 1172 (2000).

Rovira, C., Schulze, B., Echinger, M., Evanseck, J.D., Parrinello, M. *Biophys. J.* **81**, 435 (2001).

Runge, E., Gross, E.K.U. *Phys. Rev. Lett.* **52**, 997 (1984).

Seeman, N.C., Kallenbach, N.R. *Biophys. J.* **44**, 201 (1983).

Shafirovich, V., Dourandin, A., Geacintov, N.E. *J. Phys. Chem. B* **105**, 8431 (2001).

Silvestrelli, P. *Phys. Rev. Lett.* **100**, 053002 (2008).

Singh, U.C., Kollman, P.A. *J. Comp. Chem.* **7**, 718 (1986).

Sørensen, M.R., Voter, A.F., Sørensen, M.R. *J. Chem. Phys.* **112**, 9599 (2000).

Spiegel, K., Magistrato, A. *Organic Biomol. Chem.* **4**, 2507 (2006).

Sprik, M., Ciccotti, G. *J. Chem. Phys.* **109**, 7737 (1998).

Steenken, S. *Biol. Chem.* **378**, 1293 (1997).

Stock, D., Leslie, A.G., Walker, J.E. *Science* **286**, 1700 (1999).

Storer, J.W., Uchimaru, T., Tanabe, K., Uebayashi, M., Nishikawa, S., Taira, K. *J. Am. Chem. Soc.* **113**, 5216 (1991).

Sulpizi, M., Folkers, G., Röthlisberger, U., Carloni, P., Scapozza, L. *Quant. Struct.-Act. Relat. (QSAR)* **21**, 173 (2002).

Sulpizi, M., Röthlisberger, U., Laio, A. *J. Theor. Comput. Chem.* **4**, 985 (2005).

Sulpizi, M., Carloni, P., Hutter, J., Röthlisberger, U. *Phys. Chem. Chem. Phys.* **5**, 4798 (2003).

Sulpizi, M., Röhrig, U.F., Hutter, J., Röthlisberger, U. *Int. J. Quantum Chem.* **101**, 671 (2005).

Teeter, M.M., Mazer, J.A., L'Italien, J.J. *Biochemistry* **20**, 5437 (1981).

Teller, D.C., Okada, T., Behnke, C.A., Palczewski, K., Stenkamp, R.E. *Biochemistry* **40**, 7761 (2001).

Torrie, G.M., Valleau, J.P. *J. Comput. Phys.* **23**, 187 (1997).

Troullier, N., Martins, J.L. *Phys. Rev. B* **43**, 1993 (1991).

Tsukihara, T., Shimokata, K., Katayama, Y., Shimada, H., Muramoto, K., Aoyama, H., Mochizuki, M., Shinzawa-Itoh, K., Yamashita, E., Yao, M., Ishimura, Y., Yoshikawa, S. *Proc. Natl. Acad. Sci. USA* **100**, 15304 (2003).

Uchimaru, T., Tanabe, K., Nishikawa, S., Taira, K. *J. Am. Chem. Soc.* **113**, 4351 (1991).

Ullmann, G.M., Noodleman, L., Case, D.A. *J. Biol. Inorg. Chem.* **7**, 632 (2002).

Van Alsenoy, C., Yu, C.-H., Peeters, A., Martin, J.M. L., Schäfer, L. *J. Phys. Chem. A* **102**, 2246 (1998).

Vanderbilt, D. *Phys. Rev. B* **41**, 7892 (1990).

Van der Spoel, D., Lindahl, E., Hess, B., Groenhof, G., mark, A.E., Berendsen, H.J.C. *J. Comput. Chem.* **26**, 1701 (2005).

Van Doorslaer, S., Cereghetti, G.M., Glockshuber, R., Schweiger, A. *J. Phys. Chem.* **105**, 1631 (2001).

von Lilienfeld, O.A., Tavernelli, I., Röthlisberger, U., Sebastiani, D. *J. Chem. Phys.* **122**, 014133 (2005).

Vorobiev, S., Strokopytov, B., Drubin, D.G., Frieden, C., Ono, S., Condeelis, J., Rubenstein, P.A., Almo, S.C. *Proc. Natl. Acad. Sci. USA* **100**, 5760 (2003).

Vosko, S.H., Wilk, L., Nusair, M. *Can. J. Phys.* **58**, 1200 (1980).

Walker, P.D., Mezey, P.G. *J. Am. Chem. Soc.* **116**, 12022 (1994).

Wang, H., Oster, G. *Nature* **391**, 510 (1998).

Warshel, A., Levitt, M. *J. Mol. Biol.* **103**, 227 (1976).

Watson, J.D., Crick, F.H.C. *Nature* **171**, 737 (1953).

Winter, N.M., Pitzer, R.M. *J. Chem. Phys.* **89**, 446 (1988).

Yang, A.S., Gunner, M.R., Sampogna, R., Sharp, K., Honig, B. *Proteins* **15**, 252 (1993).

Zhang, Y., Seeman, N.C. *J. Am. Chem. Soc.* **116**, 1661 (1994).

Zhang, Y., Lin, H., Truhlar, D.G. *J. Chem. Theory Comput.* **3**, 1378 (2007).

Zhang, Y.K., Lee, T.S., Yang, W.T. *J. Chem. Phys.* **110**, 46 (1999).

Zhao, Q., Yang, W. *J. Chem. Phys.* **102**, 9598 (1995).

Zheng, J., Altun A., Thiel, W. *J. Comp. Chem.* **28**, 2147 (2007).

Zhou, D., Taira, K. *Chem. Rev.* **98**, 991 (1998).

Magnetoresistive phenomena in nanoscale magnetic contacts

18

J.D. Burton and E.Y. Tsymbal

18.1   Introduction

Nanomagnetic materials are playing an increasingly important role in modern technologies spanning the range from information processing and storage to communication devices, electrical power generation, and nano-biosensing (Bader 2006). A particular area of interest involves systems exhibiting the interplay between magnetism and electric transport, namely magnetoresistive properties, which are attractive for applications in magnetic field sensors and magnetic memories (Chappert *et al.* 2007). Given the worldwide technological challenge to miniaturize magnetoresistive devices, future generations of field sensors and memory elements will have to be on a length scale of a few nanometers or smaller. Magnetoresistive properties of such nanoscale objects may be very different as compared to those in respective bulk systems due to reduced dimensionality, complex surfaces and interfaces, and quantum effects. In this chapter we discuss novel features of magnetoresistive phenomena in an interesting subset of nanoscale systems, namely nano- and atomic-size ferromagnetic metal contacts.

Magnetoresistance is defined as the change in electrical resistance of a system due to a change, either in magnitude or orientation, of an applied magnetic field. There are a variety of physical mechanisms, which we refer to as *magnetoresistive phenomena*, leading to magnetoresistance. For example, magnetoresistance can arise due to direct coupling of a magnetic field with the charge of the conduction electrons via the usual electromagnetic Lorentz force. In non-magnetic metals the Lorentz force can only arise due to an externally applied magnetic field, yielding the so-called ordinary magnetoresistance (OMR). But in the case of magnetic metals the internal magnetic field generated by spontaneous magnetization also contributes, giving rise to the "anomalous" or "extraordinary" magnetoresistance. Magnetoresistive phenomena of this kind have nothing (at least directly) to do with the spin of the electron, and can be understood using semi-classical arguments (Pippard 1989).

Spin-dependent magnetoresistive phenomena, on the other hand, occur in systems containing ferromagnetic metals, like the $3d$ transition-metal ferromagnets Fe, Co, Ni or their alloys. Magnetoresistance in these systems arises due to an applied magnetic field inducing changes in the orientation of the spontaneous magnetization in either all or part of the ferromagnetic system. This leads to a corresponding change in the electrical resistance because the magnetization itself is simply a manifestation of the underlying spin-dependent electronic structure. Any changes in the electronic structure (spin-dependent or otherwise) of a system automatically lead to a corresponding change in the electrical transport properties. However, it should be emphasized that, unlike the Lorentz-force-based magnetoresistive phenomena mentioned above, the applied magnetic field is only playing an auxiliary role. It is really the changes in the magnetization and the underlying electronic structure that drive the effect.

The magnetoresistive phenomena that we will discuss in this chapter fall into two major categories. The first operates as what we will call the "spin-valve" where the flow of spin-polarized electrical current is affected by an inhomogeneous magnetization profile. The second category involves the anisotropy of electrical transport properties with respect to the orientation of the magnetization, a category collectively known as anisotropic magnetoresistance (AMR) effects.

The prototypical spin-valve magnetoresistive phenomenon is the giant magnetoresistance (GMR) effect discovered independently by two groups in 1988, earning Albert Fert and Peter Grünberg the 2007 Nobel Prize in Physics (Baibich *et al.* 1988; Binasch *et al.* 1989). GMR occurs in magnetic multilayer structures where ferromagnetic films are separated by thin non-magnetic spacer layers and the electrical resistance depends on the relative orientation of the magnetization of the ferromagnetic layers. The change in magnetization orientation is usually facilitated by the application of an external magnetic field, providing an important route to make a magnetic field sensor for data-storage applications (Chappert *et al.* 2007). The resistance in the state where all the ferromagnetic layers are parallel, R_P, differs from the resistance in the anti-parallel state, R_{AP}, by a figure of merit known as the GMR ratio: $GMR = (R_{AP} - R_P)/R_P \times 100\%$. The key to the GMR effect is that the non-magnetic layer must be thin enough so that electrons traversing it from one ferromagnetic layer to the next will retain their initial spin orientation, which we will refer to as the "spin-memory" condition. For reviews of GMR see (Levy 1994; Barthelemy *et al.* 1999; Hartmann 2000; Tsymbal and Pettifor 2001) and references therein.

A related spin-valve-like magnetoresistive phenomenon occurs due to magnetic domain walls in ferromagnetic metals. Domain walls are the regions between the uniformly magnetized magnetic domains of a magnetic material where the local magnetization rotates along a spatial dimension. In this case the electrical resistance can be altered by the presence of a magnetic domain wall, an effect known as domain-wall magnetoresistance (DWMR). This is in analogy with a trilayer GMR structure where the non-magnetic spacer layer is replaced by a magnetic domain wall that connects two anti-parallel domains acting as ferromagnetic electrodes. For DWMR to be sizable the domain wall must be thin enough that the spin of the conduction electron cannot keep track of the changing local magnetization orientation, which is analogous to the spin-memory condition on the thinness of the non-magnetic spacer layer in GMR structures. In bulk ferromagnetic materials, domain walls do not affect the resistance because the domain wall width is typically too large and the conduction electron spin can adiabatically track the inhomogeneous magnetization profile. In magnetic thin films, however, domain walls can be narrower and spin-mistracking can lead to an appreciable domain-wall resistance due to spin-dependent diffusive scattering (Viret *et al.* 1996; Levy and Zhang 1997) For reviews of DWMR see Kent *et al.* (2001) and Marrows (2005) and references therein.

Another kind of spin-valve-like magnetoresistive phenomenon occurs in magnetic tunnel junctions (MTJs) consisting of ferromagnetic metal electrodes separated by a thin insulating layer, or even vacuum. In this case current is carried from one ferromagnetic electrode to the other through the insulating layer by quantum-mechanical tunnelling. Just as in GMR, the resistance depends on the relative orientation of magnetization of the two ferromagnetic electrodes, provided the tunnelling process preserves the electron spin (i.e. the spin-memory condition must be satisfied). Tunnelling magnetoresistance (TMR) was first found by Jullière (1975). 20 years later the discovery of a large reproducible TMR at room temperature in MTJs based on amorphous Al_2O_3 (Moodera *et al.* 1995) and more recently in crystalline Fe/MgO/Fe and similar MTJs (Parkin *et al.* 2004; Yuasa *et al.* 2004) triggered tremendous activity in the experimental and theoretical investigations of the electronic, magnetic and transport properties of MTJs. (For reviews on TMR and spin-dependent tunnelling see Moodera and Mathon 1999, Zhang and Butler 2003, Tsymbal *et al.* 2003a, Tsymbal 2006 and references therein.)

AMR phenomena, on the other hand, are distinct from the spin-valve effects. Here, the electrical resistance depends on the orientation of the magnetization with respect to either the direction of current flow, orientation of the magnetization with respect to certain crystallographic orientations, or both. AMR was discovered as early as 1857 by William Thompson when he measured an "increase of resistance to the conduction of electricity along, and a diminution of resistance to the conduction of electricity across, the lines of magnetization" in a piece of ferromagnetic nickel (Thomson 1857). Even though AMR was discovered in the mid-nineteenth century, a physical explanation of the phenomenon requires concepts that were not uncovered until the twentieth century. AMR stems from the relativistic interaction of the conduction electron spin with its motion through the sample, an effect that we now call *spin-orbit coupling* (Smit 1951). Even now, more than 150 years after its discovery, researchers are continuing to uncover new regimes of AMR in nanoscale magnetic systems. For example, anisotropic conductance can occur for tunnelling from ferromagnetic electrodes into insulators or vacuum, an effect that is known as tunnelling anisotropic magnetoresistance (TAMR) (Brey *et al.*

2004; Gould *et al.* 2004). TAMR originates from the anisotropic electronic structure of the magnetic electrode(s) induced by spin-orbit coupling.

It is important to point out how AMR phenomena are different from spin-valve magnetoresistive phenomena. While the spin-valve effect is determined by the relative orientation of the magnetizations of two ferromagnetic electrodes (as in GMR and TMR) or two ferromagnetic domains (as in DWMR), AMR is intrinsic to magnetically saturated systems and can even occur in junctions geometries with only one magnetic electrode. It should also be noted that AMR does not depend on the spin polarization of the current, in contrast with the spin-valve phenomena where the spin-memory condition must be satisfied. AMR phenomena are instead determined by the angular dependence of the *total* current, regardless of its spin-polarization.

Of all the magnetoresistive phenomena described above, only OMR and AMR can be observed in bulk materials. GMR, TMR, and TAMR require small film thicknesses, either to (i) satisfy the spin-memory condition (GMR) or, (ii) tunnel through an insulating barrier (TAMR) or (iii) both (TMR). DWMR requires domain walls that are thin enough so that the conduction electron spin cannot completely track the local magnetization. All four phenomena can be observed in structures with macroscopic *lateral* dimensions. Reducing these dimensions down to the nanoscale in constrained geometries brings new physics not existent at the macroscopic scale, revealing novel features of magnetoresistive phenomena. Electronic, magnetic and transport properties of such objects are very different as compared to those in respective laterally extended systems due to reduced dimensionality, complex surfaces and interfaces, and quantum effects. In this chapter, we will discuss how three of these magnetoresistive phenomena, namely DWMR, AMR and TAMR, reveal themselves in qualitatively different ways in constrained ferromagnetic metal junctions.

The most striking difference between the laterally constrained and laterally extended systems is the very different regimes of electronic transport. First, the size of the constrained systems can be reduced below the mean-free transport path, making the transport properties ballistic rather than diffusive. In addition, the size of the constrained system can be so small that the lateral confinement of the electronic wavefunctions will play an important role in determining transport properties. The experimental and theoretical studies of these phenomena are discussed in Section 18.2. For an extensive review of this topic see Agrait *et al.* (2003), as well as Chapter 15 by van Ruitenbeek in this volume.

Constrained systems also exhibit unique magnetic properties that play an important role in magnetoresistive phenomena. For example, domain walls constrained in magnetic nanocontacts can have a width of a few nanometers and can be reduced even further to a scale comparable to the distance between atoms. In this case the magnetization may change orientation on a length scale of the order of the Fermi wavelength that can lead to much larger DWMR effects. The subject of DWMR in magnetic nanocontacts is discussed in Section 18.3.

In addition to the interesting DWMR properties of ferromagnetic contacts, novel AMR phenomena are also exhibited. Because of the reduced dimensionality of ferromagnetic contacts, which leads to strong spin-orbit effects, coupled with the very different mechanisms controlling electronic transport, the AMR effect will exhibit starkly different qualities as compared to the corresponding bulk counterparts. Recent experimental and theoretical findings in this field are discussed in Section 18.4, followed by Section 18.5, which is devoted to TAMR in broken magnetic contacts.

A review on magnetic nanocontacts and related phenomena was recently published by Doudin and Viret (2008).

18.2 Ballistic transport and conductance quantization

In bulk conducting systems, the transport properties are typically dominated by scattering processes that limit the conduction of electrons through the sample. In this regime, the electrical current is best described as a drift of electrons that travel from one scattering event to the next in the presence of an electric field. In this case the electrical conduction is generally well described by Ohm's law where the conductance G, which relates the current flowing through the sample to the electrical bias applied across it, is proportional to the area of the sample S transverse to the current and inversely proportional to the length of the sample L,

$$G = \frac{\sigma S}{L}. \tag{18.1}$$

The conductivity, σ, is determined by the properties of the conducting material, such as its electronic structure, concentration, type and distribution of defects (impurities), and temperature. While the conductance G is a useful concept for conductors of all sizes, relationships like Ohm's law, eqn (18.1), are no longer valid for conductors with dimensions smaller than a few relevant length scales.

Various transport regimes can be identified by comparing sample dimensions with different length scales. One of these length scales is known as the *phase-coherence length*, L_φ, which determines the average distance over which an electron travels between successive *inelastic* scattering events, i.e. events that destroy the quantum phase. L_φ can be of the order of a few micrometers at low temperatures. Samples where the relevant dimensions are smaller than L_φ are known as *mesoscopic*.

Another important length scale is the *elastic mean-free path*, ℓ, which measures the average distance travelled by a conduction electron between successive elastic (i.e. phase-coherent) scattering events by impurities and/or defects. When the relevant sample dimensions are much larger than both L_φ and ℓ, the transport is called *diffusive* and Ohm's law typically works well. In this regime one can picture electrons bouncing from one impurity to the next with an average flow along the direction of the current, as in Fig. 18.1. Furthermore, for mesoscopic samples where $\ell < L_\varphi$ the wave nature of the conduction electrons can lead to interesting quantum-interference phenomena between successive elastic scattering events.

Transport in samples with dimensions smaller than ℓ is described as *ballistic*. In this regime electrons retain their momentum when traversing the

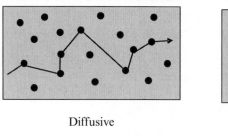
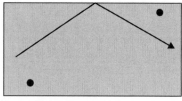

Diffusive   Ballistic

Fully quantum

Fig. 18.1 Three different transport regimes. In the diffusive regime the relevant length scale of the conductor is larger than the mean free path ℓ. In the ballistic regime, the size of the conductor is less than ℓ. In the fully quantum (and still ballistic) regime, the conductor size is comparable to the wavelength of the conduction electrons.

[1] We note that there is another important scale controlling spin-dependent electron transport. This is the *spin-diffusion length*, ℓ_{sf}, the average distance that an electron travels while retaining its spin orientation. Typically ℓ_{sf} is longer than the mean-free path ℓ. For further discussion see a recent review by Bass and Pratt (2007).

conductor and will only be altered by the scattering off the boundaries of the sample, as in Fig. 18.1. For Fe, Co and Ni ℓ is typically no larger than $\sim 10\,\text{nm}$. In this transport regime, it is sometimes appropriate to maintain a semi-classical picture of electronic transport.[1]

The *ultimate* length scale, at least where electrical transport is concerned, is the de Broglie wavelength of the conduction electrons at the Fermi level, λ_F. When one or more of the sample dimensions are of the order of λ_F, any kind of semi-classical description of electron trajectories breaks down and one must treat the system in the fully quantum-mechanical limit. In semiconductor materials λ_F can be on the scale of tens of nm, controllable by doping. In metals, however, λ_F is typically a few Å, which also happens to be the same order as the atomic spacing, and so the fully quantum limit corresponds to samples whose dimensions are only several atoms wide.

To understand the fully quantum regime it is convenient to adopt the Landauer–Büttiker scattering formalism (Büttiker 1988; Landauer 1988). In this formalism the system is modelled as a central scattering region connected to macroscopic electron reservoirs by a set of idealized leads, as in Fig. 18.2. The reservoirs are each in thermal equilibrium with their own well-defined temperature and electrochemical potential. The leads are assumed to be semi-infinite, in which case the reservoirs enter as boundary conditions at infinity. In the ideal leads electrons are carried by the Bloch states, analogous to the propagating modes of a waveguide. In fact, if the leads are treated in a free-electron picture, then the modes are precisely the same as waveguide modes whose transverse momentum (wave number) is quantized by the lateral confinement. Typically, an independent electron approach is taken in which one considers only elastic scattering, i.e. an electron with energy E coming from one of the leads will be scattered by the central region to one of the leads *with the same energy E*, so that the process is phase coherent. At a given energy E,

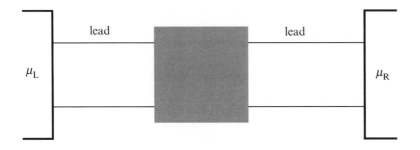

Fig. 18.2 Schematic representation of the basic two-terminal scattering device in the Landauer–Büttiker approach. The central scattering region, represented by the gray box, is connected to idealized leads that carry current via their propagating modes to and from the reservoirs. The left and right reservoirs have electrochemical potentials μ_L and μ_R, respectively.

the number and character of the propagating modes of the lead is determined by the electronic bands of the lead that cross the energy E. The interface between each lead and its respective reservoir is assumed to be reflectionless, that is, any electron travelling from the lead into the reservoir has zero probability of being reflected back into the lead.

The basic idea of the approach is to relate transport properties of the system with how the central region scatters states between the various propagating modes of the leads. For example, the probability of an electron with energy E incoming on mode n of the left lead to be scattered into the outgoing mode m of the right lead is given by $|t_{n,m}(E)|^2$. Here, $t_{n,m}(E)$ is the transmission amplitude that can be determined by the matching of the wavefunction at the interfaces of the scattering region with the propagating modes of the leads. The left reservoir injects electrons into the right-moving modes of the left lead according to the Fermi distribution $f(E - \mu_L)$.[2] Thus, the transmission of electrons through the system gives rise to a current from left to right:

$$I_{L \to R} = \frac{-2e}{h} \int_{-\infty}^{\infty} \sum_{n \in L, m \in R} |t_{n,m}(E)|^2 \, f(E - \mu_L) dE. \qquad (18.2)$$

[2]The Fermi function at a given temperature T is given by $f(x) = [\exp(x/kT) + 1]^{-1}$, where k is the Boltzmann constant.

The factor 2 accounts for spin-degeneracy of the modes in non-magnetic materials. We will discuss what happens in magnetic materials later.

The right reservoir injects electrons into the left-propagating modes of the right lead according to the Fermi distribution $f(E - \mu_R)$, giving rise to a current from right to left, $I_{R \to L}$, which can be written in a similar fashion to eqn (18.2). Because of time-reversal symmetry, the transmission probability from right to left is the same as from left to right and we find that the net current flowing from left to right in the system is given by

$$I = I_{L \to R} - I_{R \to L} = -\frac{2e}{h} \int_{-\infty}^{\infty} T(E) \left[f(E - \mu_L) - f(E - \mu_R) \right] dE. \qquad (18.3)$$

Here, we have introduced the *transmission function*

$$T(E) = \sum_{n \in L, m \in R} |t_{n,m}(E)|^2, \qquad (18.4)$$

which gives the total transmission probability over all modes at energy E. Equation (18.3) allows us to compute the total electrical current flowing between the two reservoirs for a given applied bias, $eV = \mu_L - \mu_R$, and therefore the differential conductance $G(V) = dI/dV$, which makes the Landauer–Büttiker scattering approach a powerful tool.

The most challenging part of using eqn (18.3) to determine transport properties is the calculation of $T(E)$. For an introduction to this topic, see the chapter by Datta in this volume as well as (Datta 1995). We would like to point out one interesting feature of the transmission problem though. It is always possible to describe the transport in terms of non-mixing *eigenchannels* of transmission, which correspond to the eigenvectors of the transmission matrix $\mathbf{tt}^{\dagger}$, where $\mathbf{t}$ is the matrix whose elements are $t_{n,m}(E)$ (Büttiker 1988). The corresponding transmission eigenvalues, T_i, can take any value between 0 (i.e. complete reflection), and 1 (i.e. perfect transmission), and the total transmission, eqn (18.4), reduces to a summation over all the eigenchannels: $T(E) = \sum_i T_i$.

In the zero-bias and zero-temperature limit, the conductance (3) reduces to

$$ G = \frac{2e^2}{h} T(E_F), \tag{18.5} $$

where E_F is the Fermi energy common to the whole system. Thus, in the linear response regime, the conductance is directly proportional to the electron-transmission function. The situation where perfectly transmitting eigenchannels dominate is especially interesting because, in this case, eqn (18.5) predicts *conductance quantization*. Consider a perfect ballistic channel, which would be represented in Fig. 18.2 by removing the central scattering region and connecting the two reservoirs with a single ideal lead representing a wire. In this case, there would be no reflection or scattering between the different propagating modes, and therefore for each of the active modes $|t_{n,m}(E)|^2 = \delta_{n,m}$, in which case $T_i = 1$ for all open eigenchannels. The transmission function would then be given by

$$ T(E) = \sum_i T_i = N(E), \tag{18.6} $$

where $N(E)$ is the total number of transmitting modes, determined by the number of electronic bands of the wire at energy E. Therefore, the linear response conductance, eqn (18.5), is given by $G = 2e^2/h \times N(E_F)$, i.e. the conductance is quantized in units of $G_0 = 2e^2/h$, which is known as the *quantum of conductance*.

Physically, a necessary condition to observe the quantization of conductance is that the cross-sectional radius of the ballistic channel must be comparable to the wavelength of the conduction electrons, i.e. the Fermi wavelength λ_F. However, our model of a perfect wire with non-varying cross-section connecting two reservoirs is clearly an idealization. Nevertheless, it is possible for quantization to occur in *constriction* geometries if another condition is met, namely adiabaticity. The adiabatic condition requires that the cross-section of the constriction vary slowly compared to the scale of the λ_F, in which case there would be no reflection or mixing between the modes (Glazman *et al.* 1988).

The first, and perhaps the most, convincing demonstration of ballistic conductance quantization was produced in the two-dimensional electron gas

(2DEG) formed at the GaAs/AlGaAs interface (van Wees *et al.* 1988; Wharam *et al.* 1988). A pair of metallic gates was used to manipulate the depletion region of the 2DEG to form a quantum point contact. By adjusting the gate voltage it was possible to control the effective width of the contact constriction. As the width was decreased, the conductance dropped in a stepwise fashion, where each step had a magnitude of $1G_0$, corresponding to the closing of an available mode for ballistic transport.

Can the same situation be realized in metals? In the 2DEG experiments the Fermi wavelength is $\sim$40 nm, which sets the conditions on the width and adiabaticity of the constriction cross-section. As we have already mentioned, however, for metals the Fermi wavelength is of the order of a few Å, in which case the constriction width must be on the atomic scale to observe quantization of conductance, posing a challenge to experimentally measure the effect.

There are three major experimental techniques for making atomic-scale metallic contacts, each with their own advantages and pitfalls. The first is the mechanical break junction technique, pioneered by van Ruitenbeek *et al.* (1996). Here, a thin film is deposited onto a flexible substrate, and lithographic techniques are used to define a narrow bridge in the film between two large film areas that will act as electrodes. Then, a piezo-electric actuator pushes up underneath the flexible substrate, pulling the two films apart and further narrowing the bridge down to the atomic scale. A similar situation can be realized in a scanning tunnelling microscope (STM) setup where a metallic tip in contact with a film surface is gently pulled away, forming a contact with controllable size (Gimzewski and Möller 1987). Mechanical break junction techniques are by far the most reported in the literature because of the ability to repeatedly break and reform the junction is highly favorable for performing statistical measurements such as conductance histograms, which we will discuss below. A disadvantage of this technique is that the corresponding metallic contact is usually not rigidly connected to a substrate, but instead is free-standing between two electrodes. This may lead to mechanical stability issues, especially in magnetic systems where magnetomechanical artifacts due to applied magnetic fields may play a significant role.

A second method, known as the electrical break junction technique introduced by Park *et al.* (1999), involves the migration of metallic atoms under high current densities, called *electromigration*, in lithographically patterned constrictions. This technique has a definite advantage over the mechanical break junction technique because the contacts are mechanically stabilized by contact with the substrate, and are not free-standing. However, the electrical break junction technique is inherently irreversible, and a single contact cannot be used repeatedly to gain statistical information.

The third technique differs significantly from the previous two in that the junction is not controlled by breaking, but is rather grown in an electrochemical bath (Morpurgo *et al.* 1999). Such electrochemical junctions are fabricated by first lithographically defining two metallic electrodes with a gap of $\sim$100 nm between them. The gap is then filled by electroplating metal ions from the electrochemical bath, a process that is controlled by the potential of the electrode(s) relative to the aqueous solution. The technique is clearly very different from the breaking methods described above, most notably in the "constructive" way the contacts are made, as opposed to the breaking techniques that monitor the properties of the system at its weakest point. The electrochemical method is unique in that the experiments obviously cannot be performed at cryogenic temperatures, like the breaking techniques, because of the aqueous solution. However, this can also be viewed as an advantage because future devices will need to operate at room temperature.

There are two major complications to measuring conductance quantization in atomic-size metallic contacts. First, because the constriction size cannot be controlled in a smooth fashion like the 2DEG experiments, jumps in the conductance during the breaking or reforming (or growth) of a contact may correspond to discrete changes in the atomic arrangement of the contact, not necessarily due to the opening or closing of additional conduction channels (Todorov and Sutton 1993). Second, even stable atomic arrangements are not necessarily going to correspond to a set of perfectly transmitting modes, and eigenchannels with $T_i < 1$ are expected to contribute since the adiabaticity condition may not be met.

The general experimental method for establishing conductance quantization in a given metal is the use of conductance histograms. Here, metallic contacts are broken and reformed several times (hundreds or thousands of times) and a histogram is made of all the values of conductance that are measured. (See Fig. 18.4(b) below for an example.) The resulting histogram is then a characteristic of the metal under study, and a clear indication of reproducible quantization of the conductance should yield strong peaks at integer values of the conductance quantum, G_0.

The seminal example of conductance quantization in metallic systems takes place in gold contacts (Brandbyge *et al.* 1995). It was later shown that the quantization of conductance in gold contacts is due to the formation of short atomic chains (see Fig. 18.3) (Ohnishi *et al.* 1998; Yanson *et al.* 1998).

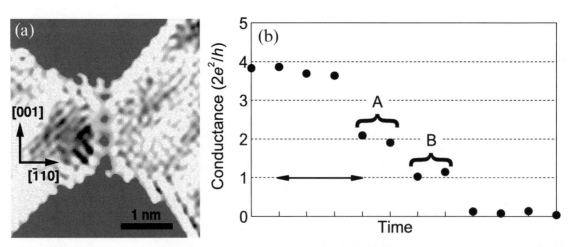

Fig. 18.3 (a) Processed electron microscope image of a linear strand of gold atoms (oriented vertically in the figure) forming a bridge between two gold films (gray areas at the top and bottom). The spacing of the four gold atoms are 0.35–0.40 nm. The strand is oriented along the [001] direction of the gold (110) film. (b) The conductance during the breaking of the Au junction. The plateaus correspond to quantized values of the conductance in double (A) and single (B) strands of gold atoms. Reprinted from Ohnishi *et al.* (1998) by permission from Macmillan Publishers Ltd: Nature, copyright 1998.

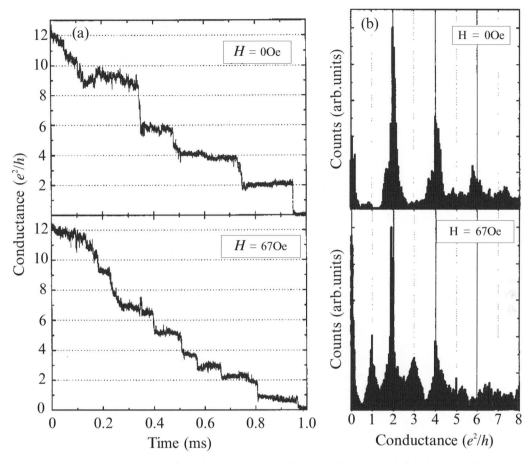

Fig. 18.4 Conductance of Ni MBJs junctions. (a) Conductance curves during breaking without magnetic field (top) and with a magnetic field of 67 Oe (bottom). (b) Conductance histograms without (top) and with a magnetic field of 67 Oe (bottom). Adapted with permission from Ono *et al.* (1999). Copyright 1999, American Institute of Physics.

For other metals, however, the situation is not so clear. The number and character of the transmission eigenchannels of an atomic-size metallic contact are essentially determined by the valence atomic orbitals available at the Fermi level. For simple metals like Au, which can be described by a single *s*-like orbital at the Fermi level, conductance quantization occurs due to low scattering. For metals with *p*- or *d*-like states at the Fermi level, like Al with 3 valence electrons or the transition-metal Nb, which has 5 valence electrons, several of the corresponding transmission eigenvalues are significantly different from perfect transmission, $T = 1$, and quantization of the conductance is not observed (Scheer *et al.* 1997; Cuevas *et al.* 1998; Ludoph *et al.* 2000).

The presence of ferromagnetism adds new features to the picture of ballistic transport in atomic contacts. In the $3d$ transition-metal ferromagnets (Fe, Co, Ni and their alloys) the electronic states are split by the exchange interaction into two sets of bands according to their spin projection along some quantization axis: one set of bands with spin parallel to the axis and another set with spin antiparallel.[3] These bands are occupied up to the Fermi level, and the

[3]This is rigorously true only in the absence of the spin-orbit interaction, which mixes the two spin channels. However, the spin-orbit interaction is small in these materials, and therefore the mixing is very weak. But, as we have already mentioned, this small spin-orbit coupling is the underlying mechanism of AMR, and will play an important role in later sections.

exchange splitting leads to unequal populations of the two spin projections and, hence, to the spontaneous magnetization. It is convenient to use the orientation of the magnetization as the spin-quantization axis and label the states that have spin magnetic moment parallel to the magnetization direction as "majority" and likewise those with spin magnetic moment antiparallel as "minority."

Due to the lifting of the spin degeneracy, and in the absence of spin-flip scattering, the conductance of uniformly magnetized ferromagnetic metals can be represented as the sum of contributions from majority- and minority-spin electrons. In the zero-bias and zero-temperature limit the conductance has the form

$$G = \frac{e^2}{h} \left[T_\uparrow (E_F) + T_\downarrow (E_F) \right], \qquad (18.7)$$

where $T_\sigma (E_F)$ is the spin-dependent transmission function different for majority- ($\sigma = \uparrow$) and minority- ($\sigma = \downarrow$) spin electrons. In the absence of scattering between different propagating modes in a ferromagnetic metal the conductance would be quantized in units of e^2/h

$$G = \frac{e^2}{h} N (E_F), \qquad (18.8)$$

where $N(E_F) = N_\uparrow(E_F) + N_\downarrow(E_F)$ is the total number of band crossings at the Fermi energy.

Signatures of spin-dependent conductance quantization in ferromagnetic metal contacts have been found in several experiments. For example, spin-dependent conductance quantization was observed in Ni nanowires electrode-posited into pores of membranes (Elhoussine *et al.* 2002), Ni atomic-size contacts made by a scanning tunnelling microscope (Komori and Nakatsuji 2001), and electrodeposited Ni nanocontacts grown by filling an opening in focused-ion-beam-milled nanowires (Yang *et al.* 2002). Experiments on ferromagnetic Co suspended atom chains evidenced electron transport of a spin-conductance quantum e^2/h, as expected for a fully polarized conduction channel (Rodrigues *et al.* 2003). Similar behavior was also observed for Pd (a quasi-magnetic 4*d* metal) and Pt (a non-magnetic 5*d* metal) atomic contacts. (Rodrigues *et al.* 2003) The latter fact was argued to indicate that the low dimensionality of the nanowire induces magnetic behavior and the lifting of spin degeneracy, even at room temperature. There were also indications that conductance quantization is sensitive to an applied magnetic field. For example, Ono *et al.* (1999) found that Ni break junctions exhibit conductance quantization in multiples of $2e^2/h$ (see top panels in Figs. 18.4(a,b)). By applying an external magnetic field they were able to achieve conductance quantization in multiples of e^2/h (bottom panels in Figs. 18.4(a,b)).

Contrary to the experiments mentioned above, however, Untiedt *et al.* (2004) reported no spin-dependent conductance quantization, even in high magnetic fields. They measured the conductance of atomic contacts made of the ferro-magnetic 3*d* metals, Fe, Co, and Ni, using mechanical break junctions under cryogenic vacuum conditions and found conductance histograms showing broad peaks above $2e^2/h$, with only little weight below. Qualitatively similar results have been obtained by other researchers (Calvo *et al.* 2008).

Theoretical studies involving model atomic structures and realistic multi orbital electronic structures of ferromagnetic atomic contacts indicate no spin-dependent conductance quantization in multiples of e^2/h and/or full spin polarization in such contacts (Bagrets *et al.* 2004, 2007; Fernández-Rossier *et al.* 2005; Jacob and Palacios 2006; Smogunov *et al.* 2006). As we have already mentioned, perfectly transmitting eigenchannels are not necessarily to be expected, especially in metals with significant *d*-electron density of states at the Fermi level where the $3d$ electrons suffer from the so-called "orbital blocking" (Jacob *et al.* 2005). Since all the $3d$ ferromagnets have a significant weight of the $3d$ electrons at the Fermi energy (at least in one of the spin channels), it is expected that this behavior would not produce a perfect transmission and conductance quantization in multiples of e^2/h.

This conclusion of the first-principles calculations is qualitatively consistent with some experiments (Untiedt *et al.* 2004), but inconsistent with the other experiments (Elhoussine *et al.* 2002; Yang *et al.* 2002; Rodrigues *et al.* 2003). The presence of adsorbates at the nanocontact was put forward to explain the presence of the histogram peak at the conductance quantum $2e^2/h$ (Untiedt *et al.* 2004). However, this does not explain the presence of other conductance steps observed experimentally in multiples of the spin-conductance quantum (see, e.g. lower panels in Fig. 18.4). Apparently, further investigations are required to clarify this issue. This is related both to experiments where a better control of the atomic structure and reproducibility of measurements are needed and to theory, where the low dimensionality of atomic-size contacts may lead to strong electron–electron correlations (e.g., Wierzbowska *et al.* 2005; Rocha *et al.* 2007) that were disregarded in the calculations mentioned above.

18.3 Domain-wall magnetoresistance at the nanoscale

A domain wall is the region separating two uniformly magnetized magnetic domains in a magnetic material, as is illustrated in Fig. 18.5(a). The study of magnetic domain walls in bulk ferromagnetic materials began in the early twentieth century with the notion of magnetic domains by Weiss (1907). In order to resolve the apparent discrepancy between the low saturation field (a few Oe) and the high internal "molecular" field responsible for the Curie temperature (tens of MOe), Weiss proposed that magnetic samples were made up of various fully magnetized regions. After the experimental confirmation of domains (Sixtus and Tonks 1931; Bitter 1932), Bloch (1932) was the first to describe the region separating two domains, the domain wall. Landau and Lifshitz (1935) showed that magnetic domains typically arise in order to reduce the extrinsic (i.e. size- and shape-dependent) magnetostatic self-interaction of a ferromagnetic sample, and refined the description of domain walls put forward by Bloch.

The most important property of domain walls, at least for the purposes of this chapter, is the domain-wall width, w, indicated in Fig. 18.5. The width of

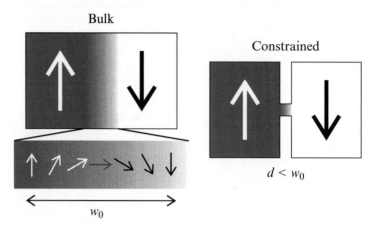

Fig. 18.5 Bulk and constrained domain walls. The intrinsic width w_0 determines the width of domain walls in the bulk. In constricted geometries, domain walls can have lower energy by squeezing their width into the length of the constriction, d.

a domain wall in a bulk magnetic material is determined by the competition between the exchange energy and the anisotropy energy. The exchange interaction tends to align neighboring regions of magnetization, whereas anisotropy prefers the magnetization to align with certain crystallographic orientations. A simple model predicts that in bulk materials the equilibrium domain-wall width is $w_0 = \pi\sqrt{A/K}$, where A is the exchange stiffness and K is the anisotropy constant (Kittel 1996).

In bulk $3d$ metal ferromagnets the domain walls are wide ($w_0 \sim 100\,\text{nm}$) on the scale of the lattice spacing, due to the stronger exchange interaction as compared to the anisotropy energy. In a constrained geometry, however, the domain-wall width w may be reduced significantly compared to the intrinsic width w_0 (Bruno 1999). The key to this phenomenon resides in the fact that it is energetically favorable for a domain wall to be compressed and fit into the constriction region. Since the energy of a domain wall grows with its area, narrowing the width of the domain wall to the constriction size (d in Fig. 18.5) reduces the total exchange energy by confining the wall to a region with smaller cross-section.

The theme of this section is to discuss how magnetic domain walls affect the electrical resistance. One of the earliest works on this subject was the theory of Cabrera and Falicov (1974) who suggested the effect of scattering due to an interaction with the electron spin. Electrons travelling in one domain experience a different potential upon entering an oppositely magnetized domain since the electronic band structure on either side of the wall differs by the exchange splitting. The reflection of the electronic wavefunction at the domain wall, represented by a potential step, gives rise to the resistance. It was found, however, that for an appreciable domain-wall resistance the width of the domain wall must be of the same order as the wavelength of the impinging electrons, otherwise the electron spin can adiabatically track the local magnetization and the reflection is very small. In bulk ferromagnetic metals, where the domain wall width is typically orders of magnitude larger than the Fermi wavelength, the reflection coefficient is exceedingly small, as was pointed out later in analogy with the propagation of microwaves in a twisted waveguide (Berger 1978).

In the mid-1990s the study of intrinsic domain-wall resistance due to the spin polarization of the current was revisited. In a semi-classical picture, the inability of the conduction-electron spin to track the local magnetization while passing through a domain wall increases the likelihood of diffusive scattering, provided the scattering rate for one spin channel was larger than for the other, which is typically the case for ferromagnetic metals (Viret *et al.* 1996). As the electron traverses a wall of thickness w, the exchange field (parallel to the local magnetization) rotates around it with angular frequency $\omega_{\text{wall}} = \pi v_{\text{F}}/w$, where v_{F} is the Fermi velocity of the electron. At the same time the electron spin precesses around the canted exchange field with Larmor frequency ω. If the domain wall is sufficiently thin, then $\omega_{\text{wall}} > \omega$ and the spin is unable to track the local magnetization quickly enough and the scattering rate is increased. This results in a relative change in resistance that scales as $\Delta R/R \propto 1/w^2$, where ΔR is the increase in resistance when a domain wall is present. A similar, however fully quantum mechanical, description of the same mistracking phenomenon also predicts a $1/w^2$ variation in the effect (Levy and Zhang 1997).

These models of Viret *et al.* (1996) and Levy and Zhang (1997) enjoyed a reasonable amount of success in describing the experimental results since the late 1990s on the measurement of the resistance due to a domain wall in thin-film structures, and to some extent mesoscopic structures as well (see a review by Marrows 2005 and references therein for further details). In all these cases the electronic transport can be considered as diffusive, a point upon which the models rely. Neither refers to the reflection of the electronic wavefunction by the domain wall, as in the Cabrera–Falicov (1974) model. In fact, using the Levy–Zhang (1997) model it is possible to compute the reflection coefficients, however, the authors did not do this since they discount the possibility of a narrow enough domain wall for such an effect to be appreciable.

The possibility of a domain-wall magnetoresistance in the ballistic transport regime, called ballistic magnetoresistance (BMR), was first put forth by García *et al.* (1999) when they observed a 200% magnetoresistance in Ni break junctions. The largest effect was measured for samples with a nominal conductance of a few times the quantum of spin conductance, e^2/h, implying that the contact was made only through a few atoms and that the transport could be considered as ballistic. The magnetoresistance was attributed to the formation of a domain wall only a few atomic layers wide that was pinned at the point contact (Tatara *et al.* 1999). An atomically thin domain wall, according to the theory of Cabrera and Falicov (1974), would lead to a sizable resistance contribution from the reflection of the electronic wavefunction. Although not directly observed in the experiments of García *et al.*, or to date any other magnetic nanojunctions where BMR has been claimed, the plausibility of atomically thin domain walls is supported by spin-polarized STM experiments by Pratzer *et al.* (2001).

The large values of magnetoresistance measured on Ni break junctions (García *et al.* 1999) stimulated further research efforts in this field. Similar behavior was found for electrodeposited Ni contacts (García *et al.* 2002) and for other ferromagnetic metals (Versluijs *et al.* 2001; Chung *et al.* 2002). These results were followed by reports of huge BMR effects in electrode-posited nanocontacts (Chopra and Hua 2002; García *et al.* 2003; Hua and Chopra 2003) which produced enormous publicity due to the prospects for a new generation of spintronic devices (Brumfiel 2003). However, a number of research groups encountered problems in reproducing these experiments. The extreme difficulty of keeping magnetic atomic-size contacts stable due to magnetostriction and magnetostatic forces convinced the researchers that the large effects observed are artifacts and that accurately performed experiments reveal only very small effects (Egelhoff *et al.* 2004; Montero *et al.* 2004; Ozatay *et al.* 2004). Currently, most researchers consider huge BMR values measured as being the consequence of lacking mechanical control and artifacts due to magnetostriction and magnetostatic effects (Doudin and Viret 2008).

These experimental results stimulated a number of theoretical studies of spin-dependent transport in constrained geometries. The simplest approach employed free-electron models. It was demonstrated that the interplay between quantized conductance and an atomic-scale domain wall results in magnetoresistance that oscillates with the cross-section of the constriction and leads to enhanced magnetoresistance values (Imamura *et al.* 2000; Tagirov *et al.* 2002). It was also shown for atomic-size constrictions that a closure of one spin-conduction channel may result in very large magnetoresistance due to "half-metallic" behavior of the electrodes (Zhuravlev *et al.* 2003).

As an instructive example of the formation of a domain wall in a magnetic nanocontact, as well as a demonstration of how this affects the ballistic transport properties, consider the model contact geometry in Fig. 18.6(a) (Burton *et al.* 2004). The difference in the shape anisotropy of the two electrodes makes the magnetization of the right-hand electrode slightly easier to switch than the left. This can be seen in the hysteresis loop shown in Fig. 18.6(c), which is calculated using micromagnetic modelling techniques (Donahue and Porter 1999). As a field is applied parallel to the junction axis, the magnetization of the right electrode switches before the left, yielding an anti-parallel alignment of the two electrodes within the switching region from 100 mT to 120 mT. When the electrode magnetizations are anti-parallel a domain wall is formed in the constriction (see Fig. 18.6(b)). The width of the domain wall is largely confined to the constriction region, and is an example of a geometrically confined domain wall, as shown in Fig. 18.5 (Bruno 1999). Indeed, the typical width of a domain wall in bulk Ni is ~50 nm, while the domain wall in Fig. 18.6(b) is confined to a wire 10 nm in length.

Taking results of the micromagnetic calculations and using a spin-split free-electron model of the constriction region, calculations of the conductance as a function of the applied field in the presence of a domain wall are presented in Fig. 18.6(e).[4] When the right electrode switches, the conductance drops abruptly due to the formation of a domain wall. When the left electrode switches at ~120 mT, the conductance abruptly jumps back up to its saturated value. However, in the intermediate region between these two switching events, there is a monotonic increase in the conductance as the domain wall spreads into the left-hand electrode, as is shown in Fig. 18.6(c). The conductance

[4]We note that this calculation assumed no defect scattering, which may play a role in real structures of that size.

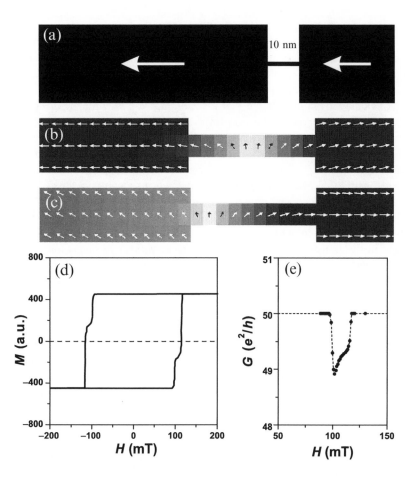

Fig. 18.6 (a) A symmetric Ni nanocontacts, with a constriction size of $2 \times 2 \times 10\,\text{nm}$. (b,c) Magnified view of the magnetic structure of the domain wall trapped in the junction's central region after the right electrode has switched for $H = 100\,\text{mT}$ (b) and $H = 110\,\text{mT}$ (c). (d) Hysteresis loop of the junction. (e) Conductance vs. applied field during the magnetic switching. Adapted with permission from Burton *et al.* (2004). Copyright 2004, American Institute of Physics.

increases due to the larger domain-wall width, indicating a regime where the domain-wall resistance is controlled by the applied magnetic field.

The spreading of the domain wall into the left electrode is favored because of the low magnetocrystalline anisotropy of Ni. The behavior changes dramatically if the left electrode is replaced with a magnetically hard material, like a CoPt multilayer. In this case, the domain wall that is formed in the contact region will not spread into the left electrode as the field is increased because of the high magnetocrystalline anisotropy, as confirmed by micromagnetic calculations (Burton *et al.* 2004). Instead, the domain wall is compressed, rather than widened, as the field increases and the corresponding domain-wall conductance will decrease with increasing field.

Although free-electron models provide a valuable insight into the domain-wall resistance, they cannot be used for quantitative comparison with experiments due to the complex spin-polarized electronic structure of the ferromagnetic metals. The band structure of transition-metal ferromagnets is dominated by d bands, which cannot be properly described by a pair of spin-split parabolic bands at the Fermi energy. Recent advances in band structure and electronic transport theory have made it possible to perform first-principles calculations of the domain-wall magnetoresistance (van Hoof *et al.* 1999;

Kudrnovský *et al.* 2000; Yavorsky *et al.* 2002). The calculations carried out for defect-free domain walls in bulk Ni, Co, and Fe found a domain-wall magnetoresistance of about 0.1% for domain-wall widths typical of bulk ferromagnets (van Hoof *et al.* 1999). These small values are consistent with a simple picture of the electron spin adiabatically tracking the local magnetization. The calculations demonstrated that a sizable domain-wall magnetoresistance can only be observed if the domain wall is abrupt, i.e. the magnetization changes orientation in one atomic layer. In particular, for an abrupt domain wall in bulk *fcc* (001) Co magnetoresistance was predicted to be as large as 250% (Kudrnovský *et al.* 2000). New features are expected for nanowires and nanoconstrictions where the lateral confinement of electronic waves plays a decisive role in electronic transport (Bagrets *et al.* 2004; Solanki *et al.* 2004; Velev and Butler 2004). In particular, it was predicted that the hybridization of the two spin subbands across the domain wall may produce pseudogaps in the transmission spectrum, leading to very large magnetoresistance (Sabirianov *et al.* 2005).

All these models assume that the domain wall is *rigid*, i.e. they neglect any spatial variation of the magnitude of the magnetic moment across the domain wall. It is well established, however, that the magnitude of the magnetic moments in itinerant magnets, like the $3d$ transition-metal ferromagnets Fe, Co and especially Ni, can depend strongly on the orientation of the neighboring moments. This effect is relatively weak in well-localized ferromagnets like Fe, but can reduce and even destroy the atomic magnetic moment of the itinerant ferromagnets, like Ni. This was demonstrated by Turzhevskii *et al.* (1990) who performed calculations in which a single atomic magnetic moment in a bulk material is rotated away from the net magnetization direction defined by the rest of the system, as shown in Fig. 18.7.

The origin of this phenomenon is the hybridization between non-collinear spin states. In the uniformly magnetized material with no spin-orbit coupling the minority- and majority-spin bands are independent. However, in a noncollinear state, such as a domain wall, this is no longer the case and the two

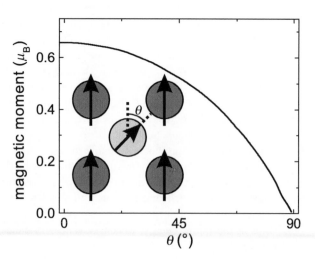

Fig. 18.7 The calculated magnetic moment on a single Ni atom whose magnetic moment is rotated by an angle θ away from the net magnetization direction defined by the rest of the moments in the system, as shown schematically in the inset. Courtesy of A.I. Lichtenstein (Turzhevskii *et al.* 1990).

spin bands are hybridized. This spin mixing leads to charge transfer and level broadening, which results in the reduction of the overall exchange splitting between majority- and minority-spin states on each atom, and hence the atomic moments are reduced. In bulk domain walls of any ferromagnetic material, this effect is small due to the slow variation of the direction of the magnetization. For the atomic-scale domain walls, however, there is a large degree of canting between neighboring magnetic moments. This leads to a significant hybridization between spin states and therefore a reduction, or *softening*, of the magnetic moments within the constrained domain wall (Burton *et al.* 2006).

It should be noted that this is different from the spatial variation of the magnetization in domain walls due to finite temperature magnetic disorder of the (fixed magnitude) atomic magnetic moments in a domain wall (Zhirnov 1959; Bulaevskii and Ginzburg 1964). In these so-called *elliptical* domain walls the magnetic moment directions in the domain wall fluctuate more rapidly due to temperature than those in the uniformly magnetized domains, and hence can be described by a locally reduced Curie temperature. This is due to the fact that the local average exchange field on an atomic moment is reduced because the non-collinearity diminishes the net exchange interaction with nearest neighbors along the domain wall (Kazantseva *et al.* 2005).

Ab-initio calculations of very narrow domain walls reveal the magnetic moment softening effect in model Ni nanowires with the 5–4 structure shown in Fig. 18.8(a) (Burton *et al.* 2006). Narrow domain walls are modelled by constraining the magnetic moment directions in a spin spiral structure between two oppositely magnetized domains, as indicated by the arrows in Fig. 18.8(a). Here, the relative angle between neighboring layers of magnetic moments are $180°/(N + 1)$, where N is the number of atomic layers in the domain wall. Figure 18.8(b) shows the calculated magnitudes of the magnetic moments for a few narrow domain-wall widths. Clearly the largest magnetic moment softening effect occurs for the narrowest domain wall, the $N = 1$ case, where the magnetic moments on the central atomic layer are reduced by ∼90% of their value in the uniformly magnetized case. The $N = 3$ and $N = 5$ cases also exhibit significant softening.

The magnetic moment softening has a profound effect on the electrical transport properties of the domain wall. This is demonstrated by calculating the transmission function, $T_{DW}(E)$, from the *ab-initio* calculations using two different methods. First, the electronic structure is taken directly from the uniformly magnetized case (i.e. the absence of a domain wall) and the local spin-dependent electronic structure parameters are then transformed according to their spin-orientation to simulate a domain wall. However, the magnetic moments are not allowed to soften, and the transmission function is computed directly. This is known as the *rigid* domain wall, and is similar to the previous studies mentioned. In the second case, however, the magnetic moments are determined self-consistently in the presence of the non-collinear structure, which we call a *soft* domain wall. The electronic structure from this calculation is then used to compute the transmission function. The results of both calculations are presented in Fig. 18.8(c) for an $N = 1$ domain wall.

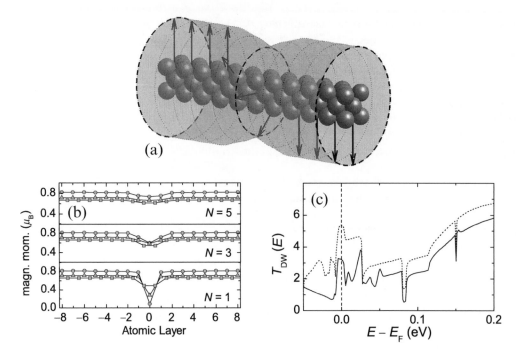

Fig. 18.8 (a) The 5–4 nanowire showing the three non-equivalent sites and a $N = 3$ domain wall. Each arrow represents the magnitude and orientation of the average magnetic moment of the plane with 5 atoms. (b) The self-consistent magnetic moments for several domain-wall widths in the 5–4 wire. The circles are for the atoms along the central axis of the wire. The triangles are for the atoms along the four corners of the wire. The squares are for the atoms along the faces of the wire. (c) A comparison of transport properties of soft and rigid domain walls. $T_{DW}(E)$ is the transmission as a function of energy for an $N = 1$ domain wall. The solid curve is for the fully self-consistent domain wall with softened magnetic moments. The dotted curve is for a rigid domain wall. Adapted with permission from Burton *et al.* (2006). Copyright 2006 from the American Physical Society.

Clearly the effect of magnetic moment softening leads to a decrease in the transmission function (and therefore the conductance) for states near the Fermi level. The origin of the enhancement of the domain wall resistance arises from the spin-memory condition common to all spin-valve magnetoresistive phenomena. For a domain-wall resistance to be appreciable, the electron spin must be *unable* to sufficiently track the local magnetization inside the domain wall. If the exchange field inside the domain wall acting on the electron spin is reduced, as it is when the effect of magnetic moment softening is present, this makes it more difficult for the electron spin to adjust to the magnetization orientation and hence there is a higher reflection off the next domain.

To conclude this section we would like to mention that experimental measurements involving domain-wall resistance and magnetoresistive effects related to reorientation of magnetic moments within the sample inevitably involve anisotropy effects. AMR may play a decisive role in experiments on domain-wall resistance because magnetic moments, apart from being non-collinear, also make different angles with respect to junction axis (see Doudin and Viret 2008 for further discussion). Anisotropy effects in magnetic nanocontacts on electron transports are discussed in the next two sections.

18.4 Anisotropic magnetoresistance in magnetic nanocontacts

Most anisotropic properties of ferromagnetic materials stem directly from spin-orbit coupling (SOC). SOC describes the relativistic effect of an electron's orbital motion on the orientation of its spin, and vice versa. Without SOC, the *orientation* of the net spin magnetic moment in ferromagnetic materials is unrelated to the geometry of the material, apart from weak macroscopic magnetostatic effects. SOC accounts for AMR, which is the central theme of this section and the next, and so it is important to understand the effect.

SOC occurs due to the interaction of the electron spin with the magnetic field that appears in the frame of the electron due to its motion through an electric potential. In a spherically symmetric potential $V(r)$, as in an atom, this introduces an additional term in the Hamiltonian of the form

$$H_{SO} = \frac{\mu_B}{e\hbar mc^2} \frac{1}{r} \frac{dV}{dr} \mathbf{L} \cdot \mathbf{S}, \qquad (18.9)$$

where $\mathbf{L}$ and $\mathbf{S}$ are the orbital and spin angular momentum operators, respectively. Even though eqn (18.9) is strictly true only for spherically symmetric potentials, it nevertheless provides a basis for the physical understanding for SOC effects in solids since the gradient of the potential is largest near the atomic positions in solids.

SOC induces changes in the electronic structure of the system that depend on the orientation of the magnetization, which is collinear with $\mathbf{S}$. First, SOC mixes minority- and majority-spin states so that spin is no longer a good quantum number. Second, SOC hybridizes different orbital states, leading to an angular-dependent shift of electronic energy levels. This orbital hybridization can also lead to the splitting of electronic states that would otherwise be degenerate in the absence of SOC. Due to this anisotropy in the electronic structure, the states at the Fermi energy that determine transport properties are now sensitive to the magnetization orientation, resulting in AMR phenomena.

Here, we give a brief overview of AMR in bulk ferromagnetic metals. For a more complete review of AMR in bulk transition metals and their alloys see (McGuire and Potter 1975). The diffusive transport in bulk materials is well described by an intrinsic resistivity tensor $\rho_{\alpha\beta}$ connecting the crystallographic components of the current density, J_β, with the components of electric field, E_α, generating the current. In ferromagnetic metals, the AMR effect manifests itself as the additional dependence of $\rho_{\alpha\beta}$ on the orientation of the magnetization. In cubic systems the resistivity is isotropic with respect to the crystallographic axes, in which case the resistivity follows a simple relationship

$$\rho(\theta) = \rho_\perp + \left(\rho_\| - \rho_\perp\right)\cos^2\theta, \qquad (18.10)$$

where θ denotes the angle of the magnetization with respect to the direction of current flow, and $\rho_\|$ and $\rho_\perp$ are the resistivities parallel and perpendicular, respectively, to the magnetization direction (see Fig. 18.9). The magnitude of the effect is quantified by the AMR ratio, which we define as

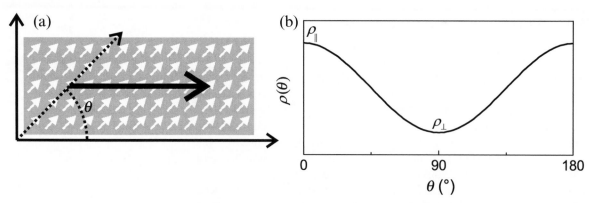

Fig. 18.9 (a) Cartoon depicting the magnetization (white arrows) and the direction of current flow (large arrow). θ is the angle between the magnetization and the current. (b) The typical $\cos^2 \theta$ angular dependence of the resistivity in bulk ferromagnets.

AMR $= (\rho_{\parallel} - \rho_{\perp})/\rho_{\perp}$. For the $3d$ transition-metal ferromagnets and their alloys, the AMR ratio is typically of the order of 1–5% at room temperature. Measured AMR ratios are generally positive (i.e. $\rho_{\parallel} > \rho_{\perp}$), but negative ratios are found for some alloys (McGuire and Potter 1975).

While eqn (18.10) provides a phenomenological expression for AMR consistent with experiment, a detailed microscopic description was not available until almost 100 years after Thompson's initial discovery (Thomson 1857). Mott's model of the transport properties of the transition metals suggests that s–d scattering dominates the conductivity of the 3d transition-metal ferromagnets (Mott and Jones 1936). Smit (1951) showed that SOC induces an anisotropy in scattering off impurities in magnetic metals that, due to the coupling of the orbital motion to the net spin-magnetization direction, breaks the otherwise crystalline symmetry. This leads to an s–d relaxation rate whose dependence on the current direction exhibits lower symmetry than the crystal, and therefore to an anisotropy of the resistivity with respect to the magnetization direction. Further refinements to the theory were made by Potter (1974), who used a more realistic band picture of the $3d$ ferromagnets. In addition to impurity scattering, phonon or magnon scattering can also lead to bulk anisotropic resistivity due to SOC. The importance of scattering phenomena in controlling bulk AMR leads to an interesting question: What happens when the scattering effects are absent? What is the manifestation of AMR in a transport regime other than the diffusive? This is where we turn to next.

While bulk AMR has been well understood for decades, it was not until recently that studies of AMR in nanoscale magnetic systems appeared in the literature. The first indications that AMR in the ballistic regime may exhibit substantially different characteristics came from theory. Velev *et al.* (2005) proposed that the origin of AMR in the ballistic transport regime must be very different compared to that in the diffusive transport regime because there is no electron scattering contributing to the conductance in the former. Specifically, the issue of conductance quantization was addressed. As we discussed previously, the conductance of narrow wires or constrictions of ferromagnetic metals may exhibit quantized values: $G = Ne^2/h$, where N is the number of open conducting channels that, in turn, is determined by the number of electronic band crossings at the Fermi energy (see eqn (18.8)). This quantity is affected by the orientation of the magnetization through SOC, which is known to be much stronger in open and constrained geometries than in bulk materials. By changing the magnetization direction one can, therefore, change the number of bands crossing the Fermi energy and thereby affect the ballistic conductance. This phenomenon was designated as *ballistic anisotropic magnetoresistance* (BAMR), and was first illustrated using *ab-initio* calculations of the band structure of a monoatomic wire of Ni atoms (Velev *et al.* 2005). As is seen from Fig. 18.10(a), near the edge of the Brillouin zone the doubly degenerate band labelled E_2 is very close to the Fermi energy when the magnetization lies perpendicular to the axis of the wire, $\mathbf{M}\perp\hat{\mathbf{z}}$. However, the degeneracy of this doublet is lifted when $\mathbf{M} \parallel \hat{\mathbf{z}}$, as is evident from Fig. 18.10(b). This spin-orbit splitting removes one band from the Fermi level, closing one conduction channel as the magnetization orientation changes from $\mathbf{M}\perp\hat{\mathbf{z}}$ to $\mathbf{M} \parallel \hat{\mathbf{z}}$. This reduces the number of bands at the Fermi level from 7 to 6, and therefore ballistic conductance is reduced by one quantum e^2/h, resulting in a positive BAMR of $1/6 \approx 17\%$.

For intermediate orientations of the magnetization between parallel ($0°$) and perpendicular ($90°$), SOC introduces splitting of the E_2 band that decreases with increasing angle between the magnetization and the axis of the wire. There is a critical angle at which the lower band of the "doublet" crosses the Fermi level, yielding a corresponding jump in conductance by e^2/h.

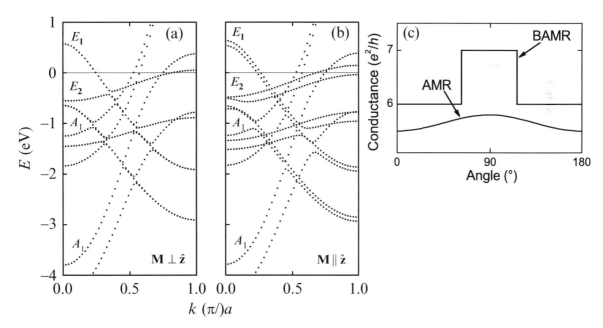

Fig. 18.10 (a, b) Calculated electronic band structure of monoatomic Ni wire with equilibrium interatomic distance in the presence of the spin-orbit interaction for magnetization lying (a) perpendicular to the wire axis $\mathbf{M}\perp\hat{\mathbf{z}}$ and (b) parallel to the wire axis $\mathbf{M} \parallel \hat{\mathbf{z}}$. The horizontal line indicates the Fermi level. (c) Conductance G of a monoatomic Ni wire as a function of angle between the magnetization and the wire axis. The sinusoidal angular dependence of AMR in bulk materials according to eqn (18.10) is schematically shown for comparison. Adapted with permission from Velev *et al.* (2005). Copyright 2005 from the American Physical Society.

This stepwise angular dependence is markedly different from the sinusoidal dependence in the bulk from eqn (18.10), as demonstrated in Fig. 18.10(c), and should be regarded as a signature of BAMR.

SOC also introduces more detailed changes to the band structure, including avoided crossings due to orbital and spin mixing. While these do not show up at the Fermi level for the monoatomic Ni wire, different materials and geometries may demonstrate a whole zoo of different BAMR signatures. This is demonstrated by examining the dependence of BAMR on the position of the Fermi level in Fig. 18.10 (see also Fig. 4 of Sokolov *et al.* 2007b). For example, it should be noted that the sign of BAMR is determined by whether bands are added to or removed from the Fermi level, which depends on the material and the details of the geometry. Indeed, calculations on Fe wires of various cross-sections have a tendency toward negative BAMR (Velev *et al.* 2005). This is in contrast to bulk, where positive AMR (i.e. $\rho_\parallel > \rho_\perp$) is observed for most materials (with a few exceptions, see McGuire and Potter 1975).

Although very revealing of the underlying physics, a perfect 1D wire is an idealization of contact geometries typical of atomic-scale metallic junctions. Due to this, the picture of transport being carried by non-scattered conduction channels may need to be modified, since the d-electrons responsible for BAMR are likely strongly scattered by the constriction (Jacob *et al.* 2005). Considering a more realistic contact geometry indicates that the angular dependence of the BAMR displays a smoothed behavior as compared to an abrupt step conductance variation shown in Fig. 18.10(c) (Burton and Tsymbal, unpublished). Nevertheless, these calculations reveal that the angular dependence of the AMR in atomic scale systems originates from the anisotropy of the electronic structure itself, as opposed to the bulk diffusive regime where AMR arises from the anisotropy in diffuse scattering. As we will see, some experimental studies of AMR in ballistic contacts do indeed demonstrate step-wise angular dependence qualitatively consistent with the picture of opening and closing conduction channels.

Early measurements performed on Ni ballistic nanocontacts found a change of sign in the magnetoresistance obtained for the fields parallel and perpendicular to the current (Viret *et al.* 2002; Yang *et al.* 2004, 2005; Keane *et al.* 2006). This behavior was interpreted as AMR in the ballistic regime, but it was unclear how this effect differed from AMR in bulk systems, despite its larger magnitude. Clear experimental indication of AMR can be obtained, instead, by monitoring the resistance while varying the angle of applied magnetic field for magnetically saturated samples. Viret *et al.* (2006) used this technique to perform AMR measurements on mechanically controlled Fe break junctions. It was found that the AMR ratio becomes quite large, up to $\sim 75\%$, for samples whose nominal conductance was of the order of e^2/h, i.e. samples that presumably have atomic size cross-section at the contact. In addition, it was found that the smallest contacts tend toward a stepwise angular dependence of the resistance, shown in Fig. 18.11, which appears to conform to the idea of opening and closing of conduction channels.

Sokolov *et al.* (2007b) reported measurements of AMR on electrodeposited Co contacts at room temperature. During the electrochemical deposition process, it was possible to monitor the increasing conductance associated

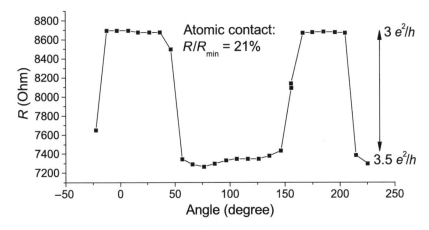

Fig. 18.11 Variation in the resistance of an atomic size contact in an Fe break junction as a 2.5-T field is rotated with respect to the contact. The AMR behavior is close to a two-level effect, reaching $3.5e^2/h$ when the magnetization is perpendicular to the contact. Adapted with permission from Viret *et al.* (2006). Copyright 2006 Springer.

with the growth of the contact region. It was found that the conductance increases in a stepwise fashion corresponding to the opening of additional conduction channels. Each plateau in conductance during growth corresponds to a different atomic configuration in the contact region, which are stable on a timescale of ~100 s. During the stable plateau, the sample was rotated in an applied magnetic field sufficient to saturate the magnetization. What was observed was remarkable: Periodic stepwise conductance jumps, close to either 1 or 2 e^2/h in amplitude, with period corresponding to the rotational frequency of the sample. Each conductance plateau, corresponding to a different atomic structure in the contact region, demonstrated a different pattern of stepwise angular dependence (see Fig. 18.12). The fact that a variety of samples exhibit quantized magnetoresistance, which was attributed to the BAMR effect predicted by Velev *et al.* (2005), points to the general nature of this phenomenon.

Other experimental measurements of AMR in magnetic nanocontacts were reported by Bolotin *et al.* (2006) who performed studies of electromigrated permalloy ($Ni_{80}Fe_{20}$) magnetic break junctions at low temperatures. Samples over a broad range of contact dimensions were studied: from $100 \times 30\,nm^2$ down to atomic-scale point contacts. For samples with zero-field resistance larger than $\sim1\,k\Omega$, AMR ratios up to 25% were observed, which is much larger than the bulk. Such samples demonstrated the angular variation in the resistance significantly different from the $\cos^2\theta$ behavior (see Fig. 18.13), but an abrupt stepwise dependence was not reported. Changes in the applied bias on the scale of a few mV also led to variations in the resistance of similar magnitude. These results were interpreted as a consequence of conductance fluctuations due to mesoscopic quantum interference. Shi *et al.* (2007) recently extended the examination of these break junctions and found that they indeed exhibit the temperature dependence expected for quantum interference of electrons in the contact region. Specifically, they found that the enhanced AMR signals, complex angular dependence, and fluctuations in applied bias are no longer significant above cryogenic temperatures. Quantum interference leads to a random dependence of the conductance on the magnetization direction due to spin-orbit scattering off a random configuration of impurities (Adam *et al.* 2006). This can be viewed as the extension of the bulk description to systems

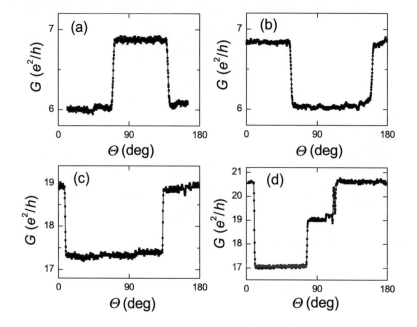

Fig. 18.12 Angular dependence of ballistic conductance of electrodeposited Co nanocontacts. (a)–(d), The angle Θ between the magnetic field and the sample plane changes from $0°$ to $180°$. Results for four different samples exhibiting different sign and magnitude of BAMR. Adapted from Sokolov *et al.* (2007b) by permission from Macmillan Publishers Ltd: Nature Nanotechnology, copyright 2007.

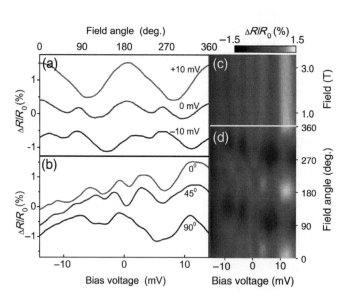

Fig. 18.13 Variations in $R = dV/dI$ at 4.2 K of an electromigrated Permalloy contact. (a) ΔR vs. field angle at different bias voltages. (b) Dependence of ΔR on V at different fixed angles of magnetic field. The curves in (a) and (b) are offset vertically. (c) Variations in $R(V)$ as a function of applied magnetic-field strength, with field directed along the current flow axis. R does not have significant dependence on the field magnitude. (d) $\Delta R(V)$ as a function of field orientation in an 800-mT magnetic field. Reprinted with permission from Bolotin *et al.* (2006). Copyright 2006 from the American Physical Society.

with relevant length scales smaller than the phase-coherence length, but still larger than the mean-free path between successive spin-orbit scattering events by impurities.

The discrepancy between the discrete AMR, observed by Viret *et al.* (2006) and Sokolov *et al.* (2007b), and the AMR observed by Bolotin *et al.* (2006) and Shi *et al.* (2007) has generated a certain level of controversy. For electromigrated Ni nanocontacts, Shi *et al.* also found abrupt changes in the angular dependence of the conductance at low temperatures, however, they

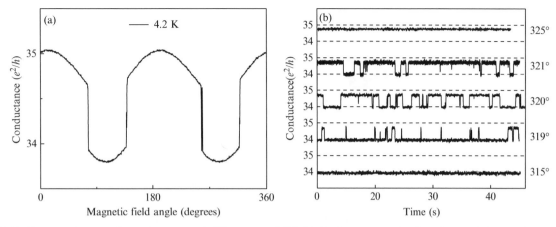

Fig. 18.14 Abrupt conductance changes in a nanoscale Ni contact at 4.2 K. (a) Conductance as a function of magnetic field angle, for a field magnitude of 800 mT. The field is rotated in the sample plane. (b) Conductance as a function of time at several fixed-field angles, for the same sample as in (a). At field angles in the vicinity of the conductance steps, in (a), two-level conductance switching is observed, which is a signature of atomic motion. Reprinted from Shi and Ralph (2007) by permission from Macmillan Publishers Ltd: Nature Nanotechnology, copyright 2007.

attributed this to a different mechanism (Shi and Ralph 2007; Shi *et al.* 2007). For field orientations close to the abrupt changes, Fig. 18.14(a), two-level fluctuations on the timescale of a few seconds were observed in time scans of the conductance shown in Fig. 18.14(b). The duty cycle of these fluctuations were very sensitive to the field orientation within the range of a few degrees, changing from being in the low-conductance state 0% of the time to being in the high-conductance state 100% of the time. The frequency of the fluctuations also appeared to increase with temperature from 4.2 K to 12 K. Based on these observations it was proposed that the abrupt angular variations in the conductance found in the electrodeposited Co nanocontacts (Sokolov *et al.* 2007b) may be due to atomic rearrangements that are sensitive to magnetic-field orientation and not due to the opening and closing of conductance channels as the result of intrinsic changes in the electronic structure.

In response, Sokolov *et al.* (2007a) pointed out that their measurements were performed on Co nanocontacts synthesized by a method different from the electromigrated Ni contacts of Fig. 18.14. Furthermore, no two-level fluctuations as a function of time were observed in the Co nanocontacts for field orientations at and around the abrupt changes in the conductance, even at room temperature. It was pointed out that electrochemically synthesized samples may have an advantage in terms of mechanical and surface stability due to being immersed in an electrolyte. If, in fact, the observed conductance steps were associated with the atomic motion induced at a given magnetic field angle, the magnitude of the field would likely affect the angle at which the conductance change occurs. However, no indications were found that the measurements were influenced by the magnitude of the saturating applied magnetic field, nor to an applied magnetic field gradient. These considerations, along with the fact that no BAMR effects were observed in electrodeposited Ni contacts (Sokolov *et al.* 2007a), indicate that the abrupt variations in the conductance in Fig. 18.14 may have a different origin as compared to those found by Sokolov *et al.* (2007a). More experiments are clearly needed to clarify the issue.

18.5 Tunnelling anisotropic magnetoresistance in broken contacts

In the previous section we discussed AMR occurring due to metallic conduction in ferromagnetic contacts. However, anisotropic conductance can also occur in the tunnelling regime of conductance when electrons quantum-mechanically tunnel from a ferromagnetic electrode through an insulator or vacuum. This phenomenon is known as tunnelling anisotropic magnetoresistance (TAMR). Just as for AMR in metal systems, TAMR originates from the anisotropic electronic structure of the ferromagnetic electrode(s) induced by SOC.

TAMR has recently been studied in a variety of planar tunnel junction heterostructures. A TAMR effect was predicted (Brey *et al.* 2004) and observed in tunnel junctions formed from semiconductor layered structures in which the ferromagnetic electrodes are formed from dilute magnetic semiconductor, Mn-doped GaAs (Gould *et al.* 2004). SOC plays an important role in this material so that the TAMR effect can be attributed to the significant anisotropy in the electronic structure linked to the magnetization direction along different crystal axes (Saito *et al.* 2005). Recently, very large TAMR effects were observed in these systems (Ruster *et al.* 2005; Gould *et al.* 2007). TAMR was also found in lithographically defined (Ga,Mn)As nanoconstrictions (Giddings *et al.* 2005). In the constriction the carrier density is depleted and the transport occurs via tunnelling, so the mechanism of AMR is not very different from that in the planar tunnel junctions, being related to the large magnetic anisotropy of (Ga,Mn)As.

Although SOC is much weaker in transition metals, a TAMR effect has also been predicted for planar tunnel junctions with transition-metal electrodes (Shick *et al.* 2006; Chantis *et al.* 2007). There have also been several recent experiments in which TAMR is observed in transition-metal systems. Although not described as such, TAMR was detected in scanning tunnelling spectroscopy measurements of thin Fe films on W(110) substrates (Bode *et al.* 2002). TAMR was demonstrated for Fe|GaAs|Au tunnel junctions with a single magnetic electrode (Fe) where the inplane magnetization of the Fe layer is varied, and the orientation of the high-resistance state could be controlled by the applied bias (Moser *et al.* 2007). A complex angular and bias dependence of the TAMR effect was also observed in magnetic tunnel junctions with CoFe electrodes, where the details were shown to be quite different depending on the barrier material used; either MgO or Al_2O_3 (Gao *et al.* 2007). Recently, TAMR in tunnel junctions with a Co/Pt multilayered electrode was observed, and it was found that the TAMR effect is much larger for a Pt-terminated electrode barrier interface, than for the Co-terminated interface (Park *et al.* 2008). TAMR was also recently reported in Co|AlO$_x$|Au tunnel junctions (Liu *et al.* 2008).

In addition to the relatively new research topic of TAMR in planar laterally extended systems, several studies of nanoscale magnetic tunnelling systems have also recently been reported. For example, a situation analogous to TAMR in planar tunnel junctions occurs in fully broken magnetic break junctions where electron transport occurs due to tunnelling through vacuum, and a couple of groups have extended their studies of AMR in magnetic contacts to the tunnelling regime.

In addition to measuring the AMR properties of Fe mechanical break junctions in the metallic contact regime, Viret *et al.* (2006) also reported AMR measurements in Fe junctions whose nominal resistance was consistent with the tunnelling regime. TAMR values reaching ∼100% were reported. There was, however, a large spread in the magnitude of the effect for various samples. It was proposed that this was due to the different geometry of atomic configurations of broken contacts, as indicated in Fig. 18.15. It was argued that for wide tunnelling gaps with flat surfaces, the dominant evanescent wavefunctions responsible for the tunnelling current would correspond to the *s* orbitals. However, for short tunnelling gaps with atomically sharper tips, *d*-electron tunnelling would also contribute significantly. Since *d*-electrons are strongly affected by SOC, the largest variation in the resistance should occur for the systems with smaller tunnelling gaps.

Measurements of TAMR in fully broken electromigrated permalloy break junctions were also reported by Bolotin *et al.* (2006). By measuring the differential resistance, $R(V) = dV/dI$, as a function of bias and magnetic-field direction at saturation, TAMR as large as 25% was observed. Similar to the metallic samples, which we discussed previously (see Fig. 18.13 in Section 18.4), the TAMR effect exhibited a complex angular dependence that was sensitive to changes in bias on a scale of several mV, as demonstrated in Fig. 18.16. However, the dependence on applied bias is much more pronounced than for the metallic contacts. In particular, there was very strong angular dependence for a few narrow ranges of applied bias superimposed on rather

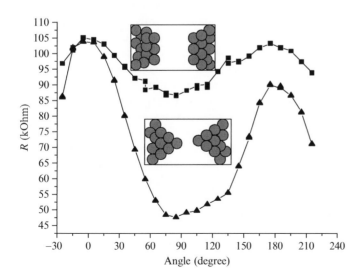

Fig. 18.15 Measured resistance variation in Fe break junctions in the tunnelling regime when a 2.5-T field is rotated with respect to the junction. The insets are schematic atomic arrangements that could explain the different magnitudes of the effect measured for different junctions. Reproduced with permission from Viret *et al.* (2006). Copyright 2006 Springer.

Fig. 18.16 Variation in differential resistance, $R(V)$, at 4.2 K of a fully broken electromigrated Permalloy break junction in the tunnelling regime. (a) ΔR vs. field angle at different bias voltages. (b) Dependence of ΔR on V at different fixed angles of magnetic field. The curves in (a) and (b) are offset vertically. (c) Variations in $R(V)$ as function of applied magnetic field strength, with field directed along the current flow axis. R does not have significant dependence on the field magnitude, indicating that the effects arise from magnetomechanical artifacts. (d) $\Delta R(V)$ as a function of field orientation in an 800-mT magnetic field. Reprinted with permission from Bolotin *et al.* (2006). Copyright 2006 from the American Physical Society.

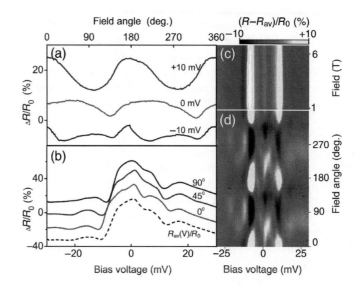

smooth (by comparison) variations. This is exemplified by the strong light-dark patterns around zero bias, as well as around both ±10 mV in the differential resistance plot in Fig. 18.16(d).

The origin of this strong sensitivity to the bias voltage and the strong dependence of conductance on the magnetization orientation was revealed in the theoretical calculations and was attributed to *tip resonant states* by Burton *et al.* (2007). In these calculations the break junction consisted of two free-standing semi-infinite Ni wire electrodes, each terminated by an apex atom and separated by a vacuum region, as shown in the top panel of Fig. 18.17. Transport properties were calculated using the Landauer–Büttiker formalism of quantum transport, as described in Section 18.2, including SOC.

The bottom panel of Fig. 18.17 shows the transmission spectrum for two orientations of the magnetization for states near the Fermi level. In addition to broad continuous bands of transmission, narrow resonant features also appear in the spectrum. These correspond to states localized in the electrodes near the junction break, which are known to show up in atomically sharp transition metal tips (Vázquez de Parga *et al.* 1998). The energy and broadening of these states are strongly affected by the magnetization orientation due to SOC. For example, the resonance seen in the bottom panel of Fig. 18.17 between 10 and 20 meV below the Fermi level changes from a narrow peak for $\theta = 0°$ to a broad resonance for $\theta = 90°$. An analysis of the symmetry of this state reveals that the spin-orbit matrix element connecting this tip state, which is localized on the apex atom, to the continuum states of the electrode varies as $\sin \theta$.

Using a simplified description in the spirit of the Bardeen approximation (Bardeen 1961) to compute the differential conductance, $G(V) = dI/dV$, calculations revealed a complex angular variation of the TAMR effect that was strongly dependent on applied bias (see Fig. 18.18(b)). This is especially true for biases corresponding to the tip resonant states. The strong light-

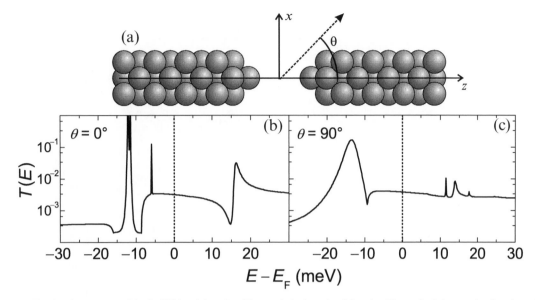

Fig. 18.17 (a) The atomic structure of the 5–4 Ni break junction. The z-axis is the axis of the wire. The angle θ denotes the direction of the spin magnetic moments in the xz-plane. (b, c) The transmission as a function of energy, $T(E)$, for $\theta = 0°$ and $\theta = 90°$. The vertical dashed line indicates the Fermi level. Adapted with permission from Burton *et al.* (2007). Copyright 2007 from the American Physical Society.

dark pattern in Fig. 18.18(c) corresponds to the resonance between -20 and -10 meV in the bottom panel of Fig. 18.17. This bears a resemblance to the experimental data of Fig. 18.16(d), indicating that the effect observed by Bolotin *et al.* (2006) may arise due to tip resonant states, which are an intrinsic property of the local electronic structure of the magnetic break junction. Calculations on Co break junctions, as well as break junctions in Ni wires with larger cross-section, also displayed tip resonant states that were sensitive to the orientation of the magnetization and that led to a similar bias-dependent TAMR effect, indicating this as a general phenomenon for magnetic break junctions.

18.6   Conclusions and outlook

Reducing the dimensions of magnetic structures leads to new magnetoresistive phenomena on the nano- and subnanometer scales. Electronic, magnetic and transport properties of such nanoscale objects are very different as compared to those in the respective bulk systems due to reduced dimensionality, complex surfaces and interfaces, and quantum effects. As was discussed in this chapter, the known magnetoresistive phenomena, such as anisotropic magnetoresistance (AMR) and domain-wall magnetoresistance (DWMR), reveal themselves in qualitatively different ways at the nanoscale.

Regarding the latter, there are indications that magnetic domain walls at the nanoscale may have a significantly reduced width as compared to bulk materials. Theory predicts that this may lead to enhanced domain-wall magnetoresistance of a few hundred per cent provided that the domain wall width is

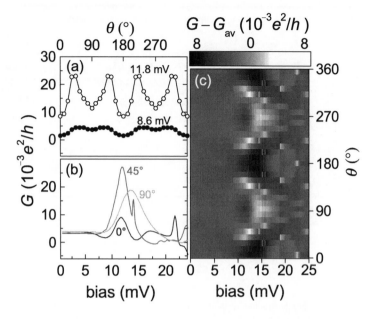

Fig. 18.18 Variation of differential conductance $G(V) = dI/dV$ at 4.2 K for the 5–4 break junction. (a) G verses θ for different bias voltages. (b) Dependence of G on bias for different θ. (c) Deviation of G from G_{av}, the conductance averaged over θ, as a function of V and θ. Adapted with permission from Burton *et al.* (2007). Copyright 2007 from the American Physical Society.

reduced down to the atomic scale. However, up to now there are no convincing experiments showing large magnetoresistance effects fully associated with the formation of a domain wall at the magnetic nanocontact. Experimental measurements involving domain-wall resistance and magnetoresistive effects related to reorientation of magnetic moments within the sample inevitably involve anisotropy effects, which may play a decisive role in these experiments (Doudin and Viret 2008). Given the extreme difficulty of the atomic-scale magnetic and structural characterization of such nanocontacts, a new approach is probably required to provide a more definitive experiment to separate the two contributions. This is especially interesting in view of the new theoretical predictions suggesting reduced magnetic moments in atomic-size domain walls (Burton *et al.* 2006) and new magnetoresistive phenomena such as ballistic anisotropic magnetoresistance (BAMR).

BAMR is a quantized change in the electrical conductance of a magnetic metal wire or a constriction that can take place in the ballistic transport regime when changing the direction of the saturation magnetic field (Velev *et al.* 2005). BAMR is a very interesting development, and recently, two groups (Viret *et al.* 2006; Sokolov *et al.* 2007b) observed experimentally a stepwise variation of ballistic conductance in a saturation magnetic field, a feature typical of BAMR. Further experiments are required to elucidate the origin of this behavior, especially in view of the possible atomic motion involved (Shi and Ralph 2007; Shi *et al.* 2007). One of the possibilities is to employ ferromagnetic materials, such as magnetic semiconductors (Ohno 1999), where a lower carrier concentration compared to metals would make it possible using larger cross-sections to observe spin-dependent conductance quantization and BAMR. Another approach would be producing a constriction by a gate voltage (van Wees *et al.* 1988; Wharam *et al.*

1988) in a spin-dependent two-dimensional electron gas (Brinkman *et al.* 2007).

An important issue in the understanding of quantum transport at the nanoscale is related to spin-dependent conductance quantization. Most theoretical calculations based on a realistic description of the electronic band structure do not predict spin-dependent conductance quantization in multiples of e^2/h, nor do they demonstrate full spin polarization in such contacts. There are, however, a number of experimental reports that do show these effects (Elhoussine *et al.* 2002; Yang *et al.* 2002; Rodrigues *et al.* 2003). Whether this behavior is an artifact of the experiments or the indication of some other factors involved, e.g. the influence of adsorbates (Untiedt *et al.* 2004) remains unclear and require further studies. The development of electronic structure methods that include electron–electron correlation effects, given the low dimensionality of atomic-size contacts, may be important for better understanding of their properties.

Throughout this chapter we have discussed how the magnetization affects the flow of electrical current in ferromagnetic materials. However, the opposite effect is also possible, i.e. the magnetization of the system being altered by electrical current. For example, electrons traversing a magnetic domain-wall experience a torque on their spin magnetic moments due to the exchange interaction that will attempt to align them with the local magnetization direction. However, Newton's third law tells us that there is an equal and opposite torque experienced by the net magnetization due to the spins of the conducting electrons, leading to current-induced motion of the domain wall (Berger 1984). These kinds of effects, which can be interpreted as a transfer of spin angular momentum from the conduction electrons to the magnetization, can be used to control magnetization without the application of magnetic fields (Slonczewski 1996). In particular the generation and sensing of magnetization dynamics in nanoscale magnetic systems (Kiselev *et al.* 2003; Krause *et al.* 2007) may offer a new route to study magnetoresistive phenomena that could help resolve some of the open questions. Further information on the topic of spin-dynamics in nanoscale magnetic conductors can be found in the chapter by Sanvito and Stamenova in this volume.

Tunnelling anisotropic magnetoresistance (TAMR) has already revealed large effects in magnetic semiconducting systems and showed an exciting possibility to produce a sizable effect in tunnel junctions with a single magnetic electrode (Gould *et al.* 2004). New features of TAMR at the nanoscale resulting from the anisotropy of resonant tunnelling driven by surface states open the possibility of studying these states in magnetic tips. This can be performed using an unconventional spin-polarized scanning tunnelling microscopy setup where, contrary to the usual convention, a magnetic tip is used to scan an already well-characterized *non-magnetic* surface, like the Cu(111) surface (Bode 2003). In this case any observed TAMR effect due the rotation of an applied magnetic field would arise solely from the variation in the electronic structure of the tip, which is the only magnetic element in the system. Careful analysis of the angular dependence of the corresponding *I–V* spectra would then reveal tip-related electronic features, including the possibility to resolve tip-resonant states.

We would also like to mention another AMR effect on the nanoscale, namely the Coulomb-blockade anisotropic magnetoresistance (CBAMR) effect in ferromagnetic single-electron transistor (SET) systems (Fernández-Rossier *et al.* 2006; Wunderlich *et al.* 2006). Here, the chemical potential and single-electron charging properties can be affected by the magnetic anisotropy of either ferromagnetic electrodes or a ferromagnetic island, or both, which opens the possibility for gate-controlled AMR effects.

Further developments in the field of magnetoresistance at the nanoscale involve the possibility of producing a "barrier" between magnetic electrodes at the nanoscale that may lead to interesting magnetoresistive behavior. For example, it was demonstrated that in magnetic tunnel junctions of small cross-section tunnelling magnetoresistance can change sign (Tsymbal *et al.* 2003b). This phenomenon occurs due to resonant tunnelling via localized states in the barrier. Depending on the position of a localized state the TMR may be positive or negative. A similar kind of resonant tunnelling effect occurs in a system comprising a carbon nanotube connecting two ferromagnetic electrodes where quantum-well states occur in the carbon nanotube due to a pure contact to the electrodes (Sahoo *et al.* 2005). These experiments found a pronounced gate-field-controlled magnetoresistance response due to the localized electronic states, the energy of which can be changed by a gate voltage. These experiments open a very interesting direction to fabricate a spin transistor at the nanoscale.

Ultimately, a single molecule may serve as a "barrier" between magnetic electrodes leading to the new field of molecular spin electronics (Rocha *et al.* 2005). New physics at the atomic scale has been revealed by demonstrating Kondo-assisted tunnelling via C-60 molecules in contact with ferromagnetic nickel electrodes (Pasupathy *et al.* 2004). Molecules themselves may be magnetic, which opens up another route for molecular spin electronics (Bogani and Wernsdorfer 2008). These research efforts demonstrate interesting opportunities for employing magnetic nanocontacts to observe novel magnetoresistive phenomena at the atomic scale.

Acknowledgments

We thank our colleagues whose research work and highly valuable insights provided conclusions presented in this chapter: Kirill Belashchenko, Bernard Doudin, Bill Egelhoff, Sitaram Jaswal, Arti Kashyap, Oleg Mryasov, Dan Ralph, Jody Redepenning, Renat Sabirianov, Andrei Sokolov, Anatoly Vedyayev, Julian Velev, Michel Viret, Chunjuan Zhang and Mikhail Zhuravlev. This work was supported by the National Science Foundation and the Nanoelectronics Research Initiative through the Materials Research Science and Engineering Center at the University of Nebraska. J. D. B. would also like to acknowledge support from the Presidential Graduate Fellowship through the University of Nebraska Foundation.

References

Adam, S., Kindermann, M., Rahav, S., Brouwer, P.W. *Phys. Rev. B* **73**, 212408 (2006).

Agrait, N., Yeyati, A.L., van Ruitenbeek, J.M. *Physics Reports* **377**, 81 (2003).

Bader, S.D. *Rev. Mod. Phys.* **78**, 1 (2006).

Bagrets, A., Papanikolaou, N., Mertig, I. *Phys. Rev. B* **70**, 064410 (2004).

Bagrets, A., Papanikolaou, N., Mertig, I. *Phys. Rev. B* **75**, 235448 (2007).

Baibich, M.N., Broto, J.M., Fert, A., Van Dau, F.N., Petroff, F., Eitenne, P., Creuzet, G., Friederich, A., Chazelas, J. *Phys. Rev. Lett.* **61**, 2472 (1988).

Bardeen, J. *Phys. Rev. Lett.* **6**, 57 (1961).

Barthelemy, A., Fert, A., Petroff, F. Giant magnetoresistance in magnetic multilayers. In Buschow, K.H.J. (ed.), *Handbook on Magnetic Materials*, Volume 12 (Elsevier, Amsterdam, 1999).

Bass, J., Pratt, W.P. *J. Physi.: Condens. Matter* **19**, 183201 (2007).

Berger, L. *J. Appl. Phys.* **49**, 2156 (1978).

Berger, L. *J. Appl. Phys.* **55**, 1954 (1984).

Binasch, G., Grünberg, P., Saurenbach, F., Zinn, W. *Phys. Rev. B* **39**, 4828 (1989).

Bitter, F. *Phys. Rev.* **41**, 507 (1932).

Bloch, F.Z. *Phys.* **74**, 295 (1932).

Bode, M. *Rep. Prog. Phys.* **66**, 523 (2003).

Bode, M., Heinze, S., Kubetzka, A., Pietzsch, O., Nie, X., Bihlmayer, G., Blügel, S., Wiesendanger, R. *Phys. Rev. Lett.* **89**, 237205 (2002).

Bogani, L., Wernsdorfer, W. *Nature Mater.* **7**, 179 (2008).

Bolotin, K.I., Kuemmeth, F., Ralph, D.C. *Phys. Rev. Lett.* **97**, 127202 (2006).

Brandbyge, M., Schiøtz, J., Sørensen, M.R., Stoltze, P., Jacobsen, K.W., Nørskov, J.K., Olesen, L., Laegsgaard, E., Stensgaard, I., Besenbacher, F. *Phys. Rev. B* **52**, 8499 (1995).

Brey, L., Tejedor, C., Fernández-Rossier, J. *Appl. Phys. Lett.* **85**, 1996 (2004).

Brinkman, A., Huijben, M., van Zalk, M., Huijben, J., Zeitler, U., Maan, J.C., van der Wiel, W.G., Rijnders, G., Blank, D.H.A., Hilgenkamp, H. *Nature Mater.* **6**, 493 (2007).

Brumfiel, G. *Nature* **426**, 110 (2003).

Bruno, P. *Phys. Rev. Lett.* **83**, 2425 (1999).

Bulaevskiĭ, L.N., Ginzburg, V.L. *Sov. Phys. JETP* **18**, 530 (1964).

Burton, J.D., Kashyap, A., Zhuravlev, M.Y., Skomski, R., Tsymbal, E.Y., Jaswal, S.S., Mryasov, O.N., Chantrell, R.W. *Appl. Phys. Lett.* **85**, 251 (2004).

Burton, J.D., Sabirianov, R.F., Jaswal, S.S., Tsymbal, E.Y., Mryasov, O.N. *Phys. Rev. Lett.* **97**, 077204 (2006).

Burton, J.D., Sabirianov, R.F., Velev, J.P., Mryasov, O.N., Tsymbal, E.Y. *Phys. Rev. B* **76**, 144430 (2007).

Büttiker, M. *IBM J. Res. Dev.* **32**, 317 (1988).

Cabrera, G.G., Falicov, L.M. *Phys. Status Solidi. B* **61**, 539 (1974).

Calvo, M.R., Caturla, M.J., Jacob, D., Untiedt, C.A.U.C., Palacios, J.J.A.P.J. *J. IEEE Trans. Nanotechnol.* **7**, 165 (2008).

Chantis, A.N., Belashchenko, K.D., Tsymbal, E.Y., van Schilfgaarde, M. *Phys. Rev. Lett.* **98**, 046601 (2007).

Chappert, C., Fert, A., Van Dau, F.N. *Nature Mater.* **6**, 813 (2007).

Chopra, H.D., Hua, S.Z. *Phys. Rev. B* **66**, 020403 (2002).

Chung, S.H., Muñoz, M., García, N., Egelhoff, W.F., Gomez, R.D. *Phys. Rev. Lett.* **89**, 287203 (2002).

Cuevas, J.C., Yeyati, A.L., Martín-Rodero, A. *Phys. Rev. Lett.* **80**, 1066 (1998).

Datta, S. *Electronic Transport in Mesoscopic Systems* (Cambridge, Cambridge University Press, 1995).

Donahue, M.J., Porter, D.G. Object Oriented MicroMagnetic Framework (OOMMF), User's Guide Version 1.0, Interagency Report NISTIR 6376. National Institute of Standards and Technology, Gaithersburg, MD; http://math.nist.gov/oommf (1999).

Doudin, B., Viret, M. *J. Phys.: Condens. Matter* **20**, 083201 (2008).

Egelhoff, J.W. F., Gan, L., Ettedgui, H., Kadmon, Y., Powell, C. J., Chen, P.J., Shapiro, A.J., McMichael, R.D., Mallett, J.J., Moffat, T.P., Stiles, M.D., Svedberg, E.B. *J. Appl. Phys.* **95**, 7554 (2004).

Elhoussine, F., Matefi-Tempfli, S., Encinas, A., Piraux, L. *Appl. Phys. Lett.* **81**, 1681 (2002).

Fernández-Rossier, J., Aguado, R., Brey, L. *Phys. Status Solidi. QC.* **3**, 4231 (2006).

Fernández-Rossier, J., Jacob, D., Untiedt, C., Palacios, J. *J. Phys. Rev. B* **72**, 224418 (2005).

Gao, L., Jiang, X., Yang, S.-H., Burton, J.D., Tsymbal, E.Y., Parkin, S.S.P. *Phys. Rev. Lett.* **99**, 226602 (2007).

García, N., Muñoz, M., Zhao, Y.W. *Phys. Rev. Lett.* **82**, 2923 (1999).

García, N., Qiang, G.G., Saveliev, I.G. *Appl. Phys. Lett.* **80**, 1785 (2002).

García, N., Wang, H., Cheng, H., Nikolic, N.D. *IEEE Trans. Magn.* **39**, 2776 (2003).

Giddings, A.D., Khalid, M.N., Jungwirth, T., Wunderlich, J., Yasin, S., Campion, R.P., Edmonds, K.W., Sinova, J., Ito, K., Wang, K.Y., Williams, D., Gallagher, B.L., Foxon, C.T. *Phys. Rev. Lett.* **94**, 127202 (2005).

Gimzewski, J.K., Möller, R. *Phys. Rev. B* **36**, 1284 (1987).

Glazman, L.I., Lesovik, G.B., Khmelnitskii, D.E., Shekhter, R.I. *JETP Lett.* **48**, 238 (1988).

Gould, C., Ruster, C., Jungwirth, T., Girgis, E., Schott, G.M., Giraud, R., Brunner, K., Schmidt, G., Molenkamp, L.W. *Phys. Rev. Lett.* **93**, 117203 (2004).

Gould, C., Schmidt, G., Molenkamp, L.W. *IEEE Trans. Electron Devices* **54**, 977 (2007).

Hartmann, U. (ed.) *Magnetic Multilayers and Giant Magnetoresistance: Fundamentals and Industrial Applications* (Springer, New York, 2000).

Hua, S.Z., Chopra, H.D. *Phys. Rev. B* **67**, 060401 (2003).

Imamura, H., Kobayashi, N., Takahashi, S., Maekawa, S. *Phys. Rev. Lett.* **84**, 1003 (2000).

Jacob, D., Fernández-Rossier, J., Palacios, J.J. *Phys. Rev. B* **71**, 220403 (2005).

Jacob, D., Palacios, J.J. *Phys. Rev. B* **73**, 075429 (2006).

Jullière, M. *Phys. Lett. A* **54**, 225 (1975).

Kazantseva, N., Wieser, R., Nowak, U. *Phys. Rev. Lett.* **94**, 037206 (2005).

Keane, Z.K., Yu, L.H., Natelson, D. *Appl. Phys. Lett.* **88**, 062514 (2006).

Kent, A.D., Yu, J., Rüdiger, U., Parkin, S.S.P. *J. Phys.: Condens. Matter* **13**, R461 (2001).

Kiselev, S.I., Sankey, J.C., Krivorotov, I.N., Emley, N.C., Schoelkopf, R.J., Buhrman, R.A., Ralph, D.C. *Nature* **425**, 380 (2003).

Kittel, C. *Introduction to Solid State Physics* (New York, Wiley, 1996).

Komori, F., Nakatsuji, K. *Mater. Sci. Eng. B-Solid State Mater. Adv. Technol.* **84**, 102 (2001).

Krause, S., Berbil-Bautista, L., Herzog, G., Bode, M., Wiesendanger, R. *Science* **317**, 1537 (2007).

Kudrnovský, J., Drchal, V., Blaas, C., Weinberger, P., Turek, I., Bruno, P. *Phys. Rev. B* **62**, 15084 (2000).

Landau, L.D., Lifshitz, E. *Phys. Z. Sowjetunion* **8**, 153 (1935).

Landauer, R. *IBM J. Res. Dev.* **32**, 306 (1988).

Levy, P.M. Giant magnetoresistance in magnetic layered and granular materials. In *Solid State Physics - Advances in Research and Applications, Vol 47* (Academic Press Inc, San Diego, 1994).

Levy, P.M., Zhang, S. *Phys. Rev. Lett.* **79**, 5110 (1997).

Liu, R.S., Michalak, L., Canali, C.M., Samuelson, L., Pettersson, H. *Nano Lett.* **8**, 848 (2008).

Ludoph, B., van der Post, N., Bratus', E.N., Bezuglyi, E.V., Shumeiko, V.S., Wendin, G. van Ruitenbeek, J.M. *Phys. Rev. B* **61**, 8561 (2000).

Marrows, C.H. *Adv. Phys.* **54**, 585 (2005).

McGuire, T., Potter, R. *IEEE Trans. Magn.* **11**, 1018 (1975).

Montero, M.I., Dumas, R.K., Liu, G., Viret, M., Stoll, O.M., Macedo, W.A.A., Schuller, I.K. *Phys. Rev. B* **70**, 184418 (2004).

Moodera, J.S., Kinder, L.R., Wong, T.M., Meservey, R. *Phys. Rev. Lett.* **74**, 3273 (1995).

Moodera, J.S., Mathon, G.J. *Magn. Magn. Mater.* **200**, 248 (1999).

Morpurgo, A.F., Marcus, C.M., Robinson, D.B. *Appl. Phys. Lett.* **74**, 2084 (1999).

Moser, J., Matos-Abiague, A., Schuh, D., Wegscheider, W., Fabian, J., Weiss, D. *Phys. Rev. Lett.* **99**, 056601 (2007).

Mott, N.F., Jones, H. *The Theory of the Properties of Metals and Alloys* (Oxford University Press, Oxford, 1936).

Ohnishi, H., Kondo, Y., Takayanagi, K. *Nature* **395**, 780 (1998).

Ohno, H.J. *Magn. Magn. Mater.* **200**, 110 (1999).

Ono, T., Ooka, Y., Miyajima, H., Otani, Y. *Appl. Phys. Lett.* **75**, 1622 (1999).

Ozatay, O., Chalsani, P., Emley, N.C., Krivorotov, I.N., Buhrman, R.A. *J. Appl. Phys.* **95**, 7315 (2004).

Park, B.G., Wunderlich, J., Williams, D.A., Joo, S.J., Jung, K.Y., Shin, K.H., Olejnik, K., Shick, A.B., Jungwirth, T. *Phys. Rev. Lett.* **100**, 087204 (2008).

Park, H., Lim, A.K.L., Alivisatos, A.P., Park, J., McEuen, P.L. *Appl. Phys. Lett.* **75**, 301 (1999).

Parkin, S.S.P., Kaiser, C., Panchula, A., Rice, P.M., Hughes, B., Samant, M., Yang, S.-H. *Nature Mater.* **3**, 862 (2004).

Pasupathy, A.N., Bialczak, R.C., Martinek, J., Grose, J.E., Donev, L.A.K., McEuen, P.L., Ralph, D.C. *Science* **306**, 86 (2004).

Pippard, A.B. *Magnetoresistance in Metals* (Cambridge, Cambridge University Press, 1989).

Potter, R.I. *Phys. Rev. B* **10**, 4626 (1974).

Pratzer, M., Elmers, H.J., Bode, M., Pietzsch, O., Kubetzka, A., Wiesendanger, R. *Phys. Rev. Lett.* **87**, 127201 (2001).

Rocha, A.R., Archer, T., Sanvito, S. *Phys. Rev. B* **76**, 054435 (2007).

Rocha, A.R., Garcia-Suarez, V.M., Bailey, S.W., Lambert, C.J., Ferrer, J., Sanvito, S. *Nature Mater.* **4**, 335 (2005).

Rodrigues, V., Bettini, J., Silva, P.C., Ugarte, D. *Phys. Rev. Lett.* **91**, 096801 (2003).

Ruster, C., Gould, C., Jungwirth, T., Sinova, J., Schott, G.M., Giraud, R., Brunner, K., Schmidt, G., Molenkamp, L.W. *Phys. Rev. Lett.* **94**, 027203 (2005).

Sabirianov, R.F., Solanki, A.K., Burton, J.D., Jaswal, S.S., Tsymbal, E.Y. *Phys. Rev. B* **72**, 054443 (2005).

Sahoo, S., Kontos, T., Furer, J., Hoffmann, C., Graber, M., Cottet, A., Schönenberger, C. *Nature Phys.* **1**, 99 (2005).

Saito, H., Yuasa, S., Ando, K. *Phys. Rev. Lett.* **95**, 086604 (2005).

Scheer, E., Joyez, P., Esteve, D., Urbina, C., Devoret, M.H. *Phys. Rev. Lett.* **78**, 3535 (1997).

Shi, S.F., Bolotin, K.I., Kuemmeth, F., Ralph, D.C. *Phys. Rev. B* **76**, 184438 (2007).

Shi, S.F., Ralph, D.C. *Nature Nano.* **2**, 522 (2007).

Shick, A.B., Maca, F., Masek, J., Jungwirth, T. *Phys. Rev. B* **73**, 024418 (2006).

Sixtus, K.J., Tonks, L. *Phys. Rev.* **37**, 930 (1931).

Slonczewski, J.C. *J. Magn. Magn. Mater.* **159**, L1 (1996).

Smit, J. *Physica* **17**, 612 (1951).

Smogunov, A., Corso, A.D., Tosatti, E. *Phys. Rev. B* **73**, 075418 (2006).

Sokolov, A., Zhang, C., Tsymbal, E.Y., Redepenning, J., Doudin, B. *Nature Nano.* **2**, 522 (2007a).

Sokolov, A., Zhang, C., Tsymbal, E.Y., Redepenning, J., Doudin, B. *Nature Nano.* **2**, 171 (2007b).

Solanki, A.K., Sabiryanov, R.F., Tsymbal, E.Y., Jaswal, S.S. *J. Magn. Magn. Mater.* **272–76**, 1730 (2004).

Tagirov, L.R., Vodopyanov, B.P., Efetov, K.B. *Phys. Rev. B* **65**, 214419 (2002).

Tatara, G., Zhao, Y.W., Muñoz, M., García, N. *Phys. Rev. Lett.* **83**, 2030 (1999).

Thomson, W. *Proc. Roy. Soc. London* **8**, 546 (1857).

Todorov, T.N., Sutton, A.P. *Phys. Rev. Lett.* **70**, 2138 (1993).

Tsymbal, E.Y. Theory of spin-dependent tunneling: Role of evanescent and resonant states. In Krohnmüller, H. and Parkin, S.S.P. (eds) *Handbook of Magnetism and Advanced Magnetic Materials* (Wiley, Chichester, 2006).

Tsymbal, E.Y., Mryasov, O.N., LeClair, P.R. *J. Phys.: Condens. Matter* **15**, R109 (2003a).

Tsymbal, E.Y., Pettifor, D.G. Perspectives of giant magnetoresistance. In *Solid State Physics* (Academic Press Inc, San Diego, 2001).

Tsymbal, E.Y., Sokolov, A., Sabirianov, I.F., Doudin, B. *Phys. Rev. Lett.* **90**, 186602 (2003b).

Turzhevskii, S.A., Likhtenshtein, A.I., Katsnel'son, M.I. *Sov. Phys. - Solid State* **32**, 1138 (1990).

Untiedt, C., Dekker, D.M.T., Djukic, D., van Ruitenbeek, J.M. *Phys. Rev. B* **69**, 081401 (2004).

van Hoof, J.B.A.N., Schep, K.M., Brataas, A., Bauer, G.E.W., Kelly, P.J. *Phys. Rev. B* **59**, 138 (1999).

van Ruitenbeek, J.M., Alvarez, A., Pineyro, I., Grahmann, C., Joyez, P., Devoret, M. H., Esteve, D., Urbina, C. *Rev. Sci. Instrum.* **67**, 108 (1996).

van Wees, B.J., van Houten, H., Beenakker, C.W.J., Williamson, J.G., Kouwenhoven, L.P., van der Marel, D., Foxon, C.T. *Phys. Rev. Lett.* **60**, 848 (1988).

Vázquez de Parga, A.L., Hernán, O.S., Miranda, R., Levy Yeyati, A., Mingo, N., Martín-Rodero, A., Flores, F. *Phys. Rev. Lett.* **80**, 357 (1998).

Velev, J., Butler, W.H. *Phys. Rev. B* **69**, 094425 (2004).

Velev, J., Sabirianov, R.F., Jaswal, S.S., Tsymbal, E.Y. *Phys. Rev. Lett.* **94**, 127203 (2005).

Versluijs, J.J., Bari, M.A., Coey, J.M.D. *Phys. Rev. Lett.* **87**, 026601 (2001).

Viret, M., Berger, S., Gabureac, M., Ott, F., Olligs, D., Petej, I., Gregg, J.F., Fermon, C., Francinet, G., Goff, G.L. *Phys. Rev. B* **66**, 220401 (2002).

Viret, M., Gabureac, M., Ott, F., Fermon, C., Barreteau, C., Autes, G., Guirado-Lopez, R. *Eur. Phys. J. B* **51**, 1 (2006).

Viret, M., Vignoles, D., Cole, D., Coey, J.M.D., Allen, W., Daniel, D.S., Gregg, J.F. *Phys. Rev. B* **53**, 8464 (1996).

Weiss, P. *J. Phys. Radium* **6**, 661 (1907).

Wharam, D.A., Thornton, T.J., Newbury, R., Pepper, M., Ahmed, H., Frost, J.E.F., Hasko, D.G., Peacock, D.C., Ritchie, D.A., Jones, G.A.C. *J. Phys. C: Solid State Phys.* **21**, L209 (1988).

Wierzbowska, M., Delin, A., Tosatti, E. *Phys. Rev. B* **72**, 035439 (2005).

Wunderlich, J., Jungwirth, T., Kaestner, B., Irvine, A.C., Shick, A.B., Stone, N., Wang, K.Y., Rana, U., Giddings, A.D., Foxon, C.T., Campion, R.P., Williams, D.A., Gallagher, B.L. *Phys. Rev. Lett.* **97**, 077201 (2006).

Yang, C.S., Thiltges, J., Doudin, B., Mark, J. *J. Phys.: Condens. Matter* **14**, L765 (2002).

Yang, C.S., Zhang, C., Redepenning, J., Doudin, B. *Appl. Phys. Lett.* **84**, 2865 (2004).

Yang, C.S., Zhang, C., Redepenning, J., Doudin, B. *J. Magn. Magn. Mater.* **286**, 186 (2005).

Yanson, A.I., Bollinger, G.R., van den Brom, H.E., Agrait, N., van Ruitenbeek, J.M. *Nature* **395**, 783 (1998).

Yavorsky, B.Y., Mertig, I., Perlov, A.Y., Yaresko, A.N., Antonov, V.N. *Phys. Rev. B* **66**, 174422 (2002).

Yuasa, S., Nagahama, T., Fukushima, A., Suzuki, Y., Ando, K. *Nature Mater.* **3**, 868 (2004).

Zhang, X.G., Butler, W.H. *J. Phys.: Condens. Matter* **15**, R1603 (2003).

Zhirnov, V.A. *Sov. Phys. JETP* **8**, 822 (1959).

Zhuravlev, M.Y., Tsymbal, E.Y., Jaswal, S.S., Vedyayev, A.V., Dieny, B. *Appl. Phys. Lett.* **83**, 3534 (2003).

Novel superconducting states in nanoscale superconductors

19

A. Kanda, Y. Ootuka, K. Kadowaki, and F.M. Peeters

19.1 Introduction

19.1.1 Characteristic lengths in superconductors

In this chapter, we deal with superconducting states in small samples. Generally, as the size of a material is decreased to a certain characteristic length scale, the character of the material drastically changes. For superconductors, there are several characteristic lengths, including the coherence length $\xi(T)$, the penetration depth $\lambda(T)$, and the size r_C where the single-particle level spacing of a particle equals the superconducting energy gap of the bulk material. For r_C, which is related to the critical size for superconductivity (Anderson 1959), the superconducting energy gap of individual nanoscale particles has been studied experimentally using single-electron tunnelling spectroscopy (Black 1996; Von Delft 2001). The characteristic length scales relevant to the present chapter are $\xi(T)$ and $\lambda(T)$, which are much larger than r_C. They have the following meanings:

(i) The coherence length $\xi(T)$ is the characteristic length over which the Cooper-pair density, $n(r)$, or the superconducting energy gap, $\Delta(r)$, can vary appreciably without undue energy increase. $\xi(T)$ was introduced by Ginzburg and Landau (1950). Note that this Ginzburg–Landau (GL) coherence length is different from the (temperature-independent) BCS (Pippard) coherence length $\xi_0 = a\hbar v_F/k_B T_C$ (Pippard 1953), which represents the size of the Cooper pair in pure materials. Here, a is a numerical constant that is estimated to be 0.18 in the BCS theory (Bardeen *et al.* 1957), v_F the Fermi velocity, and T_C the superconducting transition temperature. For Al, the BCS coherence length is about $1.6\,\mu$m, while the GL coherence length depends on the mean-free path and takes a value of at most $\xi(0) = 0.2\,\mu$m in a film deposited on a substrate.

(ii) The penetration depth, $\lambda(T)$, characterizes the exponential decay of the magnetic field in superconductors. When the sample size is smaller than $\lambda(T)$, the magnetic field penetrates into the whole sample, resulting in less energy for the (incomplete) diamagnetic Meissner effect and an increase in the critical magnetic field.

From the GL theory, which is valid for temperatures close to T_C, both $\xi(T)$ and $\lambda(T)$ diverge as $(1 - T/T_C)^{-1/2}$ when the temperature approaches T_C:

$$\xi(T) = \xi(0)\frac{1}{\sqrt{1 - T/T_C}},$$

$$\lambda(T) = \lambda(0)\frac{1}{\sqrt{1 - T/T_C}}.$$

Thus, the GL parameter $\kappa = \lambda/\xi$ is independent of temperature. At low temperatures, a two-fluid approximation (Tinkham 1996; Bertrand 2005) gives the following expressions:

$$\xi(T) = \xi(0)\frac{\sqrt{1 + (T/T_C)^2}}{\sqrt{1 - (T/T_C)^2}},$$

$$\lambda(T) = \lambda'(0)\frac{1}{\sqrt{1 - (T/T_C)^4}}.$$

We note that this leads to a temperature-dependent GL parameter.

Depending on the sign of the energy of a normal metal (N)-superconductor (S) interface, response of a superconductor to magnetic field is divided into two categories: superconductors with $\kappa = \lambda/\xi < 1/\sqrt{2}$ (type-I superconductors) show perfect diamagnetism and exhibit a sharp transition to the normal state at the critical magnetic field H_C (see Fig. 19.1). On the other hand, in superconductors with $\kappa \geq 1/\sqrt{2}$ (type-II superconductors) there exists a mixed state (a vortex state) under an applied magnetic field between the first and second critical fields (H_{C1} and H_{C2}), in which the sample is perforated by quantized units of magnetic flux with supercurrent circulating around the flux. Each magnetic flux surrounded by supercurrent is called a vortex. As the core of the vortex has a size of $\xi(T)$ and the supercurrent flows within the distance $\lambda(T)$ from the core, one can expect a strong influence of the sample size and shape on the vortex states in mesoscopic superconductors that have sizes comparable to $\xi(T)$ and $\lambda(T)$. Here, we note that vortices can nucleate in thin films made from type-I superconducting materials, because the effective penetration depth for perpendicular magnetic fields is given by $\lambda_\perp \approx \lambda^2/d$, which grows as the thickness d decreases.

19.1.2   Vortices in mesoscopic superconductors

The appearance of vortices is an intriguing phenomenon occurring at different places in nature: water in a sink, whirlwind, hurricane, typhoon and so on. They are also found in quantum systems such as superconductors (see for example, Tilley and Tilley 1990), superfluids (see for example, Vollhardt and Wolfe 1990) and Bose–Einstein condensates (Matthews *et al.* 1999). The quantum vortices are defined as a variation in the phase of an order parameter that characterizes the system; along a closed path around a vortex, the phase of

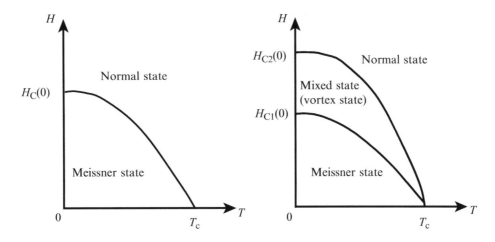

Fig. 19.1 Typical phase diagrams in H–T plane for type-I (left) and type-II (right) superconductors.

the order parameter changes by $2\pi n$ (n: integer), corresponding to a single-valued wavefunction. Due to the requirement of the lowest free energy, only singly quantized vortices with $n = 1$ are normally realized. An important breakthrough was established by the observation of doubly quantized vortex lines in superfluid ^{3}He-A (Blaauwgeers *et al.* 2000). For superconductors, expectations are even more spectacular. In mesoscopic type-II superconductors with sizes comparable to $\xi(T)$ and/or $\lambda(T)$, giant vortices with $n \geq 2$ were predicted (Deo *et al.* 1997; Moshchalkov *et al.* 1997; Schweigert and Peeters 1998; Bruyndoncx *et al.* 1999). Coalescence and reconfiguration of quantized vortices can also happen as a function of the applied magnetic field (Schweigert *et al.* 1998; Palacios 1998, 2000; Baelus *et al.* 2001). Besides, when the sample does not have a circular symmetry, it is possible to create antivortices (Chibotaru *et al.* 2000, 2001, 2004; Banča and Kabanov 2001; Baelus and Peeters 2002; Mel'nikov 2002; Mertelj and Kabanov 2003; Misko *et al.* 2003; Geurts *et al.* 2006, 2007). Furthermore, a vortex can be defined even in one-dimensional mesoscopic rings (Vodolazov *et al.* 2002; Kanda *et al.* 2007). In this chapter, we review theoretical predictions of these novel vortex states in mesoscopic superconducting films and describe some recent experimental trials that provide evidence for the existence of such novel vortex states.

19.1.3 Nanotechnology and superconductivity research

Before going into the details of the subject, we first comment on the new trend in superconductivity research, which is closely related to recent progress in nanotechnology. The studies discussed in this chapter are regarded as one of the directions of this trend. It is obvious that superconductivity research is tightly linked to applications, so that the pursuits of higher critical temperature, higher critical current, and higher critical field are the most important issues of this research. To realize higher critical current, vortex pinning is required; because of the Lorenz force exerted by the supercurrent, a vortex moves perpendicular to the current, causing a voltage along the current direction and destruction of superconductivity. Therefore, vortex pinning that prevents the vortex motion is needed to realize a higher critical current. Conventional pinning sites (called defects or pinning centers) are impurities, lattice defects, dislocations and so on. For example, in NbTi, which is the dominant material for commercial magnet applications, α-Ti is precipitated as an impurity by applying a heat treatment to a homogeneous $Nb_{0.36}Ti_{0.64}$ alloy. Because of the random distribution of positions and strengths of defects, the analysis of the pinning phenomena becomes complicated and its control is limited. On the other hand, recent development in nanofabrication techniques allows us to confine vortices in a controlled manner by introducing regularly placed pinning centers fabricated by using optical lithography, electron-beam lithography, focused ion beam, and so on. This means that a new direction in the modification of superconducting properties using nanotechnology is added to the superconductivity research that in the past was concentrated mainly on the search for new materials. Some of the advances in this direction are summarized in conference proceedings (for example, proceedings of the VORTEX conferences (Abrikosov 2004; Aliev 2006)).

19.2   Theoretical formalism

In this section, we briefly describe the theoretical methods used for the research of mesoscopic superconducting states. Here, we treat thin superconducting samples immersed in an insulating medium in the presence of a perpendicular magnetic field H_0. The two coupled GL equations are the starting point:

$$\alpha\Psi + \beta\Psi\,|\Psi|^2 + \frac{1}{4m}\left(-i\hbar\vec{\nabla} - \frac{2e}{c}\vec{A}\right)^2\Psi = 0,$$

$$\vec{j}_S = \frac{c}{4\pi}\mathrm{curl}\,\mathrm{curl}\overline{A} = -\frac{i\hbar e}{2m}(\Psi^*\vec{\nabla}\Psi - \Psi\vec{\nabla}\Psi^*) - \frac{2e^2}{mc}|\Psi|^2\,\vec{A},$$

$$(19.1)$$

where Ψ is the order parameter, α is the temperature dependent parameter, which changes the sign at the critical temperature T_C : $\alpha > 0$ at $T > T_C, \alpha = 0$ at $T = T_C, \alpha < 0$ at $T < T_C$, β a temperature-independent positive constant, m the electron mass, $\vec{A}$ the vector potential that corresponds to the microscopic magnetic field $\vec{H} = \mathrm{curl}\,\vec{A}$, $\vec{j}_S$ the supercurrent density. The boundary condition

$$i\hbar\left(\vec{\nabla}\Psi + \frac{2e}{c}\vec{A}\Psi\right)\cdot\vec{n} = 0,$$

$$(19.2)$$

where $\vec{n}$ is the unit vector normal to the surface of the superconductor, which means that the supercurrent cannot flow through the boundary. These equations are obtained by minimizing the Gibbs energy expanded in powers of Ψ:

$$G_{sH} = G_n + \int\left[\alpha\,|\Psi|^2 + \frac{\beta}{2}\,|\Psi|^4 + \frac{1}{4m}\left|-i\hbar\vec{\nabla}\Psi\right.\right.$$

$$\left.\left. - \frac{2e}{c}\vec{A}\Psi\right|^2 + \frac{H^2}{8\pi} - \frac{\vec{H}\cdot\vec{H}_0}{4\pi}\right]\mathrm{d}V.$$

$$(19.3)$$

Here, G_n is the free energy of the superconductor in the normal state, $\vec{H}_0$ the external magnetic field, and the integration is carried out over the entire space including the superconductor. From the power expansion of the free energy with respect to the order parameter, it is obvious that the GL equations are valid at temperatures close to T_C. But practically, numerical results show fairly good agreement with the experimental results even at $T \approx 0.5T_C$.

When the non-linear term in eqn (19.1) is omitted, the linearized GL equation

$$\frac{1}{4m}\left(-i\hbar\vec{\nabla} - \frac{2e}{c}\vec{A}\right)^2 \Psi = -\alpha\Psi, \tag{19.4}$$

is analogous to the Schrödinger equation for a particle:

$$\frac{1}{2m}\left(-i\hbar\vec{\nabla} - \frac{e}{c}\vec{A}\right)^2 \Psi + U\Psi = E\Psi, \tag{19.5}$$

with the potential energy $U = 0$ (Moshchalkov *et al.* 1997; Chibotaru *et al.* 2000). Note that the temperature range where eqn (19.4) is valid is limited to the close vicinity of T_C.

In the case of thin mesoscopic films in which the thickness d is assumed to be smaller than ξ and λ, it is allowed to average the macroscopic wavefunction Ψ over the disk thickness. However, the spatial magnetic field distribution should be treated in three dimensions in order to include the demagnetization effect (Schweigert *et al.* 1998):

$$\left(-i\vec{\nabla}_{2D} - \vec{A}\right)^2 \Psi = \Psi(1 - |\Psi|^2),$$
$$-\Delta_{3D}\vec{A} = \frac{d}{\kappa^2}\delta(z)\vec{j}_{2D}. \tag{19.6}$$

Here, the superconductor is placed in the xy-plane, and the external magnetic field is directed along the z-axis. The indices 2D and 3D refer to two- and three-dimensional operators, respectively. The distance is measured in units of the coherence length ξ, the vector potential in $c\hbar/2e\xi$, and the magnetic field in $H_{C2} = c\hbar/2e\xi^2$. The density of the supercurrent is given by

$$\vec{j}_{2D} = \frac{1}{2i}(\Psi^*\vec{\nabla}_{2D}\Psi - \Psi\vec{\nabla}_{2D}\Psi^*) - |\Psi|^2\,\vec{A}. \tag{19.7}$$

There are two boundary conditions: one is that the supercurrent component normal to the sample boundary is zero,

$$\vec{n} \cdot (-i\vec{\nabla}_{2D} - \vec{A})\Psi\Big|_{\text{boundary}} = 0, \tag{19.8}$$

and the other is that the magnetic field far away from the superconductor equals the applied field,

$$\vec{A}\Big|_{\vec{r}\to\infty} = \frac{1}{2}\vec{e}_\phi H_0\rho. \tag{19.9}$$

The difference between the superconducting and the normal state Gibbs free energy is given by

$$F = \frac{1}{V} \int \left[2(\vec{A} - \vec{A}_0) \cdot \vec{j} - |\Psi|^4 \right] d\vec{r}. \tag{19.10}$$

A detailed description of the theoretical model can be found in Schweigert and Peeters (1998) and Schweigert *et al.* (1998).

19.3   Theoretical predictions of vortex states in thin mesoscopic superconducting films

19.3.1   Multivortex and giant vortex states

In the mixed state of bulk type-II superconductors, vortices form triangular lattices (the Abrikosov lattices, Abrikosov 1957) due to the repulsive forces between them.[1] Because the energy associated with an interface between a normal core and surrounding superconducting parts is negative, each vortex encloses the minimum magnetic flux (flux quantum: $\Phi_0 = hc/2e \approx 2 \times 10^{-15}$ Wb) and the phase of the order parameter changes by 2π around the core. On the other hand, the vortex configuration in mesoscopic superconductors is affected not only by the vortex–vortex interaction but also by the boundary of the superconductor, leading to corruption of the Abrikosov triangular lattice and the formation of fundamentally new vortex states. The resulting vortex configuration can be divided into two categories: giant vortex states (GVSs) and multivortex states (MVSs) (Schweigert *et al.* 1998). Figure 19.2 shows the Cooper-pair density ((a) and (b)) and the phase of the order parameter ((c) and (d)) for a multivortex state and a giant vortex state corresponding to the vorticity $L = 5$ in a superconducting disk with radius $R = 6\xi$, calculated within the framework of the GL theory, described in the previous section. High Cooper-pair density is given by dark regions, and low density by light regions. The phase changes by 2π from dark to light. Here, the vorticity L is the phase change along the sample boundary divided by 2π and takes an integer value due to the fluxoid quantization:

$$\oint_C \Lambda \vec{j}_S \cdot dl + \oint_C \vec{A} \cdot dl = L\Phi_0, \tag{19.11}$$

where C is the contour along the sample boundary, $\Lambda = m/n_s e^2$ is the London coefficient (n_S: the density of superconducting electrons), and the second term of the LHS is the magnetic flux enclosed by C. In the $L = 5$ MVS, five vortices, each of which are accompanied by a phase change of 2π, are on a circular shell, reflecting the sample shape. On the other hand, in the GVS, there is only one giant vortex and the phase changes by $5 \times 2\pi = 10\pi$ around it. Notice that the Cooper-pair density near the central core does not have five small dips, but is fully axially symmetric with respect to the disk center. This GVS appears because the vortices are pushed into the disk center due to the repulsive interaction with the sample boundary. Therefore, it is stabilized in a small disk ($R \approx \xi$), or even in larger disks under high magnetic fields

[1] Square lattices are also stabilized in some anisotropic superconductors (Sakata *et al.* 2000).

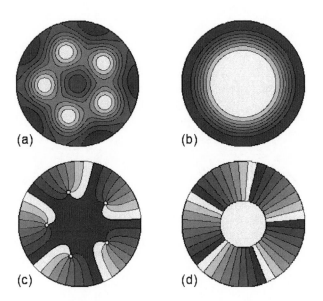

Fig. 19.2 The Cooper-pair density for the multivortex state (a) and the giant vortex state (b), and the phase of the order parameter for the multivortex state (c) and the giant vortex state (d) with vorticity $L = 5$ in a superconducting disk with radius $R = 6\xi$. High (low) Cooper-pair density is shown by black (white) regions. Phases near 2π (0) are shown in black (white).

where a strong shielding current along the boundary exerts a repulsive force on vortices. As an example, Fig. 19.3 shows the free energy of the different giant and multivortex configurations for a disk with $R = 4\,\xi$ (Schweigert *et al.* 1998), calculated within the framework of the GL theory. Every curve indicates the magnetic-field region over which the state has a local minimum in the free energy, that is, transitions to different L states occur at the ends of every curve,

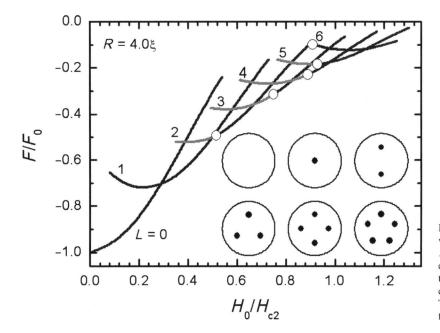

Fig. 19.3 The free energy of configurations with different number of vortices L for radius $R = 4\xi$, thickness $d \ll \xi$, and the GL parameter $\kappa = 0.28$. The open circles indicate the transitions from a multivortex state (black curve) to a giant vortex state (gray curve). The insets show the possible multivortex configurations.

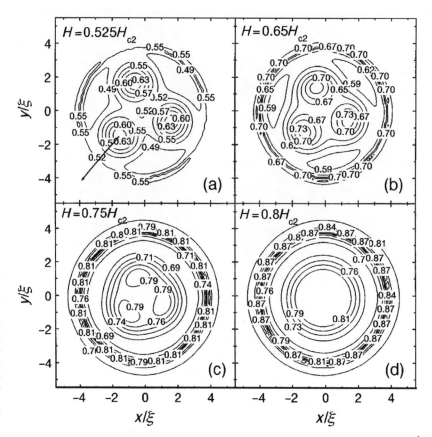

Fig. 19.4 Contour plot of the magnetic-field distribution in the disk plane for the $L = 3$ state at applied magnetic fields $H_0/H_{C2} = 0.525$, 0.65, 0.75, and 0.8 from (a) to (d) in a disk with $R = 4\xi$ and $d = 0.5\xi\kappa^2$ and $\kappa = 0.28$. For the lowest magnetic field (a) the three-vortex configuration is metastable and a small decrease in the magnetic field leads to a transition to the two vortex state due to the expulsion of a vortex through the disk boundary indicated by the arrow in (a). From Schweigert *et al.* (1998). Copyright by the American Physical Society.

when the magnetic field is swept up or down. MVSs (GVSs) are indicated by gray (black) curves. For $L \geq 2$, a MVS (a GVS) is stabilized in lower (higher) magnetic fields, and the portion of the GVS relatively increases for larger L. Notice that there is almost no MVS region for $L = 6$. The open circles indicate gradual (i.e. second-order) transitions from the MVS to the GVS with the same vorticity, as illustrated in Fig. 19.4, which shows contour plots of the magnetic-field distribution in the disk plane for the $L = 3$ state in a disk with radius $R = 4\xi$, thickness $d = 0.5\xi\kappa^2$ and $\kappa = 0.28$. Figures 19.4(a–d) correspond to different applied magnetic fields, $H_0/H_{C2} = 0.525$, 0.65, 0.75 and 0.8, respectively. For the lowest magnetic field (Fig. 19.4(a)) the $L = 3$ MVS is metastable, which corresponds to local minima in the free-energy landscape. With increasing applied magnetic field, the vortices move to the center due to the stronger shielding supercurrent along the boundary (Figs. 19.4(b) and (c)), and finally combine into one, forming a GVS (Fig. 19.4(d)). This is a second-order transition, so that there is no jump in the free energy at the MVS-GVS transition field.

For larger disks, MVSs tend to be stabilized even for larger L, because of larger distances between the sample boundary and vortices, and more shells

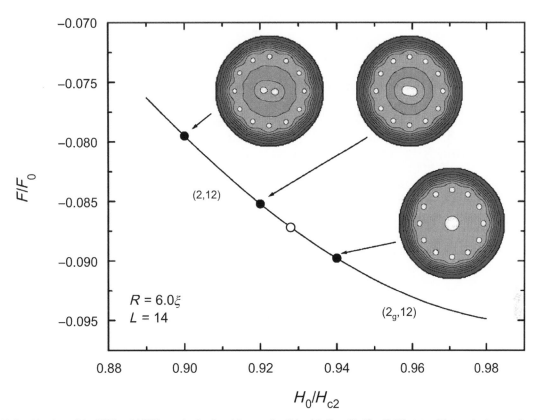

Fig. 19.5 Combination of the GVS and MVS seen in the $L = 14$ state of a disk with $R = 6\xi$. The (2,12) state with two singly quantized vortices near the center and 12 vortices on a shell transits into the (2_g,12) state with a doubly quantized vortex at the center with increasing the magnetic field. The open circles indicate the transition between the (2,12) and (2_g,12) states. From Baelus *et al.* (2004). Copyright by the American Physical Society.

and more stable vortex configurations appear for a fixed L. Figure 19.5 shows an interesting case of the MVS–GVS combination found in the $L = 14$ state of a disk with $R = 6\xi$ (Baelus *et al.* 2004). At low magnetic field, the (2, 12) state with two vortices near the center and 12 vortices on a shell is stabilized, and on increasing the magnetic field, the two vortices near the center merge into one, forming the (2_g, 12) state with a giant vortex with $L = 2$ at the center. This MVS–GVS transition, indicated by the open circle, is also of second order and does not accompany hysteresis. Figure 19.6 shows possible MVSs for $L = 16$ seen in disks with $R = 20\xi$ and their free energies as a function of the magnetic field (Baelus *et al.* 2004). In such large disks, more than one MVS is stabilized for each L, one of them has the lowest energy and the others are metastable states with slightly higher energies. Each MVS has a multiple shell structure. It is worth noting that the energy differences between these states with the same L are much smaller than that between different L states, so that even the existence of very small incompleteness such as defects, deformation, impurities, can influence the ground state, i.e. can make a different vortex

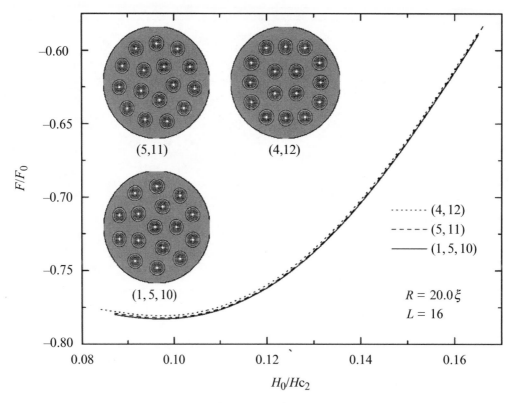

Fig. 19.6 The free energy as a function of the applied magnetic field for the (4,12)-state, the (5,11)-state, and the (1,5,10)-state in a superconducting disk with radius $R = 20\xi$. The insets show the positions of vortices for these states at $H_0/H_{C2} = 0.1$. From Baelus *et al.* (2004). Copyright by the American Physical Society.

configuration most stable. This phenomenon is seen in experiments (Kanda *et al.* 2004; Grigorieva *et al.* 2007), as will be explained in Section 19.5.2. In much larger disks, for example, with $R = 50\xi$, for small L values ($L \leq 9$), so-called *vortex molecules* are formed in which vortices are distributed in regular polygons with 0 or 1 vortex in the center of the disk, as shown in Fig. 19.7 (Cabral *et al.* 2004). This means that not many metastable states are close to the ground state. For very large values of vorticity (typically $L \geq 100$), there is a competition between the ring-like structure imposed by the disk geometry and the hexagonal lattice favored by the vortex–vortex interaction. As a result, it was found that the vortices are arranged in an almost complete Abrikosov lattice in the center of the disk, which is surrounded by at least two circular shells of vortices, as shown in Fig. 19.8 (Cabral *et al.* 2004). This situation is the same as those seen in bulk superconductors. Thus, as the size of the superconductor becomes larger, the effect of irregular vortex configuration (other than the Abrikosov triangular lattice) on physical quantities such as the free energy becomes relatively smaller. In this way, vortex states characteristic of mesoscopic superconductors are predicted to appear in samples with sizes from several ξ to tens of ξ.

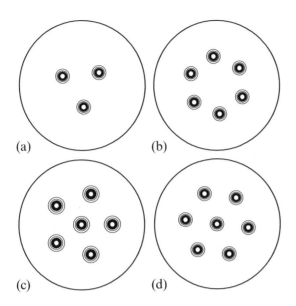

Fig. 19.7 Vortex molecules for $L = 3$ and $H_0 = 0.007H_{C2}$ (a), $L = 6$ and $H_0 = 0.01H_{C2}$ (b), $L = 6$ and $H_0 = 0.01H_{C2}$ (c), $L = 7$ and $H_0 = 0.011H_{C2}$ (d), seen in a disk with $R = 50\xi$. From Cabral *et al.* (2004). Copyright by the American Physical Society.

19.3.2   Effect of the sample geometry

Not only the sample size but also the sample geometry strongly affects the arrangement of vortices. Due to the repulsive interaction between a vortex and the sample boundary, vortices eventually tend to be situated towards the sample corners. As a result, samples with non-circular geometry favor MVSs more strongly than disks do (Baelus and Peeters 2002). Figure 19.9 shows the free energy and magnetization for the disk, the square, and the triangle with the same surface area as a function of the applied magnetic field (Baelus and Peeters 2002). In Figs. 19.9(a), (c) and (e), the free energy of the different

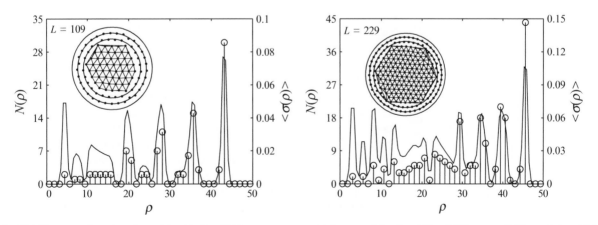

Fig. 19.8 Vortex configurations for a disk with $R = 50\xi$ in a vortex state with large vorticity, $L = 109$ and 229 seen at $H/H_{C2} = 0.1$ and 0.2, respectively. The number of vortices, N, and the average density of vortices $\langle \sigma(\rho) \rangle = N(\rho)/2\pi\rho\Delta\rho$ are plotted as a function of the radial distance ρ from the center with a step of $\Delta\rho = 1.25\xi$. From Cabral *et al.* (2004). Copyright by the American Physical Society.

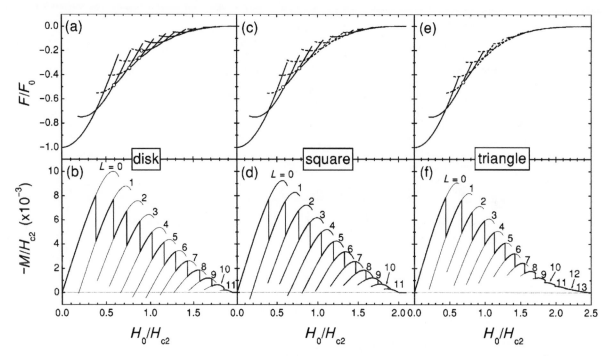

Fig. 19.9 The free energy and the magnetization for the disk, the square, and the triangle with the same surface area $S = \pi 16\xi^2$ and thickness $d = 0.1\xi$ for $\kappa = 0.28$. (a,c,e) The free energy of the GVSs (solid curves) and the MVSs (dashed curves) and the MVS–GVS transition fields (open circles). (b,d,f) The magnetization of the L states (solid curves) and the ground state (thick solid curves). From Baelus and Peeters (2002). Copyright by the American Physical Society.

GVSs (MVSs) is given by solid (dashed) curves. The transition between MVS and GVS is indicated by an open circle. In the disk, MVSs can nucleate for $L = 2, 3, 4$, and 5, while in the square and the triangle, for $L = 2, 3, 4, 5$ and 6. Moreover, for the disk, the MVSs always transits to GVSs with increasing field, while for the square and the triangle, some L states are MVSs over the whole magnetic field range. Figure 19.10 shows the Cooper-pair density for MVSs in the triangle with $L = 2, 3$, and 4 and the phase of the order parameter for $L = 5$ and 6. High (low) Cooper-pair density is indicated by dark (light) regions, and the phase near zero (2π) is given by light (dark) regions. In the $L = 2$ state, the vortices are situated on one of the perpendicular bisectors of the triangle, and for $L = 3$, all vortices are close to the corners, so that the vortex arrangement fits the sample shape. Also, for $L = 4$–6, three vortices are near the corner, while the rest are on the center. Note that for $L = 5$ and 6, the separation of the vortices at the central region becomes invisible, which shows the existence of a giant vortex at the center. That is, in both cases, there is a coexistence of a giant vortex in the center and three separated vortices placed in the direction of the corners. As shown in Figs. 19.10(d) and (e), the phase of the order parameter around the central vortex actually changes by 4π for the $L = 5$ MVS and by 6π for the $L = 6$ MVS. States with $L > 6$ are always GVS, as is seen in Fig. 19.9(e). With increasing L, this giant vortex grows, and for large vorticities, superconductivity only occurs in the corners of the sample. Further

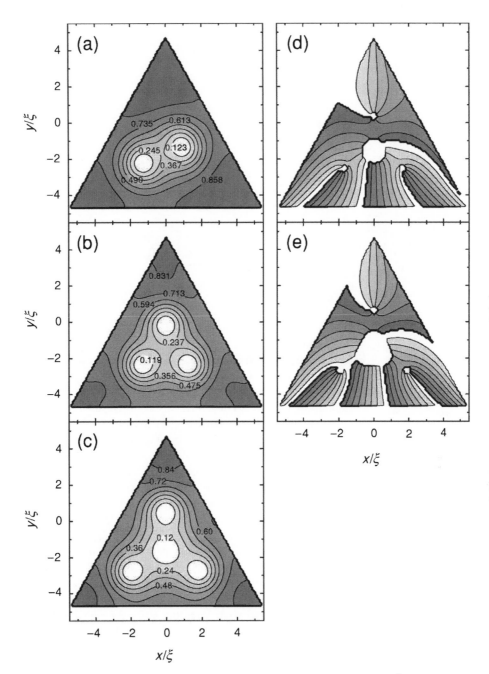

Fig. 19.10 (a–c) The Cooper-pair density for the multivortex states in a triangle with surface area $S = \pi 16\xi^2$ with $L = 2, 3$, and 4 at $H_0/H_{C2} = 0.495$, 0.82, and 0.745, respectively. High (low) Cooper-pair density is given by dark (light) regions. (d,e) The phase of the order parameter for the multivortex states with $L = 5$ at $H_0/H_{C2} = 1.27$ and with $L = 6$ at $H_0/H_{C2} = 1.345$. Phases near zero (2π) are given by light (dark) regions. From Baelus and Peeters (2002). Copyright by the American Physical Society.

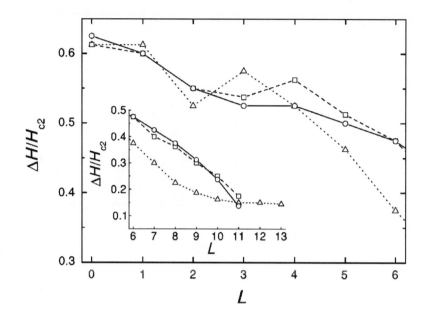

Fig. 19.11 The magnetic-field range over which the vortex states with vorticity L are (meta)stable as a function of the vorticity for the disk (open circles), the square (open squares) and the triangle (open triangles) with the same area $\pi 16\xi^2$. From Baelus and Peeters (2002). Copyright by the American Physical Society.

increasing the field pushes the superconductivity more to the corner until these corners become normal too at the superconducting/normal transition field. This kind of local superconductivity happens when the sample has sharp corners. As a consequence, the superconducting/normal transition field becomes larger in samples with sharper corners (for fixed surface area) (Schweigert and Peeters 1999; Baelus and Peeters 2002). For example, in the case of Fig. 19.9, the vortex state exists up to $L = 11$ for the disk and up to $L = 13$ for the square and the triangle. The superconducting state is destroyed at $H_{C3}/H_{C2} \approx 1.95$ for the disk, ≈ 2.0 for the square, and ≈ 2.5 for the triangle.

As mentioned above, the multivortex configuration tries to fit the sample shape. This affects the stability of the vortex states. The stability of the vortex states can be evaluated with the magnetic field region over which the state is (meta)stable, $\Delta H(L) = H_{\text{penetration}}(L) - H_{\text{expulsion}}(L)$, where $H_{\text{penetration}}(L)$ and $H_{\text{expulsion}}(L)$ are the vortex penetration and expulsion field of the L state, respectively (Baelus and Peeters 2002). Figure 19.11 shows ΔH as a function of the vorticity for the disk, the triangle, and the square with the same area $\pi 16\xi^2$. For the circular disk the stability region $\Delta H/H_{C2}$ almost uniformly decreases with increasing L. On the other hand, the triangle and the square exhibit a peak structure at $L = 3$ and $L = 4$, respectively. This is the direct correspondence of the fact that the vortex lattice tries to keep the same geometry as the sample. The following properties were pointed out (Baelus and Peeters 2002): (i) the peak structure is more pronounced for structures that fit the triangular Abrikosov lattice more closely; (ii) for $L > 4$ no clear peaks are seen; (iii) the vortex states in the square and circle geometries have almost the same stability range for $L \leq 2$ and $L \geq 6$; (iv) for $L \geq 4$ the stability range for the vortex state in the triangular geometry becomes substantially smaller than that for the other two geometries that have less sharp corners. Thus, sharp corners decrease the stability range of the vortex states.

19.3.3 Symmetry-induced vortex-antivortex molecules

It has been pointed out that the vortex state can include antivortices in order to preserve the sample symmetry (Chibotaru *et al.* 2000, 2001, 2004; Banča and Kabanov 2001; Baelus and Peeters 2002; Mel'nikov 2002; Mertelj and Kabanov 2003; Misko *et al.* 2003; Geurts *et al.* 2006, 2007). For example, for the $L = 3$ MVS state in a square, a vortex arrangement with four vortices near the corners and one antivortex at the center is possible. This vortex-antivortex (VAV) molecule was first predicted by solving the linearized GL equation, which is applicable only to the superconducting/normal phase boundary, and the theoretical superconducting/normal phase boundary was compared with the experimental one (Chibotaru *et al.* 2000, 2001). Later it turned out that the VAV molecules become rapidly unstable when moving away from the superconducting/normal phase boundary (Banča and Kabanov 2001). Due to the following properties of the VAV molecules, it is thought that the experimental detection of the VAV molecules is extremely difficult: (i) the separation between the vortex and the antivortex is very small due to the attractive interaction between them, (ii) the variation of the Cooper-pair density is extremely small ($|\Psi|^2/|\Psi_0|^2 < 10^{-6}$) in an area between vortex and antivortex, (iii) the corresponding variation of the magnetic field is also very small ($\Delta H/H_0 < 10^{-5}$), and (iv) even a tiny defect at the boundary may break the symmetry of the sample, resulting in instability of the VAV molecules (Baelus and Peeters 2002; Geurts *et al.* 2007). Therefore, even when a VAV molecule exists in a sample, the vortex configuration as a whole is very similar to a single giant vortex, and its detection by using Cooper-pair density measurement or magnetic-field measurement becomes difficult.

In order to address this issue, the VAV molecule in a superconductor with strategically placed nanoholes was investigated (Geurts *et al.* 2006, 2007). For example, in the $L = 3$ state in a square with 2×2 nanoholes the symmetry of vortex arrangement is guaranteed by nanoholes, so that the actual shape of the sample becomes less important, and the stability of the VAV configuration is substantially enhanced. Besides, vortices are pinned in the nanoholes, so that the separation between the vortices and the central antivortex is improved. Tolerance to defects was also studied (Geurt *et al.* 2007).

19.3.4 Magnetic-field distribution and demagnetization effects

When a superconductor with finite (non-zero) thickness is placed in magnetic field, the magnetic field is expelled from the sample and the magnetic-field distribution in the direction perpendicular to the sample plane changes. Some experimental techniques measure this distribution to detect the vortex states. In order to obtain sufficient contrast, the probe must be placed close to the sample surface. As an example, Fig. 19.12 shows the magnetic-field distribution for the $L = 4$ state of a square with side $W = 7.090\xi$, thickness $d = 0.1\xi$ and $\kappa = 0.28$, which is typical for Al, under a perpendicular magnetic field of $H_0/H_{C2} = 0.77$ (Baelus and Peeters 2002). High magnetic field is given by dark regions and low magnetic field by light regions. In

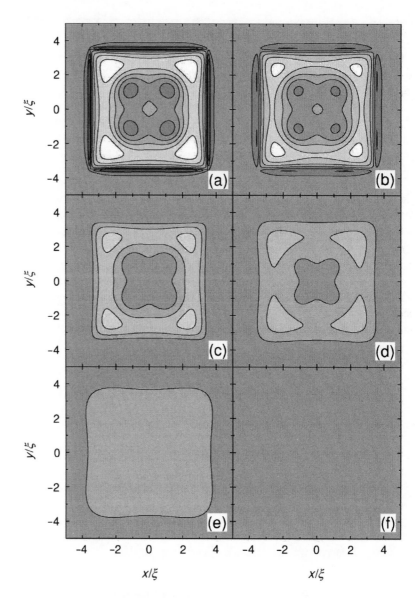

Fig. 19.12 The magnetic-field distribution for the $L = 4$ state in a square with side $W = 7.090\xi$, thickness $d = 0.1\xi$ and the GL parameter $\kappa = 0.28$ under a perpendicular magnetic field of $H_0/H_{C2} = 0.77$. The distance from the sample plane is $z/\xi = 0, 0.1, 0.3, 0.6, 1.0$ and 10.0 for (a)–(f), respectively. High magnetic field is given by dark regions and low magnetic field by light regions. From Baelus and Peeters (2002). Copyright by the American Physical Society.

the plane of the superconductor ($z = 0$, Fig. 19.12(a)), the magnetic field is compressed along the sample boundary due to the demagnetization effect and four dark spots corresponding to the vortices are clearly seen. With increasing z, the demagnetization effects decrease and the variation of the magnetic field becomes smaller. At $z = 1.0\xi$, the vortex structure in this plot is invisible, and the four vortices may be seen as one giant vortex. At $z = 10.0\xi$, the magnetic-field variation is completely washed out and the magnetic-field distribution is uniform. Although the magnetic-field distribution and the contrast of the picture depend on parameters such as the sample size, vorticity and magnetic field, we can tell that the typical distance for the detection of the vortex arrangements is of the order of the coherence length, which is at most $0.2 \, \mu$m at $T = 0$ for Al.

19.4   Experimental techniques for detection of vortices

A variety of experimental techniques have been developed for observation of vortices in superconductors. Some have been used or can be used in principle for observation of novel vortex states in mesoscopic superconductors. In the following, we briefly introduce these techniques, and discuss the possibility of application to mesoscopic superconductors.

19.4.1   Direct visualization of vortices

The most preferable technique is the direct visualization of the vortex positions. Such techniques include the classical Bitter decoration technique, several kinds of scanning probe microscopy, and magneto-optical imaging.

19.4.1.1   *Bitter decoration technique*

Bitter decoration is a classical but powerful technique that was used for the first observation of the Abrikosov triangular lattice of vortices (Essmann and Trauble 1967). This technique uses ferromagnetic nanoparticles deposited on the surface of a superconductor. The density of nanoparticles varies corresponding to the spatial distribution of the magnetic field, and the pattern of nanoparticles is observed with a scanning electron microscope. Successful observation of vortex shells and pinning-induced vortex clusters in mesoscopic Nb disks were reported (Grigorieva *et al.* 2006, 2007). The disadvantage of this method is that it cannot observe a successive transition of vortex states in an individual sample.

19.4.1.2   *Scanning SQUID microscopy*

Scanning SQUID (superconducting quantum interference device) microscope scans a small SQUID pick-up coil on the sample to obtain information on the spatial distribution of magnetic field. Vortices were successfully resolved by the scanning SQUID microscopy for large Nb disks, where vortex location is dominated by pinning (Hata *et al.* 2003). Also, it was used to determine the flux distribution in superconducting wire networks (Yoshikawa *et al.* 2004). Although a SQUID has an extremely high sensitivity, the size of the pick-up coil is 5–$10 \, \mu$m, which limits the spatial resolution and is not sufficient for the observation of mesoscopic superconducting states even in Al samples, which have the largest coherence length $\xi(0) \approx 0.2 \, \mu$m. Mathematical improvement of the images obtained by the scanning SQUID microscope has been proposed (Hayashi *et al.* 2008). An example is shown in Fig. 19.13. A more detailed description of the scanning SQUID microscopy can be found in Chapter 11, volume II of this Handbook by Kadowaki and colleagues (Nishio *et al.* 2010).

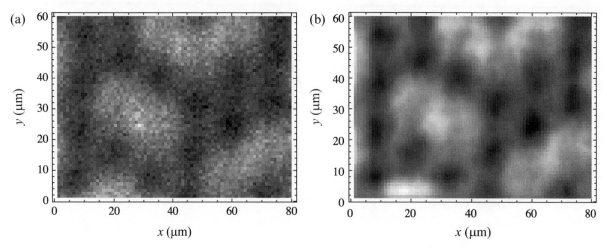

Fig. 19.13 (a) An image obtained by scanning SQUID microscopy, showing the magnetic-field distribution in a superconducting wire network, and (b) a mathematically improved image of (a). Courtesy of M. Hayashi (Akita Univ.) and T. Ishida (Osaka Prefecture Univ.).

19.4.1.3   *Scanning Hall-probe microscopy (SHPM)*

In scanning Hall-probe microscopy, a small Hall probe is scanned over the sample surface to detect the surface magnetic field (Chang *et al.* 1992). This technique is successfully used for the vortex detection in bulk superconductors (Oral *et al.* 1997; Grigorenko *et al.* 2001; Field *et al.* 2002). The spatial resolution is $> 300\,nm$. The Hall probe is fabricated on an edge of the substrate of a GaAs/GaAlAs two-dimensional electron gas. When SHPM is used for mesoscopic superconducting states, the separation of the Hall probe from the corner of the chip, which is usually several tens of μm, limits the sharpness of images, because the demagnetization effect rapidly decreases as the distance from the superconductor increases, as explained in Section 19.3.4.

19.4.1.4   *Magnetic force microscopy (MFM)*

Magnetic force microscope is a kind of non-contact atomic force microscope equipped with a magnetic material as tip. MFM reveals the best lateral resolution less than $100\,nm$ among the magnetosensitive scanning probe techniques. This technique has been used for the observation of the Abrikosov vortex lattice in Nb and $NiSe_2$ crystals (Volodin *et al.* 2000, 2002), but has not been applied to mesoscopic superconductors yet. The disadvantage of this approach is that the magnetic tip produces its own magnetic field, which may distort the vortex lattice.

19.4.1.5   *Scanning tunnelling microscopy (STM)*

The STM observation of superconductors visualizes the spatial distribution of the superconducting local density of states on the sample surface (Hess *et al.* 1989). Because it requires samples with atomically flat surfaces, sophisticated techniques are needed for the sample preparation. For mesoscopic superconductors, vortices in naturally formed Pb islands (Nishio *et al.* 2006) and artificially fabricated antidot arrays (Karapetrov *et al.* 2005) have been reported.

19.4.1.6   *Magneto-optical (MO) imaging*

The magneto-optical imaging is based on the Faraday effect, i.e. a rotation of the light polarization by a magnetic field. A magneto-optical garnet film is placed on top of a superconductor with a mirror in-between. The spatial distribution of the Faraday rotation, which corresponds to the spatial distribution of perpendicular magnetic field, is observed with an optical microscope (Koblischka and Wijngaaeden 1995). Although this method can visualize the static and dynamic magnetic behavior, it has not been applied to the study of mesoscopic superconductors.

Other direct visualization techniques include Lorentz microscopy (Harada *et al.* 1992) and low-temperature scanning electron microscopy (Straub *et al.* 2001), which have not been applied to mesoscopic superconducting states.

19.4.2   Indirect methods

Some indirect methods for the detection of the mesoscopic vortex states have been developed: cusps in the superconducting/normal phase boundary $T_C(H)$ obtained in resistance measurements have been attributed to transitions between different vortex states, corresponding to the Little–Parks effect in superconducting cylinders (Little and Parks 1962, 1964). The transition fields are in good agreement with theoretical predictions based on the linearized GL theory (Puig *et al.* 1997, 1998; Bruyndoncx *et al.* 1999; Chibotaru *et al.* 2000, 2001). Ballistic Hall magnetometry for an isolated sample that is placed on a μm-scale Hall bar shows multiple magnetization curves as a function of applied magnetic field, each of which corresponds to different vortex states (Geim *et al.* 1997a,b, 1998). The advantage of these methods is that they give macroscopic physical quantities such as the superconducting transition temperature and the magnetization, which can be compared with the theoretical values. However, it is noted that these indirect methods do not give us any information on the vortex positions, so that a numerical study (minimization of the free energy) is essential to identify the vortex states (even to determine whether the state is a GVS or an MVS) in these experiments.

19.4.2.1   *MSTJ method*

In order to distinguish experimentally the GVSs from the MVSs, Kanda *et al.* have developed the multiple-small-tunnel-junction (MSTJ) method, which is based on the classical technique for estimating the superconducting energy gap from the current–voltage ($I-V$) characteristics of a superconductor-insulator-normal metal (SIN) tunnel junction (Kanda *et al.* 1999a,b, 2004). This method gives *partial* information on the spatial distribution of the supercurrent in the sample, so that by taking into account the symmetry in the sample geometry, one can guess the vortex configuration, as will be explained in Section 19.5.1.

In the MSTJ method, several normal-metal leads are attached to a mesoscopic superconductor with an insulating layer as interface, forming highly resistive SIN tunnel junctions with tunnel resistance $R \gg h/4e^2 \approx 6.4 \, k\Omega$ (Fig. 19.14(a)). The low transparency is required to prevent the proximity-induced degradation of superconductivity. The $I-V$ characteristics of a SIN

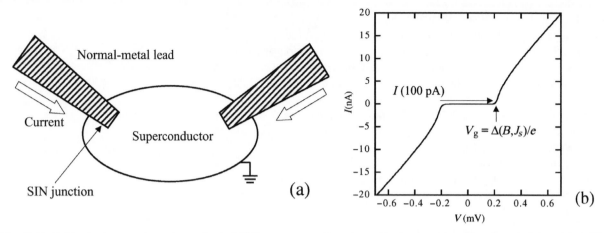

Fig. 19.14 (a) Sketch of an experimental setup for the MSTJ measurement. Normal-metal leads are connected to a mesoscopic superconductor through highly resistive small SIN tunnel junctions. A constant current I flows from each lead to the drain. (b) Typical I–V characteristics of a SIN tunnel junction at low temperatures. The current starts to flow at the threshold voltage, $V_g = \Delta/e$ (Δ: superconducting energy gap). When the current is fixed to a small value, typically 100 pA, the voltage of the junction is sensitive to change in the superconducting energy gap.

tunnel junction are highly non-linear around the origin at low temperatures, reflecting the superconducting density of states (Fig. 19.14(b)). The threshold voltage for the current onset, V_g, seen at low temperatures equals the superconducting energy gap divided by e. Therefore, when the current through the tunnel junction is fixed to a small value (typically to 0.1–2 nA), the resulting voltage is sensitive to a change in the energy gap. Moreover, if the junction size is smaller than the superconducting coherence length, the voltage reflects the local energy gap underneath the junction and also the local supercurrent density, because the energy gap decreases with increasing supercurrent density. Therefore, by employing multiple small tunnel junctions, one can obtain partial information on the supercurrent distribution and consequently the vortex structure. As an illustrative example, Fig. 19.15 shows the results for a small superconducting ring under a perpendicular magnetic field (Kanda *et al.* 2000; Kanda and Ootuka 2002). The Al ring is 0.20 μm in radius, 70 nm in linewidth, and 20 nm in thickness. Two Cu leads are attached to it with an interface layer of AlO$_x$. A constant current of 100 pA flows from one lead to the other. The ring is considered to be one-dimensional with a uniform supercurrent density because the linewidth is smaller than the coherence length $\xi(0) = 150$–190 nm. Note that the superconducting proximity effect is expected to be negligible due to the high tunnel resistance (more than several tens of kΩ). In Fig. 19.15(b), the voltage in an increasing (decreasing) magnetic field is indicated by the black (gray) curve, which shows complicated structures. The voltage variations result from two origins: (1) smearing of the energy gap due to pair breaking by the magnetic field (Tinkham 1996), and (2) a decrease of the energy gap because of the supercurrent (Bardeen 1962; Levine 1965). The former leads to a monotonic decrease in voltage as the strength of the magnetic field increases, so both the upward parabolas and the jumps in voltage come from the latter. Each parabola reflects the change in supercurrent satisfying the fluxoid quantization condition of the ring. That is,

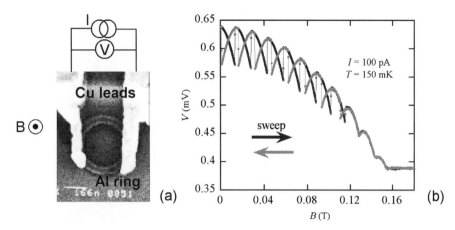

Fig. 19.15 (a) A scanning electron microscope picture of a superconducting ring with two small SIN tunnel junctions together with a schematic diagram of the measurement setup. (b) Change in the voltage between two leads under a constant current of 100 pA in increasing (gray curves) and decreasing (black curves) magnetic field.

near the maximum of each parabola, the applied magnetic flux in the ring is $n\Phi_0$ (n: integer), and no supercurrent flows. (This happens just above the field of each peak because the effect of (1) is superimposed.) Also, the voltage jumps correspond to transitions between states with different fluxoid quantum number: $n \to n \pm 1$. (In fact, the interval of the adjacent peaks is close to Φ_0/S (S: the area of the ring).) These results demonstrate that the voltage of small SIN tunnel junctions is sensitive to the change in local supercurrent flowing in a mesoscopic superconductor.

19.5   Experimental detection of mesoscopic vortex states in disks and squares

In this section, we describe the experimental study on mesoscopic super-conducting states in disks and squares by using the MSTJ method (Kanda *et al.* 2004; Baelus *et al.* 2005, 2006), which provided the first experimental distinction between MVSs and GVSs. Later, the Bitter decoration technique was also used for this purpose (Grigorieva *et al.* 2006, 2007).

19.5.1   Distinction between giant vortex states and multivortex states

Figure 19.16 shows a schematic drawing and a SEM image of the disk sample for the MSTJ measurement (Kanda *et al.* 2004). Four normal-metal (Cu) leads are connected to the periphery of the superconducting Al disk through highly resistive small tunnel junctions A, B, C, and D with area $\approx 0.01\ \mu m^2$, which is smaller than the square of the coherence length $\xi(0)$, guaranteeing the detection of the local density of states. The radius of the disk R is $0.75\ \mu m$ and the disk thickness d is 33 nm. The disk is directly connected to an Al drain lead. This sample was fabricated using e-beam lithography followed

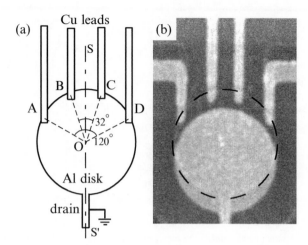

Fig. 19.16 Schematic view (a) and scanning electron micrograph (b) of a disk sample for the MSTJ measurement. From Kanda *et al.* (2004). Copyright by the American Physical Society.

by double-angle evaporation of Al and Cu. The double-angle evaporation technique is conventionally used for the fabrication of metallic small tunnel junctions, in which all of the metals (in the present case, Al and Cu) are evaporated in a single vacuum process (Dolan 1977). Most of the Al disk, indicated by the dashed circle in Fig. 19.17(b), was covered with a Cu film (bright regions). It was assumed that the Cu film did not have any influence on the superconductivity of the Al disk because of the insulating AlO_x layer in-between. In the measurement, the current flowing through each junction was fixed to a small value, and the voltages between each of the four Cu leads and the drain lead were measured, while the perpendicular magnetic field was swept.

In order to distinguish between GVSs and MVSs, the symmetry of the supercurrent distribution was taken into account. As shown in Fig. 19.2, for circular disks the Cooper-pair density and the supercurrent distribution are axially symmetric for GVSs. On the other hand, they do not have axial symmetry for MVSs. Therefore, when the voltages of the junctions attached to the disk periphery are equal to each other, the state is presumably a GVS, and otherwise it is certainly a MVS. Note that the real sample (Fig. 19.16) is symmetric only with respect to the central axis SS', when the details of the junction shapes are taken into account. Therefore, one can compare V_A and V_D or V_B and V_C in the strict sense.

Experimental results are shown in Fig. 19.17(a) for decreasing magnetic field. The overall decrease in voltage corresponds to the smearing of the energy gap due to pair breaking by the magnetic field, and the fine complex variation to a decrease of the energy gap due to the supercurrent underneath the junction. In particular, a voltage jump seen simultaneously in all junctions corresponds to a sudden change in supercurrent distribution caused by a transition between different vortex states with a vorticity change of -1. Because $L = 0$ at $B = 0$, one can determine the vorticity of each state as indicated in the figure. Note that the difference either between V_A and V_D (between V_B and V_C) at $B = 0$ mainly arises from a slight asymmetry in the junction resistance, which affects

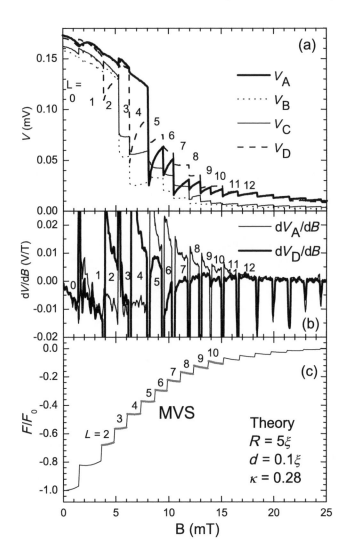

Fig. 19.17 (a) Variation of the voltages at junctions A, B, C and D in a decreasing magnetic field. The current through each junction is 100 pA. Temperature was 0.03 K. (b) Differential voltages dV_A/dB and dV_D/dB. (c) Calculated free energy F for a disk with $R = 5.0\xi$, $d = 0.1\xi$, and $\kappa = 0.28$, normalized by the $B = 0$ value F_0. Thick and thin segments indicate MVS and GVS, respectively.

the overall characteristics in the whole magnetic-field range. To make the comparison between the different curves easier, dV/dB is shown in Fig. 19.17(b). In this figure, remarkable differences are found between dV_A/dB and dV_D/dB for $L = 2$ and 4 to 11, indicating that the state is MVS for these vorticities and probably GVS otherwise.

To see whether this simple distinction between GVSs and MVSs is reliable or not, the experimental results were compared with the results of a numerical simulation. Theoretical results are shown in Fig. 19.17(c). Theoretically, MVSs nucleate for vorticity $L = 2$ to 10 for decreasing magnetic field. Thus, the theoretical calculations confirm the identification of the GVS and MVS by the MSTJ method except for $L = 3$ ($L = 11$), where theoretically the state is predicted to be a MVS (GVS), while experimentally a GVS (MVS) is inferred. The disagreement for $L = 3$ is attributed to the junction geometry ($\angle AOD = 120$ degree), which decreases significantly the voltage difference

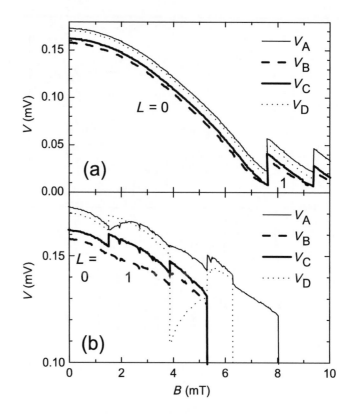

Fig. 19.18 Experimental data indicating the existence of a defect near (but not at) the disk center. Close-ups of the voltage changes near $B = 0$ for (a) increasing and (b) decreasing magnetic fields.

of the A–D junction pair for MVSs with trigonal symmetry, concealing the $L = 3$ MVS. Although this kind of symmetry-induced effect is also expected to appear in $L = 6$ and $L = 9$ MVSs, the dV/dBs are significantly different for $L = 6$ and $L = 9$ in decreasing magnetic field, as shown in Fig. 19.17(b). The difference between the experiment and theory could be attributed to the effect of defects, i.e. the stabilization of a different vortex configuration or distortions of the vortex configurations caused by defects. The experimentally observed MVS for $L = 11$ could also be attributed to the effect of defects. For increasing magnetic field, MVSs were observed for $L = 4$–6, which also agreed with the numerical simulation (MVSs for $L = 3$–6, Kanda *et al.* 2004).

19.5.1.1   *Effect of defects*

The influence of defects is noticeable as an asymmetric voltage variation for symmetric vortex states. Figure 19.18 shows the voltage variation near $B = 0$. For the $L = 0$ state both in increasing and decreasing magnetic field, all curves are parallel to each other, showing that a uniform supercurrent is flowing along the disk periphery.[2] This means that there is no crucial defect near the junctions. On the other hand, for the $L = 1$ state, where only one vortex exists in the disk, curves are not exactly parallel. This indicates that the vortex is not exactly at the center of the disk, presumably because of a defect close to (but not at) the disk center.

[2]The difference between the junction voltages are due to the slight difference in junction resistances.

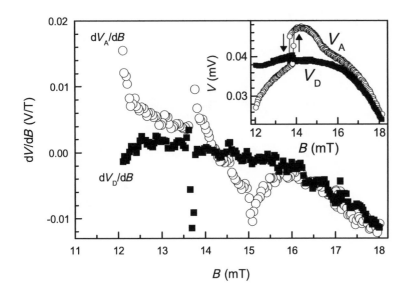

Fig. 19.19 The differential voltages dV_A/dB and dV_D/dB for the whole $L = 8$ state in decreasing magnetic field. The inset shows the measured variation of voltages V_A and V_D for the $L = 8$ state with increasing and decreasing magnetic field. The arrows show the direction of the magnetic-field sweep.

19.5.2 MVS–GVS transition and defect-induced MVS–MVS transition

The MSTJ measurement also provided the evidence for the MVS–GVS and MVS–MVS transitions (Kanda *et al.* 2004). In the above measurement, for $L = 2$ and $L = 7$ to 11, the type of vortex state (whether the state is MVS or GVS) is different for decreasing and increasing magnetic fields. This implies the existence of a MVS–GVS transition at these vorticities, which has been predicted theoretically (Palacios 1998; Schweigert *et al.* 1998) (see also Fig. 19.4), but has never been observed experimentally. Figure 19.19 shows the entire $L = 8$ state, obtained by changing the sweep direction of the magnetic field. In the main panel of Fig. 19.19, the difference between dV_A/dB and dV_D/dB is remarkable below 15.8 mT, indicating that the state is an MVS, while at larger fields dV_A/dB and dV_D/dB coincide, indicating a GVS. Thus, the MVS–GVS transition occurs at 15.8 mT. Note that the observed MVS–GVS transition is a continuous one and is not accompanied by hysteresis, corresponding to the theoretical prediction of a second-order transition (Schweigert *et al.* 1998). Moreover, as shown in the inset of Fig. 19.19, a small voltage jump with hysteresis was observed around 13.7 mT. Because there are noticeable voltage differences between V_A and V_D below and above the jump field, this jump indicates a MVS–MVS transition with different arrangements of the 8 vortices. A comparison of the experimental phase boundary with a numerical simulation provides fundamental insights in the vortex dynamics. Figure 19.20 shows the calculated free energy for the $L = 8$ state. Two possible vortex states appear: a MVS where all 8 vortices are situated on one shell (the (8) state, (i) and (ii) in the figure), and the GVS where the 8 vortices are combined in the center ((iii) in the figure). A MVS–GVS transition occurs at 14.2 mT (indicated by an open circle), without any hysteresis. The theoretical results confirm the continuous MVS–GVS transition that was observed in

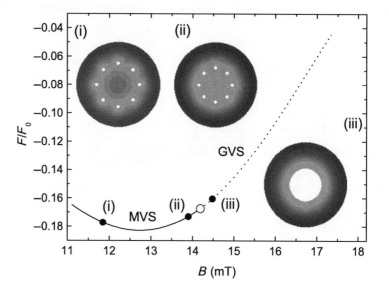

Fig. 19.20 Calculated free energy of the (8) state (solid curve) and the GVS (dashed curve) for a disk without a defect as a function of the applied magnetic field. The insets present the Cooper-pair density of the (8) state at (i) $B = 11.8$ mT, (ii) $B = 13.9$ mT, and (iii) the GVS at 14.5 mT, which are also indicated by the solid dots on the free-energy curve. The open circle indicates the MVS–GVS transition.

the experiment, although the transition field is shifted, probably due to the presence of (weak) defects in the experimental sample. On the other hand, the theory does not give any explanation for the voltage jump and the hysteresis that were observed experimentally around 13.7 mT. From the experiment it is known that near the disk center there is at least one defect that significantly affects the vortex configuration, as discussed in Section 19.5.1.1 (see also Fig. 19.18). To simulate this defect, a circular hole (with radius 0.1ξ) that is 0.2ξ shifted from the disk center, was put in the disk. The free energy including this defect is shown in Fig. 19.21. In addition to the (8) state and the GVS, we have a (1,7) state in which one vortex is near the center pinned by the defect and the other 7 vortices are on a shell. With decreasing field the GVS transits into the (8) state at $B = 14.1$ mT and then into the (1,7) state at $B = 11.7$ mT. With increasing field the (1,7) state transits into the (8) state at $B = 12.9$ mT and then into the giant vortex state at $B = 14.1$ mT. Note that the MVS–GVS transition is still a second-order one with the transition point a little bit smaller than that for a disk without a defect. This means that the presence of a defect near the disk center favors the GVS over the MVS. Also notice that the transition between the (1,7) and (8) states is of first order and is accompanied by hysteresis as well as a jump in the free energy, in agreement with the experimental observation.

19.5.3   Alternative method for the distinction between MVSs and GVSs: Temperature dependence of vortex expulsion fields

The output voltage of the MSTJ measurement is quite sensitive to the change in the supercurrent flowing underneath the junction. Therefore, transitions between different vortex states, which are accompanied by changes of the

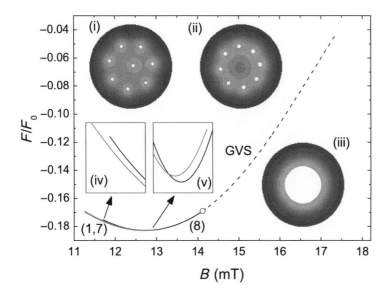

Fig. 19.21 Calculated free energy of the (1,7) state, the (8) state, and the GVS for a disk with a defect as a function of the applied magnetic field. The defect is modelled by a circular hole with radius 0.1ξ that is placed at a distance 0.2ξ from the disk center. The insets (i)–(iii) present the Cooper-pair density of the (1,7) and (8) states at $B = 12.5$ mT, and the GVS at 14.6 mT. Insets (iv) and (v) show the transition between the (1,7) state and the (8) state.

supercurrent distribution, can be detected accurately. A detailed study of the magnetic field for the transition between different vortex states in a mesoscopic disk (Baelus *et al.* 2005) showed that in increasing magnetic field, the transition field (i.e. vortex-penetration field) uniformly decreases for all vorticities with increasing temperature. On the other hand, in a decreasing magnetic field, the transition field (i.e. vortex-expulsion field) shows two kinds of behaviors that correspond to transitions from GVSs or MVSs: when a vortex-expulsion field is almost temperature independent, the transition is from an MVS, and when a vortex-expulsion field increases with temperature, the transition is from a GVS. It is worth noting that this criterion can be applicable to other geometries, for example, squares and triangles (Baelus *et al.* 2006; Kanda *et al.* 2006a). As an example, Fig. 19.22 shows the voltage of a junction in the same sample as that for Figs. 19.16–19.19 as a function of the applied magnetic field at temperatures $T = 0.1$ K (highest curve) to 0.5 K (lowest curve) for increasing (a) and decreasing (b) magnetic field (Baelus *et al.* 2005). The numbers in the figure indicate the vorticity. In increasing magnetic field (Fig. 19.22(a)), the vortex-penetration fields, given by the peaks in the voltage, always decrease with increasing temperature, as indicated by the square symbols for transitions between the states with $L = 4$ and 5, and with $L = 12$ and 13. On the other hand, in a decreasing field (Fig. 19.22(b)), the vortex-expulsion fields show several behaviors: (i) they are almost temperature independent at low fields, as indicated by open circles for the transition from $L = 5$ to 4, (ii) for intermediate fields they are almost temperature independent at low temperatures but increase at higher temperatures, as indicated by open squares for the transition from $L = 11$ to 10, (iii) at higher fields they increase with temperature, as indicated by open triangles for the transition from $L = 13$ to 12. The boundary of (ii) and (iii), i.e. the boundary where the behavior at low temperatures changes, is between $L = 11$ and 12, which corresponds

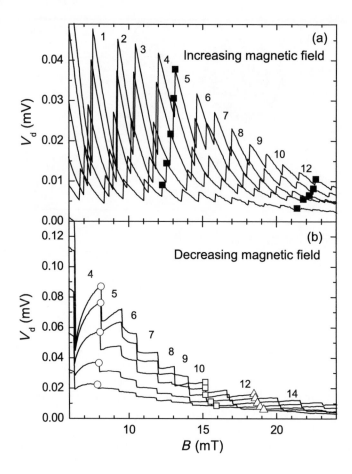

Fig. 19.22 The voltage of junction D as a function of the applied magnetic field when (a) increasing and (b) decreasing the magnetic field for several temperatures, $T = 0.1\,\text{K}$ (highest curve), 0.2, 0.3, 0.4, 0.5 K (lowest curve). The dots indicate the transition fields between the states with $L = 4$ and 5, and with $L = 12$ and 13 in increasing field (a), and the transition fields between the states with $L = 4$ and 5, with $L = 10$ and 11, and with $L = 12$ and 13 in decreasing magnetic field (b), respectively.

to the MVS–GVS boundary for 0.03 K found by the MSTJ measurement (see Fig. 19.17).

To make clear the origin of these behaviors, the temperature dependence of the free energy for the (meta)stable vortex states in a mesoscopic superconducting disk corresponding to the experimental sample was calculated (Baelus *et al.* 2005). Figure 19.23 shows theoretically obtained vortex penetration/expulsion fields together with the experimental results. The numbers in the figure indicate the vorticity of the vortex state before the transition. All the penetration fields decrease with temperature (Figs. 19.23(a) and (c)). The magnetic-field interval between the different transitions is almost constant and decreases slightly with increasing temperature. The expulsion fields (Figs. 19.23(b) and (d)) show three kinds of behaviors (i)–(iii) as discussed above. The theoretical simulation revealed that a transition from a MVS is almost temperature independent (in fact, it decreases very slightly with increasing temperature), while a transition from a GVS moves to higher magnetic field with increasing temperature. Thus, the behavior (i) corresponds to a MVS, and behavior (iii) to a GVS. Behavior (ii) agrees with the theoretical prediction that as the coherence length increases with respect to the sample size, GVSs are stabilized. It was theoretically shown

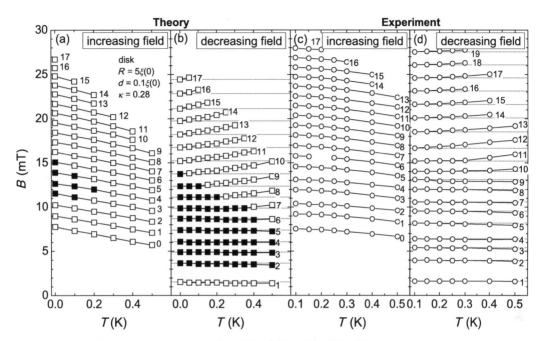

Fig. 19.23 The theoretical and experimental penetration and expulsion fields as a function of temperature; (a),(c) the theoretical and experimental $L \to L+1$ penetration fields, respectively, (b),(d) the theoretical and experimental $L \to L-1$ expulsion fields, respectively. The dashed lines correspond to the values of the expulsion field at the lowest temperature and are guides to the eye. The closed symbols in (a) and (b) correspond to a MVS and open symbols to a GVS just before the transition. From Baelus *et al.* (2005). Copyright by the American Physical Society.

that the shielding current around the sample, which is responsible for the surface barrier for vortex expulsion, has a different temperature dependence for MVSs and GVSs, i.e. in MVSs the magnitude of the shielding current is almost temperature independent, while in GVS it decreases with increasing temperature (Baelus *et al.* 2005).

19.5.4   Sample-size dependence of the vortex states

The above criterion for the distinction between MVSs and GVSs becomes valuable when applied to geometries other than disks (Baelus *et al.* 2006). In a square, for example, the MSTJ method for the distinction between MVSs and GVSs does not work well, because the Cooper-pair density and the super-current distribution for some MVSs have a tetragonal symmetry, which is the same as the sample geometry and is commensurate with the symmetry of the corresponding GVSs. On the other hand, a theoretical simulation showed that the criterion using the temperature dependence of the vortex expulsion field is still valid for squares (Baelus *et al.* 2006), that is, the vortex-expulsion field increases with temperature for transitions from GVS, while it is almost independent of temperature for transitions from MVSs. By using this criterion, the size dependence of the vortex states in squares was studied and qualitative agreement with the theoretical prediction was obtained (Kanda *et al.* 2006b).

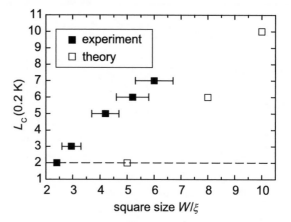

Fig. 19.24 Sample-size dependence of the smallest vorticity, L_C, for the appearance of GVSs at 0.2 K. The filled and open squares indicate the experimental and theoretical values, respectively. $L_C = 2$ means that only GVSs are found. From Kanda *et al.* (2006b), with permission from Elsevier.

Figure 19.24 shows the sample-size dependence of the smallest vorticity, L_C, for the appearance of GVSs obtained both experimentally and theoretically. In both cases, the GVSs are stabilized for smaller samples, as explained in Section 19.3.1, but the experimental L_C is always larger than the theoretical value, meaning that MVSs are more stable in the experimental samples. One of the major reasons is the effect of defects in experimental samples. Defects tend to localize vortices at the defect positions, so that multiple defects may make the MVSs more stable.

19.6   One-dimensional vortex in mesoscopic rings

In the case of a mesoscopic ring, or a mesoscopic disk with holes, vortices tend to be trapped by the hole(s), so that the vortex arrangements and their stability are significantly affected by the existence of the hole(s). Detailed analyses of the effect of holes can be found in the literature (Baelus *et al.* 2000, 2001; Geurts *et al.* 2006, 2007). Among them, one of the most peculiar phenomena is the appearance of the one-dimensional (1D) vortex in thin rings. Notice that in conventional systems, it is impossible to define a vortex in one dimension. Below, we describe the theoretical prediction and the experimental confirmation of the state.

The theoretical prediction of the unconventional behavior in mesoscopic rings has a long history. It was first derived in the 1D limit (de Gennes 1981; Fink *et al.* 1982, 1987, 1988; Straley and Visscher 1982; Berger and Rubinstein 1995, 1997, 1999; Horane *et al.* 1996; Castro and López 2005). In the 1980s it was found that in a 1D ring with an attached superconducting lead a magnetic flux Φ induces a non-uniform distribution of the order parameter, and the order parameter vanishes at one point of the ring when $\Phi = \Phi_0/2$ (Φ_0 is the magnetic flux quantum), leading to a singly connected superconducting

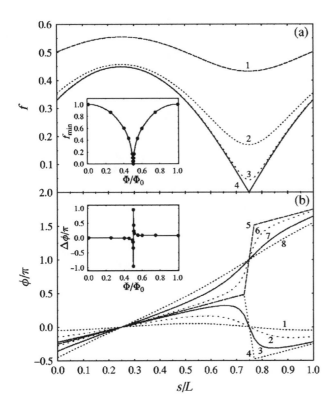

Fig. 19.25 Distribution of the amplitude f (a) and phase ϕ (b) of the order parameter in a non-uniform 1D ring having a width $w(s) = 1 + w_0 \sin(2\pi s/L)$, where $L = 3.25\xi$ is the ring periphery and $w_0 = 0.1\xi$. The magnetic flux is 1: $0.45\Phi_0$, 2: $0.49\Phi_0$, 3: $0.498\Phi_0$, 4: $(0.5-0)\Phi_0$, 5: $(0.5+0)\Phi_0$, 6: $0.502\Phi_0$, 7: $0.51\Phi_0$, 8: $0.55\Phi_0$. Note that the width is minimum at $s/L = 3/4$. In the insets, (a) the minimum amplitude $f_{\min}$ seen at $s/L = 3/4$ and (b) the phase difference $\Delta\phi$ across the point $s/L = 3/4$ are shown as a function of the magnetic flux. From Vodolazov *et al.* (2002). Copyright by the American Physical Society.

state. Later, it was found that the corresponding state appears in a 1D ring with non-uniform cross-section (Berger and Rubinstein 1995, 1997). In 2002, it was pointed out that the phase of the superconducting order parameter jumps by π at the point where the order parameter vanishes, having the property of a vortex (Vodolazov *et al.* 2002). It was called a *1D vortex*. Figure 19.25 shows the amplitude and phase of the order parameter in a thin ring with non-uniform width $w(s) = 1 + w_0 \sin(2\pi s/L)$, where $L = 3.25\xi$ is the ring periphery and $w_0 = 0.1\xi$ (Vodolazov *et al.* 2002). With increasing magnetic flux, the amplitude of the order parameter decreases around the thinnest part of the ring (at $s/L = 3/4$) and becomes zero when $\Phi/\Phi_0 = 0.5$. The phase of the order parameter jumps by π at this magnetic flux. Note that also for a conventional (2D) quantum vortex, the amplitude of the order parameter becomes zero at the core, and the phase jumps by π when passing through the core (while it changes by 2π when going around the core). In this sense, the order parameter shown in Fig. 19.25 has the property of a vortex.

The analyses were extended to rings with finite and non-uniform cross-section, allowing the variation of the order parameter in the radial direction. Although in this case the vortex core where the amplitude of the order parameter vanishes is defined as a point also in the radial direction, the properties corresponding to the above 1D case were still found (Berger and Rubinstein

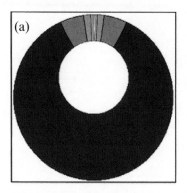 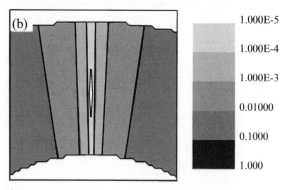

Fig. 19.26 (a) Cooper-pair density in a quasi-1D vortex state in a mesoscopic superconducting ring with finite width, plotted in logarithmic scale. (b) Close-up of the thinnest part of the ring. From Baelus *et al.* (2007), with permission from Elsevier.

1999; Vodolazov *et al.* 2002; Baelus *et al.* 2007; Kanda *et al.* 2007). The present vortex is predicted to have the following properties:

(i) This vortex is stabilized only in mesoscopic superconducting rings having diameter smaller than ξ and non-uniform cross-section and/or an attached lead. Its existence is irrespective of the type of superconductor (type I or II).

(ii) Analogous to the Abrikosov vortices in type-II superconductors, the superconducting phase changes by 2π when encircling the core. Besides, the present vortex can be located even at the sample boundary. In this case, the vortex can be identified by the phase change of π when going through the core. This is in contrast to the Abrikosov vortex that is unstable when the distance to the boundary is too short. Furthermore, the present vortex can be stabilized in a thin ring with width smaller than ξ, where the vortex becomes highly anisotropic and the region where the order parameter Ψ is close to zero extends over the whole width. This vortex corresponds to a 1D vortex derived in the 1D limit, and may be called a *quasi-1D vortex* (see Fig. 19.26, Baelus *et al.* 2007).

(iii) The vortex exists only in a narrow range of applied magnetic flux in the vicinity of a half-integer number of flux quanta: as the applied flux is increased (decreased), a vortex core appears at the outer (inner) edge of the ring, moves continuously towards the inner (outer) edge, and disappears at the inner (outer) edge. Thus, this vortex exists as a stable intermediate state during a transition between different fluxoid states. This fluxoid state transition is reversible and continuous, in contrast to the usual transitions in rings that exhibit clear hysteresis and a jump in the current circulating around the loop (Baelus *et al.* 2000). Notice that this is similar to a time-dependent phase slip center seen in narrow superconducting wires (Tinkham 1996), with the difference that here it is a stable phase slip frozen in "time" and "space". The vortex also differs from the usual Abrikosov vortex in a film: in the latter case,

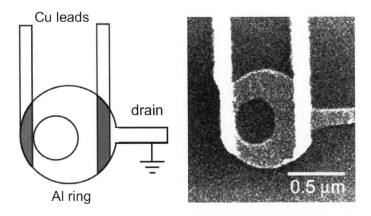

Fig. 19.27 Asymmetric ring for the MSTJ measurement. From Kanda *et al.* (2007). Copyright by the American Physical Society.

when a vortex is nucleated at the sample edge, it inevitably jumps deep inside the sample at the same magnetic field, and it does not disappear at higher magnetic fluxes.

(iv) For a mesoscopic ring with an off-centered hole, the vortex passes through the narrowest part of the ring for small magnetic fluxes, and through the widest part for large fluxes.

The existence of the 1D vortex was experimentally confirmed using the MSTJ measurement (Furugen *et al.* 2007; Kanda *et al.* 2007). The sample was a mesoscopic ring with an inside hole off-centered (Fig. 19.27). Two small tunnel junctions for the MSTJ measurement were attached to the widest part and the narrowest part of the ring, and the superconducting order parameter at these places were compared. Figure 19.28 shows the magnetic-field dependence of the junction voltage for several temperatures. At low temperatures (Fig. 19.28(a)) clear voltage jumps and hysteresis caused by fluxoid-state transitions are seen at $\Phi/\Phi_0 = \pm 0.5$ and ± 1.5, which become smaller with increasing temperature (Fig. 19.28(b)). As predicted by theories (Berger and Rubinstein 1999; Vodolazov *et al.* 2002), at temperatures close to $T_C = 1.36$ K, the voltage discontinuities and the hysteresis disappear (Figs. 19.28(c) and (d)), leading to continuous and reversible transitions. Notice that the voltage of the narrowest part at applied magnetic fluxes $\pm 0.5 \Phi_0$ is close to its normal state value seen at high magnetic flux $\Phi/\Phi_0 > 2$, while that of the widest part remains large even at $\Phi = \pm 0.5 \Phi_0$. This indicates that the superconductivity in the ring is weakened in a non-uniform manner, and that a vortex passes through the narrowest part of the ring, forming a quasi-1D vortex. This interpretation was supported by a numerical analysis within the framework of the GL theory. Besides, further comparison of experimental results with theoretical simulation revealed that for fluxoid state transitions at high magnetic fields, a vortex passes through the widest part of the ring (Kanda *et al.* 2007). These results confirmed the peculiar behaviors predicted for mesoscopic rings.

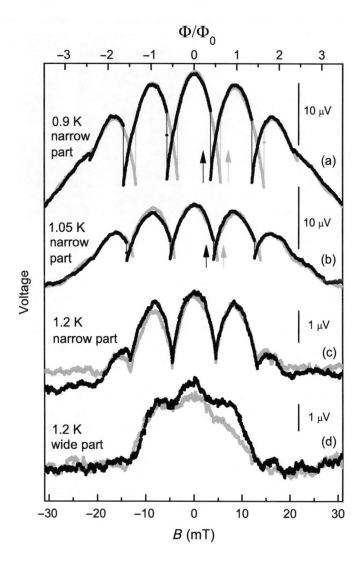

Fig. 19.28 Magnetic-field dependence of the junction voltage (current of 0.3 nA was applied) for the junction at the narrowest part of the ring at (a) 0.9 K, (b) 1.05 K, and (c) 1.2 K and (d) for the junction at the widest part at 1.2 K. The arrows indicate the sweep direction. From Kanda *et al.* (2007). Copyright by the American Physical Society.

19.7   Conclusion

In this chapter, theoretical and experimental progresses in the research of mesoscopic superconducting states, which are expected to appear in nanoscale superconductors, are reviewed. Theoretically, thanks to the development in simulation techniques, very accurate predictions of the novel vortex states in mesoscopic superconductors have been made within the framework of the non-linear Ginzburg–Landau theory, and the existence of novel vortex states such as giant vortex states, multivortex states, vortex-antivortex molecules, and a one-dimensional vortex has been foretold. Experimentally, a variety of techniques have been developed for the observation of vortices and some of them have been applied to mesoscopic superconductors. In this chapter, experimental results obtained by the MSTJ method are described and experimental evidences for mesoscopic vortex states are shown. In nanoscale superconductors, vortex states strongly depend on the geometry of the sample, so that it will be possible to control the quantum states by changing external parameters such as local field and injected currents (Milošević *et al.* 2007). The research of mesoscopic superconducting states not only gives a better understanding of the physics of quantum systems confined in a small area, but also leads to the future device application based on nanoscale superconductivity engineering.

References

Abrikosov, A.A. *Zh. Eksperim. I Theor. Fiz.* **32**, 1442 (1957); [*Sov. Phys. JETP* **5**, 1174].

Abrikosov, A.A. *Physica C* **404**, 1, and chapters in this volume (2004).

Aliev, F.G. P*hysica C* **437–438**, 1, and chapters in this volume (2006).

Anderson, P.W. *J. Phys. Chem. Solids* **11**, 28 (1959).

Baelus, B.J., Peeters, F.M., Schweigert, V.A. *Phys. Rev. B* **61**, 9734 (2000).

Baelus, B.J., Peeters, F.M., Schweigert, V.A. *Phys. Rev. B* **63**, 144517 (2001).

Baelus, B.J., Peeters, F.M. *Phys. Rev. B* **65**, 104515 (2002).

Baelus, B.J., Cabral, L.R.E., Peeters, F.M. *Phys. Rev. B* **69**, 064506 (2004).

Baelus, B.J., Kanda, A., Peeters, F.M., Ootuka, Y., Kadowaki, K. *Phys. Rev. B* **71**, 140502 (R) (2005).

Baelus, B.J., Kanda, A., Shimizu, N., Tadano, K., Ootuka, Y., Kadowaki, K., Peeters, F.M. *Phys. Rev. B* **73**, 024514 (2006).

Baelus, B.J., Kanda, A., Vodolazov, D.Y., Ootuka, Y., Peeters, F.M. *Physica C* **460–462**, 320 (2007).

Banča, J., Kabanov, V.V. *Phys. Rev. B* **65**, 012509 (2001).

Bardeen, J. *Rev. Mod. Phys.* **34**, 667 (1962).

Bardeen, J., Cooper, L.N., Schrieffer, J.R. *Phys. Rev.* **108**, 1175 (1957).

Berger, J., Rubinstein, J. *Phys. Rev. Lett.* **75**, 320 (1995).

Berger, J., Rubinstein, J. *Phys. Rev. B* **56**, 5124 (1997).

Berger, J., Rubinstein, J. *Phys. Rev. B* **59**, 8896 (1999).

Bertrand, D. Ph.D. thesis. Université Catholique de Louvain, Belgium (2005).

Blaauwgeers, R., Eltsov, V.B., Krusius, M., Ruohio, J.J., Schanen, R., Volovik, G.E. *Nature* (London) **404**, 471 (2000).

Black, C.T., Ralph, D.C., Tinkham, M. *Phys. Rev. Lett.* **76**, 688 (1996).

Bruyndoncx, V., Rodrigo, J.G., Puig, T., Van Look, L., Moshchalkov, V.V., Jonckheere, R. *Phys. Rev. B* **60**, 4285 (1999).

Cabral, L.R.E., Baelus, B.J., Peeters, F.M. *Phys. Rev. B* **70**, 144523 (2004).

Castro, J.I., López, A. *Phys. Rev. B* **72**, 224507 (2005).

Chang, A.M., Hallen, H.D., Harriot, L., Hess, H.F., Los, H.L., Kao, J., Miller, R.E., Chang, T.Y. *Appl. Phys. Lett.* **61**, 1974 (1992).

Chibotaru, L.F., Ceulemans, A., Bruyndoncx, V., Moshchalkov, V.V. *Nature* (London) **408**, 833 (2000).

Chibotaru, L.F., Ceulemans, A., Bruyndoncx, V., Moshchalkov, V.V. *Phys. Rev. Lett.* **86**, 1323 (2001).

Chibotaru, L.F., Teniers, G., Ceulemans, A., Moshchalkov, V.V. *Phys. Rev. B* **70**, 094505 (2004).

de Gennes, P.-G. *C.R. Seances Acad. Sci. Ser. 1* **292**, 279 (1981).

Deo, P.S., Schweigert, V.A., Peeters, F.M. *Phys. Rev. Lett.* **79**, 4653 (1997).

Dolan, G.J. *Appl. Phys. Lett.* **1**, 337 (1977).

Essmann, U., Traube, H. *Phys. Lett.* **24A**, 526 (1967).

Field, S.B., James, S.S., Barentine, J., Metlushko, V., Crabtree, G., Shtrikman, H., Ilic, B., Brueck, S.R. *Phys. Rev. Lett.* **88**, 067003 (2002).

Fink, H.J., López, A., Maynard, R. *Phys. Rev. B* **26**, 5237 (1982).

Fink, H.J., Grünfeld, V., López, A. *Phys. Rev. B* **35**, 35 (1987).

Fink, H.J., Loo, J., Roberts, S.M. *Phys. Rev. B* **37**, 5050 (1988).

Furugen, R., Kanda, A., Vodolazov, D.Y., Baelus, B.J., Ootuka, Y., Peeters, F.M. *Physica C* **463–465**, 251 (2007).

Geim, A.K., Dubonos, S.V., Lok, J.G.S., Grigorieva, I.V., Maan, J.C., Theil Hansen, L., Lindelof, P.E. *Appl. Phys. Lett.* **71**, 2379 (1997a).

Geim, A.K., Grigorieva, I.V., Dubonos, S.V., Lok, J.G.S., Maan, J.C., Filippov, A.E., Peeters, F.M. *Nature* (London) **390**, 259 (1997b).

Geim, A.K., Dubonos, S.V., Lok, J.G.S., Henini, M., Maan, J.C. *Nature* (London) **396**, 144 (1998).

Geurts, R., Milošević, M.V., Peeters, F.M. *Phys. Rev. Lett.* **97**, 137002 (2006).

Geurts, R., Milošević, M.V., Peeters, F.M. *Phys. Rev. B* **75**, 184511 (2007).

Ginzburg, V.L., Landau, L.D. *Zh. Eksperim. I Teor. Fiz.* **20**, 1064 (1950).

Grigorenko, A., Bending, S., Tamegai, T., Ooi, S., Henini, M. *Nature* (London) **414**, 728 (2001).

Grigorieva, I.V., Escoffier, W., Richardson, J., Vinnikov, L.Y., Dubonos, S.V., Oboznov, V. *Phys. Rev. Lett.* **96**, 077005 (2006).

Grigorieva, I.V., Escoffier, W., Misko, V.R., Baelus, B.J., Peeters, F.M., Vinnikov, L.Y., Dubonos, S.V. *Phys. Rev. Lett.* **99**, 147003 (2007).

Harada, K., Matsuda, T., Bonevich, J., Igarashi, M., Kondo, S., Pozzi, G., Kawabe, U., Tonomura, A. *Nature* (London) **360**, 51 (1992).

Hata, Y., Suzuki, J., Kakeya, I., Kadowaki, K., Odawara, A., Nagata, A., Nakayama, S., Chinone, K. *Physica C* **388**, 719 (2003).

Hayashi, M., Kaiwa, T., Ebisawa, H., Matsushima, Y., Shimizu, M., Satoh, K., Yotsuya, T., Ishida, T. *Physica C* **468**, 801 (2008).

Hess, H.F., Robinson, R.B., Dynes, R.C., Valles, J.M., Waszczak, J.V. *Phys. Rev. Lett.* **62**, 214 (1989).

Horane, E.M., Castro, J.I., Buscaglia, G.C., López, A. *Phys. Rev. B* **53**, 9296 (1996).

Kanda, A., Geisler, M.C., Ishibashi, K., Aoyagi, Y., Sugano, T. In *Quantum Coherence and Decoherence – ISQM – Tokyo '98*, (eds) Ono, Y.A. and Fujikawa, K. (Elsevier, Amsterdam, 1999a) p. 229.

Kanda, A., Ishibashi, K., Aoyagi, Y., Sugano, T. In *Physics and Applications of Mesoscopic Josephson Junctions*, (eds) Ohta, H. and Ishii, C. (The Physical Society of Japan, Tokyo, 1999b) p. 305.

Kanda, A., Geisler, M.C., Ishibashi, K., Aoyagi, Y., Sugano, T. *Physica B* **284–288**, 1870 (2000).

Kanda, A., Ootuka, Y. *Microelectron. Eng.* **63**, 313 (2002).

Kanda, A., Baelus, B.J., Peeters, F.M., Kadowaki, K., Ootuka, Y. *Phys. Rev. Lett.* **93**, 257002 (2004).

Kanda, A., Baelus, B.J., Shimizu, N., Tadano, K., Peeters, F.M., Kadowaki, K., Ootuka, Y. *Physica C* **437–438**, 122 (2006a).

Kanda, A., Baelus, B.J., Shimizu, N., Tadano, K., Peeters, F.M., Kadowaki, K., Ootuka, Y. *Physica C* **445–448**, 253 (2006b).

Kanda, A., Baelus, B.J., Vodolazov, D.Y., Berger, J., Furugen, R., Ootuka, Y., Peeters, F.M. *Phys. Rev. B* **76**, 094519 (2007).

Karapetrov, G., Fedor, J., Iavarone, M., Rosenmann, D., Kwok, W.K. *Phys. Rev. Lett.* **95**, 167002 (2005).

Koblischka, M.R., Wijngaarden, R.J. *Suprecond. Sci. Technol.* **8**, 199 (1995).

Levine, J.L. *Phys. Rev. Lett.* **15**, 154 (1965).

Little, W.A., Parks, R.D. *Phys. Rev. Lett.* **9**, 9 (1962).

Little, W.A., Parks, R.D. *Phys. Lett.* **133**, A97 (1964).

Matthews, M.R., Anderson, B.P., Haljan, P.C., Hall, D.S., Wieman, C.E., Cornell, E.A. *Phys. Rev. Lett.* **83**, 2498 (1999).

Mel'nikov, A.S., Nefedov, I.M., Ryzhov, D.A., Shereshevskii, I.A., Vinokur, V.M., Vysheslavtsev, P.P. *Phys. Rev. B* **65**, 140503 (R) (2002).

Mertelj, T., Kabanov, V.V. *Phys. Rev. B* **67**, 134527 (2003).

Milošević, M.V., Berdiyorov, G.R., Peeters, F.M. *Appl. Phys. Lett.* **91**, 212501 (2007).

Misko, V.R., Fomin, V.M., Devreese, J.T., Moshchalkov, V.V. *Phys. Rev. Lett.* **90**, 147003 (2003).

Moshchalkov, V.V., Qiu, X.G., Bruyndoncx, V. *Phys. Rev. B* **55**, 11793 (1997). Notice that the numerical results presented in Figs. 4(c) and 5(a) of this publication are incorrect, because the authors took the wrong boundary condition for the magnetic field in the center of the disk, i.e. zero magnetic field in the center of the disk, even in the case of a giant vortex state.

Nishio, T., Ono, M., Eguchi, T., Sakata, H., Hasegawa, Y. *Appl. Phys. Lett.* **88**, 113115 (2006).

Nishio, T., Hata, Y., Okayasu, S., Suzuki, J., Nakayama, S., Nagata, A., Odawara, A., Chinone, K., Kadowaki, K. Scanning SQUID microscope study of vortex states and phases in superconducting mesoscopic dots, antidots, and other structures. In *Oxford Handbook of Nanoscience and Technology*, Vol. II. (Oxford, Oxford University Press, 2010).

Oral, A., Bending, S.J., Humphreys, R.G., Henini, M. *Suprecond. Sci. Technol.* **10**, 17 (1997).

Palacios, J. J. *Phys. Rev. B* **58**, 5948 (R) (1998).

Palacios, J. J. *Phys. Rev. Lett.* **84**, 1796 (2000).

Pippard, A.B. *Proc. Roy. Soc.* (London) **A216**, 547 (1953).

Puig, T., Rosseel, E., Baert, M., Van Bael, J., Moshchalkov, V.V., Bruynseraede, Y. *Appl. Phys. Lett.* **70**, 3155 (1997).

Puig, T., Rosseel, E., Van Look, L., Van Bael, J., Moshchalkov, V.V., Bruynseraede, Y., Jonckheere, R. *Phys. Rev. B* **58**, 5744 (1998).

Sakata, H., Oosawa, M., Matsuba, K., Nishida, N., Takeya, H., Hirata, K. *Phys. Rev. Lett.* **84**, 1583 (2000).

Schweigert, V.A., Peeters, F.M. *Phys. Rev. B* **57**, 13817 (1998).

Schweigert, V.A., Peeters, F.M. *Phys. Rev. B* **60**, 3084 (1999).

Schweigert, V.A., Peeters, F.M., Deo, P.S. *Phys. Rev. Lett.* **81**, 2783 (1998).

Straley, J.P., Visscher, P.B. *Phys. Rev. B* **26**, 4922 (1982).

Straub, R., Keil, S., Kleiner, R. *Appl. Phys. Lett.* **78**, 3645 (2001).

Tinkham, M. *Introduction to Superconductivity*, 2nd edn. (McGraw-Hill, New York, 1996).

Tilley, D.R., Tilley, J. *Superfluidity and Superconductivity* (IOP, New York, 1990).

Vodolazov, D.Y., Baelus, B.J., Peeters, F.M. *Phys. Rev. B* **66**, 054531 (2002).

Vollhardt, D., Wolfe, P. *The Superfluid Phases of Helium 3* (Taylor & Francis, London, 1990).

Volodin, A., Temst, K., Van Haesendonck, C., Bruynseraede, Y. *Physica B* **284–288**, 815 (2000).

Volodin, A., Temst, K., Van Haesendonck, C., Bruynseraede, Y., Montero, M.I., Schuller, I.K. *Europhys. Lett.* **58**, 582 (2002).

Von Delft, J., Ralph, D.C. *Phys. Rep.* **345**, 61 (2001).

Yoshikawa, H., Sata, K., Nakata, S., Sato, O., Kato, M., Kasai, J., Hasegawa, T., Satoh, K., Yotsuya, T., Ishida, T. *Physica C* **412**, 552 (2004).

Left-handed metamaterials—A review

20

E. Ozbay, G. Ozkan, and K. Aydin

20.1   Introduction

Metamaterials have become a remarkable research area in recent years and received burgeoning interest due to their unprecedented properties unattainable from ordinary materials. Veselago pointed out that a material exhibiting negative values of dielectric permittivity (ε) and magnetic permeability (μ) would have a negative refractive index (Veselago 1968). Generally speaking, the dielectric permittivity (ε) and the magnetic permeability (μ) are both positive for natural materials. In fact, it is possible to obtain negative values for ε and μ by utilizing proper designs of metamaterials. To be specific, negative permittivity values at microwave frequencies are accessed by making use of thin metallic wire meshes (Pendry *et al.* 1998). It is rather difficult to obtain negative permeability due to the absence of magnetic charges. Pendry *et al.* came up with a brilliant solution and employed an array of split-ring resonators exhibiting negative effective permeability (μ_{eff}) values for frequencies close to the magnetic resonance frequency (ω_{m}) of the split-ring resonators (Pendry *et al.* 1999). The first steps to realize these novel type of materials were taken by Smith *et al.*, where they were able to observe a left-handed propagation band at frequencies where both dielectric permittivity and magnetic permeability of the composite metamaterial are negative (Smith *et al.* 2000; Shelby *et al.* 2001). Soon after, left-handed metamaterials with an effective negative index of refraction were successfully demonstrated by various groups (Shelby *et al.* 2001; Houck *et al.* 2003; Parazzoli *et al.* 2003; Aydin *et al.* 2005). Negative refraction is also achieved by using periodically modulated two-dimensional photonic crystals (Cubukcu *et al.* 2003).

One of the most exciting applications of negative-index metamaterials (NIM) is a perfect lens (Pendry 2000). Stimulated by J.B. Pendry's seminal work, superlenses that are capable of imaging subwavelength-size objects attracted a great deal of interest. Subdiffraction-free imaging is experimentally demonstrated for photonic crystals (Cubukcu *et al.* 2003), left-handed transmission lines (Grbic and Eleftheriades 2004) and left-handed metamaterials (LHM) (Lagarkov and Kissel 2004; Aydin *et al.* 2005, 2006). In the near-field regime, the electrostatic and magnetostatic limits apply and thus the electric and magnetic responses of materials can be treated as decoupled. This in turn, brings the possibility of constructing superlenses from materials with negative permittivity (Fang *et al.* 2005; Melville and Blaikie 2005; Taubner *et al.* 2006) or negative permeability (Guven and Ozbay 2006). Metamaterials offer a wide range of exciting physical phenomena that are not attainable with ubiquitous materials. To be specific, it has recently been shown that one achieves cloaking by using metamaterial coatings (Pendry *et al.* 2006; Schurig *et al.* 2006). The metamaterial structures were recently used to obtain novel mechanisms in non-linear optics (Klein *et al.* 2006), and to increase the performance of active devices (Chen *et al.* 2006; Padilla *et al.* 2006). Metamaterials are geometrically scalable, thus offering a wide range of operation frequencies including radio (Wiltshire *et al.* 2001), microwave (Bayindir *et al.* 2002; Ziolkowski 2003; Aydin *et al.* 2004; Katsarakis *et al.* 2004; Bulu *et al.* 2005; Gundogdu *et al.* 2006; Guven *et al.* 2006), millimeter-wave (Gokkavas *et al.* 2006), far infrared (IR) (Yen *et al.* 2004), mid-IR (Linden *et al.* 2004; Moser *et al.* 2005), near-IR (Shalaev *et al.* 2005; Zhang *et al.* 2005) frequencies and even visible wavelengths (Grigorenko *et al.* 2005; Dolling *et al.* 2007). Recently, an acoustic analog of electromagnetic metamaterials so-called ultrasonic metamaterials were demonstrated (Fang *et al.* 2006).

20.2   Negative-permeability metamaterials

20.2.1   Resonances of split-ring resonators

The response of materials to the incident magnetic field is determined by the magnetic permeability. Magnetic permeability is positive in normal materials. The absence of the negative values of magnetic permeability provided little motivation for studying negative-index materials. Pendry *et al.* proposed split-ring resonator (SRR) structures to obtain negative permeability values (Pendry *et al.* 1998). The resonant behavior of SRRs is due to the capacitive elements (gaps and splits), which in turn results in rather high positive and negative values of permeability near the magnetic resonance frequency (ω_m). The resonance characteristics of SRRs are studied in the literature to understand the mechanism behind negative permeability (Weiland *et al.* 2001; Gay-Balmaz and Martin 2002; Marques *et al.* 2003; Aydin *et al.* 2005; Aydin and Ozbay 2006). Split-ring resonators under investigation are built from concentric metal rings on a dielectric printed circuit board with a thickness of 1.6 mm and $\varepsilon = 3.85$. A schematic drawing of a split-ring resonator can be seen in Fig. 20.1(a). The width of the splits and the gap between the inner and outer rings are 0.2 mm, the metal width is 0.9 mm, and the outer radius is 3.6 mm. The deposited metal is copper with a thickness of 30 μm. We also fabricated a ring-resonator structure in which the splits are removed. The resulting structure is named conventionally as a closed-ring resonator (CRR) and shown schematically in Fig. 20.1(b).

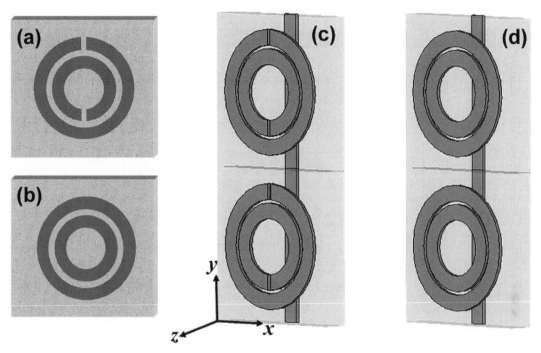

Fig. 20.1 Schematic drawing of (a) a single split-ring resonator (SRR), (b) closed-ring resonator (CRR), (c) a layer of one-dimensional left-handed metamaterial (LHM) composed of SRR and wire arrays, and (d) a layer of one-dimensional composite material (LHM) composed of CRR and wire arrays.

We first measured the frequency response of the single split-ring resonator and closed-ring resonator unit cells. Two monopole antennas were used to transmit and detect the EM waves through the single SRR unit (Aydin *et al.* 2005). Monopole antennas were then connected to the HP-8510C network analyzer for measuring the transmission coefficients. The measured frequency response of single SRR (solid line) and single CRR (dashed line) is provided in Fig. 20.2. As seen in the figure, three transmission dips were observed at frequencies of 3.82, 8.12, and 10.90 GHz throughout the transmission spectrum of a single SRR. On the other hand, a single dip was observed at 10.92 GHz for a single CRR. We also performed simulations to compare with the experimental results. Simulations were performed by using commercial software, CST Microwave Studio, which is a 3D full-wave solver, and employs the finite-integration technique. The simulated frequency responses of SRR and CRR are given in Fig. 20.2. The simulations agree well with the experiments.

The splits in the split-ring resonators structure play a key role in obtaining magnetic resonance. Removing the splits prevents the current from flowing between the inner and outer rings, and therefore, the magnetic resonance is no longer present. Based on this principle, we observed two magnetic resonances for a split-ring resonator structure at $\omega_{m1} = 3.82$ GHz and $\omega_{m2} = 8.12$ GHz. Recent studies showed that it is possible to obtain tunable metamaterials by loading SRR with varactors (Gil *et al.* 2004; Shadrivov *et al.* 2006) or capacitors (Aydin and Ozbay 2007). By changing the capacitance at the

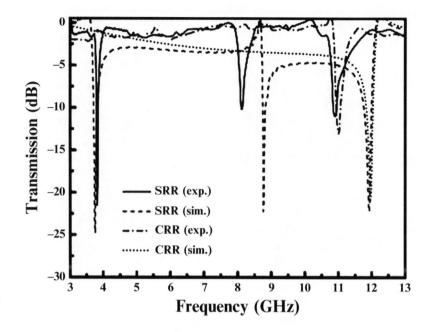

Fig. 20.2 Measured and simulated frequency responses of single SRR and CRR structures.

capacitive regions of SRRs, the magnetic resonance frequency can successfully be changed.

A split-ring resonator not only exhibits magnetic resonance induced by the splits at the rings, but electric resonance is also present via the dipole-like charge distribution along the incident electric field (Aydin *et al.* 2004; Katsarakis *et al.* 2004). Such an electric resonance behavior is observed at $\omega_e = 10.90$ GHz for the split-ring resonator and closed-ring resonator unit cells.

20.2.2   Split-ring resonator arrays

In the previous section we studied the frequency response of single SRR and CRR unit cells. If these structures are combined together, the coupling between the resonators results in bandgaps around resonance frequencies. To obtain negative permeability, we periodically arranged SRR structures. The number of unit cells along the *x*-, *y*-, and *z*-directions are $N_x = 10$, $N_y = 15$, and $N_z = 25$, with lattice spacings $a_x = a_y = 8.8$ mm and $a_z = 6.5$ mm. The directions can be seen in Fig. 20.1(c). The wave vector is along the *x*-direction, while the *E* field is along the *y*-direction, and the *H* field is along the *z*-direction. The experimental setup for measuring the transmission–amplitude and transmission–phase spectra consists of a HP 8510C network analyzer, and standard high-gain microwave horn antennas (Aydin *et al.* 2004). Figure 20.3 shows the measured and simulated transmission spectra of periodic SRRs and CRRs. The first bandgap between 3.55–4.05 GHz is observed at the transmission spectrum of the SRR array but not in the CRR array. However, the second

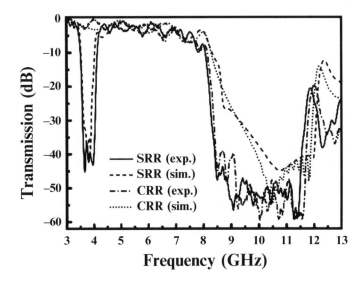

Fig. 20.3 Measured and simulated transmission spectra of SRR and CRR arrays.

bandgap between 8.15–11.95 GHz is observed for both cases. As clearly seen in the figure, the agreement between the measured and simulated data is good.

Based on the measurements and simulations, we can safely claim that the stop bands of split-ring resonator media cannot be assumed as a result of "negative μ" behavior. Some of the observed gaps (such as the second bandgap in this measurement) in the transmission spectra could also originate from the electrical response of the split-ring resonators or from Bragg gaps due to periodicity (Aydin *et al.* 2004). The bandgap between 3.55–4.05 GHz is due to the magnetic response of split-ring resonators. However, the stop band 8.15–11.95 GHz appeared due to the electrical response of the concentric rings.

20.3 Left-handed metamaterial

20.3.1 Left-handed transmission band

The LHM studied in this work consists of the periodic arrangement of SRR and thin-wire arrays. Split-ring resonator and wire patterns are fabricated on the front and back sides of FR4 printed circuit boards, respectively. A schematic drawing of a sample is provided in Fig. 20.1(c). The length and width of the continuous thin wire structures are $l = 19$ cm, and $w = 0.9$ mm. The left-handed material is composed of $N_x = 5$, $N_y = 15$, and $N_z = 32$ unit cells, with lattice spacings $a_x = a_y = 8.8$ mm and $a_z = 6.5$ mm (Aydin *et al.* 2004). Note that the lattice spacings are kept the same as the lattice spacing of only split-ring resonator medium studied in the previous section.

Figure 20.4(a) depicts the measured transmission spectra of SRR, LHM and a composite metamaterial (CMM) composed of CRR and wire arrays with 5 unit cells along the propagation direction. The bandgap in the transmission spectrum of SRR between 3.55–4.05 GHz is shown to be due to magnetic

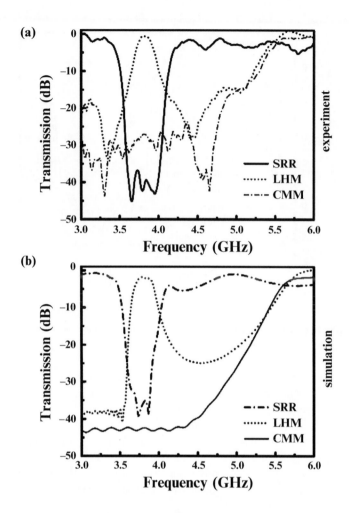

Fig. 20.4 Transmission spectra of periodic SRR, LHM and CMM arrays obtained from (a) measurements and (b) numerical simulations. The shaded regions correspond to the left-handed propagation regime.

resonance in the previous section, therefore the effective magnetic permeability is negative within this frequency range. Negative permittivity is satisfied by using thin metallic wire meshes (Pendry *et al.* 1998). The plasma frequency of the wire array that is used to construct the left-handed material in this study is shown to be at 8.0 GHz in a previous study (Aydin *et al.* 2004). Therefore, below 8.0 GHz, the effective permittivity of the wire array is negative. The condition for the formation of the left-handed transmission band is that the effective ε and μ should be simultaneously negative at a particular frequency range. As seen in Fig. 20.4(a), a transmission band is observed between 3.55–4.05 GHz. The transmission peak is measured to be -0.8 dB at 3.86 GHz. This is the highest transmission peak measured for a LHM structure. For the composite metamaterial of CRR and wire arrays, the transmission band disappeared. We indicated that the wire media has a plasma frequency around 8 GHz. SRR and CRR structures also have electric responses and therefore contribute to the total electric responses of the composite systems (LHM and CMM). The resulting LHM has a plasma frequency around 5.4 GHz, therefore the plasma frequency is reduced (Aydin *et al.* 2004). We performed numerical simulations to check the validity of the experimental results. Numerical simulation results are provided in Fig. 20.4(b) and predict the left-handed transmission band with peak value of −2.7 dB between 3.60–4.10 GHz.

20.3.2 Reflection characteristics of 1D double-negative material

In this section we present the reflection measurements of a 1D LHM structure. In the measurements, transmitter and receiver horn antennas were placed close to each other by keeping the angle between the antennas very small. The transmitter horn antenna sends the EM wave to the first surface of the structures and the receiver antenna measures the amplitude of the reflected EM waves. The number of layers along the propagation direction is $N_x = 10$ layers in this study. Figure 20.5(a) shows the measured transmission (solid line) and reflection (dashed line) spectra. The transmission peak is measured to be −9.2 dB at 3.76 GHz. As expected, as the number of layers along the propagation direction

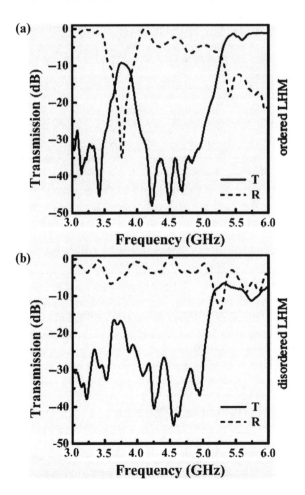

Fig. 20.5 Measured transmission (solid line) and reflection (dashed line) spectra of 10-layer (a) ordered, and (b) disordered LHM structures.

increases, the transmission within the left-handed transmission band decreases due to the losses.

In the reflection spectrum, a sharp dip with a minimum value of $-35\,\text{dB}$ is observed at 3.77 GHz. The incident EM waves with frequencies around 3.77 GHz are almost transmitted through the left-handed material without being reflected at the left-handed material air interface. The low reflection from the surface can be attributed to either matched impedance at the interface or to the thickness resonance of the slab. In a recent study, we have shown by extracting the effective parameters using a retrieval procedure, that the impedance is matched to the free space (Aydin *et al.* 2006). Impedance matching is desired for NIM structures, since it is required to achieve a perfect lens (Pendry 2000).

20.3.3   Effect of disorders on transmission and reflection

We investigated the transmission and reflection properties of ordered left-handed materials. In this section we discuss the effect of disorder on the transmission and reflection characteristics. The transmission characteristics of disordered left-handed materials have been investigated in detail (Aydin *et al.* 2004; Guven *et al.* 2005; Gorkunov *et al.* 2006). The effects of disorder need to be investigated for determining the restrictions imposed on the impedance matching of LHMs. For this purpose, we introduced disorder into the LHM system by destroying the periodicity of split-ring resonator array along x-and y-directions, randomly. The disorder is introduced as follows: Each split-ring resonator on the board with lattice point, $\mathbf{r}_n$, where $\mathbf{r} = x\mathbf{i} + y\mathbf{j}$, is displaced with $\mathbf{r}_n \pm \delta_r$. Here, δ_r is the randomness parameter and we choose it to be $|\boldsymbol{\delta}_r| = a/9$, where $a = a_x = a_y = 8.8\,\text{mm}$, is the lattice constant of the periodic SRR array. Then, the disordered LHM is obtained by arranging disordered split-ring resonator arrays with the ordered wire arrays with the same number of unit cells along all directions (Aydin *et al.* 2004).

Figure 20.5(b) plots the transmission (solid line) and reflection (dashed line) spectra of a disordered left-handed material array. There is a significant amount of reduction in the transmission peak. The peak value is measured as $-16.1\,\text{dB}$ and the transmission band became narrower. A wider reflection band is observed where the minimum value of the reflection is around $-7\,\text{dB}$. A sharp dip as in the case of ordered left-handed materials was not observed. It is obvious that the disorders affect both the reflection and transmission properties of LHMs. They affect the coupling between the SRRs and wires, therefore the resulting left-handed materials' characteristics are changed. The periodicity is important to achieve high transmission and low reflection from a left-handed metamaterial.

20.4   Negative refraction

In the previous two sections we performed transmission and reflection measurements to characterize the split-ring resonators and left-handed metamaterials. In this section we move on to the experiments to measure the refractive

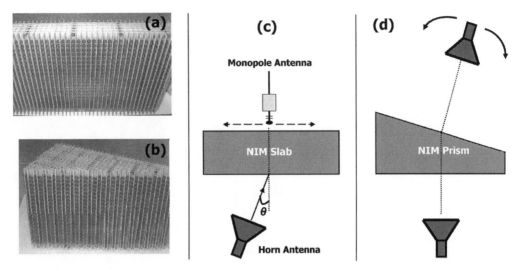

Fig. 20.6 Photographs of (a) 10-layer NIM slab and (b) wedge-shaped NIM. Scheme of experimental setups for verifying negative refraction by using (c) beam-shift method, and (d) wedge experiments.

index of our sample in the left-handed frequency region. We performed three different and independent experimental measurements to verify that the index of refraction is negative. First, we measured refraction through a 2D left-handed material slab by using the beam-shift method. We also performed refraction experiments on wedge-shaped samples. Finally, we performed a phase-shift experiment to verify and calculate the negative refractive index.

20.4.1 Refraction through slab-shaped left-handed materials

The 2D left-handed material is composed of $N_x = 10$, $N_y = 20$, and $N_z = 40$ unit cells, with lattice spacings $a_x = a_y = a_z = 9.3$ mm as seen in Fig. 20.6(a) (Aydin and Ozbay 2006). The refraction spectrum is measured by a setup consisting of a microwave horn antenna as the transmitter, and a monopole antenna as the receiver (see Fig. 20.6(c)). The incident EM wave hits the air left-handed material interface with an angle of $\theta_i = 15°$. The source is 12 cm (1.5λ) away from the first interface of the NIM slab. The spatial intensity distribution along the second NIM air interface is scanned in $\Delta x = 2.5$ mm steps.

Figure 20.7(a) displays the refracted beam profile at 3.86 GHz. The center of the refracted Gaussian beam is measured -1.25 cm away from the center of the incident Gaussian beam. Note that the incident EM wave has a Gaussian beam profile centered at $x = 0$ (not shown in the figure). One can easily find the refractive-index value by applying Snell's law, where $n_{air} \sin \theta_i = n_{LHM} \sin \theta_r$. The angle of refraction can be defined in terms of the beam shift (d_s) and width of the left-handed material slab (w) as, $\theta_r = \arctan(d_s/w)$. The effective

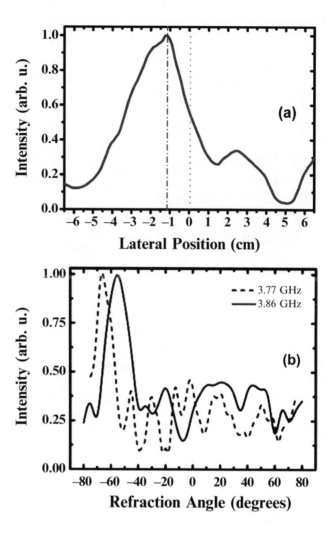

Fig. 20.7 (a) Measured refracted beam profile on the exit surface of NIM slab at 3.86 GHz. The center of the refracted beam is shifted to the left, meaning that the effective refractive index is negative. (b) The measured beam profiles from the waves refracted from a wedge-shaped 2D left-handed material at 3.77 GHz and 3.86 GHz.

refractive index of the NIM is then calculated as $n_{\text{eff}} = -1.91$ at 3.86 GHz (Aydin and Ozbay 2006).

20.4.2   Refraction through wedge-shaped left-handed materials

A typical experimental method for the observation of left-handed properties is to use wedge-shaped structures (Shelby *et al.* 2001; Houck *et al.* 2003; Parazzoli *et al.* 2003; Aydin *et al.* 2005). We constructed a wedge-shaped left-handed material with a wedge angle of $\theta = 26°$ as seen in Fig. 20.6(b). The first interface of the left-handed material is excited with EM waves emanating from the transmitter horn antenna located at a distance of 13 cm ($\sim2\lambda$) away from the first interface of the wedge. The receiver horn antenna is placed 70 cm ($\sim10\lambda$) away from the second interface of the wedge shaped NIM. It is mounted on a rotating arm to scan the angular distribution of the refracted signal. The angular refraction spectrum is scanned in $\Delta\theta = 2.5°$ step sizes. A scheme of the experimental setup is provided in Fig. 20.6(d).

The refracted-beam profiles from the NIM air interface are shown in Fig. 20.7(b) for two different frequencies, 3.77 GHz (dashed line) and 3.86 GHz (solid line). At these frequencies, the beam is refracted on the negative side of the normal, indicating that the angle of refraction is negative. The angle of the refraction at 3.77 GHz is measured to be $\theta_r = -65°$ and at 3.86 GHz $\theta_r = -60°$. We can use Snell's law for calculating the n_{eff} of the 2D NIM structure by the simple formula $n_{eff} \sin\theta_i = n_{air} \sin\theta_r$. At 3.86 GHz the refractive index is calculated as $n_{eff} = -1.98 \pm 0.05$. The angle of refraction increases when the frequency is lowered, hence the n_{eff} increases. The calculated refractive index at 3.77 GHz is $n_{eff} = -2.07 \pm 0.05$.

20.5   Negative phase velocity

For materials with a negative refractive index, the phase velocity points toward the source that is the phase velocity and energy flow are antiparallel inside a NIM (Cummer and Popa 2004; Aydin *et al.* 2005; Dolling *et al.* 2006). By measuring the transmission phases for NIMs with varying thicknesses, one can verify that the phase velocity is indeed negative. An HP 8510-C network analyzer is capable of measuring the transmission phase. We have constructed 4 different 2D NIM slabs with number of layers $N_x = 5, 6, 7$ and 8. The transmitted phases are plotted in the frequency range 3.70–4.00 GHz, which is within the left-handed transmission region.

Figure 20.8(a) shows the transmitted phase for different number of layers of NIM slabs. It is evident from the figure that increasing the number of layers of NIM results in a decrease of the phase of the transmitted EM wave. However, if the material possesses a positive refractive index, one would observe an increase in the transmitted phase with the increasing number of unit cells (Aydin *et al.* 2005).

The index of refraction in terms of wavelength, phase shift, and change in the length of left-handed material is given by Aydin *et al.* (2005):

$$n = \frac{\Delta\phi}{\Delta L} \frac{\lambda}{2\pi}. \tag{20.1}$$

At $f = 3.86$ GHz, the wavelength of the EM wave is $\lambda = 7.77$ cm. The average phase shift per unit cell ($\Delta L = 8.8$ mm) obtained from the experimental results is $\Delta\Phi = -0.45 \pm 0.04\pi$. Inserting these values in eqn (20.1), index of refraction at 3.86 GHz is found to be $n_{eff} = -1.98 \pm 0.18$. The average phase shift and calculated refractive index for the numerical simulations at 3.77 GHz are $\Delta\Phi = -0.51 \pm 0.04\pi$, and $n_{eff} = -2.31 \pm 0.18$.

In Fig. 20.8(b), we plot the refractive-index values calculated by using the phase shift between the consecutive numbers of NIM layers. The symbol (•) corresponds to the refractive indices obtained from wedge experiments at some other frequencies (data not shown here). There is a good agreement between the results obtained from two different methods. We also found the index of refraction at 3.86 GHz by using the beam-shift method as -1.91. The index of

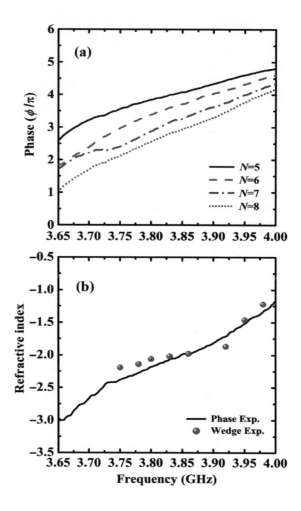

Fig. 20.8 (a) The measured transmission phase spectra of the 5 to 8 layers of LHM structures. The phase decreases with the increasing number of layers along the propagation direction. (b) Measured effective refractive indices as a function of frequency. The results obtained from the phase-shift experiments (solid line) and wedge experiments (●) are in good agreement.

refraction values obtained from the wedge and the phase-shift experiment are both −1.98 at this frequency. Therefore, we have been able to show that the results obtained from three different experiments agree extremely well.

20.6   Subwavelength imaging and resolution

A perfect lens is one of the most important applications of materials with a negative refractive index. The term, perfect lens, was coined by Pendry owing to the ability of such lenses to reconstruct a perfect image by recovering the evanescent components of EM waves (Pendry 2000). In conventional optics, the lenses are constructed from positive-index materials and require curved surfaces to bring EM waves into focus. Positive-index lenses suffer from the diffraction limit and can only focus objects with sizes on the order of a half-wavelength.

The negative-index metamaterial (NIM) slab has $40 \times 20 \times 3$ layers along the x-, y-, and z-directions with equal lattice constants in all directions,

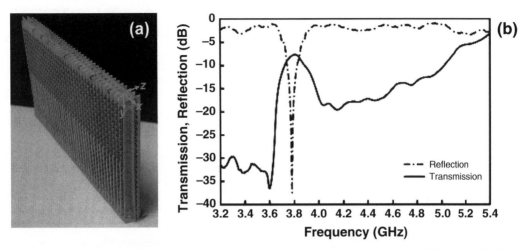

Fig. 20.9 (a) photograph of NIM slab with three unit cells along the z-direction. (b) Measured transmission (solid line) and reflection (dashed line) spectra for an NIM slab.

$a_x = a_y = a_z = 9.3$ mm (Fig. 20.9(a)). The NIM slab has a thickness of 2.56 cm ($\lambda/3$) and a length of 38 cm (4.8λ). Transmission and reflection measurements are performed to characterize 2D NIM, in which the results are plotted in Fig 20.9(b). A well-defined transmission peak is observed between 3.65–4.00 GHz, where the effective permeability and effective permittivity of NIM are simultaneously negative. A sharp dip in the reflection spectrum is observed at 3.78 GHz. The reflection is very low, -37 dB, meaning that the incident EM waves nearly do not face any reflection at the NIM surface.

The imaging measurements presented here are performed at 3.78 GHz, where the reflection is very low and the losses due to reflection are negligible. The NIM has a refractive index of $n_{\text{eff}} = -2.07 \pm 0.22$ at 3.78 GHz, which is measured by using a wedge-shaped 2D NIM (Aydin *et al.* 2005). In the imaging experiments, we employed monopole antennae to imitate the point source. The exposed center conductor acts as the transmitter and receiver and has a length of 4 cm ($\sim\lambda/2$). First, we measured the beam profile in free space that is plotted in Fig. 20.10 with a dashed line. The full width at half-maximum (FWHM) of the beam is 8.2 cm (1.03λ). Then, we inserted a NIM superlens, and measured the spot size of the beam as 0.13λ, which is well below the diffraction limit. The source is located $d_s = 1.2$ cm away from first boundary and the image forms $d_i = 0.8$ cm away from second boundary of the superlens. The intensity of the electric field at the image plane is scanned by the receiver monopole antenna with $\Delta x = 2$ mm steps. The field intensity is normalized with respect to the maximum intensity in the figure.

Since we were able to image a single point source with a subwavelength spot size, we used two point sources separated by distances smaller than a wavelength to obtain subwavelength resolution. The sources are driven by two independent signal generators and the power distribution is detected by using a microwave spectrum analyzer. The frequencies of the sources differ

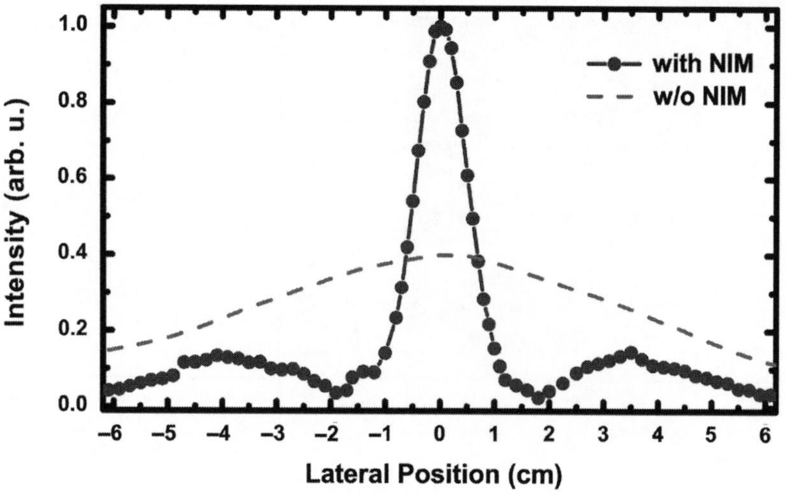

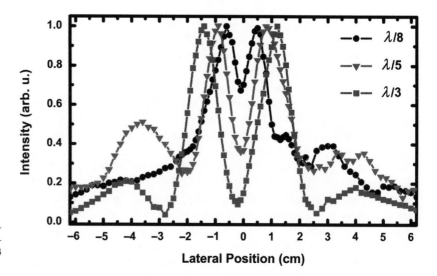

Fig. 20.10 The measured power distributions at the image plane with (solid line with circles) and without (dashed line) NIM superlens. Normalized intensity in free space is multiplied by 0.4 in the figure.

Fig. 20.11 The measured power distributions for two point sources separated by distances of $\lambda/8$ (–•–), $\lambda/5$ (–▼–), and $\lambda/3$ (–■–).

by 1 MHz to ensure that the sources are entirely incoherent. The imaging experiments are performed for three different separation distances between the sources. The measured power distribution of sources, separated by $\lambda/8$, is plotted by the solid line with circles (–•–) in Fig. 20.11. As seen in the figure, the peaks of two sources are clearly resolved. The resolution becomes better for $\lambda/5$ separation between the sources (–▼–). Finally, when the sources are $\lambda/3$ apart (–■–), two peaks are entirely resolved. In order to avoid any possible channelling effects, the sources are intentionally not placed at the line of SRR-wire boards.

In the near-field regime, the electrostatic and magnetostatic limits apply, and therefore, the electric and magnetic responses of materials can be treated as decoupled (Pendry 2000). This in turn brings the possibility of constructing superlenses from materials with negative permittivity or negative permeability. The advantage of using negative-index lenses over negative-permittivity or negative-permeability lenses is that the subwavelength resolution can be obtained for both transverse-electric (TE) and transverse-magnetic (TM) polarization of EM waves. However, single-negative lenses can only focus EM waves with one particular polarization.

20.7 Planar negative-index metamaterials

Although split-ring resonators (SRRs) were quite successful in demonstration of metamaterials, SRRs are not suitable for the planar configuration of metamaterials, since the incident EM wave has to be parallel to the SRR, provided that the magnetic field is perpendicular to the SRR. The magnetic response of split-ring resonators starts to saturate at optical frequencies (Zhou *et al.* 2005), and therefore optical magnetism is not achievable with SRRs. However, it is shown that an alternative structure of a metal–dielectric composite provides magnetic resonance at optical frequencies (Dolling *et al.* 2005; Shalaev *et al.* 2005). The structure consists of parallel metal slabs with a dielectric substrate in between, wherein the metal slabs provide the inductance and dielectric spacing that in turn provides the capacitance (Shalaev *et al.* 2005). Realizations of metamaterials using parallel metal slabs (also known as cut-wire pairs) at microwave frequencies followed soon after (Guven *et al.* 2006; Zhou *et al.* 2006). Fishnet-type metamaterials are also reported to exhibit a negative index at optical (Zhang *et al.* 2005; Dolling *et al.* 2006; Dolling *et al.* 2007) and microwave frequencies (Kafesaki *et al.* 2007; Lam *et al.* 2008; Zhou *et al.* 2008). The advantage of metamaterials with fishnet geometry is that the wires providing for negative permittivity and the slab pairs providing for negative permeability are brought together to produce a combined electromagnetic response.

In this section, we demonstrate a planar negative-index metamaterial (NIM) that operates independent of the incident polarization due to its symmetric configuration. Single-layer and multilayer metamaterials are characterized by transmission measurements and simulations. We used a low-loss Teflon substrate and obtained a high transmission from NIM structures with a peak value of -2.4 dB. The retrieved effective parameters are in good agreement with the experimental observations. A negative index appears at frequencies where both permittivity and permeability are negative. The structure size along the propagation direction is shown to affect the effective parameters. The phase shift from different sizes of the NIM samples are measured, wherein we observe a negative phase shift within the negative-index frequency regime, i.e. phase decreases with the increased NIM size. Phase measurements at the right-handed transmission band show the opposite behavior, in which the phase increases with the increasing number of NIM layers. Finally, we simulated the electric-field distribution and studied the wave propagation inside the fishnet metamaterial. Backward wave propagation is verified at the negative-index region.

20.7.1   Transmission through a fishnet-type metamaterial

The fishnet-type metamaterials were previously shown to be possible candidates for planar negative-index metamaterials (Zhang *et al.* 2005; Dolling *et al.* 2006, 2007; Kafesaki *et al.* 2007; Lam *et al.* 2008; Zhou *et al.* 2008). In the present study we employ the fishnet geometry to construct a metamaterial operating at microwave frequencies. A schematic of a fishnet unit cell is shown in Fig. 20.12(a). The highlighted areas are the metallic parts. The metal used to construct the structure is copper with a thickness of $20\,\mu$m. The substrate is a Teflon board with a thickness of $t = 1$ mm. The unit cell is repeated periodically along the x- and y-directions with a periodicity of $a_x = a_y = 14$ mm. The structure can be considered as a combination of parallel metal slab pairs with sizes $w_y \times a_x$, which provide the magnetic resonance and continuous wire pairs with sizes $w_x \times a_y$ in turn providing the negative permittivity (Kafesaki *et al.* 2007). Here, we chose $w_x = w_y = 7$ mm, and therefore the structure is symmetric in all directions. The structure will then work for TE and TM polarizations, as well as for arbitrary linear polarizations due to the symmetry of the slab and wire pairs. Therefore, the NIM will be functionally independent of the incident polarization (Kafesaki *et al.* 2007).

The numbers of layers in a single fishnet metamaterial layer are $N_x = N_y = 10$. We then stacked layers with a periodicity along the z-direction, $a_z = 2$ mm. Transmission and phase measurements were performed using an Agilent N5230A portable network analyzer, which is capable of measuring S-parameters and phase amplitudes. Transmitter and receiver antennas are standard high-gain microwave horn antennas. The wave propagates along the z-direction, with the E field parallel to the y-axis and the H field parallel to the x-axis. In these measurements, the distance between the horn antennae were kept at 35 cm. We first measured the transmission through air, and then used this data as a calibration.

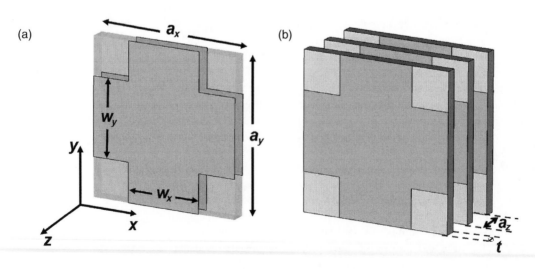

Fig. 20.12 Schematic drawings of (a) a fishnet unit cell, and (b) a multilayer fishnet metamaterial structure.

The measured transmission spectrum of a single fishnet metamaterial layer is plotted with a solid line as shown in Fig. 20.13. We observe a transmission peak between 14.2 and 14.8 GHz that is highlighted in the figure. The peak value is -2.4 dB at 14.38 GHz. This transmission band corresponds to the negative-index regime and is a left-handed transmission band, as we will show later in the following section. We performed numerical simulations in order to check the experimental results. We used the commercially available software, CST Microwave Studio, for the simulations. We simulated the unit cell as shown in Fig. 20.12(a), with periodic boundary conditions at the x- and y-axes. Along the propagation direction (z-axis), we employed open boundary conditions. We used waveguide ports to excite and detect electromagnetic waves. The dielectric constant of the Teflon board was taken as $\varepsilon = 2.16$ with a tangent loss of $\delta = 0.005$. The dashed line in Fig. 20.13 shows the simulated transmission spectrum of a single-layer fishnet structure. The agreement between the theory and experiment is quite good. The transmission peak had a value of -2.9 dB at 14.33 GHz.

Planar metamaterials, where the electromagnetic wave is incident normal to the plane of the metamaterials, were first measured at optical frequencies (Zhang *et al.* 2005; Dolling *et al.* 2006, 2007). The characterization of negative-index metamaterials was mainly performed by using a single layer. It is obvious that the basic properties of metamaterial structures, such as a negative index of refraction and the negative phase velocity cannot be verified by simply using a single layer of the metamaterial. Recently, there have been studies that characterize planar negative-index metamaterials with more than one layer along the propagation direction (Dolling *et al.* 2007; Kafesaki *et al.* 2007; Liu *et al.* 2008). Here, we report the transmission measurements of a two-layer and five-layer NIM structure. Figure 20.14(a) shows the measured (solid line) and simulated (dashed line) transmission spectra of a two-layer NIM structure. The periodicity along the propagation direction is $a_z = 2$ mm. The transmission band is present with a maximum value of -3.7 dB at 14.38 GHz that is obtained from the measurements. The simulation results

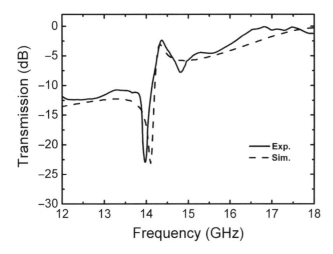

Fig. 20.13 Measured (solid line) and simulated (dashed line) transmission spectra of a single-layer fishnet metamaterial.

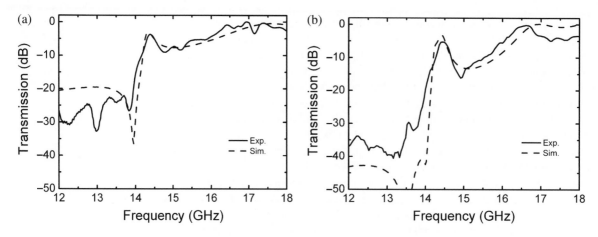

Fig. 20.14 Measured (solid line) and simulated (dashed line) transmission spectra of (a) two-layer and (b) five-layer fishnet metamaterial.

predict a similar transmission band where the peak is at 14.33 GHz with a maximum value of -3.0 dB.

Experimental and simulation results of the transmission spectra of five-layer NIM are shown in Fig. 20.14(b). The transmission peak is at 14.42 GHz, measured as -5.0 dB, and simulated as -3.2 dB. Slight differences between the simulation and experiments can be attributed to the deviation from the ideal material parameters and possible misalignments that may occur during the stacking of the layers. It is noteworthy that increasing the number of layers does not significantly change the transmission value at the left-handed transmission band. This is due to the specific choice of the substrate. The loss mechanism in the fishnet structure is mainly due to the substrate losses. Although the amount of metal is increased by a factor of five, from a one-layer structure to a five-layer structure, the transmission only decreased by -2.6 dB in the measurement. The result is more significant in the simulation results (ideal case) that the transmission peak did not change from a one-layer to a five-layer NIM.

20.7.2   Retrieved effective parameters

One of the most common characterization tools for the metamaterial structures is the retrieval procedure (Smith *et al.* 2002; Chen *et al.* 2004; Koschny *et al.* 2005; Penciu *et al.* 2006; Katsarakis *et al.* 2007). It is widely used to calculate the effective parameters of the metamaterial under investigation. The amplitude and phase of the transmission and reflection are either calculated or measured, in which the real and imaginary parts of the refractive index and wave impedance are then retrieved from the transmission and reflection coefficients. We employed the retrieval procedure to obtain effective permittivity, permeability, and a refractive index. We followed the approach as outlined in

Chen *et al.* (2004). The advantage of this procedure is that the correct branch of the effective refractive index and effective impedance was selected. The ambiguity in the determination of the correct branch was resolved by using an analytic continuation procedure.

In the retrieval procedure, we employed a single layer of NIM along the z-axis. Hence, the simulation setup coincides with a slab of NIM that consists of a single layer. The effective permittivity and permeability values were then derived from the transmission and reflection coefficients of a single layer of NIM. The real (solid line) and imaginary (dashed line) parts of the effective permittivity (Fig. 20.15(a)), permeability (Fig. 20.15(c)), and index of refraction (Fig. 20.15(e)) for a single layer of a NIM structure was consistent with the simulated and measured transmission spectra. The effective permittivity is negative within the frequency range of interest. The continuous wire arrays that are aligned along the y-axis cause plasma oscillations. The plasma frequency of an NIM structure is approx. 16.5 GHz. The resonance in the permittivity at approx. 14.25 GHz is a common characteristic of meta-materials with a negative index. The real part of the magnetic permeability has negative values between 14.23 and 14.65 GHz. The lowest value of μ is -3.38 at 14.3 GHz for a single-layer NIM. Since both permeability and permittivity are simultaneously negative, the condition needed in order to have a negative index is satisfied. As seen in Fig. 20.15(e), the NIM structure possesses negative values of the refractive index within the frequency range 13.95–14.65 GHz.

The real parts of the refractive indices of a one and two-layer structure are plotted together in Fig. 20.16(a). As was expected, the relation between ε and μ holds true for the refractive index, since the index depends on ε and μ. The minimum values of the refractive index for a one-layer NIM was $n = -4.58$ (at 14.23 GHz) and a two-layer NIM was $n = -2.66$ (at 14.19 GHz).

The frequency region between 14.23–14.65 GHz was a double-negative region, which means that the permeability and permittivity were both negative. We found that the upper edge for the negative-index regime was 14.65 GHz for a one- and two-layer NIM, where the permeability starts to take on positive values. However, the lower edge is at lower frequencies for a two-layer NIM. Although the refractive index is negative between 13.72 and 14.23 GHz for a two-layer NIM, the permeability is positive within this frequency range. Therefore, the NIM structure behaves like a single-negative material (Kafesaki *et al.* 2007). For a one-layer NIM, the single-negative frequency region was narrower, in turn appearing between 13.94 and 14.23 GHz. The figure of merit (FOM) for negative-index metamaterials was defined as a ratio of the absolute value of the real part of an index to the imaginary part of an index, FOM $= |\mathrm{Re}(n)|/\mathrm{Im}(n)$. Figures of merit for one and two-layer NIMs are shown in Fig. 20.16(b). It is obvious that the figure of merit is larger for a single-layer NIM structure. The maximum figure of merit value is 9.26 at 14.41 GHz for a one-layer NIM. For a two-layer structure we found the highest value of figure of merit to be 6.58 at 14.38 GHz. FOM decreases with the increasing number of layers due to the decrease in the refractive index.

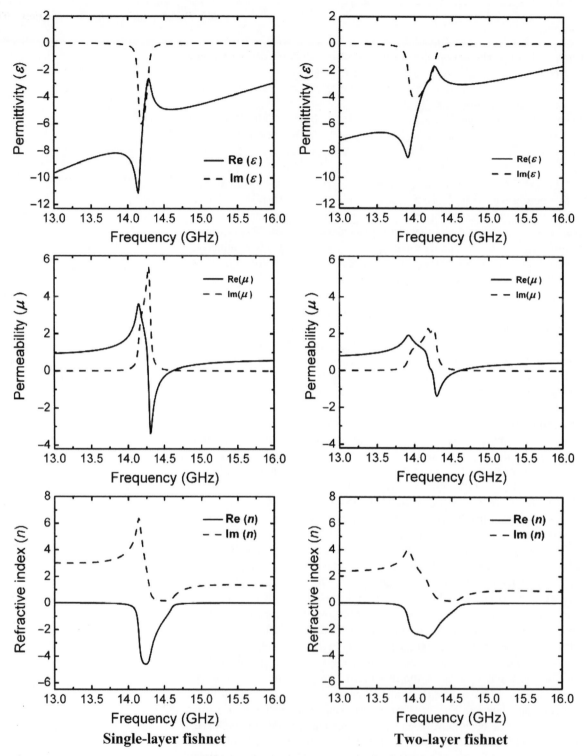

Fig. 20.15 Retrieved effective parameters of a (left) single-layer and a (right) two-layer fishnet metamaterial. The real (solid line) and imaginary (dashed line) parts of effective permittivity (top), permeability (middle), and refractive index (bottom) are shown.

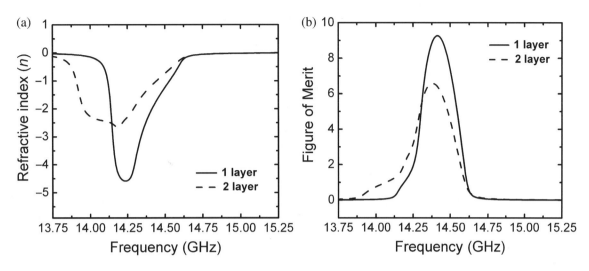

Fig. 20.16 (a) Real part of retrieved refractive index and (b) figure of merit for a single-layer (solid line) and two-layer metamaterial (dashed line).

20.7.3 Negative phase velocity and backward wave propagation

Negative-index metamaterials are of special interest for obtaining backward wave propagation, since they inherently have a negative phase velocity. A network analyzer is capable of measuring the transmitted phase. We constructed NIM structures with $a_z = 2$ mm that have $N_z = 2$, 3, 4, and 5 layers along the propagation direction. The transmitted phase of NIM structures were measured by using the same setup with the transmission measurements. In the measurements, the transmitted phase is calibrated with respect to the phase in free space. Figure 20.17(a) shows the transmission phase between 14.2 and 14.8 GHz for two (solid line), three (dashed line), four (–■–), and five (–●–) layers of the NIM structure. The phase decreases with the increasing number of layers along the propagation direction. This is typical behavior for left-handed metamaterials with a negative phase velocity (Aydin *et al.* 2005; Guven *et al.* 2006). We also performed phase measurements in a frequency range where we know that the transmission band has right-handed characteristics. The phase spectra for four different NIM structures are shown in Fig. 20.17(b) in the right-handed transmission region. We observed an increase in the phase with the increasing number of layers along the propagation direction. The negative phase shift for a NIM structure at the negative-index frequency regime means that the phase is directed towards the source. Such behavior is due to the negative phase velocity within the interested range of frequencies. These measurements clearly verify the negative phase velocity behavior of planar fishnet-type multilayer metamaterials. To our knowledge, this is the first direct observation of negative phase velocity for fishnet-type NIM structures.

Metamaterials with a negative phase velocity are also called backward wave materials, since the wave propagation is towards the source inside the

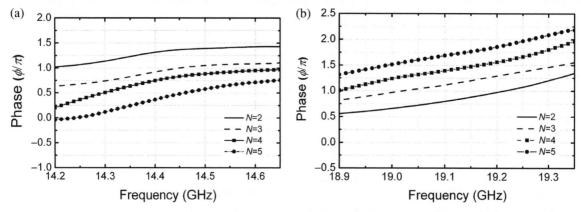

Fig. 20.17 Transmitted phase of metamaterial structure for for two (solid line), three (dashed line), four (–■–), and five (–•–) NIM (a) 14.2 and 14.8 GHz (left-handed transmission regime) and (b) 18.9 and 19.4 GHz (right-handed transmission regime).

metamaterial. We performed numerical simulations using CST Microwave Studio in order to investigate the wave propagation inside our NIM structure. Figure 20.18(a) shows the simulated NIM structure. We used the same parameters with the structure as shown in Fig. 20.12(a), but in this case with an alternative unit cell. The structure is excited with a plane wave with a wave vector along the +z-direction using CST Microwave Studio. The E field is along the y-axis and the H field is along the x-axis in the simulations. The structure had three unit cells along x and 5 unit cells along z. We calculated the electric-field profile by placing electric-field monitors along the y-axis at two different frequencies, 14.42 GHz and 17.00 GHz. 14.42 GHz was the frequency of the highest transmission within the negative-index regime for a five-layer NIM. Figure 20.18(b) shows the E-field intensity along the y-direction, where the profile was calculated at the point $y = 5$ mm. The wave propagates along the +z-direction in air. However, inside the fishnet structure, we observed backward wave propagation in the simulations. The EM wave is guided through channels as shown with a dashed rectangular box in Fig. 20.18(a). We did not observe transmission through the wire pairs. The wave propagation inside the metamaterial at 17.00 GHz is along the +z-direction, since this frequency corresponds to the right-handed transmission regime where ε and μ are both positive (Fig. 20.18(c)).

20.8   Conclusion

In conclusion, we have successfully demonstrated a left-handed transmission band for 1D double-negative material in free space with a high transmission peak. Magnetic resonances of split-ring resonator structures are verified by using a closed-ring resonator structure. The effect of disorder on the transmission and reflection characteristics of 1D double-negative materials is studied experimentally. We confirmed that 2D negative—index metamaterial has a negative refractive index at frequencies where dielectric permittivity and magnetic permeability are simultaneously negative. We have been able to observe

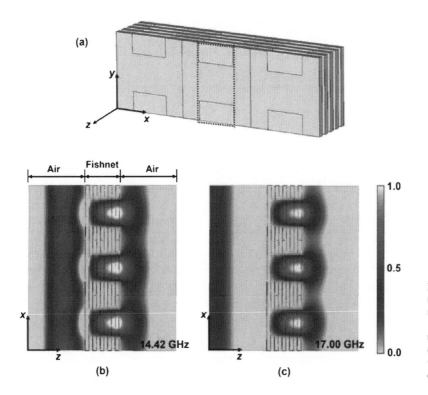

Fig. 20.18 (a) Schematic drawing of a simulated NIM structure with an alternative unit cell. (b) Simulated E-field profile at 14.42 GHz where the wave propagates along the $-z$-direction inside the fishnet structure (c) Simulated E-field profile at 17.00 GHz where the wave propagates along the $+z$-direction inside the fishnet structure.

a negative refractive index for 2D NIMs by using three different, independent methods. The results obtained from these experiments are in good agreement. The phase shift and therefore phase velocity is shown to be negative. We demonstrated an impedance-matched, low-loss negative-index metamaterial superlens that is capable of resolving subwavelength features with a record-level 0.13λ resolution, which is the highest resolution achieved by a negative-index metamaterial. We also demonstrated backward wave propagation in a planar negative-index metamaterial at microwave frequencies. The advantages of the new design are: (i) low-loss substrate in turn providing high transmission even for a five-layer NIM, (ii) negative-index behavior is present independent of the incident electromagnetic wave polarization, due to the symmetric configuration of the slab and wire pairs. We reported the highest transmission for a planar multilayer structure with -3.7 dB transmission for two layers and -5.0 dB for five layers. Negative phase velocity was experimentally verified by measuring the phase difference between those NIM structures with a different number of layers along the propagation direction. The transmitted phase decreases with an increase in the NIM size in the negative-index transmission band.

Acknowledgments

This work is supported by the European Union under the projects EU-NoE-METAMORPHOSE, EU-NoE-PHOREMOST, EU-PHOME, EU-ECONAM and TUBITAK under Projects Nos. 104E090, 105E066, 105A005, and 106A017. One of the authors (E.O.) also acknowledges partial support from the Turkish Academy of Sciences.

References

Aydin, K., Guven, K., Katsarakis, N., Soukoulis, C.M., Ozbay, E. *Opt. Exp.* **12**, 5896 (2004).

Aydin, K., Guven, K., Kafesaki, M., Zhang, L., Soukoulis, C.M., Ozbay, E. *Opt. Lett.* **29**, 2623 (2004).

Aydin, K., Bulu, I., Ozbay, E. *Opt. Exp.* **13**, 8753 (2005).

Aydin, K., Bulu, I., Guven, K., Kafesaki, M., Soukoulis, C.M., Ozbay, E. *New J. Phys.* **7**, 168 (2005).

Aydin, K., Guven, K., Soukoulis, C.M., Ozbay, E. *Appl. Phys. Lett.* **86**, 124102 (2005).

Aydin, K., Bulu, I., Ozbay, E. *New J. Phys.* **8**, 221 (2006).

Aydin, K., Bulu, I., Ozbay, E. *Microwave Opt. Tech. Lett.* **48**, 2548 (2006).

Aydin, K., Ozbay, E. *Opto-Electron. Rev.* **14**, 193 (2006).

Aydin, K., Ozbay, E. *J. Opt. Soc. Am. B* **23**, 415 (2006).

Aydin, K., Ozbay, E. *J. Appl. Phys.* **101**, 024911 (2007).

Bayindir, M., Aydin, K., Markos, P., Soukoulis, C.M., Ozbay, E. *Appl. Phys. Lett.* **81**, 120 (2002).

Bulu, I., Caglayan, H., Ozbay, E. *Opt. Express* **13**, 10238 (2005).

Bulu, I., Caglayan, H., Aydin, K., Ozbay, E. *New J. Phys.* **7**, 223 (2005).

Chen, X., Grzegorczyk, T.M., Wu, B.-I., Pacheco, J. Jr., Kong, J.A. *Phys. Rev. E* **70**, 016608 (2004).

Chen, H., Padilla, W.J., Zide, J.M.O., Gossard, A.C., Taylor, A.J., Averitt, R.D. *Nature* **444**, 597 (2006).

Cubukcu, E., Aydin, K., Ozbay, E., Foteinopoulou, S., Soukoulis, C.M. *Nature* **423**, 604 (2003).

Cubukcu, E., Aydin, K., Ozbay, E., Foteinopolou, S., Soukoulis, C.M. *Phys. Rev. Lett.* **91**, 207401 (2003).

Cummer, S.A., Popa, B. *Appl. Phys. Lett.* **85**, 4564 (2004).

Dolling, G., Enkrich, C., Wegener, M., Zhou, J.F., Soukoulis, C.M., Linden, S. *Opt. Lett.* **30**, 3198 (2005).

Dolling, G., Enkrich, C., Wegener, M., Soukoulis, C.M., Linden, S. *Science* **312**, 892 (2006).

Dolling, G., Wegener, M., Soukoulis, C.M., Linden, S. *Opt. Lett.* **32**, 53 (2007).

Dolling, G., Wegener, M., Linden, S. *Opt. Lett.* **32**, 551 (2007).

Fang, N., Lee, H., Sun, C., Zhang, X. *Science* **308**, 534 (2005).

Fang, N., Xi, D., Xu, J., Ambati, M., Srituravanich, W., Sun, C., Zhang, X. *Nature Mater.* **5**, 452 (2006).

Gay-Balmaz, P., Martin, O.J.F. *J. Appl. Phys.* **92**, 2929 (2002).

Gil, I., Garcia-Garcia, J., Bonache, J., Martin, F., Sorolla, M., Marques, R. *Electron. Lett.* **40**, 1347 (2004).

Gokkavas, M., Guven, K., Bulu, I., Aydin, K., Penciu, R.S., Kafesaki, M., Soukoulis, C.M., Ozbay, E. *Phys. Rev. B* **73**, 193103 (2006).

Gorkunov, M.V., Gredeskul, S.A., Shadrivov, I.V., Kivshar, Y.S. *Phys. Rev. E* **73**, 056605 (2006).

Grbic, A., Eleftheriades, G.V. *Phys. Rev. Lett.* **92**, 117403 (2004).

Grigorenko, A.N., Geim, A.K., Gleeson, H.F., Zhang, Y., Firsov, A.A., Khrushchev, I.Y., Petrovic, J. *Nature* **438**, 335 (2005).

Gundogdu, T.F., Tsiapa, I., Kostopoulos, A., Konstantinidis, G., Katsarakis, N., Penciu, R.S., Kafesaki, M., Economou, E.N., Koschny, T., Soukoulis, C.M. *Appl. Phys. Lett.* **89**, 084103 (2006).

Guven, K., Aydin, K., Ozbay, E. *Photon. Nanostruct.: Fund. Appl.* **3**, 75 (2005).

Guven, K., Caliskan, M.D., Ozbay, E. *Opt. Exp.* **14**, 8685 (2006).

Guven, K., Ozbay, E. *Opto-Electron. Rev.* **14**, 213 (2006).

Houck, A.A., Brock, J.B., Chuang, I.L. *Phys. Rev. Lett.* **90**, 137401 (2003).

Kafesaki, M., Tsiapa, I., Katsarakis, N., Koschny, T., Soukoulis, C.M., Economou, E.N. *Phys. Rev. B* **75**, 235114 (2007).

Katsarakis, N., Koschny, T., Kafesaki, M., Economou, E.N., Soukoulis, C.M. *Appl. Phys. Lett.* **84**, 2943 (2004).

Katsarakis, N., Kafesaki, M., Tsiapa, I., Economou, E.N., Soukoulis, C.M. *Photon. Nanostruct.: Fundam. Appl.* **5**, 149 (2007).

Klein, M.W., Enkrich, C., Wegener, M., Linden, S. *Science* **313**, 502 (2006).

Koschny, T., Markoš, P., Economou, E.N., Smith, D.R., Vier, D.C., Soukoulis, C.M. *Phys. Rev. B* **71**, 245105 (2005).

Lagarkov, A.N., Kissel, V.N. *Phys. Rev. Lett.* **92**, 077401 (2004).

Lam, V.D., Kim, J.B., Lee, S.J., Lee, Y.P. *J. Appl. Phys.* **103**, 033107 (2008).

Linden, S., Enkrich, C., Wegener, M., Zhou, J., Koschny, T., Soukoulis, C.M. *Science* **306**, 1351 (2004).

Liu, N., Guo, H., Fu, L., Kaiser, S., Schweizer, H., Giessen, H. *Nature Mater.* **7**, 31 (2008).

Marques, R., Mesa, F., Martel, J., Medina, F. *IEEE Trans. Antennas Propag.* **51**, 2572 (2003).

Melville, D.O.S., Blaikie, R. *J. Opt. Exp.* **13**, 2127 (2005).

Moser, H.O., Casse, B.D.F., Wilhelmi, O., Saw, B.T. *Phys. Rev. Lett.* **94**, 063901 (2005).

Padilla, W.J., Taylor, A.J., Highstrete, C., Lee, M., Averitt, R.D. *Phys. Rev. Lett.* **96**, 107401 (2006).

Parazzoli, C.G., Greegor, R.B., Li, K., Koltenbah, B.E., Tanielian, M. *Phys. Rev. Lett.* **90**, 107401 (2003).

Penciu, R.S., Kafesaki, M., Gundogdu, T.F., Economou, E.N., Soukoulis, C.M. *Photon. Nanostruct.: Fund. Appl.* **4**, 12 (2006).

Pendry, J.B. *Phys. Rev. Lett.* **85**, 3966 (2000).

Pendry, J.B., Holden, A.J., Robbins, D.J., Stewart, W.J. *J. Phys.: Condens. Matter* **10**, 4785 (1998).

Pendry, J.B., Holden, A.J., Robbins, D.J., Stewart, W.J. *IEEE Trans. Microwave Theory Tech.* **47**, 2075 (1999).

Pendry, J.B., Schurig, D., Smith, D.R. *Science* **312**, 1780 (2006).

Schurig, D., Mock, J.J., Justice, B.J., Cummer, S.A., Pendry, J.B., Starr, A.F., Smith, D.R. *Science* **314**, 977 (2006).

Shadrivov, I.V., Morrison, S.K., Kivshar, Y.S. *Opt. Exp.* **14**, 9344 (2006).

Shalaev, V.M., Cai, W., Chettiar, U.K., Yuan, H., Sarychev, A.K., Drachev, V.P., Kildishev, V. *Opt. Lett.* **30**, 3356 (2005).

Shelby, R.A., Smith, D.R., Nemat-Nasser, S.C., Schultz, S. *Appl. Phys. Lett.* **78**, 480 (2001).

Shelby, R.A., Smith, D.R., Schultz, S. *Science* **292**, 77 (2001).

Smith, D.R., Padilla, W.J., Vier, D.C., Nemat-Nasser, S.C., Schultz, S. *Phys. Rev. Lett.* **84**, 4184 (2000).

Smith, D.R., Schultz, S., Markos P., Soukoulis, C.M. *Phys. Rev. B* **65**, 195104 (2002).

Taubner, T., Korobkin, D., Urzhumov, Y., Shvets, G., Hillenbrand, R. *Science* **313**, 1595 (2006).

Veselago, V.G. *Sov. Phys. Usp.* **10**, 509 (1968).

Weiland, T., Schuhmann, R., Greegor, R.B., Parazzoli, C.G., Vetter, A.M., Smith, D.R., Vier, D.C., Schultz, S. *J. Appl. Phys.* **90**, 5419 (2001).

Wiltshire, M.C.K., Pendry, J.B., Young, I.R., Larkman, D.J., Gilderdale, D.J., Hajnal, J.V. *Science* **291**, 849 (2001).

Yen, T.J., Padilla, W.J., Fang, N., Vier, D.C., Smith, D.R., Pendry, J.B., Basov, D.N., Zhang, X. *Science* **303**, 1494 (2004).

Zhang, S., Fan, W., Panoiu, N.C., Malloy, K.J., Osgood, R.M., Brueck, S.R.J. *Phys. Rev. Lett.* **95**, 137404 (2005).

Zhang, S., Fan, W., Malloy, K.J., Brueck, S.R.J., Panoiu, N. C., Osgood, R.M. *Opt. Exp.* **13**, 4922 (2005).

Zhou, J., Koschny, T., Kafesaki, M., Economou, E.N., Pendry, J.B., Soukoulis, C.M. *Phys. Rev. Lett.* **95**, 223902 (2005).

Zhou, J., Zhang, L., Tuttle, G., Koschny, T., Soukoulis, C.M. *Phys. Rev. B* **73**, 041101 (2006).

Zhou, J., Economou, E.N., Koschny, T., Soukoulis, C.M. *Opt. Lett.* **31**, 3620 (2006).

Zhou, J., Koschny, T., Kafesaki, M., Soukoulis, C.M. *Photon. Nanostruct.: Fundam. Appl.* **6**, 96 (2008).

Ziolkowski, R.W. *IEEE Trans. Antennas Propag.* **51**, 1516 (2003).

2D arrays of Josephson nanocontacts and nanogranular superconductors

Sergei Sergeenkov

21.1   Introduction

Inspired by new possibilities offered by the cutting-edge nanotechnologies, the experimental and theoretical physics of increasingly sophisticated mesoscopic quantum devices, heavily based on Josephson junctions (JJs) and their arrays (JJAs), is becoming one of the most exciting and rapidly growing areas of modern science (for reviews on charge and spin effects in mesoscopic 2D JJs and quantum-state engineering with Josephson devices, see, e.g. Newrock *et al.* 2000; Makhlin *et al.* 2001; Krive *et al.* 2004; Sergeenkov 2006; Beloborodov *et al.* 2007). In particular, a remarkable increase of the measurements technique resolution made it possible to experimentally detect such interesting phenomena as flux avalanches (Altshuler and Johansen 2004) and geometric quantization (Sergeenkov and Araujo-Moreira 2004) as well as flux-dominated behavior of heat capacity (Bourgeois *et al.* 2005) both in JJs and JJAs.

Recently, it was realized that JJAs can be also used as quantum channels to transfer quantum information between distant sites (Makhlin *et al.* 2001; Wendin and Shumeiko 2007) through the implementation of the so-called superconducting qubits that take advantage of both charge and phase degrees of freedom.

Both granular superconductors and artificially prepared JJAs proved useful in studying the numerous quantum (charging) effects in these interesting systems, including Coulomb blockade of Cooper-pair tunnelling (Iansity *et al.* 1988), Bloch oscillations (Haviland *et al.* 1991), propagation of quantum ballistic vortices (van der Zant 1996), spin-tunnelling related effects using specially designed SFS-type junctions (Ryazanov *et al.* 2001; Golubov *et al.* 2002), novel Coulomb effects in SINIS-type nanoscale junctions (Ostrovsky and Feigel'man 2004), and dynamical AC re-entrance (Araujo-Moreira *et al.* 1997; Barbara *et al.* 1999; Araujo-Moreira *et al.* 2005).

At the same time, given a rather specific magnetostrictive (Sergeenkov and Ausloos 1993) and piezo-magnetic (Sergeenkov 1998b, 1999) response of Josephson systems, one can expect some non-trivial behavior of the thermal expansion (TE) coefficient in JJs as well (Sergeenkov *et al.* 2007). Of special interest are the properties of TE in applied magnetic field. For example, some superconductors like $Ba_{1-x}K_xBiO_3$, $BaPb_xBi_{1-x}O_3$ and $La_{2-x}Sr_xCuO_4$ were found (Anshukova *et al.* 2000) to exhibit anomalous temperature behavior of both magnetostriction and TE that were attributed to the field-induced suppression of the superstructural ordering in the oxygen sublattices of these systems.

The imaging of the granular structure in underdoped $Bi_2Sr_2CaCu_2O_{8+\delta}$ crystals (Lang *et al.* 2002) revealed an apparent segregation of its electronic structure into superconducting domains (of the order of a few nanometers) located in an electronically distinct background. In particular, it was found that at low levels of hole doping ($\delta < 0.2$), the holes become concentrated at certain hole-rich domains. (In this regard, it is interesting to mention a somewhat similar phenomenon of "chemical localization" that takes place in materials composed of atoms of only metallic elements, exhibiting metal–insulator transitions, see, e.g. Gantmakher 2002.) Tunnelling between such domains leads to intrinsic nanogranular superconductivity (NGS) in high-T_C superconductors (HTS). Probably one of the first examples of NGS was observed in $YBa_2Cu_3O_{7-\delta}$ single crystals in the form of the so-called "fishtail" anomaly of magnetization (Daeumling *et al.* 1990). The granular behavior has been related to the 2D clusters of oxygen defects forming twin boundaries (TBs) or dislocation walls within the CuO plane that restrict supercurrent flow and allow excess flux to enter the crystal. Indeed, there are serious arguments to consider the TB in HTS as insulating regions of the Josephson SIS-type structure. An average distance between boundaries is essentially less than the grain size. In particular, the networks of localized grain-boundary dislocations with the spacing ranged from 10 nm to 100 nm have been observed (Daeumling *et al.* 1990) which produce effectively continuous normal or insulating barriers at the grain boundaries. It was also verified that the processes of the oxygen ordering in HTS leads to the continuous change of the lattice period along TB with the change of the oxygen content. Besides, a destruction of bulk superconductivity in these non-stoichiometric materials on increasing the oxygen deficiency parameter δ was found to follow a classical percolation theory (Gantmakher *et al.* 1990).

In addition to their importance for understanding the underlying microscopic mechanisms governing HTS materials, the above experiments can provide rather versatile tools for designing chemically controlled atomic-scale JJs and JJAs with pre-selected properties needed for manufacturing the modern quantum devices (Sergeenkov 2001, 2003, 2006; Araujo-Moreira *et al.* 2002). Moreover, as we shall see below, NGS-based phenomena can shed some light on the origin and evolution of the so-called paramagnetic Meissner effect (PME) which manifests itself both in high-T_c and conventional superconductors (Geim *et al.* 1998; De Leo and Rotoli 2002; Li 2003) and is usually associated with the presence of π-junctions and/or unconventional (d-wave) pairing symmetry.

In this chapter we present numerous novel phenomena related to the magnetic, electric, elastic and transport properties of Josephson nanocontacts and NGS. The chapter is organized as follows. In Section 21.1, a realistic model of NGS is introduced that is based on 2D JJAs created by a regular network of twin-boundary dislocations with strain fields acting as an insulating barrier between hole-rich domains (like in underdoped crystals). In Section 21.2, we consider some phase-related phenomena expected to occur in NGS, such as Josephson chemomagnetism and magnetoconcentration effect. Section 21.3 is devoted to a thorough discussion of charge-related polarization phenomena in NGS, including such topics as chemomagnetoelectricity, magnetocapacitance, charge analog of the "fishtail" (magnetization) anomaly, and field-tuned weakening of the chemically induced Coulomb blockade. In Section 21.4 we present our latest results on the influence of an intrinsic chemical pressure (created by the gradient of the chemical potential due to segregation of hole-producing oxygen vacancies) on the temperature behavior of the non-linear thermal conductivity (NLTC) of NGS. In particular, our theoretical analysis (based on the inductive model of 2D JJAs) predicts a giant enhancement of NLTC reaching up to 500% when the intrinsically induced chemoelectric field $\mathbf{E}_\mu = \frac{1}{2e}\nabla\mu$ closely matches the thermoelectric field $\mathbf{E}_T = S_T\nabla T$. And finally, by introducing the concept of thermal expansion (TE) of a Josephson contact (as an elastic response of JJ to an effective stress field), in Section 21.5 we consider the temperature and magnetic-field dependence of the TE coefficient $\alpha(T, H)$ in a small single JJ and in a single plaquette (a prototype of the simplest JJA). In particular, we found that in addition to expected *field* oscillations due to a Fraunhofer-like dependence of the critical current, α of a small single junction also exhibits strong flux-driven *temperature* oscillations near T_C. The condition under which all the effects predicted here can be experimentally realized in artificially prepared JJAs and NGS are also discussed. Some important conclusions of the present study are drawn in Section 21.6.

21.2 Model of nanoscopic Josephson junction arrays

As is well known, the presence of a homogeneous chemical potential μ through a single JJ leads to the AC Josephson effect with time-dependent phase difference $\partial\phi/\partial t = \mu/\hbar$. In this section, we will consider some effects in dislocation-induced JJ caused by a local variation of excess hole concentration $c(\mathbf{x})$ under the chemical pressure (described by inhomogeneous chemical potential $\mu(\mathbf{x})$) equivalent to the presence of the strain field of 2D dislocation array $\epsilon(\mathbf{x})$ forming this Josephson contact.

To understand how NGS manifests itself in non-stoichiometric crystals, let us invoke an analogy with the previously discussed dislocation models of twinning-induced superconductivity (Khaikin and Khlyustikov 1981) and grain-boundary Josephson junctions (Sergeenkov 1999). Recall that under plastic deformation, grain boundaries (GBs) (which are the natural sources of weak links in HTS), move rather rapidly via the movement of the grain-boundary dislocations (GBDs) comprising these GBs. At the same time, there are observed (Daeumling *et al.* 1990; Moeckley *et al.* 1993; Yang *et al.* 1993; Lang *et al.* 2002) in HTS single crystals regular 2D dislocation networks of oxygen-depleted regions (generated by the dissociation of $< 110 >$ twinning dislocations) with the size d_0 of a few Burgers vectors, forming a triangular lattice with a spacing $d \geq d_0$ ranging from 10 nm to 100 nm, can provide quite a realistic possibility for the existence of 2D Josephson network within CuO plane. Recall, furthermore, that in a d-wave orthorhombic YBCO crystal TBs are represented by tetragonal regions (in which all dislocations are equally spaced by d_0 and have the same Burgers vector **a** parallel to the y-axis within the CuO plane) which produce screened strain fields (Gurevich and Pashitskii 1997) $\epsilon(\mathbf{x}) = \epsilon_0 e^{-|\mathbf{x}|/d_0}$ with $| \mathbf{x} | = \sqrt{x^2 + y^2}$.

Though in YBa$_2$Cu$_3$O$_{7-\delta}$ the ordinary oxygen diffusion $D = D_0 e^{-U_d/k_B T}$ is extremely slow even near T_C (due to a rather high value of the activation energy U_d in these materials, typically $U_d \simeq 1 eV$), in underdoped crystals (with oxygen-induced dislocations) there is a real possibility to facilitate oxygen transport via the so-called osmotic (pumping) mechanism (Girifalco 1973; Sergeenkov 1995) which relates a local value of the chemical potential (chemical pressure) $\mu(\mathbf{x}) = \mu(0) + \nabla\mu \cdot \mathbf{x}$ with a local concentration of point defects as follows $c(\mathbf{x}) = e^{-\mu(\mathbf{x})/k_B T}$. Indeed, when in such a crystal there exists a non-equilibrium concentration of vacancies, a dislocation is moved for the atomic distance a by adding excess vacancies to the extraplane edge. The produced work is simply equal to the chemical potential of added vacancies. What is important is that this mechanism allows us to explicitly incorporate the oxygen deficiency parameter δ into our model by relating it to the excess oxygen concentration of vacancies $\mathbf{c}_v \equiv c(0)$ as follows $\delta = 1 - \mathbf{c}_v$. As a result, the chemical potential of the single vacancy reads $\mu_v \equiv \mu(0) = -k_B T \log(1 - \delta) \simeq k_B T \delta$. Remarkably, the same osmotic mechanism was used by Gurevich and Pashitskii (1997) to discuss the modification of oxygen vacancies concentration in the presence of the TB strain field. In particular, they argue that the change of $\epsilon(\mathbf{x})$ under an applied or chemically induced pressure results in a significant oxygen redistribution, producing a highly inhomogeneous filamentary structure of oxygen-deficient non-superconducting regions along GB (Moeckley *et al.* 1993) (for underdoped superconductors, the vacancies tend to concentrate in the regions of compressed material). Hence, assuming the following connection between the variation of mechanical and chemical properties of planar defects, namely $\mu(\mathbf{x}) = K\Omega_0\epsilon(\mathbf{x})$ (where Ω_0 is an effective atomic volume of the vacancy and K is the bulk elastic modulus), we can study the properties of TB-induced JJs under intrinsic chemical pressure $\nabla\mu$ (created by the variation of the oxygen doping parameter δ). More specifically, a single SIS-type junction (comprising a Josephson network) is formed around TB due to a local depression of the superconducting order parameter $\Delta(\mathbf{x}) \propto \epsilon(\mathbf{x})$ over distance d_0 thus producing a weak link with (oxygen deficiency δ dependent) Josephson coupling $J(\delta) = \epsilon(\mathbf{x})J_0 = J_0(\delta)e^{-|\mathbf{x}|/d_0}$ where $J_0(\delta) = \epsilon_0 J_0 = (\mu_v/K\Omega_0)J_0$ (here $J_0 \propto \Delta_0/R_n$ with R_n being a resistance of the junction). Thus, the present model indeed describes chemically induced

NGS in underdoped systems (with $\delta \neq 0$) because, in accordance with the observations, for stoichiometric situation (when $\delta \simeq 0$), the Josephson coupling $J(\delta) \simeq 0$ and the system loses its explicitly granular signature.

To adequately describe chemomagnetic properties of an intrinsically granular superconductor, we employ a model of 2D overdamped Josephson junction array that is based on the well-known Hamiltonian

$$\mathcal{H} = \sum_{ij}^{N} J_{ij}(1 - \cos \phi_{ij}) + \sum_{ij}^{N} \frac{q_i q_j}{C_{ij}}, \tag{21.1}$$

and introduces a short-range interaction between N junctions (which are formed around oxygen-rich superconducting areas with phases $\phi_i(t)$), arranged in a two-dimensional (2D) lattice with co-ordinates $\mathbf{x_i} = (x_i, y_i)$. The areas are separated by oxygen-poor insulating boundaries (created by TB strain fields $\epsilon(\mathbf{x}_{ij})$) producing a short-range Josephson coupling $J_{ij} = J_0(\delta)e^{-|\mathbf{x}_{ij}|/d}$. Thus, typically for granular superconductors, the Josephson energy of the array varies exponentially with the distance $\mathbf{x}_{ij} = \mathbf{x}_i - \mathbf{x}_j$ between neighboring junctions (with d being an average junction size). As usual, the second term in the rhs of eqn (21.1) accounts for Coulomb effects where $q_i = -2en_i$ is the junction charge with n_i being the pair number operator. Naturally, the same strain fields $\epsilon(\mathbf{x}_{ij})$ will be responsible for dielectric properties of oxygen-depleted regions as well via the δ-dependent capacitance tensor $C_{ij}(\delta) = C[\epsilon(\mathbf{x}_{ij})]$.

If, in addition to the chemical pressure $\nabla\mu(\mathbf{x}) = K\Omega_0\nabla\epsilon(\mathbf{x})$, the network of superconducting grains is under the influence of an applied frustrating magnetic field $\mathbf{B}$, the total phase difference through the contact reads

$$\phi_{ij}(t) = \phi_{ij}^0 + \frac{\pi w}{\Phi_0}(\mathbf{x}_{ij} \wedge \mathbf{n}_{ij}) \cdot \mathbf{B} + \frac{\nabla\mu \cdot \mathbf{x}_{ij}t}{\hbar}, \tag{21.2}$$

where ϕ_{ij}^0 is the initial phase difference (see below), $\mathbf{n}_{ij} = \mathbf{X}_{ij}/|\mathbf{X}_{ij}|$ with $\mathbf{X}_{ij} = (\mathbf{x}_i + \mathbf{x}_j)/2$, and $w = 2\lambda_L(T) + l$ with λ_L being the London penetration depth of superconducting area and l an insulator thickness that, within the discussed here scenario, is simply equal to the TB thickness (Sergeenkov 1995).

To neglect the influence of the self-field effects in a real material, the corresponding Josephson penetration length $\lambda_J = \sqrt{\Phi_0/2\pi\mu_0 j_c w}$ must be larger than the junction size d. Here, j_c is the critical current density of superconducting (hole-rich) area. As we shall see below, this condition is rather well satisfied for HTS single crystals.

Within our scenario, the sheet magnetization $\mathbf{M}$ of 2D granular superconductor is defined via the average Josephson energy of the array

$$< \mathcal{H} >= \int_0^\tau \frac{dt}{\tau} \int \frac{d^2x}{s} \mathcal{H}(\mathbf{x}, t) \tag{21.3}$$

as follows

$$\mathbf{M}(\mathbf{B}, \delta) \equiv -\frac{\partial < \mathcal{H} >}{\partial \mathbf{B}}, \tag{21.4}$$

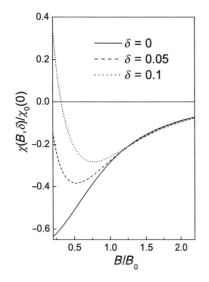

Fig. 21.1 The susceptibility as a function of applied magnetic field for different values of oxygen deficiency parameter: $\delta \simeq 0$ (solid line), $\delta = 0.05$ (dashed line), and $\delta = 0.1$ (dotted line).

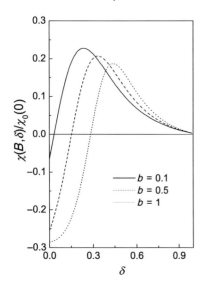

Fig. 21.2 The   oxygen-deficiency-induced susceptibility for different values of applied magnetic field (chemomagnetism).

where $s = 2\pi d^2$ is properly defined normalization area, τ is a characteristic Josephson time, and we made a usual substitution $\frac{1}{N}\sum_{ij} A_{ij}(t) \rightarrow \frac{1}{s}\int d^2 x\, A(\mathbf{x}, t)$ valid in the long-wavelength approximation (Sergeenkov 2002).

To capture the very essence of the superconducting analog of the chemomagnetic effect, in what follows we assume for simplicity that a *stoichiometric sample* (with $\delta \simeq 0$) does not possess any spontaneous magnetization at zero magnetic field (that is $\mathbf{M}(0, 0) = 0$) and that its Meissner response to a small applied field $\mathbf{B}$ is purely diamagnetic (that is $\mathbf{M}(\mathbf{B}, 0) \simeq -\mathbf{B}$). According to eqn (21.4), this condition implies $\phi_{ij}^0 = 2\pi m$ for the initial phase difference with $m = 0, \pm 1, \pm 2, \ldots$

Taking the applied magnetic field along the c-axis (and normal to the CuO plane), we obtain finally

$$\mathbf{M}(\mathbf{B}, \delta) = -\mathbf{M}_0(\delta)\frac{\mathbf{b} - \mathbf{b}_\mu}{(1 + \mathbf{b}^2)(1 + (\mathbf{b} - \mathbf{b}_\mu)^2)} \qquad (21.5)$$

for the chemically induced sheet magnetization of the 2D Josephson network. Here, $\mathbf{M}_0(\delta) = J_0(\delta)/\mathbf{B}_0$ with $J_0(\delta)$ defined earlier, $\mathbf{b} = \mathbf{B}/\mathbf{B}_0$, and $\mathbf{b}_\mu = \mathbf{B}_\mu/\mathbf{B}_0 \simeq (k_B T \tau/\hbar)\delta$, where $\mathbf{B}_\mu(\delta) = (\mu_v \tau/\hbar)\mathbf{B}_0$ is the chemically induced contribution (which disappears in optimally doped systems with $\delta \simeq 0$), and $\mathbf{B}_0 = \Phi_0/wd$ is a characteristic Josephson field.

Figure 21.1 shows changes of the initial (stoichiometric) diamagnetic susceptibility $\chi(\mathbf{B}, \delta) = \partial \mathbf{M}(\mathbf{B}, \delta)/\partial \mathbf{B}$ (solid line) with oxygen deficiency δ. As is seen, even relatively small values of δ parameter render a low-field Meissner phase strongly paramagnetic (dotted and dashed lines). Figure 21.2 presents concentration (deficiency) induced susceptibility $\chi(\mathbf{B}, \delta)/\chi_0(0)$ for different values of applied magnetic field $\mathbf{b} = \mathbf{B}/\mathbf{B}_0$ including a true *chemomagnetic* effect (solid line). According to eqn (21.5), the initially diamagnetic Meissner effect turns paramagnetic as soon as the chemomagnetic contribution $\mathbf{B}_\mu(\delta)$ exceeds an applied magnetic field $\mathbf{B}$. To see whether this can actually happen in a real material, let us estimate a magnitude of the chemomagnetic field $\mathbf{B}_\mu$. Typically (Daeumling *et al.* 1990; Gurevich and Pashitskii 1997), for HTS single crystals $\lambda_L(0) \approx 150\,\text{nm}$ and $d \simeq 10\,\text{nm}$, leading to $\mathbf{B}_0 \simeq 0.5T$. Using $\tau \simeq \hbar/\mu_v$ and $j_c = 10^{10}\,\text{A/m}^2$ as a pertinent characteristic time and the typical value of the critical current density, respectively, we arrive at the following estimate of the chemomagnetic field $\mathbf{B}_\mu(\delta) \simeq 0.5B_0$ for $\delta = 0.05$. Thus, the predicted chemically induced PME should be observable for applied magnetic fields $\mathbf{B} \simeq 0.5B_0 \simeq 0.25$ T, which are actually much higher than the fields needed to observe the previously discussed piezomagnetism and stress induced PME in high-T_c ceramics (Sergeenkov 1999). Notice that for the above set of parameters, the Josephson length $\lambda_J \simeq 1\,\mu\text{m}$, which means that the small-junction approximation assumed here (with $d \ll \lambda_J$) is valid and the so-called "self-field" effects can be safely neglected. So far, we neglected a possible field dependence of the chemical potential μ_v of oxygen vacancies. However, in high enough applied magnetic fields $\mathbf{B}$, the field-induced change of the chemical potential $\Delta\mu_v(\mathbf{B}) \equiv \mu_v(\mathbf{B}) - \mu_v(0)$ becomes tangible and should be taken into account. As is well known (Abrikosov 1988; Sergeenkov and Ausloos 1999), in a superconducting state $\Delta\mu_v(\mathbf{B}) = -\mathbf{M}(\mathbf{B})\mathbf{B}/n$, where $\mathbf{M}(\mathbf{B})$ is the corresponding magnetization, and n is the relevant carrier number density. At the same time, within our scenario, the chemical potential of a single oxygen vacancy μ_v depends on the concentration of oxygen vacancies (through the deficiency parameter δ). As a result, two different effects are possible related, respectively, to the magnetic-field dependence of $\mu_v(\mathbf{B})$ and to its dependence on magnetization $\mu_v(\mathbf{M})$. The former is simply a superconducting analog of the so-called *magnetoconcentration* effect that was predicted and observed in inhomogeneously doped semiconductors (Akopyan *et al.* 1990) with field-induced creation of oxygen vacancies $\mathbf{c}_v(\mathbf{B}) = \mathbf{c}_v(0)\exp(-\Delta\mu_v(\mathbf{B})/k_B T)$, while the latter results in a "fishtail"-like behavior of the magnetization. Let us start with the magnetoconcentration effect. Figure 21.3 depicts the predicted field-induced creation of oxygen vacancies $\mathbf{c}_v(\mathbf{B})$ using the above-obtained magnetization $\mathbf{M}(\mathbf{B}, \delta)$ (see Fig. 21.1 and eqn (21.5)). We also assumed, for simplicity, a complete stoichiometry of the system in a zero magnetic field (with $\mathbf{c}_v(0) = 1$). Notice that $\mathbf{c}_v(\mathbf{B})$ exhibits a maximum at $\mathbf{c}_m \simeq 0.23$ for applied fields $\mathbf{B} = \mathbf{B}_0$ (in agreement with the classical percolative behavior observed in non-stoichiometric YBa$_2$Cu$_3$O$_{7-\delta}$ samples (Daeumling *et al.* 1990; Gantmakher *et al.* 1990; Moeckley *et al.* 1993). Finally, let us show that in underdoped crystals the above-discussed osmotic mechanism of oxygen transport is indeed much more effective than a traditional diffusion. Using typical YBCO parameters (Gurevich and Pashitskii 1997), $\epsilon_0 = 0.01$, $\Omega_0 = a_0^3$ with $a_0 = 0.2\,\text{nm}$, and $K = 115\,\text{GPa}$, we have $\mu_v(0) = \epsilon_0 K \Omega_0 \simeq 1\,\text{meV}$ for a zero-field value of the chemical potential in HTS crystals, which leads to the creation of excess vacancies with concentration $\mathbf{c}_v(0) = e^{-\mu_v(0)/k_B T} \simeq 0.75$ (equivalent to a deficiency value of $\delta(0) \simeq 0.25$) at $T = T_C$, while the probability of oxygen diffusion in these materials (governed by a rather high activation energy $U_d \simeq 1\,\text{eV}$) is extremely low under the same conditions because $D \propto e^{-U_d/k_B T_C} \ll 1$. On the other hand, the change of the chemical potential in applied magnetic field can reach as much as (Sergeenkov and Ausloos 1999) $\Delta\mu_v(\mathbf{B}) \simeq 0.5\,\text{meV}$ for $\mathbf{B} = 0.5\,\text{T}$, which is quite comparable with the above-mentioned zero-field value of $\mu_v(0)$. Let us turn now to the second effect related to the magnetization dependence of the chemical potential $\mu_v(\mathbf{M}(\mathbf{B}))$. In this case, in view of eqn (21.2), the phase difference will acquire an extra $\mathbf{M}(\mathbf{B})$-dependent contribution and as a result the rhs of eqn (21.5) will become a non-linear functional of $\mathbf{M}(\mathbf{B})$. The numerical solution of this implicit equation for the resulting magnetization $m_f = \mathbf{M}(\mathbf{B}, \delta(\mathbf{B}))/M_0(0)$ is shown in Fig. 21.4 for three values of zero-field deficiency parameter $\delta(0)$. As is clearly seen, m_f exhibits a field-induced "fishtail"-like behavior typical for underdoped crystals with intragrain granularity. The extra extremum of the magnetization appears when the applied magnetic field $\mathbf{B}$ matches an intrinsic chemomagnetic field $\mathbf{B}_\mu(\delta(\mathbf{B}))$ (which now also depends on $\mathbf{B}$ via the above-discussed magnetoconcentration effect). Notice that a "fishtail" structure of m_f manifests itself even at zero values of field-free deficiency parameter $\delta(0)$ (solid line in Fig. 21.3) thus confirming a field-induced nature of intrinsic nanogranularity (Daeumling *et al.* 1990; Moeckley *et al.* 1993; Yang *et al.* 1993; Gurevich and Pashitskii 1997; Lang *et al.* 2002). At the same time, even a rather small deviation from the

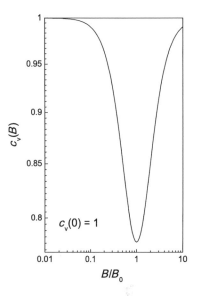

Fig. 21.3 Magnetic-field dependence of the oxygen vacancy concentration (magnetoconcentration effect).

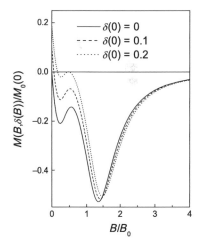

Fig. 21.4 A "fishtail"-like behavior of magnetization in an applied magnetic field in the presence of magnetoconcentration effect (with field-induced oxygen vacancies $\mathbf{c}_v(\mathbf{B})$, see Fig. 21.3) for three values of the field-free deficiency parameter: $\delta(0) \simeq 0$ (solid line), $\delta(0) = 0.1$ (dashed line), and $\delta(0) = 0.2$ (dotted line).

zero-field stoichiometry (with $\delta(0) = 0.1$) immediately brings about a para-magnetic Meissner effect at low magnetic fields. Thus, the present model predicts the appearance of two interrelated phenomena, Meissner paramagnetism at low fields and "fishtail" anomaly at high fields. It would be very interesting to verify these predictions experimentally in non-stoichiometric superconductors with pronounced networks of planar defects.

21.3   Magnetic-field-induced polarization effects in 2D JJA

In this section, within the same model of JJAs created by a regular 2D network of twin-boundary (TB) dislocations with strain fields acting as an insulating barrier between hole-rich domains in underdoped crystals, we discuss charge-related effects that are actually dual to the above-described phase-related chemomagnetic effects. Specifically, we consider the possible existence of a non-zero electric polarization $\mathbf{P}(\delta, \mathbf{B})$ (chemomagnetoelectric effect) and the related change of the charge balance in intrinsically granular non-stoichiometric material under the influence of an applied magnetic field. In particular, we predict an anomalous low-field magnetic behavior of the effective junction charge $\mathbf{Q}(\delta, \mathbf{B})$ and concomitant magnetocapacitance $\mathbf{C}(\delta, \mathbf{B})$ in paramagnetic Meissner phase and a charge analog of "fishtail"-like anomaly at high magnetic fields along with field-tuned weakening of the chemically induced Coulomb blockade (Sergeenkov 2007).

Recall that a conventional (zero-field) pair-polarization operator within the model under discussion reads (Sergeenkov 1997, 2002, 2005, 2007)

$$\mathbf{p} = \sum_{i=1}^{N} q_i \mathbf{x}_i. \tag{21.6}$$

In view of eqns (21.1), (21.2) and (21.6), and taking into account a normal "phase-number" commutation relation, $[\phi_i, n_j] = i\delta_{ij}$, it can be shown that the evolution of the pair-polarization operator is determined via the equation of motion

$$\frac{d\mathbf{p}}{dt} = \frac{1}{i\hbar} [\mathbf{p}, \mathcal{H}] = \frac{2e}{\hbar} \sum_{ij}^{N} J_{ij} \sin \phi_{ij}(t) \mathbf{x}_{ij}. \tag{21.7}$$

Resolving the above equation, we arrive at the following net value of the magnetic-field-induced longitudinal (along the x-axis) electric polarization $\mathbf{P}(\delta, \mathbf{B})$ and the corresponding effective junction charge

$$\mathbf{Q}(\delta, \mathbf{B}) = \frac{2eJ_0}{\hbar\tau d} \int_0^{\tau} dt \int_0^{t} dt' \int \frac{d^2x}{S} \sin \phi(\mathbf{x}, t') x e^{-|\mathbf{x}|/d}, \tag{21.8}$$

where $S = 2\pi d^2$ is the properly defined normalization area, τ is a characteristic time (see below), and we made a normal substitution

$\frac{1}{N}\sum_{ij}A_{ij}(t) \to \frac{1}{S}\int d^2x A(\mathbf{x},t)$ valid in the long-wavelength approximation (Sergeenkov 2002).

To capture the very essence of the superconducting analog of the chemo-magnetoelectric effect, in what follows we assume for simplicity that a *stoichiometric sample* (with $\delta \simeq 0$) does not possess any spontaneous polarization at zero magnetic field, that is $\mathbf{P}(0,0) = 0$. According to eqn (21.8), this condition implies $\phi_{ij}^0 = 2\pi m$ for the initial phase difference with $m = 0, \pm 1, \pm 2, ...$

Taking the applied magnetic field along the c-axis (and normal to the CuO plane), we obtain finally

$$\mathbf{Q}(\delta, \mathbf{B}) = \mathbf{Q}_0(\delta)\frac{2\tilde{\mathbf{b}} + \mathbf{b}(1 - \tilde{\mathbf{b}}^2)}{(1 + \mathbf{b}^2)(1 + \tilde{\mathbf{b}}^2)^2} \qquad (21.9)$$

for the magnetic-field behavior of the effective junction charge in chemically induced granular superconductors. Here, $\mathbf{Q}_0(\delta) = e\tau J_0(\delta)/\hbar$ with $J_0(\delta)$ defined earlier, $\mathbf{b} = \mathbf{B}/\mathbf{B}_0$, $\tilde{\mathbf{b}} = \mathbf{b} - \mathbf{b}_\mu$, and $\mathbf{b}_\mu = \mathbf{B}_\mu/\mathbf{B}_0 \simeq (k_B T\tau/\hbar)\delta$ where $\mathbf{B}_\mu(\delta) = (\mu_v\tau/\hbar)\mathbf{B}_0$ is the chemically induced contribution (which disappears in optimally doped systems with $\delta \simeq 0$), and $\mathbf{B}_0 = \Phi_0/wd$ is a characteristic Josephson field.

Figure 21.5 shows changes of the initial (stoichiometric) effective junction charge $\Delta\mathbf{Q}(\delta, \mathbf{B}) = \mathbf{Q}(\delta, \mathbf{B}) - \mathbf{Q}(\delta, 0)$ (solid line) with oxygen deficiency δ. According to eqn (21.9), the effective charge $\mathbf{Q}$ changes its sign at low magnetic fields (driven by non-zero values of δ) as soon as the chemomagnetic contribution $\mathbf{B}_\mu(\delta)$ exceeds an applied magnetic field $\mathbf{B}$. This is simply a charge analog of chemically induced PME. At the same time, Fig. 21.6 presents a variation of the *chemomagnetoelectric* effect with concentration (deficiency) for different values of the applied magnetic field. Notice that a zero-field contribution (which is a true *chemoelectric* effect) exhibits a maximum around $\delta_c \simeq 0.2$, in agreement with the classical percolative behavior observed in non-stoichiometric $YBa_2Cu_3O_{7-\delta}$ samples (Gantmakher *et al.* 1990).

It is of interest also to consider the magnetic-field behavior of the concomitant effective flux capacitance $\mathbf{C} \equiv \tau d\mathbf{Q}(\delta, \mathbf{B})/d\Phi$ that in view of eqn (21.9) reads

$$\mathbf{C}(\delta, \mathbf{B}) = \mathbf{C}_0(\delta)\frac{1 - 3\mathbf{b}\tilde{\mathbf{b}} - 3\tilde{\mathbf{b}}^2 + \mathbf{b}\tilde{\mathbf{b}}^3}{(1 + \mathbf{b}^2)(1 + \tilde{\mathbf{b}}^2)^3}, \qquad (21.10)$$

where $\Phi = SB$, and $\mathbf{C}_0(\delta) = \tau\mathbf{Q}_0(\delta)/\Phi_0$.

Figure 21.7 depicts the behavior of the effective flux capacitance $\Delta\mathbf{C}(\delta, \mathbf{B}) = \mathbf{C}(\delta, \mathbf{B}) - \mathbf{C}(\delta, 0)$ in an applied magnetic field for different values of oxygen deficiency parameter: $\delta \simeq 0$ (solid line), $\delta = 0.1$ (dashed line), and $\delta = 0.2$ (dotted line). Notice a decrease of magnetocapacitance amplitude and its peak shifting with increase of δ and sign change at low magnetic fields, which is another manifestation of the charge analog of chemically induced PME (cf. Fig. 21.5). Up to now, we neglected a possible field dependence of the chemical potential μ_v of oxygen vacancies. Recall, however, that in high enough applied magnetic fields $\mathbf{B}$, the field-induced change of the chemical potential $\Delta\mu_v(\mathbf{B}) \equiv \mu_v(\mathbf{B}) - \mu_v(0)$ becomes tangible and should be taken into account (Abrikosov 1988; Sergeenkov and Ausloos

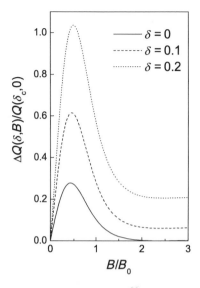

Fig. 21.5 A variation of effective junction charge with an applied magnetic field (chemomagnetoelectric effect) for different values of oxygen deficiency parameter: $\delta \simeq 0$ (solid line), $\delta = 0.1$ (dashed line), and $\delta = 0.2$ (dotted line).

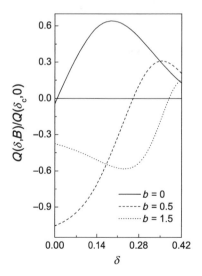

Fig. 21.6 A variation of the *chemomagnetoelectric* effect with concentration (deficiency) for different values of the applied magnetic field.

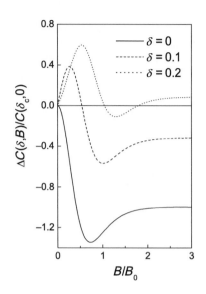

Fig. 21.7 The effective flux capacitance as a function of applied magnetic field for different values of oxygen-deficiency parameter: $\delta \simeq 0$ (solid line), $\delta = 0.1$ (dashed line), and $\delta = 0.2$ (dotted line).

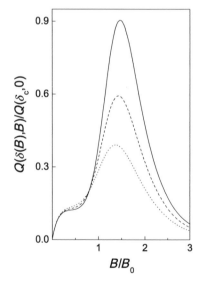

Fig. 21.8 A "fishtail"-like behavior of an effective charge in applied magnetic field in the presence of magnetoconcentration effect (with field-induced oxygen vacancies $\delta(\mathbf{B})$) for three values of field-free deficiency parameter (from top to bottom): $\delta(0) \simeq 0$ (solid line), $\delta(0) = 0.1$ (dashed line), and $\delta(0) = 0.2$ (dotted line).

1999). As a result, we end up with a superconducting analog of the so-called *magnetoconcentration* effect (Sergeenkov 2003) with field-induced creation of oxygen vacancies $\mathbf{c_v}(\mathbf{B}) = \mathbf{c_v}(0) \exp(-\Delta\mu_v(\mathbf{B})/k_B T)$ that in turn brings about a "fishtail"-like behavior of the high-field chemomagnetization (see Section 21.2 for more details). Figure 21.8 shows the field behavior of the effective junction charge in the presence of the above-mentioned magneto-concentration effect. As is clearly seen, $\mathbf{Q}(\delta(\mathbf{B}), \mathbf{B})$ exhibits a "fishtail"-like anomaly typical for previously discussed (Sergeenkov 2003) chemomagnetization in underdoped crystals with intragrain granularity. This more complex structure of the effective charge appears when the applied magnetic field $\mathbf{B}$ matches an intrinsic chemomagnetic field $\mathbf{B}_\mu(\delta(\mathbf{B}))$ (which now also depends on $\mathbf{B}$ via the magnetoconcentration effect). Notice that a "fishtail" structure of $\mathbf{Q}(\delta(\mathbf{B}), \mathbf{B})$ manifests itself even at zero values of field-free deficiency parameter $\delta(0)$ (solid line in Fig. 21.8) thus confirming a field-induced nature of intrinsic granularity. Likewise, Fig. 21.9 depicts the evolution of the effective flux capacitance $\Delta\mathbf{C}(\delta(\mathbf{B}), \mathbf{B}) = \mathbf{C}(\delta(\mathbf{B}), \mathbf{B}) - \mathbf{C}(\delta(0), 0)$ in an applied magnetic field $\mathbf{B}$ in the presence of magnetoconcentration effect (cf. Fig. 21.7).

Thus, the present model predicts the appearance of two interrelated phenomena dual to the previously discussed behavior of chemomagnetizm (see Section 21.2), namely a charge analog of Meissner paramagnetism at low fields and a charge analog of the "fishtail" anomaly at high fields. To see whether these effects can be actually observed in a real material, let us estimate an order of magnitude of the main model parameters.

Using typical for HTS single crystals values of $\lambda_L(0) \simeq 150$ nm, $d \simeq 10$ nm, and $j_c \simeq 10^{10}$ A/m^2, we arrive at the following estimates of the characteristic $B_0 \simeq 0.5$ T and chemomagnetic $B_\mu(\delta) \simeq 0.5 B_0$ fields, respectively. So, the predicted charge analog of PME should be observable for applied magnetic fields $\mathbf{B} < 0.25$ T. Notice that, for the above set of parameters, the Josephson length is of the order of $\lambda_J \simeq 1 \,\mu$m, which means that the small-junction approximation assumed in this paper is valid and the "self-field" effects can be safely neglected.

Furthermore, the characteristic frequencies $\omega \simeq \tau^{-1}$ needed to probe the effects suggested here are related to the processes governed by tunnelling relaxation times $\tau \simeq \hbar/J_0(\delta)$. Since for the oxygen-deficiency parameter $\delta = 0.1$ the chemically induced zero-temperature Josephson energy in non-stoichiometric YBCO single crystals is of the order of $J_0(\delta) \simeq k_B T_C \delta \simeq 1$ meV, we arrive at the required frequencies of $\omega \simeq 10^{13}$ Hz and at the following estimates of the effective junction charge $\mathbf{Q}_0 \simeq e = 1.6 \times 10^{-19}$ C and flux capacitance $\mathbf{C}_0 \simeq 10^{-18}$ F. Notice that the above estimates fall into the range of parameters used in typical   experiments for studying the single-electron tunnelling effects both in JJs and JJAs (van Bentum *et al.* 1988; Makhlin *et al.* 2001) suggesting thus quite an optimistic possibility to observe the above-predicted field-induced effects experimentally in non-stoichiometric superconductors with pronounced networks of planar defects or in artificially prepared JJAs. It is worth mentioning that a somewhat similar behavior of the magnetic-field-induced charge and related flux capacitance has been observed in 2D electron systems (Chen *et al.* 1994).

And finally, it can be easily verified that, in view of eqns (21.6)–(21.8), the field-induced Coulomb energy of the oxygen-depleted region within our model is given by

$$E_C(\delta, \mathbf{B}) \equiv \left\langle \sum_{ij}^{N} \frac{q_i q_j}{2C_{ij}} \right\rangle = \frac{\mathbf{Q}^2(\delta, \mathbf{B})}{2\mathbf{C}(\delta, \mathbf{B})}, \qquad (21.11)$$

with $\mathbf{Q}(\delta, \mathbf{B})$ and $\mathbf{C}(\delta, \mathbf{B})$ defined by eqns (21.9) and (21.10), respectively.

A thorough analysis of the above expression reveals that in the PME state (when $\mathbf{B} \ll \mathbf{B}_\mu$) the chemically induced granular superconductor is in the so-called Coulomb-blockade regime (with $E_C > J_0$), while in the "fishtail" state (for $\mathbf{B} \geq \mathbf{B}_\mu$) the energy balance tips in favor of tunnelling (with $E_C < J_0$). In particular, we obtain that $E_C(\delta, \mathbf{B} = 0.1\mathbf{B}_\mu) = \frac{\pi}{2} J_0(\delta)$ and $E_C(\delta, \mathbf{B} = \mathbf{B}_\mu) = \frac{\pi}{8} J_0(\delta)$. It would also be interesting to check this phenomenon of field-induced weakening of the Coulomb blockade experimentally.

21.4   Giant enhancement of thermal conductivity in 2D JJA

In this section, using a 2D model of inductive Josephson junction arrays (created by a network of twin-boundary dislocations with strain fields acting as an insulating barrier between hole-rich domains in underdoped crystals), we study the temperature, $\mathbf{T}$, and chemical pressure, $\nabla\mu$, dependence of the thermal conductivity (TC) κ of an intrinsically nanogranular superconductor. Two major effects affecting the behavior of TC under chemical pressure are predicted: decrease of the *linear* (i.e. $\nabla\mathbf{T}$-independent) TC, and giant enhancement of the *non-linear* (i.e. $\nabla\mathbf{T}$-dependent) TC with $[\kappa(\mathbf{T}, \nabla\mathbf{T}, \nabla\mu) - \kappa(\mathbf{T}, \nabla\mathbf{T}, 0)]/\kappa(\mathbf{T}, \nabla\mathbf{T}, 0)$ reaching 500% when the chemoelectric field $\mathbf{E}_\mu = \frac{1}{2e}\nabla\mu$ matches the thermoelectric field $\mathbf{E}_T = S_T\nabla\mathbf{T}$. The conditions under which these effects can be experimentally measured in non-stoichiometric high-T_C superconductors are discussed.

There are several approaches for studying the thermal response of JJs and JJAs based on the phenomenology of the Josephson effect in the presence of thermal gradients (see, e.g. van Harlingen *et al.* 1980; Deppe and Feldman 1994; Guttman *et al.* 1997; Sergeenkov 2002, 2007 and further references therein). To adequately describe transport properties of the above-described chemically induced nanogranular superconductor for all temperatures and under a simultaneous influence of intrinsic chemical pressure $\nabla\mu(\mathbf{x}) = K\Omega_0\nabla\epsilon(\mathbf{x})$ and applied thermal gradient ∇T, we employ a model of 2D overdamped Josephson junction array that is based on the following total Hamiltonian (Sergeenkov 2002)

$$\mathcal{H}(t) = \mathcal{H}_T(t) + \mathcal{H}_L(t) + \mathcal{H}_\mu(t), \qquad (21.12)$$

where

$$\mathcal{H}_T(t) = \sum_{ij}^{N} J_{ij}[1 - \cos\phi_{ij}(t)] \qquad (21.13)$$

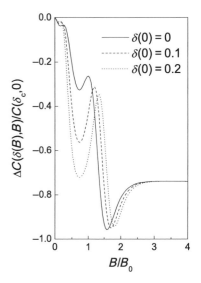

Fig. 21.9 The behavior of the effective flux capacitance in an applied magnetic field in the presence of the magnetoconcentration effect for three values of field-free deficiency parameter: $\delta(0) \simeq 0$ (solid line), $\delta(0) = 0.1$ (dashed line), and $\delta(0) = 0.2$ (dotted line).

is the well-known tunnelling Hamiltonian,

$$\mathcal{H}_{\mathrm{L}}(t) = \sum_{ij}^{N} \frac{\Phi_{ij}^2(t)}{2L_{ij}} \tag{21.14}$$

accounts for a mutual inductance L_{ij} between grains (and controls the normal-state value of the thermal conductivity, see below) with $\Phi_{ij}(t) = (\hbar/2e)\phi_{ij}(t)$ being the total magnetic flux through an array, and finally

$$\mathcal{H}_{\mu}(t) = \sum_{i=1}^{N} n_i(t)\delta\mu_i \tag{21.15}$$

describes the chemical-potential-induced contribution with $\delta\mu_i = \mathbf{x}_i \nabla\mu$, and n_i being the pair number operator.

According to the above-mentioned scenario, the tunnelling Hamiltonian $\mathcal{H}_{\mathrm{T}}(t)$ introduces a short-range (nearest-neighbor) interaction between N junctions (which are formed around oxygen-rich superconducting areas with phases $\phi_i(t)$), arranged in a two-dimensional (2D) lattice with co-ordinates $\mathbf{x_i} = (x_i, y_i)$. The areas are separated by oxygen-poor insulating boundaries (created by TB strain fields $\epsilon(\mathbf{x}_{ij})$) producing a short-range Josephson coupling $J_{ij} = J_0(\delta)e^{-|\mathbf{x}_{ij}|/d}$. Thus, typically for granular superconductors, the Josephson energy of the array varies exponentially with the distance $\mathbf{x}_{ij} = \mathbf{x}_i - \mathbf{x}_j$ between neighboring junctions (with d being an average grain size). The temperature dependence of chemically induced Josephson coupling is governed by the following expression, $J_{ij}(\mathbf{T}) = J_{ij}(0)F(\mathbf{T})$ where

$$F(\mathbf{T}) = \frac{\Delta(\mathbf{T})}{\Delta(0)} \tanh\left[\frac{\Delta(\mathbf{T})}{2k_{\mathrm{B}}\mathbf{T}}\right], \tag{21.16}$$

and $J_{ij}(0) = [\Delta(0)/2](R_0/R_{ij})$ with $\Delta(\mathbf{T})$ being the temperature-dependent gap parameter, $R_0 = h/4e^2$ is the quantum resistance, and R_{ij} is the resistance between grains in their normal state.

By analogy with a constant electric field $\mathbf{E}$, a thermal gradient $\nabla\mathbf{T}$ applied to a chemically induced JJA will cause a time evolution of the phase difference across insulating barriers as follows (Sergeenkov 2002)

$$\phi_{ij}(t) = \phi_{ij}^0 + \omega_{ij}(\nabla\mu, \nabla T)t. \tag{21.17}$$

Here, ϕ_{ij}^0 is the initial phase difference (see below), and $\omega_{ij} = 2e(\mathbf{E}_{\mu} - \mathbf{E}_{\mathrm{T}})\mathbf{x}_{ij}/\hbar$ where $\mathbf{E}_{\mu} = \frac{1}{2e}\nabla\mu$ and $\mathbf{E}_{\mathrm{T}} = S_{\mathrm{T}}\nabla\mathbf{T}$ are the induced chemoelectric and thermoelectric fields, respectively. S_{T} is the so-called thermophase coefficient (Sergeenkov 1998a) that is related to the Seebeck coefficient S_0 as follows, $S_{\mathrm{T}} = (l/d)S_0$ (where l is a relevant sample size responsible for the applied thermal gradient, that is $|\nabla\mathbf{T}| = \Delta\mathbf{T}/l$).

We start our consideration by discussing the temperature behavior of the conventional (that is *linear*) thermal conductivity of a chemically induced nanogranular superconductor paying special attention to its evolution with a mutual inductance L_{ij}. For simplicity, in what follows we limit our consideration to the longitudinal component of the total thermal flux $\mathbf{Q}(t)$ that is defined (in a q-space representation) via the total energy conservation law as follows

$$\mathbf{Q}(t) \equiv \lim_{\mathbf{q}\to 0}\left[\mathrm{i}\frac{\mathbf{q}}{\mathbf{q}^2}\dot{\mathcal{H}}_{\mathbf{q}}(t)\right], \tag{21.18}$$

where $\dot{\mathcal{H}}_{\mathbf{q}} = \partial\mathcal{H}_{\mathbf{q}}/\partial t$ with

$$\mathcal{H}_{\mathbf{q}}(t) = \frac{1}{s}\int d^2x e^{i\mathbf{q}\mathbf{x}}\mathcal{H}(\mathbf{x},t). \tag{21.19}$$

Here, $s = 2\pi d^2$ is the properly defined normalization area, and we made a usual substitution $\frac{1}{N}\sum_{ij}A_{ij}(t) \to \frac{1}{s}\int d^2x A(\mathbf{x},t)$ valid in the long-wavelength approximation ($\mathbf{q}\to 0$).

In turn, the heat flux $\mathbf{Q}(t)$ is related to the *linear* thermal conductivity (LTC) tensor $\kappa_{\alpha\beta}$ by the Fourier law as follows (hereafter, $\{\alpha,\beta\} = x,y,z$)

$$\kappa_{\alpha\beta}(\mathbf{T},\nabla\mu) \equiv -\frac{1}{V}\left[\frac{\partial \overline{<Q_\alpha>}}{\partial(\nabla_\beta \mathbf{T})}\right]_{\nabla\mathbf{T}=0}, \tag{21.20}$$

where

$$\overline{<\mathbf{Q}_\alpha>} = \frac{1}{\tau}\int_0^\tau dt <\mathbf{Q}_\alpha(t)>. \tag{21.21}$$

Here, V is the samples volume, τ is a characteristic Josephson tunnelling time for the network, and $< \ldots >$ denotes the thermodynamic averaging over the initial phase differences ϕ_{ij}^0

$$< A(\phi_{ij}^0) > = \frac{1}{Z}\int_0^{2\pi}\prod_{ij}d\phi_{ij}^0 A(\phi_{ij}^0)e^{-\beta H_0}, \tag{21.22}$$

with an effective Hamiltonian

$$H_0\left[\phi_{ij}^0\right] = \int_0^\tau \frac{dt}{\tau}\int\frac{d^2x}{s}\mathcal{H}(\mathbf{x},t). \tag{21.23}$$

Here, $\beta = 1/k_\mathrm{B}\mathbf{T}$, and $Z = \int_0^{2\pi}\prod_{ij}d\phi_{ij}^0 e^{-\beta H_0}$ is the partition function. The above-defined averaging procedure allows us to study the temperature evolution of the system.

Taking into account that in JJAs (Eichenberger *et al.* 1996) $L_{ij} \propto R_{ij}$, we obtain $L_{ij} = L_0\exp(|\mathbf{x}_{ij}|/d)$ for the explicit x dependence of the weak-link inductance in our model. Finally, in view of eqns (21.12)–(21.23), and making use of the usual "phase-number" commutation relation, $[\phi_i, n_j] = i\delta_{ij}$, we find the following analytical expression for the temperature and chemical-gradient dependence of the electronic contribution to the *linear* thermal conductivity of a granular superconductor

$$\kappa_{\alpha\beta}(\mathbf{T},\nabla\mu) = \kappa_0\left[\delta_{\alpha\beta}\eta(\mathbf{T},\epsilon) + \beta_\mathrm{L}(\mathbf{T})\nu(\mathbf{T},\epsilon)f_{\alpha\beta}(\epsilon)\right], \tag{21.24}$$

where

$$f_{\alpha\beta}(\epsilon) = \frac{1}{4}\left[\delta_{\alpha\beta}A(\epsilon) - \epsilon_\alpha\epsilon_\beta B(\epsilon)\right], \tag{21.25}$$

with

$$A(\epsilon) = \frac{5 + 3\epsilon^2}{(1 + \epsilon^2)^2} + \frac{3}{\epsilon} \tan^{-1} \epsilon, \tag{21.26}$$

and

$$B(\epsilon) = \frac{3\epsilon^4 + 8\epsilon^2 - 3}{\epsilon^2 (1 + \epsilon^2)^3} + \frac{3}{\epsilon^3} \tan^{-1} \epsilon. \tag{21.27}$$

Here, $\kappa_0 = Nd^2 S_T \Phi_0 / V L_0$, $\beta_L(\mathbf{T}) = 2\pi I_C(\mathbf{T}) L_0 / \Phi_0$ with $I_C(\mathbf{T}) = (2e/\hbar) J(\mathbf{T})$ being the critical current; $\epsilon \equiv \sqrt{\epsilon_x^2 + \epsilon_y^2 + \epsilon_z^2}$ with $\epsilon_\alpha = \mathbf{E}_\mu^\alpha / \mathbf{E}_0$, where $\mathbf{E}_0 = \hbar / 2ed\tau$ is a characteristic field. In turn, the above-introduced "order parameters" of the system, $\eta(\mathbf{T}, \epsilon) \equiv < \phi_{ij}^0 >$ and $v(\mathbf{T}, \epsilon) \equiv < \sin \phi_{ij}^0 >$, are defined as follows

$$\eta(\mathbf{T}, \epsilon) = \frac{\pi}{2} - \frac{4}{\pi} \sum_{n=0}^{\infty} \frac{1}{(2n+1)^2} \left[\frac{I_{2n+1}(\beta_\mu)}{I_0(\beta_\mu)} \right], \tag{21.28}$$

and

$$v(\mathbf{T}, \epsilon) = \frac{\sinh \beta_\mu}{\beta_\mu I_0(\beta_\mu)}, \tag{21.29}$$

where

$$\beta_\mu(\mathbf{T}, \epsilon) = \frac{\beta J(\mathbf{T})}{2} \left(\frac{1}{1 + \epsilon^2} + \frac{1}{\epsilon} \tan^{-1} \epsilon \right). \tag{21.30}$$

Here, $J(\mathbf{T})$ is given by eqn (21.17), and $I_n(x)$ stand for the modified Bessel functions.

Turning to the discussion of the obtained results, we start with a simpler zero-pressure case. The relevant parameters affecting the behavior of the LTC in this particular case include the mutual inductance L_0 and the normal-state resistance between grains R_n. For the temperature dependence of the Josephson energy (see eqn (21.17)), we used the well-known (Sergeenkov 2002) approximation for the BCS gap parameter, valid for all temperatures, $\Delta(\mathbf{T}) = \Delta(0) \tanh \left(\gamma \sqrt{\frac{\mathbf{T_C} - \mathbf{T}}{\mathbf{T}}} \right)$ with $\gamma = 2.2$. Despite the rather simplified nature of our model, it seems to quite reasonably describe the behavior of the LTC for all temperatures. Indeed, in the absence of intrinsic chemical pressure ($\nabla \mu = 0$), the LTC is isotropic (as expected), $\kappa_{\alpha\beta}(\mathbf{T}, 0) = \delta_{\alpha\beta} \kappa_L(\mathbf{T}, 0)$, where $\kappa_L(\mathbf{T}, 0) = \kappa_0 [\eta(\mathbf{T}, 0) + 2\beta_L(\mathbf{T}) v(\mathbf{T}, 0)]$ vanishes at zero temperature and reaches a normal state value $\kappa_n \equiv \kappa_L(\mathbf{T_C}, 0) = (\pi/2)\kappa_0$ at $\mathbf{T} = \mathbf{T_C}$. Figure 21.10 shows the temperature dependence of the normalized LTC $\kappa_L(\mathbf{T}, 0)/\kappa_n$ for different values of the so-called SQUID parameter $\beta_L(0) = 2\pi I_C(0) L_0 / \Phi_0$ (increasing from the bottom to the top) and for two values of the resistance ratio $r_n = R_0/R_n = 0.1$ and $r_n = R_0/R_n = 1$. First, with increasing of the SQUID parameter, the LTC evolves from a flat-like pattern (for a relatively small values of L_0) to a low-temperature maximum (for higher values of $\beta_L(0)$). Notice that the peak temperature $\mathbf{T_p}$ is practically insensitive to the variation of inductance

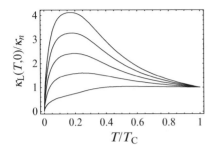 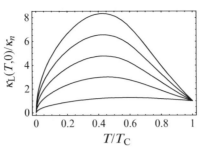

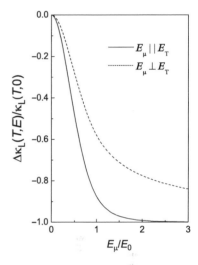

Fig. 21.10 Temperature dependence of the zero-pressure ($\nabla\mu = 0$) *linear* thermal conductivity for $r_n = 0.1$ (left) and $r_n = 1$ (right) for different values of the SQUID parameter (from bottom to top): $\beta_L(0) = 1, 3, 5, 7,$ and 9.

parameter L_0, while being at the same time strongly influenced by resistivity R_n. Indeed, as is clearly seen in Fig. 21.10, a different choice of r_n leads to quite a tangible shifting of the maximum. Namely, the smaller is the normal resistance between grains R_n (or the better is the quality of the sample) the higher is the temperature at which the peak is developed. As a matter of fact, the peak temperature T_p is related to the so-called phase-locking temperature T_J (which marks the establishment of phase coherence between the adjacent grains in the array and always lies below a single-grain superconducting temperature T_C) which is usually defined via an average (per grain) Josephson coupling energy as $J(T_J, r_n) = k_B T_J$. Indeed, it can be shown analytically that for $T_J < T < T_C$, $T_J(r_n) \simeq r_n T_C$.

Turning to the discussion of the LTC behavior under chemical pressure, let us assume, for simplicity, that $\nabla\mu = (\nabla_x\mu, 0, 0)$ with oxygen-deficiency parameter δ controlled chemical pressure $\nabla_x\mu \simeq \mu_v(\delta)/d$, and $\nabla T = (\nabla_x T, \nabla_y T, 0)$. Such a choice of the external fields allows us to consider both parallel $\kappa_{xx}(T, \nabla\mu)$ and perpendicular $\kappa_{yy}(T, \nabla\mu)$ components of the LTC corresponding to the two most interesting configurations, $\nabla\mu \| \nabla T$ and $\nabla\mu \perp \nabla T$, respectively. Figure 21.11 demonstrates the predicted chemical-pressure dependence of the normalized LTC $\Delta\kappa_L(T, \nabla\mu) = \kappa_L(T, \nabla\mu) - \kappa_L(T, 0)$ for both configurations taken at $T = 0.2 T_C$ (with $r_n = 0.1$ and $\beta_L(0) = 1$). First, we note that both components of the LTC are *decreasing* with increasing of the pressure $E_\mu/E_0 = \mu_v(\delta)\tau/\hbar$. And secondly, the normal component κ_{yy} decreases more slowly than the parallel one κ_{xx}, thus suggesting some kind of anisotropy in the system. In view of the structure of eqn (21.25), the same behavior is also expected for the temperature dependence of the chemically induced LTC, that is $\Delta\kappa_L(T, \nabla\mu)/\kappa_L(T, 0) < 0$ for all gradients and temperatures. In terms of the absolute values, for $T = 0.2 T_C$ and $E_\mu = E_0$, we obtain $[\Delta\kappa_L(T, \nabla\mu)/\kappa_L(T, 0)]_{xx} = 90\%$ and $[\Delta\kappa_L(T, \nabla\mu)/\kappa_L(T, 0)]_{yy} = 60\%$ for *attenuation* of LTC under chemical pressure. Let us turn now to the most intriguing part of this section and consider a *non-linear* generalization of the Fourier law and very unusual behavior of the resulting *non-linear* thermal conductivity (NLTC) under the influence of chemical pressure. In what follows, by the NLTC we understand a ∇T-dependent thermal conductivity $\kappa_{\alpha\beta}^{NL}(T, \nabla\mu) \equiv \kappa_{\alpha\beta}(T, \nabla\mu; \nabla T)$, which

Fig. 21.11 The dependence of the *linear* thermal conductivity on the chemical pressure for parallel ($\nabla\mu \| \nabla T$) and perpendicular ($\nabla\mu \perp \nabla T$) configurations.

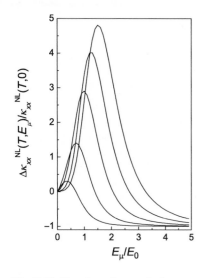

Fig. 21.12 The dependence of the *non-linear* thermal conductivity on the chemical pressure for different values of the applied thermal gradient $\epsilon_T = S_T \nabla T / E_0$ ($\epsilon_T = 0.2, 0.4, 0.6, 0.8,$ and 1.0, increasing from bottom to top).

is defined as follows

$$\kappa_{\alpha\beta}^{NL}(\mathbf{T}, \nabla\mu) \equiv -\frac{1}{V}\left[\frac{\partial \overline{<\mathbf{Q}_\alpha>}}{\partial(\nabla_\beta \mathbf{T})}\right]_{\nabla\mathbf{T}\neq 0}, \qquad (21.31)$$

with $\overline{<\mathbf{Q}_\alpha>}$ given by eqn (21.21).

Repeating the same procedure as before, we obtain finally for the relevant components of the NLTC tensor

$$\kappa_{\alpha\beta}^{NL}(\mathbf{T}, \nabla\mu) = \kappa_0 \left[\delta_{\alpha\beta}\eta(\mathbf{T}, \epsilon_{\text{eff}}) + \beta_L(\mathbf{T})\nu(\mathbf{T}, \epsilon_{\text{eff}})D_{\alpha\beta}(\epsilon_{\text{eff}})\right], \qquad (21.32)$$

where

$$D_{\alpha\beta}(\epsilon_{\text{eff}}) = f_{\alpha\beta}(\epsilon_{\text{eff}}) + \epsilon_T^\gamma g_{\alpha\beta\gamma}(\epsilon_{\text{eff}}), \qquad (21.33)$$

with

$$g_{\alpha\beta\gamma}(\epsilon) = \frac{1}{8}\left[(\delta_{\alpha\beta}\epsilon_\gamma + \delta_{\alpha\gamma}\epsilon_\beta + \delta_{\gamma\beta}\epsilon_\alpha)B(\epsilon) + 3\epsilon_\alpha\epsilon_\beta\epsilon_\gamma C(\epsilon)\right], \qquad (21.34)$$

and

$$C(\epsilon) = \frac{3 + 11\epsilon^2 - 11\epsilon^4 - 3\epsilon^6}{\epsilon^4(1+\epsilon^2)^4} - \frac{3}{\epsilon^5}\tan^{-1}\epsilon. \qquad (21.35)$$

Here, $\epsilon_{\text{eff}}^\alpha = \epsilon_\mu^\alpha - \epsilon_T^\alpha$ where $\epsilon_\mu^\alpha = \mathbf{E}_\mu^\alpha/E_0$ and $\epsilon_T^\alpha = \mathbf{E}_T^\alpha/E_0$ with $\mathbf{E}_T^\alpha = S_T\nabla_\alpha \mathbf{T}$; other parameters ($\eta$, ν, B and $f_{\alpha\beta}$) are the same as before but with $\epsilon \to \epsilon_{\text{eff}}$. As expected, in the limit $\mathbf{E}_T \to 0$ (or when $\mathbf{E}_\mu \gg \mathbf{E}_T$), from eqn (21.32) we recover all the results obtained in the previous section for the LTC. Let us see now what happens when the thermoelectric field $\mathbf{E}_T = S_T\nabla\mathbf{T}$ becomes comparable with chemoelectric field $\mathbf{E}_\mu$. Figure 21.12 depicts the resulting chemical-pressure dependence of the parallel component of the NLTC tensor $\Delta\kappa_{xx}^{NL}(\mathbf{T}, \mathbf{E}_\mu) = \kappa_{xx}^{NL}(\mathbf{T}, \mathbf{E}_\mu) - \kappa_{xx}^{NL}(\mathbf{T}, 0)$ for different values of the dimensionless parameter $\epsilon_T = \mathbf{E}_T/E_0$ (the other parameters are the same as before). As is clearly seen from this picture, in a sharp contrast with the pressure behavior of the previously considered LTC, its *non-linear* analog evolves with the chemoelectric field quite differently. Namely, NLTC strongly *increases* for small pressure values (with $\mathbf{E}_\mu < \mathbf{E}_m$), reaches a pronounced maximum at $\mathbf{E}_\mu = \mathbf{E}_m = \frac{3}{2}\mathbf{E}_T$, and eventually declines at higher values of E_μ (with $\mathbf{E}_\mu > \mathbf{E}_m$). Furthermore, as directly follows from the very structure of eqn (21.32), a similar "re-entrant-like" behavior of the *non-linear* thermal conductivity is expected for its temperature dependence as well. Even more remarkable is the absolute value of the pressure-induced enhancement. According to Fig. 21.12, it is easy to estimate that near the maximum (with $\mathbf{E}_\mu = \mathbf{E}_m$ and $\mathbf{E}_T = \mathbf{E}_0$) one gets $\Delta\kappa_{xx}^{NL}(\mathbf{T}, \mathbf{E}_\mu)/\kappa_{xx}^{NL}(\mathbf{T}, 0) \simeq 500\%$.

To understand the above-obtained rather unusual results, let us take a closer look at the chemoelectric-field-induced behavior of the Josephson voltage in our system (see eqn (21.17)). Clearly, strong heat conduction requires establishment of a quasi-stationary (that is nearly zero voltage) regime within the array. In other words, the maximum of the thermal conductivity under chemical pressure should correlate with a minimum of the total voltage in the system, $V(\nabla\mu) \equiv (\frac{\hbar}{2e}) < \frac{\partial \phi_{ij}(t)}{\partial t} > = V_0(\epsilon - \epsilon_T)$, where $\epsilon \equiv \mathbf{E}_\mu/E_0$ and

$V_0 = \mathbf{E}_0 d = \hbar/2e\tau$ is a characteristic voltage. For linear TC (which is valid only for small thermal gradients with $\epsilon_T \equiv \mathbf{E}_T/\mathbf{E}_0 \ll 1$), the average voltage through an array $V_L(\nabla\mu) \simeq V_0(\mathbf{E}_\mu/\mathbf{E}_0)$ has a minimum at zero chemoelectric field (where LTC indeed has its maximum value, see Fig. 21.11) while for non-linear TC (with $\epsilon_T \simeq 1$) we have to consider the total voltage $V(\nabla\mu)$ which becomes minimal at $\mathbf{E}_\mu = \mathbf{E}_T$ (in good agreement with the predictions for NLTC maximum that appears at $\mathbf{E}_\mu = \frac{3}{2}\mathbf{E}_T$, see Fig. 21.12).

To complete our study, let us estimate an order of magnitude of the main model parameters. Starting with chemoelectric fields $\mathbf{E}_\mu$ needed to observe the above-predicted non-linear field effects in nanogranular superconductors, we notice that according to Fig. 21.12, the most interesting behavior of NLTC takes place for $\mathbf{E}_\mu \simeq \mathbf{E}_0$. Using typical YBCO parameters, $\epsilon_0 = 0.01$, $\Omega_0 = a_0^3$ with $a_0 = 0.2\,\text{nm}$, and $K = 115\,\text{GPa}$, we have $\mu_v = \epsilon_0 K \Omega_0 \simeq 1\,\text{meV}$ for an estimate of the chemical potential in HTS crystals, which defines the characteristic Josephson tunnelling time $\tau \simeq \hbar/\mu_v \simeq 5 \times 10^{-11}\,\text{s}$ and, at the same time, leads to creation of excess vacancies with concentration $c_v = e^{-\mu_v/k_B T} \simeq 0.75$ at $\mathbf{T} = 0.2\mathbf{T}_C$ (equivalent to a deficiency value of $\delta \simeq 0.25$). Notice that in comparison with this linear defects mediated channelling (osmotic) mechanism, the probability of the conventional oxygen diffusion in these materials $D \propto e^{-U_d/k_B T}$ (governed by a rather high activation energy $U_d \simeq 1\,\text{eV}$) is extremely low under the same conditions ($D \ll 1$).

Furthermore, taking $d \simeq 10\,\text{nm}$ for typical values of the average "grain" size (created by oxygen-rich superconducting regions), we get $\mathbf{E}_0 = \hbar/2ed\tau \simeq 5 \times 10^5\,\text{V/m}$ and $|\nabla\mu| = \mu_v/d \simeq 10^6\,\text{eV/m}$ for the estimates of the characteristic field and chemical potential gradient (intrinsic chemical pressure), respectively. On the other hand, the maximum of NLTC occurs when this field nearly perfectly matches an "intrinsic" thermoelectric field $\mathbf{E}_T = S_T \nabla \mathbf{T}$ induced by an applied thermal gradient, that is when $\mathbf{E}_\mu \simeq \mathbf{E}_0 \simeq \mathbf{E}_T$. Recalling that $S_T = (l/d)S_0$ and using $S_0 \simeq 0.5\,\mu\,\text{V/K}$ and $l \simeq 0.5\,\text{mm}$ for an estimate of the *linear* Seebeck coefficient and a typical sample size, we obtain $\nabla \mathbf{T} \simeq \mathbf{E}_0/S_T \simeq 2 \times 10^6\,\text{K/m}$ for the characteristic value of the applied thermal gradient needed to observe the predicted here giant chemical pressure-induced effects. Let us estimate now the absolute value of the linear thermal conductivity governed by the intrinsic Josephson junctions. Recall that within our model the scattering of normal electrons is due to the presence of mutual inductance between the adjacent grains L_0, which is of the order of $L_0 \simeq \mu_0 d \simeq 1\,fH$ assuming $d = 10\,\text{nm}$ for an average "grain" size. In the absence of chemical-pressure effects, the temperature evolution of LTC is given by $\kappa_L(\mathbf{T}, 0) = \kappa_0[\eta(\mathbf{T}, 0) + 2\beta_L(\mathbf{T})\nu(\mathbf{T}, 0)]$ where $\kappa_0 = Nd^2 S_T \Phi_0/V L_0$. Assuming $V \simeq Nd^2 l$ for the sample volume, using the above-mentioned expression for S_T, and taking $\beta_L(0) = 5$ and $r_n = 0.1$ for the value of the SQUID parameter and the resistance ratio, we obtain $\kappa_L(0.2\mathbf{T}_C, 0) \simeq 1\,W/m\,K$ for an estimate of the maximum of the LTC (see Fig. 21.10).

And finally, it is worth comparing the above estimates for inductively coupled grains (Sergeenkov 2002) with the estimates for capacitively coupled grains (Sergeenkov 2007) where the scattering of normal electrons is governed by the Stewart–McCumber parameter $\beta_C(\mathbf{T}) = 2\pi I_C(\mathbf{T})C_0 R_n^2/\Phi_0$ due to the presence of the normal resistance R_n and mutual capacitance C_0 between the adjacent grains. The latter is estimated to be $C_0 \simeq 1\,\mathrm{aF}$ using $d = 10\,\mathrm{nm}$ for an average "grain" size. Furthermore, the critical current $I_C(0)$ can be estimated via the critical temperature $\mathbf{T_C}$ as follows, $I_C(0) \simeq 2\pi k_B \mathbf{T_C}/\Phi_0$, which gives $I_C(0) \simeq 10\,\mu\mathrm{A}$ (for $\mathbf{T_C} \simeq 90\,\mathrm{K}$) and leads to $\beta_C(0) \simeq 3$ for the value of the Stewart–McCumber parameter assuming $R_n \simeq R_0$ for the normal resistance that, in turn, results in $q \simeq \Phi_0/R_n \simeq 10^{-19}\,\mathrm{C}$ and $E_C = q^2/2C_0 \simeq 0.1\,\mathrm{eV}$ for the estimates of the "grain" charge and the Coulomb energy. Using the above-mentioned expressions for S_0 and $\beta_C(0)$, we obtain $\kappa_L \simeq 10^{-3}\,\mathrm{W/m\,K}$ for the maximum of the capacitance controlled LTC, which is actually much smaller than a similar estimate obtained above for inductance-controlled κ_L (Sergeenkov 2002) but at the same time much higher than phonon-dominated heat transport in granular systems (Deppe and Feldman 1994).

21.5   Thermal expansion of a single Josephson contact and 2D JJA

In this section, by introducing a concept of thermal expansion (TE) of a Josephson junction as an elastic response to an effective stress field, we study (both analytically and numerically) the temperature and magnetic-field dependence of TE coefficient α in a single small junction and in a square array. In particular, we found (Sergeenkov *et al.* 2007) that in addition to *field* oscillations due to Fraunhofer-like dependence of the critical current, α of a small single junction also exhibits strong flux-driven *temperature* oscillations near $\mathbf{T_C}$. We also numerically simulated the stress-induced response of a closed loop with finite self-inductance (a prototype of an array) and found that the α of a 5×5 array may still exhibit temperature oscillations if the applied magnetic field $\mathbf{H}$ is strong enough to compensate for the screening-induced effects.

Since the thermal expansion coefficient $\alpha(\mathbf{T}, \mathbf{H})$ is usually measured using mechanical dilatometers (Nagel *et al.* 2000), it is natural to introduce TE as an elastic response of the Josephson contact to an effective stress field σ (D'yachenko *et al.* 1995; Sergeenkov 1998b, 1999). Namely, we define the TE coefficient (TEC) $\alpha(\mathbf{T}, \mathbf{H})$ as follows:

$$\alpha(\mathbf{T}, \mathbf{H}) = \frac{d\epsilon}{d\mathbf{T}}, \tag{21.36}$$

where an appropriate strain field ϵ in the contact area is related to the Josephson energy E_J as follows (V is the volume of the sample):

$$\epsilon = -\frac{1}{V}\left[\frac{dE_J}{d\sigma}\right]_{\sigma=0}. \tag{21.37}$$

For simplicity and to avoid self-field effects, we start with a small Josephson contact of length $w < \lambda_J$ ($\lambda_J = \sqrt{\Phi_0/\mu_0 d j_c}$ is the Josephson penetration depth) placed in a strong enough magnetic field (which is applied normally to the contact area) such that $\mathbf{H} > \Phi_0/2\pi\lambda_J d$, where $d = 2\lambda_L + t$, λ_L is the London penetration depth, and t is an insulator thickness.

The Josephson energy of such a contact in applied magnetic field is governed by a Fraunhofer-like dependence of the critical current (Orlando and Delin 1991):

$$E_J = J \left(1 - \frac{\sin \varphi}{\varphi} \cos \varphi_0 \right), \tag{21.38}$$

where $\varphi = \pi \Phi / \Phi_0$ is the frustration parameter with $\Phi = \mathbf{H}wd$ being the flux through the contact area, φ_0 is the initial phase difference through the contact, and $J \propto e^{-t/\xi}$ is the zero-field tunnelling Josephson energy with ξ being a characteristic (decaying) length and t the thickness of the insulating layer. The self-field effects (screening), neglected here, will be considered later for an array.

Notice that in non-zero applied magnetic field $\mathbf{H}$, there are two stress-induced contributions to the Josephson energy E_J, both related to decreasing of the insulator thickness under pressure. Indeed, according to the experimental data (D'yachenko *et al.* 1995), the tunnelling-dominated critical current I_C in granular high-T_C superconductors was found to exponentially increase under compressive stress, viz. $I_C(\sigma) = I_C(0)e^{\kappa \sigma}$. More specifically, the critical current at $\sigma = 9\,\text{kbar}$ was found to be three times higher its value at $\sigma = 1.5\,\text{kbar}$, clearly indicating a weak-links-mediated origin of the phenomenon. Hence, for small enough σ we can safely assume that (Sergeenkov 1999) $t(\sigma) \simeq t(0)(1 - \beta \sigma / \sigma_0)$ with σ_0 being some characteristic value (the parameter β is related to the so-called ultimate stress σ_m as $\beta = \sigma_0 / \sigma_m$). As a result, we have the following two stress-induced effects in Josephson contacts:

(I) amplitude modulation leading to the explicit stress dependence of the zero-field energy

$$J(\mathbf{T}, \sigma) = J(\mathbf{T}, 0)e^{\gamma \sigma / \sigma_0}, \tag{21.39}$$

with $\gamma = \beta t(0) / \xi$, and

(II) phase modulation leading to the explicit stress dependence of the flux

$$\Phi(\mathbf{T}, \mathbf{H}, \sigma) = \mathbf{H}wd(\mathbf{T}, \sigma), \tag{21.40}$$

with

$$d(\mathbf{T}, \sigma) = 2\lambda_L(\mathbf{T}) + t(0)(1 - \beta \sigma / \sigma_0). \tag{21.41}$$

Finally, in view of eqns (21.36)–(21.41), the temperature and field dependence of the small single junction TEC reads (the initial phase difference is conveniently fixed at $\varphi_0 = \pi$):

$$\alpha(\mathbf{T}, \mathbf{H}) = \alpha(\mathbf{T}, 0)[1 + F(\mathbf{T}, \mathbf{H})] + \epsilon(\mathbf{T}, 0)\frac{dF(\mathbf{T}, \mathbf{H})}{d\mathbf{T}}, \tag{21.42}$$

where

$$F(\mathbf{T}, \mathbf{H}) = \left[\frac{\sin \varphi}{\varphi} + \frac{\xi}{d(\mathbf{T}, 0)} \left(\frac{\sin \varphi}{\varphi} - \cos \varphi \right) \right], \tag{21.43}$$

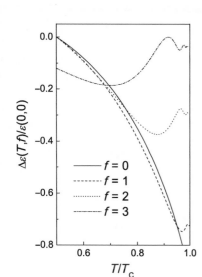

Fig. 21.13 Temperature dependence of the flux-driven strain field in a single short contact for different values of the frustration parameter **f** according to eqns (21.36)–(21.48).

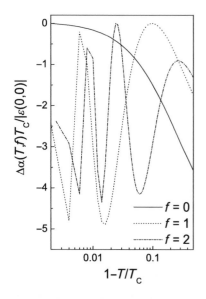

Fig. 21.14 Temperature dependence of flux-driven normalized TEC in a single small contact for different values of the frustration parameter **f** (for the same set of parameters as in Fig. 21.13) according to eqns (21.36)–(21.48).

with

$$\varphi(\mathbf{T}, \mathbf{H}) = \frac{\pi \, \Phi(\mathbf{T}, \mathbf{H}, 0)}{\Phi_0} = \frac{\mathbf{H}}{\mathbf{H}_0(\mathbf{T})}, \quad (21.44)$$

$$\alpha(\mathbf{T}, 0) = \frac{d\epsilon(\mathbf{T}, 0)}{d\mathbf{T}}, \quad (21.45)$$

and

$$\epsilon(\mathbf{T}, 0) = -\left(\frac{\Phi_0}{2\pi}\right)\left(\frac{2\gamma}{V\sigma_0}\right) I_C(\mathbf{T}). \quad (21.46)$$

Here, $\mathbf{H}_0(\mathbf{T}) = \Phi_0/\pi w d(\mathbf{T}, 0)$ with $d(\mathbf{T}, 0) = 2\lambda_L(\mathbf{T}) + t(0)$. For the explicit temperature dependence of $J(\mathbf{T}, 0) = \Phi_0 I_C(\mathbf{T})/2\pi$ we use the well-known (Meservey and Schwartz 1969; Sergeenkov 2002) analytical approximation of the BCS gap parameter (valid for all temperatures), $\Delta(\mathbf{T}) = \Delta(0)\tanh\left(2.2\sqrt{\frac{\mathbf{T_C}-\mathbf{T}}{\mathbf{T}}}\right)$ with $\Delta(0) = 1.76 k_B \mathbf{T_C}$ that governs the temperature dependence of the Josephson critical current

$$I_C(\mathbf{T}) = I_C(0)\left[\frac{\Delta(\mathbf{T})}{\Delta(0)}\right]\tanh\left[\frac{\Delta(\mathbf{T})}{2k_B\mathbf{T}}\right], \quad (21.47)$$

while the temperature dependence of the London penetration depth is governed by the two-fluid model:

$$\lambda_L(\mathbf{T}) = \frac{\lambda_L(0)}{\sqrt{1 - (\mathbf{T}/\mathbf{T_C})^2}}. \quad (21.48)$$

From the very structure of eqns (21.36)–(21.44) it is obvious that the TEC of a single contact will exhibit *field* oscillations imposed by the Fraunhofer dependence of the critical current I_C. Much less obvious is its temperature dependence. Indeed, Fig. 21.13 presents the temperature behavior of the contact area strain field $\Delta\epsilon(\mathbf{T}, \mathbf{f}) = \epsilon(\mathbf{T}, \mathbf{f}) - \epsilon(\mathbf{T}, 0)$ (with $t(0)/\xi = 1$, $\xi/\lambda_L(0) = 0.02$ and $\beta = 0.1$) for different values of the frustration parameter $\mathbf{f} = \mathbf{H}/\mathbf{H}_0(0)$. Notice the characteristic flux-driven temperature oscillations near $\mathbf{T_C}$ that are better seen on a semi-log plot shown in Fig. 21.14 that depicts the dependence of the properly normalized field-induced TEC $\Delta\alpha(\mathbf{T}, \mathbf{f}) = \alpha(\mathbf{T}, \mathbf{f}) - \alpha(\mathbf{T}, 0)$ as a function of $1 - \mathbf{T}/\mathbf{T_C}$ for the same set of parameters.

To answer an important question how the effects neglected in the previous analysis screening will affect the above-predicted oscillating behavior of the field-induced TEC, let us consider a more realistic situation with a junction embedded into an array (rather than an isolated contact) which is realized in artificially prepared arrays using a photolithographic technique that nowadays allows for controlled manipulations of the junctions parameters (Newrock *et al.* 2000). Besides, this is also a good approximation for a granular superconductor (if we consider it as a network of superconducting islands connected with each other via Josephson links). Our goal is to model and simulate the elastic response of such systems to an effective stress σ. For simplicity, we will consider an array with a regular topology and uniform parameters (such an approximation already proved useful for describing high-quality artificially prepared structures, see, e.g., Sergeenkov and Araujo-Moreira (2004).

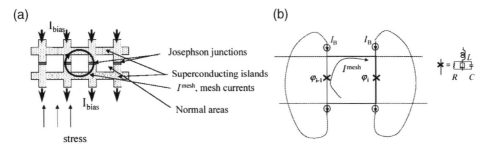

Fig. 21.15 (a) Sketch of a regular square array (a single plaquette). (b) Electrical scheme of the array with the circulating currents. The bias current is fed via virtual loops external to the array.

Let us consider a planar square array as shown in Fig. 21.15. The total current includes the bias current flowing through the vertical junctions and the induced screening currents circulating in the plaquette (Nakajima and Sawada 1981). This situation corresponds to the inclusion of screening currents only into the nearest neighbors, thus neglecting the mutual inductance terms (Phillips *et al.* 1993). Therefore, the equation for the vertical contacts will read (horizontal and vertical junctions are denoted by superscripts h and v, respectively):

$$\frac{\hbar C}{2e}\frac{\mathrm{d}^2\phi^{\mathrm{v}}_{i,j}}{\mathrm{d}t^2} + \frac{\hbar}{2eR}\frac{\mathrm{d}\phi^{\mathrm{v}}_{i,j}}{\mathrm{d}t} + I_{\mathrm{c}}\sin\phi^{\mathrm{v}}_{i,j} = \Delta I^{\mathrm{s}}_{i,j} + I_{\mathrm{b}}, \qquad (21.49)$$

where $\Delta I^{\mathrm{s}}_{i,j} = I^{\mathrm{s}}_{i,j} - I^{\mathrm{s}}_{i-1,j}$ and the screening currents I^{s} obey the fluxoid conservation condition:

$$-\phi^{\mathrm{v}}_{i,j} + \phi^{\mathrm{v}}_{i,j+1} - \phi^{\mathrm{h}}_{i,j} + \phi^{\mathrm{h}}_{i+1,j} = 2\pi\frac{\Phi^{\mathrm{ext}}}{\Phi_0} - \frac{2\pi L I^{\mathrm{s}}_{i,j}}{\Phi_0}. \qquad (21.50)$$

Recall that the total flux has two components (an external contribution and the contribution due to the screening currents in the closed loop) and it is equal to the sum of the phase differences describing the array. It is important to underline that the external flux in eqn (21.50), $\eta = 2\pi\Phi^{\mathrm{ext}}/\Phi_0$, is related to the frustration of the whole array, i.e. this is the flux across the void of the network (Grimaldi *et al.* 1996; Araujo-Moreira *et al.* 1997, 2005) and it should be distinguished from the previously introduced applied magnetic field **H** across the junction barrier that is related to the frustration of a single contact $\mathbf{f} = 2\pi\mathbf{H}dw/\Phi_0$ and that only modulates the critical current $I_{\mathrm{C}}(\mathbf{T}, \mathbf{H}, \sigma)$ of a single junction, while inducing a negligible flux into the void area of the array.

For simplicity, in what follows we will consider only the elastic effects due to a uniform (homogeneous) stress imposed on the array. With regard to the geometry of the array, the deformation of the loop is the dominant effect with its radius a deforming as follows:

$$a(\sigma) = a_0(1 - \chi\sigma/\sigma_0). \qquad (21.51)$$

As a result, the self-inductance of the loop $L(a) = \mu_0 a F(a)$ (with $F(a)$ being a geometry-dependent factor) will change accordingly:

$$L(a) = L_0(1 - \chi_{\mathrm{g}}\sigma/\sigma_0). \qquad (21.52)$$

The relationship between the coefficients χ and χ_g is given by

$$\chi_g = \left(1 + a_0 B_g\right) \chi, \qquad (21.53)$$

where $B_g = \frac{1}{F(a)} \left(\frac{dF}{da}\right)_{a_0}$. It is also reasonable to assume that in addition to the critical current, the external stress will modify the resistance of the contact:

$$R(\sigma) = \frac{\pi \Delta(0)}{2e I_C(\sigma)} = R_0 e^{-\chi \sigma/\sigma_0}, \qquad (21.54)$$

as well as capacitance (due to the change in the distance between the superconductors):

$$C(\sigma) = \frac{C_0}{1 - \chi \sigma/\sigma_0} \simeq C_0(1 + \chi \sigma/\sigma_0). \qquad (21.55)$$

To simplify the treatment of the dynamic equations of the array, it is convenient to introduce the standard normalization parameters such as the Josephson frequency:

$$\omega_J = \sqrt{\frac{2\pi I_C(0)}{C_0 \Phi_0}}, \qquad (21.56)$$

the analog of the SQUID parameter:

$$\beta_L = \frac{2\pi I_C(0) L_0}{\Phi_0}, \qquad (21.57)$$

and the dissipation parameter:

$$\beta_C = \frac{2\pi I_C(0) C_0 R_0^2}{\Phi_0}. \qquad (21.58)$$

Combining eqns (21.49) and (21.50) with the stress-induced effects described by eqns (21.54) and (21.55) and using the normalization parameters given by eqns (21.56)–(21.58), we can rewrite the equations for an array in a rather compact form. Namely, the equations for vertical junctions read:

$$\frac{1}{1 - \chi\sigma/\sigma_0} \ddot{\phi}_{i,j}^{v} + \frac{e^{-\chi\sigma/\sigma_0}}{\sqrt{\beta_C}} \dot{\phi}_{i,j}^{v} + e^{\chi\sigma/\sigma_0} \sin \phi_{i,j}^{v} = \gamma_b + \frac{1}{\beta_L \left(1 - \chi_g \sigma/\sigma_0\right)}$$
$$\times \left[\phi_{i,j-1}^{v} - 2\phi_{i,j}^{v} + \phi_{i,j+1}^{v} + \phi_{i,j}^{h} - \phi_{i-1,j}^{h} + \phi_{i+1,j-1}^{h} - \phi_{i,j-1}^{h}\right]. \quad (21.59)$$

Here, an overdot denotes the time derivative with respect to the normalized time (inverse Josephson frequency), and the bias current is normalized to the critical current without stress, $\gamma_b = I_b/I_C(0)$.

The equations for the horizontal junctions will have the same structure safe for the explicit bias-related terms:

$$\frac{1}{1 - \chi\sigma/\sigma_0} \ddot{\phi}_{i,j}^{h} + \frac{e^{-\chi\sigma/\sigma_0}}{\sqrt{\beta_C}} \dot{\phi}_{i,j}^{h} + e^{\chi\sigma/\sigma_0} \sin \phi_{i,j}^{h} = \frac{1}{\beta_L \left(1 - \chi_g \sigma/\sigma_0\right)}$$
$$\times \left[\phi_{i,j-1}^{h} - 2\phi_{i,j}^{h} + \phi_{i,j+1}^{h} + \phi_{i,j}^{v} - \phi_{i-1,j}^{v} + \phi_{i+1,j-1}^{v} - \phi_{i,j-1}^{v}\right]. \quad (21.60)$$

Finally, eqns (21.59) and (21.60) should be complemented with the appropriate boundary conditions (Binder *et al.* 2000) that will include the normalized contribution of the external flux through the plaquette area $\eta = 2\pi\,\Phi^{\text{ext}}/\Phi_0$. It is interesting to notice that eqns (21.59) and (21.60) will have the same form as their stress-free counterparts if we introduce the stress-dependent renormalization of the parameters:

$$\tilde{\omega}_J = \omega_J e^{\chi\sigma/2\sigma_0} \tag{21.61}$$

$$\tilde{\beta}_C = \beta_C e^{-3\chi\sigma/\sigma_0} \tag{21.62}$$

$$\tilde{\beta}_L = \beta_L(1 - \chi_g\sigma/\sigma_0)e^{\chi\sigma/\sigma_0} \tag{21.63}$$

$$\tilde{\eta} = \eta(1 - 2\chi\sigma/\sigma_0) \tag{21.64}$$

$$\tilde{\gamma}_b = \gamma_b e^{-\chi\sigma/\sigma_0}. \tag{21.65}$$

Turning to the discussion of the obtained numerical simulation results, it should be stressed that the main problem in dealing with an array is that the total current through the junction should be retrieved by solving self-consistently the array equations in the presence of screening currents. Recall that the Josephson energy of a single junction for an arbitrary current I through the contact reads:

$$E_J(\mathbf{T}, \mathbf{f}, I) = E_J(\mathbf{T}, \mathbf{f}, I_C)\left[1 - \sqrt{1 - \left(\frac{I}{I_C}\right)^2}\right]. \tag{21.66}$$

The important consequence of eqn (21.66) is that if no current flows in the array's junction, such a junction will not contribute to the TEC (simply because a junction disconnected from the current generator will not contribute to the energy of the system).

Below, we sketch the main steps of the numerical procedure used to simulate the stress-induced effects in the array:

(1) a bias point I_b is selected for the whole array;
(2) the parameters of the array (screening, Josephson frequency, dissipation, etc.) are selected and modified according to the intensity of the applied stress σ;
(3) the array equations are simulated to retrieve the static configuration of the phase differences for the parameters selected in step 2;
(4) the total current flowing through the individual junctions is retrieved as:

$$I_{i,j}^{v,h} = I_C \sin\phi_{i,j}^{v,h}; \tag{21.67}$$

(5) the energy dependence upon stress is numerically estimated using the value of the total current $I_{i,j}^{v,h}$ (which is not necessarily identical for all junctions) found in step 4 via eqn (21.67);
(6) the array energy E_J^A is obtained by summing up the contributions of all junctions with the above-found phase differences $\phi_{i,j}^{v,h}$;

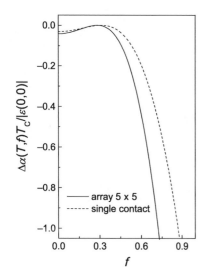

Fig. 21.16 Numerical simulation results for an array 5×5 (solid line) and a small single contact (dashed line). The dependence of the normalized TEC on the frustration parameter f (applied magnetic field $\mathbf{H}$ across the barrier) for the reduced temperature $\mathbf{T}/\mathbf{T}_C = 0.95$. The parameters used for the simulations: $\eta = 0$, $\beta = 0.1$, $t(0)/\xi = 1$, $\xi/\lambda_L = 0.02$, $\beta_L = 10$, $\gamma_b = 0.95$, and $\chi_g = \chi = 0.01$.

(7) the stress-modified screening currents $I_{i,j}^s(\mathbf{T}, \mathbf{H}, \sigma)$ are computed using eqn (21.50) and inserted into the magnetic energy of the array $E_M^A = \frac{1}{2L}\Sigma_{i,j}(I_{i,j}^s)^2$;

(8) the resulting strain field and TE coefficient of the array are computed using numerical derivatives based on the finite differences:

$$\epsilon^A \simeq \frac{1}{V}\left[\frac{\Delta\left(E_M^A + E_J^A\right)}{\Delta\sigma}\right]_{\Delta\sigma \to 0} \tag{21.68}$$

$$\alpha(\mathbf{T}, \mathbf{H}) \simeq \frac{\Delta\epsilon^A}{\Delta\mathbf{T}}. \tag{21.69}$$

The numerical simulation results show that the overall behavior of the strain field and TE coefficient in the array is qualitatively similar to the behavior of the single contact. In Fig. 21.16 we have simulated the behavior of both the small junction and the array as a function of the field across the barrier of the individual junctions in the presence of bias and screening currents. As is seen, the dependence of $\alpha(\mathbf{T}, \mathbf{f})$ is very weak up to $\mathbf{f} \simeq 0.5$, showing a strong decrease of about 50% when the frustration approaches $\mathbf{f} = 1$.

A much more profound change is obtained by varying the temperature for the fixed value of applied magnetic field. Figure 21.17 depicts the temperature behavior of $\alpha(\mathbf{T}, \mathbf{f})$ (on a semi-log scale) for different field configurations that include barrier field f frustrating a single junction and the flux across the void of the network η frustrating the whole array. First, comparing Fig. 21.17(a) and Fig. 21.14 we notice that, due to substantial modulation of the Josephson critical current $I_C(\mathbf{T}, \mathbf{H})$ given by eqn (21.38), the barrier field $\mathbf{f}$ has similar effects on the TE coefficient of both the array and the single contact including temperature oscillations. However, finite screening effects in the array result in the appearance of oscillations at higher values of the frustration $\mathbf{f}$ (in comparison with a single contact). On the other hand, Figs 21.17(b–d) represent the influence of the external field across the void η on the evolution of $\alpha(\mathbf{T}, \mathbf{f})$. As is seen, in comparison with a field-free configuration (shown in Fig. 21.17(a)), the presence of the external field η substantially reduces the magnitude of the TE coefficient of the array. Besides, with η increasing, the onset of temperature oscillations markedly shifts closer to $\mathbf{T}_C$.

21.6 Summary

In this chapter, using a realistic model of 2D Josephson junction arrays (created by 2D network of twin-boundary dislocations with strain fields acting as an insulating barrier between hole-rich domains in underdoped crystals), we considered many novel effects related to the magnetic, electric, elastic and transport properties of Josephson nanocontacts and nanogranular superconductors. Some of the topics covered here include such interesting phenomena as chemomagnetism and magnetoelectricity, electric analog of the "fishtail" anomaly and field-tuned weakening of the chemically induced Coulomb blockade as well as a giant enhancement of the non-linear thermal conductivity (reaching

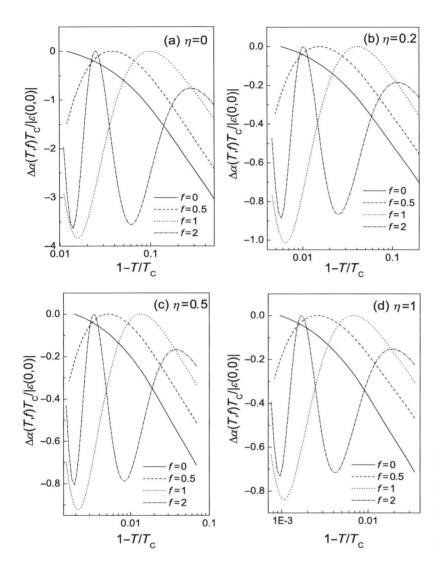

Fig. 21.17 Numerical simulation results for an array 5×5. The influence of the flux across the void of the network η frustrating the whole array on the temperature dependence of the normalized TEC for different values of the barrier field $\mathbf{f}$ frustrating a single junction for $\gamma_b = 0.5$ and the rest of the parameters are the same as in Fig. 21.16.

500% when the intrinsically induced chemoelectric field $E_\mu \propto |\nabla \mu|$, created by the gradient of the chemical potential due to segregation of hole-producing oxygen vacancies, closely matches the externally produced thermoelectric field $E_T \propto |\nabla T|$). Besides, we have investigated the influence of a homogeneous mechanical stress on a small single Josephson junction and on a plaquette (array of 5×5 junctions) and have shown how the stress-induced modulation of the parameters describing the junctions (as well as the connecting circuits) produces such an interesting phenomenon as a thermal expansion (TE) in a single contact and two-dimensional array (plaquette). We also studied the variation of the TE coefficient with an external magnetic field and temperature. In particular, near $\mathbf{T}_C$ (due to some tremendous increase of the effective "sandwich" thickness of the contact) the field-induced TE coefficient of a small junction exhibits clear *temperature* oscillations scaled with the number of flux quanta crossing the contact area. Our numerical simulations revealed that these oscillations may actually still survive in an array if the applied field is strong enough to compensate for finite screening-induced self-field effects.

The accurate estimates of the model parameters suggest quite an optimistic possibility to experimentally realize all of the predicted in this chapter, promising and important for applications effects in non-stoichiometric nanogranular superconductors and artificially prepared arrays of Josephson nanocontacts.

Acknowledgments

Some of the results presented in Section 21.5 were obtained in collaboration with Giacomo Rotoli and Giovanni Filatrella. This work was supported by the Brazilian agency CAPES.

References

Abrikosov, A.A. *Fundamentals of the Theory of Metals* (Elsevier, Amsterdam, 1988).

Akopyan, A.A., Bolgov, S.S., Savchenko, A.P. *Sov. Phys. Semicond.* **24**, 1167 (1990).

Altshuler, E., Johansen, T.H. *Rev. Mod. Phys.* **76**, 471 (2004).

Anshukova, N.V., Bulychev, B.M., Golovashkin, A.I., Ivanova, L.I., Minakov, A.A., Rusakov, A.P. *JETP Lett.* **71**, 377 (2000).

Araujo-Moreira, F.M., Barbara, P., Cawthorne, A.B., Lobb, C.J. *Phys. Rev. Lett.* **78**, 4625 (1997).

Araujo-Moreira, F.M., Barbara, P., Cawthorne, A.B., Lobb, C.J. *Studies of High Temperature Superconductors* 43, (ed.) Narlikar, A.V. (Nova Science Publishers, New York, 2002) p. 227.

Araujo-Moreira, F.M., Maluf, W., Sergeenkov, S. *Solid State Commun.* **131**, 759 (2004).

Araujo-Moreira, F.M., Maluf, W., Sergeenkov, S. *Eur. Phys. J. B* **44**, 33 (2005).

Barbara, P., Araujo-Moreira, F.M., Cawthorne, A.B., Lobb, C.J. *Phys. Rev. B* **60**, 7489 (1999).

Beloborodov, I.S., Lopatin, A.V., Vinokur, V.M., Efetov, K.B. *Rev. Mod. Phys.* **79**, 469 (2007).

Binder, P., Caputo, P., Fistul, M.V., Ustinov, A.V., Filatrella, G. *Phys. Rev. B* **62**, 8679 (2000).

Bourgeois, O, Skipetrov, S.E., Ong, F., Chaussy, J. *Phys. Rev. Lett.* **94**, 057007 (2005).

Chen, W., Smith, T.P., Buttiker, M. *et al. Phys. Rev. Lett.* **73**, 146 (1994).

Daeumling, M., Seuntjens, J.M., Larbalestier, D.C. *Nature* **346**, 332 (1990).

De Leo, C., Rotoli, G. *Phys. Rev. Lett.* **89**, 167001 (2002).

Deppe, J., Feldman, J.L. *Phys. Rev. B* **50**, 6479 (1994).

D'yachenko, A.I., Tarenkov, V.Y., Abalioshev, A.V., Lutciv, L.V., Myasoedov, Y.N., Boiko, Y.V. *Physica C* **251**, 207 (1995).

Eichenberger, A.-L., Affolter, J., Willemin, M., Mombelli, M., Beck, H., Martinoli, P., Korshunov, S.E. *Phys. Rev. Lett.* **77**, 3905 (1996).

Gantmakher, V.F., Neminskii, A.M., Shovkun, D.V. *JETP Lett.* **52**, 630 (1990).

Gantmakher, V.F. *Phys.-Uspe.* **45**, 1165 (2002).

Geim, A.K., Dubonos, S.V., Lok, J.G.S. *et al. Nature* **396**, 144 (1998).

Girifalco, L.A. *Statistical Physics of Materials* (Wiley-Interscience, New York, 1973).

Golubov, A.A., Kupriyanov, M.Yu., Fominov, Ya.V. *JETP Lett.* **75**, 588 (2002).

Grimaldi, G., Filatrella, G., Pace, S., Gambardella, U. *Phys. Lett. A* **223**, 463 (1996).

Gurevich, A., Pashitskii, E.A. *Phys. Rev. B* **56**, 6213 (1997).

Guttman, G., Nathanson, B., Ben-Jacob, E., Bergman, D.J. *Phys. Rev. B* **55**, 12691 (1997).

Haviland, D.B., Kuzmin, L.S., Delsing, P.Z. *Phys. B* **85**, 339 (1991).

Iansity, M., Johnson, A.J., Lobb, C.J. *Phys. Rev. Lett.* **60**, 2414 (1988).

Khaikin, M.S., Khlyustikov, I.N. *JETP Lett.* **33**, 158 (1981).

Krive, I.V., Kulinich, S.I., Jonson, M. *Low-Temp. Phys.* **30**, 554 (2004).

Lang, K.M., Madhavan, V., Hoffman, J.E., Hudson, E.W., Eisaki, H., Uchida, S., Davis, J.C. *Nature* **415**, 412 (2002).

Li, M.S. *Phys. Rep.* **376**, 133 (2003).

Makhlin, Yu., Schon, G., Shnirman, A. *Rev. Mod. Phys.* **73**, 357 (2001).

Meservey, R., Schwartz, B.B. *Superconductivity*, vol. 1, (ed.) Parks, R.D. (M. Dekker, New York, 1969) p. 117.

Moeckley, B.H., Lathrop, D.K., Buhrman, R.A. *Phys. Rev. B* **47**, 400 (1993).

Nagel, P., Pasler, V., Meingast, C., Rykov, R.I., Tajima, S. *Phys. Rev. Lett.* **85**, 2376 (2000).

Nakajima, K., Sawada, Y. *J. Appl. Phys.* **52**, 5732 (1981).

Newrock, R.S., Lobb, C.J., Geigenmuller, U., Octavio, M. *Solid State Phys.* **54**, 263 (2000).

Orlando, T.P., Delin, K.A. *Foundations of Applied Superconductivity* (Addison, New York, 1991).

Ostrovsky, P.M., Feigel'man, M.V. *JETP Lett.* **79**, 489 (2004).

Phillips, J.R., van der Zant, R.S.J., Orlando, T.P. *Phys. Rev. B* **47**, 5219 (1993).

Ryazanov, V.V., Oboznov, V.A., Rusanov, A.Yu. *Phys. Rev. Lett.* **86**, 2427 (2001).

Sergeenkov, S., Ausloos, M. *Phys. Rev. B* **48**, 604 (1993).

Sergeenkov, S. *J. Appl. Phys.* **78**, 1114 (1995).

Sergeenkov, S. *J. Phys. I* (France) **7**, 1175 (1997).

Sergeenkov, S. *JETP Lett.* **67**, 680 (1998a).

Sergeenkov, S. *J. Phys.: Condens. Matter* **10**, L265 (1998b).

Sergeenkov, S. *JETP Lett.* **70**, 36 (1999).

Sergeenkov, S., Ausloos, M. *JETP* **89**, 140 (1999).

Sergeenkov, S. *Studies of High Temperature Superconductors* 39, (ed.) Narlikar, A.V. (Nova Science Publishers, New York, 2001) p. 117.

Sergeenkov, S. *JETP Lett.* **76**, 170 (2002).

Sergeenkov, S. *JETP Lett.* **77**, 94 (2003).

Sergeenkov, S., Araujo-Moreira, F.M. *JETP Lett.* **80**, 580 (2004).

Sergeenkov, S. *JETP* **101**, 919 (2005).

Sergeenkov, S. *Studies of High Temperature Superconductors* 50, (ed.) Narlikar, A.V. (Nova Science Publishers, New York, 2006) p. 229.

Sergeenkov, S. *J. Appl. Phys.* **102**, 066104 (2007).

Sergeenkov, S., Rotoli, G., Filatrella, G., Araujo-Moreira, F.M. *Phys. Rev. B* **75**, 014506 (2007).

van Bentum, P.J.M., van Kempen, H., van de Leemput, L.E.C., Teunissen, P.A.A. *Phys. Rev. Lett.* **60**, 369 (1988).

van der Zant, H.S. *J. Physica B* **222**, 344 (1996).

van Harlingen, D., Heidel, D.F., Garland, J.C. *Phys. Rev. B* **21**, 1842 (1980).

Wendin, G., Shumeiko, V.S. *Low-Temp. Phys.* **33**, 724 (2007).

Yang, G., Shang, P., Sutton, S.D., Jones, I.P., Abell, J.S., Gough, C.E. *Phys. Rev. B* **48**, 4054 (1993).

Theory, experiment and applications of tubular image states

22

D. Segal, P. Král, and M. Shapiro

22.1 Introduction

Image potential states (IPS), i.e. the quantized excited states of electrons in the vicinity of metallic surfaces, have been investigated extensively in recent years (Gallagher 1994; Echenique *et al.* 2004). These states owe their stability to the attractive interaction between an electron and its image charge, given for *flat* conducting surfaces as $V(z) = -e^2/(4z)$, where z is the distance to the surface of the positive (image) charge induced by the electron below the surface.

The above one-dimensional (1D) Coulomb-like potential gives rise to the formation of a series of Rydberg-like states (Gallagher 1994). In particular, when penetration into the bulk is prohibited, the energy spectrum assumes the familiar $E_n \propto -1/(n + a)^2$ form, where $n = 1, 2, 3, \ldots$, as E_n converges to the vacuum energy. The left panel of Fig. 22.1 illustrates schematically the image potential formed above a flat surface and the resulting bound states and binding energies.

Using scanning tunnelling spectroscopy, photoemission, and two-photon photoemission (Echenique *et al.* 2004), it is possible to measure different types of electronic excitations above metals. For example, Höfer *et al.* (1997) used two-photon photoemission to populate coherent wave packets of image states above Cu(100) surfaces. In these experiments a photon is made to excite an electron out of an occupied state below the Fermi energy into an IPS close to the surface. A second photon is then used to excite the electron to an energy above the vacuum level. The kinetic energy of the emitted electron is measured, yielding useful information about the IPS energetics and lifetimes. One observes (Höfer *et al.* 1997) states with principal quantum numbers $n \sim 7$, binding energies of 15–40 meV, and lifetimes, limited by the fast collapse of the electron into the surface, of a few picoseconds. Experiments performed above molecular wires laid on surfaces (Ortega *et al.* 1994), nanoparticles (Kasperovich *et al.* 2000; Boyle *et al.* 2001), and one- and two-dimensional (2D) liquid helium (Plazman *et al.* 1999; Glasson *et al.*

(a)

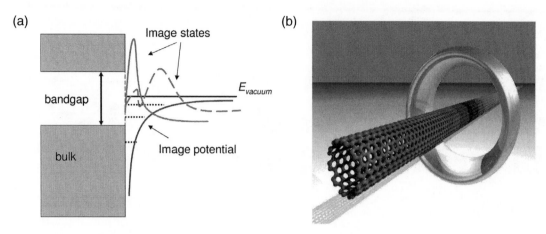

(b)

Fig. 22.1 (a) Scheme of electron image states formed in the vicinity of a flat metallic surface. (b) Scheme of electron image states formed in the vicinity of an infinitely long metallic nanotube.

2001) have demonstrated the ubiquity of extended image states in nanosystems.

Due to them being localized close to the surface, IPS can be used for surface diagnostics on the nanometric scale. In particular, IPS are extremely sensitive to adsorbates (Miller *et al.* 2002) and to changes in the dielectric susceptibility of the surface. By measuring their binding energies and lifetimes it is possible to investigate electronic structure and electronic processes, surface morphology and surface reactivity (Loly *et al.* 1983; Memmel *et al.* 1995). However, practical applications of these states are limited by their picosecond lifetimes, mostly given by their spatial overlap with the surface states of the material (Echenique *et al.* 1985; Chulkov *et al.* 1998).

The short lifetimes of IPS above flat surfaces have led us (Granger *et al.* 2002) to investigate a new class of IPS, called "tubular image states" (TIS), which are formed above *cylindrical* nanoscopic objects, such as metallic nanotubes and nanowires. As illustrated in Fig. 22.1(b), electrons occupying nanoscopic TIS are expected to hover at distances of 10–50 nm from the material surfaces. Their stability at these detached configurations is due to the "tug of war" between the attractive image potential and the repulsive centrifugal potential occurring when the electronic angular momentum l is non-zero. Because the electron is kept away from the surface and is prohibited from penetrating into it and being annihilated there, the lifetimes of the TIS are expected to be significantly longer than IPS above flat surfaces.

In subsequent publications we have shown that TIS may also be localized along the longitudinal axis of the cylinder by introducing appropriate inhomogeneities (Segal *et al.* 2004a). We have also identified mechanisms by which the nature and binding energies of TIS can be tuned by the application of external electric and magnetic fields (Segal *et al.* 2004b). Moreover, we have demonstrated the formation of *bands* of image states above 1D and 2D periodic arrays of nanotubes (Segal *et al.* 2005a), and have investigated in detail the mechanisms of the slow relaxation of TIS. This relaxation was shown to occur mainly via the excitation of circularly polarized nanotube TA phonons (Segal *et al.* 2005b). It was also suggested that TIS may serve as an interesting and new platform for investigating quantum chaos, shown by us to occur in TIS of various nanosystems (Segal *et al.* 2005c).

The system of greatest interest in conjunction with TIS are carbon nanotubes (CNT). Their mechanical, electrical, and optical properties make them useful in a wide variety of applications in nanoelectronics, nano-optics, and material science in general (Iijima 1991; Dresselhaus *et al.* 2001). CNT can be either single-walled, with a typical radius of ∼1 nm, or multiwalled, consisting of multiple layers of graphene rolled in on themselves to form a tube. In both cases the tube's length can be very large, reaching values of up to a few centimeters (Zheng *et al.* 2004).

By changing the CNT diameter and the way the graphene layers are rolled, thereby changing the chirality of the CNT, the CNT's electric properties can be made to vary vastly from those of a semiconductor to those of a nearly perfectly ballistic metallic conductor (Ando *et al.* 1998; Gomez-Navarro *et al.* 2005). Their thermal properties are also remarkable, showing a unique 1D behavior (Dresselhaus *et al.* 2001). Thus, in addition to the fundamental interest in TIS, by probing the binding energies and lifetimes of TIS we can infer about the CNT's electronic, structural properties and plasmonic and phononic collective excitation spectra (Taverna *et al.* 2002). TIS may also be used as a tool for studying surface morphology since they are sensitive to impurities, defects, and inhomogeneities in the structure beneath. In addition, TIS may be used to probe surface reactivity, growth kinetics and adsorbate reactions.

Other interesting applications of TIS to nanoscience are the development of unique "Rydberg-like", CNT-based, electronic devices: The states' high sensitivity to external fields could also be used to build tunable waveguides, mirrors, and storage devices for low-energy electrons (Segal and Shapiro 2006). TIS formed above CNT junctions can also serve as intermediate states assisting optically induced electron hopping over junction barriers, thus creating an optical switch.

Recently, using two-photon photo-emission spectroscopy (Zamkov *et al.* 2004a), the existence of low angular momenta TIS and the *prolonged lifetimes* associated with them, was confirmed experimentally in multiwalled nanotubes. Moreover, the states' binding energies and lifetimes nicely agree with theoretical predictions (Zamkov *et al.* 2004b). It is expected that similar states will soon be observed in single-walled nanotubes and nanotube bundles.

In this review we discuss our ongoing research on TIS. In Section 22.2 we discuss the theoretical background for the existence of TIS in conjunction with a single tube, with nanotube arrays, and with inhomogeneous tubular systems. We also discuss calculations of the states' lifetimes. In Section 22.3 we discuss the states' tunability by external fields and the emergence of chaos in TIS. We also describe a nanoscale electron trap based on TIS. Finally, recent experiments on TIS are presented in Section 22.4. Concluding remarks are made in Section 22.5.

22.2   Characterizing tubular image states

22.2.1   TIS formed around infinitely long homogeneous nanotubes

In this section we develop the theory of TIS using a simple model developed by Granger *et al.* (2002). We first consider an electron outside a single, infinitely long, perfectly conducting, nanotube, ignoring the atomic structure of the tube and the electronic band structure of the surface. We also ignore nuclear motions and many-body exchange and correlation interactions with other electrons in the nanotube.

Classically, an external charge q_0 approaching a metallic tube polarizes its surface, giving rise to an induced potential Φ_{ind}. Denoting as Φ_0 the direct potential due to the external charge, we require that the total potential, $\Phi_{tot} = \Phi_0 + \Phi_{ind}$ vanishes on the surface. This allows one to express the induced potential as (Arista *et al.* 2001),

$$\Phi_{ind}(\rho, \phi, z) = \frac{-2q_0}{\pi} \sum_{m=-\infty}^{\infty} \int_0^{\infty} \cos(kz) \exp(im\phi) \frac{I_m(ka)}{K_m(ka)}$$

$$K_m(k\rho_0) K_m(k\rho) dk, \tag{22.1}$$

where a is the radius of the nanotube and $(\rho, \phi, z) = (\rho_0, 0, 0)$ is the location of the external charge. The functions $(I_m(x), K_m(x), m = 0, \pm 1, \pm 2 \ldots)$ are the regular and irregular modified cylindrical Bessel functions. The interaction potential energy between the external charge and the tube is defined as

$$V_s(\rho_0) = \frac{q_0}{2} \Phi_{ind}(\rho_0, 0, 0), \tag{22.2}$$

where the factor of 1/2 is introduced to compensate for double counting. The electrostatic force between the external charge and the conducting cylinder is obtained by differentiating eqn (22.1) with respect to ρ

$$F(\rho_0) = -q \left. \frac{\partial \Phi_{ind}}{\partial \rho} \right|_{(\rho_0, 0, 0)} = \frac{2q_0^2}{\pi a^2} \int_0^{\infty} dx \left[A_0(x) + 2 \sum_{m=1}^{\infty} A_m(x) \right];$$

$$A_m(x) = \frac{I_m(x)}{K_m(x)} K_m(x\rho_0/a) x K_m'(x\rho_0/a). \tag{22.3}$$

Asymptotic analysis shows that at large distances $\rho_0/a \gg 1$ the induced force $F(\rho_0)$ is dominated by the $m = 0$ term. For large ρ the $m = 0$ term of the potential $V_s(\rho_0) = -\int^{\rho_0} F(\rho) d\rho$ of eqn (22.3) assumes the form

$$V_s(\rho_0) \sim -\frac{q_0^2}{\rho_0 \ln(\rho_0/a)}. \tag{22.4}$$

This equation agrees with the expectation that at large z distances the image potential of a conducting cylinder decays faster than the image potential of a conducting plane, $V_s \sim -1/z$, and more slowly than the image potential of a conducting sphere $V_s \sim -1/r^2$. As shown by Granger *et al.* (2002) it is possible to construct an approximate potential,

$$V_s(\rho_0) \sim \frac{2q_0^2}{\pi a} \sum_{n=1,3,5...} li\left[(a/\rho_0)^n\right];$$

$$li(x) \equiv \int_0^x \frac{dt}{\ln(t)}, \tag{22.5}$$

which is also correct near the nanotube surface, where $F(\rho_0) \sim 1/|\rho_0 - a|^2$. Thus, though approximate, eqn (22.5) interpolates well between the long-range behavior (22.4), and the behavior near the surface. Numerically, this function also reproduces the exact result, obtained by integrating eqns (22.1) and (22.3), to within a few per cent for most ρ_0/a values of interest.

We define the *effective* interaction potential $V_{\text{eff}}(\rho)$, obtained by adding the attractive induced potential (22.5) and the repulsive centrifugal potential

$$V_{\text{eff}}(\rho) = V_s(\rho) + \frac{\hbar^2(l^2 - 1/4)}{2m_e\rho^2}, \tag{22.6}$$

where m_e is the electron mass. Fig. 22.2 illustrates the effective potential $V_{\text{eff}}(\rho)$ for an electron interacting with a (10,10) metallic CNT of radius $a = 0.68$ nm. As shown there, for moderate angular momenta ($l \geq 6$), the effective potential possesses extremely long-range wells capable of supporting bound TIS. The inset in Fig. 22.2 depicts the effective potential close to the tube's surface. Clearly, for sufficiently high l the barrier in the effective potential prevents the electron from reaching the surface, thus eliminating the major decay mechanism of the image state. This point is discussed in greater detail in Section 22.2.4.

We now discuss the wavefunctions associated with the electronic motion in this effective potential. Assuming separability between the longitudinal

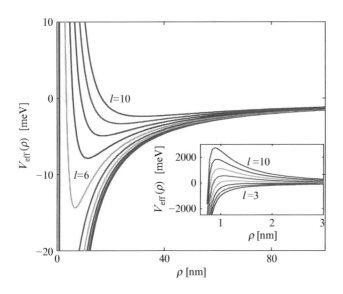

Fig. 22.2 The effective potential felt by an electron in the vicinity of a conducting nanotube at a number of angular momenta l, calculated using eqns (22.5) and (22.6) with $a = 0.68$ nm. Long-range minima appear for $l \geq 6$. The inset shows a blowup of the effective potential near the surface of the nanotube where a large potential barrier is seen to exist.

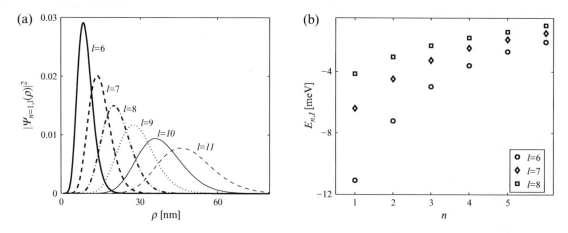

Fig. 22.3 (a) $n = 1$ wavefunctions squared for the set of potentials shown in Fig. 22.2. (b) TIS binding energies for the potentials of Fig. 22.2.

z-direction and the cylindrical co-ordinates ρ and ϕ, we obtain that

$$\Psi_{n,l,k}(\rho, \phi, z) = \psi_{n,l}(\rho)\phi_k(z)\exp(il\phi)/\sqrt{2\pi\rho},$$

$$E_{n,l,k} = E_{n,l} + E_k, \tag{22.7}$$

where $E_{n,l}$ is the eigenvalue associated with the radial motion and E_k is the eigenvalue associated with the longitudinal motion. The radial wavefunction $\psi_{n,l}(\rho)$ satisfies the (radial) Schrödinger equation

$$\left(-\frac{\hbar^2}{2m_e}\frac{d^2}{d\rho^2} + V_{\text{eff}}(\rho) - E_{n,l}\right)\psi_{n,l}(\rho) = 0. \tag{22.8}$$

Fig. 22.3(a) presents the $n = 1$ radial wavefunctions $\psi_{n=1,l}(\rho)$ for the potentials of Fig. 22.2. Due to the presence of the centrifugal barrier, wavefunctions with angular momenta $l \geq 5$ are highly detached (at distances exceeding 10 nm), from the surface. As indicated in Fig. 22.3(b), typically the binding energies range between 1–10 meV. In a similar fashion it is possible to study the interaction of *ions* with metallic tubes. One simply modifies the mass m_e and the particle charge in eqns (22.5)–(22.8).

An interesting situation that may also be realized experimentally arises when the external charge is immersed in a medium. For example, if the nanotube is embedded in GaAs, for which the dielectric constant is $\epsilon = 13.1$, the effective mass of an external electron is $m_e^* = 0.0067m_e$. In this case the delocalization of the image states exceeds that of TIS in vacuum, while the binding energies are significantly reduced to values of $\sim$0.01 meV.

TIS have also been investigated theoretically near a single semiconducting nanotube, using the random phase approximation (RPA) (Gumbs *et al.* 2005), and *between* the walls of a double-walled nanotube (Gumbs *et al.* 2006), taking into account the background dielectric constant.

We have thus far discussed the nature of TIS for an idealized case, consisting of a homogeneous, metallic, infinitely long tube. In the following sections we treat more realistic situations of inhomogeneous nanotubes of finite length. In addition, we investigate TIS associated with *arrays* of nanotubes and discuss the decay mechanisms that limit the lifetimes of TIS.

22.2.2 TIS formed around finite and inhomogeneous nanotubes—the role of heterojunctions

In Section 22.2.1 we have shown that homogeneous nanowires support stable *radially* detached TIS. We have not dealt with the confinement of these states *along* the *longitudinal* direction of the nanowire because we only considered infinitely long tubes. In this section we show (Segal *et al.* 2004a) that by systematically incorporating into the nanowire inhomogeneities that locally vary its screening ability, we can control both the radial detachment and the longitudinal confinement of the TIS. By extending this approach to periodical inhomogeneities, one can also design electronic bands of TIS.

In Fig. 22.4 we present examples of inhomogeneous nanostructures resulting from joining together two or more metallic segments, such as the (12,0) and (6,6) CNTs (Chico *et al.* 1996). In a similar fashion to the effect of impurities in one-dimensional conductors, the abrupt change in the surface electronic wavefunctions in the junction region prohibits the flow of charges across the junctions, thereby electrically insulating adjacent segments from one another. The effective barriers across junctions locally modify the ability of the system to screen external electrons, resulting in the reshaping of the TIS in the vicinity of these joints.

The electrostatic attraction of an external electron to such a complex system can no longer be calculated analytically (Jackson 1975; Granger *et al.* 2002) as done in Section 22.2.1. Instead, we develop a general numerical scheme that is capable of generating the electrostatic interaction (ignoring the atomic structure of the surface) of a charged particle with an *arbitrarily* shaped metallic surface. Fig. 22.4 displays a nanotube to which cylindrically symmetric inhomogeneities have been introduced. Due to the cylindrical symmetry, $V(\rho, z)$, the external electron potential energy expressed in cylindrical coordinates $\mathbf{r} = (\rho, \phi, z)$, is ϕ independent. It is evaluated as follows: We place

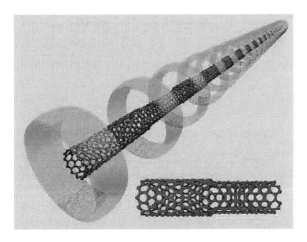

Fig. 22.4 TIS formed above junctions of electrically isolated metallic segments, such as (12,0) and (6,6) metallic CNT (see the inset). The states coalesce into electron image-state bands above periodic arrays of metallic segments.

a test charge q_0 at point $\mathbf{r}$ outside the metallic surface. Each segment of the entire surface is divided into N tiles, whose centers are placed on a grid of points $\mathbf{r}_i = (a, \phi_i, z_i)$, where a is the radius of each segment of the cylinder. The potential induced on each tile centered at $\mathbf{r}_i$ is given as,

$$V_0(\mathbf{r}_i) = \frac{k_0 q_0}{|\mathbf{r}_i - \mathbf{r}|} + \sum_{j \neq i} \frac{k_0 q_j}{|\mathbf{r}_i - \mathbf{r}_j|}, \qquad (22.9)$$

where $k_0 = 1/(4\pi\epsilon_0)$ and q_j is the charge induced on each tile. Use of charge neutrality over the entire surface of each segment, $\sum_{i=1}^{N} q_i = 0$, together with eqn (22.9), and the fact that V_0 is *constant* over the entire surface of the segment, yields a set of $N+1$ linear algebraic equations for the N tile charges q_i and the surface potential V_0. The number of tiles is then increased and the procedure repeated until convergence. After solving for the charge distribution in this manner, the potential energy of the external electron at the point $\mathbf{r}$ is given as,

$$V_s(\mathbf{r}) = \frac{q_0}{2} \sum_{i=1}^{N} \frac{k_0 q_i}{|\mathbf{r} - \mathbf{r}_i|}, \qquad (22.10)$$

where $1/2$ eliminates double counting of the energy.

This approach can be easily extended to the case of inhomogeneous structures made up of several segments. In that case we solve the above equations for each segment, while summing over contributions from *all* the induced charges in the whole system, so as to include the segment–segment Coulombic interactions. This scheme can be also generalized to the case of inhomogeneities with no radial symmetry.

Sample results of these calculations are given in Fig. 22.5. In panel (a) we present the attractive potential energy $V_s(\rho = 10\,\text{nm}, z)$ for an electron

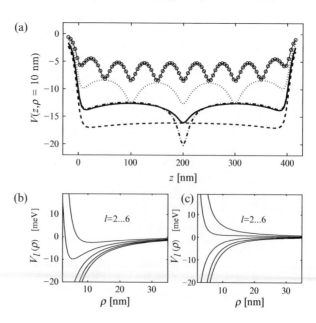

Fig. 22.5 (a) The induced potential energy $V_s(\rho = 10\,\text{nm}, z)$ for: (dashed line) a single segment of $L = 400\,\text{nm}$; (solid line) two 200-nm segments; (dashed-dotted line) the last with no interaction of the segments; (dotted line) four 100-nm segments; (circles) eight 50-nm segments. (b) The effective potential $V_l(\rho, z = 200\,\text{nm})$ with $l = 2$–6 for a system composed of two segments. (c) The same as in (b) for eight segments. We see that detached local minima exist in (b) but not in (c).

above a nanotube of total length $L = 400$ nm, made up of two, four, and eight segments. For simplicity, the radii of all the segments have been chosen as $a = 0.68$ nm, typical of the (10,10) metallic nanotube. In addition, the tube's terminal caps were ignored. We find that the average attraction becomes *weaker* as we increase the number of segments, since the screening of the external electron is reduced. We note, however, that as the number of segments is multiplied by two, the potential in the vicinity of the new junctions thus created remains practically the same, since no screening charge is needed to flow here. The localized quantum wells formed in the potential have the width $\Delta z \approx \rho$. Figs. 22.5(b) and (c) display the *effective* potential $V_1(\rho, z)$ that combines the attractive electrostatic potential of eqn (22.10) and the repulsive centrifugal potential,

$$V_1(\rho, z) = V_s(\rho, z) + \frac{\hbar^2 \left(l^2 - \frac{1}{4}\right)}{2m_e\rho^2}. \qquad (22.11)$$

It is calculated as a function of ρ at $z = 200$ nm, for $l = 2 \ldots 6$ and structures with $N = 2$ and $N = 8$ segments. For $N = 2$ and $l = 5$, the attraction gives rise to bound states, detached from the tube's surface. In the $N = 8$ case, binding is significantly reduced, resulting in the elimination of all bound states. Thus, the existence, and in particular, the shapes of the TIS above nanowires can be controlled by inserting a *variable* number of inhomogeneities into the tube.

We next examine the states formed in the potential wells of Fig. 22.5. Using the attractive potential energy $V_s(\rho, z)$ of eqn (22.10), we can evaluate the wavefunctions of the image states

$$\Psi(\rho, \phi, z) = \psi_{l,\nu}(\rho, z)\, \Phi_l(\phi), \qquad (22.12)$$

with an integer l. The combined radial-longitudinal wavefunction can be written as $\psi_{l,\nu}(\rho, z) = \chi_{l,\nu}(\rho, z)/\sqrt{\rho}$, where $\chi_{l,\nu}(\rho, z)$ is a solution of the Schrödinger equation

$$\left[\frac{-\hbar^2}{2m_e} \left(\frac{d^2}{d\rho^2} + \frac{d^2}{dz^2} \right) + V_l(\rho, z) - E_{l,\nu} \right] \chi_{l,\nu}(\rho, z) = 0. \qquad (22.13)$$

We solve eqn (22.13) numerically, using a multidimensional discrete variable representation (DVR) algorithm (Kafner and Shapiro 1984; Colbert and Miller 1992; Light and Carrington 2000).

In Fig. 22.6 (lower panel), we present the eigenenergies $E_{l,\nu}$ of the $l = 5$–7 states formed above a single segment of length $L = 400$ nm (corresponding to the dashed-line potential of Fig. 22.5(a)). The $l = 5$ eigenstates can be grouped into several clumps, characterized by an increasing number of radial (ρ) nodes. The first clump corresponds to $\nu < 40$, where the $\nu < 10$ states are localized at the segment ends, and the second clump is at $40 < \nu < 75$. As the energy is increased in each clump, the wavefunctions progressively spread toward the center regions of each segment, while acquiring additional longitudinal nodes. In the inset, we show that $E_{l,\nu\leq40}$ can be parametrized as $C\nu^2$ for the first clump of states, where $C = 2.33 \times 10^{-3}$ meV. This behavior is in excellent agreement with the level structure of a particle in a box of size L, for which

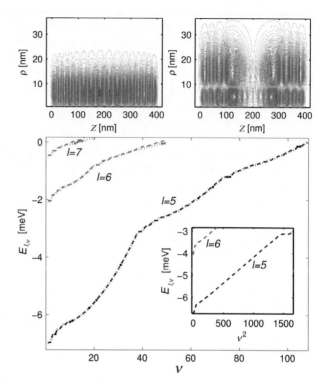

Fig. 22.6 (Upper panel) Two typical eigenstates, with $l = 5$, of the single tube with $L = 400$ nm. (left) $\nu = 16$ (right) $\nu = 48$. (Lower panel) Comparison between the exact (dashed) and adiabatic (dotted) eigenenergies. The two sets essentially coincide. (Inset) The parabolic dependence on ν of the lower-energy eigenvalues for $l = 5$ and 6; the $l = 6$ values are shifted by -2 meV for display purposes.

$E_\nu = C\nu^2$, with $C = \hbar^2\pi^2/2m_e L^2 = 2.35 \times 10^{-3}$. Eigenstates with a higher number of radial nodes follow the same behavior, but this trend is masked by the intrusion of states possessing a lower number of radial nodes. Two wavefunctions possessing $n = 0$ and $n = 1$ radial node are depicted in the left and right upper panels, respectively.

These observations can be easily explained by the fact that the motion of the image-state electron in the radial direction is much faster than its motion in the longitudinal direction. We can thus apply the *adiabatic approximation*, where one solves the one-dimensional *radial* equation in ρ, with z as a variable parameter. The adiabatic eigenvalues thus obtained are then used as potentials for the *longitudinal* eigenvalue equation. This approximation greatly reduces the calculational effort, since instead of solving the two-dimensional Schrödinger equation (eqn (22.13)), it involves solving two simple one-dimensional problems. We can thus assign every ν level two adiabatic quantum numbers n and m correlated with the number of radial and longitudinal nodes, respectively. We have verified that the adiabatic eigenenergies deviate by less than 1% from the exact eigenvalues (Segal *et al.* 2004a).

We have also studied TIS associated with an infinitely long, periodic linear array of isolated metallic segments, such as the one shown in Fig. 22.4. The attractive potential for this system is $V_s(\rho, z + R) = V_s(\rho, z)$, where $R = jL$, with j being an integer and L signifying the lattice constant. Thus, the

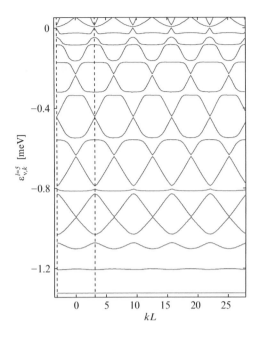

Fig. 22.7 Band structure of image states with $l = 5$ above an array of $L = 200\,\mathrm{nm}$ nanotube segments. Displayed are the $\nu = 7\text{--}25$ bands, with the flat 12th band marked, in both the reduced (vertical dashed lines) and extended zones.

wavefunctions of the system are Bloch states of the type,

$$\chi_{\nu,k}^l(\rho, z) = e^{ikz} f_{\nu,k}^l(\rho, z) = e^{ikz} f_{\nu,k}^l(\rho, z + R)\,, \qquad (22.14)$$

with band index ν, quasi-momentum $\hbar k$ and angular momentum $\hbar l$. The functions $f_{\nu,k}^l$ fulfill the equation

$$\left\{ \frac{-\hbar^2}{2m_e} \left[\frac{d^2}{d\rho^2} + \left(\frac{d}{dz} + ik \right)^2 \right] + V_l(\rho, z) \right\} f_{\nu,k}^l(\rho, z) = \epsilon_{\nu,k}^l \, f_{\nu,k}^l(\rho, z)\,, \qquad (22.15)$$

which can be solved numerically, using the DVR algorithm (Colbert *et al.* 1992).

In Fig. 22.7, we present the calculated band energies $\epsilon_{\nu,k}^l$ with $l = 5$, of a periodic system of metallic segments, whose lattice constant L is 200 nm. The low-lying bands are almost flat, since they correspond to states localized above junctions between neighboring segments, allowing electrons to slowly tunnel to neighboring segments. Many of the higher-lying bands can be smoothly connected in the extended zone to form a single nearly parabolic band. States belonging to these bands have a large number of longitudinal nodes, but are nodeless in the radial direction. These, steeply varying, bands combine with some flat bands (e.g. the 12th band, ...) of states, possessing more radial nodes and fewer longitudinal ones.

The effects outlined above in periodic systems should be observable also in *disordered* nanowires whose screening ability is smaller than that of the ordered arrays. In particular, the *average* attractive potential and image-state wavefunctions above a periodic system with segments of length L should be similar to those above a disordered nanowire with *localization length* given by L. This fact can serve as a measure of localization of conduction electrons in disordered nanowires. We can also devise a *genetic algorithm*, which for a chosen profile of the attractive potential would find an optimal profile of inhomogeneities or disorder in the system. Thus, TIS can be used in probing surface properties, or vice versa, we can shape image states above nanostructures that would control the motion of electrons and molecules in their vicinity (Král 2003) by including inhomogeneities in the structure.

22.2.3   TIS of nanotubes arrays

In Sections 22.2.1 and 22.2.2 we discuss the formation of highly extended TIS around a *single* homogeneous or inhomogeneous nanotube. An interesting issue is whether suspended arrays of parallel nanowires can support image states lacking cylindrical symmetry. In this section we show that this is indeed the case, and that highly detached TIS with *rich* band structures may form around nanotubes arrays. Such Bloch states would resemble light modes in photonic bandgap materials (Joannopoulos *et al.* 1997) or atomic matter waves formed in optical lattices (Robinson *et al.* 2000), and can be used to guide electrons inside nanowire lattice. Our discussion here closely follows Segal *et al.* (2005a). Fig. 22.8 illustrates the system consisting of an array of infinitely long metallic nanotubes (allowing us to neglect the caps) surrounded by electronic image states. The tubes are aligned along the z-direction, with their axes placed at the $x = p\,d$ and $y = q\,d$ positions (p, $q = 0, \pm1, \pm2, \ldots$; with $q = 0$, for a 1D array). Typically, the ratio of the lattice constant d to the tube radius $a \approx 1\,\text{nm}$ is $d/a = 10\text{--}100$.

The electrostatic interaction between an electron of charge e positioned at a distance ρ relative to the center of a single infinite, metallic, homogeneous nanowire is given approximately by eqn (22.5). For convenience, we copy it here

$$V_\text{s}(\rho) \approx \frac{2e^2}{\pi a} \sum_{n=1,3,5,\ldots} \text{li}\left[(a/\rho)^n\right] , \quad \text{li}(x) \equiv \int_0^x \frac{dt}{\ln(t)} . \tag{22.16}$$

Fig. 22.8 Scheme of electron image states formed in the vicinity of a 1D array of parallel metallic nanotubes.

We first examine a single-electron image state associated with *two* metallic nanotubes. The distances from the electron to the tube centers are: $\rho_{1,2} = \sqrt{(x \mp d/2)^2 + y^2}$. Without loss of generality, we approximate the electron–tube interaction by its Coulombic-like form, $V(\rho) = -e^2/4\,|\rho - a|$, valid close to the tubes. Furthermore, we assume that the total interaction is a simple sum of the electron's interaction with each nanotube,

$$V_T(\rho_1, \rho_2) = V_s(\rho_1) + V_s(\rho_2), \qquad (22.17)$$

i.e. we neglect the relatively short-range interaction between the charges induced in the two nanotubes. The wavefunctions, $\Psi(x, y, z) = \psi(x, y)\phi(z)$, are separable in the z-direction. In the x- and y-directions, they can be solved by writing the Schrödinger equation in bipolar co-ordinates, $(x, y) \to (\xi, \eta)$,

$$x = b\,\frac{\sinh(\eta)}{\cosh(\eta) - \cos(\xi)},$$

$$y = b\,\frac{\sin(\xi)}{\cosh(\eta) - \cos(\xi)}. \qquad (22.18)$$

We have chosen the free parameter to be $b = a\sinh(\eta_0)$, where $\eta_0 = \cosh^{-1}(d/2a)$, and a, the radius of the metallic (10,10) nanotube to be $a = 0.68$ nm. The tube's exterior spans the ranges $0 \le \xi \le 2\pi$ and $-\eta_0 < \eta < \eta_0$. Since the interaction potential $V_T(\rho_1, \rho_2)$ is symmetric under the reflection about the $\xi = \pi$ and $\eta = 0$ lines (the x- and y-axes), the states possesses a *two-fold reflection symmetry*. The $\psi(\xi, \eta)$ wavefunctions (their derivatives) thus vanish at $\eta = 0$ and $\xi = \pi$, points, generate odd (even) parity eigenstates. Two parity quantum numbers, $u = \pm$, $v = \pm$, associated with these reflections label the $\psi^{u,v}(\xi, \eta)$ eigenstates and the $E^{u,v}$ eigenenergies obtained from the Schrödinger equation

$$-\frac{\hbar^2}{2m_e b^2}\,(\cosh\eta - \cos\xi)^2 \left(\frac{\partial^2}{\partial\eta^2} + \frac{\partial^2}{\partial\xi^2}\right) \psi^{u,v}(\xi, \eta)$$

$$+ V_T(\xi, \eta)\,\psi^{u,v}(\xi, \eta) = E^{u,v}\,\psi^{u,v}(\xi, \eta), \qquad (22.19)$$

The calculated eigenenergies as a function of the nanotube separation d are given in Fig. 22.9. At large d, these eigenenergies correspond to energies of electronic wavefunctions localized over a single tube and can thus be numbered by the principal quantum number n and angular momentum l (Granger *et al.* 2002). As the intertube separation decreases, the higher excited single-tube states start to overlap and become gradually modified. The resulting pairs of *degenerate* states split into double-tube states, with even and odd symmetries under the tube exchange ($v = \pm$). At smaller d, single-tube states with different values of n and l mix, but in some states the number of radial and axial nodes can still be counted. Level repulsion then becomes important and avoided crossings appear, as shown in Fig. 22.9.

We have also calculated the $\psi^{u,v}(\xi, \eta)$ eigenfunctions, and have re-expressed them in Cartesian (x, y) co-ordinates. In Fig. 22.10, we present the $|\psi^{u,v}|^2$ densities for several typical wavefunctions. In the left panels, we follow the evolution of a state correlating with the single tube $l = 6$ and $n = 2$ state

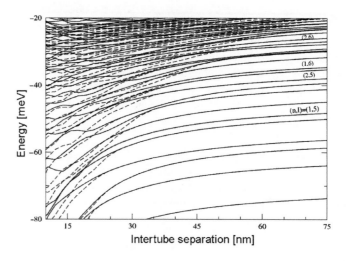

Fig. 22.9 Energies of two-tube image states with $u = +$ as a function of intertube separation d. The even ($v = +$, dashed lines) and odd ($v = -$, solid lines) parity-state energies split at small intertube separations. Selected curves are labelled by the quantum numbers (n, l) of the single-tube image potential states to which they converge as $d \to \infty$.

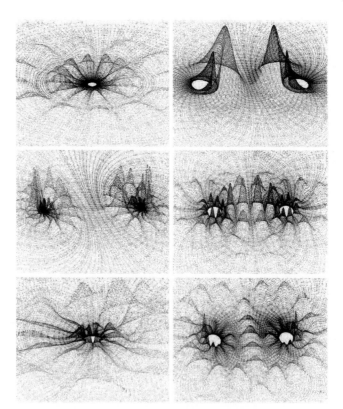

Fig. 22.10 Left panels: gradual loss of detachment of the $l = 6$, $n = 2$ wavefunction squared, as the two nanotubes get close: (top) view on one tube for $d = 200a$; (middle) the entire wavefunction for $d = 80a$; (bottom) detailed view of the right tube in the last case. Right panels: states for much smaller tube separation, $d = 10a$. (Top) The modified $l = 0$, $n = 2$ state. (Middle) An odd symmetry state with respect to reflection in the y axis. (Bottom) An even symmetry state.

as we vary the array parameters. At large tube separations, $d = 200\,a$ (upper panel), the state is detached from the tube's surface, akin to the single-tube situation (Granger *et al.* 2002). At lower but still relatively big separations, $d = 80\,a$ (middle and lower panels), this state (as other states) partially collapses on the tubes, due to *asymmetric distortion* of the attractive potential. We thus expect that the lifetime of this state would get shorter (Granger *et al.* 2002), especially, if their energies do not fall in the bandgap of the material (Höfer *et al.* 1997). In the right panel, we display more distorted states, obtained for $d = 10\,a$, showing their rather complex nodal structures.

We now investigate image states of a single electron in large periodic arrays of parallel nanowires. For a 1D array the total potential fulfills $V_T(x, y) = V_T(x + d, y)$. Thus, the transverse Bloch components of the total wavefunctions $\Psi(x, y, z) = \psi_{m,k}(x, y)\phi_{k_z}(z)$, with energies $\epsilon_{m,k} + E_{k_z}$, fulfill (Ashcroft and Mermin 1976)

$$\psi_{m,k}(x, y) = e^{ikx} f_{m,k}(x, y) = e^{ikx} f_{m,k}(x + d, y). \qquad (22.20)$$

These states can be obtained from the Schrödinger equation with the Hamiltonian parameterized by k, as,

$$\left\{ \frac{-\hbar^2}{2m_e} \left[\left(\frac{\partial}{\partial x} + ik \right)^2 + \frac{\partial^2}{\partial y^2} \right] + V_T(x, y) \right\} f_{m,k}(x, y) = \epsilon_{m,k}\, f_{m,k}(x, y). \qquad (22.21)$$

Analogous equation, with the $\mathbf{k} = (k_x, k_y)$ wave vectors, can be used for a 2D (square) lattice of nanotubes.

We solve eqn (22.21) numerically, by a multidimensional discrete-variable representation (DVR) algorithm (Colbert and Miller 1992). We use the single-tube potential (22.16), and include in V_T the interaction due to the central tube in the cell and its two (1D) or eight (2D) neighbors, while neglecting multiple reflection, as in the two-tube case. The intertube separation is varied between $d = 5$–80 nm and $a = 0.68$ nm. The calculations are performed with grid spacing proportional to d, $\Delta \approx 0.0075d$. The results at larger lattice constants ($d \geq 50$ nm) are therefore less accurate. Nevertheless, our tests show that the overall features in the wavefunctions are well converged.

In Fig. 22.11, we display the states in square 2D arrays of nanotubes, where the unit cell spans the region from $-d/2$ to $d/2$ in both axes. In the upper left panel, we show for $d = 50$ nm the probability density of a state correlated to the $l = 6$ $n = 2$ single-tube state, marked as the Γ point of Fig. 22.12. In contrast, as we move through the band to the X point, the state partially collapses on the tubes' surfaces (right panel). In the middle panels, we also display two wavefunctions for $d = 20$ nm. In the left panel, we show the state correlating with the $l = 6$, $n = 1$ single-tube state, and in the right panel we depict a state displaying quartic symmetry. In the lower-left and right panel, we display a diagonally aligned state for $d = 5$ nm and a detached state for $d = 80$ nm, respectively.

Finally, in Fig. 22.12, we present the band structure for the 2D lattice of nanotubes, for $d = 50$ nm and $\Delta/d = 0.01$, calculated between the Γ, X and M points. Some states are degenerate at the Γ point, due to their symmetry with the $\pi/2$ rotation. The *bandgaps* present at lower energies could be used for blocking the transfer of Rydberg electrons. Higher bands ($m > 60$) are broader and denser, so that bandgaps disappear, while many avoided crossings emerge. The calculated band structure is slightly dependent on the grid used. The band structure would also be modified by the inclusion of temperature-dependent

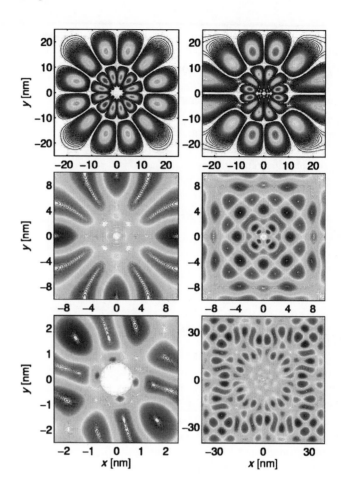

Fig. 22.11 Selected eigenstates of the periodic 2D lattice of nanotubes, described in the text.

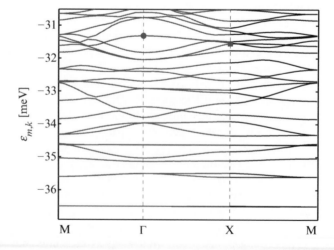

Fig. 22.12 Band structure of single-electron image states in the vicinity of a 2D array of nanotubes, with $d = 50$ nm. The boundaries are defined by the Γ ($\mathbf{k} = (k_x, k_y) = (0, 0)$), X ($\mathbf{k} = (\pi/d, 0)$) and M ($\mathbf{k} = (\frac{\pi}{d}, \frac{\pi}{d})$) points. We present the (50–73)th bands. The square (circle) denotes the X (Γ) point of the 68 (70)th band, with probability densities of the states plotted in the upper part of Fig. 22.11 right (left).

many-body effects, due to scattering of the *single* TIS electron on the electrons and phonons present in the nanotubes.

We have thus demonstrated the existence of Rydberg-like image states in *periodic arrays* of nanowires. As in the single-tube case, these Bloch states can be highly detached from the surfaces of the nanowires. This is because, an electron hovering in such a *periodic* lattice of nanowires is influenced by a Coulombic-like attraction and a centrifugal repulsion, both displaying *cylindrical symmetry* around each wire.

22.2.4   The lifetimes and stability of TIS

As already mentioned above, due to their detachment from the tube's surface, TIS are expected to have much longer lifetimes than image states on flat surfaces. In this section we explore the decay channels available to TIS, allowing us to estimate the lifetimes of these states, which we find to be in the microsecond regime. We show that the main mechanism that limits the lifetimes of TIS is the slow relaxation of their angular momentum l. In many ways this behavior is analogous to spin-orbit-coupling-mediated spin relaxation (Kikawa *et al.* 1997; Hanson *et al.* 2003).

As possible a priori routes to relaxation we list: (i) electron–electron scattering, (ii) electron–phonon scattering, and (iii) electron–defect scattering. With regard to the first mechanism, we note that scattering with surface electrons mainly causes changes in the TIS electronic linear momentum k (Pichler *et al.* 1998; Zabala *et al.* 2001; Gumbs *et al.* 2005, 2006). It is therefore the major decay process for excitations above *flat* metallic surfaces (Echenique *et al.* 2004). In contrast, the second mechanism that entails fluctuations in the effective attractive potential V_{eff} due to coupling to the nanotube phonons or impurities, can, in addition to relaxing the linear momentum k, relax the angular momenta l and the radial principal quantum number n.

The coupling with the surface phonons is highly delocalized because the centrifugal barrier prohibits the TIS electron from penetrating into the surface and sensing its detailed atomic structure. The relaxation thus proceeds via radial vibrations associated with circularly polarized (TA), angular-momenta-carrying, phonons. As illustrated in Fig. 22.13, these phonons cause "string-like" deformations of the tube (Suzuura and Ando 2002). For a (10,10) carbon nanotube these deformations, exhibiting linear dispersion, propagate at phonon velocity v_{p} of $\approx 9\,\mathrm{km/s}$ (Saito *et al.* 1998; Jiang *et al.* 2004).

We have modelled this process by considering string-like deformations of an ideally conducting infinitely long tube of radius a interacting with an electron occupying a TIS, placed a distance ρ apart from the center of the tube. The force exerted by the radially directed vibrations of the nanotube on the TIS-electron is a result of the change in the tube's equilibrium position, thus modifying the effective potential $V_{\mathrm{eff}}(\rho, z)$. For simplicity, we have used the approximate Coulombic form, $V_{\mathrm{s}}(\rho) \sim 1/|\rho - a|$, instead of the screening potential of eqn (22.5). Hence, when a nanotube oriented along the z-axis is

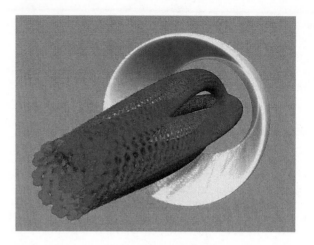

Fig. 22.13 The "string-like" deformation of a nanotube excited by the spinning TIS-electron.

displaced by $\vec{\delta}(z)$ from its equilibrium position, $\vec{\rho}_0 = (x_0, y_0)$, the potential energy can be written as

$$V_s(\rho, z) = -\frac{e^2}{4} \Big/ \left| \left[(x_e - x_0 + \delta_x(z))^2 + (y_e - y_0 + \delta_y(z))^2 \right]^{1/2} - a \right|,$$

$$(22.22)$$

where (x_e, y_e, z_e) is the external electron position, with $\rho = \sqrt{x_e^2 + y_e^2}$. The tube's displacement $\vec{\delta}(z) = \hat{x}\delta_x(z) + \hat{y}\delta_y(z)$ can be expanded in circularly polarized phonon modes co-ordinates $u_{k,\pm}$, $\vec{\delta}(z) = \sum_q \left(u_{q,+}\hat{\epsilon}_+ + u_{q,-}\hat{\epsilon}_- \right) e^{iqz}$. where q are the phonon wave vectors, and $\hat{\epsilon}_\pm \equiv \hat{x} \pm i\hat{y}$ are the circular-polarization complex directions. Expanding the potential function of eqn (22.22) in a Taylor series about the equilibrium position, and keeping only the first-order contribution, recasts the total Hamiltonian of the system as

$$H = \frac{-\hbar^2}{2m_e}\nabla^2 + V_s(\rho)|_{\vec{\delta}=0} + H_{ph} + \vec{\delta}(z) \cdot \nabla V_s(\rho)|_{\vec{\delta}=0}. \qquad (22.23)$$

In the above, the first two terms are the rigid-tube TIS-electronic Hamiltonian. H_{ph} is the free-phonon Hamiltonian, and the fourth term H_{e-ph} is the electron–phonon interaction Hamiltonian (Mahan 1990). In second quantized form, the free-phonon Hamiltonian and the electron–phonon interaction Hamiltonian can be written as

$$H_{ph} = \sum_{q,\pm} \hbar\omega_q a_{q,\pm}^\dagger a_{q,\pm}, \quad H_{e-ph} = \sum_{q,\pm,\nu,\nu'} \tilde{A}_{q,\pm}^{(\nu,\nu')} \left(a_{q,\pm}^\dagger + a_{q,\pm} \right) c_\nu^\dagger c_{\nu'},$$

$$(22.24)$$

where $c_\nu^\dagger$ and c_ν are the creation and annihilation operators of TIS-electrons in the $\psi_\nu(\rho)$ states, $\nu = (n, l, k)$, and $a_{q,\pm}^\dagger$, $a_{q,\pm}$ are creation and annihilation operators of the circularly polarized phonon modes. The matrix elements $\tilde{A}_{q,\pm}^{\nu,\nu'}$ denote the coupling elements for the $(n, l, k) \to (n', l', k')$ transitions, induced by the right and left circularly polarized phonons, respectively, with longitudinal momentum q. Energy and momentum should be conserved in the system. The linear momentum obeys $k' = k - q$, where k (k') is the initial (final) electronic momentum and q is the phononic contribution. The angular momentum conservation rule obeys $l' = l \pm 1$, and the Hamiltonian (22.24) yields in addition the energy-conservation condition,

$$\frac{\hbar^2(k^2 - k'^2)}{2m_e} + \Delta E = \pm \hbar v_p \left| k - k' \right|, \tag{22.25}$$

where $\Delta E = E_{n,l} - E_{n',l'}$, and we use the phonon-dispersion relation $\omega_q = |q| v_p$, with v_p being the phonon velocity.

In order to evaluate the zero-temperature relaxation rates for the $v = (n, l, k) \rightarrow v' = (n', l', k')$ process we use Fermi's golden rule (Jiang *et al.* 2004),

$$W_{v \rightarrow v'} = \frac{2\pi}{\hbar} \left| \tilde{A}_{q,\pm}^{(v,v')} \right|^2 \delta \left(E_k + \Delta E - E_{k'} - \hbar \omega_q \right), \tag{22.26}$$

where $q = k - k'$ is the phonon-momentum change associated with phonon emission and absorption processes. The coupling elements $\tilde{A}_{q,\pm}^{(v,v')}$ of the above were taken from Segal *et al.* (2005b) where they were calculated for different transitions associated with a (10,10) CNT of radius $a = 0.7$ nm. Because a small-l electron, which is localized closer to the tube surface, can more readily excite the phonons, these matrix elements get larger with decreasing values of l.

The *angular momentum* lifetime τ_l, which as noted above determines the relaxation rate of the TIS electron, is defined via the relation,

$$\langle L(t) \rangle = \langle L(0) \rangle e^{-t/\tau_l}, \tag{22.27}$$

where $L(t) = \sum_l \hbar l \, c_v^\dagger(t) c_v(t)$, $(v = (n, l, k))$ is the angular momentum operator in the Heisenberg representation. τ_l can be calculated using the second-order expansion of the Heisenberg equation of motion for the expectation value of $L(t)$,

$$\langle \dot{L}(t) \rangle = -\frac{1}{\hbar^2} \int_0^t d\tau \langle [H(t), [H(\tau), L(\tau)]] \rangle = -\sum_{v'} \hbar(l - l') W_{v \rightarrow v'}. \tag{22.28}$$

By combining eqns (22.27) and (22.28) we obtain the τ_l and τ_v, the dephasing lifetime, as

$$\tau_l^{-1} = \sum_{v'} (1 - l'/l) \, W_{v \rightarrow v'}, \qquad \tau_v^{-1} = \sum_{v'} W_{v \rightarrow v'}. \tag{22.29}$$

In Fig. 22.14, we display the calculated τ_l for different initial states $v = (n, l, k)$. As expected, states with large angular momenta, highly detached from the surface (see (a) inset), have long $\tau_l \approx 100$–1000 ns lifetimes (a), while states with $l \sim 1$–4 that are separated by only ~ 1 nm from the surface, have much shorter lifetimes. The effect of the proximity to the surface, causes the low-l couplings to be 10^2–10^3 times the high-l couplings, leading to a $\sim 10^4$–10^6 increase in the low-l angular momentum relaxation rates. This picture is consistent with Fig. 22.14(b), where we show that the lifetimes go up as we increase n, thereby further removing the TIS electron

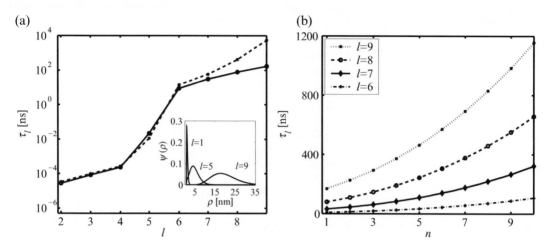

Fig. 22.14 Angular relaxation time for the (n, l, k) state at $T = 0\,\text{K}$. (a) $n = 1$, $k = 0$ (full); $n = 1$, $k = 2 \times 10^8\,\text{m}^{-1}$ (dashed). The inset shows the $n = 1$ radial states for the $l = 1$, $l = 5$ and $l = 9$ cases. (b) The n dependence of the relaxation for $l = 6, 7, 8, 9$.

from the surface. The τ_l lifetimes also get longer when the initial k is increased, allowing for excitation processes with $l \rightarrow l + 1$ for $\Delta E < 0$.

The accuracy of the above calculations for low l is not as high as for high l, because our wavefunctions do not represent the states accurately enough close to the surface. In this region, coupling to other (twisting) phonon modes might become relevant, with additional effects playing a role (Qian *et al.* 2002). The accuracy is also limited by the simplified Coulomb form used for the screening potential.

We estimate that at finite temperatures the transition rate increases relative to the $T = 0$ case by a factor of 5–50 due to emission/absorption processes induced by thermal phonons (Segal *et al.* 2005b). The slow TIS lifetimes can be tuned by changing the conductivity (Segal *et al.* 2004a) and tension in the nanotubes (stretching). As discussed in Section 22.3.3 below, the long lifetimes of TIS make them useful for a number of interesting novel applications.

22.3   Manipulating tubular image states

22.3.1   Electric- and magnetic-field control of TIS

Atomic (Gallagher 1994) and molecular (Greene *et al.* 2000) Rydberg wavefunctions can be made to vary drastically by the application of electric and magnetic fields. In particular, field-induced symmetry breaking leads to a number of interesting phenomena (Neumann *et al.* 1997), ranging from a simple levels repulsion to intriguing transitions from regular to irregular states displaying chaotic dynamics (Wunner *et al.* 1986; Gutzwiller 1990; Milczewski and Uzer 1997).

Due to their applicative potential and their long lifetimes we have investigated the effect of external fields on TIS coupled to solid-state nanosystems (Segal *et al.* 2004b), and in particular to *pairs* of parallel nanowires. In broad terms, we have found that a magnetic field directed along the nanowires long axis gives rise to "Landau-like" image states that are highly detached from the surfaces of *both* nanowires. When the magnetic field is also crossed with an electric field, generated by opposite-charging of the two nanowires, the detached TIS wavefunctions are seen to tilt towards one of the wires, displaying the increasing degree of erratic nodal patterns associated with "quantum chaos".

In what follows we consider a system comprised of two parallel, infinitely long, metallic nanotubes whose long axes are aligned along the z-direction, their centers placed at the $x = \pm d/2$, $y = 0$ points. We can positively charge one nanotube and negatively charge the other, while keeping the overall pair charge neutral, thus creating an electric field directed in the xy-plane (i.e. the two tubes have the added potentials $\pm V_a$). The total potential energy of the external electron in this configuration is given as

$$V_T(r_1, r_2) = V_s(r_1) + V_s(r_2) + V_C(r_1, r_2). \tag{22.30}$$

Here, $V_s(r_i)$ is the screening potential of eqn (22.5) of an external electron interacting with an infinitely long wire. In the above we neglect the short-range terms arising from multiple reflections of image charges belonging to different nanotubes. The charging potential V_C describes the interaction of the electron with homogeneously spread charges, due to the $\pm V_a$ potential applied on each of the tubes. Since the tubes are relatively far apart from each other, i.e. $d \gg a$, we can assume that they do not polarize each other. In this case the additional potential energy of an electron placed at r_1 and r_2 away from the centers of the two tubes, due to the charging of the tubes, is given as (Slater and Frank 1947),

$$V_C(r_1, r_2) \approx e V_a \frac{\ln(r_1/r_2)}{\ln(a/d)}. \tag{22.31}$$

This expression goes over to the correct limit of $\pm e V_a$ when the electron is placed on the surface of one of the tubes, i.e. when $r_1 = a$ and $r_2 = d$ and vice versa. We now add to the electric field a uniform magnetic field, **B**, oriented along the z-axis. With the addition of this field the total Hamiltonian is given as,

$$H = \frac{1}{2m_e} (\mathbf{p} - e\mathbf{A})^2 + V_T(x, y), \tag{22.32}$$

where $\mathbf{A} = \frac{B}{2}(-y, x, 0)$ is the vector potential of the field in the Landau gauge and $\mathbf{p}$ is the generalized momentum of the electron. The kinetic-energy term of eqn (22.32) gives rise to two additional terms in the Hamiltonian,

$$H_1 = -\frac{eB}{2m_e} L_z, \quad H_2 = \frac{e^2 B^2}{8m_e} \left(x^2 + y^2\right), \tag{22.33}$$

where $L_z = -i\hbar \left(x \frac{\partial}{\partial y} - y \frac{\partial}{\partial x}\right)$ is the angular momentum operator, and m_e is the electron mass. In what follows we ignore the electron spin. The image-state wavefunctions are separable in the z-coordinate, $\Psi(x, y, z) = \psi_\nu(x, y)\phi_{k_z}(z)$, their energies being given as, $E_\nu + \epsilon_{k_z}$. The $\psi_\nu(x, y)$ components fulfill the

Schrödinger equation

$$\left\{\frac{-\hbar^2}{2m_e}\left(\frac{\partial^2}{\partial x^2} + \frac{\partial^2}{\partial y^2}\right) + V_T(x, y) + H_1(x, y)\right.$$

$$\left. + H_2(x, y) - E_\nu\right\} \psi_\nu(x, y) = 0. \qquad (22.34)$$

Close to the surface of either tube, the screening potential, $V_s \approx 1/4|r - a|$, is equal to 1/4 the potential of a *two-dimensional* hydrogen atom (Yang *et al.* 1991). When placed in a magnetic field, a system comprised of an electron near a single tube can therefore be viewed as a highly magnified two-dimensional hydrogen atom in a magnetic field (MacDonald *et al.* 1986; Zhu *et al.* 1990). Likewise, an electron interacting with *two* tubes is the highly magnified analog of a two-dimensional H_2^+ molecule. We have solved eqn (22.34) numerically using a multidimensional discrete variable representation (DVR) algorithm (Colbert and Miller 1992).

According to the properties of TIS around two tubes in the absence of external fields described in Section 22.2.3, for large tube separations the lower excited wavefunctions are localized over each of the two tubes. As d decreases, the states start to overlap and split into double-tube states with even and odd parities (Segal *et al.* 2005a). In the limit of weak magnetic fields, for which B can be considered a small perturbation, these double-tube states become gradually modified, with the (linear) H_1 term, leading to the Zeeman effect, dominating. For a spinless hydrogen atom, H_1 yields an $eBM/2m_e$ energy term, where M, the magnetic quantum number, denotes the eigenvalues of L_z. The (non-linear) H_2 term of eqn (22.33) becomes important for strong magnetic fields or when the electronic states are significantly displaced from the tubes, as in highly excited TIS (Granger *et al.* 2002). In these cases the electrostatic $V_T(x, y)$ term constitutes a weak perturbation relative to B.

When $V_T(x, y) \to 0$ the TIS go over to the (Landau) states of a free electron in a homogeneous magnetic field,

$$\Psi_{nlk}(\mathbf{r}) = e^{ikz}e^{il\phi}u_{nl}(\rho),$$

$$E_{nlk} = \frac{\hbar^2 k^2}{2m_e} + \frac{eB\hbar}{m_e}\left(n + \frac{1 + l + |l|}{2}\right), \qquad (22.35)$$

where $u_{nl}(\xi) \propto \xi^{|l|/2}e^{-\xi/2}L_n^{|l|}(\xi)$, $\xi = eB\rho^2/2\hbar$, and $L_n^{|l|}$ are the associated Laguerre polynomials (Landau and Lifshitz 1998), with $n = 0, 1, 2, \ldots, l = 0, \pm 1, \pm 2, \ldots$. We can see that states of the same $|l|$, that differ by the sign of l, are separated in energy by $\Delta E = eB\hbar|l|/m_e$. For $B = 20$ T and $l = 6$ this gives $\Delta E \approx 12$ meV, which is about the same size as the coupling energy of the $l = 6$ state in the single-nanotube case (Granger *et al.* 2002). Thus, the $V_T(x, y)$ potential can *not* be seen here as a weak perturbation, and the two terms compete.

22.3.1.1   *Case I: $B \neq 0$, $V_a = 0$*

In Fig. 22.15, we display the dependence of the eigenenergies of an electron in the vicinity of two nanotubes on the magnetic field $B = 0$–35 T for $V_a = 0$.

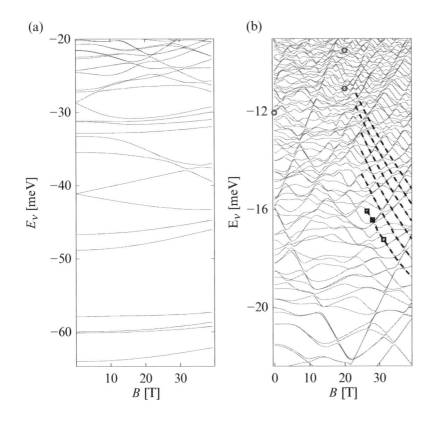

(a)

(b)

Fig. 22.15 Dependence of the eigenstates on the magnetic field B, where $V_a = 0$ and $d = 40\,$nm. (a) Eigenstates Nos. 33–80; (b) eigenstates Nos. 75–150. The probability densities of the states marked by circles (squares) are plotted in Fig. 22.16 top (bottom).

The two nanotubes of a radius $a = 0.7\,$nm are placed $d = 40\,$nm apart, at $x = \pm 20\,$nm. Most of the lower-lying eigenenergies, shown in panel (a), appear as pairs of nearly degenerate even and odd states with respect to the reflection in the $x = 0$ line. With the relatively coarse Cartesian grid used here ($\Delta(\text{grid}) = 1\,$nm), no linear Zeeman splitting appears to exist, except in very few states. This is due to an additional artificial potential of a *quadratic* symmetry, induced by the grid roughness. Then, the single-tube angular momentum states (Granger *et al.* 2002), proportional to $e^{\pm il\phi}$, combine into *split* pairs of double-tube states, proportional to the $\cos(l\phi)$ and $\sin(l\phi)$ functions, that lack the linear Zeeman term. This problem could be avoided by using a much finer grid or by going to bipolar co-ordinates (Segal *et al.* 2005a). However, the high-energy states that are of a larger interest, shown on panel (b) of Fig. 22.15, are typically highly extended and are therefore less sensitive to the grid size. These states are strongly affected by the magnetic field, appearing in the H_2 term, and at high fields $B > 20\,$T and high quantum numbers $\nu > 100$, a series of "Landau-type" levels emerges.

In Fig. 22.16, we plot probability densities for several high-energy states, denoted in Fig. 22.15 by circles (upper panels) and squares (lower panels). Because $V_a = 0$ the probability densities are symmetric with respect to $x = 0$. In the absence of external fields, the single-tube image states ($l > 6$) tend to be detached (20–40 nm) from the surface (Granger *et al.* 2002). The presence of a second nanotube breaks the cylindrical symmetry of the attractive potential around each of the tubes, and the states collapse on the tube's surface. This is

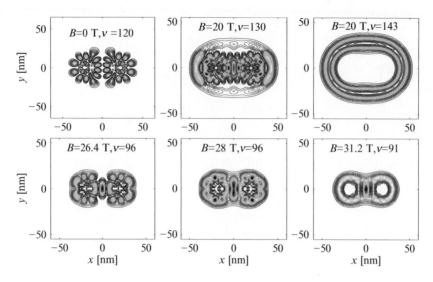

Fig. 22.16 Contour plots of the probability density of selected eigenstates for $B \neq 0$ and $V_{\mathrm{a}} = 0$, showing the formation of "Landau-like" states.

shown on the upper-left panel, where the $\nu = 120$ state at $B = 0$ resembles the single-tube $l = 6$, $n = 2$ state. (Top middle) As we increase the magnetic field, hybrid states, possessing magnetic-free and (detached) Landau-like features, emerge. The $\nu = 130$ state for $B = 20$ T displays both chaotic nodal patterns close to the tube surface, and an extended elliptic-like features ~ 20 nm away from them. (Top right) Finally, the highly excited $\nu = 143$ state for $B = 20$ T can be described as a perturbed $n = 0$, $l = 25$ Landau state, eqn (22.35), where the numbering follows the nodal pattern (not shown). Such states are only marginally affected by the presence of the tubes and tend to be completely detached from the tubes' surfaces. As a result, electrons populating these states are protected from the usual annihilation processes occurring on metallic surfaces, thus prolonging their lifetimes. Such a situation also occurs in periodic arrays of tubes without the external fields (Segal *et al.* 2005a). (Bottom) Here, we show the evolution of the $\nu = 96$ state with increasing magnetic fields, $B = 26.4$, 28, 31.2 T, with each "evolution" stage marked by a square in Fig. 22.15.

22.3.1.2   *Case II: $B \neq 0$, $V_{\mathrm{a}} \neq 0$*

We now charge the two tubes, introducing an electrostatic potential V_{a} on the right tube and $-V_{\mathrm{a}}$ potential on the left tube. In Fig. 22.17, we show the eigenenergies as a function of V_{a}, while keeping $B = 20$ T. The low-energy states (a) display nearly linear Stark splitting, $E_{\nu}^{\pm} = E_{\nu}^{0} \pm \alpha V_{\mathrm{a}}$. The electric field breaks the symmetry of the attractive double-well potential V_{T}. Thus, for each pair, the lower state localizes around the right, positively charged tube, while the higher state tends to localize around the left, negatively charged tube.

The high-energy spectrum (b) is quite complicated. Nevertheless, we can follow series of Landau-like states (marked by diamonds) corresponding to $\nu = 119$, 128, 132, 136, 138, 143. As a general rule, we find that

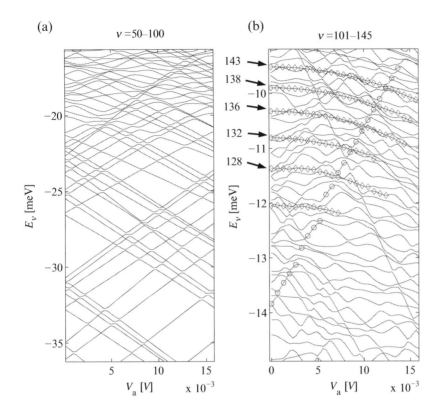

(a) $v = 50$–100

(b) $v = 101$–145

Fig. 22.17 Dependence of the eigenenergies on V_a, for $B = 20$ T and $d = 40$ nm. (a) Low-energy states. (b) High-energy states. The v index designates the positions of some of the ($V_a = 0$) Landau-type states, marked by diamonds.

as long as these states maintain their extended shape, their eigenenergies tend not to vary much with V_a. Similar behavior is observed for a classical electron colliding ballistically with an elliptically shaped quantum antidot (Kleber *et al.* 1996). As we increase V_a and the eigenstates start localizing near one of the tubes (see Fig. 22.18), their energies are seen to change precipitously. A minority of high-lying eigenenergies sharply increase/decrease with V_a, such as the state shown by circles in Fig. 22.17. As for low-lying energies, shown in the left-hand panel, these energies correspond to states that are highly localized in the vicinity of only *one* of the tubes. In the regime of high electric ($V_a = 0.04$ V) and magnetic ($B = 30$ T) fields, highly detached states form exclusively around one of the tubes, or simultaneously collapse on one tube and become detached from the other.

In Fig. 22.18 we show in detail the electric-field-induced collapse of the Landau-like states as a function of the applied electric field. Plot (a) displays a molecular-like state for the zero-field situation. As the magnetic field is turned on, eigenstates become complicated and manifest chaotic features (b). Only when the magnetic field is high enough ($B > 10$ T), do we observe the formation of regular Landau orbits (c). However, as the electric field V_a is turned on, the states first shift (d), then nest on one of the tubes (e) and finally develop chaotic nodal patterns (f). This behavior is consistent with the proliferation of avoided crossings shown in Fig. 22.17 (Chirikov 1971) and related changes of the statistics, see Fig. 22.19 below.

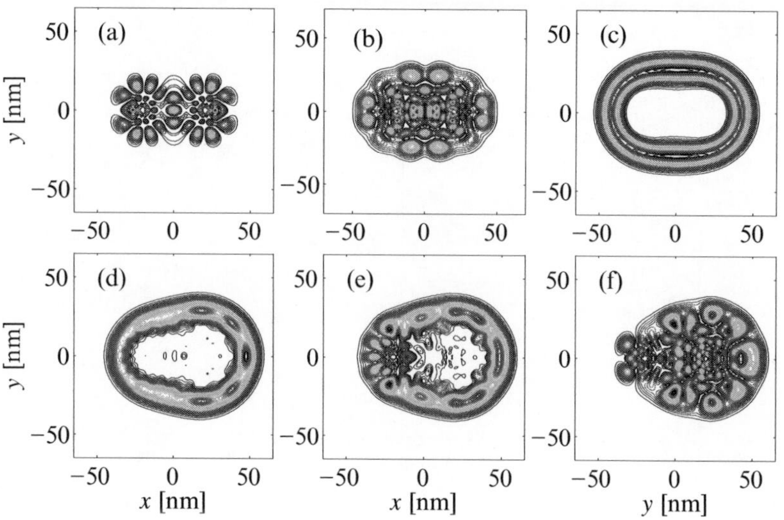

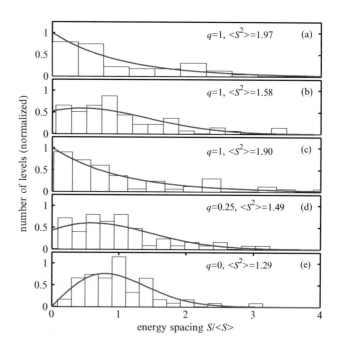

Fig. 22.18 Probability density for six representative states. In all cases $d = 40$ nm and the tubes are located at $x = \pm20$ and $y = 0$. (a) $B = 0$ T, $V_a = 0$ V; (b) $B = 6$ T, $V_a = 0$ V; (c) $B = 20$ T, $V_a = 0$ V; (d) $B = 20$ T, $V_a = 8.5 \times 10^{-3}$ V; (e) $B = 20$ T, $V_a = 1.2 \times 10^{-2}$ V; (f) $B = 20$ T, $V_a = 1.6 \times 10^{-2}$ V. The frames (c)–(f) follow the $\nu = 143$ Landau-like state (marked by "diamonds" in Fig. 22.17) as the electric field is increased.

Fig. 22.19 Nearest-neighbor energy-spacing histograms for states in the energy regime -18 meV $\leq E_\nu \leq -9$ meV. (a) $B = 0$, $V_a = 0$, (b) $B = 6$ T, $V_a = 0$; (c) $B = 20$ T, $V_a = 0$; (d) $B = 20$ T, $V_a = 0.010$ V; (e) $B = 20$ T, $V_a = 0.016$ V. The smooth curves in the histograms are the Poisson (a),(c) and Wigner (e) distributions. The curves in (b), (d) are the result of a least square fit by eqn (22.36) yielding $q = 0.31$, $q = 0.25$.

We have thus demonstrated that image states formed above parallel nanowires can be effectively controlled by the application of mutually perpendicular electric and magnetic fields. In particular, we have clearly seen that Landau-like states, formed in strong magnetic fields, can be gradually changed into chaotic states. We explore this transition in more detail in the next section.

22.3.2   Onset of chaos in TIS

In Section 22.3.1 we first encountered quantum chaos in tubular image states around nanotubes arrays. We now enlarge on this theme by systematically investigating the onset of chaos in such systems. We show below that extended image states might provide one with a new platform for investigation of quantum chaos.

In the past much of the theoretical and experimental research on quantum chaos has been performed on Rydberg atoms in strong electric and magnetic fields (Casati and Chirikov 1995). For high enough fields, these systems undergo transitions from regular to chaotic behavior. This transition is characterized by the change in the statistics of the energy spectrum; the appearance of multiple avoided crossing; and the high sensitivity of energy eigenvalues to small external perturbations (Wunner *et al.* 1986; Milczewski and Uzer 1997; Neumann *et al.* 1997; Haake 2000). Analogous studies have been performed on quantum dots, where chaos can be introduced by applying magnetic fields on regular shaped dots or due to the electron scattering with the dot's irregular boundaries (Alhassid 2000). Chaotic electron dynamics has also been explored in other nanostructures, such as antidots arrays (Weiss *et al.* 1991), resonant tunnelling diodes (Fromhold *et al.* 1994), quantum wells (Fromhold *et al.* 1995), and superlattices (Fromhold *et al.* 2001, 2004).

Here, we systematically study the onset of chaos in a simple nanosize system comprising a pair of nanotubes under crossed electric and magnetic fields, described in Section 22.3.1, with the system Hamiltonian given by eqn (22.32). The system's energy spectra under a combination of an electric field and a high magnetic field are displayed in Figs. 22.15 and 22.17.

22.3.2.1   *Level-spacing distribution*

Most of the "generic" indicators of quantum chaos, e.g. the existence of avoided crossings (Chirikov 1971), the average level density (Haake 2000), and the level-spacing distribution (Bohigas *et al.* 1984), are based on multistate properties. In particular, it is known that the level-spacing distribution should change from a Poissonian type to a Wigner type for chaotic states that obey the eigenvalue statistics of Gaussian orthogonal ensembles (GOE).

In Fig. 22.19 we demonstrate that such transitions occur for TIS as we turn on the magnetic and electric fields. We calculate the level statistics in the energy window $-18\,\text{meV} \le E_\nu \le -9\,\text{meV}$ and display it for five cases shown in Fig. 22.18 corresponding to different values of the magnetic and electric fields: In panels (a)–(c) we increase the magnetic field from 0 to 20 T, and in panels (d) and (e) we also increase the electric field. With the parameter $\langle S \rangle$ being the average energy spacing, we find that in cases (a) and (c), the nearest-neighbor energy-spacing distributions are clearly Poissonian, $P \sim \text{e}^{-S}$, indicating that the states are uncorrelated and the system is regular. In contrast, the other cases can be approximately described by the Wigner distribution $P \sim \frac{1}{2}\pi S \text{e}^{-\pi S^2/4}$, indicating chaotic dynamics. Specifically, the data can be well fitted by the Berry–Robnik distribution function (Berry and

Robnik 1984),

$$P(S,q) = \left(2q(1-q) + \pi(1-q)^3 S/2\right) e^{-qS - \frac{1}{4}\pi(1-q)^2 S^2}$$

$$+ q^2 e^{-qS} \mathrm{erfc}(\sqrt{\pi}(1-q)S/2), \qquad (22.36)$$

which interpolates well between the Poisson ($q = 1$) and the Wigner ($q = 0$) distributions. Another indicator is the mean square level spacing $\langle S^2 \rangle = \int_0^\infty P(S, q) S^2 \, dS$, which should decrease from the value of 2 in the Poissonian case to $4/\pi = 1.27$ for the Wigner distribution (Berry and Robnik 1984).

As shown in Fig. 22.19(a), in the absence of external fields the system is regular, with chaos beginning to set in as shown in panel (b) when the magnetic field is turned on. However, for stronger magnetic fields (c) the chaotic features are eliminated by the formation of the Landau-like states. Only as we increase the electric field (panels (d) and (e)) does chaos reappear. We have also verified that the fitting parameter q systematically goes to zero as V_a is increased, demonstrating the smooth reappearance ("re-entrance") and dominance of chaos. The five regimes shown in Fig. 22.19 also mark the typical states displayed in Fig. 22.18.

22.3.2.2   *Chaos and autocorrelation functions of individual states*

It is also of interest to look at chaos as a property of *individual* states. In analogy with classical chaos, which is commonly associated with randomness and the decay of correlations, one looks here at the decay of an autocorrelation function derived from the wavefunction itself. Such an autocorrelation function can be defined (Shapiro and Goelman 1984) as,

$$F(\delta) = \int \psi^*(\mathbf{q}) \, \psi(\mathbf{q} + \delta) \, d\mathbf{q} . \qquad (22.37)$$

Here, the $\mathbf{q}$ integration is performed over a self-avoiding space-filling path. In the present application, the δ displacement vector is also directed along this path. The discrete version of this expression is

$$F_n[\psi] = \frac{1}{N} \sum_{i=1}^{N} \psi^*(r_i) \, \psi(r_{i+n}), \quad n \leq N, \qquad (22.38)$$

where $r_1, r_2 \ldots r_N$ is a cyclic ($r_{N+n} = r_n$) set of ordered points belonging to the self-avoiding space-filling path. According to the indicator proposed by Shapiro and Goelman (1984), a wavefunction ψ is termed chaotic if its autocorrelation function is aperiodic and rapidly decaying.

We have calculated $F_n[\Psi]$ for the sequence of representative states plotted in Fig. 22.18, as shown in Fig. 22.20. We find that there is indeed a very strong correlation between the visual distortion of the state as viewed in Fig. 22.18, the statistical properties of the states, viewed in Fig. 22.19, and the disappearance of the periodic features of the autocorrelation function. Erratic sequences of oscillations in $F_n[\Psi]$ are obtained either when the magnetic field is weak $B = 6$ T, or when a strong magnetic field is combined with high electric field. We have also verified that different constructions of the path lead basically to the same observations.

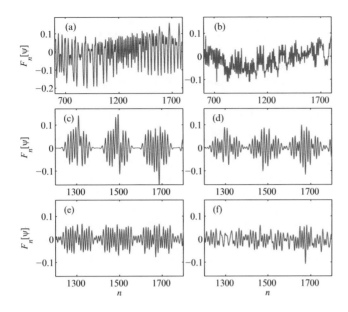

Fig. 22.20 The real part of the autocorrelation function $F_n[\Psi]$ for the states shown in panels (a)–(f) of Fig. 22.18.

22.3.3 Nano-Paul traps of single electrons

In this section we propose a novel nanoscience application of TIS formed around dc- and ac-driven carbon nanotubes, a nanoscale trap. This system, made up of four conducting nanotubes, is capable of trapping a *single* electron (Segal and Shapiro 2006). In addition to the size, the major differences between such a nanotrap and its macroscopic analog are that the electron is treated as a quantum object and that its effect ("backreaction") on the trapping device cannot be ignored. We will demonstrate focusing and trapping of an electronic *wave packet*, while fully accounting for the image charges induced by the electron on the nanotubes.

Trapping of individual charged particles in a small spatial region is a useful tool in many fields, such as, mass spectrometry, optical spectroscopy, quantum optics and metrology (Ghosh 1995). Some of the novel applications of trapped charged particles are the precise measurements of the electron magnetic moment (Dehmelt 1990) and the performance of quantum computations with a string of trapped ions (Cirac and Zoller 1995). Recent studies propose ion traps as transducers of nanowire nanomechanical oscillations (Tian and Zoller 2004; Hensinger *et al.* 2005). In the popular Penning trap the charged particles are held in a combination of electrostatic and magnetic fields (Brown 1986). In contrast, in the Paul trap, storage is achieved by means of high-frequency spatially shaped electrical fields (Paul 1990). Typical sizes of these devices are on the mm length scale, and the oscillation frequencies extend the 1–100 MHz range.

We have proposed (Segal and Shapiro 2006) a *nanosize* Paul-trap and have demonstrated computationally that it is capable of confining individual *electrons* in two or three dimensions. The nano-Paul-trap is made up of four parallel conducting nanorods, e.g., metallic CNT, spaced symmetrically about a central axis. Application of a properly tuned ac field to the nanorods can lead to two dimensional ("radial") confinement of the electrons at the trap center. Confinement along the third ("axial") direction can be attained by manipulating the structure of the tubes and/or the shapes of the end-caps.

Considering this system, some general requirements must be fulfilled. First, it should be possible to design arrays of CNT of precisely tailored dimensions. Indeed, there is a significant progress in this direction and highly ordered arrays of aligned nanotubes can be fabricated with a controlled diameter and lattice parameters (Tu *et al.* 2003). One should also be able to control the nanotubes conductive properties, i.e. if it is a metal or a semiconductor. This can be achieved utilizing new methods for non-destructive separation of semiconducting from metallic carbon nanotubes (An *et al.* 2004; Maeda *et al.* 2005). The proposed devices can be also fabricated from other metallic materials, e.g. gold (Zhang *et al.* 2001; Oshima *et al.* 2003) or palladium (Cheng *et al.* 2005), though such tube diameters are usually larger that 10 nanometers. Another basic challenge for applications is the question of populating image potential states, see discussion in Section 22.4. We proceed and discuss the design of the *nanoscale* Paul-trap.

As shown in Fig. 22.21(a), the typical dimensions of the proposed device are in the nanometric range: The distance between the tubes is $d \sim 100$ nm, each tube's radius is $a \sim 0.7$ nm, and the tube length is $L \sim 200$ nm. These requirements are within present-day technology (Tu *et al.* 2003). The proposed device has to operate at very high frequencies, 100 GHz – 1 THz, which recent studies of the ac performance of CNT suggest are realizable (Yu and Burke 2005). As discussed in previous sections an electron hovering in the vicinity of four nanotubes is electrostatically attracted to the image charges it creates on the nanotubes (Granger *et al.* 2002; Segal *et al.* 2005a). In the context of the nanotrap, the image potential has a profound effect on the electron dynamics,

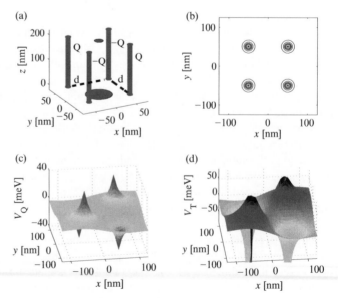

Fig. 22.21 (a) A schematic illustration of a nanotrap composed of four parallel tubes aligned in the z-direction. The large circle represents the size of the incoming electronic wave packet and the small circle the size of the trapped wave packet. (b) A contour plot of $V_s(x, y)$, the image-potential function. (c) A surface plot of $V_Q(x, y)$, the charging potential function, for $Q/L \equiv \lambda = 0.005$ e/nm. (d) The total potential energy function for $\lambda = 0.02$ e/nm. $z = 100$ nm in panels (b)–(d).

resulting, as we show below, among other things, in the *reduction* of the size of the trapping stability regions.

We have calculated numerically the image potential interaction energy $V_s(x, y, z)$, including the electrostatic interaction between the rods, using the method of Segal *et al.* (2004a). A contour plot of this potential using the typical dimensions given above is displayed in Fig. 22.21(b). The depth of the potential at the center point is ~ -0.5 meV, while near the surface of each it is as deep as -250 meV. It can be shown that far from the tubes ends, at $z \sim 100$ nm, the axial variation of the potential is weak in comparison to the radial gradient (Segal *et al.* 2004a). We can thus ignore the axial dependence of the potential, which is permissible if the tubes are long enough.

As shown in Fig. 22.21(a), when we connect the four tubes to an external (ac or dc) source we form a quadrupole in which tube no. 1 and its diagonal counterpart, tube no. 4, are charged by $+Q$ and tube no. 2 and diagonal counterpart, tube no. 3, by $-Q$. We can calculate $V_T(x, y, z)$, the total interaction energy of a charged particle with the nanoquadrupole in the presence of image charges by using a simple extension of our numerical procedure (Segal *et al.* 2004a). It is convenient to decompose V_T into two terms, $V_T = V_s + V_Q$, where V_s, which is independent of Q, is the image potential discussed above, and V_Q, the "charging" potential, is the additional potential resulting from the added quadrupolar charges. We have verified that V_Q scales linearly with the extra charge, i.e. that $V_{\alpha Q} \sim \alpha V_Q$ for distances $\rho > 5$ nm away from the tubes, and that it depends very weakly on the axial coordinate. The charging potential has an approximate quadrupolar spatial shape, as shown in Fig. 22.21(c). As shown in panel (d) the total potential V_T maintains, to a good approximation, the quadrupolar-like form. We expect that in the dc charging case, the migration of the electron to one of the tubes would lead to its eventual decay onto the surface (Segal *et al.* 2005b). In contrast, as in conventional linear Paul-traps (Paul 1990), in the ac-charging case the saddle point at the center should provide a point of stability where the electron can be confined.

We can also derive approximate analytic expressions for the image potential and the charging energy. Neglecting tube–tube polarization effects, we approximate the image potential for *infinitely* long tubes (Granger *et al.* 2002) as,

$$V_s(x, y) \approx \sum_{j=1..4} \frac{2e^2}{\pi a} \sum_{n=1,3,5,\dots} \mathrm{li}\left[(a/\rho_j)^n\right], \quad \mathrm{li}(x) \equiv \int_0^x \frac{dt}{\ln(t)}, \quad (22.39)$$

where ρ_j ($j = 1..4$) is the distance of the electron from the jth tube's center. The additional charging energy is given by (Slater and Frank 1947)

$$V_Q(x, y) \approx eV_a \frac{\ln(\rho_1\rho_4/\rho_2\rho_3)}{\ln\left(\sqrt{2}a/d\right)}, \quad (22.40)$$

where V_a ($-V_a$) is the potential applied to the 1,4 (2,3) pair. This formula goes over to the correct limit when the electron is placed on the surface of a tube. It can also be expressed in terms of the charge density on the tubes, $\lambda = Q/L$. Assuming that close to each tube surface the electric field approaches the $E_\rho = \lambda/\rho$ limiting value, valid for an infinitely long, uniformly charged wire, we find that $V_Q(x, y) \approx e\lambda \ln (\rho_2\rho_3/\rho_1\rho_4)$. This analytic expression is in good agreement with the potential obtained numerically.

We next allow for an AC charging at frequency Ω. The total potential is now a sum of a static term and an AC term,

$$V_T(x, y, t) = V_s(x, y) + V_Q(x, y) \cos(\Omega t). \qquad (22.41)$$

A more general potential could be devised by including an additional dc term, $cV_Q(x, y)$. Here, we limit our investigation to the $c = 0$ case.

The dynamics of a single electron subject to this potential is obtained by solving the time-dependent Schrödinger equation

$$i\hbar\partial_t \Psi(x, y, t) = \left[\frac{-\hbar^2}{2m_e} \left(\partial_x^2 + \partial_y^2 \right) + V_T(x, y, t) \right] \Psi(x, y, t), \qquad (22.42)$$

where m_e is the electron mass. Unlike the case of a pure quadrupole (Paul 1990), and due to the presence of the image charge potential (V_s), this equation is non-separable in the x-, y-co-ordinates and the analytical results of Brown (1991) cannot be applied. Given eqn (22.42), the propagation of an initial wave packet is done using the unitary split evolution operator (Balakrishnan *et al.* 1997). In order to eliminate artificial reflections of the wavefunction from the boundaries we introduce into the Hamiltonian an optical potential that absorbs the scattered wavelets before they reach the grid boundaries (Neuhauser and Baer 1989).

We demonstrate trapping and focusing of a single electron in the nanotrap. Fig. 22.22 displays the time evolution of an initial (highly localized) Gaussian wave packet

$$\Psi(x, y, t = 0) = \exp[-(x^2 + y^2)/2\sigma^2]/\sqrt{\pi\sigma^2},$$

of width $\sigma = 10.5$ nm, injected into the trap, subject to different charging potentials and ac modulation frequency of $\nu \equiv 1/T \equiv \Omega/2\pi = 8.3 \times 10^{11}$ Hz. The spatial parameters are as in Fig. 22.21. We see that the wave packet evolves into distinct shapes for different values of the applied potential: In the absence of the ac modulation the electron accumulates in the vicinity of the tubes, as shown in Figs. 22.22(a–d). For the case of weak charging potentials, shown in Figs. 22.22(e–h), the electron is neither trapped nor is it strongly attracted to the tubes. Rather, the wave packet slowly spreads while filling the entire intertube space. Tight focusing occurs when the charging is high enough as in Figs. 22.22(i–l), with the wave packet oscillating in the middle of the trap between its original shape and an elliptical, yet highly focused, shape. When the charging becomes too high, confinement is lost, as shown in Figs. 22.22(m–p). The electron, though avoiding the tubes, manages to leak out by following the $x = 0$ and $y = 0$ lines. Similarly, tuning the trap parameters we could also demonstrate *focusing* of electronic wave packets (Segal and Shapiro 2006).

A full characterization of the trap is presented in Fig. 22.23 where the stability diagram for different frequencies and charging potentials is provided. The diagram was generated by propagating for long times ($\sim 10T$) initially localized or unlocalized wave packets and recording whether at the end of

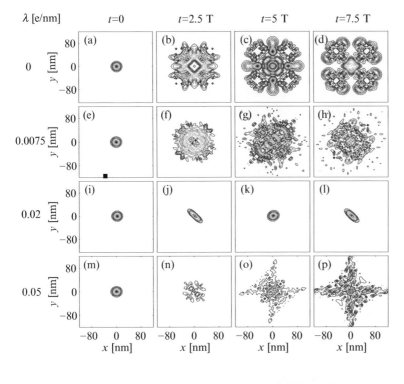

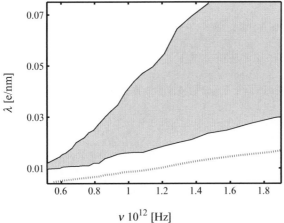

Fig. 22.22 Electron probability density for different charging potentials. (a)–(d) $\lambda = 0$; (e)–(h) $\lambda = 0.0075\,\text{e/nm}$; (i)–(l) $\lambda = 0.02\,\text{e/nm}$; (m)–(p) $\lambda = 0.05\,\text{e/nm}$. The calculation is performed in a $x - y$ $250 \times 250\,\text{nm}^2$ lattice and for $\nu = 8.3 \times 10^{11}\,\text{Hz}$.

Fig. 22.23 A stability map of the nanotrap. Trapping exists in the shaded area. The dotted line marks the lower border of the stability region when the contribution of the image potential is removed from the Hamiltonian. This removal has no effect on the upper border line.

the integration time the wavefunction remains confined at the trap center. We ascertained that the results are independent of the wavepacket's initial shape. Note that due to the image potential, the range of stability is *decreased*. This effect ultimately limits the operation of very small traps to $d > 10\,\text{nm}$.

Finally, we note that an *axial* localization of the electron in the trap results naturally from the potential barriers imposed by the tube's end-caps (Segal *et al.* 2004a). This causes the electron to oscillate in both radial and axial directions at drastically different frequencies. Whereas the radial vibrational frequency is ~500 GHz, $\nu_z \sim 10$ GHz for the axial motion (Segal *et al.* 2004a).

Strong axial confinement can also be achieved by applying a dc repulsive field on additional ring-shaped electrodes at the tubes ends (Paul 1990).

A single trapped electron is a promising candidate for a variety of quantum-information applications utilizing its spin states, the *anharmonic* quasi-energy spectrum (Mancini *et al.* 2000), and its coupling to the nanowires mechanical modes (Tian and Zoller 2004). Its dynamics can be manipulated either by magnetic fields or through coherent laser excitations.

Of even greater interest would be *arrays* of electrons (Ciaramicoli *et al.* 2003) trapped by a 2D lattice of quartets of nanotraps of the type described here. Such arrays can be used as a set of entangled qubits, where the degree of entanglement would be controlled by varying the ac charging, thereby allowing a greater or lesser portion of the electronic population to leak into neighboring cells and interact with other electrons confined there.

22.4   Experimental verifications

The first experimental evidence of an energy state within the bulk bandgap, close to the vacuum level, was reported by Johnoson *et al.* (1983) for Cu and Ni surfaces. Since then, the experimental exploration of image potential states has been mainly confined to metal surfaces, successfully characterizing the states in energy and momentum space (Echenique *et al.* 2004). A *time-domain* analysis of electron dynamics at surfaces was done by Höfer *et al.* (1997), combining photoelectron spectroscopy with ultrafast laser excitation. In this work the dynamics of image-potential states in front of clean and adsorbate-covered Cu(100) metal surfaces was studied, manifesting the coherent oscillation of the electron back and forth from the surface. Image potential states were also observed in a variety of nanoscale systems including molecular wires (Ortega *et al.* 1994), metal nanoclusters (Kasperovich *et al.* 2000; Boyle *et al.* 2001), and liquid He (Plazman and Dykman 1999; Glasson *et al.* 2001). As discussed above, the study of image potential states in carbon nanotubes is of great interest because the cylindrical symmetry, which gives rise to the repulsive centrifugal force, significantly modifies the image states from those formed above flat surfaces.

The experimental observation of high angular momenta TIS turned out to be challenging. Since these states are formed far from the surface, the nanotube has to be isolated from any source of interaction, such as the substrate or other nanotubes. Synthesis of the adequate sample with the required characteristics is complicated, especially since SWNT tend to form bundles. The experimental work therefore focused on *surface states*, i.e. image states with low *l* values.

The first experimental evidence for the existence of *low angular momentum* image-potential states in CNT was given by Zamkov *et al.* (2004a). In order to present this work we first clearly distinguish between two classes of states formed around nanotubes: (i) Tubular image states (TIS) where the electron is localized between the attractive image potential and the centrifugal barrier, and (ii) Surface states, where the states are confined between the induced potential and the repulsive potential barrier inside the surface arising from quantum correlation and exchange interactions. As discussed in Section 22.2.1, for the (10,10) single-wall nanotube with a radius of $a = 0.68$ nm, TIS have angular momentum of $\sim l > 6$, and they are located ~ 50 nm away from the surface, causing their long, microsecond scale, lifetimes. In contrast, nanotube surface states occur closer to the surface, at $\rho < 10$ nm, and have small l values (Zamkov *et al.* 2004b). Due to their weak transverse penetration into the bulk, the lifetimes of the surface states are expected to be longer than those of image states above flat surfaces, but not as long as those of the high-l TIS.

A comprehensive model describing surface states and image states in carbon nanotubes was developed by Zamkov *et al.* (2004b). The effective image-state electron–tube interaction potential combines three different terms: induced image potential, centrifugal interaction, and surface short-range potential. For an infinitely long, homogeneous, and metallic tube, the first two terms are essentially identical to the form adopted above for high-l TIS,

$$V_{\text{eff}}^{(0)}(\rho) = \frac{2}{\pi}\frac{e^2}{a} \sum_{n=1,3,5\ldots} li[(a/\rho)^n] + \frac{\hbar^2 \left(l^2 - 1/4\right)}{2m_{\text{e}}\rho^2}, \qquad (22.43)$$

where l is the angular momenta of the state, and m_{e} is the electron's mass. The main difference with high-l TIS lies in the form of the short-range potential near the surface. It is obtained by modifying the Jellium barrier model for the metallic film of Jennings *et al.* (1988), which is essentially an analytic fit of density-functional calculations, to the case of a cylindrical shell. The full effective potential then becomes (Zamkov *et al.* 2004b),

$$V_{\text{eff}}(\rho) = \begin{cases} V_{\text{eff}}^{(0)}(\rho^+)\left[1 - e^{-\lambda|\rho - \rho_0^-|^\nu}\right] & \text{if } \rho \le \rho_0^-; \\[2mm] -U_0 \Big/ \left[A_l e^{-\beta_l|\rho - \rho_0^\pm|} + 1\right] & |\rho - a| \le \Delta R/2; \qquad (22.44) \\[2mm] V_{\text{eff}}^{(0)}(\rho^-)\left[1 - e^{-\lambda|\rho - \rho_0^+|^\nu}\right] & \text{if } \rho > \rho_0^+. \end{cases}$$

Here, ΔR is the shell thickness, $\rho^\pm = \rho \pm \Delta R/2$, and the coefficients A_l and β_l are determined via the matching of the image reference planes located on either sides of the tubes walls $\rho_0^\pm = a \pm \Delta R/2$. The other parameters here are λ, which determines the transition range $(1/\lambda)$ over which the barrier saturates, and U_0, which defines the depth of the Jellium potential. The adjustable parameter ν can be determined numerically by requiring that the potential is continuous in the vicinity of ρ_0. Using this form, the total effective potential for low angular momenta values can be calculated, for SWNT as well as for MWNT. Fig. 22.24 depicts the $n = 1, 2, 3$ image potential wavefunctions for a 19-walled $d = 14.2$ nm CNT with $l = 1$ (Zamkov *et al.* 2004b), where d is the tube diameter. The lower panel displays the total effective potential in the radial direction.

The energetics and the dynamics of this system were recently measured by Zamkov *et al.* (2004a), using MWNT with $d = 10$–20 nm. First, a UV pump photon with an energy of 4.71 eV, higher than the sample work function, photoexcites an electron out of an occupied state below the Fermi energy into

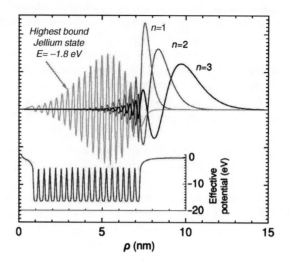

Fig. 22.24 Calculated wavefunctions for the lowest three image potential states calculated for 19-walled ($d = 14.2$ nm) MWNT with $l = 1$. The lower panel displays the total effective potential in the radial direction. Taken with permission from Zamkov *et al.* (2004b).

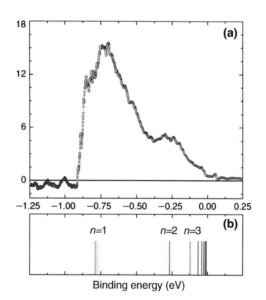

Fig. 22.25 (a) The photoelectron signal originating from image potential states plotted vs. electron-binding energies. (b) Theoretical calculations of binding energies, $l \leq 4$. Taken with permission from Zamkov *et al.* (2004a).

an image potential state. Then, a second (IR) 1.57-eV photon promotes the image state electron above the vacuum level. The kinetic-energy distribution of the electron is then measured. The output includes both binding energies and the temporal dynamics of photoexcited electrons. Figure 22.25 presents the photoelectron signal, originating from low-l image potential states, plotted vs. electron-binding energies. The experimental range of the binding energies is in excellent agreement with the model calculations (Zamkov *et al.* 2004a): The $n = 1$ states are clearly observed around -0.75 eV, while the low-energy side can be attributed to image electrons with $n = 2, 3 \ldots$. Time-domain measurements of the electron dynamics yield lifetimes of the order of 200 fs, significantly longer than the 40 fs lifetime of the $n = 1$ state above a graphite surface. This hints at a substantial difference in electron decay dynamics in tubular and planar graphene sheets.

We expect that in the future, with improvement in the nanotube synthesis and purification technology, detection of *high-l* image states around single tubes will be possible. Observation of these states around isolated double-walled nanotubes as well as in bundles of single-walled nanotubes would yield additional valuable information.

22.5 Summary

Rydberg-like tubular image states above nanostructures represent a new topic of interest with unique potential applications in optics, electronics, and surface science. The ability to manipulate these states by light fields can be exploited for quantum-information storage and retrieval. Study of the localization properties and dynamics of TIS can provide important information about processes occurring on the nanotube surface. In this review we have discussed the ongoing research in this field and have given a detailed theoretical description of the states around different tubes geometries. We have outlined methods for controlling the states, and have described the onset of chaos in the system. A detailed study of the states relaxation mechanism and resulting lifetimes has also been given. We have also described in detail the design of a nanoscale single-electron trap, made of four conducting nanotubes driven by dc and ac fields. Some ongoing experimental efforts and future directions have been described. Our methods can easily be generalized to include other nanorods structures that can be presently fabricated (Kempa *et al.* 2003), and to describe confinement of other charged particles.

Acknowledgments

A fruitful collaboration and many helpful discussions over the years with B.E. Granger and H.R. Sadeghpour are gratefully acknowledged.

References

Alhassid, Y. *Rev. Mod. Phys.* **72**, 895 (2000).

An, L., Fu, Q., Lu, C., Liu, J. *J. Am. Chem. Soc.* **126**, 10520 (2004).

Ando, T., Nakanishi, T., Saito, R. *J. Phys. Soc. Jpn.* **67**, 2857 (1998).

Arista, N.R., Fuentes, M.A. *Phys. Rev. B* **63**, 165401 (2001); Arista, N.R. *Phys. Rev. A* **64**, 032901 (2001).

Ashcroft, N.W., Mermin, N.D. *Solid State Physics* (Harcourt College Publishers, 1976).

Balakrishnan, N., Kalyanaraman, C., Sathyamurthy, N. *Phys. Rep.* **280**, 79 (1997).

Berry, M.V., Robnik, M. *J. Phys. A* **17**, 2413 (1984).

Bohigas, O., Giannoni, M.J., Schmidt, C. *Phys. Rev. Lett.* **52**, 1 (1984).

Boyle, M., Hoffmann, K., Schulz, C.P., Hertel, I.V., Levine, R.D., Campbell, E.E.B. *Phys. Rev. Lett.* **87**, 273401 (2001).

Brown, L.S. *Phys. Rev. Lett.* **66**, 527 (1991).

Brown, L.S., Gabrielse, G. *Rev. Mod. Phys.* **58**, 233 (1986).

Casati, G., Chirikov, B. *Quantum Chaos* (Cambridge University Press, New York, 1995).

Cheng, C., Gonela, R.K., Gu, Q., Haynie, D.T. *Nano Lett.* **5**, 175 (2005).

Chico, L., Benedict, L.X., Louie, S.G., Cohen, M.L. *Phys. Rev. B* **54**, 2600 (1996).

Chirikov, B.V. *J. Stat. Phys.* **3**, 307 (1971).

Ciaramicoli, G., Marzoli, I., Tombesi, P. *Phys. Rev. Lett.* **91**, 017901 (2003).

Cirac, J.I., Zoller, P. *Phys. Rev. Lett.* **74**, 4091 (1995).

Chulkov, E.V., Sarria, I., Silkin, V.M., Pitarke, J.M., Echenique, P.M. *Phys. Rev. Lett.* **80**, 4947 (1998).

Colbert, D.T., Miller, W.H. *J. Chem. Phys.* **96**, 1982 (1992).

Dehmelt, H. *Rev. Mod. Phys.* **62**, 525 (1990).

Dresselhaus, M.S., Dresselhaus, G., Avouris, Ph. *Carbon Nanotubes: Synthesis, Structure, Properties and Applications* (Springer-Verlag, Berlin, 2001).

Echenique, P.M., Berndt, R., Chulkov, E.V., Fauster, Th., Goldmann, A., Höfer, U. *Surf. Sci. Rep.* **52**, 219 (2004).

Echenique, P.M., Flores, F., Sols, F. *Phys. Rev. Lett.* **55**, 2348 (1985).

Fromhold, T.M., Eaves, L., Sheard, F.W., Leadbeater, M.L., Foster, T.J., Main, P.C. *Phys. Rev. Lett.* **72**, 2608 (1994).

Fromhold, T.M., Wilkinson, P.B., Sheard, F.W., Eaves, L., Miao, J., Edwards, G. *Phys. Rev. Lett.* **75**, 1142 (1995).

Fromhold, T.M., Krokhin, A.A., Tench, C.R., Bujkiewicz, S., Wilkinson, P.B., Sheard, F.Q., Eaves, L. *Phys. Rev. Lett.* **87**, 046803 (2001).

Fromhold, T.M., Patane, A., Bujkiewicz, S., Wilkinson, P.B., Fowler, D., Sherwood, D., Stapleton, S.P., Krokhin, A.A., Eaves, L., Henini, M., Sankeshwar, N.S., Sheard, F.W. *Nature* **428**, 726 (2004).

Gallagher, T. *Rydberg Atoms* (Cambridge University Press, New York, 1994).

Ghosh, P.K. *Ion Traps* (Clarendon Press, Oxford, 1995).

Glasson, P., Dotsenko, V., Fozooni, P., Lea, M.J., Bailey, W., Papageorgiou, G., Andresen, S.E., Kristensen, A. *Phys. Rev. Lett.* **87**, 176802 (2001).

Gomez-Navarro, C., De Pablo, P.J., Gomez-Herrero, J., Biel, B., Garcia-Vidal, F.J., Rubio, A., Flores, F. *Nat. Mater.* **4**, 534 (2005).

Granger, B.E., Král, P., Sadeghpour, H.R., Shapiro, M. *Phys. Rev. Lett.* **89**, 135506-1 (2002).

Greene, C.H., Dickinson, A.S., Sadeghpour, H.R. *Phys. Rev. Lett.* **85**, 2458 (2000).

Gumbs, G., Balassis, A. *Phys. Rev. B* **71**, 235410 (2005).

Gumbs, G., Balassis, A., Fekete, P. *Phys. Rev. B* **73**, 075411 (2006).

Gutzwiller, M.G. *Chaos in Classical and Quantum Mechanics* (Springer-Verlag, New York, 1990).

Haake, F. *Quantum Signatures of Chaos* (Springer-Verlag, Berlin, 2000).

Hanson, R., Witkamp, B., Vandersypen, L.M.K., Willems van Beveren, L.H., Elzerman, J.M., Kouwenhoven, L.P. *Phys. Rev. Lett.* **91**, 196802 (2003).

Hensinger, W.K., Utami, D.W., Goan, H.-S., Schwab, K., Monroe, C., Milburn, G.J. *Phys. Rev. A* **72**, 041405(R) (2005).

Höfer, U., Shumay, I.L., Reuβ, Ch., Thomann, U., Wallauer, W., Fauster, Th. *Science* **277**, 1480 (1997).

Iijima, S. *Nature* **354**, 56 (1991).

Jackson, J.D. *Classical Electrodynamics*, 2nd edn (John Wiley & Sons, New York, 1975).

Jennings, P.J., Jones, R.O., Weinert, M. *Phys. Rev. B* **37**, 6113 (1988).

Jiang, J., Saito, R., Grüneis, A., Dresselhaus, G., Dresselhaus, M.S. *Chem. Phys. Lett.* **392**, 383 (2004).

Joannopoulos, J.D., Villeneuve, P.R., Fan, S. *Nature* **390**, 143 (1997).

Johnson, P.D., Smith, N.V. *Phys. Rev. B* **27**, 2527 (1983).

Kanfer, S., Shapiro, M. *J. Phys. Chem.* **88**, 3964 (1984).

Kasperovich, V., Wong, K., Tikhonov, G., Kresin, V.V. *Phys. Rev. Lett.* **85**, 2729 (2000).

Kempa, K., Kimball, B., Rybczynski, J., Huang, Z.P., Wu, P.F., Steeves, D., Sennett, M., Giersig, M., Rao, D.V.G.L.N., Carnahan, D.L., Wang, D.Z., Lao, J.Y., Li, W.Z., Ren, Z.F. *Nano Lett.* **3**, 13 (2003).

Kikkawa, J.M., Smorchkova, I.P., Samarth, N., Awschalom, D.D. *Science* **277**, 1284 (1997).

Kleber, X., Gusev, G.M., Gennser, U., Maude, D.K., Portal, J.C., Lubyshev, D.I., Basmaji, P., Silva, M. de P.A., Rossi, J.C., Nastaushev, Yu.V. *Phys. Rev. B* **54**, 13859 (1996).

Král, P. *Chem. Phys. Lett.* **382**, 399 (2003).

Landau, L.D., Lifshitz, E.M. *Quantum Mechanics (Non-relativistic theory)*, (Butterworth-Heinemann, 1998).

Light, J.C., Carrington, T. *Adv. Chem. Phys.* **114**, 263 (2000).

Loly, P.D., Pendry, J.B. *J. Phys. C* **16**, 423 (1983).

MacDonald, A.H., Ritchie, D.S. *Phys. Rev. B* **33**, 8336 (1986).

Maeda, Y., Kimura, S., Kanda, M., Hirashima, Y., Hasegawa, T., Wakahara, T., Lian, Y.F., Nakahodo, T., Tsuchiya, T., Akasaka, T., Lu, J., Zhang, X.W., Gao, Z.X., Yu, Y.P., Nagase, S., Kazaoui, S., Minami, N., Shimizu, T., Tokumoto, H., Saito, R. *J. Am. Chem. Soc.* **127**, 10287 (2005).

Mahan, G.D. *Many Particle Physics* (Plenum Press, New York, 1993).

Mancini, S., Martins, A.M., Tombesi, P. *Phys. Rev. A* **61**, 012303 (2000).

Memmel, N., Bertel, E. *Phys. Rev. Lett.* **75**, 485 (1995).

Milczewski, J., Uzer, T. *Phys. Rev. E* **55**, 6540 (1997).

Miller, A.D., Bezel, I., Gaffney, K.J., Garrett-Roe, S., Liu, S.H., Szymanski, P., Harris, C.B. *Science* **297**, 1163 (2002).

Neuhasuer, D., Baer, M. *J. Chem. Phys.* **90**, 4351 (1989).

Neumann, C., Ubert, R., Freund, S., Flothmann, E., Sheehy, B., Welge, K.H., Haggerty, M.R., Delos, J.B. *Phys. Rev. Lett.* **78**, 4705 (1997).

Ortega, J.E., Himpsel, F.J., Haight, R., Peale, D.R. *Phys. Rev. B* **49**, 13859 (1994).

Oshima, Y., Onga, A., Takayanagi, K. *Phys. Rev. Lett.* **91**, 205503 (2003).

Paul, W. *Rev. Mod. Phys.* **62**, 531 (1990).

Pichler, T., Knupfer, M., Golden, M.S., Fink, J., Rinzler, A., Smalley, R.E. *Phys. Rev. Lett.* **80**, 4729 (1998).

Plazman, P.M., Dykman, M.I. *Science* **284**, 1967 (1999).

Qian, Z., Sahni, V. *Phys. Rev. B* **66**, 205103 (2002).

Robinson, M.P., Tolra, B.L., Noel, M.W., Gallagher, T.F., Pillet, P. *Phys. Rev. Lett.* **85**, 4466 (2000).

Saito, R., Takeya, T., Kimura, T., Dresselhaus, G., Dresselhaus, M.S. *Phys. Rev. B* **57**, 4145 (1998).

Segal, D., Král, P., Shapiro, M. *Phys. Rev. B* **69**, 153405 (2004a).

Segal, D., Král, P., Shapiro, M. *Chem. Phys. Lett.* **392**, 314 (2004b).

Segal, D., Granger, B.E., Sadeghpour, H.R., Král, P., Shapiro, M. *Phys. Rev. Lett.* **94**, 016402 (2005a).

Segal, D., Král, P., Shapiro, M. *Surf. Sci.* **577**, 86 (2005b).

Segal, D., Král, P., Shapiro, M. *J. Chem. Phys.* **122**, 134705 (2005c).

Segal, D., Shapiro, M. *Nano Lett.* **6**, 1622 (2006).

Shapiro, M., Goelman, G. *Phys. Rev. Lett.* **53**, 1714 (1984).

Slater, J.C., Frank, N.H. *Electromagnetism* (McGraw-Hill, NY, 1947).

Suzuura, H., Ando, T. *Phys. Rev. B* **65**, 235412 (2002).

Taverna, D., Kociak, M., Charbois, V., Henrard, L. *Phys. Rev. B* **66**, 235419 (2002).

Tian, T., Zoller, P. *Phys. Rev. Lett.* **93**, 266403-1 (2004).

Tu, Y., Lin, Y., Ren, Z.F. *Nano Lett.* **3**, 107 (2003).

Weiss, D., Roukes, M.L., Menschig, A., Grambow, P., von Klitzing, K., Weimann, G. *Phys. Rev. Lett.* **66**, 2790 (1991).

Wunner, G., Woelk, U., Zech, I., Zeller, G., Ertl, T., Geyer, F., Schweitzer, W., Ruder, H. *Phys. Rev. Lett.* **57**, 3261 (1986).

Yang, X.L., Guo, S.H., Chan, F.T., Wong, K.W., Ching, W.Y. *Phys. Rev. A* **43**, 1186 (1991).

Yu, Z., Burke, P.J. *Nano Lett.* **5**, 1403 (2005).

Zabala, N., Ogando, E., Rivacoba, A., Garcia de Abajo, F.J. *Phys. Rev. B* **64**, 205410-1 (2001).

Zamkov, M., Woody, N., Shan, B., Chakraborty, H.S., Chang, Z., Thumm, U., Richard, P. *Phys. Rev. Lett.* **93**, 156803-1 (2004a).

Zamkov, M., Chakraborty, H.S., Habib, A., Woody, N., Thumm, U., Richard, P. *Phys. Rev. B* **70**, 115419 (2004b).

Zhang, X.Y., Zhang, L.D., Chen, W., Meng, G.W., Zheng, M.J., and Zhao, L.X. *Chem. Mater.* **13**, 2511 (2001).

Zheng, L.X., O'Connell, M.J., Doorn, S.K., Liao, X.Z., Zhao, Y.H., Akhadov, E.A., Hoffbauer, M.A., Roop, B.J., Jia, Q.X., Dye, R.C., Peterson, D.E., Huang, S.M., Liu, J., Zhu, Y.T. *Nat. Mater.* **3**, 673 (2004).

Zhu, J.-L., Cheng, Y., Xiong, J.-J. *Phys. Rev. B* **41**, 10792 (1990).

Correlated electron transport in molecular junctions

23

K.S. Thygesen and A. Rubio

23.1   Introduction

The dimensions of conventional silicon-based electronics devices will soon be so small that quantum effects, such as electron tunnelling and energy quantization, will begin to influence and eventually limit the device functionality. On the other hand, molecular electronics is based on the idea of constructing electronic devices from the bottom up using organic molecules as basic building blocks and in this way integrate the quantum nature of the charge carriers directly in the design (Joachim *et al.* 2000; Cuniberti *et al.* 2005). Over the last decade it has become possible to capture individual nanostructures between metal contacts and measure the electrical properties of the resulting junction. The types of nanostructures vary all the way from a single hydrogen molecule (Smit *et al.* 2002) through organic molecules (Reichert *et al.* 2002) to metallic chains of single atoms (Yanson *et al.* 1998) to carbon nanotubes suspended over several nanometers (Nygard *et al.* 2000) and inorganic nanowires and biochromophores (Cuniberti *et al.* 2005). The physics of these systems is highly non-classical showing intriguing phenomena such as quantized conductance, conductance oscillations, strong electron–phonon coupling, Kondo physics, and Coulomb blockade. For this reason a microscopic, i.e. quantum-mechanical, understanding of nanoscale systems out of equilibrium is fundamental for the future development of molecular electronics.

The theoretical description of electron transport in molecules (we often use the term molecule to cover a general nanostructure) represents a central challenge in computational nanoscience. In principle, the problem involves an open quantum system of electrons interacting with each other and the surrounding nuclei under the influence of an external bias voltage. Fortunately, due to the large difference in mass between electrons and nuclei, it is often a good approximation to regard the nuclei as classical charges fixed in their equilibrium positions—at least for sufficiently low temperature and bias voltage. This reduces the problem to interacting electrons moving

772 Correlated electron transport in molecular junctions

through the static potential created by the frozen lattice of nuclei. Further simplification is obtained by replacing the electron–electron interactions by a mean-field potential, as is done in Hartree–Fock (HF) and Kohn–Sham (KS) theory. Within such independent-particle approximations Landauer's formula (Landauer 1970; Datta 1995) applies, giving the conductance as the (elastic) transmission probability for electrons at the Fermi level times the conductance unit $G_0 = 2e^2/h$. Landauer's formula and, in particular, its equivalent formulation in terms of non-equilibrium Green's functions (NEGF) (Meir and Wingreen 1992), has formed the basis for almost all calculations of quantum transport in nanoscale systems.[1] First-principles calculations are usually based on the KS scheme of density-functional theory (DFT) with a local exchange-correlation functional (Taylor *et al.* 2001; Brandbyge *et al.* 2002). These DFT transport schemes have been successfully applied to systems characterized by strong coupling between the molecule and the electrodes (Garci-Suarez *et al.* 2005; Thygesen and Jacobsen 2005), but systematically overestimates the conductance of weakly coupled systems (Heurich *et al.* 2002; Quek *et al.* 2007). Recently, it has been shown that the use of self-interaction corrected exchange-correlation functionals improves the agreement with experiments for such systems (Toher *et al.* 2005). However, such functionals contain parameters that basically control the size of the energy gap between the highest-occupied (HOMO) and lowest-unoccupied (LUMO) molecular orbitals, which questions the predictive power of such an approach.

Apart from the problems related to the position of molecular energy levels, there are a number of electronic effects originating from the two-body nature of the electron–electron interaction, which cannot—even in principle—be described within a single-particle picture. These include strong correlation effects like Kondo effects and Coulomb blockade (Costi *et al.* 1994; Goldhaber *et al.* 1998), renormalization of molecular levels by dynamic screening (Kubatkin *et al.* 2003; Neaton *et al.* 2006), and lifetime reduction due to quasiparticle scattering (Thygesen 2008). As we shall see in this chapter, such dynamic correlation effects can have a dramatic influence on the electrical properties of molecular junctions—in particular far from equilibrium.

In this chapter we describe how electronic correlation effects can be included in transport calculations using many-body perturbation theory within the Keldysh non-equilibrium Green's function formalism. Specifically, we use the so-called GW self-energy method (G denotes the Green's function and W is the screened interaction) which has been successfully applied to describe quasi-particle excitations in periodic solids (Hybertsen and Louie 1986; Onida *et al.* 2002). To make the problem tractable, we limit the GW description to a central region containing the nanostructure of interest and part of the leads, while the (rest of the) metallic leads are treated at a mean-field level. The rationale behind this division is that the transport properties to a large extent are determined by the narrowest part of the conductor, i.e. the molecule, while the leads mainly serve as particle reservoirs (a proper inclusion of substrate polarization effects require that a sufficiently large part of the leads are included in the central region). The use of non-equilibrium many-body perturbation theory is only one out of several methods to include correlation effects in quantum transport. In another approach the density matrix is

[1] The NEGF approach to quantum transport is also discussed in chapters 1, 3 and 24 of this volume.

obtained from a many-body wavefunction and the non-equilibrium boundary conditions are invoked by fixing the occupation numbers of left- and right going states (Delaney and Greer 2004). Exact diagonalization within the molecular subspace has been combined with rate equations to calculate tunnelling currents to first order in the lead–molecule coupling strength (Hettler *et al.* 2003). The linear response conductance of jellium quantum point contacts has been addressed on the basis of the Kubo formula that in principle allows correlation effects to be incorporated through the response function (Bokes *et al.* 2007). The time-dependent version of density-functional theory has also been used as a framework for quantum transport (Di Ventra and Todorov 2004; Stefanucci *et al.* 2007). This scheme is particularly useful for simulating transients and high-frequency ac responses and can in principle include correlations via non-adiabatic exchange-correlation kernels.

In Section 23.2 we formulate the quantum-transport problem and give a brief introduction to the non-equilibrium Green's function formalism. In Section 23.3 we present the non-equilibrium GW equations and discuss the important concept of conserving approximations. In Section 23.4.1 we obtain an expression for the current within the NEGF formalism that holds for interactions in the central region. It is demonstrated, both analytically and by numerical examples, that a self-consistent evaluation of the GW self-energy is fundamentally important for non-equilibrium transport as it—in contrast to the popular non-self-consistent approach—ensures the validity of the continuity equation. In Section 23.5, with the aim of identifying universal trends, we study a generic two-level model of a molecular junction. It is demonstrated how dynamic polarization effects renormalize the molecular levels, and a physical interpretation in terms of constrained total energy differences is provided. The application of a bias voltage is shown to enhance the dynamic polarization effects. Moreover, quasi-particle scattering becomes increasingly important at larger bias, leading to a significant broadening of the molecular resonances. These effects, which are all beyond the single-particle approximation, have a large impact on the calculated I–V curve. In Section 23.6 we combine the GW-transport scheme with DFT (for the leads) and a Wannier function basis set, and apply it to two prototypical junctions, namely a benzene molecule coupled to featureless leads and a hydrogen molecule between infinite Pt chains, and the results are analyzed using the knowledge obtained from the two-level model. It is found that non-self-consistent $G_0 W_0$ calculations depend crucially on the G_0 (whether it is the Hartree–Fock or Kohn–Sham Green's function). This together with its non-conserving nature suggests that GW-transport calculations should be performed self-consistently.

This chapter is a summary of recent work by the authors on incorporating many-body correlation effects in quantum transport, see Thygesen and Rubio (2007, 2008, 2009) and Thygesen (2008).

23.2 Formalism

In this section we formulate the quantum-transport problem and review the elements of the Keldysh–Green's function theory needed for its solution. For

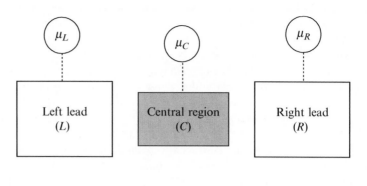

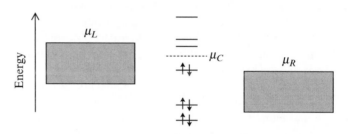

Fig. 23.1 Before the coupling between the leads and central region is established, the three subsystems are in equilibrium with chemical potentials μ_L, μ_T, and μ_C, respectively. Reprinted with permission from Thygesen and Rubio, *Phys. Rev. B* **77**, 115333 (2008). Copyright 2008 by the American Physical Society.

more detailed introductions to the subject we refer to the books by Leeuwen *et al.* (2006) and Haug and Jauho (1998). To limit the technical details we specialize to the case of an orthogonal basis set and refer to Thygesen (2006) for a generalization to the non-orthogonal case.

23.2.1   Model

We consider a quantum conductor consisting of a central region (C) connected to left (L) and right (R) leads. For times $t < t_0$ the three regions are decoupled from each other, each being in thermal equilibrium with a common temperature and chemical potentials μ_L, μ_C, and μ_R, respectively (see Fig. 23.1). At $t = t_0$ the coupling between the three subsystems is switched on and a current starts to flow as the electrode with higher chemical potential discharges through the central region into the lead with lower chemical potential. Our aim is to calculate the steady-state current that arises after the transient has died out. Notice that the duration of the steady state is determined by the size of the leads that we henceforth take to be infinite.

The single-particle state space of the electrons, $\mathcal{H}$, is spanned by the orthonormal basis set $\{\phi_i\}$. The orbitals ϕ_i are assumed to be localized such that $\mathcal{H}$ can be decomposed into a sum of orthogonal subspaces corresponding to the division of the system into leads and central region, i.e. $\mathcal{H} = \mathcal{H}_L + \mathcal{H}_C + \mathcal{H}_R$. We will use the notation $i \in \alpha$ to indicate that $\phi_i \in \mathcal{H}_\alpha$ for some $\alpha \in \{L, C, R\}$. The non-interacting part of the Hamiltonian of the *connected* system is written as

$$\hat{h} = \sum_{\substack{i,j \in \\ L,C,R}} \sum_{\sigma=\uparrow\downarrow} h_{ij} c_{i\sigma}^{\dagger} c_{j\sigma}, \tag{23.1}$$

where i, j run over all basis states of the system. For $\alpha, \beta \in \{L, C, R\}$, the operator $\hat{h}_{\alpha\beta}$ is obtained by restricting i to region α and j to region β in eqn (23.1). Occasionally, we shall write $\hat{h}_\alpha$ instead of $\hat{h}_{\alpha\alpha}$. We assume that there is no direct coupling between the two leads, i.e. $\hat{h}_{LR} = \hat{h}_{RL} = 0$ (this condition can always be fulfilled by increasing the size of the central region since the basis functions are localized). We introduce a special notation for the "diagonal" of $\hat{h}$,

$$\hat{h}_0 = \hat{h}_{LL} + \hat{h}_{CC} + \hat{h}_{RR}. \tag{23.2}$$

It is instructive to note that $\hat{h}_0$ does *not* describe the three regions in physical isolation from each other, but rather the contacted system without interregion hopping. We allow for interactions between electrons inside the central region. The most general form of such a two-body interaction is,

$$\hat{V} = \sum_{\substack{ijkl\in C \\ \sigma\sigma'}} V_{ij,kl} c_{i\sigma}^\dagger c_{j\sigma'}^\dagger c_{l\sigma'} c_{k\sigma}. \tag{23.3}$$

The full Hamiltonian describing the system at time t can then be written

$$\hat{H}(t) = \begin{cases} \hat{H}_0 = \hat{h}_0 + \hat{V} & \text{for } t < t_0 \\ \hat{H} = \hat{h} + \hat{V} & \text{for } t > t_0. \end{cases} \tag{23.4}$$

Notice, that we use small letters for non-interacting quantities while the subscript 0 refers to uncoupled quantities. At this point we shall not be concerned about the actual value of the matrix elements h_{ij} and $V_{ij,kl}$ as this is unimportant for the general formalism discussed here.

For times $t < t_0$ each of the three subsystems are assumed to be in thermal equilibrium characterized by their equilibrium density matrices. For the left lead we have

$$\hat{\varrho}_L = \frac{1}{Z_L} \exp(-\beta(\hat{h}_L - \mu_L \hat{N}_L)), \tag{23.5}$$

with

$$Z_L = \text{Tr}[\exp(-\beta(\hat{h}_L - \mu_L \hat{N}_L))]. \tag{23.6}$$

Here, β is the inverse temperature and $\hat{N}_L = \sum_{\sigma,i\in L} c_{i\sigma}^\dagger c_{i\sigma}$ is the number operator of lead L. $\hat{\varrho}_R$ and Z_R are obtained by replacing L by R. For $\hat{\varrho}_C$ and Z_C we must add $\hat{V}$ to the Hamiltonian in the exponential to account for the correlations. The initial state of the whole system is given by

$$\hat{\varrho} = \hat{\varrho}_L \hat{\varrho}_C \hat{\varrho}_R. \tag{23.7}$$

If $\hat{V}$ is *not* included in $\hat{\varrho}_C$ we obtain the uncorrelated, or non-interacting, initial state $\hat{\varrho}_{ni}$. Because $\hat{H}_0$ ($\hat{h}_0$) describes the *contacted* system in the absence of interregion hopping, $\hat{\varrho}$ ($\hat{\varrho}_{ni}$) do not describe the three regions in physical isolation. In other words the three regions are only decoupled at the *dynamic* level for times $t < t_0$.

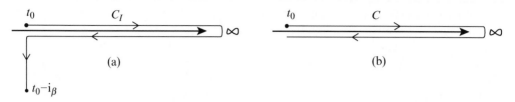

Fig. 23.2 (a) The Keldysh contour $\mathcal{C}_I$. The dynamics of the system is governed by the Hamiltonian $\hat{H}(\tau)$ on the horizontal branch, while the initial state is defined by $\hat{H}(\tau)$ on the vertical branch. (b) The Keldysh contour $\mathcal{C}$, used when correlations in the initial state are neglected.

23.2.2   Contour-ordered Green's function

To treat non-equilibrium problems it is useful to extend the time-propagation operator from the real-time axis to the so-called Keldysh contour, $\mathcal{C}_I$, depicted in Fig. 23.2(a). The contour is defined in the complex plane and runs along the real axis from t_0 to infinity, then back to t_0 and vertically down to $t_0 - i\beta$. When τ and τ' denote points on the Keldysh contour, the generalized time-propagation operator is defined by

$$\hat{U}(\tau', \tau) = T e^{-i \int_\tau^{\tau'} d\bar{\tau} \hat{H}(\bar{\tau})}, \tag{23.8}$$

where T orders a product of operators according to their time argument on the contour (later times further to the left). The integral is taken along $\mathcal{C}_I$. So far, we have defined the Hamiltonian $\hat{H}(\tau)$ only for τ on the real axis. For eqn (23.8) to make sense we must extend $\hat{H}(\tau)$ to the vertical branch of $\mathcal{C}_I$, which we will do in a moment. The contour-ordered single-particle GF relevant for our transport problem is defined by

$$G_{i\sigma, j\sigma'}(\tau, \tau') = -i \operatorname{Tr} \left\{ \hat{\varrho} T \left[c_{H,i\sigma}(\tau) c_{H,j\sigma'}^\dagger(\tau') \right] \right\}, \tag{23.9}$$

where, e.g. $c_{H,i\sigma}(\tau) = \hat{U}(t_0, \tau) c_{i\sigma} \hat{U}(\tau, t_0)$ and $\hat{\varrho}$ is the state of the system at time t_0. Notice that when evaluating $c_{H,i\sigma}(\tau)$ for τ on the real axis it does not matter whether τ is chosen on the upper or lower part of the contour. Notice also that when τ and τ' are both in the vertical branch, G is simply the Matsubara GF known from the equilibrium theory. For later reference we also note that a non-equilibrium GF is completely defined once (i) the Hamiltonian governing the dynamics and (ii) the initial state have been specified. Since the Hamiltonian contains no spin-flip processes the GF is diagonal in spin space, i.e. $G_{i\sigma, j\sigma'} = \delta_{\sigma\sigma'} G_{ij}$ and we therefore suppress the spin dependence in the following.

If we define $\hat{H}(\tau)$ along the vertical part of the contour to be

$$\hat{H}(\tau) = \sum_{\alpha=L,C,R} (\hat{h}_\alpha - \mu_\alpha \hat{N}_\alpha) + \hat{V}, \tag{23.10}$$

we see that $\hat{U}(t_0 - i\beta, t_0) = Z\hat{\varrho}$. We use this observation to write

$$G_{ij}(\tau, \tau') = -i \frac{\operatorname{Tr}\left\{ T \left[e^{-i \int_{\mathcal{C}_I} d\bar{\tau} \hat{H}(\bar{\tau})} c_i(\tau) c_j^\dagger(\tau') \right] \right\}}{\operatorname{Tr}\left\{ T e^{-i \int_{\mathcal{C}_I} d\bar{\tau} \hat{H}(\bar{\tau})} \right\}}. \tag{23.11}$$

Here, the time-argument of $c_i(\tau)$ and $c_j^\dagger(\tau')$ only serves to identify their position in the contour-ordering, i.e. they are *not* in the Heisenberg picture. In order to obtain an expansion of $G_{ij}(\tau, \tau')$ in powers of $\hat{V}$, we switch to the interaction picture defined by the non-interacting Hamiltonian $\hat{h}(\tau) = \hat{H}(\tau) - \hat{V}$. In this picture we have

$$G_{ij}(\tau, \tau') = -i \frac{\mathrm{Tr}\left\{\hat{\varrho}_{ni} T \left[e^{-i\int_{C_I} \mathrm{d}\bar{\tau}\,\hat{V}_h(\bar{\tau})} c_{h,i}(\tau)c_{h,j}^\dagger(\tau')\right]\right\}}{\mathrm{Tr}\left\{\hat{\varrho}_{ni} T e^{-i\int_{C_I} \mathrm{d}\bar{\tau}\,\hat{V}_h(\bar{\tau})}\right\}}, \tag{23.12}$$

where the time dependence of the operators is governed by the evolution operator in eqn (23.8) with $\hat{H}$ replaced by $\hat{h}$. The density matrix $\hat{\varrho}_{ni}$ is given by

$$\hat{\varrho}_{ni} = \frac{\exp(-\beta \sum_\alpha (\hat{h}_\alpha - \mu_\alpha \hat{N}_\alpha))}{\mathrm{Tr}\{\exp(-\beta \sum_\alpha (\hat{h}_\alpha - \mu_\alpha \hat{N}_\alpha))\}}, \tag{23.13}$$

which differs from $\hat{\varrho}$ in that it does not contain interactions in the central region. From the identity

$$Z\hat{\varrho} = Z_0 \hat{\varrho}_{ni} T e^{-i\int_{t_0}^{t_0 - i\beta} \mathrm{d}\bar{\tau}\,\hat{V}_h(\bar{\tau})} \tag{23.14}$$

it is clear that the integration along the vertical branch of C_I in eqn (23.12) accounts for the correlations in the initial state of region C. While it must be expected that the presence of initial correlations will influence the transient behavior of the current, it seems plausible that they will be washed out over time such that the steady-state current will not depend on whether or not correlations are present in the initial state. In practice, the neglect of initial correlations is a major simplification that allows us to work entirely on the real axis, avoiding any reference to the imaginary time. For these reasons we shall neglect initial correlations in the rest of this chapter. The GF can then be written

$$G_{ij}(\tau, \tau') = -i \frac{\mathrm{Tr}\left\{\hat{\varrho}_{ni} T \left[e^{-i\int_{C} \mathrm{d}\bar{\tau}\,\hat{V}_h(\bar{\tau})} c_{h,i}(\tau)c_{h,j}^\dagger(\tau')\right]\right\}}{\mathrm{Tr}\left\{\hat{\varrho}_{ni} T e^{-i\int_{C} \mathrm{d}\bar{\tau}\,\hat{V}_h(\bar{\tau})}\right\}}, \tag{23.15}$$

where the contour C is depicted in Fig. 23.2(b). Equation (23.15) is the starting point for a systematic series expansion of G_{ij} in powers of $\hat{V}$. Since $\hat{\varrho}_{ni}$ is a mixed state of Slater determinants and the time evolution is given by the non-interacting Hamiltonian, $\hat{h}$, Wick's theorem applies and leads to the standard Feynman rules with the exception that all time integrals are along the contour C and all Green's functions are contour-ordered. The Feynman diagrams should be constructed using the Green's function defined by $\hat{\varrho}_{ni}$ and $\hat{h}$,

$$g_{ij}(\tau, \tau') = -i\,\mathrm{Tr}\left\{\hat{\varrho}_{ni} T \left[c_{h,i}(\tau)c_{h,j}^\dagger(\tau')\right]\right\}, \tag{23.16}$$

which describes the non-interacting electrons in the contacted system.

The diagrammatic expansion leads to the identification of a self-energy, Σ, as the sum of all irreducible diagrams with no external vertices. The GF is related to the self-energy and the non-interacting GF through Dyson's equation

$$G(\tau, \tau') = g(\tau, \tau') + \int_C d\tau_1 d\tau_2 g(\tau, \tau_1) \Sigma(\tau_1, \tau_2) G(\tau_2, \tau'), \qquad (23.17)$$

where matrix multiplication is implied. As we will see in Section 23.4.2, only the Green's function of the central region is needed for the calculation of the current, and we can therefore focus on the central-region submatrix of G. Since the interactions are limited to the central region, the self-energy matrix, Σ_{ij}, will be non-zero only when both $i, j \in C$, and it should therefore be safe to use the notation Σ instead of Σ_C. Restricting eqn (23.17) to the central region we have

$$G_C(\tau, \tau') = g_C(\tau, \tau')$$
$$+ \int_C d\tau_1 d\tau_2 g_C(\tau, \tau_1) \Sigma(\tau_1, \tau_2) G_C(\tau_2, \tau'). \qquad (23.18)$$

The free propagator, $g_C(\tau, \tau')$, is a non-equilibrium GF because $\hat{\varrho}_{ni}$ is not a stationary state of $\hat{h}$, i.e. $[\hat{\varrho}_{ni}, \hat{h}] \neq 0$. It is, however, not difficult to show that g_C satisfies the following Dyson equation

$$g_C(\tau, \tau') = g_{0,C}(\tau, \tau') + \int_C d\tau_1 d\tau_2 g_{0,C}(\tau, \tau_1)$$
$$[\Sigma_L(\tau_1, \tau_2) + \Sigma_R(\tau_1, \tau_2)] g_C(\tau_2, \tau'), \qquad (23.19)$$

where g_0 is the *equilibrium* GF defined by $\hat{\varrho}_{ni}$ and $\hat{h}_0$. The coupling self-energy due to the lead $\alpha = L, R$ is given by

$$\Sigma_\alpha(\tau, \tau') = h_{C\alpha} g_{0,\alpha}(\tau, \tau') h_{\alpha C}. \qquad (23.20)$$

Notice the slight abuse of notation: Σ_α is *not* the $\alpha\alpha$ submatrix of Σ. In fact, Σ_L and Σ_R are both matrices in the central region indices only. Combining eqns (23.18) and (23.19) we can write the Dyson equation for G_C,

$$G_C(\tau, \tau') = g_{0,C}(\tau, \tau')$$
$$+ \int_C d\tau_1 d\tau_2 g_{0,C}(\tau, \tau_1) \Sigma_{tot}(\tau_1, \tau_2) G_C(\tau_2, \tau'), \quad (23.21)$$

in terms of the equilibrium propagator of the non-interacting, uncoupled system, g_0, and the total self-energy

$$\Sigma_{tot} = \Sigma + \Sigma_L + \Sigma_R. \qquad (23.22)$$

The total self-energy describes the combined effect of the interactions and the coupling to the leads.

23.2.3   Real-time Green's functions

In order to evaluate expectation values of single-particle observables we need the real-time correlation functions. We work with two correlation functions, also called the lesser and greater GFs, defined as

$$G_{ij}^{<}(t, t') = i \operatorname{Tr} \left\{ \hat{\varrho}_{ni} c_{H,j}^{\dagger}(t') c_{H,i}(t) \right\} \tag{23.23}$$

$$G_{ij}^{>}(t, t') = -i \operatorname{Tr} \left\{ \hat{\varrho}_{ni} c_{H,i}(t) c_{H,j}^{\dagger}(t') \right\}. \tag{23.24}$$

Again, the use of $\hat{\varrho}_{ni}$ instead of $\hat{\varrho}$ amounts to neglecting initial correlations. Two other important real-time GFs are the retarded and advanced GFs defined by

$$G_{ij}^{\mathrm{r}}(t, t') = \theta(t - t') \left(G_{ij}^{>}(t, t') - G_{ij}^{<}(t, t') \right) \tag{23.25}$$

$$G_{ij}^{\mathrm{a}}(t, t') = \theta(t' - t) \left(G_{ij}^{<}(t, t') - G_{ij}^{>}(t, t') \right). \tag{23.26}$$

The four GFs are related through

$$G^{>} - G^{<} = G^{\mathrm{r}} - G^{\mathrm{a}}. \tag{23.27}$$

The lesser and greater GFs are just special cases of the contour-ordered GF. For example $G^{<}(t, t') = G(\tau, \tau')$ when $\tau = t$ is on the upper branch of $\mathcal{C}$ and $\tau' = t'$ is on the lower branch. This can be used to derive a set of rules, sometimes referred to as the Langreth rules, for converting expressions involving contour-ordered quantities into equivalent expressions involving real-time quantities. We shall not list the conversion rules here, but refer to Haug and Jauho (1998) (no initial correlations) and Leeuwen *et al.* (2006) (including initial correlations). The usual procedure in non-equilibrium is then to derive the relevant equations on the contour using the standard diagrammatic techniques, and subsequently converting these equations to real time by means of the Langreth rules. An example of this procedure is given in Section 23.3.2 where the non-equilibrium GW equations are derived.

23.2.3.1 *Equilibrium*

In equilibrium, the real-time GFs depend only on the time difference $t' - t$. Fourier transforming with respect to this time difference then brings out the spectral properties of the system. In particular, the spectral function

$$A(\omega) = i[G^{\mathrm{r}}(\omega) - G^{\mathrm{a}}(\omega)] = i[G^{>}(\omega) - G^{<}(\omega)] \tag{23.28}$$

shows peaks at the *quasi-particle* energies of the system, i.e. at the energies $E_n(N + 1) - E_m(N)$ and $E_m(N) - E_n(N - 1)$, where n, m denote energy levels and N the number of electrons. We thus see that the single-particle GF carries information about the electron-addition and -removal energies. Clearly, these are the types of excitations that are relevant in a transport situation where electrons are continuously added to and removed from the central region. In Section 23.5.2 we discuss how many-body polarization effects renormalize the spectral function of a molecule adsorbed at a metal surface. In equilibrium, we, furthermore, have the important fluctuation-dissipation theorem that relates the correlation functions to the spectral function and the Fermi–Dirac distribution function, f,

$$G^{<}(\omega) = i f(\omega) A(\omega) \tag{23.29}$$

$$G^{>}(\omega) = -i(1 - f(\omega)) A(\omega). \tag{23.30}$$

The fluctuation-dissipation theorem follows from the Lehman representation that no longer holds out of equilibrium, and as a consequence one has to work explicitly with the correlation functions in non-equilibrium situations.

23.2.3.2 *Non-equilibrium steady state*

We shall make the plausible assumption that in steady state, all the real-time GFs depend only on the time-difference $t' - t$. Moreover, we take the limit $t_0 \to -\infty$. This will allow us to use the Fourier transform to turn convolutions in real time into products in frequency space. Applying the Langreth conversion rules to the Dyson equation (23.21), and Fourier transforming with respect to $t' - t$ then leads to the following expression for the retarded GF of the central region

$$G_C^r(\omega) = g_{0,C}^r(\omega) + g_{0,C}^r(\omega) \Sigma_{tot}^r(\omega) G_C^r(\omega). \tag{23.31}$$

This equation can be inverted to yield the closed form

$$G_C^r(\omega) = \left[(\omega + i\eta) I_C - h_C - \Sigma_L^r(\omega) - \Sigma_R^r(\omega) - \Sigma^r(\omega) \right]^{-1}. \tag{23.32}$$

The equation for G^a is obtained by replacing r by a and η by $-\eta$ or, alternatively, from

$$G_C^a(\omega) = \left[G_C^r(\omega) \right]^\dagger, \tag{23.33}$$

which follows from the assumption that the GFs depend on time differences only. For the correlation functions the conversion rules leads to the expression

$$G_C^{</>}(\omega) = G_C^r(\omega) \Sigma_{tot}^{</>}(\omega) G_C^a(\omega) + \Delta^{</>}(\omega), \tag{23.34}$$

where

$$\Delta^{</>}(\omega) = \left[I_C + G_C^r(\omega) \Sigma_{tot}^r(\omega) \right] g_{0,C}^{</>}(\omega) \left[I_C + \Sigma_{tot}^a(\omega) G_C^a(\omega) \right]. \tag{23.35}$$

Using that $\Sigma_{tot}^{r/a} = (g_{0,C}^{r/a})^{-1} - (G_C^{r/a})^{-1}$ together with the equilibrium relations $g_{0,C}^{<} = -f(\omega - \mu_C) \left[g_{0,C}^r - g_{0,C}^a \right]$ and $g_{0,C}^{>} = -(f(\omega - \mu_C) - 1)[g_{0,C}^r - g_{0,C}^a]$, we find that

$$\Delta^{<}(\omega) = 2i\eta f(\omega - \mu_C) G_C^r(\omega) G_C^a(\omega) \tag{23.36}$$

$$\Delta^{>}(\omega) = 2i\eta [f(\omega - \mu_C) - 1] G_C^r(\omega) G_C^a(\omega). \tag{23.37}$$

If the product $G^r G^a$ is independent of η we can conclude that $\Delta \to 0$ in the relevant limit of small η. It can in fact be shown that Δ always vanishes for interacting electrons, while for non-interacting electrons Δ vanishes only when there are no bound states (Thygesen and Rubio 2008).

23.2.3.3 *Non-interacting electrons and lead self-energy*

In the special case of non-interacting electrons, the retarded and advanced GFs are independent of the initial state of the system, i.e. of the $\hat{\varrho}$ entering the GF. Moreover, if the dynamics is governed by a time-independent Hamiltonian, g^r

and g^a depend only on the time difference $t' - t$ (even if the initial state is not a stationary state). In this case the Fourier transform of the retarded and advanced GFs with respect to $t' - t$ equals the resolvent of the Hamiltonian matrix h,

$$g^{r/a}(\omega) = [(\omega \pm i\eta)I - h]^{-1}, \qquad (23.38)$$

where I is the identity matrix and η is a positive infinitesimal. In our transport problem, the block-diagonal structure of h_0 allows us to obtain the non-interacting GF of the uncoupled system by inverting each block separately,

$$g^{r/a}_{0,\alpha}(\omega) = [(\omega \pm i\eta)I - h_\alpha]^{-1} \qquad (23.39)$$

for $\alpha \in \{L, C, R\}$. Now, it is in fact easy to show that the central region component of $g^{r/a}$ satisfies

$$g^{r/a}_C(\omega) = \left[(\omega \pm i\eta)I - h_C - \Sigma^{r/a}_L(\omega) - \Sigma^{r/a}_R(\omega)\right]^{-1}, \qquad (23.40)$$

with the retarded/advanced coupling self-energies given by

$$\Sigma^{r/a}_\alpha(\omega) = h_{C\alpha} g^{r/a}_{0,\alpha}(\omega) h_{\alpha C}. \qquad (23.41)$$

Equations (23.40) and (23.41) give the retarded and advanced components of eqns (23.19) and (23.20), respectively. Notice that $\Sigma^{r/a}_\alpha$ depends on the applied bias voltage through $g_{0,\alpha}$ because the self-consistent field in the leads follow the chemical potential to ensure charge neutrality as sketched in the lower panel of Fig. 23.1. Assuming a symmetrically applied bias ($\mu_{L/R} = \varepsilon_F \pm V/2$) we have

$$\Sigma^{r/a}_L(V; \omega) = \Sigma^{r/a}_L(0; \omega - V/2), \qquad (23.42)$$

with a similar relation for Σ_R. Since g_0 is an equilibrium GF its lesser and greater components, and thus also $\Sigma^{</>}_\alpha$, follow from the fluctuation-dissipation theorem. In contrast, the correlation functions derived from g, which is a non-equilibrium function, must be calculated using the Keldysh equation (23.34).

23.3 Many-body self-energy

In this section we discuss two specific approximations to the many-body self-energy Σ introduced in eqn (23.17), namely the GW and second Born (2B) approximations. Strictly speaking the Σ of eqn (23.17) contains the full effect of the interactions, whereas the GW and 2B self-energies only describe exchange and correlation, i.e. they do not include the Hartree potential. The GW self-energy is obtained by summing an infinite set of Feynman diagrams—one diagram at each order of the interaction—while the 2B approximation is exact to second order in the interaction (if performed self-consistently). In Section 23.3.1 we introduce an effective interaction that leads to a particularly simple form of the self-energy equations and at the same time provides a means for reducing self-interaction errors in higher-order diagrammatic expansions. In Sections 23.3.2 and 23.3.3 we discuss the non-equilibrium GW and 2B equations using the effective interaction.

23.3.1   Effective interaction

The direct use of the full interaction eqn (23.3) in a diagrammatic expansion is problematic as it introduces frequency-dependent, four-index quantities, which quickly becomes difficult to store and handle numerically. For this reason we consider instead the effective interaction defined by

$$\hat{V}_{\text{eff}} = \sum_{ij,\sigma\sigma'} \tilde{V}_{i\sigma,j\sigma'} c_{i\sigma}^{\dagger} c_{j\sigma'}^{\dagger} c_{j\sigma'} c_{i\sigma}, \qquad (23.43)$$

where

$$\tilde{V}_{i\sigma,j\sigma'} = V_{ij,ij} - \delta_{\sigma\sigma'} V_{ij,ji}. \qquad (23.44)$$

This expression follows by restricting the sum in the full interaction eqn (23.3) to terms of the form $V_{ij,ij} c_{i\sigma}^{\dagger} c_{j\sigma'}^{\dagger} c_{j\sigma'} c_{i\sigma}$ and $V_{ij,ji} c_{i\sigma}^{\dagger} c_{j\sigma}^{\dagger} c_{j\sigma} c_{i\sigma}$.

The effective interaction is local in orbital space, i.e. it is a two-point function instead of a four-point function and thus resembles the real-space representation. Note, however, that in contrast to the real-space representation $\tilde{V}_{i\sigma,j\sigma'}$ is spin dependent. In particular, the self-interactions, $\tilde{V}_{i\sigma,i\sigma}$, are zero by construction and consequently self-interaction (in the orbital basis) is avoided to all orders in a perturbation expansion in powers of $\tilde{V}$. Since the off-diagonal elements ($i \neq j$) of the exchange integrals $V_{ij,ji}$ are small for localized basis functions, the main effect of the second term in eqn (23.44) is to cancel the self-interaction in the first term.

It is not straightforward to anticipate the quality of a many-body calculation based on the effective interaction (23.43) as compared to the full interaction (23.3). Clearly, if we include all Feynman diagrams in Σ, we obtain the exact result when the full interaction (23.3) is used, while the use of the effective interaction (23.43) would yield an approximate result. The quality of this approximate result would then depend on the basis set, becoming better the more localized the basis functions and equal to the exact result in the limit of completely localized delta functions where only the direct Coulomb integrals $V_{ij,ij}$ will be non-zero. However, when only a subset of all diagrams are included in Σ the situation is different: In the GW approximation for instance, only one diagram per order (in $\hat{V}$) is included, and thus cancellation of self-interaction does not occur when the full interaction is used. On the other hand, the effective interaction (23.44) is self-interaction free (in the orbital basis) by construction. The situation can be understood by considering the lowest-order case. There are only two first-order diagrams—the Hartree and exchange diagrams—and each cancel the self-interaction in the other. More generally, the presence of self-interaction in an incomplete perturbation expansion can be seen as a violation of identities of the form $\left\langle \cdot \left| c_{k\sigma'}^{\dagger} \cdots c_{i\sigma} c_{i\sigma} \cdots c_{j\sigma''} \right| \cdot \right\rangle = 0$, when not all Wick contractions are evaluated. Such expectation values will correctly vanish when the effective interaction is used because the pre-factor of the $c_{i\sigma} c_{i\sigma}$ operator, $\tilde{V}_{i\sigma,i\sigma}$, is zero. The effect of self-interaction errors in (non-self-consistent) GW calculations was recently studied for a hydrogen atom (Nelson *et al.* 2007).

23.3.2   Non-equilibrium *GW* self-energy

It is useful to split the full many-body self-energy into its Hartree and exchange-correlation parts

$$\Sigma(\tau, \tau') = \Sigma_h(\tau, \tau') + \Sigma_{xc}(\tau, \tau').$$ (23.45)

The Hartree term is local in time and can be written $\Sigma_h(\tau, \tau') = \Sigma_h(\tau)\delta_\mathcal{C}(\tau, \tau')$, where $\delta_\mathcal{C}$ is a delta function on the Keldysh contour. The xc-self-energy is non-local in time and contains all the complicated many-body effects. In the *GW* approximation the xc-self-energy is written as a product of the Green's function, *G*, and the screened interaction, *W*. Usually, *W* is calculated in the random-phase approximation (RPA), and it has been found that improving *W* beyond RPA has little effect on the resulting *GW* self-energy (Verdozzi *et al.* 1995).

With the effective interaction (23.43) the screened interaction, *W*, and the polarization function, *P*, are reduced from four- to two-index functions. For notational simplicity we absorb the spin index into the orbital index, i.e. $(i\sigma) \to i$ (but we do not neglect it). The expressions for the contour-ordered *GW* self-energy in terms of contour-ordered quantities then read

$$\Sigma_{GW,ij}(\tau, \tau') = iG_{ij}(\tau, \tau'^{+})W_{ij}(\tau, \tau')$$ (23.46)

$$W_{ij}(\tau, \tau') = \tilde{V}_{ij}\delta_\mathcal{C}(\tau, \tau') + \sum_{kl} \int_\mathcal{C} d\tau_1 \tilde{V}_{ik} P_{kl}(\tau, \tau_1)$$

$$W_{lj}(\tau_1, \tau')$$ (23.47)

$$P_{ij}(\tau, \tau') = -iG_{ij}(\tau, \tau')G_{ji}(\tau', \tau).$$ (23.48)

It is important to notice that in contrast to the conventional real-space representation of the *GW* self-energy, the spin dependence cannot be neglected when the effective interaction is used. The reason for this is that $\tilde{V}$ is spin dependent and consequently the spin off-diagonal elements of *W* will influence the spin-diagonal elements of *G*, Σ, and *P*. A diagrammatic representation of the *GW* approximation is shown in Fig. 23.3. As they stand, eqns (23.46)–(23.48) involve quantities of the whole system (leads and central region). However, since $\tilde{V}_{ij}$ is non-zero only when $i, j \in C$, it follows from eqn (23.47), that *W* and hence Σ also have this structure. Consequently, the subscript *C* can be directly attached to each quantity in eqns (23.46)–(23.48), however, for the sake of generality and notational simplicity we shall not do so at this point. It is, however, important to realize that the GF appearing in the *GW* equations includes the self-energy due to the leads.

Using the Langreth conversion rules the retarded and lesser *GW* self-energies become (on the time axis),

$$\Sigma^r_{GW,ij}(t) = iG^r_{ij}(t)W^>_{ij}(t) + iG^<_{ij}(t)W^r_{ij}(t)$$ (23.49)

$$\Sigma^{</>}_{GW,ij}(t) = iG^{</>}_{ij}(t)W^{</>}_{ij}(t),$$ (23.50)

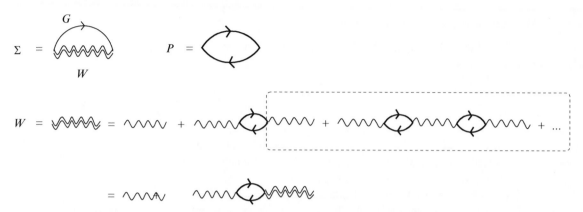

Fig. 23.3 Feynman diagrams for the GW self-energy in terms of the Green's function (G), the screened interaction (W), and the polarization bubble (P). All quantities are functions of two Keldysh times, and two basis function indices. Integration/summation over internal times/indices is implied.

where we have used the variable t instead of the time difference $t' - t$. For the screened interaction we obtain (in frequency space),

$$W^{\mathrm{r}}(\omega) = \tilde{V}[I - P^{\mathrm{r}}(\omega)\tilde{V}]^{-1} \qquad (23.51)$$

$$W^{</>}(\omega) = W^{\mathrm{r}}(\omega) P^{</>}(\omega) W^{\mathrm{a}}(\omega), \qquad (23.52)$$

where all quantities are matrices in the indices i, σ and matrix multiplication is implied. Notice that the spin off-diagonal part of $\tilde{V}$ will affect the spin-diagonal part of W^{r} through the matrix inversion. Finally, the real-time components of the irreducible polarization become

$$P^{\mathrm{r}}_{ij}(t) = -\mathrm{i}G^{\mathrm{r}}_{ij}(t)G^{<}_{ji}(-t) - \mathrm{i}G^{<}_{ij}(t)G^{\mathrm{a}}_{ji}(-t) \qquad (23.53)$$

$$P^{</>}_{ij}(t) = -\mathrm{i}G^{</>}_{ij}(t)G^{>/<}_{ji}(-t). \qquad (23.54)$$

Both the polarization and the screened interaction obey the relations $P^{\mathrm{a}}(\omega) = P^{\mathrm{r}}(\omega)^{\dagger}$ and $W^{\mathrm{a}}(\omega) = W^{\mathrm{r}}(\omega)^{\dagger}$, and similarly for the self-energy and GFs we have $\Sigma^{\mathrm{a}}_{GW}(\omega) = \Sigma^{\mathrm{r}}_{GW}(\omega)^{\dagger}$ and $G^{\mathrm{a}}(\omega) = G^{\mathrm{r}}(\omega)^{\dagger}$. In addition, all quantities fullfill the general identity $X^{>} - X^{<} = X^{\mathrm{r}} - X^{\mathrm{a}}$. We mention that the non-equilibrium GW approximation has previously been used to study bandgap renormalization in excited GaAs (Spataru *et al.* 2004).

In deriving eqns (23.51) and (23.52) we have made use of the conversion rules $\delta^{</>}_{C}(t, t') = 0$ and $\delta^{\mathrm{r/a}}_{C}(t, t') = \delta(t - t')$. With these definitions the applicability of the Langreth rules can be extended to functions containing delta functions on the contour. Notice, however, that with these definitions relation (23.25) does not hold for the delta function. The reason why the delta function requires a separate treatment is that the standard Langreth rules are derived under the assumption that all functions on the contour are well behaved, e.g. not containing delta functions. We stress that no spin symmetry has been assumed in the above GW equations. Indeed by reintroducing the spin index, i.e. $i \to (i\sigma)$ and $j \to (j\sigma')$, it is clear that spin-polarized calculations can be performed by treating $G_{\uparrow\uparrow}$ and $G_{\downarrow\downarrow}$ independently.

Within the GW approximation the full interaction self-energy is given by

$$\Sigma(\tau, \tau') = \Sigma_{\mathrm{h}}(\tau, \tau') + \Sigma_{GW}(\tau, \tau'), \qquad (23.55)$$

where the GW self-energy can be further split into an static exchange and a dynamic correlation part,

$$\Sigma_{GW}(\tau, \tau') = \Sigma_{\mathrm{x}} \delta_{\mathcal{C}}(\tau, \tau') + \Sigma_{\mathrm{corr}}(\tau, \tau'). \qquad (23.56)$$

Due to the static nature of Σ_{h} and Σ_{x} we have

$$\Sigma_{\mathrm{h}}^{</>} = \Sigma_{\mathrm{x}}^{</>} = 0. \qquad (23.57)$$

The retarded components of the Hartree and exchange self-energies become constant in frequency space, and we have (note that for Σ_h and Σ_x we do *not* use the effective interaction (23.43))

$$\Sigma_{\mathrm{h}, ij}^{\mathrm{r}} = -i \sum_{kl} G_{kl}^{<}(t = 0) V_{ik, jl} \qquad (23.58)$$

$$\Sigma_{\mathrm{x}, ij}^{\mathrm{r}} = i \sum_{kl} G_{kl}^{<}(t = 0) V_{ik, lj}. \qquad (23.59)$$

Due to eqn (23.57), it is clear that eqn (23.50) yields the lesser/greater components of Σ_{corr}. Since $\Sigma_{\mathrm{corr}}(\tau, \tau')$ does not contain delta functions its retarded component can be obtained from the relation,

$$\Sigma_{\mathrm{corr}}^{\mathrm{r}}(t) = \theta(-t) \left[\Sigma_{GW}^{>}(t) - \Sigma_{GW}^{<}(t) \right]. \qquad (23.60)$$

The separate calculation of $\Sigma_{\mathrm{x}}^{\mathrm{r}}$ and $\Sigma_{\mathrm{corr}}^{\mathrm{r}}$ from eqns (23.59) and (23.60) as opposed to calculating their sum directly from eqn (23.49), has two advantages: (i) It allows us to treat Σ_{x}, which is the dominant contribution to Σ_{GW}, at a higher level of accuracy than Σ_{corr}. (ii) We avoid numerical operations involving G^{r} and W^{r} in the time domain. For more detailed discussions of these points see Appendices A and E of Thygesen and Rubio (2008).

23.3.3 Non-equilibrium second Born approximation

When screening and/or strong correlation effects are less important, as, e.g. in the case of small, isolated molecules, the higher-order terms of the GW approximation are small and it is more important to include all second-order diagrams (Stan *et al.* 2006). The full second-order approximation, often referred to as the second Born approximation (2B), is shown diagrammatically in Fig. 23.4. As we will use the 2B for comparison with the GW results we state the relevant expressions here for completeness.

On the contour the 2B self-energy reads (with the effective interaction (23.43))

$$\Sigma_{2\mathrm{B}, ij}(\tau, \tau') = \sum_{kl} G_{ij}(\tau, \tau') G_{kl}(\tau, \tau') G_{lk}(\tau', \tau) \tilde{V}_{ik} \tilde{V}_{jl}$$

$$- \sum_{kl} G_{ik}(\tau, \tau') G_{kl}(\tau', \tau) G_{lj}(\tau, \tau') \tilde{V}_{il} \tilde{V}_{jk}. \qquad (23.61)$$

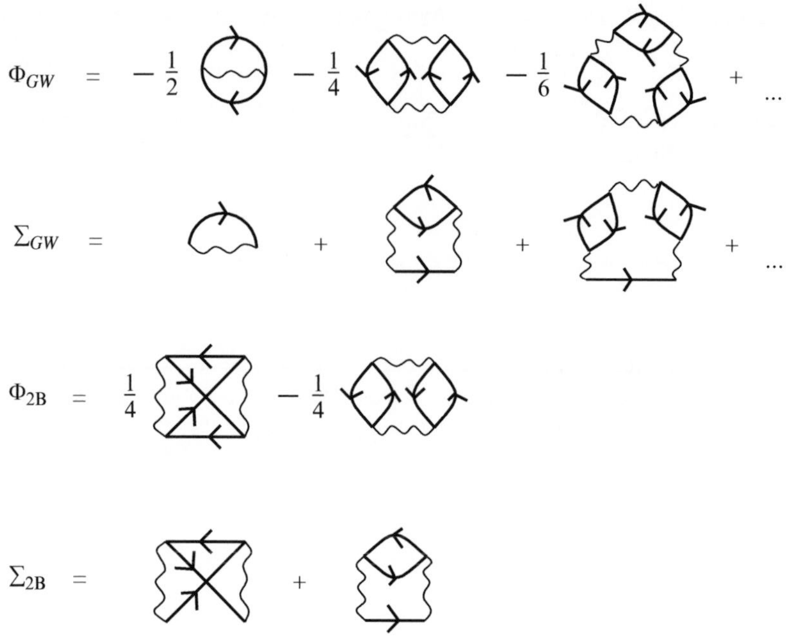

$$\Phi_{GW} \quad = \quad -\frac{1}{2} \; \bigcirc \quad - \frac{1}{4} \; \bowtie \quad - \frac{1}{6} \; \otimes \quad + \quad \ldots$$

$$\Sigma_{GW} \quad = \quad \frown \quad + \quad \otimes \quad + \quad \otimes \quad + \quad \ldots$$

Fig. 23.4 The GW and second Born self-energies, Σ_{GW} and Σ_{2B}, can be obtained as functional derivatives of their respective Φ-functionals, $\Phi_{GW}[G]$ and $\Phi_{2B}[G]$. Straight lines represent the full Green's function, G, i.e. the Green's function in the presence of coupling to the leads and interactions. Wiggly lines represent the interactions. Reprinted with permission from Thygesen and Rubio, *Phys. Rev. B* **77**, 115333 (2008). Copyright 2008 by the American Physical Society.

$$\Phi_{2B} \quad = \quad \frac{1}{4} \; \otimes \quad - \frac{1}{4} \; \bowtie$$

$$\Sigma_{2B} \quad = \quad \otimes \quad + \quad \otimes$$

Notice that the first term in Σ_{2B} is simply the second-order term of the GW self-energy. From eqn (23.61) it is easy to obtain the lesser/greater self-energies,

$$\Sigma_{2B,ij}^{</>}(t) = \sum_{kl} G_{ij}^{</>}(t) G_{kl}^{</>}(t) G_{lk}^{>/<}(-t) \tilde{V}_{ik} \tilde{V}_{jl}$$

$$- \sum_{kl} G_{ik}^{</>}(t) G_{kl}^{>/<}(-t) G_{lj}^{</>}(t) \tilde{V}_{il} \tilde{V}_{jk},$$

where t has been used instead of the time difference $t - t'$. Since these second-order contributions do not contain delta functions of the time variable, we can obtain the retarded self-energy directly from the Kramers–Kronig relation

$$\Sigma_{2B}^{r}(t) = \theta(-t) \left[\Sigma_{2B}^{>}(t) - \Sigma_{2B}^{<}(t) \right]. \tag{23.62}$$

23.4   Current formula and charge conservation

As mentioned earlier, the current flowing between the electrodes can be calculated from the Green's function of the central region. After a short introduction to the concept of conserving approximations in Section 23.4.1, we present the relevant formulas for evaluating the current under non-equilibrium conditions. In section 23.4.3 we then derive a condition on the interaction self-energy that ensures charge conservation in the sense that the current is the same at the left and right interface between the central region and the leads. Finally, in Section 23.4.4 we show that this condition is always fulfilled for the so-called conserving, or Φ-derivable, self-energies.

23.4.1 Conserving approximations

Once a self-energy has been obtained the GF follows from the Dyson and Keldysh equations (23.32) and (23.34). Any single-particle observable, such as the current or the density, can then be calculated from the GF. An important question is then whether the calculated quantities obey the fundamental conservation laws like charge and energy conservation. In the context of modelling electron transport, the condition for charge conservation as expressed by the continuity equation

$$\frac{\mathrm{d}}{\mathrm{d}t} n(\mathbf{r}, t) = -\nabla \cdot \mathbf{j}(\mathbf{r}, t) \qquad (23.63)$$

is obviously of particular interest.

Baym and Kadanoff (Baym and Kadanoff 1961; Baym 1962) showed that there exists a deep connection between the self-energy and the validity of the conservation laws. Precisely, any self-energy that can be written as a functional derivative, $\Sigma[G] = \delta\Phi[G]/\delta G$, where $\Phi[G]$ belongs to a certain class of functionals of G, produces a GF that automatically fulfills the basic conservation laws. A self-energy that can be obtained in this way is called Φ-derivable. Since a Φ-derivable self-energy depends on G, the Dyson equation must be solved self-consistently. The exact $\Phi[G]$ can be obtained by summing over all skeleton diagrams constructed using the full G as propagator. By a skeleton diagram we mean a closed diagram, i.e. no external vertices, containing no self-energy insertions. Practical approximations are then obtained by including only a subset of skeleton diagrams. Two examples of such approximations are provided by the GW and second Born Φ-functionals and associated self-energies that are illustrated in Fig. 23.4. Another example is provided by the Hartree–Fock approximation. Solving the Dyson equation self-consistently with one of these self-energies thus defines a conserving approximation in the sense of Baym.

The validity of the conservation laws for Φ-derivable self-energies follows from the invariance of Φ under certain transformations of the Green's function. For example, it follows from the closed diagramatic structure of Φ that the transformation

$$G(\mathbf{r}\tau, \mathbf{r}'\tau') \rightarrow \mathrm{e}^{\mathrm{i}\Lambda(\mathbf{r}\tau)} G(\mathbf{r}\tau, \mathbf{r}'\tau') \mathrm{e}^{-\mathrm{i}\Lambda(\mathbf{r}'\tau')}, \qquad (23.64)$$

where Λ is any scalar function, leaves $\Phi[G]$ unchanged. Using the compact notation $(\mathbf{r}_1, \tau_1) = 1$, the change in Φ when the GF is changed by δG can be written as $\delta\Phi = \int \mathrm{d}1\mathrm{d}2 \Sigma(1, 2)\delta G(2, 1^+) = 0$, where we have used that

$\Sigma = \delta\Phi[G]/\delta G$. To first order in Λ we then have

$$\delta\Phi = i \int \mathrm{d}1\mathrm{d}2\Sigma(1,2)[\Lambda(2) - \Lambda(1)]G(2,1^+)$$

$$= i \int \mathrm{d}1\mathrm{d}2[\Sigma(1,2)G(2,1^+) - G(1,2^+)\Sigma(2,1)]\Lambda(1).$$

Since this hold for all Λ (by a scaling argument) we conclude that

$$\int \mathrm{d}2[\Sigma(1,2)G(2,1^+) - G(1,2^+)\Sigma(2,1)] = 0. \qquad (23.65)$$

This condition ensures the validity of the continuity equation (on the contour) at any point in space. In the following sections we derive and discuss this result in the framework of the transport model introduced in Section 23.2.

23.4.2   Current formula

When the coupling between the leads and the central region is established, a current will start to flow. The particle current from lead α into the central region is given by the time derivative of the number operator of lead α (Meir and Wingreen 1992),

$$I_\alpha(t) = -i\langle[\hat{H}, \hat{N}_\alpha](t)\rangle$$

$$= i \sum_{\substack{i\in\alpha \\ n\in C}} G_{ni}^<(t,t)h_{in} - h_{ni}G_{in}^<(t,t). \qquad (23.66)$$

A simple diagrammatic argument shows that ($i \in \alpha, n \in C$)

$$G_{ni}(\tau, \tau') = \sum_{\substack{j\in\alpha \\ m\in C}} \int_{\mathcal{C}} \mathrm{d}\tau_1 G_{nm}(\tau, \tau_1)h_{mj}g_{0,ji}(\tau_1, \tau')$$

$$G_{in}(\tau, \tau') = \sum_{\substack{j\in\alpha \\ m\in C}} \int_{\mathcal{C}} \mathrm{d}\tau_1 g_{0,ij}(\tau, \tau_1)h_{jm}G_{mn}(\tau_1, \tau').$$

Using eqn (23.20) we notice that eqn (23.66) can be written as $i\mathrm{Tr}[A^<(t,t)]$ when A is defined as in eqn (23.92) with $B = G_C$ and $C = \Sigma_\alpha$. From the general result (23.93) it then follows that

$$I_\alpha = \int \frac{\mathrm{d}\omega}{2\pi} \mathrm{Tr}\left[\Sigma_\alpha^<(\omega)G_C^>(\omega) - \Sigma_\alpha^>(\omega)G_C^<(\omega)\right], \qquad (23.67)$$

where matrix multiplication is implied. By writing $I = (I_L - I_R)/2$ we obtain an expression that is symmetric in the L, R indices,

$$I = \frac{i}{4\pi} \int \mathrm{Tr}\left[(\Gamma_L - \Gamma_R)G_C^< + (f_L\Gamma_L - f_R\Gamma_R)(G_C^r - G_C^a)\right]\mathrm{d}\omega, \qquad (23.68)$$

where we have suppressed the ω dependence and introduced the coupling strength of lead α, $\Gamma_\alpha = i\left[\Sigma_\alpha^r - \Sigma_\alpha^a\right]$. We notice that when interactions are present, the integrals in eqns (23.68) and (23.67) will have contributions outside the bias window, $\mu_L < \omega < \mu_R$, because the conduction electrons can gain or lose energy by interacting with other electrons in the central region.

23.4.3 Charge conservation

Due to charge conservation in the steady state we expect that $I_L = -I_R = I$, i.e. the current flowing from the left lead to the molecule is the negative of the current flowing from the right lead to the molecule. Below, we derive a condition for this specific form of particle conservation.

From eqn (23.67) the difference between the currents at the left and right interface, $\Delta I = I_L + I_R$, is given by

$$\Delta I = \int \frac{d\omega}{2\pi} \text{Tr}\left[(\Sigma_L^< + \Sigma_R^<)G_C^> - (\Sigma_L^> + \Sigma_R^>)G_C^< \right]. \tag{23.69}$$

To obtain a condition for $\Delta I = 0$ in terms of Σ we start by proving the general identity

$$\int \frac{d\omega}{2\pi} \text{Tr}\left[\Sigma_{\text{tot}}^<(\omega)G_C^>(\omega) - \Sigma_{tot}^>(\omega)G_C^<(\omega) \right] = 0. \tag{23.70}$$

To prove this, we insert $G^{</>} = G_C^r \Sigma_{\text{tot}}^{</>} G_C^a + \Delta^{</>}$ (from eqn (23.34)) in the left-hand side of eqn (23.70). This results in two terms involving $G^r \Sigma_{\text{tot}}^{</>} G^a$ and two terms involving $\Delta^{</>}$. The first two terms contribute by

$$\int \frac{d\omega}{2\pi} \text{Tr}\left[\Sigma_{\text{tot}}^< G^r \Sigma_{\text{tot}}^> G^a - \Sigma_{\text{tot}}^> G^r \Sigma_{\text{tot}}^< G^a \right]. \tag{23.71}$$

Inserting $\Sigma_{\text{tot}}^> = \Sigma_{\text{tot}}^< + (G^a)^{-1} - (G^r)^{-1}$ (we use that $(G^a)^{-1} - (G^r)^{-1} = \Sigma_{\text{tot}}^r - \Sigma_{\text{tot}}^a = \Sigma_{\text{tot}}^> - \Sigma_{\text{tot}}^<$) in this expression and using the cyclic invariance of the trace, it is straightforward to show that eqn (23.71) vanishes. The two terms involving $\Delta^{</>}$ contribute to the left-hand side of eqn (23.70) by

$$\int \frac{d\omega}{2\pi} \text{Tr}\left[\Sigma_{\text{tot}}^<(\omega)\Delta^>(\omega) - \Sigma_{\text{tot}}^>(\omega)\Delta^<(\omega) \right]. \tag{23.72}$$

As mentioned in Section 23.2.3.2, $\Delta^<$ and $\Delta^>$ are always zero when interactions are present. In the case of non-interacting electrons we have $\Sigma_{\text{tot}}^{</>} = \Sigma_L^{</>} + \Sigma_R^{</>}$, which vanish outside the bandwidth of the leads. On the other hand, $\Delta^{</>}$ is only non-zero at energies corresponding to bound states, i.e. states lying outside the bands, and thus we conclude that the term (23.72) is always zero.

From eqns (23.69) and (23.70) it then follows that

$$\Delta I = \int \frac{d\omega}{2\pi} \text{Tr}\left[\Sigma^<(\omega)G_C^>(\omega) - \Sigma^>(\omega)G_C^<(\omega) \right]. \tag{23.73}$$

We notice that without any interactions particle conservation in the sense $\Delta I = 0$ is trivially fulfilled since $\Sigma = 0$. When interactions are present, particle conservation depends on the specific approximation used for the interaction self-energy, Σ.

23.4.4   Charge conservation from Φ-derivable self-energies

One expects that there should be a connection between the condition for particle conservation as expressed by $\Delta I = 0$ in eqn (23.70), and the concept of a conserving approximation in the Φ-derivable sense. Below, we show that $\Delta I = 0$ is always obeyed when the self-energy is Φ-derivable.

We start by noting that eqn (23.65) holds for *any* pair $G(1, 2)$, $\Sigma[G(1, 2)]$ provided Σ is of the Φ-derivable form. In particular eqn (23.65) does not assume that $G, \Sigma[G]$ fulfill a Dyson equation. Therefore, given any orthonormal, but not necessarily complete basis, $\{\phi_i\}$, and writing $G(1, 2) = \sum_{ij} \phi_i(\mathbf{r}_1) G_{ij}(\tau_1, \tau_2) \phi^*(\mathbf{r}_2)$ we get from eqn (23.65) after integrating over $\mathbf{r}_1$,

$$\sum_j \int_C d\tau' [\Sigma_{ij}(\tau, \tau') G_{ji}(\tau', \tau^+) - G_{ij}(\tau^-, \tau') \Sigma_{ji}(\tau', \tau)] = 0, \quad (23.74)$$

which in matrix notation takes the form

$$\int_C d\tau' \mathrm{Tr}[\Sigma(\tau, \tau') G(\tau', \tau^+) - G(\tau^-, \tau') \Sigma(\tau', \tau)] = 0. \quad (23.75)$$

Here, Σ_{ij} is the self-energy matrix obtained when the diagrams are evaluated using G_{ij} and the $V_{ij,kl}$ from eqn (23.3). The left-hand side of eqn (23.75), which is always zero for a Φ-derivable Σ, can be written as $\mathrm{Tr}[A^<(t, t)]$ when A is given by eqn (23.92) with $B = \Sigma$ and $C = G$. It then follows from the general result (23.93) and the condition (23.73) that current conservation in the sense $I_L = -I_R$ is always obeyed when Σ is Φ-derivable.

The above derivation of eqn (23.75) relied on *all* the Coulomb matrix elements, $V_{ij,kl}$, being included in the evaluation of Σ. Thus, the proof does not carry through if a general truncation scheme for the interaction matrix is used. However, in the special case of a truncated interaction of the form (23.43), i.e. when the interaction is a two-point function, eqn (23.75) remains valid. To show this, it is more appropriate to work entirely in the matrix representation and thus define $\Phi[G_{ij}(\tau, \tau')]$ as the sum of a set of skeleton diagrams evaluated directly in terms of G_{ij} and $\tilde{V}_{ij}$. With the same argument as used in eqn (23.64), it follows that Φ is invariant under the transformation

$$G_{ij}(\tau, \tau') \rightarrow e^{i\Lambda_i(\tau)} G_{ij}(\tau, \tau') e^{-i\Lambda_j(\tau')}, \quad (23.76)$$

where Λ is now a discrete vector. By adapting the arguments following eqn (23.64) to the discrete case we arrive at eqn (23.65) with the replacements $\mathbf{r}_1 \rightarrow i$ and $\mathbf{r}_2 \rightarrow j$ and with the integral replaced by a discrete sum over j. Summing also over i leads directly to eqn (23.75), which is the desired result.

To summarize, we have shown that particle conservation in the sense $I_L = -I_R$, is obeyed whenever a Φ-derivable self-energy is used *and* either (i) all Coulomb matrix elements $V_{ij,kl}$ or (ii) the truncated two-point interaction of eqn (23.43), are included in the calculation of Σ.

23.4.4.1   *Example: Transport through a single level*

As an illustrative example we consider transport through an Anderson impurity level (Anderson 1961) connected to wideband leads, i.e. the retarded lead

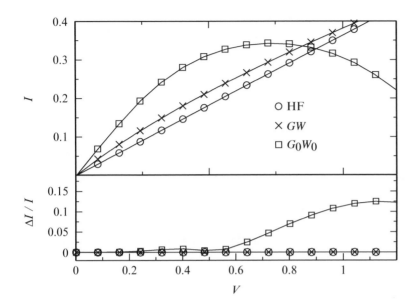

Fig. 23.5 Current–voltage characteristic for an Anderson impurity level with $\Gamma = 0.65$, $\varepsilon_c = -4$, $U = 4$. At finite bias, the non-self-consistent $G_0W_0[G_{HF}]$ approximation yields different currents at the left and right interfaces ($\Delta I \neq 0$) which means that charge conservation is violated. Moreover, G_0W_0 predicts significant negative differential conductance for $V > 0.8$. In contrast, the Φ-derivable HF and GW self-energies both conserve charge.

self-energies are taken to be frequency independent and equal to $i\Gamma$. The Hamiltonian of the central region reads

$$\hat{H}_C = \varepsilon_c c^\dagger c + U n_\uparrow n_\downarrow. \qquad (23.77)$$

where ε_c is the non-interacting energy and U is the correlation energy.

In Fig. 23.5 we show the current–voltage curve calculated from eqn (23.68) using the self-consistent HF, self-consistent GW, and non-self-consistent (NSC) GW approximations. The parameters used are given in the figure caption. In the NSC calculation (referred to as G_0W_0) we use the self-consistent Hartree–Fock GF to evaluate the GW self-energy. With this NSC self-energy we solve the Dyson's equation to obtain a NSC Green's function, which is used to calculate the current. This "one-shot" approach is not conserving, and as a result the currents calculated in the left and right leads from eqn (23.67) are not guaranteed to coincide. From the lower panel of Fig. 23.5 it can be seen that the G_0W_0 self-energy does indeed violate charge conservation, and moreover leads to unphysical negative differential conductance for $V > 0.8$. On the other hand, the self-consistent HF and GW approximations both conserve charge.

At this point we mention that the role of self-consistency in GW calculations has been much debated in the literature where it has been argued that G_0W_0, with G_0 being the DFT Green's function, produces better band structures and bandgaps than self-consistent GW. The present example clearly demonstrates that, regardless of the performance of GW for band-structure calculations, self-consistency is fundamental in non-equilibrium situations.

23.5   Two-level model

In this section we apply the general formalism presented in the preceding sections to a generic two-level model of a molecular junction. In this model the molecule is represented by its highest-occupied (HOMO) and lowest-unoccupied (LUMO) orbitals and the leads are represented by one-dimensional tight-binding chains. With the aim of identifying universal trends we compare Hartree, Hartree–Fock, and GW calculations for the spectrum and I–V characteristic. Not surprisingly the Hartree and HF results show large systematic differences due to the self-interaction errors in the Hartree potential. More interestingly, the dynamic correlations can have a large impact on both the spectrum and IV leading to significant deviations between the GW and HF results—in particular at finite bias.

In Section 23.5.1 we introduce the two-level model. In Section 23.5.2 we study how dynamic screening effects influence the equilibrium position of the HOMO and LUMO levels both in the case of weak and strong coupling between molecule and the leads. In Section 23.5.3 we consider the non-equilibrium transport properties of the model and explain the features of the I–V curves in terms of the variation of the HOMO and LUMO positions as a function of the bias voltage.

23.5.1   Hamiltonian

The model consists of two electronic levels coupled to two semi-infinite 1D tight-binding chains with nearest-neighbor hopping, see Fig. 23.6. The levels represent the HOMO and LUMO states of a molecule and the TB chains represent metallic leads. Electron–electron interactions on the molecule, and between the molecule and the first site of the chains are included. The Hamiltonian of the two-level model reads

$$\hat{H} = \hat{h}_l + \hat{h}_r + \hat{h}_{\text{mol}} + \hat{h}_{\text{coup}} + \hat{U}_{\text{mol}} + \hat{U}_{\text{ext}}. \tag{23.78}$$

Notice that we use a notation different from the canonical (L, C, R)-notation introduced in Section 23.2. This is because of the requirement that all interactions must be contained in the central region. Due to the interactions between the molecule and first sites of the leads, this implies that the central region should at least comprise the molecule *and* the first two sites of the leads. We enumerate the sites of the TB leads from $-\infty$ to -1 (left lead), and from 1 to ∞ (right lead). Thus, $\hat{h}_l$ reads

$$\hat{h}_l = \sum_{i=-\infty}^{-1} \sum_{\sigma=\uparrow,\downarrow} t \left(c_{i\sigma}^{\dagger} c_{i+1\sigma} + c_{i+1\sigma}^{\dagger} c_{i\sigma} \right) \tag{23.79}$$

with a similar expression for $\hat{h}_r$. The non-interacting part of the molecule's Hamiltonian reads

$$\hat{h}_{\text{mol}} = \sum_{\alpha=H,L} \sum_{\sigma=\uparrow,\downarrow} \xi_\alpha d_{\alpha\sigma}^{\dagger} d_{\alpha\sigma}, \tag{23.80}$$

and the coupling is given by

$$\hat{h}_{\text{coup}} = \sum_{\alpha=H,L} \sum_{\sigma=\uparrow,\downarrow} t_{\text{hyb}} \left(c_{1\sigma}^{\dagger} d_{\alpha\sigma} + d_{\alpha\sigma}^{\dagger} c_{1\sigma} + c_{-1\sigma}^{\dagger} d_{\alpha\sigma} + d_{\alpha\sigma}^{\dagger} c_{-1\sigma} \right). \tag{23.81}$$

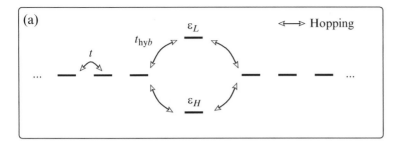

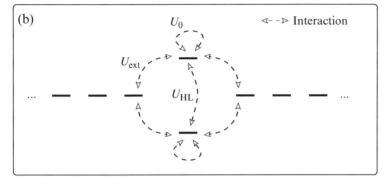

Fig. 23.6 The two-level model used to describe the HOMO and LUMO levels of a molecule couped to metallic leads. (a) The one-particle hopping matrix elements. (b) The electron–electron interactions. The interactions can be divided into intramolecule interactions (U_0 and U_{HL}) and metal–molecule interactions (U_{ext}).

For clarity we use c-operators for the lead sites and d-operators for the HOMO/LUMO levels of the molecule. The interactions are given by

$$\hat{U}_{\mathrm{mol}} = U_0(\hat{n}_{H\uparrow}\hat{n}_{H\downarrow} + \hat{n}_{L\uparrow}\hat{n}_{L\downarrow}) + U_{HL}\hat{n}_H\hat{n}_L \qquad (23.82)$$

$$\hat{U}_{\mathrm{ext}} = U_{\mathrm{ext}}(\hat{n}_1\hat{n}_H + \hat{n}_1\hat{n}_L + \hat{n}_{-1}\hat{n}_H + \hat{n}_{-1}\hat{n}_L), \qquad (23.83)$$

where, e.g. $\hat{n}_H = d_{H\uparrow}^\dagger d_{H\uparrow} + d_{H\downarrow}^\dagger d_{H\downarrow}$ is the number operator of the HOMO level, and $\hat{n}_1$ is the number operator of the first site of the right lead. We set the Fermi level to zero corresponding to half-filled bands. In general, we write $\xi_L = \xi_H + \Delta_0$, i.e. we use the non-interacting energy gap as a free parameter. The occupation of the molecule can then be controlled by adjusting ξ_H.

23.5.2   Renormalization of molecular levels by dynamic screening

In the low-bias regime, the transport properties are to a large extent determined by the position of the HOMO and LUMO levels relative to the Fermi energy of the metal electrodes. For this reason we shall first consider how the HOMO and LUMO positions are affected by the interactions in the zero-bias limit. The material presented in this section is part of ongoing work that will be published elsewhere.

When a molecule is brought into contact with a metal surface a number of different mechanisms will affect the position of the molecule's energy levels. First, the levels are shifted by the electrostatic potential outside the surface. Second, hybridization effects shift and broaden the levels into resonances with finite lifetimes. For our model, the resonance width due to coupling to the

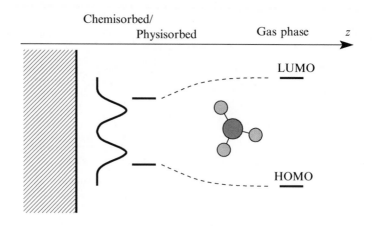

Fig. 23.7 As a molecule approaches a metal surface its HOMO–LUMO gap is reduced by image-charge formation in the metal. If the molecule–metal bond is sufficiently strong (chemisorption) dynamic charge transfer between molecule and metal can give rise to additional reduction of the gap. These renormalization effects requires a dynamic, i.e. frequency dependent, self-energy and thus cannot be described within the single-particle picture.

leads is $\Gamma \approx |t_{\text{hyb}}|^2/t$. Both the shift due to the surface potential and the hybridization are single-particle effects that can be described at a mean-field level such as Kohn–Sham (KS) or Hartree–Fock (HF) theory. On the other hand, correlation effects can renormalize the molecular spectrum in a way that cannot be described within a single-particle picture. Correlation effects are generally small for isolated molecules where HF usually yields good spectra, however, they can become significant when the molecule is in contact with a surface. An important example is the Kondo effect, where electronic interactions qualitatively changes the molecule's spectrum by introducing a narrow peak at the metal Fermi level (Costi *et al.* 1994; Goldhaber *et al.* 1998; Thygesen and Rubio *et al.* 2008). In weakly correlated systems such as molecules with a closed-shell structure adsorbed at a surface, the effect of electronic interactions is expected to be less dramatic. However, as we shall see below, dynamic screening at the molecule metal interface can introduce significant reductions of the HOMO–LUMO gap, which in turn will influence the transport through the molecule (see Fig. 23.7). Such screening effects can be observed in photoemission- and electron tunnelling spectroscopy (Johnson and Hulbert 1987; Kubatkin *et al.* 2003; Repp *et al.* 2005). Recently, first-principles GW calculations for benzene physisorbed on graphite showed a HOMO–LUMO gap reduction of more than 3 eV due to substrate polarization (Neaton *et al.* 2006). More empirical treatments of polarization/screening effects using a scissors operator on the DFT spectrum have recently been applied to transport in molecules (Quek *et al.* 2007; Mowbray *et al.* 2008) and scanning tunnelling microscopy simulations (Dubois *et al.* 2006). So far the theoretical studies have focused on weakly coupled (physisorbed) molecules where the gap renormalization is induced by substrate polarization. This is the situation studied in Section 23.5.2.2. In Section 23.5.2.3 we consider the case of strongly coupled (chemisorbed) molecules.

23.5.2.1   *The quasi-particle picture*

In interacting many-electron systems the concept of a single-particle eigenenergy becomes meaningless. However, for weakly correlated systems the concept can still be maintained due to the long lifetime of certain states of the form $c_m^\dagger|\Psi_0\rangle$ (for ϕ_m unoccupied) or $c_m|\Psi_0\rangle$ (for ϕ_m occupied). These quasi-particle (QP) states describe the many-body N-particle groundstate with an added electron (hole). The energy of the QP states relative to the ground-state energy, is given by the spectral function, $A_m(\varepsilon) = -\mathrm{Im}G^{\mathrm{r}}_{mm}(\varepsilon)$, where G^{r} is the retarded Green's function. For weakly correlated systems $A_m(\varepsilon)$ will be peaked at the QP energy, ε_m, which is equivalent to saying that the QP has a long lifetime. It is instructive to notice that the QP energies measures the cost of removing/adding an electron to the state $|\phi_m\rangle$ *in the presence* of interactions with the other electrons of the system.

For non-interacting electrons the peaks in the spectral function coincide with the eigenvalues of the single-particle Hamiltonian. Mean-field theories like KS or HF include interactions at a static level, i.e. the single-particle eigenvalues correspond to the energy cost of adding/removing an electron when the state of all other electrons is kept fixed. This is the content of Koopman's theorem that states that (for ϕ_m occupied), $\varepsilon_m^{\mathrm{HF}} = E[\Phi_0^{\mathrm{HF}}] - E[c_m\Phi_0^{\mathrm{HF}}]$, i.e. the HF eigenvalues correspond to *unrelaxed* removal/addition energies. In general, the other electrons will respond to the added electron/hole and this will shift, or renormalize, the HF energies. The size of this effect is expected to qualitatively follow the linear response function, χ, which gives the change in the particle density when the system is subject to an external field,

$$\delta n(\mathbf{r}; \omega) = \int \mathrm{d}\mathbf{r}\,\chi(\mathbf{r}, \mathbf{r}'; \omega)v_{\mathrm{ext}}(\mathbf{r}'; \omega). \qquad (23.84)$$

This suggests a direct relation between the impact of dynamic relaxations, or screening, on the QP spectrum, and the response function.

23.5.2.2 *Weak molecule–lead coupling*

We first consider the case of a weakly coupled, or physisorbed, molecule corresponding to small t_{hyb}. We use the following default parameter values: $t = 10$, $U_0 = 4$, $U_{\mathrm{HL}} = 3$, $U_{\mathrm{ext}} = 2$, $t_{\mathrm{hyb}} = 0.3$, $\Delta_0 = 4$, which yield a reso-nance width of $\Gamma \approx 0.01$. In Fig. 23.8(left) we show the HOMO and LUMO positions as function of the interaction U_{ext} as calculated using the HF and GW approximations. As U_{ext} is increased corresponding to the molecule approach-ing the surface, the GW gap decreases, while the HF gap remains unchanged. In the simplest picture the gap reduction is due to the interaction between the added/removed electron and its image charge in the metal. This effect is not present in the HF single-particle spectrum: According to Koopman's theorem the added/removed electron interacts with the HF ground state of the neutral system, which contains no image charge. For small t_{hyb} where the molecule's levels are well defined in energy and localized it is possible to include the response of the metal to the added/removed electron by performing a HF total-energy calculation with constrained HOMO/LUMO occupation numbers. Denoting by Φ_0^n the minimizing Slater determinant with the constraint of n excess electrons on the molecule, we can define constrained HF energy levels as the total energy difference

$$\bar{\varepsilon}_{L/H}^{\mathrm{HF}} = \pm\left(E\left[\Phi_0^{\pm 1}\right] - E\left[\Phi_0^0\right]\right). \qquad (23.85)$$

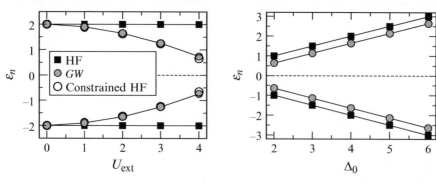

Fig. 23.8 Position of the HOMO and LUMO levels as function of interaction strength U_{ext} (left) and non-interacting gap Δ_0 (right) for a weakly coupled molecule (small Γ). The GW HOMO (LUMO) correspond to the HF energy cost of removing (adding) an electron to the molecule when the Slater determinant of the metal is allowed to relax. The gap reduction from (unrelaxed) HF to GW is thus due to polarization of the metal. This gap reduction is independent of Δ_0.

From the very good agreement between $\bar{\varepsilon}_n^{\text{HF}}$ and ε_n^{GW} seen in Fig. 23.8 (left), we conclude that GW includes the effect of relaxation or screening in the metal at the HF level. The situation is sketched in Fig. 23.10(a).

Koopman's theorem allows us to write the difference between the HF single-particle levels and the result of constrained total energies, as (in the case of the LUMO) $\varepsilon_L^{\text{HF}} - \bar{\varepsilon}_L^{\text{HF}} = E[\Phi_0^{+1}] - E[d_L^{\dagger}\Phi_0^0]$. This energy difference has two contributions: A positive one from the interaction between the added electron and the induced density in the metal, and a negative one being the cost of forming the induced density. The classical image charge approximation in contrast assumes perfect screening in the metal and zero energy cost of polarizing the metal.

From the right panel of Fig. 23.8 it can be seen that the renormalization of the gap is independent of the intrinsic gap of the molecule. This is expected since the image-charge and its interaction with the added electron/hole is independent of the HOMO–LUMO positions.

According to the above, the size of the gap reduction for fixed U_{ext} should depend on the polarizability of the metal. In Fig. 23.9 (top) we show the dependence of the levels as a function of t, i.e. the bandwidth of the metal. The effect is larger for small t corresponding to a narrow band. This is easily understood by noting that narrow bands have larger DOS at ε_F that in turn implies a larger density response function. Indeed, the right panel of Fig. 23.9 shows the diagonal elements of the static (RPA) response function for the HOMO, LUMO and terminal site of the chain. The response function of the HOMO and LUMO is negligible for all values of t, while the response of the terminal site is significant and increases as t is reduced.

23.5.2.3   *Strong molecule–lead coupling*

We now turn to the case of a strongly coupled, or chemisorbed, molecule corresponding to non-negligible t_{hyb}. In the bottom panel of Fig. 23.9 we show the center of the molecular resonances as a function of t_{hyb}. In addition to the HF and GW values, we also show the result when only $\hat{U}_{\text{mol}}$ is treated at the GW level, while $\hat{U}_{\text{ext}}$ is treated within HF. This allows us to isolate the

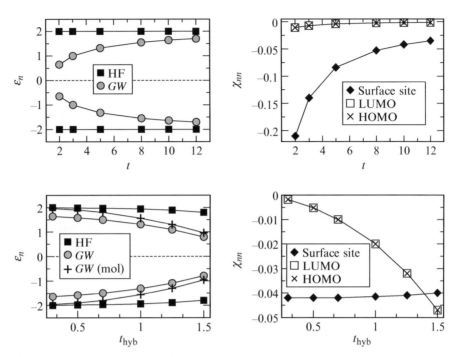

Fig. 23.9 Left panels: Position of the molecule's HOMO and LUMO levels as a function of the metal bandwidth, t and the hybridization strength, $t_{\rm hyb}$, respectively. $GW({\rm mol})$ refers to a calculation where only the interactions internally on the molecule have been treated within GW. Right panels: Static linear response functions (RPA) $\langle \phi_n | \chi (\omega = 0) | \phi_n \rangle$ for the HOMO, LUMO and terminal site of the TB chain. The reduction of the correlated gap relative to HF is due to polarization of the metal and, for large $t_{\rm hyb}$, of the molecule itself. Default parameter values are the same as in Fig. 23.8.

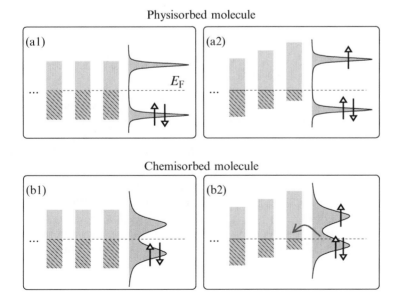

Fig. 23.10 (a1) Ground state of a physisorbed molecule at a metal surface. (a2) When an electron is added to the molecule, the metal is polarized. (b1) Ground state of a chemisorbed molecule at a metal surface. (b2) When an electron is added to the LUMO the metal is polarized, and charge is transferred from the molecule to the metal. The screening effects stabilize the system with added/removed electron and this shifts the occupied (unoccupied) quasi-particle levels up (down).

correlation effects induced by the intramolecular interactions from those of the metal–molecule interactions. Clearly, the correlated gap decreases relative to the HF gap as t_{hyb} is increased. It is also clear that the coupling-dependent part of the gap reduction comes from the interactions internally on the molecule, while the reduction due to $\hat{V}_{ext}$ is largely independent of t_{hyb}. Since $\hat{V}_{mol}$ does not produce any renormalization of the levels of the free molecule (see the $t_{hyb} \rightarrow 0$ limit), the mechanism responsible for the gap reduction must involve the metal. From the lower right panel of Fig. 23.9, we see that the response functions of the HOMO and LUMO states increase with t_{hyb} indicating the gap reduction due to $\hat{V}_{mol}$ is of a similar nature as the image-charge effect, but with the molecule itself being polarized. The effect increases with t_{hyb} because charge transfer between the molecule and the metal due to the external field from the added/removed electron, is larger when resonances are broad and have larger overlap with the metal Fermi level. The situation is sketched in Fig. 23.10(b).

23.5.3   Non-equilibrium transport

The analysis of the previous sections show that dynamic screening effects can have a large effect on the spectrum of the molecule in contact with leads. In this section we shall see that the application of a bias voltage leads to additional renormalization of the spectrum. For simplicity we limit the model to include intramolecule interactions, i.e. we set U_{ext} to zero. This means that reduction of the HOMO–LUMO gap due to image-charge formation in the leads is not included. Whereas the presence of intramolecule interactions did not have a large influence on the equilibrium positions of the HOMO and LUMO levels for small values of Γ (see the $GW(mol)$ result in Fig. 23.9), we will see that this is no longer true under finite bias conditions, where intramolecular screening is strongly enhanced and the lifetimes of the HOMO and LUMO levels can be significantly reduced due to QP scattering. Both effects lead to a reduction of the HOMO–LUMO gap as a function of the bias voltage with a large impact in the calculated I–V curve.

Throughout this section we use the following parameters: $\Delta_0 = 2$, $U_0 = 2$, $U_{HL} = 1.5$, $t = 10$. By varying the one-particle energy ξ_0, we can control the equilibrium occupation of the molecule, N_{el}. We consider the case of weak charge transfer to the molecule, i.e. N_{el} ranges from 2.0 to 2.1, corresponding to ε_F lying in the middle of the gap and slightly below the LUMO, respectively. The Fermi level is set to zero, and the bias is applied symmetrically, i.e. $\mu_L = V/2$ and $\mu_R = -V/2$. The situation is illustrated in Fig. 23.11.

In Fig. 23.12 we show the calculated dI/dV curves (obtained by numerical differentiation of $I(V)$) for different values of Γ and N_{el}. We first notice that the 2B and GW approximations yield similar results in all the cases indicating that the higher-order terms in the GW self-energy are fairly small. For $\Gamma = 1.0$, all methods yield qualitatively the same result. For even larger values of Γ (not shown), and independently of N_{el}, the results become even more similar. In this strong coupling limit, single-particle hybridization effects will dominate over the interactions and xc-effects are small.

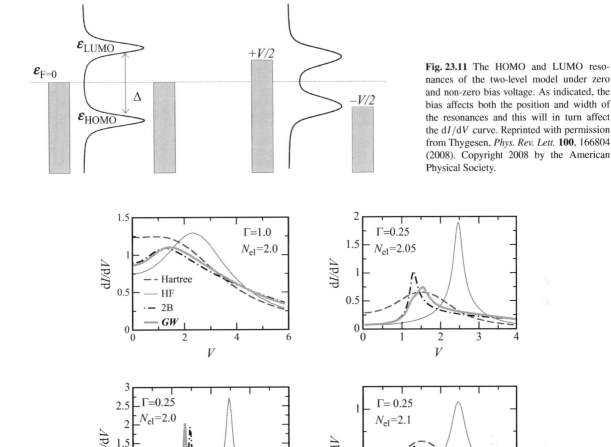

Fig. 23.11 The HOMO and LUMO resonances of the two-level model under zero and non-zero bias voltage. As indicated, the bias affects both the position and width of the resonances and this will in turn affect the dI/dV curve. Reprinted with permission from Thygesen, *Phys. Rev. Lett.* **100**, 166804 (2008). Copyright 2008 by the American Physical Society.

Fig. 23.12 dI/dV curves for different values of the tunnelling strength Γ and occupation of the molecule, N_{el}. The curves are calculated using different approximations for the xc self-energy. Reprinted with permission from Thygesen, *Phys. Rev. Lett.* **100**, 166804 (2008). Copyright 2008 by the American Physical Society.

Focusing on the $\Gamma = 0.25$ case we see that the Hartree approximation severely overestimates the low-bias conductance. This is a consequence of the self-interactions (SI) contained in the Hartree potential that leads to an underestimation of the (equilibrium) HOMO–LUMO gap, see Fig. 23.13 for $V = 0$. On the other hand, the HF, 2B, and GW methods lead to very similar conductances in the low-bias regime. This is consistent with the results of the previous section that showed that intramolecular correlations do not renormalize the equilibrium HF gap much for small Γ. For $\Gamma = 1.0$ the slightly larger conductance in GW and 2B is due to the slight reduction of the equilibrium gap.

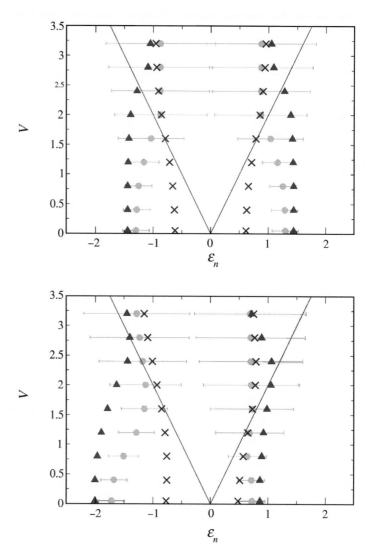

Fig. 23.13 Positions of the HOMO and LUMO levels as a function of the bias voltage for the Hartree (crosses), HF (triangles), and GW (circles) approximations. The horizontal lines show the FWHM of the GW resonances. The FWHM of the Hartree and HF resonances is 2Γ independently of V. Notice the differences in the way the levels enter the bias window: The Hartree gap opens, while the HF and GW gaps close. In the upper graph $\Gamma = 0.25$, $N_{el} = 2.0$ (symmetric case). In the lower graph $\Gamma = 0.25$, $N_{el} = 2.1$. Reprinted with permission from Thygesen, *Phys. Rev. Lett.* **100**, 166804 (2008). Copyright 2008 by the American Physical Society.

We notice that the lower left graph ($\Gamma = 0.25$, $N_{el} = 2.0$) shows an interesting feature. Namely, the HF, 2B, and GW curves all contain an anomalously strong conductance peak. Interestingly, the peak height is significantly larger than 1, which is the maximum conductance for a single level (the Anderson impurity model). Moreover, the full width at half-maximum (FWHM) of the peak is only $\sigma_{HF} = 0.27$ and $\sigma_{2B/GW} = 0.12$, respectively, which is much smaller than the tunnelling broadening of $2\Gamma = 0.5$. We note in passing that the peak loses intensity as N_{el} is increased, and that the Hartree approximation does not reproduce the narrow peak.

23.5.3.1   *Influence of bias on the HOMO and LUMO positions*

To understand the origin of the anomalous conductance peaks, we consider the evolution of the HOMO and LUMO positions as a function of the bias voltage, see Fig. 23.13 (the 2B result is left out as it is similar to GW). Focusing on the upper panel of the figure (corresponding to $N_{el} = 2.0$), we notice a qualitative difference between the Hartree and the SI-free approximations: While the Hartree gap expands as the levels move into the bias window, the HF and GW gaps shrink, leading to a dramatic increase in current around $V = 2.5$ and $V = 1.3$, respectively. This is clearly the origin of the anomalous dI/dV peaks. But why do the SI-free gaps shrink as the bias is raised?

Let us consider the change in the HOMO and LUMO positions when the bias V is increased by $2\delta V$. In general, this change must be determined self-consistently, however, a "first iteration" estimate yields a change in the HOMO and LUMO occupations of $\delta n_H \approx -A_H(-V/2)\delta V$ and $\delta n_L \approx A_L(V/2)\delta V$, respectively. Here, $A_{H/L}$ is the spectral function, or equivalently the DOS, of the HOMO/LUMO levels. At the HF level this leads to

$$\delta\varepsilon_H \approx [-U_0 A(-V/2) + 2U_{HL}A(V/2)]\delta V \qquad (23.86)$$

$$\delta\varepsilon_L \approx [U_0 A(V/2) - 2U_{HL}A(-V/2)]\delta V, \qquad (23.87)$$

where $A = H_H + A_L$ is the total DOS on the molecule and we have used that $A_H(-V/2) \approx A(-V/2)$ and $A_L(V/2) \approx A(V/2)$. The factor 2 in front of U_{HL} accounts for interactions with both spin channels. In the symmetric case ($N_{el} = 2.0$) we have $A(-V/2) = A(V/2)$. Since $U_0 < 2U_{HL}$ this means that $\delta\varepsilon_H > 0$ and $\delta\varepsilon_L < 0$, i.e. the gap is reduced as V is raised. Moreover, it follows that the gap reduction is largest when $A(\pm V/2)$ is largest, that is, just when the levels cross the bias window. In the general case ($N_{el} \neq 2.0$) the direction of the shift depends on the relative magnitude of the DOS at the two bias window edges: a level will follow the edge of the bias window if the other level does not intersect the bias edge. It will move opposite to the bias, i.e. into the bias window, if the other level is close to the bias window edge. This effect is clearly seen in the lower graph of Fig. 23.13 (triangles). Thus, the gap-closing mechanism has the largest impact on the dI/dV curve when the HOMO and LUMO levels hit the bias window simultaneously. Moreover, the effect is stronger the larger U_{HL}/U_0, and the smaller Γ (the maximum in the DOS is $\sim 1/\Gamma$). At the Hartree level, eqns (23.86) and (23.87) are modified by replacing U_0 by $2U_0$ due to self-interaction. This leads to an effective pinning of the levels to the bias window that tends to open the gap as V is increased, see Fig. 23.13 (crosses).

The above analysis shows why the HF gap is reduced as the levels hit the bias window. Interestingly, the bias-driven gap reduction is even stronger in GW and as a consequence the GW conductance peak occurs at much lower bias ($V = 1.5$) than the HF peak ($V = 2.5$). Part of the downshift of the GW conductance peak can be explained from the smaller GW equilibrium gap. Indeed, for $V = 0$ the HF gap is ~ 0.3 larger than the GW gap. However, this effect alone cannot account for the large downshift.

The reason for the bias-induced reduction of the GW gap is twofold: First, intramolecular screening effects are enhanced as the chemical potentials move closer to the molecular levels and increase the susceptibility of the levels. This is analogous to the (equilibrium) situation of increasing Γ shown in the lower panels of Fig. 23.9. The susceptibility of a molecular level is roughly given by the magnitude of the level's DOS at E_F (or chemical potentials). In the latter case this is achieved by broadening the resonances; in the former case by bringing the chemical potential(s) closer to the levels. Secondly, the rate of QP scattering, i.e. the rate at which the initial state $c_i^\dagger|\Psi_0\rangle$ is destroyed due to electron–electron interactions, increases with the bias. This follows from Fermi's golden rule by realizing that the number of available final states of the form $c_f^\dagger|\Psi_0\rangle$ having the same energy as the initial state, increases with bias. The enhanced QP scattering reduces the lifetime of the HOMO and LUMO QPs, which is equivalent to a broadening of the molecular resonances.

The full width at half-maximum (FWHM) of the GW resonances is indicated by horizontal lines in Fig. 23.13. For low bias, the width of the GW resonances is the same as the width of the Hartree and HF resonances. The latter is determined by the coupling to leads and equals $2\Gamma = 0.5$, independently of the bias. According to Fermi-liquid theory, QP scattering at the Fermi level is strongly suppressed in the ground state, i.e. $\mathrm{Im}\Sigma_{ii}(\varepsilon_F) = 0$ for $V = 0$ (recall that $\mathrm{Im}\Sigma$ is inversely proportial to the lifetime). However, as the bias is raised the phase space available for QP scattering is enlarged and $\mathrm{Im}\Sigma$ increases accordingly. As a result of the additional level broadening, $A(\pm V/2)$ increases more rapidly as a function of V. According to eqns (23.86) and (23.87), this will accelerate the gapclosing reduction already at the HF level. Finally, we notice that the long, flat tails seen in the dI/dV of the GW/2B calculations are also a result of the spectral broadening due to QP scattering.

23.6 Applications to C_6H_6 and H_2 molecular junctions

In this section we combine the GW scheme with a Wannier function (WF) basis set to study electron transport through two prototypical junctions, namely a benzene molecule coupled to featureless leads and a hydrogen molecule between two semi-infinite Pt chains. In Section 23.6.1 we briefly present the computational scheme. In the following two sections we analyze the energy spectrum and transport properties of the benzene junction. Finally, in Section 23.6.4 we present results for the I–V curve of the Pt–H_2–Pt junction.

23.6.1 Computational details

Below, we review our computational scheme for GW transport calculations in a WF basis discussed in detail in Thygesen and Rubio (2008). In a first step, periodic supercell DFT calculations are performed for the leads as well as the central region containing the molecule plus part of the leads. We use the Dacapo code (Bahn and Jacobsen 2002) that applies ultrasoft pseudopotentials (Vanderbilt 1990) for the ion cores. The KS eigenstates are expanded in plane waves with a cut-off energy of 340 Ry and the PBE xc-functional

a)   b)

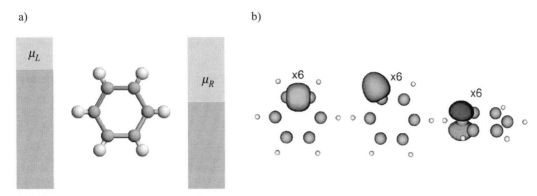

Fig. 23.14 (a) Illustration of a benzene molecule coupled to featureless electrodes with different chemical potentials. (b) Isosurfaces for the 18 partially occupied Wannier functions used as basis functions in the calculations. The WFs are linear combinations of Kohn–Sham eigenstates obtained from a DFT-PBE plane-wave calculation.

(Perdew *et al.* 1996) is used. In the second step, the KS eigenstates are transformed into maximally localized, partially occupied WFs, $\{\phi_n(\mathbf{r})\}$, and the KS Hamiltonians of the central region and the leads are subsequently evaluated in terms of the WF basis. The eighteen maximally localized WFs obtained for the benzene molecule are shown in Fig. 23.14. The xc-potential is excluded from the Hamiltonian of the central region in order to avoid double counting when the GW self-energy is added. The central region Hamiltonian reads

$$[h_C]_{ij} = \langle \phi_i | -\frac{1}{2}\nabla^2 + v_{\text{ps}} + v_{\text{h}} | \phi_j \rangle, \qquad (23.88)$$

where v_{ps} is the pseudopotential and v_{h} is the Hartree potential. Notice that v_{ps} and v_{h} contain contributions from the ion cores and electron density of the leads as they are obtained from a supercell calculation with part of the leads included.

The Coulomb integrals are evaluated for the WFs of the central region,

$$V_{ij,kl} = \int \int \mathrm{d}\mathbf{r}\mathrm{d}\mathbf{r}' \frac{\phi_i(\mathbf{r})^*\phi_j(\mathbf{r}')^*\phi_k(\mathbf{r})\phi_l(\mathbf{r}')}{|\mathbf{r}-\mathbf{r}'|}. \qquad (23.89)$$

For the correlation part of the GW self-energy, $\Sigma_{\text{corr}} = \Sigma_{GW} - \Sigma_x$, we use the effective interaction introduced in Section 23.3.1, i.e. only Coulomb integrals of the form $V_{ij,ij}$ and $V_{ij,ji}$ are included. For the Hartree and exchange self-energies, Σ_{h} and Σ_x, which are easily evaluated from eqns (23.58) and (23.59), we use all the Coulomb matrix elements. Notice, that we need Σ_{h} even though the Hartree potential from electrons in C is already contained in v_{h}. The reason is that the latter is the equilibrium Hartree potential of the DFT calculation, which might well differ from the Hartree potential of a non-equilibrium GW calculation.

The retarded Green's function is evaluated from

$$G^{\text{r}} = \left[\omega - h_C - \Sigma_L - \Sigma_R - \left(\Sigma_h^r[G] - \Sigma_h^r[g_s^{(\text{eq})}]\right) - \Sigma_{xc}^r[G]\right]^{-1}, \quad (23.90)$$

where the frequency dependence has been omitted for notational simplicity. Several comments are in order. First, Σ_L and Σ_R are the lead self-energies of eqn (23.41) (in the wideband approximation Σ_L and Σ_R are diagonal and frequency independent). The term $\Delta v_{\mathrm{h}} \equiv \Sigma_{\mathrm{h}}^{\mathrm{r}}[G] - \Sigma_{\mathrm{h}}^{\mathrm{r}}[g_s^{(\mathrm{eq})}]$ is the change in Hartree potential relative to the equilibrium DFT value. This change is due to the applied bias and the replacement of v_{xc} by Σ_{xc}. In this work Σ_{xc} can be either the exchange or the GW self-energy. Finally, we notice that the bias dependence of the various quantities entering eqn (23.90) has been suppressed for notational simplicity.

23.6.2   Equilibrium spectrum of benzene

In Fig. 23.15 we show the total density of states (DOS) of the isolated benzene molecule calculated using three different approximations: (i) DFT-PBE (ii) Hartree–Fock (iii) fully self-consistent GW. The DOS is given by

$$D(\varepsilon) = -\frac{1}{\pi} \sum_{n=1}^{N_w} \mathrm{Im} G_{nn}^{\mathrm{r}}(\varepsilon), \qquad (23.91)$$

where the sum runs over all WFs on the molecule, and the GF is obtained from eqn (23.90) using a wideband lead self-energy of $\Gamma = 0.05$. We stress that our calculations include the full dynamic dependence of the GW self-energy as well as all off-diagonal elements. Thus, no analytic continuation is performed, and we do not linearize the self-energy around the DFT eigenvalues to obtain an approximate quasi-particle equation, as is done in standard GW calculations.

The spectral peaks seen in Fig. 23.15 occurring above (below) the Fermi level correspond to electron addition (removal) energies. In particular, the HOMO level should coincide with the ionization potential of the isolated molecule, which in the case of benzene is $I_{\exp} = -9.2\,\mathrm{eV}$ (NIST Chemistry WebBook). The PBE functional overestimates this value by 3 eV, giving $I_{\mathrm{PBE}} = -6.2\,\mathrm{eV}$ in good agreement with previous calculations (Niehaus *et al.* 2005). The HF and GW calculation yields $I_{\mathrm{HF}} = -9.7\,\mathrm{eV}$ and

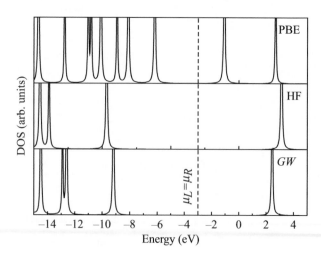

Fig. 23.15 Density of states for a benzene molecule weakly coupled to featureless leads ($\Gamma = 0.05$). The common Fermi level of the leads is indicated. Notice the characteristic opening of the bandgap when going from DFT-PBE to HF, and the subsequent (slight) reduction when correlations are included at the GW level. Reprinted with permission from Thygesen and Rubio, *Phys. Rev. B* **77**, 115333 (2008). Copyright 2008 by the American Physical Society.

$I_{GW} = -9.3\,\text{eV}$, respectively. Given the limited size of the Wannier basis, the precise values should not be taken too strictly. However, the results demonstrate the general trend: KS theory with a local xc-functional underestimates the HOMO–LUMO gap significantly due to SI errors; HF overestimates the gap slightly; GW reduces the HF gap slightly through the inclusion of dynamic screening.

In Fig. 23.16 we plot the size of the HOMO–LUMO gap as a function of the coupling strength Γ. The position of the levels has been defined as the first maximum in the DOS to the left and right of the Fermi level. Both the HF and GW gaps decrease as Γ is increased. This observation is consistent with the model calculations of Section 23.5.2.3 where it was found that the gap reduction due to the correlation part of the GW self-energy, Δ_{corr}, can be understood as a virtual charge transfer between molecule and leads. The reduction of the HF gap as a function of Γ is a consequence of the redistribution of charge from the HOMO to the LUMO when the resonances broaden and their tails start to cross the Fermi level. This is completely analogous to the bias-induced gap reduction discussed in Section 23.5.3. These results for benzene show that the conclusions obtained from the two-level model apply to more realistic systems.

23.6.3   Conductance of benzene

We consider electron transport through the benzene junction under a symmetric bias, $\mu_{L/R} = \pm V/2$, and a wideband coupling strength of $\Gamma_L = \Gamma_R = 0.25\,\text{eV}$. In Fig. 23.17 we compare the differential conductance, dI/dV, calculated from self-consistent DFT-PBE, HF, and GW, as well as non-self-consistent G_0W_0 using either the DFT-PBE or HF Green's function as G_0. The dI/dV has been obtained by numerical differentiation of the $I(V)$ curves calculated from eqn (23.68). For the DFT calculation the finite-bias effects have been included at the Hartree level, i.e. changes in the xc-potential have been neglected. We notice that the HF and $G_0W_0[G_{\text{HF}}]$ results are close to

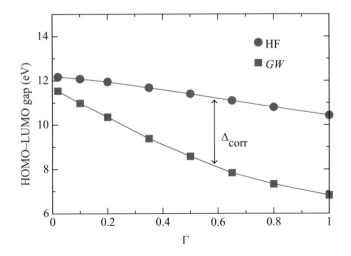

Fig. 23.16 The HF and GW HOMO–LUMO gap of the benzene molecule as a function of the coupling strength Γ. The difference between the curves represents the reduction in the gap due to the correlation part of the GW self-energy. This value increases with the coupling strength, as polarization of the molecule via dynamic charge transfer to the metal becomes possible (see Section 23.5.2.3). Reprinted with permission from Thygesen and Rubio, *Phys. Rev. B* **77**, 115333 (2008). Copyright 2008 by the American Physical Society.

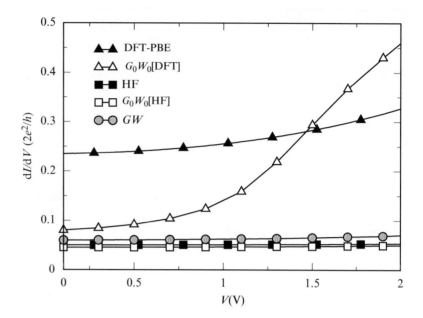

Fig. 23.17 Differential conductance of the benzene junction for $\Gamma_L = \Gamma_R = 0.25\,\mathrm{eV}$. Notice the strong G_0 dependence of the G_0W_0 result. Reprinted with permission from Thygesen and Rubio, *Phys. Rev. B* **77**, 115333 (2008). Copyright 2008 by the American Physical Society.

the self-consistent GW result. These approximations all yield a nearly linear I–V with a conductance of $\sim 0.05G_0$. In contrast, the DFT and $G_0W_0[G_{\mathrm{DFT}}]$ yield significantly larger conductances that increase with the bias voltage. We note that the violation of charge conservation in the G_0W_0 calculations is not too large in the present case ($\Delta I/I < 5\%$). This is in line with our general observation, e.g. from the Anderson model, that $\Delta I/I$ increases with I.

The trends in conductance can be understood by considering the (equilibrium) DOS of the junction shown in Fig. 23.18. As for the weakly coupled (free) benzene molecule whose spectrum is shown in Fig. 23.15, the DFT HOMO–LUMO gap is much smaller than the HF gap, and this explains the larger DFT conductance. The GW gap falls in between the DFT and HF gaps, however, the magnitude of the DOS at E_F is very similar in GW and HF, which is the reason for the similar conductances. It is interesting to note that the HOMO–LUMO gap obtained in the G_0W_0 calculations resemble the gap obtained from G_0, and that the self-consistent GW gap lies in between the $G_0W_0[G_{\mathrm{DFT}}]$ and $G_0W_0[G_{\mathrm{HF}}]$ gaps.

The increase in the $G_0W_0[G_{\mathrm{DFT}}]$ conductance as a function of bias occurs because the LUMO of the $G_0W_0[G_{\mathrm{DFT}}]$ calculation moves downwards into the bias window and becomes partly filled as the voltage is raised. In a self-consistent calculation this would lead to an increase in Hartree potential that would in turn raise the energy of the level. The latter effect is missing in the perturbative G_0W_0 approach and this can lead to uncontrolled changes in the occupations as the present example shows.

Finally, we notice that the $G_0W_0[G_{\mathrm{DFT}}]$ DOS is significantly more broadened than both the $G_0W_0[G_{\mathrm{HF}}]$ and GW DOS. The reason for this is that, as a direct consequence of the small HOMO–LUMO gap, DFT yields a larger DOS close to E_F. The larger DOS in turn enhances the QP scattering and leads

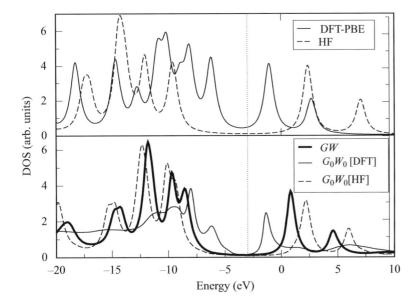

Fig. 23.18 Equilibrium DOS for the benzene molecule coupled to wideband leads with a coupling strength of $\Gamma_L = \Gamma_R = 0.25\,\text{eV}$. Upper panel shows DFT-PBE and HF single-particle approximations, while the lower panel shows the self-consistent GW result as well as one-shot G_0W_0 results based on the DFT and HF Green's functions, respectively. Reprinted with permission from Thygesen and Rubio, *Phys. Rev. B* **77**, 115333 (2008). Copyright 2008 by the American Physical Society.

to shorter lifetimes of the QP in the $G_0W_0[G_{\text{DFT}}]$ calculation. Since the QP lifetime is inversely proportional to $\text{Im}\Sigma_{GW}$ this explains the broadening of the spectrum.

23.6.4 Pt–H$_2$–Pt junction

We consider a molecular hydrogen bridge between infinite atomic Pt chains, as shown in the inset of Fig. 23.19. Experimentally, the conductance of the hydrogen junction is found to be close to the conductance quantum, $G_0 = 2e^2/h$ (Smit *et al.* 2002), and this value has been reproduced by DFT calculations (Thygesen and Jacobsen 2005). Below, we present GW-transport results for a simplified model of this system (using infinite Pt chains as leads), and refer to Thygesen and Rubio (2007) for further details on the calculations.

In the upper panel of Fig. 23.19 we show the local density of states (LDOS) at one of the two H orbitals as calculated within DFT using the PBE xc-functional, as well as self-consistent HF (in the central region). In DFT the H$_2$ bonding state is a bound state at $-7.0\,\text{eV}$ relative to E_F, while the antibonding state lies at $0.4\,\text{eV}$ and is strongly broadened by coupling to the Pt. Moving from DFT to HF the bonding state is shifted down by $\sim 8\,\text{eV}$ because for occupied states the exchange potential is more negative than the DFT xc-potential. The same effect tends to drive the half-filled antibonding state down but in this case the resulting increase in the Hartree potential (about 4 eV) stops it just below E_F.

In the lower panel of Fig. 23.19 we show the LDOS calculated in GW as well as G_0W_0 starting from either DFT or HF, i.e. G_0 is either G_{DFT} or G_{HF}. The large deviation between the two G_0W_0 results is not surprising given the large difference between G_{DFT} and G_{HF}. A closer analysis of the origin of this deviation can be found elsewhere (Thygesen and Rubio 2007). We are, however, aware that part of this large difference could be due to the limited

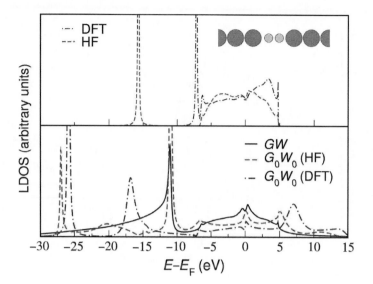

Fig. 23.19 Local density of states at one of the H orbitals of the Pt–H–H–Pt contact shown in the inset. Reprinted with permission from K.S. Thygesen and A. Rubio, J. Chem. Phys. **126**, 091101 (2007). Copyright 2007, American Institute of Physics.

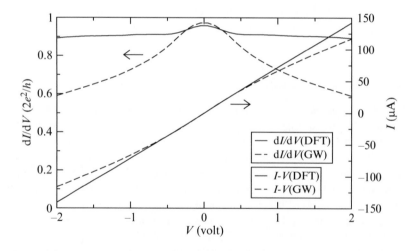

Fig. 23.20 $I–V$ and dI/dV for the hydrogen contact as calculated in DFT(PBE) and self-consistent GW. V is the source–drain bias voltage. Reprinted with permission from K.S. Thygesen and A. Rubio, J. Chem. Phys. **126**, 091101 (2007). Copyright 2007, American Institute of Physics.

size of the basis. We also mention that the LDOS results of Fig. 23.19 can be largely reproduced by including only the second-order GW diagram in the self-energy. Thus, the higher-order RPA diagrams are less important in this case.

In Fig. 23.20 we show the self-consistently calculated $I–V$ characteristics in DFT and GW. At low bias both schemes yield a conductance close to the experimental value of $1G_0$. The DFT conductance is nearly constant over the bias range, and is in fact very similar to the HF result (not shown). In contrast, the GW conductance falls off at larger bias. This is due to enhancement of quasi-particle scattering at finite bias. The QP scattering reduces the lifetime of the QPs leading to broadening of the spectral peak associated with the antibonding state of the hydrogen molecule in agreement with the results for the two-level model discussed in Section 23.5.3. Since the correlation-induced lifetime of QP at the Fermi level, $\text{Im}\Sigma_{\text{corr}}(E_F)$, vanishes identically in equilibrium, the finite-bias conductance suppression seen in Fig. 23.20 is a direct result of the non-equilibrium treatment of correlations.

23.7 Summary and perspectives

The feasibility of using many-body perturbation theory in combination with the Keldysh–Green's function formalism to address non-equilibrium quantum transport through an interacting region coupled to non-interacting leads has been demonstrated. The effect of electronic correlations was incorporated into the Green's function via the GW self-energy, and the coupling to leads was treated exactly (to all orders in the hopping between leads and central region). The important connection between self-consistency in the GW self-energy and charge conservation was emphasized, and it was demonstrated that a non-self-consistent treatment of the GW self-energy (the G_0W_0 approach) violates the continuity equation and produces unphysical results at finite bias. This, together with the arbitrariness of the G_0W_0 approximation due to its G_0 dependence, speaks in favor of the self-consistent GW approach to non-equilibrium transport.

The role of dynamic correlation effects in quantum transport was illustrated by applying the GW-transport scheme to a generic two-level model, a benzene molecule between featureless leads, and a hydrogen molecule between infinite Pt chains. It was shown that dynamic polarization of the leads as well as the molecule itself can lead to significant reduction of the molecule's HOMO–LUMO gap. The polarization effects were found to increase with the bias voltage where also quasi-particle scattering is strongly enhanced leading to broadening of the molecular resonances. As was shown, all these effects can have a large impact on the calculated I–V curve. This should always be kept in mind, when interpreting results of mean-field (DFT or Hartree-Fock) transport calculations—in particular under finite bias.

As mentioned in the introduction, the quantitative theoretical description of quantum transport in nanoscale structures from first principles is an extremely complex problem. Nevertheless, simulation methods with predictive power are required to advance the field further. It has been known for several years that the standard DFT-NEGF scheme fails to predict even the zero-bias conductance of certain classes of systems. This state of affairs makes it difficult, although not impossible, to link theory and experiments and thereby stimulate the development of nanoscale electronics.

As illustrated by the examples given in this chapter, reliable schemes for quantum transport should account for dynamic correlation effects in some way or another. The GW method discussed here includes some correlation effects, but misses others, e.g. the side peaks in the spectral function of the Anderson model are not well reproduced (Thygesen and Rubio 2007). Methods developed for strongly correlated systems, such as density matrix renormalization group theory (Schmitteckert and Evers 2008), are limited to simple models due to their inexpedient scaling with system size. The effective single-particle scheme of TDDFT makes it an attractive alternative to many-body perturbation theory, in particular for dynamical transport phenomena (Stefanucci *et al.* 2007). However, the inclusion of correlation effects requires use of xc-potentials with memory that have so far proved difficult to construct.

Another important aspect of the problem is related to the coupling between electrons and nuclei. Despite the large difference in the general time scales of electronic and nuclear motions, electronic wave packets quite often couple with the dynamics of nuclear motion (Frederiksen *et al.* 2004; Verdozzi *et al.* 2006). The proper incorporation of the electronic–nuclear interaction is crucial for describing a host of dynamical processes including Joule heating, electromigration, laser-induced electronic transport and electron transfer in molecular, biological, or electrochemical systems. Within the ground-state DFT framework, the computation of forces on the nuclei is trivial thanks to the Hellman–Feynman theorem. The situation is more complex out of equilibrium, and even more so in combination with a many-body description of the electrons, where the Hellman–Feynman theorem does not apply. However, the electron–ion dynamics must eventually be taken properly into account for a realistic description of a large class of molecular devices relevant for technological applications such as fast, integrated, optoelectronic nanodevices.

Acknowledgments

KST acknowledges support from the Danish Center for Scientific Computing through grant No. HDW-1103-06 and The Lundbeck Foundation's Center for Atomic-scale Materials Design (CAMD). AR acknowledges funding by the Spanish MEC (FIS2007-65702-C02-01), "Grupos Consolidados UPV/EHU del Gobierno Vasco" (IT-319-07), and by the European Community through NoE Nanoquanta (NMP4-CT-2004-500198), e-I3 ETSF project (INFRA-2007-1.2.2: Grant Agreement Number 211956) SANES(NMP4-CT-2006-017310), DNA-NANODEVICES (IST-2006-029192) and NANO-ERA Chemistry projects and the computer resources provided by the Barcelona Supercomputing Center, the Basque Country University UPV/EHU (SGIker Arina).

Appendix

Let $B(\tau, \tau')$ and $C(\tau, \tau')$ be two matrix-valued functions on the Keldysh contour, and consider the commutator A defined by

$$A(\tau, \tau') = \int_C [B(\tau, \tau_1)C(\tau_1, \tau') - C(\tau, \tau_1)B(\tau_1, \tau')]\mathrm{d}\tau_1, \qquad (23.92)$$

where matrix multiplication is implied. Under steady-state conditions where the real-time components of B and C can be assumed to depend only on the time difference $t' - t$, the following identity holds:

$$\text{Tr}[A^<(t, t)] = \int \frac{d\omega}{2\pi} \text{Tr}[B^<(\omega)C^>(\omega) - B^>(\omega)C^<(\omega)].$$ (23.93)

To prove this relation we first use the Langreth rules to obtain

$$A^<(t, t') = \int \Big[B^<(t, t_1)C^a(t_1, t') + B^r(t, t_1)C^<(t_1, t') $$
$$- C^<(t, t_1)B^a(t_1, t') - C^r(t, t_1)B^<(t_1, t') \Big] dt_1.$$

Since all quantities on the right-hand side depend only on the time difference we identify the integrals as convolutions that in turn become products when Fourier transformed. We thus have

$$A^<(t, t) = \int \frac{d\omega}{2\pi} A^<(\omega)$$
$$= \int \frac{d\omega}{2\pi} [B^<(\omega)C^a(\omega) + B^r(\omega)C^<(\omega)$$
$$- C^<(\omega)B^a(\omega) - C^r(\omega)B^<(\omega)].$$

Equation (23.93) now follows from the cyclic property of the trace and the identity $G^r - G^a = G^> - G^<$.

References

Anderson, P.W. *Phys. Rev.* **124**, 41 (1961).

Bahn, S.R., Jacobsen, K.W. *Comp. Sci. Eng.* **4**, 56 (2002).

Baym, G. *Phys. Rev.* **127**, 1391 (1962).

Baym, G., Kadanoff, L.P. *Phys. Rev.* **124**, 287 (1962).

Bokes, P., Jung, J., Godby, R.W. *Phys. Rev. B* **76**, 125433 (2007).

Brandbyge, M., Mozos, J.L., Ordejon, P., Taylor, J., Stokbro, K. *Phys. Rev. B* **65**, 165401 (2002).

Costi, T.A., Hewson, A.C., Zlatic V.J. *Phys. Cond. Matt.* **6**, 2519 (1994).

Cuniberti, G., Fagas, G., Richter, K. *Introducing Molecular Electronics* (Springer, 2005).

Datta, S. *Electronic Transport in Mesoscopic Systems* (Cambridge University Press, 1995).

Delaney, P., Greer, J.C. *Phys. Rev. Lett.* **93**, 036805 (2004).

Di Ventra, M., Todorov, T.N. *J. Phys.: Condens. Matt.* **16**, 8025 (2004).

Dubois, M., Latil, S., Scifo, L., Grvin, B., Rubio, A. *J. Chem. Phys.* **125**, 34708–9 (2006).

Frederiksen, T., Brandbyge, M., Lorente, N., Jauho, A.-P. *Phys. Rev. Lett.* **93**, 256601 (2004).

García-Suárez, V.M., Rocha, A.R., Bailey, S.W., Lambert, C.J., Sanvito, S., Ferrer, J. *Phys. Rev. Lett.* **95**, 256804 (2005).

Goldhaber, -G.D., Shtrikman, H., Mahalu, D., Abusch-Magder, D., Meirav, U., Kastner, M.A. *Nature* **391**, 156 (1998).

Haug, H., Jauho, A.-P. *Quantum Kinetics in Transport and Optics of Semiconductors* (Springer, 1998).

Hettler, M.H., Wenzel, W., Wegewijs, M.R., Schoeller, H. *Phys. Rev. Lett.* **90**, 076805 (2003).

Heurich, J., Cuevas, J.C., Wenzel, W., Schon, G. *Phys. Rev. Lett.* **88**, 256803 (2002).

Hybertsen, M.S., Louie, S.G. *Phys. Rev. B* **34**, 5390 (1986).

Joachim, C., Gimzewski, J.K., Aviram, A. *Nature* **408**, 541 (2000).

Johnson, P.D., Hulbert, S.L. *Phys. Rev. B* **35**, 9427 (1987).

Kubatkin, S., Danilov, A., Hjort, M., Cornil, J., Bredas, J.-L., Stuhr-Hansen, N., Hedegård, P., Bjørnholm, T. *Nature* **425**, 698 (2003).

Landauer, R. *Phil. Mag.* **21**, 863 (1970).

Leeuwen, R. van, Dahlen, N.E., Stefanucci, G., Almbladh, C.O., von Barth, U. *Time-Dependent Density Functional Theory* (Springer, 2006).

Meir, Y., Wingreen, N.S. *Phys. Rev. Lett.* **68**, 2512 (1992).

Mowbray, D.J., Jones, G., Thygesen, K.S. *J. Chem. Phys.* **128**, 111103 (2008).

Neaton, J.B., Hybertsen, M.S., Louie, S.G. *Phys. Rev. Lett.* **97**, 216405 (2006).

Nelson, W., Bokes, P., Patrick, R., Godby, R.W. *Phys. Rev. A* **75**, 032505 (2007).

Niehaus, T.A., Rohlfing, M., Della, F.S., Di Carlo, A., Frauenheim, T. *Phys. Rev. A* **71**, 022508 (2005).

NIST Chemistry WebBook, http://webbook.nist.gov/chemistry/

Nygard, J., Cobden, D.H., Lindelof, P.E. *Nature* **408**, 342–6 (2000).

Onida, G., Reining, L., Rubio, A. *Rev. Mod. Phys.* **74**, 601 (2002).

Perdew, J.P., Burke, K., Ernzerhof, M. *Phys. Rev. Lett.* **77**, 3865 (1996).

Quek, S.Y., Venkataraman, L., Choi, H.J., Louie, S.G., Hybertsen, M.S., Neaton, J.B. *Nano Lett.* **7**, 3477 (2007).

Reichert, J., Ochs, R., Beckman, D., Weber, H.B., Mayor, M., Löhneysen, H. *Phys. Rev. Lett.* **88**, 176804 (2002).

Repp, J., Meyer, G., Stojkovic, S.M., Gourdon, A., Joachim, C. *Phys. Rev. Lett.* **94**, 026803 (2005).

Smit, R.H.M., Noat, Y., Untiedt, C., Lang, N.D., Hemert, M.C., Ruitenbeek, J.M. *Nature* **419**, 906 (2002).

Schmitteckert, P., Evers, F. *Phys. Rev. Lett.* **100**, 086401 (2008).

Spataru, C.D., Benedict, L.X., Louie, S.G. *Phys. Rev. B* **69**, 205204 (2004).

Stan, A., Dahlen, N.E., Leeuwen, R. van *Europhys. Lett.* **76**, 298 (2006).

Stefanucci, G., Kurth, S., Gross, E.K.U., Rubio, A. (eds) *J. Seminario* **17**, 247–284 (2007).

Taylor, J., Guo, H., Wang, J. *Phys. Rev. B* **63**, 245407 (2001).

Thygesen, K.S. *Phys. Rev. Lett.* **100**, 166804 (2008).

Thygesen, K.S. *Phys. Rev. B* **73**, 035309 (2006).

Thygesen, K.S., Jacobsen, K.W. *Phys. Rev. Lett.* **94**, 036807 (2005).

Thygesen, K.S., Rubio, A. *J. Chem. Phys.* **126**, 091101 (2007).

Thygesen, K.S., Rubio, A. *Phys. Rev. B* **35**, 9427 (2008).

Thygesen, K.S., Rubio, A. *Phys. Rev. Lett.* **102**, 046802 (2009).

Toher, C., Filippetti, A., Sanvito, S., Burke, K. *Phys. Rev. Lett.* **95**, 146402 (2005).

Vanderbilt, D. *Phys. Rev. B* **41**, 7892 (1990).

Verdozzi, C., Godby, R.W., Holloway, S. *Phys. Rev. Lett.* **74**, 2327 (1995).

Verdozzi, C., Stefanucci, G., Almbladh, C.-O. *Phys. Rev. Lett.* **97**, 046603 (2006).

24

Spin currents in semiconductor nanostructures: A non-equilibrium Green-function approach

Branislav K. Nikolić, Liviu P. Zârbo, and Satofumi Souma

24.1 Introduction

Over the past two decades, *spintronics* (Žutić *et al.* 2004) has emerged as one of the most vigorously pursued areas of condensed-matter physics, materials science, and nanotechnology. A glimpse at the Oxford English Dictionary reveals the following attempt to define the field succinctly: *"spintronics is a branch of physics concerned with the storage and transfer of information by means of electron spins in addition to electron charge as in conventional electronics."* The rise of spintronics was ignited by basic research on ferromagnet/normal-metal multilayers in late 1980s (Maekawa and Shinjo 2002), as recognized by the Nobel Prize in Physics for 2007 being awarded to A. Fert and P. Grünberg for the discovery of giant magnetoresistance (GMR). The GMR phenomenon also exemplifies one of the fastest transfers of basic research results in condensed-matter physics into applications, where in less then ten years since its discovery it has revolutionized information-storage technologies by enabling a 100-fold increase in hard-disk storage capacity.

In recent years the frontiers of spintronics have been reshaped through several intertwined lines of research: (*i*) ferromagnetic metal devices where the main theme is manipulation of magnetization via electric currents and vice versa (Ralph and Stiles 2008); (*ii*) ferromagnetic semiconductors that, unlike metal ferromagnets, offer additional possibilities to manipulate their magnetic ordering (such as Curie temperature, coercive fields, and magnetic dopants), but are still below optimal operating temperature (Jungwirth *et al.* 2006);

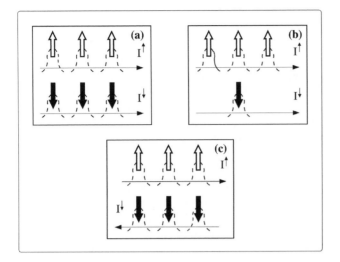

Fig. 24.1 The classification of spin I^S and charge I currents in metal and semiconductor spintronic systems corresponding to spatial propagation of spin-↑ and spin-↓ electronic wave packets carrying spin-resolved currents $I^\uparrow$ and $I^\downarrow$: (a) conventional charge current $I = I^\uparrow + I^\downarrow \neq 0$ is spin-unpolarized $I^S = \frac{\hbar}{2e}(I^\uparrow - I^\downarrow) \equiv 0$; (b) spin-polarized charge current $I \neq 0$ is accompanied also by spin current $I^S \neq 0$; and (c) *pure* spin current $I^S = \frac{\hbar}{2e}(I^\uparrow - I^\downarrow) \neq 0$ arising when spin-↑ electrons move in one direction, while an equal number of spin-↓ electrons move in the opposite direction, so that total charge current is $I \equiv 0$.

(*iii*) paramagnetic semiconductor spintronics (Awschalom and Flatté 2007) largely focused on *all-electrical* manipulation of spins via spin-orbit (SO) coupling effects in solids (Fabian *et al.* 2007); and (*iv*) spins in semiconductors as building blocks of futuristic solid-state-based quantum computers (Hanson *et al.* 2007). Unlike early "non-coherent" spintronics phenomena (such as GMR), the major themes of the "second-generation" spintronics are moving toward the spin-coherent realm where the spin component persists in the direction transverse to external or effective internal magnetic fields. Recent experiments exploring such phenomena include: spin-transfer torque where spin current of large enough density injected into a ferromagnetic layer either switches its magnetization from one static configuration to another or generates a dynamical situation with steady-state precessing magnetization (Ralph and Stiles 2008); transport of coherent spins (able to precess in the external magnetic fields) across ∼100-µm thick silicon wafers (Huang *et al.* 2007); and the direct and inverse spin-Hall effects (SHE) in bulk (Kato *et al.* 2004a) and low-dimensional (Sih *et al.* 2005; Wunderlich *et al.* 2005) semiconductors and metals (Saitoh *et al.* 2006; Valenzuela and Tinkham 2006; Kimura *et al.* 2007) (SHE in both metals and semiconductors has been observed even at room temperature (Stern *et al.* 2006; Kimura *et al.* 2007)). A closely related effort that permeates these subfields is the generation and detection of *pure spin currents* (Nagaosa 2008) which do not transport any net charge, as illustrated in Fig. 24.1. Their harnessing is expected to offer both new functionality and greatly reduced power dissipation.[1]

In this chapter we discuss how different tools of quantum-transport theory, based on the non-equilibrium Green-function (NEGF) techniques,[2] can be extended to treat spin currents and spin densities in *realistic* open paramagnetic semiconductor devices out of equilibrium. In such devices, the most important spin-dependent interaction is the SO coupling stemming from relativistic corrections to the Pauli–Schrödinger dynamics of spin-$\frac{1}{2}$ electrons. Recent theoretical efforts to understand spin transport in the presence of SO couplings

[1] The Joule heat losses induced by the current flow set the most important limits (Keyes 2005) for conventional electronics, as well as for hybrid electronic–spintronic or purely spintronic devices envisioned to perform both storage and information processing on a single chip.

[2] The NEGF techniques for finite-size devices have been developed over the past three decades mainly through the studies of charge currents of non-interacting quasi-particles in mesoscopic semiconductor (Datta 1995) and nanoscopic molecular (Koentopp *et al.* 2008) where quantum-coherent effects on electron transport are the dominant mechanism because of the smallness of their size. Current frontiers of the NEGF theory are also concerned with the inclusion of many-particle interaction effects responsible for dephasing (Okamoto 2007; Thygesen and Rubio 2008). For a lucid introduction to the general scope of NEGF formalism applied to finite-size devices see Chapter 1 by S. Datta in this volume, as well as Chapter 23 by K.S. Thygesen and A. Rubio in this volume focusing on the inclusion of electronic correlations in NEGF applied to molecular junctions.

[4] After learning about the NEGF techniques for spin transport discussed in this chapter, an interested reader might enter the field by trying to reproduce many other examples treated within this framework in our journal articles (Souma and Nikolić 2004, 2005; Nikolić and Souma 2005a; Nikolić *et al.*, 2005a, 2005b, 2006; Dragomirova and Nikolić 2007; Nikolić and Zârbo 2007; Zârbo and Nikolić 2007; Nikolić and Dragomirova 2009).

by using conventional approaches, such as the bulk conductivity of infinite homogeneous systems (computed via the Kubo formula (Murakami *et al.* 2003; Sinova *et al.* 2004) or the kinetic equation (Mishchenko *et al.* 2004) or spin-density[3] diffusion equations for bounded systems (Burkov *et al.* 2004; Bleibaum 2006; Galitski *et al.* 2006), have encountered enormous challenges even when treating non-interacting quasi-particles. Such intricacies in systems with the intrinsic SO couplings, that act homogeneously throughout the sample, can be traced to spin non-conservation due to spin precession that leads to ambiguity in defining spin currents (Shi *et al.* 2006; Sugimoto *et al.* 2006) in the bulk or ambiguity in supplying the boundary conditions for the diffusion equations (Bleibaum 2006; Galitski *et al.* 2006).

On the other hand, spin-resolved NEGF techniques discussed in this chapter offer a consistent description of both phase-coherent (at low temperatures) and semi-classical (at finite temperatures where dephasing takes place) coupled spin and charge transport in both clean and disordered realistic finite-size devices attached to external current and voltage probes, as encountered in experiments. The physical quantities that can be computed within this framework yield experimentally testable predictions for outflowing spin currents, induced voltages by their flow, and spin densities within the device. The presentation is tailored to be mostly of a tutorial style, introducing the essential theoretical formalism and practical computational techniques at an accessible level that should make it possible for graduate students and non-specialists in physics and engineering to engaged in theoretical and computational modelling of nanospintronic devices. We illustrate formal developments with examples drawn from the filed of the *mesoscopic* SHE (Hankiewicz *et al.* 2004; Nikolić *et al.* 2005b; Ren *et al.* 2006; Sheng and Ting 2006; Bardarson *et al.* 2007) in low-dimensional SO-coupled semiconductor nanostructures.[4]

24.2   What is pure spin current?

Pure spin current represents flow of spin angular momentum that is not accompanied by any net charge transport (Fig. 24.1(c)). These circuits can be contrasted with traditional electronic circuits where equal number of spin-↑ and spin-↓ electrons propagate in the same direction, so that total charge current in that direction $I = I^\uparrow + I^\downarrow$ is unpolarized $I^S = 0$ (Fig. 24.1(a)). Spin currents are substantially different from familiar charge currents in two key aspects: they are time-reversal invariant and they transport a vector quantity. In metal spintronic devices, ferromagnetic elements polarize electron spin, thereby leading to a difference in charge currents of spin-↑ and spin-↓ electrons (Fig. 24.1(b)). Such spin-polarized charge currents are accompanied by a net spin current $I^S = \frac{\hbar}{2e}(I^\uparrow - I^\downarrow) \neq 0$, as created and detected in magnetic multilayers (Maekawa and Shinjo 2002). Figure 24.1 also provides a transparent illustration of one of the major advantages of pure spin currents over the spin-polarized charge currents employed by the "first-generation" spintronic devices. For example, to transport information via $2 \times \hbar/2$ spin angular momenta, Fig. 24.1(b) utilizes four electrons. In Fig. 24.1(c) the same transport of $2 \times \hbar/2$ is achieved using only two electrons moving in opposite direction.

The Joule heat loss in the latter situation is only 25% of the dissipative losses in the former case.

24.3   How can pure spin currents be generated and detected?

Among a plethora of imaginative theoretical proposals (Sharma 2005; Tserkovnyak *et al.* 2005; Tang 2006; Nagaosa 2008) to generate pure spin currents using quantum effects in ferromagnet, semiconductor, and superconductor systems and their hybrids, only a few have received continuous experimental attention. These include: non-local spin injection in lateral spin-valves (Valenzuela and Tinkham 2006; Kimura *et al.* 2007); adiabatic quantum spin-pumps based on semiconductor quantum dots (Watson *et al.* 2003); spin pumping by rotating magnetization of a ferromagnetic layer driven by microwaves (Saitoh *et al.* 2006); optical pump-probe experiments on semiconductors (Stevens *et al.* 2003); and the spin-Hall effect (Sih *et al.* 2006).

Even if spin currents are induced easily, their detection can be quite challenging since transport of electron spin between two locations in real space is alien to Maxwell electrodynamics and no "spin-current ammeter" exists (Adagideli *et al.* 2006). In metal spintronic devices spin currents can be converted into voltage signal (Jung and Lee 2005) by injection into ferromagnetic electrode (as achieved in lateral spin-valves). On the other hand, for semiconductor spintronic devices, which do not couple well to metallic ferromagnets (Fabian *et al.* 2007), it is important to avoid ferromagnetic elements (and their stray fields) in both the spin-injection and the spin-detection processes. Multifarious theoretical ideas have been contemplated to solve this fundamental problem, ranging from the detection of tiny electric fields induced by the flow of magnetic dipoles associated with spins (Meier and Loss 2003) to nanomechanical detection of oscillations induced by spin currents in suspended rods (Mal'shukov *et al.* 2005). Desirable schemes to detect pure spin current in semiconductors should exploit fundamental quantum-mechanical effects that can transform its flux into conventionally measurable voltage drops and charge currents within the same circuit through which the spin current is flowing. The recently discovered SHE holds great promise to revolutionize generation, control, and detection of pure spin fluxes within the setting of all-electrical circuits.

24.4   What is the spin-Hall effect?

The SHE actually denotes a *collection* of phenomena manifesting as transverse separation of spin-↑ and spin-↓ states driven by longitudinally injected standard unpolarized charge current or longitudinal external electric field (Murakami 2006; Schliemann 2006; Sinova *et al.* 2006; Engel *et al.* 2007; Nagaosa 2008). The spins separated in this fashion comprise either a pure spin current or accumulate at the lateral sample boundaries. Its Onsager reciprocal phenomenon—the inverse spin-Hall effect (Hirsch 1999; Hankiewicz *et al.*

Fig. 24.2 Basic phenomenology of the direct and inverse SHEs: (a) conventional (unpolarized) charge current flowing longitudinally through the sample experiences transverse deflection of opposite spins in opposite direction due to SO-coupling-induced "forces". This generates pure spin current in the transverse direction or spin accumulation (when transverse electrodes are removed) of opposite sign at the lateral sample edges; (b) pure spin current flowing through the same sample governed by SO interactions will induce transverse charge current or voltage drop $\Delta V = V_2 - V_3$ (when transverse leads are removed). Note that to ensure purity ($I_2 = I_3 \equiv 0$) of the transverse spin-Hall current in (a), employed to define manifestations of the mesoscopic SHE in ballistic or disordered SO-coupled multiterminal nanostructures, one has to apply proper voltages V_2 and V_3.

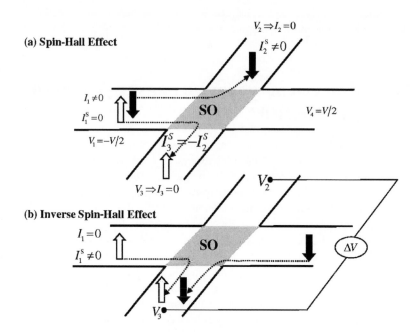

(a) Spin-Hall Effect

(b) Inverse Spin-Hall Effect

2004, 2005; Li *et al.* 2006) where longitudinal pure spin current generates transverse charge current or voltage between the lateral boundaries—offers one of the most efficient schemes to detect elusive pure spin currents by converting them into electrical quantities (Saitoh *et al.* 2006; Valenzuela and Tinkham 2006; Kimura *et al.* 2007). For an illustration of possible experimental manifestations of the direct and inverse SHE in multiterminal nanostructures see Fig. 24.2.

While SHE is analogous to the classical Hall effect of charges, it occurs in the absence of any externally applied magnetic fields or magnetic ordering in the equilibrium state. Instead, both the direct and the inverse SHE essentially require the presence of some type of SO interactions in solids. Although SO couplings are a tiny relativistic effect for electrons in vacuum, they can be enhanced by several orders of magnitude for itinerant electrons in semiconductors due to the interplay of crystal symmetry and strong electric fields of atom cores (Winkler 2003). They have recently emerged as one of the central paradigms (Fabian *et al.* 2007) of semiconductor spintronics—unlike cumbersome magnetic fields, they make possible spin control on very short length and timescales via electric fields, and could, therefore, enable smooth integration with conventional electronics.

Some of the observed SHE manifestations (Kato *et al.* 2004a) have been explained (Engel *et al.* 2007) by the SO-coupling effects localized to the region around impurities that bring the interplay of skew-scattering (asymmetric SO-dependent scattering, which deflects spin-↑ and spin-↓ electrons of an unpolarized flux in opposite directions transverse to the flux) and side jump (due to the non-canonical nature of the physical position and velocity operators in the presence of the SO coupling around an impurity (Sinitsyn 2008)). These impurity-driven mechanisms were a crucial ingredient of the seminal arguments in early 1970s (D'yakonov and Perel' 1971) predicting theoretically the existence of (in modern terminology (Hirsch 1999)) the *extrinsic* SHE. The extrinsic SO effects are fixed by the material properties and the corresponding SHE is hardly controllable, except through charge density and mobility (Awschalom and Flatté 2007).

A strong impetus for the revival of interest into the realm of SHE has arisen from speculations (Murakami *et al.* 2003; Sinova *et al.* 2004) that transverse pure spin currents, several orders of magnitude larger than in the case of extrinsic SHE, can be driven by longitudinal electric fields in systems with *intrinsic* SO couplings. Such SO couplings manifest in materials with bulk inversion asymmetry or semiconductor heterostructures where inversion symmetry is broken structurally. They act homogeneously throughout the sample inducing the spin splitting of the quasiparticle energy bands. The *intrinsic* (or band-structure-driven) SHE could account for large spin-current signals observed in 2D hole gases (Wunderlich *et al.* 2005) or huge SHE response in some metals (Guo *et al.* 2008). In addition, since the strength of the Rashba SO coupling (Winkler 2003) in two-dimensional electron gases (2DEGs) within heterostructures with strong structural inversion asymmetry can be controlled experimentally by a gate electrode (Nitta *et al.* 1997; Grundler 2000), intrinsically driven SHEs are amenable to easy all-electrical manipulation in realistic nanoscale multiterminal devices (Nikolić *et al.* 2005b).

The magnitude of both the extrinsic and intrinsic SHE also depends on the impurities, charge density, geometry, and dimensionality. Such a variety of SHE manifestations poses immense challenge for attempts at a unified theoretical description of spin transport in the presence of relativistic effects. This has not been resolved by early hopes (Murakami *et al.* 2003; Sinova *et al.* 2004) that auxiliary spin-current density computed within infinite homogeneous systems could be elevated to a universally applicable and experimentally measurable quantity (for more technical discussion of these issues see Section 24.7). Thus, theoretical analysis has increasingly been shifted toward experimentally relevant quantities in confined geometries and predictions on how to control parameters that can enhance them (Onoda and Nagaosa 2005a). Examples of such quantities are edge spin accumulation (Nikolić *et al.* 2005c; Nomura *et al.* 2005; Onoda and Nagaosa 2005a; Zyuzin *et al.* 2007) and bulk spin density (Nikolić *et al.* 2006; Reynoso *et al.* 2006; Chen *et al.* 2007; Finkler *et al.* 2007; Liu *et al.* 2007), or outflowing spin currents driven by them (Nikolić *et al.* 2006).

24.5 What is the mesoscopic spin-Hall effect?

Realistic devices on which SHE experiments are performed are always in contact with external electrodes and circuits that typically inject the charge current (rather than applying a "longitudinal electric field") or perform measurement of the resulting voltage drops and spatial distribution of spins and charges. In the seminal arguments (Sinova *et al.* 2004) for the intrinsic SHE in infinite Rashba spin-split 2DEG (in the "clean" limit), electric-field-driven acceleration of electron momenta and associated precession of spins plays a crucial role. On the other hand, mesoscopic SHE was introduced (Nikolić *et al.*

2005b) for ballistic finite-size 2DEGs attached to multiple current and voltage probes where an electric field is absent in the SO-coupled central sample (on the proviso that the surrounding leads are reflectionless). Another stunning difference (Sheng and Ting 2006) between intrinsically driven SHE in the bulk and finite-size 2DEGs is extreme sensitivity to disorder in the former case (Inoue *et al.* 2004) which, for linear in momentum SO couplings (such as the Rashba one), is able to completely destroy the spin-Hall current density in unbounded systems (Mishchenko *et al.* 2004; Adagideli and Bauer 2005). Unlike in three-dimensional semiconductor and metallic devices, which are always disordered and where extrinsic contribution to the SHE is therefore present or dominant, ballistic conditions for the mesoscopic SHE can be achieved in low-dimensional semiconductor systems. In fact the very recent experiment on nanoscale H-shaped structures realized using high mobility HgTe/HgCdTe quantum wells has reported for the first time the detection of mesoscopic SHE via non-local and purely electrical measurements (Brüne *et al.* 2008).

The magnitude of pure spin currents flowing out of mesoscopic SHE device (illustrated in Fig. 24.2) through ideal (spin and charge interaction free) electrodes is governed by the spin precession length L_{SO}. This mesoscopic length scale (e.g. $L_{SO} \sim 100$ nm in recently fabricated 2DEGs), on which the vector of the expectation values of spin precesses by an angle π, has been identified through intuitive physical arguments (Engel *et al.* 2007) as an important parameter for spin distributions (e.g. in clean systems the spin response to inhomogeneous field diverges at the wave vector $q = 2/L_{SO}$). In fact, the mesoscopic SHE analysis predicts (Nikolić *et al.* 2005b) via numerically exact calculations (see Fig. 24.8) that the optimal device size for achieving large spin currents is indeed $L \simeq L_{SO}$. This is further confirmed by alternative analyses of the SHE response in finite-size systems (Moca and Marinescu 2007). In the general cases (Sih *et al.* 2005; Hankiewicz and Vignale 2008), where both the extrinsic and intrinsic SO interaction effects are present, the intrinsically driven contribution to SHE in finite-size devices dominates (Nikolić and Zârbo 2007) when the ratio of characteristic energy scales (Nagaosa 2008) for the disorder and SO coupling effects satisfies $\Delta_{SO}\tau/\hbar \gtrsim 1$ (Δ_{SO} is the spin splitting of quasi-particle energies and $\hbar/\tau$ is the disorder-induced broadening of energy levels due to the transport scattering time τ).

For mesoscopic SHE devices in the phase-coherent transport regime (device smaller than the dephasing length), one can also observe the effects of quantum confinement and quantum interferences in spin-related quantities that counterpart familiar examples from mesoscopic charge transport (Datta 1995). They include: SHE conductance fluctuations (Ren *et al.* 2006; Bardarson *et al.* 2007); resonances in SHE conductance due to opening of new conducting channels (Nikolić *et al.* 2005b; Sheng and Ting 2006) or mixing of bound (Bulgakov *et al.* 1999) and propagating states due to SO couplings; and constructive or destructive quantum interference-based control (Souma and Nikolić 2005) of spin-Hall current in multiterminal Aharonov–Casher rings (as the electromagnetic dual of Aharonov–Bohm rings where SO coupling, rather than magnetic field, permeates the ring). The charge and spin dephasing can be included (Golizadeh-Mojarad and Datta 2007a) within the same NEGF transport formalism to allow for comparison with experiments performed at finite temperatures where quantum coherence effects are smeared out (Golizadeh-Mojarad and Datta 2007b).

In general, the presence of SO couplings requires to treat the whole device geometry when studying the dynamics of transported spin densities. For example, the decay of non-equilibrium spin polarizations in ballistic or disordered quantum wires is highly dependent on the transverse confinement effects (Nikolić and Souma 2005a; Holleitner *et al.* 2006) or chaotic vs. regular boundaries of quantum dots (Chang *et al.* 2004). Since SO couplings in SHE devices manifest through both of their aspects—creation of spin currents and concurrently relaxation of spins—it is a non-trivial task to understand how spin currents and edge spin accumulations scale with increasing the strength of the SO couplings (Onoda and Nagaosa 2005a).

The analysis of the whole device setup, where a central SO-coupled sample is treated together with the surrounding electrodes, also simplifies the discussion of esoteric SHE concepts, such as the SHE in insulators (Onoda and Nagaosa 2005b) (where electrodes introduce dissipation necessary to obtain a non-zero value of spin accumulation as time-reversal odd quantity) or quantum SHE (König *et al.* 2008) whose quantized spin Hall conductance is due to chiral spin-filtered (or helical) edge states (i.e. Kramers doublets of states forcing electrons of opposite spin to flow in opposite directions along the edges of the sample) in a multiterminal SO-coupled bridge with energy gap in the central sample (Sheng *et al.* 2005). For example, recent direct experimental evidence (Roth 2009) for nonlocal transport in HgTe quantum wells in the quantum spin-Hall regime, which shows how non-dissipative quantum transport occurs through chiral spin-filtered edge states while the contacts lead to equilibration between the counter-propagating spin states at the edge, can be analyzed only by the multiterminal-device-oriented quantum transport techniques discussed in this chapter.

24.6 SO couplings in low-dimensional semiconductors

The coupling between the orbital and the spin degree of freedom of electrons is a relativistic effect described formally by the non-relativistic expansion of the Dirac equation in external electric and magnetic fields (for which exact solutions do not exist) in powers of the inverse speed of light c. In the second order v^2/c^2, one identifies (Zawadzki 2005) the SO term[5]

$$\hat{H}_{SO} = \frac{\hbar}{4m_0^2 c^2} \hat{\mathbf{p}} \cdot (\hat{\boldsymbol{\sigma}} \times \nabla V(\mathbf{r})), \qquad (24.1)$$

responsible for the entanglement of the spin and orbital degrees of freedom in the two-component non-relativistic Pauli Hamiltonian for spin-$\frac{1}{2}$ electron. Here, m_0 is the free-electron mass, $\hat{\boldsymbol{\sigma}} = (\hat{\sigma}_x, \hat{\sigma}_y, \hat{\sigma}_z)$ is the vector of the Pauli matrices, and $V(\mathbf{r})$ is the electric potential. The SO coupling term can also be extracted from the semi-classical analysis that usually invokes the interaction of the electron magnetic dipole moment (associated with spin) with

[5] Although a topic of numerous textbooks on relativistic quantum mechanics and quantum field theory, the v^2/c^2 expansion has recently been carefully re-examined (Zawadzki 2005) to find all terms at this order of approximation in manifestly gauge invariant form, thereby revealing various inconsistencies in the textbook literature.

magnetic field in the frame moving with the electron (Jackson 1998). In the instantaneous rest frame of an electron, a magnetic field is obtained by Lorentz transforming the electric field from the laboratory frame. It is actually more efficient for intuitive analysis of different experimental situations to remain (Fisher 1971) in the laboratory frame where a magnetic dipole $\boldsymbol{\mu}$ moving with velocity $\mathbf{v}$ generates an electric dipole moment

$$\mathbf{P}_{\text{lab}} = \mathbf{v} \times \boldsymbol{\mu}/c^2. \tag{24.2}$$

[6]The Thomas precession takes into account change in rotational kinetic energy due to the precession of the accelerated electron as seen by a laboratory observer in the "extended" special relativity of accelerated objects.

Here, the right-hand side is evaluated in the electron rest frame and $\mathbf{P}_{\text{lab}}$ is measured in the lab (both sides can be evaluated in the lab yielding the same result to first order in v/c). The potential energy of the interaction of the electric dipole with the external electric field $\mathbf{E}_{\text{lab}}$ in the lab frame, $U_{\text{dipole}} = -\mathbf{P}_{\text{lab}} \cdot \mathbf{E}_{\text{lab}}$, corrected to the Thomas precession[6] $U_{\text{Thomas}} = -U_{\text{dipole}}/2$ leads to the SO-coupling term $U_{\text{SO}} = U_{\text{dipole}} + U_{\text{Thomas}} = -\mathbf{P}_{\text{lab}} \cdot \mathbf{E}_{\text{lab}}/2$.

The lab frame analysis allows one to quickly depict the Mott skew-scattering off a target whose Coulomb field deflects a beam of spin-↑ and spin-↓ particles in opposite directions, thereby, e.g. polarizing the beam of neutrons (Fisher 1971) or generating skew-scattering contribution (Engel *et al.* 2007; Hankiewicz and Vignale 2008) to the extrinsic SHE. For example, if we look at spin-↑ electron from behind moving along the y-axis, whose expectation value of the spin vector is oriented along the positive z-axis so that the corresponding magnetic dipole moment lies along the negative z-axis, then in the lab frame we also see its Lorentz-transformed electric dipole moment $\mathbf{P}_{\text{lab}}$ oriented along the negative x-axis. The electric dipole feels the force $\mathbf{F} = (\mathbf{P}_{\text{lab}} \cdot \nabla)\mathbf{E}_{\text{lab}}$, oriented in this case along the positive x-axis (right transverse direction with respect to the motion of the incoming electron) since the gradient of the electric field $\mathbf{E}_{\text{lab}}$ generated by the target is always negative outside of it. Note that this simple-minded classical picture only explains one aspect of the SO-dependent interaction with impurity. The other one—the so-called side jump (i.e. sideways shift of the scattering wave packet)—requires more quantum-mechanical analysis (Sinitsyn 2008) to extract additional contribution to the velocity operator due to impurity potential $V_{\text{disorder}}(\mathbf{r})$ in the SO Hamiltonian eqn (24.1).

The heuristic discussion based on the Lorentz transformations gives only a minuscule effect and the influence of electronic band structure is essential (Engel *et al.* 2007) to make these effects experimentally observable in solids. In the case of atoms, SO coupling is due to interaction of the electron spin with the average Coulomb field of the nuclei and other electrons. In solids, $V(\mathbf{r})$ is the sum of a periodic crystalline potential and an aperiodic part containing potentials due to impurities, confinement, boundaries and external electric fields. The non-relativistic expansion of the Dirac equation can be viewed as a method of systematically including the effects of the negative-energy solutions on the positive energy states starting from their non-relativistic limit (Zawadzki 2005). This effect in vacuum is small due to the huge gap $2m_0c^2$ between positive and negative energy states. In solids, strong nuclear potential competes with this huge denominator in eqn (24.1), so that the much smaller bandgap between the conduction and valence band (playing the role of electron positive energy sea and positron negative energy sea, respectively) replaces $2m_0c^2$, thereby illustrating the origin of strong enhancement of the SO couplings in solids (Winkler 2003).

Although intrinsic SO couplings can always be written in the Zeeman form $\hat{\sigma} \cdot \mathbf{B}_{SO}(\mathbf{p})$, their effective magnetic field $\mathbf{B}_{SO}(\mathbf{p})$ is momentum dependent and, therefore, does not break the time-reversal invariance. The Kramers theorem (Ballentine 1998) for time-reversal invariant quantum systems requires that the energy bands $\varepsilon_n(\mathbf{k})$ of an electron in a periodic potential satisfy $\varepsilon_n(\mathbf{k}, \uparrow) = \varepsilon_n(-\mathbf{k}, \downarrow)$, since $\mathbf{k} \mapsto -\mathbf{k}$ and $\sigma = \uparrow \mapsto \sigma = \downarrow$ upon time reversal ($\hbar \mathbf{k}$ is crystal momentum). Therefore, in semiconductors invariant under spatial inversion $\mathbf{k} \mapsto -\mathbf{k}$ (such as silicon) the Kramers theorem gives double degenerate spin states for any $\mathbf{k}$ value, $\varepsilon_n(\mathbf{k}, \uparrow) = \varepsilon_n(\mathbf{k}, \downarrow)$. To obtain a non-zero $\mathbf{B}_{SO}(\mathbf{p})$ that breaks the spin degeneracy in three-dimensional crystals $\varepsilon_n(\mathbf{k}, \uparrow) \neq \varepsilon_n(\mathbf{k}, \downarrow)$ the host crystal has to be inversion asymmetric. In bulk semiconductors with zinc-blende symmetry, the conduction band of III-V compounds will split into two subbands where anisotropic spin splitting is proportional to k^3. Such SO-induced splitting is termed cubic Dresselhaus SO coupling, and it is associated with bulk inversion asymmetry (BIA). In semiconductor heterostructures (Winkler 2003), the ideal symmetry of the 3D host crystal is broken by the interface where 2DEG or two-dimensional hole gas (2DHG) is confined within a quantum well. In such reduced effective dimensionality, the symmetry of the underlying crystal lattice is lowered, so that an additional linear in k Dresselhaus term becomes relevant. Besides microscopic crystalline potential $V_{\text{crystal}}(\mathbf{r})$ as the source of the electric field in the SO coupling term in eqn (24.1), an interface electric field accompanying the quantum-well structural inversion asymmetry (SIA) gives rise to the Rashba spin splitting of conduction-band electrons

$$\hat{H}_{\text{Rashba}} = \frac{\alpha}{\hbar}(\hat{\sigma} \times \hat{\mathbf{p}}) \cdot \hat{\mathbf{e}}_z, \tag{24.3}$$

for 2DEG in the xy-plane and $\hat{\mathbf{e}}_z$ as the unit vector along the z-axis. In narrow-gap semiconductors the Rashba effect linear in k should dominate over the bulk k^3 Dresselhaus term. Moreover, it has been experimentally demonstrated that Rashba coupling can be changed by as much as 50% by external gate electrode (Nitta *et al.* 1997; Grundler 2000) covering 2DEG, which has become one of the key concepts in semiconductor spintronics. Nevertheless, there is a lengthy theoretical debate on the importance of different electric-field contributions to the value of experimentally observed α and the parameters that are effectively manipulated via the gate electrode to cause its increase (Grundler 2000)—we refer to a comprehensive overview of these issues by Fabian *et al.* (2007) and Winkler (2003). Note that impurity-determined $V_{\text{disorder}}(\mathbf{r})$ contribution to eqn (24.1) does not require broken inversion asymmetry of the pure crystal or of the structure.

The coupling between electron momentum and spin correlates charge currents and spin densities in SO-coupled semiconductors (Silsbee 2004) leading to highly non-trivial effects in non-equilibrium situations. Some of these have been observed in recent magnetoelectric experiments (Kato *et al.* 2004b; Silov *et al.* 2004; Ganichev *et al.* 2006), where charge current induces spin density, as well as in the spin galvanic experiments (Ganichev and Prettl 2003),

[7]The same analysis applies to linear Dresselhaus coupling since they can be transformed
into each other by a unitary matrix.

where non-equilibrium spin density drives a charge current. It is also the key ingredient of intrinsically driven SHEs (where the induced spin density is oriented out-of-plane, rather than inplane as in the magnetoelectric effects (Silsbee 2004)).

In this section we focus on the description of the Rashba SO coupling[7] in the form suitable for spin-dependent NEGF calculations using a local orbital basis and the lattice Hamiltonian defined by it. The Rashba Hamiltonian prepared in this fashion will be used as the starting point for illustrating SHE-related spin and charge-transport calculations in Section 24.8. We also add the treatment of the extrinsic SO coupling in 2DEG within the same local orbital basis framework in Section 24.6.3.2 to enable the description of the most general experimental situations (Sih *et al.* 2005) in low-dimensional devices where both extrinsic and intrinsic mechanisms can act concurrently (Nikolić and Zârbo 2007; Hankiewicz and Vignale 2008).

24.6.1   Rashba coupling in bulk 2DEG

The effective single-particle SO Hamiltonian for a clean infinite homogeneous 2DEG with the Rashba coupling eqn (24.3) can be formally rewritten as

$$\hat{H}_R^{2D} = \frac{\hat{\mathbf{p}}^2}{2m^*} \otimes \mathbf{I}_S + \frac{\alpha}{\hbar}\left(\hat{p}_y \otimes \hat{\sigma}_x - \hat{p}_x \otimes \hat{\sigma}_y\right). \tag{24.4}$$

Here, $\otimes$ stands for the tensor product of two operators acting in the tensor product $\mathcal{H}_O \otimes \mathcal{H}_S$ of the orbital and spin Hilbert spaces, and m^* is the effective mass. To enforce pedagogical notation, we also use $\mathbf{I}_O$ as the unit operator in $\mathcal{H}_O$ and $\mathbf{I}_S$ for the unit operator in $\mathcal{H}_S$. The internal *momentum-dependent* magnetic field corresponding to the Rashba coupling is extracted from eqn (24.3), recast in the form of the Zeeman term $-g\mu_B\hat{\boldsymbol{\sigma}} \cdot \mathbf{B}_R(\mathbf{p})/2$, as $\mathbf{B}_R(\mathbf{p}) = (2\alpha/g\mu_B)(\mathbf{p} \times \mathbf{e}_z)$. The Hamiltonian commutes with the momentum operator $\hat{\mathbf{p}}$, time-reversal operator $\hat{T}$, and the chirality operator $(\hat{\boldsymbol{\sigma}} \times \hat{\mathbf{p}}/|\mathbf{p}|) \cdot \mathbf{e}_z$. The zero commutator $\left[\hat{H}, \hat{\mathbf{p}}\right] = 0$ due to the translation invariance of infinite 2DEG implies that solutions of the Schrödinger equation are of the form

$$\Psi(\mathbf{r}) = Ce^{i\mathbf{k}\mathbf{r}}\left|\chi\right\rangle_s, \tag{24.5}$$

where $\left|\chi\right\rangle_s$ is a two-component spinor. Due to time-reversal invariance $\left[\hat{H}, \hat{T}\right] = 0$, the two eigenstates of opposite momenta and spins are degenerate, $E(\mathbf{k}, \sigma) = E(-\mathbf{k}, -\sigma)$. The time-reversal operator can be written as $\hat{T} = \hat{K} \otimes \exp(i\pi\hat{\sigma}_y/2)$, where $\hat{K}$ is the complex conjugation operator (Ballentine 1998). If the Hamiltonian is invariant to chirality transformation, the spin state is momentum dependent. In the Rashba case the spin and momentum of an eigenstate are always perpendicular to each other.

The Rashba Hamiltonian eqn (24.4) in the case of a 1D electron gas (1DEG) reduces to

$$\hat{H}_R^{1D}(k_x) = \frac{\hbar^2 k_x^2}{2m^*} - \alpha k_x\hat{\sigma}_y. \tag{24.6}$$

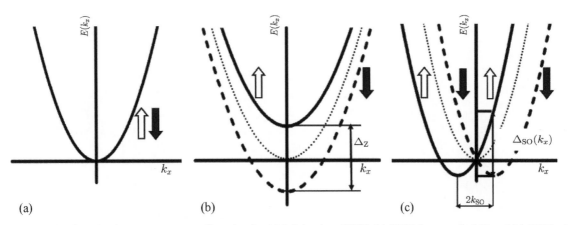

Fig. 24.3 The one-dimensional energy-momentum dispersion for: (a) infinite clean 1DEG, (b) 1DEG in magnetic field, and (c) 1DEG with the Rashba SO coupling. The SO-induced spin splitting in (c) is signified by the energy separation Δ_{SO} at a given k_x. The complementary description of the SO-induced spin splitting is k_{SO} difference between the momenta of two electrons of opposite spins and chiralities in panel (c).

Its eigenvalues define the energy-momentum dispersion

$$E_\pm(k_x) = \frac{\hbar^2 k_x^2}{2m^*} \pm \alpha |k_x|, \qquad (24.7)$$

with the corresponding eigenvectors

$$\psi_{k_x,\pm} = \frac{e^{ik_x x}}{\sqrt{2\pi\hbar}} \cdot \frac{1}{\sqrt{2}} \begin{pmatrix} 1 \\ \mp i \dfrac{k_x}{|k_x|} \end{pmatrix}. \qquad (24.8)$$

These eigenstates are labelled with the momentum operator eigenvalue $\hbar k_x$ and the chirality operator eigenvalue $\lambda = \pm 1$. Equation (24.8) shows that

$$\langle \psi_{k_x,\pm} | \hat{\sigma}_y | \psi_{k_x,\pm} \rangle = \mp \frac{k_x}{|k_x|}, \qquad (24.9)$$

meaning that each spin-split parabolic subband has a well-defined spin.

To elucidate the physical meaning of the SO-induced spin splitting, we compare in Fig. 24.3 the Rashba dispersion with the more familiar Zeeman splitting of 1DEG placed in the external (momentum-independent) magnetic field. In the case of the Zeeman splitting, the spin-↑ and spin-↓ subbands are shifted vertically with respect to each other by the Zeeman energy Δ_Z. On the other hand, the Rashba spin-splitting depends on momentum

$$\Delta_{SO}(k_x) = 2\alpha k_x, \qquad (24.10)$$

so that $E_+(k_x)$ and $E_-(k_x)$ are shifted horizontally along the momentum axis rather than along the energy axis. This ensures that the system remains spin unpolarized in equilibrium, as dictated by the time-reversal invariance and the fact that spin density is a time-reversal odd quantity. The spin splitting can also be described using k_{SO} as the difference between the momenta of two electrons of opposite spins and chiralities, and different experiments (such as Shubnikov–de Haas, Raman scattering, or spin precession) probe either Δ_{SO} or k_{SO} illustrated in Fig. 24.3(c).

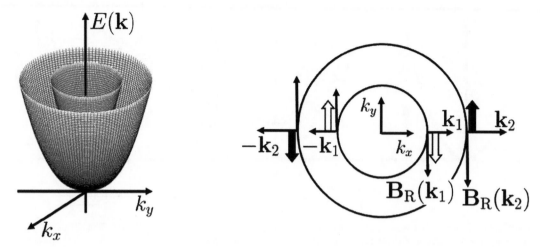

Fig. 24.4 The energy-momentum dispersion $E(k_x, k_y)$ of the Rashba spin-split 2DEG. The states at the Fermi level lie on two concentric Fermi circles, $E_+(\mathbf{k}) = E_F$ for the inner circle and $E_-(\mathbf{k}) = E_F$ for the outer circle, as shown in the right panel. The radius of the inner circle is k_1 and the radius of the outer one is denoted by k_2. The Fermi momentum k_F of the free particle is $k_1 \leq k_F \leq k_2$. For a given momentum $\mathbf{k}$, the spin of an electron is oriented either parallel [$E_+(\mathbf{k})$ branch] or antiparallel [$E_-(\mathbf{k})$ branch] to the momentum-dependent effective magnetic field $\mathbf{B}_R(\mathbf{k})$. The states of opposite momentum and spin on each Fermi circle are Kramers degenerate.

From eqn (24.9) we see that the y-component of spin of the eigenstate $\left|\psi_{k_x,\lambda}\right\rangle$ is $\uparrow$ for $\lambda = 1$ and k_x negative, and $\downarrow$ for $\lambda = 1$ and k_x positive. The opposite is true for the other branch $E_{\lambda=-1}(k_x)$. Thus, the eigenvalue of spin $\hat{\sigma}_y$ is a good quantum number since in 1D $\hat{\sigma}_y$ commutes with the Hamiltonian. Following this argument, it is easy to understand the relative spin orientations of the states along each of the dispersion branches.

As in the 1D case, the Rashba Hamiltonian in 2D commutes with momentum, chirality, and the time-reversal operator. Its eigenenergies

$$E_\pm(\mathbf{k}) = \frac{\hbar^2 \mathbf{k}^2}{2m^*} \pm \alpha \left|\mathbf{k}\right|, \qquad (24.11)$$

are plotted in Fig. 24.4. They are labelled by a 2D wave vector $\left|\mathbf{k}\right| = \sqrt{k_x^2 + k_y^2}$ and the chirality eigenvalues $\lambda = \pm 1$ (i.e. the spin projection perpendicular to both $\mathbf{k}$ and growth direction along the z-axis). The corresponding eigenstates are

$$\psi_{\mathbf{k},\pm} = \frac{e^{i\mathbf{k}\mathbf{r}}}{2\pi\hbar} \cdot \frac{1}{\sqrt{2}} \left(\begin{array}{c} 1 \\ \pm \dfrac{k_y - ik_x}{\left|\mathbf{k}\right|} \end{array} \right). \qquad (24.12)$$

Finding the expectation value of the spin operator $\hbar\hat{\boldsymbol{\sigma}}/2$ in eigenstate eqn (24.12)

$$\left\langle \psi_{\mathbf{k},\pm} \left| \hat{\boldsymbol{\sigma}} \right| \psi_{\mathbf{k},\pm} \right\rangle = \pm \frac{k_y}{\left|\mathbf{k}\right|} \mathbf{e}_x \mp \frac{k_x}{\left|\mathbf{k}\right|} \mathbf{e}_y, \qquad (24.13)$$

demonstrates (Winkler 2003) that no common spin quantization axis can be found for all eigenstates of the Rashba spin-split 2DEG ($\mathbf{e}_x$ and $\mathbf{e}_y$ are the unit vectors within the xy-plane of the 2DEG).

24.6.2   Rashba coupling in quantum wires

24.6.2.1   *Energy dispersion of Rashba-split transverse propagating subbands*

The Rashba Hamiltonian for a quantum wire patterned within 2DEG (along the *x*-axis)

$$\hat{H}_{\mathrm{R}}^{\mathrm{Q1D}} = \frac{\hat{\mathbf{p}}^2}{2m^*} + \frac{\alpha}{\hbar}\left(\hat{\boldsymbol{\sigma}} \times \hat{\mathbf{p}}\right) \cdot \mathbf{z} + V_{\mathrm{conf}}(y), \qquad (24.14)$$

describes a quasi-one-dimensional electron gas (Q1DEG) of width d whose lateral confinement is accounted by the potential $V_{\mathrm{conf}}(y)$. The motion along the confinement direction is quantized, such that the energy-momentum dispersion in the absence of SO coupling is split into spin-degenerate quasi-1D subbands with quadratic dispersion $E_{\pm}(k_x) = E_n + \hbar^2 k_x^2/2m^*$ in a 1D wave vector k_x labelling an eigenstate of the subband with index n. The effective-mass Rashba Hamiltonian for a quantum wire is translationally invariant along the wire direction x, and can be separated into three terms $\hat{H} = \hat{H}_{\mathrm{sb}} + \hat{H}_{\mathrm{mix}} + \hat{H}_{\mathrm{1D}}$ (Governale and Zülicke 2004):

$$\hat{H}_{\mathrm{sb}} = \frac{\hat{p}_y^2}{2m^*} + V_{\mathrm{conf}}(y), \qquad (24.15a)$$

$$\hat{H}_{\mathrm{mix}} = -\frac{\hbar k_{\mathrm{SO}}}{m^*}\hat{\sigma}_x \hat{p}_y, \qquad (24.15b)$$

$$\hat{H}_{\mathrm{1D}} = \frac{\hbar^2}{2m^*}(k_x + k_{\mathrm{SO}}\hat{\sigma}_y)^2 + \frac{\hbar^2 k_{\mathrm{SO}}^2}{2m^*}. \qquad (24.15c)$$

The term $\hat{H}_{\mathrm{sb}}$ defines the eigenenergies E_n of the transverse confining potential, while $\hat{H}_{\mathrm{1D}}$ is the 1D translationally invariant Rashba term for the quantum wire. Therefore, in the hypothetical case where the second term $\hat{H}_{\mathrm{mix}}$ is absent, the eigenenergies of the Hamiltonian eqn (24.15) show only k_x-dependent splitting

$$E_{\pm}(k_x) = E_n + \frac{\hbar^2 k_x^2}{2m^*} \pm \alpha|k_x|. \qquad (24.16)$$

When the expectation values E_{mix} of $\hat{H}_{\mathrm{mix}}$ between the eigenstates of $\hat{H}_{\mathrm{sb}} + \hat{H}_{\mathrm{1D}}$ are of the order of $\Delta E_n = E_{n+1} - E_n$, the mixing of the spin-split subbands becomes important. Thus, the ratio

$$\frac{E_{\mathrm{mix}}}{\Delta E_n} \simeq \frac{\hbar k_{\mathrm{SO}}}{m^*}\frac{\pi\hbar}{d}\left(\frac{\pi^2\hbar^2}{m^*d^2}\right)^{-1} = \frac{d}{L_{\mathrm{SO}}} = \frac{k_{\mathrm{SO}}d}{\pi}, \qquad (24.17)$$

gives a simple criterion to separate the strong and weak SO coupling regimes in Q1DEGs. In the weak coupling regime $L_{\mathrm{SO}} \gg d$ the subband mixing is negligible, as illustrated in Fig. 24.5(a). In quantum wires with a strong Rashba effect, realized for $L_{\mathrm{SO}} \lesssim d$ (or $k_{\mathrm{SO}}d \gtrsim \pi$) hybridization of quasi-1D subbands becomes important and eqn (24.16) can not capture the non-parabolic energy dispersions in Fig. 24.5(b),(c). Unlike the weak coupling regime where the Hamiltonian eigenstates are also eigenstates of $\hat{\sigma}_y$ (i.e. their

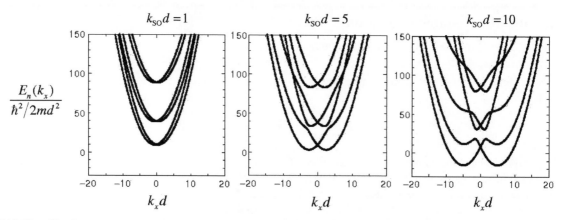

Fig. 24.5 The subband energy-momentum dispersion in the Rashba spin-split clean quantum wire. The parameter $k_{SO}d$ (d is the width of the wire) defines the weak and strong SO coupling regimes. For large values of $k_{SO}d$ the mixing of subbands is non-trivial, giving rise to a strong energy dependence of the wire conductance.

spins are polarized inplane along the y-axis), in the strong coupling regime no common spin quantum number can be assigned to states within a given subband (Governale and Zülicke 2004).

24.6.2.2   *Spin precession in Rashba quantum wires*

The lifting of spin degeneracy of each transverse subband of a quantum wire by the Rashba coupling means that an electron injected at the Fermi energy E_F can have two different wave vectors within the wire, $E_+(k_{x1}) = E_-(k_{x2}) = E_F$. The wave vector k_{x1} labels the eigenstates of the E_+ spin-split subband whose spinor part is $\frac{1}{\sqrt{2}}\binom{1}{i}$ (describing spin-↑ along the y-axis), while k_{x2} labels the eigenstates of E_- subband associated with the eigenspinor $\frac{1}{\sqrt{2}}\binom{1}{-i}$ (describing spin-↓ along the y-axis). Thus, when an electron with spin-↑ along the z-axis is injected from a half-metallic ferromagnet into the Rashba quantum wire, as proposed in the Datta–Das spin-FET device (Datta and Das 1990), the outgoing wavefunction from a wire of length L will be

$$\psi(L) \sim e^{ik_{x1}L}\binom{1}{i} + e^{ik_{x2}L}\binom{1}{-i}. \tag{24.18}$$

The probability to detect a particle with spin-↑ or with spin-↓ along the z-axis is given by

$$|\langle (1\ 0) \mid \psi(L)\rangle| = 4\cos^2\frac{(k_{x1} - k_{x2})L}{2}, \tag{24.19a}$$

and

$$|\langle (0\ 1) \mid \psi(L)\rangle| = 4\sin^2\frac{(k_{x1} - k_{x2})L}{2}, \tag{24.19b}$$

respectively. Such modulation of current by spin precession with angle $\Delta\theta = \Delta k_x L$, where $\Delta k_x = k_{x2} - k_{x1} = 2m^*\alpha/\hbar^2$ stems from $E_+(k_{x2}) - E_-(k_{x1}) = 0$, would provide the basis for the operation of the envisioned spin-FET device (Datta and Das 1990). It also introduces in a transparent fashion the concept of the *spin precession length*

$$L_{SO} = \frac{\pi \hbar^2}{2m^* \alpha}, \tag{24.20}$$

along which the vector of the expectation values of spin precesses by an angle $\Delta\theta = \pi$ while propagating along the Rashba wire.

We emphasize that the energy independence of L_{SO} holds for weak SO coupling and narrow wires. As discussed in Section 24.6.2.1, for strong SO coupling $d > L_{SO}$, intersubband mixing becomes important, so that eqn (24.20) is inapplicable. The L_{SO} scale appears as the characteristic length scales for various processes in semiconductor spintronic systems. For example, in weakly disordered systems L_{SO} also plays the role of a characteristic scale for the D'yakonov–Perel' spin dephasing where non-equilibrium spin density decays due to randomization of $\mathbf{B}_R(\mathbf{p})$ in each scattering event changing $\mathbf{p}$ and, therefore, the direction of magnetic field around which spin precesses (Chang *et al.* 2004; Fabian *et al.* 2007). It also sets a mesoscopic scale ($\sim$100 nm in heterostructures with large α (Nitta *et al.* 1997; Grundler 2000)) at which one can expect the largest SHE response (Nikolić *et al.* 2005b, 2006; Moca and Marinescu 2007).

24.6.3 Discrete representation of effective SO Hamiltonian

In order to perform numerical calculations on finite-size systems of arbitrary shape attached to external electrodes it is highly advantageous to discretize the effective SO Hamiltonian. Here, we introduce a discretization scheme for the Rashba Hamiltonian. In addition, we also discuss the discretization scheme for the extrinsic SO Hamiltonian that makes it possible to treat intrinsic and extrinsic SO coupling effects on an equal footing in nanostructures (Nikolić and Zârbo 2007).

The grid used for discretization is a collection of points $\mathbf{m} = (m_x, m_y)$ on a square lattice of constant a, where m_x, m_y are integers. We use the following notation: $\psi_{\mathbf{m}} \equiv \langle \mathbf{m} \mid \psi \rangle$ for the wavefunction evaluated at point $\mathbf{m}$; and $A_{\mathbf{mm'}} \equiv \langle \mathbf{m} | \hat{A} | \mathbf{m'} \rangle$ for the matrix element of an operator $\hat{A}$. In finite-difference methods, one has to evaluate the derivatives of a function $f(x)$ on a grid $\{\ldots, x_{m-1}, x_m, x_{m+1}, \ldots\} \equiv \{\ldots, (m-1)a, ma, (m+1)a, \ldots\}$:

$$\left(\frac{df}{dx}\right)_m = \frac{f_{m+1} - f_{m-1}}{2a}, \tag{24.21}$$

where $f_m \equiv f(x_m)$. The second derivative can be computed as

$$\left(\frac{d^2 f}{dx^2}\right)_m = \frac{f_{m+1} - 2f_m + f_{m-1}}{a^2}. \tag{24.22}$$

For a particular discretization scheme, the matrix elements of the derivative operators can be expressed in the local orbital basis $|\mathbf{m}\rangle$, which can be interpreted as each site hosting a single *s*-orbital as in the tight-binding models of solids. For instance, using eqn (24.21) we get

$$\left\langle m \left| \frac{d}{dx} \right| n \right\rangle = \frac{\langle m+1 \mid n \rangle - \langle m-1 \mid n \rangle}{2a} = \frac{\delta_{n,m+1} - \delta_{n,m-1}}{2a}, \quad (24.23)$$

where δ is the Kronecker symbol. The same can be done for the operator d^2/dx^2 using eqn (24.22)

$$\left\langle m \left| \frac{d^2}{dx^2} \right| n \right\rangle = \frac{\langle m+1 \mid n \rangle - 2 \langle m \mid n \rangle + \langle m-1 \mid n \rangle}{a^2}$$
$$= \frac{\delta_{n,m+1} - 2\delta_{n,m} + \delta_{n,m-1}}{a^2}. \quad (24.24)$$

It is worth mentioning that for a given grid, the differentiation operators matrix elements are dependent on the discretization scheme used. For example, if eqn (24.23) is used to get $d^2/dx^2 = d/dx \cdot d/dx$ it would lead to a different result from the one in eqn (24.24). This is due to the fact that these matrix elements are not computed but approximated, and this depends on the selected discretization scheme.

The discrete one-particle operators are written in the second quantized notation as

$$\hat{A} = \sum_{m,n} \left\langle m \left| \hat{A} \right| n \right\rangle \hat{c}_m^\dagger \hat{c}_n, \quad (24.25)$$

which, together with eqns (24.23) and (24.24), leads to

$$\frac{d}{dx} = \sum_m \frac{1}{2a} \left(\hat{c}_m^\dagger \hat{c}_{m+1} - \hat{c}_m^\dagger \hat{c}_{m-1} \right), \quad (24.26a)$$

$$\frac{d^2}{dx^2} = \sum_m \frac{1}{a^2} \left(\hat{c}_m^\dagger \hat{c}_{m+1} - 2\hat{c}_m^\dagger \hat{c}_m + \hat{c}_m^\dagger \hat{c}_{m-1} \right). \quad (24.26b)$$

Equations (24.26a) and (24.26b) can be used to discretize the effective-mass Hamiltonian of a quasielectron in the clean spin and charge interaction-free 2DEG

$$\hat{H}_{\text{free}} = \frac{\hat{p}^2}{2m^*} \otimes \mathbf{I}_S = -\frac{\hbar^2}{2m^*} \left(\frac{\partial^2}{\partial x^2} + \frac{\partial^2}{\partial y^2} \right) \otimes \mathbf{I}_S, \quad (24.27)$$

where the discretized version of the same Hamiltonian is

$$\hat{H}_{\text{free}} = \sum_{\mathbf{mm}'\sigma} t_{\mathbf{mm}'} \hat{c}_{\mathbf{m}\sigma}^\dagger \hat{c}_{\mathbf{m}'\sigma}. \quad (24.28)$$

This is the familiar tight-binding Hamiltonian on the square lattice with single s-orbital $\langle \mathbf{r}|\mathbf{m} \rangle = \phi(\mathbf{r} - \mathbf{m})$ per site. Here, $\hat{c}_{\mathbf{m}\sigma}^\dagger$ ($\hat{c}_{\mathbf{m}\sigma}$) is the creation (annihilation) operator of an electron at site $\mathbf{m} = (m_x, m_y)$. The spin-independent hopping matrix element $t_{\mathbf{mm}'}$ is given by

$$t_{\mathbf{mm}'} = \left\langle \mathbf{m} \left| -\frac{\hbar^2}{2m^*} \frac{\partial^2}{\partial \mathbf{r}^2} \right| \mathbf{m}' \right\rangle = \begin{cases} -t_O & \text{if } \mathbf{m} = \mathbf{m}' \pm a\mathbf{e}_{x,y} \\ 0 & \text{otherwise}, \end{cases} \quad (24.29)$$

where $t_O = \hbar^2/(2m^*a^2)$ is the orbital hopping parameter.

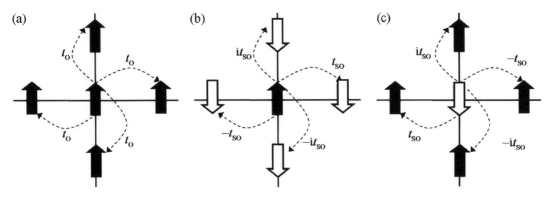

Fig. 24.6 An illustration of the physical meaning of the orbital hopping t_O (a) and the Rashba SO hopping t_{SO} for spin-↑ (b) and spin-↓ (c) electrons on the tight-binding lattice. As shown in panel (a), the probability amplitude for a spin to hop between two sites without flipping is proportional to t_O. The z-spin $|\!\uparrow\rangle_z$ at the central site **m** in panels (b), (c) flips by jumping to any of the four neighboring sites with probability amplitude proportional to t_{SO}.

24.6.3.1   *Discretization of the Rashba SO Hamiltonian*

Following the general discretization procedure outlined above, we recast the Rashba Hamiltonian in the local orbital basis representation as

$$\hat{H} = \sum_{\mathbf{m}\sigma} \varepsilon_{\mathbf{m}} \hat{c}^{\dagger}_{\mathbf{m}\sigma} \hat{c}_{\mathbf{m}\sigma} + \sum_{\mathbf{mm'}\sigma\sigma'} \hat{c}^{\dagger}_{\mathbf{m}\sigma} t^{\sigma\sigma'}_{\mathbf{mm'}} \hat{c}_{\mathbf{m'}\sigma'}. \tag{24.30}$$

While this Hamiltonian is of the tight-binding type, its hopping parameters are non-trivial 2×2 Hermitian matrices $\mathbf{t_{m'm}} = (\mathbf{t_{mm'}})^{\dagger}$ in the spin space. The on-site potential $\varepsilon_{\mathbf{m}}$ describes any static local potential, such as the electrostatic potential due to the applied voltage or the disorder that is usually simulated via a uniform random variable $\varepsilon_{\mathbf{m}} \in [-W/2, W/2]$ modelling short-range isotropic scattering off spin-independent impurities. The generalized nearest-neighbor hopping $t^{\sigma\sigma'}_{\mathbf{mm'}} = (\mathbf{t_{mm'}})_{\sigma\sigma'}$ accounts for the Rashba coupling

$$\mathbf{t'_{mm}} = \begin{cases} -t_O\mathbf{I}_S - it_{SO}\hat{\sigma}_y & (\mathbf{m} = \mathbf{m'} + \mathbf{e}_x) \\ -t_O\mathbf{I}_S + it_{SO}\hat{\sigma}_x & (\mathbf{m} = \mathbf{m'} + \mathbf{e}_y) \end{cases}, \tag{24.31}$$

through the SO hopping energy scale $t_{SO} = \alpha/2a$ whose physical meaning is illustrated in Fig. 24.6. It is also useful to express the spin precession length eqn (24.20)

$$L_{SO} = \frac{\pi t_O}{2t_{SO}} a, \tag{24.32}$$

in terms of the lattice spacing a.

A direct correspondence between the continuous effective Rashba Hamiltonian eqn (24.14) (with quadratic and isotropic energy-momentum dispersion) and its lattice version eqn (24.30) (with tight-binding dispersion) is established by selecting the Fermi energy (e.g. $E_F = -3.8t_O$) of the injected electrons to be close to the bottom of the band $E_b = -4.0t_O$ (so that tight-binding dispersion reduces to the quadratic one), and by using $t_O = \hbar^2/(2m^*a^2)$ for the orbital hopping, which yields the effective mass m^* in the continuum limit. For example, the InGaAs/InAlAs heterostructure employed in the experiments of Nitta *et al.* (1997) is characterized by the effective mass $m^* = 0.05m_0$ and the width of the conduction band $\Delta_b = 0.9\,\text{eV}$, which sets $t_O = \Delta_b/8 = 112\,\text{meV}$ for the orbital hopping parameter on a square lattice (with four nearest neighbors of each site) and $a \simeq 2.6\,\text{nm}$ for its lattice spacing. Thus, the Rashba SO coupling of 2DEG formed in this heterostructure, tuned to a maximum value (Nitta *et al.* 1997) $\alpha = 0.93 \times 10^{-11}\,\text{eV m}$ by the gate electrode, corresponds to the SO hopping $t_{SO}/t_O \simeq 0.016$ in the lattice Hamiltonian eqn (24.30).

24.6.3.2   *Discretization of the extrinsic SO Hamiltonian*

The extrinsic SO coupling in disordered 2DEG is described by the Pauli Hamiltonian

$$\hat{H} = \frac{\hat{p}_x^2 + \hat{p}_y^2}{2m^*} + V_{\text{conf}}(y) + V_{\text{disorder}}(x, y) + \lambda \left(\hat{\sigma} \times \hat{\mathbf{p}} \right) \cdot \nabla V_{\text{disorder}}(x, y),$$

$$(24.33)$$

where λ is the extrinsic SO coupling strength. The fourth term in eqn (24.33) can be rewritten as

$$\lambda \left(\hat{\sigma} \times \hat{\mathbf{p}} \right) \cdot \nabla V_{\text{disorder}}(x, y) = -i\lambda \left[\left(\partial_y V_{\text{disorder}} \right) \partial_x - \left(\partial_x V_{\text{disorder}} \right) \partial_y \right] \hat{\sigma}_z.$$

$$(24.34)$$

Equation (24.23) can then be used to find the discrete version of $\partial_x V_{\text{disorder}}$

$$\langle \mathbf{m} \,|\partial_x V_{\text{disorder}}|\, \mathbf{m}' \rangle = \sum_{\mathbf{m}''} \langle \mathbf{m} \,|\partial_x|\, \mathbf{m}'' \rangle \langle \mathbf{m}'' \,|V_{\text{disorder}}|\, \mathbf{m}' \rangle$$

$$= \frac{V_{\text{disorder}}(\mathbf{m} + a\mathbf{e}_x)\delta_{\mathbf{m}',\mathbf{m}+a\mathbf{e}_x} - V_{\text{disorder}}(\mathbf{m} - a\mathbf{e}_x)\delta_{\mathbf{m}',\mathbf{m}-a\mathbf{e}_x}}{2a}.$$

$$(24.35)$$

The final expression (Nikolić and Zârbo 2007) of the discretized version of eqn (24.33)

$$\hat{H} = \sum_{\mathbf{m},\sigma} \varepsilon_{\mathbf{m}} \hat{c}_{\mathbf{m}\sigma}^\dagger \hat{c}_{\mathbf{m}\sigma} - t_O \sum_{\mathbf{mm}'\sigma} \hat{c}_{\mathbf{m}\sigma}^\dagger \hat{c}_{\mathbf{m}'\sigma} - i\lambda_{SO} \sum_{\mathbf{m},\alpha\beta} \sum_{ij} \sum_{\nu\gamma} \epsilon_{ijz}\nu\gamma$$

$$\times (\varepsilon_{\mathbf{m}+\gamma\mathbf{e}_j} - \varepsilon_{\mathbf{m}+\nu\mathbf{e}_i})\hat{c}_{\mathbf{m},\alpha}^\dagger \hat{\sigma}_{\alpha\beta}^z \hat{c}_{\mathbf{m}+\nu\mathbf{e}_i+\gamma\mathbf{e}_j,\beta}, \qquad (24.36)$$

introduces second-neighbor hopping in the third term (in addition to usual nearest-neighbor hopping in the second term). Here, λ_{SO} is the dimensionless extrinsic SO coupling strength $\lambda_{SO} = \lambda\hbar/4a^2$ and ϵ_{ijz} stands for the Levi-Civita totally antisymmetric tensor (i, j denote the inplane coordinate axes and dummy indices ν, γ take values ± 1). For example, for the 2DEG in SHE experiments conducted by Sih *et al.* (2005), the SO parameters are set to $\lambda_{SO} \simeq 0.005$ and $t_{SO} \simeq 0.003 t_O$ for an additional Rashba term, assuming a conduction bandwidth $\simeq 1\,\text{eV}$ and using the effective mass $m^* = 0.074m_0$.

24.7 Spin-current operator, spin density, and spin accumulation in the presence of intrinsic SO couplings

The theory of the intrinsic SHE in infinite homogeneous systems is formulated in terms of the spin-current density that is not conserved in media with SO coupling and, therefore, does not have a well-defined experimental measurement procedure associated with it. Moreover, these spin-current densities can be non-zero even in thermodynamic equilibrium (Rashba 2003; Nikolić *et al.* 2006). The conservation of charge implies the continuity equation in quantum mechanics for the charge density $\rho = e|\Psi(\mathbf{r})|^2$

$$\frac{\partial \rho}{\partial t} + \nabla \cdot \mathbf{j} = 0, \tag{24.37}$$

associated with a given wavefunction $\Psi(\mathbf{r})$. From here one can extract the charge-current density $\mathbf{j} = e\mathrm{Re}\,[\Psi^\dagger(\mathbf{r})\hat{\mathbf{v}}\Psi(\mathbf{r})]$ viewed as the quantum-mechanical expectation value (in the state $\Psi(\mathbf{r})$) of the charge current density operator

$$\hat{\mathbf{j}} = e\frac{\hat{n}(\mathbf{r})\hat{\mathbf{v}} + \hat{\mathbf{v}}\hat{n}(\mathbf{r})}{2}. \tag{24.38}$$

This operator can also be obtained heuristically from the classical charge current density $\mathbf{j} = en(\mathbf{r})\mathbf{v}$ via quantization procedure where the particle density $n(\mathbf{r})$ and the velocity $\mathbf{v}$ are replaced by the corresponding operators and symmetrized to ensure that $\hat{\mathbf{j}}$ is a Hermitian operator.

In SO-coupled systems $\hat{\mathbf{j}}$ acquires extra terms since the velocity operator $i\hbar\hat{\mathbf{v}} = [\hat{\mathbf{r}}, \hat{H}]$ is modified by the presence of SO terms in the Hamiltonian $\hat{H}$. For example, for the Rashba SO Hamiltonian eqn (24.3) the velocity operator is $\hat{\mathbf{v}} = \hat{\mathbf{p}}/m^* - (\alpha/\hbar)(\hat{\sigma}_y\mathbf{e}_x - \hat{\sigma}_x\mathbf{e}_y)$. The spin density $S^i = \frac{\hbar}{2}[\Psi^\dagger(\mathbf{r})\hat{\sigma}_i\Psi(\mathbf{r})]$ then satisfies the following continuity equation

$$\frac{\partial S^i}{\partial t} + \nabla \cdot \mathbf{j}^i = F_S^i. \tag{24.39}$$

In contrast to the charge continuity equation (eqn (24.37)), this contains the spin-current density

$$\mathbf{j}^i = \frac{\hbar}{2}\Psi^\dagger(\mathbf{r})\frac{\hat{\sigma}_i\hat{\mathbf{v}} + \hat{\mathbf{v}}\hat{\sigma}_i}{2}\Psi(\mathbf{r}), \tag{24.40}$$

as well as a non-zero spin source term

$$F_S^i = \frac{\hbar}{2}\mathrm{Re}\left(\Psi^\dagger(\mathbf{r})\frac{i}{\hbar}[\hat{H}, \hat{\sigma}_i]\Psi(\mathbf{r})\right). \tag{24.41}$$

The non-zero $F_S^i \neq 0$ term reflects non-conservation of spin in the presence of intrinsic SO couplings that act as internal momentum-dependent magnetic field forcing spin into precession. Thus, the plausible Hermitian operator of the spin current density

$$\hat{j}_k^i = \frac{\hbar}{2} \frac{\hat{\sigma}_i \hat{v}_k + \hat{v}_k \hat{\sigma}_i}{2}, \tag{24.42}$$

is a well-defined quantity (a tensor with nine components, where $\hat{j}_k^i$ describes transport of spin S_i in the k-direction, $i, k = x, y, z$) only when $\hat{\mathbf{v}}$ is spin independent.

The lack of the usual physical justification for eqn (24.42) in systems with intrinsic SO couplings leads to an arbitrariness in the definition of the spin current (Shi *et al.* 2006). Thus, different definitions lead to ambiguities (Sugimoto *et al.* 2006) in the value of the intrinsic spin-Hall conductivity $\sigma_{\text{sH}} = j_y^z / E_x$ computed as the linear response to the applied longitudinal electric field E_x penetrating an infinite SO-coupled (perfect) crystal. It also yields qualitatively different conclusions about the effect of impurities on SHE (Nagaosa 2008), and does not allow us to connect (Nomura *et al.* 2005) directly the value of σ_{sH} to measured edge-spin accumulation of opposite signs along opposite lateral edges.

Under the time-reversal transformation, the mass, charge, and energy do not change sign, while the velocity operator and the Pauli matrices change sign, $t \to -t \Rightarrow \hat{\mathbf{v}} \to -\hat{\mathbf{v}}$ and $t \to -t \Rightarrow \hat{\sigma} \to -\hat{\sigma}$. Since the charge-current density operator eqn (24.38) contains velocity, it changes sign under the time reversal $t \to -t \Rightarrow \hat{\mathbf{j}} \to -\hat{\mathbf{j}}$ and, therefore, has to vanish in the thermodynamic equilibrium (except in the presence of an external magnetic field that breaks time-reversal invariance, thereby allowing for circulating or diamagnetic charge currents even in thermodynamic equilibrium (Baranger and Stone 1989)). On the other hand, the spin-current density operator eqn (24.42) is the time-reversal invariant (or even) quantity $t \to -t \Rightarrow \hat{j}_k^i \to \hat{j}_k^i$: if the clock ran backward, spin current would continue to flow in the same direction. Thus, $\hat{j}_k^i$ can have non-zero expectation values even in thermodynamic equilibrium. This has been explicitly demonstrated (Rashba 2003) for the case of an infinite clean Rashba spin-split 2DEG where such *equilibrium* spin currents are polarized inside the plane. Therefore, the out-of-plane polarized spin-current density has been considered as a genuine non-equilibrium SHE-induced response (Sinova *et al.* 2004; Hankiewicz and Vignale 2008). However, within multiterminal finite-size Rashba coupled devices even out-of-plane polarized spin currents can have an equilibrium nature (Nikolić *et al.* 2006). This is essential information for the development of a consistent theory for transport (non-equilibrium) spin currents where the contributions from background (equilibrium) currents must be eliminated.

A plausible solution to these issues appears to be in defining a conserved quantity where the non-conserved part eqn (24.41) is moved to the right-hand side of the continuity equation eqn (24.39) and incorporated in the definition of a new spin current operator (Shi *et al.* 2006). However, this solution is of limited value since it cannot be used for realistic devices with arbitrary boundary conditions due to sample edges and attached electrodes (Nagaosa 2008). In Section 24.8 we discuss how to by-pass these issues altogether by employing a NEGF-based description of SHE in realistic finite-size devices where spin transport is quantified through: (*i*) total spin currents flowing out of the sample through attached leads in which they are conserved due to leads being ideal (i.e. spin and charge interaction free); (*ii*) local spin currents whose sum in the leads is equal to the total currents of (*i*); and (*iii*) non-equilibrium spin density near the boundaries (i.e. edge spin-Hall accumulation that is part of the definition of SHE) or within the SO-coupled sample (which is not typically associated with SHE experimentally measured quantities, but represents a natural ingredient of theoretical modelling).

24.8 NEGF approach to spin transport in multiterminal SO-coupled nanostructures

The NEGF theory (Leuween *et al.* 2006; Haug and Jauho 2007; Rammer 2007) provides a powerful conceptual framework, as well as computational tools, to deal with a variety of out-of-equilibrium situations in quantum systems. This includes a steady-state transport regime and more general transient responses. Over the past three decades, it has become one of the major tools to study steady-state quantum transport of charge currents through small devices in the phase-coherent regime where an electron is described by a single wavefunction throughout the device. In this respect, NEGF offers efficient realization of Landauer's seminal ideas to account for phase-coherent conduction in terms of the scattering matrix of the device (Datta 1995). Furthermore, NEGF is a much more complete framework, making it possible to go beyond the paradigms of the Landauer–Büttiker scattering approach by including many-body effects in transport where electron–electron interactions, electronic correlations, and electron–phonon interactions introduce dephasing, leading to incoherent electron propagation (Datta 1995, 2005; Thygesen and Rubio 2008).

The central concepts in the NEGF description of quantum transport are: the sample Hamiltonian $\hat{H}$ and its matrix representation $\mathbf{H}$ in a chosen local orbital basis (such as the discrete versions of the effective SO Hamiltonian discussed in Section 24.6.3); self-energy matrices due to the interaction of the sample with the electrodes $\boldsymbol{\Sigma}_{\text{leads}}$, $\boldsymbol{\Sigma}_{\text{leads}}^{<}$; self-energy matrices due to many-body interactions within the sample $\boldsymbol{\Sigma}_{\text{int}}$, $\boldsymbol{\Sigma}_{\text{int}}^{<}$; and two independent Green functions—the retarded $\mathbf{G}^{r}$ and the lesser $\mathbf{G}^{<}$ one. The retarded Green function describes the density of available quantum-mechanical states, while the lesser one determines how electrons occupy those quantum states.[8]

In the case of phase-coherent calculations of total spin and charge currents, we only need to find the retarded spin-resolved self-energies due to the electrodes $\boldsymbol{\Sigma}_{\text{leads}}^{\sigma}$ (determining the escape rates of spin-σ electrons into the electrodes) and compute the retarded Green-function elements connecting sites between interfaces where leads are attached to the sample. For convenience, we term this usage of NEGF the "Landauer–Büttiker approach" in Section 24.8.1 since it allows us to describe transport in multiterminal structures by constructing the transmission block of the scattering matrix in terms of the (portion of) retarded Green-function matrix.

For the computation of local quantities within the sample, we also need to find matrix elements of the retarded Green function between any two sites within the sample as well as $\mathbf{G}^{<}$. This requires solving the Keldysh (integral

[8] Both Green functions can be obtained from the contour-ordered Green function defined for any two time values that lie along the Kadanoff–Baym–Keldysh time contour (Leuween *et al.* 2006; Haug and Jauho 2007; Rammer 2007).

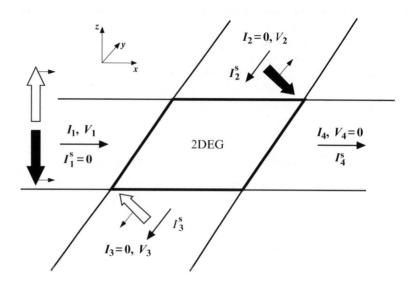

Fig. 24.7 The four-terminal bridge for the detection of the mesoscopic SHE. The central region is 2DEG, where electrons are confined within a semiconductor heterostructure grown along the z-axis whose SIA induces the Rashba SO coupling. The four attached leads are clean, non-magnetic, and without any SO coupling. The unpolarized ($I_1^S = 0$) charge current ($I_1 \neq 0$) injected through the longitudinal leads induces a spin-Hall current $I_2^{S_z} = -I_3^{S_z}$ in the transverse leads that act as the voltage probes $V_2 = V_3 \neq 0$, $I_2 = I_3 = 0$.

or matrix) equation for $\mathbf{G}^<$. The same is true for incoherent transport where dephasing takes place in the sample requiring $\mathbf{\Sigma}_{\text{int}}$, $\mathbf{\Sigma}_{\text{int}}^<$ self-energies to be computed (typically in the self-consistent fashion together with $\mathbf{G}^r$ and $\mathbf{G}^<$), or for non-linear transport where local charge density and the corresponding electric potential profile within the sample play a crucial role (Christen and Büttiker 1996) in understanding the device current–voltage characteristics. We denote this procedure in Sections 24.8.2 and 24.8.3 as the "Landauer–Keldysh" approach since here the full NEGF theory is applied to finite-size (clean or disordered) devices attached to semi-infinite electrodes, i.e. to the Landauer setup where such electrodes simplify boundary conditions for electrons assumed to escape to infinity through them to be thermalized in the macroscopic reservoirs (ensuring steady-state transport).[9]

[9]The NEGF theory, which is often called "Keldysh formalism," (Leeuwen *et al.* 2006) has been initiated by the pioneering works of Schwinger, Baym, Kadanoff and Keldysh who considered infinite homogeneous systems out of equilibrium (Haug and Jauho 2007; Rammer 2007). Its present applications (Datta 1995, 2005) to quantum transport in finite-size systems attached to semi-infinite electrodes can be traced to an early analysis of Caroli *et al.* (1971).

24.8.1   Landauer–Büttiker approach to total spin currents in multiterminal nanostructures

The experiments on quantum Hall bridges in the early 1980s were posing a challenge for theoretical interpretation of multiterminal transport measurements in the mesoscopic transport regime (Ando 2003). By viewing the current and voltage probes on an equal footing, Büttiker (1986) has provided an elegant solution to these puzzles in the form of multiprobe formulas

$$I_p = \sum_q (G_{qp} V_p - G_{pq} V_q) = \sum_q G_{pq}(V_p - V_q). \qquad (24.43)$$

They relate charge current $I_p = I_p^\uparrow + I_p^\downarrow$ in lead p to the voltages V_q in all other leads attached to the sample via the conductance coefficients G_{pq}. To study the spin-resolved charge currents I_p^σ ($\sigma = \uparrow, \downarrow$) of individual spin species $\uparrow, \downarrow$ we imagine that each non-magnetic lead in Fig. 24.7 consists of the two leads allowing only one spin species to propagate (as realized by, e.g. half-metallic ferromagnetic leads). Upon replacement $I_p \to I_p^\sigma$ and

$G_{pq} \to G_{pq}^{\sigma\sigma'}$, this viewpoint allows us to extract the multiterminal formulas for the spin-resolved charge currents I_p^σ, thereby obtaining the linear response relation for spin current

$$I_p^S = \frac{\hbar}{2e}(I_p^\uparrow - I_p^\downarrow), \qquad (24.44)$$

flowing through the lead p

$$I_p^S = \frac{\hbar}{2e} \sum_q \left[\left(G_{qp}^{\uparrow\uparrow} + G_{qp}^{\downarrow\uparrow} - G_{qp}^{\uparrow\downarrow} - G_{qp}^{\downarrow\downarrow} \right) V_p \right.$$

$$\left. - \left(G_{pq}^{\uparrow\uparrow} + G_{pq}^{\uparrow\downarrow} - G_{pq}^{\downarrow\uparrow} - G_{pq}^{\downarrow\downarrow} \right) V_q \right]. \qquad (24.45)$$

The spin-Hall conductance of the four-terminal bridge sketched in Fig. 24.7 is then defined as

$$G_{\text{sH}} = \frac{\hbar}{2e} \frac{I_2^S}{\Delta V} = \frac{\hbar}{2e} \frac{I_2^\uparrow - I_2^\downarrow}{V_1 - V_4}. \qquad (24.46)$$

Below, we simplify the notation by introducing the labels

$$G_{pq}^{\text{in}} = G_{pq}^{\uparrow\uparrow} + G_{pq}^{\uparrow\downarrow} - G_{pq}^{\downarrow\uparrow} - G_{pq}^{\downarrow\downarrow}, \qquad (24.47\text{a})$$

$$G_{pq}^{\text{out}} = G_{pq}^{\uparrow\uparrow} + G_{pq}^{\downarrow\uparrow} - G_{pq}^{\uparrow\downarrow} - G_{pq}^{\downarrow\downarrow}. \qquad (24.47\text{b})$$

In fact, these coefficients have a transparent physical interpretation: $\frac{\hbar}{2e} G_{qp}^{\text{out}} V_p$ is the spin current flowing from the lead p with voltage V_p into other leads q whose voltages are V_q, while $\frac{\hbar}{2e} G_{pq}^{\text{in}} V_q$ is the spin current flowing from the leads $q \neq p$ into the lead p.

The standard charge conductance coefficients (Büttiker 1986; Baranger and Stone 1989; Datta 1995) in the multiprobe Landauer–Büttiker formalism eqn (24.43) are expressed in terms of the spin-resolved conductances as $G_{pq} = G_{pq}^{\uparrow\uparrow} + G_{pq}^{\uparrow\downarrow} + G_{pq}^{\downarrow\uparrow} + G_{pq}^{\downarrow\downarrow}$. Their introduction in 1980s was prompted by the need to describe the linear transport properties of a single sample (with specific impurity arrangements and attached to a specific probe configuration) by using measurable quantities (instead of the bulk conductivity, which is inapplicable to mesoscopic conductors (Baranger and Stone 1989)). They describe the total charge current flowing in and out of the system in response to voltages applied at its boundaries.

Regardless of the detailed microscopic physics of transport, conductance coefficients must satisfy the sum rule $\sum_q G_{qp} = \sum_q G_{pq}$ in order to ensure the second equality in eqn (24.43), i.e. the charge current must be zero $V_q = \text{const.} \Rightarrow I_p \equiv 0$ in equilibrium. On the other hand, the multiprobe spin-current formulas (24.45) apparently possess a non-trivial equilibrium solution $V_q = \text{const.} \Rightarrow I_p^S \neq 0$ (found by Pareek (2004)) that would contradict the Landauer–Büttiker paradigm demanding usage of only measurable quantities. However, when all leads are at the same potential, a purely equilibrium non-zero term, $\frac{\hbar}{2e}(G_{pp}^{\text{out}} V_p - G_{pp}^{\text{in}} V_p) = \frac{\hbar}{e}(G_{pp}^{\downarrow\uparrow} - G_{pp}^{\uparrow\downarrow})V_p$, becomes relevant for I_p^S, cancelling all other terms in eqn (24.45) to ensure that no *unphysical* total spin current $I_p^S \neq 0$ can appear in the leads of an unbiased ($V_q = \text{const.}$)

multiterminal device (Kiselev and Kim 2005; Souma and Nikolić 2005; Scheid *et al.* 2007).

At zero temperature, the spin-resolved conductance coefficients

$$G_{pq}^{\sigma\sigma'} = \frac{e^2}{h} \sum_{ij} \left| t_{ij,\sigma\sigma'}^{pq} \right|^2, \qquad (24.48)$$

where summation is over the conducting channels in the leads, are obtained from the Landauer-type formula as the probability for spin-σ' electron incident in lead q to be transmitted to lead p as spin-σ electron. The quantum-mechanical probability amplitude for this process is given by the matrix elements of the transmission matrix t^{pq}, which is determined only by the wavefunctions (or Green functions) at the Fermi energy (Baranger and Stone 1989). The stationary states of the structure 2DEG + two leads supporting one or two conducting channels can be found exactly by matching the wavefunctions in the leads to the eigenstates of the Hamiltonian eqn (24.14), thereby allowing one to obtain the charge conductance from the Landauer transmission formula (Governale and Zülicke 2004). However, modelling of the full bridge geometry with two extra leads attached in the transverse direction, the existence of many open transverse propagating modes ("conducting channels"), and possibly a strong SO coupling regime when the sample is bigger than the spin-precession length ($L > L_{SO}$), is handled much more efficiently through the NEGF formalism.[10]

For a non-interacting particle that propagates through a finite-size sample of arbitrary shape, the transmission matrices

$$t^{pq} = \sqrt{-\mathrm{Im}\,\hat{\Sigma}_p \otimes \mathbf{I}_S} \cdot \hat{G}_{pq}^r \cdot \sqrt{-\mathrm{Im}\,\hat{\Sigma}_q \otimes \mathbf{I}_S},$$

$$\mathrm{Im}\,\hat{\Sigma}_p = \frac{1}{2i}\left(\hat{\Sigma}_p^r - \hat{\Sigma}_p^a \right), \qquad (24.49)$$

between different leads can be evaluated in a numerically exact fashion using the real⊗spin-space Green functions $G^r(\mathbf{m}'\sigma';\mathbf{m},\sigma)$ defined in the Hilbert space $\mathcal{H}_O \otimes \mathcal{H}_S$. This requires computation of a single object, the retarded Green operator

$$\hat{G}^r = \frac{1}{E\hat{I}_O \otimes \hat{I}_S - \hat{H} - \sum_{p=1}^{4} \hat{\Sigma}_p^r \otimes \hat{I}_S}, \qquad (24.50)$$

which becomes a matrix (i.e. the Green function) when represented in a basis $|\mathbf{m}\rangle \otimes |\sigma\rangle \in \mathcal{H}_O \otimes \mathcal{H}_S$ introduced by the Hamiltonian eqn (24.30). Here, $|\sigma\rangle$ are the eigenstates of the spin operator for the chosen spin-quantization axis. The matrix elements $G^r(\mathbf{m}'\sigma';\mathbf{m},\sigma) = \langle \mathbf{m}',\sigma'|\hat{G}^r|\mathbf{m},\sigma \rangle$ yield the probability amplitude for an electron to propagate between two arbitrary locations $\mathbf{m}$ and $\mathbf{m}'$ (with or without flipping its spin σ during the motion) inside an open conductor in the absence of inelastic processes. Its submatrix $\hat{G}_{pq}^r$, which is required in eqn (24.49), consists of those matrix elements that connect the layer of the sample attached to the lead p to the layer of the sample attached to the lead q. The self-energy $\sum_{p=1}^{4} \hat{\Sigma}_p^r \otimes \mathbf{I}_S$ (r-retarded, a-advanced, $\hat{\Sigma}_p^a = [\hat{\Sigma}_p^r]^\dagger$)

[10]The multiterminal and multichannel finite-size device can be also modelled using the random matrix theory for its scattering matrix (Bardarson *et al.* 2007). However, this method is strictly applicable only for weak SO coupling regime $L \ll L_{SO}$ (Aleiner and Fal'ko 2001).

account for the "interaction" of an open system with the attached four ideal semi-infinite leads p (Datta 1995).

24.8.1.1 *General expression for spin-Hall conductance*

Since the total charge current $I_p = I_p^\uparrow + I_p^\downarrow$ depends only on the voltage difference between the leads in Fig. 24.7, we set one of them to zero (e.g., $V_4 = 0$ is chosen as the reference potential) and apply voltage V_1 to the structure. Imposing the requirement $I_2 = I_3 = 0$ for the voltage probes 2 and 3 allows us to get the voltages V_2/V_1 and V_3/V_1 by inverting the multiprobe charge current formulas (24.43). Finally, by solving eqn (24.45) for I_2^S we obtain the most general expression for the spin-Hall conductance defined by eqn (24.46)

$$G_{\text{sH}} = \frac{\hbar}{2e} \left[\left(G_{12}^{\text{out}} + G_{32}^{\text{out}} + G_{42}^{\text{out}} \right) \frac{V_2}{V_1} - G_{23}^{\text{in}} \frac{V_3}{V_1} - G_{21}^{\text{in}} \right]. \qquad (24.51)$$

This quantity is measured in units of the spin conductance quantum $e/4\pi$ (as the largest possible G_{sH} when transverse leads support only one open conducting channel), which is the counterpart[11] of the familiar charge conductance quantum e^2/h.

In contrast to charge current, which is a scalar quantity, spin current has three components because of the vector nature of spin (i.e. different "directions" of spin correspond to different quantum-mechanical superpositions of $|\uparrow\rangle$ and $|\downarrow\rangle$ states). Therefore, we can expect that, in general, the detection of spin transported through the transverse leads of mesoscopic devices will find its expectation values to be non-zero for all three axes. However, their flow properties

$$I_2^{S_z} = -I_3^{S_z}, \qquad (24.52a)$$

$$I_2^{S_x} = -I_3^{S_x}, \qquad (24.52b)$$

$$I_2^{S_y} = I_3^{S_y}, \qquad (24.52c)$$

show that only the z- and the x-components can represent the SHE response for the Rashba SO-coupled four-terminal bridge. That is, if we connect the transverse leads 2 and 3 to each other (thereby connecting the lateral edges of 2DEG by a wire), only the spin-current-carrying z- and x-polarized spins will flow through them, as expected from the general SHE phenomenology where non-equilibrium spin-Hall accumulation detected in experiments (Kato *et al.* 2004a) has opposite sign on the lateral edges of 2DEG.

Therefore, to quantify all non-zero components of the vector of transverse spin current in the linear response regime, we can introduce three spin conductances (Nikolić *et al.* 2005b)

$$G_{\text{sH}}^x = I_2^{S_x}/V_1, \qquad (24.53a)$$

$$G_{\text{sp}}^y = I_2^{S_y}/V_1, \qquad (24.53b)$$

$$G_{\text{sH}}^z = I_2^{S_z}/V_1, \qquad (24.53c)$$

(assuming $V_4 = 0$). They can be evaluated using the same general formula (24.51) where the spin-quantization axis for $\uparrow$, $\downarrow$ in spin-resolved charge

[11] Note that $(\hbar/2e)(e^2/h) = e/4\pi$, where (e^2/h) is the natural unit for spin-resolved charge conductance coefficients $G_{pq}^{\sigma\sigma'}$.

conductance coefficients is chosen to be the x-, y-, or z-axis, respectively. For example, selecting $\hat{\sigma}_z| \uparrow\rangle = +| \uparrow\rangle$ and $\hat{\sigma}_z| \downarrow\rangle = -| \downarrow\rangle$ for the basis in which the Green operator eqn (24.50) is represented allows one to compute the z-component of the spin current $I_p^{S_z}$. In accord with their origin revealed by eqn (24.52), we denote G_{sH}^z and G_{sH}^x as the spin-Hall conductances, while G_{sp}^y is labelled as the "spin-polarization" conductance since it stems from the polarization of 2DEG by the flow of unpolarized charge current in the presence of SO couplings (Silsbee 2004).

24.8.1.2   *Symmetry properties of spin conductances*

Symmetry properties of the conductance coefficients with respect to the reversal of a bias voltage or the direction of an external magnetic field play an essential role in our understanding of linear response electron transport in macroscopic and mesoscopic conductors (Büttiker 1986; Datta 1995). For example, in the absence of a magnetic field they satisfy $G_{pq} = G_{qp}$ (which can be proved assuming a particular model for charge transport (Datta 1995)). Moreover, since the effective magnetic field $\mathbf{B}_R(\mathbf{p})$ of the Rashba SO coupling depends on momentum, it does not break the time-reversal invariance, thereby imposing the following property of the spin-resolved conductance coefficients $G_{pq}^{\sigma\sigma'} = G_{qp}^{-\sigma'-\sigma}$ in multiterminal SO-coupled bridges.

In addition, the ballistic four-terminal bridge in Fig. 24.7 with no impurities possesses various geometrical symmetries. It is invariant under rotations and reflections that interchange the leads, such as: (i) rotation C_4 (C_2) by an angle $\pi/2$ (π) around the z-axis for a square (rectangular) 2DEG central region; (ii) reflection σ_{vx} in the xz-plane; and (iii) reflection σ_{vy} in the yz-plane. These geometrical symmetries, together with the $G_{pq} = G_{qp}$ property, specify $V_2/V_1 = V_3/V_1 \equiv 0.5$ solution for the voltages of the transverse leads when the $I_2 = I_3 = 0$ condition is imposed on their charge currents.

The device Hamiltonian containing the Rashba SO term commutes with the unitary transformations that represent these symmetry operations in the Hilbert space $\mathcal{H}_O \otimes \mathcal{H}_S$: (i) $\hat{U}(C_2) \otimes \exp(i\frac{\pi}{2}\hat{\sigma}_z)$, which performs the transformation $\hat{\sigma}_x \to -\hat{\sigma}_x, \hat{\sigma}_y \to -\hat{\sigma}_y, \hat{\sigma}_z \to \hat{\sigma}_z$ and interchanges the leads 1 and 4 as well as the leads 2 and 3; (ii) $\hat{U}(\sigma_{vx}) \otimes \exp(i\frac{\pi}{2}\hat{\sigma}_y)$, which transforms the Pauli matrices $\hat{\sigma}_x \to -\hat{\sigma}_x, \hat{\sigma}_y \to \hat{\sigma}_y, \hat{\sigma}_z \to -\hat{\sigma}_z$ and interchanges leads 2 and 3; and (iii) $\hat{U}(\sigma_{vy}) \otimes \exp(i\frac{\pi}{2}\hat{\sigma}_x)$ that transforms $\hat{\sigma}_x \to \hat{\sigma}_x, \hat{\sigma}_y \to -\hat{\sigma}_y, \hat{\sigma}_z \to -\hat{\sigma}_z$ and exchanges lead 1 with lead 4. The Hamiltonian also commutes with the time-reversal operator.

The effect of these symmetries on the spin-resolved charge-conductance coefficients, and the corresponding spin conductances G_{sH}^z, G_{sH}^x and G_{sp}^y expressed in terms of them through eqn (24.51), is as follows. The change in the sign of the spin operator means that spin-$\uparrow$ becomes spin-$\downarrow$ so that, e.g. G_{pq}^{in} will be transformed into $-G_{qp}^{\text{in}}$. Also, the time reversal implies changing the signs of all spin operators and all momenta so that $G_{pq}^{\sigma\sigma'} = G_{qp}^{-\sigma'-\sigma}$ is equivalent to $G_{pq}^{\text{in}} = -G_{qp}^{\text{out}}$. Thus, invariance with respect to $\hat{U}(\sigma_{vy}) \otimes \exp(i\frac{\pi}{2}\hat{\sigma}_x)$ yields the identities $G_{21}^{\text{in},x} = G_{24}^{\text{in},x}$, $G_{21}^{\text{in},y} = -G_{24}^{\text{in},y}$, and $G_{21}^{\text{in},z} = -G_{24}^{\text{in},z}$. These symmetries do not imply cancellation of $G_{23}^{\text{in},x}$. However, invariance with respect to $\hat{U}(\sigma_{vx}) \otimes \exp{(i\frac{\pi}{2}\hat{\sigma}_y)}$ and $\hat{U}(C_2) \otimes \exp{(i\frac{\pi}{4}\hat{\sigma}_z)}$ implies that $G_{23}^{\text{in},y} \equiv 0$ and $G_{23}^{\text{in},z} \equiv 0$.

These symmetry-imposed conditions simplify the general formula (24.51) for spin conductances of a perfectly clean Rashba SO-coupled four-terminal bridge to

$$G_{\text{sH}}^x = \frac{\hbar}{2e}\left(2G_{12}^{\text{out},x} + G_{32}^{\text{out},x}\right), \tag{24.54a}$$

$$G_{\text{sp}}^y = \frac{\hbar}{2e}G_{12}^{\text{out},y}, \tag{24.54b}$$

$$G_{\text{sH}}^z = \frac{\hbar}{2e}G_{12}^{\text{out},z}, \tag{24.54c}$$

where we employ the result $V_2/V_1 = V_3/V_1 \equiv 0.5$ valid for a geometrically symmetric clean bridge. Because this solution for the transverse terminal voltages is violated in disordered bridges, its sample specific (for given impurity configuration) spin conductance cannot be computed from simplified formulas (24.54), as is often assumed in the recent mesoscopic SHE studies (Ren *et al.* 2006; Sheng and Ting 2006).

It is insightful to apply the same symmetry analysis to the bridges with other types of SO couplings. For example, if the Rashba term in the Hamiltonian eqn (24.14) is replaced by the linear Dresselhaus SO term $\frac{\beta}{\hbar}\left(\hat{p}_x\hat{\sigma}_x - \hat{p}_y\hat{\sigma}_y\right)$ due to BIA (Žutić *et al.* 2004), no qualitative change in our analysis ensues since the two SO couplings can be transformed into each other by a unitary matrix $(\hat{\sigma}_x + \hat{\sigma}_y)/\sqrt{2}$. In this case, the spin-Hall response is signified by $I_2^{S_z} = -I_3^{S_z}$ and $I_2^{S_y} = -I_3^{S_y}$ components of the transverse spin current, while $I_2^{S_x} = I_3^{S_x}$. For the Dresselhaus SO-coupled bridge, the general expression (24.51) simplifies to

$$G_{\text{sp}}^x = \frac{\hbar}{2e}G_{12}^{\text{out},x}, \tag{24.55a}$$

$$G_{\text{sH}}^y = \frac{\hbar}{2e}\left(2G_{12}^{\text{out},y} + G_{32}^{\text{out},y}\right), \tag{24.55b}$$

$$G_{\text{sH}}^z = \frac{\hbar}{2e}G_{12}^{\text{out},z}. \tag{24.55c}$$

The qualitatively different situation emerges when both the Rashba and the linear Dresselhaus SO couplings become relevant in the central region of the bridge since in this case it is impossible to find spin rotation that, combined with the spatial symmetry, would keep the Hamiltonian invariant, while only transforming the signs of its spin matrices. Moreover, for such a ballistic bridge the condition $I_2 = I_3 = 0$ leads to the $V_2/V_1 = 1 - V_3/V_1$ solution for the voltages, whereas imposing the alternative condition $V_2 = V_3$ generates non-zero charge currents flowing through the transverse leads 2 and 3 together with the spin currents (no simple relations akin to eqn (24.52) can be written in either of these cases).

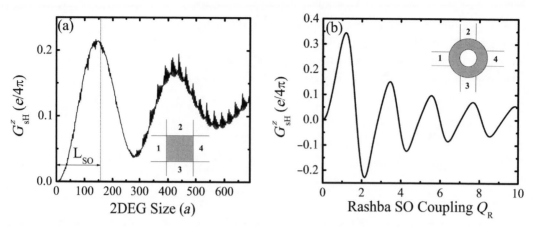

Fig. 24.8 The spin-Hall conductance G_{sH}^z of: (a) Rashba SO-coupled square-shaped 2DEG as a function its size L (in the units of the lattice spacing $a \simeq 3$ nm) for $t_{SO} = 0.01 t_O$ setting the spin precession length $L_{SO} \approx 157a$; and (b) single-channel Aharonov–Casher ring as the function of the dimensionless Rashba SO coupling $Q_R = (t_{SO}/t_O)N/\pi$ where the number of lattice sites discretizing the ring is $N = 100$.

24.8.1.3   *Example: Transverse spin currents in mesoscopic SHE nanostructures*

Figure 24.8 plots two examples of G_{sH}^z computed for simply connected (square-shaped) 2DEG and multiply connected ring realized within 2DEG. In both cases, the Rashba SO coupling is present within the device (gray area in the insets of Fig. 24.8). Figure 24.8 confirms that the optimal device size to observe large mesoscopic SHE in multiterminal structures is indeed governed by L_{SO} (Nikolić *et al.* 2005b). The computation of SHE conductance for very large square lattices in Fig. 24.8(a) is made possible by the usage of the recursive Green-function algorithm for four-terminal nanostructures discussed in Section 24.9.2.

The spin conductance of the single-channel ring attached to four single-channel ideal leads in Fig. 24.8(b) illustrates the *quantum-interference-driven SHE* introduced by (Souma and Nikolić 2005). The ring represents a solid-state realization of the two-slit experiment since an electron entering the ring can propagate in two possible directions—clockwise and counterclockwise. The superpositions of corresponding quantum states are sensitive to the acquired topological phases in a magnetic (Aharonov–Bohm effect) or an electric (Aharonov–Casher effect for particles with spin) external field whose changing generates an oscillatory pattern of the ring conductance. Thus, unlike commonly discussed extrinsic and intrinsic SHE, whose essence can be understood using semi-classical arguments and wave-packet propagation (Sinitsyn 2008), here the Aharonov–Casher phase difference acquired by opposite spin states during their cyclic evolution around the ring plays a crucial role. The spin conductance G_{sH}^z becomes zero at specific values of the Rashba coupling when the destructive interference of opposite spins travelling in opposite directions around the ring takes place (Souma and Nikolić 2005; Tserkovnyak and Brataas 2007). The amplitude of such quasi-periodic oscillations of G_{sH}^z, which are absent in simply connected mesoscopic SHE devices of Fig. 24.8(a), gradually decreases at large Rashba coupling because of the reflection at the ring lead interface.

24.8.2 Landauer–Keldysh approach to local spin currents in multiterminal nanostructures

The theory of imaging of charge flow on nanoscale can be constructed efficiently within the framework of lattice models of mesoscopic devices and the corresponding bond charge currents (Baranger and Stone 1989; Todorov 2002). This makes it possible to obtain a detailed picture of charge propagation between two arbitrary sites of the lattice (Nonoyama and Oguri 1998; Cresti *et al.* 2003; Metalidis and Bruno 2005; Nikolić *et al.* 2006; Zârbo and Nikolić 2007), thereby providing a way to interpret recent scanning probe experiments. These experimental advances (Topinka *et al.* 2003) have brought new insights into quantum transport by imaging its local features within a single sample, rather than performing conventional measurement of macroscopically averaged quantities. At the same time, scanning-probe techniques are becoming increasingly important in the quest for smaller electronic devices. For example, recent imaging of charge flow in conventional $p - n$ junctions suggests that in structures shrunk below 50 nm individual positions of scarce dopants will affect their function, thereby requiring to know precisely how charge carriers propagate on the nanoscale (Yoshida *et al.* 2007). In the case of spin transport, Kerr rotation microscopy has made possible imaging of steady-state spin density driven by charge flow through various SO-coupled semiconductor structures (Crooker and Smith 2005; Kato *et al.* 2005).

In this section, we discuss NEGF-based tools that allow us to compute the spatial details of spin flow on the scale of a few nanometers by introducing the *bond spin currents* (Nikolić *et al.* 2006). They represent the analog of bond charge currents, as well as a lattice version of the spin-current density conventionally employed in the studies of the intrinsic SHE in macroscopic systems (Murakami *et al.* 2003; Sinova *et al.* 2004). Even though their sums over the cross-sections within the sample change as we move from the bottom to the top transverse electrode due to spin-current non-conservation, they illustrate propagation of precessing spins and within the ideal transverse leads their sums over the cross-section reproduce (Nikolić *et al.* 2006) conserved total spin currents discussed in Section 24.8.1 using the Landauer–Büttiker approach.

24.8.2.1 *Bond charge-current operator in SO-coupled systems*

The charge conservation expressed through the familiar continuity equation eqn (24.37) yields a uniquely determined bond charge-current operator for quantum systems described on a lattice by a tight-binding-type of Hamiltonian eqn (24.30). That is, the Heisenberg equation of motion

$$\frac{d\hat{N}_{\mathbf{m}}}{dt} = \frac{1}{i\hbar}\left[\hat{N}_{\mathbf{m}}, \hat{H}\right], \tag{24.56}$$

for the electron number operator $\hat{N}_{\mathbf{m}}$ on site $\mathbf{m}$, $\hat{N}_{\mathbf{m}} \equiv \sum_{\sigma=\uparrow,\downarrow} \hat{c}^{\dagger}_{\mathbf{m}\sigma} \hat{c}_{\mathbf{m}\sigma}$, leads to the charge continuity equation on the lattice

$$e\frac{d\hat{N}_{\mathbf{m}}}{dt} + \sum_{k=x,y}\left(\hat{J}_{\mathbf{m},\mathbf{m}+\mathbf{e}_k} - \hat{J}_{\mathbf{m}-\mathbf{e}_k,\mathbf{m}}\right) = 0. \tag{24.57}$$

This equation introduces the bond charge-current operator (Todorov 2002) $\hat{J}_{\mathbf{mm}'}$ that describes the particle current from site $\mathbf{m}$ to its nearest-neighbor site $\mathbf{m}'$. The "bond" terminology is supported by a picture where current between two sites is represented by a bundle of flow lines bunched together along a line joining the two sites.

Thus, the spin-dependent Hamiltonian eqn (24.30) containing 2×2 hopping matrix defines the bond charge-current operator $\hat{J}_{\mathbf{mm}'} = \sum_{\sigma\sigma'} \hat{J}_{\mathbf{mm}'}^{\sigma\sigma'}$, which can be viewed as the sum of four different *spin-resolved* bond charge-current operators

$$\hat{j}_{\mathbf{mm}'}^{\sigma\sigma'} = \frac{e}{i\hbar} \left[\hat{c}_{\mathbf{m}'\sigma'}^{\dagger} t_{\mathbf{m}'\mathbf{m}}^{\sigma'\sigma} \hat{c}_{\mathbf{m}\sigma} - \text{H.c.} \right], \tag{24.58}$$

where H.c. stands for the Hermitian conjugate of the first term. In particular, for the case of $t_{\mathbf{mm}'}^{\sigma\sigma'}$ being determined by the Rashba SO interaction eqn (24.31), we can decompose the bond charge-current operator into two terms, $\hat{J}_{\mathbf{mm}'} = \hat{J}_{\mathbf{mm}'}^{\text{kin}} + \hat{J}_{\mathbf{mm}'}^{\text{SO}}$, having a transparent physical interpretation. The first term

$$\hat{j}_{\mathbf{mm}'}^{\text{kin}} = \frac{eit_O}{\hbar} \sum_{\sigma} \left[\hat{c}_{\mathbf{m}'\sigma}^{\dagger} \hat{c}_{\mathbf{m}\sigma} - \text{H.c.} \right], \tag{24.59}$$

can be denoted as "kinetic" since it originates only from the kinetic energy t_O and does not depend on the SO coupling energy t_{SO}. On the other hand, the second term

$$\hat{j}_{\mathbf{mm}'}^{\text{SO}} = \begin{cases} -\dfrac{4et_{\text{SO}}}{\hbar^2} \hat{S}_{\mathbf{mm}'}^{y} & (\mathbf{m} = \mathbf{m}' + \mathbf{e}_x) \\[2mm] +\dfrac{4et_{\text{SO}}}{\hbar^2} \hat{S}_{\mathbf{mm}'}^{x} & (\mathbf{m} = \mathbf{m}' + \mathbf{e}_y) \end{cases}$$
$$= \frac{4et_{\text{SO}}}{\hbar^2} \left((\mathbf{m}' - \mathbf{m}) \times \hat{\mathbf{S}}_{\mathbf{mm}'} \right)_z \tag{24.60}$$

represents an additional contribution to the intersite charge current flow due to non-zero Rashba SO hopping t_{SO}. Here, we also introduce the "bond spin-density" operator

$$\hat{\mathbf{S}}_{\mathbf{mm}'} = \frac{\hbar}{4} \sum_{\alpha\beta} \left[\hat{c}_{\mathbf{m}'\alpha}^{\dagger} \hat{\sigma}_{\alpha\beta} \hat{c}_{\mathbf{m}\beta} + \text{H.c.} \right], \tag{24.61}$$

defined for the bond connecting the sites $\mathbf{m}$ and $\mathbf{m}'$, which reduces to the usual definition of the local spin-density operator for $\mathbf{m} = \mathbf{m}'$ (see eqn (24.74)).

24.8.2.2   *Non-equilibrium bond charge current in SO-coupled systems*

The formalism of bond charge current makes it possible to compute physically measurable (Topinka *et al.* 2003) spatial profiles of local charge-current density within the sample as the quantum-statistical average[12] of the bond charge-current operator in the non-equilibrium state (Caroli *et al.* 1971; Nonoyama

[12]The quantum statistical average $\langle \hat{A} \rangle = \text{Tr}[\hat{\rho}\hat{A}]$ is taken with respect to the density matrix $\hat{\rho}$ that has evolved over sufficiently long time, so that non-equilibrium state and all relevant interactions are fully established.

and Oguri 1998; Cresti *et al.* 2003),

$$\left\langle \hat{J}_{\mathbf{mm'}} \right\rangle = \sum_{\sigma\sigma'} \left\langle \hat{j}_{\mathbf{mm'}}^{\sigma\sigma'} \right\rangle, \tag{24.62}$$

$$\left\langle \hat{j}_{\mathbf{mm'}}^{\sigma\sigma'} \right\rangle = \frac{-e}{\hbar} \int_{-\infty}^{\infty} \frac{\mathrm{d}E}{2\pi} \left[t_{\mathbf{m'm}}^{\sigma'\sigma} G_{\mathbf{mm'},\sigma\sigma'}^{<}(E) - t_{\mathbf{mm'}}^{\sigma\sigma'} G_{\mathbf{m'm},\sigma'\sigma}^{<}(E) \right]. \tag{24.63}$$

Here, the local charge current is expressed in terms of the non-equilibrium lesser Green function $G_{\mathbf{m'm},\sigma'\sigma}^{<}(E)$.

The usage of the second quantized notation in eqn (24.30) facilitates the introduction of NEGF expressions for the non-equilibrium expectation values (Caroli *et al.* 1971; Nonoyama and Oguri 1998). We imagine that at time $t' = -\infty$ the sample and the leads are not connected, while the left and the right longitudinal lead of a four-probe device are in their own thermal equilibrium with the chemical potentials μ_L and μ_R, respectively, where $\mu_L = \mu_R + eV$. The adiabatic switching of the hopping parameter connecting the leads and the sample generates time evolution of the density matrix of the structure (Caroli *et al.* 1971). The physical quantities are obtained as the non-equilibrium statistical average $\langle\ldots\rangle$ [with respect to the density matrix (Keldysh 1965) at time $t' = 0$] of the corresponding quantum-mechanical operators expressed in terms of $\hat{c}_{\mathbf{m}\sigma}^{\dagger}$ and $\hat{c}_{\mathbf{m}\sigma}$. This will lead to the expressions of the type $\left\langle \hat{c}_{\mathbf{m}\sigma}^{\dagger} \hat{c}_{\mathbf{m}\sigma'} \right\rangle$, which define the lesser Green function (Caroli *et al.* 1971; Nonoyama and Oguri 1998)

$$\left\langle \hat{c}_{\mathbf{m}\sigma}^{\dagger} \hat{c}_{\mathbf{m'}\sigma'} \right\rangle = \frac{\hbar}{\mathrm{i}} G_{\mathbf{m'm},\sigma'\sigma}^{<}(\tau = 0) = \frac{1}{2\pi\mathrm{i}} \int_{-\infty}^{\infty} \mathrm{d}E\, G_{\mathbf{m'm},\sigma'\sigma}^{<}(E). \tag{24.64}$$

Here, we utilize the fact that the two-time correlation function $[\hat{c}_{\mathbf{m}\sigma}(t) = \mathrm{e}^{\mathrm{i}\hat{H}t/\hbar} \hat{c}_{\mathbf{m}\sigma} \mathrm{e}^{-\mathrm{i}\hat{H}t/\hbar}]$

$$G_{\mathbf{mm'},\sigma\sigma'}^{<}(t, t') \equiv \frac{\mathrm{i}}{\hbar} \left\langle \hat{c}_{\mathbf{m'}\sigma'}^{\dagger}(t') \hat{c}_{\mathbf{m}\sigma}(t) \right\rangle, \tag{24.65}$$

depends only on $\tau = t - t'$ in stationary situations, so the time difference τ can be Fourier transformed to energy

$$G_{\mathbf{mm'},\sigma\sigma'}^{<}(\tau) = \frac{1}{2\pi\hbar} \int_{-\infty}^{\infty} \mathrm{d}E\, G_{\mathbf{mm'},\sigma\sigma'}^{<}(E) \mathrm{e}^{\mathrm{i}E\tau/\hbar}, \tag{24.66}$$

which will be utilized for steady-state transport studied here. We use the notation where $\mathbf{G}_{\mathbf{mm'}}^{<}$ is a 2×2 matrix in the spin space whose $\sigma\sigma'$ element is $G_{\mathbf{mm'},\sigma\sigma'}^{<}$.

The spin-resolved bond charge current in eqn (24.63) describes the flow of charges, which start as spin σ electrons at the site $\mathbf{m}$ and end up as a spin σ' electrons at the site $\mathbf{m'}$ where possible spin-flips $\sigma \neq \sigma'$ (instantaneous or due to precession) are caused by spin-dependent interactions. The decomposition of the bond charge-current operator into the kinetic and SO terms leads to a Green function expression for the corresponding non-equilibrium bond charge currents

$$\left\langle \hat{J}_{\mathbf{mm'}} \right\rangle = \left\langle \hat{J}_{\mathbf{mm'}}^{\mathrm{kin}} \right\rangle + \left\langle \hat{J}_{\mathbf{mm'}}^{\mathrm{SO}} \right\rangle, \tag{24.67}$$

with kinetic and SO terms given by

$$\left\langle \hat{J}_{\mathbf{mm'}}^{\text{kin}} \right\rangle = \frac{et_O}{\hbar} \int_{-\infty}^{\infty} \frac{dE}{2\pi} \text{Tr}_S \left[\mathbf{G}_{\mathbf{mm'}}^{<}(E) - \mathbf{G}_{\mathbf{m'm}}^{<}(E) \right], \qquad (24.68)$$

$$\left\langle \hat{J}_{\mathbf{mm'}}^{\text{SO}} \right\rangle = \frac{et_{\text{SO}}}{\hbar} \int_{-\infty}^{\infty} \frac{dE}{2\pi i} \text{Tr}_S \left\{ \left[(\mathbf{m'} - \mathbf{m}) \times \hat{\boldsymbol{\sigma}} \right]_z \right.$$
$$\left. \times \left[\mathbf{G}_{\mathbf{mm'}}^{<}(E) + \mathbf{G}_{\mathbf{m'm}}^{<}(E) \right] \right\}. \qquad (24.69)$$

Note, however, that the "kinetic" term is also influenced by the SO coupling through $\mathbf{G}^{<}$. In the absence of the SO coupling, eqn (24.69) vanishes and the bond charge current reduces to the standard expression (Caroli *et al.* 1971; Nonoyama and Oguri 1998; Cresti *et al.* 2003). The trace Tr_S is performed in the spin Hilbert space. Similarly, we can also obtain the non-equilibrium local charge density in terms of $\mathbf{G}^{<}$

$$e\left\langle \hat{N}_{\mathbf{m}} \right\rangle = e \sum_{\sigma=\uparrow,\downarrow} \left\langle \hat{c}_{\mathbf{m}\sigma}^{\dagger} \hat{c}_{\mathbf{m}\sigma} \right\rangle = \frac{e}{2\pi i} \int_{-\infty}^{\infty} dE \sum_{\sigma} G_{\mathbf{mm},\sigma\sigma}^{<}(E)$$

$$= \frac{e}{2\pi i} \int_{-\infty}^{\infty} dE \, \text{Tr}_S[\mathbf{G}_{\mathbf{mm}}^{<}(E)], \qquad (24.70)$$

which is the statistical average value of the corresponding operator.

24.8.2.3 *Bond spin current operator in SO-coupled systems*

To mimic the plausible definition of the spin-current density operator j_k^i in eqn (24.42), we introduce the bond spin-current operator for the spin-S_i component as the symmetrized product of the spin-$\frac{1}{2}$ operator $\hbar\hat{\sigma}_i/2$ ($i = x, y, z$) and the bond charge-current operator from eqn (24.57)

$$\hat{j}_{\mathbf{mm'}}^{S_i} \equiv \frac{1}{4i} \sum_{\alpha\beta} \left[\hat{c}_{\mathbf{m'}\beta}^{\dagger} \left\{ \hat{\sigma}_i, \mathbf{t}_{\mathbf{m'm}} \right\}_{\beta\alpha} \hat{c}_{\mathbf{m}\alpha} - \text{H.c.} \right]. \qquad (24.71)$$

By inserting the hopping matrix $\mathbf{t}_{\mathbf{m'm}}$ eqn (24.31) of the lattice SO Hamiltonian into this expression we obtain its explicit form for the case of the Rashba coupled system

$$\hat{j}_{\mathbf{mm'}}^{S_i} = \frac{it_O}{2} \sum_{\alpha\beta} \left(\hat{c}_{\mathbf{m'}\beta}^{\dagger} \left(\hat{\sigma}_i \right)_{\beta\alpha} \hat{c}_{\mathbf{m}\alpha} - \text{H.c.} \right) + t_{\text{SO}} \hat{N}_{\mathbf{mm'}} \left[\mathbf{e}_i \times (\mathbf{m'} - \mathbf{m}) \right]_z,$$

$$(24.72)$$

which can be considered as the lattice version of eqn (24.42). Here, we simplify the notation by using the "bond electron-number operator" $\hat{N}_{\mathbf{mm'}} \equiv \frac{1}{2} \sum_{\sigma} \left(\hat{c}_{\mathbf{m'}\sigma}^{\dagger} \hat{c}_{\mathbf{m}\sigma} + \text{H.c.} \right)$, which reduces to the standard electron-number operator for $\mathbf{m} = \mathbf{m'}$.

24.8.2.4 *Non-equilibrium bond spin currents in SO-coupled systems*

Similarly to the case of the non-equilibrium bond charge current in Section 24.8.2.2, the non-equilibrium statistical average of the bond spin-current operator eqn (24.72) can be expressed using the lesser Green function $\mathbf{G}^<$ as

$$\left\langle \hat{J}^{S_i}_{\mathbf{mm'}} \right\rangle = \left\langle \hat{J}^{S_i(\text{kin})}_{\mathbf{mm'}} \right\rangle + \left\langle \hat{J}^{S_i(\text{so})}_{\mathbf{mm'}} \right\rangle, \tag{24.73a}$$

$$\left\langle \hat{J}^{S_i(\text{kin})}_{\mathbf{mm'}} \right\rangle = \frac{t_{\mathrm{O}}}{2} \int_{-\infty}^{\infty} \frac{\mathrm{d}E}{2\pi} \mathrm{Tr}_S \left[\hat{\sigma}_i \left(\mathbf{G}^<_{\mathbf{m'm}}(E) - \mathbf{G}^<_{\mathbf{mm'}}(E) \right) \right], \tag{24.73b}$$

$$\left\langle \hat{J}^{S_i(\text{SO})}_{\mathbf{mm'}} \right\rangle = \left[\mathbf{e}_i \times (\mathbf{m'} - \mathbf{m}) \right]_z \frac{t_{\mathrm{SO}}}{2} \int_{-\infty}^{\infty} \frac{\mathrm{d}E}{2\pi \mathrm{i}} \mathrm{Tr}_S \left[\mathbf{G}^<_{\mathbf{mm'}}(E) + \mathbf{G}^<_{\mathbf{m'm}}(E) \right]. \tag{24.73c}$$

Here, we also encounter two terms that can be interpreted as the kinetic and the SO contribution to the bond spin current crossing from site $\mathbf{m}$ to site $\mathbf{m'}$. However, we emphasize that such SO contribution to the spin-S_z bond current is identically equal to zero, which simplifies the expression for this component to eqn (24.73b) as the primary spin-current response in the SHE.

24.8.3   Landauer–Keldysh approach to local spin densities

Motivated by recent advances in Kerr rotation microscopy, which have made possible experimental imaging of steady-state spin polarization in various SO-coupled semiconductor structures (Crooker and Smith 2005; Kato *et al.* 2005), we discuss in this section a NEGF-based approach to computation of the spatial profiles of $\left\langle \hat{\mathbf{S}}_{\mathbf{m}} \right\rangle$. Unlike local spin-current density, such *local flowing spin density* is a well-defined and measurable quantity that offers insight into the spin flow in the non-equilibrium steady transport state. For $\left\langle \hat{\mathbf{S}}_{\mathbf{m}} \right\rangle$ computed at the lateral edges of the sample, we use the term "spin accumulation" that was directly measured in the seminal spin-Hall experiments (Kato *et al.* 2004a; Wunderlich *et al.* 2005).

24.8.3.1   *Local spin density and its continuity equation*
The local spin density in the lattice models is determined by the local spin operator $\hat{\mathbf{S}}_{\mathbf{m}} = (\hat{S}^x_{\mathbf{m}}, \hat{S}^y_{\mathbf{m}}, \hat{S}^z_{\mathbf{m}})$ at site $\mathbf{m}$ defined by

$$\hat{\mathbf{S}}_{\mathbf{m}} = \frac{\hbar}{2} \sum_{\alpha\beta} \hat{c}^\dagger_{\mathbf{m}\alpha} \hat{\boldsymbol{\sigma}}_{\alpha\beta} \hat{c}_{\mathbf{m}\beta}. \tag{24.74}$$

The Heisenberg equation of motion for each component $\hat{\mathbf{S}}^i$ ($i = x, y, z$) of the spin-density operator

$$\frac{\mathrm{d}\hat{S}^i_{\mathbf{m}}}{\mathrm{d}t} = \frac{1}{\mathrm{i}\hbar} \left[\hat{S}^i_{\mathbf{m}}, \hat{H} \right], \tag{24.75}$$

can be written in the form

$$\frac{\mathrm{d}\hat{S}^i_{\mathbf{m}}}{\mathrm{d}t} + \sum_{k=x,y} \left(\hat{J}^{S_i}_{\mathbf{m},\mathbf{m}+\mathbf{e}_k} - \hat{J}^{S_i}_{\mathbf{m}-\mathbf{e}_k,\mathbf{m}} \right) = \hat{F}^{S_i}_{\mathbf{m}}, \tag{24.76}$$

where $\hat{J}^{S_i}_{\mathbf{mm'}}$ is the bond spin-current operator given by eqn (24.71) so that the second term on the left-hand side of eqn (24.76) corresponds to the "divergence" of the bond spin current on site **m**. Here, in analogy with eqn (24.41), we also find the lattice version of the spin source operator $\hat{F}_{\mathbf{m}}^{S_i}$ whose explicit form is

$$\hat{F}_{\mathbf{m}}^{S_x} = -\frac{t_{SO}}{t_O} \left(\hat{J}_{\mathbf{m,m+e}_x}^{S_z} + \hat{J}_{\mathbf{m-e}_x,\mathbf{m}}^{S_z} \right), \tag{24.77a}$$

$$\hat{F}_{\mathbf{m}}^{S_y} = -\frac{t_{SO}}{t_O} \left(\hat{J}_{\mathbf{m,m+e}_y}^{S_z} + \hat{J}_{\mathbf{m-e}_y,\mathbf{m}}^{S_z} \right), \tag{24.77b}$$

$$\hat{F}_{\mathbf{m}}^{S_z} = \frac{t_{SO}}{t_O} \left(\hat{J}_{\mathbf{m,m+e}_x}^{S_x} + \hat{J}_{\mathbf{m-e}_x,\mathbf{m}}^{S_x} + \hat{J}_{\mathbf{m,m+e}_y}^{S_y} + \hat{J}_{\mathbf{m-e}_y,\mathbf{m}}^{S_y} \right). \tag{24.77c}$$

The presence of the non-zero term $\hat{F}_{\mathbf{m}}^{S_i}$ on the right-hand side of the spin continuity eqn (24.76) signifies, within the framework of bond spin currents, the fact that spin is not conserved in SO-coupled systems. The fact that the bond spin-current operator eqn (24.71) appears in the spin continuity equation eqn (24.76) as its divergence implies that its definition in eqn (24.71) is plausible. However, the presence of the spin source operator $\hat{F}_{\mathbf{m}}^{S_i}$ reminds us that such a definition cannot be made unique (Shi *et al.* 2006), in sharp contrast to bond charge current that is uniquely determined by the charge continuity eqn (24.57).

Evaluating the statistical average of eqn (24.76) in a steady state (which can be either equilibrium or non-equilibrium), leads to the identity

$$\sum_{k=x,y} \left(\left\langle \hat{J}_{\mathbf{m,m+e}_k}^{S_i} \right\rangle - \left\langle \hat{J}_{\mathbf{m-e}_k,\mathbf{m}}^{S_i} \right\rangle \right) = \left\langle \hat{F}_{\mathbf{m}}^{S_i} \right\rangle. \tag{24.78}$$

In particular, for the spin-S_z component we get

$$\sum_{k=x,y} \left(\left\langle \hat{J}_{\mathbf{m,m+e}_k}^{S_z} \right\rangle - \left\langle \hat{J}_{\mathbf{m-e}_k,\mathbf{m}}^{S_z} \right\rangle \right) = \frac{t_{SO}}{t_O} \sum_{k=x,y} \left(\left\langle \hat{J}_{\mathbf{m,m+e}_k}^{S_k} \right\rangle + \left\langle \hat{J}_{\mathbf{m-e}_k,\mathbf{m}}^{S_k} \right\rangle \right), \tag{24.79}$$

which relates the divergence of the spin-S_z current (left-hand side) to the spin-source (right-hand side) determined by the sum of the longitudinal component of the spin-S_x current and the transverse component of the spin-S_y current.

Since no experiment has been proposed to measure local spin current density within the SO-coupled sample, defined through eqn (24.40) or its lattice equivalent eqn (24.73), we can obtain additional information about the spin fluxes within the sample by computing

$$\left\langle \hat{\mathbf{S}}_{\mathbf{m}} \right\rangle = \frac{\hbar}{2} \sum_{\alpha,\beta=\uparrow,\downarrow} \hat{\sigma}_{\alpha\beta} \left\langle \hat{c}_{\mathbf{m}\alpha}^\dagger \hat{c}_{\mathbf{m}\beta} \right\rangle = \frac{\hbar}{4\pi i} \int_{-\infty}^{\infty} dE \sum_{\alpha,\beta=\uparrow,\downarrow} \hat{\sigma}_{\alpha\beta} G_{\mathbf{mm},\beta\alpha}^{<}(E)$$

$$= \frac{\hbar}{4\pi i} \int_{-\infty}^{\infty} dE \, \mathrm{Tr}_S \left[\hat{\boldsymbol{\sigma}} \mathbf{G}_{\mathbf{mm}}^{<}(E) \right], \tag{24.80}$$

as the non-equilibrium spin density driven by charge transport in the presence of SO couplings.

24.8.4 Spin-resolved NEGFs for finite-size multiterminal devices

The spin-dependent NEGF formalism discussed in Section 24.8.2 does not actually depend on the details of the external driving force that brings the system into a non-equilibrium state. That is, the system can be driven by either the homogeneous electric field applied to an infinite homogeneous 2DEG or the voltage (i.e. electrochemical potential) difference between the electrodes attached to a *finite-size* sample. For example, in the latter case, the external bias voltage only shifts the relative chemical potentials of the reservoirs into which the longitudinal leads (employed to simplify the boundary conditions) eventually terminate, so that the electrons do not feel any electric field in the course of ballistic propagation through clean 2DEG central region. The information about these different situations is encoded into the lesser Green function $\mathbf{G}^<$.

Here, we focus on experimentally relevant spin Hall devices where the finite-size central region (C), defined on the $L \times L$ lattice, is attached to four external semi-infinite leads of the same width L. The leads at infinity terminate into the reservoirs where electrons are brought into thermal equilibrium, characterized by the Fermi–Dirac distribution function $f(E - eV_\mathrm{p})$, to ensure the steady-state transport. In such a multiterminal Landauer setup (Büttiker 1986; Baranger and Stone 1989; Datta 1995), current is limited by quantum transmission through a potential profile, while power is dissipated non-locally in the reservoirs. The voltage in each lead of the four-terminal spin-Hall bridge is V_p ($p = 1, \ldots, 4$), so that the on-site potential $\varepsilon_\mathbf{m}$ within the leads has to be shifted by eV_p.

The spin-dependent lesser Green function $\mathbf{G}^<$ defined in eqn (24.65) is evaluated within the finite-size sample region as a $2L^2 \times 2L^2$ matrix in the site⊗spin space through the spin-resolved matrix Keldysh equation (Keldysh 1965)

$$\mathbf{G}^<(E) = \mathbf{G}^r(E)\mathbf{\Sigma}^<(E)\mathbf{G}^a(E), \qquad (24.81)$$

which is valid for steady-state transport (i.e., when transients have died away). Within the effective single-particle picture, the retarded Green function is computed by inverting the Hamiltonian of an open system sample + leads

$$\mathbf{G}^r(E) = \left[E\mathbf{I}_C - \mathbf{H}_C - eU_\mathbf{m} - \sum_p \mathbf{\Sigma}_p(E - eV_p) \right]^{-1}, \qquad (24.82)$$

and the advanced Green function matrix is $\mathbf{G}^a(E) = [\mathbf{G}^r(E)]^\dagger$. Here, the retarded $\mathbf{\Sigma}_p(E)$ and the lesser $\mathbf{\Sigma}^<(E)$ self-energy matrices

$$\mathbf{\Sigma}_p(E) = \mathbf{H}_{pC}^\dagger \left[(E + \mathrm{i}0_+)\,\mathbf{I}_p - \mathbf{H}_p^{\mathrm{lead}} \right]^{-1} \mathbf{H}_{pC}, \qquad (24.83)$$

$$\mathbf{\Gamma}_p(E) = \mathrm{i} \left[\mathbf{\Sigma}_p(E) - \mathbf{\Sigma}_p^\dagger(E) \right], \qquad (24.84)$$

$$\mathbf{\Sigma}^<(E) = \mathrm{i} \sum_p \mathbf{\Gamma}_p(E - eV_p) f(E - eV_p), \qquad (24.85)$$

are exactly computable in the non-interacting electron approximation and without any inelastic processes taking place within the sample. They account for the "interaction" of the SO-coupled sample with the attached leads, thereby generating a finite lifetime that the electron spends within the 2DEG before escaping through the leads toward the macroscopic thermalizing reservoirs. Here, $\mathbf{I}_C$ is the $2L^2 \times 2L^2$ identity matrix and $\mathbf{I}_p$ is the identity matrix in the infinite site⊗spin space of the lead p. We use the following Hamiltonian matrices

$$(\mathbf{H}_C)_{\mathbf{mm'},\sigma\sigma'} = \langle 1_{\mathbf{m}\sigma}| \hat{H} |1_{\mathbf{m'}\sigma'}\rangle, \quad (\mathbf{m}, \mathbf{m'} \in C), \qquad (24.86a)$$

$$\left(\mathbf{H}_p^{\text{lead}}\right)_{\mathbf{mm'},\sigma\sigma'} = \langle 1_{\mathbf{m}\sigma}| \hat{H} |1_{\mathbf{m'}\sigma'}\rangle, \quad (\mathbf{m}, \mathbf{m'} \in p), \qquad (24.86b)$$

$$\left(\mathbf{H}_{pC}\right)_{\mathbf{mm'},\sigma\sigma'} = \langle 1_{\mathbf{m}\sigma}| \hat{H} |1_{\mathbf{m'}\sigma'}\rangle, \quad (\mathbf{m} \in p, \mathbf{m'} \in C), \qquad (24.86c)$$

where $|1_{\mathbf{m}\sigma}\rangle$ is a vector in the Fock space (meaning that the occupation number is one for the single-particle state $|\mathbf{m}\sigma\rangle$ and zero otherwise) and $\hat{H}$ is the Hamiltonian given in eqn (24.30).

In the general case of arbitrary applied bias voltage, the gauge invariance of measurable quantities (such as the current–voltage characteristics) with respect to the shift of electric potential everywhere by a constant V_c, $eV_p \rightarrow eV_p + eV_c$ and $eU_{\mathbf{m}} \rightarrow eU_{\mathbf{m}} + eV_c$, is satisfied, with the proviso that the retarded self-energies $\mathbf{\Sigma}_p(E - eV_p)$ introduced by each lead depend explicitly on the applied voltages at the sample boundary, while the computation of the retarded Green function $\mathbf{G}^r(E)$ has to include the electric potential landscape[13] $U_{\mathbf{m}}$ within the sample (Christen and Büttiker 1996). However, when the applied bias is low, so that linear-response zero-temperature quantum transport takes place through the sample (as determined by $\mathbf{G}^r(E_F)$), the exact profile of the internal potential becomes irrelevant (Baranger and Stone 1989; Nikolić and Allen 1999).

[13]The potential landscape $U_{\mathbf{m}}$ can be obtained from the Poisson equation with charge density eqn (24.70) as the source.

24.8.4.1   *How to introduce dephasing into mesoscopic SHE nanostructures*

Much of the applications of NEGF techniques to mesoscopic semiconductor and nanoscopic molecular systems treat phase-coherent transport by relying on some type of Landauer transmission formula (Rocha *et al.* 2006). On the other hand, applications demand devices operating at room temperature where device size is typically much bigger than the dephasing length so that one seldom observes quantum interference effects in the transmission due to multiple scattering off impurities and boundaries. Since such quantum-coherence effects are also seen in the linear response spin density in SO-coupled quantum wires, typically computed at zero-temperature (Nikolić *et al.* 2005c; Reynoso *et al.* 2006), to compare such calculations with experiments conducted at finite temperature it is desirable to introduce dephasing into NEGF formalism in a simple yet controllable fashion (which leaves the charge-current conservation intact).

The most widely used example of the simple approach to inclusion of dephasing are fictitious Büttiker voltage probes (Büttiker 1986), such as one-dimensional electrodes attached at each site $\mathbf{m}$ of the lattice and with their electrochemical potential adjusted to ensure that the current drawn by each probe is zero.[14] In NEGF formalism this introduces additional self-energy whose matrix elements are $(\boldsymbol{\Sigma}_{\text{deph}})_{\mathbf{mm},\sigma\sigma} = -i\eta$ for all sites $\mathbf{m}$ within the sample except for those where the external electrodes are attached (Golizadeh-Mojarad and Datta 2007a). However, while Büttiker probes coupled to each site can remove sharp features in the transmission of the device, they also introduce additional scattering events thereby artificially enhancing the resistance.

A "momentum-conserving" phenomenological model for dephasing, that is easily implemented within the NEGF formalism, has recently been proposed by Golizadeh-Mojarad and Datta (2007a). Here, the self-energies are taken as $\boldsymbol{\Sigma}_{\text{deph}}(E) = d_i \mathbf{G}^r(E)$ and $\boldsymbol{\Sigma}_{\text{deph}}^<(E) = d_i \mathbf{G}^<(E)$, with parameter d_i controlling the strength of the elastic dephasing processes. This choice is motivated by the general expressions for the first-order self-consistent Born approximation for the self-energies that ensure the current conservation (Leeuwen *et al.* 2006). The additional choice of the function multiplying matrix elements of NEGFs is taken to guarantee momentum conservation, so that dephasing does not artifactually enhance the resistance (Golizadeh-Mojarad and Datta 2007a). This form of the retarded and lesser self-energies requires that eqn (24.82) be solved self-consistently (with the initial guess for $\mathbf{G}^r(E)$ being the solution with no dephasing) and then use such $\mathbf{G}^r(E)$ to solve the Keldysh equation eqn (24.81) directly as the Sylvester equation of matrix algebra. This allows one to compare NEGF calculations (Golizadeh-Mojarad and Datta 2007b) to SHE experiments all the way to room temperature. Another simple way to smear sharp features in the local spin density due to quantum coherence is averaging over an energy interval (Nikolić *et al.* 2005c), as shown in Fig. 24.9. For more details about different NEGF-based models of dephasing in quantum transport and their effect on momentum and spin relaxation see Chapter 3 by R. Golizadeh-Mojarad and S. Datta in this volume.

[14]Since current through Büttiker probes is fixed to be zero, for every electron that enters the lead and is absorbed by the corresponding reservoir another one will come with phase memory washed out due to equilibration by inelastic effects assumed to take place in the reservoir. Thus, this method introduces inelastic effects is a simple phenomenological way.

24.8.5   Example: Local spin density and spin currents in mesoscopic SHE nanostructures

Figure 24.9(a) shows NEGF computation (Nikolić *et al.* 2005c) of spin density within the Rashba coupled 2DEG where all essential features of the experimentally observed SHE in disordered semiconductor devices (Kato *et al.* 2004a) are obtained but for a perfectly clean sample, such as: the out-of-plane $\langle S_{\mathbf{m}}^z \rangle$ component of the spin accumulation develops two peaks of opposite signs at the lateral edges of the 2DEG; upon reversing the bias voltage, the edge peaks flip their sign; around the left-lead-2DEG interface $\langle S_{\mathbf{m}}^z \rangle$ is suppressed since unpolarized electrons are injected at this contact. By attaching two additional transverse leads at the lateral edges in Fig. 24.9(b), we show how spin-↑ and spin-↓ densities will flow through those leads in opposite directions to generate spin-Hall current, thereby providing an all-electrical semiconductor-based spin injector using ballistic nanostructures with intrinsic SO couplings. No discussion of controversial spin currents within the sample is necessary to reach this conclusion. Note that the simple picture of opposite spin densities

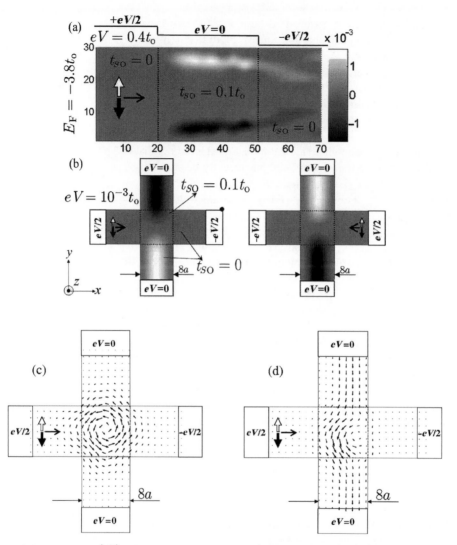

Fig. 24.9 (a) The out-of-plane component $\langle S_{\mathbf{m}}^z \rangle$ of the non-equilibrium spin accumulation induced by non-linear ($eV = 0.4\,t_O$) quantum transport of unpolarized charge current injected from the left lead into a two-terminal clean 2DEG nanostructure (of size $L = 30a > L_{SO}$, $a \simeq 3\,\mathrm{nm}$) with the Rashba SO coupling $t_{SO} = 0.1t_O$ and spin precession length $L_{SO} \approx 15.7a$. (b) Lateral spin-↑ and spin-↓ densities will propagate in opposite directions through the attached transverse ideal ($t_{SO} = 0$) electrodes to yield a linear response spin-Hall current $I_y^{S_z}$ flowing out of a 2DEG ($L = 8a < L_{SO}$), which changes sign upon reversing the bias voltage. (c), (d) The spatial distribution of local spin currents in ballistic four-terminal 2DEG bridges with the Rashba SO coupling $t_{SO} = 0.1\,t_O$ setting $L_{SO} \approx 15.7a$ and linear response bias voltage $eV = 10^{-3}t_O$. The local spin current is the sum of the equilibrium (persistent) spin current in (c), carried by the fully occupied states from $-4\,t_O$ to $E_F - eV/2$, and the non-equilibrium (transport) spin current in (d) carried by the partially occupied states around the Fermi energy from $\mu_R = E_F - eV/2$ (electrochemical potential of the right reservoir) to $\mu_L = E_F + eV/2$ (electrochemical potential of the left reservoir).

flowing in opposite directions in Fig. 24.9(b) is obtained in samples smaller than L_{SO}, while in larger samples these patterns are more complicated (Nikolić *et al.* 2006) due to multiple spin precession within the sample before spin has a chance to exit through the leads.

The spatial profile of local spin currents corresponding to Fig. 24.9(b) is shown in Figs. 24.9(c) and (d) where we separate the integration in eqn (24.73b) for S_z bond spin current into two parts $\left\langle \hat{J}_{\mathbf{mm'}}^{S_z} \right\rangle = \left\langle \hat{J}_{\mathbf{mm'}}^{S_z(\mathrm{eq})} \right\rangle +$ $\left\langle \hat{J}_{\mathbf{mm'}}^{S_z(\mathrm{neq})} \right\rangle$:

$$\left\langle \hat{J}_{\mathbf{mm'}}^{S_z} \right\rangle = \frac{t_{\mathrm{O}}}{2} \int\limits_{E_b}^{E_{\mathrm{F}}-eV/2} \frac{\mathrm{d}E}{2\pi} \mathrm{Trs} \left[\hat{\sigma}_z \left(\mathbf{G}_{\mathbf{m'm}}^{<}(E) - \mathbf{G}_{\mathbf{mm'}}^{<}(E) \right) \right]$$

$$+ \frac{t_{\mathrm{O}}}{2} \int\limits_{E_{\mathrm{F}}-eV/2}^{E_{\mathrm{F}}+eV/2} \frac{\mathrm{d}E}{2\pi} \mathrm{Trs} \left[\hat{\sigma}_z \left(\mathbf{G}_{\mathbf{m'm}}^{<}(E) - \mathbf{G}_{\mathbf{mm'}}^{<}(E) \right) \right], \quad (24.87)$$

plotted as panels (c) and (d), respectively. The states from the band bottom E_b to $E_{\mathrm{F}} - eV/2$ are fully occupied, while states in the energy interval from the electrochemical potential $E_{\mathrm{F}} - eV/2$ ($eV > 0$) of the right reservoir to the electrochemical potential $E_{\mathrm{F}} + eV/2$ of the left reservoir are partially occupied because of the competition between the left reservoir that tries to fill them and the right reservoir that tries to deplete them. The spatial distribution of the microscopic spin currents in Fig. 24.9(c) is akin to the vortex-like pattern of bond spin currents within the device that would exist in equilibrium $eV_p = \mathrm{const.}$ (Nikolić *et al.* 2006). These do not transport any spin between two points in real space since their sum over any cross-section is zero. Thus, Fig. 24.9(d) demonstrates that non-zero spin-Hall flux through the transverse cross-sections is due to only the wavefunctions (or Green functions) around the Fermi energy. The same sum of $\left\langle \hat{J}_{\mathbf{mm'}}^{S_z(\mathrm{neq})} \right\rangle$ in (d) over an arbitrary transverse cross-section (orthogonal to the y-axis) defines the total spin current that, although not the same on different cross-sections within the sample, flows into the leads where it becomes conserved and it is measurable (e.g. via the inverse SHE (Hankiewicz *et al.* 2004)).

24.9 Computational algorithms for real⊗spin space NEGFs in multiterminal devices

While all formulas in Section 24.8.4 for the core NEGF quantities, $\mathbf{G}^r(E)$ and $\mathbf{G}^<(E)$, can be implemented by brute force operations on full matrices, this is typically restricted to small lattices of a few thousands sites due to computational complexity of matrix operations. For example, for a system of size L in d dimensions, the computing time scales as L^{3d}, while the memory needed scales as L^{2d}. By separating the system into slices described by much smaller Hamiltonian matrices, and by using the recursive Green function algorithm in serial or parallel implementation (Drouvelis *et al.* 2006), the complexity can be reduced drastically to L^{3d-2} scaling of the required computing time.

It is often considered (Kazymyrenko and Waintal 2008) that the recursive algorithm can be applied only to two-terminal devices and for the computation of total transport quantities in the leads. Some alternative algorithms with reduced computational complexity tailored for multiterminal structures have been proposed recently (Kazymyrenko and Waintal 2008), and the recursive algorithm has also been extended to get local charge quantities within the sample (Cresti *et al.* 2003; Metalidis and Bruno 2005). We discuss in Section 24.9.2 how to extend the recursive scheme to four-terminal mesoscopic spin-Hall bridges in Fig. 24.7 that makes possible computation of the spin-Hall conductance on large lattice sizes shown in Fig. 24.8(a). In addition, Section 24.9.3 shows how to obtain local spin densities within large two-terminal structures (Nikolić *et al.* 2005c) using the recursive-type approach. Since the starting point of any algorithm to obtain $\mathbf{G}^r(E)$ and $\mathbf{G}^<(E)$ for open finite-size system is computation of self-energies for the infinite part (ensuring a continuous spectrum and dissipation) of the system, we briefly review recent developments in numerical techniques used to compute the surface Green function of semi-infinite leads that generates matrices for $\mathbf{\Sigma}(E)$ and $\mathbf{\Sigma}^<(E)$.

24.9.1   Numerical algorithms for computing self-energy matrices

Several different numerical techniques are available to evaluate single-particle Green function of a semi-infinite electrode using a localized basis (Velev and Butler 2004). For example, the so-called Ando method (Ando 1991) computes the surface Green function of a homogeneous lead, consisting of a repeating supercell described by the Hamiltonian $\mathbf{H}_{i,i}$ where supercells are coupled by the Hamiltonian $\mathbf{H}_{i,i-1}$, from its exact Bloch propagating modes at energy E. This method offers high-precision numerical evaluation of the self-energy and can be generalized (Khomyakov *et al.* 2005; Rocha *et al.* 2006) or accelerated (Sørensen *et al.* 2008) to complicated homogeneous periodic systems, where $\mathbf{H}_{i,i-1}$ is not invertible (required in the original Ando algorithm) due to the atomistic structure of the leads that are more complicated than a square tight-binding lattice (such as the honeycomb one (Chen *et al.* 2007; Zârbo and Nikolić 2007)).

An alternative and approximate method, applicable also to inhomogeneous systems (when the lead does not consist of repeating supercells with identical $\mathbf{H}_{i,i}$), is the recursive (or continued fraction) algorithm for the surface Green function (Velev and Butler 2004). It connects the Green function of a given layer to the Green functions of neighboring layers, so that starting from the surface layer and repeating this process until the effect of all other layers on the surface layer is taken account yields the surface Green function (in practice, the recursive method is executed to finite depth into the lead where the number of considered neighboring layers affecting the surface layer is set by the convergence criterion).

The Ando method can be executed analytically for a simple square lattice lead with no SO coupling (Datta 1995). For comparing mesoscopic SHE calculations to the bulk SHE description, it is also useful to have a system where SO-coupled leads are attached to an identical SO-coupled sample to form an infinite spin-split wire, while two additional ideal (with no spin interactions) leads are attached in the transverse direction (Sheng and Ting 2006). The self-energies for semi-infinite electrodes with the Rashba SO coupling can

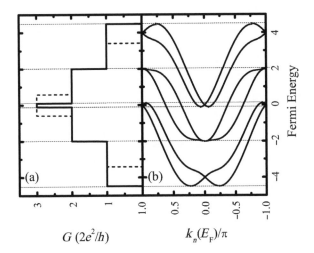

$G\ (2e^2/h)$ $k_n(E_F)/\pi$

Fig. 24.10 The conductance quantization (a) in the Rashba spin-split infinite clean quantum wire modelled on the lattice with three sites per cross-section. The wire is decomposed into the finite-size central sample attached to two semi-infinite leads where the Rashba hopping $t_{SO} = t_O$ exists in all three segments. The dashed line in (a) corresponds to quantized conductance of the same wire in the absence of the SO coupling $t_{SO} = 0$. Panel (b) plots the Rashba-split subband dispersion responsible for this unusual conductance quantization.

be obtained via the Ando method, with double-size matrices $\mathbf{H}_{i,i}$ and $\mathbf{H}_{i,i-1}$ to include spin. We employ this method in Fig. 24.10 to provide a simple example of the conductance quantization through a three-channel infinite spin-split quantum wire that can be used as a building block of the four-terminal calculations. The Rashba-split subbands define new conducting channels whose spin-polarization properties can be highly non-trivial for strong SO coupling (Governale and Zülicke 2004), as discussed in Section 24.6.2.1. This is in contrast to standard wires where conducting channels are always separable states $|n\rangle \otimes |\sigma\rangle$, which simplifies the analysis of spin injection (Nikolić and Souma 2005a) while also introducing the scattering at the lead sample interface that can reduce the spin-Hall conductance (Nikolić *et al.* 2005b; Sheng and Ting 2006).

24.9.2   Recursive Green function algorithm for total spin and charge currents in multiterminal nanostructures

For many transport calculations the full retarded Green function, eqn (24.82), is not required. For instance, to compute the transmission matrix (24.49) from lead q to lead p, one only needs the Green function matrix elements $(\mathbf{G}^r)_{\mathbf{mm}',\sigma\sigma'}$ where the sites $\mathbf{m}$ belong to the lead p edge supercell and the sites $\mathbf{m}'$ belong to the first supercell of the lead q. We first discuss the recursive algorithm (Drouvelis *et al.* 2006) that evaluates only this portion $\mathbf{G}_{N+1,0}$ of the full retarded Green function matrix for the two-probe setup in Fig. 24.11(a). The sample is divided into N layers, and to account for the effect of the leads, their self-energies are added onto the layers 0 and $N + 1$. These two layers are the first principal layers of the two leads. It is important to emphasize that adding the self-energies directly onto the sample in more complicated cases than a simple square lattice can lead to an incorrect result for the transmission eigenvalues.

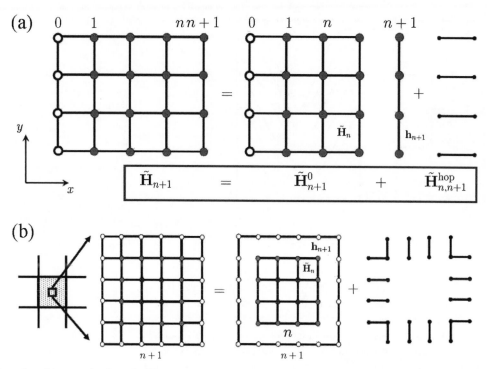

Fig. 24.11 Illustration of the recursive Green function computational algorithm for: (a) two-terminal setup; and (b) four-terminal setup defined on the square tight-binding lattice.

The recursive scheme starts by adding the self-energy of the left lead $\mathbf{\Sigma}_L$ onto its principal layer directly connected to the sample. The Hamiltonian of any of the $N + 2$ blocks is labelled by $\mathbf{h}_n$. The retarded Green function matrix of the nth isolated block is $\mathbf{g}_n = (E - \mathbf{h}_n - i\eta)^{-1}$. The hopping matrix connecting blocks n and $n + 1$ is

$$\mathbf{H}^{\mathrm{hop}}_{n,n+1} = \begin{pmatrix} 0 & \mathbf{H}_{n,n+1} \\ \mathbf{H}_{n+1,n} & 0 \end{pmatrix}. \tag{24.88}$$

We also introduce $\tilde{\mathbf{H}}_n$ as the matrix that is the sum of the Hamiltonian of $n + 1$ blocks and the lead self-energy added to the 0th block. The corresponding retarded Green function matrix is $\mathbf{G}_n = (E - \tilde{\mathbf{H}}_n + i\eta)^{-1}$. One can add the next isolated block $\mathbf{h}_{n+1}$ to $\tilde{\mathbf{H}}_n$ and obtain $\tilde{\mathbf{H}}^{(0)}_{n+1} = \tilde{\mathbf{H}}_n \oplus \mathbf{h}_{n+1}$.

Assuming that the Green function of n blocks is known, we add the $(n + 1)$th block to $\tilde{\mathbf{H}}_n$ without turning on the coupling $\mathbf{H}^{\mathrm{hop}}_{n,n+1}$ between them. The new Hamiltonian, is block diagonal

$$\tilde{\mathbf{H}}^{(0)}_{n+1} = \begin{pmatrix} \tilde{\mathbf{H}}_n & 0 \\ 0 & \mathbf{h}_{n+1} \end{pmatrix}, \tag{24.89}$$

whose inversion leads to

$$\mathbf{G}^{(0)}_{n+1} = \begin{pmatrix} \mathbf{G}_n & 0 \\ 0 & \mathbf{g}_{n+1} \end{pmatrix}. \tag{24.90}$$

To find the Green function $\mathbf{G}_{n+1}$ it is necessary to add the matrix elements connecting the $(n + 1)$th block to $\tilde{\mathbf{H}}_n$

$$\tilde{\mathbf{H}}_{n+1} = \begin{pmatrix} \tilde{\mathbf{H}}_n & \mathbf{H}_{n,n+1} \\ \mathbf{H}_{n+1,n} & \mathbf{h}_{n+1} \end{pmatrix}. \tag{24.91}$$

The Green function of $n + 2$ blocks can be found by using the Dyson equation

$$\mathbf{G}_{n+1} = \mathbf{G}_{n+1}^{(0)} + \mathbf{G}_{n+1}^{(0)} \mathbf{H}_{n,n+1}^{\text{hop}} \mathbf{G}_{n+1}. \tag{24.92}$$

All matrices in this equation are of the dimension $n + 2$ in the block co-ordinate space. Equation (24.92) can be used to obtain the recurrence relationships for the Green function blocks we are interested in. To get the conductance of the two-terminal device we need to compute $\langle N + 1 | \mathbf{G}_{N+1} | 0 \rangle \equiv \mathbf{G}_{N+1}(N + 1, 0)$, which connects the right and the left lead.

In the multiterminal case, we choose to start from the Hamiltonian of the block of sites in the center of the sample and then add one-by-one additional blocks of the same shape until we reach the attached electrodes, as sketched in Fig. 24.11(b). Therefore, in this case we need the Green function elements $\mathbf{G}_{N+1}(N + 1, N + 1)$. From eqn (24.92) we find

$$\mathbf{G}_{n+1}(n + 1, n + 1) = \mathbf{G}_{n+1}^{(0)}(n + 1, n + 1) + \left\langle n+1 \left| \mathbf{G}_{n+1}^{(0)} \mathbf{H}_{n,n+1}^{\text{hop}} \mathbf{G}_{n+1} \right| n+1 \right\rangle. \tag{24.93}$$

By inserting the identity matrix in the subspace of block coordinates into eqn (24.93)

$$\mathbf{I} = \sum_m |m\rangle \langle m|, \tag{24.94}$$

the second term in eqn (24.93) becomes

$$\left\langle n + 1 \left| \mathbf{G}_{n+1}^{(0)} \sum_m |m\rangle \langle m| \mathbf{H}_{n,n+1}^{\text{hop}} \sum_m |l\rangle \langle l| \mathbf{G}_{n+1} \right| n + 1 \right\rangle. \tag{24.95}$$

It is easy to see from eqns (24.88), (24.90), and (24.91) that

$$\langle n + 1 | \mathbf{G}_{n+1}^{(0)} | m \rangle = \delta_{n+1,m} \mathbf{g}_{n+1}, \tag{24.96a}$$

$$\langle m | \mathbf{H}_{n,n+1}^{\text{hop}} | l \rangle = \delta_{m,n} \delta_{l,n+1} \mathbf{H}_{n,n+1} + \delta_{m,n+1} \delta_{l,n} \mathbf{H}_{n+1,n}. \tag{24.96b}$$

Equation (24.96) can be substituted into eqn (24.95) to yield the Dyson equation (24.92) for $\mathbf{G}_{n+1}(n + 1, n + 1)$

$$\mathbf{G}_{n+1}(n + 1, n + 1) = \mathbf{g}_{n+1} + \mathbf{g}_{n+1} \mathbf{H}_{n+1,n} \mathbf{G}_{n+1}(n, n + 1), \tag{24.97}$$

in which $\mathbf{G}_{n+1}(n, n + 1)$ is an unknown quantity. By computing the matrix elements of the terms in the Dyson equation (24.92) between $|n\rangle$ and $|n + 1\rangle$ we find that

$$\mathbf{G}_{n+1}(n, n + 1) = \mathbf{G}_n(n, n) \mathbf{H}_{n,n+1} \mathbf{G}_{n+1}(n + 1, n + 1). \tag{24.98}$$

Finally, combining eqns (24.97) and (24.98) leads to the following recurrence relationship

$$\mathbf{G}_{n+1}(n+1, n+1) = \frac{\mathbf{I}_{n+1}}{\mathbf{g}_{n+1}^{-1} - \mathbf{H}_{n+1,n}\mathbf{G}_n(n, n)\mathbf{H}_{n,n+1}}. \tag{24.99}$$

Here, the matrix $\mathbf{I}_{n+1}$ is the identity matrix in the subspace of the $(n+1)$th block and the denominator is the matrix to be inverted.

If the recursive scheme in Fig. 24.11(b) is used, we need to obtain $\mathbf{G}_{N+1}(N+1, N+1)$ from which the transmission matrix can be computed. For the two-terminal case, the recursive scheme sketched in Fig. 24.11(a) is more efficient computationally. Computing the matrix elements of the Dyson equation (24.92) between the states $|n+1\rangle$ and $|m\rangle$ gives

$$\mathbf{G}_{n+1}(n+1, m) = \mathbf{G}_{n+1}(n+1, n+1)\mathbf{H}_{n+1,n}\mathbf{G}_n(n, m), \tag{24.100}$$

where $m < n + 1$.

We summarize the general recursive Green function algorithm for the computation of (spin-resolved) transmissions, and therefore spin or charge conductances, in the form of the following steps:

1. Calculate the self-energies for the leads, as described in Section 24.9.1.
2. Identify the $N + 2$ blocks for the recursive method as shown in Fig. 24.11 (the number of sites of each block might be different).
3. Initialize the recurrence by computing $\mathbf{G}_0(0, 0) \equiv \mathbf{g}_0$, where $\mathbf{g}_0 = (E - \tilde{\mathbf{H}}_0 - i\eta)^{-1}$ is the Green function of the first block. This block has the self-energy added to it in the two-terminal case. In the multiterminal case, the first block can be chosen to be a rectangle of the size $N_x^{(0)} \times N_y^{(0)}$ sites, as illustrated in Fig. 24.11(b). A difficulty may arise in the multi-terminal cases where the matrix $E - \mathbf{h}_0$ can turn out to be singular for certain choices of the shape of the sample or its parameters.
4. Using the formulas (24.99) and (24.100) compute the Green functions $\mathbf{G}_{N+1}(N+1, N+1)$ and $\mathbf{G}_{N+1}(N+1, 0)$.
5. Obtain the retarded Green function matrix elements $\mathbf{G}_{pq}$ connecting each pair of leads p and q, either from $\mathbf{G}_{N+1}(N+1, 0)$ in the two-terminal case, or $\mathbf{G}_{N+1}(N+1, N+1)$ in the multiterminal case.
6. Compute the transmission matrix using eqn (24.49).

24.9.3   Recursive Green function algorithm for local spin and charge densities

The computation of non-equilibrium spin and charge densities using eqns (24.80) and (24.70), respectively, requires us to obtain $\mathbf{G}^<$. This has the highest computational complexity since one has to perform the matrix multiplication in eqn (24.81). Moreover, this matrix multiplication has to be done for all the energy points in eqn (24.80). Here, we introduce a method for recursive calculation of the spin and charge non-equilibrium densities within the NEGF formalism, which makes it possible to greatly reduce this computational complexity and, thereby, analyze spin densities in sizable samples (Nikolić *et al.* 2005c).

In order to illustrate the essential steps of this method, it is enough to consider a simple two-probe setup in Fig. 24.11(a). The two-probe system considered here is made of $N + 2$ blocks, where the 0th block is the first left lead supercell onto which the left lead self-energy is added, and $(N + 1)$th block is the first right lead supercell onto which the right lead self-energy is added. The voltage applied onto the left and right lead is V_L and V_R, respectively. The lesser self-energy matrix in eqn (24.81) can be rewritten as

$$\Sigma^<(E) = i\Gamma_L(E - eV_L)f(E - eV_L) + i\Gamma_R(E - eV_R)f(E - eV_R)$$

$$= \begin{pmatrix} i\Gamma_L(E - eV_L)f(E - eV_L) & \cdots & 0 \\ \cdots & \cdots & \cdots \\ 0 & \cdots & i\Gamma_R(E - eV_R)f(E - eV_R) \end{pmatrix},$$

$$(24.101)$$

where the only non-zero elements are the first and the last on the diagonal. This simplifies greatly the matrix multiplication in eqn (24.81), since only multiplication by the non-zero matrix elements of $\Sigma^<$ is required

$$\Sigma^<(0, 0) = i\Gamma_L(E - eV_L)f(E - eV_L), \quad (24.102a)$$

$$\Sigma^<(N + 1, N + 1) = i\Gamma_R(E - eV_R)f(E - eV_R). \quad (24.102b)$$

In order to calculate the non-equilibrium spin density eqn (24.80) or charge density eqn (24.70), only the diagonal elements of $G^<(E)$ are required. Using eqns (24.101) and (24.102) in eqn (24.81), leads to

$$G^<(m, m) = G(m, N + 1)\Sigma^<(N + 1, N + 1)G^\dagger(N + 1, m)$$

$$+ G(m, 0)\Sigma^<(0, 0)G^\dagger(0, m), \quad (24.103)$$

where m labels the coordinates of the mth block.

The retarded Green function matrices $G(m, N + 1)$ and $G^r(m, 0)$ can be found using the Dyson equation (24.92) introduced in Section 24.9.2. To find $G^r(m, N + 1)$, one starts the recurrence from the left lead and finds

$$G_{n+1}(m, n + 1) = G_n(m, n)H_{n,n+1}G_{n+1}(n + 1, n + 1), \quad (24.104)$$

where $m \leq n$. The Green function block $G_{n+1}(n + 1, n + 1)$ can be obtained from eqn (24.99). To find $G(m, 0)$, the recurrence has to start from the right lead and we get

$$G_{n-1}(m, n - 1) = G_n(m, n)H_{n,n-1}G_{n-1}(n - 1, n - 1), \quad (24.105)$$

where $m \geq n$. The system of equations is finally closed by using

$$G_{n-1}(n - 1, n - 1) = \frac{I_{n-1}}{g_{n-1}^{-1} - H_{n-1,n}G_n(n, n)H_{n,n-1}}. \quad (24.106)$$

In this method, one has to compute recursively the retarded Green functions $G(m, N + 1)$ and $G(m, 0)$ for all $m = \overline{0, N + 1}$. It is not as computationally costly as the full Green function matrix inversion, and it can be extended to compute bond spin and charge currents. In that case, we also have to find $\mathbf{G}(m \pm 1, N + 1)$ and $\mathbf{G}(m \pm 1, 0)$.

24.10   Concluding remarks

The pure spin currents, which arise when equal numbers of spin-↑ and spin-↓ electrons propagate in opposite directions so that the net charge current carried by them is zero, have recently emerged as one of the major topics of both metal- and semiconductor-based spintronics. This is due to the fact that they offer a new realm to explore fundamental spin transport phenomena in solids, as well as to construct a new generation of spintronic devices with greatly reduced power dissipation and new functionality in transporting either classical or quantum (such as flying spin qubits) information. Their generation (which is currently experimentally achieved mostly through spin pumping, non-local spin injection using lateral spin valves, and SHE), control, and demanding detection typically involves quantum-coherent spin dynamics. In particular, the recently discovered SHE and the inverse SHE have offered some of the most efficient schemes to generate and detect pure spin currents, respectively. However, they have also posed numerous challenges for their theoretical description and device modelling since a conventionally defined spin current operator does not satisfy standard continuity equation for spin when intrinsic SO couplings (which act as homogeneous internal momentum-dependent magnetic fields affecting the band structure and causing spin precession) are present. Even with modified spin-current definitions, this fundamental problem makes difficult easy connection between theoretically computed spin conductivities of infinite homogeneous systems and usually experimentally measured spin densities for devices with arbitrary boundaries and attached external electrodes. An attempt to give up on the theoretical description of spin flow through spin current altogether, by using spin density within the diffusion equation formalism (which is expected to provide a complete and physically meaningful description of spin and charge diffusive transport in terms of position- and time-dependent spin and charge densities), leads to another set of obstacles. That is, the explicit solution of the diffusion equation strongly depends on the boundary conditions (extracted for charge diffusion from conservation laws) where spin non-conservation, manifested as ballistic spin precession at the edges, cannot be unambiguously matched with the diffusive dynamics in the bulk captured by the diffusion equation for length scales much longer than the mean-free path.

This chapter provides a summary of such fundamental problems in spin transport through SO-coupled semiconductors and possible resolution through spin-dependent NEGF formalism that can handle *both* ballistic and diffusive regimes, as well as phase-coherent and dephasing effects, in realistic multiterminal nanostructures operating at different mobilities and temperatures. The central quantities of this approach—total spin currents flowing out of the device through ideal (interaction-free) electrodes, local spin densities within the sample and along the edges, and local spin currents (whose sum within the leads gives total spin currents)—can be used to understand experiments or model novel spintronics devices. For example, NEGF approach is indispensable (Hankiewicz *et al.* 2004) for computational modelling of generator-detector experimental setups (Brüne *et al.* 2008; Roth *et al.* 2009) where direct and inverse SHE multiterminal bridges are joined into a single circuit to sense the pure spin-Hall current through voltages induced by its flow through regions with SO interactions. The chapter also provides extensive coverage of relevant technical and computational details, such as: the construction of retarded and lesser non-equilibrium Green functions for SO-coupled nanostructures attached to many electrodes; computation of relevant self-energies introduced by the lead–sample interaction; and accelerated algorithms that make possible spin-transport modelling in a device whose size is comparable to the spin-precession length of a few hundred nm where the SHE response to injected unpolarized charge current is expected to be optimal (Nikolić *et al.* 2005b; Moca and Marinescu 2007).

Acknowledgments

We acknowledge numerous insightful discussions with İ. Adagideli, S. Datta, J. Fabian, J. Inoue, S. Murakami, N. Nagaosa, J. Nitta, K. Nomura, M. Onoda, E.I. Rashba, and J. Sinova. This work was supported by NSF Grant No. ECCS 0725566, DOE Grant No. DE-FG02-07ER46374, and the Center for Spintronics and Biodetection at the University of Delaware.

References

Adagideli, İ., Bauer, G.E.W. *Phys. Rev. Lett.* **95**, 256602 (2005).

Adagideli, İ., Bauer, G.E.W., Halperin, B.I. *Phys. Rev. Lett.* **97**, 256601 (2006).

Aleiner, I.L., Fal'ko, V.I. *Phys. Rev. Lett.* **87**, 256801 (2001).

Ando, T. *Phys. Rev. B* **44**, 8017 (1991).

Ando, T. *Physica E* **20**, 24 (2003).

Awschalom, D.D., Flatté, M.E. *Nature Phys.* **3**, 153 (2007).

Ballentine, L.E. *Quantum Mechanics: A Modern Development* (World Scientific, Singapore, 1998).

Baranger, H.U., Stone, A.D. *Phys. Rev. B* **40**, 8169 (1989).

Bardarson, J.H., Adagideli, İ., Jacquod, P. *Phys. Rev. Lett.* **98**, 196601 (2007).

Bleibaum, O. *Phys. Rev. B* **74**, 113309 (2006).

Brüne, C., Roth, A., Novik, E.G., König, M., Buhmann, H., Hanklewicz, E.M., Hanke, W., Sinova, J., Molenkamp, L.W. Preprint arXiv:0812.3768 (2008).

Bulgakov, E.N., Pichugin, K.N., Sadreev, A.F., Středa, P., Šeba, P. *Phys. Rev. Lett.* **83**, 376 (1999).

Burkov, A.A., Núñez, A.S., MacDonald, A.H. *Phys. Rev. B* **70**, 155308 (2004).

Büttiker, M. *Phys. Rev. Lett.* **57**, 1761 (1986).

Caroli, C., Combescot, R., Nozieres, P., Saint-James, D.J. *Phys. C: Solid State Phys.* **4**, 916 (1971).

Chang, C.H., Mal'shukov, A.G., Chao, K.A. *Phys. Rev. B* **70**, 245309 (2004).

Chen, S.-H., Liu, M.-H., Chang, C.-R. *Phys. Rev. B* **76**, 075322 (2007).

Christen, T., Büttiker, M. *Europhys. Lett.* **35**, 523 (1996).

Cresti, A., Farchioni, R., Grosso, G., Parravicini, G.P. *Phys. Rev. B* **68**, 075306 (2003).

Crooker, S.A., Smith, D.L. *Phys. Rev. Lett.* **94**, 236601 (2005).

Datta, S., Das, B. *Appl. Phys. Lett.* **56**, 665 (1990).

Datta, S. *Electronic Transport in Mesoscopic Systems* (Cambridge University Press, Cambridge, 1995).

Datta, S. *Quantum Transport: Atom to Transistor* (Cambridge University Press, Cambridge, 2005).

Dragomirova, R.L., Nikolić, B.K. *Phys. Rev. B* **75**, 085328 (2007).

Drouvelis, P.S., Schmelcher, P., Bastian, P. *J. Comp. Phys.* **215**, 741 (2006).

D'yakonov, M.I., Perel', V.I. *Sov. Phys. JETP* **13**, 467 (1971).

Engel, H.A., Rashba, E.I., Halperin, B.I. *Handbook of Magnetism and Advanced Magnetic Materials*, (eds) H. Kronmüller and S. Parkin (Wiley, Hoboken, NJ, 2007).

Fabian, J., Matos-Abiaguea, A., Ertlera, C., Stano, P., Žutić, I. *Acta Phys. Slov.* **57**, 565 (2007).

Finkler, I.G., Engel, H.A., Rashba, E.I., Halperin, B.I. *Phys. Rev. B* **75**, 241202 (2007).

Fisher, G.P. *Am. J. Phys.* **39**, 1528 (1971).

Galitski, V.M., Burkov, A.A., Das Sarma, S. *Phys. Rev. B* **74**, 115331 (2006).

Ganichev, S.D., Prettl, W. *J. Phys.: Condens. Matter* **15**, R935 (2003).

Ganichev, S.D., Danilov, S.N., Schneider, P., Bel'kov, V.V., Golub, L.E., Wegscheider, W., Weiss, D., Prettl, W. *J. Magn. Magn. Mater.* **300**, 127 (2006).

Golizadeh-Mojarad, R., Datta, S. *Phys. Rev. B* **75**, 081301 (2007a).

Golizadeh-Mojarad, R., Datta, S. preprint arXiv:cond-mat/0703280 (2007b).

Governale, M., Zülicke, U. *Solid State Commun.* **153**, 581 (2004).

Grundler, D. *Phys. Rev. Lett.* **84**, 6074 (2000).

Guo, G.Y., Murakami, S., Chen, T.W., Nagaosa, N. *Phys. Rev. Lett.* **100**, 096401 (2008).

Hankiewicz, E.M., Molenkamp, L.W., Jungwirth, T., Sinova, J. *Phys. Rev. B* **70**, 241301 (2004).

Hankiewicz, E.M., Li, J., Jungwirth, T., Niu, Q., Shen, S., Sinova, J. *Phys. Rev. B* **72**, 155305 (2005).

Hankiewicz, E.M., Vignale, G. *Phys. Rev. Lett.* **100**, 026602 (2008).

Hanson, R., Kouwenhoven, L.P., Petta, J.R., Tarucha, S., Vandersypen, L.M.K. *Rev. Mod. Phys.* **79**, 1217 (2007).

Haug, H., Jauho, A.P. *Quantum Kinetics in Transport and Optics of Semiconductors*, 2nd edn (Springer, Berlin, 2007).

Hirsch, J.E. *Phys. Rev. Lett.* **83**, 1834 (1999).

Holleitner, A.W., Sih, V., Myers, R.C., Gossard, A.C., Awschalom, D.D. *Phys. Rev. Lett.* **97**, 036805 (2006).

Huang, B., Monsma, D.J., Appelbaum, I. *Phys. Rev. Lett.* **99**, 177209 (2007).

Inoue, J., Bauer, G.E.W., Molenkamp, L.W. *Phys. Rev. B* **70**, 041303 (2004).

Jackson, J.D. *Classical Electrodynamics*, 3rd edn (Wiley, Hoboken, NJ, 1998).

Jung, S., Lee, H. *Phys. Rev. B* **71**, 125341 (2005).

Jungwirth, T., Sinova, J., Mašek J., Kučera, J., MacDonald, A.H. *Rev. Mod. Phys.* **78**, 809 (2006).

Kato, Y.K., Myers, R.C., Gossard, A.C., Awschalom, D.D. *Science* **306**, 1910 (2004a).

Kato, Y., Myers, R.C., Gossard, A.C., Awschalom, D.D. *Nature* **427**, 50 (2004b).

Kato, Y.K., Myers, R.C., Gossard, A.C., Awschalom, D.D. *Appl. Phys. Lett.* **87**, 022503 (2005).

Kazymyrenko, K., Waintal, X. *Phys. Rev. B* **77**, 115119 (2008).

Keldysh, L. *Sov. Phys. JETP* **20**, 1018 (1965).

Keyes, R.W. *Rep. Prog. Phys.* **68**, 2701 (2005).

Khomyakov, P.A., Brocks, G., Karpan, V., Zwierzycki, M., Kelly, P.J. *Phys. Rev. B* **72**, 035450 (2005).

Kimura, T., Otani, Y., Sato, T., Takahashi, S., Maekawa, S., *Phys. Rev. Lett.* **98**, 156601 (2007).

Kiselev, A.A., Kim, K.W. *Phys. Rev. B* **71**, 153315 (2005).

Koentopp, M., Chang, C., Burke, K., Car, R. *J. Phys.: Condens. Matter* **20**, 083203 (2008).

König, M., Buhmann, H., Molenkamp, L.W., Hughes, T., Liu, C.X., Qi, X.L., Zhang, S.C. *J. Phys. Soc. Jpn.* **77**, 031007 (2008).

van Leeuwen, R., Dahlen, N.E., Stefanucci, G., Almbladh, C.-O., von Barth, U. *Lecture Notes in Physics 706*, 33 (Springer, Berlin, 2006).

Li, J., Dai, X., Shen, S., Zhang, F. *Appl. Phys. Lett.* **88**, 162105 (2006).

Liu, M.-H., Bihlmayer, G., Blügel, S., Chang, C.-R. *Phys. Rev. B* **76**, 121301(R) (2007).

Maekawa, S., Shinjo, T. (eds) *Spin Dependent Transport in Magnetic Nanostructures* (Taylor and Francis, London, 2002).

Mal'shukov, A.G., Tang, C.S., Chu, C.S., Chao, K.A. *Phys. Rev. Lett.* **95**, 107203 (2005).

Meier, F., Loss, D. *Phys. Rev. Lett.* **90**, 167204 (2003).

Metalidis, G., Bruno, P. *Phys. Rev. B* **72**, 235304 (2005).

Mishchenko, E.G., Shytov, A.V., Halperin, B.I. *Phys. Rev. Lett.* **93**, 226602 (2004).

Moca, C.P., Marinescu, D.C. *Phys. Rev. B* **75**, 035325 (2007).

Murakami, S., Nagaosa, N., Zhang, S.C. *Science* **301**, 1348 (2003).

Murakami, S. *Advances in Solid State Physics 45*, 197 (Springer, Berlin, 2006).

Nagaosa, N. *J. Phys. Soc. Jpn.* **77**, 031010 (2008).

Nikolić, B., Allen, P.B. *Phys. Rev. B* **60**, 3963 (1999).

Nikolić, B.K., Dragomirova, R.L. *Semicond. Sci. Tech.* **24**, 064006 (2009).

Nikolić, B.K., Souma, S. *Phys. Rev. B* **71**, 195328 (2005a).

Nikolić, B.K., Zârbo, L.P., Souma, S. *Phys. Rev. B* **72**, 075361 (2005b).

Nikolić, B.K., Souma, S., Zârbo, L.P., Sinova, J. *Phys. Rev. Lett.* **95**, 046601 (2005c).

Nikolić, B.K., Zârbo, L.P., Souma, S. *Phys. Rev. B* **73**, 075303 (2006).

Nikolić, B.K., Zârbo, L.P. *Europhys. Lett.* **77**, 47004 (2007).

Nitta, J., Akazaki, T., Takayanagi, H., Enoki, T. *Phys. Rev. Lett.* **78**, 1335 (1997).

Nomura, K., Wunderlich, J., Sinova, J., Kaestner, B., MacDonald, A.H., Jungwirth, T. *Phys. Rev. B* **72**, 245330 (2005).

Nonoyama, S., Oguri, A. *Phys. Rev. B* **57**, 8797 (1998).

Okamoto, S. *Phys. Rev. B* **76**, 035105 (2007).

Onoda, M., Nagaosa, N. *Phys. Rev. B* **72**, 081301 (2005*a*).

Onoda, M., Nagaosa, N. *Phys. Rev. Lett.* **95**, 106601 (2005*b*).

Pareek, T.P. *Phys. Rev. Lett.* **92**, 076601 (2004).

Ralph, D., Stiles, M. *J. Magn. Magn. Mater.* **320**, 1190 (2008).

Rammer, J. *Quantum Field Theory of Non-Equilibrium States* (Cambridge University Press, Cambridge, 2007).

Rashba, E.I. *Phys. Rev. B* **68**, 241315 (2003).

Ren, W., Qiao, Z., Wang, J., Sun, Q., Guo, H. *Phys. Rev. Lett.* **97**, 066603 (2006).

Reynoso, A., Usaj, G., Balseiro, C.A. *Phys. Rev. B* **73**, 115342 (2006).

Rocha, A.R., García-Suárez, V.M., Bailey, S., Lambert, C., Ferrer, J., Sanvito, S. *Phys. Rev. B* **73**, 085414 (2006).

Roth, A., Brüne, C., Buhmann, H., Molenkamp, L.W., Maciejko, J., Qi, X.-L., Zhang, S.-C. preprint arXiv: 0905.0365 (2009).

Saitoh, E., Ueda, M., Miyajima, H., Tatara, G. *Appl. Phys. Lett.* **88**, 182509 (2006).

Scheid, M., Bercioux, D., Richter, K. *New J. Phys.* **9**, 401 (2007).

Schliemann, J. *Int. J. Mod. Phys. B* **20**, 1015 (2006).

Sharma, P. *Science* **307**, 531 (2005).

Sheng, L., Sheng, D.N., Ting, C.S., Haldane, F.D.M. *Phys. Rev. Lett.* **95**, 136602 (2005).

Sheng, L., Ting, C.S. *Int. J. Mod. Phys. B* **20**, 2339 (2006).

Shi, J., Zhang, P., Xiao, D., Niu, Q. *Phys. Rev. Lett.* **96**, 076604 (2006).

Sih, V., Myers, R.C., Kato, Y.K., Lau, W.H., Gossard, A.C., Awschalom, D.D. *Nature Phys.* **1**, 31 (2005).

Sih, V., Lau, W.H., Myers, R.C., Horowitz, V.R., Gossard, A.C., Awschalom, D.D. *Phys. Rev. Lett.* **97**, 096605 (2006).

Silov, A.Y., Blajnov, P.A., Wolter, J.H., Hey, R., Ploog, K.H., Averkiev, N.S. *Appl. Phys. Lett.* **85**, 5929 (2004).

Silsbee, R.H. *J. Phys.: Condens. Matter* **16**, 179 (2004).

Sinitsyn, N.A. *J. Phys.: Condens. Matter* **20**, 023201 (2008).

Sinova, J., Culcer, D., Niu, Q., Sinitsyn, N.A., Jungwirth, T., MacDonald, A.H. *Phys. Rev. Lett.* **92**, 126603 (2004).

Sinova, J., Murakami, S., Shen, S.-Q., Choi, M.-S. *Solid State Commun.* **138**, 214 (2006).

Sørensen, H.H.B., Hansen, P.C., Petersen, D.E., Skelboe, S., Stokbro, K. *Phys. Rev. B* **77**, 155301 (2008).

Souma, S., Nikolić, B.K. *Phys. Rev. B* **70**, 195346 (2004).

Souma, S., Nikolić, B.K. *Phys. Rev. Lett.* **94**, 106602 (2005).

Stern, N.P., Ghosh, S., Xiang, G., Zhu, M., Samarth, N., Awschalom, D.D. *Phys. Rev. Lett.* **97**, 126603 (2006).

Stevens, M.J., Smirl, A.L., Bhat, R.D.R., Najmaie, A., Sipe, J.E., van Driel, H.M. *Phys. Rev. Lett.* **90**, 136603 (2003).

Sugimoto, N., Onoda, S., Murakami, S., Nagaosa, N. *Phys. Rev. B* **73**, 113305 (2006).

Tang, C.S. *Int. J. Mod. Phys. B* **20**, 869 (2006).

Thygesen, K.S., Rubio, A. *Phys. Rev. B* **77**, 115333 (2008).

Todorov, T.N. *J. Phys.: Condens. Matter* **14**, 3049 (2002).

Topinka, M.A., Westervelt, R.M., Heller, E. *J. Phys. Today* **56**(12), 47 (2003).

Tserkovnyak, Y., Brataas, A., Bauer, G.E.W., Halperin, B.I. *Rev. Mod. Phys.* **77**, 1375 (2005).

Tserkovnyak, Y., Brataas, A. *Phys. Rev. B* **76**, 155326 (2007).

Valenzuela, S.O., Tinkham, M. *Nature* **442**, 176 (2006).

Velev, J., Butler, W. *J. Phys: Condens. Matter* **16**, R637 (2004).

Watson, S.K., Potok, R.M., Marcus, C.M., Umansky, V. *Phys. Rev. Lett.* **91**, 258301 (2003).

Winkler, R. *Spin-Orbit Coupling Effects in Two-Dimensional Electron and Hole Systems* (Springer, Berlin, 2003).

Wunderlich, J., Kaestner, B., Sinova, J., Jungwirth, T. *Phys. Rev. Lett.* **94**, 047204 (2005).

Yoshida, S., Kanitani, Y., Oshima, R., Okada, Y., Takeuchi, O., Shigekawa, H. *Phys. Rev. Lett.* **98**, 026802 (2007).

Zârbo, L.P., Nikolić, B.K. *Europhys. Lett.* **80**, 47001 (2007).

Zawadzki, W. *Am. J. Phys.* **73**, 756 (2005).

Žutić, I., Fabian, J., Das Sarma, S. *Rev. Mod. Phys.* **76**, 323 (2004).

Zyuzin, V.A., Silvestrov, P.G., Mishchenko, E.G. *Phys. Rev. Lett.* **99**, 106601 (2007).

Disorder-induced electron localization in molecular-based materials

Sven Stafström and Mikael Unge

25.1 Introduction

A lot of attention has recently been focused on the issue of charge transport in different carbon-based organic systems and materials that are built up from either molecular or polymers constituents. Ideally, these constituents are fully conjugated with an extended π-electronic system. Due to the inter-molecular interactions, there is also the possibility of inter-molecular electron delocalization. The role played by order/disorder in this context is of special importance since disorder can prevent delocalization, which is the particular feature that makes this class of materials of interest for electronic applications. In the ordered regime, however, molecular or polymer-based materials can exhibit electron and hole mobilities comparable with that of amorphous silicon, indicating that these novel materials are possible substitutes in technical applications.

Despite a large amount of experimental and theoretical work, there are still unanswered questions, in particular related to the interplay between the structure of the materials at the nanometer scale and their transport properties. Also, the extent of electron–phonon coupling and how this coupling effects electron transport is not yet fully understood. It is well known that in the limit of localized states, for which charge transport is non-adiabatic and described in terms of hopping (Miller and Abrahams 1960) (or charge transfer) between such localized states, the reorganization energy plays an important role for the charge-transfer rates according to the Marcus formula (Marcus 1993). However, for more well-ordered systems in which the donor and acceptor states couple more strongly, perturbation theory, which is the basis for both the Miller and Abrahams, and the Marcus rate expressions, can no longer be applied. Instead the electron–phonon coupling has to be treated in a direct way by solving the time-dependent Schrödinger equation together with the lattice equation of motion (Johansson and Stafström 2001; Hultell and Stafström 2007).

In this review we will focus on the problem of electron localization and how the extent of localization depends on the type of system we are dealing with. We will use DNA (deoxyribos nucleic acid) and a pentacene molecular crystals as examples of two different classes of materials. In DNA, on the one hand, the disorder is intrinsic and strong, which results in very short localization lengths. The pentacene crystal, on the other hand, is intrinsically homogeneous and the disorder is extrinsic and weak, which makes a metal–insulator transition (MIT) possible.

25.1.1 Background, carbon-based materials for electronic applications

In the 1960s, organic molecular crystals became of large interest due to the observation if high electron mobilities in acene-based crystals (Karl 1989). Molecular crystals could also be made in the form of charge-transfer salts (Bechgaard salts (Jerome *et al.* 1980)) with superconducting properties. Following this development, a number of breakthroughs have been made that advanced the research field both from the point of view of fundamental science and in terms of applications and technological breakthroughs.

In 1976 MacDiarmid, Heeger, and Shirakawa observed that a conjugated polymer, polyacetylene, can become highly conducting following the treatment of this polymer with iodine (Chiang *et al.* 1978). Iodine acts as a dopant and charges (holes) are transferred to the polyacetylene chains. This process is made possible due to the low ionization potential of polyacetylene. Based on the seminal finding of MacDiarmid, Heeger and Shirakawa, which gave them the Nobel prize in Chemistry for the year 2000, a number of other conjugated polymers and dopants have been investigated (see e.g. Skotheim and Reynolds 2006). Since the discoveries of luminescence (Burroughes *et al.* 1990), photovoltaic properties (Sariciftci *et al.* 1992), and the possibility to make field effect transistor devices from organic materials such as molecular crystals and polymers (Tsumura *et al.* 1986; Horowitz *et al.* 1989; Horowitz 1998) the focus of this research area has changed from studies of how charge carriers can be created by means of doping to studies of the mobility of the charge carriers and the excited-state dynamics that are related to luminescence and photovoltaic properties.

Parallel with the developments of molecular crystals and conjugated polymers, another molecule and molecular-based material was discovered in 1985. In that year, Smalley, together with Kroto and Curl made the discovery of a particularly stable all-carbon molecule, the icosahedral C_{60} molecule (Kroto *et al.* 1985). The discovery of this new form of pure carbon, and the insight that led to the interpretation of the magic number features in the observed mass spectra in terms of hollow carbon molecules, led to the award of the Nobel Prize in Chemistry to Curl, Kroto and Smalley in 1996.

The interest in electronic properties, and charge transport in particular, has shifted from fullerenes to carbon nanotubes (Iijima 1991). Carbon nanotubes are today one of the most interesting candidates to be used in truly nanoelectronic devices. In this context, the electronic robustness of the tubes with respect to both topological and external disorder has to be fully understood. It is therefore of fundamental interest to study disorder-induced (Anderson) electron localization also in carbon nanotubes (White and Todorov 1998; Hjort and Stafström 2001a). This lies, however, outside the scope of this presentation in which we limit the discussion to localization properties of molecular-based organic systems.

In this review we focus the attention to the effects of correlation in the disorder. It is well known that such correlations can lead to the existence of delocalized states (Dunlap *et al.* 1990; de Moura and Lyra 1998; Liu *et al.* 2003; Xiong and Zhang 2003; Shima *et al.* 2004). Such correlations can be found in many systems and are represented here by DNA and pentacene crystals. These two molecular-based systems, the types of disorder that apply to them, and how correlation enters into the disorder distribution are introduced below.

25.1.2   DNA

The conducting properties of the double-helix structure of DNA (see Fig. 25.1) have been investigated and debated intensively during the last 4–5 years. Electron conduction is of course of interest for physicists, in particular in the context of nanotechnology, but also for biological reasons. The latter for understanding the repair process of oxidative-damaged DNA (Dandliker *et al.* 1998). In the context of molecular electronics, DNA has shown properties that might make it suitable to use as conducting wire or as a building block in construction of biomolecular logical devices (Braun *et al.* 1998; Ben-Jacob *et al.* 1999). However, there is no unified description of the conducting properties of DNA. Some reports on the subject classify DNA as a conductor (Fink and Schönenberger 1999; Porath *et al.* 2000; Yoo *et al.* 2001; Hwang *et al.* 2002), while others state that DNA is an insulator (de Pablo *et al.* 2000; Zhang *et al.*

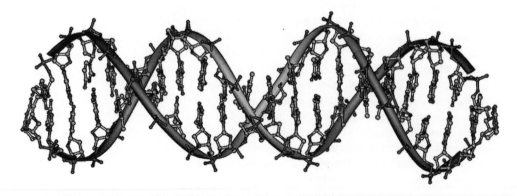

Fig. 25.1 Base-pair stacking geometry in DNA (picture created by Michael Ströck, Wikipedia).

2002). To a large extent, the different outcomes of the studies are due to the fact that the experiments have been performed on different types of DNA. Clearly, there is a large difference between synthetic poly(G)-poly(C) DNA, which is an ordered homogeneous systems in which one strand contains guanine only and the other strand cytosine only (Porath *et al.* 2000; Hjort and Stafström 2001b), and the more or less random sequences of all four base molecules as in λ-DNA (de Pablo *et al.* 2000; Zhang *et al.* 2002). Furthermore, the environment plays an important role. Experiments on charge transport have been performed on dry DNA (Porath *et al.* 2000; Yoo *et al.* 2001), as well as in solution (Fink and Schönenberger 1999; Hwang *et al.* 2002), and in vivo (Boudaïffa *et al.* 2000). Recently, a theoretical study was presented for DNA trapped inside a carbon nanotube (Gao *et al.* 2003), thus yet another environment for the DNA molecule. As the results from these investigations show, the environment in which the DNA molecule exists has a major effect on the result of the conductivity measurements/calculations (Basko and Conwell 2002).

The double-helix structure of DNA results in an essentially one-dimensional system. The electronic properties of this system as a whole result from interactions of the π-orbitals along this system. Similar to most organic electroactive materials, the majority-charge carriers in DNA are holes (Lewis *et al.* 2000; Giese *et al.* 2001). Therefore, by only considering the highest-occupied molecular orbitals (HOMO) of each base in the DNA molecule we get a physically relevant but less complex system. It should be noted that such a system only treats the static electron-conduction mechanism. Effects from electron–phonon coupling (Basko and Conwell 2002) or from ionic motions are not included. Nevertheless, it is important to obtain knowledge of the different interactions separately before we can understand the full system. Furthermore, our model relates directly to studies of dry DNA at low temperatures for which we will be able to draw some important conclusions.

Based on this simple model for DNA we calculate the localization length of electronic states around the Fermi energy. For a periodic and perfectly ordered one-dimensional system of this type all eigenstates are extended (Bloch 1928). Introduction of uncorrelated disorder leads to localization of all states (Anderson 1958). Recently, one-dimensional systems with correlated disorder were investigated theoretically and based on the Anderson type of model it was concluded that under specific conditions, such systems can in fact contain extended states (Carpena *et al.* 2002). In this case the disorder is not completely random but has some underlying structure that also affects the eigenstates of the system. The correlations that exist in the sequence of base pairs in DNA (Holste *et al.* 2001) might apply to such a model and thus lead to electron delocalization in DNA (Carpena *et al.* 2002). However, we have shown (Unge and Stafström 2003a) together with others (Bagci and Krokhin 2007; Daz *et al.* 2007) that the disorder in DNA is so strong that, even though such correlations exist, they are of no importance for the electronic properties of the system, i.e. DNA lacks the existence of extended electronic states. The results that give this conclusion will be presented and discussed in Section 25.3.1.

25.1.3   Pentacene molecular crystals

One particularly interesting example of nanotechnology is molecular electronics in which molecules are used as building blocks in an electroactive material. This so-called bottom-up approach is based on self-assembly of molecular nanostructures. Molecular crystals as active materials in electronic devices have gained considerable interest during recent years. Pentacene is reported to have the highest field-effect mobility values for organic field-effect transistors (OFETs), and also to have good environmental stability (Lin *et al.* 1997; Dimitrakopoulos *et al.* 1999; Jurchescu *et al.* 2004; Sundar *et al.* 2004). Pentacene OFETs with low (2.5 V) operating voltage have also been demonstrated (Halik *et al.* 2004). This feature increases the possibility to use pentacene OFETs in real applications. Organic field-effect transistors (FET) with high mobility have been studied for some different organic materials. Pentacene thin-film FETs have shown mobilities similar to amorphous silicon (Nelson *et al.* 1998). These mobility measurements were performed at low temperatures, at which the mobility and the mean-free path is directly related to the extent of delocalization of the electronic wavefunction.

The pentacene crystal is build up from a triclinic unitcell consisting of two pentacene molecules (Campbell *et al.* 1961). The crystal can be considered as a 2D material (see Fig. 25.2), since the electron interaction in the third dimension is significantly smaller than in the two others (Cornil *et al.* 2000). In the Anderson model (Anderson 1958), which includes diagonal (on-site) disorder, random (uncorrelated) disorder results in localized wavefunctions in 1D (one dimension) and 2D. In 3D there are localized states in the band tails and extended states otherwise (Souillard 1987).

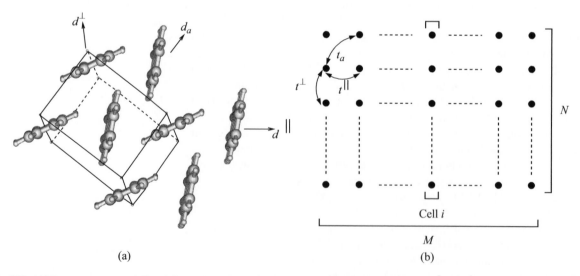

(a)   (b)

Fig. 25.2 (a) The pentacene crystal. The triclinic unit cell is indicated by the parallelepiped. The directions $d^{\parallel}$ and $d^{\perp}$ correspond to the directions with the strongest intermolecular interaction. Also, the direction d_a has non-negligible intermolecular interaction. (b) The rectangular lattice, where each point corresponds to a molecule and M and N are the length and width, respectively, of the stripes used in the calculations of the localization lengths. From Unge and Stafström (2006) *Phys Rev. B* **74**, 235403 ©2006 The American Physical Society, reproduced with permission.

Thus, band motion in pentacene crystals should not be possible in the presence of random disorder, which we have shown earlier (Unge and Stafström 2003b). However, as discussed above, both short-range (Dunlap *et al.* 1990) and long-range (de Moura and Lyra 1998; Liu *et al.* 2003; Xiong and Zhang 2003; Shima *et al.* 2004) correlations in the disorder can yield delocalized states in both one and two dimensions. Furthermore, even though the states are localized, the localization length itself can be of interest. In particular in the context of nanotechnology since in this case the size of the device can be of the same order of magnitude as the localization length. Systems that exhibit this behavior remain highly conducting even in the presence of disorder.

The relatively weak interactions in a van der Waals bonded molecular solid, compared to, e.g. a covalent-bonded crystal, yield temperature-induced disorder due to displacements and rotations of the pentacene molecules. This type of disorder affects both the on-site potential (diagonal disorder) and the intermolecular interaction (off-diagonal disorder) (Bussac *et al.* 2004; Unge and Stafström 2006). Furthermore, it is highly likely that the dominating disorder in molecular crystals is not completely random but contains long-range correlations. The reason is that charges present in solids of large polarizable molecules yields a polarization cloud around the carrier. With structural disorder in the form of displacements and rotations of the individual molecules with respect to the perfect lattice structure this polarization cloud gives rise to long-range correlated disorder in both the on-site and off-diagonal matrix elements as well as a coupling between these two elements (Flambaum and Sokolov 1999; Bussac *et al.* 2004). The charge that induces the polarization can be the carrier of the electric current itself, in this case the carrier is dressed by a polarization cloud and referred to as an electronic polaron or Coulomb polaron (Toyozawa 1954). Another possible source of this type of disorder is charged impurity ions located in the molecular crystal or in the substrate on which the crystal is grown (Flambaum and Sokolov 1999).

In this work we present results for a 2D rectangular lattice. This lattice can be considered as a model system for a layered organic crystal, in which each molecule is represented by one site. We calculate the localization length numerically for this system, using transfer matrices and Lyapunov exponents. We use finite-size scaling to study the effect of on-site and off-diagonal disorder with long-range correlation on the metal–insulator transition (MIT) in a two-dimensional (2D) system. The results are discussed in the context of pentacene molecular crystals, but can be applied to any 2D system described by a tight-binding model with both on-site and off-diagonal disorder with long-range correlation. We also study the effect on the localization length by the introduction of molecular interaction between next-nearest-neighbor in addition to nearest-neighbor interaction.

25.2   Methodology

The localization properties of the two different systems included in this review, a DNA double strand and a pentacene molecular crystal, are studied using the same type of methodology. DNA is a one-dimensional system for which the localization properties can be obtained directly using the method of transfer matrices and Lyapunov exponents. When the dimensionality is increased, it is no longer possible to calculate the localization properties in a direct way. Instead we have to use finite-size scaling and extrapolate to the properties of the infinite system.

The presentation below is divided into three Sections, Sections 25.2.1, 25.2.2, and 25.2.3, which describe the steps to calculate the localization properties of the systems in question.

25.2.1   Hamiltonian

All calculations presented here are based on a tight-binding Hamiltonian that couples nearest-neighbor sites via the transfer integral t. In DNA a site corresponds to one of the nucleobases, whereas in the pentacene crystal each pentacene molecule constitutes a site. The electronic basis function on each site is the highest-occupied molecular orbital (HOMO). We have chosen to study the localization properties in the HOMO manifold of stated since both DNA and pentacene are p-type with holes as majority charge carriers.

In the case of DNA, the double-strand topology has to be considered, which results in the following tight-binding Hamiltonian,

$$\mathbf{H}_{\mathrm{DNA}} = \sum_{i}^{M} \left(\sum_{j=1}^{2} \epsilon_i^j |j, i\rangle\langle j, i| \right.$$

$$\left. + \sum_{j \neq j'}^{2} \left(t_{i,i+1}^{j,j'} |j, i\rangle\langle j', i+1| + t_{i+1,i}^{j',j} |j', i+1\rangle\langle j, i| \right) \right), \quad (25.1)$$

where ϵ_i^j is the on-site energy (in this case the eigenenergy of the HOMO) of the base molecule i on each of the two strands ($j = 1, 2$), and M is the number of base pairs in the system. The intermolecular electron-transfer integral between neighboring base molecule in the strand is denoted by $t_{i,i+1}^{j,j}$. The term $t_{i,i+1}^{j,j'}$, $j \neq j'$, describes the interstrand interactions between the HOMO of base i in strand j and the HOMO of base $i + 1$ in strand j'. The numerical values taken for the on-site energies were obtained from the HOMO energies of guanine (G), adenine (A), cytosine (C) and thymine (T) as calculated from *ab-initio* density-functional theory (DFT) on each of the individual base molecules. The calculations were performed with a double-zeta basis set with polarization functions for the valence orbitals. The values are: $\epsilon_G = 0.0\,\mathrm{eV}$ (reference), $\epsilon_A = -0.33\,\mathrm{eV}$, $\epsilon_C = -0.54\,\mathrm{eV}$, and $\epsilon_T = -0.91\,\mathrm{eV}$ (Carlsson 2002). The values for the intermolecular transfer integrals are taken from Voityuk *et al.* (2000) and Voityuk *et al.* (2001). The intrastrand transfer integrals range from 0.029–0.158 eV and the interstrand transfer integrals from 0.0007–0.062 eV. As a result of these different energy values, the Hamiltonian describing the electronic structure of DNA contains disorder in both the on-site and the off-diagonal terms. Note that the differences in the on-site energies are large compared to the intermolecular transfer integrals, which indicates that delocalization will be strongly hindered in a random sequence of these base pairs. Other values for the intrastrand transfer integral have been used, e.g., in Caetano and Schulz (2005) the authors used a value of 1.0 eV. The effect of these different values for the transfer integrals will be commented on below.

The degree of hybridization of the π-orbitals of the base-pair stack to the sugar and phosphate groups forming the backbone of the double strand has also been discussed (Cuniberti *et al.* 2002). We have calculated the molecular orbitals from *ab-initio* DFT of the combined system of the base molecule and the neighboring sugar and phosphate groups. For guanine and adenine, the hybridization is practically non-existent and the HOMO energies are very close to those obtained for the base molecule itself. This can be easily understood from the shape of the HOMO, which for these two molecules have a node close to the nitrogen atom that connects the base molecule to the sugar moiety. Due to the absence of coupling between the guanine HOMO and the backbone the guanine band is electronically unaffected by the backbone. Since most charge carriers originate from this band we conclude that our model is sufficient to describe the static scattering mechanism of the electronic charge carriers in DNA.

In the case of cytosine and thymine, the hybridization is stronger and there is a shift of the HOMO energies towards higher energies. Again, the explanation for this behavior is obvious from the shape of the HOMO, cytosine and thymine have large amplitudes at the nitrogen site attached to the sugar. As a result of this hybridization the HOMO energy values are now: $\epsilon_G' = 0.0$ eV (reference), $\epsilon_A' = -0.31$ eV, $\epsilon_C' = -0.42$ eV, and $\epsilon_T' = -0.77$ eV. In comparison with the corresponding values presented above, this gives on-site energies that are closer to each other. The effect of this difference on the localization length will be commented on below.

The treatment of the other system included in this review, the pentacene molecular crystal, is similar, except that the geometrical structure is very different. Pentacene crystallizes in a two-dimensional rectangular lattice, see Fig. 25.2. To model this system, we use a finite supercell of this crystal that has a width N and length M. Similar to the case of DNA, we only include nearest-neighbor inter-molecular interactions. The on-site energy corresponds to the HOMO energies of the pentacene molecule in its particular environment (see below). The pentacene crystal is described by using the Hamiltonian derived in Bussac *et al.* (2004)

$$
\mathbf{H}_{\text{pentacene}} = \sum_{i=1}^{M} \left(\sum_{j=1}^{N} \left(\epsilon_i^j |j,i\rangle\langle j,i| + t_{i,i}^{j,j+1} |j,i\rangle\langle j+1,i| \right. \right.
$$

$$
\left. + t_{i,i}^{j+1,j} |j+1,i\rangle\langle j,i| \right) + \sum_{j,j'=1}^{N} \left(t_{i,i+1}^{j,j'} |j,i\rangle\langle j',i+1| \right.
$$

$$
\left. \left. + t_{i+1,i}^{j',j} |j',i+1\rangle\langle j,i| \right) \right), \tag{25.2}
$$

where M and N are the lengths and the widths of the system, respectively. Note that the transfer integrals within cell i appear here, as opposed to in DNA (see eqn (25.1)) for which the transfer integrals between the two bases within the same base pair is assumed to be zero. The on-site energies represent the potential due to the induced dipoles, i.e. the polarization cloud discussed above. The off-diagonal terms are the effective transfer integrals corresponding to the bare tight-binding parameters scaled (amplified) to account for the disorder due to molecular polarization effects.

$$t_{i,i'}^{j,j'} = \tilde{t}_{i,i'}^{j,j'} \exp\left[\frac{\epsilon_i^j + \epsilon_{i'}^{j'}}{\delta}\right]. \tag{25.3}$$

These effects also introduce a coupling between the on-site and off-diagonal terms in the Hamiltonian (Bussac *et al.* 2004).

The first summation in eqn (25.2) includes nearest-neighbor interactions within the unit cell, i.e. in the $d^{\perp}$ direction (see Fig. 25.2), whereas the second summation describes the interaction in the $d^{\parallel}$ and the d_a directions.

Similar to the case of DNA, the electronic band considered is the HOMO of pentacene. Values of the intermolecular transfer integrals of the HOMO orbitals are obtained from Cheng *et al.* (2003) and renormalized according to Bussac *et al.* (2004). The same renormalization constant is used for the three different transfer integrals $t_{i,i'}^{j,j'}$ resulting in the following values: $t^{\parallel} = 66$ meV, $t^{\perp} = 47$ meV, and $t_a = 32$ meV (see Fig. 25.2). The strength of the coupling between on-site and off-diagonal elements, δ, is set to 0.4 eV from Bussac *et al.* (2004).

The electric field produced by fluctuations in the polarization of the system (randomly oriented dipole moments) will not result in completely random on-site energies. Instead, as discussed above, there is a certain long-range correlation in the potential produced by the variations in this field from site to site. The correlation function for the on-site energies, ε_i, is $< \varepsilon_i(r_i)\varepsilon_j(r_j) > \propto 1/|r_i - r_j|$. The modified Fourier filtering method is used to generate the appropriate correlated on-site energies (Makse *et al.* 1996). A set of numbers $\{u_i\}_{i=1...M\times N}$ with a Gaussian distribution are generated and distributed over a rectangular lattice of size $M \times N$. The mean square of the Gaussian distribution is set to 15.8 meV corresponding to typical molecular rotation angles due to thermal disorder at room temperature (Bussac *et al.* 2004). The two-dimensional Fourier transform of these numbers, $u_{\mathbf{q}}$, is then weighted/filtered by the spectral density, $S(\mathbf{q})$, according to $\varepsilon_{\mathbf{q}} = \sqrt{S(\mathbf{q})}u_{\mathbf{q}}$. The inverse Fourier transform of $\{\varepsilon_{\mathbf{q}}\}$ yields a set of numbers $\{\varepsilon_i\}$ with long-range correlation (Unge and Stafström 2006).

In the modified Fourier filtering method the spectral density is defined as

$$S(\mathbf{q}) = \frac{2\pi}{\Gamma(\beta + 1)} \left(\frac{|\mathbf{q}|}{2}\right)^{\beta} K_{\beta}(|\mathbf{q}|), \tag{25.4}$$

where $\mathbf{q}$ is a vector in reciprocal space, $\beta = (\gamma - 2)/2$, K_{β} is the modified Bessel function of the second kind and of order β, and Γ is the gamma function. The strength of the correlation is determined by γ, $\gamma = 2$ ($\beta = 0$)

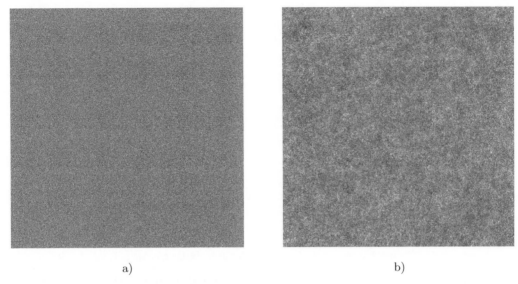

a)   b)

Fig. 25.3 On-site energy plots of square lattices with 1024 × 1024 sites, (a) uncorrelated numbers, (b) correlated numbers with $\gamma = 1.0$. From Unge, M. and Stafström (2006) *Phys Rev. B* **74**, 235403 ©2006 The American Physical Society, reproduced with permission.

corresponds to a very weakly correlated set with contribution from $K_0(|\mathbf{q}|)$ only. The long-range correlation resulting from randomly distributed dipoles correspond to $\gamma = 1$ (Flambaum and Sokolov 1999). All results for pentacene presented in this work are based on calculations performed for this value of γ. It should be noted, however, that deviations from this situation occur if, for instance, the distribution of the rotational angles is somewhat anisotropic as a result of the anisotropy of the crystal structure. In this case, the value of γ would be less than unity. On the other hand, additional screening of the electric field of the induced dipoles leads to a reduction in the correlation and a corresponding increase in the value of γ. The implications of such deviations are discussed in connection to the results presented below.

Figure 25.3 shows a set of 1024 × 1024 correlated uncorrelated (left) and correlated (right) numbers with $\gamma = 1.0$ (Unge and Stafström 2006). In the uncorrelated set the variations in the numbers are such that they almost average out at the level of resolution of the picture, whereas the correlated set shows much more structure with more or less well-defined regions with close-lying on-site energies. The largest set of correlated numbers generated here is $N \times M = 512 \times 2^{22} \approx 2.1 \times 10^9$.

25.2.2 Transfer matrix

With the two different expressions for the Hamiltonian in the case of DNA (eqn (25.1)) and pentacene (eqn (25.2)) in mind we can now proceed to describe the methodology for calculating the localization properties.

In matrix form the Hamiltonians in eqns (25.1) and (25.2) have a block-tridiagonal structure

$$
\mathbf{H} = \begin{pmatrix}
\mathbf{H}_1 & \mathbf{T}_1 & & & & \\
\mathbf{T}_1^\dagger & \mathbf{H}_2 & \mathbf{T}_2 & & & 0 \\
& \ddots & \ddots & \ddots & & \\
& & \mathbf{T}_{i-1}^\dagger & \mathbf{H}_i & \mathbf{T}_i & \\
0 & & & \mathbf{T}_i^\dagger & \mathbf{H}_{i+1} & \mathbf{T}_{i+1} \\
& & & & \ddots & \ddots & \ddots
\end{pmatrix},
\tag{25.5}
$$

where $\mathbf{H}_i$ is a $N \times N$ matrix that describe the on-site and transfer integrals within cell i, see Fig. 25.2(b), and $\mathbf{T}_i$ is a $N \times N$ matrix describing the interaction between cell i and cell $i + 1$.

The secular equation for the systems is:

$$
\mathbf{T}_{i-1}^\dagger \mathbf{A}_{i-1} + (\mathbf{H}_i - E\mathbf{I})\mathbf{A}_i + \mathbf{T}_i \mathbf{A}_{i+1} = \mathbf{0}.
\tag{25.6}
$$

The coefficients of the wavefunction in unit cell i (base pair in the case of DNA) at energy E are described by $\mathbf{A}_i$ and $\mathbf{I}$ is the identity matrix.

We can now introduce the transfer matrix, $\tau_i(E)$, acting as

$$
\begin{pmatrix} \mathbf{A}_{i+1} \\ \mathbf{A}_i \end{pmatrix} = \tau_i \begin{pmatrix} \mathbf{A}_i \\ \mathbf{A}_{i-1} \end{pmatrix}.
\tag{25.7}
$$

From eqn (25.6) the transfer matrix is thus defined as

$$
\tau_i(E) = \begin{pmatrix} \mathbf{T}_i^{-1}(E\mathbf{I} - \mathbf{H}_i) & -\mathbf{T}_i^{-1}\mathbf{T}_{i-1}^\dagger \\ \mathbf{I} & \mathbf{0} \end{pmatrix}.
\tag{25.8}
$$

The transfer matrix gives a connection between the coefficients of the wavefunction along the system and therefore contains information about the exponential decay of the eigenstates. To determine the localization lengths of the eigenstates we use the concept of Lyapunov exponents, that describe the exponential evolution of the eigenstates (Pichard and Sarma 1981). The evolution of the eigenstate is described by the product of the transfer matrices:

$$
\mathbf{Q}_M = \prod_{i=1}^{M} \tau_i(E).
\tag{25.9}
$$

If the determinant of each $\tau_i(E)$ is non-zero and finite the following matrix exists (Oseledec 1968)

$$
\boldsymbol{\Gamma} = \lim_{M \to \infty} (\mathbf{Q}_M^\dagger \mathbf{Q}_M)^{1/2M}.
\tag{25.10}
$$

The eigenvalues of $\boldsymbol{\Gamma}$ are of the type $\exp(\gamma_i)$, where γ_i are the Lyapunov characteristic exponents (LCEs) of $\mathbf{Q}_M$. The LCEs describe the rate of the exponential decay of the eigenstates. The localization length, λ_N, is the inverse of the smallest LCE (Pichard and Sarma 1981), which are calculated numerically using an orthogonalization process described by Benettin and Galgani (1979).

The convergence criterion is set to $\Delta\lambda < 0.001$ unit cells. In the calculations with long-range correlation this requires a value of M up to $M = 2^{22}$ (see above).

It is, however, not straightforward to obtain the transfer matrix from eqn (25.8) since the **T**-matrices in general are singular. We show here how this apparent problem in many cases can be overcome (Hjort and Stafström 2000, 2001a). By multiplying two consecutive secular matrix equations by the $M \times M$ matrices $\mathbf{X}_{i1}$ and $\mathbf{X}_{i2}$, respectively, we get

$$\mathbf{X}_{i1}\mathbf{T}_{i-1}^{\dagger}\mathbf{A}_{i-1} + \mathbf{X}_{i1}\hat{\mathbf{H}}_i\mathbf{A}_i + \mathbf{X}_{i1}\mathbf{T}_i\mathbf{A}_{i+1} = \mathbf{0} \qquad (25.11)$$

$$\mathbf{X}_{i2}\mathbf{T}_i^{\dagger}\mathbf{A}_i + \mathbf{X}_{i2}\hat{\mathbf{H}}_{i+1}\mathbf{A}_{i+1} + \mathbf{X}_{i2}\mathbf{T}_{i+1}\mathbf{A}_{i+2} = \mathbf{0}, \qquad (25.12)$$

where we have used the notation $\hat{\mathbf{H}} = \mathbf{H}_i - E\mathbf{I}$. Combining eqns (25.11) and (25.12) writing $\mathbf{X}_i = (\mathbf{X}_{i1}, \mathbf{X}_{i2})$ then gives

$$\mathbf{X}_i \begin{pmatrix} \mathbf{T}_{i-1}^{\dagger} \\ \mathbf{0} \end{pmatrix} \mathbf{A}_{i-1} + \mathbf{X}_i \begin{pmatrix} \hat{\mathbf{H}}_i \\ \mathbf{T}_i^{\dagger} \end{pmatrix} \mathbf{A}_i + \mathbf{X}_i \begin{pmatrix} \mathbf{T}_i \\ \hat{\mathbf{H}}_{i+1} \end{pmatrix} \mathbf{A}_{i+1}$$

$$+\mathbf{X}_i \begin{pmatrix} \mathbf{0} \\ \mathbf{T}_{i+1}^{\dagger} \end{pmatrix} \mathbf{A}_{i+2} = \mathbf{0}. \qquad (25.13)$$

If $\mathbf{X}_i$ is chosen so that (Hjort 2001)

$$\mathbf{X}_i \begin{pmatrix} \mathbf{T}_i & \mathbf{0} \\ \hat{\mathbf{H}}_{i+1} & \mathbf{T}_{i+1}^{\dagger} \end{pmatrix} = \begin{pmatrix} \mathbf{I} & \mathbf{0} \end{pmatrix}, \qquad (25.14)$$

then

$$\mathbf{A}_{i+1} = -\mathbf{X}_i \begin{pmatrix} \hat{\mathbf{H}}_i \\ \mathbf{T}_i^{\dagger} \end{pmatrix} \mathbf{A}_i - \mathbf{X}_i \begin{pmatrix} \mathbf{T}_{i-1}^{\dagger} \\ \mathbf{0} \end{pmatrix} \mathbf{A}_{i-1}, \qquad (25.15)$$

and the transfer matrix can be constructed in the same way as in eqn (25.12). Note that the only assumption made about the system is the tridiagonal-block structure of its Hamiltonian. The internal structure of the cells may be varying as well as the intermolecular transfer integrals. Equation (25.14) may be rewritten in the more familiar form $\mathbf{CX} = \mathbf{D}$

$$\begin{pmatrix} \mathbf{T}_i^{\dagger} & \hat{\mathbf{H}}_{i+1} \\ \mathbf{0} & \mathbf{T}_{i+1}^{\dagger} \end{pmatrix} \mathbf{X}_i = \begin{pmatrix} \mathbf{I} \\ \mathbf{0} \end{pmatrix}. \qquad (25.16)$$

A unique solution to eqn (25.16) exists only when the $2M \times 2M$ **C** matrix has a non-zero determinant. This is often not the case and there may then be either no solution or infinitely many. The criterion for the existence of a solution is that every column in **D** is a linear combination of the columns in **C**. This also implies that the transfer matrix in general is not uniquely determined. In general, **C** will be rank deficient and from the infinite set of solutions (if they exist!) to eqn (25.16), the minimum norm solution ($\|\mathbf{Cx} - \mathbf{d}\|_2 = \min$; **x** and **d** are column vectors in **X** and **D**) can be obtained from, e.g. singular-value decomposition (SVD) or complete orthogonal factorization (Golub and van Loan 1996). Any other solution can then be acquired from the null space vectors of eqn (25.16). However, we emphasize that the existence of a solution to eqn (25.16) may depend on the choice of cell structure.

A problem with the transfer matrices obtained from eqns (25.15) and (25.16) is that their determinant most often is equal to zero. Hence, they cannot be used to determine the Lyapunov exponent (see eqn (25.10)). However, it always seems possible to extract a submatrix (the reduced transfer matrix (Hjort 2001)), having a non-zero determinant. The reduced transfer matrix describes the coefficients of a certain subset of the atoms in the unit cell. All other coefficients in the unit cell can be obtained from a linear combination of this subset of coefficients. Thus, the exponential behavior of any wavefunction in the molecular crystal is totally determined by this submatix.

25.2.3   Finite-size scaling

The numerical calculation of the localization length of 1D and quasi-1D systems such as DNA, for which the parameter N in eqn (25.2) is equal to 2, was described above. To be able to calculate the localization lengths in systems for which $N \to \infty$, i.e. 2D systems, we have used the finite-size scaling technique introduced by MacKinnon and Kramer (MacKinnon *et al.* 1981) and Pichard and Sarma (1981). The localization length, λ_N, for a particular energy will increase with increasing values of N. In the limit of $N \to \infty$ the localization length converges to λ_∞, which is the localization length for the 2D system. Consequently, the renormalized localization length, $\Lambda_N = \lambda_N/N$, decreases with increasing N if the states are localized, whereas a constant or increasing Λ_N with increasing N indicates extended states.. According to finite-size scaling theory (MacKinnon *et al.* 1981; Pichard and Sarma 1981), Λ_N can be expressed as

$$\Lambda_N = f(\xi(w)/N), \tag{25.17}$$

where $\xi(w)$ is a function of a variable, w, that depends on energy and the amount of disorder. By fitting the calculated values of Λ_N to a polynomial of the form:

$$f(\xi(w)/N) = \frac{a\xi(w)}{N} + b\left(\frac{a\xi(w)}{N}\right)^2, \tag{25.18}$$

where a is the lattice constant and b is a parameter (MacKinnon *et al.* 1983) the localization length of the infinite system can be obtained since eqns (25.17) and (25.18) yield that

$$\lambda_\infty = \lim_{N \to \infty} Nf(\xi(w)/N) = \lim_{N \to \infty} (a\xi + b\frac{(a\xi(w)^2}{N}) = a\xi. \tag{25.19}$$

Thus, by calculating the localization length for different sizes, N, of a system, the renormalized localization length, Λ_N, is obtained and λ_∞ is determined by fitting the scaling function to the Λ_N values.

In all cases studied here that include long-range correlations in the disorder, the localization lengths are very long and consequently the value of N has to be even larger to allow for this kind of fitting procedure. Since the computations become very extensive for large values of N this limits the application of this procedure. The maximum value that we have been able to treat is $N = 512$. We therefore use finite-size scaling as an indicator of extended states and a possible MIT and not to calculate localization lengths for the infinite systems.

25.3 Results and discussion

25.3.1 DNA

We present here the electronic properties of different types of base-pair stacking of the DNA double-helix structure. The discussion is restricted to the electron localization properties from which we can make some important conclusion regarding the electron-transport properties of this type of organic system. As is clear from the discussion above, the results depend crucially on the values for the on-site energies and the transfer integral for electron transfer between neighboring nucleobases. The work that is summarized here is performed with values for these quantities that have been obtained from *ab-initio* calculations (Voityuk *et al.* 2000, 2001; Carlsson 2002). The values for the transfer integral can be compared to what has been reported for other systems. In the context of the content in this chapter, it is natural to make comparison between the transfer integral for the system of coupled nucleobases and those of the pentacene molecular crystal. We find from the literature that the nearest-neighbor guanine–guanine transfer integral is 84 meV (Voityuk *et al.* 2000, 2001) whereas the largest value for pentacene (see Fig. 25.2) is 66 meV (Cheng *et al.* 2003; Bussac *et al.* 2004). Thus, even though the pentacene molecules are slightly larger, the herringbone structure results in a reduced overlap. It should be pointed out that in some reports of the electron-localization properties of DNA, values of 1.0 eV have been used for the transfer integral (Carpena *et al.* 2002; Caetano and Schulz 2005). Since this value exceeds that of the difference in on-site energies, it is not surprising that the calculated localizations lengths are much larger than reported here.

Besides the relation between the on-site energies and the nearest-neighbor electron-transfer energies, the resulting values of the localization lengths are also dependent on the sequence of the base pairs. We calculate here the localization properties of computer-generated G–C and T–A sequences at different base-pair concentrations, both for completely random sequences and for different short-range correlated distributions. The purpose of introducing short-range correlation is to study the effect of increasing sequences of identical bases. For the values of the on-site energies considered here, this is the only type of correlation that can affect the localization length (see below). The correlated sequence is generated from a random distribution of base pairs. To each such base pair we generate a number that controls how many such identical pairs that will appear in a row. The length of this subsequence is generated from a normal distribution with standard deviation σ and expectation value 1. Only numbers larger than or equal to 1 in this distribution are used. For comparison, we also report the localization length of different known DNA sequences, e.g. the human chromosome 22 (NT_011520) and λ-DNA (NC_001416)(NCBI).

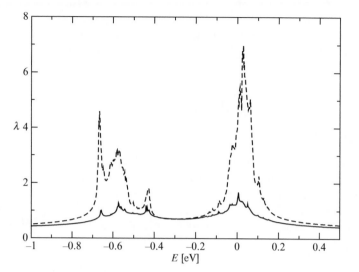

Fig. 25.4 Localization lengths (in units of the base-pair separation distance) as a function of energy for G–C DNA with (dashed) and without (solid) interstrand coupling. From Unge and Stafström (2003) *Nano. Lett.* **3**, 1417. ©2003 ACS Publications, reproduced with permission.

First, we study the effect of the interstrand electron transfer. The localization length of a disordered G–C sequence is calculated. The appearance of G and C along a single strand is completely random in this sequence. The result is shown in Fig. 25.4. Without the interstrand interactions included, the electronic system consists of two uncoupled single strands with a random distribution of two different on-site energies $\epsilon_G = 0.0$ eV and $\epsilon_C = -0.537$ eV.

With the interstrand interaction included, there is always the possibility to hop from one G (or C) site to another. The localization length in the G- and C-band is therefore increased significantly, the peak in the G-band reaches $\lambda \approx 7$ base pairs. The localization is in this case caused primarily by disorder in the electron-transfer interactions, the intrastrand G–G transfer integral is $t_{i,i+1}^{j,j} = 0.084$ eV and the two interstrand G–G transfer integral are: $t_{i,i+1}^{1,2} = 0.019$ eV and $t_{i,i+1}^{2,1} = 0.043$ eV, respectively (Voityuk *et al.* 2000, 2001). Furthermore, these values are considerably larger than the corresponding C–C transfer integral, which explains the shorter localization lengths in the C-band.

As discussed above (Cuniberti *et al.* 2002), there is also a small but non-zero hybridization of the C molecule with the neighboring sugar group. This hybridization shifts the energy of the HOMO level of C by approximately 0.12 eV towards the HOMO level of G, which results in a small increase in the localization length. The peak in the G-band reaches $\lambda \approx 8$ base pairs instead of ≈ 7. For the results presented below, this difference is even smaller and we can conclude that the nucleobase–sugar hybridization can be neglected in relation to the electron localization properties of DNA.

In Fig. 25.5 is shown the localization length for G–C DNA generated randomly, but with increased probability that a G(C) is followed by a G(C). We generate two types of DNA sequences with different standard deviations, one in which $\sigma = 1$ and another in which $\sigma = 8$, where larger values of σ means longer uniform sequences. The maximum localization length for $\sigma = 8$ reaches approximately 50 base pairs even though the average length of a

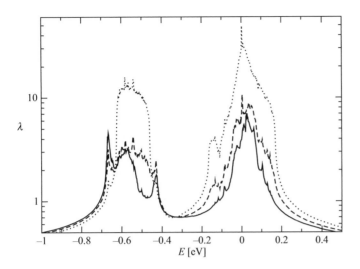

Fig. 25.5 Localization lengths (in units of the base-pair separation distance) as a function of energy for G–C DNA for a completely random rectangular distribution (solid) and with different σ: $\sigma = 1$ (dashed), $\sigma = 8$ (dotted). From Unge and Stafström (2003) *Nano. Lett.* **3**, 1417. ©2003 ACS Publications, reproduced with permission.

uniform G–C sequence is in this case 14 base pairs only. In these calculations G and C appear with equal probability in each strand. If the concentration of G(C) in strand 1(2) is increased, the amount of disorder is reduced and the localization length consequently increases. With $\sigma = 8$ and 80% G in strand 1 the maximal localization length is approximately 70 base pairs. The average length of a uniform G–C sequence is in this case 35 base pairs.

As shown in Fig. 25.5, the localization length decreases rapidly away from the center of the G-band. Thus, most charge carriers in this band are very localized. This result points to the fact that electron localization will occur even for synthesized structures with quite long sequences of uniform stacking. This is in agreement with experimental observations of band-like transport through a maximum of 30–60 base pairs of poly(G)-poly(C) DNA (Porath *et al.* 2000; Hwang *et al.* 2002).

Calculations of the localization lengths of A–T DNA do not give as long localization lengths as in G–C DNA. With $\sigma = 8$, the electronic states at the center of the A and T bands are extended over approximately 12 base pairs. The shorter length in these bands compared to the G-band occur because of the weaker intra- and interstrand transfer integral between nearest-neighbor adenine and nearest-neighbor thymine, respectively.

The localization lengths in a DNA sequence containing all four bases are considerably shorter than in G–C DNA since the disorder in on-site energies is increased. In Fig. 25.6 is shown the maximum localization lengths in the G-band for different concentrations of G–C pairs. In all these calculations the G–C (T–A) and C–G (A–T) ordering appear with equal probability. A G–C concentration of 50% corresponds to equal concentrations of G–C and T–A base pairs. We note that the localization lengths are very short in this regime. Below 90% G–C pairs, the localization lengths are shorter than 5 and it is not until 95–100% that we see a significant increase in the localization length. DNA sequences containing all four bases will therefore act as insulators, e.g. λ-DNA (NC_001416), which has a G–C concentration of approximately 50%, (NCBI), has been reported to have insulating behavior

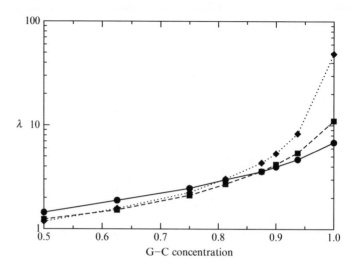

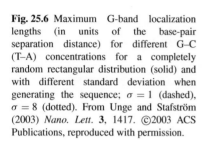

Fig. 25.6 Maximum G-band localization lengths (in units of the base-pair separation distance) for different G–C (T–A) concentrations for a completely random rectangular distribution (solid) and with different standard deviation when generating the sequence; $\sigma = 1$ (dashed), $\sigma = 8$ (dotted). From Unge and Stafström (2003) *Nano. Lett.* **3**, 1417. ©2003 ACS Publications, reproduced with permission.

(Zhang *et al.* 2002). Another example is the human chromosome 22 (NT_011520) that has a G–C concentration of approximately 47% (NCBI). Our results of the localization lengths in the G-band of this system do not even reach the separation length between two adjacent base pairs. This shows that the long-range correlation that exists in the human chromosome 22 has no effect on the electronic properties of this system, a result that is in disagreement with the results presented by other authors (Carpena *et al.* 2002; Caetano and Schulz 2005). We can conclude, in agreement with the discussion above, that this difference is due to the fact we use values of the parameters based on *ab-initio* results that are considerably smaller than those used in the references mentioned above. This behavior is a natural result of the large variations in the individual on-site energies of the four bases and the different transfer integrals between the different base combinations.

As can be seen in Fig. 25.6, the localization lengths for the completely random sequence exceed those of the correlated sequences below G–C concentrations of about 87%. In this case correlation results in longer potential (A–T) barriers between the (short) uniform G–C sequences. In the completely random sequence the barriers are in general shorter, which allows for extension of the orbitals across such barriers. As the concentration increases, the A–T barriers in the correlated sequences become shorter and the localization lengths in the G-band increase and become substantially longer than the average length of a uniform G–C sequence.

From the results presented above we can make one definite conclusion, the electronic states in any DNA sequence in which the concentration of G–C and A–T pairs is about the same are strongly localized and charge transport along the DNA strand can only occur via a thermally activated electron-transfer process. This is a direct consequence of the large variations in the on-site energies, variations that cause a high amount of disorder in comparison with the electron-transfer energy of the system (Unge and Stafström 2003a; Bagci and Krokhin 2007; Daz *et al.* 2007). In this situation, even though there is long-range correlation in the distribution of the nucleobases, this correlation is insufficient to affect the electronic properties. Below, we will see that this situation changes drastically in a uniform pentacene molecular crystal, a systems in which the intrinsic disorder is absent and in which the disorder is coming from extrinsic imperfections such as impurity ions, etc.

25.3.2 Pentacene

The exceptional high mobilities that have been observed in field effect transistors (FET) based on pentacene molecular crystals (Lin *et al.* 1997; Dimitrakopoulos *et al.* 1999; Jurchescu *et al.* 2004) indicate that the electron mean-free paths are long. This can only occur in the case of extended states or if the localization length is substantially longer than the average distance between phonon-scattering events. The primary reason for this is of course that the material forms pure crystals with very little disorder. The calculations were performed based on the Hamiltonian described above and using finite-size scaling to look for a possible metal–insulator transition (MIT). Furthermore, we stress the importance of both the correlation between the on-site and the off-diagonal elements in the Hamiltonian in eqn (25.3).

Before we go into the detailed results of the model, we study the effect of including the inter-molecular interaction, t_a, in the d_a direction of the pentacene molecular crystal (see Fig. 25.2). These studies were performed on systems without long-range correlations in the disorder and with the coupling between on-site and off-diagonal elements excluded. The off-diagonal elements are in this case generated from eqn (25.3) but using an auxiliary set of on-site energies that is completely independent of the energies used in the on-site matrix element.

The results from a set of calculations with N (see Fig. 25.2) in the range from 8 to 128 are shown in the left panels in Fig. 25.7. In the lower left panel the intermolecular t_a interaction is absent, which results in moderate localization lengths, $\sim$16 unit cells for the largest system. The introduction of

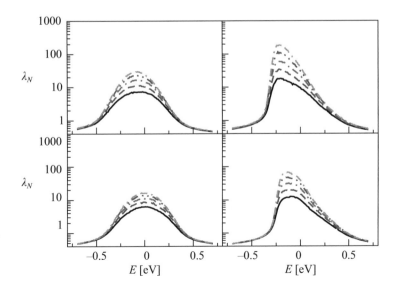

Fig. 25.7 Localization length, λ_N, vs. energy, E, for different N; 8 (lowest curve in each panel), 16, 32, 64 and 128 (uppermost curve). The two left and the two right panels correspond to uncorrelated disorder and uncorrelated disorder with coupling between the on-site and off-diagonal terms, respectively. The upper and lower panels are results with and without t_a included, respectively. From Unge and Stafström (2006) *Phys. Rev. B* **74**, 235403. ©2006 The American Physical Society, reproduced with permission.

t_a (top left panel) yields an increase in λ_N with a factor of ~ 1.4 and a shift of the maximum localization lengths to negative energies.

In the right panels in Fig. 25.7 are shown the results including the coupling between the on-site and off-diagonal matrix elements. These results are still with the long-range correlations excluded. The localization lengths increase quite dramatically compared to the system discussed above with λ_N reaching close to 68 lattice units in the case with t_a excluded (lower right panel) and above 185 lattice units with t_a included. Note that the mean values and standard deviation of the distribution of both on-site and off-diagonal matrix elements are the same with and without coupling included. Thus, the increase in the localization length is due to the qualitative difference between the two distributions. This difference also results in a shift in the maximum of λ_N towards negative energies by $\sim 0.2\,\text{eV}$ in the right panels (with coupling between on-site and off-diagonal matrix elements) compared to the left panels. To some extent this shift is also present in the density of states (DOS) that we have calculated for a finite system of size 64×150. The DOS plots are shown in Fig. 25.8 for the four different types of systems discussed above. The difference in shape in the DOS can be explained by the fact that the largest values of the intermolecular interaction strength occur between sites with positive on-site energies (see eqn (25.2)). The bonding linear combination of atomic orbitals (LCAO) form eigenstates at energies slightly below $\varepsilon = 0$. States with large bonding LCAO contributions on these strongly coupled sites are also the most extended states observed in Fig. 25.7. Since the coupling between large positive energies and large intermolecular interaction strengths is missing in the uncoupled model, the localization lengths become shorter in this case.

We now proceed by introducing long-range correlations into the on-site energies. The result for these calculations are shown in Fig. 25.9 without (left panel) and with (right panel) intermolecular interaction in the d_a direction

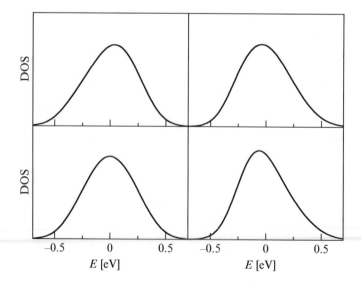

Fig. 25.8 Density of states, DOS (arb. units), vs. energy, E, for system of size 64×150. The two left and the two right panels correspond to uncorrelated disorder and uncorrelated disorder with coupling between the on-site and off-diagonal terms, respectively. The upper and lower panels are results with and without t_a included, respectively. From Unge and Stafström (2006) *Phys. Rev. B* **74**, 235403. ©2006 The American Physical Society, reproduced with permission.

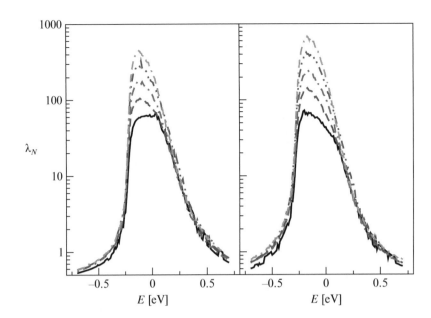

Fig. 25.9 Localization length, λ_N, vs. energy, E, with long-range correlation, $\gamma = 1.0$ for different N; 8, 16, 32, 64 and 128. The right plot includes t_a. From Unge and Stafström (2006) *Phys Rev. B* **74**, 235403. ©2006 The American Physical Society, reproduced with permission.

included. The coupling between the on-site and off-diagonal terms is included in both graphs, hence the intermolecular interaction term includes long-range correlation. In these systems the localization length reaches ~446 unit cells and ~683 unit cells without and with t_a, respectively. Thus, the localization length increases by a factor of 4–5 compared to the case with uncorrelated disorder. This is purely an effect of the long-range correlation since the mean values and standard deviation of the distribution of both on-site and off-diagonal matrix elements are the same in both cases. The maximum values of λ_N are shifted towards negative energies for the same reason as discussed above.

It is clear from Fig. 25.9 that the localization lengths of the states in the band tails are much shorter than in the center of the band. Furthermore, these localization lengths do not scale with the width of the system but remain approximately constant for increasing N. This is a clear indication that the region of the band tails contain localized states and with the Fermi level positioned in this region, the system should behave as an insulator. Note, however, that the DOS is small in these regions and even moderate (field effect) charging of the molecular layer can shift the Fermi energy into the region where the localization lengths scale with N.

We now turn to investigate the scaling properties of the localization length. Λ_N is calculated as a function of N for both systems presented in Fig. 25.9 as well as for three of the four systems presented in Fig. 25.7 (the case with no coupling and t_a is excluded). Due to the different energies at which the peak in the localization length occur, we choose to study the scaling behavior at energies corresponding to the maximum localization length in the spectra shown in Figs. 25.7 and 25.9. These energies are listed in Table 25.1. The results are shown in Fig. 25.10 for N up to 256 in all cases and also for $N = 512$ for the two cases of correlated disorder, i.e. the cases with the largest values of λ_N.

Table 25.1 Energies used in the finite-size scaling analysis corresponding to the maximum localization lengths for the different models

Model	$E_{\max\lambda}$ [eV]
Uncorr	0.00
Coupling	−0.13
Coupling with t_a	−0.22
Long-range corr.	−0.12
Long-range corr. with t_a	−0.17

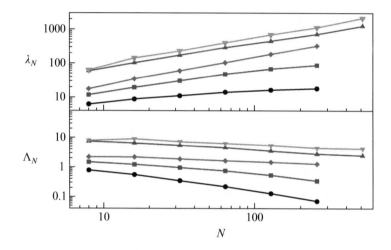

Fig. 25.10 Localization length, λ_N, and renormalized localization length, Λ_N, vs. N. Uncorrelated (circles), coupling between the on-site and off-diagonal terms (squares), coupling between the on-site and off-diagonal terms with t_a (diamonds), long-range correlation $\gamma = 1.0$ without (triangle up) and with (triangle down) t_a. The solid lines are a guide to the eye. The localization lengths are calculated at the energies specified in Table 25.1. The deviation for each calculated value are within the size of the symbols. From Unge and Stafström (2006) *Phys Rev. B* **74**, 235403. ©2006 The American Physical Society, reproduced with permission.

Clearly there is a large difference between the results of the systems with long-range correlation presented in Fig. 25.10 and those of the uncorrelated model. The uncorrelated system yields localization lengths for the two-dimensional system of 17 unit cells at $N = 256$, corresponding to 7 nm, assuming an intermolecular distance of 4 Å. With long-range correlations included the localization length reaches 678 unit cells without t_a included and 1074 unit cells with t_a included. The finite-size scaling results up to $N = 256$ still show a localized behavior, i.e. Λ_N is decreasing with increasing N. Since the systems with the most extended wavefunctions could be close to a MIT we have performed calculations with $N = 512$ for these cases in order to look for a possible change in the behavior of Λ_N. The maximum localization length obtained is 2115 unit cells, which corresponds to 0.85 μm. Compared to the system with uncorrelated disorder and no coupling between the diagonal and off-diagonal disorder this is an increase of the localization length by more than two orders of magnitude.

The renormalized localization lengths for the system with t_a included show a tendency to saturate at a value of $\Lambda_N \sim 4$ (topmost curve in the lower panel in Fig. 25.10). According to finite-size scaling this indicates that the system is reaching the transition from an insulating to a metallic state (Pichard and Sarma 1981). Unfortunately, at present we are unable to perform calculations for even larger values of N to give a more conclusive result concerning the MIT. Note that the particular type and strength of the disorder used in our calculations represent the behavior of a typical molecular crystal such as pentacene. It is of course possible to drive the model system further towards the metallic state by increasing the correlation factor below the present value of $\gamma = 1$ (or alternatively, reduce the width of the Gaussian distribution of on-site energies or increase the intermolecular interaction strength). This would, however, require another type of correlated disorder than that obtained from the random dipole model. It will be the subject of further studies to include conductivity in the model and to relate the results obtained here to the experimental result of band-like transport of single crystals of pentacene at room temperature (Jurchescu *et al.* 2004).

In conclusion, the results presented here show that molecular crystals such as pentacene can exhibit extended states only if there exists correlation in the static disorder. The existence of this type of disorder is motivated by the model of randomly oriented induced dipole moments. Our results show a substantial increase in the localization length of the electronic state with correlated disordered as compared to the case of uncorrelated disorder. The localization length exceeds the lattice constant by more than three orders of magnitude when calculated for a stripe of the molecular crystal with a width of 512 lattice sites. The renormalized localization length indicates a transition into an extended state, which can explain the experimental observation of band-like transport of single crystals of pentacene at room temperature (Jurchescu *et al.* 2004).

25.4 Summary

Fifty years after the first results of "Absence of Diffusion in Certain Random Lattices" (Anderson 1958), the problem of electron localization still attracts a lot of attention. This interest involves novel materials, such as organic materials for electronic applications, fullerenes, carbon nanotubes, graphene, etc.

In this work we have limited the discussion to molecular-based systems, i.e. materials that are not covalently bonded but in which the cohesive energy comes from van der Waals forces between the molecules. Such systems are inherently more sensitive to disorder for two reasons. First, the intermolecular transfer integrals are small, usually of the order of 100 meV or less, which results in narrow electronic bands and thus a higher sensitivity to disorder. Second, the disorder can be larger since the molecules experience a weaker potential well for, for instance, displacements and rotations.

In DNA we have an additional very important contribution to the disorder, namely, the distribution of the different constituent molecules adenine, cytocine, guanine, and thymine. One of the main conclusions from the work presented here is that the difference in ionization potential between these nucleobases results in a disorder strength that is enough to localize the wave-functions over segments of the systems that contain identical nucleobase molecules. Unless the DNA system is built up from guanine–cytocine base pairs only, charge transport through such a system occurs via a thermally activated electron-transfer process. We have also included the effects of interstrand coupling and coupling to the sugar–phosphate backbone and found that these interactions have no qualitative effect on the localization properties of DNA. Addition disorder due to fluctuations in the intermolecular transfer integrals and in the on-site energies, i.e. effects that have been neglected in our model, could momentarily act to reduce disorder but on the average, such an effect will further increase electron localization in the segments that are homogeneous (Renger and Marcus 2003). One important interaction that is missing in our model is electron–phonon interaction. This interaction plays an important role in the electron-transfer process but the strength of the interaction is of the same order as or smaller than the intermolecular transfer integral and thus cannot compete with the disorder in the on-site potential. Thus, in models dealing with transport in DNA this interaction has to be included, but such a transport model has to be based on the fact that the states are delocalized at most over segments of identical nucleobases (Renger and Marcus 2003).

The situation in pentacene is quite different. Even though the forms of the Hamiltonians are very similar (see eqns (25.1) and (25.2)) the main difference comes from the fact that the fluctuations in the on-site energies in the homomolecular pentacene system are much smaller than in DNA. Thus, in the absence of extrinsic disorder, this system would have been metallic. Such disorder is, however, unavoidable since the molecules can relatively easily reorient in the crystal and with the molecules polarized due to the presence of charges (charge carriers and/or impurities) this will create a correlated set of on-site energies and transfer integrals. The electron-localization properties in this system show that, given the type of disorder that is realistic in a molecular crystal of this type (Bussac *et al.* 2004), the electron-localization lengths can be very large, of the order of μm, but for the system size we have been able to study we see no clear evidence for a MIT. The correlation in the disorder that arises from the random distribution of molecular dipole moments is indeed important for the system to exhibit such long localization lengths. What limits further studies in this context is the large size of the system. It will certainly, with the next generation of high-performance computing facilities, be possible to further extend the width (the parameter N) of the system and, using finite-size scaling theory, reach the system size that is needed for the scaling theory to produce a conclusive result concerning the possible MIT.

Finally, we would like to point out the importance of the dimensionality. Both the systems discussed here are highly anisotropic and are either quasi-1D (DNA) or quasi-2D (pentacene). In a 3D system there can exist a mobility edge, i.e. a threshold energy separating localized and delocalized states if the disorder strength is less than 5 times the electronic bandwidth (Anderson 1958). Thus, from the point of view of charge-transport properties over longer distances, a 3D system is preferred. Such properties have been observed, for instance, in doped polyacetylene and doped (alkali-metal intercalated) fullerenes in which the dopant both transfers charge carriers to the system and acts as a bridge that improves the intermolecular interactions in such a way that a metallic state can be achieved. It would be very interesting to find similar properties in a pristine molecular crystal. Such a material would in that case be ideal for electronic applications.

Acknowledgments

Financial support from the Swedish Research Council (VR) is gratefully acknowledged. We also thank the National Supercomputer Centre in Linköping for providing code-development support and computational facilities.

References

Anderson, P.W. *Phys. Rev.* **109**, 1492 (1958).

Bagci, V.M.K., Krokhin, A.A. *Phys. Rev. B* **76**, 134202 (2007).

Basko, D.M., Conwell, E.M. *Phys. Rev. Lett.* **88**, 098102 (2002).

Ben-Jacob, E., Hermon, Z., Caspi, S. *Phys. Lett. A* **263**, 199 (1999).

Benettin, G., Galgani, L. In G. Laval, D. Grésillon, (eds) *Intrinsic Stochasticity in Plasmas* (Editions de Physique, Orsay, 1979) p. 93.

Bloch, F. *Z. Physik* **52**, 555 (1928).

Boudaïffa, B., Cloutier, P., Hunting, D., Huels, M.A., Sanche, L. *Science* **287**, 1658 (2000).

Braun, E., Eichen, Y., Siva, U., Ben-Yoseph, G. *Nature* **391**, 775 (1998).

Burroughes, J.H., Bradley, D.D.C., Brown, A.R., Marks, R.N., Mackay, K., Friend, R.H., Burns, P.L., Holmes, A.B. *Nature* **347**, 539 (1990).

Bussac, M.N., Piconand, J.D., Zuppiroli, L. *Europhys. Lett.* **66**, 392 (2004).

Caetano, A., Schulz, P.A. *Phys. Rev. Lett.* **95**, 126601 (2005).

Campbell, R., Robertsson, J.M., Trotter, J. *Acta Cryst.* **14**, 705 (1961).

Carlsson, M. *Studies of the Electronic Structure of DNA*. Master's thesis, Linkoping University (2002).

Carpena, P., Bernaola-Galván, P., Ivanov, P.C., Stanley, H.E. *Nature* **418**, 955 (2002).

Cheng, Y.C., Silbey, R.J., da Silva Filho, D.A., Calbert, J.P., Cornil, J., Brédas, J.L. *J. Chem. Phys.* **118**, 3764 (2003).

Chiang, C.K., Druy, M.A., Gau, S.C., Heeger, A.J., Louis, E.J., MacDiarmid, A.G., Park, Y.W., Shirakawa, H. *J. Am. Chem. Soc.* **100**, 1013 (1978).

Cornil, J., Calbert, J.P., Brédas, J.L. *J. Am. Chem. Soc.* **123**, 1250 (2000).

Cuniberti, G., Porath, D., Dekker, C. *Phys. Rev. B* **65**, 241314(R) (2002).

Dandliker, P.J., Núñes, M.E., Barton, J.K. *Biochemistry* **37**, 6491 (1998).

Dimitrakopoulos, C.D., Purushothaman, S., Kymissis, J., Callegari, A., Shaw, J.M. *Science* **283**, 822 (1999).

Dunlap, D.H., Wu, H.L., Phillips, P.W. *Phys. Rev. Lett.* **65**, 88 (1990).

Daz, E., Sedrakyan, A., Sedrakyan, D., Domnguez-Adame, F. *Phys. Rev. B* **75**, 014201 (2007).

Fink, H.W., Schönenberger, C. *Nature* **398**, 407 (1999).

Flambaum, V.V., Sokolov, V.V. *Phys. Rev. B* **60**, 4529 (1999).

Gao, H., Kong, Y., Cui, D., Ozkan, C.S. *Nano Lett.* **3**, 471 (2003).

Giese, B., Amaudrut, J., Köhler, A.K., Spormann, M., Wessely, S. *Nature* **412**, 318 (2001).

Golub, G.H., van Loan, C.F. *Matrix Computations* (The Johns Hopkins University Press, London, 1996).

Halik, M., Klauk, H., Zschieschang, U., Schmid, G., Dehm, C., Schütz, M., Maisch, S., Effenberger, F., Brunnbauer, M., Stellacci. F. *Nature* **43**, 963 (2004).

Hjort, M. *Electronic Structure Studies of Carbon Based Nanomaterials*. Ph.D. thesis, Linkoping University (2001).

Hjort, M., Stafström, S. *Phys. Rev. B* **61**, 14089 (2000).

Hjort, M., Stafström, S. *Phys. Rev. B* **63**, 113406 (2001a).

Hjort, M., Stafström, S. *Phys. Rev. Lett.* **87**, 228101 (2001b).

Holste, D., Grosse, I., Herzel, H. *Phys. Rev. E* **64**, 041917 (2001).

Horowitz, G. *Adv. Mat.* **10**, 365 (1998).

Horowitz, G., Fichou, D., Peng, X., Xu, Z., Garnieret, F. *Solid State Commun.* **72**, 381 (1989).

Hultell, M., Stafström, S. *Phys. Rev. B* **75**, 104304 (2007).

Hwang, J.S., Kong, K.J., Ahn, D., Lee, G.S., Ahn, D.J., Hwang, S.W. *Appl. Phys. Lett.* **81**, 1134 (2002).

Iijima, S. *Nature* **354**, 56 (1991).

Jerome, D., Mazaud, A., Ribault, M., Bechgaard, K. *J. Phys. Lett.* (Paris) **41**, L95 (1980).

Johansson, A., Stafström, S. *Phys. Rev. Lett.* **86**, 3602 (2001).

Jurchescu, O.D., Baas, J., Palstra, T.T.M. *Appl. Phys. Lett.* **84**, 3061 (2004).

Karl, N. *Molec. Cryst. Liquid Crys.* **171**, 31 (1989).

Kroto, H.W., Heath, J.R., O'Brien, S.C., Curl, R.F., Smalley, R.E. *Nature* **318**, 162 (1985).

Lewis, F.D., Liu, X., Liu, J., Miler, S.E., Hayes, R.T., Wsiewski, M.R. *Nature* **406**, 51 (2000).

Lin, Y.Y., Gundlach, D.J., Nelson, S.F., Jackson, T.N. *IEEE Electron Device Lett.* **18**, 606 (1997).

Liu, W.S., Liu, S.Y., L. Lei, X. *Eur. Phys. J. B* **33**, 293 (2003).

MacKinnon, A., Kramer, B. *Phys. Rev. Lett.* **47**, 1546 (1981).

MacKinnon, A., Kramer, B.Z. *Phys. B* **53**, 1 (1983).

Makse, H.A., Havlin, S., Schwartz, M., Stanley, H.E. *Phys. Rev. E* **53**, 5445 (1996).

Marcus, R.A. *Rev. Mod. Phys.* **65**, 599 (1993).

Miller, A., Abrahams, E. *Phys. Rev.* **120**, 745 (1960).

de Moura, F.A.B.F. Lyra, L.M. *Phys. Rev. Lett.* **81**, 3735 (1998).

NCBI, National Center for Biotechnology Information. http://www.ncbi.nlm.nih.gov/.

Nelson, S.F., Lin, Y.Y., Gundlach, D.J., Jackson, T.N. *Appl. Phys. Lett.* **72**, 1854 (1998).

Oseledec, V.I. *Trans. Moscow Math. Soc.* **19**, 197 (1968).

de Pablo, P.J., Moreno-Herrero, F., Colchero, J., Herrero, J.G., Herrero, P., Baró, A., Ordejón, P., Soler, J.M., Artacho, E. *Phys. Rev. Lett.* **85**, 4992 (2000).

Pichard, J.L., Sarma, G. *J. Phys. C* **14**, 127 (1981).

Porath, D., Bezryadin, A., Vries, S.D., Dekker, C. *Nature* **403**, 635 (2000).

Renger, T., Marcus, R.A. *J. Phys. Chem. A* **107**, 8404 (2003).

Sariciftci, N., Smilowitz, L., Heeger, A., Wudl, F. *Science* **258**, 1474 (1992).

Shima, H., Nomura, T., Nakayama, T. *Phys. Rev. B* **70**, 075116 (2004).

Skotheim, T.A., Reynolds, J.R., (eds) *Handbook of Conducting Polymers*, 3rd edn (CRC Press, 2006).

Souillard, B. In *Les Houches XLVI*. Chance and Matter, Amsterdam, **305** (1987).

Sundar, V.C., Zaumseil, J., Podzorov, V., Menard, E., Willett, R.L., Someya, T., Gershenson, M.E., Rogrers, J.A. *Science* **303**, 1644 (2004).

Toyozowa, Y. *Prog. Theor. Phys.* **12**, 421 (1954).

Tsumura, A., Koezuka, H., Ando, T. *Appl. Phys. Lett.* **49**, 1210 (1986).

Unge, M., Stafström, S. *Nano Lett.* **3**, 1417 (2003a).

Unge, M., Stafström, S. *Synth. Met.* **139**, 239 (2003b).

Unge, M., Stafström, S. *Phys. Rev. B* **74**, 235403 (2006).

Voityuk, A.A., Bixon, M., Rösch, N. *J. Chem. Phys.* **114**, 5614 (2001).

Voityuk, A.A., Rösch, N., Bixon, M., Jortner, J. *J. Phys. Chem. B* **104**, 9740 (2000).

White, C.T., Todorov, T.N. *Nature* **393**, 240 (1998).

Xiong, S.J., Zhang, G.P. *Phys. Rev. B* **68**, 174201 (2003).

Yoo, K.H., Ha, D.H., Lee, J.O., Park, J.W., Kim, J., Kim, J.J., Lee, H.Y., Kawai, T., Choi, H.Y. *Phys. Rev. Lett.* **87**, 198102 (2001).

Zhang, Y., Austin, R.H., Kraeft, J., Cox, E.C., Ong, N.P. *Phys. Rev. Lett.* **91**, 198102 (2002).

Subject Index

Printed and bound by CPI Group (UK) Ltd, Croydon, CR0 4YY